Manual
of
Agricultural
Nematology

Manual
of
Agricultural
Nematology

edited by
William R. Nickle

Beltsville Agricultural Research Center
Agricultural Research Service
United States Department of Agriculture
Beltsville, Maryland

 **CRC Press**
Taylor & Francis Group
Boca Raton London New York

CRC Press is an imprint of the
Taylor & Francis Group, an **informa** business

CRC Press
Taylor & Francis Group
6000 Broken Sound Parkway NW, Suite 300
Boca Raton, FL 33487-2742

First issued in paperback 2019

© 1991 by Taylor & Francis Group, LLC
CRC Press is an imprint of Taylor & Francis Group, an Informa business

No claim to original U.S. Government works

ISBN-13: 978-0-8247-8397-6 (hbk)
ISBN-13: 978-0-367-40297-6 (pbk)

Visit the Taylor & Francis Web site at
http://www.taylorandfrancis.com

and the CRC Press Web site at
http://www.crcpress.com

Library of Congress Cataloging-in-Publication Data

Manual of agricultural nematology / edited by William R. Nickle.
    p.   cm.
  Includes bibliographical references and index.
  ISBN 0-8247-8397-2
  1. Plant nematodes--Handbooks, manuals, etc. 2. Nematoda--Handbooks, manuals, etc. I. Nickle, William R.
  SB998.N4M33   1991
  632'.65182--dc20   91-6775
                                 CIP

Foreword

I. N. Filipjev's seminal *Manual of Agricultural Helminthology* has served nematologists and agriculturalists for many years; however, an updating of the data has been badly needed. For example, this 1941 text cited 19 genera and 185 species of plant parasitic nematodes which has now grown to 207 genera and over 4832 species of plant parasites as of 1990. The present *Manual of Agricultural Nematology* is an attempt to provide a contemporary exposition of the 1941 *Manual of Agricultural Helminthology*.

To achieve this task, W. R. Nickle has assembled an international group of specialists in the field of nematology, each with particular expertise in one or more groups of nematodes or in a pertinent area of nematology such as anatomy, morphology, or systematics. As one would expect, each author has introduced into the text not only expert commentary but also style and a little personality.

This book will neither solve nor amend all the terrible problems incumbent on modern nematology. It will, however, come as close to achieving that task as one can hope for, and it should certainly prove an indispensable tool in any nematological workplace. It has an undeniable international, individualistic flavor, and at the same time is instilled with a logical, well-coordinated sequence of data. Reviewing this book has been both an education and a pleasure.

R. P. Esser
Nematologist
Division of Plant Industry
Florida Department of Agriculture and Consumer Services
Gainesville, Florida

Preface

This book is an attempt to update the *Manual of Agricultural Helminthology* written by the Russian I. N. Filipjev in 1934 and translated and revised by Schuurmans Stekhoven, Jr. in 1941. The format is the same; however, because of the current complexity of the science of nematology, 29 carefully selected, internationally eminent contributors have been called upon to prepare this volume. The contributors were asked to present biology and a conservative taxonomic treatment of the main groups of plant and insect parasitic nematodes.

This book has been designed as a reference text and guide for higher level students and workers in nematology, plant pathology, zoology, horticulture, agronomy, and insect pathology. The most comprehensive treatment of nematode races available, it stands as a companion volume to *Plant and Insect Nematodes* (Marcel Dekker, Inc., 1984).

Plant parasitic nematodes cause over $6 billion in losses to agricultural production in the United States each year. Insect parasitic nematodes are beginning to take a more prominent place in biological control of pest insects as we try to reduce the amount of pesticides in our groundwater. Separate chapters are devoted to nematode morphology, biology and ecology, collection and preparation, and the newest methods of nematode identification including DNA fingerprinting and other techniques. Next is a very learned evolutionary treatise on higher level classification. The major groups of nematodes then follow, with the most important groups taken up first. The root-knot nematodes and cyst nematodes, which are the most serious pests, are presented in detail. Other important groups of plant parasitic nematodes follow with aids to their identification and biology. The last three chapters deal with insect parasitic nematodes.

I am indebted to the Agricultural Research Service, United States Department of Agriculture, for the opportunity to work in the science of nematology since 1965 and to the contributors to this book and colleagues around the world who offered their support and encouragement. Several journals and publishers have generously approved the reproduction of some of the illustrations and descriptions presented in the book, and this cooperation is appreciated. I wish to thank my wife, Cathy, for helping with some of the typing.

This book is dedicated to the 29 contributors who took the time to give us their best efforts in the pages that follow.

William R. Nickle

Contents

V.   INSECT PARASITIC NEMATODES

Contributors

R. V. Anderson   Biosystematics Research Centre, Agriculture Canada, Ottawa, Ontario, Canada

James G. Baldwin   Department of Nematology, University of California, Riverside, Riverside, California

Michał W. Brzeski   Nematology Laboratory, Research Institute of Vegetable Crops, Instytut Warzywnictwa, Skierniewice, Poland

Lynn K. Carta   Division of Biology, California Institute of Technology, Pasadena, California

John Curran   Division of Entomology, Commonwealth Scientific and Industrial Research Organization, Canberra, Australia

Wilfrida Decraemer   Department of Invertebrates, Section of Recent Invertebrates, Koninklijk Belgisch Instituut voor Natuurwetenschappen, Brussels, Belgium

Jonathan D. Eisenback   Department of Plant Pathology, Physiology and Weed Science, Virginia Polytechnic Institute and State University, Blacksburg, Virginia

Renaud Fortuner   Analysis and Identification Branch, California Department of Food and Agriculture, Sacramento, California

Etienne Geraert   Laboratorium voor Morfologie en Systematiek der Dieren, Rijksuniversiteit Gent, Ghent, Belgium

David J. Hooper   Entomology and Nematology Department, Institute of Arable Crops Research, Rothamsted Experimental Station, Harpenden, Hertfordshire, England

Parviz Jatala   Department of Nematology and Entomology, International Potato Center, Lima, Peru

Helmut Kaiser   Institut für Zoologie, Karl-Franzens-Universität Graz, Graz, Austria

Eino L. Krall   Laboratory of Entomology and Nematology, Institute of Zoology and Botany, Estonian Academy of Sciences, Tartu, Estonia, USSR

Christain Laumond   Institut National de la Recherche Agronomique, Antibes, France

Pieter A. A. Loof   Department of Nematology, Wageningen Agricultural University, Wageningen, The Netherlands

Armand R. Maggenti   Department of Nematology, University of California, Davis, Davis, California

Manuel Mundo-Ocampo   Department of Nematology, University of California, Riverside, Riverside, California

Khuong B. Nguyen   Entomology and Nematology Department, University of Florida, Gainesville, Florida

Terry L. Niblack   Department of Plant Pathology, University of Missouri—Columbia, Columbia, Missouri

William R. Nickle   Nematology Laboratory, Beltsville Agricultural Research Center, Agricultural Research Service, United States Department of Agriculture, Beltsville, Maryland

Don C. Norton   Department of Plant Pathology, Iowa State University, Ames, Iowa

John W. Potter   Nematology Section, Research Station, Agriculture Canada, Vineland Station, Ontario, Canada

Dewey J. Raski   Department of Nematology, University of California, Davis, Davis, California

Michel Remillet*   Biological Control Laboratory, Faculty of Agriculture, University of Cairo, Cairo, Egypt

Robert D. Riggs   Department of Plant Pathology, University of Arkansas, Fayetteville, Arkansas

Grover C. Smart, Jr.   Entomology and Nematology Department, University of Florida, Gainesville, Florida

Dieter Sturhan   Institute for Nematology and Vertebrate Research, Federal Biological Research Center for Agriculture and Forestry, Münster, Germany

Hedwig Hirschmann Triantaphyllou   Department of Plant Pathology, North Carolina State University, Raleigh, North Carolina

Valerie M. Williamson   Department of Nematology, University of California, Davis, Davis, California

Wilhelmus M. Wouts   Department of Scientific and Industrial Research Plant Protection, Mt. Albert Research Centre, Aukland, New Zealand

* *Present affiliation*: Institut Français de Recherche Scientifique pour le Développement en Coopération, Paris, France

Manual
of
Agricultural
Nematology

I
GENERAL MORPHOLOGY
AND BIOLOGY OF
NEMATODES

1

General Nematode Morphology

ARMAND R. MAGGENTI *University of California, Davis, Davis, California*

I. INTRODUCTION

Understanding of organismal biology advances only where there is substantial knowledge of the form, structure, and function of the embodied taxa. A knowledge of morphology is essential if we are intent on not only entering but contributing to modern biology in aspects seemingly unrelated to nematode morphology. Without morphology there would be no taxonomy and no systematic organization of nematodes in a classification. Such a vacuum would negate studies in population ecology, pathology, physiology, and, to a large extent, genetics where processes could be studied but application would be limited. Molecular diagnostics might show relationships among individuals or populations but we could not comprehend the broader significance of such relationships.

The more that is known about the organisms one is working with the more significant will be any findings, for knowledge leads to confidence, insight, and to the ultimate objective of almost any research—predictability. Predictions have enhanced value when they are consistent with observable facts. Predictions not founded on fact are conjectures that through repetition become obstacles to the advancement of knowledge.

One of the most important contributions morphologists can make to their respective science is the determination and testing of homologous and analogous structures. Comparative morphology reveals the true nature of structures and thus can lead to significant modifications in our understanding of nemic relationships. Elucidating the analogous nature of the postcorporal valves of Plectidae and Rhabditidae led to a revision of the higher classification of Nemata (Maggenti, 1963). Inglis's (1964) comparative study of the head structures of Enoplida resulted in a clearer understanding of the origin and homologies of stomatal structures throughout Nemata. This knowledge along with the work of Baldwin and Hirschmann (1976) resulted in a new interpretation of the stomatal structures in Tylenchina (Maggenti, 1981). Discussions over whether or not a structure is analogous or homologous will be inevitable wherever morphologists gather in conversation; however, comparative morphology remains the bastion which when consistent with observed facts sustains all biological studies.

II. GAMETOGENESIS AND REPRODUCTION

Genetics elucidates the laws by which some aspects of life may be understood. This is the discipline that very often is able to offer explanations for observed biology. Through genetics the mechanisms of inherent patterns in living things are revealed; disclosure of the mechanism established laws such as the laws of inheritance. Biological observation precedes our understanding of the nature and properties of things. Observation forms the basis of biological theories that are, without knowledge of the mechanisms, logical abstractions of reality.

Mendel (1865) formulated the first laws of inheritance and established the basis on which the later developments of genetics have taken place. The work of Mendel lay dormant and unappreciated for more than 30 years; however, the biological observations on the mechanisms underlying Mendel's work were known. Van Benden's work (1883) with the eggs and spermatozoa of *Parascaris equorum* was basic to the doctrine of the genetic continuity of chromosomes. In this same report he described the process of meiosis and the fact that gametes contained only one-half the chromosome complement of somatic cells.

A. Spermatogenesis

The formation of spermatozoa is accomplished by two consecutive reduction divisions (meiosis) (Fig. 1). The process of spermatogenesis begins when the nuclear reticulum of the primary spermatocyte is resolved into the diploid number of chromosomes. During this process the chromatin material is threadlike and may be in the process of doubling. The chromatin threads then come to lie side by side, forming a monoploid set of conjugating homologous pairs of male- and female-derived chromosomes. The paired chromosome threads at this stage are clearly doubled and held together by a centromere. Each original chromosome now becomes four strands associated in a tetrad of four chromatids. In each tetrad two chromatids separate from the other two and thus each chromatid has a pairing partner. At this time, when the chromosomes are shortened and thickened, they can easily be counted and characterized. At the next step in the process the tetrads align at the spermatocytes' equatorial plate. The paired chromatids separate from the other pair of the tetrad and as dyads move to opposite poles of the cell. The cell now divides and the cell so formed has its own complement of dyads. Therefore, each cell has $1n$ chromosome but $2n$ chromatids. The cell is now a secondary spermatocyte.

The process of spermatogenesis proceeds through the second meiotic division. The dyads of the preceeding program align along the equatorial plate. Upon the separation of the chromatids of each dyad the integrity of the complete chromosome is returned. The now distinct chromosomes move to opposite poles of the cell which then divides and the resulting haploid progeny are called spermatids. Thus four spermatids have been derived from the primary spermatocyte.

The formation of spermatozoa involves mitosis, maturation, and morphogenesis. Among Nemata spermatozoa are variable in shape and motility. The mechanism of spermatozoan motility in *Caenorhabditis* has been studied by Roberts and Ward (1982a,b). Their findings indicate that ameboid spermatozoa propel themselves by tip to base flow of membrane over the pseudopod surface i.e., a new membrane component is inserted at the tip of the pseudopod and these components flow backward to be taken up at the base of the pseudopod. Because the pseudopod membrane is continuously being rebuilt at the tip, new attachment sites are created at the leading edge of the cell which is thus propelled forward.

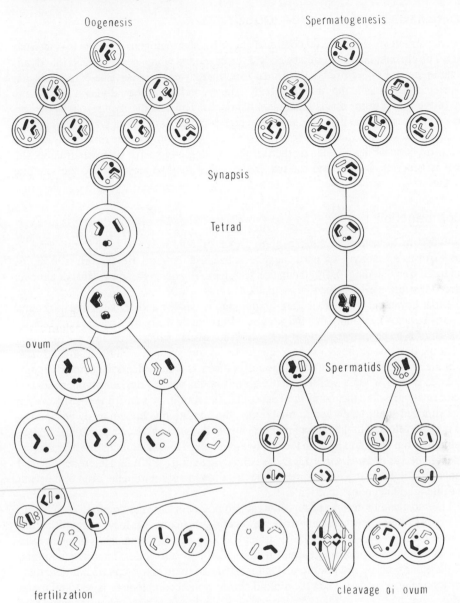

FIGURE 1 Schematic of oogenesis, spermatogensis, fertilization, and cleavage. (Maggenti/Springer-Verlag, 1981; adapted from Huettner.)

B. Oogenesis

A sequence of events similar to that outlined above for spermatogensis occurs in oogenesis (Fig. 1). The chromatin of the primary oocyte forms tetrads which are separated into dyads in the same manner as occurred in the primary spermatocyte. However, one of the cells that is constricted off is called the first polar body. The second maturation division follows the pattern of the first, i.e., the dyads separate into monads. The next division results in the formation of the second polar body and the ovum (egg pronucleus), both of which are haploid.

In nematodes the sperm generally enters the egg prior to or at the time of the second reduction division of the oocyte. At the time of the egg and sperm pronuclei fusion, the eggshell is generally present and the two polar bodies may be seen outside the vitelline membrane.

C. Reproduction

Reproduction in Nemata is sexual. The term asexual should never be used in reference to nematodes or as a synonym for parthenogenesis because the term asexual refers to reproduction by fission, budding, or fragmentation which usually occurs among plants or animals considered evolutionarily lower than nematodes.

Most commonly nematodes are amphigonous (males and females separate) and oviparous (exotokia), meaning that the eggs develop outside the body. Ova are fertilized by sperm stored in a sperm receptacle at the end of the ovary, from sperm in a spermatheca located at the junction of the oviduct and uterus, or, in some monovarial forms, from sperm stored in the vestigial posterior uterus. Though exotoky is common among nematodes, endotoky does occur. When a well-defined eggshell is not seen during embryogenesis and larvae are retained within the uterus until deposited, the condition is called viviparity. When the eggs (normal shell) hatch within the female, the condition is known as ovoviviparity. Larvae may be released in the normal manner; however, larvae may (when females are senile) remain in the body and feed on the decaying female. This phenomenon is often incorrectly designated as *endotokia matricida*. As coined by Seurat (1920) for Heteroderidae and Oxyuridae, endotokia matricida referred to the retention of eggs within the dead female and was not intended to connote matricide.

The most common form of reproduction among nematodes is amphimixis, i.e., the union of two gametes. Parthenogenesis is the second most common form of reproduction and is most often seen as a reproductive strategy among nematode parasites of plants and animals. Three types of parthenogenesis are known among nematodes: meiotic automixis, meiotic parthenogenesis, and mitotic parthenogenesis. Meiotic automixis (facultative meiotic parthenogenesis) occurs when the diploid chromosomal complement is restored in the reduced oocytes by fusion of the second polar body with the egg pronucleus. Meiotic parthenogenesis refers to a reproductive process whereby synapsis takes place and the reduced chromosome number appears at first maturation prophase. The somatic chromosome number (diploid, tetraploid, etc.) is restored by chromosome duplication at anaphase 1; there is no second maturation division. Mitotic parthenogenesis occurs when there is no pairing of homologous chromosomes during prophase and therefore the somatic chromosome number is maintained. The result is the formation of the first polar body and diploid egg pronucleus; from this point embryology proceeds normally.

What may be a precursor step in the development of parthenogenesis is a modified form of parthenogenetic reproduction called pseudogamy or pseudofertilization. In this type of reproduction the sperm enters the oocyte but does not fuse with the pronucleus; it merely activates further development. The sperm nucleus degenerates and development proceeds by parthenogenesis.

The least common form of reproduction among nematodes is hermaphroditism. It occurs in *Caenorhabditis* (bacterial feeders in nature) (Wood, 1988) and may occur in *Heterogonema* (mermithid parasite of nitidulid beetles) (Van Waerbeke and Remillet, 1973). In both of these genera normal amphimictic reproduction occurs, although less commonly in the former than the latter.

The hermaphrodites of these genera differ: *Caenorhabditis* has two tubular ovotestes, whereas the hermaphrodite of *Heterogonema* has an anterior testis, a posterior ovary, and the secondary sexual characteristics of males (spicules) and normal females occur in the breeding population. *Caenorhabditis* hermaphrodites are self-fertilizing except on those rare occasions when normal males are present and cross-fertilization occurs. On the other hand, in *Heterogonema* cross-fertilization is the rule; self-fertilization, if it occurs, happens only at the end of the hermaphrodite's life when eggs are seen in the posterior female gonad (this may happen by parthenogenesis). Only cross-fertilized females are infective to new coleopteran hosts.

In both genera the form of hermaphroditism seems to serve a senseless purpose. Among other invertebrates, including those considered evolutionarily primitive as compared to nematodes, such as Gastrotrichs, hermaphrodites cross-fertilize each other. Cross-fertilization enhances an organism's reproductive potential in a discontinuous environment and furthermore it assures gene flow as opposed to clonal stagnation. It is interesting that among hermaphrodites gene flow is assured by periodic production of normal males (*Caenorhabditis*) or by cross-fertilization of normal females by hermaphroditic males (*Heterogonema*). Other reports of hermaphroditism among nematodes need to be investigated; the presence of sperm in the spermatheca and the rarity of males does not indicate hermaphroditism. It may only indicate that males are seasonal and sperm can be stored for prolonged periods.

III. EMBRYOGENESIS

Studies of nemic embryology and cell lineage began more than 100 years ago; the most notable effort was that of Boveri in 1892. His study of *Parascaris equorum* is the foundation of modern nematode embryology. Following Boveri were several rather detailed studies and among these the most memorable are the studies of Martini (1903, 1909) and Pai (1927, 1928). Interest in comprehensive studies of embryonic cell lineage waned over the next 50 years and a revitalization did not occur until scientific inquisitiveness was stimulated by the manipulability of *Caenorhabditis elegans* and the availability of modern techniques and equipment. In 1983 Sulston et al. published the embryonic cell lineage of *Caenorhabditis elegans* from zygote to newly hatched larva.

Throughout the years there have been differences of opinion as to what particular cells give rise to in the completed larva. These differences were conceptual rather than based on actual conditions in the given nematode species. This should not be taken as a criticism of early workers. It merely points out that misinterpretations are to be expected when one must follow the development of hundreds of cells.

Caenorhabditis elegans has proven to be an ideal animal for these and other studies because of certain inherent features: it has a short life cycle, it can be easily cultured, it is small enough that large numbers are not a hindrance to laboratory procedures, and it has relatively few cells in the juvenile and mature stages. Embryogenesis takes about 14 hrs. at 22°C (Wood, 1988), as opposed to 14 days for *Plectus parietinus* (Maggenti, 1961).

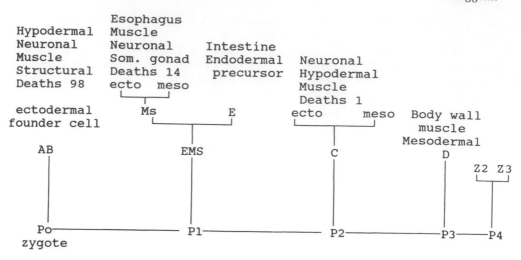

FIGURE 2   Lineage pattern of cleavages in the *Caenorhabditis elegans* embryo. (Modified from Wood, 1988.)

The following summarization of the embryonic development of a *C. elegans* hermaphrodite is based on Wood (1988) and Fig. 2 is a generalized outline of blastomere lineage.

The first cleavage of the ovum (P0) results in two cells of unequal size. The larger anterior ectodermal founder cell is designated AB (S1 in traditional nemic embryology); the smaller posterior cell (P1)is the posterior germline cell whose potential is to form the remainder of the embryo. From this point on the cleavages are nonsynchronous. The derivative lineages of cell AB are 254 neuronal cells, 72 hypodermal cells, 23 muscle cells, and 40 other structural cells. In the production of the forgoing there are 98 cell deaths and 389 survivors.

The second cleavage (about 10 min. after the first) results in the second somatic daughter cell designated EMS (ethylmethanesulfonate) and P2. The EMS cell divides unequally and produces the MS founder cell and the slightly small E founder cell of the mesenteron (20 cells). The MS cell produces 48 muscle cells, 13 neuronal cells, 9 gland cells, 2 somatic gonad cells, and 8 other cells; there are 14 cell deaths and 80 survivors.

Shortly after the formation of EMS, P2 divides to produce somatic founder cell C and germline cell P3. C produces 32 muscle cells, 13 hypodermal cells, and 2 neuronal cells; there is 1 death and 47 survivors. P3 produces founder cell D that gives rise to 20 muscle cells and P4 that produces the two germinal cells Z2 and Z3.

At eclosion the hermaphroditic larva consists of 558 cells. During embryogenesis 671 cells were formed and of these, 113 underwent programmed death. These cell deaths are specific, autonomous, and invariant from one embryo to the next.

As of now there is no one satisfactory explanation for programmed cell death. The responsible factors could be intrinsic or extrinsic, i.e., cell death could be suicide or murder. The advocates of murder point to the phagocytic activity of surrounding cells. However, phagocytosis can be blocked by mutation and cell death still occurs. There is some evidence that cell death is sex-linked because certain cells die in males that do not die in females and vice versa. Another view is that it is because cell lineage is dichotymous that unnecessary cells or excess cells are produced. Sulston et al. (1983) put forth the idea that cell–cell interactions that were originally essential for developmental decisions have been replaced by autonomous programs that are fast and reliable. In other words, the loss of flexibility was outweighed by efficiency. With this in mind, cell death is a feature that could seemingly be eliminated by a more efficient design—cell death represents a developmental fossil. There may be some circumstantial evidence to support the latter view. Most cell deaths are recorded in the development of the nervous system, and from ancestral to derived forms we see a reduction in the system i.e., the noticeable reduction occurs with tactile sensory hairs from marine to terrestrial nematodes. There is also a simplification of male supplementary organs and a reduction in somatic glands both of which are often associated with simple sensory cells. This circumstantial observation has some support in that most deaths occur in the AB founder line and in the MS founder line, the greatest contributors to the nervous system, somatic musculature, and glands (Fig. 2).

Postembryonically nematodes undergo four molts and pass through five stages, the fifth stage being the adult. Among Adenophorea it is typically the first-stage larva that emerges from the egg. Among Secernentea it is often the second-stage larva that emerges, there being many exceptions among free-living and parasitic Rhabditia; an exception is noted in *Caenorhabditis*, which emerges as a first-stage larva.

IV. INTEGUMENT

The exoskeleton of the nematode consists of the cuticle and its underlying producer, the hypodermis. Together they form a complex organ that protects the animal from detrimental external conditions and functions, in part to maintain the delicate chemical balance internally.

The advent of the electron microscope revived interest in nematode cuticle because it was more easily fixed and less subject to distortion than other tissues of the body. However, it was soon evident that the universal model of nematode cuticle, *Ascaris,* was misleading. As attempts were made to incorporate the two prevalent nomenclatures to the strata of the cuticle, it became apparent that the most studied feature of nematodes was also the least understood. A basic problem is the attempt to apply the same nomenclature to cuticular layers, ignoring the variation occurring throughout Nemata. The number and nature of the layering in *Enoplus* and *Pratylenchus* are not the same; yet attempts are made to apply the same nomenclature and to seek the same number of layers.

The simplistic nomenclature applied to ascarid cuticle simply labels the layers: cortical, matrix (median), and basal. The most comprehensive divides the cuticle into nine layers (Chitwood and Chitwood, 1974): (1) and external cortical layer; (2) an internal cortical layer; (3) a fibrillar layer; (4) a matrix layer (homogeneous layer); (5) a boundary layer; (6–8) external middle and internal fiber layers; and (9) a basal lamella.

Shepherd (1972) proposed a separate nomenclature for Heteroderidae and this scheme was followed by Baldwin (1983). In the Shepherd system the layering is labeled A–D. What needs to be recognized is that this layering has no relationship with the cuticle of *Ascaris* other than the fact that the entire cuticle of Heteroderidae, indeed all Tylenchina, is cortical. There are no fibrillar, matrix, or fiber layers in Tylenchina.

After reviewing the available literature and studying electron micrographs of the cuticle, Maggenti (1979) proposed nomenclature that avoided the traditional designations that had become confused in application. The components of the system were designed after the system applied to other invertebrates. Only those components (strata) that are present are used in a discussion of cuticle. One need not try to apply all nine layers or four strata to every nematode and the system is applicable throughout Nemata. The ancestral cuticle as seen in *Deontostoma* (Fig. 3) is complete in its strata: epicuticle, exocuticle, mesocuticle, and endocuticle. The crossed-fiber mesocuticlular layer is sometimes replaced by structural struts. Note also that *Ascaris* has no endocuticle and that Tylenchina has neither meso- nor endocuticle. However, there is evidence of sublayering in the tylench exocuticle and this is where the Shepherd–Baldwin system should be employed.

There are few studies on the cuticle that lines the esophagus, vagina, cloacal pouch, and other minor intrusions into the body such as glands or sensory structures. In gross features it appears similar to epi- and exocuticle; however, differential staining shows that there are different isoelectric points between external and internal cuticle. Through differential staining it can also be demonstrated that the cuticle of the cheilostome is more closely related to that of the external body than the esophageal lining (including the esophastome). Other evidence of the differences in external and internal cuticle can be inferred from the virus vectors where the receptor sites vary internally and, as far as is known, are never external.

A. Cuticular Structures

Cuticular structures can be divided into two categories: ornamental and sensory. Those that are classified as ornamentation are not to be interpreted as nonfunctional. Many of these are important taxonomically because they are conservative and biologically because they contribute to the animals' lifestyle, e.g., anterior retrorse spines on animal parasites to give purchase (Fig. 4). Sensory structures are also important taxonomically and they offer a great deal of information about a given taxa's evolutionary state.

B. Ornamentation

There are numerous forms of external cuticular ornamentation and a few that are internal. A common ornamentation seen among Chromadoria and Secernentea is somatic transverse striae; the interstices between striae are called annuli (s. annulus). (The use of annule and annules is incorrect and should be avoided; the word annule does not exist in the English language and is not used in zoology outside of nematology.) In addition to transverse striae there may be longitudinal striae or ridges. These longitudinal ornamentations may be present along with the transverse striae in which case the body facia looks like a corn cob. Striae, either longitudinal or transverse, are rarely seen among Enoplia; however, electron micrographs show that sub-light microscopic striae do occur.

Laterally there may be special longitudinal body striae called the lateral field; these are found among Chromadoria and Secernentea. The number and nature of these striae are commonly used in generic and species characterizations. These lateral longitudinal striae may be broken, diagonal, or joined asymmetrically by other striae (a condition known as areolation). Lateral fields often take on dramatic forms such as extended alae that may be as wide as the body diameter. Lateral alae occur as special structures cervically and caudally. The latter are found only on males.

For the most part male caudal alae are restricted to Secernentea; however, they do occur rarely in Enoplia. Among the animal parasites, especially those in Rhabditia, the caudal ala and its supporting musculosensory rays have great taxonomic significance. In these groups and especially among the strongyles the caudal ala because of the sensory rays and musculature is called bursa copulatrix (Fig. 5B,C). This term should not be used to describe caudal alae outside these specialized groups of nematodes. Though commonly associated with the tylenchid plant parasites, the significance of the form and shape of the caudal alae are only occasionally important in taxonomy. When the caudal alae are restricted to the two

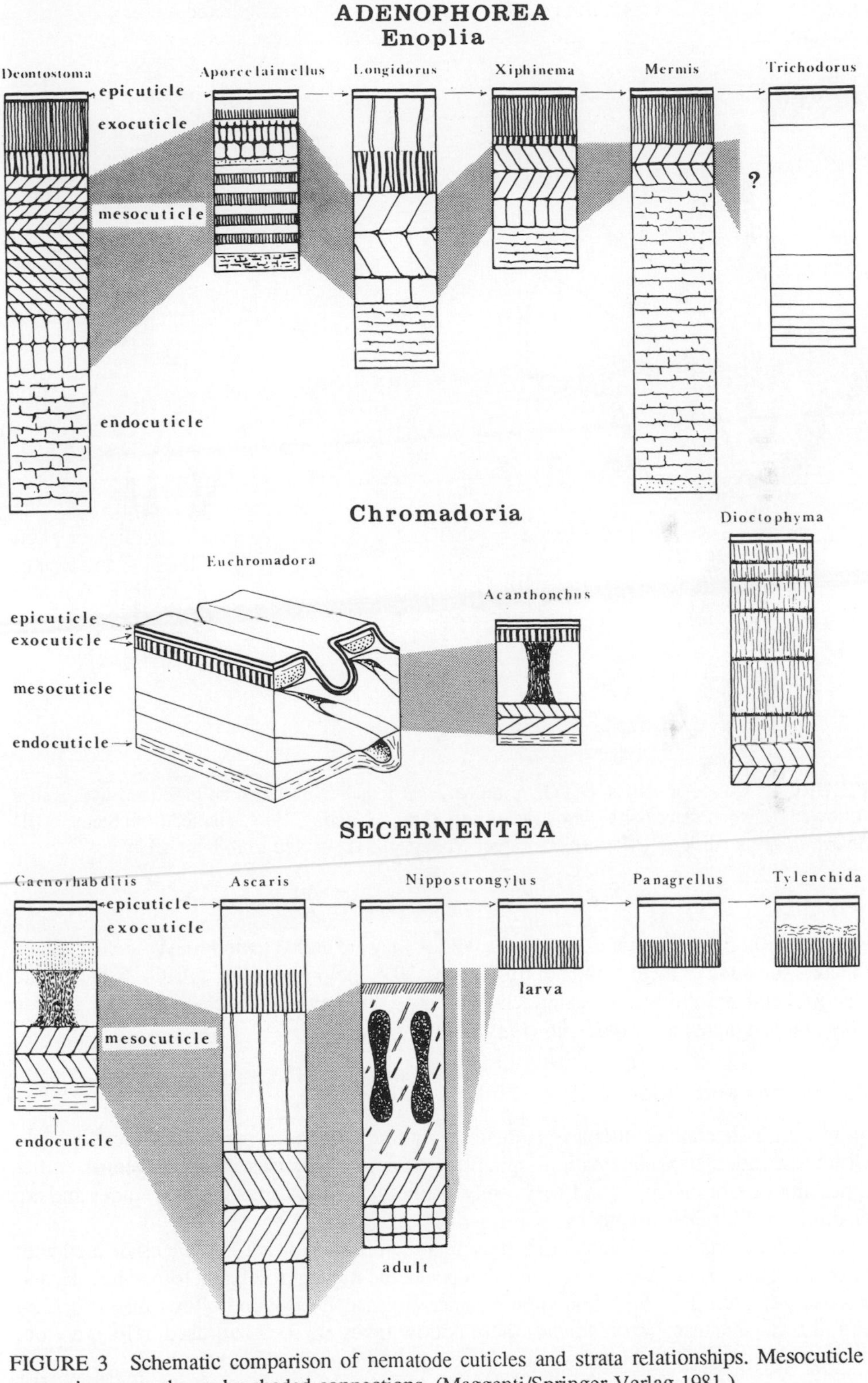

FIGURE 3   Schematic comparison of nematode cuticles and strata relationships. Mesocuticle comparisons are shown by shaded connections. (Maggenti/Springer-Verlag,1981.)

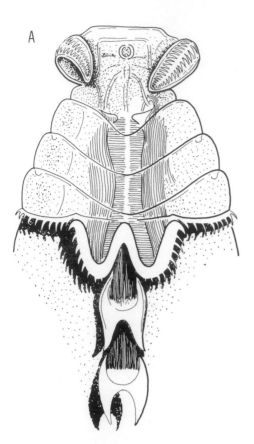

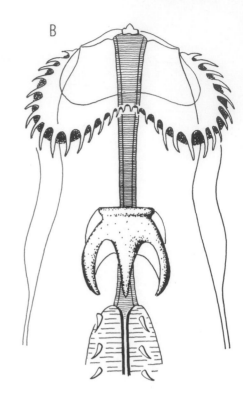

FIGURE 4   Cuticular spines. (A) Oxyuroidea: *Heth* female with lobed collar edged with spines followed by two massive body spines. (Maggenti/Springer-Verlag, 1981; adapted from Steiner.) (B) *Seuratia*, spiny cordons. (Maggenti/Springer-Verlag, 1981; adapted from Seurat.)

sides of the body and do not extend to the tail tip, they are called leptoderan (Fig. 5E). When the alae surround or meet at the tail tip they are called peloderan (Fig. 5A–C). If the caudal alae join both anteriorly and posteriorly (forming a bowl around the cloacal opening), then they are designated as arakoderan (Fig. 5D,F).

C. Spines and Setae

Spines are noncellular cuticular protrusions without muscle or nervous connections. As with most cuticular ornamentations spines are commonly seen among Secernentea, sometimes among Chromadoria, and very rarely in Enoplia. Scales are similar to spines and are distinguished by their bluntness.

Setae and their derivatives are sensory structures. These sensilla are associated with three cells: the tormogen cell that forms the socket, the trichogen cell that forms the seta, and a sensory neuron (Fig. 6A). Sometimes glands are associated with hollow setae (Fig. 6B) and in some instances (*Draconema*) these hollow tubes are long and used in locomotion; these are then called ambulatory setae (Allen and Noffsinger, 1978).

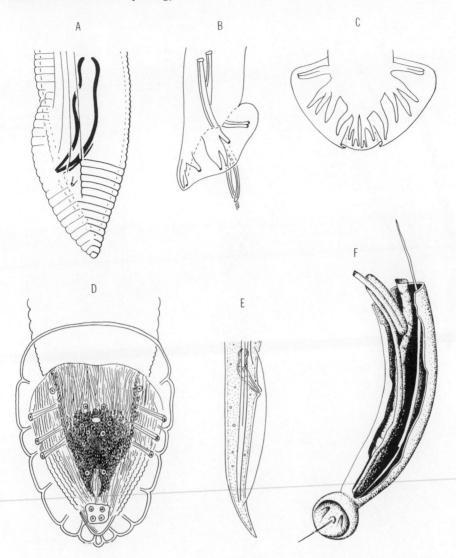

FIGURE 5   Types of caudal alae; Secernentea. (A) Peloderan, Tylenchida: *Rotylenchus*. (Maggenti/Springer-Verlag, 1981; adapted from Thorne.) (B) Peloderan, lateral view with well developed (bursa/caudal alae) bursal rays; Strongylida: *Chabertia*. (Maggenti/Springer-Verlag, 1981; adapted from Yorke and Maplestone.) (C) Peloderan, ventral view with bursal rays; Strongylida: *Delafondia*. (Maggenti/Springer-Verlag, 1981; adapted from Loos.) (D) Arakoderan, ventral view; Strongylida: *Physaloptera turgida*. (Maggenti/Springer-Verlag, 1981; adapted from Chitwood.) (E) Leptoderan, lateral view; Tylenchida: *Ditylenchus*. (Maggenti/Springer-Verlag, 1981; adapted from Thorne.) (F) Arakoderan (bowlshaped); Ascaridida: Dioctophymatoidea. (Maggenti/Springer-Verlag, 1981; adapted from Goeze.)

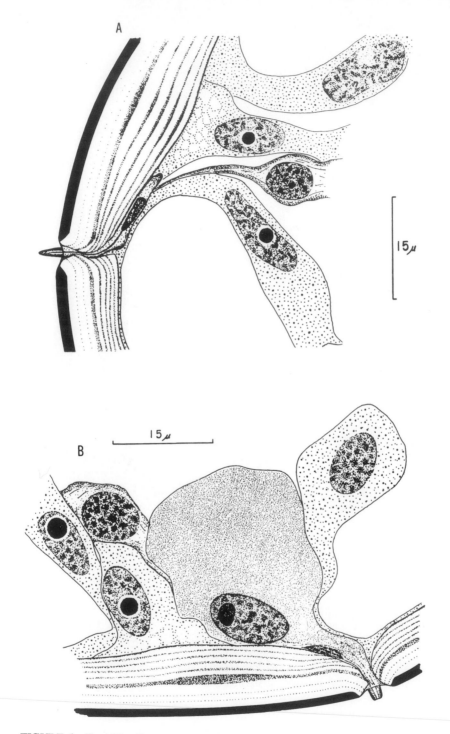

FIGURE 6   Enoplida: *Deontostoma californicum*. (A) Transverse section through sublateral somatic seta showing relationship of hypodermal cells and sensory neurons. (B) Tranverse section
through a sublateral hypodermal gland. (Maggenti/Springer-Verlag, 1981.)

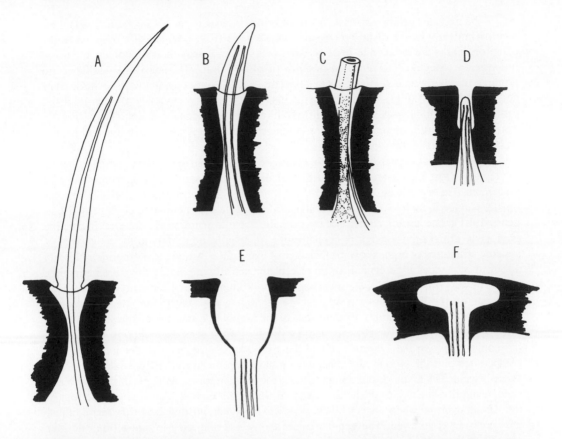

FIGURE 7 Some forms of sensilla found among Nemata. (A) Sensillum trichodeum. (B) Sensillum basiconicum. (C) Sensillum basiconicum. (D) Sensillum coeloconicum. (E) Sensillum ampullastomum. (F) Sensillum insiticii. (Maggenti/Springer-Verlag, 1981.)

Other cuticular ornamentations are cephalic helmets; these are characteristic and prominent in some chromadorids. Similar but only occasionally very prominent is the double cephalic cuticle of enoplids. However, in some groups, such as *Deontostoma*, this cephalic capsule and its fenestrae are highly significant in leptosomatid taxonomy. Tail shape is often an expression of cuticular ornamentation, e.g., long and filamentous, adorned with digits, mucros, etc. Preanal suckers are another feature classed under ornamentation; they are known only among animal parasites and their function is not understood. Suckers are only verified among the Seuratoidea. In this group the sucker is a combination of muscle and glandular tissue.

D. Sensory Structures

There are a variety of morphological manifestations of the sensilla of nematodes (Fig. 7). For the most part the sensilla of nematodes are primary sense cells (Fig. 6A). A primary sense cell receives stimuli either directly or through a distal process. A secondary sense cell is an ectodermal cell that receives stimuli indirectly by way of distal branches of a sensory neuron.

Nematodes exhibit six basic forms of primary sense organs (Maggenti, 1981): sensilla trichodea, with the distal processes elongate setae (Fig. 7A); sensilla basiconica, the distal processes are reduced to pegs or cones (Figs. 7B,C); sensilla coeloconica, the distal processes are peglike and recessed in pores (papillae) (Fig. 7D); sensilla ampullacea are large, or relatively so, deep flask-shaped pouches, such as amphids, that lead to the sensory cell endings (cilia) (Figs. 7D, 8B); sensilla insitica, there is no evidence of external distal processes, the ciliary processes or modified cilia are embedded in the cuticle (Figs. 7F, 8A); sensilla colei, similar to sensilla insitica but the ciliary endings are beneath the cuticle, not embedded in it (Fig. 8B).

These sensory structures of nematodes are important to our understanding of nemic phylogeny. Most notable are the so-called cephalic sensilla that differ by form, number, and placement. There are 16 cephalic sensilla that follow the circlet formula of 6–6–4. The combination and reduction of these sensilla signals derived states in nematode phylogeny (Maggenti, 1981). The use of contemporary cephalic sensilla patterns for interpreting nemic phylogeny utilizes the concept of cephalization. (Cephalization refers to the movement of somatic structures or organs toward the anterior extremity.) Among contemporary Nemata a logical sequence of the cephalization of cephalic sensilla, including the paired amphids, can be developed. Currently, the most ancestral condition of three separate whorls of sensilla, one on the lips and two postlabial, exists among the Oxystominoidea (superorder Marenoplica; order Enoplida). In some taxa of the family Oxystominidae the sensilla are in three separate whorls, are setiform, and are of two forms: sensilla trichodea and sensilla basiconica. Among the Leptosomatidae the first whorl is reduced to six labial sensilla coeloconica and the second and third whorls have combined into 10 postlabial setiform sense organs. The amphids are coincidentally moving anteriorly as well. In the combined state, generally, the organs of the first external whorl are sensilla trichodea and the four of the second external whorl are sensilla basiconica. However, in some taxa (Prismatolaimus) the condition is reversed. The genus *Plectus* shows a further step in that all but the four sensilla of the second external whorl (sensilla basiconica) are on the lip region as sensilla coeloconica. Among Adenophorea the final step is for all sensilla to be on the lips as sensilla coelonconica; however, the amphids always remain postlabial. In Secernentea the sensilla are labial and sensilla coeloconica; the amphids are, with rare exceptions (some larval diplogasterids), also labial. Cephalization among Secernentea is moving toward reduction in sensory organs on the labia to either sensilla insitica or sensilla colei. Among animal parasitic nematodes, especially in Spiruria, it is not uncommon to find only the four sensilla of the second external whorl and the amphids visible. Plant parasitic nematodes vary from all present to 10 or six or only four visible; the amphids remain labial.

E. Amphids

The paired amphids are presumed to be chemoreceptors. Their function has never been verified; nor is it known if they have the same function in all taxa.

Amphids are sensilla ampullacea (Figs. 7E, 8B). The aperture regardless of its external manifestation is followed by a pouch or in some instances a tube in which the ciliary sensors lie. These ciliary sensors are contiguous with the dendritic ending of the amphidial nerve whose neurocytes are clustered posterior to the nerve ring in the laterally located amphidial ganglia. This forms the basic morphology of all amphids.

Phylogenies have been proposed on the basis of amphid aperture morphology (Schuurmans Stekhoven and DeConinck, 1933). These have proven less than satisfactory because the morphological countenance of the external aperture reveals no logical sequential development. Among marine enoplids the amphids are often large and vesiculate but in

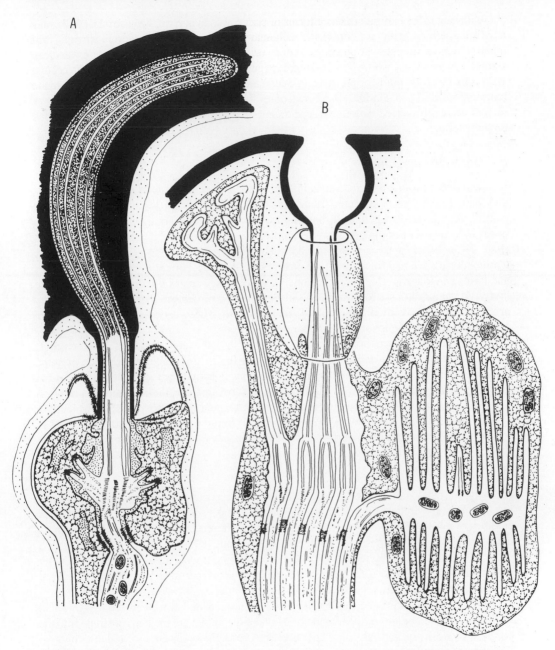

FIGURE 8   Cephalic sensilla of Tylenchida. (A) Sensillum insiticii. (Maggenti/Springer-Verlag, 1981; adapted from DeGrisse.) (B) Sensillum ampullastomum (amphid) with an associated sheath sensillum: sensillum colei. (Maggenti/Springer-Verlag, 1981; adapted from Coomans.)

many instances they are seen as small, hardly discernible oval slits. Terrestrial Enoplia may have slit apertures with large internal pouches so that they appear as inverted stirrups. Still other taxa have simple oval openings with small pouches. The greatest variations are seen among Chromadoria where the apertures range from large and disklike circles through unispirals to multispiral forms. Among Secernentea they vary from porelike, to elongate longitudinal clefts, to relatively large oval pores. It would seem that some extrinsic factor or factors have influenced aperture shape rather than an intrinsic factor that is reflected phylogenetically.

F. Somatic Sensilla

In Adenophorea somatic sensilla are common. This is especially true of marine Enoplia that they have sensory setae scattered along the body sublaterally. In some instances they are combined with glands that open through hollow setae (Fig. 6B). Though somatic sensilla do occur in terrestrial Enoplia, they are generally most obvious on the tail region of females (Fig. 14C). When present along the body they occur most often as papillae or pores. Chromadoria are noted for possessing somatic sensilla, most commonly on the tail of males and females. However, there are some notable exceptions such as *Epsilonema* whose entire body is covered with sensory hairs. It is also among Chromadoria that deirids occur (they are not known from any other group of Adenophorea). These cervical sensory organs are generally found at the level of the nerve ring.

Secernentea are notable for their paucity of somatic sensory organs. The two most commonly observed are deirids and phasmids. Phasmids are similar to amphids or some deirids but are generally confined to the tail region. Phasmids are generally porelike but they may be large and disklike. In addition to their form and size, their location is also variable; in some taxa they are found anterior to the vulva and at different positions on either side of the body. Unusually large phasmids are also found on some vertebrate parasites in Rhabditia and on some earthworm parasites in Spiruiria (Drilonematoidea).

G. Hypodermis

Most hypodermal studies have dealt with Secernentea. As a result a syncytial hypodermis has been misinterpreted as the norm among Nemata. This is not true, as there are no verified findings of multinucleate cells in any taxon of Adenophorea. It may occur to some extent in mermithids but this has not been confirmed (Chitwood, 1974).

In Adenophorea increased body size is accommodated by an increase in the number of individual hypodermal cells. Secernentea accommodate increased body size by syncytial growth of the hypodermis, i.e., nuclear increase without an increase in cell number.

The hypodermis underlies and secretes the cuticle whether it be external or internal. In general, the cell bodies of the hypodermis (where the nuclei are located) protrude into the body cavity dorsally, ventrally, and laterally. The protruding cells are referred to as chords. This nomenclature (and spelling) comes from early morphological studies in which the chords marked the segment of the secant between the muscle intersections with the body curvature. Dorsally the single hypodermal chord normally extends posteriorly only for the length of the esophagus. The ventral chords, in cross-section, may consist of one or two rows of cells. The lateral chords are generally the most prominent and generally they consist of three rows of cells.

From the dorsal, dorsolateral, ventrolateral, and ventral hypodermal cell bodies thin sheets of tissue extend between the cuticle and somatic musculature. The cell bodies that are directly lateral (between the lateral subdorsal/subventral cell bodies) have no sheet like extension and are called seam cells. The number of cells that make up an individual hypodermal chord longitudinally varies according to the size and species of nematode.

What has here been described as a generalized hypodermis is typical of Adenophorea except for parasitic forms. Among the animal parasitic forms the trend is for an increase in the number of chords, especially in the anterior body. These variations are most commonly seen among mermithids. One variant is to have uninucleate chords as is normal but in addition there are four submedian chords, also uninucleate. In another example there are three rows in each lateral chord, three rows in the dorsal chord, two rows in the ventral chord with two additional chords subventrally; once again all chordal cells are uninucleate. Some chromadorids do show pseudochords in the submedian sectors. This results from the meeting of the subcuticular hypodermal cellular sheets that are then pushed or protruded into the body cavity. Another modification seen among the parasitic adenophores is the bacillary band, which in reality is no more than an "excess" of hypodermal glands mixed with the regular hypodermal cells (Chitwood and Chitwood, 1974). Bands are found on either side or both sides of the body in some Stichosomida.

Among Secernentea multinucleate (syncytial) hypodermal cells are very common. How the syncythium (multinucleate condition) arises has become a matter of disagreement. The majority of observations report that the multinucleate condition results from coenocytic processes, i.e., nuclear division without cell membrane formation. This pheonomenon is commonly seen among animals and occurs in the production of cartilage and connective tissue. The opposing view holds that the multinucleate condition is formed by the amalgamation of separate cells by cell membrane dissolution. This is reportedly the mechanism that occurs in *Caenorhabditis* (Sulston and Horvitz, 1977). Amalgamated cells are not common among plants or animals; among plants certain latex cells are of this type of syncytial development; and some pathological animal tissues are formed in this manner. The most commonly observed syncytial development results from nuclear division without cell membrane formation.

It is disquieting that in the literature on *Caenorhabditis* cell lineage, the initial reports are of nuclei destined to become a part of the lateral syncytia. At this stage there is no reference to cells or cell membranes. It is only after the multinucleate condition is seen that there is reference to formation by cell membrane fusion (Wood, 1988). No mechanistic explanation is given that explains directional cell membrane fusion with immediate cell membrane dissolution! Also there is no explanation given for the fact that the number of nuclei in the multinucleate cell are more reflective of a geometric progression than an arithmetic one: 110 nuclei are reported. Since all divisions are not synchronous, this is still on the order of the expected nuclei number as produced by synchronous geometric progression, i.e., 128. If multinucleate cells are derived from the amalgamation of uninucleate cells, why are the nuclei smaller than those found in nonamalgamated cells? This is not to say dogmatically that the "syncytium" is not formed by cell fusion but that uniqueness among any animal group deserves a very critical viewing. In either event multinucleate hypodermal cells are common among Secernentea. This is one reason put forth for the placement of dioctophymatids in Secernentea: all chords and submedial chords are multinucleate.

H. Molting

Postembryonically nematodes undergo four molts prior to achieving adulthood. The adults do not molt unless one considers *Deladenus siricidicola* as an exception. However, it is not a true molt but a very specialized adaptation to their parasitism of woodwasps. The adult female sheds the final cuticle and the exposed hypodermis develops microvilli whose function is direct absorption of nutrients from the host hemocele (Riding, 1970). The lack of cuticle also allows the females to undergo inordinate growth.

All known or verified observations of Adenophorea indicate that the first-stage larva emerges from the egg. Secernenteans, on the other hand, may emerge as either a first- or second-stage larva from the egg. In Diplogasteria the emergence of the second-stage larva at hatching is common and apparently it is universal in Tylenchida.

Though frequently observed, molting is among the most poorly understood phenomena in Nemata. Two differing processes of molting are reported for nematodes (Bird and Rogers, 1965; Lee, 1970). In one form of molting the entire cuticle with all layers intact is shed; in the other process only the epicuticle is shed while other "layers" are dissolved. In addition to the above we also know that as the molting process begins there are significant changes in the hypodermis. Increased cellular activity is manifested by a thickening of the hypodermis underlying the cuticle with a concomitant increase in mitochondria and enlargement of the hypodermal cell nuclei and nucleoli.

Prior to the time of the loosening or dissolution of specific cuticular layers, the new epicuticle is layed down completely. This could mean that the molting fluid does not activate until the underlying new cuticle is protected by the epicuticle.

V. SOMATIC MUSCULATURE

The longitudinal obliquely striated somatic muscles are located peripherally and attached to the hypodermis as a single layer of spindle-shaped cells. Most often the muscle cells are separated into four muscle fields by the hypodermal chords: two dorso-and two ventrosubmedian. When there are submedian protrusions of the hypodermis into the body cavity there may be six or eight fields.

Unusual, though not unique to nematodes, is the phenomenon of the muscle sending an elongate "innervation" process from the noncontractile portion of the cell to the central nervous system rather than having nerves extend from the central nervous system to the muscle cell (Fig. 9A). This unusual situation has also been recognized among Echinodermata and Cephalochordata. These extensions are seen throughout the body of the nematode from the nerve ring posteriorly. Anteriorly, because there are no longitudinal nerve cords, the muscle innervation processes enter directly into the nerve ring.

Transmission electron microscopy is antiquating many terms that in the past were applied to the somatic musculature. Some of the terms remain useful as points of reference but they have lost value in the development of phylogenies. Among these are the descriptive morphological terms for the position of the contractile elements in muscle cells: platymyarian, coelomyarian, and, in some restricted instances, circomyarian. The configurations seen in the light microscope are actually formed by the intrusion of the sarcolemma, sarcoplasmic reticulum, or sarcoplasma among the contractile elements of the muscle fiber. *Xiphinema* (Dorylaimid) is described as having coelomyarian musculature, as are *Ascaris* (Ascaridida) and *Deontostoma* (Enoplida), yet each is different. The appearance of coelomyarian muscle "bundles" is created in *Xiphinema* because the contractile elements are separated by fingers of sarcoplasmic reticulum (Fig. 10). *Ascaris*, on the other hand, has discrete elements apically and basally. The same differences occur in so-called platymyarian muscles. In *Desmoscolex* (Desmoscolecida) the contractile fibers are separated by sarcoplasmic reticulum and sarcoplasma. Tylenchida are also described as having platymyarian musculature. However, the contractile elements are not clearly separated but have sarcoplasmic reticulum penetrating the contractile elements irregularly. Two other terms of limited phylogenetic value are meromyarian and polymyarian. These terms refer to

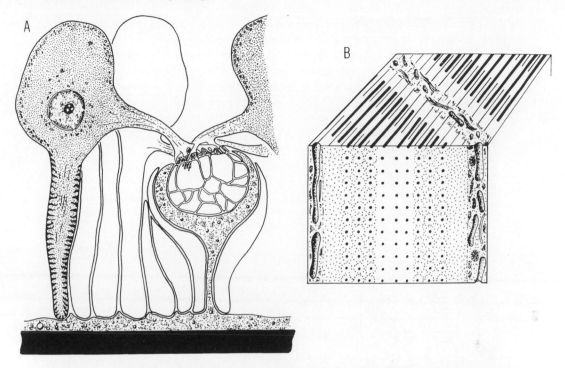

FIGURE 9   (A) Schematic transverse section of muscle cells and their armlike myoneural extensions; the ventral nerve cord is embedded in troughlike hypodermis. (B) Schematic of a muscle fiber showing striation patterns in two planes. The obliqueness is greatly exaggerated in this diagram; in reality, the angle of the striations is less than 6°. (Maggenti/Springer-Verlag, 1981; adapted from Rosenbluth.)

the number of muscle cells in a quadrant; mero- refers generally to less than six and poly- to six or more. These are arbitrary terms. Some authors have made attempts to utilize these conditions as markers for the ancestral and derived states (Chitwood and Chitwood, 1974). The premise was that the ancestral nema was an areolaim, which is meromyarian. However, the nemas now considered as ancestral representatives, oxystominids, are polymyarian-coelomyarian as adults. This proposal breaks down because all known nemas are platymyarian-meromyarian in at least the first juvenile stage. This was pointed out by Filipjev (1934) where he referred to this pheonomenon among derived forms as pedogenesis.

Throughout the body there are specialized somatic muscles that serve special functions or are associated with the secondary sexual organs. Among the better known are the somatointestinal, somatoesophageal, copulatory, bursal, spicular, gubernacular, and vulval muscles. These are generally converted somatic muscles that are recognized as such because the noncontractile portion containing the nucleus is located on the body wall.

The somatic muscles of nematodes are obliquely striated at about 6°. It should be noted that the obliqueness as illustrated by Rosenbluth (1967) is 60° in the figure for convenience of illustration (Fig. 9B). Oblique musculature is known to occur in other invertebrate groups and is typical of the longitudinal tentaclular muscles found in cephalopods. In squids the obliquely striated muscles are capable of a much greater range of lengthening and shortening than the cross-striated perpendicular muscles and they have a faster reaction time.   The versatility of movement attributed to tentacles, tongues, and elephant trunks is

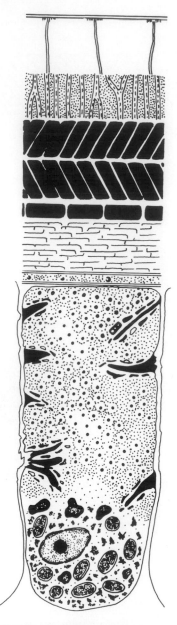

FIGURE 10   Portion of a transverse section through *Longidorus* showing the cuticle, underlying hypodermis, and a transverse section through a muscle cell showing fingerlike extensions of the sarcolemma. (Maggenti/Springer-Verlag, 1981; adapted from Aboul-eid.)

credited to the peripherally arranged longitudinal muscles, and internally perpendicular and transverse muscles that form a muscular hydrostatic skeleton (Smith and Kier, 1989).

Much emphasis has been placed on the role of the hydrostatic skeleton (fluid-filled pseudocoelom) in nematode movement since the introduction of the terminology into nematological literature by Harris and Crofton (1957). In nematodes it is true that a part of the skeleton support system is provided by the liquid-filled pseudocoelom as well as the enteron but such a hydrostatic skeleton alone does not explain the diversity of movement exhibited by nematodes. Perhaps the latter can only be understood when nematode hydrostatics are viewed as intermediate between hydrostat movement as exemplified by polyps along with other wormlike invertebrates with large, fluid-filled cavities/"coeloms," and muscular hydrostats as seen in the organs of cephalods and many mammals. In a muscular hydrostat (Kier and Smith, 1985) the musculature itself effects movement and provides skeletal support for that movement. Skeletal support is provided because the muscle is composed primarily of an incompressible liquid and is thus constant in volume. This mechanism offers advantages over the hydrostatic skeleton, which provides support through large, liquid-filled cavities and thus allows only unlocalized movement.

In muscular hydrostats where structures are capable of complex bending, there are peripherally arranged muscle cells parallel to the long axis (longitudinal muscles). These are arranged helically around the long axis by a slightly offset orientation or by a combination of obliquely striated muscle cells and staggered arrangement that creates a helix. (The somatic musculature of nematodes is peripherally located, longitudinally arranged, obliquely striated, and helically arranged by alternation of muscle cells and by not being oriented on the direct longitudinal axis.)

The most important biomechanical feature of a muscular hydrostat is its constant volume (incompressibility at physiological pressures). Without some means of resisting longitudinal compression, unilateral shortening will not produce bending. Nematodes lack antagonistic muscles and hence bring into play the hydrostatic skeleton and the cuticle to prevent shortening by maintaining a constant diameter. If a constant diameter is not maintained, then the muscular hydrostat organ will shorten but not bend (Smith and Kier, 1989). The helical arrangement allows, through directed contraction along the helix, the animal to twist. The muscle cell arrangement diagrams (Fig. 11) of Ohmori and Ohbayashi (1975) indicate that the nematode could contract a helical in either a left-handed or right-handed direction that would result in a left or right twist.

The foregoing is still speculative for nematodes but does offer an alternate explanation for the diverse movements of nematodes not allowed by a high-pressure hydrostatic skeleton that functions as an antagonist to the longitudinal somatic musculature.

VI. SYSTEMS OF EXCRETION

The cells of the excretory system in nematodes are hypodermal (ectoderm) in origin. The system consists of a sinus cell (= excretory cell, rennete, ventral gland), a pore or socket cell (tormogen); and, depending on the class, collecting tubules may be present. However, within Nemata there are alternate systems of excretion that may or may not involve tissue of hypodermal origin. Therefore, an excretory system is not universal among nematodes but a system of excretion is.

In an orthodox discrete excretory system there are never less than two cells contributing to the form of the organ: the excretory cell and the pore-forming cell. This is the basic system as seen among those Adenophorea that exhibit an excretory system. The system is not universally present among Adenophorea but appears to be constantly present in Secer-

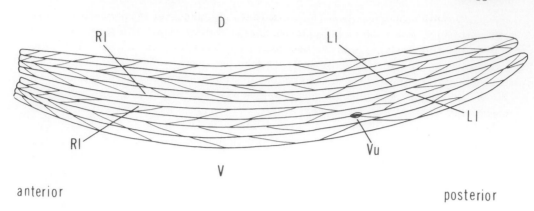

FIGURE 11   Diagram of the muscle cell arrangement in *Decrusia additicta*. The nematode is split through the lateral line. The lower sector represents the ventral portion of the body and the upper section the dorsal sector of the body. Anterior is to the left, posterior to the right. Abbreviations: V, ventromedial line; D, dorsolateral line; Ll, left lateral line; Rl, right lateral line; Vu, vulva. (Maggenti/Springer-Verlag, 1981; adapted from Ohmori and Ohbayashi.)

nentrea where the organ is more complex and may consist of as many as six cells: the pore cell, the duct cell, the excretory cell, tubule cell(s), and fused coelomocytes in the preadult and adult.

Collecting tubules and elongate cuticularized ducts are unknown among Enoplia and in this subclass more taxa possess an excretory system in Marenoplica than among Terrenoplica. In Chromadoria some taxa possess an elongate cuticularized duct (*Plectus*) and some exhibit pseudo-collecting tubules (nonhollow) as are found in *Anonchus*.

It is axiomatic that when an excretory system is absent, as it is in many taxa of Adenophorea, the vital function of excretion must be assumed by some other organ or tissue. There are several candidates among the Adenophorea: hypodermal glands, caudal glands, tubular gland setae, coelomocytes, and the prerectum. It may be coincidental but as the excretory system becomes smaller relative to body size the number of hypodermal glands increases. Numerous hypodermal cells would seem more efficient that a single anteriorly restricted excretory cell.

As previously mentioned, the system in Secernentea appears to be universal but its form is variable. The basic system, possibly the ancestral secernentean system, is found in Rhabditia and is referred to as the "H" system, i.e., consisting of two anteriorly directed and two posteriorly directed collection tubules. All the systems seen in Secernentea are modifications of this condition. All taxa have one or two posteriorly directed collection tubules and these then may or may not be associated with anterior tubules.

In Rhabditia an unusual and unexplained condition is seen; it was first described in depth by Waddell (1968) in the kidney worm *Stephanurus dentatus*. Juveniles, stages 1–3, have two separate subventral coelomocytes that are located near the excretory cell. In the preadult and adult these coelomcytes join the excretory system and are the basis for the observation that the system in rhabdits has three cells. The function of these cells in the early juveniles is unknown; perhaps they have some juvenile hormone role that ceases to function in the pre- and adult stage. Cellular changes in function are known to occur with age in Arthropoda.

Taxa in the order Tylenchida most often have a single collecting tubule that has both an anteriorly and posteriorly directed branch. This tubule may be located on the right or left side of the body. Among the taxa in Tylenchulinae that produce a gelatinous matrix, it is the excretory cell that is the source of the gelatin (Maggenti, 1962). It is not unusual for the excretory cell in these nematodes to occupy one-half or more of the body volume.

VII. ALIMENTARY CANAL

The alimentary canal or enteron of a nematode is divisible into three major sections: the stomodeum, mesenteron, and proctodeum. The stomodeum and the proctodeum are lined with cuticle which is absorbed and/or shed at each molt.

The stomodeum can be subdivided into the stoma, esophagus, and esophagointestinal valve. The lips, even though they are not strictly a part of the stomodeum, will be discussed here. "Lips" is somewhat of a misnomer inasmuch as they are seldom movable except in some marine taxa and animal parasites.

A. Lip Region

Among Adenophorea the labial region is most often distinctly or indistinctly hexaradiate and, as discussed earlier, the lip region may bear from one to all three whorls of cephalic sensilla but never the amphids. The lip region may be smoothly rounded or each sector may be prominent and pyramidal or conical. The shape appears to have more to do with the biology of the taxa than its phylogenetic placement.

Amphid apertures among Adenophorea are highly variable. They are laterally placed and they may be simple ovals, laterally elongate slits, huge dorsoventral ellipses that occupy much of the width of the cervical region, large circles, uni- or multiple spirals, or simple pores. Sometimes the amphid aperture is guarded by a tonguelike or flaplike cuticular accessory piece (Maggenti et al., 1983; Hope, 1988).

There is far more variability in the lip regions of Secernentrea; the basic plan is hexaradiate but in many taxa this is not discernible. Though the lips may not be discernible the cephalic sensilla are, and they do occupy that region designated as labial and in almost all cases the amphidial apertures are also components of the region. Ascarids have three lips, spirurids two, and Diplogasteria exhibit almost all available combinations from hexaradiate to little more than an oral plate.

The plant parasites in Tylenchida display a wide range of en face views. Rarely is a full complement of sensilla visible; most often the second whorl of six sensilla (surrounding the oral opening) and the outer whorl of four sensilla are visible. In many taxa only the four sensilla of the third whorl are seen. DeGrisse et al. (1974) demonstrated that the first and second whorls may subside into the anterior stoma.

Lips as such do not exist in Tylenchida. The lip region is often reduced to an unlobed labial plate or there may be two or six lobes with an undivided oral plate (*Meloidodera*). Terms that have been applied to the various structures associated with the labial region are lobes, pseudolips, liplets, etc. What needs to be done is a systematic morphological comparison of the "labial region" so that homologies and analogies can be determined.

The amphidial apertures are highly variable among the Tylenchida. In the family Tylenchidae the apertures may be elongated sinuous slits extending posterolaterally, or arc-shaped to rounded pits on the labial plate; rarely are they seen as oblique slits on the labial plate. In Anguinidae the apertures are elliptical and directed toward the oral opening. In the remainder of Tylenchina they are round, oval, elliptical, (the latter two are dorsoventrally

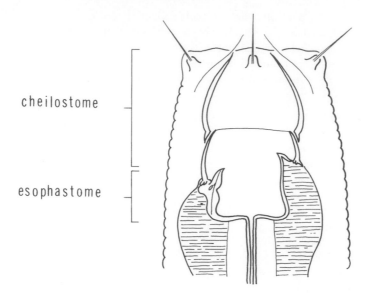

cheilostome

esophastome

FIGURE 12   Stomatal division in Nemata. (Maggenti/Springer-Verlag, 1981; *Butlerius* head adapted from Goodey.)

directed), and located laterally. Often the amphids are abutting the labial disk and in some instances they are on the disk. The obvious difference among these apertures is their relative dorsoventral length.

B.   Stoma

The nematode stoma, though highly variable, has limited phylogenetic value when overviewing the phylum. However, because the form, shape, and armature of the stoma is an indication of the animal's biological habits, its form is often valuable for the separation of lower taxa.

In ancestral nematodes the stoma is part of the primary invagination that forms the esophagus (= pharynx) and is called the esophastome. Generally, these stomas are described as collapsed or undeveloped. However, in some more derived nematodes the esophastome may be highly developed, bulbous, and armed or guarded by teeth (*Bullbodacnitis*; Seuratoidea; Spirurida).

In most nematodes the stoma is constructed of two parts that originate from the primary invagination (esophastome) and a secondary invagination (cheilstome) (Fig. 12). One or both of these invaginations form the stoma throughout Nemata and they do allow comparative homologies to be made. Among derived taxa the cheilostome is generally well developed and contributes to stomatal armature such as teeth, denticles, plates, and stylets (Fig. 13). The foundation of movable armature (though the armature is cheilostomal) is part of the esophastome and the operating musculature is most commonly esophageal; retractor muscles are generally converted somatic muscles. It is for this reason that all movable armature is dorsal or subventral.

The adjudication of which portion of the stoma is cheilostome or esophastome is determined by the anterior-most extent of esophageal tissue. The rhabdion designations of the stoma by Steiner (1933) are not useful in determining homologies because they are ap-

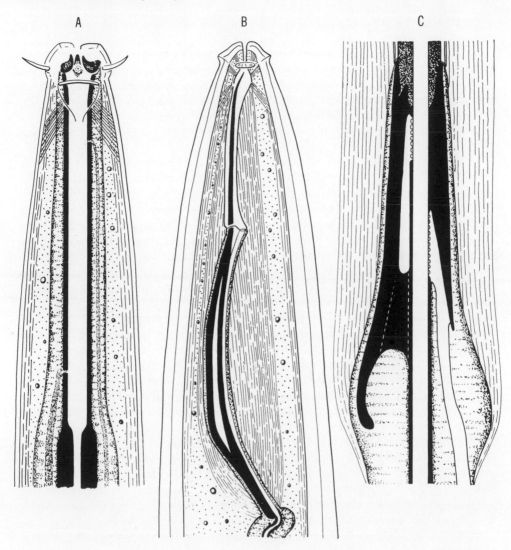

FIGURE 13 (A) Ironidae: *Ironus*; anterior movable teeth, cheilostomal; elongated cylindrical portion of stoma, esophastome. (Maggenti/Springer-Verlag, 1981, adapted from Van der Heiden.) (B) Diphtherophoroidea: *Trichodorus*; the anterior slender portion of the tooth (extending from where the esophageal lumen attaches to the cheilostome) is cheilostomal in origin; the basal portion of the tooth is esophastomal. (Maggenti/Springer-Verlag, 1981, adapted from Allen and Noffsinger.) (C) Esophastomal flanges (odontophore) of *Xiphinema* showing the internal sensory "gustatory" organs. (Maggenti/Springer-Verlag, 1981, adapted from Robertson.)

plied without regard for the rhabdion origin, i.e., whether they are of cheilostomal or esophastomal origin.

Stomatal structure alone cannot determine the feeding habits of a nematode. For example, the variety of stomas found among vertebrate parasites encompasses the array of types seen in predators and bacterial feeders whereas plant parasitic nematodes share stomal characteristics with the cellular (algae and fungi) feeders and predators whose commonality is a hollow (or derivative therefrom) axial spear capable of piercing cells of both lower and higher plants. Such armature occurs independently in both the subclass Enoplia and Diplogasteria.

In Adenophorea plant parasitic nematodes are known only from the order Dorylaimida in the suborders Dorylaimina and Diphtherophorina. The suborder Diptherophorina contains plant parasites that utilize a modified solid "tooth" derived from a hollow spear for feeding on plant roots (Fig. 13B). Features common to all Adenophorean plant parasites are: all are below ground root parasites, all are ectoparasitic, and they all are capable of transmitting specific plant viruses. The hollow axial spear of dorylaims is the basic model for those seen in the plant parasitic Longidoridae (Fig. 14A). The spear is derived from the two stomal sections: the anterior odontostyle is cheilostomal and the posterior odontophore is esophastomal. The specialized cell in the anterior esophagus is hypodermal and belongs to the somatic cheilostomal hypodermis, not to that of the esophagus. The cell is embedded in the anterior esophagus but its opening is at the junction of the cheilostome and esophastome. The ancestral cheilostome remains as the "guiding rings." In reality this structure is the membranous walls of the stoma that extend from the oral opening (may include heavily sclerotized structures as in *Actinolaimus* or *Carcharolaimus*) posteriorly to where it is attached at the base of the odontostyle at the ferrule junction. The illusion of rings is created by the folding of the membrane to accommodate the protrusion of the retractable stylet. Longidoridae differ from this model by great elongation of the odontostyle and odontophore. The latter in some taxa is ornamented by flanges (apodemes for muscle attachment). Within the odontophore of Longidoridae are three sinuses (Fig. 13C) that contain sensory organs that have been designated as gustatory organs (Robertson, 1976).

The "stylet" of the trichodorids is derived from a hollow axial spear that can be seen in the ancestral taxa of Diphtherophorina. Much of the original spear is atrophied back to the membranous cheilstome. However, the dorsal segment (originating from the dorsal odontophore) and the extended cheilostomal "mural tooth" remain to form the feeding apparatus. The ancestral membranous stoma projects posteriorly from the oral opening to the junction of the anterior "tooth" and the dorsal, mural odontophore. From this juncture the lumen of the esophagus is visible running down the ventral surface of the odontophore (Fig. 13B). As in all Dorylaimida the odontostyle (in this instance, the anterior mural tooth) is formed by a specialized cheilostomal hypodermal cell embedded in the anterior esophagus.

The hollow axial spear of Tylenchida does not differ significantly in architecture from that found in Dorylaimida. The tylench spear is thought to be derived from the diplogasterid fossoria. A much closer link is seen in *Neodiplogaster* and *Tylopharynx*; both have stomatal armature reminiscent of the tylench spear (Maggenti, 1963). The cheilostomal structure is somewhat more complex than that described in Longidoridae; in tylenchs it includes the cephalic framework, the stomatal cavity (vestibule), and the spear cone (Fig. 15). The cephalic framework is the anterior modification of the cheilostome that extends like an umbrella over the modified anterior esophagus. The framework has two supposed functions: it supports the labial region as an internal skeleton, and it provides apodemal structure to the protractor muscles of the spear. The so-called guiding apparatus is the ancestral stoma and it extends from the oral opening to the ferrule junction of the ante-

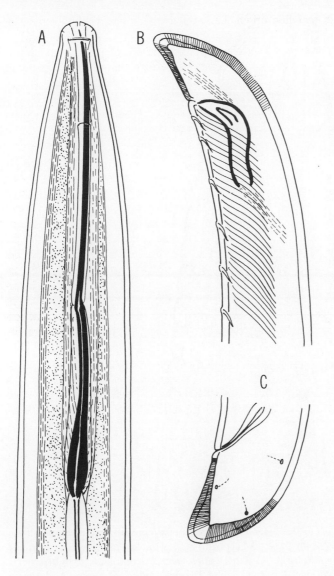

FIGURE 14   *Longidorus* spp. (A) Female anterior. (B) Male tail. (C) Female tail. (Maggenti/
Springer-Verlag, 1981.)

rior cone and the esophastomal shaft. The opening of the spear cone is subterminal and ventral. The esophastome consists of the spear shaft and accompanying apodemal knobs (one
dorsal and two subventral) (Fig. 15); these are not well developed or seen at the base of the
spear in all taxa. The protractor muscles are modified from the three anterior esophageal
muscles. The contractile portion of the muscles have their origin on the knobs and their
insertion on cephalic framework and body wall. The noncontractile cell bodies are located
in the anterior esophagus. There are no retractor muscles as described for dorylaims.

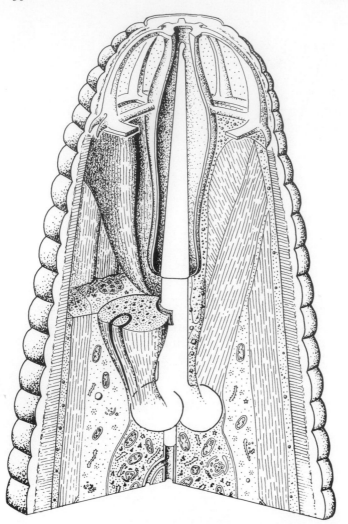

FIGURE 15   A three-dimensional reconstruction of the anterior extremity of Tylenchina illustrating the cheilostome as consisting of the basketlike cephalic framework and the tubelike extension attached to the base of the spear cone (also cheilostomal). These parts are contributed to by the external hypodermis seen lining the right side of the central tube. The shaft and knobs composing the base of the spear are esophageal in origin as are the protractor muscles that are conversions of the three anterior esophageal muscles. (Maggenti/Springer-Verlag, 1981; adapted from Baldwin and Hirschmann.)

C. Esophagus

The second region of the stomodeum is the esophagus or pharnyx. The latter term has had ambiguous use in nematology. Among helminthologists it is used correctly and refers to the esophastome. It also seems misleading to equate the esophagus of nematodes with the mastax (pharnyx, trophi) of rotifers that is designated posteriorly as an esophagus. There is no proof of any homology of the structures other than functionally, i.e., the esophagus transports food from the stoma to the intestine. [The original hypothesis of homology was based on a *Rhabditis/Plectus* ancestor as compared to rotifers. This view of nemic phylogeny was discarded a quarter of a century ago (Maggenti, 1963).] The pharnyx in other invertebrates is a recognizable division of the esophagus that links the stoma and esophagus proper. As such helminthologists are correct when they refer to the esophastome as the pharynx. Confusion was engendered by Hyman's (1951) unsubstantiated claim that the pharynx (mastax) of rotifers is homologous to the esophagus of nematodes and this interpretation was perpetuated by Roggen (1973). Terms that are used to support unfounded evolutionary relationships, such as "pharynx" sensu Hyman–Roggen, purporting homology with rotifers and other so-called pseudocoelomates are misleading and detrimental to the advancement of our knowledge of nematode evolution. Esophagus refers to the primary embryological invagination of the stomodeum and does not imply a single- or multiple-tissue organ. By definition pharynx is, when recognizable, a subdivision of an esophagus. In the absence of any proof of homology with rotifers the term as a substitute for esophagus should be avoided.

The esophagus is the most complex organ in the nematode body. This one organ has nerve, muscle, gland, and hypodermal constituents. Nemic esophagi are highly variable in shape and functional parts. Esophageal form is indicative of the animal's trophic behavior and an important structure in the reconstruction of nemic phylogeny.

In ancestral adenophorean taxa the constituent cells and their nuclei are evenly distributed throughout the esophagus; there is no congregation noted anteriorly or posteriorly (Chitwood and Chitwood, 1974). Externally the esophagus is either cylindrical or long and tapering to a broad posterior that houses the five glands, all of which open posterior to the nerve ring. Derived forms may have the same external shape; however, internally the cell nuclei are aggregated anteriorly and posteriorly. The area that is devoid of nuclei is, in derived taxa, called the isthmus. When the nuclei are so aggregated the anterior region is called the corpus and the posterior region the postcorpus. These two regions are, by rough estimation, marked by the nerve ring. Figure 16 compares these regions in Adenophorea and Secernentea.

Esophageal musculature has been primarily reported on the number of radial nuclei observed. As has been shown with the detailed studies of *Caenorhabditis* (Wood, 1988), nuclear studies can be misleading as to the number of individual cells, since some cells have multiple nuclei. In the ancestral state (Enoplia) there appears to be a total of 36 muscle nuclei in the esophagus. This is a recurring number throughout Adenophorea though the distribution varies. Dorylaims have 24 radial nuclei in the corpus and 12 in the postcorpus. Among chromadorids the distribution of radial nuclei is commonly 12 in the procorpus, 12 in the metacorpus, and 12 in the postcorporal bulb. In Secernentea the number varies with the subclass: Spiruria in general have 36 radial nuclei (18 corpus, 18 postcorpus); Rhabditia 24 (6 procorpus, 6 metacorpus, and 12 in the postcorporal bulb) (20 radial nuclei are reported for *Caenorhabditis*); and Diplogasteria/Tylenchida 12, divided between the procorpus and the metacorpus, and rarely are even remnants of the radial nuclei seen in the glandular postcorpus.

The esophageal nervous system as described in *Caenorhabditis* is fundamentally bilateral in symmetry in contrast to the triradiate symmetry of the esophagus. The neuropile

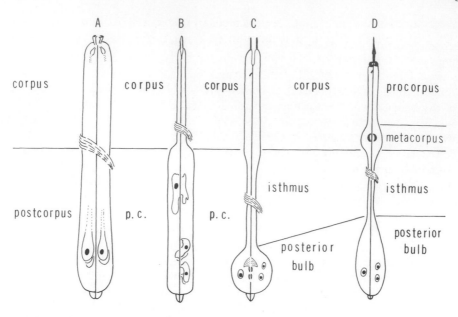

FIGURE 16   Diagramatic comparison of homologous regions of nemic esophagi. (A) Enoplida. (B) Dorylamida. (C) Rhabditida. (D) Tylenchina. (Maggenti/Springer-Verlag, 1981.)

is for the most part organized into two nerve rings, one in the metacorpus and the other in the postcorporal bulb. Connection with the central nervous system is by way of two interneurons that penetrate the basement membrane of the esophagus in the procorpus. White (1988) suggests that the postsynaptic interactions with the nonesophageal nervous system indicates that information flows unidirectionally from the central nervous system to the esophagus. White further reports that with the exception of one motor neuron, "M4," the esophageal nervous system is not required for pumping. Therefore, the system may be for sensory-mediated modulation and inhibition of pumping.

The vertebrate parasites show a similar degree of variation in the external morphological diversity of esophageal form as the so-called free-living taxa. However, plant parasitic nematodes have two basic types of esophagi according to class: a two-part esophagus (Adenophorea) or a three-part esophagus (Secernentea).

The two-part adenophorean esophagus consists of a cylindrical corpus followed by either an elongated cylindrical muscular/glandular region (Longidoridae) or a short pyriform muscular/glandular region (Trichodoridae). (The postcorpus is here designated as muscular/glandular rather than simply glandular as is seen commonly throughout the literature. The distinction is being made here because of the nature of the postcorpus in secernentean plant parasites where the postcorpus is truly glandular.) The generalized esophagus of Dorylaimida contains 24 muscles in the corpus and 12 muscles in the glandular postcorpus. Six of the muscles of the corpus are the protractor muscles of the stomatal armature. The number of glands in the postcorpus varies from three to five; indications are that the plant parasitic Dorylaimida may have five glands: one dorsal and four subventral. In all known instances the orifices for these glands are located near the gland body posterior to the nerve ring.

The three-part esophagus of secernentean plant parasites finds it ancestral form in diplogasteroids. In these forms the esophagus can be subdivided into the corpus (procorpus and metacorpus), isthmus, and glandular postcorpus (= posterior bulb). The postcorporal glands are variable from three to five, rarely more. Among the derived Tylenchida the metacorporal valve may be present or absent, and the absence does not necessarily denote taxonomic divergence beyond the species or generic level with the exception of Sphaerulariina. Musculature is limited to the 12 muscles of the corpus; there are no muscles in the postcorpus. A recurring phylogenetic development is seen in the enlargement and overlap of the glands constituting the postcorpus. The glands may overlap the anterior intestine slightly or they may extend posteriorly for a considerable distance. This developmental trend is seen in many families of Tylenchina. The assumption is that overlapping glands are indicative of the more derived forms (Luc et al., 1987). In Tylenchina when the stylet is elongated there may be an amalgamation of the procorpus and the metacorpus with the lumen of the procorpus becoming tortuous (*Dolichodorus* in Tylenchoidea and in many taxa of Criconematoidea).

Tylenchoidea and Aphelenchoidea are divergent in some features of their esophagi; these differences along with others are used in separation of the suborders Tylenchina and Aphelenchina. In aphelenchs the dorsal esophageal gland orifice is located in the metacorpus anterior to the valve, whereas in tylenchs the orifice is in the procorpus and generally near the base of the stylet. A "typical" three-part esophagus is seen among aphelenchs only in Paraphelenchidae. The other families in Aphelenchina are distinguished not only by having the postcorpus glands overlapping the intestine but by the glands taking the form of an appendage coming directly from the metacorporal bulb (Aphelenchoidea) or as a distinct cecalike appendage of the columnar isthmus.

D. Mesenteron (Intestine)

The single-layered mesenteron is derived from embryonic endoderm and is the first tissue invaginated during gastrulation. The two parent intestinal cells lie ventrally and posteriorly on the blastula just prior to gastrulation. At the start of gastrulation (Wood, 1988) the two parent intestinal cells sink inward. Later, after the somatic musculature and sex cell have invaginated, the esophageal precursors sink into the interior. Through further divisions of the endoderm and esophageal precursor cells, a central cylinder is formed that in the completed embryo becomes the esophagus and intestine.

Some taxa show subdivisions of the mesenteron and these are designated as the ventricular region, midgut, and prerectum. In totomounts the three regions are seldom recognized. However, the prerectum in Dorylaimida is easily recognized by the conspicuous change in the nature of the constituent cells and their long microvilli. The ventricular region is somewhat arbitrary and is distinguished by packed-cell inclusions and insoluble spherocrystals. Among the plant parasites in the order Tylenchida subdivisions of the mesenteron are not recognized. Animal parasitic nematodes may have the mesenteron separated from the stomodeum and proctodeum, in which case it is a food storage organ and is called a "trophosome."

The ventricular region is in some parasitic forms described as having diverticula (cecae) projecting anteriorly, posteriorly, or in both directions. In all known instances anteriorly directed cecae are ventricular in origin and posteriorly directed cecae may be either ventricular or esophageal.

In all regions of the intestine the internal border of the cells are covered by microvilli which in older literature is referred to as the bacillary layer or brush border. Below the microvilli there may be an area of dense fibrils known as the terminal web which is perforated by cytoplasmic connections to the remainder of the cell cytoplasm. The terminal web may extend into the base of the microvilli. This condition, though unexplained, was observed with light microscopy and was called the subbacillary layer (Fig. 17A).

The cells throughout the intestine may have the same or differing characteristics, and such conditions are designated homocytous and heterocytous, respectively. If all the cells in cross section are of equal height, then the intestine is isocytous; if different in height the condition is called anisocyty. Nomenclature is also applied to the total number of cells in the intestine: oligocytous, up to 128 cells; polycytous, 256 to 8192 cells; and myriocytous, 16,384 cells or more. These figures are based on theoretical divisions of the endoderm and the actual numbers are seldom achieved. Cell shape is also affected by number (Fig. 18): Oligocytous intestinal cells are longitudinally elongate and rectangular (64) (Fig. 18B, C, E, F) to hexagonal (128); polycytous intestines have cuboidal cells (Fig. 18A) and myrioytous intestines have tall columnar cells (Fig. 18D). The shape of the intestinal lumen is also dictated by the number of intestinal cells: oligocytous, cylindrical/rounded lumen; polycytous, subpolygonal lumen; myriocytous, the lumen is a multifolded or flattened tube. Intestinal cell descriptions are further complicated by being uninucleate to polynucleate. The latter condition is only known among mermithids in Adenophorea but is common among Secernentea. Syncytial intestines are reported for the highly derived secernentean plant parasites such as *Meloidogyne*.

E. Proctodeum

The proctodeum is the posterior ectodermal invagination counterpart to the anterior ectodermal invagination of the stomodeum (Fig. 17B). In cross-section the rectum (proctodeum) is flattened subtriangular, or an irregular tube that anteriorly is surrounded by a sphincter muscle and posteriorly ends at the surface orifice, the anus. In addition to the sphincter muscle, internally there may be an intestinorectal valve (= pylorus), formed from intestinal epithelium. Two muscles may be associated with the rectum, the uncommon dilator ani and the universal depressor ani (the H-shaped muscle). In association there may be rectal glands, six in the male and three in the female. An exception to this occurs in mature females of *Meloidogyne* (Tylenchina) there are six rectal glands that produce the gelatinous matrix. The proctodeum in males is complicated by the secondary sexual organs and will be discussed with the reproductive system. Suffice it to say here that the proctodeum of males joins the reproductive system to form a cloaca.

VIII. REPRODUCTIVE SYSTEM

Nematodes are mostly dioeceous i.e., ordinarily only one sex is represented in any one individual. A few rare instances are known of hermaphroditic nematodes in which both sexes are represented. The best known example is *Caenorhabditis elegans*. The reproductive system of nematodes is quite similar in both sexes (Figs. 19 and 20) and is not unlike a single ovariole or testicular tubule of arthropods. Generally the reproductive system is composed of one or two (rarely multiple) tubular gonads.

The complete reproductive system consists of the primary sex organs and the secondary sex organs. The primary sex organs are mesodermal in embryonic origin. The mesodermal parts of the genitalia not only house the germ cells but also provide for their development and nutrition. The secondary sex organs are ectodermal in origin and are produced by invaginations of the body wall.

Sexual dimorphism, other than the secondary sex structures, is not a common feature of nematodes. When it does occur it is most often evident among parasitic groups and

A

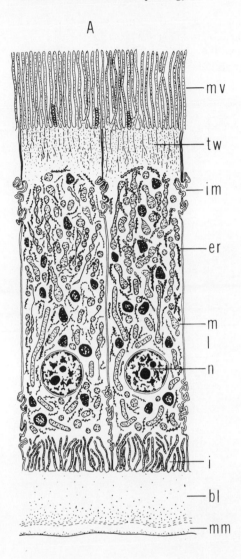

B

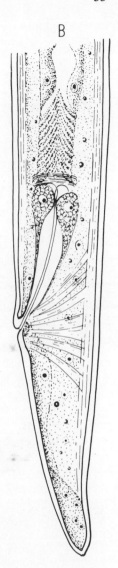

FIGURE 17   (A) Intestinal cells of *Ascaris*. Abbreviations: mv, microvilli; tw, terminal web; im, infolding of cell membrane; er, endoplasmic reticulum; m, mitochrondria; l, lipid inclusions; n, nucleus; i, infoldings of plasma membrane; bl, basal lamella; mm, mesenterial membrane. (Maggenti/Springer-Verlag 1981; adapted from Kessel et al.) (B) Female tail of *Bulbodacntis* showing intestinorectal valve, sphincter muscle, rectal glands, and depressor ani muscle. (Maggenti/Springer-Verlag, 1981.)

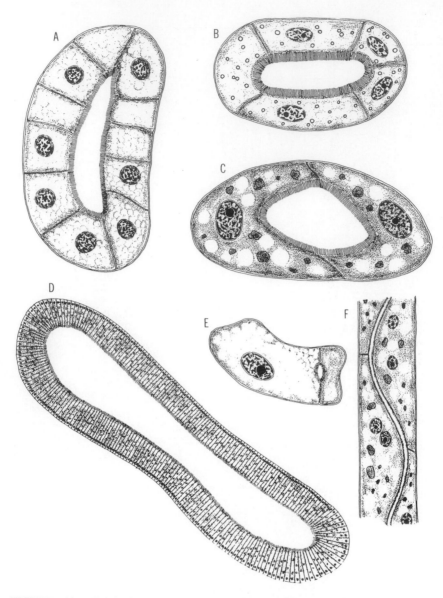

FIGURE 18 Histological sections of nemic intestines. (A) Enoplia: *Deontostoma.* (B) Chromadoria: *Axonolaimus.* (C) Rhabditia: *Rhabditis.* (D) Spiruria: *Ascaris.* (E) Diplogasteria: *Ditylenchus.* (F) Diplogasteria: *Ditylenchus* (longitudinal section). (Maggenti/Springer-Verlag, 1981; adapted from Chitwood and Chitwood.)

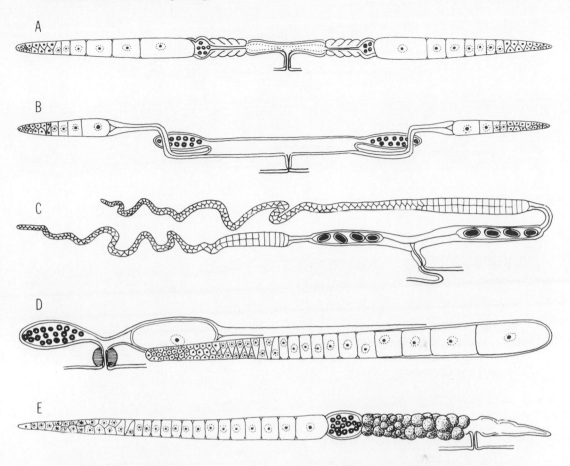

FIGURE 19 Amphidelphic female reproductive systems. (A) Ovaries outstretched. (B) Specialized spermatheca between oviduct and uterus. (C) Anterior ovary outstretched with flexures; posterior ovary reflexed and anteriorly directed with flexures. (D) Amphidelphic postvulval reproductive system (postpudenal). (E) Amphidelphic prevulval reproductive system (antepudendal). (Maggenti/Springer-Verlag, 1981; adapted from Chitwood and Chitwood.)

rarely in freeliving, most often marine, taxa. The commonest example is the swollen saccate female in contrast to the vermiform male. Other illustrations of sexual dimorphism among nematodes include atrophy of the male feeding apparatus, notable differentiation of cuticular ornamentation or sense organs (amphids), and the degeneration of the female into a reproductive sac or the prolapse and growth of the reproductive system independent of the female body (*Sphaerularia bombi*).

Aberrant individuals that are gynandromorphs (individuals in which male and female characteristics or structures occur) are not uncommon. Generally, the female gonad is complete and only portions of the male secondary sexual characteristics are present. These individuals are not hermaphrodites.

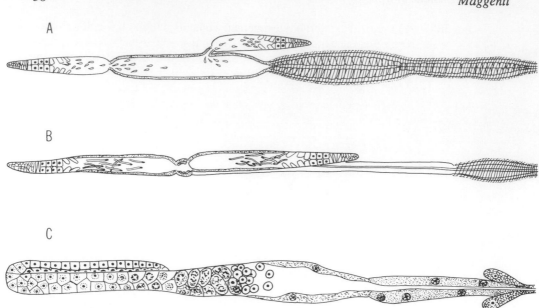

FIGURE 20   Male reproductive systems. (A) Enoplia: *Enoplus*; seminal vesicle and vas deferens heavily muscled. (B) Chromadoria: *Tobrilus*; musculature limited to vas deferens. (C) Secernentea: *Rhabditis*; musculature lacking but glands present. (Maggenti/Springer-Verlag, 1981; adapted from Chitwood and Chitwood.)

A. Female Reproductive System

The typical system consists of two ovaries, one anterior and one posterior (Fig. 19A–C). The ovaries are proximally connected to their respective oviduct and uterus, which in turn are connected proximally to the cuticularly lined vagina that terminates at the vulva. The vulva is generally located at or near the midbody. There are many variants from this typical plan; the ovaries may be outstretched or reflexed; there may be one, two, or more ovaries (32 in *Placentonema gigantissima*). A spermatheca may be present at the junction of the ovary and oviduct or at the junction of oviduct and uterus, the former being the most common. The uterus may be divided into columned uterus (= tricolumella, quadricollumella) and uterus vera; the vagina may also be divided, generally in parasitic taxa, into a vagina uterina (mesodermal) and vagina vera (ectodermal). When only one complete gonad is present, then the vestigial uterus of the second gonad may act as a seminal receptacle for sperm storage (*Ditylenchus, Anguina*) even though a functional spermatheca may be present. The ovaries of nematodes are of three types: panoistic, teleotrophic, and hologonic. The most common ovarial type is the panoistic ovary in which new germ cells originate at the blind distal end from one cell. *Caenorhabditis* and *Ascaris* are examples of teleotropic ovaries. New germ cells are produced in the same manner as in the panoistic ovary; however, the developing oocytes receive nutrients from the central rachis rather than through the single-layered epithelial covering that absorbs nutrients from the body cavity. The attachment of the rachis persists until the oocytes are ready for fertilization. In a hologonic ovary (*Trichuris*) the germinal cells reportedly are proliferated from a series of germinal areas extending the length of the ovary either on one or on both sides.

Seurat (1920) ascribed explicit phylogenetic designations to the types of morphological variability seen among nematode reproductive systems. The essence of Seurat's nomenclature is the position and direction of the uteri (not the ovaries) and how this can then be applied to studies of nematode phylogeny. Many nematologists have lost sight of the original purpose of the terms, their contributory significance to understanding nematode phylogeny, and the inherent genetic significance imparted by the correct application of this terminology. Seurat's nomenclature and definitions are as follows:

Amphidelphic: uteri opposed
Opistodelphic: uteri parallel and posteriorly directed
Prodelphic: uteri parallel and anteriorly directed

All plant parasitic nematodes are amphidelphic whether they have one or two ovaries anteriorly or posteriorly directed. It is interesting to note that when adenophorean plant parasites have a single ovary it is posteriorly directed, and if there is a shift in vulval position from midbody, it is toward the anterior. When secernentean plant parasites have one ovary it is anteriorly directed and any shift of vulva position is toward the posterior.

To apply Seurat's terminology incorrectly is to ignore the inherent knowledge of the system and to miss the entire phylogenetic message contained in the reproductive system under consideration. For the examples *Pratylenchus vulnus* and *Xiphinema bakeri*, some nematologists would teach that *Pratylenchus* is prodelphic and that *Xiphinema bakeri* is opisthodelphic; this is erroneous, misleading, and devoid of factual information. Both examples are amphidelphic. By recognizing this we know the ancestral condition, that there has been a reduction from the diovarial condition to monovarial, that this was followed by a reduction of one uterus (the extent of reduction is variable), and that the gonad reduction occurred anteriorly or posteriorly.

The value of the information gained by the correct application of terminology was recently demonstrated in the genus *Helicotylenchus* (two ovaries, amphidelphic) and its synonym *Rotylenchoides* with a well-developed anterior ovary and an atrophied posterior ovary. The cline of ovarial reduction throughout the genus *Helicotylenchus* and the recognition of the amphidelphic condition strongly supports the synonomy of these once separate genera (Fortuner, 1984). Designation of *Rotylenchoides* as prodelphic would by definition mean that the genetic histories and compositions of the two genera are different and therefore they could not be synonymized.

Many have questioned the condition in *Meloidogyne* which is not an example of prodelphy but the victim of secondary prodelphy induced by the swelling of the female with the coincident posterior shift of the vulval position. In the Heteroderidae, the swelling of the body is genetically controlled, whereas the condition of the reproductive system is an architectural accommodation. This is proven by such ancestral heteroderid genera as *Meloinema*, *Nacobbodera*, and *Bursadera*, where the young adult female is vermiform and amphidelphic but later becomes swollen, sedentary, and secondarily prodelphic. Hopefully, these examples clarify how biased misinterpretations of established terminology obstructs enhanced communication and the advancement of factual knowledge.

B. Male Reproductive System

In general organization the male reproductive system is similar to that of the female in that it consists of one or two testes, associated ducts, and sperm reservoirs and outlets to the outside of the body (Fig. 20). The typical system consists of three parts: the testis, the seminal receptical, and the vas deferens. Rarely, there is a vas efferens between the testis and the seminal receptical. The testes are of two types: panoistic or hologonic. The hologonic testes are known only from the same parasitic taxa (*Trichuris*) that had females with hologonic ovaries.

There are differences between the male reproductive systems of Adenophorea and Secernentea. As a rule adenophorean males have two testes (diorchic) (exception: *Trichodorus* and some taxa in Chromadoria). Secernentean males have a single testis (monorchic) (exception: sex-reversed males in *Meloidogyne*). Other differences are evident in the musculature associated with the proximal end of the system (Fig. 20). In males of Enoplia the ejaculatory duct is heavily muscled and this is easily detected (Fig. 20A). Chromodorids also have a muscle layer surrounding the ejaculatory duct but it is weak and difficult to detect (Fig. 20B). In Secernentea the males lack ejaculatory muscles but there are well-developed ejaculatory glands that are often mistaken for muscle (Fig. 20C).

The secondary sexual organs of the male are more prominent than those associated with the female reproductive system. The secondary sex organs referred to are the cloaca, spicular pouch, spicules, gubernaculum, copulatory muscles, supplements, and caudal alae. Caudal alae have already been discussed.

The presence of a cloaca in male nematodes and the absence of it in females is one of the characteristics that separates nematodes from other pseudocoelomates. A cloaca is known in females of two genera, *Lauratonema* and *Rondonia*, neither of which is considered ancestral. There is a difference in the cloaca between adenophoreans and secernenteans. The difference adds evidence to the assumption that Adenophorea is ancestral and Secernentea derived. In the adenophorean cloaca the confluence of the rectum and vas deferens is posterior as a result a distinguishable rectum persists. (This could indicate that the progenitors of Nemata as males and females had separate gonopore openings.) In males of Secernentea the cloaca is formed by the entrance of the vas deferens into the hindgut either at or just posterior to the intestinorectal valve: thus no distinguishable rectum exists.

C. Spicules

On the dorsal wall of the cloaca there are specialized cells called the spicula primordia that by a quasi-evagination form the spicular pouch and by invagination form the spicules. As such the spicules do not lie directly within the cloacal pouch but in an offset pouch. The spicules are extruded to the exterior by way of the cloacaspicular orifice.

Spicules are not, as often supposed, flat bladelike structures (Fig. 21A). Each spicule is in cross-section crescentic or tubelike, and in all instances with a cytoplasmic core in which sensory nerves may be embedded (Fig. 21C). Like most nematode structures, terminology of parts is not consistent because proposals were made independently in the various fields of helminthology, plant nematology, marine nematology, etc. The proximal end of the spicule modified for muscle attachment and contiguous with the spicular pouch is called the manubium (= capitulum, head). Beyond the head the spicule may narrow to a section known as the calomus (= shaft). The main portion of the spicule is the lamina (= blade). The blade may have a longitudinal, winglike, membranous extension called the velum.

The basic number of spicules is two but there may be only one or none. In addition to variability in number, the paired spicules may be very unequal in length. For example, males of *Viguiera hawaiiensis* del Prado Vera, Maggenti, and Van Riper, 1985 have extremely unequal spicules. The short right spicule is 0.14–0.17 mm and the long left spicule 3.35–4.10 mm long. The long spicule in this genus is nearly 60% of the total body length (average length of male is, 6.1 mm). Each spicule has both a protractor and a retractor muscle.

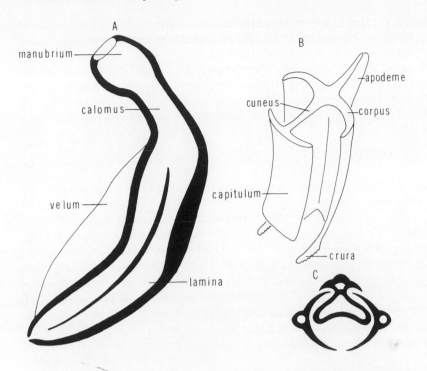

FIGURE 21 (A) Spicule. (B) Gubernaculum. (C) Transverse section through spicules and gubernaculum showing relationship to each other. (Maggenti/Springer-Verlag, 1981.)

D. Gubernaculum

The gubernaculum is a cuticular thickening of the dorsal wall of the spicular pouch and acts as a guide during spicular protrusion. The simplest gubernaculum is merely a thickened plate or trough called the corpus. In its most complex form protruding from the corpus and at right angles to it, is an additional longitudinal plate that keeps the spicules separate; this plate is termed the cuneus (Fig. 21B). In turn the cuneus may be tipped by a transverse plate called the capitulum. Toothed lateral extensions at the bottom of the corpus are called crura. Extending posteriorly from the corpus there may be apodemes for muscle attachment.

Most commonly there are two pairs of muscles that operate the gubernaculum: the protractor gubernaculi that extend from the ventral body wall anteriorly to the gubernaculum and the retractor gubernaculi that extend from the gubernaculum to the dorsal body wall. In addition, there may be paired seductor gubernaculi muscles that extend from the lateral body wall to the gubernaculum.

E. Telamon

The telamon is a specialized structure known only from strongyles. In strongyles it is an immovable thickening of the ventral cloacal wall; it is independent of the spicular pouch. This structure is essential in strongyles because the spicular pouch orifice is not directly opposite the cloacal (anal) orifice as in other Nemata. Therefore, the telamon turns the spicules to the exterior as they are being protruded.

The term should not be applied to the spicular accessories associated with males of tylenchid plant parasites. The so-called "telamon" of hoplolaims is merely the capitulum of the gubernaculum.

Additional secondary sexual characters often useful in identifying taxa are the preanal and postanal male supplements. In Adenophorea preanal supplements are in a single medioventral row; in Secernentea they are subventral and paired. When the supplements are large and extensible, as they are in some marine Adenophorea, they are called appendicules.

IX. NERVOUS SYSTEM

One of the responsibilities of the nervous system is to transmit stimuli received externally by way of somatic sensory organs to the central nervous system and then to the internal tissues where they are translated into a "proper" response. The nervous system acts to mediate all of the animal's activity through stimulation, coordination, and responsive actions.

The basic operative unit of the nervous system is the neuron. The neuron has a large cell body (neurocyte) with a conspicuous nucleus and two or more protoplasmic processes. The process transmitting stimuli to the neurocyte is called the dendrite and the protoplasmic process carrying impulses away from the neurocyte is the axon. There are two basic types of neurons, bipolar and multipolar. Bipolar neurons have one dendrite and one axon. Multipolar neurons have multiple dentrites and one axon. The axon which is generally unbranched may have a collateral branch near the neurocyte. A group of neurocytes constitute a ganglion.

Neurons fall into three categories: sensory neurons (afferent), motor neurons (efferent), and adjustor neurons (internuncial or associative). Sensory neurons transmit impulses to the central nervous system; motor neurons conduct impulses to the effectors and internuncial neurons interconnect with sensory and motor neurons so that more than one effector may be activated.

The central nervous system of nematodes (Figs. 22 and 23) consists of a large aggregate of ganglia that are situated dorsally (two subdorsal ganglia), laterally (six ganglia), and ventrally (a bilobed ganglion) around the esophagus and are anteriorly connected to the major nerve bundle surrounding the esophagus, i.e., the circumesophageal commissure or nerve ring (Chitwood and Chitwood, 1974; Wood, 1988). Anterior to the "brain" (nerve ring and associated ganglia) are six small ganglia that receive nerves (dendritic fibrils) from the cephalic sensory organs; axonic nerves transmit the impulses through the nerve ring to internuncial neurons in the lateral ganglia. The amphids and deirids are innervated from separate ganglia that are located laterally and posteriorly to the nerve ring.

The area under the medial ventral body wall where the two arms of the nerve ring amalgamate before proceeding posteriorly as the ventral nerve cord is called the hemizonid. In lateral view the hemizonid stands out as a refractive body. All the refractive structures called "cephalids" are really subcuticular commissures. Just posterior to the hemizonid is the anterior-most and largest ganglion of the ventral nerve cord: the retrovesicular ganglion. The ventral nerve trunk proceeds posteriorly passing to the right of the excretory pore and to the right of the vagina. Two loose aggregates of neurocytes anterior and posterior to the vagina are designated by some authors as the pre- and postvulvar ganglia. At the rectum (Fig. 23) the ventral nerve trunk sends off two commissures. Each one is directed dorsally around the rectum or cloaca and laterally enters the laterorectal ganglion. The commissure then proceeds dorsally where the two arms now merge in the dorsorectal ganglion. The nerve originating from this ganglion innervates the tail.

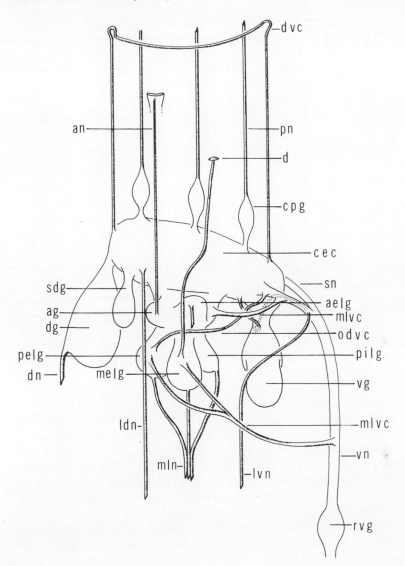

FIGURE 22 Generalized anterior central nervous system (right side) illustrating nerve ring, associated ganglia, nerves, and commissures. Abbreviations: dvc, dorsoventral commissure: an, amphidial nerve; pn, papillary nerve; d, deirid; cpg, cephalic papillary ganglia, cec, circumesophageal commissure; sn, subventral nerve; sdg, subdorsal ganglion; ag, amphidial ganglion; aelg, anterior externolateral ganglion; mlvc, major lateroventral commissure; dg, dorsal ganglion; odvc, oblique dorsoventral commissure; pelg, posterior externolateral ganglion; dn, dorsal nerve; melg, median externolateral ganglion; pilg, posterior internolateral ganglion; vg, ventral ganglia; mlvc, minor lateroventral commissure; ldn, laterodorsal nerve; mln, mediolateral nerve; lvn, lateroventral nerve; vn, ventral nerve; rvg, retrovesicular ganglion. (Maggenti/Springer-Verlag, 1981.)

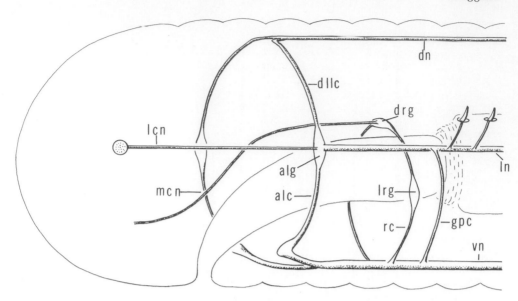

FIGURE 23   Generalized central nervous system, caudal region. Abbreviations: dn, dorsal nerve; dllc, dorsolateral lumbar commissure; drg, dorsorectal ganglion; lrg, laterorectal ganglion; mcn, median caudal nerve; ln, lateral nerve; rc, rectal commissure; alg, anolumbar ganglion; lcn, laterocaudal nerve; alc, anolumbar commissure; rc, rectal commissure; gpc, genitopapillary commissure; vn, ventral nerve. (Maggenti/Springer-Verlag, 1981.)

      Caudal sensory organs (including the phasmids) are innervated from the nerve that emanates from the anolumber ganglion which terminates the lateral nerves and receives commissures that connect the ventral nerve with the dorsal nerve.

      The subsidiary somatic nerves are the dorsal nerve cord and the lateral nerve cord. Just as the muscles of the subventral sectors send processes to synapse with the ventral nerve cord, so also do the muscles of the subdorsal sectors send processes to the dorsal nerve, which is primarily motor. Muscles anterior to the nerve ring connect directly into it. The lateral and ventral nerves are composed of both motor and sensory nerves. The entire system is connected by commissures that connect the dorsal nerve with the lateral and ventral nerves and vice versa.

      Among Adenophorea there appears to be a peripheral nerve net that extends from the anterior extremity to the posterior extremity forming plexuses at the lip region, vulva, and in males along the entire caudal region. A notable feature is the linking of the somatic setae and papillae, which in turn are individually connected to sensory neurons of the peripheral nervous system.

      The system is controversial but has been seen several times and in several genera of Enoplia and Chromadoria when silver-stained as described by Croll and Maggenti (1968). They postulated that the network coordinates impulses from seta to seta and thence to the central nervous system, thereby compensating for the paucity of sensory organs over the body. A network has not been demonstrated in Secernentea; however, free nerve endings under and in the cuticle (sensilla insitica and sensilla colei) have been and these could act as proprioceptors.

REFERENCES

Allen, M. W. and Noffsinger, E. M. 1978. A revision of the marine nematodes of the superfamily *Draconematoidea* Filipjev, 1918 (Nematoda: Draconematina). University Calif. Pub. Zool. Vol. 109 Univ. Calif. Press, Berkeley.

Baldwin, J. G. 1983. Fine structure of bodywall cuticle of females of *Meloidodera charis, Atalodera lonicerae* and *Sarisodera hydrophila* (Heteroderidae). *J. Nematol.* 15: 370–381.

Baldwin, J. G., and Hirschmann, H. 1976. Comparative fine structure of the stomatal region of males of *Meloidogyne incognita* and *Heterodera glycines. J. Nematol.* 8:1–17.

Bird, A. F., and Rogers, G. E. 1965. Ultrastructure of the cuticle and its formation in *Meloidogyne javanica. Nematologica* 11:224–230.

Boveri, T. 1892. Ueber die Entstehung des Gegensatzes zwischen den Geschlechtszellen und der somatischen Zellen bei *Ascaris megalocephala,* nebst Bermekungen zur Entwickelungsgeschichte der Nematoden. Sitsungsb. *Gellsch. Morphol. Physiol.* 8:114–125.

Chitwood, B. G., and Chitwood, M. B. 1974. *Introduction to Nematology.* University Park Press, Baltimore.

Croll, N. A., and Maggenti, A. R. 1968. A peripheral nervous system in Nematoda with a discussion of its functional and phylogenetic significance. *Proc. Helminthol. Soc. Wash.* 35:108–115.

DeGrisse, A. T., Lippens, P. L., and Coomans, A. 1974. The cephalic sensory system of *Rotylenchus robustus* and a comparison with some other tylenchids. *Nematologica* 20:88–95.

Filipjev, I. N. 1934. The classification of the free-living nematodes and their relation to the parasitic nematodes. *Smithsonian Misc. Coll.* 89:1–63.

Harris, J. E., and Crofton, H. D. 1957. Structure and function in the nematodes: Internal pressure and the cuticular structure in *Ascaris. J. Exp. Biol.* 34:116–130.

Hope, W. D. 1988. *Syringonomus dactyltus,* a new species of bathyal marine nematode (Enoplida: Leptosomatidae) and a supplementary description of *Syringonomus typicus* Hope and Murphy, 1969. *Proc. Biol. Soc. Wash.* 101: 717–729.

Hyman, L. H. 1951. The Invertebrates: Acanthocephala, Aschelminthes and Entoprocta. The Pseudoscoelomate Bilateria, *Vol. 3. McGraw-Hill, New York.*

Inglis, W. G. 1964. The marine Enoplida (Nematoda): A comparative study of the head. *Bull. Br. Mus. (Nat. Hist.) Zool.* 2:263–376.

Kier, W. M., and Smith, K. K. 1985. Tongues, tentacles and trunks: The biomechanics of movement in muscular hydrostats. *Zool. J. Linn. Soc.* 83:307–324.

Lee, D. L. 1970. Moulting in nematodes: The formation of the adult cuticle during the final moult of *Nippostrongylus brasiliensis. Tissue and Cell* 2:139–153.

Luc, M., Maggenti, A. R., Fortuner, R., Raski, D. J., and Geraert, E. 1987. A reappraisal of Tylenchina (Nemata). 1. For a new approach to the taxonomy of Tylenchina. *Revue Nématol.* 10:127– 134.

Maggenti, A. R. 1961. Morphology and biology of the genus *Plectus* (Nematoda: Plectidae). *Proc. Helminthol. Soc. Wash.* 28:118–130.

Maggenti, A. R. 1962. The production of the gelatinous matrix and its taxonomic significance in *Tylenchulus* (Nematoda: Tylenchulinae). *Proc. Helminthol. Soc. Wash.* 29:139–144.

Maggenti, A. R. 1963. Comparative morphology in nemic phylogeny. In *The Lower Metazoa Comparative Biology and Phylogeny,* E. C. Doughtery, eds. Univ. Calif. Press, Berkeley, pp. 273–282.

Maggenti, A. R. 1979. The role of the cuticular strata nomenclature in the systemaics of Nemata. *J. Nematol.* 11:94–98.

Maggenti, A. R. 1981. *General Nematology.* Springer-Verlag, New York.

Maggenti, A. R., Raski, D. J., Koshy, P. K., and Sosamma, V. K. 1983. A new species of *Chronogaster* Cobb, 1913 (Nemata: Plectidae) with an amended diagnosis of the genus and discussion of cuticular ornamentation. *Revue Nématol.* 6:257–263

Martini, E. 1903. Ueber Furchung und Gastrulation bei *Cucullanus elegans* Zed., *Ztschr. Wissensch. Zool.* 74:501–556.

Martini, E. 1909. Ueber Subcuticula und Seitenfelder einiger Nematoden. *Ztschr. Wisssensch. Zool.* 93:535–624.

Mendel, G. 1865. Versuche uber Pflanzen Hybriden. *Verh. naturf. Ver. in Brunn, Abhandlungen, IV*: 3–47. [English Trans.] 1956. Harvard Univ. Press, Cambridge, MA.

Ohmori, Y., and Ohbayashi, M. 1975. Arrangement of the somatic muscle cells of meromyarian nematodes. *Jap. J. Nematol.* 24:294– 299.

Pai, S. 1927. Lebenzyklus der *Anguillula aceti* Ehrbg., *Zool. Anz., Leipzig* 74:257–270.

Pai, S. 1928. Die Phasen des lebenscyclus der *Anguillula aceti* Ehrbg. und ihre experimentell-morphologische Beeingflussung. *Ztschr. Wissensch. Zool.* 131:293–344.

Riding, I. L. 1970. Microvilli on the outside of a nematode. *Nature* 226:179–180.

Roberts, T. M., and Ward, S. 1982a. Centripetal flow of pseudopodial surface components could propel the amoeboid movement of *Caenorhabditis elegans* spermatozoa. *J. Cell Biol.* 92: 132–138.

Roberts, T. M., and Ward, S. 1982b. Directed membrane flow on the pseudopods of *Caenorhabditis elegans* Spermatozoa. *Cold Spring Harbor Symp. Quant. Biol.* 46:695–702.

Robertson, W. M. 1976. A possible gustatory organ associated with the odontophore in *Longidorus leptocephalus* and *Xiphinema diversicaudatum*. *Nematologica* 21:443–448.

Roggen, D. R. 1973. Functional morphology of the nematode pharynx. 1. Theory of the soft-walled cylindrical pharynx. *Nematologica* 19:349–365.

Rosenbluth, J. 1967. Ultrastructural organization of obliquely striated muscle fibers in *Ascaris lumbricoides*. *J. Cell Biol.* 25:495–510.

Schuurmans Stekhoven, J. H., and DeConinck, L. A. 1933. Morphologische Fragen zur Systematik der freilebenden Nematoden. *Verhandl. Deutsch. Zool. Gesellsch.* 35:138–143.

Seurat, L. G. 1920. Histoire Naturelle des Nematodes de la Berbérie, Alger.

Shepherd, A. M., Clark, S. A., and Dart, P. J. 1972. Cuticule structure in the genus *Heterodera*. *Nematologica* 18:1–17.

Smith, K. K., and Kier, W. M. 1989. Trunks, tongues, and tentacles: Moving with skeletons of muscle. *Am. Scientist* 77:28–35.

Steiner, G. 1933. The nematode *Cylindrogaster longistoma* (Stefanski) Goodey, and its relationship. *J. Parasitol.* 20:66–68.

Sulston, J. E., and Horvitz, H. R. 1977. Postembryonic cell lineages of the nematode *Caenorhabditis elegans*. *Dev. Biol.* 56:110–156.

Sulston, J. E., Schierenberg, E., White, J. G., and Thomson, J. N. 1983. The embryonic cell lineages of the nematode *Caenorhabditis elegans*. *Dev. Biol.* 100:64–119.

Van Beneden, E. 1883. *Recherches sur la maturation de l'oeuf, la fecondation et la division cellulaire*. Gand and Leipzig, Paris.

Van Waerbeke, D., and Remillet, M. 1973. Morphologie et biologie de *Heterogonema ovomasculis* n. sp. (Nematoda: Tetradonematidae) parasite de nitidulidae (Coleoptera). *Nematologica*, 19:80–92.

Waddell, A. H. 1968. The excretory system of the kidney worm *Stephanurus dentatus* (Nematoda). *Parasitology* 58:907–919.

Wood, W. B. (ed.) 1988. The nematode *Caenorhabditis elegans*. Cold Spring Harbor Laboratory, Cold Spring Harbor, NY.

2
Biology and Ecology of Nematodes

DON C. NORTON   *Iowa State University, Ames, Iowa*

TERRY L. NIBLACK   *University of Missouri—Columbia, Columbia, Missouri*

I. INTRODUCTION

Nematodes are the most numerous multicellular organisms present in agroecosystems and can be found at densities up to 30 million/m^2. They are found exploiting every niche provided by the vegetation, soil, and other biota as resources. Filipjev and Schuurmans Stekhoven (1941) limited their view of nematodes that are important for agriculture to plant and insect parasitic species; hence the emphasis in the current tome. It is clear, however, that taxa belonging to other trophic groups or guilds have an impact on agricultural production that exceeds the direct effects of parasitism (Anderson, 1987; Yeates and Coleman, 1982; Yeates, 1987). Thus, there are historical perceptual limits on what taxa or guilds of nematodes should be studied or included in a manual of agricultural nematology. Nematology has suffered because of these perceptions (Stone et al., 1983). Nonetheless, our emphasis here will be on plant parasitic nematodes; it is the acute interaction between plant pathogenic nematodes and their economically important hosts that directs our attention to the plant parasites.

Taxonomic limits are also imposed on the study of nematode ecology: May and Seger (1986) deplored the unsatisfactory state of systematics, especially with invertebrates, and they recognized the importance of knowing what is where as a basis for understanding and managing the ecosystem. Problems exist even with extant nematode identifications let alone the myriad of species yet to be described. The limitations on studies of nematode biology and ecology due to sampling and extraction procedures are well documented but not inimical to valuable research (Barker and Noe, 1988; Ferris and Noling, 1987; McSorley, 1987).

All the limitations notwithstanding, research into nematode ecology has progressed increasingly rapidly in the past decade. Movement in mathematical and molecular ecology should enhance rather than replace studies in the classical and traditional areas. In this treatment, we do not wish merely to present a collation of facts but to give examples as a basis for understanding these organisms, an inadequate as our understanding may be. Nor are these examples presented to elucidate a "principle." Although one cannot ignore the human factor in discussions of agriculture, nematodes are not inclined to behave according to human-contrived principles. We do search for common patterns, but the student, matriculating or life-long, should not be conditioned to think or search in learning from principles to cases. Thus, our examples, but we hope that they are pertinent.

Our knowledge and ability to manage nematode populations when desired is still much too meager. This chapter can only be a preliminary statement concerning a phase of nematology in which most questions are yet to be answered. Even so, the literature is too vast to be exhaustively treated here, so the reader is referred to review articles and books wherever possible.

II. GENERAL BIOLOGY

A. Life Cycles

Life cycles of most plant nematodes are not complex. There are six stages, including the undifferentiated egg, four juvenile stages (J1–J4), and the adult. In the Secernentea, the "tylenchid" plant parasites, the J1 forms in the egg where it molts into the J2, which is the stage that hatches. In the Adenophorea, the "dorylaimids," the hatching stage is the J1.

Hatching, along with various stimulants and depressants of the process, has received attention as a possible target of control measures aimed at preventing it or inducing it in the absence of a host (Perry, 1987). Preceded by induced changes in eggshell permeability, hatching may be either a physical or an enzymatic process. Plant nematodes, such as *Rotylenchulus reniformis* and *Meloidogyne incognita*, may hatch freely in water at appropriate temperatures with some increase in percentage hatch in the presence of a host. Others, such as *Globodera rostochiensis*, may require a specific hatch-inducing signal from a host. Some may undergo diapause in the egg (e.g., *Heterodera avenae, Meloidogyne naasi*; Evans, 1987); other forms of arrested development will be discussed in Section III. D. A naturally occurring chemical isolated from red kidney bean induced hatch of *Heterodera gylcines* (Masamune et al., 1982), but specific stimuli operative in soils have not been identified.

Molting separates each of the J1–adult stages in all nematodes, involving a sequence of events similar to that of the insects: (a) apolysis, separation of the cuticle from the hypodermis; (b) cuticle formation; and (c) ecdysis (Wharton, 1986). Molting may represent an evolutionary relict, a mechanism in which the cuticle is modified between stages, or in which the cuticle is modified between stages, or in which excess nitrogen is excreted. It may simply be an accommodation to growth proceeds; for example, Bird (1983) found that the volume growth of adult *Rotylenchulus reniformis* was 17–19% lower than that of the J2 stage. *Paratylenchus projectus* is dependent on a host stimulus for molting (Ishibashi et al., 1975).

Dorylaimid plant parasites remain vermiform throughout their life cycles, but among the tylenchids there is some variation. Most species remain vermiform, but the females of some important plant parasitic species become sedentary in the J2 stage and increasingly saccate to globose as they mature, e.g., *Meloidogyne* and *Heterodera* spp. Sexual primordia are visible in the J3 stages of the most plant nematodes.

B. Reproduction

Plant nematodes reproduce amphimictically or autokonously, never asexually. In species that have males, unbalanced sex ratios are common and show a general trend toward an increased proportion of males when a population is subjected to environmental stress, even when sex is genetically determined (Yeates, 1987). Life cycles vary from less than 5 days, as with some neotylenchids, to a year or more, as with some longidorids, and within the limits imposed by biology are influenced primarily by temperature and substrate quality. Population cycles of most soil- and root-inhabiting nematodes have overlapping generations. Some nematodes, such as the anguinids *Anguina tritici* and *Subanguina calamagrostis*, begin their life cycles as a distinct cohort and progress separately through each of their developmental stages with little overlapping of generations.

Egg production is high (300+) for species whose eggs are confined, as in plant tissue (*Ditylenchus dipsaci*), an egg sac (*Meloidogyne* spp.), or the body of the female (*Heterodera* spp.). It is more difficult to obtain information with nematodes where eggs are deposited in the soil because of difficulty in recovery and identification. Because of the rapidity with which some migratory nematodes increase in a season (e.g., *Pratylenchus* spp.), egg production must occur rapidly. Fecundity is usually seasonal in temperature zones. A question remaining is whether females keep reproducing as long as the environment is favorable or there is a period in the migratory species in which the female can live after fecundity ceases. Female cyst nematodes, *Heterodera* spp., and species of related genera die with the maturation of the adult. See Wharton (1986) for a discussion of reproductive energetics.

C. Feeding

The feeding processes of dorylaimid and tylenchid plant parasites have been documented on videotape and described in detail by Wyss (1981, 1987). Feeding is by means of a stylet or spear, usually containing a hollow tube through which secretions from pharyngeal glands are injected into plant cells, and cell contents are ingested. The secretions may partly digest cell cytoplasm before ingestion, induce the formation of specialized feeding sites, or have other activity (Hussey, 1989).

Most plant nematodes parasitize underground parts, but some feed on above-ground parts. Nematode feeding habits are often described as ecto- or endoparasitic, with each category subdivided into migratory and sedentary, or sessile, habits. Sometimes an additional category, semiendoparasitic, is included in the scheme. Galling may be induced by certain species in each group. True ectoparasites can feed on epidermal or deeper tissues, depending on stylet length and other factors, their bodies remaining outside the plant tissue. Endoparasites penetrate tissues completely. Feeding habits are not always easily delimited because some species fit into different categories at different life stages. *Heterodera glycines* females are sedentary, endoparasitic during development until sexual maturity when their swollen bodies erupt through the root cortex; they continue feeding semiendoparasitically. *Hoplolaimus galeatus* individuals can be endoparasitic or semiendoparasitic during the same development stage. *Pratylenchus agilis*, a member of a genus of endoparasites, can feed ectoparasitically in vitro, but it is unknown whether *P. agilis* exhibits this behavior in nature (Rebois and Huettel, 1986).

In the root, most migratory endoparasites feed in the cortical parenchyma, but a few species, such as *Pratylenchus vulnus*, penetrate beyond the endodermis and even into the more lignified tissues. Migratory endoparasites may cause extensive necrotic lesions in the tissues in which they feed, leaf-feeding nematodes usually feed on mesophyll tissue, and may feed either ecto- or endoparasitically (*Aphelenchoides* spp.). Sedentary root endoparasites and some deep-feeding ectoparasites induce development of specialized feeding sites that act as metabolic sinks in or adjacent to stelar tissue (e.g., giant cells, syncytia, nurse cells). The pinewood nematode *Bursaphelenchus xylophilus* invades the resin canals of many coniferous trees and feeds on the epithelial lining of the canals. Knowledge of feeding habits is essential in choosing appropriate extraction methods.

D. Host Reaction

The overall effects of nematode parasitism on plants range from stimulatory to lethal, even within single plant–nematode interactions (Nickle, 1984; Oostenbrink, 1966). Host reactions are characterized along a continuum from susceptible to resistant (even "immune"), which designations may or may not include reference to the nematode's ability to reproduce on the host (Cook and Evans, 1987). Yield responses of susceptible annual hosts are often a function of initial, or preplant, densities of the parasite and all of the conditions affecting the interaction. In deleterious (to the host) associations, above-ground parasites may cause characteristic malformations of various plant organs. Root-knot (*Meloidogyne* spp.), stubby root (*Paratrichodorus* and *Trichodorus* spp.), and some other root parasites cause characteristic root symptoms, but most below-ground parasites cause nonspecific shoot symptoms.

For many economically important plant nematodes and their hosts, genetic variability in the interaction results in a situation in which there is some degree of host cultivar–nematode isolate specificity. The nomenclature applied to the nematode varies, including race, strain, ecotype, and pathotype, but there are many examples, including the interactions between *Ditylenchus dipsaci* and various hosts, *Globodera rostochiensis* and potato, *Heterodera glycines* and soybean, and *Bursaphelenchus xylophilus* and pine (Dropkin, 1988). This variability causes a number of problems, not the least of which are the difficulties presented to plant breeders attempting to provide growers with nematode-resistant cultivars, but it is characteristic of nematode–host relations.

E. Locomotion and Dissemination

All nematodes have at least one motile stage. In agroecosystems, motile stages most often occur in the soil. Locomotion is by out-of-phase waves of muscle contraction in the dorsoventral plane, resulting in draconic (rather than snakelike) serpentine undulations. In contrast, in the Criconematidae the waves are in phase, so that movement is earthwormlike. The repertoire of body movements exhibited by nematodes is somewhat limited (Crofton, 1971), and the plant parasites in soil probably move actively only a few to several centimeters per year. Plant nematodes tend to be more sluggish than microbivorous nematodes. Soil nematodes in general are capable of their most rapid movement when body length is about three times the average diameter of soil particles (Nicholas, 1984). Because nematodes are essentially aquatic creatures, their movement in soil is in the water phase and is affected by soil characteristics affecting moisture, e.g., texture, structure, slope position, rainfall, compaction, and so on.

Plant nematodes are probably attracted by metabolic products or other factors emanating from roots, but few specific attractants have been identified. Root feeders may be attracted to root tips as invasion sites (*Meloidogyne* spp.), to young tissues farther back (*Pratylenchus* spp.), or to older tissue (*Helicotylenchus dihystera*). Temperature and CO_2 have been shown to attract plant parasitic nematodes in increasing gradients (Dusenberry, 1987). Often roots contain or produce feeding deterrents or toxins (Anderson, 1987).

Long-distance movement can be by any means that transports soil or infected plant parts, such as farm machinery, animals (Fig. 1), wind, water, root crops, seed, soil peds in seed lots, or nursery stock. There may be a striking similarity in the species comprising plant nematode communities on crops growing in a region exhibiting similar macroclimatic char-

FIGURE 1   Two common ways of spreading nematodes: farm machinery and animals. (Iowa Agricultural Experiment Station.)

acteristics (Ferris et al., 1971a,b), but this region/crop similarity is exceeded by the efficiency of humans in distributing plant nematodes (Niblack, 1989; Norton, 1978). Nematodes can be carried in guts of rodents, birds, and probably other animals. Dissemination is a natural phenomenon and is difficult to stop, although it can be inhibited is some situations by quarantines and sanitation practices.

III. POPULATION ECOLOGY

We are fortunate in nematology to work with organisms of unitary structure. We are not confronted with modular organisms such as plants with continuous growth and clonal reproduction. Although nematodes can be counted as unitary organisms, there are problems involved in obtaining reliable estimates of populations, and this should be kept in mind when reading generalizations about nematode ecology.

"Ecological" studies of plant parasitic nematodes most often involve single nematode species or populations. This is a natural consequence of the impact of some plant nematodes on their very economically important hosts. Population ecology in this chapter is meant to include species populations and their interaction with biotic and abiotic influences, and subjects independent of a population–community (synecological–autecological) dichotomy. Nematode community studies range from the purely qualitative, i.e., lists of species found in association with a given host, to more quantitative analyses. The community ecology section in this chapter will deal with subjects involving more than one nematode species population at a time.

TABLE 1   Some Factors that Affect Populations of Plant Parasitic Nematodes

I. Abiotic	II. Biotic

A. Topographic
  1. Elevation
  2. Slope
  3. Exposure
  4. Surface
 	
B. Soil environment
  1. Moisture
    Rainfall
    Snowfall
    Runoff
    Internal drainage
    Frost
  2. Temperature
    Mean
    Extremes
    Duration of extremes
    Cumulative heat units
  3. Aeration
  4. Texture
  5. Structure
  6. pH and fertility
  7. Organic matter
  8. Gas exchange
	A. Host
	  1. Suitability
	  2. Availability of feeding site
	
	B. Parasite
	  1. Life cycle
	  2. Reproductive rate
	  3. Survival mechanisms
	  4. Sex ratios
	  5. Infectivity
	
	C. Human
	  1. Cultural practices
	    Plowing
	    Rotation
	    Resistant varieties
	    Pesticides
	  2. Conservation practices
	
	D. Other biota
	  1. Fungi
	  2. Bacteria
	  3. Nematodes
	  4. Viruses and related forms
	  5. Insects and mites
	  6. Other fauna

A. Biotic and Abiotic Factors Affecting Populations

Some factors that affect distributions and dynamics of populations of plant parasitic nematodes are outlined in Table 1 and many were discussed in reviews (Norton, 1978; Wallace, 1971; Yeates, 1981, 1987). A little thought should illustrate possible ways in which these factors, alone or in combination, can or could act on nematode populations. Naturally, some factors are more important than others and must be determined for individual species under different circumstances. Any combination of factors can change the carrying capacity of a habitat for a species or act as screens to limit the number of species present in a community of parasites.

The host plant, of course, has primacy in the biotic factors affecting nematodes. Its myriad effects on every facet of nematode biology and ecology were reviewed by Yeates (1987). Other biotic factors, including pathogens, parasites, and predators, represent a field of study that has and will provide wide scope for study (Poinar and Jansson, 1988) and perhaps application.

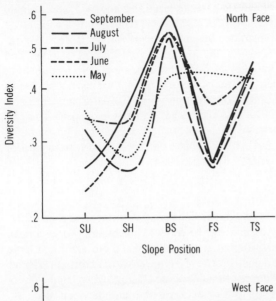

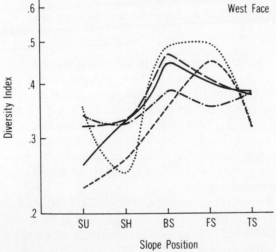

FIGURE 2   Diversity (*H'*) indices of nematodes along a maize toposequence. SU = summit; SH = shoulder; BS = backslope; FS = footslope; TS = toeslope. (Reproduced by permission from Norton and Oard, 1981.)

Among the abiotic factors that govern nematode populations, moisture and temperature are generally considered to be the most important. Nematode densities vary considerably with time, edaphic conditions, and slope aspect (Fig. 2). The effects of soil properties on nematode populations may be direct or indirect.

B.   Distribution of Populations

1.   Geographic Distribution

Many nematodes have a wide host range, occur in a wide range of environments, and are cosmopolitan. These are apt to be rather primitive types within a taxon in that they are not especially specialized morphologically or in host–parasite reactions; however, some highly specialized parasites, such as *Meloidogyne* spp., can also be characterized by a wide range and wide distribution. Some species are restricted by environment, but not necessarily exclusively. For example, *Meloidogyne incognita*, the "southern root-knot nematode," and *Meloidogyne hapla*, the "northern root-knot nematode," tend to be most common in North America in the ranges to which their common names refer (Anonymous, 1984). *Belonolaimus longicaudatus* has a widespread distribution in the United States but is restricted to soils containing 80–90% sand (Robbins and Barker, 1974). Other species may be distributed in other patterns and for reasons that are not entirely clear. For example, *Hoplolaimus galeatus* is more common and better known in the middle and eastern parts of the United States than in the western part (Anonymous, 1984).

2. Local Distribution

Nematode populations can vary in three spatial dimensions and over time, and changes affecting variation in one dimension may or may not be reflected in others. For example, *Belonolaimus longicaudatus* populations sampled at a depth of 5–15 cm exhibited large fluctuations in densities over time, while densities at 25–50 cm were fairly constant (Barker et al., 1969). In row crops, plant nematodes are often distributed "lengthwise," in the direction of tillage. Horizontal distribution refers to distribution within and between rows.

a. Horizontal distribution

It is generally agreed that of the basic distribution patterns,—uniform, random, and clustered,—the last is by far the most common for plant and soil nematodes. Among the causes of clustering are (a) occurrence of qualitative differences within hosts, resulting in some parts attracting and nourishing some nematodes more than others; (b) production of eggs in clumps by sedentary females; (c) production of several generations by nematodes with short life spans that flourish in some habitats and not in others; (d) competition among various fauna and microflora for nutrients or space; (e) inhibition by local environmental factors such as toxic substances; and (f) crop management practices in agroecosystems that may reduce numbers of some species and increase others.

Horizontal spatial patterns will vary with temporal ones. Within a field species and numbers of nematodes will vary widely among plants or even from one side of a plant to another (Alby et al., 1983; Barker and Nusbaum, 1971). Frequency distributions are often positively skewed (Fig. 3) in that large populations occur in relatively few samples while most samples contain few nematodes. It is this high degree of variability that makes measuring of nematode populations so difficult and imprecise. This skewing has serious implications for damage forecasting when the imprecision is associated with overestimates of crop loss (Noe and Barker, 1985; Seinhorst, 1973). A number of distribution functions have been used to describe horizontal distribution of nematode populations and can help reduce the risk involved in basing crop damage estimates on nematode population estimates (Barker and Noe, 1988; McSorley, 1987).

b. Temporal distribution

Populations of nematodes rarely remain constant for long. Some populations peak early in the season and then decline, abruptly or steadily, for the remainder of the season, while others increase throughout the season only to be limited by a reduction in resource, by the physical-chemical environment, or competition, or predation by other biota. Some plant nematodes, such as *Xiphinema* and *Longidorus* spp., are long-lived, and their densities may vary little within a year (Flegg, 1968). Differences in temporal distributions may reflect

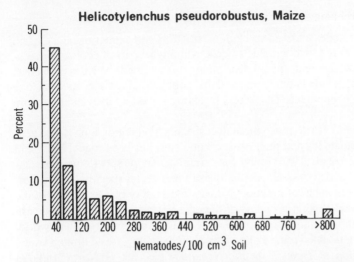

FIGURE 3 Typical nematode frequency distributions when many samples are collected. (Iowa Agriculture Experiment Station.)

inherent differences among nematode species, or be associated with seasonal changes in the quantity or quality of plant material, or both. This allows cohabitation of roots by pathogenic species and complicates studies of competition. Yeates et al. (1985) found sequential, complementary distributions of species of *Meloidogyne, Heterodera*, and *Pratylenchus* on white clover sampled over several months at 3-week intervals. Kraus-Schmidt and Lewis (1979) found similar relationships among *Hoplolaimus, Meloidogyne*, and *Scutellonema* on cotton.

c. Vertical distribution

In agroecosystems, the largest numbers of plant nematodes are found in the top 15–20 cm of soil, but some may be found at depths of 240 cm (Raski et al., 1965). Nematodes of many species are stratified in soil. *Criconemella xenoplax*, for example, is found to a depth of 1 m on peach roots, whereas *C. ornata* is found mostly in the upper 15 cm on peanut (Barker, 1982). Root distribution may control distribution of plant nematodes but is not the only factor. Shaping of citrus trees will change soil temperatures, compared with nonshaded ones, with the result that *Radopholus similus* will be closer to the surface in shaded areas than in nonshaded areas, even though abundant roots are present in both levels (Reynolds and O'Bannon, 1963).

Vertical migration probably is largely controlled by temperature, moisture tensions, and root distribution (O'Bannon et al., 1972; Schmitt, 1973), but often evidence of migration is circumstantial. The possibility of preferential colonization at various levels as a response to environmental changes at various levels over time must be considered. Prot (1980) reviewed the literature on migration. In any study of migration, however, techniques must be examined carefully to ensure that conclusions are not based on artifacts of methodology. Doubtless short distance migration occurs, but documentation of the distances migrated in soils are difficult to obtain.

C. Biogeography

As with all organisms, nematodes undergo speciation, radiation, and extinction, and as such these processes are pertinent in biogeographic studies. Although a part of ecology biogeographic studies usually examine widespread distribution as patterned over time, whereas traditional ecology studies include habitat relationships, population increase, survival, age structure, and similar local phenomena. These local facets are important in evaluating more extensive distributions.

It probably is premature to do much more than speculate on regional or global geographic distributions of nematodes and their causes, but a start has been made. Boag and Topham (1985) and Topham et al. (1985) found that certain nematode species were associated with and could be used to detect small populations of virus vector species, demonstrating the potential usefulness of such information. Nematologists in only a few states of the United States have published general accounts of the occurrences of plant nematodes in their respective states, even in what is probably the most surveyed country in the world. Mere listing with minimal annotated remarks and infrequent updating are of limited value. The publication *Distribution of Plant Parasitic Nematode Species in North America* (Anonymous, 1984) was an attempt to collate the known occurrences up to 1984. Although it accomplished much, periodic revisions are necessary.

In studies of nematode distributions and their causes, caution should be made to avoid introductions, i.e., agricultural settings. Patterns of distribution can best be shown where natural forces, barriers such as oceans or climatic forces, limit long-range spread of species. Ferris et al. (1986) related the distributions of some dorylaimids to plate tectonics. As might be expected, there is a tropical fauna in which circumstantial evidence indicates that certain species would not become established in higher latitudes even if introduced. Some examples are *Rotylenchulus reniformis* and many species of *Xiphinema*, among others.

Taxonomic problems further confuse the problems in studying wide distributions of plant parasitic nematodes. For example, *Helicotylenchus dihystera* collected from Iowa prairies are not the same as those collected from maize in South America (Norton, unpublished). Geographic isolation, eliminating gene flow among related populations, is one of the most important causes of speciation. Local abiotic conditions exert selection pressures and, combined with the well-documented effects of host on morphometric variation (Yeates, 1987), result in morphological changes through evolutionary processes such as natural selection and genetic drift. Small morphological changes result in ecotypes or geographic variants. Eventually, some of these variants may evolve into new species.

Although in science we often state that the best experiments are the simplest ones, we must keep in mind that in a holistic view of ecology, as well as other phases of biology, simplistic explanations of phenomena are usually to be avoided.

D. Survival

Doubtless many species of nematodes have become extinct, and doubtless many are now endangered as habitats are being destroyed or chemicals applied. Purposeful local eradication has evidently been successful in a few instances, as with *Globodera rostochiensis* in Delaware and upstate New York in the United States. Small local populations of nematodes are in greater danger of extinction than large populations unless survival capabilities of the former permit persistence. Planting of resistant or nonhost crops may so reduce a population that severe climatic and edaphic conditions may eliminate populations locally. Some plants nematodes have greater capabilities for survival than others. Nematodes with long life cycles frequently survive, although numbers may be few.

Plant parasitic nematodes are able to survive unfavorable conditions by entering dormant states or states of arrested development classified as quiescent or diapause (Evans, 1987; Antoniou, 1989). Quiescent states are induced by unfavorable environmental conditions; facultative quiescence is a readily reversible response to sudden environmental changes, and obligate quiescence is life stage-specific, requiring specific environmental signals to end. Either form may be induced by lack of water (anhydrobiosis), high salt concentration (osmobiosis), lack of oxygen (anoxybiosis), low temperature (cryobiosis), and high temperature (thermobiosis) (Antoniou, 1989). Anhydrobiosis is the best characterized and documented (Antoniou, 1989; Demeure and Freckman, 1981), and may allow nematode survival for a few months up to 39 years.

Quiescence is not confined to one stage of development. In an Iowa native prairie, evidence indicated that *Helicotylenchus pseudorobustus, Merlinius joctus*, and *Xiphinema americanum* overwintered mainly as eggs, while many vermiform individuals of *H. leiocephalus, Tylenchorhynchus maximus, T. nudus*, and *T. silvaticus* survived the winter (Schmitt, 1973). The host may also affect survival. Koenning et al. (1985) found that survival of *Pratylenchus brachyurus* was less in a winter cover of wheat than in fallow soil, and that winter survival was generally density-independent.

Diapause, like quiescence, may be either a facultative or an obligate state (Evans, 1987), but differs from quiescence in that endogenous factors are responsible for the arrest in development. Most of the plant nematodes known to exhibit diapause do so in the egg (*Meloidogyne, Heterodera*, and related species).

E. Population Dynamics

1. Parameters of Populations

Population dynamics, according to Ferris and Wilson (1987), is a term "used to convey changes in the numbers, age class distribution, sex ratio, and behavior of a population through time and space, determined by inherent characteristics of the individuals and mediated by environmental condition, food resources, and interacting biotic agents." Plant nematode populations have often been characterized as r or K strategists (Ferris and Wilson, 1987; Nicholas, 1984; Wharton, 1986), but these categories are not mutually exclusive and tend to most useful in comparisons within rather than between taxonomic groups (Yeates, 1987).

Densities of nematodes, totals or broken down into stages, are the usual data collected in population studies. Feedbacks that decrease density by decreasing births, survivorship, growth rates, or immigration are negative feedbacks; the opposite are positive feedbacks. Density-vague populations are characterized by high variances that can only be weakly explained by density (Strong, 1984). Often variances are so high that any density effects on population regulation are either absent or not discernable. Except in extreme pathological situations that may occur with plant pathogenic nematodes (e.g., *Meloidogyne*), plant parasitic nematodes do not deplete their resources. Their populations are density-vague or density-independent, governed by other factors such as abiotic ones, host compatibilities, or combinations of such.

Although numbers are used most commonly in analyzing nematode abundance in populations and communities, the use of biomass is often intuitively more satisfying. Biomass is generally defined as the amount of living matter in a given volume or area of habitat. (It seems counterintuitive when abundance and biomass data on nematodes are given per square meter, despite their existence in three dimensions; however, there are historical arguments in favor of retaining the two-dimensional mindset.) Nematode biomass is frequently calculated by the Andrássy equation:

$$\text{Biomass (}\mu g) = W^2 \times L \div (16 \times 100{,}000)$$

where W = greatest body width (μm) and L = body length (μm) (Yeates, 1988). The biomasses of adults of several plant parasitic species are listed by Waliullah (1983).

One measurement that probably would give the best information on the biology of an organism is body weight (Brown and Gibson, 1983). Body weight, however, must be used in conjunction with feeding habits and many ecological parameters. Different expressions of population change can result depending on whether actual nematode densities (counts) or biomass is used (Fig. 4). Similarly, Yeates (1988) showed how a related parameter, biovolume, does not vary linearly with abundance. Duncan and Freckman (in Freckman, 1982) found alarming disparity between biomass calculated by the Andrássy formula given above and by a more laborious method, itself based on several assumptions. Because other estimates are based on the calculation of biomass (e.g., respiration, production), they suggested that this area needs research attention.

2. Modeling

Studies on the seasonal fluctuations of populations of plant nematodes and the influences thereon by soil characteristics, management practices, host suitability, community structure, and other factors are legion. They have demonstrated a wonderful variety of interactions among nematode reproduction, host response, and environmental influences. There are several comprehensive reviews of the theories and various implementations of population modeling as an end in itself or as a basis for nematode management (Barker and Noe, 1988; Duncan and McSorley, 1987; Ferris and Noling, 1987; Ferris and Wilson, 1987). For economically important plant parasites, empirical models can easily be constructed and applied for predictive purposes using field data and regression analyses. Recently, work has increased on more complex and biologically descriptive simulation models whose parameters do not have to be redefined for changes in environmental conditions, for example.

The most important factor affecting populations of plant nematodes is the presence of a suitable host. For annual crops, the critical point for measuring nematode population density is at planting. This reflects the biological as well as economic reality that crop yields are related to initial nematode population densities (P_i) and that currently available nematode management strategies must be applied at planting. A general, well-known model for relating nematode P_i to crop yield was proposed by Seinhorst (1965) (Fig. 5). The model describes several characteristics of a given nematode–host interaction: the sigmoid relationship between P_i due to intraspecific competition. Duncan and Ferris (1982) expanded the model to describe the effects of multispecies infestations. Economic threshold concepts can easily be applied to this and similar models to be used as a basis for optimizing management decisions. While useful for modeling interactions in annual crops, a different approach must be taken for perennials (Duncan and McSorley, 1987; Ferris and Noling, 1987).

Evolving technology has and will have a profound effect on the development and implementation of predictive modeling for nematode management (Bird and Thomason, 1980), e.g., in the development of expert systems. The sensitivity of the annual crop–pathogenic nematode interaction to initial conditions would seem to make their long-term interactions a suitable system for description using chaos theory.

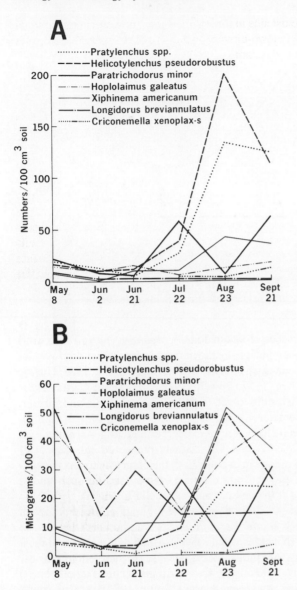

FIGURE 4   Comparison of the same nematode populations in a maize field on the basis of densities (A) and biomass (B). (Iowa Agriculture Experiment Station.)

F.   Interactions with Other Microorganisms

Although known for some time but largely ignored until the 1950s, that nematodes influence other organisms and their effects on plants has now been widely accepted. The comparative ease of working with single species may have obscured more complex interactions. While often true that if one species of nematode is controlled, a severe disease situation may be mitigated so that it is no longer economically important, other complexes may not be solved so readily. Atkinson (1892) was perhaps the first to find that a disease, fusarium wilt

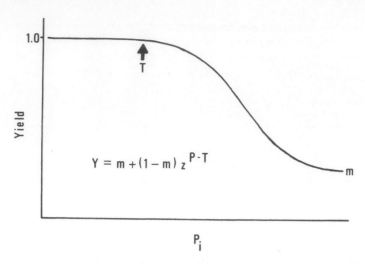

FIGURE 5   Generalized relationship between plant parasitic nematode densities at planting (P_i) and relative yield of a susceptible host. T = tolerance limit; m = minimum yield; z = a constant representing the amount of root not attacked by the nematodes. (From Seinhorst, 1965.)

of cotton, was more serious in the presence of a nematode (a *Meloidogyne* sp.) than in its absence. The range of interactions among plant parasitic nematodes and other microorganisms as they affect plant growth have been extensively reviewed (e.g., Huang, 1987; Hussey and McGuire, 1987; Lamberti and Roca, 1987; Powell, 1979; Sikora and Carter, 1987; Smith, 1987). The "other microorganisms" involved include fungi, bacteria, and viruses. Nematodes also interact with larger creatures (e.g., earthworms, mites, insects, and rodents), but these interactions are not well characterized except for nematode–nematode interactions, discussed below. Powell (1971) and his students pioneered this aspect of nematology. Any change caused by an organism in one part of the plant affects the physiology of other parts of the plant and thus may act some distance from the source; such studies now provide a basis for investigations at the cellular and molecular level.

Nematodes may be involved in disease caused by fungi (*Cylindrocladium crotalariae, Fusarium* spp., *Rhizoctonia solani, Verticillium* spp., and others) that invade the roots, as well as foliage pathogens (Nicholson et al., 1985). Much of the early and most current work was with *Meloidogyne* spp., but fungal–nematode disease interactions are now known with *Belonolaimus, Pratylenchus*, and others. A range of interactions among hosts, nematode species, and mycorrhizal fungi have been reported (Smith, 1987). As with fungi, a range of interactions with bacterial plant pathogens exists. Nematodes may be involved in or required for development of certain plant diseases caused by bacteria (see below), interact positively or negatively with rhizobial symbionts of legumes (Huang, 1987), or feed on plant pathogenic bacteria (Nicholas, 1984).

As with other areas of nematode ecology, there are a number of problems that complicate interaction studies (Sikora and Carter, 1987; Wallace, 1983) because of the number of variables involved, especially in soil systems. The task is easier, but not without complications, with interactions of nematodes with other organisms in above-ground plant parts. Examples are the association of *Anguina tritici* and *Corynebacterium tritici* that results in a different symptom alone from that in combination (Gupta and Swarup, 1972); the association of *Subanguina calamagrostis* and the fungus *Dilosphora alopecuri* (Norton et al., 1987); and *Anguina agrostis* and *Corynebacterium rathayi* (Bird, 1981). In the latter, the nematode is a vector for the bacterium as a surface contaminant, the galls induced by the nematode become toxic to animals only when the galls are colonized by the bacteria.

Since the classic work of Hewitt et al., (1958), the transmission of plant viruses by nematodes has received much attention (Lamberti and Roca, 1987). About 20 nepoviruses are known to be transmitted by species of *Longidorus* or *Xiphinema*. At least two tobraviruses, tobacco rattle and pea early browning, are transmitted by species of *Paratrichodorus* or *Trichodorus*. Transmission of viruses by members of the Tylenchida has not been established. Nematodes ingest viruses when they feed on virus-infested plants, and evidently act mainly as a mechanical transfer for viruses, as the viruses do not multiply in the vector. However, viruses are retained in specific sites in the nematode, due partly at least to the nature of the protein surface of the virus particles, and perhaps the surface charge (Raski et al., 1973; Taylor and Robertson, 1977). It appears that many viruses are transmitted by a few nematodes, but because many viruses have wide plant hosts ranges, and some of the vectors are common and widespread, it is easy to comprehend that virus transmission is common.

G. Bioenergetics

Bioenergetics is the study of energy transformations and energy exchanges within and between living things and their environments. The scope of ecological energetics (versus biochemical-molecular energetics) and a derivation of the generalized formula used to describe energy flow through a population or community was given by Phillipson (1975) as:

$$C = P + R + F + U$$

where C (consumption) = total intake of food energy during a specified time interval; P (production) = energy content of the biomass of food digested less that respired or rejected; R (respiration) = energy converted to heat and loss in life processes; F (egesta) = energy content of food not digested; and U (excreta) = energy content of digested material passed from the body. The preferred energy unit used is the kilojoule (1 calorie = 4.816 J).

Each of these components requires a number of estimates (total densities, growth rates, reproduction, etc.) and laboratory determinations (oxygen consumption, biomass, energy content), and though the generalized formula appears quite simple, doing the actual calculations is complex. Yet some have attempted it, to allow comparisons to be made among populations, trophic groups, or habitats. Sohlenius (1980) and Yeates (1979) reviewed the literature on the contribution of the total nematode component to energy flow in terrestrial ecosystems. Sohlenius (1980) estimated that nematodes contribute only about 1% to total soil respiration, perhaps 10–15% of animal respiration. Yet bacterial-feeding nematodes can consume as much as 50% of the annual production of microfloral biomass (Paul and Clark, 1988); thus, as regulators of the primary decomposers combined with plant parasites as primary consumers, the contribution nematodes make to energy flow of a system is substantial, as their numbers would indicate. With respect to plant parasitic nematodes, analysis of energetics can also provide insight into their pathogenic effects on plants, and the interrelatedness of the biology of a parasite and its host (Atkinson, 1985).

IV. COMMUNITY ECOLOGY

A. Populations and Communities

A community, in the sense of bioenergetics, must be composed of primary producers, herbivores, and carnivores. However, the term "animal community" is used to describe assemblages of animals in a given habitat, assumed to have food web dependencies as well as mutualistic interactions (Boughey, 1973). A nematode community is an assemblage of nematodes, including primary consumers to predators. It can be studied from several aspects: by numbers of species, numbers of individuals, biomass, physiological or ecological activity, trophic groups, and so on. Techniques applied in population ecology can be applied to communities; community changes can result from normal cyclic patterns, competition or other density-dependent factors, environmental constraints, and many other events. Their interactions may be weak or tightly knit. If species become established, they fill a niche, but niches tend to complement each other and do not work in direct competition (Whittaker, 1975). Usually an increase in niches means an increase in productivity (Boughey, 1973).

A question often asked about communities is whether additional parasitic species could be introduced so that they persist and reproduce without causing extinction of other species. Most species capable of attacking a host species are not locally available and therefore we do not know how much species packing can occur. If an annual monocropping system could be perpetuated for hundreds of years, and if enough different nematode species could eventually be available, would there be greater species packing than there is now? There is evidence of increased species packing in the short term in that there was a trend toward an increase in numbers of obligatory parasites and total nematodes in separate alfalfa plots monitored during 1–3 and 3–5 years (Wasilewska, 1967, 1979). Longer term experiments are needed to allow small populations low and undetectable species to increase (Sohlenius et al., 1987).

The makeup of a community is not entirely a fortuitous one, although chance is important. The following factors in community formation, modified from Mueller-Dombois and Ellenberg (1974), seem to be pertinent for nematodes.

1. The species of an area provide the basic materials for a community.
2. Because of limited short-term dispersal and barriers of various kinds, most plant parasitic nematodes do not come in contact with a given host (see also Price, 1986; Holmes, 1986).
3. Properties of the nematodes themselves, especially their life forms, life cycles, and physiological requirements, allow different species to coexist, persist, and reproduce.
4. The niche is the total of factors operative in a given habitat.
5. Communities change over time; the time elapsed from the initial occupancy to any desired point will affect the community.
6. In addition is the biology, including the ecology, of the host. Although the host is part of a nematode's habitat, it is so important that a separate category is justified.

B. Habitats

The number of nematode species in a community varies with habitat. Prairies and woodlands generally are richer in all nematode species than are cultivated fields; and prairies are richer than woodlands (Burkhalter, 1928; Egunjobi, 1971; Wasilewska, 1979; Weaver and Smolik, 1987; Yeates, 1979), but the number of species in a cultivated field can be as high as

74 (Baird and Bernard, 1984). Plant parasites generally compose a higher percentage of the total nematode population in agricultural settings than in natural or lightly managed areas (Ferris and Ferris, 1974; Niblack and Bernard, 1985), with up to 16 species in a site.

In agroecosystems, many nematode species are an integral part of the community through parasitism of the host or interactions with other nematodes. Others are residual from a preceding crop or associated weeds. Probably most species are rare most of the time (Caughley and Lawton, 1981). Around a given crop, only a limited number of nematode species occur regularly. Other will be erratic, but most species will be absent. For example, over 170 plant parasitic nematode species have been associated with maize, but usually only three to eight species are found around a plant at any time. Similar number of species per site have been shown for other crops (Niblack, 1989; Yeates, 1987); Yeates (1987) suggested that around seven species per guild or trophic group per habitat is "normal" for nematodes.

Humankind is a great dispenser of nematodes, noted above, but most species in cultivated fields are probably residuals from natural areas before cultivated agriculture appeared. As is true with insects (Brown and Gibson, 1983; MacArthur, 1972), most introduced contemporary nematode species became established in disturbed areas. Agricultural systems generally develop toward monoculture, and associated management practices restrict herbage diversity. Cultivation, or compaction in no-till regimes, results in soil structural changes which in turn causes more moisture and temperature fluctuations than are found in noncultivated areas. These changes result in conditions often exceeding the ecological amplitude of many nematodes; the resulting unstable habitats inhibit or prevent many nematodes from becoming established or persisting. Agriculture favors some nematodes whose genetic makeup allows survival and reproduction under frequent environmental changes, including nutritional resources. Thus, nematodes in agricultural systems generally have lower richness and diversities than in natural areas.

C. Trophic Groups

During the long course of evolution, morphological and biochemical modifications result in a diversity of nematode feeding types and habits; thus trophic groups make a useful classification for community studies. Trophic groups have been variously categorized (e.g., Freckman and Caswell, 1985; Niblack, 1989; Overgaard-Nielsen, 1960; Wasilewska, 1971a,b; Yeates, 1971, 1979), and usually the categories are based on known feeding habits or pharyngeal morphology. They usually include those that feed primarily on bacteria (bacterivores, microbivores), fungi (mycophages, fungivores), higher plants (phytophages, plant parasites), small animals (carnivores, predators), and those that feed on a variety of substrates (omnivores). Freckman and Caswell (1985) provided descriptions of each category and presented a model of how interactions among them might occur (Fig. 6). Demarcations are not always clear and may overlap. For example, some of the primitive tylenchids feed on fungi as well as on higher plants. Several species of rhabditid genera (e.g., *Acrobeloides, Cephalobus, Panagrolaimus*) may be able to obtain nourishment from plant tissue (Poinar, 1983). A system of trophic classification that could be applied to nematodes and would reflect the importance of size among soil biota was proposed by Heal and Dighton (1985) to include microtrophic, mesotrophic, and macrotrophic groups.

Our main interest is with plant parasites, but they are only one component of a community, and thus may be influenced directly or indirectly by other biota, including other nematodes. Whether the nematode component of the soil biota represents a community of interacting members, or interactions are minimal among groups exploiting different resources, is a matter of question (Niblack, 1989; Yeates, 1984). Yeates (1987) reviewed the studies showing positive correlations between total nematode abundance and primary pro-

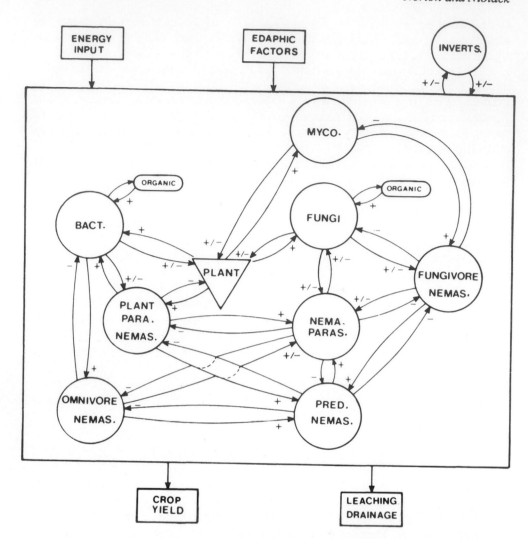

FIGURE 6   Interactions among some components of an agricultural food web. Arrows indicate
the direction of the influence; + and − indicate a positive or negative effect. Bact. = bacteria; Fun-
givore Nemas. = fungivorous nematodes; Inverts. = invertebrates; Myco. = mycorrhizae; Nema.
Paras. = parasites and predators of nematodes; Omnivore Nemas. = omnivorous nematodes; Organic = organic matter; Plant Para. Nemas. = plant parasitic nematodes; Pred. Nemas. = predatory
nematodes. (Reproduced with permission from Freckman and Caswell, 1985.)

duction in natural or lightly managed ecosystems. Obviously, the nematode community has
an impact on the vegetation that is not limited to the deleterious effects of parasitism, but
few researchers have looked into this question in agroecosystems. Investigations of nematode community structure in agroecosystems can help explain whether the correlation is
coincidental or, if not, determine the basis for the relationship.

D. Structure

Community structure means different things to different people. Most agree that community structure involves patterns of species occurrence, their relative abundance, and resource use (Holmes, 1986). Pielou (1972) takes the position that there has to be interdependence among the species of a community to have structure and that a haphazard assemblage of species with no interaction is devoid of structure. In studies of a complete nematode fauna there is more likely to be evidence of structure if the community is analyzed in terms of its trophic group (or some group other than taxonomic) composition, but this is a matter of controversy (Niblack, 1989).

The appearance of structure will depend on the level of resolution of the study. In agroecosystems, studies often focus on the plant parasitic nematodes only. Take, for example, two papers mentioned earlier. First, the work of Boag and Topham (1985) on species associations demonstrated a useful structure that might allow prediction of the occurrence of virus vector nematodes. Second, Yeates et al. (1985) investigated a temporal structure allowing coexistence of three potentially pathogenic nematode populations. In the first instance, the structure involves particular species, and in the second, the particular species are not important to the conclusion of temporal structure.

The most important first step in studying a nematode community is to make a faunistic list or to classify the community in some way (e.g., Baird and Bernard, 1985; Bostrom and Sohlenius, 1986; Sohlenius et al., 1987). Except in rare or unusual circumstances, nematode communities are not fully censused but rather are estimated as we sample the total population, recovering only those species that are at a detectable level based on the extraction technique used. Thus, estimates are nearly always conservative. Different lists can be obtained within the same area depending on the thoroughness of sampling. The rare species that might be missed are usually not important unless they have a large biomass, such as *Longidorus* spp., or carry a virus.

Once the classification and enumeration of the community is complete, the numbers can be investigated in a variety of ways. We use whatever expression is appropriate to the question, such as diversity or concentration of dominance, prominence values, dispersion, and so on, combined with proper analytical techniques (Niblack, 1989).

E. Diversity

It is a common belief that natural ecosystems are more diverse and thus more stable and resistant to perturbations than habitats disturbed by humans (Mindermann, 1956; Wasilewska, 1979). Although it is often stated that community diversity leads to stability, modern thought and evidence support the contention that environmental stability leads to community stability, which permits high diversity (Pielou, 1975). Diversity can be expressed in several ways (Pielou, 1975), most simply by counting up the number of species and comparing their proportions in a community. However, a single expression is useful, allowing for statistical analyses and comparisons among communities. The expression most commonly used in some modification of the Shannon function, usually the Shannon–Wiener index:

$$H' = - \sum_{i=1}^{s} \log_2 p_i$$

where s = number of species, p_i = proportion belonging to the ith species, and H' estimates the probability of correctly predicting the species of an individual randomly drawn from the population. H' confounds the number of species and their evenness, and it is desirable to keep them distinct (Pielou, 1975). Therefore, a measurement of evenness (often called J or J') can be used:

$$J' = \frac{H'}{H'_{max}}$$

where $H'_{max} = \log_2 s$.

What have diversity studies shown in plant nematology? We know that there are differences in community composition when β diversity is measured along a toposequence or over time (Fig 2; Norton and Oard, 1981). Yeates (1984) found in his study of nematode populations in seven soils with grazed pastures that diversity was not related to pasture productivity, and evenness was negatively correlated with the abundance of the abundance of the dominant species. Richness of species and diversity can vary greatly in a generalized biome. For example, the diversity of plant parasitic nematodes in forests of the Adirondack Mountains of New York State decreases as elevation increases due greatly to the more rigorous climate at the higher elevations (Norton and Oard, 1982). H' was lower in a maize field when biomass rather than numbers was used to calculate the statistic (Fig.4; Norton and Edwards, 1988). This is because a few large nematodes, including *Longidorus brevian-nulatus*, dominated the community biomass whereas they constituted a small numerical part of the community. Similarly, H' averaged over 15 sites in an Iowa prairie was lower when nematode biomass was used rather than numbers (Norton and Schmitt, 1978). That H' can vary with crops and soils can be demonstrated by calculating it from data given by Ferris and Bernard (1971b, Table 1) from rotation crops in Illinois; H' tended to be higher around corn than with soybeans. Niblack and Bernard (1985) reported that H' was higher around maple than peach or dogwood in Tennessee nurseries. H' was positively correlated with age of tree in dogwood, but not with maple sites, and only weakly correlated with degree of weed cover or number of weed species present; thus increasing diversity of herbage did not increase herbivorous nematode diversity. Species richness and diversity were not affected by soil cultivation in an annual compared with a perennial cropping system in Sweden (Bostrom and Sohlenius, 1986; Sohlenius et al., 1987).

F. Interaction Among Plant Parasitic Nematodes

It is common for three or more species of plant parasitic nematodes to occupy the same general area and feed on the same host at the same time, and these nematodes may each have measurable effects on their hosts. Thus, it is reasonable to expect that interactions between or among the nematodes will affect either the nematodes or plant growth, or both. Indeed, the interactions documented range from neutral (no measurable interaction) to stimulatory to deleterious to one or more of the participants. Most research on interactions has been under confined conditions such as in pots in the greenhouse and often with numbers much larger than those found in nature. Nonetheless, nematodes do interact in reality, with consequences for agriculture. The literature has been reviewed by Eisenback and Griffin (1987).

Oostenbrink (1966) attributed the polyspecific nature of plant parasitic nematode communities to four factors, three of which have been mentioned in previous sections of this chapter: (a) the effect of humans in distributing nematodes in soil and plant parts; (b) the wide host ranges of many plant nematodes; (c) their ability to survive; and (d) the low incidence of interspecific competition among them. Opinions on the existence and importance of competition, difficult to study in any event, have varied from almost complete disregard to those of the "competitionists," who consider competition to be a major characteristic of interactions among species (Lawton, 1984; Schoener, 1982). In studies of population changes, it is sometimes tempting to ascribe inverse peaks between them. This pattern has lead to the assumption that one nematode can outcompete another, but population changes may merely be a matter of differences in niche dimensions and have nothing to do with interactions among species. For instance: (a) nematodes are highly aggregated, and their distribution patterns may not overlap; (b) life cycles vary so that some nematodes are common or rare at different times; (c) several kinds of substrates occur in a root system, such as root hairs and other epidermal cells, cortical parenchyma, stelar tissue, and so on, that may be preferred by different nematodes; (d) even when parasitized by large numbers of nematodes, usually there is a lot of tissue that is not colonized. A clear distinction should be made between true competition and population decline of a species due to natural cyclic patterns or indirect interactions.

Competition is not likely to occur when many niches occur, when there are vacant niches, when the substrate is rapidly growing, when there is slow colonization, or when the substrate is ephemeral (Price, 1986). Competition is most likely to occur when reproduction of the parasite is rapid and the host tissue damage is great. At the beginning of the season, at least with annual crops in temperate regions, the resources are growing faster than most plant parasitic nematodes can multiply. Thus, there is apt to be no competition unless the initial population is so large that root growth is markedly reduced or there are physiology changes in the host that may favor the nematode.

Competition theory is difficult to test with nematodes for at least three reasons: (a) suitable controls are difficult to establish; (b) growing seasons are often too short for long-term interactions to be evident; and (c) the change in substrate by nematode destruction is too rapid to allow continuous "healthy" nutritional resource. Theoretically, competition should increase diversity by forcing some nematodes to feed on a less favorable substrate. Because of a diversity of substrates and physical and chemical factors, and thus many niches, competition is more apt to occur intraspecifically than interspecifically. The fact remains that, in general, plants are not being eliminated by nematodes, and a plant is capable of supporting a far greater nematode population than it usually does.

V. CONCLUSIONS

In this chapter, we have often emphasized how many problems there are with studies of nematode ecology, even at the most prosaic level. We cannot begin to suggest that all questions have been addressed, much less answered. Our discussions have also frequently included such statements as, "A range of interactions exist, from one extreme to another." One cannot assume that the "average" nematode falls somewhere in the middle, and over-generalizations are a constant hazard. Every level of resolution gives a new perspective on the complexity of the nematodes' milieu. Because of space constraints, we have dealt with a number of subjects in a cursory way. Even so, it should be obvious that the major theme in the biology and ecology of nematodes is interrelatedness. Changing even the smallest component of a community will have far-reaching effects, whether or not they are immediately obvious or even measurable. Yet communities retain remarkable stability in the absence of catastrophe. There is increasing emphasis on developing sustainable agricultural production systems (i.e., with less harsh environmental effects) given the increasing demands of human populations on productivity. This is the best reason for plant nematologists to take a broad view of the role nematodes play in agroecosystems and still satisfy what Bird (1981) called our "understandable tendency to study organisms only when they impinge on [our] economy."

REFERENCES

Alby, T., Ferris, J. M., and Ferris, V. R. 1983. Dispersion and distribution of *Pratylenchus scribneri* and *Hoplolaimus galeatus* in soybean fields. *J. Nematol.* 15: 418–426.

Anderson, D. C. 1987. Below-ground herbivory in natural communities: a review emphasizing fossorial animals. *Quart. Rev. Biol.* 62: 261–286.

Anonymous. 1984. *Distribution of Plant-Parasitic Nematode Species in North America.* Society of Nematologists, Hyattsville, MD.

Antoniou, M. 1989. Arrested development in plant parasitic nematodes. *Helminthol. Abstr. Ser. B.* 58: 1–19.

Atkinson, H. J. 1985. The energetics of plant parasitic nematodes: A review. *Nematologica* 31: 62–71.

Atkinson, G. F. 1892. Some diseases of cotton. *Alabama Agr. Expt. Sta. Bull.* 41: 1–65.

Baird, S. M., and Bernard, E. C. 1984. Nematode population and community dynamics in soybean-wheat cropping and tillage regimes. *J. Nematol.* 16: 379–386.

Barker, K. R. 1982. *Criconemella* in the Southeastern United States. In *Nematology in the Southern United States*, R. D. Riggs, ed. Southern Cooperative Series Bulletin 276, pp. 150–156.

Barker, K. R., and Noe, J. P. 1988. Techniques in quantitative nematology. In *Experimental Techniques in Plant Diseases Epidemiology*, J. Kranz and J. Rotem, eds. Springer-Verlag, New York, pp. 223–236.

Barker, K. R., and Nusbaum, C. J. 1971. Diagnostic and advisory programs. In *Plant Parasitic Nematodes*, Vol. 1, B. M. Zuckerman, W. F. Mai, and R. A. Rohde, eds. Academic Press, New York, pp. 281–302.

Barker, K. R., Nusbaum, C. J., and Nelson, L. A. 1969. Seasonal population dynamics of selected plant-parasitic nematodes as measured by three extraction procedures. *J. Nematol.* 1: 232–239.

Bird, A. F. 1981 The *Anguina-Corynebacterium* association. In *Plant Parasitic Nematodes*, Vol. 3, B. M. Zuckerman, W. F. Mai, and R. A. Rohde, eds. Academic Press, New York, pp. 303–323.

Bird, A. F. 1983. Growth and moulting in nematodes: Changes in the dimensions and morphology of *Rotylenchulus reniformis* from start to finish of moulting. *Int. J. Parasitol.* 13: 201–206.

Bird, G. W., and Thompson, I. J. 1980. Integrated pest management: The role of nematology. *Bioscience* 30: 670–674.

Boag, B., and Topham, P. B. 1985. The use of association of nematode species to aid the detection of small numbers of virus-vector nematodes. *Plant Pathol.* 34: 20–24.

Bostrom, S., and Sohlenius, B. 1986. Short-term dynamics of nematode communities in arable soil. Influence of a perennial and an annual cropping system. *Pedobiologia* 29: 345–357.

Boughey, A. S. 1973. *Ecology of Populations.* Macmillan, New York.

Brown, J. H., and Gibson, A. C. 1983. *Biogeography.* Mosby, St. Louis.

Burkhalter, M. 1928. Die Verbreitung der freilebenden Erdnematoden in verschiedenen Gelandarten in Massif der Rochers de Naya (2045 m.) *Rev. Suisse Zool.* 35: 389–437.

Caughley, G., and Lawton, J. H. 1986. Plant-herbivore systems. In *Theoretical Ecology: Principles and Applications*, R. M. May Ed. Sinauer, Sunderland, MA.

Cook, R., and Evans, K. 1987. Resistance and tolerance. In *Principles and Practices of Nematode Control in Crops*, R. H. Brown and B. R. Kerry, eds. Academic Press, New York, pp. 179–222.

Crofton, H. D. 1971. Form, function, and behavior. In *Plant Parasitic Nematodes*, Vol. 1, B. M. Zuckerman, W. F. Mai, and R. A. Rohde, eds. Academic Press, New York, pp. 83–116.

Demeure, Y., and D. W. Freckman. 1981. Recent advances in the study of anydrobiotic nematodes. In *Plant Parasitic Nematodes*, Vol. 3, B. M. Zuckerman and R. A. Rohde, eds. Academic Press, New York, pp. 205–226.

Dropkin, V. H. 1988. The concept of race in phytonematology. *Ann Rev. Phytopathol.* 26: 145–161.

Duncan, L. W., and Ferris, H. 1982. Interactions between phytophagus nematodes. In *Nematodes in Soil Ecosystems*, D. W. Freckman, ed. University of Texas Press, Austin, pp. 29–54.

Duncan, L. W., and McSorley, R. 1987. Modeling nematode populations. In *Vistas on Nematology*, J. A. Veech and D. W. Dickson, eds. Society of Nematologists, Hyattsville, MD, pp. 377–389.

Dusenberry, D. B. 1987. Prospects for exploring sensory stimuli in nematode control. In *Vistas on Nematology*, J. A. Veech and D. W. Dickson, eds. Society of Nematologists, Hyattsville, MD, pp. 131–135.

Egunjobi, O. A. 1971. Soil and litter nematodes in some New Zealand forests and pastures. *N. Zealand J. Sci.* 14: 568–579.

Eisenback, J. D., and Griffin, G. D. 1987. Interactions with other nematodes. In *Vistas on Nematology*, J. A. Veech and D. W. Dickson, eds. Society of Nematologists, Hyattsville, MD, pp. 313–320.

Evans, A. A. F. 1987. Diapause in nematodes as a survival strategy. In *Vistas on Nematology*, J. A. Veech and D. W. Dickson, eds. Society of Nematologists, Hyattsville, MD, pp.180–187.

Ferris, H., and J. W. Noling. 1987. Analysis and prediction as a basis for management decisions. In *Principles and Practices of Nematode Control in Crops*, R. H. Brown and B. R. Kerry, eds. Academic Press, New York, pp. 49–85.

Ferris, H., and Wilson, L. T. 1987. Concepts and principles of population dynamics. In *Vistas on Nematology*, J. A. Veech and D. W. Dickson, eds. Society of Nematologists, Hyattsville, MD, pp. 373–376.

Ferris, V. R., and Bernard, R. L. 1971a. Crop rotation effects on population densities of ectoparasitic nematodes. *J. Nematol.* 3: 119–122.

Ferris, V. R., and Bernard, R. L. 1971b. Effect of soil type on population densities of nematodes in soybean rotation fields. *J. Nematol.* 3: 123–128.

Ferris, V. R., and Ferris, J. M. 1974. Interrelationships between nematode and plant communities in agricultural ecosystems. *Agro-Ecosystems* 1: 275–299.

Ferris, V. R., Goseco, C. G., and Ferris, J. M. 1976. Biogeography of free-living soil nematodes from the perspective of plate tectonics. *Science* 193: 508–510.

Filipjev, I. N., and Schuurmans Stekhoven, J. H., Jr. 1941. *A Manual of Agricultural Helminthology*. E. J. Brill, Leiden.

Flegg, J. J. M. 1968. The occurrence and distribution of *Xiphinema* and *Longidorus* species in southeastern England. Nematologica 14: 189–196.

Freckman, D. W. 1982. Parameters of the nematode contribution to ecosystems. In *Nematodes in Soil Ecosystems*, D. W. Freckman, ed. University of Texas Press, Austin, pp. 81–97.

Freckman, D. W., and Caswell, E. P. 1985. The ecology of nematodes in agroecosystems. *Ann. Rev. Phytopathol.* 23: 275–296.

Gupta, P., and Swarup, G. 1972. Ear-cockle and yellow ear-rot diseases of wheat. II. Nematode bacterial association. *Nematologica* 18: 320–324.

Heal, O. W., and Dighton, J. 1985. Resource quality and trophic structure in the soil system. In *Ecological Interactions in Soil: Plants, Microbes, and Animals*, A. H. Fitter, D. Atkinson, D. J. Read, and M. B. Usher, eds. Blackwell Scientific, Oxford, pp. 339–354.

Hewitt, W. B., Raski, D. J., and Goheen, A. C. 1958. Nematode vector of soil-borne fanleaf virus of grapevines. *Phytopathology* 48: 586–595.

Holmes, J. C. (1986). The structure of helminth communities. In *Parasitology: Quo Vadit? Proceedings of the Sixth International Congress of Parasitology*, M. J. Howell, ed. Australian Academy of Science, Canberra, pp. 203–208.

Huang, J.-S. 1987. Interactions of nematodes with rhizobia. In *Vistas on Nematology*, J. A. Veech and D. W. Dickson, eds. Society of Nematologists, Hyattsville, MD, pp. 301–306.

Hussey, R. S. 1989. Disease-inducing secretions of plant parasitic nematodes. *Ann. Rev. Phytopathol.* 27: 123–141.

Hussey, R. S., and McGuire, J. M. 1987. Interaction with other organisms. In *Principles and Practices of Nematode Control in Crops*, R. H. Brown and B. R. Kerry, eds. Academic Press, New York, pp. 292–328.

Ishibashi, N., Kondo, E., and Kashio, T. 1975. The induced molting of 4th stage larvae of pin nematode, *Paratylenchus aciculus* Brown (Nematoda: Paratylenchidae) by root exudate of host plant. *Appl. Ent. Zool.* 10: 275–283.

Koenning, S. R., Schmitt, D. P., and Barker, K. R. 1985. Influence of selected cultural practices on survival of *Pratylenchus brachyurus* and subsequent effects on soybean yield. *J. Nematol.* 17: 464–469.

Kraus-Schmidt, H., and Lewis, S. A. 1979. Seasonal fluctuations of various nematode populations in cotton fields in South Carolina. *Plant Dis. Reptr.* 63: 859–863.

Lamberti, F., and Roca, F. 1987. Present status of nematodes as vectors of plant viruses. In *Vistas on Nematology*, J. A. Veech and D. W. Dickson, eds. Society of Nematologists, Hyattsville, MD, pp. 321–328.

Lawton, J. H. 1984. Herbivore community organization: General models and specific tests with phytophagous insects. In *A New Ecology*, P. W. Price, C. N. Slobodchikoff, and W. S. Gaud, eds. John Wiley and Sons, New York, pp. 329–352.

MacArthur, R. H. 1972. *Geographical Ecology, Patterns in the Distribution of Species*. Harper and Row, New York.

May, R. M., and Seger, J. 1986. Ideas in ecology. *Am. Scientist* 74: 256–267.

Masamune, T., Anetai, M., Takasuge, M., and Katsui, N. 1982. Isolation of a natural hatching stimulus, glycinoeclepin A, for the soybean cyst nematode. *Nature* 297: 495–496.

McSorley, R. 1987. Extraction of nematodes and sampling methods. In *Principles and Practices of Nematode Control in Crops*, R. H. Brown and B. R. Kerry, eds. Academic Press, New York, pp. 13–41.

Mindermann, G. 1956. Aims and methods in population researches on soil-inhabiting nematodes. *Nematologica* 1: 47–50.

Mueller-Dombois, D., and Ellenberg, H. 1974. *Aims and Methods of Vegetation Ecology*. John Wiley and Sons, New York.

Niblack, T. L. 1989. Applications of community structure research to agricultural production and habitat disturbance. *J. Nematol.* 21:437–443 .

Niblack, T. L., and Bernard, E. C. 1985. Plant-parasitic nematode communities in dogwood, maple, and peach nurseries in Tennessee. *J. Nematol.* 17: 132–139.

Nicholas, W. L. 1984. *The Biology of Free-Living Nematodes*, 2nd ed. Clarendon Press, Oxford.

Nicholson, R. L., Bergeson, G. B., DeGennaro, F. P., and Viveiros, D. M. 1985. Single and combined effects of the lesion nematode and *Colletotrichum graminicola* on growth and anthracnose leaf blight of corn. *Phytopathology* 75: 654–661.

Nickle, W. R. (Ed.). 1984. *Plant and Insect Nematodes*. Marcel Dekker, New York.

Noe, J. P., and Barker, K. R. 1985. Overestimation of yield loss of tobacco caused by the aggregated spatial pattern of *Meloidogyne incognita*. *J. Nematol.* 17: 245–251.

Norton, D. C. 1978. *Ecology of Plant-Parasitic Nematodes*. John Wiley and Sons, New York.

Norton, D. C. and Edwards, J. 1988. Age structure and community diversity of nematodes associated with maize in Iowa sandy soils. *J. Nematol.* 20:340–350.

Norton, D. C., and Oard, M. 1981. Plant-parasitic nematodes in loess toposequences planted with corn. *J. Nematol.* 13: 314–321.

Norton, D. C., and Oard, M. 1982. Occurrence and diversity of some Criconematidae and Paratylenchidae in the Adirondack Mountains of New York State. *Proc. Iowa Acad. Sci.* 89: 11–14.

Norton, D. C., and Schmitt, D. P. 1978. Community analyses of plant-parasitic nematodes in the Kalsow Prairie, Iowa. *J. Nematol.* 10: 171–176.

Norton, D. C., Cody, A. M., and Gabel, A. W. 1987. *Subanguina calamagrostis* and its biology in *Calamagrostis* spp. in Iowa, Ohio, and Wisconsin. *J. Nematol.* 19: 260–262.

O'Bannon, J. H., Radewald, J. D., and Tomerlin, A. T. 1972. Population fluctuation of three parasitic nematodes in Florida citrus. *J. Nematol.* 4: 194–199.

Oostenbrink, M. 1966. Major characteristics of the relation between nematodes and plants. *Meded. Landouwhogeschool Wageningen* 66–4.

Overgaard-Nielsen, C. 1949. Studies on the soil microfauna. II. The soil inhabiting nematodes. *Natura Jutland.* 2: 1–131.

Paul, E. A., and Clark, F. E. 1988. *Soil Microbiology and Biochemistry.* Academic Press, New York.

Perry, R. N. 1987. Host–induced hatching of phytoparasitic nematode eggs. In *Vistas on Nematology*, J. A. Veech and D. W. Dickson, eds. Society of Nematologists, Hyattsville, MD, pp. 159–164.

Phillipson, J. 1975. Introduction to ecological energetics. In *Methods for Ecological Bioenergetics*, W. Grodzinski, R. Z. Klekowski, and A. Duncan, eds. Blackwell Scientific, Oxford, pp. 3–13.

Pielou, E. C. 1972. Measurement of structure in animal communities. In *Ecosystem Structure and Function*, J. A. Weins, ed. Oregon State University Press, pp. 113–135.

Pielou, E. C. 1975. *Ecological Diversity.* Wiley–Interscience, New York.

Poinar, G. O., Jr. 1983. *The Natural History of Nematodes.* Prentice-Hall, Englewood Cliffs, N.J.

Poinar, G. O., Jr., and Jansson, H.-B. 1988. *Diseases of Nematodes.* CRC Press, Boca Raton.

Powell, N. T. 1971. Interactions between nematodes and fungi disease complexes. *Ann. Rev. Phytopathol.* 9: 253–273.

Powell, N. T. 1979. Internal synergisms among organisms inducing disease. In *Plant Disease, Vol. 4*, J. G. Horsfall and E. B. Cowling, eds. Academic Press, New York, pp. 113–133.

Price, P. W. 1986. Evolution in parasitic communities. In *Parasitology: Quo Vadit? Proc. 6th Inter, Congr. Parasitol.*, M. J. Howell, ed. Australian Acad. Sci., Canberra, pp. 209–214.

Prot, J.-C. 1980. Migration of plant parasitic nematodes towards plant roots. *Rev. Nematol.* 3: 305–318.

Raski, D. J., Hewitt, W. B., Goheen, A. C., Taylor, C. E., and Taylor, R. H. 1965. Survival of *Xiphinema index* and reservoirs of fanleaf virus in fallowed vineyard soil. *Nematologica* 11: 349–352.

Raski, D. J., Maggenti, A. R., and Jones, N. O. 1973. Location of grapevine fanleaf and yellow mosaic virus particles in *Xiphinema index.* *J. Nematol.* 5: 208–211.

Rebois, R. V., and Huettel, R. N. 1986. Population dynamics, root penetration, and feeding behavior of *Pratylenchus agilis* in monoxenic root cultures of corn, tomato, and soybean. *J. Nematol.* 18: 392–397.

Reynolds, H. W., and O'Bannon, J. H. 1963. Factors influencing the citrus nematode and its control on citrus replants in Arizona. *Nematologica* 9: 337–340.

Robbins, R. T., and Barker, K. R. 1974. The effects of soil type, particle size, temperature, and moisture on reproduction of *Belonolaimus longicaudatus.* *J. Nematol.* 6: 1–6.

Schmitt, D. P. 1973. Population fluctuations of some plant-parasitic nematodes in the Kalsow Prairie, Iowa. *Proc. Iowa Acad. Sci.* 80: 69–71.

Schoener, T. W. 1982. The controversy over interspecific competition. *Am. Scientist* 70: 586–595.

Seinhorst, J. W. 1965. The relation between nematode density and damage to plants. *Nematologica* 11: 137–154.

Seinhorst, J. W. 1973. The relation between nematode distribution in a field and loss in yield at different average nematode densities. *Nematologica* 19: 421–427.

Sikora, R. A., and Carter, W. W. 1987. Nematode interactions with fungi and bacterial plant pathogens: Fact or fancy. In *Vistas on Nematology*, J. A. Veech and D. W. Dickson, eds. Society of Nematologists, Hyattsville, MD, pp. 307–312.

Smith, G. S. 1987. Interactions of nematodes with mycorrhizal fungi. In *Vistas on Nematology*, J. A. veech and D. W. Dickson, eds. Society of Nematologists, pp. 292–300.

Sohlenius, B. 1980. Abundance, biomass, and contribution to energy flow by soil nematodes in terrestrial ecosystems. *Oikos* 34:186–194.

Sohlenius, B., Bostrom, S., and Sandor, A. 1987. Long-term dynamics of nematode communities in arable soil under four cropping systems. *J. Appl. Ecol.* 24: 131–144.

Stone, A. R., Platt, H. M., and Khalil, L. F., eds. 1983. *Concepts in Nematode Systematics*. Academic Press, New York.

Strong, D. R. 1984. Density-vague ecology. In *A New Ecology*, P. W. Price, C. N. Slobodchikoff, and W. S. Gaud, eds. Wiley-Interscience, New York, pp. 313–327.

Taylor, C. E., and Robertson, W. M. 1977. Virus vector relationships and mechanics of transmission. *Proc. Am. Phytopathol. Soc.* 4: 20–29.

Topham, P. B., Alphey, T. J. W., Boag, B., and De Waele, D. 1985. Comparison between plant-parasitic nematode species association in Great Britain and in Belgium. *Nematologica* 31: 458–467.

Waliullah, M. I. S. 1983. Biomass of some selected plant-parasitic and soil nematodes. *Ind. J. Nematol.* 13: 32–47.

Wallace, H. R. 1971. Abiotic influences in the soil environment. In *Plant Parasitic Nematodes*, Vol. 1, B. M. Zuckerman, W. F. Mai, and R. A. Rohde, eds. Academic Press, New York, pp. 257–280.

Wallace, H. R. 1983. Interactions between nematodes and other factors on plants. *J. Nematol.* 15: 221–227.

Wasilewska, L. 1967. Analysis of the occurrence of nematodes in alfalfa crops. II. Abundance and quantitative relations between species and ecological groups of species. *Ekol. Pol. Ser. A.* 15: 347–371.

Wasilewska, L. 1971a. Nematodes of the dunes of the Kampinos Forest. II. Community structure based on numbers of individuals, state of biomass, and respiratory metabolism. *Ekol. Pol.* 19: 651–688.

Wasilewska, L. 1971b. Trophic classification of soil and plant nematodes. *Wiadm. Ekol.* 17: 379–388.

Wasilewska, L. 1979. The structure and function of soil nematode communities in natural ecosystems and agrocenoses. *Ekol Pol.* 27: 97–146.

Weaver, T., and Smolik, J. 1987. Soil nematodes of northern Rocky Mountain ecosystems: Genera and biomass. *Great Basin Nat.* 47: 473–479.

Wharton, D. A. 1986. *A Funcional Biology of Nematodes*, Johns Hopkins University Press, Baltimore.

Whittaker, R. H. 1975. *Communities and Ecosystems*. Macmillan, New York.

Wyss, U. 1981. Ectoparasitic root nematodes: Feeding behavior and plant cell responses. In *Plant Parasitic Nematodes*, Vol. 3, B. M. Zuckerman and R. A. Rohde, eds. Academic Press, New York, pp. 325–354.

Wyss, U. 1987. Video assessment of root cell responses to dorylaimid and tylenchid nematodes. In *Vistas on Nematology*, J. A. Veech and D. W. Dickson, eds. Society of Nematologists, Hyattsville, MD., pp. 211–220.

Yeates, G. W. 1971. Feeding types and feeding groups in plant and soil nematodes. *Pedobiologia* 11: 173–179.

Yeates, G. W. 1979. Soil nematodes in terrestrial ecosystems. *J. Nematol.* 11: 213–229.

Yeates, G. W. 1981. Nematode populations in relation to soil environmental factors: A review. *Pedobiologia* 22: 312–338.

Yeates, G. W. 1984. Variation in soil nematode diversity under pasture with soil and year. *Soil Biol. Biochem.* 16: 95–102.

Yeates, G. W. 1987. How plants affect nematodes. *Adv. Ecol. Res.* 17: 61–113.

Yeates, G. W. 1988. Contribution of size classes to biovolume, with special reference to nematodes. *Soil Biol. Biochem.* 20: 771–773.

Yeates, G. W., and Coleman, D. C. 1982. Role of nematodes in decomposition. In *Nematodes in Soil Ecosystems*, D. W. Freckman, ed. University of Texas Press, Austin, pp. 55–80.

Yeates, G. W., Watson, R. N., and Steele, K. W. 1985. Complementary distribution of *Meloidogyne, Heterodera*, and *Pratylenchus* (Nematoda: Tylenchida) in roots of white clover. *Proc. 4th Aust. Conf. Grassland Invert. Ecol.*, pp. 71–79.

II
TECHNICAL METHODS
FOR COLLECTION
AND PREPARATION
OF NEMATODES

3

Methods for Collection and Preparation of Nematodes

Part 1. Field Sampling and Preparation of Nematodes for Optic Microscopy

RENAUD FORTUNER   *California Department of Food and Agriculture, Sacramento, California*

I. INTRODUCTION

This chapter reviews the techniques used for studies of nematode systematics and identification: collection, fixation, mounting, and related studies.

The excellent book *Laboratory Methods for Work with Plant and Soil Nematodes* has been in wide use since it was first published by T. Goodey in 1949. It was recently updated for the fifth time (Southey, 1986a) to include detailed descriptions of all published techniques up to 1984. The present chapter will not duplicate information available from Southey's book, but it will review the pros and cons of the various techniques.

II. COLLECTION OF NEMATODE SAMPLES

Southey (1986b) and Barker (1985a) gave accounts of the problems and errors attached to nematode sampling due to the patchy distribution of most nematode species. Southey discusses mostly cyst nematodes, and Barker is interested in *Meloidogyne*, but their comments are true for other plant parasitic nematodes. Nematode distribution is aggregated, and the distribution pattern varies depending on seasonal fluctuation, crop and the species considered. Vertical distribution patterns also vary.

Nematode populations most often are fitted to a negative binomial distribution, described by two parameters: the mean and an aggregated index k, that reflects clumping. Aggregation is also measured by Taylor's power law where the aggregation index also increases with clumping.

Merny and Déjardin (1970) sampled the nematode population levels in two 1-hectare fields in Ivory Coast by taking 100 samples per hectare. The populations followed Taylor's law with an aggregation index of 1.65. This level of sampling allows a reasonable estimate of the population after log transform of the raw data, but only if the mean population is high enough. If only 10 samples per hectare are taken, it is reasonable to estimate the population levels within the following five classes:

Very low: Very high variability; or four samples or more have no nematodes,
Low: Lower end of the confidence interval of the mean < 25 nematodes per liter of soil,
Average: Lower end of confidence interval between 25 and 99 nem/liter soil,
High: Lower end of confidence interval between 100 and 399 nem/liter soil,
Very High: Lower end of confidence interval > 400 nem/liter soil.

The sampling method must be adapted to circumstances and to the purpose of sampling. Vertical distribution should be considered, as most plant parasitic nematodes follow the root distribution. Sampling for nematodes associated with trees should be done on the drip line where the actively growing rootlets can be found. Quénéhervé and Cadet (1986) described a sampling technique for banana roots separated into roots attached to the mother plant, and those attached to first- and second-generation shoots. Barker (1985a) reviews several patterns for sample collection in the field. Southey (1986b) discusses the problem of sampling for regulatory purposes, and he gives a table with the percentage chances of detection and failure to detect various population levels. Sampling should be done across the rows, but a deliberate sampling bias may be introduced. For example, in California strawberry nurseries the strawberries are planted by a machine that does four rows at a time. There is one plant container per row, so the first container will do rows 1, 5, 9, etc. Field sampling at the end of the growing season is biased so that plants that came from all four containers have the same chance to be sampled.

When investigating the eventuality that nematodes are responsible for a patch of poor growth, sampling should be done from the center of the diseased patch toward a place outside the patch where the plants are still in good condition. If a nematode is responsible for the damage, the largest populations will probably be found at the boundary of the patch, where the plants are still able to provide an abundant source of food to the parasites. Wallace (1971) gave the results of such a sampling for damages of *Helicotylenchus dihystera* on turf in Australia.

The biology of the nematodes should be known, particularly if a particular species is targeted. Searching for an infestation by *Xiphinema americanum* by placing carefully scrubbed roots in a mist extractor will surely fail because all *Xiphinema* are ectoparasites.

Knowing the life cycle of the nematodes can help sampling. In South Dakota, gravid females are found only from late April to early June, when new roots are produced by the hosts. In samples collected during August not a single gravid female was found and very few adult were present (Thorne and Malek, 1968). The life span of the species sampled should also be considered (Barker, 1985a).

Obviously, the field samples should be kept cool and moist, and be processed as soon as possible. Some species are very fragile and special precaution must be taken for their recovery. Trichodorids may be killed in a soil sample dropped from a height. The mortality may be very high when the sample is dropped repeatedly (Brown and Boag, 1988). In Africa, the trichodorids often disappear completely when samples are brought back from the field via dirt roads with washer-board surface. It was found that the best way to recover trichodorids was to force an aluminum can into the soil and carefully remove the can with the soil core inside. The can protected the nematodes until extraction.

Hooper (1986a) gave some general advice: recovery depends on host plant, sampling depth, soil type, and type of nematode. Samples should be kept in plastic bags as survival decreases in dry soil. However, saturated soils also are adverse to survival. Direct sunlight and excessive heat should be avoided. The samples should be stored at 5–20°C and processed as soon as possible. A few days of storage may increase recovery probably due to hatching of eggs. This increases the chances for detecting a species present in low numbers, but newly hatched individuals should not be counted when estimating populations.

Air photography may be useful for evaluating damages of some species. Barker (1985a) reviewed historical attempts. Shesteperov (1986) described trials held in the USSR for evaluation of damages caused by *Globodera rostochiensis* on potato from aerial photographs. The method was fairly successful from the time of bud formation to the flowering of potatoes, but overall success was only 50% against 95% accuracy from examination of soil samples.

III. EXTRACTION

Numerous methods exist that take advantage of various characteristics of the nematodes for separating them from the substrate: difference in size, in density, in motility, etc. Some methods are more suited for soil extractions, others for extractions of nematodes from vegetal tissues. Any method is rarely used alone, but rather in combination with other methods. For example, the residue on the sieve in the sieving method can be cleaned by Baermann funnel, migration, or centrifugation.

A. Direct Examination

Root material can be cleaned from most debris. They can be dissected under the dissecting microscope for direct observation of the nematodes within the tissues. This method is particularly suited when juvenile forms of species with obese mature females have been found in a soil sample. Roots from nearby plants may be dissected for recovery of the adult females. Staining can help by selectively coloring the nematodes within the roots.

B. Sieving

Sieving is mostly used for soil, but also for cleaning nematode suspensions after recovery from soil and roots by other methods.

Sieving discards the particles either smaller or larger than the nematodes. Soil and nematodes are suspended in water. After a short wait to allow heavy particles to sink, the supernatant is poured through a sieve. Cobb recommends using a series of sieves, with smaller and smaller mesh size.

Nematode recovery is improved by briefly (less than 5 min) soaking the soil in water before the extraction. The clay particles can be dispersed using a mixer or automatically shaking the sieves, or by various chemicals, such as sodium oxalate, or detergents containing sodium (Seinhorst, 1956; Wehunt, 1973). Meerzainudeen et al. (1984) put small stones in the sieves to prevent clogging.

Specimens can be lost by passing through sieves with too large mesh size. Sieve openings should not be greater than one-tenth of the nematode length, and even then there should be repeated sieving for reasonably accurate results. However, repeated sieving has its own faults, when specimens are trapped on the screen and cannot be washed out.

C. Migration

Nematode suspension contaminated by debris that hinders observation can be cleaned by active migration through a filter, generally a tissue paper filter. Inactive nematodes and many nematodes with body cuticular appendages and ornamentations (criconematids) are unable to go through the filter. There is also a risk of toxicity from components of the tissue paper.

Ryss (1987) described a filter composed of a 1-cm layer of coarse sand placed in a 1-mm mesh sieve partially submerged in water. Active nematodes, including large-size nematodes, migrated through the sand in one-half to 12 hr. Results were better than with tissue paper filters, but criconematids were lost.

Sudakova et al. (1986) report a technique for extracting *Aphelenchoides avenae* from thick fungal cultures using the nematodes' ability to migrate toward water-filled chambers. It is conceivable to improve this technique for the selective recovery of plant parasitic nematodes by using a solution of nematode attractant (root extract) instead of pure water.

D. Flotation

1. General Principles

Flotation methods take advantage of the difference of gravity between nematodes and debris. They process either by elutriation in a stream of water (including the Seinhorst two-flask method) or by centrifugation.

2. Seinhorst Two-Flask Technique

This type of extraction from soil can be done in the field with very simple material. It is remarkably efficient for small and medium-sized nematodes. It was found to be slower and less efficient than sieving for large nematodes such as longidorids (Brown and Boag, 1988).

3. Elutriation

In the elutriation technique, the nematode and soil particles are poured at the top of a column where is maintained an upward current of water. The current allows the heavier particles to settle while retaining the nematodes and the smaller particles in the column. Glass elutriators are fragile and easily broken, but there exist metallic models. The upward flow of water must be monitored and adjusted. The size of the soil sample is limited by the size of settling contained at the bottom of the apparatus. An advantage of the method is that it can be automatized (Byrd et al., 1976).

Winfield et al. (1987) describe a new soil elutriator, the Wye Washer, which they claim achieves extractions as good as or better than those from existing techniques, as well as easier operation.

4. Centrifugation

During centrifugation the nematodes float in a solution with a density greater than the average density of the nematode. The method is very good for extracting sluggish forms such as criconematids. It is generally more efficient than the other methods, and it may be used to clean extracts obtained from sieving or elutriation. The density depends on the species of nematode: *Pratylenchus vulnus* and *Meloidogyne incognita* seem to be recovered at lower densities than *Criconemella xenoplax*, while *Xiphinema index* requires even higher densities (Viglierchio and Yamashita, 1983). Recovery also depends on the solute used. *Pratylenchus vulnus* is recovered at 1.060 with "Percoll" (a colloidal silica with polyvinyl-pyrrolidone), but at 1.100 with zinc sulfate (Viglierchio and Yamashita, 1983). Sugar is the most used solute because it is cheap. Sulfate of magnesium does not have the stickiness of sucrose. Sulfate of zinc has fewer osmotic effects but is more acid and toxic. Other manu-factured solutes (Ludox, Ficoll, Percoll) have advantages over the simple chemicals but are more expensive. It is important to verify the specific gravity of the solution after mixing the solute. In hypertonic solutions, the nematode shrinks longitudinally, and the body assumes an accordion appearance, distorts, and collapses. It may or may not recover after being transferred back to water. In extreme cases, membrane functions collapse, allowing free inflow of the hypertonic solution through the body wall. Dead nematodes precipitate and they are eliminated in the pellet. Use of sucrose is not recommended for extraction of speci-mens to be studied with SEM (Eisenback, Part 2 of this chapter).

5. Settling Methods

Gravity is also used in "settling" methods (Barker, 1985a), where nematodes are allowed to settle at the bottom of a beaker or a test tube. Settling also occurs as the final step of most extraction processes, when extra water is siphoned out of the test tube containing the nema-todes. Enough time should be allowed for all nematodes to settle (see "Mist Extraction," below). The rate of sedimentation of the nematodes is not as critical as the time it takes for the water to come to a complete rest after the initial nematode suspension has been poured into the tube. Differences of temperature may increase this time by allowing convection currents to further disturb the water.

E. Maceration

1. General Principles

Maceration (called root incubation by Hooper, 1986b) is used mostly for nematodes inside vegetal tissues because nematodes tend to leave roots immersed in water. Leaves and stems are generally not suitable for maceration techniques; however, the method was successful for *Aphelenchoides ritzemabosi* in chrysanthemum leaves (Cranston and Newton, 1965). Most nematodes emerge within 4–7 days. After about 2 weeks, a new generation appears that has developed within the plant tissues. The roots or shoots and leaves can be stored moist or immersed in water. The water can be either still (jars, plastic bags, Baermann fun-nel with closed stem) or flowing (mist extractor with open stem). Enzymes or chemicals can be added for speeding the decomposition of plant material.

2. Baermann Funnel

The Baermann funnel uses little labor and simple equipment: a funnel, a piece of rubber tube, and a clamp. It is used mostly for nematode extraction from plant material, but also from soil finely crumbled, or to clean nematode suspensions extracted using another tech-nique. Active nematodes leave the decaying tissues, go through the cloth, and sink to the bottom of the funnel. Lack of oxygenation can kill or immobilize some nematodes. To limit this risk, it is better to use a polyethylene tube through which oxygen can diffuse, or add H_2O_2 to water, or use an air stream. This method allows recovery of active nematodes only, and specimens can be trapped by the tissue and the sides of the funnel. The tissue material can be toxic and kill the nematodes. There is also a possibility of contamination of the water by bacteria or fungi that attack nematodes.

3. Mist Extraction

With this method, there is no risk of nematode death by lack of oxygen. A mist sprayer send a mist of water to several funnels, usually arranged in a square below. There may not be enough water in the corners for proper humidification of the plant material, or there may be too much water under the spray cone, creating a strong current that may carry individuals away, particularly good swimmers such as *Aphelenchoides* spp. Viglierchio and Schmitt (1983a) found that the sedimentation rate for various nematode species ranges from about 0.004 cm/sec for small *Meloidogyne* juveniles to 0.1 cm/sec and more for large nematodes such as *Xiphinema index*. Losses due to overflow or adherence to the funnel are negligible except with extremely low (10-min cycle with less than 10% spray time) or high (more than 50% spray time) mist cycles. The water pressure may not remain constant; the pipes and spray nozzles may have deposits that diminish the water flow and/or modify the spatial distribution of the spray. The system should be regularly cleaned and retested. To achieve more uniform spray, the mist system at the ORSTOM lab in Abidjan was changed to a battery of individual mist chambers, each chamber with its own sprayer and a single funnel. The funnel chambers were arranged in a single row to avoid water spillage and contamination that may occur when a funnel is removed from the back of a large chamber with several rows of funnels. Viglierchio and Schmitt (1983a) linked poor recovery to the quality of the paper tissue used. Some tissues may have a low permeability to nematodes. Permeability varies with each brand of paper tissue and with each lot of the same brand. Recovery of *Meloidogyne incognita* varied from 37 to 90% depending on the tissue used.

F. Flocculation

In soil samples, flocculating agents such as Separan or ferric chloride ($FeCl_3$) help with the separation of nematodes from flocculated soil particles. This method cannot really be used alone, but it is a first step to other methods (Byrd et al., 1966). There is a risk that the flocculate traps the smaller nematodes.

G. Shredding

With the shredding method (often called maceration) a blender or mixer is used to shred or lacerate roots or other plant material. The shedding time should be adjusted for the blender used and for the plant material. It should not be too long as to damage the nematodes. About 5 sec at full speed is generally adequate. This technique would not be practical for processing large numbers of samples because of the time wasted between each extraction to thoroughly wash the blender. As for soil flocculation, shredding should be completed by another method, such as migration, centrifugation, flocculation, or sedimentation.

H. Comparison of the Various Techniques

Many authors have compared the methods for nematode extraction as listed by Viglierchio and Schmitt (1983b) and, more recently, McSorley et al. (1984), Shesteperov et al. (1984), and Clayden et al. (1985). Results depend on the nematode, the soil, and the host. Most methods or combinations of methods do not even achieve a 50% recovery of the inoculum, and results are extremely variable.

Generally speaking, routine is the enemy of accuracy. It is recommended to check the efficiency of the extraction techniques at least once when they are first implemented. Then, the procedure followed by the technical staff should be checked at least once a year. Checking should be done by running known numbers of specimens through the system with at least 10 replicates.

I. Extraction of Heteroderid Cysts

Cyst extraction requires special techniques and procedures, as recently reviewed by Shepherd (1986). The cysts are first extracted from the soil by a gravity method, either flotation (Fenwick can and its various modification) or centrifugation. The cysts must then be separated from the debris, either directly (wet debris) or after drying. The cysts also can be floated away from thoroughly dried debris by ethanol or glycero-ethanol mixtures. An improved machine for the rapid separation of cysts from dried root debris is described by Faulkner and Greet (1984).

The reliability of cyst extraction appears to be higher than that of soil and root nematodes. Miller (1983) studying the variation of results of *Heterodera schachtii* cyst extraction in different labs found that collecting the cysts from debris is the most critical phase. Cyst collection is helped by separation in ethanol. Intensive training of the staff and standardized procedures are necessary to attain uniform results. Cooke et al. (1983) compared several methods (Fenwick and flotation) for cyst recovery. All results were consistent and the methods were equally good. Caswell et al. (1985) described a Fenwick technique and separation of cysts by an ethanol-glycerine mixture. Recovery depended on soil type and on the number of cysts present. It was almost 90% in the best cases. Reilly and Grant (1985) found that recovery depended on cyst density. Centrifugal method is to be preferred at or above 400 cysts/100 cm³ of soil. Flotation method gave better results when cyst density was under this figure. Rajan and Swarup (1985) compared several techniques for extraction of cysts of *Heterodera cajani*. Sieving and the Fenwick can are best. For separation of cysts from debris, acetone or acetone carbon tetrachloride is the most effective, but all evaluated chemicals affect hatching and root penetration in biological studies.

J. Waste Disposal

Waste is composed of solid material (part of the sample not used for the extraction; bulk of the material left at the end of the extraction procedure) and liquid material (overflow of water used during the extraction; water suspension with the nematodes after examination). It is recommended to dispose of this waste to avoid contamination of the environment with exotic nematodes. This is particularly true for regulatory and quarantine labs. Treatment of residues can be by dry heat, steam, or fumigation, under proper conditions. For example, the California Department of Food and Agriculture recommends various ranges of temperature and time: dry heat from 230–249°F for 16 hr to 430–450°F for 2 min; steam heat, 15 lb pressure for 30 min. The water used to process samples may be boiled, chemically treated, or filtered. The residues left on the filter should be burned. Contaminated shipping containers may also be burned.

IV. IDENTIFICATION AND COUNTING

A. Observation

At the end of the extraction process, the nematodes are typically recovered in a test tube, in about 25 cm³ of water. Observation and counting of nematodes present is usually done in a smaller container with only 5 cm³ of water. Identification and eventually counting of the forms present can be to genus only (species to be determined later) or immediately to species after making temporary slides (nematodes heat-killed in a drop of water on a glass slide, coverslip with a temporary seal made with a mixture of eight parts paraffin wax to three parts petroleum jelly). Immediate identification is required for regulatory action.

Nematode populations can be expressed (Hooper, 1986a) as (a) number of nematodes per unit volume of soil (per liter, per 100 cm³), but this causes large operator errors due to difficulty of packing a volume of moist soil; (b) number of nematodes per volume of soil, determined by displacement of water, but results are affected by compaction and moisture content of soil; (c) number of nematodes per unit weight of soil, preferably dry weight. Hooper recommends that the number N of nematodes recovered from Z g of moist soil with a moisture content of Y g of water/100 g of dry soil be reported to:

$$N \times \frac{100 + Y}{100} \times \frac{200}{Z} \text{ nematodes/200 g dry soil}$$

B. Reporting Results

The absolute density of a species (also called abundance) is the number of specimens of this species per unit of volume or weight of soil independent of other species that may have been present.

For comparing the populations of different species in one sample, the various absolute densities can be compared, or the relative densities can be calculated as the number of individuals of each species divided by total number of individuals in this sample, in percent.

When several samples have been collected, e.g., from several fields during a general survey, the frequency (also called constancy) indicates how widely distributed is a species, regardless of its density. The absolute frequency is the percentage of samples where the species is observed (number of samples containing a species divided by total number of samples collected, in percent). Comparison between several species is facilitated by computing their relative frequency, which is the absolute frequency of a species divided by the sum of the absolute frequencies of all the species present, in percent.

The importance of each species depends on both its absolute density (population level in the fields where the species has been found). This can be shown graphically (e.g., Fortuner, 1976) or by computing a prominence value for this species equal to the absolute density multiplied by the square root of the absolute frequency of this species.

Finally, the importance of a species in a community also depends on its size expressed as its biomass. Biomass is computed by dividing the volume of the species (body length multiplied by square of body width) by a correction factor equal to 1.6×10^6 (Andrássy, 1956). Robinson (1984) proposed a computer program for calculating nematode volume from its dimension obtained with a digitizing tablet. The importance value is equal to the sum of relative frequency, relative density, and relative biomass.

V. CULTIVATION OF NEMATODES

A. Interest of Cultivation for Systematic Studies

It may be necessary to establish a lab culture of the specimens to obtain an accurate identification. For example, *Deladenus siricidicola* cultured on a young fungus is a free-living, mycetophagous species, but old and brown cultures (usually after a month at 22°C) produce insect parasitic forms that are morphologically very different from the mycetophagous ones (Bedding, 1973). Until recently, the mycetophagous form and the insect parasitic form were classified into different genera, even different families (Fortuner and Raski, 1987). Another example is the development of the esophageal glands that varies with age in *Ditylenchus myceliophagus* where the long overlap observed in second-stage juveniles later regresses until it almost disappears in the oldest females (Fortuner, 1982).

Another interest of nematode lab cultures is that field populations are often represented by a small number of specimens, which makes it impossible to give a good account of variability. Also, specimens obtained from a single host do not give a good account of the large environment-induced variability. Lab culture allows the placement of limits on the extent of this variability.

B. Culture Techniques

1. Agnotobiotic or Xenic Cultures

The culture is said to be agnotobiotic or xenic when the nematode is cultivated with an unknown number of associated organisms, e.g., a mixture of fungi and/or bacteria. Greenhouse culture on a whole plant belongs in this category.

2. Gnotobiotic or Monoxenic Cultures

In gnotobiotic cultures, the nematodes are cultivated with known associated organisms. When there is only one such organism, the culture may be called monoxenic. Monoxenic cultures include cultures on callus tissues or excised roots.

3. Axenic Cultures

In axenic cultures, there are no associated organisms and the nematodes are cultivated on a chemical nutritive medium that contains no living organisms, or part of organisms, other than the nematodes themselves. Bolla (1987) gave a comprehensive account of the current problems for axenic culture of plant parasitic nematodes. Success was achieved only for aphelenchids. Mechanical problems (need for solid substrate, feeding tube, stylet size and action) and biochemical problems (host attraction, nutrient concentration) have so far prevented the establishment of truly axenic cultures for tylenchs.

4. Surface Sterilization

In addition to the sterilization techniques reviewed by Hooper (1986e), Krusberg and Sardanelli (1984) report the use of a glass chromatography column filled with small glass beads and wrapped in aluminum foil. The column is sterilized, then filled with a solution of streptomycin sulfate, penicillin G, and potassium salt in sterile distilled water. Xenic nematodes are rinsed several times in the above solution then transferred to the top of the column. As they work their way down the column the nematodes are surface-sterilized while the contaminating microorganisms are left behind on the top of the column. The glass beads prevent convection currents that would carry the microorganisms to the bottom of the sedimentation column.

C. Nematode Banks

As an alternative to lab culture of nematodes in individual labs, Plant Genetics, Inc. in California is offering "Nematest," large populations of several nematode species (*Meloidogyne hapla, M. incognita, M. chitwoodi, Pratylenchus penetrans, Ditylenchus dipsaci*) to nematologists studying these species.

Mai and Riedel (1987) call for the creation of a large germplasm bank for nematodes that would supply large quantities of well-defined plant parasitic species for research and teaching. An alternative to maintaining a large collection at a single location would be a network of universities and research centers maintaining local collections of nematodes species. Bridge and Ham (1985) describe a technique for cryopreservation of living specimens of *Meloidogyne graminicola* that could be used for long-term preservation in a germplasm bank.

D. Staining Nematodes

Hooper (1986b) gave a review of the traditional stains used for nematodes: acetic orcein, nile blue B and toluidine blue, gold chloride, silver nitrate, and, for vital staining, methyl red and neutral red pH indicator dyes. Premachadran et al. (1988) used Coomassie brilliant blue G to stain secretions from amphids, phasmids, excretory system of live nematodes.

Meyer et al. (1988) compared seven stains for distinguishing live and dead eggs of *Heterodera glycines*. With bright-field microscopy, chrysoidin, eosin Y, new blue R, and nile blue A gave the best results, even better with added DMSO (dimethylsulfoxide). No single stain was consistently better with fluorescence microscopy, and the results depended on the combination filter/stain.

VI. PREPARATION OF SLIDES

Recent reviews on methods for killing, fixing, and mounting nematodes were given by Hooper (1986c) and Santos and Almeida (1989). It must be stressed that there exists no technique for eternally preserving lifelike dead nematodes (Maggenti and Viglierchio, 1965), and that fixation and mounting always result in a certain amount of distortion that tends to increase with age of the slide. After studying the effect of various methods, Brown and Topham (1984) concluded that some of the differences in published morphometrics of *Xiphinema diversicaudatum* are due to differences in processing methods. It is important that methods used be recorded and published with each description.

A. Killing

Killing is generally done by heat, with or without simultaneous fixation. If only a few specimens have to be killed, they are placed in a drop of water on a glass slide, and the slide is heated over an alcohol lamp. Heating should stop as soon as the nematodes are dead. As a rule of thumb, the specimens are often dead when condensation droplets that appear when the slide is first placed over the flame have evaporated. The specimens should be checked with the dissecting microscope and if some are still twitching the slide should be put back a few more seconds over the flame.

Mass killing is often done with hot fixative following the methods of Seinhorst (1966), Netscher and Seinhorst (1969), or Netscher (1971). Safe mass killing of nematodes can be done by slowly heating the test tube with the nematode suspension in a beaker with hot or boiling water. The temperature of the suspension should be monitored with a thermometer placed in the test tube. The nematodes die after a few seconds at 60°C.

Other killing methods have been proposed (e,g., vapor phase perfusion, Maggenti and Viglierchio, 1965) but they are not in common use.

B. Fixing

TAF was widely used in the 1950s as a fixative because of the remarkable lifelike appearance of the specimens, but it was soon discovered that long-term storage in TAF prior to mounting resulted in distortion of the specimens. For this reason its use has decreased dramatically. However, specimens fixed in TAF and mounted lactophenol or glycerin within a year remain in good condition (Hooper, 1987).

Mixtures of formalin (40% formaldehyde) and either glacial acetic acid or propionic acid (FA 4:1 of Seinhorst, 1954; FP 4:1 of Netscher and Seinhorst, 1969) currently are the most widely used fixatives. Olowe and Corbett (1983) tested several fixatives for *Pratylenchus brachyurus* and *P. zeae*. None were perfect but F4 and FP 4:1 were the most satisfactory.

It is best to conserve part of the nematodes in mass collection in case the mounted specimens deteriorate. Long-term storage runs the risk of slow evaporation, but MacGowan (1986) solves this problem by storing preserved nematodes in heat-sealed glass ampules.

C. Mounting

Lactophenol mounts allow an excellent preservation of the specimens, even after 30 years, but it is difficult to obtain a good seal and many slides dry out (Hooper, 1987).

Glycerine mounts are the favorite. A number of techniques exist that allow processing the specimens through alcohol to glycerine with minimum time and efforts (Hooper, 1986c, 1987).

Nematodes are often mounted on Cobb's slides that allow observations at high magnifications (oil immersion) from either side of the specimens. The coverslips must always be supported to avoid squashing the specimens. Glass rods or beads are often used, but it is difficult to select supports of the same diameter than the specimens. ORSTOM labs uses tungsten filaments of calibrated diameter, by 5-µm increments. Huang et al. (1984) use polyester base adhesive transparent tapes 45–50 µm thick with a 9-mm-diameter opening where the glycerol and nematodes are placed, then covered with the usual coverslip.

Taxonomy and identification of *Meloidogyne* rest in part on characteristics seen in lip region profile, excretory pore region, and perineal pattern in mature globose females. Gerber and Taylor (1988) describe a method which by the removal of the contents of the posterior half of the body followed by cutting away a quarter of the cuticle close to the perineal pattern allows mounting of whole specimens clearly showing all three regions.

Esser (1988) describes how whole cysts, freshly taken from the root, can be set on water-agar on a microscope slide and the cone area examined directly without having to be cut and trimmed. This method is said to give excellent results, even though the light is passing lengthwise through the cyst.

D. Labeling

It is important to record the origin of the specimens and to maintain proper records of the slides and their contents. The system advocated by Thorne for numbering slides by genus name, with each species in a genus identified by a number, and each slide with a particular species distinguished from the others by a letter, is still used in many labs. The inconvenience of such a system is obvious because of the continuous changes in nomenclature. Also, it is impractical for computerized records.

Several collections have already stored, or are in the process of storing, their records on computer. It would be preferable if all nematode collections would follow the same format. It would then be conceivable to regroup all records in a central computer for easier search and use of the records. A comprehensive format has been proposed by the present author, and it is implemented at UC–Riverside and UC–Davis, while curators of several other collection have expressed interest in this concept.

VII. COLLECTING DATA

A. Microscope

A top-quality research microscope is the most indispensable piece of equipment for systematic studies. It should be properly installed and maintained. In bright-field microscopy, Köhler illumination gives the best resolution and least glare (Winfield and Southey, 1986). Most details are seen only with oil immersion objectives of 100x magnification. Immersion oil can also be placed between the condenser and the bottom of slide to achieve maximum resolution.

Oil-smeared slides must be cleaned after use. Placing some alcohol on the slide makes the oil gather in a small drop that is then easily picked up by a brush. What is left of the oil on the slide can then be wiped by a clean brush, a piece of cotton, or a tissue while avoiding crushing the specimens. Xylene is a good solvent for immersion oil but it releases toxic fumes.

Interference microscopy, and particularly the differential interference contrast of Nomarsky, reveals morphological details invisible with bright-field microscopy. It is necessary to describe the observation methods employed (microscope, magnification, bright-field or interference microscopy) in the "Materials and Methods" paragraph of taxonomical papers because the same feature may appear differently in different setups.

B. Measurements

Some authors take measurements directly from the ocular micrometer, but it may be difficult to exactly position the micrometer in reference to the feature to be measured. Error becomes prohibitive when the feature is longer than the graduated bar of the micrometer.

The outline or the axis of the feature can be traced on a piece of paper using a traditional camera lucida or, better, a drawing tube available with most microscopes. Most standard objectives are subject to field curvature resulting in differences of magnification between the center and the edges of the field.

Measurements can be taken from the drawing with a ruler and a map measurer for curved lines. Actual measurements in millimeters are written on a piece of paper, then converted to micrometers. Cheap hand-held calculators allow easy computation of mean and standard deviation. Any published description should state these basis statistics (Fortuner, 1984).

Semiautomatic measurement systems have been available for some time on microcomputers (Boag, 1981). Current systems allow electronic magnification and enhancement of morphological details by selection of grey levels. Measuring, converting, and computing data is done automatically and accurately by the computer after the operator has marked the beginning and the end of the feature to be measured. The results can be obtained as square matrix, ready to be loaded into a statistical analysis package. Cost is about $4000 for the software, $4000–$5000 for specialized hardware (digitizing pad, video camera, frame grabber, graphic card, high-resolution monitor, etc.), plus the cost of the computer itself.

Statistical packages such as SAS PC are now available on microcomputers and they make using elaborate statistical procedures such as discriminant function analyses a relatively easy task.

C. Descriptions

Accurate illustrations, made from feature outlines taken with drawing tube or camera lucida, should accompany every description. All observed shapes of each feature should be fully illustrated.

Photographs are rarely used for illustrating shapes because the narrow depth of field at high magnifications keeps most of the subject out of focus. Eisenback (1988) described a technique whereby a single exposure of, e.g., 40 sec is replaced by four exposures of 10 sec each, at four different levels of focus.

Part 2.   Preparation of Nematodes for Scanning Electron Microscopy

JONATHAN D. EISENBACK   *Virginia Polytechnic Institute and State University, Blacksburg, Virginia*

I. INTRODUCTION

The scanning electron microscope (SEM) is a useful tool to study nematode morphology and taxonomy (Baldwin and Powers, 1987; Eisenback, 1985; Hirschmann 1983). This instrument has been used to detect differences among populations or races of species, to group species within a genus or genera within a family, and has helped in some studies to reconstruct proposed phylogenies (Hirschmann, 1983). Clearly, the SEM has clarified many difficult taxonomic questions in several nematode groups, and descriptions of new species increasingly include scanning electron micrographs. The value of the SEM to nematode morphology and taxonomy will probably increase in the future.

The most useful taxonomic characters revealed by SEM include morphology of the anterior end, in particular lip patterns, and details of the posterior end and general body region (Hirschmann, 1983). In addition, the SEM is useful to visualize surface morphology of nematode eggs and dissected body parts including stylets (Eisenback and Rammah, 1987), spicules (Rammah and Hirschmann, 1987), sperm (Eisenback, 1985), and other tissues (Abrantes and Santos, 1989). Details revealed by SEM often aid in a more precise interpretation of morphology as seen in the light microscope (LM); as a result, the LM is made more useful (Fig. 1).

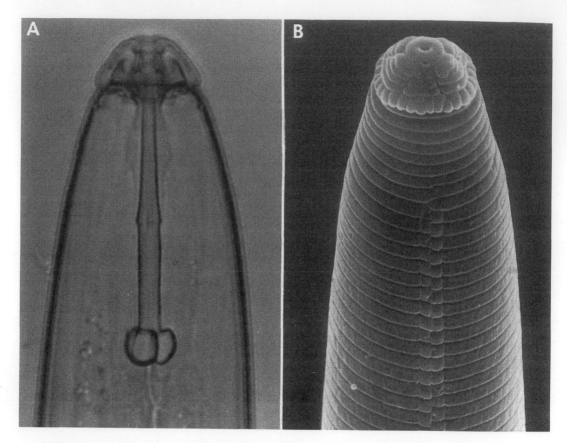

FIGURE 1   Anterior end of a female of *Hoplolaimus galeatus* (Cobb) Filip'ev and Schuurmans Stekhoven. (A) Light micrograph showing that the specimen is transparent to light and revealing that the depth of focus is low. (B) Scanning electron micrograph showing the three-dimensional image and large depth of focus. (From Eisenback, 1985.)

Proper preparation of the specimens is important for SEM observations. Poor preparation may obscure small details and often produces artifacts that interfere with the correct interpretation of the scanned image. At times the utility of information obtained by SEM is limited by adequate specimen preparation (Eisenback, 1986).

Preparing nematodes for SEM usually involves numerous steps. The success of the preparation is dependent on the success of each individual step. Shrinkage, swelling, and surface precipitation that occurs during the initial stages of preparation are likely to worsen in the final stages.

Techniques for preparing nematodes for SEM are now available that produce specimens that are stable in the microscope and relatively free of artifacts (Eisenback, 1986). The description of the technique that follows is adequate for most genera of plant parasitic nematodes, although slight modifications of the procedure may be required for the optimum preservation of detail, depending on the genus or nematode tissues. In cases where these modifications are known, they will be described for each individual genus.

II. SELECTION AND HANDLING NEMATODES

A. Handling

Careful selection and handling of specimens are necessary for proper preparation (Eisenback, 1985). Extraction techniques that are harsh may cause artifacts and give poor results even though the remaining procedures are adequately performed. Harsh extraction techniques include those that use bleach, sucrose, or flocculating agents, and those that allow the increase of contaminating organisms or anaerobic conditions.

Hand-picking individual specimens with a small, fine wire, such as a dental pulp canal file, or other suitable pick ensures that each specimen will be of adequate quality, and of the same species and developmental stage (Eisenback, 1985). Micropipetting is a useful alternative method of handling specimens provided that specimens are in good physiological condition, in a relatively clean monospecific solution, and in good supply. This technique is particularly useful for large, saccate forms.

B. Processing Containers

Processing nematodes for SEM usually involves placing the specimens in many different types of solutions and equipment. A chamber that allows the bulk transfer of many specimens from one liquid to another is necessary to facilitate the preparation procedures. An ideal container allows for rapid exchange of fluids, prevents loss of few specimens, and minimizes additional artifacts (Eisenback, 1985).

Several containers have been described; however, a modified epoxy embedding (BEEM) capsule is perhaps the most widely used and is easily assembled (Fig. 2) (Eisenback, 1985). The container is made by cutting the conical end off the capsule. A perforated cap is fashioned with a small hole punch. A small piece of fine-mesh nylon screen is held in place over the hollow cylinder as the cap is snapped into position (Fig. 2C). Specimens are placed in the modified BEEM container inside a small glass Stendor dish. Fluids are exchanged by withdrawing one solution from the Stendor dish and pipetting in a new one.

III. KILLING AND FIXATION

A. General Precepts

Proper fixation is perhaps the most important step in preparing good specimens for SEM (Fig. 3). In most cases, fixatives and techniques commonly used for preparing nematodes for light microscopy are inadequate for SEM (Fig. 3C) (Eisenback, 1986). Therefore hot formalin-based fixatives are not useful; instead, a sequential fixation with cold glutaraldehyde-based fixative or mixtures of glutaraldehyde and formalin are usually adequate for most species and genera (Fig. 3D).

Fixation is usually a two-part process. Primary fixation of protein molecules occurs by an aldehyde or mixture of aldehydes, and the unsaturated fats are postfixed with osmium tetroxide. Glutaraldehyde is the aldehyde of choice because each molecule of the fixative has two reactive groups that are capable of crosslinking proteins (Hopwood, 1969, 1972).

The fixatives can greatly upset the tonicity of the tissues and must be used with a suitable buffering system (Schiff and Gennaro, 1979). Sodium cacodylate and phosphate buffers are most commonly used for biological material. The selection of the buffer is dependent on the preference of the user. Sodium cacodylate contains arsenic and is thus toxic to contaminating organisms, but may be hazardous to use; whereas the phosphate buffers

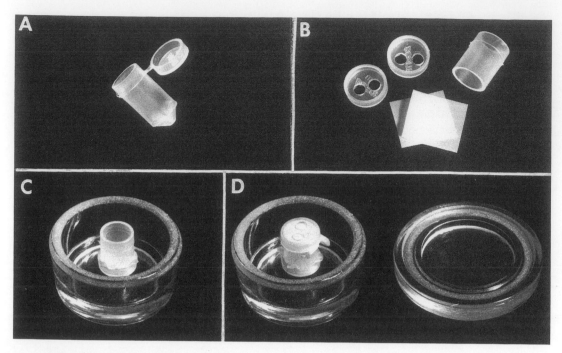

FIGURE 2   BEEM capsule container. (A) An unmodified BEEM capsule. (B) Component parts of container including two caps with holes, two squares of nylon cloth, and the hollow cylindrical portion of a modified capsule. (C) Container in a Stendor dish, container without cap. (D) Container with cap. (From Eisenback, 1985.)

are less hazardous to the user but easily become contaminated with other organisms, particularly fungi. Additionally, the tonicity of the buffer can be changed by the incorporation of nonelectrolytes or electrolytes (Hopwood, 1969; Pentilla et al., 1974).

B. Buffers

To prepare sodium cacodylate buffer (0.1 M), mix 21.4 g of sodium cacodylate or 15.5 g of anhydrous sodium cacodylate with distilled water for a final volume of 500 ml. The pH of the buffer is adjusted with concentrated hydrochloric acid to approximately 7.0.

To prepare sodium phosphate buffer, mix a solution of monobasic sodium phosphate and dibasic sodium phosphate in different combinations to adjust the pH. For the monobasic solution, 26.85 g of monobasic sodium phosphate is added to tap water for a final volume of 500 ml. The dibasic solution is prepared by adding 13.80 g of dibasic sodium phosphate and distilled water for a final volume of 500 ml. To make a 0.2 M sodium phosphate buffer with a 7.0 pH, 19.5 ml of the monobasic solution is added to 30.5 ml of the dibasic solution.

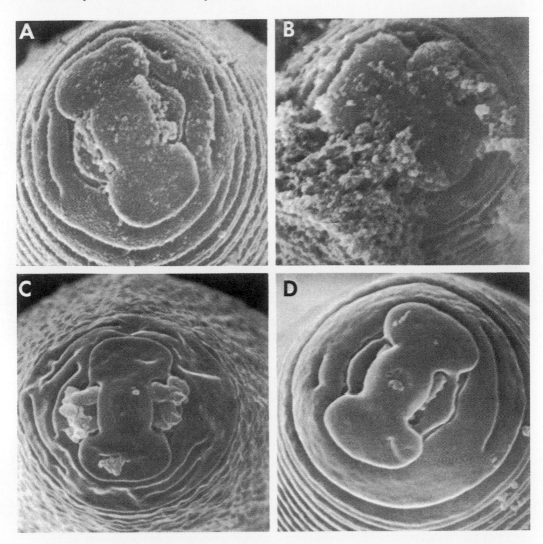

FIGURE 3 *Meloidogyne incognita* second-stage juveniles prepared for scanning electron microscopy. (A) Rapid fixation at room temperature with a mixture of 2% glutaraldehyde/1% formalin. Preservation is good but a surface precipitate obscures details. (B) Slow sequential fixation in the cold with a mixture of 2% glutaraldehyde/1% formalin. Preservation is poor and a surface precipitate greatly obscures details. (C) Sequential fixation with cold 2% formalin. Preservation is poor as seen by many wrinkles even though precipitate is absent. Formalin alone is not a suitable fixative for SEM. (D) Sequential fixation with cold 2% glutaraldehyde. Preservation is excellent and the specimen is relatively free from artifacts. (From Eisenback, 1986, courtesy *J. Nematol.*)

C. Fixatives

Glutaraldehyde is usually available commercially in 2-ml ampules at a 70% concentration. The 2-ml ampules are added to the necessary amount of buffer to make the desired concentration. For a 4% solution, the 2 ml of 70% glutaraldehyde is added to 33.0 ml of buffer.

D. Fixation

Strong fixatives often cause shrinkage because they are hypertonic (Fig. 3A). Weak fixatives initially preserve cell morphology but do not harden tissues enough to prevent deleterious effects caused by other preparative procedures of dehydration and drying. Sequential fixation begins with a low concentration of fixative which is gradually increased until it is quite strong. For most nematode specimens, sequential fixation in the cold is superior to rapid fixation at room temperature (Fig. 3). Routinely, specimens are placed in a known amount of cold (4°C) tap water and a 4% solution of buffered (0.1 M) glutaraldehyde is dropped onto the specimens every 15 min until the concentration of the solution with the specimens is increased to 2%. After the final concentration is reached, fixation continues for an additional 24–48 hr at 4°C.

The fixed material is washed with several changes of cold buffer to completely rinse unbound glutaraldehyde from the tissues. The rinse must be complete (usually four to six changes within 10–15 min); however, if it is too lengthy, unfixed unsaturated fats may begin to leach out of the tissue. For this reason the specimens are postfixed with osmium tetroxide.

E. Postfixation

A 2% solution of osmium tetroxide is prepared with distilled water several hours before needed so that the crystals will have time to go into solution. Osmium is extremely hazardous and caution must be used.

Specimens are postfixed in cold (4°C) osmium for 4–8 hr or longer (overnight). After postfixation, the specimens are rinsed with several changes of cold buffer within a 15-min period and the specimens are allowed to gradually return to room temperature.

IV. DEHYDRATION AND INTERMEDIATE FLUIDS

Dehydration after fixation is necessary because the most common methods of drying use fluids that are not miscible with water. Ethanol is the dehydrating agent of choice. It is miscible with carbon dioxide for specimens that will be critical point-dried from CO_2, and miscible with Freon 13 for specimens dried from Freon 13. Differences in preservation between critical point drying from carbon dioxide and Freon 13 are minimal to undetectable (Fig. 4); the chemical of choice is usually dependent on standard lab operating procedures. Likewise, ethanol is a useful media for freeze drying because it forms much smaller ice crystals than water.

Gradual dehydration prevents osmotic shock to the specimens and ensures a complete exchange with the water. An eight-step graded series of ethanol in water is usually adequate. Exchanges with gradual increases of 5, 10, 25, 50, 75, 90, 95, and 100% ethanol are 15–20 min each in duration, and 100% ethanol is exchanged three times to ensure that it has replaced all of the water. Usually a newly opened bottle of ethanol is used in these last steps to ensure that it is completely free of water.

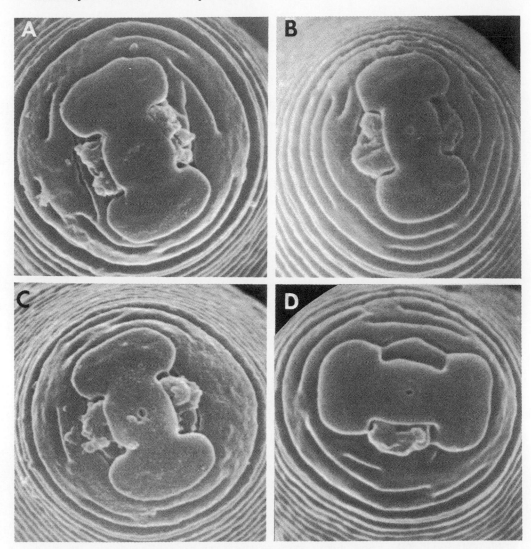

FIGURE 4 *Meloidogyne incognita* second-stage juveniles prepared for scanning electron microscopy. (A) Critical point-dried from Freon. Preservation is good and artifacts are minimal. (B) Critical point-dried from carbon dioxide. Preservation is good and artifacts are minimal. (C) Air-dried from Freon. Preservation is fair and only a small amount of surface residue is evident. (D) Freeze-dried from 100% ethanol. Preservation is good and artifacts are few. (From Eisenback, 1986, courtesy *J. Nematol.*)

V. DRYING

A. General Precepts

Drying the specimens is necessary to remove the inter- and intracellular fluids within the fixed and dehydrated nematodes. The two most common methods used for drying nematodes, critical point and freeze drying, produce nearly the same results (Fig. 4); the method selected is dependent on the equipment available and standard operating procedures of each lab. Usually, freeze-drying equipment is more readily available, more economical to use, and has fewer operational difficulties than critical point drying. Likewise, air drying from a highly volatile low-surface-tension liquid produces nearly as good results as the more complicated drying techniques (Fig.4C). The major problem with air drying from a volatile liquid is the possible formation of a slight surface precipitate.

B. Critical Point Drying

At a certain temperature and pressure, a solution reaches a "critical point" where the phases are indistinguishable (Eisenback, 1986). Unfortunately, the critical point of water is extremely high and more suitable liquid must be substituted. Usually the liquid of choice is either Freon 13 or carbon dioxide. Details of critical point drying will not be presented here; the reader is encouraged to follow the directions prescribed by the manufacturer of the critical point drying apparatus.

C. Freeze Drying

This technique involves two step: (a) rapid freezing of fluids in the specimen and (b) sublimation of the ice under a high vacuum (MacKenzie, 1976). Ice crystals of water are very large and usually damage the cells and tissues of the nematodes; therefore, the specimens are dehydrated with ethanol, a nonpolar solvent that forms much smaller ice crystals. Rapid freezing is necessary to minimize the size of the crystals that are formed during the process. A Freon liquefier is utilized to liquefy propane obtained from a bottle distributed for use with a small torch. The specimens are rapidly immersed in the liquified propane cooled by liquid nitrogen. The nematodes are evacuated to 10^{-1} Torr at liquid nitrogen temperatures ($-80°C$) until all of the ice sublimes (approximately 2 hr).

Dried nematodes are always stored in a desiccator to prevent them from absorbing moisture from the atmosphere. They may be stored in a desiccator for several weeks to months, unmounted inside the BEEM capsule container, or mounted on SEM stubs ready for coating and viewing.

VI. MOUNTING

Dried whole nematodes are mounted on a SEM viewing stub with double-coated adhesive tape. The specimens are transferred from the container to the stub with a fine metal wire, such as a dental pulp canal file. The individual nematodes are propped against a short hair which is centered on the adhesive tape. The nematodes are placed on the tape so that their anterior ends are projected beyond the hair at a 45° angle to the surface of the stub. Other mounting procedures have been described that may also be satisfactory (Eisenback, 1986).

VII. COATING AND VIEWING

Mounted specimens are coated with a thin layer of gold or gold/palladium alloy. Sputter coating is usually superior to evaporative coating. The thickness of the coating is critical for optimal results; too little coating may cause charging artifacts, whereas too much coating obscures surface details. Usually, approximately 20 nm of metal is adequate. Additional metal may be deposited on the specimens if the first coat is too thin.

VIII. ALTERNATIVE TECHNIQUES FOR SPECIFIC GENERA OR SPECIMENS

A. *Heterodera* spp. Second-Stage Juveniles

Nematodes are chilled to 4°C in 0.5 ml of 0.1 M phosphate buffer at pH 7.2. One drop of cold 4% phosphate-buffered glutaraldehyde is added two time a day until a 2% final solution of fixative is obtained. The specimens are rinsed with phosphate buffer containing 7% sucrose and dehydrated with cold (4°C) ethanol. They are dried, mounted, and viewed as previously described in the standard preparation procedures (Noel and Stanger, 1986).

The main differences with this procedure and the standard procedure are the prolonged sequential fixation period, the use of sucrose in the rinses, and dehydration in the cold. These modifications appear to eliminate the shrinkage and subsequent wrinkles caused by other preparations techniques.

B. Glycerin-Infiltrated Material

Infiltrated specimens are drained on a small piece of filter paper and mounted on a small sliver of glass. The nematodes are attached to the glass with a minute amount of nail polish or thinned Zut (Sher and Bell, 1975).

The success of this method is often dependent on the methods that were used to preserve and infiltrate the specimens with glycerin. Sometimes specimens are prepared for routine examination with the light microscope and subsequently viewed with the SEM. The nematodes are fixed with 5% formalin and slowly infiltrated with glycerin (Sher and Bell, 1975). Problems with this technique include excess glycerin obscuring small morphological details, and difficulties of using glycerin in the microscope. Specimens are likely to be sensitive to damage by the electron beam and can only be viewed at low operating voltages, at low magnifications, and for short periods.

C. Stylets

Nematodes are transferred to a drop of 45% lactic acid on a coverslip dissecting chamber (Eisenback, 1985) and the head is cut off with an eye knife. The stylet is pushed out through the cut opening with a sharpened dental pulp canal file. The stylet is cleaned by swishing in the lactic acid and gently attached to the coverslip by applying slight pressure with the file. Two percent formalin is dropped into the lactic acid every few minutes until the coverslip is flooded and the lactic acid is washed away. The formalin is drained from the coverslip and the stylet is allowed to air-dry and marked with a small triangular piece of translucent tape. The coverslip is attached to a SEM stub and sputter-coated with 20 nm of gold/palladium.

This technique has been shown to be useful for dissecting stylets of several genera of plant parasitic nematodes including *Belonolaimus, Criconemella, Hemicycliophora, Hoplolaimus, Meloidogyne, Pratylenchus, Scutellonema, Sphaeronema,* and *Xiphinema*

(Eisenback and Rammah, 1987). The stylets of some genera cannot be successfully dissected with this technique because parts of the stylets are soluble in lactic acid; therefore, other dissecting solutions are necessary.

D. Stylets of Cyst and Other Nematodes

Dissection of stylets is similar to that described previously; however, the 45% lactic acid is replaced with 0.01% sodium hypochlorite (NaOCl) (Mota and Eisenback, 1988). Stylets of several genera have been successfully removed with this dilute solution of household bleach including *Criconemella, Heterodera, Globodera, Longidoroides, Longidorus, Paralongidorus,* and *Xiphinema* (Eisenback, 1985: Mota and Eisenback, 1988; Swart and Heyns, 1987).

E. Spicules

Nematode spicules can be dissected in lactic acid similar to that of stylets (Eisenback, 1985). This technique has been used to dissect spicules from males of several genera including *Aphelenchoides, Aporcelaimellus, Belonolaimus, Dolichodorus, Globodera, Heterorhabditis, Heterodera, Hoplolaimus, Meloidogyne, Mesorhabditis, Panagrellus, Tylenchorhynchus,* and *Xiphinema* (Rammah and Hirschmann, 1987).

F. Vulval Cyst Cones

Mature cysts are sonicated in water and fixed in 5% formalin for 48 hr. They are placed in 45% lactic acid and the vulval cones are cut off with a sharp eye knife. The dissected specimens are transferred to anhydrous glycerin and stored overnight in a desiccator. They are removed and the excess glycerin is drained from them on a narrow strip of filter paper in an oven at 40°C for 1 hr. The drained cones are mounted on double-coated tape on SEM stubs and sputter-coated with approximately 20 nm of gold/palladium (Hirschmann and Triantaphyllou, 1979).

G. Perineal Patterns

Perineal patterns are prepared for SEM with a modification of the technique useful for dissecting stylets from nematodes. Patterns are cut and cleaned in 45% lactic acid and transferred to a fresh drop of lactic acid on a coverslip dissecting chamber. The lactic acid is washed from the specimens by flooding the coverslip with drops of 2% formalin; the patterns are fixed for an additional 10 min in the formalin. The fixative is drained from the coverslip and the patterns are allowed to air-dry. The coverslip is mounted onto a SEM stub, sputter-coated, and viewed with the surface of the stub perpendicular with the electron beam (Abrantes and Santos, 1989).

Part 3.   Preparation of Nematodes for Transmission Electron Microscopy

LYNN K. CARTA   *California Institute of Technology, Pasadena, California*

I. INTRODUCTION

The basic steps for processing nematodes for transmission electron microscopy (TEM) are the same as for other biological specimens. These steps include fixation, dehydration, infiltration and embedding, sectioning, and staining. The microscopic size of most nematodes and the fact that it is often difficult for fixatives to penetrate the cuticle may create some additional problems.

General TEM references (Hayat, 1981, 1986; Lee, 1983; Aldrich and Mollenauer, 1986; Robinson et al., 1987; Crang and Klomparens, 1988) should be consulted for selecting methods for specific needs not well represented in the nematological literature. An excellent summary of TEM methods with one protocol for nematode processing was recently published (Shepherd and Clark, 1988). A detailed presentation of TEM histochemistry (McClure, 1981) is available and will not be addressed in this discussion.

This portion of the chapter provides an overview of and comment on various TEM methods for resolving morphology in different nematode systems. Different methods are often needed for different nematode genera, life stages, and parts of the body. The best possible fine-structural morphology is especially important for organ reconstructions, detecting phenotypic variations in mutant nematodes, and describing host–parasite associations.

II. FIXATION

For best general preservation, glutaraldehyde (sometimes with other aldehydes) is followed by osmium tetroxide after thorough rinsing to avoid undesirable fixative complexes. Glutaraldehyde fixes proteins and nucleic acids. Osmium tetroxide fixes many unsaturated lipids, acts as an electron stain, and enhances lead poststaining. Note that aldehydes can penetrate latex gloves, whereas gloves containing neoprene give better protection.

A. Chemical Combinations Used in Nematode Fixation

The basic fixation strategy involves fast and nearly simultaneous killing and fixation to minimize autolytic changes. This goal is not easily achieved with many terrestrial parasitic nematodes. In these nematodes chemical fixation may be difficult due to a relatively impermeable cuticle and high turgor pressure.

Glutaraldehyde followed by osmium, always with puncture in glutaraldehyde, is a standard fixation method that works well in many nematodes (Bird, 1971; Baldwin and Hirschmann, 1973; Ward et al., 1975; Perkins et al., 1986). The author has found this method to work fairly well in the head region of heteroderid plant parasites, perhaps because of the many sensory organ openings, but detail was poor in the phasmidial sheath cell in tails (Fig. 5).

Some of the earliest fixation methods employed osmium alone with *Xiphinema index* (Roggen et al., 1965), *Heterodera schachtii* (Wisse and Daems, 1968), and *Caenorhabditis elegans* (Ward et al., 1975; Sulston et al., 1980). Although it penetrates more slowly

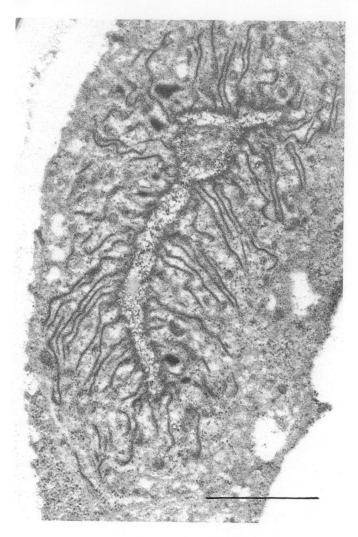

FIGURE 5   Glutaraldehyde: 2%, 25°C; osmium tetroxide: 2%, 25°C. *Meloidodera floridensis*, tail: phasmid sheath cell enclosing neuron, x.s. Bar = 1 μm.

than aldehydes, osmium fixes faster. As a consequence, it takes less time before puncture can be completed than with aldehydes. Unlike glutaraldehyde, osmium tetroxide destroys the semipermeability of membranes, but it allows leaching of proteins. The sensillar matrix material in nematodes is not fixed by osmium alone (Perkins et al., 1986).

Osmium followed by glutaraldehyde was effective in fixing tails of *Scutellonema brachyurus* (Wang and Chen, 1985). It was also fairly effective in juvenile and male tails of heteroderids, and very good for *C. elegans*, but some leaching of cytoplasmic ground substance can occur (Carta, unpublished observations). The method was not effective in fixing female *Heterodera schachtii* (Cordero, pers. commun.).

Acrolein is well known for its rapid penetration of tissues, but it gave very poor resolution when used alone and followed by osmium (Zeikus and Aldrich, 1975). However, acceptable results are reported when 1 or 10% acrolein is added to glutaraldehyde and followed by osmium (Wright and Jones, 1965). This method works better in marine than soil nematodes. Very short (1-min) exposure of acrolein with glutaraldehyde and propylene phenoxetol allowed earlier puncture and better overall preservation than standard methods in *C. elegans* (Chalfie and Thomson, 1979). This method gave acceptable preservation of *H. schachtii* second-stage juveniles. However, greater exposure to acrolein for 2 hr with 1.5% acrolein, 1.5% paraformaldehyde, and 3% glutaraldehyde (Hayat, 1981) followed by osmium distorted the cuticle, inflated membranes, and generally obscured cellular detail so that only nuclei were discernible (Carta, unpublished observations). A similar fixation mix (3% acrolein, 3% glutaraldehyde) was used on *Paratrichodorus porosus* (G. W. Bird, 1970) with better results. Fixation in this relatively permeable plant parasite was generally good except for poorly preserved muscle fibers.

Simultaneous double fixation with glutaraldehyde and osmium at 4°C worked well in preserving the insects *Periplaneta americana* and *Musca domestica* larvae (Singh et al., 1985). The same method was ineffective with the nematodes *Meloidodera floridensis* (Carta, unpublished observation.) and *Romanomermis culicivorax* (S. El-Din, pers. commun.). Tissues were fixed but very poorly resolved.

Cryofixation was first demonstrated with *Xiphinema index* and *Caenorhabditis briggsae* and compared with a standard glutaraldehyde-formaldehyde-osmium fixation. It gave relatively poor resolution of myofilaments and uterine double membranes and slightly better cuticle preservation. Mitochondrial cristae were poorly resolved with both techniques (Himmelhoch and Zuckerman, 1982). Poor membrane preservation and "gummy" phasmid secretion were noted in nematodes immersed in liquid nitrogen and slowly thawed in glutaraldehyde followed by osmium (Fig. 6) (Carta, unpublished observation).

Formaldehyde when added to glutaraldehyde followed by osmium has given superior results over other methods for a number of authors, either without heat (Wharton and Barrett, 1985; Himmelhoch and Zuckerman, 1982) or with 70°C heat (Zeikus and Aldrich, 1975; Byard et al., 1986). This method worked extremely well with hydrogen peroxide (Byard et al., 1986) and 60°C heat in heteroderid tails (Carta and Baldwin, 1989) and females of *H. schachtii* (Cordero, pers. commun.). Membranes and ground cytoplasm were especially well fixed in phasmidial sense organs (Fig. 7).

B. Fixation Additives

Hydrogen peroxide has been reported to improve the fixation of glutaraldehyde, presumably by increasing the rate of cross-linkage to proteins. Note that it must not be added to fixatives with paraformaldehyde or an explosive mixture may result (Byard et al., 1986).

Tannic acid (TA) (1% in glutaraldehyde) has been very effective in resolving microtubules in nematodes where total size and number of protofilaments is critical to accurate interpretation (Chalfie and Thompson, 1982). It should be remembered that apparent microtubule size may increase in proportion to TA concentration (Hayat, 1981).

En block staining with aqueous or alcoholic uranyl acetate is often used to enhance contrast, stabilize membranes, and reduce shrinkage from dehydration (Lee, 1983).

C. Timing

Nematodes killing can be very fast, but it may take much longer before the specimen body is well enough fixed that it can be cut to allow further penetration. Plant parasitic nematodes can be killed in 5–6 min at 4°C and 2 sec at 70°C with 3% formaldehyde (Zeikus and

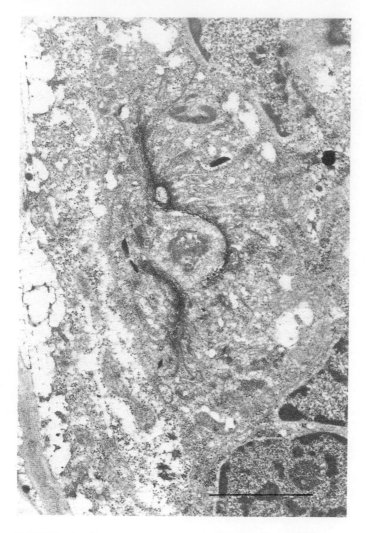

FIGURE 6   Glutaraldehyde: 2%, –60°C thawed to 25°C; osmium tetroxide: 2%, 25°C. *Meloidodera floridensis*, tail: phasmid sheath cell enclosing neuron, x.s. Bar = 1 μm.

Aldrich, 1975). Yet it may take 30 min to 2 hr before they are sufficiently fixed for puncture. Complete aldehyde fixation may take hours beyond that.

Free-living nematodes are penetrated by fixatives more readily than most plant or animal parasites. When free-living *C. elegans* is placed in 2% glutaraldehyde in M9 buffer, the majority of worms cease movement, and explosive release of internal tissue does not occur when punctured after 3–5 min. With Heteroderidae, the time increases to between 70 and 90 min (Carta, unpublished observations). Zeikus and Aldrich (1975) advise that tylenchid plant parasites being fixed by aldehyde should not be punctured before 2 hr. Puncture time is about 30 min for second-stage *Meloidogyne javanica* at 4°C (Bird, 1967) and fourth-stage *Ditylenchus dipsaci* (Wharton and Barrett, 1985). Third-stage trichostrongyle juveniles survive at least 30 hr in 6.5% glutaraldehyde (Eckert and Schwarz, 1965). Initial

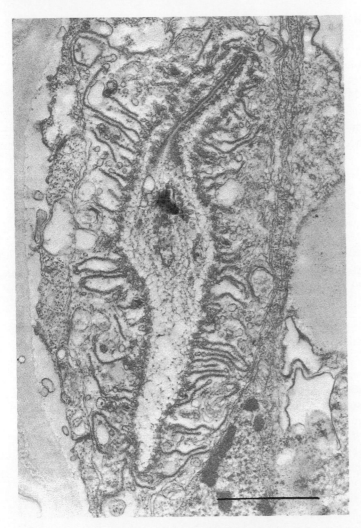

FIGURE 7 Glutaraldehyde: 2%, and paraformaldehyde: 1.5%, 60°C cooled to 25°C; osmium tetroxide: 2%, 25°C. *Meloidodera floridensis*, tail: phasmid sheath cell enclosing neuron, x.s. Bar = 1 μm.

fixative penetration sufficient to inactivate turgor pressure, and total fixation time varies among genera. Males of *Heterodera glycines* needed up to 8 hr more glutaraldehyde fixation than *Meloidogyne incognita* (Baldwin and Hirschmann, 1975).

Nematodes may have different osmolarity requirements for good fixation results, but no systematic studies have been published to determine optimum ranges. It is especially important to determine the effective osmolarity in different buffer systems on organisms with different membrane properties during early fixation when the greatest amount of distortion can occur to surface cells (Hayat, 1981).

Empirical evaluation from osmometer readings of osmolarities can sometimes be found in the nematological literature. Otherwise an approximation can be derived from osmolarity tables (Hayat, 1981) to determine the range of osmolarities (for different nematode fixation protocols). Osmometer readings were used to show that *Xiphinema americanum* is well fixed in 150 mosm glutaraldehyde in cacodylate (Wright and Carter, 1980) while animal parasitic trichuroid nematodes were fixed in 350 mosm (Wright, 1971). Plant parasitic *Ditylenchus dipsaci* was fixed in about 1800 mosm aldehyde fixative (derived from Wharton and Barrett, 1985), and *Dolichodorus heterocephalus* was fixed (Zeikus and Aldrich, 1975) in approximately 1100 mosm.

A given hypertonic solution may alter the body volume of different stages of *Panagrellus silusiae* to different degrees (Prencepe et al., 1984). It might be necessary to select the osmolarity of buffer and fixative for different nematode stages as was done with insect stages (Beaulaton and Gras, 1980).

Prehatch nematodes show better resolution and less shrinkage from the cuticle than posthatch, but many structures are developmentally different. The cuticle, hypodermis, and sensory organs are especially variable just before hatch (Carta and Baldwin, 1990). Therefore fixation quality may have to be sacrificed to developmental accuracy. It is very important to indicate whether nematodes have been naturally hatched or artificially hatched by physical or chemical release from the egg. Even hatching induction by $ZnCl_2$ in heteroderids seems to have an adverse effect on the physiological health and perhaps tissue preservation of newly hatched juveniles.

D. Relaxation

Nematodes will appear coiled or twisted unless properly relaxed and straightened before or during fixation. This makes orientation of nematodes easier for sectioning. Cooling to 4 or 5°C for 30 min is done most commonly before or during initial fixation (Bird, 1971; Wisse and Daems, 1968; Shepherd and Clark, 1988).

Propylene phenoxetol (1% for 1 min) has been used with acrolein and glutaraldehyde for 1 min to straighten nematodes at the beginning of fixation (Chalfie and Thompson, 1979). It also facilitates entrance of aldehydes and preserves flexibility of tissues (Steedman, 1976). This compound is more difficult to obtain than sodium azide and has mildly toxic effects during recovery. Sodium azide has generally replaced propylene phenoxetol in narcotization of *C. elegans* for light microscopy. No reports of successful nematode fixation with sodium azide in comparison with other means of narcotization are available to evaluate. However, in other organisms sodium azide also improves fixation of mitochondria in deeper regions of tissue (Minassian and Huang, 1979).

Fifteen minutes of 0.001 N iodine at room temperature was used to relax third- and fourth-stage *Haemonchus contortus* before fixation with osmium (Ross, 1967).

Mild heat of molten agar (45–60°C) straightens nematodes before fixation without any noticeable impairment of fine structure (Endo, 1979). Heat also has other properties important for nematode preservation. These properties to be discussed include chemical fixative penetration, physical fixation, and cuticle permeabilization.

E. Puncture and Penetration

Many investigations have shown that puncture improves penetration of fixatives and the quality of fine structure (Bird, 1971; Zeikus and Aldrich, 1975). This is especially true with terrestrial parasites. Cuticle puncture may under some circumstances cause the pulling of body contents from cuticle at some distance from the cut. However, it is this author's experience that cold temperature without puncture showed considerable tissue pulling compared to standard fixation with puncture. If tissue distortion occurs, osmolarity during fixation and dehydration should be considered as well as puncture.

Eye knives are popular instruments for cuticle puncture (Baldwin and Hirschmann, 1973; Zeikus and Aldrich, 1975). A lancet was used to cut nematodes in buffer on slides during osmium fixation (Wisse and Daems, 1968). A stone-ground, three-sided sewing needle point (A. H. Bell, pers. commun.), and electrochemically sharpened tungsten needle (Hubel, 1957) have given the author reliable results during glutaraldehyde fixation.

Dimethylsulfoxide (DMSO) can increase the rate of penetration of fixatives, but its potential for artifacts has not been well assessed (Hayat, 1981). It was first used in *Xiphinema index* to aid in the penetration of osmium tetroxide as a primary fixative (Roggen et al., 1966), and was added to speed penetration of glutaraldehyde after treatment with hot formaldehyde-glutaraldehyde (Zeikus and Aldrich, 1975).

Using 60–70°C heat, mechanical puncture may be unnecessary when hot formaldehyde-glutaraldehyde is followed by glutaraldehyde–hydrogen peroxide and osmium in *C. elegans*. Puncture was not required for good fixation under these conditions (Byard and Sigurdson, 1986). When considering the need for puncture, it must be remembered that this free-living nematode loses its capacity for internal displacement very quickly after fixation compared to many plant and animal parasites.

The use of heat to aid in penetration of chemical fixatives as well as fixation itself is not recommended for routine electron microscopy (Hayat, 1981). However, when fixing nematodes, 60–70°C heat seems to provide a significant improvement in fine structure where other methods have failed (Zeikus and Aldrich, 1975; Byard and Sigurdson, 1986; Carta and Baldwin, 1990).

When heat was used for straightening nematodes, 70°C was the most reliable temperature for doing this without noticeably affecting the fine structure in tylenchids (Zeikus and Aldrich, 1975). However, it was suggested that this temperature may be too high for obtaining the best preservation in all nematodes or stages. First- and second-stage juveniles of *C. elegans* were not well fixed compared to good fixation in third through fifth (adult) stages, perhaps because of differences in heat resistance (Byard and Sigurdson, 1986). However, osmolarity problems had not been ruled out as a contributing cause.

Mild heat (50°C) has been helpful in fixing eggs of *Meloidogyne javanica* (Bird and McClure, 1976). The use of mild heat as a protein fixative in other systems has been carefully examined, indicating that tissue shrinkage is no greater than with aldehydes (Hopwood et al., 1984).

It seems that heat may affect more than simply the activity of the fixative, or reduction in turgor from relaxation or death. A recent study on cuticular permeability to water and osmium showed that heat between 40 and 50°C applied to *Ditylenchus dipsaci* caused a phase change in the cuticular lipid. A further increase in permeability between 55 and 70°C was believed to be due to the death of the nematode (Wharton et al., 1988). Whether this phase change could be translated into faster penetration of glutaraldehyde into other nematodes was recently tested with *C. elegans* and *Verutus volvingentis*, a heteroderid plant parasite. Juveniles and adults were subjected to 2% glutaraldehyde in M9 buffer at 45°C. This allowed puncture after 10 min rather than the usual 90 min for *Verutus*. No noticeable improvement was seen within the 3-min control time for *C. elegans* (Carta, unpublished observations). Any potential artifacts from heat as part of the fixation process must be weighed against generally more severe postmortem artifacts (Sjostrand, 1967).

Various potential cuticle solvents have been tried with variable success. A brief (4-min) immersion in 10% bleach (0.5% NaOCl) before a standard fixation showed poor resolution of structures with the TEM in *Heterodera schachtii*. However, 2% bleach for 15 min enhanced fixative penetration in *C. elegans* (Sulston et al., 1983) and *Romanomermis culicivorax* (Platzer and Platzer, 1988).

Difficulties in penetrating eggs with needles and knives have been overcome with finer tools made from broken coverslips (Bird and McClure, 1976), and more easily controlled pieces of cellulose acetate sheeting (Perry et al., 1982). Lasers (von Ehrenstein et al., 1981) and chitinase enzymes (Albertson and Thomson, 1982) have also been effective in puncturing eggs for enhanced penetration of fixatives.

F. Buffers

Phosphate buffer is one of the most physiologically compatible buffers available (Hayat, 1981). It has been effective at pH 6.8 (Endo, 1979), pH 7.2 (Baldwin and Hirschmann, 1973), pH 7.3 (Bird, 1971), to pH 7.4 (Ward et al., 1975). Artifactual deposits may sometimes result with phosphate buffer, particularly when used above 0.1 M concentrations (Hayat, 1986).

For optimal physiological functioning nematodes require certain salts in the surrounding medium (Wright and Newall, 1980). The M9 buffer which contains these salts is commonly used with *C. elegans* (Sulston and Hodgkin, 1988) and has been used successfully in nematode fixation (Byard and Sigurdson, 1986). This buffer comes closest to the physiological ideal but has not been compared to stronger phosphate buffers for general preservation or artifactual deposits.

Cacodylate buffer used by various authors (Webster, Wharton, Carta, Shepherd, Wright) participates with the fixative in the killing of nematodes. It does not lead to artifactual precipitates and it may also reduce osmotic shrinkage with glutaraldehyde as compared with phosphate buffer (Lee, 1984).

PIPES buffer has been used with good results in some plant and animal tissue (Hayat, 1981). However, even with careful buffer washes, this buffer badly obscured the cytoplasmic ground substance in *Meloidodera floridensis* (Carta, unpublished observation). HEPES buffer has been used for fixing *C. elegans* (Sulston and Hodgkins, 1988).

III. TRANSFER

Transfer of specimens is important in fixation as well as dehydration. Nematodes may be transferred through solutions either in agar or unenveloped.

One of the most common methods of safe transfer of individual specimens involves tucking them into a shallow agar groove and overlaying with a drop of 45°C agar. The agar should be made up in salt solution of similar osmolarity to fixatives or shrinkage may result (Wright and Jones, 1965). Commonly, the agar is cut into oblong sheets tapered at one end, or in blocks. Timing of placement in agar may vary. Specimens may be immobilized or killed in agar just before fixation (Endo, 1979), after puncture in the middle of glutaraldehyde fixation (Shepherd and Clark, 1988), before osmium (Zeikus and Aldrich, 1975), or after osmium (Baldwin and Hirschmann, 1973).

Transfer of unenveloped individuals generally saves time during the osmotically fragile steps of fixation. Unenveloped processing would also be necessary when using flat embedding procedures (Reymond and Pickett-Heaps, 1982) for precise longitudinal orientation or correlative LM and EM.

Vehicles for transfer of unenveloped specimens include standard-sized centrifuge tubes (Wisse and Daems, 1968), small polypropylene microcentrifuge tubes (Hong and Barta, 1986), and Bureau of Plant Industry (BPI) dishes for transfer of specimens individually (Zeikus and Aldrich, 1975). Alternatively, small sieves or baskets made from 5-100-μm fabric mesh fitted to modified Beem R* capsules (Eisenback, 1985) or glass tubes and Teflon rings (De Grisse, 1977) may be used. If the mesh is appropriate for the size and activity of the specimens, they will not be lost from the open end unless pipetting or immersion is very rough. The sieves rely on diffusion of solvent through the mesh, which may be slow and require blotting. A more elegant apparatus employs a syringe connected to the longer of two opposed, 10–µm mesh-covered Teflon tubes. The Teflon tubes create a chamber when concentrically shoved within a glass tube. The syringe allows for complete and rapid solution transfers (McClure and Stowell, 1978).

IV. ORIENTATION

The positioning of agar sheets (Wright and Jones, 1965) or blocks containing carefully oriented specimens within a gelatin capsule or other mold is one of the most common methods of orientation of nematodes. However, manipulations in the molds can be difficult and imprecise. Sandwiching nematodes between sheets of epoxy (Wisse and Daems, 1968) or on a prepolymerized plate below small epoxy drops (DeGrisse, 1977) provides more precise orientation within both the agar and the mold. However, using a jeweler's saw to shape specimen blocks is tedious, and the epoxy dust may be dangerous. A safer and more precise method involves placing worms in a thin layer of epoxy on a slide treated with a releasing agent (McClure, pers. commun.). Individual cells may be observed with the LM, scored with a diamond scriber, and glued to resin stubs for sectioning (Reymond and Pickett-Heaps, 1982).

Eggs or larvae can be oriented into a distal, compact pellet at the end of a small rod. This can be done by microcentrifuging in a microcapillary containing a tiny polyethylene tube with specimens in unpolymerized epoxy. After polymerization the specimens in the rod can be positioned close to the tip of a larger mold (Hong and Barta, 1986).

V. EMBEDDING

The cuticle represents a barrier to good epoxy resin infiltration (Wright and Jones, 1965). As a result, a relatively long time (2–3 days) is required for infiltration of nematode tissue compared to tissue such as rat liver (hrs). Slow dropwise infiltration of 50% epoxy-acetone over 1 1/2 hr was effective for vermiform plant parasites and *C. elegans*; 1/2 hour was insufficient in the author's experience. The possibility that puncture of the cuticle may significantly aid in the penetration of embedding media seems to be a logical prediction. However, even the tougher plant parasites did not require puncture for good infiltration of low-viscosity Spurr's epoxy resin with a slow infiltration schedule (Endo, 1979). No puncture was necessary for *C. elegans* with either Spurr's or Epon. Both epoxies showed equally good infiltration and staining (Byard et al., 1986). A number of undescribed methods were tested for the especially difficult infiltration of egg shells. The final reported method involved infiltration of unaccelerated Epon CY212 resin for 3 days at room temperature and in accelerated resin for 12 hr at 37°C (Preston and Jenkins, 1984). Very fresh epoxy is of some importance in obtaining the best quality sections possible, and should be kept no longer than three months to a year (Endo, pers. commun.).

Ultra low-viscosity LR white acrylic resin shows promise for nematode embedding (Hussey, pers. commun.). However, curing the resin to the desirable consistency for sectioning is not always reliable, and some report UV radiation to be better than heat. Because of the sensitivity to oxygen, a gelatin capsule must be used rather than flat embedding molds. In this capsule, orientation in agar can be done by the pointed block configuration of Wright and Jones (1965).

VI. SECTIONING

The process of trimming a block in preparation for sectioning may be more exacting with microscopic nematodes than with larger organisms and tissues. This is an advantage since the smaller the block face, the greater the ease of sectioning. However, the consistency of the epoxy-infiltrated tissue and that of the surrounding epoxy are generally different. Fibrous, heterogeneous tissues such as cuticle are difficult to section (Hayat, 1986). This can create problems with nematodes, especially swollen parasitic females, falling out of the surrounding sections.

Grids coated with formvar or parlodion are necessary for obtaining serial sections. Although image contrast is lowered and formvar may tear, the specimen itself may be protected from falling out of the section. This is not uncommon even with good tissue infiltration (Baldwin, pers. commun.). When tissue of different consistency is sectioned, the consistency of the embedding media should approach the hardest cuticular tissue for easiest sectioning. Adjustment in the epoxy components, orientation of specimens near block surfaces, and extra polymerization time will help to create the desired hardness (Wright and Jones, 1965).

Although Epon may section with difficulty on glass knives (Wright and Jones, 1965), softer Spurr's resin may have a greater tendency to hold a static charge making sectioning difficult on either glass or diamond.

Excessive static electricity may occasionally make the cutting of serial ribbons an impossible task. Antistatic guns, ionizing polonium bars, a grounded copper wire in the boat, and humidifiers are sometimes helpful in reducing static charge which can build up on a block during sectioning. However, a diamond knife may also contribute to a problem. In our lab a drop of metal paint was used to connect and ground the diamond to its conductive metal boat. Diamond knives are generally mounted in nonconductive epoxy (Koen, pers. commun.).

VII. STAINING

Staining with uranyl acetate and lead citrate is one procedure for which there is the least variation from standard methods. En block staining with uranyl acetate (aqueous or alcoholic) is commonly used in nematode studies to improve preservation and increase contrast when combined with poststaining of uranyl acetate. When this is done it is important that any cacodylate or phosphate buffer be well washed out before addition of uranyl acetate to avoid precipitation (Hayat, 1986). Sometimes heat is used with uranyl acetate to enhance staining after sectioning (Ward et al., 1975; Hope, 1988).

The likelihood of lead contamination by carbon dioxide can be reduced by inserting grids at the base of the drop rather than on its surface (Aldrich and Mollenhauer, 1986).

Staining times for uranyl acetate and lead citrate vary somewhat with nematodes. In our lab *Meloidogyne* requires less staining time than *Heterodera*.

Part 4. Molecular Techniques for Nematode Species Identification

VALERIE M. WILLIAMSON   *University of California, Davis, Davis, California*

I. INTRODUCTION

Molecular characters are becoming increasingly useful for species identification and systematics in nematology. Recent developments in methodology resulting in more convenient, rapid, and specific assays have opened the door to greater use of these tests for detecting and identifying phytopathogenic nematodes. Proteins and DNA are particularly useful characters for species identification because they are less subject to environmental effects and investigator interpretation than are many other characters. DNA is the genetic code of the organism and thus its properties are a direct reflection of the genetic identity of that organism. Proteins are products of the genetic code and, because of the availability of techniques to directly measure their properties, are also particularly useful as characters.

This section will concentrate on the use of proteins and DNA as characters in identification of plant pathogenic nematodes. Emphasis will be placed on describing techniques currently used or techniques potentially valuable, for rapid, routine identification. In general, molecular techniques are most useful for identification at the lower taxonomic levels (genus, species, and subspecies levels). Currently, molecular characters are used in conjunction with traditional morphological characters. In a few instances, specific molecular techniques have been extensively validated and are appropriate for routine identification. One such case is the use of isozyme electrophoresis for identification of root-knot nematode species. As more molecular characters become validated and as they become more routine, these characters are likely to find increasing applications in identification. Advantages and disadvantages of various techniques will be discussed with emphasis on protein characters for use in identification. DNA techniques will be outlined and discussed only briefly as they described in more detail elsewhere in this volume (Chapter 4). Immunological techniques and their potential use will also be introduced.

II. PROTEIN ELECTROPHORESIS

A. Background

Electrophoretic separation of molecules is the foundation of many biochemical analyses. In general, the materials to be analyzed (e.g., extracts of nematodes or their eggs) are subjected to electrophoresis in a matrix such as starch, agar, acrylamide, or cellulose acetate. Proteins separate in such matrices because of their net electrical charge and/or size. Protein charge reflects amino acid composition (and sometimes chemical modification). The amino acid composition is determined by the sequence of DNA (the gene) that encodes the protein. Thus mobility of proteins is determined in large part by the genetic material, the DNA, of the organism. Total protein can be analyzed by staining the matrix containing the electrophoresed protein extract with a protein-specific stain. Specific enzymes or isozymes can be localized using activity stains. Both analysis of total protein and analysis of specific isozymes have been useful in identification. For an excellent review of the history of protein electrophoresis in nematode systematics, see Hussey (1979). Here we will try to emphasize more recent developments in techniques and implementation.

B. Total Protein, 1-D and 2-D Acrylamide Gels

1. Uses of the Technique

Separation of soluble proteins in whole nematode extracts by acrylamide gel electrophoresis followed by staining the separated proteins yields species-specific patterns. Dickson et al. (1971a) found that some protein band patterns were useful for distinguishing selected species of *Meloidogyne* and *Ditylenchus* as well as the genera *Heterodera* and *Aphelenchus*. Hussey and Krusberg (1970) found distinct differences in soluble protein patterns between *D. dipsaci* and *D. triformis*. *Heterodera* species were found to produce distinguishable protein patterns (Pozdol and Noel, 1984). Huettel et al. (1983b) found a major protein band differing between two races of *Radopholus similis*. Bakker et al. (1988) saw reproducible differences between *Globodera rostochiensis* and *G. pallida*. The complexity of total protein patterns makes interpretation of results difficult. A single nematode comprises thousands of different proteins differing in size and other physical properties as well as in abundance and temporal regulation. In *Meloidogyne* species 10–12 abundant protein bands are seen above the background proteins but not all are seen in all species (Dalmasso and Bergé, 1978). It is difficult to tell whether differences in band patterns indicate different proteins or changes due to environmental or developmental conditions.

Two-dimensional polyacrylamide gel electrophoresis (2-D PAGE) can be used for better separation of total soluble proteins. For 2-D PAGE, proteins are separated in the first dimension using isoelectric focusing, a procedure that distinguishes proteins by their isoelectric point. The isoelectric point of each protein is determined mainly by its amino acid composition. A strip of matrix carrying the separated proteins is applied to a second gel and subjected to electrophoresis in the presence of a detergent (SDS). The second gel (second dimension) results in the separation of proteins by size. 2-D gels have been valuable for phylogenetic studies on Dorylaimida (Ferris and Ferris, 1988; Ferris et al., 1986) and have been used to compare *H. glycines* isolates and root-knot nematode species (Ferris et al., 1986; Premachandran et al., 1984). However, 2-D PAGE is labor-intensive and thus more useful for systematics than for routine identification.

2. Extraction Methods

Generally, female root-knot or cyst nematodes are prepared by grinding in buffer in an ice bath with an ice-cooled glass homogenizer (Dickson et al., 1971a). Nematodes can be stored frozen (–80°C) before and after extraction. Usually 30–200 young females are used per analysis. Vermiform nematodes are harder to break apart and extract. One method is to freeze nematodes in liquid nitrogen, then grind in a porcelain mortar (Dickson et al., 1971a). *Ditylenchus* were broken by passing through a cold French pressure cell (Hussey and Krusberg, 1971).

3. Electrophoresis and Analysis

Comparison of samples is best done after separation by polyacrylamide slab gel electrophoresis (see Dalmasso and Bergé, 1978 for procedures). 2-D gel analysis usually relies on the technique of O'Farrell (1975) (Ferris et al., 1985, 1986). Proteins are visualized after electrophoresis by using a protein-specific stain (Poehling et al., 1980; Premachandran et al., 1984). Another approach has been to label the proteins with a radioactive isotope before electrophoresis then to use autofluorography to visualize the separated proteins (Ferris et al., 1987).

C. Isozyme Electrophoresis

1. Uses of the Technique

Total protein patterns are complex due to the presence of thousands of different proteins and thus difficult to interpret. Often it is preferable to examine a single protein or a small subset of proteins. Enzymes catalyzing a defined reaction can in many cases be easily visualized after electrophoresis by application of a specific activity stain to the matrix. Isozymes are enzymes catalyzing the same reaction, but differing in a measurable property such as electrophoresic mobility. When these variant enzymes correspond to products of analogous genes in different organisms, they are often referred to as allozymes. A major advantage of isozyme staining is that it is generally much more sensitive than total protein staining and one can examine single individuals in groups such as cyst or root-knot nematodes that have large females. This sensitivity permits testing several individual nematodes in a population to detect mixed infections.

Extensive characterizations of isozymes have been carried out in root-knot nematode species (Dalmasso and Bergé, 1978, 1983; Esbenshade and Triantaphyllou, 1985a,b, 1987; 1990). Analysis of a number of different isozymes in individual root-knot females by polyacrylamide gel electrophoresis has shown that characteristic isozyme patterns can be used to reliably identify species. Esbenshade and Triantaphyllou (1985a,b) examined 291 populations from 16 species of *Meloidogyne* originating from 65 countries. Patterns of 27 enzymes were analyzed to construct a phylogenetic tree of the genus *Meloidogyne* (Esbenshade and Triantaphyllou, 1987). The technique is sufficiently sensitive that the isoenzyme pattern of a single adult female can be observed using acrylamide gel electrophoresis (or five females using starch gel electrophoresis). Enzyme phenotypes have been given designations corresponding to the relative mobility of the activity bands after electrophoresis, the species with each phenotype, and the number of bands present (see Esbenshade and Triantaphyllou, 1985a). For example, the esterase pattern of *M. javanica* is called J3 (javanica, three bands) because of its distinctive three activity bands. Isozyme electrophoresis has not been shown to distinguish races of root-knot nematodes.

A number of investigators have observed that esterase isozyme patterns are different for each of the major root-knot nematode species and thus are particularly useful (Dalmasso and Bergé, 1978; Esbenshade and Triantaphyllou 1985a,b; Fargette, 1987a,b; Pais and Abrantes, 1989). Fargette (1987b) analyzed the electrophoretic pattern of the b esterases (b esterases hydrolyze both α- and β-naphthyl acetate) in 57 populations of *Meloidogyne* collected in West Africa and concluded that electrophoretic identification is more accurate and objective than the other available criteria including analysis of perineal patterns. Isozyme patterns within a species are consistent even for isolates from diverse geographic locations, making identification very reliable. Patterns of nematodes grown on different hosts were the same (Dickson et al., 1971b; Fargette, 1987a; Pais and Abrantes, 1989). Differences were seen between isozyme patterns of adults, juveniles, and eggs (Dickson et al., 1971b). Esterase patterns obtained by different groups from different parts of the world are essentially the same and can be easily correlated.

Other isozymes have been useful for identifying root-knot nematodes. For example, *M. hapla* and *M. incognita* have similar esterase patterns (Esbenshade and Triantaphyllou, 1990) but can easily be distinguished by comparing malate dehydrogenase (MDH) patterns. In addition, esterase activity can be very low in older females, and some isolates of *M. hapla* appear to lack b esterase activity. In these cases other isozymes must be used (Dalmasso and Bergé, 1978). Glucose phosphate isomerase is useful for distinguishing some species of root-knot nematodes (see Esbenshade and Triantaphyllou, 1985a).

Isozyme analysis has also been applied to other phytopathogenic as well as free-living nematode species (reviewed in Platzer, 1981). Analysis of single individuals has been limited to cyst and root-knot nematodes because of the large size of adult females. Analysis of vermiform nematodes is more difficult because of their smaller size and the difficulty of extracting proteins from them. Analysis of single individuals of *Heterodera glycines* has shown that isozyme polymorphisms do exist within populations (Radice et al., 1988a; Esbenshade and Triantaphyllou, 1988). Correlation of isozymes analysis has also been applied to identify species and pathotypes of potato cyst nematode (Fox and Atkinson, 1984, 1985, 1988), to distinguish races of *Radopholus similis* (Huettel et al., 1983a), and to delineate morphologically indistinguishable *Anguina* species (Riley et al., 1988).

Despite its many advantages, electrophoretic determination of nematode species is not yet widely used for routine identification. Reasons for this include the difficulty of preparation of acrylamide gels and the toxicity of the acrylamide as well as a lack of familiarity and some skepticism on the part of identification labs about the accuracy of the new technique. The recent availability of an automated electrophoresis system (PhastSystem by Pharmacia) has greatly simplified electrophoretic analysis and eliminated the need for acrylamide gel preparation (Esbenshade and Triantaphyllou, 1990). This development is likely to increase the use of protein electrophoresis in routine identification.

2.   Sample Preparation and Storage

Adult root-knot nematode females dissected from root tissue can be stored in 0.9% saline solution at room temperature for up to 3 weeks, a property very useful in field situations. Prior to electrophoresis individual root-knot nematode females are crushed in a small volume of extraction solution (20% sucrose, 2% Triton X-100). Crushed nematodes can also be stored frozen (–15°C) until ready for electrophoretic analysis (Esbenshade and Triantaphyllou, 1985c). Esterase activity is stable for several months in nematodes prepared in this way. *Anguina* species have been lysed by sonication then stored frozen in capillary tubes (Riley et al., 1988).

3.   Electrophoresis

Thin (0.7-mm) acrylamide slab gels provide the highest resolution matrix for separation of isozymes. A detailed description of the apparatus and techniques for miniaturized acrylamide gel electrophoresis analysis have been presented by others (Esbenshade and Triantaphyllou, 1985b, c; Dalmasso and Bergé, 1978). For some enzymes, separation by isoelectric focusing in acrylamide or agarose gels has been useful (Dalmasso and Bergé, 1978; Radice et al., 1988a; Fox and Atkinson, 1988).

With the PhastSystem automated electrophoresis system the nematode extract is applied to a precast, commercially available, small (4.3 × 5 cm), very thin (0.4-mm) polyacrylamide gel. Two gels with 12 samples each can be electrophoresed at a time. Electrophoresis is carried out for about 30 min. Conditions are preprogrammed to give highly reproducible results from run to run. For esterase analysis native gel electrophoresis has been carried out using 10–15% gradient gels.

Other, less expensive electrophoresis systems using other matrices including starch gels, agarose gels, and cellulose acetate paper can also be used. Starch and agarose gels are less sensitive and require more material than other matrices, so in general they cannot be used to look at single nematodes. These techniques could be used if large numbers of uniform nematode populations are available. Cellulose acetate gels have been used for distinguishing acid phosphate alleles linked to nematode resistance in tomato (Bolkan et al.,

TABLE 1 Isozyme Stains for MDH and Esterase

Malate dehydrogenase stain:
    38 ml H_2O
    5.0 ml solution A (10.6 g Na_2CO_3, 1.34 g L-malic acid, H_2O to 100 ml)
    7.5 ml solution B (0.5 M Tris-HCL, pH 7.1)
    25 mg NAD^+
    15 mg nitroblue tetrazolium
    1.0 mg phenazine methosulfate

Esterase stain:
    25 ml 0.1 M sodium phosphate buffer, pH 7.4
    7.5 mg EDTA
    15 mg fast blue RR salt
    10 mg α-naphthyl acetate in 0.5 ml acetone

Note: These solutions are derived from Esbenshade and Triantaphyllou (1985c) and should be prepared immediately before use.

1987). The technique is sensitive enough to detect isozymes in single nematodes but no data have accumulated on nematode isozymes by this method.

4. Activity Staining and Analysis

After electrophoresis, the gel is stained for isozyme activity. Directions for staining with esterase and malate dehydrogenase (MDH) are presented in Table 1. For preparation of other enzyme stains, see Esbenshade and Triantaphyllou (1985c) and Huettel et al. (1983). It is possible to stain a single gel run on an automated gel apparatus for both MDH and esterase activities (Esbenshade and Triantaphyllou, 1990). The gel is first stained for 5 min with MDH activity stain, then rinsed with esterase stain buffer and stained for an additional 45 min with esterase activity stain. A diagram of the esterase and malate dehydrogenase phenotypes for common root-knot species can be seen elsewhere in this volume (Eisenback). Species identification is made by comparison to known samples. For root-knot nematode identification, *M. javanica* makes an excellent standard because of the distinctive esterase and MDH patterns.

III. DNA ANALYSIS

A. Background

DNA is a biopolymer of two complementary strands of nucleotides. There are four nucleotides in DNA and their order in the genome is the code which determines the identity of the individual. The DNA of an organism is not generally subject to the environmental influences or developmental changes that influence the utility of many other characters. The genome of *M. incognita* comprise about 51 million nucleotides of DNA sequence (Pableo and Triantaphyllou, 1989), one of the smallest genomes of any multicellular animal. DNA analysis has been useful for identification of other plant pathogens (reviewed by Miller and Martin, 1988). There is enormous potential for the use of DNA sequence analysis in establishing and testing phylogenetic relationships of nematodes (reviewed in Ferris and Ferris, 1987; Baldwin and Powers, 1987; Curran and Webster, 1987; and Curran, this volume).

Besides the DNA present in the nuclear chromosomes, DNA is also present in the cytoplasm inside mitochondria. Mitochondrial DNA (mtDNA) is a relatively short DNA sequence (14,000–36,000 nucleotides in nematodes) compared to nuclear DNA (Hyman, 1988). It is present in hundreds of copies per cell, so less material is required for analysis. In addition, mtDNA sequences evolve more rapidly than single-copy nuclear DNA. This makes mtDNA particularly useful for observing differences between closely related organisms (Brown et al., 1979). For a recent review of nematode mitochondrial DNA biology and applications to nematological problems, see Hyman (1988).

The current expansion in the use of DNA for identification is the consequence of the recent availability of a number of tools for study of DNA sequence and properties. One of these tools is a large number of commercially available restriction endonucleases. These enzymes recognize short sequences of DNA (usually four to six nucleotides long) and cleave the DNA at these sequences. The resulting fragments are then subjected to electrophoretic analysis. This technique has been invaluable for the characterization of particular DNA sequences. A second technique, hybridization, is based on the ability of complementary strands of DNA to find each other and reanneal.

B. Restriction Enzyme Digestion Patterns

Patterns of nematode DNA digested with restriction endonucleases and subjected to agarose gel electrophoresis have been analyzed (reviewed in Curran and Webster, 1987). Differences in DNA sequence result in differences in the number and size of fragments produced (restriction fragment length polymorphisms or RFLPs). Fragments are separated by size using agarose gel electrophoresis and stained, usually with ethidium bromide, a dye which binds to DNA and fluoresces when exposed to UV light. Because of the complexity of genomic DNA it is usually not possible to directly observe or distinguish sequences present in a single copy. However, some sequences of DNA, such as ribosomal DNA sequences, are present in hundreds of copies in the genome and can be viewed directly as distinct bands visible above a background smear of single-copy DNA. Curran and Webster (1989) found differences in restriction patterns of genomic DNA digests of *Heterorhabditis* isolates; Kalinski and Huettel (1988) saw differences between races of *Heterodera glycines*.

Mitochondrial DNA can be separated fairly easily from genomic DNA. Restriction endonuclease-digested mitochondrial DNAs from *M. incognita, M. javanica, M. hapla, M. chitwoodi,* and *M. arenaria* all produce different RFLP patterns (Powers et al., 1986; Hyman, 1988). Polymorphisms were not observed within *M. incognita* races (Powers et al., 1986). When mtDNA from the sibling species *Heterodera glycines* and *H. schachtii* were compared, cut separately with 12 different restriction endonucleases, only nine of 90 fragments that could be scored were of the same size, indicating that these species had diverged a long time ago (Radice et al., 1988b).

C. Hybridization Analysis

1. Uses of the Technique

Specific sequences in a complex electrophoretic pattern of DNA can be detected by a method developed by Southern (1975). The DNA in the agarose gel is denatured, the complimentary strands are separated, then transferred to a membrane. Specific DNA probes are labeled, usually with the isotope ^{32}P, and hybridized with the membrane. Specific sequences, including RFLPs, complementary to the added probe are detected in this way. Species-specific probes, i.e., fragments of DNA that can hybridize to one species but not to related species, allow identification without electrophoresis. Species can be identified using relatively crude DNA samples by simply spotting the DNA onto a matrix then hybridizing with species-specific probes. Hybridization signals can be obtained from a single egg, larva, or adult using DNA probes that are present in multiple copies, such as certain mtDNA fragments. Several hundred samples can be simultaneously processed. Currently, only a few species-specific probes are available. *Meloidogyne* mtDNA has been detected in the presence of host tissue and other organisms using a simple spot-blot procedure (Powers et al., 1986). A cloned DNA fragment that can hybridize to *Globodera pallida* but not *G. rostochiensis* has been identified (Burrows and Perry, 1988). A DNA probe cloned from mitochondrial DNA has been developed for the specific detection of *Heterodera glycines* (Besal et al., 1988). This probe can detect a single *H. glycines* female nematode that has been applied as an unpurified homogenate to nitrocellulose filters. However, some weak reaction with *H. schachtii* was observed, indicating that the probe is not completely species-specific.

2. Sample Preparation

Analysis of nucleic acid sequence requires many specialized techniques. There are excellent manuals now available which describe in detail techniques such as agarose gel electrophoresis, DNA isolation and hybridization, and restriction enzyme digestion (e.g., Maniatis et al., 1982; Berger and Kimmel, 1987). For specific applications such as preparation of total DNA from nematodes, see, for example, Curran et al., 1985; or for mtDNA isolation and DNA preparation, Radice et al., 1988 or Powers et al., 1986.

D. Future Developments

Molecular identification using DNA is developing rapidly and the use of DNA based techniques in nematology certainly will continue to increase. The need to use radioactively labeled probes has slowed the expansion of practical applications of DNA hybridization because of the hazards associated with handling and disposing of radioactive material. Nonradioactive probes are now being tested and should enhance the applicability of nucleic acid probes as identification tools (Miller and Martin, 1988; Burrows, 1989). Another technique showing promise in nematode species identification is the polymerase chain reaction (PCR). The amplification in number of copies of specific fragments of DNA by PCR allows their analysis without the use of isotopes. This technique should make it possible to amplify a single-copy DNA sequence using a single nematode or even an egg (Harris et al., 1989; White et al., 1989).

IV. SEROLOGICAL TECHNIQUES

A. Background

Use of immunological methods has not been extensive in phytonematology (reviewed in Hussey, 1979). Classical techniques have involved generating antisera by injecting rabbits with whole nematodes or fractions of nematodes. Antibodies recognizing a number of different determinants, i.e., polyclonal antibodies, are produced in the rabbit. Serum from the rabbit is placed in a well in an agar plate and allowed to diffuse toward the antigen (in this case, nematode extract). Immunoprecipitin bands are observed in the agar where the antibody and antigen interact. One problem has been the difficulty of obtaining species-specific antibodies. Most polyclonal antibodies recognize several antigenic determinants in the nematode, many of which are present on broad groups of nematodes such as family or genus. Specificity can be improved by combining antibody-antigen reactivity with electrophoresis. Immunoelectrophoresis has been used to compare potato cyst nematode species and pathotypes (Wharton, 1983). Even then discrimination of all species and pathotypes was not possible.

B. ELISA

Immunological methods for detecting other plant pathogens have been used for many years. Recent developments in immunodiagnostic methods have expanded the scope of application to the diagnosis of plant diseases caused by many viruses, bacteria, spiroplasms, mycoplasm-like organisms, and fungi (reviewed in Miller and Martin, 1988). Advantages include high sensitivity and the ability to detect pathogen antigens in complex mixtures such as plant material or soil. ELISA (enzyme linked immunoabsorbent assay) and related assays have been particularly useful. For this assay, the antibody is attached to a solid support, then a solution with the antigen (e.g., a plant extract or soil sample) is added to allow binding of antigen to the attached antibody and unbound material is washed away. Additional antibody linked to an enzyme capable of producing a color reaction is added. Substrates required to produce the color reaction are added and the intensity of the color is proportional to the amount of antigen present. One adaptation of this type of assay for clinical or field use is the "dipstick" test which has been used to detect fungal plant pathogens in turfgrass (Lankow et al., 1987). In this technique disposable strips with antibody are incubated with extracted plant material then transferred to a substrate solution for color development. Some quantitation can be made by monitoring the intensity of the color development. For applications where identification of broad groups of nematodes, at family or genus level, is required, such as quarantine facilities, rapid dipstick tests for the presence of these nematodes should be of great value and are likely to be developed commercially in the future.

C. Monoclonal Antibodies

Recent advances in the techniques available for production of monoclonal antibodies may allow development of species or even race-specific probes. Monoclonal antibodies are produced by fusion of antibody-secreting cells from an immunized mouse to myeloma cells, producing hybridomas which secrete a specific antibody. Individual hybridomas each produce a monospecific antibody and a panel of hybridomas can be tested to identify those producing the most useful antibodies. Hybridomas can be cultured to produce large amounts of the specific antibody or can be stored frozen in liquid nitrogen for later use. A monoclonal antibody can be a very specific and standardized reagent because it recognizes a single determinant. For plant pathogenic fungi production of highly specific monoclonal antibodies has been difficult but has worked very well in some cases (Miller and Martin, 1988). Monoclonal antibodies specific to the genus *Globodera* and one specific to *G. pallida* have been identified and produced (Schots et al., 1990). Production of monoclonal antibodies is a complex procedure requiring specialized techniques. Such product development is probably best undertaken by commercial organizations because they can use the antibodies they have developed as standardized reagents for identification kits. It is likely that commercial production of such kits for economically important species will occur in the future and will be of much value for routine identification.

V. CONCLUSIONS: IDENTIFICATION TODAY AND TOMORROW

The use of molecular techniques is just beginning to show its impact on nematode identification. From this review it should be clear that useful techniques are now being developed. In the next 10 years these techniques will assume a major role in identification of nematodes and for establishment of taxonomic species definitions. Several cases are described above where isozymes and protein patterns have been used to determine species boundaries. Because DNA is a primary determinant of genetic composition, it is likely DNA sequence analysis will have an even more important role in species determination (see Chapter 4, this volume).

This section has discussed mainly the techniques that are or have the potential to be used for rapid, routine identification. Currently, isozyme electrophoresis is the most broadly applicable and validated of the molecular techniques. It is highly sensitive, allowing analysis of individual nematodes, and adaptable to use with many different groups of nematodes. The newly available automated gel electrophoresis system makes isozyme analysis even easier and more reproducible. Use of DNA hybridization in identification is likely to increase as species-specific probes become available and as new technological developments are applied to nematology. In cases where appropriate antibodies have been produced, the development of serological kits for identification will allow rapid and inexpensive identification, and therefore become the method of choice. Finally, it is important to remember that these molecular characters are mostly useful at lower taxonomical levels and should be used in conjunction with more classical characters such as morphological and ecological factors.

ACKNOWLEDGMENTS

I thank P. Esbenshade, A. C. Triantaphyllou, and B. Hyman for unpublished preprints. Also thanks to B. Westerdahl for helpful comments on the manuscript.

REFERENCES

Abrantes, I.O., and Santos, S.M. 1989. A technique for preparing perineal patterns of root-knot nematodes for scanning electron microscopy. *J. Nematol.* 21: 138–139.

Albertson, P.G., and Thomson, J.N. 1982. The kinetochores of *Caenorhabditis elegans*. *Chromosoma* 86: 409–428.

Aldrich, H.C., and Mollenhauer, H.H. 1986. Secrets of successful embedding, sectioning, and imagining. In *Ultrastructure Techniques for Microorganisms*, H.C. Aldrich and W.J. Todd, eds. Plenum Press, New York, pp. 101–132.

Andrássy, I. 1956. Die Rauminhalts-und Gewichtsbestimmung der Fadenwürmer (Nematoden). *Acta Zool.* 2: 1–15.

Bakker, J., Schots, A., Bouwman-Smits, L., and Gommers, F.J. 1988. Species-specific and thermostable proteins from second-stage larvae *Globodera rostochiensis* and *G. pallida*. *Phytopathology* 78: 300–305.

Baldwin, J.G., and Hirschmann, H. 1973. Fine structure of cephalic sense organs in *Meloidogyne incognita* males. *J. Nematol.* 5: 285–302.

Baldwin, J.G., and Hirschmann, H. 1975. Fine structure of cephalic sense organs in *Heterodera glycines* males. *J. Nematol.* 7: 40–53.

Baldwin, J.G., and Powers, T.O. 1987. Use of fine structure and nucleic acid analysis in systematics. In *Vistas on Nematology*, J.A. Vecch and D.W. Dickson, eds., Society of Nematologists, Hyattsville, MD, pp. 336–345.

Barker, K.R. 1985a. Sampling nematode communities. In *An Advanced Treatise on Meloidogyne, Vol. 2, Methodology*, K.R. Barker, C.C. Carter, and J.N. Sasser, eds., North Carolina State University, Raleigh, pp. 3–17.

Barker, K.R. 1985b. Nematode extraction and bioassays. In *An Advanced Treatise on Meloidogyne, Vol. 2, Methodology*, K.R. Barker, C.C. Carter, and J.N. Sasser, eds., North Carolina State University, Raleigh, pp. 19–35.

Beaulaton, J., and Gras, R. 1980. Influence of aldehyde fixatives on the ultrastructure of the prothoracic gland cells in *Rhodnius prolixus* (Insecta: Heteroptera) with special reference to their osmotic effects. *Microsc. Acta* 82: 351.

Bedding, R.A. 1973. Biology of *Deladenus siricidicola* (Neotylenchidae) an entomophagous-mycetophagous nematode parasitic in siricid woodwasps. *Nematologica* 18 (1972): 482–493.

Berger, S.L., and Kimmel, A.R., eds. 1987. *Guide to Molecular Cloning Techniques*, Methods in Enzymology, Vol. 152. Academic Press, San Diego.

Besal, E.A., Powers, T.O., Radice, A.D., and Sandall, L.J. 1988. A DNA hybridization probe for detection of soybean cyst nematode. *Phytopathology* 78: 1136–1139.

Bird, A. 1967. Changes associated with parasitism in nematodes. Morphology and physiology of preparasitic and parasitic larvae of *Meloidogyne javanica. J. Parasitol.* 53: 768–776.

Bird, A.F. 1971. *The Structure of Nematodes*. Academic Press, New York.

Bird, A., and McClure, M.A. 1976. The tylenchid (Nematoda) egg shell: Structure, composition and permeability. *Parasitology* 72: 19–28.

Bird, G.W. 1971. Digestive system of *Trichodorus porosus. J. Nematol.* 3: 50–57.

Boag, B. 1981. Measuring nematodes using a digitising tablet and microcomputer. *Syst. Parasitol.* 2: 145–147.

Bolkan, H.A., Williamson, V.M., and Waters, C.M. 1987. Use of cellulose acetate electrophoresis as an alternative to starch gel electrophoresis for detecting root-knot nematode resistance in tomato. *Plant Dis.* 71: 1001–1003.

Bolla, R.I. 1987. Axenic culture of plant-parasitic nematodes: Problems and perspectives. In *Vistas on Nematology* (J.A. Veech and D.W. Dickson, eds.), Society of Nematologists, Hyattsville, MD, pp. 401–407.

Bridge, J., and Ham, P.J. 1985. A technique for the cryopreservation of viable juveniles of *Meloidogyne graminicola. Nematologica* 31: 185–189.

Brown, D.J.F. and Boag, B. 1988. An examination of methods used to extract virus-vector nematodes (Nematoda: Longidoridae and Trichodoridae) from soil samples. *Nematol. Medit.* 16: 93–99.

Brown, D.J.F., and Topham, P.B. 1984. A comparison of reported variation in the morphometrics of *Xiphinema diversicaudatum* (Nematoda: Dorylaimida) and the effect of some methods of preparing specimens for examination by optical microscopy. *Nematol. Medit.* 12: 169–186.

Brown, W.M., Gcorge, M., and Wilson, A.C. 1979. Rapid evolution of animal mitochondrial DNA. *Proc. Natl. Acad. Sci. USA* 77: 1967–1971.

Burrows, P.C. 1989. The identification of plant-parasitic nematodes using biotic labeled DNA probes. *J. Nematol.* (abstract) 21:553.

Burrows, P.R., and Perry, R.N. 1988. Two cloned DNA fragments which differentiate *Globodera pallida* from *G. rostochiensis. Revue Nématol.* 11: 441–445.

Byrd, D.W., Barker, K.R., Ferris, H., Nusbaum, C.J., Griffin, W.E., Small, R.H., and Stone, C.A. 1976. Two semi-automatic elutriators for extracting nematodes and certain fungi from soil. *J. Nematol.* 8: 206–212.

Byrd, D.W., Nusbaum, C.J., and Barker, K.R. 1966. A rapid flotation-sieving technique for extracting nematodes from soil. *Plant Dis. Reptr.* 50:954–957.

Byard, E.H., Sigurdson, W.J., and Woods, R.A. 1986. A hot aldehyde-peroxide fixation method for electron microscopy of the free-living nematode *Caenorhabditis elegans. Stain Technol.* 61: 33–38.

Carta, L.K., and Baldwin, J.G. 1989. The ultrastructure of phasmid development in *Meloidodera floridensis* and *M. charis* (Heteroderinae). *J. Nematol.* (in press).

Caswell, E.P., Thomason, I.J., and McKinney, H.E. 1985. Extraction of cysts and eggs of *Heterodera schachtii* from soil with an assessment of extraction efficiency. *J. Nematol.* 17: 337–340.

Chalfie, M., and Thomson, J.N. 1979. Organization of neuronal microtubules in the nematode *Caenorhabditis elegans. J. Cell. Biol.* 82: 278–289.

Clayden, I.J., Turner, S.J., and Marks, R.J. 1985. Comparison of the Fenwick can and Schuiling centrifuge methods for the extraction of potato cysts nematodes from soil. *Bull. OEPP* 15: 285–287.

Cohen, A.L. 1979. Critical point drying: Principles and procedures. *Scanning Electron Microsc.* II: 303–324.

Cooke, D.A., Mathias, P.L., Chwarszczynska, D.M., and Coppock, L.J. 1983. Comparison of methods of assessing field populations of *Heterodera schachtii. Plant Pathol.* 32: 339–343.

Crang, R.F.E., and Klomparens, K.L., eds. 1988. *Artifacts in Biological Electron Microscopy.* Plenum Press, New York.

Cranston, D.M., and Newton, P. 1965. A trouble-free test for the presence of eelworms on chrysanthemum. *Plant Pathol.* 14: 74.

Curran, J., Baillie, D.L., and Webster, J.M. 1985. Use of restriction fragment length differences in genomic DNA to identify nematode species. *Parasitology* 90: 137–144.

Curran, J., and Webster, J.M. 1987. Identification of nematodes using restriction fragment length differences and species-specific DNA probes. *Can. J. Plant Pathol.* 9: 162–166.

Curran, J., and Webster, J.M. 1989. Genotypic analysis of *Heterorhabditis* isolates from North Carolina. *J. Nematol.* 21: 140–145.

Dalmasso, A., and Bergé, J.B. 1978. Molecular polymorphism and phylogenetic relationship in some *Meloidogyne* spp.: Applications to the taxonomy of *Meloidogyne. J. Nematol.* 10: 323–332.

Dalmasso, A., and Bergé, J.B. 1983. Enzyme polymorphism and the concept of parthenogenetic species, exemplified by *Meloidogyne*. In *Concepts in Nematode Systematics*, A.R. Stone, H.M. Platt, and L.F. Khalil, eds. Academic Press, New York, pp. 187–196.

Day, J.W. 1974. A Beem R* capsule chamber pipette for handling small specimens for electron microscopy. *Stain Technol.* 49: 408–410.

De Grisse, A. 1977. Modification of the mini-sieve and prepolymerized plate techniques for use in electron microscopy, *J. Nematol.* 9: 196–199.

Dickson, D.W., Sasser, J.N., and Huisingh, D. 1971a. Comparative disc-electrophoretic protein analyses of selected *Meloidogyne, Ditylenchus, Heterodera* and *Aphelenchus* spp. *J. Nematol.* 2: 286–293.

Dickson, D.W., Huisingh, D., and Sasser, J.N. 1971b. Dehydrogenases, acid and alkaline phosphatases, and esterases for chemotaxonomy of selected *Meloidogyne, Ditylenchus, Heterodera* and *Aphelenchus* spp. *J. Nematol.* 3: 1–16.

Eckert, J., and Schwarz, R.Z. 1965. Zur Struktur der Cuticula Invasionsfahiger Larven einiger Nematoden. *Z. Parasitenkunde* 26: 116.

Eisenback, J.D. 1985. Techniques for preparing nematodes for scanning electron microscopy. In *An Advanced Treatise on Meloidogyne, Vol. 2, Methodology*, K.R. Barker, C.C. Carter, and J.N. Sasser, eds. North Carolina State University Graphics, Raleigh.

Eisenback, J.D. 1985. Techniques for preparing nematodes for scanning electron microscopy. In *An Advanced Treatise on Meloidogyne, Vol. 2, Methodology*, K.R. Barker, C.C. Carter, and J.N. Sasser, eds. North Carolina State University Graphics, Raleigh, pp. 79–105.

Eisenback, J.D. 1986. A comparison of techniques useful for preparing nematodes for scanning electron microscopy. *J. Nematol.* 18: 479–487.

Eisenback, J.D. 1988. Multiple focus and exposure photomicroscopy of nematodes for increased depth of filed. *J. Nematol.* 20: 333–334.

Eisenback, J.D., and Rammah, A. 1987. Evaluation of the utility of a stylet extraction technique for understanding morphological diversity of several genera of plant-parasitic nematodes. *J. Nematol.* 19: 384–386.

Endo, B.Y. 1980. Ultrastructure of the anterior neurosensory organs of the larvae of the soybean cyst nematode, *Heterodera glycines. J. Ult. Res.* 72: 349–366.

Esbenshade, P.R., and Triantaphyllou, A.C. 1985a. Use of enzyme phenotypes for identification of *Meloidogyne* species. *J. Nematol.* 17: 6–20.

Esbenshade, P.R., and Triantaphyllou, A. 1985b. Identification of major *Meloidogyne* species employing enzyme phenotypes as differentiating characters. In *An Advanced Treatise on Meloidogyne, Vol. 1, Biology and Control*, J.N. Sasser and C.C. Carter, eds. North Carolina State University Press, pp. 135–140.

Esbenshade, P.R., and Triantaphyllou, A. 1985c. Electrophoretic methods for the study of root-knot nematode enzymes. In *An Advanced Treatise on Meloidogyne, Vol. 2, Methodology*, K.R. Barker, C.C. Carter, and J.N. Sasser, eds. North Carolina State University Press, Raleigh, pp. 115–123.

Esbenshade, P.R. and Triantaphyllou, A.C. 1987. Enzymatic relationships and evolution in the genus *Meloidogyne* (Nematoda: Tylenchida). *J. Nematol.* 19: 8–18.

Esbenshade, P.R., and Triantaphyllou, A.C. 1988. Genetic analysis of esterase polymorphism in the soybean cyst nematode. *J. Nematol.* 20: 486–492.

Esbenshade, P.R., and Triantaphyllou, A.C. 1990. Isozymes phenotypes for the identification of *Meloidogyne* species. *J. Nematol.* 22: 10–15.

Esser, R.P 1988. A simple method for examination of the vulva area of mature cyst of *Heterodera* spp. *J. Nematol.* 20: 497–498.

Fargette, M. 1987a. Use of the esterase phenotype in the taxonomy of the genus *Meloidogyne*. 1. Stability of the esterase phenotype. *Revue Nématol.* 10: 39–43.

Fargette, M. 1987b. Use of the esterase phenotype in the taxonomy of the genus *Meloidogyne*. 2. Esterase phenotypes observed in West African populations and their characterization. *Revue Nématol.* 10: 45–56.

Faulkner, G.J.N., and Greet, D.N. 1984. A machine for the rapid and efficient separation of nematode cysts from dried root debris. *Nematologica* 30: 99–102.

Ferris, V.R., Ferris, J.M., and Murdock, L.L. 1985. Two-dimensional protein patterns in *Heterodera glycines. J. Nematol.* 17: 422–427.

Ferris, V.R., Ferris, J.M., Murdock, L.L. and Faghihi, J. 1986. *Heterodera glycines* in Indiana. III. 2-D protein patterns of geographical isolates. *J. Nematol.* 18: 177–182.

Ferris, V.R., and Ferris, J.M. 1987. Phylogenetic concepts and methods. In *Vistas on Nematology*, J.A. Veech and D.W. Dickson, eds. Society of Nematologists, Hyattsville, MD, pp. 346–353.

Ferris, V.R., Ferris, J.M., Murdock, L.L., and Faghihi. 1987. Two-dimensional protein patterns in *Labronema, Aporcelaimellus* and *Eudorylaimus* (Nematoda: Dorylaimida). *J. Nematol.* 19: 431–440.

Ferris, V.R., and Ferris, J.M. 1988. Phylogenetic analyses in Dorylaimida using data from 2-D protein patterns. *J. Nematol.* 20: 102–108.

Fortuner, R. 1976. Les nématodes parasites des racines associés au riz au Sénégal (Haute-Casamance et régions Centre et Nord) et en Mauritanie. *Cah. ORSTOM, sér. Biol.* 10 (1975): 147–159.

Fortuner, R. 1982. On the genus *Ditylenchus* Filipjev, 1936 (Nematoda: Tylenchida). *Revue Nématol.* 5: 17–38.

Fortuner, R. 1984. Statistics in taxonomic descriptions. *Nematologica* 30: 187–192.

Fortuner, R., and Raski, D.J. 1987. A review of Neotylenchoidea Thorne, 1941 (Nemata: Tylenchida). *Revue Nématol.* 10: 257–267.

Fox, P.C., and Atkinson, H.J. 1984. Glucose phosphate isomerase polymorphism in field populations of the potato cyst nematodes, *Globodera rostochiensis* and *G. pallida. Ann. App. Biol.* 104: 503–509.

Fox, P.C., and Atkinson, H.J. 1985. Enzyme variation in pathotypes of the potato cyst nematodes, *Globodera rostochiensis and G. pallida. Parasitology* 91: 499–506.

Fox, P.C., and Atkinson, H.J. 1988. Non-specific esterase variation in field populations of the potato cyst nematodes *Globodera rostochiensis* and *G. pallida. Nematologica* 34: 156–163.

Gerber, K., and Taylor, A.L. 1988. A simple technique for mounting whole root-knot nematode females. *J. Nematol.* 20: 502–503.

Harris, T.S., Sandall, L.J., and Powers, T.O. 1989. Enhanced molecular diagnostics using polymerase chain reaction. *J. Nematol.* (Abstract)21:564.

Hayat, M.A. 1986. *Basic Techniques for Transmission Electron Microscopy.* Academic Press, New York.

Hayat, M.A. 1981. *Fixation for Electron Microscopy.* Academic Press, New York.

Himmelhoch, S., and Zuckerman, B.M. 1982. *Xiphinema index* and *Caenorhabditis elegans*: preparation and molecular labeling of ultrathin frozen sections. *Exp. Parasitol.* 54: 250–259.

Hirschmann, H. 1983. Scanning electron microscopy as a tool in nematode taxonomy. In *Concepts in Nematode Systematics*, A.R. Stone, H.M. Platt, and L.F. Khalil, eds., Systematics Association Special Volume No. 22, Academic Press, New York, pp. 95–111.

Hirschmann, H., and Triantaphyllou, A.C. 1979. Morphological comparison of members of the *Heterodera trifolii* species complex. *Nematologica* 25: 458–481.

Hooper, D.J. 1986a. Extraction of free-living stages from soil. In *Laboratory Methods for Work with Plant and Soil Nematodes*, J.F. Southey, ed., Ministry of Agriculture, Fisheries and Food, London, HMSO, pp. 5–30.

Hooper, D.J. 1986b. Extraction of nematodes from plant material. In *Laboratory Methods for Work with Plants and Soil Nematodes*, J.F. Southey, ed., Ministry of Agriculture, Fisheries and Food, London, HMSO, pp. 51–58.

Hooper, D.J. 1986c. Handling, fixing, staining, and mounting nematodes. In *Laboratory Methods for Work with Plants and Soil Nematodes*, J.F. Southey, ed., Ministry of Agriculture, Fisheries and Food, London, HMSO, pp. 59–80.

Hooper, D.J. 1986d. Preserving and staining nematodes in plant tissues. In *Laboratory Methods for Work with Plants and Soil Nematodes*, J.F. Southey, ed., Ministry of Agriculture, Fisheries and Food, London, HMSO, pp. 81–85.

Hooper, D.J. 1986e. Culturing nematodes and related experimental techniques. In *Laboratory Methods for Work with Plants and Soil Nematodes*, J.F. Southey, ed., Ministry of Agriculture, Fisheries and Food, London, HMSO, pp. 133–157.

Hooper, D.J. 1987. Observations on the curation of plant and soil nematodes. *Nematologica* 32 (1986): 312–321.

Hope, W.D. 1988. Ultrastructure of the feeding apparatus of *Rhabdodemania minima* Chitwood, 1936 (Enoplida: Rhabdodemaniidae). *J. Nematol.* 20: 118–140.

Hopwood, D. 1969. Fixatives and fixation: A review. *Histochem. J.* 1: 323–360.

Hopwood, D. 1972. Theoretical aspects of glutaraldehyde fixation. *Histochem. J.* 4: 267–303.

Hopwood, D., Coghill, G., Ramsay, J., Milne, G., and Kerr, M. 1984. Microwave fixation: its potential for routine techniques, histochemistry, immunocytochemistry and electron microscopy. *Histochem. J.* 16: 1171–1191.

Huang, C.S., Bettencourt, C., and Mota-Silva, E.F.S. 1984. Preparing nematode permanent mounts with adhesive tapes. *J. Nematol.* 16: 341–342.

Hubel, D.H. 1957. Tungsten microelectrode for recording from single units. *Science* 5: 185–195.

Huettel, R.N., Dickson, D.W., and Kaplan, D.T. 1983a. Biochemical identification of the two races of *Radopholus similis* by starch gel electrophoresis. *J. Nematol.* 15: 338–344.

Huettel, R.N., Dickson, D.W., and Kaplan, D.T. 1983b. Biochemical identification of the two races of *Radopholus similis* by polyacrylamide gel electrophoresis. *J. Nematol.* 15: 345–348.

Hussey, R.S. 1979. Biochemical systematics of nematodes: A review. *Helminth. abstr. (ser. B)* 48: 141–148.

Hussey, R.S., and Krusberg, L.R. 1971. Disc-electrophoresis patterns of enzymes and soluble proteins of *Ditylenchus dipsaci* and *D. triformis. J. Nematol.* 3: 79–84.

Hyman, B.C. 1988. Nematode mitochondrial DNA: anomalies and applications. *J. Nematol.* 20: 523–531.

Kalinski, A., and Huettel, R.N 1988. DNA restriction fragment length polymorphism in races of the soybean cyst nematode, *Heterodera glycines. J. Nematol.* 20: 532–538.

Krusberg, L.R., and Sardanelli, S. 1984. Techniques for axenizing nematodes. *J. Nematol.* 16: 348.

Lankow, R.K., Grothaus, G.D., Miller, S.A. 1987. Immunoassays for crop management systems and agricultural chemistry. In *Biotechnology in Agriculture, ACS Syp. Ser.*, H.M. LeBaron, R.O. Mumma, R.C. Honeycutt, J. H. Duesing, eds. Am. Chem. Soc., Washington, D.C. 334: 228–252.

MacGowan, J.B. 1986. A new, efficient technique for permanent nematode storage. *J. Nematol.* 18: 419–420.

MacKenzie, A.P. 1976. Principles of freeze-drying. *Transplant Proceedings* 8(2) Supplement 1: 181–188.

McClure, M.A., and Stowell, L.J. 1978. A simple method of processing nematodes for electron microscopy. *J. Nematol.* 10: 376–377.

McClure, M.A. 1981. Electron Microscopy Histochemistry. In *Plant Parasitic Nematodes*, Vol. 3, B.M. Zuckerman and R.A. Rohde, eds. Academic Press, New York, pp. 89–124.

McSorley, R., Parrado, J.L., and Dankers, W.H. 1984. A quantitative comparison of some methods for the extraction of nematodes from roots. *Nematropica* 14: 72–84.

Maggenti, A.R., and Viglierchio, D.R. 1965. Preparation of nematodes for microscopic study: Perfusion by vapor phase in killing and fixing. *Hilgardia* 36: 435–463.

Mai, W.F., and Riedel, R.M. 1987. Culture of plant-parasitic nematodes: Importance of a germplasm bank. In *Vistas on Nematology*, J.A. Veech and D.W. Dickson, eds., Society of Nematologists, Hyattsville, MD, pp. 398–400.

Maniatis, T., Fritsch, E.F., and Sambrook, J. 1982. *Molecular Cloning: A Laboratory Manual.* Cold Spring Harbor Laboratory, New York.

Meerzainudeen, M., Vijayaraghavan, S., Ganesaraja, V., and Nirmal Jhonson, S.B. 1984. A quicker method of washing and extraction of nematodes than the conventional method. *Ind. J. Nematol.* 14: 196–197.

Merny, G., and Déjardin, J. 1970. Les nématodes phytoparasites des rizières inondées de Côte d'Ivoire. II. Essai d'estimation de l'importance des populations. *Cah. ORSTOM, Sér. Biol.* No. 11: 45–67.

Meyer, S.L.F., Sayre, R.M., and Huettel, R.N. 1988. Comparisons of selected stains for distinguishing between live and dead eggs of the plant-parasitic nematode *Heterodera glycines. Proc. Helminthol. Soc. Wash.* 55: 132–139.

Miller, S.A., and Martin, R.R. 1988. Molecular diagnosis of plant disease. *Ann. Rev. Phytopathol.* 26: 409–432.

Mota, M.M., and Eisenback, J.D. 1988. Optimization of stylet extraction in cyst nematodes. *Virginia J. Sci.* 39: 97. (abstr.)

Müller, J. 1983. Zur Problematik der quantitativen Erfassung von *Heterodera schachtii* mit Hilfe von Bodenuntersuchungen. III. Einfluss von Bearbeiter und Extraktionsmethodik. *Nachrichtenblatt des Deutschen Pflanzenschutzdienstes, Braunschweig* GFR 35: 168–172.

Netscher, C. 1971. A rapid technique for mass-killing of nematodes. *Nematologica* 16 (1970): 603.

Netscher, C., and Seinhorst, J.W. 1969. Propionic acid better than acetic acid for killing nematodes. *Nematologica* 15: 286.

Noel, G.R., and Stanger, B.A. 1986. Scanning electron microscopy of second-stage juvenile cephalic morphology in *Heterodera glycines* races. *J. Nematol.* 18: 475–478.

O'Farrell, P.H. 1975. High resolution two-dimensional electrophoresis of proteins. *J. Biol. Chem.* 250: 4007–4021.

Olowe, T., and Corbett, D.C.M. 1983. Morphology and morphometrics of *Pratylenchus brachyurus* and *P. zeae.* I. Effect of fixative and processing. *Ind. J. Nematol.* 13: 141–154.

Pableo, E.C., and Triantaphyllou, A.C. 1989. DNA complexity of the root-knot nematode (*Meloidogyne* spp.) genome. *J. Nematol.* 21: 260–263.

Pais, C.S., and Abrantes, I.M.O. 1989. Esterase and malate dehydrogenase phenotypes in Portuguese populations of *Meloidogyne* species. *J. Nematol.* 21: 342–346.

Pentilla, A., Kalimo, H., and Trump, B.F. 1974. Influence of glutaraldehyde and/or osmium tetroxide on cell volume, ion content, mechanical stability, and membrane permeability of Ehrlich ascites tumor cells. *J. Cell Biol.* 63: 197–214.

Perkins, L.A., Hedgecock, E.M., Nichol Thomson, J., and Culotti, J.G. 1986. Mutant sensory cilia in the nematode *Caenorhabditis elegans*. *Dev. Biol.* 117: 456–487.

Perry, R.N., Wharton, D.A., and Clarke, A.J. 1982. The structure of the egg shell of *Globodera rostochiensis* (Nematoda: Tylenchida). *Int. J. Parasitol.* 12: 481–485.

Platzer, E.G., 1981. Potential use of protein patterns and DNA nucleotide sequences in nematode taxonomy. In: *Plant Parasitic Nematodes*, Vol. 3, B.M. Zuckerman and R.A. Rohde, eds. Academic Press, New York, pp. 3–21.

Poehling, H.M., Wyss, U., and Neufoff, V. 1980. Two dimensional microelectrophoresis of proteins from plant parasitic nematodes: increased sensitivity of protein detection by silver staining. *Electrophoresis* 1: 198–200.

Powers, T.O., Platzer, E.G., and Hyman, B.C. 1986. Species-specific restriction site polymorphism in root-knot nematode mitochondrial DNA. *J. Nematol.* 18: 288–293.

Powers, T.O., and Sandall, L.J. 1988. Estimation of genetic divergence in *Meloidogyne* mitochondrial DNA. *J. Nematol.* 20: 505–511.

Pozdol, R.F., and Noel, G.R. 1984. Comparative electrophoretic analyses of soluble proteins from *Heterodera glycines* races 1–4 and three other *Heterodera* species. *J. Nematol.* 16: 332–340.

Premachandran, D., Bergé, J.B., and Bride, J.M. 1984. Two dimensional electrophoresis of proteins from root-knot nematodes. *Revue Nématol.* 7: 205–207.

Premachandran, D., Mende, N. von, Hussey, R.S., and McClure, M.A. 1988. A method for staining nematode secretions and structures. *J. Nematol.* 20: 70–78.

Prencepe, A., Bianco, M., Viglierchio, D.R., and Scognamiglio, A. 1984. Response of the nematode *Panagrellus silusiae* to hypertonic solutions. *Proc. Helminthol. Soc. Wash.* 51: 36–41.

Preston, C.M., and Jenkins, T. 1984. *Trichuris muris*: Structure and Formation of the egg-shell. *Parasitology* 89: 263–273.

Quénéhervé, P., and Cadet, P. 1986. Une nouvelle technique d'échantillonnage pour l'étude des nématodes endoparasites du bananier. *Revue Nématol.* 9: 95–97.

Radice, A.D., Riggs, R.D., and Huang, F.H. 1988a. Detection of intraspecific diversity of *Heterodera glycines* using isozymes phenotypes. *J. Nematol.* 20: 29–39.

Radice, A.D., Powers, T.O., Sandall, L.J., and Riggs, R.D. 1988b. Comparison of mitochondrial DNA from the sibling species *Heterodera glycines* and *H. schachtii*. *J. Nematol.* 20: 443–450.

Rajan, and Swarup, G. 1985. Evaluation of cyst extraction techniques and their effect on biology of *Heterodera cajani*. *Ind. J. Nematol.* 15: 75–82.

Rammah, A., and Hirschmann, H. 1987. Morphological comparison and taxonomic utility of copulatory structures of selected nematode species. *J. Nematol.* 19: 314–323.

Reilly, J.J., and Grant, C.E. 1985. Seasonal fluctuations of *Globodera tabacum solanacearum* as estimated by two soil extraction techniques. *J. Nematol.* 17: 354–360.

Reymond, O.L., and Pickett-Heaps, J.D. 1982. A routine flat embedding method for electron microscopy of microorganisms allowing selection and precisely orientated sectioning of single cells by light microscopy. *J. Microsc.* 130: 79–84.

Riley, I.T., Reardon, T.B., and McKay, A.C. 1988. Electrophoretic resolution of species boundaries in seed-gall nematodes, *Anguina* spp. (Nematoda: Anguinidae), from some graminaceous hosts in Australia and New Zealand. *Nematologica* 34: 401–411.

Robinson, A.F. 1984. Comparison of five methods for measuring nematode volume. *J. Nematol.* 16: 343–347.

Robinson, D.G., Ehlers, U., Herken, R., Herrmann, B., Mayer, F., and Schurmann, F.W. 1987. *Methods of Preparation for Electron Microscopy*. Springer-Verlag, New York.

Roggen, D.R., Raski, D.J., and Jones, N.O. 1966. Cilia in nematode sensory organs. *Science* 152: 515–516.

Ross, M.M.R. 1967. Modified cilia in sensory organs of juvenile stages of a parasitic nematode. *Science* 156: 1494–1495.

Ryss, A.Yu. 1987. [Effective field method for the extraction of nematodes from soil and plant roots.] *Trudy Zool. Inst. Akad. Nauk SSSR* 161: 107–109.

Santos, M.S.N. de A., and Abrantes, I.M. de O. 1989. Morphological characters and methods for preparing nematodes. In *Nematode Identification and Expert System Technology*, R. Fortuner, ed. Plenum Press, New York, pp. 201–215.

Schiff, R., and Gennaro, J.F. 1979. The role of the buffer in the fixation of biological specimens for transmission and scanning electron microscopy. *Scanning* 2: 135–148.

Schots, A., Gommmers, F.J., Egberts, E., and Bakker, J. 1989. Serodiagnosis of nematode species. *J. Nematol.*, abstract 21:587.

Seinhorst, J.W. 1954. On *Trichodorus pachydermus* n.sp. (nematoda: Enoplida). *J. Helminth.* 28: 111–114.

Seinhorst, J.W. 1956. The quantitative extraction of nematodes from soil. *Nematologica* 1: 249–267.

Seinhorst, J.W. 1966. Killing nematodes for taxonomic study with hot F.A. 4:1. *Nematologica* 12: 178.

Shepherd, A.M. 1986. Extraction and estimation of cyst nematodes. In *Laboratory Methods for Work with Plant and Soil Nematodes*, J.F. Southey, ed. Ministry of Agriculture, Fisheries and Food, London, HMSO, pp. 31–49.

Shepherd A.M., and Clark, S.A. 1988. Preparation of nematodes for electron microscopy. In *Laboratory Methods for Work with Plant and Soil Nematodes*, J.F. Southey, ed. HMSO, London, pp. 121–128.

Sher, S.A., and Bell, A.H. 1975. Scanning electron micrographs of the anterior region of some species of Tylenchoidea (Tylenchida: Nematoda). *J. Nematol.* 7: 69–83.

Shesteperov, A.A. 1986. [Use of air photography as one of the methods for detecting foci of *Globodera* infection on potatoes.] *Byll. Vses. Inst. Gel'mint. K.I. Skryabina*, No.45: 74–89 (see ha57: 1680).

Shesteperov, A.A., Hernandes, U., Valdes, Z., and Pérez, U. 1984. [Effectiveness of different methods for extraction of nematodes from roots and rhizosphere of banana.] *Byll. Vses. Inst. Gel'mint. K.I. Skryabina* No. 36: 59–64.

Singh, G.J.P., and Singh, B. 1984. Action of dieldrin and trans-aldrin diol upon the ultrastructure of the sixth abdominal ganglion of *Periplaneta americana* in relation to their electrophysiological effects. *Pest. Biochem. Physiol.* 21: 102–126.

Sjostrand, F.S. 1967. *Electron Microscopy of Cells and Tissues*, Vol. 1. Academic Press, New York.

Southern, E.M., 1975. Detection of specific sequences among DNA fragments separated by gel electrophoresis. *J. Mol. Biol.* 98: 503–517.

Southey, J.F., 1986a. *Laboratory Methods for Work with Plant and Soil Nematodes*. Ministry of Agriculture, Fisheries and Food, London, HMSO.

Southey, J.F. 1986b. Principles of Sampling for Nematodes. In J.F. Southey, ed. (ED.), *Laboratory Methods for Work with Plant and Soil Nematodes*, Ministry of Agriculture, Fisheries and Food, London, HMSO: 1–4.

Sudakova, I.M., Mikulina, R.V., and Pugachev, V.V. 1986. [Obtaining a clean culture of plant nematodes from a nutrient substrate by mean of active migration.] *Zool. Zh.* 65: 444–447.

Steedman, H.F. 1976. *Zooplankton Fixation and Preservation*. Unesco Press, Paris.

Swart, A., and Heyns, J. 1987. Morphological study of Longidorid nematode stylets using the SEM. *Phytophylactica* 19: 103–106.

Sulston, J.E., Schierenberg, E., White, J.G., and Thomson, J.N. 1983. The embryonic cell lineage of the nematode *Caenorhabditis elegans*. *Dev. Biol.* 100: 64–119.

Sulston, J.E., and Hodgkin, J. 1988. Methods. In *The Nematode* Caenorhabditis elegans, W.B. Wood, and the Community of *C. elegans* Researchers, eds. Cold Spring Harbor Laboratory, New York, pp. 587–606.

Sulston, J.E., Schiernberg, E., White, J. G., and Thomsen, J. N. 1980. The embryonic cell lineage of the nematode *Caenorhabditis elegans. Dev. Biol.* 100: 64–119.

Thorne, G., and Malek, R.B. 1968. Nematodes of the Northern Great Plains. Part I. Tylenchida (Nemata: Secernentea). *South Dakota Agr. Exp. Stn Tech. Bull.* No.31.

Townshend, J.L. 1983. Anaesthesia of three nematode species with propylene phenoxetol. *Nematologica* 29: 357–360.

Viglierchio, D.R., and Schmitt, R.V. 1983a. On the methodology of nematode extraction from field samples: Baermann funnel modifications. *J. Nematol.* 15: 438–444.

Viglierchio, D.R., and Schmitt, R.V. 1983b. On the methodology of nematode extraction from field samples: Comparison of methods for soil extraction. *J. Nematol.* 15:450–454.

Viglierchio, D.R., and Yamashita, T.T. 1983c. On the methodology of nematode extraction from field samples: density flotation techniques. *J. Nematol.* 15: 444–449.

von Ehrenstein, G., Sulston, J.E., Schierenberg, E., Laufer, J.S., and Cole, T. 1981. Embryonic cell lineages and segregation of developmental potential in *Caenorhabditis elegans*. In *International Cell Biology*, 1980–1981, Springer-Verlag, New York.

Wallace, H.R. 1971. The influence of the density of nematode populations on plants. *Nematologica* 17: 154–166.

Wang K.C., and Chen, T.A. 1985. Ultrastructure of the phasmids of *Scutellonema brachyurum. J. Nematol.* 17: 175–186.

Wehunt, E.J. 1973. Sodium-containing detergents enhance the extraction of nematodes. *J. Nematol.* 5: 79–80.

Wharton, D.A., and Barrett, J. 1985. Ultrastructural changes during recovery from anabiosis in the plant-parasitic nematode *Ditylenchus. Tissue and Cell* 17: 79–96.

Wharton, D.A., Preston, C.M., Barrett, J., and Perry, R.N. 1988. Changes in cuticular permeability associated with recovery from anhydrobiosis in the plant parasitic nematode, *Ditylenchus dipsaci. Parasitology.* 97: 317–330.

Wharton, R.J., Storey, R.M.J., and Fox, P.C. 1983. The potential of some immunochemical and biochemical approaches to the taxonomy of potato cyst nematodes. In *Concepts in Nematode Systematics*, A.R. Stone, H.M. Platt, and L.F. Khalil, eds. Academic Press, New York

White, T.J., Arnheim, N., and Erlich, H.A. 1989. The polymerase chain reaction. *Trends Genet.* 5: 185–189.

Winfield, A.L., Enfield, M.A., and Foreman, J.H. 1987. A column elutriator for extracting cyst nematodes and other small invertebrates from soil samples. *Ann. Appl. Biol.* 111: 223–231.

Winfield, A.L., and Southey, J.F. 1986. Use of the optical microscope in nematology. In J.F. Southey, ed., *Laboratory Methods for Work with Plant and Soil Nematodes*, Ministry of Agriculture, Fisheries and Food, London, HMSO, pp. 95–106.

Wisse, E., and Daems, W.Th. 1968. Electron microscopic observations on second stage larvae of the potato root eelworm *Heterodera rostochiensis. J. Ult. Res.* 24: 210–231.

Wright, D.J., and Newall, D.R. 1980. Osmotic and ionic regulation in nematodes. In *Nematodes as Biological Models*, Vol. 2 B.M. Zuckerman, ed. Academic Press, New York, pp. 143–164.

Wright, K.A., and Jones N.O. 1965. Some techniques for the orientation and embedding of nematodes for electron microscopy. *Nematologica* 11: 125–130.

Wright, K.A., and Carter 1980. Cephalic sense organs and body pores of *Xiphinema americanum* (Nematoda: Dorylaimoidea). *Can. J. Zool.* 58: 1439–1451.

Zeikus, J.A., and Aldrich, H.C. 1975. Use of hot formaldehyde fixative in processing plant-parasitic nematodes for electron microscopy. *Stain Technol.* 50: 220–225.

Zuckerman, B.M., Himmelhoch, S., and Kisiel, M. 1973. Fine structure changes in the cuticle of adult *Caenorhabditis briggsae* with age. *Nematologica* 19: 109–112.

4

Application of DNA Analysis to Nematode Taxonomy

JOHN CURRAN *Commonwealth Scientific and Industrial Research Organization, Canberra, Australia*

I. INTRODUCTION

Morphology has been and will continue to be the mainstay of nematode taxonomy. It is an essential component of any higher level classification of nematodes and in many cases morphology provides a rapid and unambiguous diagnosis to species. However, the identification of intrasubspecific groupings requires the application of other taxonomic techniques; to date, available methods include host range testing, protein electrophoresis, immunological techniques; and, most recently, DNA sequence analysis. Fortuner et al. examined the application of protein and immunological techniques to the resolution of taxonomic problems in plant nematology. The focus of this chapter will be to examine the uses of DNA-based taxonomic methods in nematology with particular reference to the identification of nematode species, subspecies, and other intrasubspecific forms such as races, pathotypes, and strains. The general principles of DNA analysis will be presented along with detailed protocols for basic procedures. In addition, the potential use of DNA-based data in the classification of nematodes will be discussed and an overview of available methods given. Reference will be made to polymerase chain reaction techniques which, though currently not widely used in nematode taxonomy, are likely to contribute to significant advances in the near future.

The direct examination of an organism's genotype by analysis of DNA sequence is a taxonomic tool whose applications range from differentiation of intrasubspecific groupings of nematodes (Curran et al., 1986; Powers et al., 1986; Curran and Webster, 1987; Bolla et al., 1988; Kalinski and Huettel, 1988) to studies into the evolutionary relationships between pro-and eukaryotic organisms (Vahidi et al., 1988). Because the genotype of the nematode is examined directly, problems associated with phenotypic variation in taxonomic characters are avoided. Furthermore, greater discrimination is possible because the entire genome is available for study, including the 75–80% of the genome which is noncoding and contains many highly variable sequences. This DNA sequence variability provides a large pool of information that can provide useful diagnostic characters for separation of taxa and infrasubspecific categories or for the unraveling of evolutionary relationships between nematode groups. Within this great range of application, it is important that the appropriate molecular approach be used for the taxonomic problem under study.

A. Choice of DNA Sequence for Analysis

In diagnosis, it is necessary only to discriminate between a limited number of nematode species or infrasubspecific categories (e.g., to separate *Globodera rostochiensis* from *G. pallida*; *Bursaphelenchus xylophilus* from *B. mucronatus*). Furthermore, it is not necessary to determine the evolutionary relationships between the taxa merely to distinguish them. Therefore, the choice of DNA sequence is not limited because any data that display a difference between the organisms, e.g., restriction fragment length differences (RFLDs), DNA probe, or sequence data (see below), for any gene or region of the genome is useful.

In classification, rather than identification, the choice of DNA sequences or genes to be studied is important and knowledge of the evolutionary relationship between sequences is crucial. There are differences in the degree of DNA sequence conservation both between different genes and between different regions within genes (e.g., introns and exons). This degree of sequence conservation determines the taxonomic level at which the sequence is useful. The more highly conserved sequences are most useful at higher taxonomic levels (genus to phylum, e.g., heat shock genes, histone genes), while variable sequences have greatest utility at lower taxonomic levels (species or below, e.g., nontranscribed spacer regions in rDNA).

A useful approach is to study genes which exhibit different degrees of conservation along the sequence. One such gene, rDNA, has a general trend of sequence conservation tending to greater sequence divergence from 5′ to 3′ in transcribed regions. Furthermore, there is great variability within the internal transcribed spacer regions and nontranscribed spacer regions (rDNA also has the advantage of being repetitive and its abundance in the genome makes its cloning and visualization easier). Thus, by studying one gene it is possible to cover all taxonomic levels (see Dutta, 1986, for other examples).

However, a danger of studying a single gene (or region of the genome) is that the derived phylogeny may reflect the evolution of the gene rather than the organism. Care is needed also to ensure that only homologous sequences are compared (i.e., those with an inferred common ancestry) and that the problems of pseudogenes, gene conversion, and duplication are considered. These points are thoroughly reviewed by Paterson (1987, 1988).

B. Genomic and Mitochondrial DNA Analyses

The majority of DNA analyses conducted on nematodes have used total DNA (i.e., a mixture of genomic and mtDNA) or mtDNA and have predominantly looked at RFLDs between reference populations, detecting some intraspecific variation. It is generally held that because of the higher rates of evolution in mtDNA compared to chromosomal (genomic) DNA, mtDNA sequence data are more likely to display taxonomically useful variation at lower taxonomic levels (genus or below). However, this assumption of differences in rates of evolution between mtDNA and chromosomal DNA does not hold true for all organisms and has not been tested in the Nematoda (see review by Birley and Croft, 1986). It will be necessary to determine the rates of mtDNA and genomic DNA sequence divergence in nematodes before their comparative taxonomic values can be assessed.

There are a large number of molecular methods available but I will emphasize those strategies and techniques I feel most efficiently answer two types of taxonomic questions: (1) diagnosis of species and infraspecific groupings (e.g., races, pathotypes) and (2) higher level classifications. There are often a number of alternative procedures available and molecular methods are being continuously updated. I have chosen to present simple methods which have been tried and tested in the study of nematodes. Confidence in these basic procedures would facilitate the application of more complex and more elegant procedures. Materials and methods for basic techniques will be provided in detail. For more complex methodologies reference should be made to the several excellent molecular technique manuals that are available (e.g., Sambrook et al., 1989). Another valuable resource is the technical information and specific protocols provided by the major suppliers of biochemicals and equipment used in DNA analyses. The availability of various types of "kits" for standard techniques greatly simplifies the use of DNA methods and frequently is associated with improved reliability.

II. OVERVIEW OF METHODS

The techniques to visualize DNA sequence (genotype) differences between organisms can be divided into three basic approaches: (1) detection of restriction fragment length differences between nematode DNA samples, (2) DNA probes/dot blots, and (3) DNA sequencing (Table 1).

A. Detection of Restriction Fragment Length Differences

Detection of RFLDs relies on the digestion of DNA with site-specific restriction endonucleases. Since the distribution of sites recognized by any particular restriction endonuclease is determined by the nucleotide sequence (genotype) of the isolate, the size distribution of the DNA restriction fragments generated is characteristic for that genotype. The nematode genome contains some 8×10^7 base pairs. This relatively small genome size means that RFLDs in repetitive DNA are easily visualized by examination of the size distribution of DNA fragments in ethidium bromide-stained agarose gels (see Curran et al., 1985).

RFLDs in low or single copy and/or repetitive DNA sequence can be detected by the more sensitive technique of hybridization to Southern blots of labeled cloned DNA fragments (Fig. 1).

The use of RFLD data derived from cloned DNA fragments or from genomic digests in combination with DNA probes can be used to construct restriction maps of regions of the genome. The organization and size of these regions provides additional information which is potentially taxonomically useful (Vahidi et al., 1988; Wheeler, 1989; see Fig. 2).

B. DNA Probes–Dot Blots

DNA probes can be used to detect as little as a 5% sequence divergence. The probe, a single-stranded DNA sequence, is applied to a filter, to which is bound single-stranded target DNA under defined temperature and salt concentration conditions. If the sequences of the target DNA and probe diverge by less than 5% the probe anneals to the target. After removal of the unbound probe, the extent of the annealing can be assessed. By judicious choice of probe this method can be used to distinguish nematode taxa. As detailed in methods, the most common application is in "dot blot" procedures were DNA samples from unknown populations can be tested with a probe specific for a given taxon. Dot blot procedures are the most cost-effective and simple procedure for nematode identification.

Species-specific probes have been identified for species within the genera *Heterodera, Globodera,* and *Bursaphelechus* (Besal et al., 1987; Marshall and Crawford,

TABLE 1 Comparison of Available DNA-Based Molecular Methods for the Taxonomy of Nematodes

Taxonomic problem	Technique	Sensitivity	Comment
Identification	RFLDs	Species and intraspecific identification, may not be able to identify individual nematodes	Easiest of techniques to implement at a moderate cost per sample
	DNA probes (dot blots)	Species only at present, may not be able to identify individual nematodes	Allows rapid, low-cost analysis of large numbers of samples
	PCR	Species and intraspecific identification, single nematodes can be identified	Most accurate method but cost may preclude its use on large numbers of samples
Classification	RFLDs	Any taxonomic level but lacks resolution at higher levels	Easy technique for both mtDNA and genomic DNA but lacks the fine resolution of sequencing methods
	DNA sequencing	Any taxonomic level, choice of gene for study is important	DNA library construction and subcloning are time consuming
	PCR–DNA sequencing	Any taxonomic level, requires little nematode DNA	Rapid sequence data generation, preserved nematode specimens could be used

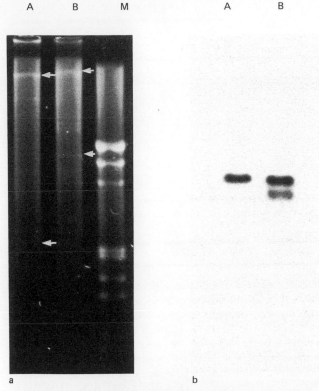

FIGURE 1   Identification of *Meloidogyne hapla* cytological races A and B by detection of restriction fragment length differences in repetitive DNA: (a) EcoRI digest of total genomic DNA, (b) 28S, 18S rDNA probe hybridized to EcoRI-digested total DNA. (a) Photograph of an ethidium bromide-stained agarose gel viewed under 260 nm transmitted irradiation showing the fragment size distribution of EcoRI-cut total DNA. Diagnostic restriction fragment length differences are arrowed. (b) Autoradiograph of a Southern blot of Eco-Ri-digested total DNA hybridized with a 32P-labeled 28S, 18S ribosomal DNA repeat Charon 4 phage probe cloned from *Caenorhabditis elegans*. Legend: A, *M. hapla* cytological race A(NCSU#86); B,*M. hapla* cytological race B (NCSU#48); M, EcoRI/Hind III-cut C1857 λ DNA size marker, fragment sizes top to bottom: 21.7, 5.1/5.0, 4.3, 3.5, 1.98, 1.9, 1.6 and 1.4 kilobases. Hybridization conditions: 5X SSPE, 0.3% SDS at 62°C, washed in 2X SSPE, 0.3% SDS at 62°C. (From Curran and Webster, 1987.)

1987; Burrows and Perry, 1988; Webster et al., 1990) (Fig. 3). However, to date no isolate-specific hybridization probes are available for positive/negative determinations in dot blot procedures, but given sufficient sequence divergence oligonucleotide probes might be constructed.

Identification of single nematodes is possible using highly repetitive DNA sequences as probes (Rollinson et al., 1986; Curran and Webster, 1987; Burrows, 1989). However, there may be insufficient DNA present in an individual nematode to be detected by current labeling technology (Curran and Webster, 1987). This is not a problem in the taxonomy of nematodes which can be cultured in vivo or in vitro because there would be a virtually unlimited amount of nematodes and thus DNA available. However, for ecological, population, and pest management studies the identification of individual nematodes is highly desirable.

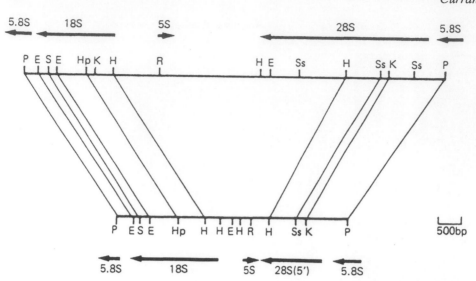

FIGURE 2 Mapping of cloned rDNA repeats from *M. arenaria*. *Meloidogyne arenaria* DNA was digested with Pst I and subcloned into pUC13. Two distinct repeats of 5 and 9 kb were then isolated and restriction mapped. Individual rRNAs were localized roughly by Southern blotting of digested *M. arenaria* repeats with *C. elegans* DNA probes for 5S or rRNA genes. The 5.8S RNA position and polarity were derived from DNA sequencing of the ends of the Pst I fragments (data not shown). In the 5-kb repeat, the position and polarity of the 5S and 5′-28S rRNA were assigned from DNA sequence data. Sequence data also localized and oriented the 5S RNA sequence within the 9-kb fragment. Enzyme sites: P, Pst I; E, Eco RI; S, Sma I; Hp, Hpa I; H, Hind III; Ss, Sst I; K, Kpn I; R, Rsa I. Not all Rsa I sites are shown, only the site within the 5S RNA sequence. For clarity, there is also an Sst I site not indicated on the 5-kb fragment, near the Hind III site at the boundary between the 5S and 28S sequences. (From Vahidi et al., 1988.)

An alternative approach might be the use of in situ hybridization techniques to detect a unique mRNA if this was expressed in sufficient quantity. This technique has been used to visualize mRNA of moderately expressed genes within individual *C. elegans*. This approach may be of value in identifying species and higher taxa. However, given the generally higher rate of sequence conservation in coding regions, it is unlikely that sufficient sequence divergence would be found between naturally occurring mRNAs of different isolates to allow diagnosis of infrasubspecific groupings (though this may be dependent on the gene used).

The DNA extracted from a single infective stage nematode can be amplified using polymerase chain reaction (PCR) technology (in essence DNA is replicated in vitro; see below for details) to generate sufficient DNA for detection by DNA-DNA hybridization techniques, restriction fragment length differences, or DNA base sequencing. At present the cost involved may make PCR technology too expensive for routine processing of large numbers of samples, e.g., in diagnostic services, though automation of processing may change this (Kocher et al., 1989).

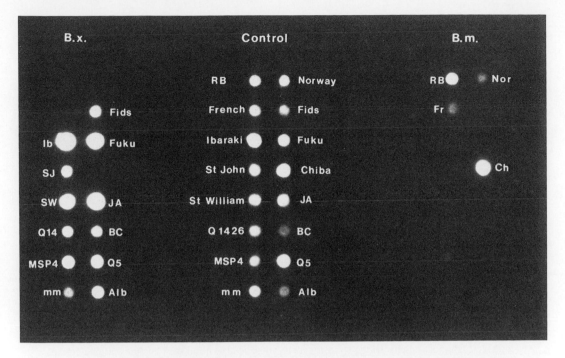

FIGURE 3   DNA from 16 isolates was spotted in triplicate onto nitrocellulose filter so that they could be tested against the probe (pBx6) for *Bursaphelenchus xylophilus* (MSP-4), a probe (pBm4) for *B. mucronatus* (RB), and a control probe (pBm3). The isolates RB (Japan); French; Ibaraki (Japan); St. John (Canada); St. William (Canada); Q1426 (Canada); MSP-4 (U.S.); mm (Canada); Norway; Fids (Canada); Fukushima (Japan); Chiba (Japan); BxUJA (Japan); British Columbia (Canada); Q52A (Canada); Alb (Canada). The hybridization conditions were 70°C and wash at 68°C 0.2X SSPE. (From Webster et al., 1990.)

C.   DNA Sequencing

The Sanger dideoxy sequencing method is the most frequently used method for obtaining DNA sequence data from cloned DNA fragments. The use of M13 phage, a single-stranded DNA bacteriophage, as the cloning vector to provide a single-stranded DNA template for sequencing has made the Sanger dideoxy method a rapid procedure which can provide 300–1000 bp of sequence data from a cloned DNA fragment. Detailed protocols are usually provided by the suppliers of reagents, kits for M13 cloning (an a number of alternative cloning vectors which are available) and dideoxy sequencing. In outline, the cloned DNA fragment (previously identified by colony/plaque hybridization from a library of DNA fragments) is digested with restriction endonuclease(s) and ligated into suitably prepared M13 phage DNA and transfected into *E. coli*. Recombinant phage clones are then identified. These recombinant phage clones synthesize single-stranded phage plus insert DNA which can be purified as a template for use in dideoxy sequencing.

A short primer sequence is added to the template and in the presence of DNA polymerase (Klenow fragment) and the four DNA bases (dNTPs) a complementary DNA

<pre>
                            10   20   30   40   50   60
Mel. 5 kb   GATTACGACCATACCGCGTTGAAAGCACGCCATCCCGTCCGATCTGGCAAGTTAAGCAAC-

C. elegans GCTTACGACCATATCACGTTGAATGCACGCCATCCCGTCCGATCTGGCAAGTTAAGCAAC-

Mel. 9 kb   GATTACGACCATACCGCGTTGAAAGCACGCCATCCCGTCCGATCTGGCAAGTTAAGCAAC-

-GCTGGGCTTCCTTAGTACTTGGAACGGAGACGTCCTGGGAATCGGGCCAGATACTGGGGCTCAT Mel. 5 kb

-GTTGAGTCCAGTTAGTACTTGGATCGGAGACGGCCTGGGAATCCTGGATGTTGGTAAGCTTTTT C. el.

-GCTGGGCTTCCTTAGTACTTGGAACGGAGACGTCCTGGGAATTGAGTCTCTATCAAAAAAATAT Mel. 9 kb
       70   80   90   100   110   120
</pre>

FIGURE 4   The 5S RNA-like sequence within the 5-and 9-kb *M. arenaria* rDNA repeats. For the 5-kb fragment, the central 800-bp Hind III fragment containing 5S RNA homology was sequenced. For the 9-kb fragment, DNA sequencing of subcloned Rsa I, Eco RI, and Hind III fragments localized and oriented the 5S RNA sequence. Both strands were sequenced for all data presented in this figure. Sequences are aligned above and below the *C. elegans* 5S RNA gene with homologies underlined. Numbers indicate nucleotide position within the 120-bp 5S RNA sequences. The putative RNA polymerase III termination sequence TTTT (sequence after nucleotide 120) in *C. elegans* and the corresponding bases in *M. arenaria* are also shown. For the 5-kb fragment, there is no available termination sequence (four or more T residues further downstream of the *Meloidogyne* 5S RNA sequence (data not shown). In the 9-kb fragment, there is also a run of T residues at nucleotides 136–140. (From Vahidi et al., 1988.)

sequence is synthesized. The substitution of 2′,3′-dideoxynucleotides (ddNTPs) in place of the appropriate dNTPs terminates chain elongation in a random manner. Thus, adding each of the ddNTPs in turn (= four reactions) generates fragments of differing lengths, each randomly terminating as a complementary dNTP is encountered along the template DNA. These fragments can then be size-fractionated in a polyacrylamide gel and the DNA sequence read directly from the autoradiograph (either a ^{35}S-or ^{32}P-labeled dNTP is used in each reaction) (Fig. 4). The major limiting steps in this procedure are the construction of DNA libraries, the initial identification of clones carrying the desired sequence for each comparison, and the subsequent subcloning of appropriately sized fragments for M13 cloning and dideoxy sequencing.

D. Polymerase Chain Reaction

Any portion of the genome can be amplified and sequenced. The DNA can be derived from a single nematode and even from preserved specimens, which is not possible with other techniques (Kocher et al., 1989). Two DNA primers which hydridize to opposite strands of the DNA and flank the DNA sequence to be amplified are required. Successive rounds of denaturation of the DNA (by heating), then cooling and annealing of primers, and sequence extension by *Taq* DNA polymerase in the presence of dNTPs effectively doubles the amount of DNA present after each cycle. This amplified DNA can then be used in RFLD or dot blot analysis, or in the final cycles it is possible to alter primer ratios to generate single-stranded DNA for sequencing (see Saiki et al., 1988; Kocher et al., 1989).

III. RESEARCH STRATEGIES

As discussed in the introduction, the choice of detailed strategies for DNA analysis is dependent on the taxonomic problem, i.e., identification or classification, the taxonomic level at which diagnosis is necessary and the required sensitivity (i.e., individual nematodes or not). The following outlines the strategies that have been successfully used in identifying nematodes and discusses the available alternatives for DNA analysis in nematode classification.

A. Diagnosis of Species and Intraspecific Groupings

A general research strategy for species and race/pathotype separation that has proved useful for several taxonomic problems uses the RFLD analysis and DNA probes dot blots. The first step is the comparison of RFLDs in total DNA [genomic + mtDNA (mitochondrial DNA)] of unknown nematode populations against species and race standards. RFLDs in repetitive DNA, mtDNA, and rDNA (ribosomal DNA) (representative clones are available for these sequences, e.g., from *Drosophila, Caenorhabditis elegans*) as well as randomly cloned sequences (see Curran et al., 1985, 1986; Powers et al., 1986; Curran and Webster, 1987, 1989; Vahidi et at, 1988) can be detected by agarose gel electrophoresis and Southern blot analysis. These data are used to place the nematode isolates into species groups or define their taxonomy in conjunction with other data, e.g., morphology, isozymes, etc.

Once the taxonomy is, or at least informal species groupings of isolates are, established the focus can shift to the development of a rapid diagnostic system. The "yes-no" hybridization of DNA probes to dot blots of total DNA is very useful in this regard (Besal et al., 1987; Curran and Webster, 1987; Marshall, 1987; Webster et al., 1990). Randomly cloned fragments from a representative nematode population can then be screened against all populations for (a) species specificity and (b) isolate specificity. Data indicates that a screen of ~30–100 random clones will provide the necessary discrimination (Besal et al., 1987; Curran and Webster, 1987; Marshall and Crawford, 1987; Curran, unpublished).

If required diagnosis is not possible by this simple and rapid method, then probes can be obtained from specific DNA sequences (e.g., the level of sequence divergence in nontranscribed spacer regions from rDNA can be an effective and sensitive discriminator of nematode populations; Curran and Webster, 1987; Webster et al., 1990). Specific probes can then be constructed by subcloning the often short, highly specific DNA sequence.

The probes identified have then to be subjected to exhaustive validation against characterized populations to confirm their diagnostic utility. Since the ability to detect a specific DNA sequence in a DNA sample is in part dependent on the amount of DNA present, it is vital that appropriate positive controls (e.g., using a probe for transcribed rDNA sequences; see Webster et al., 1990) be incorporated to control for differences in efficiency of extraction of DNA from nematode samples and for nonspecific binding.

With current labeling techniques and dot blot procedures it is technically feasible to obtain sufficient copies of the target highly repetitive DNA from a single female/male. Identification of single (infective) juvenile stage nematodes has been reported (Burrows, 1989) but usually 3–10 individuals are required. If identification of single (infective)-juvenile-stage nematodes is necessary then polymerase chain reaction technology can be used to amplify the target sequence from individual juveniles for RFLD or dot blot analysis.

B.   Higher Level Classification

Analysis of DNA sequence differences by construction of restriction maps has been little used in comparing different nematode taxa (see Powers and Sandall, 1988). It is a time-consuming methodology requiring the cloning, restriction mapping, and then alignment of maps for each taxonomic unit.

A simpler though less informative technique is available if sufficient mtDNA can be isolated for labeling and direct visualization (see Powers and Sandall, 1988). Following restriction enzyme digestion of the labeled mtDNA and its separation by agarose gel electrophorsesis (or acrylamide gels for greater sensitivity), the proportion of shared fragments can be calculated as follows:

$$F = \frac{2Nxy}{(Nx + Ny)}$$

Nx = number of fragments in genotype x, Ny = number of fragments in genotype y, Nxy = number of common fragments. The F value is then substituted in the following formula to calculate sequence divergence p (Nei and Li, 1979):

$$p = 1 - \left[\frac{-F + (F^2 + 8F)}{2} \right] 1/n$$

n = the number of base pairs per cleavage site (see Powers and Sandall, 1988 for an application).

Despite the potential wealth of information provided by analysis of RFLDs, it only provides a small, possibly biased sample of the total DNA sequence information available. An alternative approach is to use sequence data derived by sequencing a complete region (i.e., cloned DNA fragment). This would allow finer detail and greater accuracy in the determination of relationships between taxa. There are a number of algorithms that have been developed for analysis and construction of phylogenies from DNA sequence data and these are available as computer programs in most common computer languages (see Hobish, 1986 for further details). However, despite the obvious advantages of full sequence data, the time taken and expertise required to prepare and screen genomic DNA libraries, subclone appropriate fragments and sequence them has meant that this approach has not been widely used in general taxonomy. It has not been applied to the taxonomy of plant parasitic nematodes.

Polymerase chain reaction technology may rapidly alter this lack of sequence data. Although research effort would be needed to construct the specific primers for most genes (sequences that are highly conserved across the whole nematode taxon have to be determined), several universal primers are available, e.g., for mtDNA or rDNA sequences, which bind to highly conserved regions present in all taxa (Kocher et al., 1989, Wheeler, 1989). With such universal primers the data required to construct phylogenies at almost any taxonomic level are rapidly generated. PCR technology is becoming generally available and has been applied in plant nematology in a study on the mtDNA of *Meloidogyne* for diagnostic purposes (Harris et al., 1989). It has almost unlimited potential for the study of taxonomic relationships between organisms (see Kocher et al., 1989) and is sure to find many applications in nematode taxonomy.

IV. MATERIALS AND METHODS

The following section provides details of specific equipment and protocols to enable a general nematology laboratory to complete RFLD analysis, cloning of DNA fragments, and dot blot procedures. These techniques should allow the establishment of a DNA-based means of identification for any nematode group. The equipment and procedures to acquire DNA sequence data, necessary for higher level classification, are more complex and are not given; instead reference should be made to molecular biology manuals.

A. Equipment Requirements

In addition to general laboratory equipment, e.g., pH meter, balances, glassware, the following are required:

Microcentrifuge—indispensable, to hold 1.5-ml polypropylene tubes
Adjustable pipettes, 0–20 µl, 20–200 µl, 200–1000 µl plus disposable tips
Electrophoresis power supply
Flat-bed gel box(es) for agarose gels
Camera equipment to photograph gels
UV transilluminator (short-wavelength)
Culture shaker—to be run at 37°C
Vortex mixer
Plate incubator—37°C
Water bath(s)—to run at 37–90°C
Vacuum oven—not essential, depends on type of filter used in Southerns, etc.
Access to cool room (4°C) if cooled centrifuges not available
Access to –20°C freezer for enzyme and biochemical storage
Access to liquid nitrogen source—not essential

B. DNA Extraction

The aim is to extract high molecular weight DNA (greater than 100 kb average fragment size) free of contaminants which might inhibit subsequent enzymatic reactions. The most common procedure uses the proteolytic enzyme proteinase K in the presence of sodium dodecyl sulfate at 65°C. This combination ensures the rapid lysis of cells and denaturation of nucleases which would rapidly degrade the liberated DNA. The remaining steps in the procedure involve the separation of the DNA, which is retained in the aqueous phase, from proteins, lipids, and carbohydrates which are extracted into organic solvents.

Although DNA can be detected in pg–ng amounts in the initial stages of an investigation it is desirable to have sufficient DNA to be readily visualized by simple techniques such as agarose gel electrophoresis (which requires 0.1–1 µg of DNA per sample). As a rule of thumb approximately 1 ug of DNA can be obtained from 1 mg wet weight of nematodes. Given the need to establish basic information about the nematode's genome, including suitable restriction sites for cloning, 50–100 µg of DNA (50–100 mg wet weight of nematodes) should be prepared from the reference population of a given nematode group (once sufficient information is obtained, considerably fewer nematodes are required for analysis, with good diagnostic results possible from as few as 1–10 nematodes).

The nematodes should be washed free of surface contaminants with water and either used immediately or stored under liquid nitrogen. Nematodes differ in their response to lysis in proteinase K; for example, the infective stage juveniles of *Meloidogyne* and *Romanomermis* spp. lyse readily while the infective stage juveniles of *Steinernema* and *Heterorhabditis* tend to be more resistant, and must be physically broken open (e.g., by grinding in a mortar and pestle). If liquid nitrogen is available the most satisfactory method is to freeze the nematodes in a precooled mortar, grind to a fine powder, and then add the 5–10 volumes of proteinase K solution (1–10 mg/ml proteinase K, in 0.1 M Tris–Cl, pH 8.5, 0.05 M EDTA, 0.2 M NaCl, and 1% w/v sodium dodecyl sulfate butter) to 1 volume of nematode powder, bringing the temperature to 65°C as fast as possible. Individual nematodes can be cut with a syringe or scalpel blade under proteinase K solution. The suspension of nematodes is placed in a water bath at 65°C with occasional gentle swirling for 15–30 min.

C. DNA Purification

The liberated DNA is extracted against an equal volume of phenol saturated with TE (0.01 M Tris-Cl pH 8, 0.001 M EDTA). The phenol should be colorless; any with straw pink coloration should be redistilled or discarded. The aqueous phase is transferred to a clean container after each extraction using a wide-bored pipette (high molecular weight DNA is susceptible to mechanical shearing) avoiding transfer of the denatured proteins, etc., present at the water-phenol interface. After the third phenol extraction, the DNA solution is mixed with an equal volume of chloroform-isoamyl alcohol (24:1). For each extraction of 15–30 min, the phases are mixed by occasional gentle inversion of the container and then, prior to transfer of the aqueous phase, briefly centrifuged (12,000g for 30 sec) to improve separation of the phases.

The DNA is precipitated from the final aqueous phase by addition of 2.5 volumes of ice cold 95% ethanol. The DNA is pelleted in a centrifuge (5–10 min in a microcentrifuge or at 12,000g), washed with 70% ethanol, air-dried, and then resuspended in TE buffer. RNA is removed by addition of 1 µl of a 10 mg/ml solution of DNase-free RNase, incubated at room temperature for 15–30 min. The DNA is then extracted against an equal volume of chloroform-isoamyl alcohol (24:1), precipitated, and resuspended as above. At this stage the DNA should be sufficiently pure for restriction endonuclease digestion or cloning procedures. The considerable time and expense of further purifying DNA in CsCl gradients in an ultracentrifuge is unnecessary. The quality of the DNA can be checked by electrophoresing a small sample (1–2 µg) in an agarose gel stained with ethidium bromide; the DNA should appear as a tight band near the origin with little smearing (RNA appears as an intense band at 450 bp).

D. Restriction Endonuclease Digestion

Restriction endonuclease digestions are normally performed on 0.1–1 µg aliqouts of DNA in a volume of 10–30 µl under the reaction conditions recommended by the enzyme's manufacturer. Although there is an ever-increasing choice of restriction endonucleases, useful information can be obtained from fewer than 10–15 different restriction endonucleases. An initial screen might include EcoRI, HindIII, and BamH1. Generally restriction endonucleases with 6–bp recognition sequence are sufficient; it is rare that 4–bp restriction endonucleases offer any advantage.

A common cause of partial or complete failure of the restriction endonuclease to digest the DNA is a carryover of a high salt concentration, organic solvents, or other contaminants from the DNA extraction and purification steps. Frequently this problem can be solved by simple reprecipitating the DNA and resuspending in fresh TE buffer or repeating the phenol-chlorofom steps. However, there are reports of specific inhibitors (e.g., associated with egg debris of *Globodera*), which may require modification of the basic procedure (Burrows and Boffey, 1986).

E. Agarose Gel Electrophoresis

Restriction endonuclease-digested DNA samples are mixed with loading buffer (to a final concentration of 5% glycerol, 0.025% bromophenol blue) and placed in the slots of an agarose gel containing 0.5 µg/ml ethidium bromide. The usual gel composition ranges from 0.5 to 1% agarose w/v in 0.089 M Tris-borate, 0.089 M boric acid, and 0.002 M EDTA running buffer. The lower percentage gels provide greater resolution of large DNA restriction fragments, while the higher percentage gels provide increased resolution of small fragments; the most frequently used gel is 0.7%. Larger gels and lower voltage give greater resolution than smaller gels and higher voltage but there is a trade-off in time and material costs. For most routine purposes "minigels" (100 x 65 mm) run at 30–70 V for 6–2 hr provide adequate results. C1857 λ DNA digested with EcoRI and HindIII is used as size marker. The gel, illuminated by 260 nm transmitted irradiation, can be photographed with appropriate film and filters (e.g., Polaroid type 57 or 667, Kodak 22A Wratten filter).

F. Southern Transfer

To increase the sensitivity of analysis (up to 1000–fold lower limit of detection) and to visualize low/single-copy DNA sequences, labeled DNA is hybridized to the size fractionated DNA obtained after agarose gel electrophoresis. A convenient means of achieving this is to transfer the DNA from the agarose gel into a solid support (e.g., a nitrocellulose filter). During this process the DNA is single-stranded and thus ready to hybridize to a complementary single-stranded, labeled DNA probe.

The agarose gel is washed (2 × 5 min, in several times the volume of the gel) with 0.25 M HCl in a glass dish. This partially hydrolyzes the DNA, breaking the larger DNA fragments into smaller fragments and increasing the efficiency of their transfer; longer acid treatment can be deleterious—sufficient acid depurination has usually occurred by the time the bromothymol blue in the loading buffer has changed from blue to yellow. The agarose gel is then washed (2 × 15 min, several volumes) in 1 M Tris-Cl (pH 8), 1.5 M NaCl at room temperature. This neutralizes the gel and denatures the DNA. This is followed by 2 × 15 min washes (several volumes) in a 1 M ammonium acetate, 0.02 M NaOH; two pieces of nitrocellulose and six pieces of Whatman 3M filter paper, each slightly larger than the gel, are also thoroughly wetted in this solution. The transfer is set up as a sandwich comprising a 5– to 7–cm layer of paper toweling followed by three sheets of presoaked filter paper, then one of nitrocellulose, the gel, then nitrocellulose, three sheets of filter paper, and a final 5– to 7– cm layer of paper toweling (care should be taken not to trap air bubbles between the various layers). A glass plate is placed on top and the transfer allowed to proceed for 2–16 hr, the single-stranded DNA diffusing from the gel and binding to the surface of the nitrocellulose filter. After transfer the position of the gel lanes is marked on the nitrocellulose filters and gel discarded. The nitrocellulose filters are air-dried and then baked at 80°C for 2 hr in a vacuum oven. Gloves should be worn throughout these procedures to protect the worker as well as reduce the surface contamination of the nitrocellulose filers.

G. Dot Blots

To determine only the presence or absence of a particular DNA sequence in a sample, it is unnecessary to size-fractionate the DNA. Instead, dot blots, which allow processing of numbers of samples, can be used. Denatured DNA samples can be directly spotted (10– to 30–μl samples) onto the surface of nitrocellulose. However, more uniform dots are obtained using a commercially available dot blot apparatus. This consists of a multiwelled manifold in which a nitrocellulose filter is clamped; denatured DNA solutions are added to the wells and drawn under suction through the filter. The test DNA sample can be denatured by the addition of NaOH to 0.3 M (e.g., 20 μl of the DNA solution plus 4 μl of 2 M NaOH), and NH4OAc to 1 M is added after 10 min. An aliquot of this solution is then placed in the well of the dot blot apparatus under suction and drawn down onto the nitrocellulose, the single-stranded DNA binding to the filter. The filter is rinsed with 1 M NH4OAc and baked as above.

H. Hybridization of Probes to Nitrocellulose Filters

The nitrocellulose filter is prehybridized by placing the filter in a heat-sealable plastic bag and adding a 5X SSPE, 0.3% SDS solution (approximately 0.1–0.5 ml/cm^2 of filter area; a stock solution of 20X SSPE contains 174 g NaCl, 27.6 g NaH$_2$PO$_4$.H$_2$O, and 7.4 g EDTA per liter, adjusted to pH 7.4 with 10 M NaOH). As much air as possible is removed; the bag is sealed with a heat sealer and then immersed in a waterbath at 65–68°C for 2 hr taking care to ensure that air bubbles do not prevent the even treatment of the filter surface.

The corner of the heat seal bag is cut off, the prehybridization solution discarded, and fresh 5X SSPE, 0.3% SDS solution added. To this the labeled DNA probe, which has been denatured (single-stranded) by placing it in boiling water for 15 min (and then rapidly cooled on ice to slow the rate of reannealing), is added by Pasteur pipette. Excess air is removed, the corner of the bag resealed as above, and the bag is then placed at the appropriate hybridization temperature (e.g., 65–68°C overnight or approximately 16 hr). The filter(s) is then removed and excess unbound probe removed by washing a minimum of three times in 0.2 or 2X SSPE, 0.3% SDS at the desired temperature (e.g., 68°C, 50°C, or room temperature). The desirable hybridization and wash conditions are dependent on the degree of homology between the DNA probe and the target sequence on the filter. The less exact the match, the less stringent the hybridization conditions should be and stringency is a function of temperature and salt conditions (see Sambrook et al. 1989). In general, the conditions used above work well with most probes, the less stringent conditions (low temperature, high salt) being used for heterologous probes, the more stringent (high temperature, low salt) for homologous and total DNA probes.

I. Visualization of Probe Bound to the Target Sequence

Radiolabeled dNTP–^{32}P DNA probes are visualized by exposing the hybridized filter to X-ray film, having marked the nitrocellulose filter with gel origin, etc., using radioactive ink. Length of exposure will depend on activity of the probe and quantity bound to the target sequence—typically it is 24–48 hr, but can be up to 1 week. Nonradioactive probes (e.g., biotinylated probes) can be visualized more rapidly (frequently in hours) but are generally less sensitive and require greater care in probe and filter preparation as nonspecific binding of probe to contaminants can pose problems. Biotinylated probes are visualized by the addition of a biotin-specific antibody linked to a fluorescent moiety or enzyme (e.g., peroxidase) capable of causing a color change on its substrate.

There are many cost-effective commercially available kits for labeling DNA fragments and the specific protocols provided by the manufacturer should be followed. Two common methods of labeling are (1) by nick translation, which incorporates labeled dNTP at random along the DNA fragment (using the excision/repair functions of DNase and DNA polymerase I) and (2) endlabeling (e.g., by using Klenow fragment of DNA polymerase I), which labels by the addition of labeled dNTP complementary to the first unpaired nucleotide of the overhanging terminus of the restriction fragment. Nick translation has the advantage of potentially adding greater quantities of label to the probe (particularly if the DNA fragment is large) while end labeling adds the same amount of label to each DNA fragment irrespective of size. This latter approach is useful in labeling restricted total DNA (see below).

In choosing between radioactive vs. nonradioactive labeling methods a number of factors must be considered. dNTP–^{32}P labeling is generally less sensitive to contaminants, it has a lower limit of detection, and its visualization requires considerable fewer steps than nonradioactive methods. It does require extra safety equipment (e.g., shielding, monitors) and, because of the relatively short half-life of dNTP–^{32}P, batches of probe must be prepared at regular intervals. Nonradioactive labeling is a less hazardous procedure and does not require additional safety equipment, etc. The labeled probe is stable (a given batch may last many months) and results can be visualized more rapidly. It is generally less sensitive in its ability to detect small amounts of target sequence and is more sensitive to contaminants, and processing is more time consuming, but Burrows (1989) indicates that these difficulties can be overcome and reports the detection of a species-specific DNA sequence from one invasive stage nematode of *Globodera pallida* using biotinylated probes. On balance, if there is not ready access to established radioisotope facilities, it may prove preferable to adopt the use of nonradioactive probes.

J. Probes

A number of different types of probe (here defined as any labeled DNA sequence which can hybridize to specific target DNA sequence) have been used in the study of plant parasitic nematodes. The simplest is the use of labeled total DNA (e.g., Bolla et al., 1988), which is useful in detection of RFLDs by identifying medium-high copy number sequences and by increasing the sensitivity. Unrestricted DNA can be used as a probe but frequently better resolution (particularly of smaller fragments) is achieved by end-labeling total DNA that has been digested using the same restriction endonuclease used on the target DNA bound to the Southern transfer filter.

Cloned DNA fragments are the most frequently used probes for the detection of RFLDs and as "yes-no" probes for detecting the presence or absence of a particular sequence in the target DNA (in this context their greatest utility is as species-specific probes). The cloned DNA fragments can be from another taxon and of known or unknown gene function. There are many cloned DNA sequences of known genes derived from *Caenorhabditis elegans* (the *Wormbreeder's Gazette*, the newsletter of *C. elegans* research, is a valuable resource; contact Mark Edgley, Caenorhabditis Genetics Center, Division of Biological Sciences, 110 Tucker Hall, Columbia, Missouri 65211). These genes are often readily available from other research groups and have sufficient homology to be useful in any nematode taxonomic group. Highly conserved sequences such as rDNA or histone genes from diverse organisms, e.g., *Drosophila* and *Xenopus*, are also useful for nematode taxonomy.

K. Randomly Cloned DNA Fragments as Taxonomic Probes

Cloning is achieved by digesting both donor (= nematode) and vector DNA (a DNA sequence capable of replicating in a bacterial cell) with the same restriction endonuclease, thus generating DNA fragments with compatible termini. These DNA fragments are then ligated together and the recombinant molecule transformed into a host bacteria which allows the replication of the cloned DNA sequences. There are numerous cloning vectors and strategies for their use (e.g., plasmid, phage, cosmid vectors, expression vectors, total DNA, mtDNA or cDNA libraries, etc.). However, to answer clearly defined taxonomic questions, such as species diagnosis, experience indicates that it is necessary only to screen small numbers of randomly cloned fragments (Curran and Webster, 1987; Besal et al., 1987; Marshall and Crawford, 1987; Burrows and Perry, 1988). Simple cloning strategies using plasmid vectors (even at their most inefficient) can provide sufficient clones for a successful screen (~30–100 cloned fragments). Plasmid vectors include pUC 18/19, pTZ 18/19 and puc 118/119.

The detailed protocols for cloning are dependent on the vector-host system employed and are provided by the supplier. A particularly useful vector-host for these purposes is pUC19 in *E. coli* JM83 in which those bacterial colonies transformed by recombinant (i.e., nematode DNA-containing) plasmids are readily selected. The following is a specific protocol that has provided consistently good results. The previously restricted total DNA from a representative nematode population is mixed with similarly digested pUC19 plasmid DNA at a ratio of 3–5:1 insert DNA to vector DNA, with a final DNA concentration of ~0.02–0.05 µg/µl in a 20–µl ligation reaction.To this is added 2 µl of a 10X DNA ligase reaction buffer (final concentration 50 mM Tris-Cl, pH 7.8, 20 mM DTT, 1 mM ATP, 50 µg/ml BSA, bovine serum albumin) and distilled water added to a final volume of 20 µl. T4 DNA ligase is then added (0.01 U) and the reaction incubated at 4°C for 12–16 hr.

The ligated DNA can then be transformed into competent *E. coli* JM 83 (i.e., host bacterial cells which have been treated with $CaCl_2$ to allow the passage of extracellular DNA into the bacterial cytoplasm). Bacterial cells can be prepared by inoculating from a single colony into 10 ml of LB, Luria-Bertani broth (10 g tryptone, 5 g yeast extract, 6 g NaCl, 1 liter distilled water) and grown overnight at 37°C with shaking (240 rpm). One hundred microliters of this culture is then transferred to 50 ml of fresh LB and grown until $OD590 = 0.2$. The bacteria are then pelleted by centrifugation (preferably at 4°C) and resuspended in a half-volume (~25 ml) of ice cold 50 mM $CaCl_2$ and incubated on ice for 30 min. The cells are then repelleted and gently resuspended in one-tenth volume (~5 ml) ice cold 50 mM $CaCl_2$ and stored on ice until used for transformation.

One hundred microliters of competent cells can then be added to the 20-µl ligation reaction and incubated on ice for 1 hr with occasional gentle mixing. The cells are heat-shocked at 42°C for 2 min and 1 ml of fresh LB is added. The cells are then incubated at 37°C for 1 hr. Aliquots (in the range 25–100 µl) are plated on selective medium (15 g agar, 10 g tryptone, 5 g yeast extract, 6 g NaCl, 1 liter distilled water), autoclaved, and then cooled to 40–45°C prior to the addition of 30 µg/ml ampicillin, 40 µg/ml X-gal (5–bromo–4–chloro–3–indolyl–B–galactopyranoside) and 160 µg/ml IPTG (isopropyl–B–D–thlogalactopyranoside). Plasmid pUC19 has an ampicillin resistance gene so that bacterial cells successfully transformed with plasmid DNA will grow in the presence of ampicillin. Recombinant (i.e., nematode DNA-containing) plasmids have DNA inserted into pUC19 such that the expression of B– galactosidase is inhibited. Therefore in the presence of IPTG (inducer) and X-gal (chromogenic substrate) bacterial colonies containing pUC19 without insert DNA have an active B-galactosidase and appear blue; bacterial colonies containing pUC19 with a nematode DNA insert have an inactive B-galactosidase and are white. The white colonies are then picked and streaked/spotted on LB + ampicillin plates. Hundreds to thousands of colonies are routinely obtained from one such ligation reaction, and restreaked single colonies can be stored at 4°C on plates or in tube slants or stabs.

The plasmid DNA can then be isolated by growing the bacterial clone in 1 to 1000-ml volumes of LB overnight and extracting the plasmid DNA. A rapid small-scale procedure is to pellet 1.5 ml of cell culture in a microcentrifuge tube, remove the spent medium, and resuspend the pellet in 100 μl of ice cold 50 mM glucose, 10 mM EDTA, 25 mM Tris-Cl (pH 8) solution. Keep at room temperature for 5–10 min, then add 200 μl of a freshly prepared solution of 0.2 M NaOH–1% SDS. Mix by gentle inversion (do not shake vigorously or vortex; the aim is to release the small plasmid molecules into the solution without releasing large quantities of chromosomal DNA) and store on ice for 5 min. Then add 150 μl of ice cold potassium acetate solution (60 ml 5 M potassium acetate, 11.5 ml glacial acetic acid, and 28.5 ml water), vortex the inverted tube gently for 10 sec and store on ice for 5 min. The tube is then microcentrifuged for 5 min at 4°C. The plasmid-containing supernatant is transferred to a fresh tube and extracted against a 1:1 phenol-chloroform mix. The plasmid DNA is then precipitated with ethanol and resuspended in TE as described under DNA extraction. Yields of 10–50 μg of plasmid DNA can be expected.

The plasmid DNA can then be digested with the appropriate restriction endonuclease and the presence and size of the inserted nematode DNA determined by agarose gel electrophoresis. The recombinant plasmids obtained can be labeled and used as hybridization probes.

Random fragments cloned from a representative nematode population can then be screened against dot blots of DNA extracted form the test nematode populations for (a) species specificity and (b) isolate specificity under defined hybridization conditions.

V. CONCLUSION

The use of molecular techniques in plant nematology has been centered on the detection of sequence differences between limited numbers of test populations and usually at the species level. It is now necessary to go beyond the demonstration that DNA technology can detect sequence variation within a nematode grouping. The taxonomic significance of this type of variation can only be assessed by the examination of many individual nematodes (or clonal lines) and geographically distant populations of a given nematode species. In essence it will become necessary to move toward a more "population genetic" approach rather than a "typological" methodology. Progress in this direction was recently made in the study of inter– and intra–isolate variation in DNA sequence of species of *Steinernema* and *Bursaphelenchus*; and, as might be expected, considerable variation was detected (Reid and Hominick, 1989; Webster et al., 1990). Given that ultimately it is possible to detect a single nucleotide change in the millions of base pairs of a nematode's genome, the assignment to a particular taxonomic grouping will be dependent on the acceptance that an arbitrary degree of sequence divergence is representative of a given taxonomic level.

The recent literature documents the increased application of DNA-based taxonomic techniques in agricultural nematology. At present the most useful approach for nematode identification is the development of taxon-specific DNA probes which can be used in dot blot procedures. There is little doubt that current research applications will soon transfer to practical use in diagnostics services with perhaps the work on *Globodera rostochiensis* and *G. pallida* identification being the closest to fruition (Marshall and Crawford, 1987; Burrows and Perry, 1988). In the near future, the ease of use and rapidity of PCR-based techniques is set to revolutionize molecular taxonomy (Kocher et al., 1989) in both diagnostics and classification and see its widespread application in nematode taxonomy. The application of DNA-based molecular methods is in its infancy in plant parasitic nematode taxonomy. However, the power of DNA-based approaches is sure to prompt their wider use.

LITERATURE CITED AND OTHER REFERENCES

Beckenbach, K., Xue, B., Kachinka, B., and D. Baillie. 1989. Use of Tc-1 related sequences to identify races in the Phyla nematoda. *J. Nematol.* 21: 551.

Besal, E. A., Powers, T. O., and Sandall, L. J. 1987. A DNA hybridization probe for the detection of the soybean cyst nematode. *J. Nematol.* 19: 513.

Birley, A. J., and Croft, J. H. 1986. Mitochondrial DNAs and phylognetic relationships, in *DNA Systematics, Vol. 1, Evolution*, Dutta, S. K., ed. CRC Press, Boca Raton.

Bolla, R. I., Weaver, C., and Winter, R. E. K. 1988. Genomic differences among pathotypes of *Bursaphelechus xylophilus*. *J. Nematol.* 20: 309–316.

Burrows, P. R. 1989. The identification of plant-parasitic nematodes using biotin labelled DNA probes. *J. Nematol.* 21: 553– 554.

Burrows, P. R., and Perry, R. N. 1988. Two cloned DNA fragments which differentiate *Globodera pallida* from *G. rostochiensis*. *Rev. Nematol.* 11: 441–445.

Burrows, P. R., and Boffey, S. A. 1986. A technique for the extraction and restriction endonuclease digestion of total DNA from *Globodera rostochiensis* and *Globodera pallida* second stage juveniles. *Rev. Nematol.* 9: 199–200.

Curran, J. 1990. Molecular techniques in entomopathogenic nematode taxonomy. In *Entomopathogenic Nematodes in Biological Control*, CRC Press (in press).

Curran, J., Baillie, D. L. and Webster, J. M. 1985. Use of restriction fragment length differences in genomic DNA to identify nematode species. *Parasitology* 90: 137–144.

Curran, J., McClure, M. A., and Webster, J. M. 1986. Genotypic analysis of *Meloidogyne* populations by detection of restriction fragment length differences in total DNA. *J. Nematol.* 18: 83–86.

Curran, J., and Webster, J. M. 1987. Identification of nematodes using restriction fragment length differences and species-specific DNA probes. *Can. J. Plant Pathol.* 9: 162–166.

Curran, J., and Webster, J. M. 1989. Genotypic analysis of *Heterorhabditis* isolates from North Carolina, USA. *J. Nematol.* 21: 140–145.

Dutta, S. K. 1986. *DNA Systematics, Vol. 1, Evolution*, Dutta, S. K., ed. CRC Press, Boca Raton.

Harris, T. S., Sandall, L. J., and T. O. Powers. 1989. Enhanced molecular diagnostics using polymerase chain reaction. *J. Nematol.* 21: 564.

Hobish, M. K. 1986. The role of the computer in estimates of DNA nucleotide sequence divergence. In *DNA Systematics, Vol. 1, Evolution*, Dutta, S. K., ed. CRC Press, Boca Raton.

Hyman, B. C. 1988. Nematode mitochondrial DNA: anomalies and applications. *J. Nematol.* 20: 523–531.

Hyman, B. C. 1989. Molecular diagnosis of *Meloidogyne* species. *J. Nematol.* 21: 567.

Hyman, B. C., Peloquin, J. J., and Platzer, E. G. 1987. Physical constitution of the *Meloidogyne chitwoodi* mitochondrial genome: Application for molecular diagnostics. *J. Nematol.* 19: 528.

Kalinski, A., and Huettel, R. N. 1988. DNA restriction fragment length polymorphism in races of the soy bean cyst nematode *Heterodera glycines*. *J. Nematol.* 20: 532–538.

Kocher, T. D., Thomas, W. K., Meyer, A., Edwards, S. V., Paabo, S., Villablanca, F. X., and Wilson, A. C. 1989. Dynamics of mitochondrial DNA evolution in animals: Amplification and sequencing with conserved primers. *Proc. Natl. Acad. Sci. USA*, 86: 6196–6200.

Marshall, J. W., and Crawford, A. M. 1987. A cloned DNA fragment that can be used as a sensitive probe to distinguish *Globodera pallida* from *Globodera rostochiensis* and other cyst forming nematodes. *J. Nematol.* 19: 541.

Nei, M., and Li, W.–H. 1979. Mathematical model for studying genetic variation in terms of restriction endonucleases. *Proc. Nat. Acad. Sci. USA* 76: 5269–5273.

Paterson, C. 1987. *Molecules and Morphology in Evolution: Conflict or Compromise*. Cambridge University Press, Cambridge, pp. 1–22.

Paterson, C. 1988. Homology in classical and molecular biology. *Mol. Biol. Evol.* 5: 603–625.

Peloquin, J. J., Bird, D., McK. and Platzer, E. G. 1989. Morphologically homogeneous populations of M. hapla have two distinct mitochondrial genome sizes. *J. Nematol.* 21: 579–580.

Platzer, E. G. 1981. Potential use of protein patterns and DNA nucleotide sequences in nematode taxonomy, in *Plant Parasitic Nematodes*, Vol. 3. Zuckerman, B. M. and Rohde, R. A., eds. Academic Press, New York, 1981, p. 3.

Powers, T. O., and Sandall, L. J. 1988. Estimation of genetic divergence in *Meloidogyne* mitochondrial DNA. *J. Nematol.* 20: 505– 511.

Powers, T. O., and Sandall, L. J. 1989, Mitochondrial DNA and species boundaries in *Meloidogyne*. *J. Nematol.* 21: 581.

Powers, T. O., Platzer, E. G., and Hyman, B. C. 1986. Species-specific restriction site polymorphism in root-knot nematode mitochondrial DNA. *J. Nematol.* 18: 288–293.

Reid, A. P., and W. M. Hominick. 1989. Characterization of entomophilic nematode species and strains using recombinant DNA technology. *J. Nematol.* 21: 582.

Rollinson, D., Walker, T. K., and Simpson A. J. G. 1986. The application of recombinant DNA technology to problems of helminth identification. *Parasitology* 91: S53–S71.

Saiki, R. K., Gelfand, D. H., Stoffel, S., Scharf, S. J., Higuchi, R., Horn, G. T., Mullis, K. B., and Erlich, H. A. 1988. Primer-directed enzymatic amplification of DNA with a thermostable DNA polymerase. *Science* 239: 487–491.

Sambrook, J., Fritsch, E. F., and Maniatis, T. 1989. *Molecular cloning*: *A laboratory manual*. 2nd ed. Cold Spring Harbor, NY. Cold Spring Harbor Laboratory 1989, 3 vols.

Vahidi, H., Curran, J., Nelson, D. W., Webster, J. M., McClure, M. A., and Honda, B. M. 1988. Unusual sequences, homologous to 5s RNA, in ribosomal DNA repeats of the nematode *Meloidogyne arenaria. J. Mol. Evolution* 27: 222–227.

Vrain, T. C., and D. Wakarchuk. 1989. Construction of a set of cloned rDNA probes of the specific differentiation of the *Xiphinema americanum* group. *J. Nematol.* 21: 593–594.

Webster, J. M., Anderson, R. V., Baillie, D. L., Beckenbach, R., Curran, J., and Rutherford, T. A. 1990. DNA probes for differentiating isolates of the pinewood nematode species complex. *Rev. Nematol.* (in press).

Wheeler, W. C. 1989. Hierarchy of life: molecules and morphology in phylogenetic analysis. In *Proc. Nobel Symposium 70* held at Alfred Nobles Bjorkborn, Karlskoga, Sweden, Aug. 29–Sept. 2, 1988. Amsterdam, New York. Excerpta Medica, New York, NY. Elsevier Science Publ. Co. 1989.

Xue, B. G., Baillie, D. L., Beckenbach, K., and Webster, J. M. 1989. DNA hybridization probes for the genotypic differentiation of *Meloidogyne* spp. *J. Nematol.* 21: 595–596.

III
SYSTEMATICS

5

Nemata: Higher Classification

ARMAND R. MAGGENTI   *University of California, Davis, Davis, California*

I. INTRODUCTION

Classifications are constructed from the foundation rank (species) up, not from the top (phylum or class) down; however, systematists are obliged to write classifications from the top down. Understanding a classification is dependent on our ability to keep the perspective of development from the base (species) to the apex (phylum). It is imperative that before proceeding to the next higher category all available information has been exhausted from the formation of the relevant lower category. Enhancement of any classification is dependent on forming higher taxa which suggest relationships that can only be reflected by the formation of a higher category. It is redundant to use the category superfamily for a single family. The present classification deletes all such examples. Likewise, in the orders Mononchida and Desmodorida the subordinal rank has been replaced by the superfamily category that was bypassed in the formation of the order (suborder in this instance is superfluous). Other than cases of monotypic ranks, any given group of taxa utilizing the same definition in the absence of enhancing knowledge of relationships should be subject to severe scrutiny and possible rejection.

     The history of the higher classification of nematodes spans just a little more than 100 years when Orley (1880) proposed grouping similar nematode genera into families. However, the first major attempt to organize nematodes (marine, freshwater, terrestrial, and parasitic) into a classification above the family rank was that of Filipjev (1934). It is creditable that many of his orders stand today. Filipjev did more than just list his proposed orders; for the first time in nematode classification an attempt was made to show phylogeny and to hypothesize nematode evolution from marine forms within his order Enoplata. Filipjev stated in 1934 that because classes at that time could not be distinguished within "Nematoda," he had little choice but to recognize nematodes as a class in the phylum Aschelminthes. However, he noted that if classes could be distinguished he would accept phylum status as proposed by Cobb (1919); he was aware of a lack of proof of relationships among the Aschelminthes or any other conglomerate phylum.

Chitwood (1937) expanded on the classification of Filipjev and proposed the classes Phasmidia (Secernentea) and Aphasmidia (Adenophorea) as well as recognizing nematodes as a phylum, as had Cobb (1919). However, against his own judgment he used the Grobben (1909) class designation: Nematoda. Chitwood (1958), recognizing his earlier error, accepted the phylum designation Nemates as proposed by Cobb (1919) but amended the spelling to Nemata in order to conform with the Pearse (1936) system for nomenclatorial endings; however, his classification essentially remained the same. Hyman (1951) rejected Chitwood's 1950 classification and reassigned nematodes to the phylum Aschelminthes Grobben, 1909. Her classification adopted the format of Filipjev, i.e., below the class Nematoda only ordinal divisions were recognized. This most inappropriate and uninformed decision to place nematodes in Aschelminthes (or Nemathelminthes) continues to plague nematology, but oddly enough not invertebrate zoology, which accepted nematodes as a phylum and rejected Hyman's and other such proposals more than 30 years ago.

Neither Chitwood (1950) nor Hyman (1951) agreed with Filipjev's proposal that the most ancestral nematodes were to be found among the marine enoplids. They both believed in a chromadorid ancestor, Hyman believing it to be a marine form and Chitwood believing in a freshwater/terrestrial form that superficially resembled *Rhabditis*, namely, *Plectus*. Maggenti (1961) studied the morphology and taxonomy of the genus *Plectus* and in 1963 utilized the 1961 study, along with comparative morphology of the major groups of nematodes, to reexamine the classifications proposed to that date. Maggenti's proposal maintained the basic framework of Chitwood's 1950 proposal; however, there were some significant philosophical differences. The evolution of Secernentea from Araeolaimida was emphasized as was the ancestral origin of nematodes from an enoplid-like ancestral form thought to be found among the Leptosomatidae as was proposed by Filipjev (1934). It now seems more likely that the contemporary form closest to the ancestral candidate will be from among the Oxystominoidea that manifest a more ancestral state of cephalization, i.e., the 6–6–4 cephalic setal pattern of some taxa is preserved in three separate whorls of setae.

De Coninck et al. (1965) proposed two important changes in the hierarchy of Adenophorea. They proposed that Adenophorea be subdivided into two infraclasses: Enoplia and Chromadoria. This move reflected the phylogenetic divergence of these major groupings.

Gadea (1973) proposed that Adenophorea and Secernentea be abandoned and replaced by Enoplimorpha and Chromadorimorpha. If this proposal and the remaining placement of taxa had been accepted, it would have effectively destroyed the relationships reflected in the classification that had been evolving over the previous 40 years. Andrássy (1974, 1976) reinstated the basic format that had been emerging prior to Gadea; however, he discarded the dichotomous division of Adenophorea subdivided into Enoplia and Chromadoria. These names were replaced with the synonyms Torquentia (= Chromadoria) and Penetrantia (= Enoplia). He retained Secernentea so that now all categories were at the same rank; once again relationships and phylogeny were obscured.

Maggenti (1982) in a review of the phylum to family level noted discrepancies in the hierarchy that did not reflect relationships. As a result, within Secernentea three subclasses were proposed: Rhabditia, Spiruria, and Diplogasteria. Within the Diplogasteria, Tylenchina was reduced from more than 20 families to seven families. This work led to the proposal (Maggenti, 1983) that within Adenophorea the subclasses Enoplia be subdivided into two superorders: Marenoplica and Terrenoplica. These ranks were proposed in order to clarify the major evolutionary trends within Enoplia, namely, a marine development and a terrestrial development. In addition, the orders Trichosyringida, Trichocephalida, and Mermithida were combined in the single order: Stichosomida. This proposal recognizes the distinctiveness of the stichosome structure as well as the biological nature of these parasites of invertebrates and vertebrates, an evolutionary direction that cannot be ignored.

Lorenzen (1983) proposed that Mermithoidea, Trichuroidea, and Dictophymatoidea be removed from Adenophorea and placed within Secernentea. Among the arguments presented was his acceptance of Richter's (1971) claim that the stylet in these groups was formed from "the ventral wall of the buccal cavity," whereas in some taxa within Dorylaimida the stylet is formed from the subventral stomatal wall. However, in Diphtherophorina the stylet is in part formed from the dorsal wall. The formation of the stylet from the ventral wall is highly unlikely in that a ventral wall within the stoma or esophagus of Nemata is unknown. Embryologically, in the formation of the stomodeum, the primary invagination forms the esophagus and the secondary invagination the stoma; both are composed of a dorsal wall, a left subventral wall, and a right subventral wall. The hinge points for these plates are subdorsal and ventral; the latter, though mislabeled, are evident in Richter's (1971) electron micrographs (Abb. 1b-d; Abb. 2). The evidence indicates that Richter's plate orientation is off by 180° and that the stylet is a product of the dorsal wall (that does exist in nematodes) and is similar to the formation as seen in the family Trichodoridae. Poinar and Hess (1974) claim that whether the stylet is formed by the dorsal plate or the "ventral" plate is dependent on the genus of mermithid under study. I cannot accept that in the same group the stylet is formed either dorsally or "ventrally"; not even if a ventral wall existed. Therefore, Lorenzen's proposal is rejected. Accepted is the proposal of Maggenti (1983) that Mermithoidea and Trichuroidea be combined (minus Dioctophymatoidea) in the order Stichosomida. The removal of Dioctophymatoidea, as proposed by Lorenzen (1983) and Maggenti (1983), is accepted and Dioctophymatoidea is here placed in the class Secernentea, subclass Spiruria, order Ascarida.

In Nemata the phytoparasitic nematodes of the orders Dorylaimida and Tylenchida are among the most taxonomically abused groups in the phylum. The animal parasitic groups have experienced a great deal of taxonomic scrutiny; however, in recent decades this has been directed toward seeking the correct placement of taxa and groups rather than an inflation of genera and species, which did exist historically. To a large extent the approach to the systematics of the free-living nematodes in marine, freshwater, and soil habitats has been the subject of the same type of review as the animal parasitic forms. The inflation among the phytoparasites has been largely due to a complete disregard for the principles of systematics, regardless of the taxonomic school followed.

Classification in the latter half of the 19th century, under the force of Darwin's perceptions, shifted from utilitarian classifications that were logic-based but designed for the sake of identification to the broader concept that diversity among organisms resulted from evolutionary divergence. An intent of classification thus became recognized as being a reflection of relationship and divergence. The application of the concept of diversity from evolutionary divergence has resulted in a somewhat stable understanding of the free-living and, to a slightly lesser extent, the animal parasitic nematodes. The differing concepts of a classification, i.e., either identification or divergence in evolution, were for the most part seemingly abandoned by many phytonematologists through the 1960's and into the early 1980's. Certainly, most of the prevailing classifications of plant parasitic nematodes in those years could not be interpreted as instruments of identification or illustrative of divergence. The number of genera and species being proposed (Fig. 1) was beyond reason: in 1967, 1239 species, 55 genera; in 1979, 2535 species, 129 genera; in 1987, 4305 species, 197 genera (Esser, 1979, 1987); minutia were being used to differentiate species. Inflation at the species level in turn stimulates an increase in the number of proposed genera (Fig. 1) for apparently no better reason than the perception that there are too many species in a par-

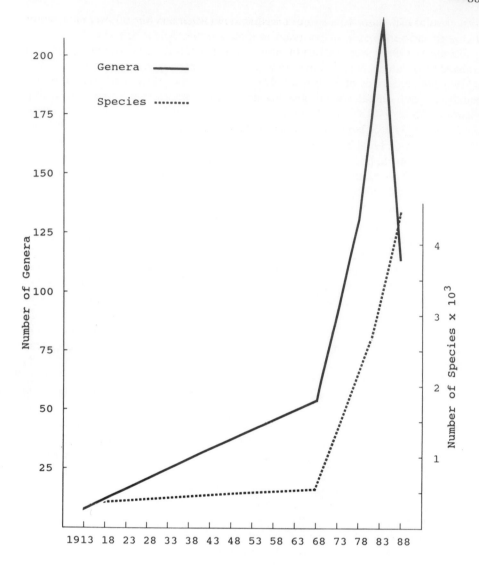

FIGURE 1 Cumulative annual number of genera and species of phytoparasitic nematodes from 1913 to 1988.

ticular genus. (The classification presented here recognizes 107 genera of phytoparasitic nematodes.) Reliance on minutia played a significant role in confusing the classification of plant parasitic nematodes. Each level of expansion begat further inflation in subfamilies, families, and superfamilies; this was especially true among the Tylenchina. This structuring of the hierarchy to accommodate the inflation of species and generic proposals resulted in a disastrous misunderstanding of plant parasitic nematodes. The function of a hierarchy was. ignored (Goodey, 1963; Golden, 1971; Fotedar and Handoo, 1978; Siddiqi, 1980, 1986) and diversity from evolutionary divergence was submerged by an inability to any longer recognize natural groupings. There was opposition to this unbridled inflation and attempts were made to hold to Darwinian concepts of classification (Allen and Sher, 1967, Paramanov, 1967, 1968; Andrássy, 1976; Inglis, 1983; Maggenti, 1971, 1978, 1981, 1982, 1983).

The above efforts did manage to insert into nematological literature on phytoparasites evolutionary concepts and terminology. Andrássy's 1976 classification attempted to utilize evolution as a basis of classifying nematodes. Unfortunately, it was founded on unsubstantiated morphology and an absence of biological data; however, it was an encouraging effort.

Siddiqi (1980, 1986) offered a chimerical classification and phylogeny of Tylenchida: suborders Tylenchina, Criconematina, Hexatylina, and Myenchina (Siddiqi's origin of Tylenchina is from an oxyurid-drilonematid!). Siddiqi ignores the fact that oxyurids are rhabdits and drilonematids are spirurids. Note also that the aphelenchs are treated by Siddiqi as a separate order: Aphelenchida whose origin is from among Diplogasteria. The latter is most likely correct as Tylenchida (sensu Tylenchina, Aphelenchina, and Sphaerulariina) are generally accepted as having their origin among the mycetophagous Diplogasteria. To accept Siddiqi's unfounded and confused concept is to destroy the knowledge of the evolution of plant and animal parasitism that has been unfolding for more than 50 years from morphological diversity and biology. An example is Siddiqi's totally unfounded proposal that the origin of Tylenchida (origination: the two separate subclasses Rhabditia and Spiruria) is separate from Aphelenchina (= Siddiqi's Aphelenchida) and to be found among Oxyurida, as well as Drilonematoidea (two different subclasses), and originating from three unrelated species of nematode parasites reported from frogs and leeches (Annelida; therefore, their parasites are always drilonematids ???) that have not been seen or studied since their descriptions by Eberth (1863), Schuberg and Schroder (1904), and Pereira (1931, unpublished thesis). Studying or recollecting these nematodes would only help to place them correctly among the animal parasites; it would not add to our knowledge of the history of plant parasitic nematode evolution. Modern morphology and biology of secernentean nematodes supports the evolution of animal parasitic nematodes from bacterial feeding forms and that of plant parasitic nematodes from mycetophagus forms (Maggenti, 1981). Siddiqi (1986) bodes disaster for anyone interested in the evolution of plant parasitism among Tylenchida.

In 1983 an international team (Maggenti, Luc, Raski, Fortuner, and Geraert) agreed to work together on a revision of the suborder Tylenchina. The results of this cooperative effort were recently published (Luc et al, 1987; Maggenti et al., 1988). The section of this chapter on Tylenchina condenses their taxonomic conclusions. Likewise the section on Trichodoridae reflects Decraemer and Aphelenchina reflects the decisions of Nickle and Hooper as contained in the present book.

This disquisition will deal for the most part with the classification of Nemata to the level of family. Exceptions to this will be found among the plant parasitic nematodes in Adenophorea and Secernentea, which will be dealt with to the level of genus.

II. CLASSIFICATION

Phylum: Nemata

The phylum Nemata is sometimes listed as a class in the phylum Nemathelminthes or Aschelminthes, groups whose membership or status is continuously being challenged. In recent years the majority of workers in invertebrate zoology and nematology have treated nematodes as the separate phylum Nemata (= Nematoda) with two classes: Adenophorea and Secernentea.

Nematodes are bilaterally symmetrical "pseudocoelomates" most often with elongate, cylindrical, unsegmented, wormlike bodies covered by a cuticle secreted by the underlying hypodermis. In specialized parasites the body may be spindle-shaped, pear-shaped, lemon-shaped, or other versions of saccate. The anterior extremity is generally rounded and the oral opening is normally terminal. characteristically the anterior extremity bears 16 sensilla that may be setiform, papilliform, or, in the more advanced forms, sensilla insitica or colea. Paired chemoreceptors (amphids) are located laterally and posteriorly to or intermingled with the four posterior cephalic sensilla. The oral opening is followed by the alimentary canal that is divisible into the stomodeum (stoma, esophagus, esophagointestinal valve), mesenteron (intestine), and proctodeum (rectum), which opens through a subterminal anus. When present the excretory system opens through a ventromedial pore, generally anterior. The sexes are usually separate and the reproductive system in males and females are most often simple and panoistic. Teleotrophic forms with a nutritive rachis do occur e.g., *Caenorhabditis*. In females the genital and anal openings are separate. In males the reproductive system forms a cloacal junction with the proctodeum. Males generally have cuticularized secondary sex organs (spicules) that may be accompanied by a guiding gubernaculum. The central nervous system is composed of the ganglionated circumesophageal commissure (nerve ring) and longitudinal nerves, the main nerve being ventral and ganglionated. The somatic musculature is largely limited to longitudinally oriented fibers.

The eggs undergo determinate cleavage. The life cycle is normally direct: egg, four juvenile or larval stages, and adult. Molts occur between each juvenile stage.

Nematodes are mycetophagus, herbivorous, and carnivorous; they are predators of micrometazoans and parasites of plants and animals.

Class I: Adenophorea

The oral opening is generally guarded by six lip-like structures. The stoma or buccal cavity may be armed with movable jaws or teeth, fixed teeth, a hollow stylet, or it may be unarmed. The amphids are postlabial and the orifice is variable in shape being porelike, pocketlike, circular, or spiral. The cephalic sensilla (16) are setose to papilloid and postlabial to labial in position. Generally, somatic setae, hypodermal glands, and somatic papillae are present. The cells of the hypodermis are uninucleate. The body covering often lacks visible transverse or longitudinal striae. When present the excretory system is single celled and usually lacks a cuticularized duct. Generally there are six or more coelomocytes in the pseudocoelom. Caudal glands (three) are present in select groups. Males generally have two testes and paired spicules with an accompanying gubernaculum. Males also have a single, medioventral series of papilloid or tubular preanal supplements. Rarely do the males have caudal alae. At eclosion most often it is the first-stage juvenile that emerges from the egg.

The class is divided into two subclasses: Enoplia and Chromadoria.

Subclass A: Enoplia

The subclass is characterized by a diversity of characters. When the amphid is pouchlike the aperture is transverse or cyathiform. When the internal structure is tubiform the aperture may be porelike, ellipsoid, or greatly elongated. The cephalic sensilla are papilliform, setiform, or a mixture of both. Caudal glands are present or absent; however, they are generally present in marine nematodes. The esophageal glands open either into the anterior esophagus, through stomatal teeth, or posterior to the nerve ring. The typical number of esophageal glands appears to be five; however, in animal parasitic forms the esophageal glands are duplicated exterior to the esophagus and are referred to as a stichosome. The shape of the esophagus varies from cylindrical to wine bottle-shaped. The external cuticle is generally devoid of transverse or longitudinal striae; however, both may occur.

The subclass is divided into two superorders: Marenoplica and Terrenoplica.

Superorder 1: Marenoplica

This superorder is composed primarily of the marine nematodes but does include some soil and freshwater forms. The group is characterized by having six papillate cephalic sensilla on the lip region and 10 sensilla generally posterior to the labial region. These latter ones are generally setiform. In all the forms examined the esophageal glands (generally five) open anteriorly into the stoma or into the anterior portion of the esophagus, i.e., anterior to the nerve ring. Commonly a single excretory cell is evident as well as three caudal glands that open through a terminal cuticularized spinneret. The ductus ejaculatorius is heavily muscled and males generally have elaborate or prominent preanal supplements in a single ventromedial row.

There are three orders in the superorder: Enoplida, Oncholaimida, and Tripylida.

Order 1: Enoplida

The order is characterized by having the cuticle of the cephalic region doubled or formed into a cap or helmet. The helmet results from the infolding of the cuticle over the extreme anterior end of the esophagus. The stoma may or may not possess armature in the form of teeth or movable jaws. Even though the stoma is surrounded by esophageal tissue, the anterior portion of the movable jaws is of cheilostomal origin. The well-developed esophagus is nearly cylindrical and the esophagointestinal valve is prominent. The five esophageal glands are located in the posterior region of the esophagus; however, the orifices for these glands are located anterior to the nerve ring. The dorsal and two anterior subventral glands open at the base of the stoma or through the stomatal teeth. The amphid apertures may be transverse slits or oval elongate. The cephalic sensilla are most often in two whorls: an anterior circumoral whorl of six and a posterior whorl of 10 (6 + 4); in some taxa this whorl is clearly separated into the ancestral state of a circlet of six and one of four. A single medioventral excretory cell is usually present; the orifice is generally at the level of or anterior to the nerve ring. Medioventral supplementary organs may or may not be present on the males. The male spicules are paired and accompanied by a gubernaculum. Caudal glands and cuticular spinneret are found in males and females.

There are two superfamilies: Oxystominoidea and Enoploidea.

Superfamily 1: Oxystominoidea

The included taxa differ from other members of the order by not having the labial region set off by a distinct groove. The cephalic region is narrowed and within it the stoma is weakly developed and completely surrounded by esophageal tissue. Most characteristic is the separation of the cephalic sensilla into three distinct whorls with the most anterior sometimes being setose. The distinct separation of the whorls is unusual and considered to represent the ancestral state. The amphidial pouch is generally elongate and the external aperture varies from a transverse slit to a small oval; in some taxa the aperture is longitudinally elongate. The esophagus is not distinctive; it is elongate and gradually broadens posteriorly. Females with one or two ovaries and the males may have one or two testes. Male supplementary organs are generally absent but when present they vary from conspicuous to minute and setose.

There are three families: Paraoxystominidae, Oxystominidae, and Alaimidae.

Superfamily 2: Enoploidea

The anterior helmet is punctate, smooth, or penetrated by numerous fenestrae. The stoma is completely surrounded by esophageal tissue and when armed with jaws or teeth there are usually three; sometimes only one tooth is present. The nearly cylindrical esophagus may have posterior crenations. The cephalic sensilla are in the typical two-whorl pattern. The pouchlike amphids have transverse slits as apertures. The esophageal glands empty through the teeth or into the stoma. One or two tubiform supplementary organs are found on the male.

There are five families in the superfamily: Enoplidae, Lauratonematidae, Leptosomatidae, Phanodermatidae, and Thoracostomopsidae.

Order 2: Oncholaimida

The amphids are pocketlike with oval or elliptical apertures. The stoma is generally vaselike with heavily cuticularized walls. Stomatal armature consists of one dorsal tooth and two subventral teeth; the walls may additionally be armed with rows of small denticles. The stoma is divisible into a cheilostome and an esophastome. The male stoma in some taxa is indistinct or collapsed. There are cephalic sensilla in the typical two-whorl pattern. In some taxa the sensilla are all papilliform. Generally, the esophagus is conoid to cylindrical; in some taxa the posterior portion is composed of a series of bulbs.

The order contains three families: Oncholaimidae, Eurystominidae, and Symplocostomatidae.

Order 3: Tripylida

The cephalic cuticle is simple and not duplicated; there is no helmet. The body cuticle is smooth or sometimes superficially annulated. Cepahalic sensilla follow the typical pattern of one whorl circumoral and the second whorl is often the combination of circlets two and three. The pouchlike amphids have apertures that are inconspicuous or transversally oval. The stoma is variable being simple, collapsed, funnel-shaped or cylindrical, and armed or unarmed. In most taxa the stoma is surrounded by esophageal tissue, i.e., it is entirely esophastome. When the stoma is expanded both the cheilostome and esophastome are evident. Esophagi are cylindrical-conoid. Esophageal glands open anterior to the nerve ring. Males generally have three supplementary organs; more in some taxa. A gubernaculum accompanies the spicules. Caudal glands are generally present.
The order has two suborders: Tripylina and Ironina.

Suborder 1: Tripylina

Taxa in this group are the only forms that have well-developed cuticular annulation. Two types of stomas occur: collapsed (esophastome), or globular and composed of two distinct parts. Esophageal glands open anterior to the nerve ring; when a dorsal tooth is present the gland opens through it. Cephalic sensilla pattern is typical of the order. Amphidial apertures are porelike or oval. The esophagus is nearly cylindrical and the esophagointestinal valve is well developed and may be denticulated. Posteriorly the males have either papilloid or vesiculate preanal supplements. Caudal glands and spinneret are usually present.

There are two families: Tripylidae and Prismatolaimidae.

Suborder 2: Ironina *inquirenda*

There are cephalic sensilla in the usual two circlets. The stoma is cylindrical and elongate, and is armed with three teeth either anteriorly or posteriorly. The three anterior teeth are eversible. The walls of the cylindrical stoma are heavily cuticularized. All esophageal glands open anteriorly. The esophagointestinal valve is small. Males have papillose or setose preanal supplementary organs. The paired spicules are accompanied by a weakly developed gubernaculum.

Currently only one family is recognized: Ironidae.

Superorder 2: Terrenoplica

The superorder represents the terrestrial Enoplia and is distinguished by the cephalization of the cephalic sensilla. The sensilla are all found on the lip region and are of two basic types: porelike (sensilla coeloconica) and conelike (sensilla basiconica). In all included taxa the three to five esophageal glands open posterior to the nerve ring. A discrete excretory cell is seldom seen. Male preanal supplements, though often numerous, are generally weak and tubiform.

There are four orders in the superorder: Isolaimida, Mononchida, Dorylaimida, and Stichosomida.

Order 1: Isolaimida *inquirenda*

This is an order of uncertain position because its morphology is poorly understood; it is represented by the single genus *Isolaimium*. They are elongate and cylindrical nematodes that range from 3 to 6 mm. Anteriorly and posteriorly the cuticle may be annulated or ornamented with transverse and longitudinal punctuations that are most obvious on the tail. The oral opening is surrounded by six prominent cuticularized tubes and two whorls of six sensilla. The amphids are presumed to be the dorsolateral papillae of the second whorl. The stoma is elongate and has thickened walls anteriorly. The esophagus is clavate. The females are amphidelphic, sometimes with flexures in the ovaries. Male spicules are well developed and the weakly developed gubernaculum has two dorsal apophyses. Male preanal supplements are papilloid; in addition, there are paired caudal papillae on the tail of both sexes.

There is one monotypic family: Isolaimiidae.

Order 2: Mononchida

A full complement of cephalic sensilla is present on the lips in two circlets of six and 10. The amphids are small, cuplike, and located just posterior to the lateral lips; the amphidial aperture is either slitlike or ellipsoidal. The stoma is globular, heavily cuticularized, and derived primarily from the cheilostome. The stoma bears one or more massive teeth that may be opposed by denticles in either transverse or longitudinal rows. The esophagus is cylindrical-conoid, with a heavily cuticularized lumenal lining. The excretory system is atrophied. Males have ventromedial supplements and paired spicules. A gubernaculum that may possess lateral accessory pieces may be present or absent. Females have one or two ovaries. Caudal glands and a spinneret are common; however, they may be degenerate or absent.

There are two superfamilies: Mononchoidea and Bathyodontoidea.

Superfamily 1: Mononchoidea

The labial region is expanded and the labia are generally angular and distinct. The outer circle of cephalic sensilla is terminal on each labium. The stoma is barrel-shaped and there is a large immovable tooth on the dorsal wall. Subventral teeth may be present as denticles or as teeth as massive as the dorsal tooth. The esophagus is nearly cylindrical with a heavily cuticularized lumen. The esophagointestinal valve is well developed. Females have one or two reflexed ovaries. The male spicules are slender and the gubernaculum is well developed and has prominent lateral accessory pieces; preanal supplements are papilloid. Caudal glands and spinneret are present in both sexes.

There are five families: Mononchidae, Mylonchulidae, Cobbonchidae, Anatonchidae, and Iotonchulidae.

Superfamily 2: Bathyodontoidea

The lip region is rounded or only slightly expanded. The cephalic sensilla are conelike. The stoma comprises mostly esophastome; the cheilstome is shortened and hexaradiate. The stomatal cavity is tubular posteriorly. The esophastome is armed at the posterior end of the cheilostome with a ventrosublateral tooth of varying size; sometimes there is an accompanying small denticle. The esophagus is nearly cylindrical and posteriorly houses five esophageal glands. The esophagointestinal valve is well developed. Females are amphidelphic and with the ovaries reflexed; in some there is only one gonad. Males are rare; when described the spicules are paired, simple, unaccompanied by a gubernaculum, and the preanal supplements are papilloid. In both sexes caudal glands and a spinneret are present.

The suborder is divided into two families: Bathyodontidae and Mononchulidae.

Order 3: Dorylaimida

The labia are generally well developed; however, many taxa exhibit a smoothly rounded anterior. The labial region is often set off from the general body contour by a constriction. The cephalic sensilla are all located on the labial region. When there is no constriction the labial region is defined as that region anterior to the amphids. The amphidial pouch is shaped like an inverted stirrup and the aperture is ellipsoidal or a transverse slit. The stoma is armed with a movable mural tooth or a hollow axial spear. The anterior portion of the tooth or spear is produced by a special cell in the anterior esophagus. The esophagus is divided into a slender, muscular anterior region and an elongated or pyriform glandular/muscular posterior region. There are generally five esophageal glands with orifices posterior to the nerve ring. In some taxa there are three glands and in others seven have been reported. The esophagointestinal valve is well developed. The mesenteron is often clearly divided into an anterior intestine and a prerectum. Females have one or two reflexed ovaries; when only one the vulva may shift anteriorly. Males have paired equal spicules that are rarely accompanied by a gubernaculum. The males often have the ventromedial preanal supplements preceded by paired adanal supplements.

The order is divided into three suborders: Dorylaimina, Diptherophorina, and Nygolaimina.

Suborder 1: Dorylaimina

This suborder is distinguished by its hollow axial spear or stylet. The lips bear the full complement of cephalic sensilla; the amphids are postlabial with slitlike or ellipsoidal apertures.

The protrusible spear is composed of two parts: the odontostyle (tip) and the odontophore (posterior shaft). Anterior to the odontostyle and guarding the oral opening there may be denticles or other cuticular accessories. The anterior esophagus is narrow and the posterior muscular/glandular portion is expanded. In some taxa the posterior esophagus is surrounded by spiral muscles. Usually there are five esophageal glands; three is not uncommon but seven is rare. The excretory cell and pore are absent or atrophied. A prerectum is usually present. Males have paired spicules with lateral accessory pieces; seldom is the gubernaculum present.

There are four superfamilies in the suborder: Dorylaimoidea, Belonolaimoidea, Actinolaimoidea, and Belondiroidea.

Superfamily 1: Dorylaimoidea

The stoma (cheilostome) is tubular; into the stoma projects the axial spear (odontostyle) whose opening is obliquely dorsal. Posteriorly the odontophore may be modified into three flanges (apodemes). The body cuticle is smooth; in some taxa it is ornamented with longitudinal ridges. The posterior portion of the esophagus is one-third the length of the esophagus. The esophagointestinal valve is well developed. Females generally have paired gonads; when one is present it is posteriorly directed. Male supplements consist of an adanal pair preceded by a ventromedial series.

The superfamily contains seven families: Dorylaimidae, Encholaimidae, Tylencholaimidae, Tylencholaimellidae, Leptonchidae, Belonenchidae, and Longidoridae. Longidoridae contains important plant parasites that vector viruses.

Family: Longidoridae

The stylets are greatly elongated; the odontostyle may be 10 times longer and the odontophore five times longer than the width of the lip region. Basally there may be knobs or flanges on the odontophore. The lip region is smoothly rounded and the postlabial ampidial apertures are porelike or slitlike. The posterior expanded portion of the esophagus is one-third to one-fifth the length of the entire esophagus. Generally, only three esophageal glands are evident. Females have paired reflexed gonads; when only one the vulva usually shifts anteriorly.

Family Longidoridae
Genera: *Xiphinema* Cobb, 1913
     Longidorus Micoletzky, 1922
     Paralongidorus Siddiqi, Hooper, and Khan, 1963
     Longidorusoides Jacobs and Heyns, 1982
     Californidorus Robbins and Weiner, 1978
     Xiphidorus Monteiro, 1976

Superfamily 2: Actinolaimoidea

This superfamily is readily recognized by the heavily cuticularized stomatal walls or the cuticular basketlike vestibule. These stomatal modifications may also take the form of large teeth or denticles. The axial spear is always present. The labial region, which is generally rounded, may be separated from the general body contour by a constriction; then the labia are distinct. The esophagus is nearly equally divided between the narrow anterior portion and the expanded muscular/glandular posterior region. Females have paired reflexed gonads.

The superfamily contains four families: Actinolaimidae, Brittonematidae, Carcharolaimidae, and Trachypleurosidae.

Superfamily 3: Belondiroidea

The common characteristics of this superfamily is the thick sheath of spiral muscle surrounding the basal enlarged portion of the esophagus. The labial region is narrow and may be rounded or angular. Amphid apertures are very large, often as wide as the head width. The odontostyle is normally shorter than the width of the labial region; the odontophore is rarely flanged. Females have one or two gonads; when single it is posteriorly directed.

This superfamily contains five families: Belondiridae, Roqueidae, Dorylaimellidae, Oxydiridae, and Mydonomidae.

Suborder 2: Diphtherophodrina

The labial region is only slightly expanded and bears the cephalic sensilla. The amphids have an inverted stirrup shape and the aperture is ellipsoidal. When the stomatal spear is complete it is comprised of both cheilorhabdions and esophorhabdions, most strongly developed on the dorsal side. In some taxa the subventral walls are membranous. The thickened dorsal wall than functions as a movable tooth. The elongate narrow esophagus expands posteriorly into a short pyriform bulb. An excretory pore and cell are present ventromedially in the anterior body. Females have paired ovaries but males have a single testis. The male has paired spicules, sometimes accompanied by a gubernaculum. In addition, a weakly developed caudal alae may be present on the tail. The preanal supplements are papilloid.

There are two families in the suborder: Diphtherophoridae and Trichodoridae. The latter contains important plant parasites that vector plant viruses.

Family: Trichodoridae

Sensu
W. Decraemer

Members are distinguished by the stomatal armature, which is a modification of the stylet found in Diphterophoridae. The now curved dorsal tooth acts as a stylet but is not hollow. The anterior dorsal tooth is produced within the esophagus. The posterior portion is the thickened dorsal esophageal wall. The esophagus is narrow, cylindrical anteriorly and expanded posteriorly. Females have one or two gonads; the male has a single testis. The spicules are paired and accompanied by a gubernaculum. Preanal supplements are present.

Genera: *Trichodorus* Cobb, 1913
        Paratrichodorus Siddiqi, 1974
        Monotrichodorus Andrassy, 1976
        Allotrichodorus Rodriquez–M, Sher, and Siddiqi, 1978

Suborder 3: Nygolaimina

The principle distinguishing character is the eversible stoma that is armed with a protrusible subventral mural tooth. The esophagus varies from a dorylaim two-part, to bulbular, to having a distinct posterior bulb that appears to be valved. Three cardiac glands are characteristically present at the junction of the esophagus and intestine. Prerectum is present or absent.

There are four families: Nygolaimidae, Nygolaimellidae, Aetholaimidae, and Campydoridae.

Order 4: Stichosomida

The order is composed of taxa that are parasitic in either vertebrates or invertebrates. The most distinguishing characteristics is the modification of the posterior esophagus into a stichosome, i.e., a series of glands exterior to the esophagus proper. The stichosome may be in one or two rows. The early larval stages possess a protrusible stylet that is absent in the adults. Amphids are postlabial.

There are three superfamilies relegated to this order, one of which requires confirmation: Trichocephaloidea, Mermithoidea, and Echinomermelloidea *incertae sedis*.

Superfamily 1: Trichocephaloidea

The stichosome is generally in a double row; however, in the family Trichinellidae it is formed as a single short row. The orifices are posterior to the nerve ring. A unique feature of the superfamily is the form of the female gonad: the germinal zone extends along the length of the gonad but no rachis is formed. Males and females possess a single gonad. Males have one or two spicules. The eggs are distinguished by being operculate. There are three families: Trichuridae, Trichinellidae, and Trichosyringidae.

Superfamily 2: Mermithoidea

The stichosome in this family is always formed of two rows of stichocytes with orifices posterior to the nerve ring. The gonads are unmodified i.e., the germinal zone is distal. Both sexes generally have two gonads but a single gonad does occur in some taxa . Males may have a single spicule, or the spicules paired and fused or separate. The intestine in adults forms a trophosome. The eggs are not operculate.

There are two families: Mermithidae and Tetradonematidae.

Superfamily 3: Echinomermelloidea

Superfamily *Incertae sedis*

The superfamily comprises three families, all *incertae sedis*: Echinomermellidae, Marimermithidae, and Benthimermithidae.

Subclass B: Chromadoria

This subclass represents the transitional taxa between the marine and terrestrial nematodes. They are in all environments but are frequently encountered in the freshwaters of the world. The subclass contains some of the most elaborately ornamented forms in Nemata.

The cephalic and/or labial sensilla are generally in three whorls; however, the first circlet is always papilliform whereas the outer circlets may in varying combinations be setiform, elongate coniform, or papilliform. The manifestation of the amphids and their apertures are the most varied seen throughout the phylum. They may be spiral, circular, vesicular, or forms derivable therefrom. When well developed the stoma may contain a large dorsal tooth, three jaws, or six inwardly acting teeth. There are three uninucleate esophageal glands. The dorsal gland orifice is anterior to the nerve ring while the subventral glands empty into the posterior corpus. The esophageal shape varies from near-cylindrical to one clearly divisible into three parts: corpus, isthmus, and posterior bulb; the latter is often valved. Usually the esophagointestinal valve is well developed. In both sexes there may be one or two gonads. Males in the subclass are distinguished by having the muscles of the ductus ejaculatorius only moderately developed. Caudal glands are nearly ubiguitous.

The subclass encompasses five orders: Chromadorida, Desmoscolecida, Desmodorida, Monhysterida, and Araeolaimida.

Order 1: Chromadorida

The amphid manifestation is variable but within superfamilies some constancy is apparent. The various amphids are reniform, transverse elongate loops, simple spirals, or multiple spirals not seen in any other orders or subclasses. The cephalic sensilla are in one or two whorls at the extreme anterior. In all taxa the cuticle shows some form of ornamentation, usually punctuations that are apparent whether the cuticle is smooth or annulated. When developed the stoma is primarily esophastome and is usually armed with a dorsal tooth, jaws, or protrusible rugae. The corpus of the esophagus is cylindrical, the isthmus is not seen, and the postcorpus is distinctly expanded in which the heavily cuticularized lumen forms the cresentic valve. The esophagointestinal valve is triradiate or flattened. The females usually have paired reflexed ovaries.

The order is divided into two superfamilies: Chromadoroidea and Choanolaimoidea.

Superfamily 1: Chromadoroidea

The amphids are transversely elongate ovals to loop-shaped, circularly spiral, or tightly coiled multispirals. The amphids, normally located just posterior to the third circlet of cephalic sensilla, may sometimes be located among the four sensilla of the third whorl. The second and third circlets may be combined. The vestibule leading to the stoma proper (esophastome) generally possesses weakly to well-developed rugae. The stoma proper is cylindrical, funnel-shaped, or bowl-shaped, and may be armed with a dorsal tooth or the main chamber may have three well-developed teeth that are protrusible; in some taxa there are small denticles at the anterior of the esopastome. The postcorpus bulb may be slightly expanded or well developed. The cuticle is usually ornamented with transverse rows of punctuations or other intracuticular designs. Males may have knoblike preanal supplements. In some taxa the male spicules may be elongate and accompanied by a gubernaculum with or without a caudal apophysis.

There are six families: Chromadoridae, Hypodontolaimidae, Microlaimidae, Spirinidae, Cyatholaimidae, and Comesomatidae.

Superfamily 2: Choanolaimoidea

The structure of the complex stoma is not well understood. The cheilostome is composed of flaps or rugae. In some taxa there are jaws or movable teeth. The esophastome is sometimes divided into two sections with immobile teeth between the sections. The cephalic region is blunt and not distinctly set off. The cephalic sensilla are in the form of short cones or setae. The amphids are circular, spiral, or multispiral. The cuticle is ornamented with heterogeneous punctuations and pores. The preanal male supplements may be large and cuticularized or simply a series of papillae.

There are three families in the superfamily: Choanolaimidae, Selachinematidae, and Ethmolaimidae.

Order 2: Desmoscolecida

The taxa in this order are readily recognized by their conspicuous body annulation. The annuli may be covered with concretion rings or the cuticle may be ornamented with scales, warts, or bristles. The cephalic sensilla reportedly are reduced in number. The internal whorl is absent, the second whorl is papilliform, and the four sensilla of the third circlet are setiform. The vesiculate amphids are oval to circular and occupy much of the cephalic region. The somatic setae are tubular and the open distal ends are often elaborate. The posterior esophagus is only slightly expanded. When pigment spots or ocelli are present they are just posterior to the esophagus. Females are amphidelphic and the ovaries are generally outstretched. The male spicules are generally accompanied by a gubernaculum. Both sexes have three caudal glands.

There are two families: Desmoscolecidae and Greeffiellidae.

Order 3: Desmodorida

The stoma among the included taxa is variable in that it may or may not be armed with a dorsal tooth which may or may not be opposed by subventral denticles. The variable amphids range from reniform to elongate loops to simple or multiple spiral. The cephalic sensilla are generally in three whorls; however, the second and third whorls may be combined. Distinguishing the order is the cephalic capsule or helmet and the conspicuous somatic annuli. In some groups anterior and posterior adhesion tubes are utilized in locomotion.

There are two superfamilies: Desmodoroidea and Draconematoidea.

Superfamily 1: Desmodoroidea

The external amphid varies from an elongate hook to a simple spiral or multiple spiral. The cephalic sensilla are generally in three circlets but in some taxa the two outermost whorls are combined. Usually the esophagus ends in a valved bulb. The body annulation varies from fine to coarse. Longitudinal rows of somatic setae are not uncommon and numerous setae often are present on the cervical region. Males may have one or two spicules.

There are three families in the suborder: Desmodoridae, Ceramonematidae, and Monoposthiidae.

Superfamily 2: Draconematoidea

The labia range from obscure to well developed. In most genera the presence of numerous anteriorly placed cervical setae obscures the symmetry of the outer circlets of cephalic sensilla. The external amphid varies from a simple elongate loop to one that intersects itself and appears a single spiral. Cephalic adhesion tubes may be present. The somatic annuli may be ornamented with spines, ridges, granules, or internal inflations. Elongate setae are scattered over the body. Adhesion tubes are generally present posteriorly. The females are amphidelphic and the ovaries are reflexed. In males the spicules are accompanied by a gubernaculum.

There are three families in the group: Draconematidae, Epsilonematidae, and Prochaetosomatidae.

Order 4: Monhysterida

In general the stoma is funnel-shaped and lightly cuticularized; however, in some taxa the stoma is spacious, heavily cuticularized, and armed with protrusible teeth. The amphids vary from simple spirals to circular. Usually the second and third circlets of cephalic sensilla are combined but in some taxa the third circlet of four distinct setae is separate. The normal pattern of distribution is often disrupted by numerous cervical setae. The near-cylindrical esophagus is sometimes swollen posteriorly. The cuticle may be smooth or may have annuli or ornamentation. When the annuli are distinct the somatic setae may be long and in four to eight longitudinal rows. The female gonads are outstretched and either single or paired.

There are seven families: Linhomoeidae, Siphonolaimidae, Monhysteridae, Scaptrellidae, Sphaerolaimidae, Xyalinidae, and Meyliidae.

Order 5: Araeolaimida

The amphids are simple spirals that appear as elongate loops, shepherd's crooks, question marks, or circular. The cephalic sensilla are often separated into three circlets: the first two papilliform or the second coniform, the third circlet is usually setiform; rarely are the second and third whorls combined. Body annulation is simple. The stoma is anteriorly funnel-shaped and posteriorly tubular; rarely is it armed. Usually the esophagus ends in a bulb that may be valved. In all but a few taxa the females have paired gonads. Male preanal supplements are generally tubular, rarely papilloid.

There are two suborders: Araeolaimina and Tripylina.

Suborder 1: Araeolaimina

Cephalic sensilla are in three circlets. The circumoral and the first external circle are generally papilliform. The third circlet consists of four long or short setae. The stoma is most often a simple tube, slightly funnel-shaped anteriorly. The diverse amphids are always derivable from simple spirals, sometimes appearing circular. The esophagus has tuboid endings on the esophageal radii of the anterior corpus. The basal esophageal bulb may be valved.

There are three unassigned families: Axonolaimidae, Camacolaimidae, Tripyloididae, and two superfamilies (Araeolaimoidea, Plectoidea) make up the suborder.

Superfamily 1: Araeolaimoidea

The amphids are generally simple spirals, elongate loops, or hooks. In some taxa the spiral nature of the amphid is obscure and it appears to be circular. The cephalic sensilla are in three whorls; generally the four sensilla of the third whorl are setiform. The stoma is not well developed; anteriorly it is thinly cuticularized and the esophastome is tubular or funnel-shaped. The esophagus is cylindrical or externally divisible into a corpus, an isthmus, and a swollen valveless terminal bulb. The females have paired outstretched ovaries. The males lack preanal tubular supplements.

There are four families: Araeolaimidae, Cylindrolaimidae, Diplopeltidae, and Rhabdolaimidae.

Superfamily 2: Plectoidea

These nematodes either have papilloid cephalic sensilla or the outer circlet of four may be setiform or elaborated into winglike lamellae. The amphids are simple circles, ovals, or shepherd's crook. The stoma is tubular or expanded anteriorly and followed by a shortened tubular portion. The esophagus is cylindrical anteriorly, with or without an isthmus, and followed by an expanded muscular bulb with a bellows-type valve. In the family Bastianidae there is no valve. In all known taxa there are tuboid endings on the radii of the anterior corpus. Females have paired reflexed ovaries. Male preanal supplements vary from none to several and tuboid.

There are four families in the superfamily: Plectidae, Leptolaimidae, Haliplectidae, and Bastianidae.

Class II: Secernentea

The amphidial aperture shows little variation throughout the class. Most commonly it is porelike and located dorsolaterally on the lateral labia or anterior extremity. In some instances they are oval or cleftlike or located postlabially (generally in immatures). The cephalic sensilla are located on the labial region and are porelike or papilliform. In some parasitic groups the number of sensilla are reduced from 16 to four. At the level of the nerve ring there are generally a pair of sensilla called deirids. Generally there is a pair of sensilla located on the caudal region called phasmids. The cells of the hypodermis may be multinucleate. The cuticle varies from four layers to two layers and is generally transversely striated. Laterally, along most of the body length, there is a lateral wing area marked by longitudinal striae or ridges. In some parasitic forms the lateral alae may extend out from the body contour for a distance equal to the body diameter.

The esophagus generally has three glands. The dorsal gland always opens anteriorly, either in the procorpus or in the anterior metacorpus or posterior corpus. The two subventral glands open posterior to the metacorpus. The excretory system opens through a ventromedian cuticularized duct. Collecting tubules extend from the excretory cell either on both sides of the body or on one side. Females are devoid of somatic setae or papillae. The second-stage juvenile, with a few exceptions among parasitic forms, emerges from the egg. Caudal papillae may be found on the male. When present, male preanal supplements are paired; sometimes there is a ventromedial preanal papilloid supplement. Commonly males possess caudal alae; in some taxa these are called the bursa or bursa copulatrix.

There are three subclasses: Rhabditia, Spiruria, and Diplogasteria.

Subclass A: Rhabditia

One of the most distinguishing characteristics of the subclass is the form of the esophagus that, at least in immatures, is divisible into a corpus, isthmus, and posterior bulb. In adult parasitic forms the esophageal form may be clavate or cylindrical; however, the second-stage juveniles have the rhabditoid valved bulb or a form reminiscent of it. The valve within the posterior bulb acts as a rolling valve, not as a bellows as in Plectidae. The two-part stoma generally lacks armature; the stoma may be further subdivided into two or more sections. Males generally have well-developed caudal alae that may be supported by rays either papilliform or cuticularized.

Within the subclass the "family" Myenchidae that contains parasites of frogs and leeches is considered *incertae sedis* as to position and *familia dubium* as to ranking.

Two orders are recognized: Rhabditida and Strongylida.

Order 1: Rhabditida

The number of labia varies from a full complement of six to three or two or none. The tubular stoma may be composed of five or more sections called rhabdions. The three-part esophagus always ends in a muscular bulb that is invariably valved. The excretory tube is cuticularly lined and paired lateral collecting tubes generally run posteriorly from the excretory cell; some taxa have anterior tubules also. Females have one or two ovaries; when only one is present the vulva shifts posteriorly. The cells of the intestine may be uninucleate, binucleate, or tetranucleate. The hypodermal cells may also be multinucleate. Caudal alae, when present, contain papillae.

There are two suborders: Rhabditina and Cephalobina.

Suborder 1: Rhabditina

The stoma is commonly cylindrical and devoid of distinct rhabdions. The stoma is generally two or more times as long as it is wide. In most taxa the labia are distinct and bear the papilloid cephalic sensilla and porelike amphids. The esophagus is three part and ends in a valved muscular bulb. Females have one or two ovaries. Males usually have paired spicules accompanied by a gubernaculum. Caudal alae are common on males.

The suborder encompasses one family *insertae sedis*: Alloionematidae; five superfamilies are recognized: Rhabditoidea, Bunonematoidea, Cosmocercoidea, Oxyuroidea, and Heterakoidea.

Superfamily 1: Rhabditoidea

The superfamily is distinguished by the presence of a well-developed cylindrical stoma. The number of labia varies from two to six. The esophagus, at least in the immature stages, has a muscular posterior bulb with a roller-type valve. The caudal alae of males are supported by five to nine papilloid rays.

The superfamily contains 12 families: Rhabditidae, Rhabditonematidae, Odontorhabditidae, Steinernematidae, Rhabdiasidae, Angiostomatidae, Agfidae, Strongyloididae, Syrphonematidae, Heterorhabditidae, Carabonematidae, and Pseudodiplogasteroididae.

Superfamily 2: Bunonematoidea

The taxa in this superfamily are all distinguished by their strikingly asymmetrical bodies in terms of both the cephalic sensilla and labial distribution. Furthermore, the asymmetry extends to the elaborate ornamentation on the right side of the body; the left side of the body appears near-normal. The male caudal alae are often asymmetrical. Asymmetry is limited to the external covering of the body. The stoma and esophagus are rhabditoid. Females have two gonads.

There are two families: Bunonematidae and Pterygorhabditidae.

Superfamily 3: Cosmocercoidea

The labial region bears three or six labia and ventrolateral papillae are always present. The stoma is weakly developed and for the most part is surrounded by esophageal tissue. The posterior bulb of the three-part esophagus is always valved. Esophageal glands are always uninucleate and no caeca are associated with the esophagus or intestine. A precloacal sucker may be present on males when present the spicules are equal and in some taxa are functionally replaced by the gubernaculum. Two families are placed in the superfamily: Cosmocercidae and Atractidae.

Superfamily 4: Oxyuroidea

Characteristically the ventrolateral papillae are absent and the cepahlic sensilla occur in a circle of eight or fused as four. The labia are reduced or absent. The stoma is primarily composed of the cheilostome and the esophastome is not distinguished. A valve is always discernible in the posterior bulb. Precloacal suckers may occur on males and the spicules may be greatly reduced.

There are three families in the superfamily: Oxyuridae, Thelastomatidae, and Rhigonematidae.

Superfamily 5: Heterakoidea

Ventrolateral papillae are present and the eight labial sensilla are paired in a whorl of four. The labia are well defined but small. The major development of the stoma is esophastomal in origin. The esophagus is three-parted and the posterior bulb is valved; rarely is the esophagus cylindrical. The subventral esophageal glands are duplicated. Males have a precloacal sucker surrounded by a cuticular ring. The spicules are always paired.

There are only two families in the superfamily: Heterakidae and Ascaridiidae.

Suborder 2: Cephalobina

The characteristic thin-walled vestibular stoma is always expanded and followed by a heavy-walled chamber of about the same width. The funnel-shaped or tubular posterior stoma (esophastome) is always narrower than the cheilostome and composed of four separate rhabdions. The esophagus is of the three-part rhabditoid type and the posterior bulb is always valved. Females have, with rare exceptions, a single anteriorly directed ovary, often with several flexures. Caudal alae are absent on the male tail; however, the tail does bear papillae. The male spicules are paired and a gubernaculum is present.

The superfamily contains four unassigned families: Cephalobidae, Robertiidae, Chambersiellidae, Elaphonematidae; and one superfamily: Panagrolaimoidea.

Superfamily 1: Panagrolaimoidea

The broad, thick-walled cheilostome is as long as it is broad or only slightly longer than broad. The posterior stoma is funnel-shaped, short, and lined with small rhabdions. The esophagus may have a distinctly muscular metacorpus. Females have one gonad that first is anteriorly directed but then reflexes posteriorly to extend beyond the vulva, usually without further flexures. The posterior uterus is vestigial and functions as a seminal receptacle.

The superfamily embraces three families: Panagrolaimidae, Alirhabditidae, and Brevibuccidae.

Order 2: Strongylida

The cephalic region may be adorned with either three or six labia or the labia may be replaced by a corona radiata. The variable stoma may be well developed or rudimentary but is never collapsed or inconspicuous. The capacious stoma consists primarily of cheilostome; only the base of the cup is surrounded by esophageal tissue. The stoma even when reduced is represented by the cheilostomal vestibule. The esophagus in larval forms is rhabdiform and the posterior bulb is valved. The adult esophagus is cylindrical to clavate. In the fourth juvenile stage and the adult the excretory system consists of paired lateral collecting tubules, an excretory cell, and paired subventral glands. Other immatures have a single excretory cell. The females have one or two gonads and the uterus is heavily muscled. The males have unusual caudal alae; in this group it is referred to as a bursa copulatrix. The bursa differs from other caudal alae in that it contains muscles and rays. Males have paired and equal spicules.

There are three unassigned families: Diaphanocephalidae, Metastrongylidae, Maupasinidae; and three superfamilies: Strongyloidea, Ancylostomatoidea, and Trichostrongyloidea.

Superfamily 1: Strongyloidea

Though the stoma is variable, it is always well developed and spacious. It may be hexagonal in cross-section, globular, cylindrical, or infundibuliform; it is never armed with teeth or cutting plates. The oral opening may be surrounded by six small labia or a corona radiata. In newly hatched larvae the esophagus is rabditiform but the adult esophagus is clavate or near-cylindrical. Males have well-developed bursae and the bursal rays are never fused.

The superfamily embraces three families: Strongylidae, Cloacinidae, and Syngamidae.

Superfamily 2: Ancylostomatoidea

This group is distinguished by the characteristic capacious, thick-walled, globose stoma that is unarmed or armed anteriorly with teeth or cutting plates. The branches of the dorsal bursal rays in the male bursa are greatly reduced.

The superfamily contains three families: Ancylostomatidae, Uncinariidae, and Globocephalidae.

Superfamily 3: Trichostrongyloidea

Three or six inconspicuous labia surround the oral opening; in some taxa the labia are absent. The stoma is reduced or collapsed and there is no evidence of a corona radiata. The sometimes inflated cuticle of the cephalic region is unusual among strongyles. The thick somatic cuticle often has longitudinal ridges causing these worms to be wiry rather than limp. In second-stage juveniles the esophagus is rhabditoid but the posterior bulb may be valveless. In the bursa the lateral rays are well developed even though the dorsal rays may be atrophied.

There are six families in the superfamily: Trichostrongylidae, Amidostomatidae, Strongylacanthidae, Heligmosomatidae, Ollulanidae, and Dictyocaulidae.

Subclass B: Spiruria

This is the only subclass in Nemata that is entirely parasitic and the only group for which no free-living counterparts can be identified.

The oral opening is surrounded by three or six apical lobes, often called pseudolabia, or the labia are modified into two lateral lobes or "labia." The complement of cephalic sensilla is variable throughout the group and ventrolateral papillae are present only among some taxa in the Ascarida. The stoma may be globose or long and cylindrical; when armed the armature is cheilostomal in origin. There are taxa where the stoma is not distinguishable; in these taxa esophageal tissue extends to the much reduced cheilostome. The esophagus is variable, it may be clavate, nearly cylindrical, or divided into a narrow anterior region and an elongate swollen glandular region (wine bottle-shaped). The esophagus is never in any stage rhabditoid in character or valved posteriorly. The collecting tubules of the excretory system may be in the form of an "H," or an inverted tuning fork; in a few instances the system is limited to one side of the body, usually the left. The vagina may be greatly elongate and tortuous. The generally paired male spicules may be greatly disproportionate in length. Caudal alae with embedded papillae may occur. Eggshells are often highly ornamented.

Three orders are recognized: Ascaridida, Cammallanida, and Spirurida.

Order 1: Ascaridida

Generally, the oral opening is surrounded by three or six labia, but in some taxa labia are absent; however, the cephalic sensilla are always evident. Usually there are eight cephalic or labial sensilla; the submedians may be fused then only four sensilla are seen. The stoma varies from being completely reduced to spacious or globose. The esophagus varies from club-shaped to nearly cylindrical, never rhabditoid. There may be posterior esophageal or anterior intestinal cecae. The collecting tubules of the excretory system may extend posteriorly and anteriorly. Males generally have two spicules; however, in some taxa there may be none or only one. The gubernaculum may also be present or absent. Though females generally have two ovaries, multiple ovaries do occur. The number of uteri is also variable; there may be two, three, four, or six. Phasmids are sometimes large and pocketlike. Reportedly the larvae lack a stomatal hook or barb.

The order encompasses five superfamilies: Ascaridoidea, Seuratoidea, Camallanoidea, Dioctophymatoidea, and Muspiceoidea.

Superfamily 1: Ascaridoidea

Body length varies from 1 to 40 cm. Cuticle marked by superficial annuli. The oral opening is generally surrounded by three well-developed lips: one dorsal and two subventrals that bear the porelike amphids. The stoma is primarily esophastome followed by a cylindrical to clavate esophagus. Cecae may extend anteriorly or posteriorly from the esophagointestinal junction. Females have paired ovaries or multiples of three, four, or six. Males have paired spicules rarely accompanied by a gubernaculum.

The superfamily contains seven families: Ascarididae, Toxocaridae, Anisakidae, Acanthocheilidae, Goeziidae, Crossophoridae, and Heterocheilidae.

Superfamily 2: Seuratoidea

Lip region and lips are greatly reduced or absent. Ventrolateral sensilla are present or absent, and the submedian sensilla are doubled or single. The stoma may be small and poorly developed, or strongly developed and provided posteriorly with teeth. The esophagus is cylindrical or slightly clavate. Intestinal cecae are rarely present. In males there may or may not be a precloacal sucker. Male spicules paired and often accompanied by a well-developed gubernaculum. Caudal alae, if present, are very narrow.

The superfamily contains five families: Seuratidae, Schneidernematidae, Quimperiidae, Subuluridae, and Cucullanidae.

Superfamily 3: Camallanoidea

The stoma, the main chamber of which is always cheilostomal, is variable in size but always well developed. It may be either globose or transversely rectangular and internally supported by numerous longitudinal or oblique ridges. The esophastome forms the basal stomatal plates. The internal circlet of cephalic sensilla are minute and the external circlet of eight is partly fused. The esophagus is short and generally clavate.

Two families are recognized in this superfamily: Camallanidae and Anguillicolidae.

Superfamily 4: Dioctophymatoidea

This superfamily was placed in Adenophorea prior to 1981 when it was moved to the Ascaridida within Secernentea/Rhabditia. It is here retained with Ascardida but moved to Secernentea/Spiruria.

The cuticle, with a well-developed oblique fiber layer, lacks the endocuticular layer. Externally the cuticle may be annulated or ornamented with distinct hooklike spines. When well developed the stoma is primarily esophastomal. The esophagus is externally cylindrical but internally divisible into the corpus and the postcorpus with three ramifying glands. Both males and females have a single gonad. Male tail is expanded into a thickened cuplike bursa copulatrix. Males have a single, elongate spicule.

There are two families in the superfamily: Dioctophymatidae and Soboliphymidae.

Superfamily 5: Muspiceoidea *incertae sedis*

These nematodes are arbitrarily relegated not only to superfamily level but to inclusion among the Ascaridida. The known taxa are parasites of vertebrates where they cause tumorous and cancerlike damage.

The cephalic sensory organs are greatly reduced. In only one species have amphids or cephalic papillae been reported. Males are unknown. In females the digestive tract is reduced. The females are reportedly amphidelphic, diovarial, and viviparous. However, the vulva may or may not be present; when absent larvae are released by the rupturing of the female body; thus reminiscent of *Dioctophyma*.

There are three families all *insertae cedis*: Muspiceidae, Robertdollfusidae, and Phlyctainophoridae.

Order 2: Spirurida

The labial region is most frequently provided with two lateral labia or pseudolabia; in some taxa there are four or more lips; rarely lips are absent. Because of the variability in lip number there is a corresponding variation in the shape of the oral opening which may or not be surrounded by teeth. The amphids are most often laterally located; however, in some taxa they may be located immediately posterior to the labia or pseudolabia. The stoma may be cylindrical and elongate or rudimentary. The esophagus is generally divisible into an anterior muscular portion and an elongate swollen posterior glandular region where the multinucleate glands are located. Eclosion larvae are usually provided with a cephalic spine or hook and the presence of a porelike phasmid on the tail.

All known taxa utilize an invertebrate in their life cycle; the definitive hosts are mammals, birds, reptiles, and, rarely, amphibians.

There are six superfamilies in the order: Spiruroidea, Drilonematoidea, Physalopteroidea, Dracunculoidea, Diplotriaenoidea, and Filarioidea.

Superfamily 1: Spiruroidea

Usually the lateral lips are well developed; some taxa have four lips and in a few taxa lips are absent. There are either four or eight cephalic sensilla in the external circlet. In some taxa the sensilla are doubled or fused. The cephalic and cervical regions may manifest cordons, collarettes, or cuticular rings. The oral opening is variable, being circular, hexagonal, or dorsoventrally elongated. The stoma, composed almost entirely of the cheilostome, is always well developed and anteriorly it may be armed with teeth. The vulva is generally at midbody and in only a few taxa is it located posteriorly or anteriorly near the esophagus. In males the ventral body surface, anterior to the cloacal opening, is often decorated with incomplete longitudinal ridges or lines.

The superfamily encompasses five families: Spiruridae, Thelaziidae, Acuariidae, Hedruridae, and Tetrameridae.

Superfamily 2: Drilonematoidea

The stoma, which may be surrounded by rudimentary lips (often not visible), is greatly reduced. The amphids may be postlabial and enlarged. The esophagus in mature adults may be three parted, i.e., with a corpus, isthmus, and a pyriform glandular bulb that is never valved or the esophagus may be short and clavate. Females have a single elongate ovary. Male spicules are either absent or paired, and a gubernaculum may be present or absent. The phasmids, sometimes called caudal suckers, are greatly enlarged and may occupy the full width of the tail.

The placement of these nematodes in the hierarchy is uncertain; however, their biology and known morphology are characteristic of Spiruria and not Rhabditia.

The superfamily contains six families: Drilonematidae, Ungellidae, Scolecophilidae, Creagrocericidae, Mesidionematidae, and Homungellidae.

Superfamily 3: Physalopteroidea

The inner surface of the two large lateral labia (pseudolabia) are generally provided with teeth; interlabia are not present between the pseudolabia. Of the cephalic sensilla the inner circlet (circumoral) is reduced or absent and the outer circle consists of four fused sensilla. The anterior cuticle in some taxa is reflexed and forms a collar that may partly cover the pseudolabia. The cephalic region may be ornamented with a very prominent spined bulbous structure; however, there are no cordons, collarettes, or rings. On males the caudal alae are well developed when the caudal papillae are pedunculate; when the caudal papillae are nonpedunculate the caudal alae are absent. In females the ovaries may number four or more and the vulva may be either anterior or posterior to the midbody.

The superfamily contains three families: Physalopteridae, Megalobatrachonematidae, and Gnathostomatidae.

Superfamily 4: Dracunculoidea

The stoma is generally very reduced and most often is only vestibular. The internal circle of six sensilla are well developed; the external circle consists of eight separate and well-developed sensilla. In immature females the vulva is at midbody; however, in gravid females it is atrophied as is the female's posterior intestine. If caudal alae are present on the male tail, they are small and postcloacal.

There are three families in the superfamily: Dracunculidae, Philometridae, and Micropleuridae.

Superfamily 5: Diplotriaenoidea

The pattern of sensilla surrounding the dorsoventrally elongate oral opening is variable. There may be four small sensilla in a circle around the oral opening, followed by a circle of six and a third circle of four, or there may be but two circlets of four sensilla each or these may join as paired or fused sensilla. Within the anterior stomatal region are protrusible cuticular structures called tridents located on either side of the anterior esophagus. The esophagus is divided into a narrow anterior portion and a wide, elongate, glandular portion. The vulva is anterior and females are oviparous or ovoviviparous. Males have unequal spicules and lack caudal alae.

There are two families: Diplotriaenidae and Oswaldofilariidae.

Superfamily 6: Filarioidea

The internal circle of sensilla may be absent or consist of two or four papillae surrounding the circular or oval oral opening. The external circle consists of eight sensilla. The stoma is rudimentary. The esophagus shows little difference between the corpus and the postcorpus with its multinucleate glands. The vulva is generally located in the anterior half of the body. In males the spicules are either equal or unequal and caudal alae may be present or absent. The gubernaculum is always absent.

The superfamily contains five families: Filariidae, Aproctidae, Setariidae, Desmidocercidae, and Onchocercidae.

Subclass C: Diplogasteria

Mostly small-to medium-sized nematodes, seldom exceeding 3–4 mm. The cuticle is ornamented by annuli that are sometimes transversed by longitudinal striae; the cuticle may be punctated. The labia may be well developed but hexaradiate symmetry is almost always evident. The full complement of cephalic sensilla is often present, especially on males. In derived taxa the inner circle of six sensilla may be lacking. In the external circle of 10 sensilla the externolaterals are ventrolateral on the lateral labia. The amphids, when labia are evident, are located dorsolaterally on the lateral lips and the apertures are porelike, small ovals, slits, or clefts. In some larval forms, particularly dauerlarvae, the amphids may be postlabial. The variable stoma is primarily composed of cheilostomal tissue and is most often armed. Except for rare instances, movable stomatal armature in the form of large teeth, opposable fossores, or axial spears is limited to this subclass. As in other nematodes the movable armature is controlled by the three anterior-most muscles of the esophagus. The ferrule (shaft and apodemes) of the axial spear or fossores are the product of the esophastome. The esophagus has a muscular corpus divided into an almost cylindrical procorpus and a muscular, almost always valved metacorpus followed by an isthmus and a glandular postcorpus. A valve is never present in the postcorporal bulb. Females have one or two ovaries and males have one testis with the ductus ejaculatoris almost devoid of musculature. Males have paired spicules but may or may not have a gubernaculum or caudal alae.

The subclass is divided into two orders: Diplogasterida and Tylenchida.

Order 1: Diplogasterida

The labia are seldom well developed; however, a hexaradiate symmetry is distinct. The external circle of labial sensilla may appear setose but they are always short, never long and hairlike. The stoma may be slender and elongate or spacious or any gradation between these two. The stoma may be armed or unarmed; the armature may be movable teeth, fossores, or a pseudostylet. The corpus is always muscled and distinct from the postcorpus that is divisible into an isthmus and glandular posterior bulb. The metacorpus is almost always valved. The female reproductive system may have one or two ovaries and males may or may not have caudal alae; however, a gubernaculum is always present. The male tail most commonly has nine pairs of caudal papillae; three are preanal and six are caudal. The order has four families: Diplogasteridae, Odontopharyngidae, Diplogasteroididae, and Cylindrocorporidae *incertae sedis*.

Order 2: Tylenchida

The labial region in Tylenchida is variable; it may be distinctly set off or smoothly rounded and well developed; the hexaradiate symmetry is most often retained or discernible. The amphids are porelike, oval, slitlike, or clefts located on the lips. The internal circlet of six sensilla may be lacking. The external circle of 10 sensilla is often evident; however, these may be reduced to a visible four or some may be doubled. The hollow stylet is the product of the cheilostome (conus, "guiding apparatus," and framework) and the esophastome (shaft and knobs). Throughout the order and its suborders the stylet may be present or absent and may or may not be adorned with knobs. The variable esophagus is most often divisible into the corpus, isthmus, and glandular posterior bulb. The corpus is further divisible into the procorpus and metacorpus. The metacorpus is generally valved but may not be in some females and males, and the absence is characteristic of some taxa. The orifice of the dorsal esophageal gland opens either into the anterior procorpus or just anterior to the metacorporal valve. The excretory system is asymmetrical and there is but one longitudinal collecting tubule. Females have one or two genital branches; when only one branch is present it is anteriorly directed. Except for sex-reversed males there is only one genital branch. Males may have one (= phasmid) or more caudal papillae. The spicules are always paired and variable in shape; they may or may not be accompanied by a gubenaculum.

The order contains four suborders: Tylenchina, Aphelenchina, Sphaerulariina, and Hexatylina *taxon dubium*.

Suborder 1: Tylenchina

Sensu
Armand R. Maggenti, Michel Luc, Dewey J. Raski,
Renaud Fortuner, and Etienne Geraert

The female esophagus is a distinctive characteristic of the suborder; it is composed of a procorpus and a generally valved matacorpus and a glandular postcorpus with an isthmus between the metacorpus and the posterior glandular region. The glandular postcorpus manifests varying degrees of development and enlargement; it may join the intestine directly or overlap the anterior intestine extensively. Among some taxa the male feeding apparatus is atrophied. In some groups the metacorporal valve is lacking in both males and females but the esophagus is still functional. The hollow feeding stylet in males and females generally has three basal knobs, one dorsal and two subventral. The stylet may be absent in males of some taxa or during specific stages of the life cycle. The labial region may be distinct or undifferentiated from the general anterior body outline. Transverse body annulation is interrupted by lateral longitudinal incisures. In some taxa the entire circumference of the body is marked by longitudinal incisures as well as transverse annuli. The females have one or two ovaries and the oviduct has two rows of seven cells. The males generally have simple caudal alae and paired spicules that may or may not be accompanied by a gubernaculum.

The suborder embraces two superfamilies: Tylenchoidea and Criconematoidea.

Superfamily 1: Tylenchoidea

The labial region generally portrays the hexaradiate condition and is often distinguishable from the general body profile. Internally the labial region may be supported by a cuticularized skeleton which may or may not be well developed. The generally slender and cylindrical procorpus is set off from the metacorpus which is followed by the narrow isthmus that leads to the expanded glandular region. The latter may consist of a true bulb enclosing the glands or the glands may form a lobe or lobes that extend posteriorly past the anterior intestine. The glandular region generally contains three glands but through duplication there may be five. Females have one or two ovaries and the males generally have caudal alae. The phasmids are commonly adanal, on the tail, or erratically on the body.

The superfamily contains seven families: Tylenchidae, Anguinidae, Dolichodoridae, Belonolaimidae, Pratylenchidae, Hoploaimidae, and Heteroderidae.

Family: Tylenchidae

Body is slender, vermiform; lateral fields are variable: none to multiple lines. The transverse body annuli may be transected by longitudinal ridges. Labial region generally elevated, rounded, and annulated. Labial framework generally weakly developed; stylet usually small and delicate. Amphids variable: small oblique slits to long, sinuous, longitudinal clefts. Deirids present or absent, phasmidlike structures present or absent, usually advulval sometimes on tail. Esophagus divided into slender procorpus, elliptical metacorpus most often valved, long slender isthmus followed by symmetrical pyriform glandular region. Females generally with a single anteriorly directed genital branch; 12-celled spermatheca often offset; columned uterus with four rows of cells; postuterine sac (PUS) length less than one vulval body diameter. Male caudal alae leptoderan. Sperm cells with little cytoplasm. Tails elongate-conoid, generally narrowing to filiform outline. Free-living algal and fungus feeders, rarely parasites of higher plants. There are five subfamilies:

Subfamily Tylenchinae
    Genera: *Tylenchus* Bastian, 1865
            Miculenchus Andrássy, 1951
            Filenchus Andrássy, 1954
            Malenchus Andrássy, 1968
            Irantylenchus Kheiri, 1972
            Polenchus Andrássy, 1980
            Allotylenchus Andrássy, 1984
            Cucullitylenchus Huang and Raski, 1986
            Mukazia Siddiqi, 1980

Subfamily Ecphyadorphorinae
   Genera: *Ecphyadorphora* de Man, 1921
            Lelenchus Andrássy, 1954
            Ecphyadorphoroides Corbett, 1964
            Epicarinema Raski, Maggenti, Koshy, and Sosamma, 1982
            Mitranema Siddiqi, 1986

Subfamily Tylodorinae
   Genera: *Eutylenchus* Cobb, 1913
            Macrotrophurus Loof, 1958
            Cephalenchus Goodey, 1962
            Tylodorus Meager, 1963
            Campbellenchus Wouts, 1977

Subfamily Atylenchinae
   Genera: *Atylenchus* Cobb, 1913
            Aglencus Andrássy, 1954
            Pleurotylenchus Szczgiel, 1969
            Antarctenchus Spaull, 1972
            Gracilancea Siddiqi, 1976
            Coslenchus Siddiqi, 1978

Subfamily Boleodorinae
   Genera: *Psilenchus* de Man, 1921
            Boleodorus Thorne, 1941
            Basiria Siddiqi, 1959
            Neopsilenchus Throne and Malek, 1968
            Atetylenchus Khan, 1973
            Neothada Khan, 1973
            Duotylenchus Saha and Khan, 1982
            Basirienchus Geraert and Raski, 1986

   Genus   *incertae sedis*:
           Luella Massey, 1974
   Genera *dubia*:
           Sakia Khan, 1964
           Basilophora Husain and Khan, 1965

Family: Anguinidae

Body is elongate slender, somewhat swollen in mature females. Lateral field with four or six or more lines. Low and flattened labial region. Small delicate stylet; labial framework lightly sclerotized. Amphid apertures small, lateral slits. Female genital system with 16-celled tubular spermatheca, in line with genital tract; columned uterus with either four rows of cells or multicelled. PUS length varies from very long to none at all. Female tails conoid not elongated. Male caudal alae short, leptoderan, sometimes long and peloderan.

Sperm cells most often with large amount of cytoplasm. Deirids and phasmids generally absent. Family contains fungus feeders, as well as facultative and obligate higher plant parasites, generally parasites of the above-ground parts of plants.

Genera: *Anguina* Scopoli, 1777
Halenchus Cobb, 1933
Ditylenchus Filipjev, 1936
Thada Thorne, 1941
Sychnotylenchus Ruhm, 1956
Pseudhalenchus Tarjan, 1958
Subanguina Paramonov, 1967
Cynipanguina Maggenti, Hart, and Paxman, 1974
Pterotylenchus Siddiqi and Lenne, 1984

Genus *incertae sedis*:
Chitinotylenchus Micoletsky, 1922

Family: Dolichodoridae

Large, slender, cylindroid nematodes. Cuticle distinctly annulated, lateral field with three to four lines. No deirids. Labial region distinctly offset and annulated. Amphid apertures small slits laterally or dorsoventrally directed. Stylet well developed up to 150 μm long. Esophagus with amalgamated prometacorpus, short isthmus, and a nonoverlapping pyriform glandular region. Female tail rounded to hemispherical with spikelike extension, rarely elongate-conoid. Females with two genital branches, columned uterus with four rows of cells. Male caudal alae winglike and lobed. Obligate migratory ectoparasites of plant roots.

Genera: *Dolichodorus* Cobb, 1914
Brachydorus de Guiran and Germani, 1968
Neodolichodorus Andrássy, 1976

Family: Belonolaimidae

Body is slender to robust. Labial region is high, rounded, ogival, or slightly flattened. Labial framework variable from poorly to very well developed. Amphid apertures dorsoventral slits located at the edge of the labial disk. Esophagus usually with a slender procorpus, rounded metacorpus, slim isthmus, and a glandular postcorpus that may overlap the intestine. Lateral field: two to six lines. External cuticle sometimes with longitudinal ridges. Female tails cylindroid to conoid, more than twice as long as wide, often thickened terminal cuticle. Phasmids confined to the tail. Females with two genital branches (exception: *Trophurus*); columned uterus with three rows of cells. Male caudal alae peloderan. Deirids present or absent. Obligate, generally migratory ectoparasites. Some taxa are capable of feeding as endoparasites of higher plants. There are two subfamilies.

Subfamily Belonolaiminae
Genera: *Belonolaimus* Steiner, 1949
Carphodorus Colbran, 1965
Morulaimus Sauer, 1966
Geocenamus Thorne and Malek, 1968

        Sauertylenchus Sher, 1974

Subfamily Telotylenchinae
  Genera: *Tylenchorhynchus* Cobb, 1913
         Trophurus Loof, 1956
         Trichotylenchus Whitehead, 1960
         Nagelus Throne and Malek, 1968
         Paratrophurus Arias, 1970
         Merlinius Siddiqi, 1970
         Triversus Sher, 1974
         Amphimerlinius Siddiqi, 1976

Genus *dubium*:
         Tetylenchus Filipjev, 1936

Family: Pratylenchidae

Body elongate, slender to greatly swollen. Labial region low, height less than 0.5 the diameter of the basal lip annulus, and generally with fewer than five annuli. Stylet strong but short, less than 2.5 times longer than the diameter of basal lip region annulus. Labial framework well developed, especially the cheilostomal cylinder and the basal plate. Esophageal glands overlap the intestine (exception: some *Pratylenchoides*). Deirids rare. Phasmids located on the tail. Tails generally more than two anal body diameters long. Male caudal alae peloderan. Females with one or two genital branches; when one the posterior branch forms a PUS. Columned uterus with three rows of cells. Sexual dimorphism may occur; either male feeding apparatus atrophied or females may be saccate. Obligate (migratory or sedentary) endoparasites of higher plants. There are two subfamilies.

Subfamily Pratylenchinae
  Genera: *Pratylenchus* Filipjev, 1936
         Radopholus Thorne, 1949
         Pratylenchoides Winslow, 1958
         Hoplotylus s'Jacob, 1960
         Zygotylenchus Siddiqi, 1963
         Hirschmanniella Luc and Goodey, 1964
         Apratylenchoides Sher, 1973

Subfamily Nacobbinae
  Genus: *Nacobbus* Thorne and Allen, 1944

Family: Hoplolaimidae

Females vermiform to kidney-shaped. Labial region higher than 0.5 the diameter of the basal lip annulus, with rounded or trapezoidal outline in lateral profile. Stylet strong, 2.5–3 times longer than the diameter of the basal lip annulus. Esophageal glands overlap the intestine (exception: *Pararotylenchus*). Females with two genital branches; posterior branch may be reduced to a PUS. Columned uterus with three rows of four cells. Lateral field generally with four lines. Phasmids generally located anterior to the anus; rarely are they on the tail. Tail two anal body diameters long or less. Tail profile generally exhibits pronounced dorsal curvature, sometimes hemispherical. Among the sedentary taxa the eggs are laid in a gelatinous matrix. In some taxa the males show sexual dimorphism manifested by a non-functional feeding apparatus. Caudal alae of male leptoderan. Gubernaculum with titillae. Obligate migratory ectoendoparasites (sometimes semiendosedentary parasites) of higher plants. There are two subfamilies.

Subfamily Hoplolaiminae
    Genera: *Hoplolaimus* von Daday, 1905
            Rotylenchus Filipjev, 1936
            Helicotylenchus Steiner, 1945
            Scutellonema Andrássy, 1958
            Aorolaimus Sher, 1963
            Aphasmatylenchus Sher, 1965
            Antarctylus Sher, 1973
            Pararotylenchus Baldwin and Bell, 1981

Subfamily Rotylenchulinae
    Genera: *Rotylenchulus* Lindford and Oliveira, 1940
            Acontylus Meagher, 1968
            Senegalonema Germani, Luc, and Baldwin, 1984

Family: Heteroderidae

Body vermiform and slender in juveniles, robust-vermiform in males, and always swollen in mature females. Labial framework usually well developed, especially in juveniles and males. Valved metacorpus generally large, postcorporal glands always overlap the anterior intestine. Somatic cuticle of males and juveniles annulated; female body ornamentation varies from annuli to reticulations. Females swollen, with two amphidelphic "prodelphic" genital branches; columned uterus with three rows of cells. Eggs are either laid in a gelatinous matrix or retained inside the female body (cyst: tanned female). Males lack caudal alae and the cloacal opening is nearly terminal (exception: *Bursadera*). Sedentary obligate parasites of roots, forming galls in some cases. There are three subfamilies.

Subfamily Heteroderinae
    Genera: *Heterodera* Schmidt, 1871
            Meloidodera Chitwood, Hannon, and Esser, 1956
            Globodera Skarbilovich, 1959
            Cryphodera Colbran, 1966
            Atalodera Wouts and Sher, 1971
            Sarisodera Wouts and Sher, 1971
            Punctodera Mulvey and Stone, 1976
            Cactodera Krall' and Krall', 1978
            Hylonema Luc, Taylor, and Cadet, 1978
            Thecavermiculatus Robbins, 1978
            Dolichodera Mulvey and Ebsary, 1980
            Verutus Esser, 1981
            Rhizonema Cid del Prado Vera, Lownsbery, and Maggenti, 1983

Afenestrata Baldwin and Bell, 1985
Bellodera Wouts, 1985

Subfamily Meloidogyninae
  Genus: *Meloidogyne* Goeldi, 1892

  Genus *dubium*: Meloidoderella

Subfamily Nacobboderinae
  Genera: *Meloinema* Choi and Geraert, 1974
    Nacobbodera Golden and Jensen, 1974
    Bursadera Ivanova and Krall', 1985

Superfamily 2: Criconematoidea

Labial region poorly developed; represented by a labial disk often with four submedian lobes. Prometacorpus generally amalgamated, postcorpus pyriform, and clearly set off from the intestine (exception: *Sphaeronema whittoni*). Females have a single anteriorly directed genital branch and there is no evidence of a postuterine sac. Columned uterus not defined. Males and in some taxa specific juvenile stages have the stylet reduced or nonfunctional. Rarely do males have a caudal alae. Deirids mostly absent, phasmids unknown.

Family: Criconematidae

Female body vermiform to sausage-shaped. Cuticle thick, in some cases double, lacking a lateral field in females. Annuli vary from rounded with or without extracuticular layer to retrorse with lobation, crenation, scales, fringe, or spines. Deirids absent. Labial region with or without submedian lobes; internal framework well developed. Stylet massive, conus (cone) much longer than the shaft and knobs. Esophagus mostly with amalgamated prometacorpus, followed by a short isthmus and a small pyriform glandular region. The stylet is lacking in mature males; caudal alae uncommon but when present may be well developed. The lateral field may consist of two, three, or four lines. Obligate migratory to nearly sedentary ectoparasites of plant roots; some taxa induce terminal root galls. There are two subfamilies.

Subfamily Criconematinae
  Genera: *Criconema* Hofmanner and Menzel, 1914
    Ogma Southern, 1914
    Hemicriconemoides Chitwood and Birchfield, 1957
    Bakernema Wu, 1964
    Criconemella De Grisse and Loof, 1965
    Discocriconemella De Grisse and Loof, 1965
    Nothocriconemoides Maas, Loof, and De Grisse, 1971
    Blandicephalanema Mehta and Raski, 1971
    Pateracephalanema Mehta and Raski, 1971

  Genera *dubia*:
    Macroposthonia de Man, 1921
    Criconemoides Taylor, 1936

Subfamily Hemicycliophorinae
  Genera: *Hemicycliophora* de Man, 1921
            Caloosia Siddiqi and Goodey, 1964

Family: Tylenchulidae

Female body slender, swollen, or globose. Cuticle thin and annulated except on swollen forms, which may have fine punctations or minute tubercles. Lateral field present but not visible on swollen females. Internal labial framework weak and the stylets are most often delicate but may be extremely long. Esophagus with an amalgamated prometacorpus or the procorpus may be slightly swollen but the metacorpus is distinct and followed by the isthmus and posterior glandular bulb. The feeding apparatus of males is degenerate; in some taxa this is also true of specific juveniles. Males do not have caudal alae. Deirids rarely present. These nematodes are near sedentary to sedentary obligate parasites of plant roots. The sedentary forms induce the plant root to form nutritive cells. There are three subfamilies.

Subfamily Tylenchulinae
  Genera: *Tylenchulus* Cobb, 1913
            Sphaeronema Raski and Sher, 1952
            Trophotylenchulus Raski, 1957
            Trophonema Raski, 1957
            Meloidoderita Pogosyan, 1966

Subfamily Paratylenchinae
  Genera: *Paratylenchus* Micoletzky, 1922
            Cacopaurus Thorne, 1943
            Gracilacus Raski, 1962

Subfamily Tylenchocriconematinae
  Genus: *Tylenchocriconema* Raski and Siddiqi, 1975

Suborder 2: Aphelenchina

Sensu
William R. Nickle and David J. Hooper

The labial cap is usually distinct and often set off by a definite constriction. The hollow axial spear is seldom strongly developed and basally there may or may not be knobs or thickenings. Esophagus almost always with a large valved metacorpus, often squarish in outline. All glands open into the metacorpus; the dorsal gland opens anterior to the valve and the subventrals posterior to the valve. With the exception of one genus, the glands overlap the anterior intestine. Females have a single, anteriorly directed ovary; a postuterine sac may or may not be present. When present it functions as a spermatheca. Vulva always located posteriorly. Males sometimes possess caudal alae; when present the genital papillae form alal rays; there are always two or more pairs of caudal papillae. The spicules are most commonly thornlike in shape; however, in a few taxa they may be slender and slightly arcuate. When a gubernaculum is present it is reportedly forked.

Included in the suborder are mycetophagous forms, higher plant parasites (generally they attack the above ground parts of plants), predators, and obligate insect parasites.

The suborder contains five families: Aphelenchidae, Paraphelenchidae, Aphelenchoididae, Seinuridae, Entaphelenchidae. Parasites of higher plants are found only in the family Aphelenchoididae.

Family: Aphelenchoididae

The labial cap is evident and the lip region is often expanded and set off by a constriction. The lips are amalgamated but distinguished by the faint cephalic framework. The stylet is usually slight and basal knobs may or may not be present. The esophagus lacks an isthmus; the esophagus joins the intestine immediately posterior to the metacorporal bulb and the elongate glands overlap the intestine dorsally from this juncture. The spicules are robust and due to the development of the rostrum they are characteristically rose thorn-shaped; a gubernaculum is generally absent. Bursa, when present, an insignificant lobe on the tail tip. There are three subfamilies.

Subfamily Aphelenchoidinae
    Genus: *Aphelenchoides* Fischer, 1894

Subfamily Bursaphelenchinae
    Genus: *Bursaphelenchus* Fuchs, 1937

Subfamily Rhadinaphelenchinae
    Genus: *Rhadinaphelenchus* J. B. Goodey, 1960

Suborder 3: Sphaerulariina

Generally there are three distinct life cycles; two free living and one parasitic. The free-living cycles take place in the environs of the immature hosts; the parasitic cycle occurs in the host's hemocele. The free-living female has a stylet but the metacorpus lacks a valve. The dorsal gland orifice opens into the corpus; however, it may be some distance posterior to the stylet base. The elongated glands overlap the anterior intestine. In the free-living cycle the female reproductive system consists of a small and fingerlike anteriorly directed ovary with few developing oocytes. The short oviduct is followed by a prominent uterus, filled with sperm in the infective stage adult female. Within the host the parasitic female either becomes grossly enlarged and degenerates to a reproductive sac, or the uterus prolapses and gonadal development takes place outside the female's body. The oocytes may be arranged around a rachis. In some instances the prolapsed uterus/gonad is 30 times longer than and 300 times the volume of the female's body. Males are always free living. The males are usually longer than the freeliving females and the stylet is often absent or obscure as is the degenerate esophagus. The spicules are not always typically tylenchoid; they may be elaborate and accompanied by a gubernaculum. When present the caudal alae is peloderan.

There are four families in the suborder: Sphaerulariidae, Allantonematidae, Iotonchiidae, and Fergusobiidae. All are hemocele parasites of insects and mites. Fergusobiidae contains insect/plant parasites.

Family: Fergusobiidae

The labial cap is set off from the general body contour. The stylet is well developed and has three basal knobs. The dorsal esophageal gland opens near the base of the stylet. The sausagelike female is dorsally curved. Male spicules are robust-knobbed and arcuate. There is a narrow, peloderan caudal ala.

These nematodes have two life cycles, one gametogenic and one parthenogenic. Impregnated gametogenic females are the infective stage.

Genus: *Fergusobia* Currie, 1937

III. CLASSIFICATION OF NEMATA

Phylum Nemata

Class 1: Adenophorea
  Subclass A: Enoplia
    Superorder 1: Marenoplica
      Order 1: Enoplida
        Superfamily 1: Oxystominoidea
          Families: Paraoxystominidae
                     Oxystominidae
                     Alaimidae
        Superfamily 2: Enoploidea
          Families: Enoplidae
                     Lauratonematidae
                     Leptosomatidae
                     Phanodermatidae
                     Thoracostomopsidae
      Order 2: Oncholaimida
          Families: Oncholaimidae
                     Eurystominidae
                     Symplocostomatidae
      Order 3: Tripylida
        Suborder 1: Tripylina
          Families: Tripylidae
                     Prismatolaimidae
        Suborder 2: Ironina *inquirenda*
          Family: Ironidae
    Superorder 2: Terrenoplica
      Order 1: Isolaimida *inquirenda*
          Family: Isolaiimidae
      Order 2: Mononchida
        Superfamily 1: Mononchoidea
          Families: Mononchidae
                     Mylonchulidae
                     Cobbonchidae
                     Anatonchidae
                     Iotonchulidae

          Superfamily 2: Bathyodontoidea
            Families: Bathyodontidae
                    Mononchulidae
Order 3: Dorylaimida
   Suborder 1: Dorylaimina
      Superfamily 1: Dorylaimoidea
         Families: Dorylaimidae
               Encholaimidae
               Tylencholaimidae
               Tylencholaimellidae
               Leptonchidae
               Belonenchidae
               Longidoridae
           Genera: see text
      Superfamily 2: Actinolaimoidea
         Families: Actinolaimidae
               Brittonematidae
               Carcharolaimidae
               Trachypleurosidae
      Superfamily 3: Belondiroidea
         Families: Belondiridae
               Roqueidae
               Dorylaimellidae
               Oxydiridae
               Mydonomidae
   Suborder 2: Diptherophorina
         Families: Diphtherophoridae
               Trichodoridae
   Suborder 3: Nygolaimina
         Families: Nygolaimidae
               Nygolaimellidae
               Aetholaimidae
               Campydoridae
Order 4: Stichosomida
      Superfamily 1: Trichocepaloidea
         Families: Trichuridae
               Trichinellidae
               Trichosyringidae
      Superfamily 2: Mermithoidea
         Families: Mermithidae
               Tetradonematidae
      Superfamily 3: Echinomermelloidea *incertae sedis*
         Families: *incertae sedis*
               Enchinomermellidae
               Marimermithidae
               Benthimermithidae
Subclass B: Chromadoria
   Order 1: Chromadorida
      Superfamily 1: Chromadoroidea

     Families: Chromadoridae
         Hypodontolaimidae
         Microlaimidae
         Spirinidae
         Cyatholaimidae
         Comesomatidae
    Superfamily 2: Choanolaimoidea
     Families: Choanolaimidae
         Selachinematidae
         Ethmolaimidae
   Order 2: Desmoscolecida
     Families: Desmoscolecidae
         Greeffiellidae
   Order 3: Desmodorida
    Superfamily 1: Desmodoroidea
     Families: Desmodoridae
         Ceramonematidae
         Monoposthiidae
    Superfamily 2: Draconematoidea
     Families: Draconematidae
         Epsilonematidae
         Prochaetosomatidae
   Order 4: Monhysterida
     Families: Linhomoeidae
         Siphonolaimidae
         Monhysteridae
         Scaptrellidae
         Sphaerolaimidae
         Xyalinidae
         Meyliidae
   Order 5: Araeolaimida
    Suborder 1: Araeolaimina
     Families: Axonolaimidae
         Camacolaimidae
         Tripyloididae
    Superfamily 1: Araeolaimoidea
     Families: Araeolaimidae
         Cylindrolaimidae
         Diplopeltidae
         Rhabdolaimidae
    Superfamily 2: Plectoidea
     Families: Plectidae
         Leptolaimidae
         Haliplectidae
         Bastianidae

Class 2: Secernentea
 Subclass A: Rhabditia
     Myenchidae: *Familium incertae sedis*; *dubium*

Order 1: Rhabditida
   Suborder 1: Rhabditina
      Superfamily 1: Rhabditoidea
         Families: Rhabditidae
                   Rhabditonematidae
                   Odontorhabditidae
                   Steinernematidae
                   Rhabdiasidae
                   Angiostomatidae
                   Agfidae
                   Strongyloididae
                   Syrphonematidae
                   Heterorhabditidae
                   Carabonematidae
                   Pseudodiplogasteroididae
      Superfamily 2: Bunonematoidea
         Families: Bunonematidae
                   Pterygorhabditidae
      Superfamily 3: Cosmocercoidea
         Families: Cosmocercidae
                   Atractidae
      Superfamily 4: Oxyuroidea
         Families: Oxyuridae
                   Thelastomatidae
                   Rhigonematidae
      Superfamily 5: Heterakoidea
         Families: Heterakidae
                   Ascaridiidae
   Suborder 2: Cephalobina
         Families: Cephalobidae
                   Robertiidae
                   Chambersiellidae
                   Elaphonematidae
      Superfamily 1: Panagrolaimoidea
         Families: Panagrolaimidae
                   Alirhabditidae
                   Brevibuccidae
Order 2: Strongylida
         Families: Diaphanocephalidae
                   Metastrongylidae
                   Maupasinidae
      Superfamily 1: Strongyloidea
         Families: Strongylidae
                   Cloacinidae
                   Syngamidae
      Superfamily 2: Ancylostomatoidea
         Families: Ancylostomatidae
                   Uncinariidae
                   Globocephalidae

Superfamily 3: Trichostrongyloidea
    Families: Trichostrongylidae
                 Amidostomatidae
                 Strongylacanthidae
                 Heligmosomatidae
                 Ollulanidae
                 Dictyocaulidae

Subclass B: Spiruria
  Order 1: Ascaridida
    Superfamily 1: Ascaridoidea
        Families: Ascarididae
                 Toxocaridae
                 Anisakidae
                 Acanthocheilidae
                 Goeziidae
                 Crossophoridae
                 Heterocheilidae
    Superfamily 2: Seuratoidea
        Families: Seuratidae
                 Schneidernematidae
                 Quimperiidae
                 Subuluridae
                 Cucullanidae
    Superfamily 3: Camallanoidea
        Families: Camallanidae
                 Anguillicolidae
    Superfamily 4: Dioctophymatoidea
        Families: Dioctophymatidae
                 Soboliphymidae
    Superfamily 5: Muspiceoidea *incertae sedis*
        Families: Muspiceidae
                 Robertdollfusidae
                 Phlyctainophoridae
  Order 2: Spirurida
    Superfamily 1: Spiruroidea
        Families: Spiruridae
                 Thelaziidae
                 Acuariidae
                 Hedruridae
                 Tetrameridae
    Superfamily 2: Drilonematoidea
        Families: Drilonematidae
                 Ungellidae
                 Scolecophilidae
                 Creagrocericidae
                 Mesidionematidae
                 Homungellidae
    Superfamily 3: Physalopteroidea
        Families: Physalopteridae

                Megalobatrachonematidae
                Gnathostomatidae
      Superfamily 4: Dracunculoidea
        Families: Dracunculidae
                Philometridae
                Micropleuridae
      Superfamily 5: Diplotriaenoidea
        Families: Diplotriaenidae
                Oswaldofilariidae
      Superfamily 6: Filarioidea
        Families: Filariidae
                Aproctidae
                Setariidae
                Desmidocercidae
                Onchocercidae
Subclass C: Diplogasteria
    Order 1: Diplogasterida
        Families: Diplogasteridae
                Odontopharyngidae
                Diplogasteroididae
                Cylindrocorporidae *insertae sedis*
    Order 2: Tylenchida
      Suborder 1: Tylenchina
        Superfamily 1: Tylenchoidea
          Families: see text for genera
                 Tylenchidae
                 Anguinidae
                 Dolichodoridae
                 Belonolaimidae
                 Pratylenchidae
                 Hoplolaimidae
                 Heteroderidae
        Superfamily 2: Criconematoidea
          Families: see text for genera
                 Criconematidae
                 Tylenchulidae
      Suborder 2: Aphelenchina
        Families: see text for genera
               Aphelenchidae
               Paraphelenchidae
               Aphelenchoididae
               Seinuridae
               Entaphelenchidae
      Suborder 3: Sphaerulariina
        Families: Sphaerulariidae
               Allantonematidae
               Iotonchidae
               Fergusobiidae

REFERENCES

Allen, M.W., and Sher, S. 1967. Taxonomic problems concerning the phytoparasitic nematodes. *Ann. Rev. Phytopathol.* 5: 247–264.

Andrássy, I. 1974. A nematodák evolúciója és rendszerezése. *MTA Biol. Oszt. Köal.* 17: 13–58.

Andrássy, I 1976. *Evolution as a Basis for the Systematization of Nematodes.* Pitman, London.

Chitwood, B. G. 1937. A revised classification of the Nematoda. *Papers in helminthology, 30 year Jubileum K. I. Skrjabin, Moscou,* pp. 69–80.

Chitwood, B. G. 1950. General structure of nematodes. In *Introduction to Nematology,* B. G. Chitwood and M. B. Chitwood, eds. Monumental Printing, Baltimore, Chap. 2, pp. 7–27.

Chitwood, B. G. 1958. The designation of official names for higher taxa of invertebrates. *Bull. Zool. Nomencl.* 15: 860–895.

Cobb, N. A. 1919. The orders and classes of nemas. *Contrib. Sc. Nematol.* 8: 213–216.

De Coninck, L., Theodorides, J., Roman, E., Ritter, M., and Chabaud, A. G. 1965. Systématique des nématodes. In *Traité de Zoologie,* P.-P. Grassé, ed. Masson, Paris, pp. 586–1200.

Eberth, C. J. 1863. Über *Myoryctes weissmanni* einen neuen Parasiten des Froschmuskels. *Z. Wiss. Zool.* 12: 530–535.

Esser, R.P. 1979. Phytoparasitic genera and species. *Nematol. News Lett.* 25(4): 13–14.

Esser, R.P. 1987. Checklist of phytoparasitic nematode genera, subgenera, and the number of species described in each taxon. *Nematol. News Lett.* 33(2): 7–8.

Filipjev, I. N. 1934. The classification of the free-living nematodes and their relation to the parasitic nematodes. *Smithson. Misc. Coll.* 89: 1–63.

Fotedar, D. N., and Handoo, Z. A. 1978. A revised scheme of classification to order Tylenchida Thorne, 1949 (Nematoda). *J. Sci. Univ. Kashmir.* 3: 55–82.

Gadea, E. (1973). Sobre la filogenia interna de los nematodos. *P. Inst. Biol. Apll., Barcelona.* 15: 87–92.

Golden, A. M. 1971. Classification of the genera and higher categories of the order Tylenchida (Nematoda). In *Plant Parasitic Nematodes. Vol. 1, Morphology, Anatomy, Taxonomy, and Ecology,* Zuckerman, Mai, and Rohde, eds. Academic Press, New York, pp. 191–232.

Goodey, J. B. 1963. *Soil and Freshwater Nematodes,* by T. Goodey, rewritten. Methuen and Co., London.

Grobben, K. 1909. Die systematische Einteilung des Thierreichs Verhandl. *K. K. Zool. Bot. Gesellsch. Wien (1908)* 58: 491–511.

Hyman, L. H. 1951. The Invertebrates: Acanthocephala, Aschelminthes and Entoprocta. The Pseudocoelomate Bilateria. *Vol. 3. McGraw-Hill, New York.*

Inglis, W. G. 1983. An outline of the phylum Nematoda. *Aust. J. Zool.* 31: 243–255.

Lorenzen, S. 1983. Phylogenetic systematics: Problems, achievements and its application to the Nematoda. In *Concepts in Nematode Systematics,* A. Stone, H. Platt, and L. Khalil, eds. Academic Press, London, pp. 11–23.

Luc, M., Maggenti, A. R., Fortuner, R., Raski, D. J., and Geraert, E. 1987. A Reappraisal of Tylenchina (Nemata). 1. For a new approach to the taxonomy of Tylenchina. *Revue Nématol.* 10: 127–134.

Maggenti, A. R. 1961. Morphology and biology of the genus *Plectus. Proc. Helminthol. Soc. Wash.* 28: 118–130.

Maggenti, A. R. 1963. Comparative morphology in nemic phylogeny. In *The Lower Metazoa: Comparative Biology and Phylogeny,* E. C. Dougherty, ed. University of California Press, Berkeley, pp. 273– 282.

Maggenti, A. R. 1971. Nemic Relationships and origins of plant parasitic nematodes. In *Plant Parasitic Nematodes, Vol. 1, Morphology, Anatomy, Taxonomy, and Ecology,* Zuckerman, Mai, and Rohde, eds. Academic Press, New York, pp. 65–81.

Maggenti, A. R. 1978. Influence of morphology, biology, and ecology on evolution of parasitism in Nematoda. In *Biosystematics in Agriculture,* Romberger, ed. John Wiley and Sons, New York, pp. 173–191.

Maggenti, A. R. 1981. Nematodes: Development as plant parasites. *Ann. Rev. Microbiol.* 35: 135–154.

Maggenti, A. R. 1982. Nemata. In *Synopsis and Classification of Living Organisms*, S. P. Parker, McGraw-Hill, New York, pp. 878–923.

Maggenti, A. R. 1983. Nematode higher classification as influenced by species and family concepts. In *Concepts in Nematode Systematics*, A. Stone, H. Platt, and L. Khalil, eds. Academic Press, London, pp. 25–40.

Maggenti, A. R., Luc, M., Raski, D. J., Fortuner, R., and Geraert, E. 1987. A reappraisal of Tylenchina (Nemata). 2. Classification of the suborder Tylenchina (Nemata: Diplogasteria). *Revue Nématol.* 10: 135–142.

Orley, L. 1880. Az anguillulidak maganrajza. A kir. m. termeszettudom. tersulat altal a bugatdijjal jutalmazott palymii. *Termeszetr. Fuz.* 4: 16–150.

Paramanov, A. A. 1967. A critical review of the suborder Tylenchina (Filipjev, 1934) (Nematoda: Secernentea). *Akad. Nauk SSSR Trudy gel'mint. Lab.* 18: 78–101.

Paramanov, A. A. 1968. Principles of ecological and morphological analysis of the classification of Tylenchida. *Izv. Acad. Nauk SSSR, Ser. Biol.* 6: 793–801.

Pearse, A. S. 1936. *Zoological names, a list of phyla, classes, and orders: Sect. F. A. A. A. S.*, Duke Univ. Press, Durham, North Carolina, pp. 1–24.

Pereira, C. 1931. *Myenchus bothelhoi* n. sp., curioso nematoide parasito de *Limnobdella brasiliensis* Pinto (Hirudinea). *Thèse Fac. Màd. São Paulo*, p. 29.

Poinar, G. O. Jr. and Hess, R. 1974. Structure of the preparasitic juveniles of *Filipjevimermis leipsandra* and some other Mermithidae (Nematoda). *Nematologica* 20: 163–173.

Richter, S. 1971. Zum Feinbau von Mermithiden (Nematoda). I. Der Bohrappart der vorparasitischen Larven von *Hydromeris contorta* (Linstow, 1889) Hagmeier, 1912. *Z. Parasitkde.* 36: 32–50.

Siddiqi, M. R. 1980. The origin and phylogeny of the nematode orders Tylenchida Thorne, 1949 and Aphelenchida n. ord. *Helminth. Abstr.* 49: 143–170.

Siddiqi, M. R. 1986. *Tylenchida Parasites of Plants and Insects*. Slough, U.K., Commonw. Inst. Parasit., p. 645.

IV
PLANT PARASITIC NEMATODES

6

Root-Knot Nematodes: *Meloidogyne* Species and Races

Jonathan D. Eisenback   *Virginia Polytechnic Institute and State University, Blacksburg, Virginia*

Hedwig Hirschmann Triantaphyllou   *North Carolina State University, Raleigh, North Carolina*

I. INTRODUCTION

A. Agricultural Importance

Root-knot nematodes, *Meloidogyne* spp., are economically important plant pathogens and distributed worldwide. They are obligate parasites and parasitize thousands of different plant species including monocotyledons, dicotyledons, and herbaceous and woody plants. Species of *Meloidogyne* are pests of major food crops, vegetables, fruit, and ornamental plants grown in tropical, subtropical, and temperate climates. They reduce the yields as well as the quality of the produce. Populations of certain species occur as physiological races (e.g., *M. incognita*, Table 1) with varied abilities to reproduce on key differential host plants (Sasser, 1972; Hartman and Sasser, 1985).

At the end of the calendar year 1988, the genus included more than 60 species (61 species and two subspecies). Four species are major pests and are distributed widely in agricultural areas around the world (Table 2). Another seven species are important in general but are more limited in their distribution and host range. Several other species may cause damage to a few crops or are prevalent in specific agricultural regions. The remaining species may be more specialized and attack only one or a few hosts and are limited in their distribution.

B. Life Cycle

Root-knot nematodes display marked sexual dimorphism, i.e., the females are pyriform or saccate, the males vermiform (Figs. 1A,B, 2). These general differences in body shape between female and male become established during the postembryonic development of *Meloidogyne* (Fig. 3). The embryonic development results in the first-stage juvenile which molts once in the egg and hatches as a second-stage juvenile. This motile, vermiform, infective stage migrates through the soil and enters the root of a suitable host plant.

TABLE 1   Differential Host Test Reactions for the Identification
of the Widely Recognized Host Races of *M. incognita*

Meloidogyne incognita race	Cotton Deltapine 61	Tobacco NC95
Race 1	– a	–
Race 2	–	+
Race 3	+	–
Race 4	+	+

a +indicates reproduction by the nematode; – indicates no or little
reproduction.

It moves through the plant tissue to a preferred feeding site and establishes a complex host–parasite relationship with the plant. The second-stage juvenile becomes sedentary and as it feeds on special nurse cells (giant cells), it undergoes more morphological changes. It becomes flask-shaped and without further feeding molts three times into the third- and fourth-stage juvenile, and finally becomes an adult. Shortly after the last molt the saccate adult female resumes feeding and continues to do so for the remainder of her life. During this postembryonic development, the reproductive system develops and grows into functional gonads (Figs. 4 and 5) (Triantaphyllou and Hirschmann, 1960). The sexes can be distinguished on the basis of the number of gonads: females (Fig. 4) always have two gonads; males (Fig. 5) usually have only one. The change in shape from the saccate male juvenile to the vermiform adult male takes place during the fourth juvenile stage. At that time, the juvenile male undergoes a form of metamorphosis in which the body elongates from the saccate to a vermiform shape (Fig. 5f). The fourth-stage male is enclosed within the second- and third-stage cuticles and, after the final molt is completed, emerges as a fully developed male (Fig. 5f). The adult male does not feed. It will leave the root and move freely through the soil (Fig. 3). Depending on the type of mode of reproduction of the particular species, amphimixis or parthenogenesis, the male may search for a female and mate, or remain in the soil and finally die. The length of the life cycle of root-knot nematodes is greatly influenced by temperature. For *M. incognita* on tomato at approximately 29°C, the first adult females appear 13–15 days after root penetration; the first egg-laying females are found 19–21 days after penetration (Triantaphyllou and Hirschmann, 1960). The life span of egg-producing females may extend from 2 to 3 months, but that of males may be much shorter.

C. Host–Parasite Relationships

Root-knot nematodes usually cause the formation of knots or galls on roots of susceptible host plants (Figs. 1D, 6, 7A). The second-stage infective juveniles are attracted to host roots (Fig. 3). They accumulate in the regions of the apical meristem, cell elongation, and near points of emergence of lateral roots. The juveniles usually enter the roots behind the root cap. Penetration involves mechanical action of the stylet (Lindford, 1942) as well as enzymatic action (cellulolytic or pectolytic) through certain esophageal gland secretions (Bird et al., 1975). After penetration, the second-stage juveniles migrate intercellularly in the cortex to the region of cell differentiation where they settle and begin feeding. Their heads lie at the periphery of the vascular parenchyma tissue, and the remainder of their bodies in the cortex

TABLE 2 Distribution of Root-Knot Nematodes, *Meloidogyne* Species, by Continent and Order of Economic Importance

Cont.	North America	South America	Africa	Europe	Asia	Australia
Major pests	*M. incognita* *M. javanica* *M. arenaria* *M. hapla*	*M. incognita* *M. javanica* *M. arenaria* *M. hapla*	*M. incognita* *M. javanica* *M. arenaria* *M. hapla* *M. acronea* *M. artiellia*	*M. incognita* *M. javanica* *M. arenaria* *M. hapla* *M. naasi*	*M. incognita* *M. javanica* *M. arenaria* *M. hapla* *M. graminicola*	*M. incognita* *M. javanica* *M. arenaria* *M. hapla* *M. naasi*
Important pests	*M. chitwoodi* *M. graminicola*	*M. exigua* *M. kikuyensis*				
Important pests in some locations	*M. microtyla* *M. graminis* *M. naasi*	*M. coffeicola* *M. oryzae* *M. salasi*	*M. africana* *M. decalineata* *M. litoralis*	*M. ardenensis*	*M. brevicauda* *M. mali* *M. camelliae*	

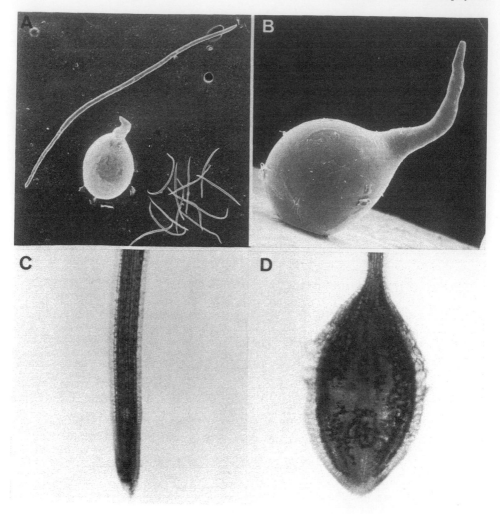

FIGURE 1   Micrographs of signs and symptoms of root-knot nematodes (*Meloidogyne* spp.). (A) Scanning electron micrograph (SEM) of male, female, and second-stage juveniles, respectively. (B) SEM of a *Meloidogyne* female. (C) Light micrograph (LM) of a normal tomato root tip. (D) LM of a tomato root tip 4 days after infection with root-knot nematode second-stage juveniles.

parallel to the long axis of the root. Preferred feeding sites are primary phloem or adjacent undifferentiated parenchymalike cells of the pericycle. In response to the feeding of the second-stage juvenile, the host tissue undergoes pronounced morphological and physiological changes (Fig. 1C,D). Some parenchyma cells develop into permanent nurse cells for the nematode. They become hypertrophied, multinucleate "giant cells" (Fig. 7D), possibly as a result of the introduction of secretions produced by the subventral esophageal gland cells of the feeding second-stage juvenile (Dropkin, 1972; Hussey, 1987). The nematodes drive nourishment from these specialized nutritive cells and cannot continue their development to adulthood without them. Males apparently do not feed as adults and leave the roots, but

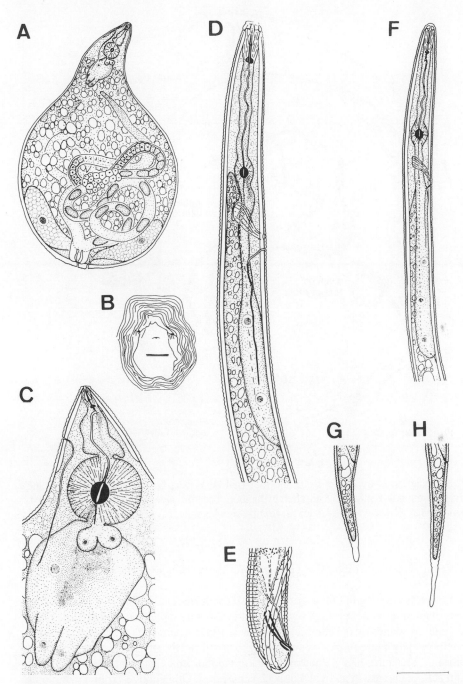

FIGURE 2 Line drawings of females, males, and second-stage juveniles of *Meloidogyne* spp. (A) Entire female. (B) Perineal pattern. (C) Anterior end including the esophagus. (D) Anterior end of male. (E) Tail of male. (F) Anterior end of second-stage juvenile. (G) Tail of second-stage juvenile. (H) Tail of second-stage juvenile of "graminis group." Bar = 100 μm for A and 30 μm for B–H. (After Eisenback, 1989.)

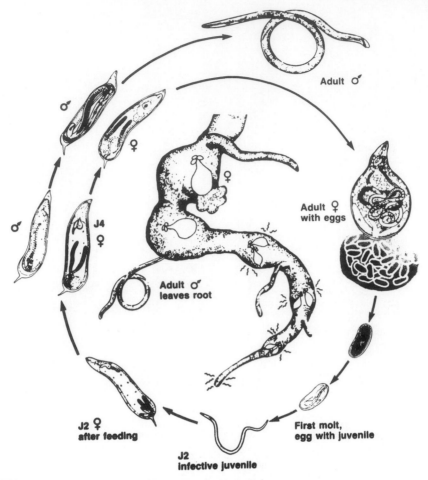

FIGURE 3   Diagram of the life cycle of the root-knot nematode, *Meliodogyne*. The third and fourth juvenile stages do not feed and are short in duration. The third juvenile stage is not illustrated. Abbreviations: J2, second-stage juvenile; J4, fourth-stage juvenile. (Courtesy of H.D. Shew.)

females continue feeding at the same giant cells for the remainder of their life spans (Fig. 7B–D).

Giant cells are essential for a successful host–parasite relationship. Their active metabolism is maintained either through secretions of the dorsal esophageal gland (Bird, 1968) or the removal of solutes by the adult female (Jones, 1981). Giant cells are essentially food transfer cells passing nutrients to the nematode (Jones and Northcote, 1972), which acts as a metabolic sink (McClure, 1977). Photosynthates are mobilized to the giant cells in the roots and, as a result, plant growth and yield may be suppressed. Other above-ground symptoms of infected plants include chlorosis of foliage and temporary wilting during periods of water stress. Nutrient and water absorption are greatly reduced by the damaged, galled root system (Fig. 7A).

The mechanism of giant cell formation has been intensely debated. Based on chromosome numbers in giant cells, it was suggested that they are formed through repeated en-

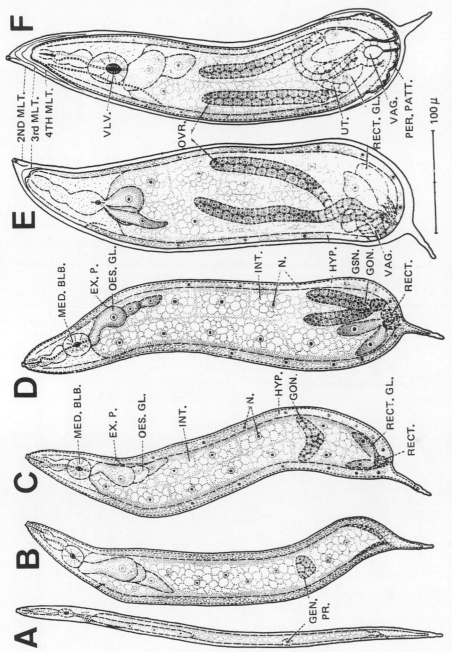

FIGURE 4 Development of a female of *M. incognita* from second-stage juvenile to adult. (A) Second-stage infective juvenile. (B) Swollen, sexually undifferentiated second-stage juvenile. (C) Early second-stage juvenile differentiating into a female. (D) Second-stage female juvenile shortly before second molt. (E) Fourth-stage female juvenile. (F) Adult female shortly after fourth molt. Abbreviations: EX. P., excretory pore; GEN. PR., genital primordium; GON., gonad; HYP., hypodemis; INT., intestine; MED. BLB., median bulb; 2nd MLT., second molt; 3rd MLT., third molt; 4th MLT., fourth molt; N, nucleus; OES. GL., esophageal glands; OVR., ovary; PER. PATT., perineal pattern; RECT. GL., rectal glands; RECT., rectum; UT., uterus; VAG., vagina; VLV., valve. (After Triantaphyllou and Hirschmann, 1960.)

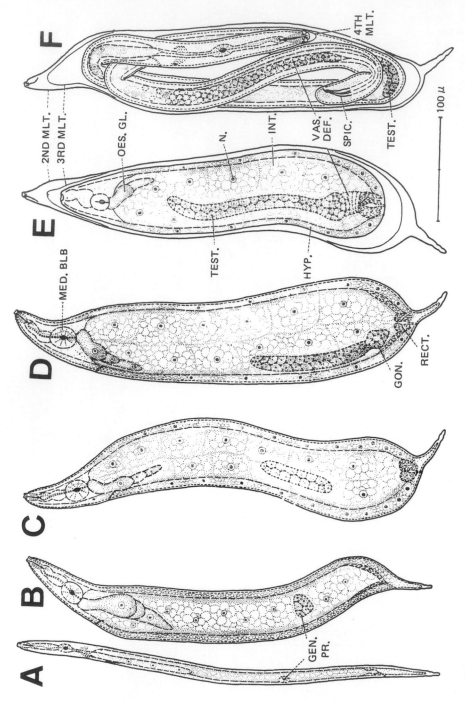

FIGURE 5   Development of a male of *Meloidogyne incognita* from second-stage juvenile to adult. (A) Second-stage infective juvenile. (B) Swollen, sexually undifferentiated second-stage juvenile. (C) Early second-stage juvenile differentiating into a male. (D) Second-stage juvenile male shortly before second molt. (E) Early fourth-stage juvenile male. (F) Adult male shortly after fourth molt. Abbreviations: GEN. PR., genital primordium; GON., gonad; HYP., hypodermis; INT., intestine; MED. BLB., median bulb; 2nd MLT., second molt; 3rd MLT., third molt; 4th MLT., fourth molt; N., nucleus; OES. GL., esophageal glands; RECT., rectum; SPIC., spicule; TEST., testis; VAS. DEF., vas deferens. (After Triantaphyllou and Hirschmann, 1960.)

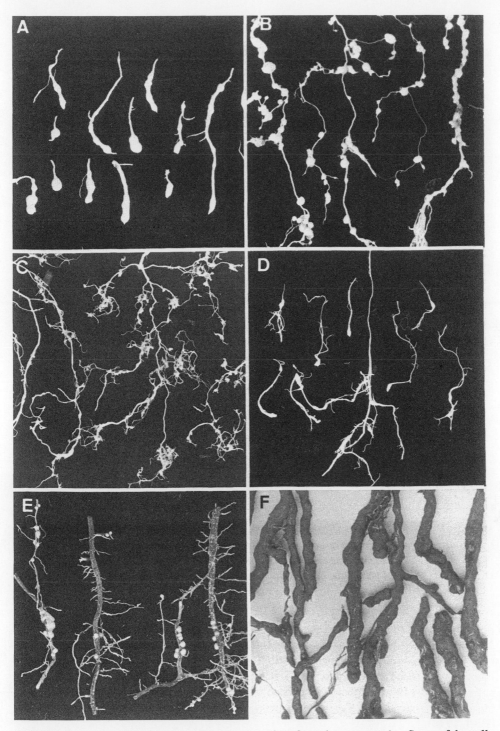

FIGURE 6   Roots with galls caused by various species of root-knot nematodes. Some of the galls are typical and may be helpful for species identification. (A) *M. exigua* on coffee: large terminal galls. (B) *M. arenaria* on tomato: smooth, round, bead-like galls. (C) *M. hapla* on tomato: small, discrete galls associated with proliferation of rootlets. (D) *M. naasi* on barley: elongated galls merging gradually with root; sometimes these galls affect one side of the cortex more than the other, causing the root to appear hooked. (E) *M. kikuyensis* on sugarcane: galls resembling *Rhizobium* nodules. (F) *M. brevicauda* on tea: many large, elongate, coalesced galls. (F courtesy of N. Gnanapragasam.)

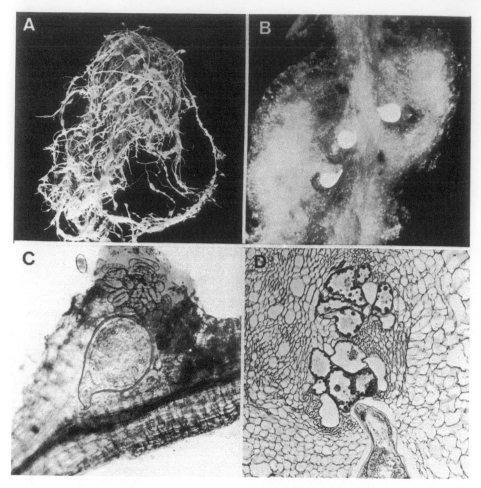

FIGURE 7   Roots infected with root-knot nematodes, *Meloidogyne* spp. (A) Entire root system of tomato. (B) Gall teased apart to reveal three root-knot nematode females. (C) Gall seen with transmitted light to show root–knot nematode female and egg mass in vivo. (D) Light micrograph of a cross-section of a root-gall showing several giant cells (specialized feeding sites).

domitoses with cytokineses, i.e., they are considered expansions of single cells (Huang and Maggenti, 1969). With the use of advanced electron microscope techniques, no evidence of cell wall dissolution and subsequent fusion of giant cells was demonstrated (Jones and Payne, 1978; Jones and Dropkin, 1976). In spite of these findings, however, the possibility still exists that cell wall dissolution and cell fusion could be involved in giant cell formation in some host–parasite relationships depending on the nematode and host species involved (Bird, 1974).

Root tissues around the nematode and the giant cells undergo hyperplasia and hypertrophy resulting in the characteristic root gall (Figs. 1D, 6, 7A,D). Galls usually develop 1–2 days after juvenile penetration. Gall size is commonly related to the number of nematodes present in the tissue but may also depend on the plant species parasitized. Galls induced by most *Meloidogyne* species are similar in their morphology. Some species, however, such as

M. exigua, *M. hapla*, and *M. kikuyensis*, produce characteristic galls which may be useful in species identification (Fig. 6).

II. GENERAL MORPHOLOGY

A. Females

Adult females of the genus *Meloidogyne* (Eisenback, 1985) have swollen, saccate bodies that range in median length from about 0.44 to 1.30 mm and in median width from about 0.325 to 0.700 mm. The neck protrudes anteriorly; the vulva and anus are located terminally, flush with the perineal region (Figs. 2A, 8A–C, E–F, 9) or they may be slightly raised (Fig. 8D). In most species, the females have symmetrical bodies and the neck and perineal region are in a straight line (Fig. 9). In some species, however, the neck projects from the longitudinal axis at an angle of 15–90° to one side (Fig. 8D).

The pearly white body has a moderately thick cuticle (Fig. 9) which remains soft throughout the life of the female. The cuticularized hexaradiate cephalic framework surrounds the stomatal cavity, which holds a hollow protrusible stylet (Fig. 10). The delicate stylet consists of cone, shaft, and knobs and is 10–24 µm in length among the species. Most species have average stylet lengths of 14.0–16.0µm. The stylet is moved by protractor muscles and functions like a hypodermic needle. The morphology of the stylet is a good supplemental character for species identification (Tables 3 and 4). The shape of the cone, shaft, and knobs appears to be species-specific in may cases. The shape of the knobs and the type of junction made with the shaft are the most important diagnostic features. Important mor-

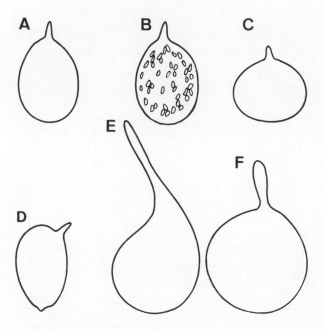

FIGURE 8   (A–F) Body shapes of females of *Meloidogyne*. (A) "Javanica group." (B) "Acronea group." (C) "Exigua group." (D) "Graminis group." (E) "Brevicauda group." (F) "Nataliei group." (Figures are not drawn to scale.) (After Eisenback, 1989.)

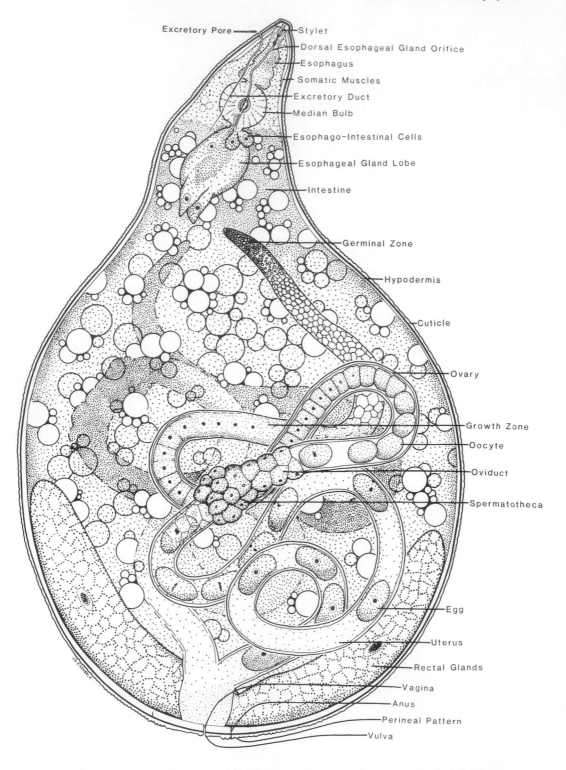

FIGURE 9 Gross morphology and anatomy of a female root-knot nematode, *Meloidogyne* sp. (After Eisenback, 1985.)

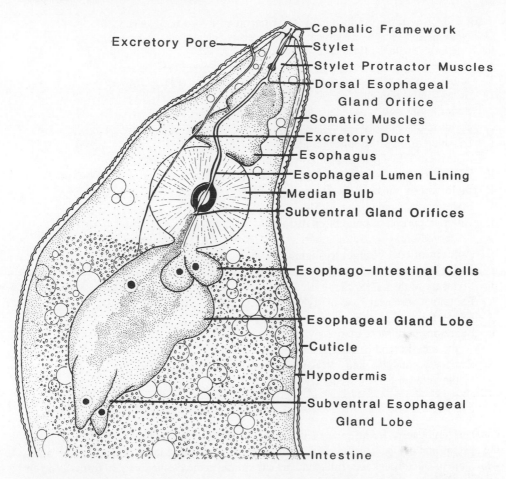

FIGURE 10   Line drawing of the anterior end of a female root-knot nematode, *Meloidogyne* sp. (After Eisenback, 1985.)

phometric features are stylet length and cone length in relation to stylet length, stylet knob height and width, and the width/height ratio. The esophagus has a large muscular median bulb with conspicuous valve plates and three ventrally overlapping esophageal glands that empty their secretions at three points into the digestive system (Fig. 10). The dorsal esophageal gland orifice (DEGO) is located posteriorly to the stylet knobs; the two subventral gland orifices open into the esophagus lumen immediately behind the valve plates. The DEGO distance has a broad range among the species (2–10 μm) and seems variable within populations and species. This character, however, can have limited value in identification of species groups (Jepson, 1987). Two esophagointestinal cells connect the esophagus with the syncytial intestine (Fig. 10). The excretory pore is always located anterior to the median bulb, but its position varies much within and among the species and is not a good diagnostic character. Females are didelphic and the two gonads are very long and greatly convoluted, occupying a major part of the total body content (Figs. 9 and 11). The various gonad parts are well defined. Each gonad is composed of an ovary with germinal zone and growth zone, narrow oviduct, globular spermatotheca, and long uterus. The number of oviduct cells is

TABLE 3   Key to the Agriculturally Most Important Root-Knot Nematodes (*Meloidogyne* spp.) Based on Morphology of Stylets of Females and Distance of the Excretory Pore to the Head End

1.	Stylet knobs irregular in outline	*M. chitwoodi*
1′	Stylet knobs smooth	2
2.	Stylet knobs small, rounded; set off from shaft	3
2′	Stylet knobs broadly elongate or sloping posteriorly	4
3.	Distance of dorsal esophageal gland orifice (DEGO) to knobs more than 6.5 μm	*M. exigua*
3′	DEGO to knobs less than 6.5 μm	*M. hapla*
4.	Stylet knobs transversely elongate	5
4′	Stylet knobs sloping posteriorly	7
5.	Anterior portion of cone often distinctly curved dorsally	*M. incognita*
5′	Anterior of cone straight to slightly curved dorsally	6
6.	Stylet less than 12 μm long . . . . , ...	*M. graminicola*
6′	Stylet more than 12 μm long	*M. javanica*
7.	Excretory pore near base of stylet	*M. naasi*
7′	Excretory pore posterior to base of stylet	8
8.	Excretory pore more than 2 stylet lengths from the anterior end of the female head	*M. arenaria*
8′	Excretory pore less than 2 stylet lengths from the anterior end of the female head	*M. artiellia*

constant (eight cells) for all species, whereas that of the spermatotheca cells differs among the species and can be a useful diagnostic character. Both uteri unite in a common vagina which opens out through the slitlike vulva. The cuticle in the perineal region forms a finger-print-like pattern, the perineal pattern. It comprises the tail terminus, phasmids, lateral lines, anus, and vulva surrounded by cuticular striae or folds (Fig. 12). It is the most characteristic feature of females (Tables 5 and 6). Although variability of the pattern occurs within species and populations, basic characteristics do not change significantly. Pattern features, including general shape, presence or absence of markings in the lateral field areas, punctations on the tail terminus, and form of striae, remain relatively stable among females within species. Morphometric features, such as interphasmidial distance, distance of anus to tail terminus, and anus to center of vulva may be helpful differentiating characters. Six large, unicellular rectal glands in the posterior body region (Fig. 9) are connected to the rectum and produce a very large amount of gelatinous matrix material which is excreted through the rectum. All eggs are usually deposited in this protective egg sac (Figs. 3, 7C).

B. Males

The vermiform males of root-knot nematode species (Eisenback, 1985) (Fig. 13) vary greatly in body size from 700 to 2,000 μm due to the varying environmental conditions existing during their development. Morphometric characters such as body length and ratios involving body length, including esophagus length, tail length, and body width are nearly useless. The male head, composed of head cap and head region, provides many good diagnostic features (Tables 7 and 8; Fig. 14). The head cap includes a labial disk surrounded by

TABLE 4   Summary of Important Diagnostic Characters of Stylet of Females of the Agriculturally Most Important Root-Knot Nematodes (*Meloidogyne* spp.)

Species	Stylet cone	Stylet shaft	Stylet knobs	Length (μm)	DEGO[a] (μm)
M. exigua	Straight to slightly curved dorsally	Cylindrical, occasionally narrows near junctions with knobs	Small, rounded; slightly indented anteriorly	12–14	4–8
M. incognita	Anterior half distinctly curved dorsally	Slightly wider posteriorly	Set off, rounded to transversely elongate, some-times indented anteriorly	15–17	2–4
M. javanica	Slightly curved dorsally	Cylindrical	Set off, transversely elongate	14–18	2–5
M. arenaria	Straight, broad and robust	Wider posteriorly	Not set off, sloping posteriorly, merging with shaft	13–17	3–7
M. hapla	Slightly curved dor-sally, narrow and delicate	Slightly wider posteriorly	Set off, small and rounded	14–17	5–6
M. chitwoodi	Slightly curved dorsally	Cylindrical to slightly wider posteriorly	Not set off, irregular in outline, indented medially	11–12.5	4–5.5
M. graminicola	Slightly curved dorsally	Cylindrical to slightly wider posteriorly	Set off, transversely elongate	10.5–11	3–4
M. naasi	Slightly curved dorsally	Slightly wider posteriorly	Not set off, large and rounded; sloping posteriorly	11–15	2–4
M. artiellia	Slightly curved dorsally	Slightly wider posteriorly	Not set off, large and rounded; sloping posteriorly	12–16	4–7

[a]DEGO is the distance from the base of the stylet to the dorsal esophageal gland orifice.

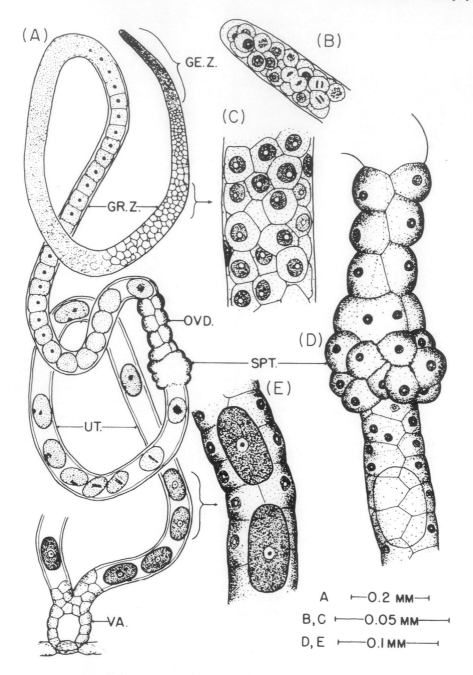

FIGURE 11 Reproductive system of a female of *Meloidogyne*. (A) One of the two gonads. (B) Distal tip of germinal zone of ovary. (C) Growth zone of ovary. (D) Oviduct and spermatotheca. (E) Uterus with eggs. Abbreviations: GE.Z., germinal zone; GR. Z., growth zone; OVD., oviduct. SPT., spermatotheca; UT., uterus; VA., vagina. (After Triantaphyllou, 1962.)

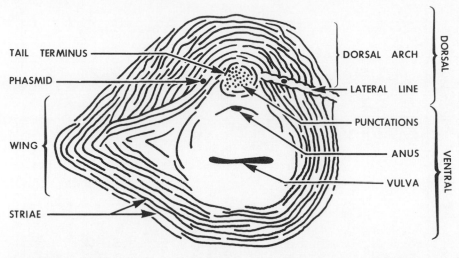

FIGURE 12   Diagram of the perineal pattern of a root-knot nematode, *Meloidogyne* sp. (After Eisenback et al., 1981.)

TABLE 5   Key to the Agriculturally Most Important Root-Knot Nematodes (*Meloidogyne* spp.) Based on Morphology of the Perineal Pattern

1.   Punctations present in tail terminal area . . . . *M. hapla*
1′   Punctations absent in tail terminal area 2
2.   Lateral lines clearly marked by deep incisures;
       usually extending well beyond perineum . . . . *M. javanica*

2′   Lateral lines not clearly marked or
       ending near perineum 3
3.   Striae in dorsal arch twisted and fused . . . . *M. chitwoodi*
3′   Striae in dorsal arch not twisted and fused 4
4.   Lateral fields near perineum marked
       by coarse, raised, looped, and folded striae . . . . *M. exigua*
4′   Lateral fields not marked by looped and folded striae 5
5.   Dorsal arch high and square, striae smooth to wavy . . . *M. incognita*
5′   Dorsal arch low and rounded 6
6.   Striae near perineum in dorsal arch coarse and thick . . . *M. artiellia*
6′   Striae near perineum in dorsal arch not coarse and thick 7
7.   Striae in dorsal arch rounded and forming shoulders . . . *M. arenaria*
7′   Dorsal arch without distinct shoulders 8
8.   Phasmids large . . . . *M. naasi*
8′   Phasmids very small . . . . *M. graminicola*

TABLE 6 Summary of Important Diagnostic Characters of Perineal Patterns of the Agriculturally Most Important Root-Knot Nematodes (*Meloidogyne* spp.)

Species	Dorsal arch	Lateral field	Striae	Tail terminus
M. exigua	Low, rounded to high and squarish	Inner regions with coarse, raised, looped and folded striae	Coarse, smooth	Whorl absent
M. incognita	High, squarish	Distinct lateral lines smooth to wavy, absent, marked by breaks and forks in striae	Fine to coarse, distinct, sometimes zigzaggy	Often with whorl
M. javanica	Moderately high, rounded	Distinct lateral lines	Coarse, smooth to slightly wavy	Often with distinct whorl
M. arenaria	Low, rounded to high and squarish	Lateral lines absent, marked by short, irregular, forked striae	Coarse, smooth to slightly wavy	Usually without distinct whorl
M. hapla	Low, rounded	Lateral lines inconspicuous	Fine, smooth to slightly wavy	Whorl absent marked by subcuticular punctations
M. chitwoodi	Low, oval to rounded	Indistinct of forking dorsal and ventral striae	Coarse, smooth to wavy and twisted near perineum, often fused	Whorl present
M. graminicola	High, squarish to low and rounded	Not clearly defined; in-dicated by breaks in striae	Smooth and continuous around pattern	Large terminus, usually free of striae
M. naasi	Low, rounded	Absent	Coarse, smooth; broken and irregular near phasmids	Large terminus, usually free of striae
M. artiellia	High, angular	Absent, or marked by irregularities in thick striae	Fine, smooth, to wavy, very coarse and thick near perineum	Whorl absent

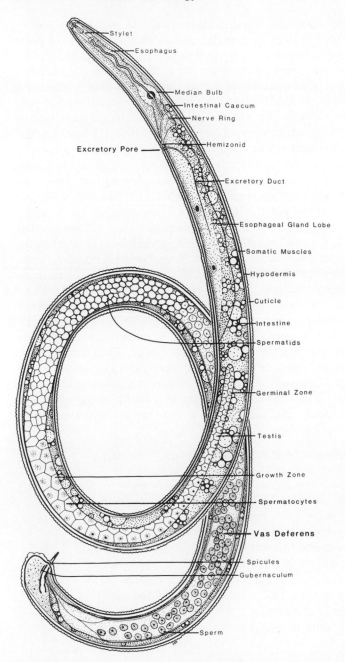

FIGURE 13   Line drawing of the gross morphology and anatomy of a male *Meloidogyne* sp. (After Eisenback, 1985.)

lateral and medial lips. A centrally located prestoma leads to a slitlike stoma. Four sensory organs terminate on the medial lips (cephalic sensilla), and six others surround the stoma area (labial sensilla). The head region may or may not be set off from the remainder of the body. The size, height, shape, and slope of the head cap, the shape and proportion of labial disk, and lips; the expression of labial and cephalic sensilla; and the presence or absence of annulations in the head region can be used to distinguish species and populations. The male stylet length has a broad range within the genus (13–30 µm); although most species have an average stylet length of 18–24 µm, with a low coefficient of variability (CV = 4%), which makes it a good differentiating character. Size and shape of the stylet cone, shaft, and knobs are also excellent supporting characters for species identification (Tables 7 and 8) (Eisenback and Hirschmann, 1981). The DEGO is located 2–13 µm posterior to the stylet knob base. In general, this morphometric character exhibits much variation, although some species can be distinguished on the basis of DEGO distance. The esophagus has a slender procorpus and oval-shaped median bulb with distinct valve plates (Figs. 2D and 13). The isthmus is short, and in most species the ventrally overlapping gland lobe has two instead of the usual three esophageal nuclei. The obscure esophagointestinal junction is situated at the height of the nerve ring. The position of excretory pore exhibits large intraspecific variation and is limited value as a differential character. The long excretory duct leads to an excretory gland cell with a large nucleus. The hemizonid is usually located anterior to the excretory pore and can be helpful taxonomically only in a few species in which it is located posterior to the excretory pore. One gonad is present in normal males, whereas sex-reversed males have two gonads. Germinal and growth zones of the testis, which is usually outstretched, are well defined (Fig. 13). Most of the gonad consists of a long vas deferens (seminal vesicle) packed with developing sperm. The vas deferens ends posteriorly in a glandular region and forms a cloaca with the intestine in the region of the spicules (Fig. 15). Spicule length ranges from 19 to 40 µm among the species, and there is much overlap in spicule measurements.

TABLE 7   Key to the Agriculturally Most Important Root-Knot Nematodes (*Meloidogyne* spp.) Based on Morphology of Heads and Stylets of Males

1.	Labial disk usually raised above medial lips .... *M. incognita*
1′	Labial disk and medial lips form one smooth head cap 2
2.	Head annule not in contour with the first body annule ... *M. hapla*
2′	Head annule in contour with the first body annule 3
3.	Stylet knobs irregular in outline, indented anteriorly . *M. chitwoodi*
3′	Stylet knobs smooth 4
4.	Stylet knobs transversely elongate .... *M. javanica*
4′	Stylet knobs not transversely elongate 5
5.	Dorsal esophageal gland orifice to base of stylet more than 5 µm 6
5′	Dorsal esophageal gland orifice to base of stylet less than 5 µm 7
6.	Stylet robust .... *M. arenaria*
6′	Stylet narrow, delicate; knobs taper posteriorly .... *M. artiellia*
7.	Stylet length 18–20 µm .... *M. exigua*
7′	Stylet length usually less than 18 µm 8
8.	Stylet shaft often narrows near its base .... *M. graminicola*
8′	Stylet shaft cylindrical .... *M. nassi*

TABLE 8 Summary of Important Diagnostic Characters of Head Shapes and Stylets of Males of the Agriculturally Most Important Root-Knot Nematodes (*Meloidogyne* spp.)

Species	Head cap	Head region	Stylet cone	Stylet shaft	Stylet knobs	lgth (μm)	DEGO[a] (μm)
M. exigua	High, rounded, set off from head annule	Not set off, smooth, lateral lips present	Bluntly pointed	Cylindrical, narrows at distinctly angular its base	Small rounded; in some	18–20	3–5
M. incognita	Flat to concave, labial disk raised above the medial lips	Not set off, usually marked by 2–3 incomplete annulations	Tip blunt, bladelike	Usually cylindrical, often narrows near knobs	Set off, rounded to transversely elongate, sometimes indented anteriorly	23–25	2–4
M. javanica	High rounded, set off from head annule	Set off, smooth or marked by 2–3 incomplete annulations	Tip pointed, cone straight	Usually cylindrical	Set off, low, and transversely very elongated broad	18–22	2–4
M. arenaria	Low, to moderately raised, sloping posteriorly	Not set off, smooth or marked by 1–2 incomplete annulations	Tip pointed, cone broad and robust	Usually cylindrical, often broadens near knobs	Not set off, sloping posteriorly merging with shaft	20–28	4–8

TABLE 8 (Continued)

Species	Head cap	Head region	Stylet cone	Stylet shaft	Stylet knobs	lgth (μm)	DEGO[a] (μm)
M. hapla	High and narrow	Set off, smooth, larger in diameter than first body annule	Tip pointed, cone narrow, delicate	Cylindrical, often wider or narrower at its base	Set off, small, and round	17–23	4–5
M. chitwoodi	High, rounded, set off from head annule present	Not set off, smooth, large lateral lips	Tip pointed, cone narrow, delicate	Cylindrical to conical	Anteriorly indented, irregular in outline	16–19	3
M. grami-nicola	High, rounded, set off from head annule	Not set off, smooth	Tip pointed often narrows near base	Cylindrical, to distinctly angular	Set off, rounded	16–17	3–4
M. naasi	High, rounded, set off	Not set off, smooth or marked by few annulations	Tip often pointed,cone straight	Cylindrical may narrow near base	Set off, rounded	16–19	2–4
M. artiellia	High, rounded, set off	Lateral lip present, not set off	Straight, narrow	Cylindrical, wider at its base	Rounded, set off to sloping posteriorly	17–27	5–7

[a]DEGO is the distance from the base of the stylet to the dorsal esophageal gland orifice.

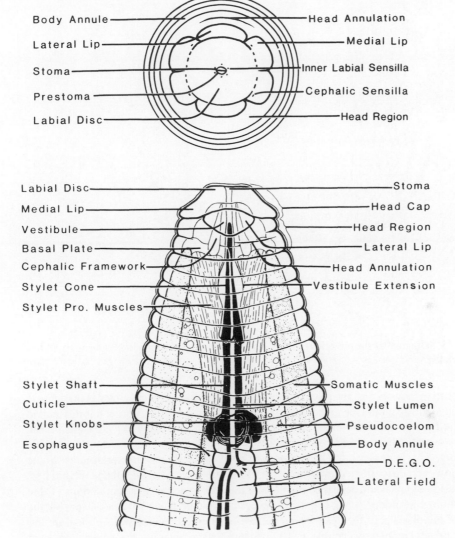

Body Annule ──── Head Annulation
Lateral Lip ──── Medial Lip
Stoma ──── Inner Labial Sensilla
Prestoma ──── Cephalic Sensilla
Labial Disc ──── Head Region

Labial Disc ──── Stoma
Medial Lip ──── Head Cap
Vestibule ──── Head Region
Basal Plate ──── Lateral Lip
Cephalic Framework ──── Head Annulation
Stylet Cone ──── Vestibule Extension
Stylet Pro. Muscles

Stylet Shaft ──── Somatic Muscles
Cuticle ──── Stylet Lumen
Stylet Knobs ──── Pseudocoelom
Esophagus ──── Body Annule
──── D.E.G.O.
──── Lateral Field

FIGURE 14 Diagrams of the anterior end of a male root-knot nematode, *Meloidogyne* sp., in face and lateral views, respectively. (After Eisenback, 1985.)

Slight differences in spicule structure have been described in some species, but in general spicule morphology is not diagnostic. The male tail is bluntly rounded and short without caudal alae (Figs. 2E, 15). There is little variation in tail shape among the species.

C. Second-Stage Juveniles

Infective second-stage juveniles of root-knot nematodes (Eisenback, 1985) vary in body length from 290 to 912 μm within the genus (Fig. 16). Except for second-stage juveniles of *M. spartinae*, which are 600–912 μm long, the range among the species is approximately

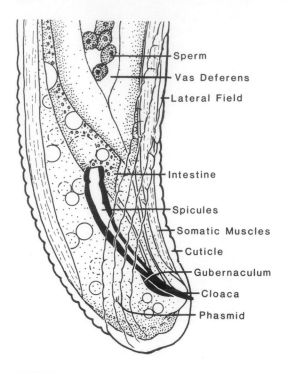

FIGURE 15   Diagram of the posterior end of a male root-knot nematode, *Meloidogyne* sp. (After Eisenback, 1985.)

300–500 μm. Many species overlap in body length; therefore, this character is inadequate for species identification. Due to the small size of the second-stage juvenile, it is difficult to discern the head morphology precisely. Second-stage juveniles have the same basic head characters as males, but they can be seen clearly only with the scanning electron microscope (SEM) (Fig. 17). In general, the head morphology is quite similar among species. Some species, however, differ distinctly in the shape of the labial disk, the lateral and medial lips, the expression of labial and cephalic sensilla, and the occurrence of head annulations (Table 9). In the light microscope (LM), slight differences may occur in general head shape, but the number of annulations in the head region cannot be counted precisely since they are too small and do not completely encircle the head. Second-stage juveniles have a delicate stylet (Fig. 16) that ranges in length from 8 to 18 μm within the genus. The stylet tip is very fine and difficult to see with LM. Measurements from the stylet knob base to the head end are more reliable in properly fixed specimens. Stylet length shows generally low variability and may be a useful supplemental character in certain exotic species (Table 9). Stylet morphology can be distinctive as differences in size and shape of the stylet knobs occur among species and populations. The DEGO distance 2–8 μm seems to be a good differentiating feature, and groups of species can be distinguished by this measurement (Table 9). The esophagus (Fig. 16) has a narrow, faintly outlined procorpus, a well-defined median bulb with large valve plates, and three long ventrally overlapping glands that are instrumental in molting and feeding. The esophagointestinal junction in the vicinity of the nerve ring is obscure. The position of the excretory pore is variable. The hemizonid can be a fairly useful diagnostic feature in those species in which it is located posteriorly to the pore. Tail length varies considerable among the species from 15 to 100 μm. Due to its small intraspecific variation

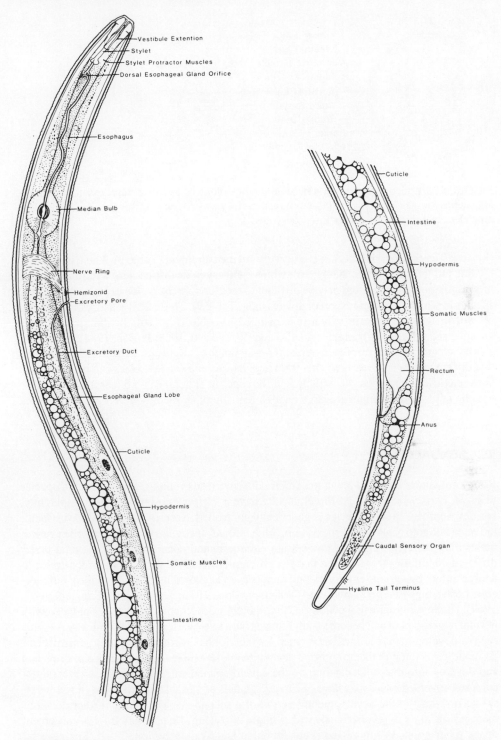

FIGURE 16   Gross morphology and anatomy of the anterior and posterior ends of a second-stage juvenile of a root-knot nematode, *Meloidogyne* sp. (After Eisenback, 1985.)

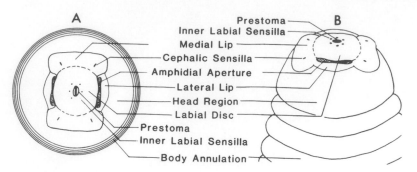

FIGURE 17   Line drawing of the surface head morphology of a second-stage juvenile of a root-knot nematode, *Meloidogyne* sp., in face (A) and lateral views (B), respectively, as seen in the scanning electron microscope. (After Eisenback, 1985.)

(low coefficient of variability), it is a very useful measurement (Table 9). The tail ends in a hyaline tail terminus (Figs. 2G–H, 16), which can vary in its distinctness within specimens of a population and therefore may be difficult to measure precisely. It can be useful in those species in which it is always short or always long (Table 9). The caudal sensory organ can be helpful in those species in which it is distinctly visible (Fig. 16). Juveniles have been grouped according to tail length and tail shape (Whitehead, 1968; Jepson, 1984). Some species are clearly distinct from each other in overall range of tail length. Jepson (1984) showed that differences in tail measurements from populations of a single species can be larger than between different species. Nevertheless, differences in either mean tail length and/or mean hyaline tail terminus length are large enough to distinguish species within groups (Jepson, 1987).

III. SEXUALITY

Root-knot nematodes are known to exhibit unbalanced sex ratios. Cross-fertilizing species (e.g., *M. carolinensis*, *M. spartinae*) usually have a 1:1 male-to-female ratio. Apparently, sex development in such species is genetically controlled. Species that reproduce be facultative or obligatory parthenogenesis (e.g., *M. hapla*, *M. incognita*) have variable sex ratios; males may be rare or absent under certain environmental conditions and abundant under different conditions. It is believed that sex differentiation in the latter group is controlled to a large extent by various environmental factors such as crowding of the juveniles, unfavorable hosts, high temperatures, etc. (Triantaphyllou, 1973; Davide and Triantaphyllou, 1967). Under environmental conditions favorable for development, juveniles proceed with normal development and become adult females with two ovaries (Fig. 18, pathway A). This type of development is normally expected for thelytokous parthenogenetic organisms. Under conditions unfavorable for normal development, female juveniles undergo sex reversal and develop into males. Depending on the developmental stage at which sex reversal occurs, sex-reversed males may have one to two gonads of variable length. Thus, if sex reversal occurs early in the developmental period of a second-stage juvenile, further development results in a sex-reversed male with a single testis (Fig. 18, pathway Bc). Development of the other gonad is suppressed (Papadopoulou and Triantaphyllou, 1982). Sex reversal near the end of the developmental period of a second-stage juvenile results in a sex-reversed male with two testes which correspond to the two ovaries of a female (Fig. 18, pathway Ba).

TABLE 9 Summary of Important Diagnostic Characters of Second-Stage Juveniles of the Agriculturally Most Important Root-Knot Nematodes (*Meloidogyne* spp.)

Species	Head cap	Head region	Stylet width	Stylet knobs	Stylet (μm)	DEGO (μm)	Tail (μm)	Terminus (μm)	Body (μm)
M. exigua	Anteriorly flattened, elongate	Usually smooth	Moderately sized cone and shaft	Set off, small, and rounded	9–10	2.5–4	44–46	12–14	334–358
M. incognita	Anteriorly flattened, elongate	Usually marked by 1–3 incomplete annulations	Moderately sized cone and shaft	Set off, posteriorly rounded, sloping backward	10–12	2–3	42–63	6–13.5	346–463
M. javanica	Anteriorly flattened, elongate	Usually smooth	Moderately sized cone and shaft	Set off, posteriorly rounded, sloping backward, transversely elongate	10–12	3.5	51–63	9–18	402–560
M. arenaria	Anteriorly flattened, elongate	Usually smooth	Broad cone and shaft	Not set off, posteriorly rounded, merging with shaft	10–12	3–5	44–69	6–13	398–605

TABLE 9 (Continued)

Species	Head cap	Head region	Stylet width	Stylet knobs	Stylet (μm)	DEGO (μm)	Tail (μm)	Terminus (μm)	Body (μm)
M. hapla	Rounded, narrow	Rounded, usually smooth	Narrow cone and shaft	Set off, small, and rounded	10–12	3–4	46–69	12–19	357–517
M. chitwoodi	Anteriorly flattened, elongate	Rounded, usually smooth	Narrow cone and shaft	Not set off, outline irregular	10	3–4	39–47	9–14	336–417
M. graminicola	Anteriorly flattened, elongate	Rounded, usually smooth	Narrow cone and shaft	Set off, small, and rounded	11–12	2.8–3.4	67–76	14–21	415–484
M. naasi	Anteriorly flattened, elongate	Usually smooth	Narrow cone and shaft	Not set off, tapering onto shaft	13–15	2–3	52–78	17–27	418–435
M. artiellia	Anteriorly flattened, elongate	Usually smooth	Moderately sized cone and shaft	Not set off, tapering onto shaft	14–16	2.5–4.5	18–26	2–7	334–370

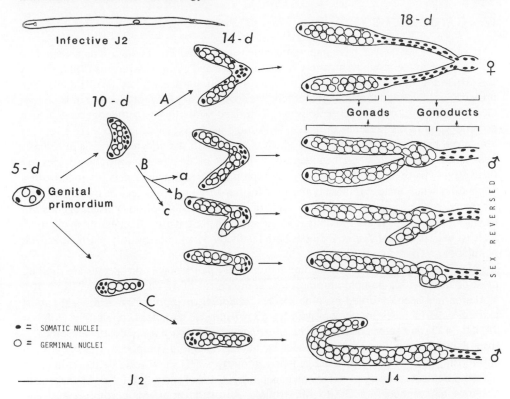

FIGURE 18 Development of the genital primordium (diagrammatic) of *M. incognita* juveniles 5, 10, 14, and 18 days after initiation of feeding illustrating the anatomical changes that lead to the development of females (pathway A), sex-reversed males (pathway B), or normal males (pathway C). Adverse environmental conditions imposed during an early (Ba), a late (Bc), or an intermediate (Bb) developmental period on second-stage juveniles (J2) lead to sex reversal and the development of males with variable numbers of testes. For further explanations, see Papadopoulou and Triantaphyllou (1982). (Illustration provided by A. C. Triantaphyllou.)

It may also result in the development of an intersexual individual, male or female, with some secondary sexual characteristics of the opposite sex. Sex reversal at an intermediate period often results in a sex-reversed male with a well-developed testis and a rudimentary one of the variable length (Fig. 18, pathway Bb). A small number of juveniles follow a developmental pattern that leads to the development of males with one testis even under unfavorable environmental conditions (Fig.18, pathway C). Sex determination in these juveniles appears to be genetically controlled.

These developmental peculiarities have been revealed by anatomical studies of the gonad primordium of juveniles of *M. incognita* developing under different conditions of crowding (Papadopoulou and Triantaphyllou, 1982). Other parthenogenetic species probably follow a similar pattern. A practical way to induce sex reversal and thus control root-knot nematode populations by changing them to populations of males with few or no females is still a far-reaching objective (Triantaphyllou, 1973).

IV. CYTOGENETICS

The root-knot nematodes have undergone extensive cytogenetic evolution as revealed by the presence in the group of a variety of reproductive and cytological forms (Table 10). A few *Meloidogyne* species reproduce obligatory by amphimixis (cross-fertilization). The presence of males and insemination of females are necessary for reproduction in such species. Most *Meloidogyne* species, however, reproduce primarily by parthenogenesis, which does not require the presence of males. Within the parthenogenetic group there are many obligatory parthenogenetic species in which maturation of the oocytes consists of a single mitotic division (mitotic parthenogenesis). The rest of the parthenogenetic species are facultatively parthenogenetic, i.e., they can reproduce by parthenogenesis (meiotic parthenogenesis) when males are absent and females remain uninseminated, or by amphimixis when males are present and females become inseminated. Thus, mitotic parthenogenesis in root-knot nematodes is always obligatory, whereas meiotic parthenogenesis is facultative. At least one species, *M. hapla*, includes both facultatively and obligatorily parthenogenetic populations (Triantaphyllou, 1985).

The obligatory amphimictic and facultatively parthenogenetic species are diploid, most of them with a haploid chromosomal complement of 18 chromosomes (Table 10). Variation has been observed in some species. *Meloidogyne hapla*, representing the most extreme case, includes populations with $N = 13-17$ and also tetraploid populations with $N = 28$ and 34 chromosomes. *Meloidogyne spartinae* and *M. kikuyensis* deviate substantially from the rest of the root-knot nematodes by having a haploid complement of only seven chromosomes. Obligatory parthenogenetic (mitotic) species show wide variation in chromosome numbers due to polyploidy an aneuploidy. Diploid aneuploid forms have somatic chromosome numbers of 30 to 40, triploid forms have 50–56 chromosomes, and hypotriploid forms have intermediate somatic chromosome numbers. More specific cytological information is presented with the description of individual species of *Meloidogyne*.

V. BIOCHEMISTRY

Several enzyme phenotypes, especially those of esterases, are species-specific and can be used as reliable taxonomic characters for identification of most major and several minor *Meloidogyne* species (Esbenshade and Triantaphyllou, 1985). Recently, the use of such enzyme phenotypes in *Meloidogyne* identification became easy and practical. Advanced technology made available automated electrophoretic apparati that can run very thin polyacrylamide slab gels on which the phenotypes of two or more enzymes can be revealed from the protein extract of a single *Meloidogyne* female. Such apparati are now used routinely in some laboratories as a first step in the species identification of *Meloidogyne* populations (Esbenshade and Triantaphyllou, 1989). In most cases, the esterase phenotypes of a small number of females of an unknown population are compared to the phenotype of an *M. javanica* female, run in a central lane of the same electrophoretic gel. Such a comparison often reveals whether the population is pure (all females have the same esterase phenotype), or the population is mixed, consisting of two or more species (two or more species-specific phenotypes present). In most cases, when the species involved are among those with characteristic esterase phenotype, identification of the species is easy and reliable on the basis of esterase phenotype alone. In other cases, when two or more species have the same or very similar esterase phenotypes, staining the same gel for a second enzyme (usually for malate dehydrogenase) helps differentiate between such species (Fig. 19).

TABLE 10   Summary of Cytogenetic Information Related to Root-Knot Nematodes (*Meloidogyne* spp.)

Meloidogyne species	Populations studied (number)	Countries of origin (number)	Chromosome number		Mode of reproduction
			n	2*n*	
M. kikuyensis	1	1	7		
M. spartinae	4	1	7		
M. carolinensis	2	1	18		
M. megatyla	1	1	18		Amphimixis
M. microtyla	2	1	18–19		
M. subartica	1	1	18		
M. exigua	6	1	18		
M. graminicola	1	1	18		
M. graminis	10	5	18		Facultative
M. naasi	1	1	18		
M. ottersoni	1	1	18		meiotic
M. hapla (race A)	48	24	13–17		parthenogenesis
(Polyploid)	1	1	28		
(Polyploid)	2	2	34		
M. chitwoodi	6	3	14–18		
M. arenaria	18	13		30–38	
	34	21		40–48	
	68	32		51–56	
M. criciani	1	1		42–44	
M. enterolobii	1	1		46–	
M. hapla (race B)	6	3		30–32	
	11	8		43–48	Obligatory
M. hispanica	4	4		33–36	
M. incognita	6	6		32–38	
	215	64		41–46	mitotic
M. javanica	126	45		42–48	
M. microcephala	3	2		36–38	
M. oryzae	2	1		51–55	parthenogenesis
M. plantani	1	1		42–44	
M. querciana	1	1		30–32	
Total	584				

Source: Triantaphyllou, 1985; updated by same author.

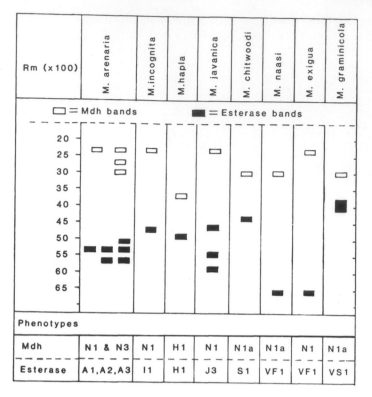

FIGURE 19   Species-specific phenotypes of malate dehydrogenase (Mdh) and esterases helpful in differentiating eight species of *Meloidogyne*. On the same polyacrylamide gel, malate dehydrogenase bands are bluish, those of esterases black. Phenotype designations correspond to those of Esbenshade and Triantaphyllou (1989). (Illustration provided by A. C. Triantaphyllou.)

VI. TAXONOMY

A. Diagnosis of the Subfamily Meloidogyninae Skarbilovich, 1959 (Wouts, 1973)

Heteroderidae. Cuticle usually not abnormally thick, annulated in all female and male stages. Cephalic framework moderately sclerotized; subdivides head into six sectors with lateral sectors equal to wider-than-submedian sectors.

 Female: Sedentary, globose with protruding neck. Cuticle moderately thick. No cyst stage. Preadult, vermiform female stage absent. Labial disk fused with medial and lateral lips. Cephalic framework delicate. Stylet slender, less than 25 µm long, knobs well developed. Excretory pore opposite or anterior to median bulb of esophagus, often slightly posterior or at stylet knob level. Vulva and anus terminal. Perineal region flush or slightly raised; fingerprintlike pattern of cuticular annulation around vulva and anus (perineal pattern). Eggs usually not retained in body; deposited in gelatinous matrix formed by specialized rectal glands. Oviduct and spermatotheca of reproductive system of characteristic structure.

Male: Vermiform, migratory. Body twisted through 180°. Labial disk and medial lips fused. Stylet and framework well developed. Tail short, without caudal alae. One or two testes.

Second-stage juvenile: Infective juvenile vermiform, migratory. Stylet and cephalic framework delicate. Shortly after infection, juvenile becomes swollen and sedentary with tail spike.

Third-stage juvenile: Swollen, sedentary with short, blunt tail. Stylet absent. Enclosed in shed second-stage cuticle. Very short lived.

Fourth-stage juvenile: Swollen, sedentary with terminal anus. Stylet absent. Enclosed in second-and third-stage shed cuticles.

B. Nominal Species

GENUS
  Meloidogyne Goeldi, 1889
    syn. *Caconema* Cobb, 1924
    syn. *Hypsoperine* Sledge and Golden, 1964
        Hypsoperine (Hypsoperine) Siddiqi, 1985
        Hypsoperine (Spartonema) Siddiqi, 1985
TYPE SPECIES
  M. exigua Goeldi, 1889
    syn. *Heterodera exigua* (Goeldi, 1889)
            Marcinowski, 1909
SPECIES
  M. acronea Coetzee, 1956
    syn. *Hypsoperine acronea* (Coetzee, 1956) Sledge and Golden, 1964
        Hypsoperine (Hypsoperine) acronea (Coetzee, 1956) Siddiqi, 1985
  M. africana Whitehead, 1960
  M. aquatilis Ebsary and Eveleigh, 1983
  M. ardenensis Santos, 1968
  M. arenaria (Neal, 1889) Chitwood, 1949
    syn. *Anguillula arenaria* Neal, 1889
        Tylenchus arenarius (Neal, 1889) Cobb, 1890
        Heterodera arenaria (Neal, 1889) Marcinowski, 1909
        M. arenaria arenaria (Neal, 1889) Chitwood, 1946
        M. arenaria thamesi Chitwood in Chitwood, Specht, and Havis, 1952
        M. thamesi (Chitwood in Chitwood et al., 1952) Goodey, 1963
  M. artiellia Franklin, 1961
  M. brevicauda Loos, 1953
  M. californiensis Abdel-Rahman and Maggenti, 1987
  M. camelliae Golden, 1979
  M. caraganae Shagalina, Ivanova, and Krall, 1985
  M. carolinensis, Eisenback, 1982
  M. chitwoodi Golden, O'Bannon, Santo, and Finley, 1980
  M. christiei Golden and Kaplan, 1986
  M. coffeicola Lordello and Zamith, 1960
    syn. *Meloidodera coffeicola* (Lordello and Zamith, 1960) Kirjanova, 1963
  M. cruciani Garcia-Martinez, Taylor, and Smart, 1982
  M. decalineata Whitehead, 1968

M. deconincki Elmiligy, 1968

M. enterolobii Yang and Eisenback, 1983

M. ethiopica Whitehead, 1968

M. fanzhiensis Chen, Liang, and Wu, 1988

M. fujianensis Pan, 1985

M. graminicola Golden and Birchfield, 1965

M. graminis (Sledge and Golden, 1964) Whitehead, 1968
  syn. *Hypsoperine graminis* Sledge and Golden, 1964
        Hypsoperine (*Hyposoperine*) *graminis* (Sledge and Golden, 1964) Siddiqi, 1985

M. hapla Chitwood, 1949

M. hispanica Hirschmann, 1986

M. incognita (Kofoid and White, 1919) Chitwood, 1949
  syn. *Oxyuris incognita* Kofoid and White, 1919
        Heterodera incognita (Kofoid and White, 1919) Sandground, 1923
        M. incognita incognita (Kofoid and White, 1919) Chitwood, 1949
        M. incognita acrita Chitwood, 1949
        M. acrita (Chitwood, 1949) Esser, Perry, and Taylor, 1976
        M. incognita inornata Lordello, 1956
        M. elegans da Ponte, 1977
        M. grahami Golden and Slana, 1978
        M. incognita wartellei Golden and Birchfield, 1978
        M. inornata Lordello, 1956

M. indica Whitehead, 1968

M. javanica (Treub, 1885) Chitwood, 1949
  syn. *Heterodera javanica* Treub, 1885
              Cobb, 1890
        Anguillula javanica (Treub, 1885) Lavergne, 1901
        M. javanica javanica (Treub, 1885) Chitwood, 1949
        M. javanica bauruensis Lordello, 1956
        M. bauruensis (Lordello, 1956) Esser, Perry, and Taylor, 1976
        M. lucknowica Singh, 1969
        M. lordelloi da Ponte, 1969

M. jinanensis Zhang and Su, 1986

M. kikuyensis de Grisse, 1961

M. kirjanovae Terenteva, 1965

M. kralli Jepson, 1984

M. litoralis Elmiligy, 1968

M. mali Itoh, Oshima and Ichinohe, 1969

M. maritima Jepson, 1987

M. marylandi Jepson and Golden, 1987

M. mayaguensis Rammah and Hirschmann, 1988

M. megadora Whitehead, 1968

M. megatyla Baldwin and Sasser, 1979

M. microcephala Cliff and Hirschmann, 1984

M. microtyla Mulvey, Townshend, and Potter, 1975

M. naasi Franklin, 1965

M. nataliei Golden, Rose, and Bird, 1981

M. oryzae Maas, Sanders, and Dede, 1978

M. oteifae Elmiligy, 1968

M. ottersoni (Thorne, 1969) Franklin, 1971

syn. *Hypsoperine ottersoni* Thorne, 1969

   Hypsoperine (Hypsoperine) ottersoni Thorne, 1969 Siddiqi, 1985

M. ovalis Riffle, 1963

M. partityla Kleynhans, 1986

M. pini Eisenback, Yang and Hartman, 1985

M. plantani Hirschmann, 1982

M. propora Spaull, 1977

syn. *Hypsoperine (Hypsoperine) propora* Spaull, 1967; Siddiqi, 1985

M. querciana Golden, 1979

M. salasi Lopez, 1984

M. sewelli Mulvey and Anderson, 1980

M. sinensis Zhang, 1983

M. spartinae (Rau and Fassuliotis, 1965) Whitehead, 1968

syn. *Hypsoperine spartinae* Rau and Fassuliotis, 1965

   Hypsoperine (Spartonema) spartinae (Rau and Fassuliotis, 1965) Siddiqi, 1985

M. subartica Bernard, 1981

M. suginamiensis Toida and Yaegashi, 1984

M. tadshikistanica Kirjanova and Ivanova, 1965

M. turkestanica Shagalina, Ivanova, and Krall, 1985

M. vandervegtei Kleynhans, 1988

SPECIES *INQUIRENDAE*

M. marioni (Cornu, 1879) Chitwood and Oteifa, 1952

syn. *Anguillula marioni* Cornu, 1879

   Heterodera marioni (Cornu, 1879) Marcinowski, 1909

   Heterodera goeldi Lordello, 1951 = nom. nov. for *M. marioni*

M. vialae (Lavergne, 1901) Chitwood and Oteifa, 1952

syn. *Anquillula vialae* Lavergne, 1901

   Heterodera vialae (Lavergne, 1901) Marcinowski, 1909

M. poghossianae Kirjanova, 1963

syn. *M. acronea* apud Poghossian, 1961

NOMEN NUDUM

M. californiensis Abdel-Rahman, 1981

M. carolinensis Fox, 1967

M. megriensis (Poghossian, 1971) Esser, Perry, and Taylor, 1976

syn. *Hypsoperine megriensis* Poghossian, 1971

   Hypsoperine (Hypsoperine) megriensis (Poghossian, 1971) Siddiqi, 1985

C. Description and Figures of Important Species

1. *Meloidogyne exigua,* Type Species

Female: The mature female body is small and rounded, and the distinct neck is in line with the tail end (Fig. 8A,C). Females and egg masses are usually completely embedded within root tissue.

   The perineal pattern (Fig. 20C–F) is round to hexagonal, and the dorsal arch varies from low and rounded to somewhat high and squarish (Chitwood, 1949; Lordello and Zamith, 1958; Cain, 1974; Jepson, 1987). The striae are coarse and widely spaced. The lateral fields are usually inconspicuous and only indistinctly forked; however, the inner lateral line regions may have coarse, raised, looped, and folded striae which also cover the anus. The phasmids are widely separated. Sometimes the patterns resemble those of *M. hapla* in

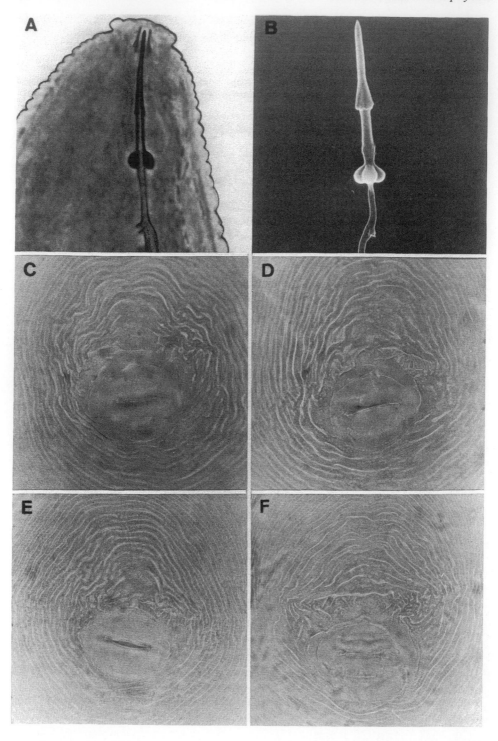

FIGURE 20   Micrographs of females of *M. exigua*. (A) Light micrograph of the anterior end. (B) Scanning electron micrograph of an excised stylet. (C–F) LM of perineal patterns.

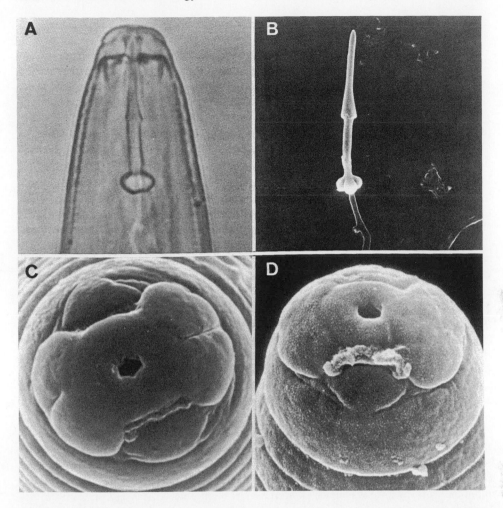

FIGURE 21   Micrographs of males of *M. exigua*. (A) Light micrograph of the head and stylet in
lateral view. (B) Scanning electron micrograph of excised stylet. (C) SEM of the anterior end in face
view. (D) SEM of the anterior end in lateral view.

overall shape but the striae are coarser in *M. exigua*, and *M. hapla* females almost always
have punctations in the tail terminal area.

The stylet of the female (Fig. 20A,B) is 12–14 μm long (Eisenback and Hirschmann,
unpublished). The shaft is cylindrical but occasionally narrows at the junction with the
knobs. Set off from the shaft, the three stylet knobs are small and rounded; each knob may
be slightly indented anteriorly. The DEGO is far from the base of the knobs, usually 4–8 μm.

Male: The shape of the head (Fig. 21A) is unique for the species as seen in lateral
position in the LM. The labial disk and medial lips are fused and form one smooth continuous head cap. In the SEM, the head region is smooth and the lateral lips are incompletely to
completely formed (Lopez, 1985; Eisenback, unpublished) (Fig. 21C,D). The medial lips
are often divided medially by a shallow groove. The head region and body annules are in the

same contour. Several other species are similar in overall appearance: *M. graminis, M. naasi, M. graminicola, M. kralli,* and others.

Stylets of males of *M. exigua* are 18–20 μm long (Eisenback and Hirschmann, unpublished). The cone is bluntly pointed. The opening of the sytlet is 2–3 μm posterior to the tip and appears as a longitudinal slit on the dorsal side of the cone. Although it narrows at the junction with the knobs, the shaft is straight and cylindrical (Fig. 21A,B). The stylet knobs are rounded, but in some specimens distinctly angular, and set off from the shaft. The distance of the DEGO to the base of the knobs is variable and moderately long, 3–5 μm.

Second-stage juvenile: Body length of preparasitic juveniles is 334–358 μm (Lordello and Zamith, 1958), and the length of the tail is 44–46 μm (Fig. 22). The hyaline tail terminus is 12– 14 μm.

Juvenile head shape is characteristic for the species as revealed by SEM; however, the details are too small to be seen clearly by LM (Eisenback and Hirschmann, unpublished) (Fig. 22). The labial disk and medial lips are fused and dumbbell-shaped in face view. Slightly raised above the lips, the labial disk is rounded; however, the lateral edges are nearly parallel with each other. The rounded medial lips are slightly raised above the elongate lateral lips. The head region is smooth and in contour with the body annules.

The stylet of second-stage juveniles of *M. exigua* is 9–10 μm long and the distance of the DEGO to the base of the knobs is 2.5–4 μm (Lordello and Zamith, 1958). The stylet is sharply pointed and the orifice is very near the tip and often is marked by a small, bumplike projection (Eisenback and Hirschmann, unpublished) (Fig. 22B). The shaft and cone gradually widen throughout their length. The knobs are small and rounded to transversely elongate.

The moderately long tail has a smooth tail terminus that ends in a bluntly rounded tip. A few narrow constricting annulations at the beginning of the terminus are quite characteristic.

Morphological variants: Populations of *M. exigua* appear to be very similar to each other morphologically (Lima and Ferraz, 1985), and variants within this species have not been widely reported. Lordello and Zamith (1958) described variants of males based on the distance of the DEGO to the base of the stylet. In one population the distance was 3 μm, and in another this distance was much shorter, almost negligible. Likewise, Lopez (1985) described variants of males of *M. exigua* populations on the basis of minor differences in the head morphology as seen in the SEM. In one population the medial lips were deeply divided into lip pairs, whereas the division was less pronounced in another population.

Host races: Several physiological strains of *M. exigua* have been described, but they have not been widely recognized as distinct host races. Curi et al., (1970) reported three physiological races in Brazil; Chitwood and Berger (1960) detected two strains in Guatemala; and Morera and Lopez (1987) reported variation in virulence among populations from islands of the Caribbean. The reported variability of the virulence of *M. exigua,* however, may be due to the involvement of additional undescribed species that parasitize coffee (Triantaphyllou, 1985).

Symptoms: *Meloidogyne exigua* galls on *Coffea* spp. are usually elongated, approximately 4 mm in diameter. Often galls located at the root tip are terminal (Cain, 1974) (Fig. 6A). Mature females with their eggs are completely embedded in root tissue.

Cytology: Populations of *M. exigua* reproduce by facultative meiotic parthenogenesis and have a haploid chromosome number of $n = 18$ (Fig. 23) (Triantaphyllou, 1985).

Biochemistry: The esterase phenotype of *M. exigua,* namely VF1, is shared with *M. naasi;* however, *M. exigua* can be differentiated from *M. naasi* by the malate dehydrogenase phenotype, N1 (Fig. 19) (Esbenshade and Trantaphyllou, 1989).

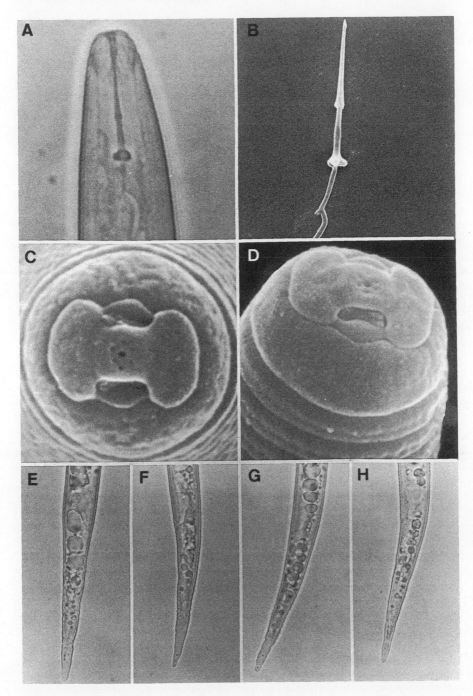

FIGURE 22 Micrographs of second-stage juveniles of *Meloidogyne exigua*. (A) Light micrograph of the head and stylet in lateral view. (B) Scanning electron micrograph of excised stylet. (C) SEM of the anterior end in face view. (D) SEM of the anterior end in lateral view. (E–H) LM of tails. (C and D Eisenback and Hirschmann, 1979.)

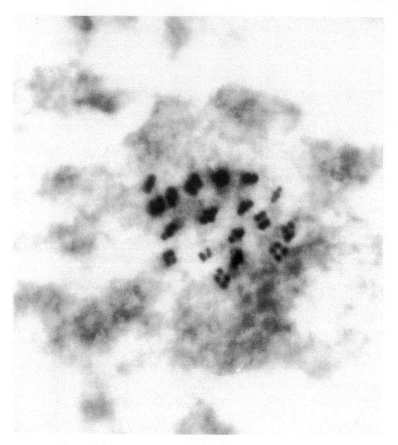

FIGURE 23   Light micrograph of chromosomes of *M. exigua*. (Courtesy of A. C. Triantaphyllou.)

Distribution: *Meloidogyne exigua* occurs in the major coffee-producing areas of the world, mainly in South and Central America, and islands in the Caribbean. Populations of this species have been reported in Brazil, Peru, Venezuela, Colombia, Guatemala, Dominican Republic, Costa Rica, El Salvador, Martinique, Trinidad (Cain, 1974), Paraguay, Bolivia, and Surinam (Jepson, 1987). The coffee root-knot nematode may also occur on the Indian subcontinent and in Thailand, Greece, Java (Jepson, 1987), and Italy (Scognamiglio et al., 1985).

2.   *Meloidogyne incognita*

Female: The body is characteristically pear-shaped; the posterior end is globose and the neck projects anteriorly, usually in line with the tail end (Fig. 8A). Females are often completely embedded in plant tissue, whereas the egg masses break through and protrude from the gall.

The perineal pattern (Fig. 24C–F) characteristically has a high dorsal arch with smooth to waxy striae. However, some patterns have a lower dorsal arch and finely spaced, wavy striae; others have indications of lateral wings and distinct forking near the lateral lines. Often some striae bend toward the vulva. Two morphological variants have been described (Chitwood, 1949; Triantaphyllou and Sasser, 1960). A population derived from a

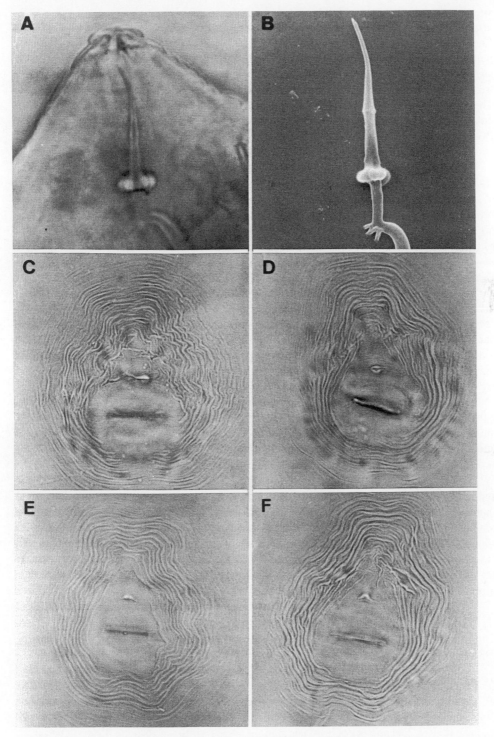

FIGURE 24   Micrographs of females of *M. incognita.* (A) Light micrograph of the anterior end. (B) Scanning electron micrograph of an excised stylet. (C–F) LM of perineal patterns. (After Eisenback et al., 1980.)

single egg mass isolate produced offspring with both types of perineal patterns. *M. incognita* var. *incognita* type of patterns are characterized by a high squarish dorsal arch with fine, wavy striae in the subdorsal sectors (Fig. 24C). Likewise, *M. incognita* var. *acrita*-type patterns (Fig. 24E) have a high squarish dorsal arch, but the striae in the subdorsal sectors are not wavy and they are coarser than in *M. incognita* var. *incognita*. Both variants have a triangular postanal whorl and often have transverse striae that tend to bend toward the vulva. The lateral lines are faintly marked by breaks and forks in the striae, but distinct lateral incisures are not present. There is no relationship between morphology and parasitism in these two forms of *M. incognita* (Canto-Saenz and Brodie, 1987), and both the species *M. acrita* and the subspecies *M. incognita acrita* are not widely accepted as valid taxa.

The stylet of the female (Fig. 24A,B) is 15–17 µm long (Eisenback et al., 1980). The stylet cone is sharply pointed and distinctly curved dorsally near its middle; the shaft is slightly wider at its base. The stylet knobs are strongly set off from the shaft; they are broadly elongate and anteriorly indented. In some specimens the indentations are so pronounced that each knob almost appears as two. The opening of the DEGO is relatively close to the base of the knobs, generally 2–4 µm.

Male: The shape of the head (Fig. 25A,C,D) is very distinct and shared by all of the named subspecies of *M. incognita* (Jepson, 1987) and *M. kirjanovae*, *M. tadshikistanica*, and one of the morphological variants of *M. californiensis*. The large rounded labial disk is distinctly raised above the medial lips and may be centrally concave (Eisenback et al., 1981). In some populations, the labial disk is consistently not as much rounded and raised, and is more continuous with the medial lips. The medial lips are crescent-shaped in face view and extend for some distance posteriorly onto the head region. In some populations they seem to be more elongate and extend further onto the head region. Lateral lips are generally absent (Fig. 25D), although in a few populations they are well developed (Fig. 25C), and in others only remnants are visible on the head region (Eisenback and Hirschmann, unpublished). The head region is usually subdivided by incomplete head annulations, whose number varies within and among populations. Usually, up to five rows of short, broken annulations are common; but some specimens or populations have very few or no annulations at all. Head annule and body annules are usually in the same contour.

The stylet of males of *M. incognita* (Fig. 25A,B) varies in length from 23 to 25 µm and has a characteristic morphology (Eisenback, 1981). The stylet tip is bluntly rounded and flattened laterally. A ventral projection approximately 4 µm from the tip marks the posterior of the opening of the stylet lumen. The shaft is cylindrical but often narrows near the knobs which are set off from the shaft. The knobs vary in size and shape among different populations from small and rounded to large and broadly elongated. Often their anterior surfaces are slightly indented. The distance from the base of the knobs to the DEGO is relatively short, 2–4 µm.

Second-stage juvenile: The total body length of infective second-stage juveniles of *M. incognita* is 346–463 (405) µm, range (average), and the tail length is 42–63 (52) µm (Eisenback et al., 1981) (Fig. 26). The hyaline tail terminus is 6– 13.5 (9) µm (Jepson, 1987).

The head morphology as revealed by SEM is very characteristic for the species but the various features are too small to be seen in the LM (Eisenback, 1982) (Fig. 26). The labial disk and medial lips are dumbbell-shaped in face view, and the labial disk is rounded and raised above the medial lips. The lateral lips are rounded to triangular and may fuse with the head region. The number of head annulations varies within and among different populations. Usually the head region is marked by two to three irregular, incomplete annulations,

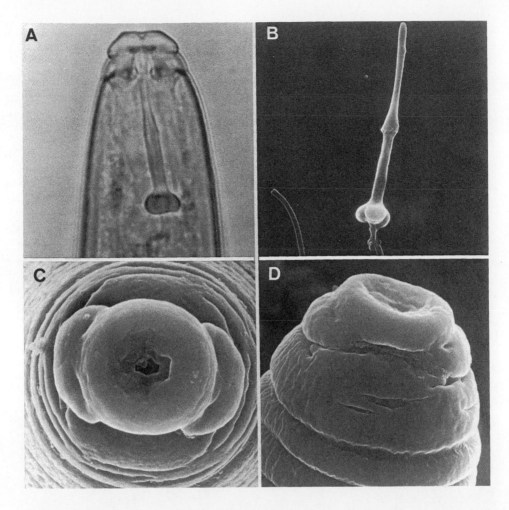

FIGURE 25   Micrographs of males of *M. incognita*. (A) Light micrograph of the head and stylet in lateral view. (B) Scanning electron micrograph of excised stylet. (C) SEM of the anterior end in face view. (D) SEM of the anterior end in lateral view. (After Eisenback et al., 1981.)

but may have up to seven annulations, or may be smooth in some specimens or populations. Both the head region and the body annules are in the same contour.

The stylet of the second-stage juveniles of *M. incognita* is 10–12 μm long (Fig. 26A,B). The knobs are rounded and set off from the shaft (Eisenback, 1982). They generally slope slightly posteriorly, and the distance from the base of the knobs to the DEGO is relatively short, (2.0–3.0 μm). The tail is usually narrow and conical, tapering to a finely rounded tip (Fig. 26E–H), but some populations have wider tails ending in broadly rounded tips (Eisenback and Hirschmann, unpublished).

Morphological variants: Detailed morphological investigations of several populations from the two cytological and four physiological host races demonstrated that all populations of *M. incognita* were sufficiently morphologically similar to be considered as one taxonomic unit (Hirschmann, 1984). Populations from several described subspecies, in-

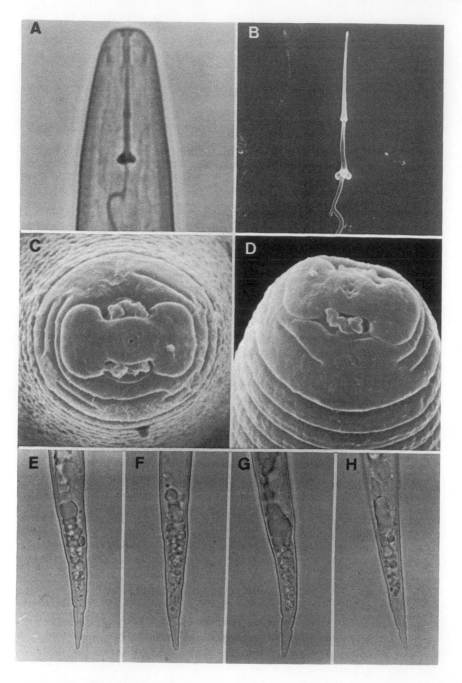

FIGURE 26   Micrographs of second-stage juveniles of *M. incognita*. (A) Light micrograph of the head and stylet in lateral view. (B) Scanning electron micrograph of excised stylet. (C) SEM of the anterior end in face view. (D) SEM of the anterior end in lateral view. (E–H) LM of tails. (A and B after Eisenback, 1982; C and D after Eisenback et al., 1981.)

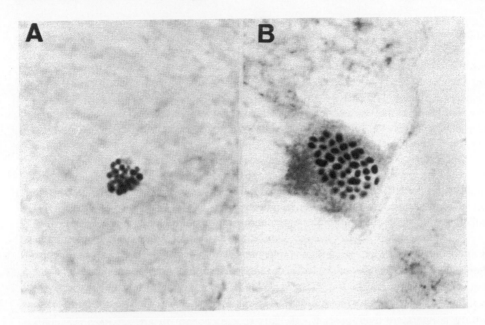

FIGURE 27   Light micrographs of prophase and metaphase chromosomes in maturing oocytes of *M. incognita*. (A) Most of the chromosomes in oocytes present in the uteri are clumped together at prophase. (B) Discrete, countable chromosomes are present in the posterior part of the uteri at metaphase in only one or two oocytes. (After Eisenback et al., 1981.)

cluding *M. incognita grahami*, *M. incognita acrita*, and *M. incognita wartellei*, were found to belong within the morphological variability of the species. Physiological differences of the four host races were not expressed morphologically, although some populations of each cytological race were distinct. Also, when considered as a group, the populations of the diploid race were slightly different from those of the triploid race; however, all populations were morphologically similar in many respects, and recognition of new taxa was not recommended.

Host races: Four host races of *M. incognita* have been widely recognized and accepted (Hartmann and Sasser, 1985). Populations of all four races reproduce on pepper, tomato, and watermelon; they vary in their response to tobacco NC95 and cotton (Table 1). Race 1 populations do not reproduce on tobacco NC95 or on cotton, Race 2 populations reproduce on tobacco but not on cotton, race 3 populations reproduce on both. Race 1 populations are most frequent and race 2 and 3 populations are less common, whereas race 4 populations are rare. Additional physiological races have been described from the reactions of additional host plants (Southards and Priest, 1971), but they have not been widely accepted.

Symptoms: Galls produced by *M. incognita* are not characteristic for the species. Although galls may occur singly, when populations are high, many galls coalesce to form large and sometimes massive knots. Secondary roots may emerge from the galls but they are not numerous as with *M. hapla*, and the galls are not terminal as with *M. exigua*.

Cytology: Populations of *M. incognita* reproduce exclusively by mitotic parthenogenesis and occur as two cytological forms (Fig. 27) (Triantaphyllou, 1985). The triploid form, 3N = 40– 46, is the most common and is widely distributed around the world. The diploid form 2N = 32–36, is less common. All populations of *M. incognita* have a unique cytological feature in the maturation of oocytes that easily distinguishes them from all other species of *Meloidogyne*. As the oocytes pass through the spermatotheca, they remain in prophase until they have traveled to the posterior portion of the uterus. There they rapidly proceed to metaphase. The chromosomes are closely grouped together during the prolonged prophase and cannot be easily counted as individual chromosomes (Fig. 27). The chromosomes of all other root-nematode species advance to metaphase when they pass through the spermatotheca into the uterus. They are discrete, spread out in a large area, and can be counted as individual units.

Biochemistry: Most populations of *M. incognita* have a unique esterase phenotype, namely I1 (Fig. 19) (Esbenshade and Triantaphyllou, 1985). Variability of esterase activity in populations of *M. incognita* from around the world is small; only one of many populations does not have the typical phenotype. The malate dehydrogenase phenotype, N1, is similar to that of *M. javanica*, *M. exigua*, and some populations of *M. arenaria*.

Distribution: *Meloidogyne incognita* is one of the most ubiquitous species in the genus. It occurs over a wider geographic range than any other species, from approximately 40° N latitude to 33° S (Taylor et al., 1982), and has an extremely large host range. The annual average temperature range for *M. incognita* is between 18 and 30°C, but the majority of populations are found in an annual temperature range of 24–30°C. The optimum warmest monthly temperature average is 27°C. This species is rare in areas where the coldest monthly average temperature is below 3°C. *Meloidogyne incognita* is commonly found concomitantly with *M. javanica*.

3.   *Meloidogyne javanica*

Female: The body is pear-shaped and the distinct neck is in line with the posterior end (Fig. 28A). Egg masses often protrude from the gall, but the females are usually completely embedded within plant tissue.

The perineal pattern (Fig. 28C–F) is often very diagnostic for the species because many specimens have distinct lateral lines that clearly delineate the dorsal and ventral regions of the pattern (Chitwood, 1949; Eisenback, 1981). Usually few, if any, striae cross the lateral lines which often extend for some distance away from the perineal area. Patterns that resemble those of *M. arenaria* and *M. incognita* can be easily identified as *M. javanica* by the presence of distinct lateral lines. The overall shape of the perineal pattern is round or oval to slightly squarish. The dorsal arch is moderately high and narrow and has smooth to coarse, broken striae that may be forked near the lateral lines. Some striae may bend toward the vulva.

The stylet of the female (Fig. 28A,B) is similar to that of *M. incognita*, except that the cone is only slightly curved dorsally (Eisenback et al., 1980). It is 14–18 (16) μm long. The opening is near the tip (0.5 μm) and is marked by a slight protuberance posteriorly. The shaft is cylindrical and broadens near the base. The large, transversely elongate knobs are set off from the shaft and are often market anteriorly by a shallow indentation. The distance from the DEGO to the stylet knob base is variable (2–5 μm).

Male: The shape of the head of *M. javanica* is unique for the species in lateral view in the LM (Eisenback et al., 1981) (Fig. 29A). The labial disk and medial lips form one smooth continuous head cap that is high and rounded and distinctly set off from the head region. In face view, the lateral edges of the oral disk are nearly parallel with each other and overlap the amphidial openings (Fig. 29C,D). The medial lips are rounded and their lateral edges are slightly wider than the labial disk. Lateral lips are not expressed. The head region

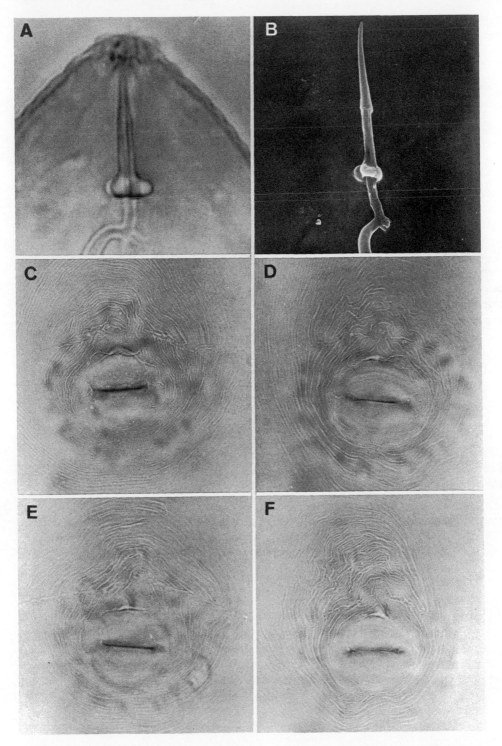

FIGURE 28   Micrographs of females of *M. javanica*. (A) Light micrograph of the anterior end. (B) Scanning electron micrograph of an excised stylet. (C–F) LM of perineal patterns. (After Eisenback et al., 1980.)

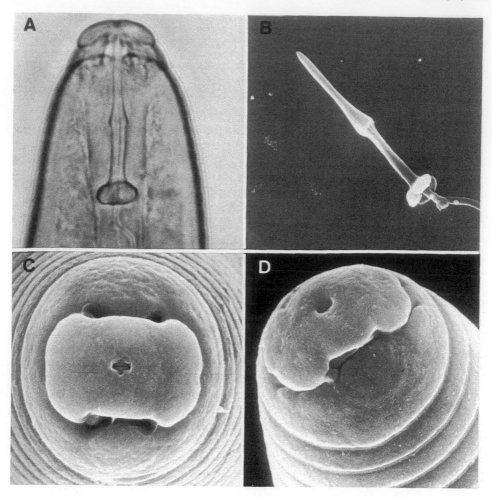

FIGURE 29   Micrographs of males of *M. javanica*. (A) Light micrograph of the head and stylet in
lateral view. (B) Scanning electron micrograph of excised stylet. (C) SEM of the anterior end in face
view. (D) SEM of the anterior end in lateral view. (After Eisenback et al., 1981.)

is usually smooth, but may be marked by two to three irregular and incomplete annulations
in some populations.

The head shape of males of *M. javanica* is similar to that of several other species,
namely, *M. exigua*, *M. graminis*, *M. naasi*, *M. graminicola*, *M. krali*, and others. The head
cap of *M. javanica* is slightly higher and more rounded, and the area around the prestoma is
slightly elevated as well. Also, the anterior body diameter is wider than the head region and
appears set off in *M. javanica*; but in the other species the head and body are in the same
contour.

The stylet of the male is unique for the species and is useful for separating *M.
javanica* males from the other species with similar head shapes (Eisenback et al., 1981)
(Fig. 29A,B). The cone is bluntly pointed. In many specimens the stylet opening is long and
slitlike and lies some distance (3–4 μm) from the point. The transversely elongate, broad

knobs are distinctly set off from the shaft. The distance of the dorsal gland orifice to the base of the knobs is relatively short (2–4 μm).

Some populations of *M. javanica* produce male intersexes that exhibit varying degrees of female secondary sex characteristics ranging from a small ventral protuberance anterior to the cloacal opening to a large protuberance marked with a rudimentary vulva (Chitwood, 1949; Eisenback et al., 1981). In some populations nearly all of the males are intersexes; in others, intersexes are rare of absent. The occurrence of intersexes can be useful in species identification.

Second-stage juvenile: The body length of preparasitic juveniles of *M. javanica* is 402–560 μm (488 μm), and the tail is 51– 63 μm (56) μm (Eisenback et al., 1981) (Fig. 30). The hyaline tail terminus is 9–18 (13.7)μm long (Jepson, 1987).

The shape of the head of *M. javanica* second-stage juveniles as seen by SEM is characteristic for the species, but the differences are too small to be seen in the LM (Eisenback, 1982) (Fig. 30). The labial disk and medial lips are often dumbbell-shaped as in *M. arenaria* and *M. incognita*, but in some specimens they may be bowtie-shaped (Eisenback et al., 1981). The lateral lips are generally triangular in shape, and sometimes fused with the head region, but in some specimens they may be more elongate, similar to those of *M. arenaria* and *M. incognita* (Cliff and Hirschmann, 1985; Hirschmann, 1984). The head may be marked by a few short incomplete annulations but is generally smooth.

The stylet of *M. javanica* second-stage juveniles is 10–12 μm long, and the distance of the DEGO to the base of the knobs is 3.5 μm (Eisenback, 1982). The shape of the stylet is similar to that of *M. incognita*, but the knobs are more transversely elongate (Fig. 30A,B).

The conical tail tapers to a fine, rounded tip (Fig. 30 E–H). The hyaline terminus is distinct and may have constricting annules. Second-stage juveniles of *M. javanica* exhibit few useful differentiating characters; they resemble juveniles of the other major species in many aspects.

Morphological variants: A morphological and morphometric comparison by LM and SEM examined the variability of certain morphological characters within and among six populations of *M. javanica* belonging to three host races (reproducing on pepper, peanut, or not reproducing on either) (Rammah and Hirschmann, 1989). No morphological differences were observed between the populations of the three host races. The most useful diagnostic characters of *M. javanica* were found to be head and stylet morphology of males, and stylet and perineal pattern morphology of females.

Some populations may produce intersexes that appear as normal males, but they have rudimentary vaginas in various states of development (Eisenback and Hirschmann, unpublished). Other populations do not produce any males.

Host races: Host races of *M. javanica* have not been widely recognized, although a few populations vary from the normal host response in the standard differential host test (Hartman and Sasser, 1985). Usually populations of *M. javanica* reproduce on watermelon, tomato, and *M. incognita*-resistance tobacco, but do not reproduce on cotton, pepper, or peanut (Hartman and Sasser, 1985). Occasionally, however, some populations may reproduce on pepper and more rarely on peanut. These populations have been informally referred to as the "pepper race" and the "peanut race" of *M. javanica*.

Symptoms: Galls produced by this species are similar to those of *M. incognita*, *M. arenaria*, and many other species, and are thus not diagnostic.

Cytology: *Meloidogyne javanica* populations reproduce exclusively by mitotic parthenogenesis (Triantaphyllou, 1985). All populations examined belong to one chromosomal form that may represent a triploid with a chromosome number of 2N = 43–48. The chromosomes of *M. javanica* populations are univalents during metaphase of the single-

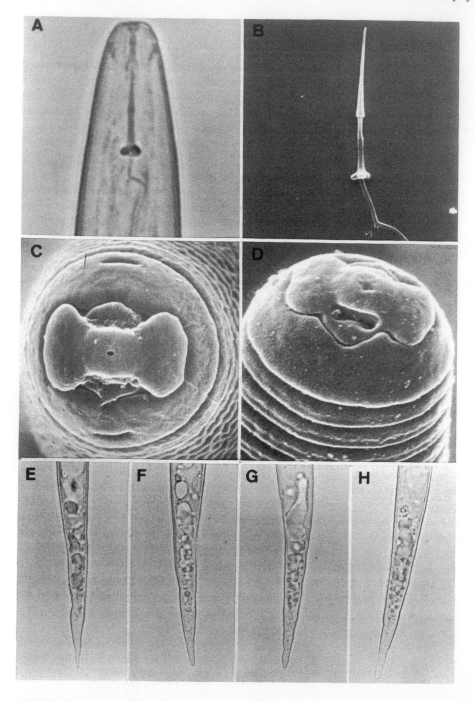

FIGURE 30   Micrographs of second-stage juveniles of *M. javanica*. (A) Light micrograph of the head and stylet in lateral view. (B) Scanning electron micrograph of excised stylet. (C) SEM of the anterior end in face view. (D) SEM of the anterior end in lateral view. (E–H) LM of tails. (A and B after Eisenback, 1982; C and D after Eisenback et al., 1981.)

maturation division (Fig. 31). They are usually spread in a large metaphase plate and are easier to count than in the other species of *Meloidogyne*. Commonly, two to four oocytes are at metaphase in the uterus near the spermatotheca and can be studied. All other oocytes in the uterus are usually in anaphase and telophase, and are of limited value for cytological investigations.

Biochemistry: One distinctive esterase phenotype characterizes all populations of *M. javanica* examined, namely J3 (Fig. 19) (Esbenshade and Triantaphyllou, 1985). Little variation of esterase phenotypes was found in populations of this species from around the world, and no other population of the other species examined had the J3 phenotype. Most populations of *M. javanica* have a malate dehydrogenase phenotype N1, similar to that of *M. incognita*, *M. exigua*, and some populations of *M. arenaria*. One population of *M. javanica* from Bangladesh and one from Korea have the N3 phenotype similar to that of a few populations of *M. arenaria*.

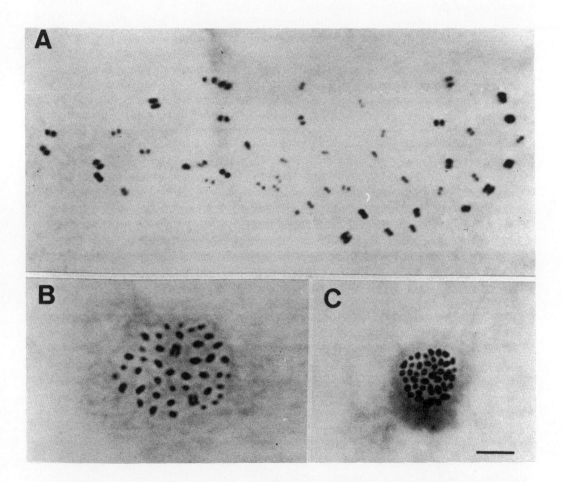

FIGURE 31   Micrographs of chromosomes during the single maturation division of oocytes of *M. javanica*. The univalent chromosomes (dyads) indicate that the species reproduces by mitotic parthenogenesis. (A) Prometaphase with 46 univalents in an oocyte that just entered the uterus. (B) Metaphase or anaphase (polar view). (C) Telophase in an advanced oocyte in the posterior part of the uterus. (After Triantaphyllou, 1985.)

Distribution: *Meloidogyne javanica* occurs throughout the world. It is the second most common species. Like *M. incognita*, *M. arenaria*, and *M. hapla*, it has an extensive host range. The latitude range for *M. javanica* is from approximately 33 °N to 33 °S, about 3° less than that of *M. incognita* (Taylor et al., 1982). This species is rare in areas where the coldest monthly temperature averages below 3°C. The distribution seems to be limited by cold temperatures and enhanced by periods of drought. *Meloidogyne javanica* is the predominant species in areas with a distinct dry season with less than 5 mm of precipitation per month for three or more successive months. Commonly, *M. javanica* occurs concomitantly with *M. incognita* and sometimes with *M. arenaria*; sometimes all three species occur in the same field.

4.   *Meloidogyne arenaria*

Female: The body is pear-shaped and the neck is in line with the posterior body end (Fig. 28A). Females often are embedded completely in the plant tissue whereas the egg masses usually break through the gall and protrude externally.

The perineal pattern is often characteristic for the species (Chitwood, 1949; Eisenback et al., 1981; Cliff and Hirschmann, 1985) (Fig 32C–F) but is highly variable and not very useful for identification. The dorsal arch is low and rounded to high and squarish; the striae are coarse and smooth to wavy. Some striae may bend toward the vulva. In the lateral areas, the lines in the dorsal arch tend to curve sharply toward the tail terminus and meet the ventral striae at an angle. Also these striae become forked and wider near the lateral areas, which are often delineated, although distinct lateral incisures are generally absent. If lateral incisures are present, they extend only a short distance anteriorly. Patterns may also have striae that extend laterally and form one or two "wings." Sometimes the perineal patterns of *M. arenaria* may be reminiscent of *M. incognita*. *Meloidogyne arenaria* patterns with lateral lines may also be confused with those of *M. javanica*, which have longer lateral lines. Some rounded to hexagonal patterns may resemble those of *M. hapla*, but almost all specimens of *M. hapla* have punctations in the tail terminus and the striae are generally finer.

The stylet length of females of *M. arenaria* is 13–17 (16) μm and the distance of the DEGO to the base of the stylet is 3–7 (5) μm (Eisenback et al., 1981; Cliff and Hirschmann, 1985). The opening of the stylet, which is market posteriorly by a small protuberance, is close (0.5 μm) to the pointed stylet tip. The robust stylet is very characteristic for the species (Fig. 32A,B). The cone is slightly curved dorsally and gradually increases in width posteriorly. Likewise, the shaft gradually widens posteriorly to near the junction with the stylet knobs which are rounded or tear-dropped shaped and gradually taper onto the shaft. The lumen of the stylet is very wide.

Male: The shape of the head of *M. arenaria* is a useful diagnostic character; however, some specimens of other species such as *M. enterolobii*, *M. partityla*, and *M. pini* may have very similar head shapes. In face view the labial disk is large and round and fused with the crescent-shaped medial lips (Eisenback et al., 1981; Cliff and Hirschmann, 1985) (Fig. 33). Lateral lips are usually absent and only occasionally marked by short remnant lines. In lateral view the large, smooth head cap is variable in height but it is usually low to moderately raised, and extends a short distance onto the head region. The head region is usually smooth but rarely may be marked by one to two incomplete annulations. Both the head region and the body annules are in the same contour.

The stylet of the male of *M. arenaria* is 20–28 (23) long (Eisenback et al., 1981). The stylet opening is 3–4 μm from the bluntly pointed tip (Fig. 33A,B). The anterior portion of the cone is cylindrical and the posterior one-third gradually increases in width. The wide, cylindrical shaft merges with the large rounded to slightly tear-drop-shaped knobs. The dis-

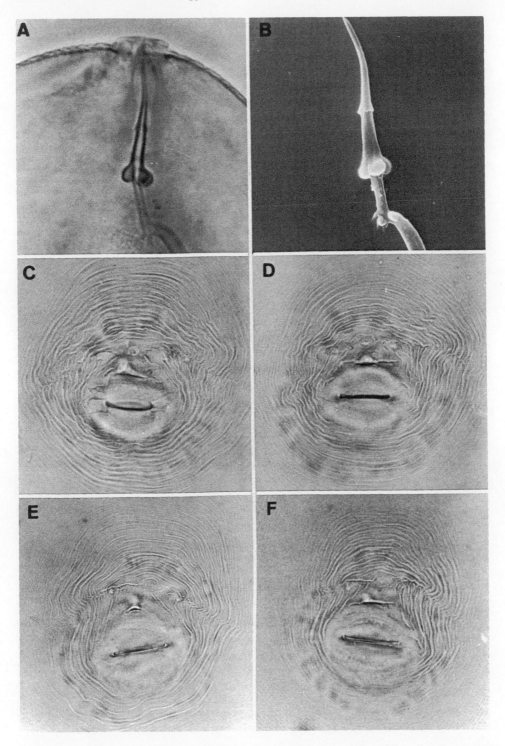

FIGURE 32 Micrographs of females of *M. arenaria*. (A) Light micrographs of the anterior end. (B) Scanning electron micrograph of an excised stylet. (C–F) LM of perineal patterns. (After Eisenback et al., 1980.)

tance of the DEGO to the base of the stylet knobs is usually long to very long (4–8 μm, average 6 μm).

Second-stage juvenile: The total body length of preparasitic juveniles of *M. arenaria* is 398–605 (504) μm (Eisenback et al., 1981; Cliff and Hirschmann, 1985). The tail length is 44–69 (56) μm, and the hyaline tail terminus is 6–13 (9)μm long (Fig. 34).

The head shape is very similar to that of several other species, namely, *M. graminis*, *M. graminicola*, *M. marylandi*, *M. ardenensis*, *M. microcephala*, *M. oryzae*, *M. naasi*, *M. querciana*, and others. Also, sometimes the heads of *M. incognita* and *M. javanica* may be difficult to distinguish from those of *M. arenaria*. The labial disk and medial lips are fused and form a dumbbell-shaped head cap (Eisenback et al., 1981) (Fig. 34C,D). The rounded to

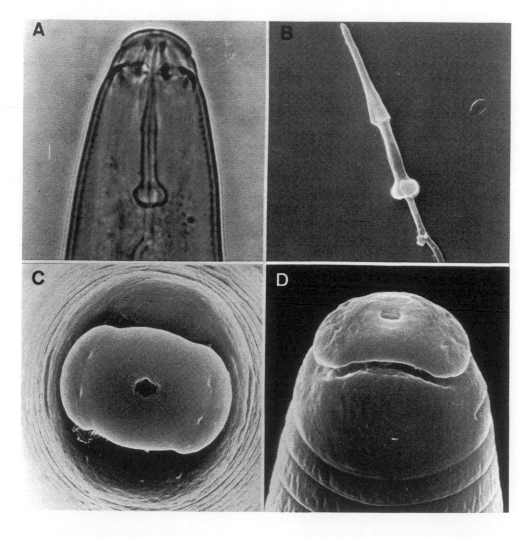

FIGURE 33   Micrographs of males of *M. arenaria*. (A) Light micrograph of the head and stylet in lateral view. (B) Scanning electron micrograph of excised stylet. (C) SEM of the anterior end in face view. (D) SEM of the anterior end in lateral view. (After Eisenback et al., 1981.)

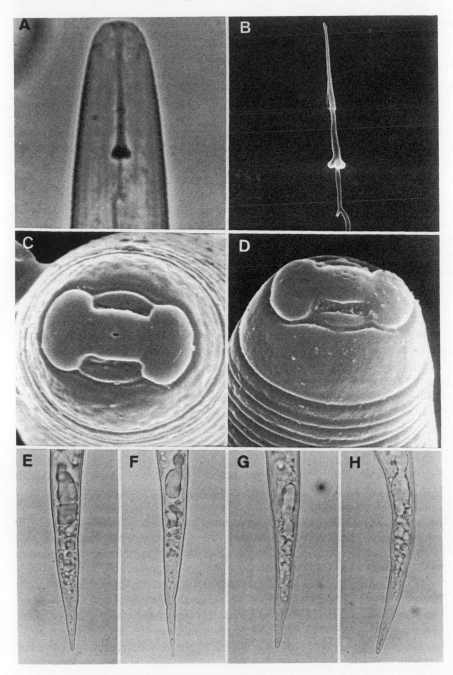

FIGURE 34   Micrographs of second-stage juveniles of *M. arenaria*. (A) Light micrograph of the head and stylet in lateral view. (B) Scanning electron micrograph of excised stylet. (C) SEM of the anterior end in face view. (D) SEM of the anterior end in lateral view. (E–H) LM of tails. (A and B after Eisenback, 1982; C and D after Eisenback et al., 1981.)

triangular lateral lips are lower than the medial lips and sometimes fused with the head region. The head region is usually smooth, and only occasionally has one or two incomplete annulations. The head region is in contour with the body annules.

The stylet of second-stage juveniles of *M. arenaria* is 10–12 (11) μm long. It has fairly large, distinctly separate knobs that are round to triangular in profile and gradually taper onto the shaft (Eisenback, 1982) (Fig. 34A,B). The distance of the base of the knobs to the DEGO is relatively long (3–5 μm, average 4 μm) (Cliff and Hirschmann, 1985).

The long, slender tail has a narrow, tapering terminus which may bear several distinct annulations (Fig. 34E–H). The tail tip is finely rounded to pointed. The hyaline tail terminus is not well defined.

Morphological variants: The morphological variability of seven populations representing the triploid form (2N = 51–56) and the two host races of *M. arenaria* was evaluated recently by Cliff and Hirschmann (1985). Populations of the two host races could not be distinguished on the basis of morphology. Two of the seven populations were slightly different from the typical populations of *M. arenaria*; however, they were not divergent enough to be considered as new species. The variants were similar to each other, and differed from the typical populations in perineal pattern morphology, head shape, and stylet morphology of males and second-stage juveniles, stylet morphology of females, and tail shape of second-stage juveniles.

A morphological comparison of seven hypotriploid populations (2N = 40–48) of *M. arenaria* showed that these populations differed from each other and the typical *M. arenaria* by certain characteristic features (Rammah and Hirschmann, 1984). The differences, however, were insufficient to recognize the variant populations as new species. The most useful species-specific characters of *M. arenaria* were stylet and head morphology of males and stylet morphology of females. Perineal patterns were highly variable and constituted the least useful differentiating character.

Host races: Two host races of *M. arenaria* have been widely recognized and accepted (Hartman and Sasser, 1985). All populations reproduce on tomato, watermelon, and *M. incognita*-resistance tobacco. They vary in their ability to parasitize peanut and pepper. Some populations readily reproduce on peanut and are referred to as race 1 or the peanut race; whereas populations of race 2 cannot reproduce on peanut. Frequently populations of race 2 will not reproduce on pepper, and neither race reproduces on cotton or strawberry.

Symptoms: Galls produced by populations of *M. arenaria* are large and similar to those caused by several other species (Fig. 6). Sometimes, however, a population may produce many small, beadlike galls that lack lateral roots and are regularly spaced along the root simulating a string of pearls.

Cytology: Populations of *M. arenaria* reproduce exclusively by mitotic parthenogenesis (Triantaphyllou, 1985). Cytologically, *M. arenaria* is a very complex species that includes a triploid form (2N = 30–38) and a hypotriploid form (2N = 40–48), both of which occur less frequently (Fig. 35). The behavior and morphology of the chromosomes of *M. arenaria* are similar to those of *M. javanica*. They are univalents during metaphase of the single-maturation division and spread out in a large plate. These two species differ only in chromosome number and accurate counts are essential for their differentiation. Populations of *M. javanica* have a somatic chromosome number of 42–48.

Biochemistry: Populations of *M. arenaria* are more variable than those of the other three most common species (Esbenshade and Triantaphyllou, 1985). Three phenotypes of esterase activity commonly occur namely, A1, A2, and A3, and several other phenotypes occur less commonly (Fig. 19). The phenotypes of A1 and A2 include several cytological forms of *M. arenaria*, whereas phenotype A3 includes only the most typical cytological

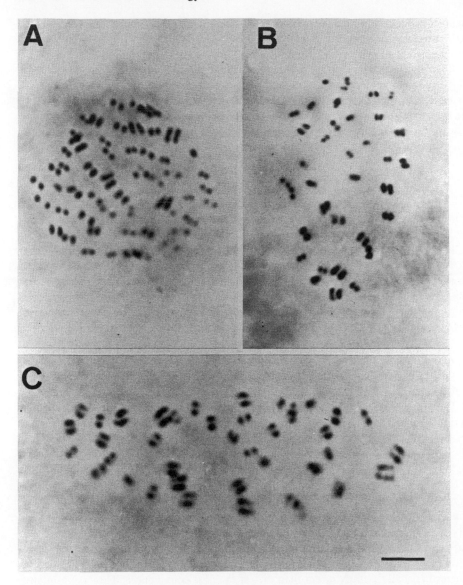

FIGURE 35   Micrographs of chromosomes of *M. arenaria* in oocytes at prometaphase of the single maturation division. (A) Triploid form with 54 chromosomes. (B) Diploid form with 36 chromosomes. (C) Hypotriploid form with 42 chromosomes. All populations reproduce by mitotic parthenogenesis as evidenced by the univalent chromosomes (dyads). (After Triantaphyllou, 1985.)

form with a somatic chromosome number of 51–56. Some populations of *M. arenaria* have atypical esterase phenotypes, namely, S1–M1, S2–M1, and M3–F1. The malate dehydrogenase phenotype varies from N1 to N3 according to population.

Distribution: *Meloidogyne arenaria* is widely distributed around the world and is the third most common species. Although it is not as common as *M. incognita* and *M. javanica*, its distribution is nearly the same as that of *M. incognita*, approximately 40 °N latitude to 33 °S latitude (Taylor et al., 1982). The optimum warmest monthly average temperature is approximately 24 °C. This species is rare in areas where the coldest monthly average temperature is below 3 °C.

5.   *Meloidogyne hapla*

Female: The body is pear-shaped and the neck is in line with the posterior end of the body (Fig. 8). The egg masses normally break through the gall and protrude externally, but the females are usually contained inside the gall.

The perineal pattern is usually characteristic for the species (Chitwood, 1949; Eisenback et al., 1981) (Fig. 36 C–F). The striae are fine and smooth to wavy. In general, patterns vary from rounded hexagons to flattened ovals. The dorsal arch is usually rounded and flattened, but in some specimens it may be higher and squarish. Lateral lines are inconspicuous; however, they may be marked by slight irregularities in the striae or by dorsal and ventral striae that meet at an angle. Often the lateral areas are marked by a deep groove dividing the dorsal and ventral sectors of the pattern. Sometimes striae may extend laterally and form one or two wings. The most diagnostic feature for this species is the occurrence of punctations in the tail terminal area. These punctations may not be readily apparent in poorly preserved specimens.

The length of the stylet of females of *M. hapla* is 14–17 (15.5) μm and the distance of the DEGO to the base of the stylet is 5–6 μm (Eisenback et al., 1980). The stylet is more delicate than that of the other three most common species (Fig. 36A,B). The cone is sharply pointed and gradually increases in width along its entire length. The shaft is cylindrical to slightly tapered posteriorly. The knobs are rather small and rounded, set off from the shaft.

Male: The shape of the head of the male of *M. hapla* is diagnostic and useful for species identification (Eisenback et al., 1981) (Fig 37). The labial disk and medial lips are fused into a high head cap that is slightly elongate medially; however, it does not extend posteriorly onto the head region as in the other three most common species. In lateral view the head cap is rounded. Lateral lips may be present in some populations or in some specimens. The head region is smooth; it is not in the same contour with the body annules because its diameter is usually slightly larger than that of the first body annule. The body annules also decrease in width as they near the head region. Based on head shape alone, *M. hapla*, *M. suginamiensis*, *M. jinanensis*, and *M. microtyla* are difficult to distinguish; however, the stylet of *M. microtyla* is much more delicate and the distance of the DEGO to the base of the knobs is shorter. *Meloidogyne jinanensis* may eventually become a synonym of *M. hapla*.

The stylet length of *M. hapla* males is 17–23 (20.5)μm, and the distance of the DEGO to the base of the stylet is 4–5 (4.5) μm (Eisenback et al., 1981). The stylet of males of *M. hapla* is small and thin; the cone gradually increases in width posteriorly. The shaft is variable in its morphology (Fig. 37A,B); it may be cylindrical, often narrowing at its junction with the knobs, or it may gradually increase in diameter along its entire length. In some specimens the stylet opening is marked by a slight protuberance. posteriorly. The knobs are small and rounded, set off from the shaft.

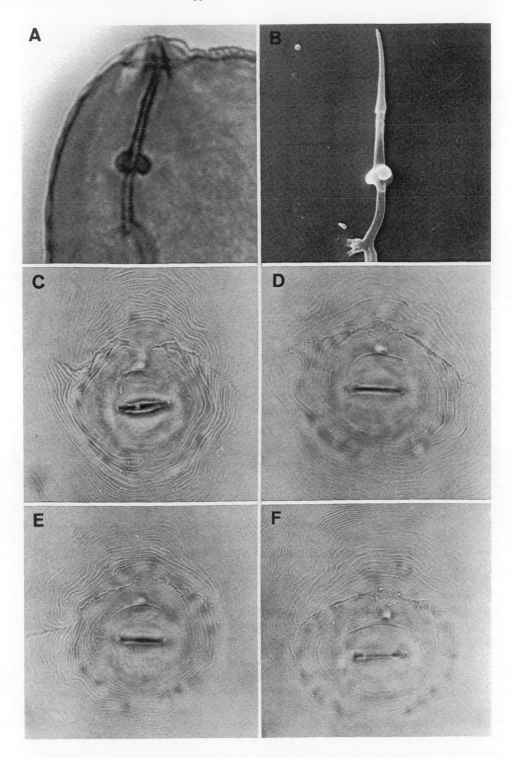

FIGURE 36   Micrographs of females of *M. hapla*. (A) Light micrograph of the anterior end. (B) Scanning electron micrograph of an excised stylet. (C–F) LM of perineal patterns. (C–F after Eisenback et al., 1980.)

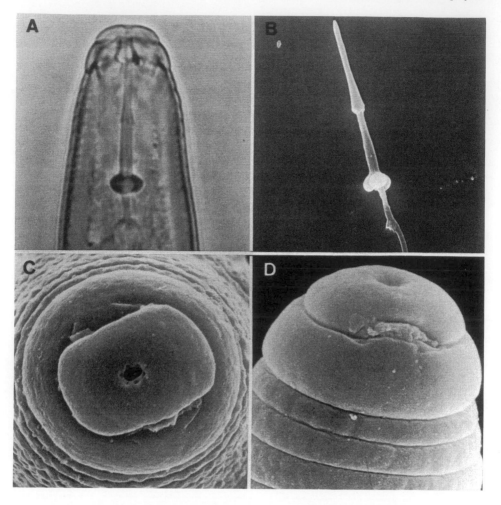

FIGURE 37   Micrographs of males of *M. hapla*. (A) Light micrograph of the head and stylet in lateral view. (B) Scanning electron micrograph of excised stylet. (C) SEM of the anterior end in face view. (D) SEM of the anterior end in lateral view. (A after Eisenback et al., 1981.)

357–467 (413)μm long, whereas populations of race B are longer, 410–517 (474)μm. The tail length of race A populations is 46–58 (53)μm and 54–69 (62) μm in race B populations. The hyaline tail terminus is 12–19 (16) μm long (Fig. 38).

The shape of the head is characteristic for the species (Eisenback, 1982) (Fig. 38). The labial disk and medial lips are fused into a head cap. In some populations the labial disk may by slightly raised above the medial lips, but in others it is in the same contour. Lateral lips are usually present, but in some populations they may be absent. The posterior edges of the medial lips are usually rounded to slightly indented medially; however, in some populations they may be pointed medially. The head region is large and smooth and has no annulations. In lateral view in the LM, the head region is more rounded and the head cap is smaller than in the three other common species.

The stylet of juveniles of *M. hapla* is 10–12 (11) μm long, and the distance from the base of the stylet to the DEGO is 3–4 (3.5) μm (Eisenback, 1982). The stylet is delicate, and the rounded knobs are small and set off from the shaft (Fig. 38A,B).

The long, slender tail has a narrow tapering terminus that may bear several distinct annulations (Fig. 38E–H). The tail tip is finely rounded to pointed and the hyaline tail terminus is not clearly demarcated.

Morphological variants: Differences occur in the shape of the male head and stylet, head shape of second-stage juveniles, and in several morphometric values of various stages (Eisenback and Hirschmann, 1979). Generally populations of race A are smaller than populations of race B.

Differences in head shape, other than those already described between the cytological races, are usually small except for one population from Canada (Eisenback and Hirschmann, 1979) in which the head cap is elongate and the medial lips are pointed rather than rounded. Also, the head region is not set off from the body annules but is in the same contour with the body annules. In the second-stage juvenile, the medial lips are also pointed and the lateral lips are fused with the head region. Only the differences in the morphology of the male can be seen clearly in the LM.

Host races: Populations of *M. hapla* reproduce on peanut, pepper, NC95 tobacco, and tomato, but do not reproduce on cotton or watermelon (Hartman and Sasser, 1985). Populations of race A and race B varied in their ability to reproduce on *Tagetes patula* (Eisenback, 1987). Race B populations readily reproduced on *T. patula* but race A populations did not. Only one population of race A with 17 chromosomes from Virginia reproduced on *T. erecta*. *Meloidogyne hapla* is a serious pest on strawberry, peanut, potato, carrot, rose, lettuce, celery, and other cool climate crops. This species does not reproduce on any of the grasses or grains such as wheat, oats, barley, rye, and corn.

Symptoms: The galls produced by *M. hapla* on susceptible plants are often diagnostic of the species. They are small and have numerous secondary roots emerging from them (Fig. 6C). Heavily galled root systems generally appear thick and matted; however, some populations may produce galls that are similar to those of the other three most common species.

Cytology: *Meloidogyne hapla* occurs as two distinct cytological forms (Triantaphyllou, 1985). Populations of race A are most common. They reproduce by facultative meiotic parthenogenesis and usually have haploid chromosome numbers of 16 or 17, although some populations have chromosome numbers of 14 or 15 (Fig. 39A,B).

Race B populations are less common and reproduce exclusively by mitotic parthenogenesis (Triantaphyllou, 1985). Some race B populations are diploid with chromosome numbers of 2N = 30–31, but most populations are polyploids and have chromosome numbers of 3N = 43–48 (Fig. 39C,D).

The chromosomes of race A populations are bivalent (tetrads) during metaphase of the first maturation division and are easy to count (Triantaphyllou, 1885). Thus, race A population of *M. hapla* can easily be distinguished from the other three common species by the presence of 14–17 bivalent chromosomes. Race B populations, on the other hand, cannot be distinguished from the other three common species on the basis of cytology alone. The chromosomes of race B populations are univalent (dyads) and appear similar and behave like the chromosomes of *M. javanica* and *M. arenaria*. Even the number of chromosomes overlaps between race B populations of *M. hapla* and *M. javanica*.

Biochemistry: Most populations of *M. hapla* have a common esterase phenotype, namely H1 (Fig. 19) (Esbenshade and Triantaphyllou, 1985). However, one population from France does not have any major esterase activity and another population from Minne-

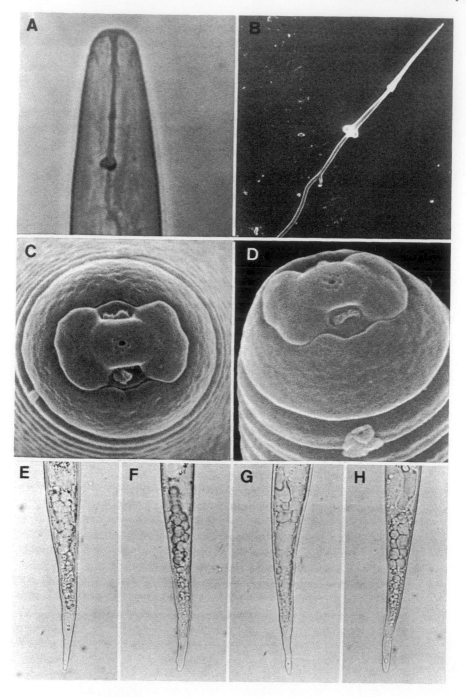

FIGURE 38   Micrographs of second-stage juveniles of *M. hapla*. (A) Light micrograph of the head and stylet in lateral view. (B) Scanning electron micrograph of excised stylet. (C) SEM of the anterior end in face view. (D) SEM of the anterior end in lateral view. (E–H) LM of tails. (A and B after Eisenback, 1982.)

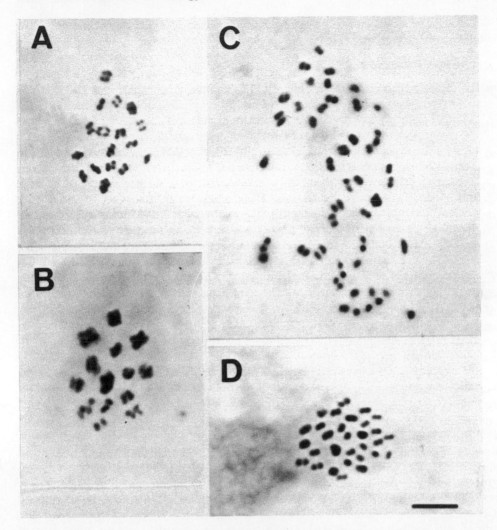

FIGURE 39   Micrographs of chromosomes of *M. hapla* during the first maturation division. Cytological race A chromosomes are bivalent (tetrads) as a result of the pairing of homologous chromosomes. (A) Seventeen bivalent chromosomes. (B) Fourteen bivalent chromosomes. (C) Prometaphase of oocytes of *M. hapla* race B with 45 univalent chromosomes. (D) Race B with 32 chromosomes. Cytological race B chromosomes are univalents (dyads) and are similar to those of *M. arenaria*. (After Triantaphyllou, 1985.)

sota has an esterase phenotype of A1 which is identical that of *M. arenaria*. The malate dehydrogenase phenotype H1, however, is unique for the species.

     Distribution: *Meloidogyne hapla* is less prevalent than *M. incognita* and *M. javanica*, but is nearly as common or more so than *M. arenaria*. Given the appropriate common name of "northern root-knot nematode," *M. hapla* generally occurs in the cooler regions of the world. In the northern hemisphere, it is most common between latitudes 34 °N and 43 °N (Taylor et al., 1982). In subtropical or tropical areas, it is usually found at high altitudes (more than 1000 m). In the southern hemisphere, *M. hapla* may be present at low altitudes, south of latitude 45 °S.

6. *Meloidogyne chitwoodi*

Female: The posterior end of the body is in line with the neck of the pear-shaped female (Fig. 8). The vulva sometimes appears to be located on a slight posterior protuberance. The females are generally contained in the gall and the egg masses protrude from the root tissue.

The overall shape of the perineal pattern of *M. chitwoodi* is round to oval (Golden et al., 1980; Eisenback and Hirschmann, unpublished) (Fig. 40C–F). The dorsal arch is low and rounded to high and sometimes squarish, similar to that of *M. incognita*. The striae are coarse and smooth to wavy, depending on their distance from the perineal area. The striae near the perineal area are twisted and curved, and often overlap each other. The striae in the dorsal area are more twisted than those in the ventral area. Striae further away from the perineal area are smoother. Faint outlines of the lateral field are formed by indistinct forking of the dorsal and ventral striae. Lateral striae may bend toward the vulva.

The stylets of females of *M. chitwoodi* are morphologically unique and useful for species identification (Fig. 40A,B) (Jepson, 1987; Eisenback and Hirschmann, unpublished). The length of the stylet is 11–12.5 (12) μm and the distance of the DEGO to the base of the stylet is 3.5–5.5 (4.2) μm. The anterior end of the cone is sharply pointed and slightly curved dorsally, whereas the shaft is cylindrical to slightly tapered posteriorly. The morphology of the knobs is characteristic for the species; they are small, irregular in outline, indented medially, and taper onto the shaft.

Male: The shape of the male head as seen in the SEM is characteristic for the species, but similar to several other species as seen in the LM (Jepson, 1987) (Fig. 41). In SEM the labial disk and medial lips are fused and form the head cap; however, the medial lips are somewhat lower than the labial disk, and the posterior edges are irregular in outline (Golden et al., 1980; Eisenback and Hirschmann, unpublished). The cephalic sensilla on each medial lip appear as small, elongate depressions in the cuticle. Large, clearly delineated lateral lips are present which may also have irregular outlines and are marked by additional short grooves. In lateral view in the LM, the head shape of *M. chitwoodi* is similar to that of several other species, including *M. graminis*, *M. graminicola*, *M. kralli*, *M. oryzae*, *M. exigua*, and others. The smooth and rounded head cap is set off from the head region, which is usually without annulations. Both the head region and the body annules are in the same contour. The distinct, large lateral lips are helpful in the identification of this species in the LM. Other species with similar head shape may have lateral lips but usually they are not large and distinct.

The length of the stylet of males is 16–19 (18) μm and the distance of the DEGO to the base of the knobs is 3 μm long (Golden et al., 1980). The stylet of males of *M. chitwoodi* is long and thin (Eisenback and Hirschmann, 1982; Jepson, 1987) (Fig. 41A,B). The opening of the stylet is several (3)μm from the tip. The anterior two-thirds of the cone is cylindrical and the posterior one-third gradually increases in width. The shaft is cylindrical to conical, and the knobs are small and taper onto the shaft. As in the female, the anterior margins of the knobs are usually intended medially and are irregular in outline. Often the anterior margins of the knobs appear to be stippled.

Second-stage juvenile: The length of preparasitic juveniles of *M. chitwoodi* is 336–417 (390) μm, the tail length is 39–47 (43) μm, and the hyaline tail terminus length is 9–14 (11) μm (Golden et al., 1980) (Fig. 42).

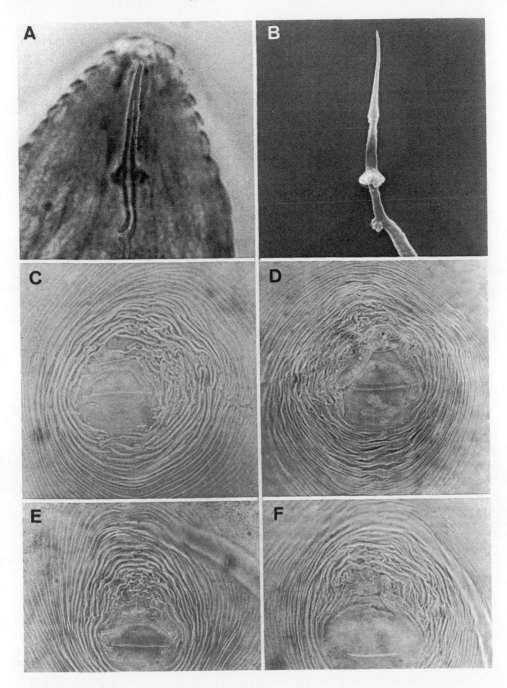

FIGURE 40   Micrographs of females of *M. chitwoodi*. (A) Light micrograph of the anterior end.
(B) Scanning electron micrograph of an excised stylet. (C–F) LM of perineal patterns.

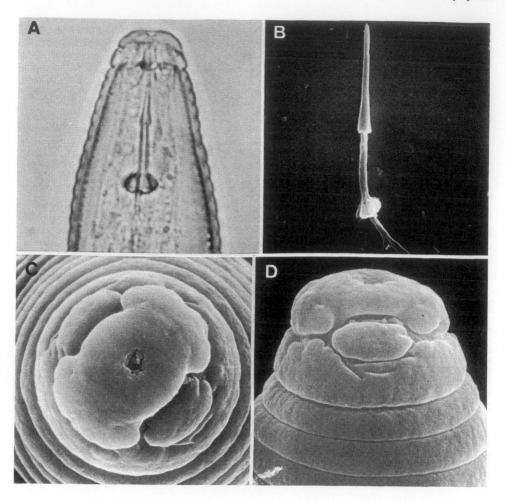

FIGURE 41   Micrographs of males of *M. chitwoodi*. (A) Light micrograph of the head and stylet in lateral view. (B) Scanning electron micrograph of excised stylet. (C) SEM of the anterior end in face view. (D) SEM of the anterior end in lateral view.

The shape of the head of *M. chitwoodi* is not diagnostic for the species because several species have the same or similar head morphology, namely, *M. graminis*, *M. graminicola*, *M. oryzae*, *M. naasi*, and others. The labial disk and medial lips are fused to form the head cap (Golden et al., 1980; Eisenback and Hirschmann, unpublished) (Fig. 42). The head cap appears anchor-shaped. Sometimes a slight median indentation marks each medial lip. The lateral lips are small and elongate, and in contour with the head region, which is usually smooth except for a rare, short, incomplete annulation posterior to the lateral lip.

The stylet of juveniles of *M. chitwoodi* is 10 μm long and the distance of the DEGO to the base of the stylet is 3–4 (3.5) μm (Golden et al., 1980). The morphology of the stylet of *M. chitwoodi* is useful for species identification (Eisenback and Hirschmann, unpublished) (Fig. 42A,B). As in the stylet of the male and female, the anterior margins of the knobs are

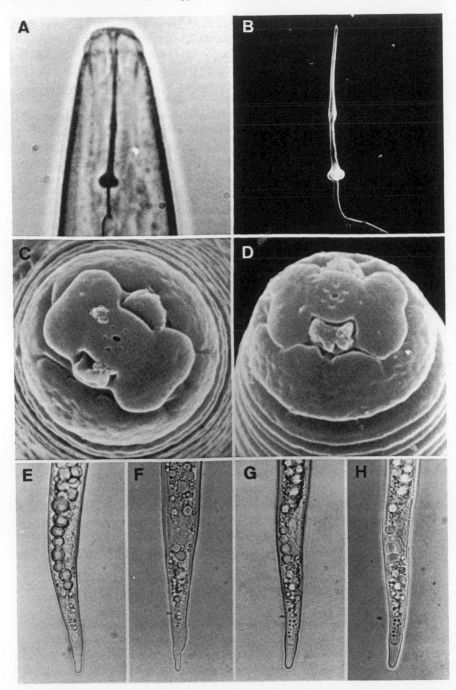

FIGURE 42   Micrographs of second-stage juveniles of *M. chitwoodi.* (A) Light micrograph of the head and stylet in lateral view. (B) Scanning electron micrograph of excised stylet. (C) SEM of the anterior end in face view. (D) SEM of the anterior end in lateral view. (E–H) LM of tails.

indented, irregular, and appear to be stippled. Although these details are too small to be resolved in the LM, they are indicated by the poor definition of the knobs.

The shape of the tail is somewhat diagnostic for the species. The tail, which ends in a broad, blunt tip, is much shorter than that of most other species that are similar in head morphology. Likewise, the hyaline tail terminus is short and clearly delineated (Nyzcepir et al., 1982) (Fig. 42E–H).

Morphological variants: Certain morphological differences have been reported for various populations of *M. chitwoodi*, and a correlation was made between morphology and host race (Pinkerton 35 al., 1985). Differences occurred in the nature of the striae in the perineal pattern, shape of the hyaline tail terminus in the second-stage juvenile, and induced root proliferation in the host.

Host races: Two host races have been described for populations of *M. chitwoodi* (Santo and Pinkerton, 1985). Populations of race 1 cannot reproduce on alfalfa and most carrot cultivars are good to moderate hosts; whereas race 2 populations readily reproduce on alfalfa but not on most cultivars of carrot (Santo et al., 1988).

Symptoms: The galls produced by *M. chitwoodi* ar similar to those produced by several other species. On potato, the symptoms of infected tubers may be useful in distinguishing *M. chitwoodi* from *M. hapla*. Tubers infected by *M. chitwoodi* have numerous small, pimplelike raised areas on the surface, whereas in *M. hapla* these swellings are not evident.

Cytology: The haploid chromosome number of populations of *M. chitwoodi* varies from N = 14 to N = 17 (Fig. 43) (Triantaphyllou, 1985). This species reproduces by facultative meiotic parthenogenesis.

Biochemistry: The esterase phenotype of *M. chitwoodi* has one slow band and is designated as S1 (Fig. 19) (Esbenshade and Triantaphyllou, 1985). All populations tested had the same phenotype; however, one population each of *M. incognita* and *M. plantani* have the S1 phenotype. The S1 phenotype expressed by *M. chitwoodi* populations was slightly different in that the band was less dense and retained less stain. The malate dehydrogenase phenotype N1a is similar to that of *M. naasi* and *M. graminicola*.

Distribution: *Meloidogyne chitwoodi* is a major pest of economic importance in the Pacific Northwest of the United States (Golden et al., 1980). It has been reported in the states of Washington, Oregon, California, Utah, Nevada, Idaho, and Virginia. This species has also been reported incidentally from Argentina, Holland, and Mexico (Esbenshade and Triantaphyllou, 1985; Sosa-Moss, 1985). In Mexico, *M. chitwoodi* appears to be a major economic pest on several crops including potato and sunflower (L. I. Miller, pers. commun.). This species has also been found in South Africa and has been detected on seed potatoes imported into South Africa from Australia (Kleynhans, 1989).

7. *Meloidogyne graminicola*

Female: The body shape is elongate and the vulva is located on a slight posterior protuberance (Golden and Birchfield, 1965) (Fig. 8D). The anterior end of the body is usually in line with the posterior end, but in some specimens it may be at an angle with the body axis.

The perineal pattern of *M. graminicola* is dorsoventrally ovoid (Golden and Birchfield, 1965; Eisenback and Hirschmann, unpublished) (Fig. 44C–F). The dorsal arch is usually high and squarish but may be low and rounded. The tail arch is usually high and squarish but may be low and rounded. The tail terminal area may be flat with shallow transverse striae, or it may be raised and surrounded by numerous irregular striae. Lateral fields are not clearly defined. The striae are generally smooth and continuous around the entire pattern. Short, broken, and irregular striae may occur in the dorsal arch, around the tail terminus, and in the lateral sectors adjacent to the perineum. The phasmids are tiny and very

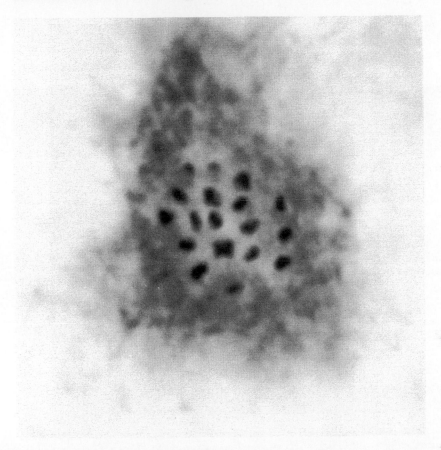

FIGURE 43   Light micrograph of 18 bivalent chromosomes (tetrads) during metaphase in *M. chit-woodi*. (Courtesy of A. C. Triantaphyllou.)

close together, approximately two-thirds the width of the vulva. In some specimens a few shallow striae may bend towards the vulva.

The stylet of females of *M. graminicola* is 10.5–11 μm long and the distance of the DEGO to the base of the stylet is 3–4 μm (Golden and Birchfield, 1965). The cone is slightly curved dorsally and the shaft is cylindrical but sometimes widens posteriorly (Jepson, 1987; Eisenback and Hirschmann, unpublished) (Fig. 44A,B). The knobs are set off from the shaft, broadly elongate, and very similar to those of *M. javanica*.

Male: The shape of the head is similar to that of several other species, including *M. exigua, M. graminis, M. kralli, M. oryzae*, and *M. naasi* (Jepson, 1987). In the SEM, the labial disk and medial lips are fused and form one smooth, continuous head cap. The posterior edges of the medial lips are rounded and extend onto the head region. The lateral edges of the labial disk and medial lips meet at a sharp angle, nearly 90 μm, and each medial lip appears anchor-shaped. Cephalic sensilla appear as small, elongate, shallow depressions on each medial lip pair. Lateral lips may be demarcated on the head region by shallow grooves, or they may be reduced or completely lacking. The head region is usually not annulated.

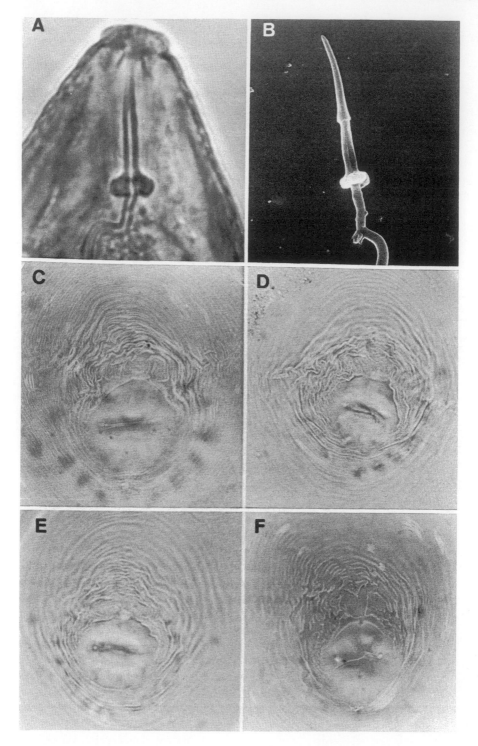

FIGURE 44   Micrographs of females of *M. graminicola*. (A) Light micrograph of the anterior
end. (B) Scanning electron micrograph of an excised stylet. (C–F) LM of perineal patterns.

appears anchor-shaped. Cephalic sensilla appear as small, elongate, shallow depressions on each medial lip pair. Lateral lips may be demarcated on the head region by shallow grooves, or they may be reduced or completely lacking. The head region is usually not annulated.

In the LM the head cap of *M. graminicola* is smooth, rounded, and set off from the head region (Jepson, 1987; Eisenback and Hirschmann, unpublished) (Fig. 45A). The head region and the body annules are in the same contour.

The length of the stylet of males of *M. graminicola* is 16–17 (16.8) μm, and the distance of the DEGO to the base of the stylet is 3–4 μm (Golden and Birchfield, 1965). The shaft is cylindrical but often narrows near its junction with the knobs (Jepson, 1987; Eisenback and Hirschmann, unpublished) (Fig. 45A,B), which are small rounded, and set off from the shaft. The stylet of the male of *M. graminicola* is very similar to that of *M. hapla*, *M. exigua*, *M. naasi*, and *M. oryzae*.

Second-stage juvenile: Preparasitic second-stage juveniles of *M. graminicola* are 415–484 (441)μm long (Golden and Birchfield 1965). Their tail length is 67–76 (71) μm, and the hyaline tail terminus is 14–21 (18) μm long (Jepson, 1987) (Fig. 46).

As revealed by SEM, the labial disk and medial lips of *M. graminicola* are fused and form one smooth, continuous head cap (Jepson, 1987). The posterior edges of the medial lips are rounded and extend onto the head region. The lateral edges of the labial disk and the medial lips meet at a sharp angle. Sometimes each medial lip may be slightly indented medially indicating its derivation from a medial lip pair. The elongate lateral lips are in the same contour with the head region, which usually is smooth. The shape of the head is not diagnostic since it is similar to other species, such as *M. graminis*, *M. oryzae*, and *M. naasi*.

The stylet of second-stage juveniles of *M. graminicola* is 11–12 (11.4) μm long, and the distance from the DEGO to the base of the stylet is 2.8–3.4 (2.8) μm (Golden and Birchfield, 1965). The stylet is thin and the knobs are small, rounded, set off from the shaft, and slightly sloping posteriorly (Eisenback and Hirschmann, unpublished) (Fig. 46A,B).

The tail of second-stage juveniles of *M. graminicola* is very long and thin, and tapers to a long, narrow hyaline terminus (Fig. 46C–F). The tail tip is sharply pointed and may

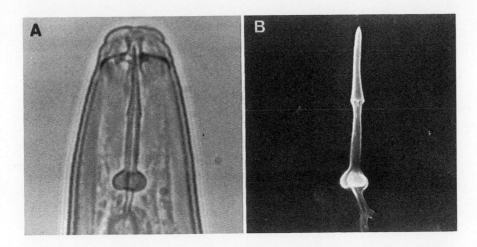

FIGURE 45 Micrographs of males of *M. graminicola*. (A) Light micrograph of head and stylet in lateral view. (B) Scanning electron micrograph of excised stylet.

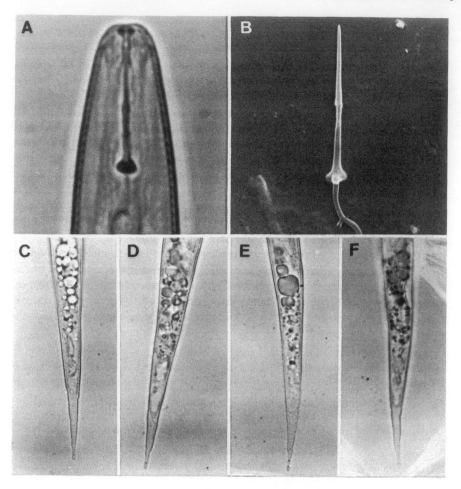

FIGURE 46   Micrographs of second-stage juveniles of *M. graminicola*. (A) Light micrograph of
the head and stylet in lateral view. (B) Scanning electron micrograph of excised stylet. (C–F) LM of
tails.

Host races: Although differences in pathogenicity of populations have been reported, races of this species have not been widely recognized. Populations from different
geographic areas vary in their ability to reproduce on various rice cultivars (Rao et al., 1977;
Sahu and Chawla, 1986).

Symptoms: *Meloidogyne graminicola* females often remain embedded within the
root tissue and the eggs are deposited within the cortex. Secondary infections may take
place by second-stage juveniles migrating in the plant tissue without leaving the root system. Galls produced by *M. graminicola* are large and usually associated with secondary
rootlets and root hairs. Galls at the root tips are often terminal and sharply curved, almost
ringlike.

Cytology: Populations of *M. graminicola* reproduce by facultative meiotic parthenogenesis (Triantaphyllou, 1985). The haploid chromosome number of this species is N =

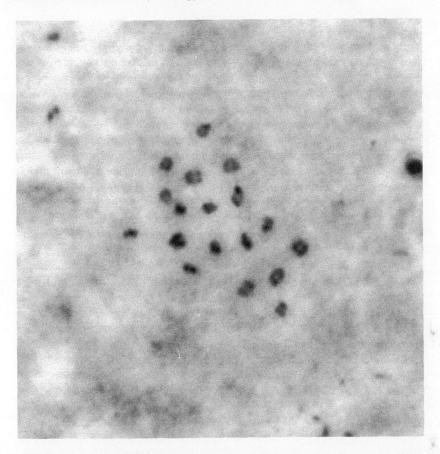

FIGURE 47   Micrograph of the 18 bivalent chromosomes (tetrads) during prometaphase in *M. graminicola*. (After Triantaphyllou, 1969.)

18 (Fig. 47). Cytologically, *M. graminicola* is similar to *M. exigua*, *M. graminis*, and *M. naasi*.

Biochemistry: Populations of *M. graminicola* have one slow band with a very large drawn-out area of enzymatic activity, namely VS1 (Fig 19) (Esbenshade and Triantaphyllou, 1985). A very similar but slightly faster phenotype occurs in *M. oryzae* and *M. salasi*. The malate dehydrogenase phenotype N1a is similar to that of *M. chitwoodi* and *M. naasi*.

Distribution: *Meloidogyne graminicola* is widespread around the world and common in the subtropics and tropics. Populations of this species have been reported in the United States, Bangladesh, Burma, India, Laos, Libya, Thailand, and Vietnam (Jepson, 1987). *Meloidogyne graminicola* has also been found in South Africa (Kleynhans, 1989). Rice is the most important host, and nearly all varieties and selections tested are susceptible to *M. graminicola*, which causes much damage to this economically important crop. Numerous other grasses, broad-leaf weeds, and water weeds serve as alternative hosts (Manser, 1971).

8. *Meloidogyne naasi*

Female: The body of females of *M. naasi* is almost spherical in mature specimens and the neck is not in line with the posterior end (Fig. 8D). The vulva is located on a slight posterior protuberance.

The perineal pattern of *M. naasi* is dorsoventrally ovoid (Franklin, 1965; Eisenback and Hirschmann, unpublished). The dorsal arch is generally low and rounded (Fig. 48C–F). The coarse striae in the dorsal arch are broken and irregular around and between the phasmids where they run in right angles to the dorsal arch. The tail terminus area is free of striae and a prominent fold covers the anus. The large, round phasmids are slightly closer together than the vulval slit length. The striae in the ventral pattern area are fine and continuous.

The stylet of females of *M. naasi* is 11–15 (13) μm long, and the DEGO is 2–4 (3.5) μm from the base of the stylet (Franklin, 1965). The cone is curved dorsally and the shaft widens slightly posteriorly (Eisenback and Hirschmann, unpublished) (Fig. 48A,B). The knobs are very large and rounded, and taper onto the shaft, very similar to those of *M. arenaria* and *M. oryzae*.

Male: The shape of the head resembles that of *M. exigua*, *M. graminicola*, *M. graminis*, *M. kralli*, and *M. oryzae*. The labial disk and medial lips are fused and form a smooth and continuous head cap (Jepson, 1987) (Fig. 49A). The rounded posterior edges of the head cap extend onto the usually smooth head region. The lateral edges of the labial disk and medial lips meet at a sharp angle. Cephalic sensilla are expressed on the medial lips as small, elongate, shallow depressions. The lateral lips may be reduced or lacking.

In lateral view in the LM, the head cap of males is high, rounded, and distinctly set off from the head region, which may be smooth or have a few annulations (Jepson, 1987; Eisenback and Hirschmann, unpublished) (Fig. 49C). The head region is in the same contour with the body annules.

The stylet of the male of *M. naasi* is 16–19 (18) μm long, and the distance of the DEGO to the base of the stylet is 2–4 (3) μm (Franklin, 1965). The cone is straight and the cylindrical shaft often narrows near its junction with the knobs (Jepson, 1987; Eisenback and Hirschmann, unpublished) (Fig. 49B). The small, rounded knobs are set off from the shaft. The stylet of the male of *M. naasi* is very similar to that of several other species including *M. graminicola*, *M. exigua*, and *M. oryzae*.

Second-stage juvenile: Juveniles of *M. naasi* are 418–465 (435)μm long (Franklin, 1965). Their tails are 52–78 (70) μm long and the hyaline tail terminus measures 17–27 (22.5) μm (Fig. 50F–I).

In the SEM, the medial lips of second-stage juveniles of *M. naasi* are fused with the labial disk and form a smooth, continuous head cap (Jepson, 1987) (Fig 50). The rounded posterior edges of the medial lips, which may be slightly indented medially, extend onto the head region. The lateral edges of the labial disk join with the lateral edges of the medial lips at a sharp angle. The lateral lips are lower than the labial disk and medial lips, and in the same contour with the head region. The head region is usually smooth. The head shape is similar to that of several other species including *M. graminis*, *M. oryzae*, and *M. graminicola*. Second-stage juveniles of *M. naasi* usually have several (3–9), vesicles associated with the lumen lining of the anterior metacorpus which are useful diagnostically (Fig. 50E).

The stylet of second-stage juveniles of *M. naasi* is 13–15 (14) μm long, and the DEGO is 2–3 (2.5) μm from the base of the stylet (Franklin, 1965). The very slender stylet has large knobs that gradually taper onto the shaft (Eisenback and Hirschmann, unpub-

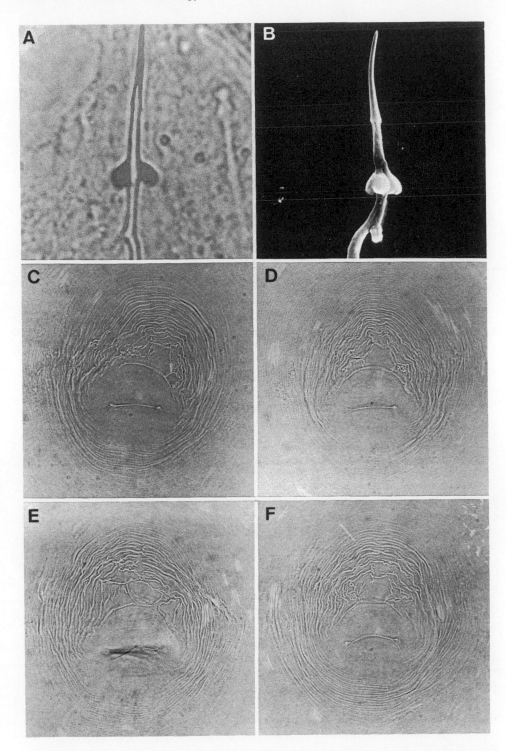

FIGURE 48   Micrograph of females of *M. naasi*. (A) Light micrograph of the anterior end with stylet. (B) Scanning electron micrograph of an excised stylet. (C–F) LM of perineal patterns.

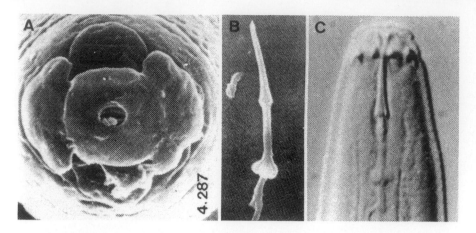

FIGURE 49   Micrographs of males of *M. naasi*. (A) Scanning electron micrograph of the anterior end in face view. (B) SEM of excised stylet. (C) Light micrograph of the head and stylet in lateral view. (After Jepson, 1987).

lished) (Fig. 50A,B). The stylet is similar in shape to that of *M. arenaria* but is several micrometers longer.

The tails of second-stage juveniles of *M. naasi* are long and narrow. They taper to a long, thin hyaline portion often marked with a clavate terminus (Jepson, 1987) (Fig. 50F–I). Tails of juveniles of *M. naasi* are similar to those of *M. graminicola*, *M. oryzae*, and *M. ottersoni*.

Morphological variants: None described.

Symptoms: The females are generally contained in the root tissue and the egg masses are enclosed within the gall. Secondary infections may occur by juveniles migrating inside the plant tissue. Galls produced by *M. naasi* may be small and spindle-shaped or large and terminal, often hooked or spiraled, almost ringlike (Franklin, 1965) (Fig. 6D).

Host races: Five host races have been described for *M. naasi* on the basis of the response of four key differential host species, namely, curly dock, sorghum, creeping bentgrass, and common chickweed (Mitchell et al., 1973).

Cytology: Populations of *M. naasi* reproduce by facultative meiotic parthenogenesis and have a haploid chromosome number of $N = 18$ (Fig. 51). Cytologically, *M. naasi* is similar to *M. graminicola*, *M. exigua*, and *M. graminis*.

Biochemistry: *Meloidogyne naasi* has the esterase phenotype, VF1, which is characterized by one very fast band (Fig. 19) (Esbenshade and Triantaphyllou, 1985). The only other species with the same esterase phenotype is *M. exigua*. *Meloidogyne naasi* can be differentiated from *M. exigua* by the malate dehydrogenase band N1a.

Distribution: *Meloidogyne naasi* is widely distributed in Europe including England, Wales, France, the Netherlands, Yugoslavia, Belgium, Germany, Italy, and Malta (Jepson, 1987). It has also been reported in North and South America and in Asia (Franklin, 1965; Mitchell et al., 1973).

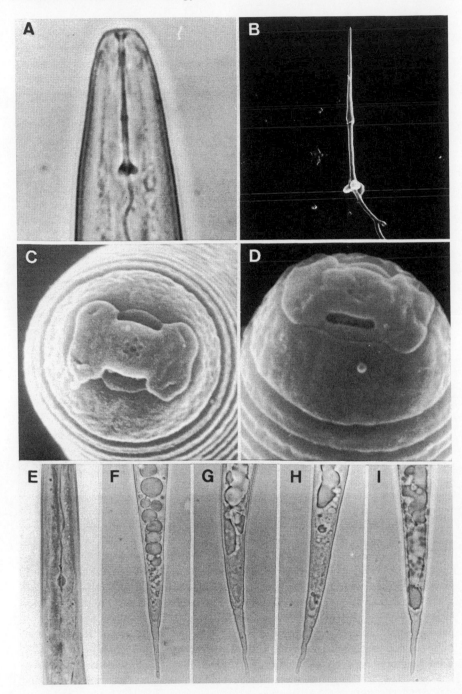

FIGURE 50 Micrographs of second-stage juveniles of *Meloidogyne naasi*. (A) Light micrograph of the head and stylet in lateral view. (B) Scanning electron micrograph of excised stylet. (C) SEM of the anterior end in face view. (D) SEM of the anterior end in lateral view. (E) LM of esophagus showing vesicles in the median bulb. (F–I) LM of tails.

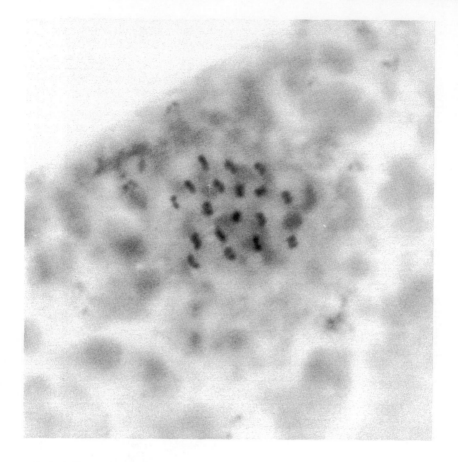

FIGURE 51   Micrograph of the 18 bivalent chromosomes (tetrads) during prometaphase in *M. naasi*. (After Triantaphyllou, 1969.)

9. *Meloidogyne artiellia*

Female: The body of females of *M. artiellia* is pear-shaped and the neck is in line with the posterior end. Normally the posterior end of the female protrudes through the slightly swollen gall and the egg mass is deposited completely outside the root (Franklin, 1961).

The perineal pattern of *M. artiellia* is very unique and diagnostic for the species (Franklin, 1961; Jepson, 1987; Eisenback and Hirschmann, unpublished) (Fig. 52C–F). Overall, the pattern is rounded and made up of smooth, fine striae; however, the cuticle in the lateral sectors of the pattern, often including the tail terminus, is thicker and forms large, distinct ridges. Lateral fields may be distinguished in some specimens by irregularities in the striae or depressions within the large ridges.

The stylet of females of *M. artiellia* measures 12–16 (14) μm and the distance of the DEGO to the base of the stylet is 4–7 (5.5) μm (Franklin, 1961). The stylet is slightly curved dorsally, and the cone and the shaft gradually widen posteriorly (Jepson, 1987; Eisenback

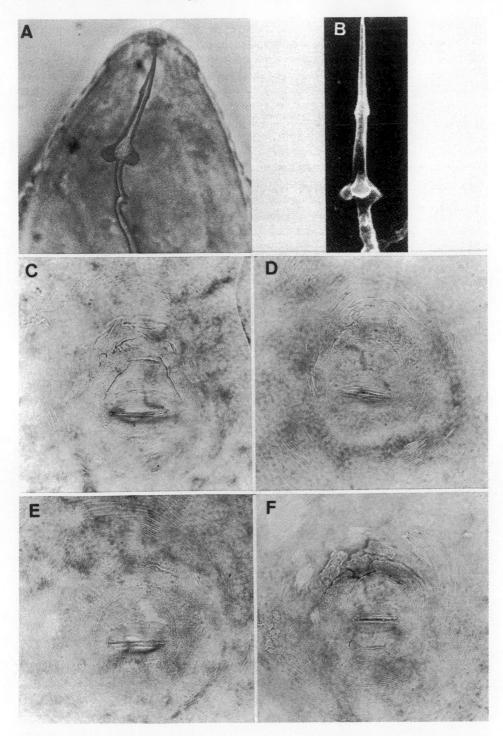

FIGURE 52   Micrographs of females of *M. artiellia*. (A) Light micrograph of the anterior end. (B) Scanning electron micrograph of an excised stylet. (C–F) LM of perineal patterns. (B after Jepson, 1987.)

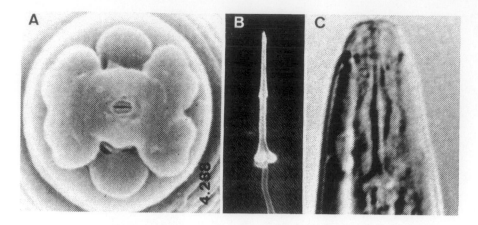

FIGURE 53   Micrographs of males of *M. artiellia*. (A) Scanning electron micrograph of the anterior end in face view. (B) SEM of excised stylet. (C) Light micrograph of the head and stylet in lateral view. (After Jepson, 1987.)

and Hirschmann, unpublished) (Fig. 52A,B). The knobs are large and gradually taper onto the shaft, similar to those of *M. arenaria*, *M. naasi*, and *M. oryzae*.

    Male: The head morphology of *M. artiellia* as revealed by SEM is unique for the species (Jepson, 1987). The labial disk is fused with the medial lips forming one smooth and continuous head cap (Fig. 53A). The medial lips are distinctly divided medially and almost appear as separate lip pairs. The cephalic sensilla form shallow, elongate depressions in the cuticle of each lip. The elongate lateral lips are clearly defined on the head region by deep grooves; otherwise the head region is smooth and free of additional annulations.

    In lateral view in the LM, the rounded head cap of males of *M. artiellia* is high and set off from the head region (Jepson, 1987) (Fig. 53C), which is in the same contour with the body. The head shape is very similar to that of several other species including *M. graminis*, *M. graminicola*, *M. naasi*, and *M. oryzae*.

    The length of the stylet of males of *M. artiellia* is 17– 27 (19) μm, and the distance from the base of the stylet to the DEGO is 5–7 (6) μm (Franklin, 1961). The stylet is thin and straight (Jepson, 1987) (Fig. 53B). The cone gradually increases in width posteriorly and the shaft is almost cylindrical. The small rounded knobs are set off from the shaft. The stylet shape and DEGO distance to the base of stylet are similar to those of *M. hapla*.

    Second-stage juvenile: Juveniles of *M. artiellia* are 334–370 (354) μm long, and their tails measure 18–26 (22) μm with a hyaline tail terminus length of 2–7 (5) μm (Franklin, 1961; Jepson, 1987) (Fig. 54C).

    As revealed by SEM, the labial disk of *M. artiellia* second-stage juveniles is small and round and partially fused with the medial lips (Jepson, 1987) (Fig. 54A). The medial lips are indented medially, and cephalic sensilla are indicated on each lip by shallow, elongate depressions in the cuticle. Elongate lateral lips are also demarcated on the head region, but this region is otherwise smooth. The head morphology of *M. artiellia* is very similar to that of *M. microtyla* second-stage juveniles.

    The stylet of second-stage juveniles of *M. artiellia* is 14–16 (14.5) μm long, and the distance from the base of the knobs to the DEGO is 2.5–4.5 (3.5)μm (Franklin, 1961). The stylet is long and thin with large knobs that gradually taper onto the shaft and appear to slope posteriorly (Jepson, 1987) (Fig. 54B).

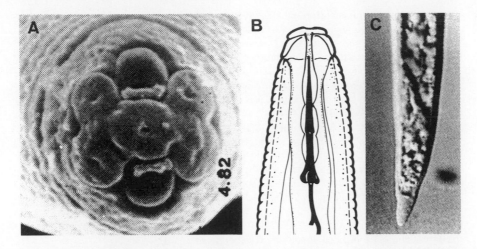

FIGURE 54   Micrographs of second-stage juveniles of *M. artiellia*. (A) Scanning electron micrograph of the anterior end in face view. (B) Drawing of the head and stylet in lateral view. (C) Light micrograph of the tail in lateral view. (After Jepson, 1987.)

Symptoms: The galls are very small, and lateral root proliferation near the site of infection is stimulated (Franklin, 1961).

Host races: None reported.

Cytology: Unavailable.

Biochemistry: Unavailable.

Distribution: Reports of populations of *M. artiellia* are mainly from Europe, namely, England, France, Syria, Italy, and Greece (Greco and di Vito, 1987; di Vito and Zaccheo, 1985). It is an important pest on wheat, chick-pea, and vetch. In experimental host tests, nearly all of the species of Leguminosae and Graminaceae tested were good to very good hosts. Species of Solanaceae, Umbelliferae, Chenopodiaceae, Malvaceae, Compositae, Liliaceae, Linaceae, and Rosaceae were poor hosts or nonhosts.

ACKNOWLEDGMENTS

The authors especially thank Dr. A. C. Triantaphyllou for his contributions of text, tables, and figures to the sections on sexuality, cytogenetics, and biochemistry. Thanks are also extended to Drs. P. R. Esbenshade, L. I. Miller, K. P. N. Kleynhans, H. D. Shew, J. N. Sasser, F. Lamberti, and M. di Vito; and to all the nematologists who have contributed directly or indirectly to the information presented herein.

We sincerely appreciate the unique contributions to this manuscript by Mr. Eugene McCabe.

REFERENCES

Bird, A. F. 1968. Changes associated with parasitism in nematodes. IV. Cytochemical studies on the ampulla of the dorsal esophageal gland of *Meloidogyne javanica* and on exudations from the buccal stylet. *J. Parasitol.* 54: 879–890.

Bird, A. F. 1974. Plant response to root-knot nematodes. *Annu. Rev. Phytopathol* 12: 69–85.

Bird, A. F., Downton, W. S. J., and Hawker, J. S. 1975. Cellulase secretion by second stage larvae of the root-knot nematode (*Meloidogyne javanica*). *Marcellia* 38: 165–169.

Cain, S. C. 1974. *Meloidogyne exigua. C. I. H. Descriptions of Plant-Parasitic Nematodes, Set 4, No. 49*, Commonwealth Inst. Helminthology, St. Albans, Herts., England.

Canto-Saenz, M., and Brodie, B. B. 1987. Relationship between morphology and parasitism in two populations of *Meloidogyne incognita. J. Nematol.* 19: 19–22.

Chitwood, B. G. 1949. Root-knot nematodes. I. A revision of the genus *Meloidogyne* Goeldi, 1887. *Proc. Helm. Soc. Wash.* 16: 90–104.

Chitwood, B. G., and Berger, C. A. 1960. Preliminary report on nemic parasites of coffee in Guatemala, with suggested ad interim control measures. *Plant Dis. Rep.* 44: 841–847.

Cliff, G., and Hirschmann, H. 1985. Evaluation of morphological variability in *Meloidogyne arenaria. J. Nematol.* 17: 445–459.

Curi, S. M., Lordello, L. G. E., deBona, A., and Cintra, A. F. 1970. Atual distribuição geographica dos nematoides do cafeeiro, (*Meloidogyne coffeicola* e *M. exigua*), no Estado de São Paulo. *Biologico* 36: 26–28.

Davide, R. G., and Triantaphyllou, A. C. 1967. Influence of environment on development and sex differentiation of root-knot nematodes. *Nematologica* 13: 102–110.

Dropkin, V. H. 1972. Pathology of *Meloidogyne*—galling, giant cell formation effects on host physiology. *EPPO Bull. No.* 6: 23–32.

Eisenback, J. D. 1982. Morphological comparison of head shape and stylet morphology of second-stage juveniles of *Meloidogyne* species. *J. Nematol.* 14: 339–343.

Eisenback, J. D. 1985. Detailed morphology and anatomy of second-stage juveniles, males, and females of the genus *Meloidogyne* (root-knot nematodes). In *An advanced treatise on Meloidogyne, Vol. 1, Biology and control*, J. N. Sasser and C. C. Carter, eds., North Carolina State Univ. Graphics, Raleigh.

Eisenback, J. D. 1987. Reproduction of northern root-knot nematode (*Meloidogyne hapla*) on marigolds. *Plant Dis.* 71: 281.

Eisenback, J. D. 1988. Identification of Meloidogynids. In *Nematode Identification and Expert Systems Technology*, R. Fortuner, ed. Plenum Press, New York.

Eisenback, J. D., and Hirschmann, H. 1979. Morphological comparison of second-stage juveniles of several *Meloidogyne* species (root-knot nematodes) by scanning electron microscopy. *Scanning Electron Micros.* 1979/3: 223–230.

Eisenback, J. D., and Hirschmann, H. 1981. Identification of *Meloidogyne* species on the basis of head shape and stylet morphology of the male. *J. Nematol.* 13: 413–521.

Eisenback, J. D., and Hirschmann, H. 1982. Morphological comparison of stylets of male root-knot nematodes (*Meloidogyne* spp.). *Scanning Electron Micros.* 3:837–843.

Eisenback, J. D., Hirschmann, H., Sasser, J. N., and Triantaphyllou, A. C. 1981. *A Guide to the Four Most Common Species of Root-knot Nematodes, (Meloidogyne species) with a pictorial key.* A Coop. Publ. Depts. Plant Pathol. and Genetics and U.S. Agency International Dev., Raleigh, NC.

Eisenback, J. D., Hirschmann, H., and Triantaphyllou, A. C. 1980. Morphological comparisons of *Meloidogyne* female head structures, perineal patterns, and stylets. *J. Nematol.* 12: 300–313.

Esbenshade, P. R., and Triantaphyllou, A. C. 1985. Use of enzyme phenotypes for identification of *Meloidogyne* species. *J. Nematol.* 17: 6–20.

Esbenshade, P. R., and Triantaphyllou, A. C. 1989. Isozyme phenotypes for the identification of *Meloidogyne* species. *J. Nematol.* 22:10–15.

Franklin, M. T. 1961. A British root-knot nematode, *Meloidogyne artiellia*, n. sp. *J. Helminthol., Suppl.*, pp. 85–92.

Franklin, M. T. 1965. A root-knot nematode, *Meloidogyne naasi* n. sp., on field crops in England and Wales. *Nematologica* 11: 79–86.

Golden, A. M., O'Bannon, J. H., Santo, G. S., and Finley, A. M. 1980. Description and SEM observations of *Meloidogyne chitwoodi* n. sp. (Meloidogynidae), a root-knot nematode on potato in the Pacific Northwest. *J. Nematol.* 12:319–327.

Golden, A. M., and Birchfield, W. 1965. *Meloidogyne graminicola* (Heteroderidae) a new species of root-knot nematode from grass. *Proc. Helm. Soc. Wash.* 32: 228–231.

Greco, N., and di Vito, M. 1987. The importance of plant parasitic nematodes in food legume production in the Mediterranean Region. In *Nematodes Parasitic to Cereals and Legumes in Temperature and Semi-arid Regions*, M. C. Saxena, R. A. Sikora, and J. P. Srivastava, eds. Proc. Workshop at Larnaca, Cyprus, 1–5 March 1987. Int. Cencer Agr. Res. in Dry Areas, Aleppo, Syria.

Hartman, K. M., and Sasser, J. N. 1985. Identification of *Meloidogyne* species on the basis of differential host test and perineal pattern morphology. In *An Advanced Treatise on Meloidogyne, Vol. 2, Methodology*, K. R. Barker, C. C. Carter, and J. N. Sasser, eds. North Carolina State Univ. Graphics, Raleigh.

Hirschmann, H. 1984. Morphological variability of *Meloidogyne incognita* revealed by light and scanning electron microscopy. *Proc. 1st Int. Congress Nematol.*, Guelph, Ontario, Canada, p. 35.

Hussey, R. S. 1987. Secretions of esophageal glands of Tylenchida nematodes. In *Vistas on Nematology*, J. A. Veech and D. W. Dickson, eds. Soc. Nematologists, Hyattsville, MD.

Huang, C. S., and Maggenti, A. R. 1969. Mitotic aberrations and nuclear changes of developing giant cells in *Vicia faba* caused by root-knot nematode, *Meloidogyne javanica*. *Phytopathology* 59: 447–455.

Jepson, S. B. 1984. The use of second-stage juvenile tails as an aid in the identification of *Meloidogyne* species. *Nematologica* 29: 11–28.

Jepson, S. B. 1987. *Identification of the Root-Knot Nematodes (Meloidogyne species)*. Commonwealth Agr. Bureaux, Farnham Royal.

Jones, M. G. K. 1981. Host cell responses to endoparasitic nematode attack: Structure and function of giant cells and syncytia. *Ann. Appl. Biol.* 97: 353–372.

Jones, M. G. K., and Dropkin, V. H. 1976. Scanning electron microscopy of nematode-induced giant transfer cells. *Cytobios* 15: 149–161.

Jones, M. G. K., and Northcote, D. J. 1972. Multinucleate transfer cells induced in coleus roots by the root-knot nematode, *Meloidogyne arenaria*. *Protoplasma* 75: 381–395.

Jones, J. G. K., and Payne, H. L. 1978. Early stages of nematode-induced giant-cell formation in roots of *Impatiens balsamina*. *J. Nematol.* 10: 70–81.

Kleynhans, K. P. N. 1989. *The root-knot nematodes of South Africa*. Plant Protection Research Institute, Pretoria (in press).

Lima, R. D'A. de, and Ferraz, S. 1985. [Morphometric character analysis among diferent populations of *Meloidogyne exigua* (Nematoda: Meloidogynidae).] Análise comparativa das variações morfométricas entre diferentes populações de *Meloidogyne exigua*. *Revista Ceres* 32: 362–372.

Lindford, M. B. 1942. The transient feeding of root-knot nematode larvae. *Phytopathology* 32: 580–589.

Lopez, C. R. 1985. [Observations on the morphology of *Meloidogyne exigua* with the scanning electron microscope.] Observaciones sobre la morfologia de *Meloidogyne exigua* con el microscopio electronico de rastreo. *Nematropica* 15: 27–36.

Lordello, L. G. E., and Zamith, A. P. L. 1958. On the morphology of the coffee root knot nematode, *Meloidogyne exigua* Goeldi, 1887. *Proc. Helmin. Soc. Wash.* 25: 133–137.

Manser, P. D. 1971. Notes on the rice root-knot nematode in Laos. *FAO Plant Protect. Bull.* 19: 138–138.

McClure, M. A. 1977. *Meloidogyne incognita*: A metabolic sink. *J. Nematol.* 9: 88–90.

Michell, R. E., Malek, R. B., Taylor, D. P., and Edwards, D. I. 1973. Races of barley root-knot nematode, *Meloidogyne naasi*. I. Characterization by host preference. *J. Nematol.* 5: 41–44.

Morera, N., and Lopez, R. 1987. [Evaluation of the virulence of three *Meloidogyne exigua* Goeldi, 1887, populations in coffee and of the resistance of 6 coffee cultivars to one of these populations.] Evaluacion de la virulencia de tres poblaciones de *Meloidogyne exigua* Goeldi, 1887, en el cafeto y de la resistencia de seis lineas do dicho cultivo a una base de esas poblaciones. *Boletin de PROMECAFE No.* 35: 9–15.

Nyczepir, A. P., O'Bannon, J. H., Santo, G. S., and Finley, A. M. 1982. Incidence and distinguishing characteristics of *Meloidogyne chitwoodi* and *M. hapla* in potato from the Northwestern United States. *J. Nematol.* 14: 347–353.

Papadopoulou, J., and Triantaphyllou, A. C. 1982. Sex differentiation in *Meloidogyne incognita* and anatomical evidence of sex reversal. *J. Nematol.* 14: 549–566.

Pinkerton, J. N., Mojtahedi, H., and Santo, G. S. 1985. Intraspecific variation between populations of *Meloidogyne chitwoodi* from Washington. *J. Nematol.* 17: 509 (abstr.).

Rammah, A., and Hirschmann, H. 1990. Morphological comparison of three host races of *Meloidogyne javanica. J. Nematol.* 22.

Rao, Y. S., Jena, R. N., and Prasad, K. S. K. 1977. Infectivity of two isolates of *Meloidogyne graminicola* in rice. *Indian J. Nematol.* 7: 98–99.

Sahu, S. C., and Chawla, M. L. 1986. A new virulent strain of rice root-knot nematode from Agartala, India. *Int. Rice Res. Newslett.* 11: 40.

Santo, G. S., Mojtahedi, H., and Wilson, J. H. 1988. Host-parasite relationship of carrot cultivars and *Meloidogyne chitwoodi* races and *M. hapla. J. Nematol.* 20: 555–564.

Santo, G. S., and Pinkerton, J. N. 1985. A second race of *Meloidogyne chitwoodi* discovered in Washington. *Plant Dis.* 69: 361.

Sasser, J. N. 1972. Physiological variation in the genus *Meloidogyne* as determined by differential hosts. *OEPP/EPPO Bull.* 6: 41–48.

Scognamiglio, A., Bianco, M., and Marullo, R. 1985. *Meloidogyne exigua* Goeldi su radici di Bougainvillea glabra (Sanderiana). Osservatorio per le Malattie delle Piante per la Campania, Napoli. *Estratto da: (L'Informatore Agrario) Verona,* XLI.

Sosa-Moss, C. 1985. Report on the status of *Meloidogyne* research in Mexico, Central America, and the Caribbean countries. In *An advanced treatise on Meloidogyne. Volume I. Biology and control.* J. N. Sasser and C. C. Carter, eds. North Carolina State Univ. Graphics, Raleigh.

Southards, C. J., and Priest, M. F. 1971. Physiologic variation of seventeen isolates of *Meloidogyne incognita. J. Nematol.* 3: 330 (Abstr.)

Taylor, A. L., Sasser, J. N., and Nelson, L. A. 1982. *Relationship of climate and soil characteristics to geographical distribution of Meloidogyne species in agricultural soils.* A Coop. Public. Depart. Plant Path., North Carolina State Univ. and U. S. Agency Int. Devel., Raleigh.

Triantaphyllou, A. C. 1962. Oogenesis in the root-knot nematode *Meloidogyne javanica. Nematologica* 7: 105–113.

Triantaphyllou, A. C. 1969. Gametogenesis and the chromosomes of two root-knot nematodes, *Meloidogyne graminicola* and *M. naasi. J. Nematol.* 1: 62–714.

Triantaphyllou, A. C. 1973. Environmental sex differentiation of nematodes in relation to pest management. *Ann. Rev. Phytopathol.* 11: 441–462.

Triantaphyllou, A. C. 1985. Cytogenetics, cytotaxonomy and phylogeny of root-knot nematodes. In *An Advanced Treatise on Meloidogyne, Vol. 1, Biology and Control,* J. N. Sasser and C. C. Carter, eds. North Carolina State Univ. Graphics, Raleigh.

Triantaphyllou, A. C., and Hirschmann, H. 1960. Post-infection development of *Meloidogyne incognita* Chitwood 1949 (Nematoda: Heteroderidae). *Ann. Inst. Phytopathol., Benaki* 3: 3–11.

Triantaphyllou, A. C., and Sasser, J. N. 1960. Variations in perineal patterns and host specificity of *Meloidogyne incognita. Phytopathology* 50: 724–735.

di Vito, M., and Zaccheo, G. 1985. On the host range of *Meloidogyne artiellia. Nematol. Medit.* 13: 207–212.

Whitehead, A. G. 1968. Taxonomy of *Meloidogyne* (Nematodea: Heteroderidae) with descriptions of four new species. *Trans. Zoo. Soc. London* 31: 263–401.

7

Heteroderinae, Cyst- and Non-Cyst-Forming Nematodes

JAMES G. BALDWIN and MANUEL MUNDO-OCAMPO *University of California, Riverside, Riverside, California*

I. INTRODUCTION

A. Biology

1. Distribution

Heteroderinae Filipjev and Schuurmans Stekhoven, *sensu* Luc et al. 1988, are of worldwide concern because they include some of the most important nematode pathogens of agriculture, the cyst nematodes. New reports including discovery of new species underscore the worldwide distribution of Heteroderinae, challenging earlier views that the group is nearly restricted to cool climates. Indeed, the first descriptions were of pathogens in temperate Europe and North America and these continue to be among the greatest menaces to agriculture. *Heterodera* was first recognized as *H. schachtii* on sugar beet in Europe in the late 1800s; today the range extends to wherever sugar beets are grown and beyond, including Europe, USSR, USA, Mexico, and Africa, even becoming established in hot deserts and tropics. In the early years following description of *H. schachtii* the diversity of the cyst nematodes was recognized by many "races." Today nearly 60 species of *Heterodera* extend throughout the world. These include species limited to hot climates such as *H. sacchari* on sugar cane and rice and *H. oryzae* on rice and banana. *Heterodera zeae* while occurring on corn in both India and Maryland seems to be better adapted and more pathogenic in warm climates of India. About one-third of the species, including *H. amygdali* on almonds and other *Prunus* in Tajik, USSR, seem highly restricted in distribution, some with only a single known site. Others, including *H. avenae*, *H. cruciferae*, *H. glycines*, and *H. trifolii*, are widely distributed. *Heterodera avenae* occurs in Europe, USSR, North America, South America, the Middle East, Africa, Asia, and Australia; although it generally attacks cool weather cereals, it persists in the tropics where cultivars, seasons, or altitudes allow suitable hosts to be grown. *Heterodera cruciferae* is widely distributed in Europe and USSR; it also extends to California and South Australia. *Heterodera glycines* occurs from Asia to the eastern USA with reports in Egypt and South America; it is not known in Europe.

The golden potato cyst nematode was first considered a race of *H. schachtii*. It is now recognized as *Globodera rostochiensis*, the type species of a genus with about 12 species including a second potato cyst nematode, *G. pallida*. Several species, including *G. zelandica* on *Fuchsia* in New Zealand, are restricted in distribution. However, the potato cyst nematodes *G. rostochiensis* and *G. pallida* occur worldwide, extending from South and Central America to Europe, USSR, Mediterranean region, Asia, India, Africa, Australia, New Zealand, Iceland, Philippines, and restricted regions of Canada and the USA. Although the potato cyst nematodes have been reported in tropics, they are typically found at high elevations. However, rare reports suggest occurrences in hot tropics of Mozambique and southern India.

Heterodera (= *Punctodera*) *punctata* was among the first cyst nematodes known in North America and was described only 5 years after the golden potato cyst nematode was recognized as a species distinct from *H. schachtii*. This parasite of cereals is of little economic concern, although it is widespread in Europe, USSR, and USA. In contrast, *Punctodera chalcoensis* is highly destructive to *Zea mays*, and it is apparently restricted to Mexico. *Cactodera* is abundant in the warm lower elevations of Mexico and Cuba as well as the hot desert regions of the western USA. The type species *C. cacti* occurs on cacti and *Euphorbia*; widespread reports outside the New World may be the result of exported ornamental plants. While some species are subtropical, others, such as *C. weissi*, extend from southern Virginia to the cooler climates of the USA and Canada, and *C. aquatica* and *C. estonica* are described from the USSR.

Afenestrata is a cyst nematode with only a single known species, *A. africana*, of very limited distribution, on a woody host in tropical Africa. Records suggest that species and most genera of non-cyst-forming heteroderines have restricted distributions. However, they are often not known to be economically important, so that their presence may go unnoticed and the breadth of their distribution may yet be unknown. Collectively, the six species of *Meloidodera* are particularly widely distributed including the USSR, Alaska, and mild climates of the southeastern USA. *Verutus* spp. occur at restricted sites in the southeastern and southwestern USA as well as Japan and Europe. *Cryphodera* apparently occurs only in Australia and New Zealand, *Hylonema* in Africa, and *Camelodera* in central Asia of the USSR. Most sarisoderines (*Sarisodera, Rhizonema, Bellodera, Ekphymatodera*) are limited to the western USA and perhaps subtropical and tropical Mexico and South America (unpublished observations). Ataloderines (*Atalodera, Thecavermiculatus*) extend from Alaska through the western USA to South America. The potential of *Meloidodera charis* on corn (*Zea mays*) in Nebraska and *Thecavermiculatus* n. sp. on potato (*Solanum tuberosum*) in Alaska suggests that non-cyst-forming heteroderines may yet prove important to agriculture.

The distribution of Heteroderinae is largely determined by the evolutionary history of the taxa and the history of the areas of the earth in which they live (Ferris, 1979). However, superimposed on these patterns is the effect of dispersal, particularly of agricultural pests, by man. Several authors have speculated on the introduction of potato cyst nematodes from the Andean region of South America to Europe and then throughout the world (Mia, 1977); new infestations can sometimes be traced to the introduction of contaminated potato seed pieces. Containment in the case of the potato cyst nematode in the USA is often attributed to rigorous quarantine enforcement. Conversely, spread suggests failure of quarantine, as in the case of soybean cyst nematode in the USA. Occurrence of the soybean cyst nematode in the USA has been linked to importation from Japan (Riggs, 1977), a hypothesis which may be testable by molecular biology methods which analyze geneological relationships among geographic isolates (Ferris et al., 1985; Radice et al.,

1988; Sandall and Powers, 1988). Whereas interpretation of results of these methods are not yet conclusive, they nevertheless point to the promising new approach to understanding dipersal of Heteroderinae and other nematodes.

2. Life History

The basic life history is consistent within Heteroderinae, including four molts and four juvenile stages plus adults (Fig. 1). The first and part of the second stage occur within the egg, and the emerging infective second-stage juvenile establishes a sedentary feeding site in a host where the final three molts occur.

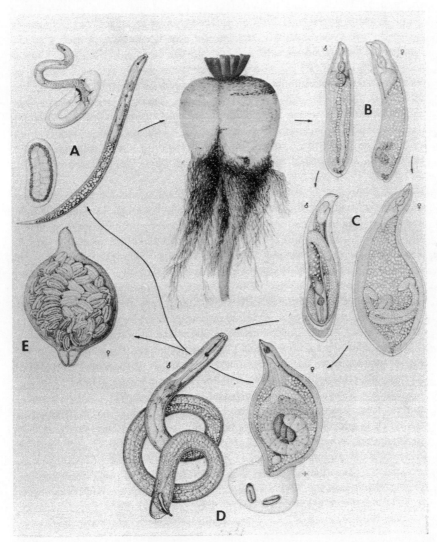

FIGURE 1   Life cycle of *Heterodera schachtii* on sugar beet. (A) Egg and second-stage juvenile. (B) Third-stage juveniles. (C) Fourth-stage juveniles. (D) Adults. (E) Cyst. (Illustrated by C. Papp.)

In most *Heterodera* spp. egg hatching as well as attraction to hosts is stimulated by host root exudates (Stone, 1974; Williams and Siddiqi, 1972; Brown, 1958; Perry and Clarke, 1981; Perry, 1986; Perry et al., 1989). However, other factors such as soil moisture, aeration, and temperature as well as nematode diapause are also responsible for seasonal emergence of juveniles in many species of *Heterodera* (Oostenbrink, 1967, Mulvey and Anderson, 1974; Mankau and Lindford, 1960; Franklin, 1972; Stone, 1974). In the absence of a host, cysts usually decompose slowly. Even where a host is present, hatching is often delayed over a period of years, with most second-stage juveniles released by the fifth or sixth year (e. g., Thorne, 1961). A few cyst nematode species, including *H. schachtii* and *H. trifolii*, apparently do not have a strong and consistent response to host exudates. Although there are many exceptions, little response to host exudates might be expected in species with relatively low host specificity and a wide host range.

Freshly hatched second-stage juveniles move toward roots of a potential host searching with movements of the head and lips and continuous stylet probing (Doncaster and Seymour, 1973). Once the nematode reaches the root tip surface, it makes quick, intermittent probes on the epidermis until a favorable site is found (Fig. 2B). Then it penetrates the cell wall (Fig. 2C). Penetration of the host occurs mostly at the growing tip of roots or at sites where lateral roots emerge; however, in some cases penetration may occur at any available site including wounds (Wyss and Zunke, 1986). Once the nematode penetrates epidermal cells, it moves intracellularly in the cortex toward the vascular cylinder, frequently in tissue adjacent to the protoxylem poles (Jones, 1981). Here probing occurs on cells that surround the head region until a feeding site is selected, stylet penetration takes place, and material from the esophageal gland lobe is injected (Endo, 1987).

After the infective juvenile establishes a suitable feeding site and nurse cells, the nematode enlarges rapidly and the final molts are completed (Figs. 2D, 3). Adults include sedentary obese egg-producing females and in most cases vermiform migratory males. Unbalanced sex ratios in amphimitic cyst-forming heteroderines are probably due to differential rates of death in unfavorable conditions. Females often die where there is reduced food availability and competition for feeding sites is high. Conversely, males require less food than females and are more tolerant of crowding (Koliopanos and Triantaphyllou, 1972). Some dispute that differential death rate alone can account for unbalanced sex ratios and consider sex reversal, as occurs in *Meloidogyne*, to also occur in some heteroderines (Müller, 1986). Environmentally controlled sex expression has indeed been demonstrated for *Meloidodera floridensis* (Triantaphyllou and Hirschmann, 1973).

Males of amphimitic species are apparently attracted to pheromones which may occur in exudates secreted through the vulva of females often into a clear gelatinous matrix (Green, 1971a; Cordero, 1989) (Fig. 4). Several males may be attracted to a single female, and copulation may occur more than once. After copulation, the clear matrix changes in color and intensity; subsequently fertilized eggs begin embryogenesis and may be laid or retained within the uterus. In many genera of heteroderines, females, full of eggs, may be transformed into tanned cysts upon death. These cysts typically have posterior fenestration, areas in the cuticle of loose fiberous mesh (see "Morphology" below). Fenestrae are eventually ruptured by second-stage juveniles hatching from eggs retained within the cysts (Fig. 2A). The first juveniles to hatch thrust their stylets against the fibrous areas creating rows of perforations. With pressure of these first juveniles against the fenestrae, the fibrous tissue is broken and infective juveniles emerge from the cyst (Cordero, 1989). Non-cyst-forming heteroderines, including *Thecavermiculatus*, may also retain eggs which hatch within the female, and they emerge as second-stage juveniles (Robbins, 1978) (Fig.

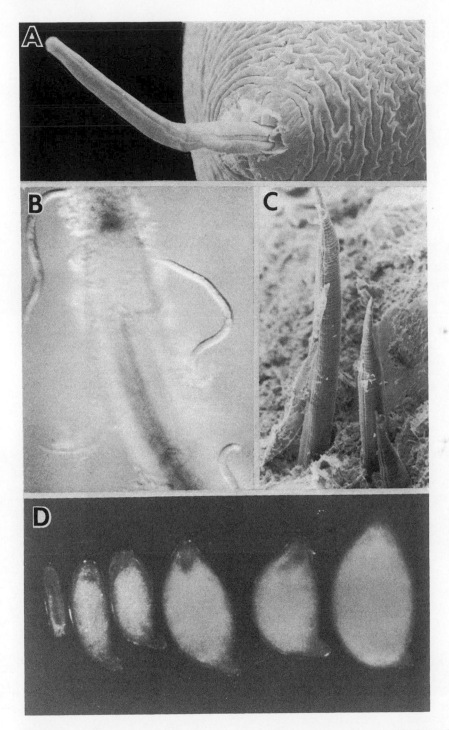

FIGURE 2 Biology of *Heterodera schachtii*. (A) Second-stage juvenile (J2) emerging from fenestrae of cyst (SEM). (B) J2s probing surface of sugar beet root prior to penetration. (C) Tail end of J2s protruding from root following penetration (SEM). Development of the female from the saccate second (left), third, fourth stages, as well as the young adult, mature adult, and cyst (right). (A, B, C courtesy D. Cordero Clark.)

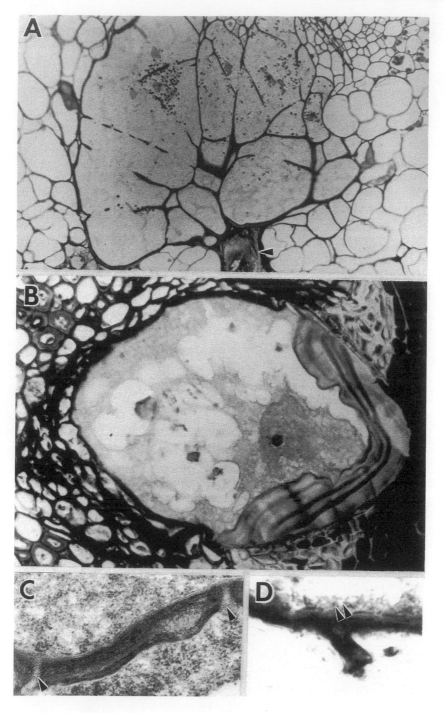

FIGURE 3   Host responses to Heteroderinae. (A) Syncytium. Arrowhead indicates head of female. (B) Single uniculate giant cell (SUGC). (C) Cell wall of SUGC with plasmodesmata (arrowheads). (D) Cell wall with wall ingrowths (double arrowheads). (A–C after Mundo-Ocampo and
Baldwin, 1983a,b; reprinted by permission *J. Nematol.*)

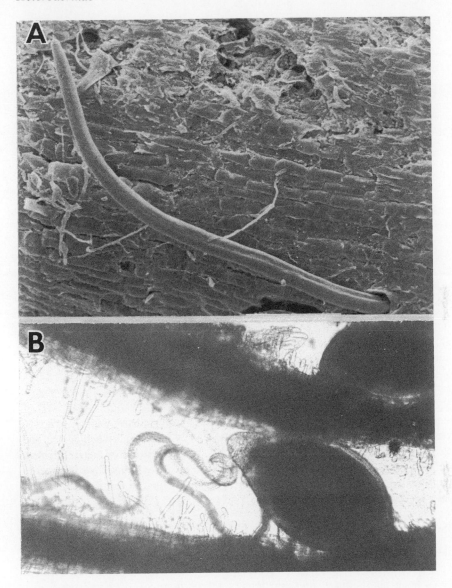

FIGURE 4   Males of *Heterodera schachtii*. (A) Male emerging from host root (SEM) (B) Males attracted to posterior terminus of newly hatched female. (B courtesy of D. Cordero Clark.)

26A). Comparisons among genera of cyst nematodes suggest that whether or not eggs are retained or laid is correlated with the degree of development of vaginal musculature in the adult (Cordero, 1989).

Although the entire life cycle of a heteroderine may be completed in less than 30 days, generation time varies with the heteroderine species and is greatly affected by environmental conditions.

Cytogenetically the heteroderines constitute a uniform group (Triantaphyllou, 1975a, 1983). Most of them are bisexual, cross-fertilizing species with a haploid

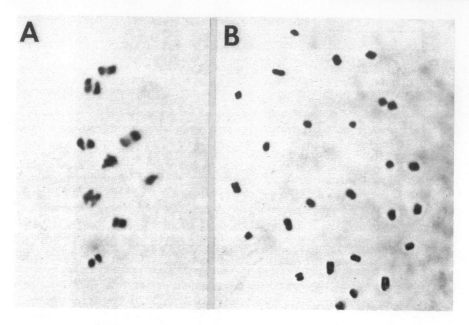

FIGURE 5   Chromosomes of oocytes of *Heterodera* spp. (A) Late prophase I of *Heterodera trifolii* with nine chromosomes. (B) Metaphase I of *H. trifolii* with 26 chromosomes. (Courtesy A. C. Triantaphyllou; after Triantaphyllou and Hirschmann, 1962, 1978.)

complement of nine chromosomes (Fig. 5A). *Cactodera betulae* deviates slightly from the normal cytogenetic status of the group in that it reproduces by meiotic parthenogenesis and has $n = 12$ or 13 chromosomes (Triantaphyllou, 1970). Also, members of the *Heterodera trifolii* species complex have no males and reproduce exclusively by mitotic parthenogenesis (Triantaphyllou and Hirschmann, 1978). They represent polyploid forms with somatic chromosome numbers of $24 - 35$ (Fig. 5B).

3.   Morphology

The specialized morphology of Heteroderinae reflects a complex life history including sedentary parasitism. The morphology is varied, accommodating diverse adaptations to a wide range of hosts and habitats. Investigations of morphology are the most reasonable place to begin unraveling questions of biology and physiology of heteroderines. Comparative morphology is presently the most important approach for investigating phylogenetic relationships and for identification. At the most practical level morphological identification is used to predict pathogenicity and host specificity; this is an essential first step for efficient application of every control strategy.

Those morphological characters that diagnose heteroderines from other Tylenchida have become more complex, yielding a broader understanding of the group. In 1934, Filipjev diagnosed *Heterodera*, which then included all cyst and root-knot nematodes, on the basis of sexual dimorphism and infective vermiform juveniles. Chitwood (1949) reinstated root-knot nematodes as a separate genus and distinguished *Heterodera* as distinct from root-knot nematodes primarily by the presence of a cyst with a complex nonstriated cuticular surface pattern as well as at least partial retention of eggs. Subsequent description of non-cyst-forming as well as striated heteroderines has required a more complex diagnosis to reflect the great morphological diversity of this group. Heteroderine females are generally nearly spherical but exceptions are sausage-shaped; the vulva is terminal or subterminal but exceptions are equatorial (subterminal) (Fig. 6); the excretory pore is near the level of the median bulb, but in exceptions it is far anterior; eggs may be laid in a gelatinous matrix, or they may not. Eggs may be retained in the body of the female and a cyst may form, or they may not (e.g., Luc et al., 1988). Regardless of heteroderine variabilities, in practice only a little experience is needed to distinguish them from root-knot nematodes. The perineal region of heteroderines is unlike the distinctive fingerprint-like perineal pattern in root-knot nematode females (Fig. 7A,B). Second-stage juveniles are relatively robust with a hyaline terminal region, unlike the delicate appearance, weak stylet, and nonhyaline tail terminus of root-knot nematode juvenile (Fig. 7C,D). Heteroderine males lack the distinctive lateral amphidial cheeks of root-knot males; they also have a long, slender esophageal isthmus in contrast to the very short, broad isthmus of root-knot nematodes (Chitwood, 1949; Baldwin and Hirschmann, 1973, 1975; Baldwin et al., 1977) (Fig. 7E,F). The first heteroderines described were cyst forming. Since the cyst is the life stage most readily available from the soil, taxonomic characters are primarily related to the cyst (Hesling, 1978; Stone, 1986) (Fig. 8). Yet, there is a disagreement about the definition and homology of the cyst and even whether or not genera which acquire some color but do not retain eggs after death of the female have a cyst (Baldwin and Bell, 1985; Luc et al., 1986; Wouts, 1985). Luc et al. (1986) defined a heteroderine cysts as a persistent tanned sac which retains eggs and is derived from some or all components of the mature female body

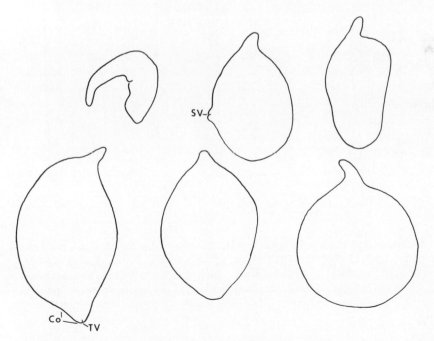

FIGURE 6 Shapes of females of Heteroderinae genera ranging from reniform to saccate and with a rounded terminus or cone (Co). Vulva may be subterminal (SV) or terminal (TV). (After Baldwin and Schouest, 1990, reprinted by permission Kluwer Academic.)

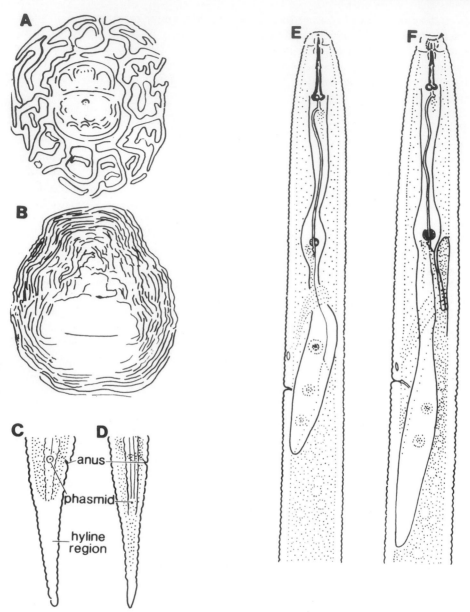

FIGURE 7   Morphological features distinguishing Heteroderinae from *Meloidogyne*. (A) Terminal region of female of Heteroderinae. (B) Terminal region of female of *Meloidogyne*. (C) Tail region of second-stage juvenile (J2) of Heteroderinae with lens-like phasmid and hyaline terminus. (D) Tail region of J2 of *Meloidogyne* with pore-like phasmid. (E) Anterior region of male of Heteroderinae. (F) Anterior region of male of *Meloidogyne* (arrowhead indicates position of amphidial cheek).

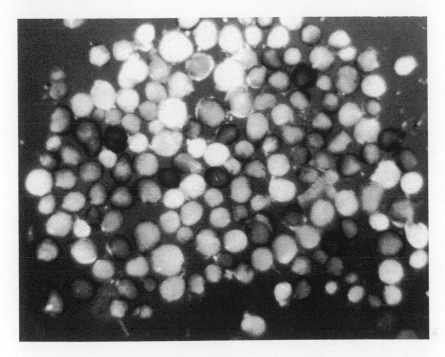

FIGURE 8   White females and pigmented cysts of *Punctodera chalcoensis*.

wall. By this definition genera with cysts include *Afenestrata, Heterodera, Cactodera, Globodera, Punctodera*, and *Dolichodera*; the majority of heteroderine genera lack cysts (Table 1). Genera with females which become cysts and genera with females that do not share similar features of shape, cuticle layering, surface patterns, and digestive, reproductive, and sensory systems.

All Heteroderinae females have a narrow anterior protuberance or "neck" and swollen body, but genera vary in details for shape (Fig. 6). *Verutus* is readily recognized by its reniform shape, whereas other heteroderines are more swollen and overall shape is particularly affected by the presence or absence of a terminal cone (Fig. 11). The code apparently forms in response to differential elasticity of the body wall as the female enlarges, i.e., posteriorly the body wall is least elastic for girth expansion (Cordero, 1989). It is uncertain if reduced elasticity is a function of cuticle thickness, cuticle layering and the physical nature of layers, attachment points of vaginal muscles, or there parameters. Some genera lacking a cone, such as *Meloidodera*, do not have cysts; other genera with globose females and no cone, including *Globodera*, do form cysts. Many genera have distinct terminal cones so that the female is lemon-shaped as in non-cyst-forming *Atalodera* or cyst-forming *Heterodera*. The genus *Thecavermiculatus* shares many specialized characters with Atalodera but the cone is reduced, as is also the case in certain species of *Cactodera* and *Heterodera*. In some cases fine details of female/cyst shape or size can be characteristic of species and of some limited value for identification as noted for *Heterodera cardiolata* or *Cactodera estonica* (Golden, 1986). Although intraspecific variability is usually great, taxonomic reliability of shape and size of females/cysts can sometimes be improved by considering the largest individuals in the population (Hirschmann, 1956; Thorne, 1961).

TABLE 1   Genera of Heteroderinae and Numbers of Valid Species[a]

Genera	Authority	Number of species
Heterodera[b]	Schmidt, 1871	57
Afenestrata	Baldwin and Bell, 1985	1
Cactodera	Krall and Krall, 1978	9
Globodera[b]	Skarbilovich, 1959	12
Punctodera[b]	Mulvey and Stone, 1976	3
Dolichodera	Mulvey and Ebsary, 1980	1
Atalodera	Wouts and Sher, 1971	4
Thecavermiculatus	Robbins, 1978	4
Camelodera	Krall, Shagalina, and Ivanova, 1988	1
Bellodera	Wouts, 1985	1
Sarisodera	Wouts and Sher, 1971	1
Rhizonema	Cid Del Prado Vera, Lownsbery, and Maggenti, 1983	1
Hylonema	Luc, Taylor, and Cadet, 1978	1
Ekphymatodera	Baldwin, Bernard, and Mundo, 1989	1
Cryphodera	Colbran, 1966	4
Meloidodera	Chitwood, Hannon, and Esser, 1956	6
Verutus	Esser, 1981	3

[a]Groupings generally correspond to tribes of Baldwin and Schouest (1990).
[b]Genera with economically most important species.

The body wall cuticle of Heteroderinae females is modified from the A and B layers present in vermiform Tylenchida to include a broad C layer; in some genera, including non-cyst-forming *Atalodera* and cyst-forming *Globodera*, an additional D layer occurs (Shepherd et al., 1972; Cliff and Baldwin, 1985) (Fig.9). The presence or absence of the D layer among cyst nematodes was used to support separation of round and lemon-shaped cysts into distinct genera (Shepherd et al., 1972). However, *Heterodera*, previously reported to lack a D layer, was recently found to have a very thin D layer 4 weeks after the final molt (Cordero, 1989). Overall thickness of cuticle does not seem to relate to number of layers. For example, *Punctodera*, with a D layer, may have a very thin cuticle. Homology of layers in the female and layers in the cyst have not been established, nor is it known if particular cuticular layers or other parts of the body are especially responsible for the color that occurs in many dead females and cysts (Fig. 8). Changes in color are attributed to action of phenoloxidase on substrates in the cyst wall (Awan and Hominick, 1982; Ellenby, 1946a, 1955), but the precise process and nature of variation in color are unknown. Cyst color varies among species from light tan, as in *Heterodera zeae*, to shades of brown and nearly black in other species (Golden, 1986). In some cases the sequence of color change as a female matures is distinctive, as in *G. rostochiensis* which forms a golden, followed by a brown cyst, versus *G. pallida* with no golden stage. Although color may be helpful in identification, it must be used cautiously because color variations may be associated with age, translucence, reflection, surface texture, and variable environmental influences on the cyst (Hesling, 1978; Golden, 1986).

The color of a young cyst may be masked by the presence of a subcrystalline layer, a layer external to the cuticle of some species (Fig. 10A,B). The material may occur on molted fourth-stage cuticles but is most conspicuous in adults and young cysts where it varies greatly in thickness from a uniform film to a thick layer apparently cracked into a pattern of blocks. The subcrystalline layer occurs in both cyst-and non-cyst-forming taxa, and its presence or absence and thickness, while variable, may be somewhat species-specific. Brown et al. (1971) proposed that the subcrystalline layer was partially composed of a fatty acid generated by a symbiotic fungus feeding on metabolic products of the nematode. However, subsequent investigations show that the subcrystalline layer persists in aseptic culture (Zunke, 1986; Cordero, 1989).

Overall cuticular surface patterns, and particularly those of the midbody region, occur in diverse forms among females and cysts of Heteroderinac. These particular patterns may characterize specific taxa (Fig. 10). These midbody patterns broadly include striated and zig-zag. However, particular patterns may be further specified as fine or coarse, ridged, reticulate, lacelike, or punctate. Striated patterns are common among some non-cyst-form-ing genera including *Meloidodera* and *Cryphodera*. Wouts (1985) questions homology of this pattern with the similar parallel ridges of *Bellodera*, and cyst-forming *Dolichodera*; questions of homology with the parallel ridges or broken wavy lines of some *Cactodera* must also be considered. Zig-zag patterns occur in noncyst genera such as *Atalodera* as well as most cyst nematodes including *Heterodera*. Zig-zag patterns range from coarse to fine. They also vary in details of anastomosing and reticulation, and they may be associated with punctations (Fig. 10G). Punctations especially characterize *Punctodera* where they occur as distinct rows beneath the zig-zag surface, but they also occur with varying reliability and clarity in many other genera (Golden, 1986). Previously punctations were interpreted as pores or pits (Franklin, 1939; Thorne, 1928), but they are more likely homologous with optically and electron dense deposits in the A layer of the cuticle. In some taxa these

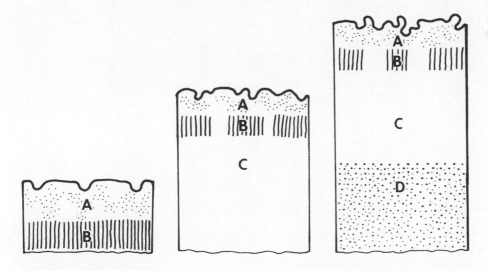

FIGURE 9   Diagramatic cross-sections of female cuticles showing A and B layers alone as they occur in males and second-stage juveniles, as well as presence of layers C and C+D as they occur in adult females of Heteroderinae. (After Baldwin and Schouest, 1990; reprinted by permission Kluwer Academic.)

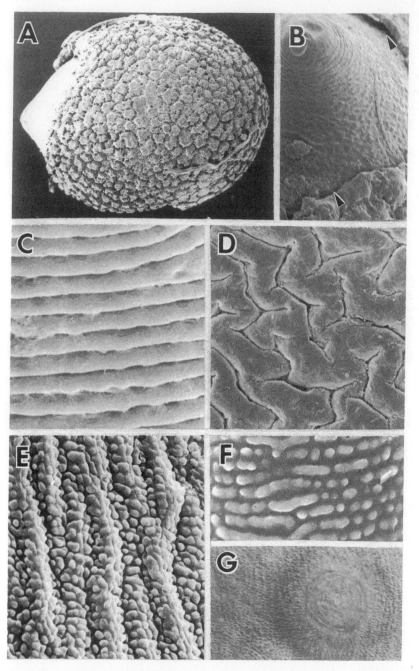

FIGURE 10   Surface patterns of females and cysts of Heteroderinae (SEM unless indicated otherwise). (A) Subcrystalline layer, lateral view of cyst of *Heterodera avenae*. (B) Enlargement of terminal region from A showing boundary of subcrystalline layer (arrowheads). (C) Striated cuticle, *Rhizonema sequoiae*. (D) Zig-zag cuticle, *Atalodera lonicerae*. (E) Tuberculate cuticle with longitudinal striae, *Ekphymatodera thomasoni*. (F) Cuticle of tuberculate neck region of *Globodera rostochiensis*, (G) subsurface cuticular punctations, *Punctodera chalcoensis* (light microscopy).

deposits occur in periodic globules (= punctations?), whereas in others the material is diffuse (Cliff and Baldwin, 1985; Baldwin, unpublished) and may not be apparent with light microscopy (= no punctations?).

The derivation of the surface patterns of heteroderine cysts and females is not clear. Franklin (1939) and Wieser (1953) considered them to be derived from the cuticular annulations of second-stage juveniles. In *H. schachtii*, video-recorded development indicates that the striated pattern persists only a short time after infection; by early fourth stage the midbody has irregular longitudinal ridges. The zig-zag pattern first forms posteriorly in the late fourth stage and spreads to the midbody prior to the last molt (Cordero, 1989). There is not yet an explanation as to how the various patterns develop differently among genera and species.

Cuticular surface patterns of the midbody of females and cysts arc helpful in species identification as first shown by Taylor (1957) but have become less useful with awareness of the large number of heteroderine species and discovery of intermediate patterns which make precise terminology and characterization more difficult to apply (Hesling, 1978). Midbody cuticular patterns vary within a population, from one region of the body to another, with age, and between females and cysts of a given species. Improved understanding of these variables throughout Heteroderinae may allow further standardization of conditions for comparing patterns and the use of finer details and more precise descriptive terminology for taxonomic characterization (Othman et al., 1988).

The surface pattern anteriorly is similar among heteroderine females and cysts with striations throughout the neck region, but some taxa, such as *Globodera*, are distinctive by abundant neck tubercles (Othman et al., 1988))(Fig. 10F). The en face region typically includes a squarish labial disc just anterior to a plate consisting of six fused lips (Fig. 13G), but in some genera, such as *Verutus*, the lateral lips are partially separate from the rest of the plate (Othman and Baldwin, 1985). The excretory pore of heteroderine females generally occurs near the base of the neck at the level of the medium bulb or further posteriorly; however, in *Bellodera* it occurs far anterior, near the level of the stylet knobs (Fig. 25C).

The surface pattern of the posterior region of females and cysts varies among heteroderines and is usually modified from that of the midbody. However, in *Meloidodera*, the striae are unchanged at the terminus, forming concentric rings around the anus and phasmid openings; posteriorly the position of lateral lines is indicated by interruptions in the striae, not unlike the lateral lines of some *Meloidogyne* females. In *Verutus* the striae are interrupted or continue around the anus (Othman and Baldwin, 1985; Baldwin et al., 1989) (Fig. 14A). In *Meloidodera* and *Verutus*, the pattern of striae is barely interrupted at the equatorial vulva. In other heteroderines the vulva is terminal in both cyst-and non-cyst-forming species the surrounding terminal pattern is coarser and more irregular than at midbody; there is little or no evidence of lateral lines and no evidence of phasmid openings (Fig. 14). The anus is relatively dorsal to the terminal vulva and distance between the two varies from 150 µm in some cyst-forming nematodes to only a few micrometers in *Atalodera* (Fig. 14B), and finally *Sarisodera* where the anus occurs on the lip of the vagina. The intervening perineal area (*sensu* Green, 1971b) often has distinctive cuticular markings. For example, in *Globodera* the pattern may be mazelike or have parallel ridges, and the number of ridges between vulva and anus may vary with the species (Figs. 14E, 21D, E). In the immediate area of the vulva of most *Globodera* the vulva spans a basin or slight depression which has a crescent-shaped structure on both the dorsal and ventral rims. The crescents are covered with nonsensory cuticular protuberances which may be tightly packed, scattered, or, in some species, absent (Stone, 1986) (Fig. 14E, F). In *Heterodera* cysts a region analogous to the rim of the basin may have wrinkles (rivulets) or other patterns distinctive among certain species (Golden, 1986) (Figs. 11, 14B–D).

In cyst-forming nematodes, excluding *Afenestrata*, the cuticle in the area surrounding the vulva is thin-walled consisting of loose mesh fibers (Cordero, 1989). The area eventually ruptures in the mature cyst. This ruptured area is the fenestra, and the size and shape of fenestrae is variable among genera and species (Figs. 11 and 12). Fenestrae occur in two basic patterns: with two openings (semifenestrae), one on each side of the vulva and surrounding tissue, or circumfenestrate with a single hole which deletes the vulva and surrounding tissue (Figs. 12C, 23A). Circumfenestrae encompass most or all of the basin, but exclude the crescents (Fig. 14E). If the semifenestrae are separated by a wide bridge enclosing a short vulva and surrounding tissue, and if the holes are each more than one-half a circle, the pattern is bifenestrate (Fig. 12B). If the bridge is narrow and the semifenestrae are flattened next to the bridge, the pattern is ambifenestrate (Stone, 1986) (Fig. 12A). In *Punctodera* the circumfenestrae is accompanied by a distinct anal fenestra (Fig. 12D). Circumfenestrae without anal fenestration are associated with round cyst nematodes including *Globodera* and *Dolichodera*, as well as in the cyst of *Cactodera* which has a small cone. Both bifenestrae and ambifenestrae are common in *Heterodera*, but bifenestrae are variable within species and may be difficult to apply in a diagnosis (Stone, 1986). In other cases the narrow bridge of some species such as *H. carotae* may break giving a false appearance of a circumfenestra and thereby confound diagnosis for the inexperienced (Golden, 1986).

Taxonomic characters associated with fenestrae include the distance from the anus to the nearest edge of fenestrae. Dividing this measurement by length of fenestrae gives Granek's ratio as defined by Hesling (1973), a helpful but variable tool for species diagnosis. Other commonly used characters include width and length of vulval slit and surrounding bridge.

Some surface patterns of the posterior region of heteroderine females and cysts are associated with the reproductive system. The vulva and associated lips lead to a dorsoventrally flattened vagina which connects to two long, coiled reproductive tracts (Fig. 1B–D). Each tract consists of an elongate uterus which connects to a short oviduct and spermatheca; the spermatheca connects to a long growth zone and relatively short germinal zone of the ovary (Triantaphyllou and Hirschmann, 1962). The cuticle-lined vagina is enclosed by a circular sheath of muscles near the uterus, the sphincter vagina (Fig. 11). Additional bundles of muscle fibers, the dilatores vagina, extend from the cuticle lining of the vagina to the body wall cuticle on the dorsal and ventral side of the vulva. In *H. schachtii* the dilatores vagina occur as six muscles on each side at four levels resulting in 48 muscles. In other heteroderine genera, particularly those that do not lay eggs, musculature is diminutive (Cordero, 1989). In many genera the cuticle thickens at the proximal end of the vagina in mature females and forms an underbridge (Fig. 11). The underbridge approaches the adjacent body wall cuticle, sometimes forking at the junction. The underbridge persists in the cyst and the thinner cuticle of the vagina may also persist in the cyst as a "sheaf" (Golden, 1986). The hypodermis of the posterior region of cyst-forming heteroderines includes regions of crystalline cuticular material which are deposited unevenly on the inner body wall cuticle as the female becomes senescent (Cordero, 1989). Large projections of the material extend into the body cavity and they become dark-colored, persisting in the cyst as bullae (Fig. 11) which are characteristic of many *Heterodera*, *Dolichodera*, and some *Punctodera* cysts. Bullae are particularly associated with species that have a pronounced underbridge, and they are predominant just internal from the underbridge. Their presence, size, and pattern may be useful for diagnosis. In *Cactodera* and some *Heterodera* small

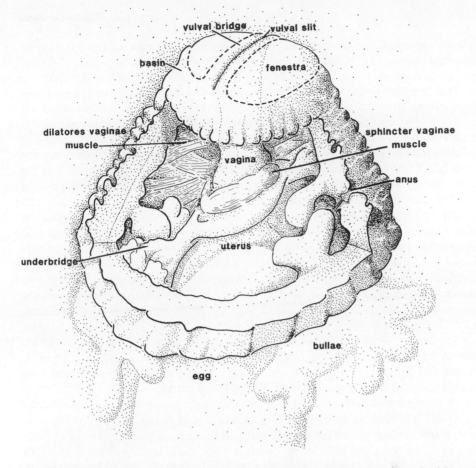

FIGURE 11 Illustration of terminal cone region of female of *Heterodera schachtii* with cutaway showing internal structures.

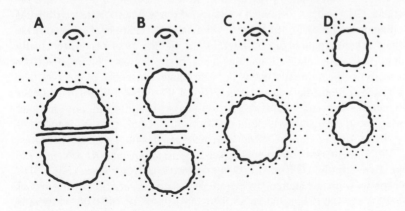

FIGURE 12 Fenestral patterns of terminal region of Heteroderinae. (A) Semifenestrate-ambifenestrate vulval region; anal region not fenestrate. (B) Semifenestrate-bifenestrate vulval region, anal region not fenestrate. (C) Circumfenestrate vulval region anal region not fenestrate. (D) Circumfenestrate vulval region; separate anal fenestra.

pigmented structures, vulval denticles, are apparent near the apex of the vulval cone (Golden and Raski, 1977). Although they may resemble small bullae, Cordero (1989) demonstrated homology with remnants of vaginal muscles which persist in the cyst.

Poorly understood structure of the cyst include Mulvey's bridge, which may occur in some species at right angles to the underbridge. Some have speculated that Mulvey's bridge may be a hardened remnant of muscle (Mulvey, 1959), but recent studies suggest it is more likely of cuticular origin (Cordero, 1989). Cells specifically involved in secretion of the gelatinous matrix of eggs have not been identified although tissue surrounding the uterus is filled with rough endoplasmic reticulum and seems a likely source of exudate (Cordero, 1989). Absence or presence of matrix and the size of egg mass may vary among species, but it is not usually of diagnostic value.

The digestive system of heteroderine females consists of a stylet, esophagus, intestine, rectum, and anus apparently with little variation among species, except in the size and shape of the stylet and distance of the dorsal gland orifice to the stylet knobs. The anus is important in identification with respect to its distance from the vulva as well as its fenestration in cysts of *Punctodera*. Little is known about the fine structure of the digestive system of the female heteroderine, whereas males and juveniles have been thoroughly investigated (Baldwin, 1982; Baldwin and Hirschmann, 1976; Baldwin et al., 1977; Endo, 1984, 1987, 1988).

The sensory system of female heteroderines has not been adequately investigated. Apparently anterior sensory organs, including amphids and labial papillae, are well developed and active, whereas phasmids are only visible in females of *Verutus* and *Meloidodera* (Othman and Baldwin, 1985, 1986; Othman et al., 1986, 1988).

Males of heteroderines are vermiform, with a short, rounded tail lacking caudal alae (Fig. 13L, M). The body of males of most species is twisted as much as 90°, although in *Verutus*, males are not twisted. Males of most species develop from sedentary globose fourth-stage juveniles, although in *Meloidodera* they may develop from nonfeeding fourth-stage vermiform juveniles (Hirschmann and Triantaphyllou, 1973). The male length varies from 450 to 1700 µm. However, length may be affected by environmental considerations such as nutrition before the final molt; thus, intraspecific variation is so great that male length is generally not useful as a diagnostic character. In contrast to females, the cuticle of males consists solely of an A (exocuticle) and B (endocuticle) layer (Fig. 9).

Surface structures have been examined in males of many heteroderines with SEM (Baldwin, 1986; Othman and Baldwin, 1985, 1986; Othman et al., 1986, 1988). In each case the body in annulated. There are three or four lateral lines and the outer bands are usually areolated. The lateral lines originate 7–10 annules from the lip region. The basic surface pattern of the lip region of heteroderine males is hexaradiate consisting of a labial disc surrounded by six lips, four submedial and two lateral (Fig. 13D–F, J, K). Males of a number of genera have lip patterns with longitudinal blocks on the head region similar to those in many Hoplolaimidae; these are visible with light microscopy and can assist in diagnosis of species, including some *Meloidodera*, *Cryphodera*, *Atalodera*, and *Thecavermiculatus*. Similarly, narrow lips characterize some genera such as *Sarisodera* and *Rhizonema*, the fusion of the labial disc with the submedial lips is distinctive in *Heterodera* (Fig. 13D). Nevertheless, lip patterns of males tend to have more intraspecific variation than those of second-stage juveniles, and the diagnostic value of many aspects of male lip patterns is limited. Posteriorly, the tail length of males may be diagnostic, and a sheath (tubus) that extends from the cloaca and partially encloses retracted spicules may be distinctively large in some genera including *Sarisodera* and *Afenestrata*. The tail terminus may be smooth or

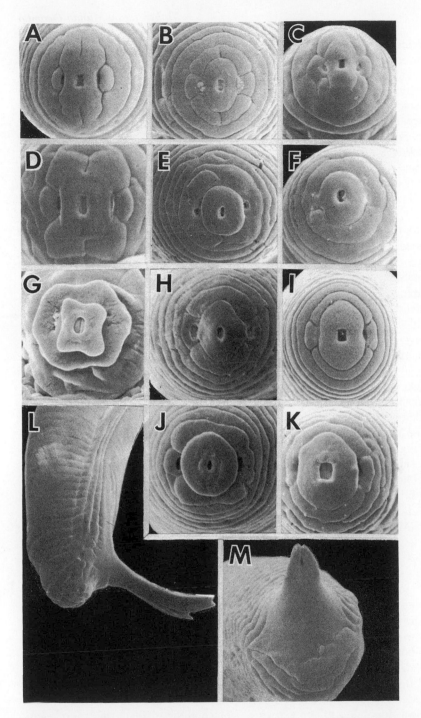

FIGURE 13 En face lip patterns and the terminal region of the male of Heteroderinae. (A) *Heterodera glycines*, second-stage juvenile (J2). (B) *Globodera rostochiensis*, J2. (C) *Punctodera chalcoensis*, J2. (D) *H. glycines*, male. (E) *G. rostochiensis*, male. (F) *P. chalcoensis*, male. (G) *H. trifolii*, female. (H) *Verutus californicum*, J2. (I) *Rhizonema sequoiae*, J2. (J) *V. californicus*, male. (K) *R. sequoiae*, male. (L) Terminal region of male of *H. schachtii* with bifid spicule tip. (M) End view of terminal region of *R. sequoiae* without bifid tip. (B, C, E, F after Othman et al., 1988; reprinted by permission, *Revue de Nematologie;* J after Baldwin et al., 1989, reprinted by permission *J. Nematol.)*

have complex ridges, or tubercles, but the patterns are highly variable, even within a given population (Fig. 13L, M).

The reproductive system of heteroderine males is composed of a pair of retractable copulatory spicules which protrude from a cloaca (Figs. 1, 13L, M). The cloaca leads to the vas deferens, which in turn connects with a single testes. The spicules have a broad proximal end, a cylindrical shaft, and a flattened pointed blade with incurved edges; the broad end is attached to two protractor muscles (Clark et al., 1973). The pair of blades interlock along part of their length to form a conduit for sperm transmission from the cloaca. The spicules are innervated and spicule neurons include those which terminate in a minute channel which opens near the tip. The tip may be rounded as in *Rhizonema* or *Globodera*. Conversely, it may be bifid as in *Heterodera*, *Cactodera*, and *Verutus* (Fig. 13L, M). Spicule length varies from 20 to 45 μm, and reportedly it is 2.5–4% of male length (Geraert and DeGrisse, 1981); almost the entire range of length is expressed within some genera (e.g., *Heterodera* spp). The cuticle-lined tract in which spicules move is thickened dorsally, forming a gubernaculum.

The reproductive system of heteroderine males is linked to the digestive system through the common terminal duct, the cloaca. Although the stylet, dorsal esophageal gland orifice, esophagus, and intestine generally appear well developed and robust relative to infective juveniles, males apparently do not feed (Fig. 7E). This lack of feeding might be reflected, however, by diminutive esophageal musculature and dorsal gland (Baldwin et al., 1977; Baldwin, 1982). The excretory pore generally occurs at the level of the esophageal gland lobe.

The sensory system of heteroderine males includes well-developed amphids and six inner labial sensillae which open to the exterior and are probably chemoreceptive. In addition, four outer labial and four cephalic receptors terminate beneath the cuticle surface. Phasmids, while present in *Meloidodera*, are diminutive or absent in males of most heteroderines (Fig. 13L, M) (Carta and Baldwin, 1990a, b).

Second-stage juveniles of heteroderines are slender with a rounded anterior and tapering tail, with total body length varying from about 330 to 700 μm (Figs. 1, 15A, D). Generally length is sufficiently stable within species to be a useful diagnostic character (Wouts and Weischer, 1977), although some rare populations include aberrant giant or dwarf juveniles (Mulvey, 1959, 1960a). As in males, the cuticle consists only of an A and B layer, although it is modified posteriorly to include a fiberous (collagenous?) hyaline region (Figs. 7C, 9, 15A, D). The length of this hyaline region varies among species and may be diagnostic.

Surface structures have been examined in second-stage juveniles of many heteroderines with SEM (Baldwin, 1986; Othman and Baldwin, 1985, 1986; Othman et al., 1986, 1988). The body is annulated and possesses lateral lines similar to that of males, except that they terminate relatively anteriorly on the tail, often only a few annules posterior to the phasmid opening (Fig. 7C). The basic surface pattern of the lip region of heteroderine juveniles is hexaradiate consisting of a labial disc surrounded by six lips, four submedial and two lateral (Fig. 13A–C, H, I). Posterior to the disc and lips, head annules vary in number among species and may be useful in some diagnoses. Contrary to some males, there are no longitudinal blocks. In addition to narrow submedial lips distinguishing *Sarisodera* or *Rhizonema* (Fig. 13I) and fusion of submedial lips with the labial disc characteristic of *Heterodera* (Fig. 13A), other features of lip patterns of juveniles are useful systematic characters. For example, *Punctodera* has submedial lips fused with head annules (Fig. 13C), and in *Atalodera*, *Thecavermiculatus*, and some cyst-forming genera, lateral lips are

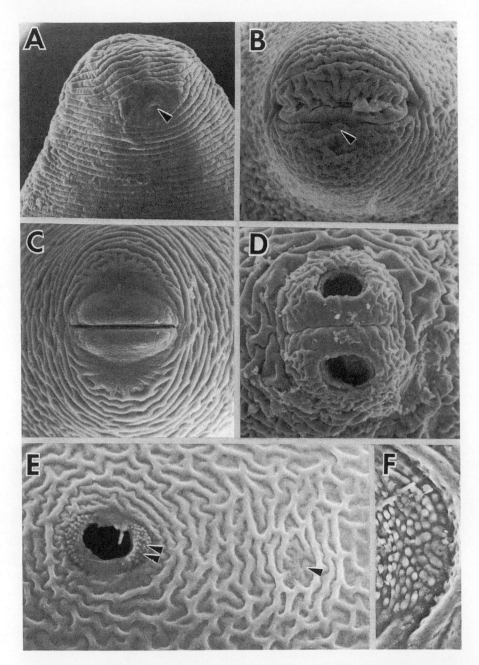

FIGURE 14   SEM of posterior region of Heteroderinae. (A) Female of *Verutus californicus* showing anus (arrowhead). (B) Female of *Atalodera lonicerae* showing vulval lips and anus (arrowhead). (C) Female of *Heterodera trifolii* showing vulval lips. (D) Cyst of *Heterodera fici* with semifenestra. (E) Cyst of *Globodera* sp. with circumfenestrae. Arrowhead indicates position of anus, double arrowheads indicate perineal tubercles on vulval crescent. (F) Enlargement of tubercles from E.

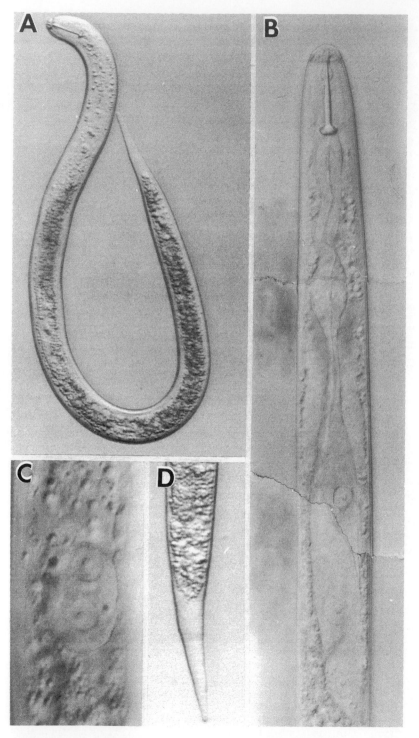

FIGURE 15   Morphology of second-stage juveniles (J2) of Heteroderinae. (A) Entire J2. (B) Anterior region showing stylet and ventral overlap of esophageal glands. (C) Genital premordium. (D) Hyaline region of tail.

fused with the labial disc. These distinctive lip patterns are relatively stable and useful in identification and phylogenetic analysis (Baldwin and Schouest, 1990).

The reproductive system of second-stage juveniles, including both males and females, is an oval primoridium with four cells (Fig. 15C). The primordium occurs slightly posterior to the center of the body in the region of the intestine. In heteroderines with a terminal vulva, the primordium doesn't migrate posteriorly until the third stage and third molt (Fig. 1B).

The digestive system of second-stage juveniles includes a stylet whose length and knob morphology is particularly stable so that even minute differences may be diagnostic for species (Hirschmann and Triantaphyllou, 1979; Wouts and Weischer, 1977) (Figs. 15, 17A–C). The esophagus of juveniles resembles that of males, except that the musculature or dorsal gland are relatively well developed (Fig. 15B). The gland region fills the body width in nearly all heteroderines, except that in *Atalodera* and *Thecavermiculatus* it is particularly narrow, and it may be intermediate in width in *Hylonema* and *Verutus* (Figs. 15B, 25A, B). The position of the excretory pore is near the level of the esophageal gland lobe. The intestine, which may be highly vacuolated in starved juveniles, joins a muscular rectum and opens posteriorly through a cuticle-lined anus (Fig. 7C). The position of the anus is important because it marks the beginning of the tail, and tail length is often a useful diagnostic character.

The nervous system of heteroderine juveniles includes highly developed amphids as well as a full complement of 16 anterior sensillae (Endo, 1980). Posteriorly, the phasmid may terminate as a pore as in cyst-forming nematodes, or the opening may lead to a broad lenslike ampulla (Baldwin, 1985) (Fig. 7C, D). In many heteroderines, the internal structure of the phasmid may begin to deteriorate soon after hatching (Carta and Baldwin, 1990a, b).

4.   Feeding Habits and Parasitism

Heteroderinae are obligate sedentary parasites that feed as mature females, generally with most of their enlarged body protruding from the root surface, and their elongate anterior region embedded in the host root (Fig. 16). As second-stage juveniles, they induce highly specialized nurse cells which are sustained throughout the life of the parasite (Fig. 3).

Details of feeding by heteroderines has been documented throughout the life history with inverted interference contrast microscopy and high-resolution video-enhanced recording, by examining nematodes in vitro culture. After an infective juvenile settles in the region of the vascular cylinder, the stylet probes, and finally penetrates an adjacent cell. Ingestion of cell contents is associated with alternate stylet protrusion, intermittent pumping of the metacorpus, and synthesis in esophageal glands followed by release of contents of the dorsal gland ampulla through the stylet and into the host cell (Wyss and Zunke, 1986). During this initial contact the cell hypertropies and a nurse cell begins to form.

Nurse cells may take many forms; the best known are multinucleate syncytia which are characteristic of cyst-forming genera (Fig. 3A). However, syncytia also occur in non-cyst-forming heteroderines including *Atalodera* and *Verutus*. Many non-cyst-forming genera, such as *Meloidodera* and *Sarisodera*, induce a single giant cell with one large nucleus (Fig. 3B). Nonheteroderines such as *Rotylenchulus* and *Meloidogyne* also induce nurse cells, but their similarity to those of heteroderines may be superficial. For example, the multinucleate nurse cells of *Meloidogyne* and *Rotylenchulus reniformis* are coenocytes because they originate by hypertrophy and karyokinesis without cytokinesis. This is in contrast to the heteroderine syncytium which forms from merging hypertrophied adjacent cells, apparently through enlargement of plasmodesmata openings between cells and putative dissolution of cell walls (Jones, 1981a, b; Mundo-Ocampo and Baldwin, 1983).

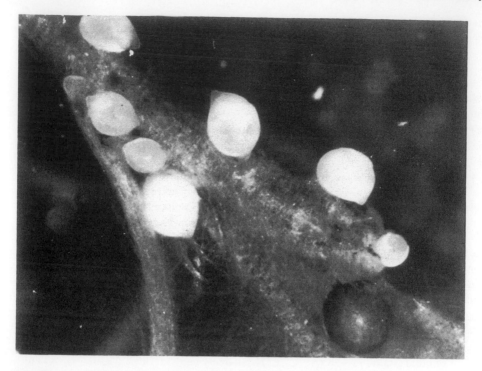

FIGURE 16   Females of *Heterodera schachtii* feeding on roots.

Nurse cells allow the parasite to sequester basic nutrients from adjacent plant tissues for
nematode development and reproduction. The host, in response, continuously replaces the
lost cell nutrients. The flow of nutrients from adjacent plant tissue to the nurse cell may be
enhanced by increased numbers of plasmodesmata or by increased surface area associated
with wall ingrowths, particularly in regions in contact with xylem (Fig. 3C, D). During
prolonged feeding, both single uninucleate giant cells and syncytia increase in size, first
expanding toward the root stele, and then extending several millimeters longitudinally. As
the nurse cell increases in size, adjacent cells may be crushed and hyperplasia may occur in
surrounding tissues. Nurse cells typically have a feeding tube and a plug or thickening of the
cell wall around the point of stylet penetration. Their dense cytoplasm appears to be the
result of loss of a central vacuole and replacement with numerous small vacuoles, as well as
abundant organelles including mitochondria, plastids, and Golgi (Endo, 1987; Jones,
1981a, b). Whether the nurse cell has one or many nuclei, the nuclear and nucleolar volume
is greatly increased, and multilobed (Fig. 3A, B). Conditions unfavorable for development
of the nematode such as a resistant host may be associated with a "hypersensitive response,"
i.e., the nurse cell is not adequately established and is surrounded by localized lignification
and necrotic brown tissue which, at least in potatoes, is autofluorescent (Robinson et al.,
1988).

B.   Agricultural Importance

The worldwide agricultural impact of cyst nematodes is probably second only to root-knot
nematodes. Cyst nematodes are set apart by the number of highly pathogenic species
specialized on major agricultural commodities including grains, root crops, and most

legumes. Many of the most serious cyst nematode pests are spread worldwide, in part because eggs in cysts remain viable under conditions of dispersal that would be fatal to most nematodes. Protective cysts often confound management practices which would be effective against other nematodes; they may escape nematicides and only long rotations may be effective. Furthermore the value of many resistant varieties is sometimes limited by "resistance-breaking" pathotypes or races.

Although worldwide losses to cyst nematodes are considerable, they are difficult to quantify, in part because many costs are indirect. Losses are indirect through interactions with other biotic factors including pathogenic bacteria, fungi, and probably viruses. Miller (1986) citing the work of Evans and Brodie (1980) notes, for example, that losses to potato cyst nematode are particularly high in the presence of *Verticillium*. These interactions may be facilitated by wounds associated with penetration of juveniles as well as physiological changes of the host which result in increased susceptibility to secondary pathogens. Losses of soybean to *H. glycines* are increased by inhibition of root nodulation of nitrogen-fixing bacteria (Lehman et al., 1971). Similarly, nitrogen fixation on peas is inhibited by *H. goettingiana* (Fig. 19B, C). Losses to cyst nematodes may also be compounded by abiotic factors. In Mexico, *Punctodera chalcoensis* may reduce yields on corn (*Zea mays*) by nearly 100%, but losses reportedly can often be offset by increased water and fertilizer. On the other hand, Mai (1977) notes that losses to potato cyst nematode may result in a harvest less than the seed pieces, even with heavy application of fertilizer. Symptoms similar to nutrient deficiencies are common, but their expression varies with the crop and depends on soil texture, nutrient makeup, and pH. Greater susceptibility to damage by heat and draught is widely reported as a result of heteroderine infection (Barker and Lucas, 1984; Steele, 1984), but expression of these symptoms is clearly affected by condition such as soil type. Cost of additional fertilizer and water must be considered in assessing damage by cyst nematodes. Other losses include expenses of management including nematicides. Other subtle losses are due to rotation with crops of lesser value or use of resistant varieties which sometimes require compromises in desirable qualities or yield. Since cultivars resistant to heteroderines typically have a hypersensitive reaction to infection, the plants may be adversely affected by invasion of juveniles and unsuccessful nurse cells. Quarantine against cyst nematodes adds to economic losses, with regulation of the potato cyst nematode in the USA being one of the most extensive and expensive (Mai, 1977; Miller, 1986). In some cases losses to cyst nematodes have been exacerbated through closed markets, with clean produce being rejected from an area wider than the actual region of infestation; in other cases markets may be closed to "harmless" species, subspecies, or races which are not readily distinguished from a related pathogen. Indirect losses to cyst nematodes measured in the expense of research programs, such as breeding for resistance, are nearly impossible to estimate but undoubtedly add greatly to the expense of coping with heteroderines.

Losses to cyst nematodes may be dramatic including deformed roots (e.g., carrots, potatoes), necrosis, and plant death. More often damage is subtle, with unevenly distributed nematodes apparent in a field as patches of stunted growth, vulnerable to water stress and wilt (Figs. 18–20, 22 , 23). Stunted plants may be chlorotic, sometimes with browning of leaf margins or read veining. Often symptoms can be confused with other problems of crop production including nutrient deficiencies, soil compaction, or toxicity from agricultural chemicals (Moore, 1984). The means by which cyst nematodes induce poor growth is not fully understood. Often it is suggested that feeding interferes with water and nutrient absorption, perhaps through blocked vessels or mechanical damage related to the expansion of nurse cells. Conversely, Seinhorst (1986) notes that plant weights may be greatly reduced without visible signs of root damage; he suggests that the primary effect of cyst nematodes, even at small densities, is reduced growth by retarded development, and that secondary effects may be decreased water consumption, early senescence, and perhaps reduced potassium content. Seinhorst's view of retarded development may be consistent with observations of Norton (1984) that corn plants infected with *H. avenae* were only 2 feet tall in mid-August with no inflorescences, whereas unaffected plants were 6 feet tall with fully formed tassels. Similar observations are made on corn infected with *H. zeae* (Koshy et al., 1970b) and *P. chalcoensis* (Jeronimo, 1988).

Although the heteroderines include some of the most destructive pests of agriculture, investigation of their unusual characteristics may be a valuable source of insight into basic biology. The usual host–parasite relationship in which the nematode alters the development of host cells, after years of investigation, continues to confound understanding. However, new tools of specific labeling and monoclonal antibodies promise insight into these mechanisms and perhaps into processes of host differentiation in general (Hussey, 1988). Heteroderines have contributed to pioneering investigations of pheromones and other attractants of nematodes, although there are many opportunities for further investigation of communication between nematodes as well as host specialization. Heteroderines, with their remarkable persistence in a cyst, may prove to be good tools for investigation of diapause in general. Heteroderines may also contribute to our general understanding of evolution. They are well suited to phylogenetic investigation because they can often be obtained in large numbers, often can be grown under controlled conditions on tissue cultured hosts, and they have a range of characters which can be scrutinized from living cultures with respect to comparative development, fine structure, and molecular biology.

II. TAXONOMY

A. Genera

1. Relationships

Insight into phylogenetic relationships of Heteroderinae is essential to a useful, stable taxonomic system which readily accommodates discoveries of new species. Yet phylogenetic relationships both of Heteroderinae to other Tylenchida and within Heteroderinae are controversial. Difficulties in determining phylogenetic relationships are due to the large number of genera and paucity of known reliable characters.

The taxonomic relationship of Heteroderinae to other Tylenchida historically has implies a unique common ancestor between cyst nematodes and root-knot nematodes. From 1909, these nematodes generally shared the same genus until Chitwood (1949) clearly described root-knot nematodes and recognized the genus *Meloidogyne* as separate from *Heterodera*. The relationship between the two genera was believed to be further strengthened by description of *Meloidodera* as an "intermediate noncyst forming link" between *Heterodera* and *Meloidogyne* (Chitwood el al., 1956). The proposed close relationship between *Heterodera* and *Meloidogyne* was believed to be supported not only by this intermediate, but by the shared characters of enlarged sedentary females and sexual dimorphism, males lacking caudal alae, and perhaps inducement of multinucleate nurse cells for feeding. Many systematic schemes, while recognizing a close relationship between heteroderines and meloidogynines, have also suggested close phylogenetic ties with other Tylenchida with globose females including *Tylenchulus*, *Rotylenchulus*, and *Naccobus*. In their reappraisal, Luc et al. (1988) underscore the historical view of a shared ancestor between heteroderines and root-knot nematodes, considering Heteroderidae as including

Heteroderinae (e.g., *Heterodera*: cyst nematodes, *Meloidodera*: noncyst nematodes) and Meloidogyninae (e.g., *Meloidogyne*: root-knot nematodes); they also include Nacobboderinae, an unusual group of three rare species in *Nacobbodera*, *Meloinema*, and *Bursadera*.

An alternate view of phylogenetic relationships is that the heteroderines and root-knot nematodes are "widely separated" (Wouts and Sher, 1971) and that their shared character is the enlarged female, and considered the shape of the male and loss of caudal alae as secondary adaptations to this shape. Therefore Wouts (1973a) proposed separate families, Heteroderidae (e.g., *Heterodera*, *Meloidodera*) and Meloidogynidae (*Meloidogyne*). Wouts (1973a) suggests that the Heteroderidae arose from vermiform, didelphic Hoplolaiminae with annulations, a short tail, large phasmids, and longitudinal lip striae. Ultrastructural evidence on host responses, the esophagus, cuticle, lip patterns, and other characters underscore the disparity between heteroderines and meloidogynines, and strengthens arguments of common ancestry between heteroderines and hoplolaimids.

Hypotheses of phylogenetic relationships within Heteroderinae based primarily on classical characters were proposed by Ferris (1979) and Wouts (1985). Baldwin and Schouest (1990) introduced a number of new characters and a more detailed analysis of classical characters for a parsimonious computer-generated phylogenetic analysis. Results supported three distinct groups of genera, designated as tribes within Heteroderinae: Sarisoderini, Ataloderini, and Heteroderini. Verutini, Cryphoderini, and Meloidoderini, while basically monogeneric, are also included to indicate equivalent sister group rank (Table 1). Each of these three groups of genera is supported as a line originating from a unique common ancestor; in addition, Ataloderini and Heteroderini are proposed as collectively having a unique common ancestor. New evidence to test hypotheses of relationships among heteroderines, meloidogynines, and hoploliamids as well as within Heteroderinae will undoubtedly be forthcoming through ultrastructural, developmental, and molecular biological approaches.

Although Heteroderinae includes 17 genera, major agricultural pathogens primarily occur in three cyst-forming genera: *Heterodera*, *Globodera*, *Punctodera* (Table 1).

Heterodera

Heterodera Schmidt, 1871
      syn. *Tylenchus* (*Heterodera*) Schmidt, 1871
            Heterodera (*Heterodera*) Schmidt, 1871
            Heterobolbus Raillet, 1896
            Bidera Krall and Krall, 1981
            Ephippiodera Shagalina and Krall, 1981

Diagnosis: Heteroderinae Filipjev and Schuurmans Stekhoven
    Females. Cyst present. Body more or less lemon-shaped, with posterior cone. Cuticle surface with zig-zag pattern of ridges; D layer diminutive; absent in young females and in cone of mature females.* Vulva terminal. Vulva area of cyst with two fenestrae (ambifenestrate or bifenestrate); no anal fenestration. Underbridge generally present; perineal tubercles absent. Bullae present or absent. Eggs retained in body; in some cases egg masses also present; egg surface smooth.

---

* Based on previous reports (Shepherd et al., 1972), Luc et al. (1988) considered absence of a D layer a character of *Heterodera*. Recent investigations suggest a rudimentary D layer anterior to the cone in mature females (Cordero, 1989).

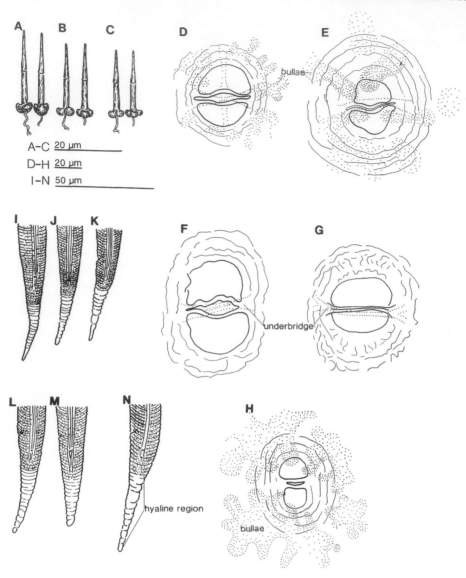

FIGURE 17   Characteristics for identification of some species of *Heterodera*. (A) *Heterodera tri-folii*, second–stage juvenile (J2) stylet (left dorsal, right lateral view). (B) *Heterodera glycines*, J2 stylet. (C) *Heterodera zeae*, J2 stylet. (D) *H. glycines*, terminal region of cyst; (E) *H. zeae*, terminal region of cyst. (F) *Heterodera cruciferea*, terminal region of cyst. (G) *H. carotae*, terminal region of cyst. (H) *Heterodera avenae*, terminal region of cyst. (I) *H. trifolii*, J2 tail. (J) *H. glycines*, J2 tail. (K) *H. zeae*, J2 tail. (L) *Heterodera goettingiana*, J2 tail. (M) *H. cruciferea*, J2 tail. (N) *H. avena*, J2 tail. (A, B, I, J redrawn after Hirschmann, 1956.)

FIGURE 18   Field damage by *Heterodera* spp. (A) *Heterodera schachtii* on cabbage. (Courtesy H. L. Rhoades.) (B) *H. schachtii* on red table beet. (C) *Heterodera glycines* on soybean. (A, courtesy H. L. Rhoades; B, courtesy J. L. Townshend; C, courtesy R. S. Hussey.)

FIGURE 19   Field damage by *Heterodera* spp. (A) *Heterodera goettingiana* on peas. (B) Control lacking *H. goettingiana* and with extensive nodulation. (C) *H. goettingiana* inhibition of nitrogen-fixing nodulation on pea. (D) Damage by *Heterodera carotae* on carrot. (A–C courtesy C. Green; D, courtesy N. Greco.)

FIGURE 20   Field damage by *Heterodera avenae*. (A) Wheat crop (left) is severely chlorotic following attack by *H. avenae*. Rye (right) is tolerant, remaining green. (B) Field damage on oats. (A and B courtesy F. Green.)

Males. Body twisted. Lateral field with four (rarely three) lines. Spicules > 30 μm long, distal extremity with single or more often bifid distal tip.

Second-stage juveniles. Lateral field with four (rarely three) lines. Esophageal glands filling body cavity. Tail conical, pointed; hyaline part variable, generally half tail length. Phasmids opening porelike. En face pattern with fusion of submedial lips with labial disc.

Nurse cell. Syncytium generally with wall ingrowths.

Species. (Table 2)

Identification. Although *Heterodera* includes nearly 60 species, many of these are only found in limited regions or restricted hosts, so that routine identification of the economically most important species in a given region can be simplified. In addition, three groupings of related species are recognized within *Heterodera*, so that familiarity with characteristics of each species group can quickly narrow choices. Comparison and keys to all the species of *Heterodera* is beyond the scope of this chapter and nearly comprehensive keys are published elsewhere (e.g., Golden, 1986). Instead we have selected representatives of the *schachtii*, *goettingiana*, and *avenae* groups based on broad distribution, economic importance, and recent awareness and interest (Table 3; Fig. 17). The *schachtii* group has a long vulval slit (>35 μm), is ambifenestrate, and bullae and underbridge are well developed. The *goettingiana* group has a long vulva slit (>30 μm), underbridge is absent or very slender and weakly developed, ambifenestrate or bifenestrate, and bullae are absent. The *avenae* group has a short vulval slit (<16 μm long), is bifenestrate, and bullae and underbridge may or may not be present (Mulvey, 1972). Morphometrics of many populations of a given species has made it possible to identify a few reliable characters as aids to preliminary identification (Table 3; Fig. 17). In addition, increasing recognition is being given to the value of second-stage juveniles for identification (Wouts and Weischer, 1977). Confirmation of identity requires more thorough examination of descriptions and comparison with closely related species of the three basic groups (Table 2).

Heterodera schachtii A. Schmidt 1871 (the sugar beet cyst nematode)

Synonymns and history. The sugar beet nematode, *H. schachtii*, devastated sugar beet production in Europe during the second half of the nineteenth century (Thorne, 1961). Previously, poor growth of sugar beets was associated with "exhausted soils" resulting from continuous planting without rotation, but in 1850 Schmidt observed that stunted beets were covered with fine hairs full of white bodies. He subsequently described the white bodies as a nematode pathogen and identified common cultural practices contributing to the dispersal of the nematode throughout production areas in Europe (Filipjev and Schuurmans Stekhoven, 1941; Thorne, 1961).

The sugar beet cyst nematode was taxonomically described from the host *Beta vulgaris* L. and from a locality near Aschersleben, East Germany by A. Schmidt (1871) and maned *Heterodera schachtii* in honor of Schacht (Raski, 1950). shortly after its description, Orley placed *H. schachtii* in a different genus, *Tylenchus schachtii* (Schmidt, 1871) Orley, 1880. During a period of confusion concerning races of cyst nematode (see history of *G. rostochiensis* below), the sugar beet cyst nematode was designated *H. schachtii minor* by O. Schmidt (1930) to distinguish it from the oat cyst nematode. However, Franklin (1940) designated the "beet strain" as a separate species which regained the original name of *H. schachtii* A. Schmidt.

Hosts. Although *H. schachtii* is primarily associated with *B. vulgaris*, its host range extends to *Spinacea* (spinach), *Brassica* (cabbage, brussels sprouts, cauliflower, broccoli, turnip), *Raphanus* (radish), and a diversity of common weeds of the Brassicaceae

TABLE 2   Species of *Heterodera*

Schachtii Group

    Heterodera schachtii A. Schmidt, 1871
       syn. *Tylenchus schachtii* (A. Schmidt, 1871) Orley, 1880
       syn. *Heterodera schachtii minor* O. Schmidt, 1930

    H. amygdali Kirjanova and Ivanova, 1975

    H. cajani Koshy, 1967
       syn. *H. vigini* Edward and Misra, 1968

    H. ciceri Vovlas, Greco, and Di Vito, 1985

    H. daverti Wouts and Sturhan, 1979

    H. elachista Ohshima, 1974

    H. fici Kirjanova, 1954

    H. galeopsidis Goffart, 1936
       syn. *H. schachtii galeopsidis* Goffart, 1936

    H. gambiensis Merny and Netscher, 1976

    H. glycines Ichinohe, 1952

    H. lespedezae Golden and Cobb, 1963

    H. leuceilyma Di Edwardo and Perry, 1964

    H. limonii Cooper, 1955

    H. medicaginis Kirjanova in Kirjanova and Krall, 1971

    H. oryzae Luc and Berdon Brizuela, 1961

    H. oryzicola Rao and Jayaprakash, 1978

    H. oxiana Kirjanova, 1962

    H. rosii Duggan and Brennan, 1966

    H. sacchari Luc and Merny, 1963

    H. salixophila Kirjanova, 1969

    H. sonchophila Kirjanova, Krall, and Krall, 1976

    H. sorghi Jain, Sethi, Swarup, and Srivastava, 1982

    H. tadshikistanica Kirjanova and Ivanova, 1966

    H. trifolii, Goffart, 1932
       syn. *H. schachtii* var *trifolii* Goffart, 1932
           H. paratrifolii Kirjanova, 1961
           H. rumicis Poghossian, 1961
           H. scleranthii Kaktina, 1957

    H. zeae Koshy, Swarup, and Sethi, 1971

Goettingiana Group

    H. bergeniae Maqbool and Shahina, 1988

    H. canadensis Mulvey, 1979

TABLE 2   (Continued)

H. cardiolata Kirjanova and Ivanova, 1969

H. cruciferae Franklin, 1945

H. cyperi Golden, Rau, and Cobb, 1962

H. delvii Jairajpuri, Khan, Setty, and Govindu, 1979

H. goettingiana Liebscher, 1892

H. graminis Stynes, 1971

H. graminophila Golden and Birchfield, 1972

H. humuli Filip'ev, 1934

H. longicolla Golden and Dickerson, 1973

H. mediterranea Volvas, Inserra, and Stone, 1981

H. menthae Kirjanova and Narbaer, 1977

H. methwoldensis Cooper, 1955

H. mothi Khan and Husain, 1965

H. pakistanensis Maqbool and Shahina, 1986

H. phragmitidis Kazachenko, 1986

H. plantaginis Narbaev and Sidikov, 1987

H. polygoni Cooper, 1955

H. graduni Kirjanova, 1971

H. raskii Basnet and Jayaprakash, 1984

H. urticae Cooper, 1955

H. uzbekistanica Narbaev, 1980

Avenae Group

H. arenaria Cooper, 1955
    syn. *Bidera arenaria* (Cooper, 1955) Krall and Krall, 1978

H. avenae Wollenweber, 1924
    syn. *H. schachtii* var. *avenae* Wollenweber, 1924
        H. Bidera avenae (Wollenweber, 1924) Krall and Krall, 1978
        H. schachtii major O. Schmidt, 1930
        H. major O. Schmidt, 1930
        H. ustinovi Kirjanova, 1969; Krall and Krall, 1978

H. bifenestra Cooper, 1955
    syn. *H. Bidera bifenestra* (Cooper, 1955) Krall and Krall, 1978
        H. longicaudata Seidel, 1972
        H. Bidera longicaudata (Seidel, 1972) Krall and Krall, 1978

H. filipjevi (Madzhidov, 1981) Stone, 1985
    syn. *Bidera filipjevi* Madzhidov, 1981

H. hordecalis Andersson, 1975
    syn. *Bidera hordecalis* (Andersson, 1975) Krall and Krall, 1978.

TABLE 2   (Continued)

H. iri Mathews, 1971
    syn. *Bidera iri* (Mathews, 1971) Krall and Krall, 1978

H. latipons Franklin, 1969
    syn. *Bidera latipons* (Franklin, 1969) Krall and Krall 1978
        Ephippiodera latipons (Franklin, 1969) Shagalina and Krall, 1981

H. mani Mathews, 1971
    syn. *Bidera mani* (Mathews, 1971) Krall and Krall, 1978

H. turcomanica Kirjanova and Shagalina, 1965
    syn. *Bidera turcomanica* (Kirjanova and Shagalina, 1965) Krall and Krall, 1978
        Ephippiodera turcomanica (Kirjanova and Shagalina, 1965) Shagalina and
        Krall, 1981

Source: Adapted from Luc et al., (1988).

(Cruciferae) and Chenopodiaceae families. Although Solanaceae are generally not hosts, some *H. schachtii* isolates have been reported to develop on tomato. Discrepancies occur among the large number of host range tests (Filipjev and Schuurmans Stekhoven, 1941; Franklin, 1972; Griffin, 1982; Raski, 1952; Steele, 1965; Winslow, 1954), which could reflect differences among isolates including physiological races (Miller, 1983). Other discrepancies may occur because juveniles penetrate the roots of some plants which are unable to sustain feeding, development, and reproduction; such plants might be included in host lists but their inclusion could be misleading. Although *H. schachtii* has a broad host range, damage to the sugar beet production is generally of greatest importance (Graney and Miller, 1982). Franklin (1972) suggested that damage to other brassicas occurs rarely and is associated with very high initial inoculum but in the USA, Canada, and Poland a number of brassicas are severely affected including table beets (Brzeski, 1969; Miller, 1986) (Fig. 18A, B). Griffin and Waite (1982) suggest that *H. schachtii* is a potential threat to tomato production, particularly when it occurs in combination with root-knot nematodes. However, in California, tomatoes in rotation with sugar beets for many years are not affected by *H. schachtii*, regardless of whether or not root-knot nematodes are present (Thomason, pers. commun.).

   Distribution. *Heterodera schachtii* favors temperate regions but apparently tolerates a broad range of climates; it is widespread in Europe, the United States, and Canada. Thorne (1961) believed that some populations of *H. schachtii* were introduced to the USA on sugar beets, but considered other populations to be indiginous to weed hosts in the southwestern deserts of the USA including those of Utah and California. *Heterodera schachtii* also occurs in parts of the Middle East and USSR, western and southern Africa, Australia, Chile, and Mexico. It is readily dispersed with soil and plant parts, an important consideration in control (see below).

   Biology. Lemon-shaped cysts remain dormant in the soil from one host season to the next with only a percentage hatching and emerging each year. Consequently, large populations accumulate with continuous planting of sugar beet. Fully developed eggs containing second-stage juveniles hatch in response to the stimulus of root exudates and emerge from the cyst. Hatching is favored at 25°C and can be stimulated by host and nonhost root exudates and may even occur in the absence of exudates. Second-stage

TABLE 3 Compendium for Identification Among Select Heteroderidae of Particular Economic Significance (Fig. 17).

Schachtii Group: Bullae and underbridge well developed; vulval slit > 35 μm, ambi-fenestrate

Species	Bullae	J2 stylet length (μm)	J2 tail length (μm)	J2 stylet knobs shape	Hosts
H. schachtii	Scattered	25–27	48–55	Anchor	Wide
H. glycines	Scattered	22–25	40–49	Subventral rounded	Wide
H. trifolii	Scattered	> 27	> 55	Anchor	Wide
H. zeae	Four distinct "fingers"	< 22	40–49	Concave on top	Gramineae

Goettingiana Group: No bullae, underbridge poorly developed or absent; vulval slit > 30 μm; ambifenestrate or bifenestrate

Species	Underbridge	Egg mass	J2 hyline part tail μm	Hosts
H. goettingiana	Slender but present	Small	> 34	Legumes
H. crucifera	Slender but present	Small	< 26	Crucifers (especially Brassicae)
H. carotae	Very thin; rarely persists in cysts	Large, often size of female	26–32	Carrots

Avenae group: Bullae and underbridge present or absent; short vulaval slit (< 16 μm), bifenestrate

Species	Bullae	Underbridge	J2 lateral field	Width vulval bridge (μm)
H. avenae	Well-developed	present	Four lines outer two faint	Wide; 18–39

juveniles have greatest motility in soil at 15°C in soil that is less than saturated. They penetrate the epidermal host root tissue behind secondary root tips and become sedentary in the vicinity of the vascular tissue where they induce formation of syncytial nurse cells. Rate of development is optimal at 18–28°C, with adults occurring about 18 days after penetration and brown cysts developing in about 38 days (Franklin, 1972; Griffin, 1988; Raski, 1950). The white females protrude from host roots and attract males for copulation and fertilization. Gravid females secrete a gelatinous matrix into which they may deposit more than 100 eggs (Raski, 1950); but as many as 600 eggs may persist in the female body and young cyst (Raski, 1950; Thorne, 1961). Early in the season when hosts are vigorous, some eggs may hatch almost immediately and multiple generations occur in a single growing season. But late in the season, as plant roots deteriorate, cysts full of eggs are released into the soil until conditions are suitable for hatch and infection.

Control. The first line of defense against *H. schachtii* is to exercise extreme caution to prevent its spread to noninfested areas. The previous practice of moving tare soil from factories back into fields resulted in spread in many growing areas in the USA (Thomason, pers. commun.). Similarly, cysts may be spread by canal water (Faulkner and Bolander, 1966) and farm machinery. They may also be dispersed by survival in the digestive tract of cattle gleaning postharvest sugar beet residue (Kontaxis et al., 1971).

Attempts to develop commercially acceptable cultivars of sugar beet with resistance to *H. schachtii* have not yet been successful, so that most practices aimed at management of this pest include rotations with nonhosts. Rotation with grains or leguminous crops is effective, particularly if sugar beets are limited to one year in five, but this approach is costly. Chemical control is difficult because eggs and juveniles in the cyst may be somewhat protected; it is also expensive and may be environmentally hazardous. Nevertheless, nematicides, including aldicarb and 1, 3-dichloropropene, have been important in reducing yield loss where long rotations are not practical (Miller, 1986). Müller (1986) considers trap crops risky, but suggests that rotation combining green manure with resistant crucifers may effectively reduce populations. Since natural antagonists of the sugar beet cyst nematode are abundant, biological control may eventually prove useful in managing these nematodes.

Heterodera trifolii Goffart, 1932 (the clover cyst nematode)

Synonyms and history. White clover (*Trifolium repens* L.) is frequently unthrifty and wilted during warm weather in Europe and the causal agent was recognized as clover cyst nematode. The nematode was first described as *H. schachtii* var. *trifolii* Goffart, 1932 but no type material was designated. A cyst-infected piece of clover root from Goffart's collection could have provided type specimens but the tissue was lost during an attempt at remounting. Therefore, a cyst with eggs from a new collection on *T. repens* near Rendsburg, Schleswig-Holstein, Germany, believed to approximate the type locality, was designated a neotype (Wouts and Sturhan, 1978) [It is noteworthy, however, that Mulvey and Golden (1983) and Mulvey and Anderson (1974) indicate the type host as *Trifolium pratense*.] Goffart's original collection, like the neotype, apparently came from a population without males; in this respect *H. trifolii*, a mitotic parthenogenetic species, is distinct from *H. diverti*, a cyst nematode of clover which has abundant males and is cross-fertilizing (Wouts and Sturhan, 1978). In the years following the original description of *H. trifolii*, its distinctiveness from *H. glycines* with respect to morphology and host range was not clear until Hirschmann's (1956) detailed comparisons clearly defined the two species. The monosexual species *Heterodera lespedezae* (described above in relation to *H. glycines*) and *H. galeopsidis* are regarded as closely related to *H. trifolii*, and it is suggested that these and certain other species may be polyploid forms ($2n$ = about 27) that have evolved from diploid amphimitic relatives of the *H. schachtii* group (n = 9) (Fig. 5). As such they may best be represented as subspecies within a species complex (Triantaphyllou and Hirschmann, 1978; Maas et al., 1982; Sikora and Maas, 1986). Other species have been proposed as distinct from *H. trifolii* but are included as junior synonyms by Luc et al. (1988); these are *H. paratrifolii* Kirjanova and Ivanova, 1966; *H. rumicis* Poghossian, 1961; and *H. scleranthii* Kaktina, 1957.

Host range. Heterodera trifolii has a broad host range within the pea family *Fabaceae* (Leguminosae) including *Trifolium* (clover), *Medicago* (alfalfa), *Lespedeza*, *Glycine* (soybean, some cultivars), *Pisum* (pea), and *Phaseolus* (greenbean). It also attacks many nonlegumes representing diverse families; genera including *Dianthus* (carnation), *Vinca* (vetch), *Lotus*, *Melilotus*, *Spinacea* (spinach), *Polygonum*, *Rumex*, and *Chenopodium* (Thorne, 1961; Norton and Isely, 1976; Mulvey and Anderson, 1974). Individual populations vary in host range (Maas et al., 1982), an important consideration in pest management strategies. In addition, the broad host range of *H. trifolii* indicates the importance of considering reservoirs of the pathogen in control.

Distribution. Heterodera trifolii is present throughout Europe as well as the USSR; it has also been reported in the Mediterranean region including Israel and Libya. It is present in Australia, New Zealand, and India. *Heterodera trifolii* may be the most widely dispersed cyst nematode in North America, reported primarily in pastures and turf with white clover in 30 of the United States including Hawaii, and throughout southern Canada (Mulvey and Anderson, 1974; Mulvey and Golden, 1983). Although the nematode may be indigenous to both North America and the Old World, some speculate that the nematode was introduced to North America. Indeed, some suggest that cysts could be dispersed long distance by bird migrations (Sikora and Maas, 1986) as has also been reported for *H. glycines* (Epps, 1971).

Biology. Young females lay eggs in a gelatinous matrix but as the female matures eggs are retained and persist in the cyst (Mulvey, 1959). Hatching of second-juveniles from eggs in the cyst takes place over a period of years whenever moisture requirements are favorable and requirements for breaking diapause are met. Hatch occurs over a temperature range of 4–31°C, but optimum temperature is 17.2°C (Oostenbrink, 1967). Hatch may also be favored by the presence of certain host leachetes, but not all hosts seem to affect hatch (Winslow, 1955). Following the penetration of the root tips and establishment of a feeding site near the stele (Mankau and Lindford, 1960), the rate of development is primarily affected by soil temperature. At an optimum of about 25°C the life cycle is completed in 17 days (Norton, 1967; Mankau and Linford, 1960), but at cooler temperatures of 10–20°C egg laying does not occur until about 1 month after host penetration (Mulvey, 1959). Several generations may take place within a season, so that populations may increase rapidly (Mulvey, 1959; Yeates and Risk, 1976).

In many cases clover is only damaged by very high numbers of *H. trifolii* and damage is most apparent in mature plants that have been cut (Sikora and Maas, 1986; Seinhorst and Sen, 1966). With large numbers of infective juveniles, clover, and even nonhosts, may first be damaged by penetration because multiple invasions may kill a root tip (Sikora and Maas, 1986). Hosts are also damaged by diversion of nutrients from the plant, interaction with other pathogens (Sikora and Maas, 1986; Ennik et al., 1964; Yeates, 1973), and in some cases inhibition of nitrogen fixation (Yeates et al., 1977). High population levels may reduce water uptake, as reflected by wilting in warm weather (Thorne, 1961).

Control. Control of *H. trifolii* by rotation or use of alternative crops is confounded by the broad host range and susceptibility, which may vary among nematode populations and races (Sikora and Maas, 1986). Although resistance may also be difficult to maintain against such a heterogeneous pathogen, some lines of white clover are reportedly resistant; resistance is inherited by more than one gene (Dijkstra, 1971; Kuiper, 1960). Chemical control of *H. trifolii* is feasible for ornamentals such as carnation, and solarization, as discussed for *H. carotae*, may also be effective (Sikora and Maas, 1986).

Heterodera glycines Ichinohe, 1952 (the soybean cyst nematode)

Synonyms and history. Cyst nematodes may have been known to occur on soybeans in north central Japan since the late 1800s, although Ichinohe (1961) notes that a report by Hori (1915) was apparently the first to bring widespread attention to the soybean cyst nematode and its causal role in "yellow dwarf" disease. Subsequent reports in Japan by Ito (1921), Tanaka (1921), and Fujita and Miura (1934) compared morphology and host range of the soybean cyst nematodes with sugar beet cyst nematodes; they concluded that cysts of soybean were probably a race of *H. schachtii*. It was also reported that the distribution of this soybean race extended to Korea and Manchuria (Yokoo, 1936; Nakata and Asuyana, 1938). During this period, cyst nematodes, including the soybean cyst nematode, were generally considered races of *H. schachtii*, but in 1940 Franklin's comparative morphological descriptions led to many cyst "races" being elevated to species. It was taxonomically consistent that the morphological distinctions of the soybean cyst nematode and its economic importance led Ichinohe (1952) to elevate the putative soybean race to a new species, *H. glycines*, with the type host soybean (*Glycine max*) and the type locality of Obihiro-shi, Hokkaido, Japan. About the same time the soybean nematode was detected and found to be widely distributed in three counties of North Carolina (Winstead et al., 1955; Thorne, 1961). Although the distribution and host range of *H. glycines* and *H. trifolii*, the clover cyst nematode, overlap, Hirschmann (1956) clarified morphological differences between them. *Heterodera glycines* is now known throughout most of eastern USA, and also occurs in Taiwan, Colombia, and the Middle East (Moore, 1984). It has considerable intraspecific variability with respect to its capacity to infect different hosts and cultivars (Golden et al., 1970; Inagaki, 1979; Ross, 1962; Triantaphyllou, 1975b) and there are reports of morphological distinctions among isolates (Faghihi et al., 1986b; Golden and Epps, 1965; Koliopanos and Triantaphyllou, 1971; Miller and Duke, 1967; Riggs et al., 1982). Some isolates, previously believed to be *H. glycines* which heavily attacked Kobe lespedeza (*Lespedza striata*) in North Carolina, were subsequently recognized as morphologically distinct from both *H. glycines* and *H. trifolii*, and were described as a new species, *H. lespedezae* (Bhatti et al., 1972; Golden and Cobb, 1963). This species may occur in the same locality with *H. glycines*, and some isolates may even attack soybean (Fagbenle et al., 1986), so that the two species are often confused.

Hosts. *Heterodera glycines* is of major economic importance on soybeans (Fig. 18C). It has a broad host range, especially among legumes, including *Phaseolus* (e.g., greenbeans), *Lespedeza*, *Vicia* (vetch), *Trifolium* (clover), and *Pisum* (pea), but it also attacks a number of nonlegumes including ornamentals, such as *Geranium*, and *Papaver* (poppy) as well as many weeds (Moore, 1984). The host range, and particularly weed hosts, is an important consideration, particularly for management strategies involving sanitation, quarantine regulation, and crop rotation. Host range on soybean cultivars varies with putative "races," some cultivars being resistant to certain races and susceptible to others (Golden et al., 1970). However, races of *H. glycines* are quantitative and determined by gene frequencies, which may shift by selection within a given population (Triantaphyllou, 1975b). Thus, host range tests often are not reliable for characterizing races (Riggs et al., 1988). Yet new races are being described to accommodate the great range of variation (Golden et al., 1970; Riggs et al., 1981; Riggs and Schmitt, 1988). Although isolates may be distinct with respect to morphology, serology, isozymes, or 2-D protein patterns, these have not been linked to host variants (Faghihi et al., 1986b; Ferris et al., 1986; Koliopanos and Triantaphyllou, 1971; Radice et al., 1988; Riggs et al., 1982). However, the potential of restriction fragment length polymorphisms for identifying races of soybean cyst nematode is being explored (Kalinski and Huettel, 1988).

Distribution. *Heterodera glycines* may have originated in the orient including Japan, Korea, China (e.g., Manchuria), where soybeans are historically an important crop. It has been suggested that *H. glycines* was introduced to the United States on bulbs from Japan and that it subsequently spread to 26 central and southeastern states (Riggs et al., 1988; Wrather et al., 1984). An alternative hypothesis is that populations of the soybean cyst nematode have been endemic on a variety of weed hosts, and only become apparent as soybean culture became widespread in the United States. The origin and geneology of populations of *H. glycines* might prove to be traceable through ongoing investigations of molecular biology of geographic isolates (Ferris et al., 1985, 1986; Radice et al., 1988; Sandall and Powers, 1988). Some reports suggest that cysts may be ingested and dispersed by birds (Epps, 1971) or mammals (Smart, 1964), and undoubtedly man has dispersed nematodes through moving water or soil, as well as soil peds on seed, vehicles, and agricultural products (Wrather et al., 1984). Soybean cyst nematode is now reported in Taiwan, Egypt, Colombia, Indonesia, and parts of Canada and the Soviet Union.

Biology. A female of *H. glycines* produces a gelatinous matrix in which one or more males is likely to be attracted. Once fertilized the diploid amphimitic female produces 200–600 viable eggs (Moore, 1984). Some eggs are exuded into the gelatinous matrix and may hatch quickly, so that several generations may occur in a single growing season. Other eggs, perhaps the majority, are retained in the cyst after death of the female and continue to hatch at a slow rate, so that 50–90% are released within a year and some eggs may survive in the cysts up to 8 years under favorable cool, moist conditions (Moore, 1984). Esser and Langdon (1967) report that cysts held at –40°C for 7 months still contain viable eggs. However, at 36°C, development in the egg is halted at the four-cell stage and eggs die (Alston and Schmitt, 1988). Hatching is greatest at 26 and 22°C (day and night) and is also enhanced by pod-producing rather than vegetative soybeans (Hill and Schmitt, 1989). Root diffusates are believed to play a role in hatching of second-stage juveniles (Tsatsum and Sakurai, 1966). Diffusates may also be involved in attraction and penetration of juveniles to host roots. Second-stage juveniles generally penetrate host roots near the tip. They become sedentary and establish a syncytial feeding site. Time required for the nematode to complete molts to adults is usually from 21 to 24 days at ideal soil temperatures of 21–24°C. Cooler soil temperatures increase length of the cycle but outside a range of 10–34°C development is not completed (Moore, 1984; Slack et al., 1972; Wrather et al., 1984).

Susceptible reactions to *H. glycines* can result in yield losses in individual fields exceeding 70% with overall losses of soybean to this nematode in the southern USA estimated at 2–4% (Wrather et al., 1984). Symptoms include stunting, yellowing, browning of leaf margins, and loss of leaves with reduced flower and seed production. Lateral roots may be increased in number and there may be fewer *Rhizobium* nodules for nitrogen fixation. The most severe losses are associated with sandy soils (Koenning et al., 1988; Todd and Pearson, 1988).

Control. Since the soybean cyst nematode can greatly limit production of soybeans, past management strategies must be considered. A primary defense has been to attempt to limit spread to noninfested areas by sanitation, including cleaning of contaminated farm vehicles and bean seed, and enforcing quarantine regulations. since many agricultural crops are nonhosts of *H. glycines*, crop rotation is an effective approach to management. In warm climates persistence of cysts with viable eggs may be relatively short, many eggs hatch, and juveniles may senesce in the soil prior to planting. In such cases rotations of 1 year may be effective. However, in cooler climates where eggs are likely dormant through winter, nearly synchronous hatching may occur near the time of planting; thus rotations as long as 3 years may be needed (Moore, 1984). One year in a nonhost may reduce populations by 75%, and 2 years by 92% (Moore, 1984). Rotation programs must be specific to the region and must consider possible weed hosts. Moreover, rotations may include, in addition to a nonhost, both a resistant and a susceptible soybean cultivar (Riggs, 1977) to reduce pressure toward shifts in gene frequencies to resistance-breaking races (Moore, 1984).

A number of soybean cultivars with resistance to *H. glycines* are available, but their resistance is generally specific to particular putative races as determined by soybean differential tests (Moore, 1984; Riggs et al., 1988). In some cases resistant genes are linked to undesirable characteristics (e.g., black seed coat in Peking), and suitable resistant cultivars are not necessarily available in lines adapted to all the geographic areas in which soybeans are grown (Wrather et al., 1984). Resistant varieties must be managed carefully to avoid shifts in gene frequencies of the nematode and thus buildup of new races which may occur during 2–4 years of continuous cropping. Although Moore (1984) suggests that a combination of resistant cultivars and rotation is likely to be more economical than use of nematicides, both fumigant and nonfumigant nematicides may be effective in reducing populations of soybean cyst nematode. Nematicides, however, are environmentally and economically costly and, due to protection of eggs in the cyst and a preliminary increase in root feeding sites, they only reduce initial populations (Wrather et al., 1984). There is some evidence that nematicides are more effective when planting is late, as activity of infective juveniles in the soil may be greatest (Hussey and Boerma, 1983).

Several fungi have been recognized as parasites of eggs of *H. glycines*, and natural predators may act in conjunction with other management practices to reduce nematode populations. Fields in continuous cropping with susceptible cultivars may have a decline in populations of soybean cyst nematodes due to increase in natural antagonists (Hartwig, 1981). However, these biological control factors generally are understood insufficiently to manipulate them as part of management strategies.

Heterodera zeae Koshy, Swarup, and Sethi, 1971 (the corn cyst nematode)

Synonyms and history. While collecting the oat cyst nematode, *H. avenae*, in Rajasthan, India during 1969, a sample mixed with an unknown second species was discovered on *Zea mays* L. and taxonomically described as *Heterodera zeae* (Koshy et al., 1970). Additional samples of *H. zeae* were discovered in Egypt and Pakistan, and in 1981 it was reported in Maryland (Sardanelli et al., 1981). Shortly thereafter, Golden and Mulvey (1983) redescribed *H. zeae*, with greater morphological and SEM detail. They also compared populations from India and the USA.

Hosts. The corn cyst nematode was first recognized as distinct from *H. avenae* by its reproduction on maize and barley but not on a number of other grains susceptible to *H. avenae* (Bhargava and Yadav, 1978; Koshy et al., 1970). Ringer et al. (1987) carried out extensive host range tests on cyst populations from Maryland. All 22 cultivars of *Zea mays* tested, as well as *Zea mexicana* (teosinte), were susceptible. Other economic hosts include some cultivars of barley (*Hordeum vulgare*), oat (*Avena sativa*), rice (*Oryza sativa*), sorghum (*Sorghum bicolor*, with poor nematode reproduction), sugar cane (*Saccharum* sp.), and wheat (*Triticum aestivum*). Several grass weed hosts have been identified in India and the USA (Ringer et al., 1987; Srivastava and Swarup, 1975; Vermaand Yadau, 1978). Populations from India and the USA apparently differ some in their host ranges, although these have not yet been compared under identical conditions. *Heterodera zeae* has been associated with a number of dicots including legumes, solanaceous plants, cucurbits, crucifers, and many ornamentals, but reproduction has not been detected on any dicot (Krusberg, pers. commun.). Curiously, populations from Pakistan are reported to be widespread on citrus, pear, and garlic (Maqbool, 1981), but these also may be field associations and reproduction on these hosts has not been established.

Distribution. Heterodera zeae is widespread in India but also occurs in Pakistan, the Nile Valley, Egypt, and Maryland (Krusberg and Sardanelli, 1982; Sardanelli et al., 1981). In Maryland it is found primarily in heavy silty-clay soils and population densities are usually sparse with 1–10 cysts/250 cm^3 soil (Krusberg, pers. commun.). Although the corn cyst nematode could be indigenous to all of these areas, its apparently narrow distribution in the western hemisphere suggests recent introduction.

Biology. The life cycle of *H. zeae* is completed in about 22 days (Lauritis et al., 1983; Verma and Yadav, 1975) if temperatures are optimally warm (about 30°C). Second-stage juveniles hatch readily and penetrate both the main root and lateral roots of the host, and a syncytium is initiated. Postembryogenesis is typical for the genus (Shahina and Maqbool, 1989). the final molt to mature females with a gelatinous egg mass is completed in about 10 days, but eggs are not laid into the mass until about 14 days after penetration (Lauritis et al., 1983). Males, which are rare, apparently are not required for reproduction (Hutzell, 1984). Under condition in Maryland, cysts with infective juveniles apparently only persist in fallow soil for about 2 years (Krusberg and Sardanelli, 1989). The corn cyst nematode is considered an economic pest in India, but pathogenicity to corn in the field has not been demonstrated in Maryland (Ringer et al., 1987). The nematode is however, a potential threat to corn production in the USA. However, its spread and pathogenicity could be restricted by high-temperature requirements (>30°C) which are more likely to be sustained in soils of the Southeast than in the corn belt of the USA. Preliminary microplot tests in Maryland suggest that high densities of the nematode reduce yields of corn in sandy soils, particularly where other stresses such as low soil fertility occur (Krusberg, pers. commun.)

Control. Since May 1984, a quarantine has been imposed on the areas of Maryland infested with *H. zeae* but only a few tests have been carried out on its pathogenicity and strategies for management. Preliminary trials show that several fumigants reduce soil population density, but without an increase in corn yield (Krusberg, pers. commun.). Crop rotation may be confounded by the wide host range including a number of weed hosts. Sources of resistance in *Zea mays* are unknown.

Heterodera goettingiana Liebscher, 1892 (the pea cyst nematode)

Synonyms and history. Liebscher (1890) was the first to attribute yellowish stunted pea plants (*Pisum sativum* L.) at the agriculture institute in Göttingen, Germany to a cyst nematode which he at first considered a strain of *H. schachtii*. His investigations of this nematode indicated morphological and biological distinctions, which let him to consider the pea cyst nematode as a new species, *H. goettingiana* (Di Vito and Greco, 1986). Liebscher's recognition of a distinct species was remarkable in 1892, considering that most other cyst nematodes were viewed as strains of *H. schachtii* for another 50 years, and little was known about biology and pathogenicity of *H. goettingiana* until the mid-1900s (Franklin, 1951; Goffart, 1941).

Host range. Winslow (1954) and Jones (1950b) indicate several hosts of *H. goettingiana* in the Fabaceae (Leguminosae), including peas (*Pisum sativum* L.), broadbean (*Vicia faba*), vetch (*Vicia* spp.), soybean (*Glycines max*), lentils (*Lens sculenta*), and weed hosts including *Pisum, Vicia*, and perhaps *Lathyrus*. Conflicting reports suggest that populations may differ in their response to ornamental sweet pea or grosspea (*Lathyrus odoratus*) (Di Vito et al., 1980; Jones, 1950b; Stone and Course, 1986; Thorne, 1961; Winslow, 1954). Di Vito et al. (1980) found that many legumes susceptible to other cyst species are resistant to *H. goettingiana*, including various clovers (*Trifolium* spp.), lupine (*Lupinus* sp.), and chickpea (*Cicer arietinum*). In the case of chickpea, tissues are invaded but become necrotic and only males develop (Varo Alcala et al., 1970). Attempts to find resistance in garden peas have failed (Oostenbrink, 1951; Stone and Course, 1986). Di Vito and Greco (1986) report a few families and cysts on the nonleguminose plant *Asperula arvensis* L. Symptoms are typically of cysts nematodes, including patches of stunted and chlorotic plants in the field (Fig. 19A).

Distribution. Shortly after *H. goettingiana* was found in Germany, it was reported in Great Britain (1912) and France (1917) (Di Vito and Greco, 1986). Although it occurs throughout the world, it is most widespread in Europe including Germany (Goffart, 1941), the Netherlands (Oostenbrink, 1951), Great Britain (Jones, 1965), and Belgium (D'Herde, 1966). It occurs in the USSR and in the Mediterranean region including Spain, Portugal, Italy, Sicily, Malta, Israel, and Algeria. It has been reported in the USA (Thorne, 1961), but it is not widespread, and Stone and Course (1986) suggest that U.S. populations are chance introductions. Undoubtedly *H. goettingiana*, like other cyst nematodes, is spread locally by water and farm vehicles, but possible natural means of long-distance dispersal are unknown.

Biology. The life history of *H. goettingiana*, from egg through four molts, adult and cyst, is illustrated by Macara (1963) and thoroughly discussed by Stone and Course (1986) and Di Vito and Greco (1986). Cysts with eggs are highly persistent and in cooler climates may remain viable in the absence of a host for 12 years (Brown, 1958; Di Vito and Greco, 1986). Development takes 3–15 weeks depending on soil temperature and moisture as well as host species. Thus, only one or two generations are completed during the growing season for peas in England, but three generations may occur in southern Italy (Di Vito and Greco, 1986). Under ideal conditions, including cool soil (10–13°C), females mate and produce a gelatinous matrix which may have more than 100 eggs; however, under adverse conditions eggs are not usually exuded but are retained within the cyst (Greco et al., 1986). Development is inhibited below 4.4°C or above 25°C, even where soil moisture is adequate (Beane and Perry, 1984; Di Vito and Greco, 1986; Jones, 1975). Unfavorable conditions, including poor hosts, favor development of males (Guevara-Benite et al., 1970; Jones, 1965).

Hatching is greatest at about 15°C. Although age of host was not considered in early tests of response to root exudates, more recent investigations indicate that exudates are required for a high percentage of hatch and that exudates from older hosts (e.g., 16-week-old broadbeans) are apparently most effective in stimulating hatch (Beane and Perry, 1983; Shepherd, 1963). Infective juveniles penetrate host roots and induce a syncytium (Volvas and Inserra, 1978), causing stunting and yellowing which is most evident at the flowering stage. In peas, *Rhizobium* nodulation and nitrogen fixation is inhibited (Fig. 19B, C), seed production is greatly reduced, and nematode-infected plants are vulnerable to root invasion and death by soil fungi.

Control. The pea cyst nematode may be highly persistent in some soils; nevertheless, the host range is relatively narrow and 3 to 6 year rotations with nonhosts generally reduce populations to levels that are not damaging. However, possible weed hosts must be carefully controlled in such rotations (Di Vito and Greco, 1986). In some warm regions, such as those with Mediterranean climates (where irrigation can be provided), late varieties of peas can be used which tolerate soil conditions that are too warm for rapid development of the pea cyst nematode (Greco et al., 1986b). Cultivars are not yet available with resistance to *H. goettingiana*, although some species of *Pisum* may show some promise as sources of resistance (Di Vito and Perrino, 1978; Di Vito and Greco, 1986). Chemical control of the pea cyst nematode is confounded by its slow rate of hatch; nevertheless a number of nematicides are show to be useful in management programs (Di Vito and Greco, 1986; Stone and Course, 1986).

Heterodera cruciferae. Franklin, 1945 (the Brassica or cabbage cyst nematode)

Synonyms and history. Franklin recognized that what were commonly considered biological strains of *H. schachtii* until 1940 were four morphologically distinct species which are now recognized as *H. schachtii*, *G. rostochiensis*, *H. goettingiana*, and *H. avenae*. However, even within the newly defined *H. schachtii*, she recognized a distinct strain in specimens from a plot of old cabbage plants. The nematodes did not infect sugar beet and cysts were smaller than those of *H. schachtii*. She described the cyst nematode of cabbage as *Heterodera cruciferae* from the type host, *Brassica oleracea*; and type locality at St. Albans, England (Franklin, 1945). Although there were some initial criticism that the characters diagnosing *H. cruciferae* were difficult to use in identification and may not be reliable (Miles, 1951), the species became clearly established with subsequent morphological investigations (Stone and Rowe, 1976).

Hosts. *Heterodera cruciferae* has a narrower host range than *H. schachtii* but apparently attacks all species of *Brassica* including common agricultural crops of cabbage, brussels sprouts, cauliflower, broccoli, turnips, and radish (Franklin, 1945, 1951; Jones, 1950b). Cruciferea (Brassicaceae) other than *Brassica* may be less susceptible to the nematode, and a few crucifers including *Aethionema*, *Hesperis*, and *Matthiola* are not hosts at all (Winslow, 1954). Some members of Labiatae, the mint family, may be hosts, and certain species could be important as weed reservoirs (Stone and Rowe, 1976).

Distribution. *Heterodera cruciferea* is widespread in England, where it was first described, but it also occurs throughout Europe including Ireland, Netherlands, Belgium, Germany, Switzerland, France, Portugal, Turkey, Yugoslavia, Bulgaria, Hungary, Poland, and the USSR. In the United States, *H. cruciferae* is only known to occur in California where Raski (1952) described it as widespread and well established; it is not known if the nematode is indigenous to California of if it was introduced from Europe. Curiously, in California it is rarely if ever found separate from *H. schachtii*. *Heterodera cruciferae* was described from South Australia where it is speculated that contaminated wooden barrels or crates from Europe were responsible for its introduction (Stirling and Wicks, 1975).

Biology. Unlike most temperate heteroderines, *H. cruciferae* parasitizes cool-weather or winter-grown crops, so that the number of generations completed in a season depends on the growing period; in northwestern Europe this is one generation on transplanted summer cauliflowers and cabbage, but as many as three generations on late cultivars of brussels sprouts, spring cauliflower, and spring cabbage (Stone and Rowe, 1976). Juveniles of *H. cruciferae* are stimulated to hatch from eggs of cysts in the presence of diffusates from *Brassica* spp., but not by diffusates of other hosts (Shepherd, 1965; Winslow, 1953). Second-stage juveniles explore and probe the root tip (Doncaster and Seymour, 1973) and at temperatures above 4°C they invade roots and establish a syncytial feeding site typical of cyst nematodes. Mackintosh (1960) speculates that secondary roots compensate for some potential damage of the nematode to the host, and agreed with Franklin (1951) that some crop failures attributed to *H. cruciferae* actually may be due to secondary fungal invaders. Lear et al. (1966) found that the nematode may interact synergistically with the fungus causing club root disease. Severe crop damage in Europe seems to be primarily on transplanted seedlings; conversely, in California and southern Australia damage may be more widespread and severe (Lear, 1971; Stirling and Wicks, 1975; Stone and Rowe, 1976).

Control. The limited host range of *H. cruciferae* makes it subject to excellent management by crop rotation. Several fumigants are effective in reducing initial populations (Lear et al., 1965a, b, 1966; Stirling and Wicks, 1975; Stone and Rowe, 1976).

Heterodera carotae Jones, 1950 (the carrot cyst nematode)

Synonyms and history. Prior to knowledge of the carrot cyst nematode "carrot sickness" had been attributed to pathogenic fungi. In 1931, when virtually all cyst nematodes were considered strains of *H. schachtii*, Triffit (1931) reported females on carrots (*Daucus carota*) in England which were atypical of *H. schachtii* because they were small with a strikingly large egg sac (Greco, 1986). By 1944, when Jones discovered the carrot cyst nematode, many "strains" of *H. schachtii sensu lato* were being described as morphologically distinct species. Perhaps this context invited closer morphological inspection of the carrot cyst nematode which was then described as *H. carotae* on the type host, *Daucus carota* L., from the Isle of Ely, England (Jones, 1950).

Hosts. The host range of *H. carotae* is restricted to *Daucus* including *D. carota* ssp. *sativus*, *D. carota* ssp. *sarota*, *D. pulcherrimus* (Jones 1950a; Mathews, 1975; Winslow, 1954), and the wild Umbelliferae, *Torilis* spp. Vallotton, 1980). However, the life cycle reportedly is only completed on *Daucus carota* L. and *Torilis leptophylla* L. (Mugniery and Bossis, 1988). The weed host *Torilis* is significant as an important source of infection of cultivated carrot (Greco, 1986). Infected fields of carrots show poor stands, with chlorotic plants and poor yields (Fig. 19D).

Distribution. *Heterodera carotae* occurs throughout the carrot-growing areas of Europe and has been reported from England, Ireland, Netherlands, Scotland, France, Italy, Switzerland, Germany, Sweden, Poland, Sweden, Czechoslovakia, Hungary (Mathews, 1975; Greco, 1986). It has also been reported from the USSR, Cyprus, and India (Greco, 1986), and from Michigan (Graney, 1985). *Heterodera carotae* may be dispersed locally on farm vehicles, but export of cysts adhering to tap roots or seeds could also account for more distant spread (Greco, 1986).

Biology. Temperature requirements for hatch and development of *H. carotae* are highly specific and frequently conditions are only suitable for one life cycle per season. However, with a long growing season, repeated crops, and cool temperatures, as many as four cycles may occur (Greco, 1986; Jones, 1950a; Mugniery and Bossis, 1988; Stelter, 1969). The source of repeated infections is eggs (typically about 160) exuded into the large egg sac. Second-stage juveniles hatch quickly from these eggs if root exudates from young carrots are present, if soil moisture is favorable, and if soil temperature is 15–20°C. Hatching occurs as low as 5°C but is repressed at 25°C (Greco et al., 1982, 1986; Winslow, 1955). In addition to eggs in the egg mass, many eggs are protected in the cyst. Eggs from cysts less than 2 months old rarely hatch, but second-stage juveniles hatch from mature cysts when conditions of soil temperature and moisture are suitable (Aubert, 1986; Greco, 1981, 1986). Curiously, eggs in the cyst are apparently unaffected by host leachates (Greco, 1986). Second-stage juveniles invade the tips of host feeder roots at 5–30°C but do not develop below 10°C; they establish a typical syncytial feeding site and at 20°C adult females and fertilization occurs at 26 days (Greco, 1986;; Volvas, 1978). Heavily infected plants are stunted and the tops become yellowish red and necrotic; tap roots are small and unmarketable and lateral roots may proliferate (Greco, 1986).

Control. Management strategies for *H. carotae* including rotation might seem promising because of the narrow host range of this species. However, eggs protected in cysts are persistent, and rotations of 4–5 years are apparently required for adequate reduction of inoculum (Bossis, 1986; Bossis and Mugniery, 1988; Greco, 1986; Vollotton, 1983). Although no resistant varieties are known, early cultivars appear less susceptible than late ones (Greco, 1986), perhaps because they escape field conditions ideal for hatch and development of the nematode. Greco (1986) suggests that because *H. carotae* is also inactive at high soil temperatures, yield losses could be reduced in warm countries by planting in the late summer. Various nematicides are economically feasible on early crops (Bossis, 1986; Greco, 1986; Greco et al., 1986b; Mathews, 1975). Because of the sensitivity of *H. carotae* to high temperatures, solarization of soil in summer with polyethylene sheets or mulch may result in temperatures that are lethal to eggs if populations are low and soil temperatures can be elevated at an adequate depth. Solarization may create additional beneficial conditions, such as increased antagonists to the nematode (Greco, 1986; Greco et al., 1985).

Heterodera avenae Wollenweber, 1924 (oat cyst nematode)

Synonyms and History. Heterodera schachtii was reported as attacking cereals in Germany by Kühn in 1874 and in the years following there were several references to the "oat strain" of *H. schachtii* (Franklin, 1957; Hansen, 1904; Rostrup, 1896). Moortensen et al. (1908) designated *H. schachtii* var. *avenae* but only recorded the occurrence of the nematode, and no morphological description was included. Wollenweber in 1924 used the name *H. schachtii* var. *avenae* for a population from oats (*Avena sativa* L.) in Aschersleben, Germany; he gave a more complete description including diagnostic characters on the cyst, which, combined with knowledge of the host, could be used for identification (Thorne, 1961). Schmidt (1930) also observed morphological differences between the oat and beet strain of cyst nematode, and particularly noted that the second-stage juveniles were longer in *H. avenae*; he adopted the name *H. schachtii* ssp. *major*. However, Filipjev (1934) elevated the older variety name to species *Heterdera avenae*, which after some controversy (Franklin, 1957) has been adapted as consistent with the international zoological rules of priority. Although original type material was not designated, a neotype from Aschersleben has been proposed (Franklin et al., 1959).

Heterodera avenae, together with other bifenestrate cyst nematodes having a short vulva, were placed in a genus *Bidera* Krall and Krall (1978), but Mulvey and Golden (1983) synonymized *Bidera* with *Heterodera*. Although Wouts (1985) retained *Bidera*, it was rejected by Stone (1986) and Luc et al. (1988). Other synonyms of *H. avenae* include *H. ustinovi* Kirjanova, 1969 and *Bidera ustinvoni* (Kirjanova, 1969) Krall and Krall, 1978.

Hosts. Hosts of *H. avenae* are primarily restricted to Poaceae (Gramineae) including economically important cereals, oats (*Avena*), wheat (*Triticum*), barley (*Hordeum*), and rye (*Secale*). Other hosts are noncereal grasses including *Agropyrum, Agrostis, Alopecurus, Brachypodium, Bromus, Cynosurus, Echinochloa, Festuca, Koeleria, Lolium, Phalaris, Phleum, Poa, Polypogun, Setaria, Sorghum, Vulpia, Zerna*, and *Zea* (Andersen, 1961; Bovien, 1953; Filipjev and Schuurmans Stekhoven, 1941; Stoyanov, 1982; Thorne, 1961; Williams and Siddiqi, 1972). Many grasses, such as *Zea* (corn), are poor hosts (Johnson and Fushtey, 1966), but even a low rate of reproduction must be considered in rotations. In Canada corn is severely injured by an intolerant reaction to invading juveniles which do not complete development (Miller, 1986). There are reports of *H. avenae* occurring worldwide in a variety of habitats, *Pisum, Sonchus*, and *Trifolium*, but they are limited to certain restricted areas or could be results of contamination of samples (Thorne, 1961), or cyst species other than *H. avenae. Heterodera avenae* consists of as many as 20 pathotypes (Andersen and Andersen, 1982, 1986; Brown, 1982a, b; Stone and Hill, 1982; Stone and Williams, 1974), and populations are also biochemically diverse (Dalmasso et al., 1982). In addition, several species are morphologically similar to *H. avenae*. Thus some populations of *H. avenae* may vary in host range from others, and closely related species may also differ in host range (e.g., *H. mani* Mathews, 1971 and *H. hordecalis* Anderson, 1975).

Distribution. Heterodera avenae is widespread throughout a variety of climates including northwestern Europe, the Mediterranean region (Italy, Spain, Greece, Israel, North Africa), North America (particularly Canada but also California and Michigan), India, New Zealand, Australia, and perhaps Japan (Grandison, 1982; Graney, 1985; Hackney, 1981; Williams and Siddiqi, 1972). Nevertheless, the species seems to be diverse so that populations from a given region may be particularly adapted to the environmental conditions of that region, and may also differ in pathogenicity on host species and cultivars. This diversity and occurrence in nonagricultural habitats (e.g., in California) could reflect wide dispersal aside from the activities of man. McLeod (1968) notes that *H. avenae*, like other cyst nematodes, spreads through soil very slowly unless distribution is promoted by rainfall runoff, wind, and activities of man. Wide distribution in dry southern Australia may be attributed to transport of cysts during turbulent dust storms (Brown, 1984; Meagher, 1982a). Some regions of southern Australia may lack the nematode primarily because heavy soils do not favor its development (Brown, 1984).

Biology. Although *H. avenae* occurs worldwide in a variety of habitats, it completes only one or two generations per year, depending on the population (Valdeolivas and Romero, 1985). Whether or not hatching of second-stage juveniles from eggs within cysts is stimulated by root exudates is controversial, and even roots of poor hosts may be attractive for penetration (Andersen and Andersen, 1986; Johnson and Fushtey, 1966). The hatching rate from a cyst may vary from 40 to 90%. Eggs can survive within cysts for years, particularly under cool, dry conditions (Meagher, 1982a). Infective juveniles enter behind the root cap and move intracellularly to the growth zone, establishing the feeding site near the endodermis or pericycle. Sedentary females mature from 6 to 9 weeks after penetration, whereas males leave the roots in about 3 weeks. Although a gelatinous material is secreted through the vulva of white females of *H. avenae*, eggs are not laid through the small vulval slit but rather escape through the fenestrae of the cyst, usually at the beginning of the subsequent growing season. In field conditions, populations develop to higher numbers of individuals after moist and cool periods. Although temperature requirements appear to be complex, cool temperatures (2–10°C) are optimal for egg hatching, whereas dry conditions during growth of the host seem to favor development of the parasite (Williams and Siddiqi, 1972). Populations of different areas vary in response to moisture contents of soil. Indian and Australian populations are better adapted to arid conditions, whereas European populations seem to be less resistant to dry conditions (Williams and Siddiqi, 1972). Populations may also vary in response to soil type. Reproduction is typically favored in sandy and calciferous soils, but there are exceptions (Filipjev and Schuurmans Stekhoven, 1941; Williams and Siddiqi, 1972).

Cereals infected with *H. avenae* generally are stunted (Fig. 20). Leaves yellow and may even be tinged with purple. Root systems are reduced in size and highly branched with a proliferation of short, thick side roots. Seminal root growth is especially impaired in intolerant cultivars (Volkmar, 1989). Stunted, chlorotic plants may appear in patches within a field of otherwise healthy plants (Fig. 20) (McLeod, 1968; Williams and Siddiqi, 1972), and some symptoms may be enhanced by interaction with secondary pathogens (e.g., *Rhizoctonia solani*; Brown, 1984) as well as physiological stresses.

Control. Damage of *H. avenae* is most severe when hosts such as wheat or barley are grown continuously, but populations can be greatly reduced by 3-year rotations with noncrops such as legumes, where such rotations are economically feasible. In Australia, rotation with legumes has the added benefit of improving nitrogen-depleted soils which are common in the southern wheat belt (Brown, 1982a, 1984). In addition to rotation, in some regions crops can be planted sufficiently early to establish a good root system prior to heavy rainfall. Heavy rain maximized hatch and population densities of infective juveniles (Brown, 1984; Meagher, 1982b). Although resistant varieties are useful for control of *H. avenae* on barley or oats, their value may be limited by specificity to certain pathotypes, possible effectiveness only under certain environmental conditions, and limited availability of resistant varieties (Andersen and Andersen, 1982, 1986; Brown, 1982b, 1984; Hansen,

1986). Yet resistance in oats and barley generally is not linked to undesirable horticultural characteristics, and resistant cultivars tolerate invasion and hypersensitive reaction to the nematode with little adverse reaction. Resistance in wheat is limited, although some lines are available in duram wheat. Brown (1984) indicates that resistance in other types of wheat is linked to low yield potential and therefore is not likely to be accepted by growers.

As with many cyst nematodes, resistant quiescent juveniles in cysts may be difficult to control by fumigation, although reports indicate a striking response to preplant fumigants such as ethylene dibromide (EDB) on the sandy soils of southern Australia (Brown, 1984; Gurner, 1982). Generally, however, fumigation may only be economically worthwhile for certain high-value crops, in small areas where infestation is very heavy, or in small isolated introductions where there may be hope of eradication (McLeod, 1968). There is some promise that biological control agents, particularly certain fungi, may prove useful for suppressing populations of *H. avenae* (Andersson, 1982; Kerry, 1975; Kerry and Crump, 1977). Recently, chitin was reported to be as effective as EDB or aldicarb in increasing wheat yield and controlling the nematode (Spiegel et al., 1989). Restriction of spread, regulation by quarantine, and sanitation (e.g., care not to move contaminated soil on farm equipment) play an important role in limiting infection by *H. avenae*.

Globodera

Globodera Skarbilovich, 1959
    syn. *Heterodera* (*Globodera*) Skarbilovich, 1959
Diagnosis. Heteroderinae Filipjev and Schuurmans Stekhoven

Females. Cyst present. Body spheroidal, without terminal cone. Cuticle surface with zig-zag pattern of ridges; distinct D layer present. Vulva terminal; perineal tubercles on crescents near vulva. Anus and vulva both in a vulval basin. Underbridge and bullae rarely present; circumfenestrae develops around vulva in cyst; no fenestration around anus. All eggs retained in body (no egg mass); egg surface smooth.

Males. Body twisted. Lateral field with four lines. Spicules >30 µm, distally pointed.

Second-stage juveniles. Lateral field with four lines. Esophageal glands filling width of body cavity. Tail conical, pointed, with terminal region half hyaline. Phasmid openings porelike. En face pattern typically with six separate lips; variants with fusion of adjacent submedial lips.

Nurse cell. Syncytium, generally with wall ingrowths.

Species. (Table 4)

Identification. *Globodera* includes less than a dozen species, many of which are considered unimportant because their distribution is narrow or they are not known on economic hosts. Nevertheless the genus is especially significant because of the worldwide economic impact of two species, *G. rostochiensis* and *G. pallida*, the potato cyst nematodes. Identification between the two species is essential to some management strategies such as use of resistant cultivars. A species complex, the tobacco cyst nematodes, *G. tabacum* (*G. tabacum tabacum*; *G. tabacum solanacearum*, *G. tabacum virginiae*) *sensu* Stone, 1983, while globally less important than potato cyst nematodes, is included in this review. Subspecies *G. t. tabacum* is damaging on tobacco in Connecticut and Massachusetts. *Globodera t. solanacearum* damages tobacco and *G. t. virginiae* occurs on solanaceous weeds (and at least experimentally on tobacco) in the southern USA. *Globodera* is widespread on weed hosts in Mexico and Central America and these are putatively *G. t. virginiae*. It is not clear if additional subspecies of *G. tabacum* occur in Mexico, or if species

TABLE 4   Species of *Globodera*

Globodera Skarbilovich, 1959
   syn. *Heterodera (Globodera)* Skarbilovich, 1959

Globodera rostochiensis (Wollenweber, 1923) Behrens, 1975
   syn. *Heterodera schachtii rostochiensis* Wollenweber 1923
      H. schachtii solani Zimmermann, 1927

G. achilleae (Golden and Klindic', 1973) Behrens, 1975
   syn. *Heterodera achilleae* Golden and Klindic', 1973

G. artemisiae (Eroshenko and Kazachenko, 1972) Behrens, 1975
   syn. *H. artemisiae* Eroshenko and Kazachenko, 1972

G. hypolysi Ogawa, Ohshima, and Ichinohe, 1983

G. leptonepia (Cobb and Taylor, 1953) Behrens, 1975
   syn. *H. leptonepia* Cobb and Taylor, 1953

G. millefolii (Kirjanova and Krall, 1965) Behrens, 1975
   syn. *H. millefolii* Kirjanova and Krall, 1965

G. mirabilis (Kirjanova, 1971) Mulvey and Stone, 1976
   syn. *H. mirabilis* Kirjanova, 1971

G. pallida (Stone, 1973) Behrens, 1975
   syn. *H. pallida* Stone, 1973

G. pseudorostochiensis (Kirjanova, 1963) Mulvey and Stone, 1976
   H. pseudorostochiensis Kirjanova, 1963

G. tabacum tabacum (Lownsbery and Lownsbery, 1954) Behrens, 1975
   syn. *H. tabacum* Lownsbery and Lownsbery, 1954

G. tabacum solanacearum (Miller and Gray, 1972) Behrens, 1975
   syn. *H. solanacearum* (Miller and Gray, 1972)
      G. solanacearum (Miller and Gray, 1972)

G. tabacum virginiae (Miller and Gray, 1968) Behrens, 1975
   syn. *H. virginiae* (Miller and Gray, 1968)
      G. virginiae (Miller and Gray, 1968)

G. zelandica Wouts, 1984

Source: Adapted from Luc et al., (1988).

not included in the *G. tabacum* complex are present. Indeed some isolates from weed hosts in Mexico which do not infect potato are morphologically similar or even identical to *G. pallida*, a similarity that undoubtedly confounds management strategies for the potato cyst nematodes (Miller, pers. commun.). Taxonomic relationships which reflect phylogeny among some groups of *Globodera* and particularly those of the *G. tabacum* complex are elusive. One hypothesis is that a continuum of morphological variants occurs within some members of the genus because interspecific and intrasubspecific hybridization occurs not only under experimental conditions (e.g., Miller, 1983) but in nature. Such fertile hybrids may give rise to isolates with recombination of morphological characters. Because of the current taxonomic dilemmas concerning *Globodera*, we consider identification of subspecies of the *G. tabacum* beyond the scope of this review, and instead emphasize

TABLE 5   Compendium for Identification Among Select *Globoderas* (Fig.21)

Species	J2 stylet knob shape	No. cuticular ridges between vulva-anus	Granek's ratio of cysts	Host
G. rosto-chiensis	Rounded: dorsal may slope posteriad	16–31 ridges (about 22)[a]	1.3–9.5 (4.5)[a]	No tobacco potato
G. pallida	Pointed anteriorly; anterior surface of dorsal slightly concave	8–20 ridges (about 12)[a]	1.2–3.5 (2.3)[a]	No tobacco potato[c]
G. tabacum complex	Pointed to rounded anteriorly; dorsal slightly concave	10–14 ridges	1–4.2 (2.8 or less)[a,b]	Tobacco[d] no potato[c]

	J2 stylet length (μm)	Female stylet length (μm)		
G. rostochiensis	(21–23) 21.8–22[a]	(21–25) 23[a]		
G. pallida	(21–26) 22.8–24[a]	(23–29) 27[a]		
G. tabacum	(19–28) 22.8–24[a]	(18–30) 25[a]		

[a]Measurement which are most typical (Golden, pers. commun.).
[b]Generally less than 2.1 in *G. t. tabacum.*
[c]See text for reports of exceptions.
[d]*G. t. virginiae* only occurs on tobacco experimentally. Also many Mexican populations do not reproduce on tobacco or potato (Miller, pers. commun.).

morphological characters reported to be useful in distinguishing the combined subspecies of *G. tabacum* with *G. pallida* and *G. rostochiensis* (Table 5). Identification of the *G. tabacum* complex is often aided by the slightly more pointed anterior surface of stylet knobs and concave surface of the dorsal knob on second-stage juveniles relative to *G. rostochiensis* in which the knobs are more rounded and the dorsal knob slopes posteriorly (Fig. 21A–C). Miller and Gray (1968, 1972) illustrate a concave posterior-sloping dorsal knob in females of *G. t. virginiae* and *G. t. solanacearum*. These knobs apparently differ from the more rounded knobs in females of *G. rostochiensis*. However, J2 and female stylet knobs of the *G. tabacum* complex are apparently not different from populations of *G. pallida*. The pattern between the vulva and anus of the *G. tabacum* complex may form a zig-zag maze rather than more or less parallel ridges of *G. rostochiensis* Miller and Gray, 1968, 1972), but the pattern in *G. tabacum* may overlap with the highly variable patterns of *G. pallida*. Granek's ratio, the distance from the anus to the nearest edge of the vulva

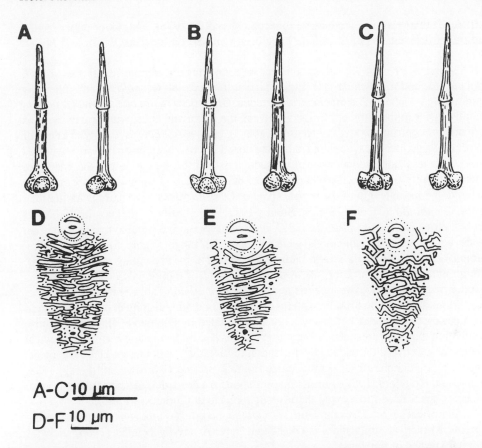

FIGURE 21 Characteristics for identification of some species of *Globodera*. (A) *Globodera rostochiensis*, stylet of second-stage juvenile (J2) (left dorsal and lateral view). *Globodera pallida*, stylet of J2. (C) *Globodera tabacum tabacum*, stylet of J2. (D) *G. rostochiensis*, anal-vulval region of female. (E) *G. pallida*, anal-vulval region of female; (F) *G. tabacum*, anal-vulval region of female. (A, B, D, E, redrawn after Stone, 1973.)

opening divided by the diameter of the vulva (Granek, 1955) or length of the fenestra (Hesling, 1973), may be useful. Granek's ratio in the *G. tabacum* complex may overlap with *G. pallida*, particularly in subspecies *G. t. virginiae*, but in other subspecies it is distinctive in that the mean of a number of individuals in a population in generally less than 2.1 (Fig. 21D–F). Because of the morphological distinctiveness of the *G. tabacum* complex is subtle, and variation among subspecies and populations is great, a bioassay including certain potato cultivars susceptible to other cyst nematodes may prove to be the most reliable approach to identification by the inexperienced. Subspecies of *G. tabacum* rarely reproduce on potato, and *G. rostochiensis* and *G. pallida* rarely reproduce on tobacco. Nevertheless, there are several reports of isolates that behave to the contrary (Lownsbery and Lownsbery, 1954; Miller and Gray, 1968, 1972). Stone (1972) reports infection by subspecies of *G. tabacum* on *S. tuberosum* "Arran Banner." However, subsequent tests did not support Stone's initial findings except in rare cases where a few females were found on a test plant (Miller, pers. commun.; Miller et al., 1974; Stone and Miller, 1974). Some isolates of *G. rostochiensis* from Venezuela readily reproduce on tobacco (Meredith, 1976), and under experimental condition a few females of *G. pallida* have been reported on tobacco (Parrot and Miller, 1977).

Stone (1972; 1973) gives the most complete differential diagnoses of *G. rostochiensis* and *G. pallida*. He suggests that females of *G. pallida* are internally cream-colored whereas *G. rostochiensis* is golden. *Globodera rostochiensis* and *G. pallida* generally differ in the number of ridges between the vulva and anus, with nearly twice as many ridges occurring in *G. rostochiensis* (Table 5). In *G. rostochinesis*, the stylet knobs of the second-stage juveniles are rounded but the dorsal knob may slope posteriorly, whereas in *G. pallida* it is elongated along the longitudinal axis of the nematode and the knobs are pointed anteriorly (Fig. 21A, B). When a number of individuals of a population are carefully measured, the mean length of the female stylet of *G. rostochiensis* typically is longer than that of *G. pallida* (27 vs. 23 μm; Stone, 1972, 1973). Many diagnostically useful morphometrics for *G. rostochiensis* and *G. pallida* have relatively stable means, but often include a few unusual individuals which extend the range (Table 5). Golden (pers. commun.) considers Granek's ratio particularly valuable for separating the two species; although unusual variants occur, the mean of a number of individuals of a population is generally near 4.5 for *G. rostochiensis* and 2.3 for *G. pallida* (Table 5).

Stone suggests the SEM lip patterns are useful for distinguishing between *H. pallida* from other cyst nematodes but this was not substantiated by Othman et al. (1988), whose investigation indicated considerable variation within species. Promising nonmorphological methods for identification include hybridization of unknown with known DNA fragments to separate populations of *G. rostochiensis* from *G. pallida* (Burrows and Perry, 1988; Marshall and Crawford, 1987). Further investigations are needed to identify fragments that are species-specific and to expand the investigations to additional species.

Identification of *G. rostochiensis*, *G. pallida*, and the *G. tabacum* complex, having the greatest practical application, is considered here. A key by Golden (1986) includes additional species of the genus.

Globodera rostochiensis (Wollenweber, 1923) Behrens, 1975
Globodera pallida (Stone, 1973) Behrens, 1975 (potato cyst nematodes)

Synonyms and history. Kühn was the first to report potato as a host of *H. schachtii* in 1881 when this species encompassed all cyst nematodes. Perhaps potato cysts nematodes were not widespread at the time, since through the 1800s other authors considered solanaceous plants as not susceptible to *H. schachtii sensu lato* (Franklin, 1951). In the early 1900s the cyst nematode of potato became more widely known throughout Europe and many concluded that at least one form of *H. schachtii* was specialized on this host, causing "soil sickness of potatoes" (Spears, 1968). Wollenweber (1923) recognized morphological differences between the potato cyst nematodes and cyst nematodes of sugar beet and proposed the separate species, *Heterodera rostochiensis*. Others continued to consider potato cysts as subspecific within *H. schachtii* until 1940 when Franklin published a more complete morphological description and diagnosis of *H. rostochiensis*. Interest in the potato cyst nematode intensified as its wide distribution was recognized including discovery in 1941 of infested fields in Long Island (USA). Description of additional species of cyst nematodes prompted Golden and Ellington (1972) to redescribe *H. rostochiensis* to allow more precise comparison with related species; about the same time Stone (1972) recognized a second species of potato cyst nematode, *H. pallida*. These several "related" species were primarily set apart by the absence of a terminal cone resulting in round cysts. The group was first represented taxonomically by a subgenus, *Globodera* Skarbilovich. Subsequently,

Mulvey and Stone (1976) confirmed Behrens (1975) in elevating the round cyst nematodes, including *H. rostochiensis* and *H. pallida*, to generic rank resulting in *Globodera rostochiensis* and *Globodera pallida*.

Hosts. The potato cyst nematodes parasitize about 90 species of the large genus *Solanum*. Many hosts are wild species of South America (Southey, 1965) including close relatives of potato, *S. tuberosum*, with varying levels of resistance, *S. tuberosum andigena*, *S. vernei*, and *S. sucrense* (Spears, 1968). However, other weed hosts such as *S. sarachoides*, *S. dulcamara*, and *Datura stramonium* in Europe may exacerbate persistence of this nematode in certain agricultural areas (Goodey et al., 1965, Stone, 1973). Although the potato cyst nematodes are not typically found on tobacco, we have noted exceptions (see "Identification"; Meredith, 1976; Parrott and Miller, 1977). *Solanum tuberosum* is by far the most important agricultural host of potato cyst nematodes; other agricultural hosts are *S. melongena* (eggplant) and *Lycopersicon esculentum* (tomato) (Stone, 1973).

Distribution. For many years the potato cyst nematode was believed to be restricted in distribution to Europe and was thought to have originated there as a mutation from other cyst species (De Segura, 1952). However, discovery in the 1950s of the nematode on a ship from Peru led to surveys in South America and recognition that potato cyst nematode occurred on native plants in the Andean region of Peru and adjacent countries. The nematode was likely a pest of pre-Inca potato agriculture, with *G. pallida* still present on *Solanum acaule* in terraces uncultivated since ancient times (Jatala and Garzon, 1987). Today, a widely accepted hypothesis is that the potato cyst nematodes, like their potato host, originated in the mountains of South America and that they may have been introduced to Europe, with the potato, in the 1600s. Lack of awareness of these nematodes for more than 200 years is interpreted as evidence of a slow buildup of the nematodes in Europe (Spears, 1968). In addition, old potato cultivars may have been relatively tolerant to cyst nematodes, compared to the selected lines introduced following the Irish potato famine in the mid-1800s (Spears, 1968). From Europe, it is believed that the potato cyst nematodes were spread, primarily on seed pieces, throughout the potato-growing regions of the world including more than 50 countries (Evans and Stone, 1977; Mai, 1977). It is noteworthy, however, that the genus *Globodera* is widespread and that it includes apparently indigenous nonagricultural forms in Europe, Asia, America, and perhaps New Zealand. The range of the potato cyst nematodes on weed hosts prior to movement by commerce is unknown and commerce may not entirely account for its modern distribution (Ferris, 1979). Seinhorst (1986) suggests that containment or restricted spread in some regions, such as warmer climates and the USA, probably cannon be wholly attributed to successful quarantine, but rather adverse conditions in areas of introduction. Similarly, *G. rostochiensis* in northern Mexico probably has not moved into Texas on vehicles and become established in the potato-growing region of Texas, perhaps because of the high soil temperature (Miller, pers. commun.). In many regions *G. rostochiensis* and *G. pallida* occur together, but often temperature seems to restrict distribution of one or the other species. In cooler areas such as north of 15.6 S, *G. pallida* occurs exclusively, whereas the range of *G. rostochiensis* extends into regions with slightly warmer climates than are tolerated by *G. pallida* (Evans et al., 1975; Evans and Stone, 1977; Mai, 1977).

Biology. The life history of potato cyst nematodes follows the general pattern of other cyst-forming heteroderines. However, details of hatching, establishing feeding, mating, and cyst formation include adaptations peculiar to these nematodes, with some distinctions even between *G. rostochiensis* and *G. pallida*.

Tanned cysts, often with about 500 eggs, are especially persistent in soil; eggs within cysts may survive 28 years where soil type and temperature are ideal (Grainger,

1964), but even under ideal conditions the percentage of hatch decreases steadily over time. Hatching of freshly developed eggs apparently can occur in water, but for eggs in diapause, hatching is stimulated by host root diffusates which effect a change in the permeability of lipoprotein membranes of the eggshell. These diffusates allow leakage of solutes and increase in O_2 uptake (Atkinson and Ballantyne, 1977a, b) and hydration of juveniles (Ellenby and Perry, 1976; Perry et al., 1982; Clark and Perry, 1985; Perry and Feil, 1986). Hydration of juveniles is also associated with secretory granule accumulation in subventral glands in preparation for feeding after hatching (Perry et al., 1989). These changes may be accompanied by greater activity of juveniles within the egg including stylet thrusting (Doncaster and Shepherd, 1967). Root diffusates from both susceptible and resistant cultivars elicit similar effects (Mägi, 1971), but the effectiveness of diffusates depends on temperature; and some substances, such as exudates of the fungus, *Rhizoctonia solani*, inhibit hatching of one or sometimes both nematode species. *Globodera pallida* and *G. rostochiensis* differ in their response to various hatching agents and their concentrations (Clark and Hennessy, 1987). Furthermore, *G. pallida* hatches less freely and is less persistent at high temperatures than *G. rostochiensis* (Clarke and Perry, 1977; Robinson et al., 1987). The importance of root diffusates to hatching and completion of the life cycle of the potato cyst nematode has generated speculation on possible control measures that would interfere with the role of diffusates. Indeed, Perry and Beane (1988) demonstrated that activated charcoal in greenhouse pots temporarily limits hatching by interfering with diffusates. Hatched second-stage juveniles are attracted to penetrate the host just behind the root tip or lateral root. They move through the root and feed on the pericycle, cortex or endodermis, and become sedentary, inducing formation of a large syncytial transfer cell at the feeding site. (Spears, 1968). In resistant reactions, the host response typically involves localized necrosis and thickening of the syncytial walls, perhaps in response to accumulation of lignin which might act as a barrier to the flow of nutrients (Rice et al., 1986; Robinson et al., 1988).

Development from hatching to adult requires 38–45 days (Spears, 1968) and females mate within 50 days of root invasion (Evans, 1970). As the sedentary female enlarges, the tail end bursts through the root, becoming exposed for mating. Whereas females of *Globodera* produce little if any gelatinous matrix, they nevertheless exude a pheromone which attracts males for copulation (Green et al., 1970; Green and Greet, 1972; Greet et al., 1968). Males are motile and develop from sedentary saccate juvenile stages; some suggest that males are more abundant under environmental conditions which may be stressful to the population (Trudgill et al., 1967). Males do not feed and Evans (1970) reports they live up to 10 days inseminating up to 10 females, and each may be inseminated by more than one male. A degree of mating compatibility is conserved between *G. rostochiensis*, *G. pallida*, and many cyst nematodes, so that under laboratory condition males of one species may inseminate females of another, and in many cases hybrids are produced (Miller, 1983). We have noted that the possible occurrence of such hybrids in nature may account for the high variability in morphology and host specificity within some species and groups of species (Miller, pers. commun.). The potato cyst nematode is amphimitic with the basic n = 9 chromosomes although some variants in chromosome numbers have been reported (Triantaphyllou, 1975a).

Eggs are retained within the body of the potato cyst female and the second-stage juvenile develops within the egg. Upon death, the body of the female becomes a cyst, changing in color from white to brown, also passing through a golden or orange phase in *G. rostochiensis* (Spears, 1968). Juveniles may hatch so that more than one generation is completed per year on a crop if temperatures remain favorable or, in the absence of favorable conditions including a suitable host, they may persist within the cyst for years.

At low levels potato cyst nematodes do little damage, but after years of repeated potato culture, these cyst nematodes may increase in number to a point of limiting production, and we have noted that in some extreme cases the yield may be less than the seed pieces planted (Mai, 1977). Field symptoms of heavy infestations are similar among cyst-forming nematodes; first they induce poor growth in spots, followed by increase in size and number of the spots. Damage to the root system is typically associated with wilting of plants and stunting (Spears, 1968). Tomato plants yield symptoms similar to potatoes except that roots may have slight swelling resembling root-knot nematode nodules (Spears, 1968).

We have noted that potato cyst nematodes are heterogeneous, consisting of two morphologically and physiologically distinct species, *G. rostochiensis* and *G. pallida*. With the introduction of resistant potato cultivars, however, a greater diversity was recognized in response to these cultivars, leading to identification of a number of pathotypes and various schemes for their nomenclature (Canto and de Scurrah, 1977; Kort et al., 1977). Seinhorst (1986) notes that five pathotypes of *G. rostochiensis* and three of *G. pallida* are recognized but that the diversity of these nematodes is probably greater than our ability to distinguish the pathotypes; one limitation is that host differential tests typically indicate "susceptibility" or "resistance" of an isolate but do not easily accommodate intermediate responses. Diversity of pathotypes beyond the present nomenclature is also underscored at the level of molecular biology (Bakkar and Gommers, 1986; Fox and Atkinson, 1985a, b).

Control. It has been noted (see "Hosts") that closely related wild forms of potato— *S. tuberosum andigena*, *S. vernei*, *S. sucrense*— provide sources of resistance to potato cyst nematode which have been incorporated into agricultural cultivars. Although resistant cultivars stimulate hatching, roots respond to infection by necrosis and hypersensitive reaction. In resistant cultivars, juveniles often fail to mature to females, whereas production of males apparently is unrestrained (Seinhorst, 1986). Resistance is based on dominant gene, but females that are homozygously recessive for the virulence gene are capable of reproduction on resistant plants (Jones et al., 1981; Seinhorst, 1986). The effectiveness of resistant cultivars is confounded by pathotypes which overcome resistance, and more such pathotypes may be selected by repeated use of resistant cultivars (Seinhorst, 1986). Resistant cultivars are primarily useful as a means to maintain an already low population density of the potato cyst nematode below damaging levels, when the cultivars meet other commercial requirements. However, resistant cultivars are not available for every purpose, and those resistant to a range of pathotypes are in short supply (Seinhorst, 1986).

Crop rotation might be a promising method of control for a nematode with a narrow host range such as potato cyst nematode, except that the infective state is so persistent in the absence of a host. In most of Europe, where populations decrease only about 35%/year in the absence of a host, a 5-7 year rotation is needed to minimize damage, but in warmer climates, such as the Mediterranean area, decline is apparently faster and shorter rotations may be effective (Grainger, 1964; Seinhorst, 1986; Spear, 1968). Volunteer plants and weed hosts may interfere with effective rotations.

Trap crops, if implemented with caution and careful timing, are a promising method of reducing populations (Balandras, 1986; Carlsson and Videgard, 1971). Sprouted small tubers (with only a small reserve of nutrients) of a susceptible late cultivar are planted and invaded by juveniles. The plants are carefully sampled so that they can be destroyed just before the females produce eggs. Optimally this method can result in a 75% decrease in populations and combined with chemical treatment of soil reduction of 90% has been achieved (Mugniery and Balandras, 1984).

Chemical control has the disadvantages of environmental hazard and high cost, but is often effective in managing the potato cyst nematode, particularly when integrated with other strategies. Whitehead (1986) notes that withdrawal of DBCP and EDB from general use may be discouraging investment in discovery and registration of new nematicides, yet nematicides are likely to be needed for potato production in the foreseeable future.

In some cases, as in the Netherlands, chemical treatment of the soil may be required by the government to keep nematode infestations at a low level nationally and thus to protect the international trade of seed potatoes (Seinhorst, 1986). Import of potatoes is widely regulated by quarantine laws as a first line of defense in managing potato cyst nematodes. In the USA the potato cyst nematode has successfully been limited to a small area, but under different regulatory conditions containment has been more difficult. For example, in Mexico the potato cyst nematode seems to be a relatively new introduction and spread within the mountainous regions of the country is rapid. Quarantine is confounded by a number of undescribed wild *Globodera* forms on weed hosts, which are morphologically similar to *G. rostochiensis* and *G. pallida*, but which do not reproduce on potato.

The potential for biological control of potato cyst nematodes has been considered by only a few investigators (Roessner, 1986). The Andes, being the putative origin of potatoes and *G. rostochiensis* and *G. pallida*, is deemed a possible source of natural enemies to the nematode, and several associated fungi were found in Peru (Jones and Rodriguez Kabana, 1985). Although antagonists of practical importance for management have not been found (Rossner, 1986), biological control may nevertheless be feasible in the future (Whitehead, 1986).

Globodera tabacum (Lownsbery and Lownsbery, 1954) Behrens, 1975 *sensu* Stone, 1983 (tobacco cyst nematode complex) including *G. tabacum tabacum* (Lownsbery and Lownsbery, 1954); *G. tabacum virginiae* (Miller and Gray, 1968); *G. tabacum solanacearum* (Miller and Gray, 1972)

Synonyms and history. The round cyst nematodes; previously known only as potato cyst nematodes, have long been recognized as heterogeneous in morphology and host range, so it was not surprising when a population with round cysts was found in Connecticut parasitizing shade tobacco. Reproduction on tobacco was contrary to previous host range tests of the potato cyst nematodes in Europe (Lownsbery and Lownsbery, 1954). In addition, the new population of cyst nematode did not develop on cultivars of potato tested (Lownsbery, 1953; Mai, 1952). Closer examination of the population indicated morphological differences from *G. rostochiensis* including patterns between the vulva and anus, and led to the description of the new species *Heterodera tabacum* Lownsbery and Lownsbery, 1954. A second cyst nematode which parasitizes tobacco, but which was recovered from horsenettle (*Solanum carolinese* L.), a weed in Virginia, was described as distinct from *H. tabacum* Miller and Gray, 1968, in part on the basis of the shape of female stylet knobs, as well as a higher value of Granek's ratio, and differential host response on certain species of *Nicotiana* (Miller and Gray, 1968). A nematode similar to *H. virginiae* had also been described in Mexico, but never published as a new species (Campos Vela, 1967); in subsequent investigations this was considered conspecific with *H. virginiae* (Golden and Ellington, 1972). Yet many populations, including that described by Campos Vela, do not reproduce on tobacco (Miller, pers. commun.). Furthermore, some populations of the Mexican cyst nematode are morphologically indistinguishable from *G. pallida*, although they do not reproduce on potato. A few years prior to the description of *H.*

virginiae, Osborne and Holmes recovered a cyst nematode parasitizing flue-cured tobacco and horsenettle on a farm in Virginia, but the nematode was not described as a new species and for several years was referred to as the "Osborne cyst." Miller and Gray (1972) noted that this nematode differed from *H. tabacum* and *H. virginiae* in its response to *Nicotiana* hosts and that morphological differences from *H. tabacum* and *H. virginiae* included the pattern of the vulva anal region and Granek's ratio. Shortly thereafter, the round cyst nematodes were increasingly recognized as distinctive from other cysts; these differences were first reflected by placement in a separate subgenus *Globodera* Skarbilovich, 1959, and later a separate genus *Globodera* including *G. tabacum*, *G. virginiea,* and *G. solanacearum* (Behrens, 1975; Mulvey and Stone, 1976).

Morphological characters are highly conserved among many species of *Globodera*. Stone (1983) believed that only *G. rostochiensis*, *G. pallida*, and *G. tabacum* could be consistently recognized morphologically. Furthermore, he suggested that *G. virginiae*, *G. solanacearum*, and *G. tabacum*, while apparently allopatric, were difficult to separate. This view was supported by multivariate analyses. Furthermore, Stone reported that the latter three species hybridize with viable offspring more readily than other species (i.e., *G. rostochiensis* and *G. pallida*; see also Miller, 1983). Stone considered that only *G. tabacum* was significant on agricultural hosts, and it was his opinion that this species alone needed to be recognized for practical taxonomy. Based on these views, he proposed that *G. tabacum*, *G. solanacearum*, and *G. virginiae* be given a new rank of subspecies within *G. tabacum* (Stone, 1983). Unfortunately, Stone did not propose a diagnosis for the new inclusive *G. tabacum*.

Distribution. Globodera t. tabacum is apparently restricted in distribution to Connecticut and Massachusetts, whereas *G. t. solanacearum* is only known to occur in Virginia and North Carolina (Barker and Lucas, 1984). *Globodera t. virginiae* also occurs in Virginia but is reported to be conspecific with populations of cyst nematodes which occur throughout Mexico and Central America.

Hosts. The three subspecies of the *G. tabacum* complex are primarily limited to host range to *Nicotiana* and *Solanum*, although host range tests including other plant genera are limited. All three subspecies of the *G. tabacum* complex reproduce on *Nicotiana tabacum* L. However, *N. tabacum* is a poor host for *G. t. virginiae* (Nusbaum, 1969; Barker and Lucas, 1984) and not observed to be a host under field conditions (Miller, pers. commun.). Differential responses separate the subspecies of *G. tabacum* on other *Nicotiana* species (Miller and Gray, 1972). Isolates of the potato cyst nematodes have been reported not to reproduce on *N. tabacum*. Conversely, *G. t. tabacum* was originally believed not to reproduce on potato, *Solanum tuberosum* (specifically *S. tuberosum* "Katahdin"; Miller and Gray, 1972) but subsequent reports indicate formation of some cysts on at least one potato cultivar, *S. tuberosum* "Arran Banner" (Stone, 1972). Lownsbery and Lownsbery (1954) reported *S. nigrum* as a host of *G. t. tabacum* but a nonhost of *G. rostochiensis*. Roberts and Stone (1981) tested eight isolates including the three subspecies of the *G. tabacum* complex on 19 other species of *Solanum*. Although isolates varied in their host range, the variation weakly confirmed subspecies, particularly when the data were corrected for a mislabeled population (Roberts; Miller, pers. commun.). About five host species were not susceptible to most nematode isolates. Nusbaum (1969) reported that *G. t. solanacearum* reproduces on tobacco, tomato, and eggplant but not on potato. Clearly, isolates of all three subspecies of *G. tabacum* are highly variable in response to many host isolates and cultivars, leading further support to the physiological heterogeneity of this group. *Globodera t. solanacearum* is particularly destructive on many cultivars of flue-cured tobacco (Fig. 22).

FIGURE 22   *Globodera tabacum solanacearum* of flue-cured tobacco. (A) Stunted infected plants adjacent to large noninoculated controls. (B) Cysts on roots (A, B courtesy L. I Miller.)

Biology. Knowledge of the biology of the *G. tabacum* complex is primarily limited to reports in original descriptions and a few brief reviews (e.g., Barker and Lucas, 1984). Mature females of the *G. tabacum* complex retain eggs in the cyst, which may survive for years in the absence of a suitable host. Host exudates are believed to play a role in hatch followed by infection of seedlings. Females are fertilized and about 20 days is required for *G. t. tabacum* to produce eggs under field conditions, so that in the USA it may have four to five generations on a single tobacco crop (Barker and Lucas, 1984). Conversely, *G. t. solanacearum* is reported to require 32–82 days to complete a generation depending on conditions (34 days at 24–29°C; Adams et al., 1982; Miller and Gray, 1972), so that as few as two generations may be completed per year; *G. t. virginiae* apparently has similar generation time and temperature requirements (Miller and Gray, 1968).

Subspecies of *G. tabacum* may cause poor growth of tobacco and losses may exceed 50% (Barker and Lucas, 1984). *Globodera t. virginiae*, while reproducing poorly on tobacco, may cause considerable stunting on K-16 burley tobacco because of intolerance to invasion of infective juveniles (Barker and Lucas, 1984; Miller, pers. commun.).

Control. Tobacco cultivars with some resistance to the *G. tabacum* complex are available, and resistance in other *Nicotiana* species holds some promise for developing additional suitable lines (Baalawy and Fox, 1971; Barker and Lucas, 1984; LaMondia, 1988; Miller et al., 1972). Currently available lines are not tolerant of invasion and may be damaged by penetration of high nematode populations; nevertheless these lines may be effective in combination with nematicides (Johnson et al., 1989). Use of resistant cultivars may also be confounded by pathogenic variation of the nematode and the potential selection for pathogenicity (Elliott et al., 1986). The relatively narrow host range of these cyst nematodes also suggests that effective rotation programs could be developed as part of management programs (Barker and Lucas, 1984), although the relatively low economic value of rotation crops has discouraged their use (Johnson et al., 1989). Several nematicides give effective control, although in some cases their value could be limited by resistance of eggs retained within cysts (Miller, 1969, 1970; Barker and Lucas, 1984).

Cactodera

Cactodera Krall and Krall, 1978
  syn. *Heterodera* Schmidt, 1959

Diagnosis: Heteroderinae Filipjev and Schuurmans Stekhoven
Females. Cyst present. Spheroidal to lemon-shaped with small posterior cone. Cuticle surface with irregular transverse ridges; distinct D layer present (except in *C. betulae*). Vulva terminal. In cysts, circumfenestrae develop around vulval region; no anal fenestration. Underbridge, perineal papillae and bullae absent, but denticles may be present. All eggs retained in body (no egg mass); egg surface smooth or tuberculate.
Males. Body twisted. Lateral field with four lines. Spicules >30 µm, bifid distal tip.
Second-stage juveniles. Lateral field with four lines. Esophageal glands filling body cavity. Tail pointed with terminal half hyaline. Phasmid openings porelike. En face pattern commonly with six separate lips; sometimes fusion of adjacent submedial lips, without fusion of labial disc and submedial lips (except *C. betulae*).
Nurse cell. Syncytium, generally with wall ingrowths.
Species. (Table 6)
Identification. *Cactodera* includes nine species, although the status of *C. betulae* and *C. chaubattia* in the genus is controversial (Baldwin and Schouest, 1990; Stone, 1986; Luc et al., 1988). *Cactodera* is distinctive from other heteroderines with the combination of

TABLE 6   Species of *Cactodera*

Cactodera cacti (Filipjev and Schuurmans Stekhoven, 1941) Krall and Krall, 1978
   syn. *Heterodera cacti* Filipjev and Schuurmans Stekhoven, 1941

C. acnidae (Schuster and Brezina, 1979) Wouts, 1985
   syn. *H. acnidae* Schuster and Brezina, 1979

C. amaranthi (Stoyanov, 1972) Krall and Krall, 1978
   syn. *H. amaranthi* Stoyanov, 1972

C. aquatica (Kirjanova, 1971) Krall and Krall, 1978
   syn. *H. aquatica* Kirjanova, 1971

C. betulae (Hirschmann and Riggs, 1969) Krall and Krall, 1978
   syn. *H. betulae* Hirschmann and Riggs, 1969

C. chaubattia (Gupta and Edward, 1973) Stone, 1986
   syn. *G. chaubattia* (Gupta and Edward, 1973) Wouts, 1924
     H. chaubattia Gupta and Edward, 1973
     H. mali Kirjanova and Borisenko, 1975
     G. mali (Kirjanova and Borisenko, 1975) Behrens, 1975

C. estonica (Kirjanova and Krall, 1963) Krall and Krall, 1978
   syn. *H. estonica* Kirjanova and Krall, 1963

C. eremica Baldwin and Bell, 1985

C. thornei (Golden and Raski, 1977) Krall and Krall, 1978
   syn. *H. thornei* Golden and Raski, 1977

C. weissi (Steiner, 1949) Krall and Krall, 1978
   syn. *H. weissi* Steiner, 1949

---

a cyst and vulval cone with a circumfenestrae (Fig. 23A). The pattern on the cuticle of females and cysts is often ambiguous with some zig-zag patterns in the cones becoming more parallel and horizontal at midbody, forming striations. Cuticular striations are unusual in cyst nematodes, occurring only in *Cactodera* and *Dolichodera*. Bullae are absent in *Cactodera* but vulval denticles, which are residual vaginal muscles (described as small, toothlike structures) within the vulval cone, are characteristically distinct in the genus. Golden and Raski (1977) and Cordero (1989) suggest that these denticles occur in other cyst nematodes, sometimes in combination with bullae (see "Morphology" above). Species of *Cactodera* are distinctive and generally easily identified by second-stage juveniles on the basis of length of the stylet, tail, and hyaline tail terminus. The shape of cysts and presence or absence of tubercles on eggs is also useful for species identification. Although the name may suggest that *Cactodera* is limited to cacti as a host, other hosts include *Euphorbia*, *Montia*, *Polygonum*, *Atriplex*, and certain grasses. Although affinity with *Cactodera* is uncertain, *C. butulae* attacks *Betulae* and a range of unrelated hosts (Golden and Raski, 1977), and *C. chaubattia* is reported from soil around *Malus* (Gupta and Edward, 1973). Golden (1986) provides a key including most species of *Cactodera*.

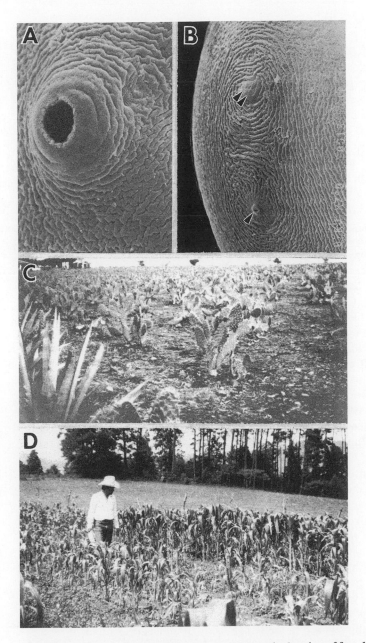

FIGURE 23 *Cactodera* and *Punctodera*. (A) Terminal region of female of *Cactodera cacti* showing cone (SEM). (B) Terminal region of female of *Punctodera chalcoensis*, lacking cone (SEM). Arrowhead indicates vulva; double arrow, anus. (C) Commercially grown cactus in Mexico, commonly infected with *C. cacti*. (D) Field plots in Mexico showing rows of stunted corn infected with *P. chalcoensis*, adjacent to taller controls in soil treated with Nemacure® and Furadan®.

Cactodera cacti (Filipjev and Schuurmans Stekhoven, 1941) Krall and Krall, 1978 (the cactus cyst nematode)

Synonyms and history. The cactus cyst nematode was first described as *H. schachtii* on cacti ([*Phyllocactus* (= *Epiphyllum*), *Cereus* (= *Heliocerus*)] in the Netherlands by Adam (1932) but Filipjev and Schuurmans Stekhoven (1941) redescribed the nematode as *Heterodera cacti*. Goffart (1936) believed the species to be closely related to *H. schachtii* because root exudates of sugar beet, a nonhost, nevertheless "activated" juveniles. Krall and Krall (1978) created the new genus *Cactodera* to include *C. cacti* and several additional species. The cactus cyst nematode may have originated on cacti in the New World but it has become widely distributed on ornamentals throughout the world.

Hosts. The cactus cyst nematode is apparently nearly restricted in host range to succulents of Cactaceae and Euphorbiaceae. As the widespread distribution of this nematode became apparent in the late 1950s a number of reports indicated a wide range of host species within the two families (Kumar, 1964; Southey, 1957; Shmal'ko, 1959). Since then records in the USA indicate a host range of 76 species of Cactaceae and two Euphorbiaceae (Langdon and Esser, 1969). Little is known of the pathogenicity of *C. cacti* on ornamental plants, although it is likely that they reduce plant vigor and interact with root-rotting organisms. They also affect the salability of cacti, since infected plants are often subjected to quarantine regulations targeting cyst nematodes in general.

Distribution. The cactus cyst nematode is probably native of Mexico (Krall and Krall, 1978) and perhaps Central and South America where it occurs widely in native habitats. Although it may be best adapted to arid subtropics, it has been widely dispersed on ornamentals and in cooler climates is commonly found in greenhouses. For example, Shamal'ko (1959) reports the cactus cyst nematode in the botanical garden of the Academy of Sciences in the USSR and indicates that the nematode was introduced on a large collection of succulent plants from foreign countries in 1946–1948. In the western hemisphere it is reported from Canada, Mexico, Colombia, Argentina, and Brazil, and in the USA from Florida, Colorado, Georgia, Minnesota, and New York. *Cactodera cacti* occurs in Europe in the Netherlands, England, France, Germany, Sweden, Switzerland, Austria, Belgium, Yugoslavia, Hungary, Czechoslovakia, and Italy. It is present in other Mediterranean regions of Malta, Algeria, and Israel and extends to the USSR, India, Japan, Korea, and Vietnam (Johnson, 1968; Mulvey and Golden, 1983; Krall and Krall, 1978).

Biology. Although the life history of *C. cacti* is generally similar to that of other cyst nematodes, it has not been investigated in detail. Under greenhouse conditions a generation takes 29–34 days (Shamal'ko, 1959), and in monoxenic cultures at 22°C a generation takes about 30 days (Cordero, 1989). In the Sonoran deserts of Mexico, it is not known if the nematode persists during unfavorable periods of cold or draught as a cyst, although Krall and Krall (1978) speculate that only one generation a year may occur under field conditions. Conversely, roots of long-lived succulents may provide adequate protection to sustain the nematodes throughout the year and the cyst may function primarily for dispersal among widely spaced hosts. Eggs are not deposited in a gelatinous matrix as for many other Heteroderinae; they are fully retained in the female body until death (Cordero, 1989).

The cactus cyst nematode occurs on cacti grown for food as well as on ornamentals (Fig. 23C). Shaml'ko (1959) reports that a number of species of infected ornamental cacti were stunted, had reduced turgor, and were subject to decay by secondary invaders. Langdon and Esser (1969) note that infected commercial plantings of *Schlumbergera* Lem (= *Zygocactus*) were wilted with an abnormal reddish color. In Mexico, cacti (*Opuntia* sp.) are extensively cultivated in large land holdings in several states for food, including fruit and cladodes (Fig. 23C). We recently detected *C. cacti* in commercial plantings in the states of Zacatecas and Durango where cactus is one of the most important crops. Although pathogenicity of the nematode on agricultural *Opuntia* has not been established, it is likely that the nematode contributes to secondary root-rotting fungi and bacteria common in cacti. Similarly, pathogenicity and interactions are likely in commercial plantations of *Amaranth*, a high-protein seed crop of increasing agricultural importance, which is infected with *Cactodera amaranthi*, a species closely related to *C. cacti* and widespread throughout Mexico. In most of the world *C. cacti* is limited to ornamental plants and economic losses may relate to quarantine regulations and markets which reject plants infected with cyst nematodes.

Control. Quarantine regulations against cyst nematodes in general may be effective in containing *C. cacti*. Sanitation is undoubtedly important in control of cactus cyst nematodes, with care not to transport infected plants and soil. Since infection is limited to roots and the collar of the host, clean plants can generally be propagated from seed or vegetative parts in sterilized soil or other clean planting material. Where available, systemic nematicides may provide adequate control in ornamentals, and many fumigant nematicides are likely to be effective for controlling soil stages. Granular nematicides have been used with little success (Schneider, 1961; Shmal'ko, 1959).

Punctodera

Punctodera Mulvey and Stone (1976)
   syn. *Heterodera* Skarbilovich, 1947
Diagnosis: Heteroderinae Filipjev and Schuurmans Stekhoven

Females. Cyst present. Spheroidal, ovoid, or pear-shaped, without posterior cone. Cuticle surface with reticulate pattern, pronounced subsurface pattern of parallel rows of punctations. Punctations most pronounced on cyst. D layer present. Vulva terminal. In cyst, circumfenestrae develops around vulval region, separate fenestrae around anal region. Underbridge and perineal tubercles absent. Bullae present or absent. Eggs retained in body (no egg mass); egg surface smooth.

Males. Body twisted. Lateral field with four lines. Spicules >30 μm, distally with single point.

Second-stage juveniles. Stylet 30 μm. Lateral field with four lines. Esophageal glands filling body cavity. Tail conical with long hyaline terminal portion. Phasmid openings porelike. En face pattern with adjacent submedial lips fused with each other and partially with labial disc.

Nurse cell. Syncytium.

Species. (Table 7)

Identification. Cyst nematodes which are now considered *Punctodera* have been recognized for 60 years, primarily on the basis of a cyst without a cone but with both a vulval circumfenestra and an anal fenestra (Figs. 12D, 23B). Subsurface punctations, while

TABLE 7   Species of *Punctodera*

P. punctata (Thorne, 1928) Mulvey and Stone, 1976

   syn. *Heterodera punctata* Thorne, 1928

P. chalcoensis Stone, Sosa Moss and Mulvey, 1976

P. matadorensis Mulvey and Stone, 1976

---

indicated by the genus name, also occur in other heteroderines, although their intensity and occurrence in parallel rows is striking in *Punctodera* (Fig. 10G). *Punctodera punctata* is distributed worldwide on turf, wheat, and weed grasses and *P. matadorensis* is apparently limited to weed grasses at the type locality in Canada. Only with discovery of a third species, *Punctodera chalcoensis* in Mexico, has the economic menace of the genus and the importance of accurate species identification been understood. *Punctodera chalcoensis* and *P. matadorensis* can be separated from *P. punctata* by the spherical to subspherical shape of the females and cysts in the former vs. more elongate, pear-shaped females and cysts in the latter. Cysts of *P. chalcoensis* differ from those of *P. matadorensis* by small, scattered bullae or no bullae vs. massive and consistently present bullae in *P. matadorensis*. *Punctodera chalcoensis* also differs slightly by the shape of the stylet knobs of the second-stage juveniles, which are flat to slightly concave anteriorly in *P. chalcoensis* vs. strongly concave and anchor-shaped in *P. matadorensis*, and rounded in *P. punctata*. The esophagus of second-stage juveniles of *P. chalcoensis* and *P. punctata* occupies about 30% of the body length whereas in *P. matadorensis* it includes about 50% of the body length. Original diagnoses suggested that juveniles of *P. matadorensis* are larger than those of *P. punctata* but reexamination of types and additional populations indicate that these measurements are not reliable and that the size of juveniles overlaps the two species (Franklin, 1938; Horne, 1966; Mundo-Ocampo and Baldwin, unpublished; Wouts, 1985). Since *P. chalcoensis* is the only species known to reproduce on *Zea mays*, a bioassay may be useful in identification.

Punctodera chalcoensis Stone, Sosa-Moss, and Mulvey, 1976 (the Mexican corn cyst nematode)

  Synonyms and history. The Mexican corn cyst nematode was first observed by Vazquez (1976) in corn (*Zea mays* L.) fields in Huamantla, Tlaxcala, Mexico in the late 1950s, and was considered *H. punctata* (= *P. punctata*; Becerra and Sosa Moss, 1978). In the early 1960s a cyst nematode was observed attacking corn in the Valley of Mexico at Chalco and subsequently in the states of Puebla and Tlaxcala which was also identified as *H. punctata* (Sosa-Moss and Gonzalez, 1973; Vazquez, 1976). Sosa-Moss (1965)believed that the Chalco population of the Mexican corn cyst nematode, unlike *H. punctata*, was limited to corn as a host. In 1976, a new genus, *Punctodera*, was proposed to accommodate *P. punctata* and a new species from Canada, *P. matadorensis* (Mulvey and Stone, 1976). Meanwhile, morphological differences were noted when comparing the Mexican corn cyst nematode with *H. punctata* populations from other countries (Sosa-Moss, 1965; Villanueva, 1974) and with *P. matadorensis* (Stone et al., 1976). These observations led to description of the Mexican corn cyst nematode as a distinct species, *P. chalcoensis*, with the type locality in Chalco, Mexico and *Zea mays* as the type host (Stone et al., 1976).

  Hosts. Although *P. chalcoensis* has been tested on a variety of potential hosts including many Poaceae (Gramineae), reproduction only occurs on cultivated maize (*Zea mays* L.) and a wild type of maize, "teozinte" (*Zea mexicana* [Schrd.] Kuntz); Stone et al., 1976). Tests of five nematode populations on 13 corn hybrids (commonly grown in the USA), eight cultivars (from Mexico), six breeding lines (from CIMMYT in Mexico) did not indicate any resistant *Zea* (unpublished). Although there are informal reports of *P. chalcoensis* on weed hosts including *Bidens*, *Tithonia*, and *Wedelia*, our tests (unpublished) failed to demonstrate nematode reproduction.

  Distribution. Punctodera chalcoensis is only reported from Mexico, where it is believed to be indigenous based on its narrow host range on *Zea*, a genus which is historically Mexican (Stone et al., 1976). Earlier reports of wide distribution in Mexico often have not been recognized because identifications confused *P. chalcoensis* with *P. punctata*. However, recent surveys indicate broad distribution at elevations between 1500 and 2800 m throughout major corn-growing areas of Central Mexico from the Pacific Ocean to the Gulf of Mexico. These states include Mexico, Tlaxcala, Puebla, Veracruz, Michoacan, and Jalisco (Mundo et al., 1987; Santacruz, 1982; Vazquez, 1976). The temperate conditions which favor development of this nematode and its pathogenicity on corn, an important crop worldwide, may indicate potential for this nematode to become a widespread economic threat.

Biology. Becerra and Sosa Moss (1978) and Munoz (1983) concluded that the life cycle of *P. chalcoensis* spans the growing season of corn, so that only one generation occurs per year and that the cysts must remain in the soil during winter to initiate the next cycle. However, Jeronimo (1988) carried out a particularly thorough study and reported a typical heteroderine life cycle of about 30 days. Apparently fresh eggs may hatch readily, but once diapause is established within cysts, only a low percentage of hatch occurs each season. Villanueva (1974) reported that root exudates of most plants induce egg hatching of the Mexican populations, but that exudates from corn roots resulted in a significantly greater degree of hatching.

The environmental requirements of *P. chalcoensis* are unknown although the distribution at high elevations in Mexico, and its absence in corn fields in warmer subtropical regions suggests that it does not tolerate continuous warm, humid conditions.

Punctodera chalcoensis is highly damaging on corn in Mexico, causing severe yellowing, stunting, and even death of young seedlings (Fig. 23D). Sosa-Moss and Gonzalez (1973) observed that plants without supplementary fertilization showed more damage and less tolerance to the nematode. However, low inoculum levels (10 cysts/kg soil) induced a physiological effect that increased tolerance to further nematode attack. Greenhouse tests conducted by Hernandez (1965) suggest that improved varieties of corn were more tolerant to different levels of inoculum of the corn cyst nematode than local varieties.

Control. Too little is known of the biology of *P. chalcoensis* to develop precise control procedures. Whereas some corn cultivars may be tolerant of the cyst nematode, none are known to be resistant or to reduce population levels. Many farmers in Mexico recognize that continuous planting to corn results in reduced productivity and fallow 1 out of 3 years is a common practice which might reduce cyst populations. Although subsistence farmers have few acceptable crops for rotation with corn, large farms may be able to reduce populations by rotation with other grains such as wheat or barley. The persistence of the Mexican corn cyst nematode in the absence of a suitable host is not known. In some cases chemical control may be feasible and good control has been achieved with Furadan and Nemacure (Jeronimo, 1988). Becerra and Sosa Moss (1978) noted that early sowing prior to cool rains may allow the seedlings to get a foothold in advance of nematode hatch and thus result in less damage. Damage is minimized by good cultural practices including adequate fertilization and water.

Since *P. chalcoensis* is not likely to be dispersed with seed or harvested corn, containment is more feasible than for cyst nematodes of root crops and root propagules. Nevertheless, sanitation with respect to farm vehicles, water, and transport of contaminated soil is an important aspect of containment.

Genera of Lesser Economic Importance

Heterodera, Cactodera, Globodera, and *Punctodera* are the most widely recognized Heteroderinae, with more than 70% of the known species of the subfamily. These four genera are cyst forming and include nearly all species of economic importance. With the exception of *Afenestrata* and *Dolichodera,* the remaining 13 genera of the subfamily do not form cysts. (Table 1). The importance of these 13 genera, while not directly apparent as *Heterodera* or *Globodera,* cannot be overlooked. Some lesser known, non-cyst-forming genera, including species of *Meloidodera* and *Thecavermiculatus,* have potential to attack agricultural crops. The history of agriculture suggests that indigenous nematodes may adapt from weed hosts to become important agricultural pests as new crops are introduced and agriculture extends to previously uncultivated areas. Furthermore, nematodes of restricted distribution and little economic importance may become a threat when introduced to new habitats. Some heroderine genera restricted to natural ecosystems may have a significant impact on habitats such as forests, e.g., species of *Rhizonema, Sarisodera, Hylonema, Cryphodera,* and *Meloidodera* attack trees. In light of the physiological effects of more thoroughly studied heteroderines, it is plausible that these less familiar parasites of trees increase vulnerability of their hosts to stresses such as pollution, drought, and other pathogens.

Knowledge of all genera of Heteroderinae is essential to understanding evolution of the subfamily as a basis for stable, predictive taxonomy. Lesser known genera from natural habitats are particularly important to such investigations, in part because their biogeography, being little affected by man, often can be attributed to plate tectonics, which is useful for testing hypotheses of phylogeny (Ferris, 1979). Evaluation of processes such as diapause, cyst formation, and induction of nurse cells by heteroderines may contribute to understanding basic biological phenomena. However, these processes can best be understood in the context of evolution and variation within the subfamily as a whole. For example, insight into cyst formation may be gained through considering a transformation series including heteroderines that do not form cysts, as well as intermediates. Similarly, the process of induction of nurse cells may best be understood by considering not only syncytia of *Heterodera* and *Globodera,* but ancestral genera which do not induce syncytia, as well as genera which induce diminutive syncytia. Lesser known heteroderines are considered in four groups of genera to facilitate comparison. The groups do not fully represent taxonomic categories, but are largely based on the phylogenetic analysis of Baldwin and Schouest (1990): (a) cyst-forming genera include *Afenestrata* and *Dolichodera*; (b) ataloderines are *Atalodera, Thecavermiculatus,* and *Camelodera*; (c) sarisoderines are *Sarisodera, Rhizonema, Bellodera, Hylonema,* and *Ekphymatodera*; (d) ancestral species include *Verutus, Meloidodera,* and *Cryphodera.*

Cyst-forming

Afenestrata Baldwin and Bell, 1985
  syn. *Afrodera* Wouts, 1985
      Dolichodera Mulvey and Ebsary, 1980

Afenestrata africana and *D. fulvialis* are the only species of these cyst-forming genera; neither is reported outside their localities, respectively, in Ivory Coast and Quebec, Canada, and both are probably parasites of grasses. *Afenestrata* is similar to *Heterodera,* the cyst having a pronounced cone and zig-zag cuticular pattern, but it is distinguished from other cyst nematodes in that it lacks fenestration (Fig. 24C, D). *Dolichodera* is similar to *Globodera,* the cyst lacking a cone and having circumfenestrae (Fig. 24A, B). Unlike

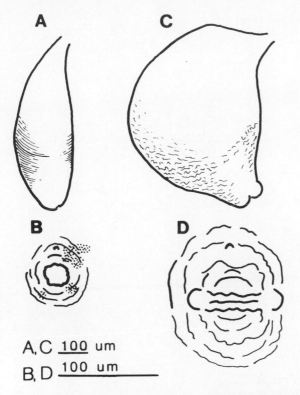

A, C 100 um

B, D 100 um

FIGURE 24   Characteristics for identification of *Dolichodera* and *Afenestrata*. (A) *Dolichodera fluvialis* cyst. (B) *D. fluvialis* cyst, terminal region showing circumfenestrae. (C) *Afenestrata africana* cyst. (D) *A. africana,* terminal region lacks fenestrae. (A and B redrawn after Mulvey and Ebsary, 1980.)

Globodera, the female body wall cuticle of *Dolichodera* is thin with fine striations. *Dolichodera* is unusual among heteroderines in that it was recovered from bottom sediment of a river; it is likely that the cysts were washed from plants on the shore. Luc et al. (1988) review the history and synonyms of *Afenestrata* and *Dolichodera* and include a complete diagnosis.

Ataloderines

Atalodera Wouts and Sher, 1971
   syn. *Sherodera* Wouts, 1974
       Thecavermiculatus Robbins, 1978
       Camelodera Krall, Sagalina, and Ivanova, 1988

Non-cyst-forming *Atalodera* and *Thecavermiculatus* are similar in that they both have second-stage juveniles with very narrow esophageal glands (Fig. 25A, B), unlike other heteroderines. Contrary to most other non-cyst-forming heteroderines, the nurse cell induced by *Atalodera* and *Thecavermiculatus* is a syncytium. The syncytium is unlike that of most cyst-forming heteroderines, however, because it lacks wall ingrowths (Mundo and Baldwin, 1983). Females of *Atalodera* have a pronounced posterior cone whereas those of *Thecavermiculatus* have a greatly reduced cone (Fig. 25D, E; 26A). *Atalodera* includes four

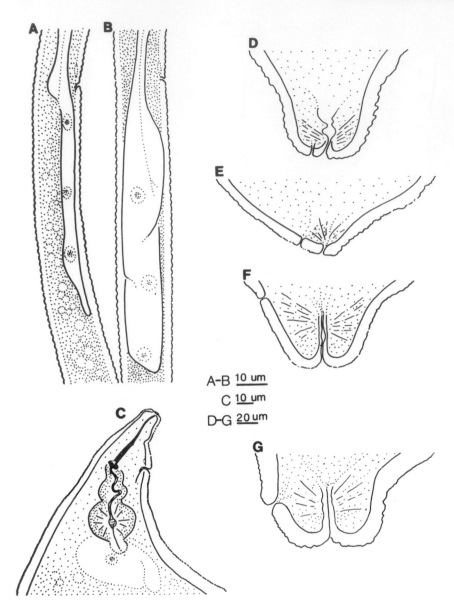

A–B <u>10 um</u>
C <u>10 um</u>
D–G <u>20 um</u>

FIGURE 25   Characteristics for identification of ataloderines and sarisoderines. (A) Narrow esophageal gland region of second-stage juveniles (J2) typical of ataloderines. (B) Broad esophageal gland region of J2, typical of sarisoderines. (C) Head region of female of *Bellodera utahensis* showing anterior position of excretory pore, typical of this genus. (Redrawn after Baldwin et al., 1983.) (D) *Atalodera ucri*, lateral view of terminal region of the female. (E) *Thecavermiculatus gracililancea*, lateral view of terminal region of female. (F) *Camellodera eremophila*, lateral view of terminal region of female. (Redrawn from Krall et al., 1988.) (G) *B. utahensis*, lateral view of terminal region of female.

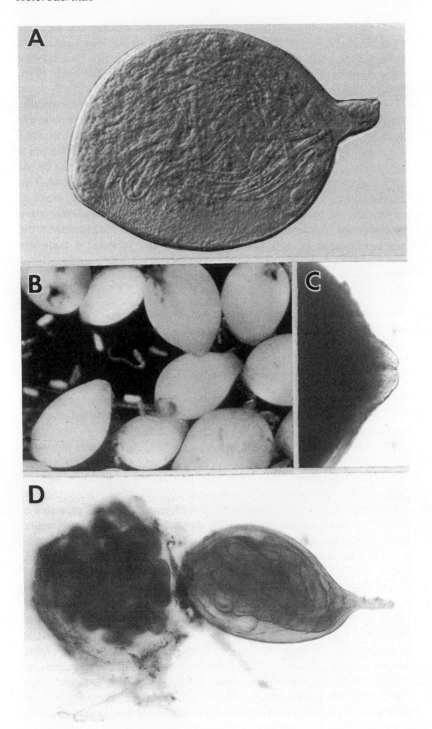

FIGURE 26 Females of ataloderines and sarisoderines. (A) *Thecavermiculatus gracililancea*, female full of second-stage juveniles. (B) *Sarisodera hydrophila* females. (C) *S. hydrophila* female, terminal cone region. (D) *Ekphymatodera thomasoni* with egg mass.

species, all described form California, and most species are only known from one or a few localities (Baldwin et al., 1989; Wouts, 1973c; Wouts and Sher, 1971). Hosts are native plants including grasses, sage (*Haplopappus*), and honeysuckle (*Lonicerea*). Four species of *Thecavermiculatus* have been described. These occur in California (Robbins, 1978, 1986), Alaska (Bernard, 1981), and Peru (Golden et al., 1983). The species are all from grasses with the exception of *T. andinus* from Peru, which is described from *Oxalis*. However, the taxonomic position of *T. andinus* is controversial (Wouts, 1985; Luc et al., 1988) and may prove to be a separate genus, with many characteristics ancestral to other ataloderines (Baldwin and Schouest, 1990). *Camelodera* was recently described from a woody shrub in the deserts of Turkmenian, USSR and closely resembles *T. andinus* by the relatively narrow esophagus in second-stage juveniles, the broad vulval–anal distance (Fig. 25F), and obscure surface pattern of females. A population of *Thecavermiculatus*, which is an undescribed new species, was recovered from cultivated potato in Alaska and may prove to be agriculturally important (Bernard, pers. commun.). Luc et al. (1988) reviewed the history and synonyms of *Atalodera* and *Thecavermiculatus* and included a complete diagnosis.

Sarisoderines

Sarisodera Wouts and Sher, 1971
Rhizonema Cid Del Prado Vera, Lownsbery, and Maggenti, 1983
Bellodera Wouts, 1985
Hylonema Luc, Taylor, and Cadet, 1978
Ekphymatodera Baldwin, Bernard, and Mundo, 1988

Sarisoderines are a diverse group of non-cyst-forming heteroderines which are distinct form other heteroderines by the lip patterns of juveniles and males as viewed with the scanning electron microscope; in each case the submedial lips are very thin in relation to the labial disc (Fig. 13I) (Othman and Baldwin, 1986; Baldwin et al., 1989; Baldwin and Schouest, 1990). Each genus is monospecific and all the species have a terminal cone (Fig. 26B, C). All the genera, except *Hylonema* described from Ivory Coast, are known only in the western USA. Contrary to ataloderines, all (with the possible exception of *Ekphymatodera*) induce a single uninucleate giant cell. *Sarisodera*, *Rhizonema*, and *Hylonema* are all parasites of trees, *Bellodera* occurs on a woody shrub, whereas *Ekphymatodera* is a parasite of rushes (Fig. 26D). An undescribed sarisoderine genus which is a parasite of trees in Brazil (Hirschmann, pers. commun.) suggests that these nematodes are more widely distributed than was previously indicated. *Sarisodera* is distinctive by the zig-zag cuticle pattern and very close vulval–anal distance, with the anus located on the inner lip of the vulva. *Rhizonema* and *Bellodera* both have annulated cuticles in the females, but the vulval–anal distance in the former is 15–33 μm, whereas in *Bellodera* it is 53–88 μm. Unlike other Heteroderinae in which the excretory pore is near the level of the esophageal glands, in *Bellodera* the opening is near the level of the stylet knobs (Fig. 25C). Unlike other sarisoderines, *Hylonema* and *Ekphymatodera* both have lip patterns in males and juveniles, in which the submedial lips are fused with the labial disc (Othman and Baldwin, 1986; Baldwin et al., 1989). Also the hyaline portion of the tail of second-stage juveniles is unusually long in both genera. The body wall pattern of females of *Ekphymatodera* is full of uneven protuberances (Fig. 10E), whereas the cuticle of *Hylonema* is relatively smooth or faintly striated. Luc et al. (1988) reviewed the history and synonyms of sarisoderines and included a complete diagnosis of each except *Ekphymatodera* (Baldwin et al., 1989).

Ancestral heteroderines

Cryphodera Colbran, 1966
  syn. *Zelandodera* Wouts, 1973
Verutus Esser, 1981
Meloidodera Chitwood, Hannon, and Esser, 1956

Although *Cryphodera*, *Verutus*, and *Meloidodera* apparently do not form a cohesive taxonomic group (Baldwin and Schouest, 1989), they nevertheless are considered together because they share a number of ancestral similarities. All have striated cuticles and lack a distinct posterior protuberance. *Verutus* and *Meloidodera* have a subequatorial vulva and terminal anus (Figs. 6; 14A); in *Cryphodera* the vulva and anus are also widely separated. The four species of *Cryphodera* are all from trees in Australia and New Zealand. The uniquely large vulval lips and elongate-reniform shape of females of *Verutus* readily distinguish it from *Meloidodera* with smaller vulval lips and more ovoid females. Two species of *Verutus* occur in the USA, one in Japan, and a fourth undescribed species is present in West Germany; apparently the genus favors marshy habitats and grass or herbaceous hosts (Baldwin et al., 1989). *Meloidodera*, a diverse genus with six species, is widely distributed in the USSR as well as the United States, including Alaska. Most are parasites of woody plants including forest trees. However, *M. eurytyla* from Alaska parasitizes dunegrass, and *M. charis*, while often found on woody desert hosts such as *Mesquite*, also parasitizes a number of herbaceous plants including corn (*Zea mays*), sorghum, okra (*Abelmoschus esculentus*), and turf grass (Heald and Golden, 1969; Heald, 1984). *Meloidodera charis* is apparently widely distributed on field corn in Nebraska, where in recent years it appears to be pathogenic (Minton and Golden, 1971; Dappen and Powers, pers. commun.). Luc et al. (1988) reviewed the history and synonyms of *Cryphodera*, *Verutus*, and *Meloidodera* and included a complete diagnosis of each.

KEY TO GENERA OF HETERODERINAE

1.	Vulval subequatorial	2
	Vulva terminal	3
2.	Female reniform	*Verutus*
	Female not reniform	*Meloidodera*
3.	Cyst absent; eggs not retained in body of dead female	4
	Cyst present; at least some eggs retained in tanned body of dead female	13
4.	Vulval anal distance 15–35 μm	5
	Vulval anal distance greater than 35 μm	10
5.	Esophageal gland lobe of J2 nearly fills diameter of body cavity	6
	Esophageal gland lobe of J2 narrow; about one-third body diameter	7
6.	Female with rounded terminus	*Hylonema*
	Female with distinct cone	8

7.	Little or no terminal cone	*Thecav-miculatus*
	Distinct terminal cone	*Atalodera*
8.	Cuticle pattern of female striated at midbody	*Rhizonema*
	Cuticle pattern of female not striated at midbody	9
9.	Cuticle pattern of female zig-zag	*Sarisodera*
	Cuticle pattern of female not zig-zag but with rough surface and longitudinal furrows	*Ekphy-matodera*
10.	Female cuticle with prominent striae at midbody	11
	Female cuticle without prominent striae at midbody	12
11.	Mature female lemon-shaped; prominent vulval cone	*Bellodera*
	Mature female nearly spherical little or no vulval cone	*Cryphodera*
12.	Mature female nearly spherical, little or no terminal prominence	*Thecaver-miculatus andinus*
	Mature female with distinct terminal prominence	*Camelodera*
13.	Cyst lacking fenestration of vulval region	*Afenestrata*
	Cyst with fenestration of vulval region	14
14.	Cyst with fenestration of anal region	*Punctodera*
	Cyst lacking fenestration of anal region	15
15.	Cyst with two semifenestrae in vulval region	*Heterodera*
	Cyst with one fenestrae (circumfenestrae) in vulval region	16
16.	Terminal cone of female prominent to slightly reduced	*Cactodera*
	Terminal cone of female absent, rounded terminus	17
17.	Cuticle of female zig-zag	*Globodera*
	Cuticle of female striated	*Dolichodera*

ACKNOWLEDGMENTS

The authors especially thank G. E. Dappen for his valuable comments.

REFERENCES

Adam, W. 1932. Note sur *Heterodera schachtii* Schm. parasite des cactus. *Bull. Mus. Rol. Hist. Nat. Belg.* 8: 1–10.

Adams, H. S., Osborne, W. W., and Webber, Jr., A. J. 1982. Effect of temperature on development and reproduction of *Globodera solanacearum*. *Nematropica* 12: 305–311.

Alston, D. G., and Schmitt, D. P. 1988. Development of *Heterodera glycines* life stages as influenced by temperature. *J. Nematol.*, 20: 366–372.

Andersen, S. 1961. Resistens mod havreal, *Heterodera avenae*. Med. Kgl. Vet. Landbohjsk. No. 68.

Andersen, S., and Andersen, K. 1982. Suggestions for determination and terminology of pathotypes and genes for resistance in cyst-forming nematodes, especially *Heterodera avenae*. *OEPP/ EPPO Bull.* 12: 379–386.

Andersen, K., and Andersen, S. 1986. *Heterodera avenae*: Virulence and resistance. In *Cyst Nematodes*. Lamberti, F., and Taylor, C. E., eds. Plenum Press, New York, pp. 277–285.

Andersson, S. 1982. Population dynamics and control of *Heterodera avenae*, a review with some original results. *OEPP/EPPO Bull.*, 12: 463–475.

Atkinson, H. J., and Ballantyne, A. J. 1977a. Changes in the oxygen consumption of cysts of *Globodera rostochiensis* associated with the hatching of juveniles. *Ann. Appl. Biol.*, 87: 159–166.

Atkinson, H. J., and Ballantyne, A. J. 1977b. Changes in the adenine nucleotide content of cysts of *Globodera rostocheiensis* associated with the hatching of juveniles. *Ann. Appl. Biol.*, 87: 167–174.

Aubert, V., 1986. Hatching of the carrot cyst nematode. In *Cyst Nematodes*. Lamberti, F., and Taylor, C. E., eds. Plenum Press, New York, pp. 347–353.

Awan, F. A., and Hominick, W. H. 1982. Observation on tanning of the potato cyst nematode, *Globodera rostochiensis*. *Parasitology*, 85: 61–71.

Baalawy, Halima A., and Fox, J. A. 1971. Resistance to Osborne's cyst nematode in selected *Nicotiana* species. *J. Nematol.*, 3: 395–398.

Bakker, J., and Gommers, F. J. 1986. Genotyping European potato cyst nematode populations with two dimensional gel electrophoresis of total protein extracts (abstract). *Rev. Nematol.*, 9: 287.

Balandras, C. M. 1986. Limits to methods for management of potato cyst nematodes in western France. In *Cyst Nematodes*. Lamberti, F., and Taylor, C. E., eds. Plenum Press, New York, pp. 229–234.

Baldwin, J. G., 1982. Fine structure of the esophagus of *Sarisodera hydrophila* (Heteroderoidea). *J. Nematol.*, 14: 279–291.

Baldwin, J. G. 1985. Fine structure of the phasmid of second-stage juveniles of *Heterodera schachtii* Tylenchida: Nematoda). *Can. J. Zool.* 63: 534–542.

Baldwin, J. G. 1986. Testing hypotheses of phylogeny of Heteroderidae. In *Cyst Nematodes*. Lamberti, F., and Taylor, C. E., eds. Plenum Press, New York, pp. 75–99.

Baldwin, J. G., and Bell, A. H. 1985. *Cactodera eremica* n. sp., *Afenestrata africana* (Luc et al., 1973) n. gen., n. comb., and an emended diagnosis of *Sarisodera* Wouts and Sher, 1971 (Heteroderidae). *J. Nematol.*, 17:187–201.

Baldwin, J. G., and Hirschmann, H. 1973. Fine structure of cephalic sense organs in *Meloidogyne incognita* males, *J. Nematol.*, 5: 285–302.

Baldwin, J. G., and Hirschmann, H. 1975. Fine structure of cephalic sense organs in *Heterodera glycines* males. *J. Nematol.*, 7: 40–53.

Baldwin, J. G., and Hirschmann, H. 1976. Comparative fine structure of the stomatal region of males of *Meloidogyne incognita* and *Heterodera glycines*. *J. Nematol.*, 8: 1–17.

Baldwin, J. G., and Schouest, L. 1990. Comparative detailed morphology of Heteroderinae Filip'ev and Schuurmans Stekhoven, 1941 sensu Luc et al., 1988 for phylogenetic systematics and revised classification. *Systematic Parasitol.* 15: 81–106.

Baldwin, J. G., Bernard, E. C., and Mundo-Ocampo, M. 1989. Four new species of Heteroderidae including *Ekphymatodera* n. gen. from California, *J. Nematol.*, 21: 48–68.

Baldwin, J. G., Hirschmann, H., and Triantaphyllou, A. C. 1977. Comparative fine structure of the esophagus of males of *Heterodera glycines* and *Meloidogyne incognita. Nematologica* 23: 239–252.

Barker, K. R., and Lucas, G. B. 1984. Nematode parasites of tobacco. In *Plant and Insect Nematodes*. Nickle, W. R., ed. Marcel Dekker, New York, pp. 212–242.

Beane, J., and Perry, R. N. 1983. Hatching of the cyst nematode *Heterodera goettingiana* in response to root diffusate from bean (*Vicia faba*). *Nematologica*, 29: 360–362.

Beane, J., and Perry, R. N. 1984. Observations on the invasion and development of the pea cyst -nematode *Heterodera goettingiana* in winter field beans (*Vicia faba*). *Ann Appl. Biol.*, 105: 503–508.

Becerra, L. E. N., and Sosa-Moss, C. 1978. Relacion del Nematodo enquistado del maiz *Punctodera chalcoensis* (Stone, Sosa-Moss y Mulvey) con otros microorganismos fitopatogenos del suelo. Tesis de Maestria en ciencias agricolas del primer autor. Colegio de Postgraduados. Chapingo, Mexico.

Behrens, E. 1975. *Globodera* Skarbilovic, 1959. Eine eelbatandige Gattung in der Unterfamilie Heteroderinae Skarbilovic, 1947 (Nematoda: Heteroderidae). *Vortragstagung zu Aktuellen Problemem der Phytonematologie* 29: 12–26.

Bernard, E. C. 1981. Three new species of Heteroderoidea (Nematoda) from the Aleutian Islands. *J. Nematol.*, 13: 499–513.

Bhargava, S., and Yadav, B. S. 1978. Host range study and evaluation of certain barley varieties to the maize cyst nematode, *Heterodera zeae* (abstract). *Ind. J. Mycol. Plant Pathol.* 8: 72.

Bhatti, D. S., Hirschmann, H., and Sassar, J. N. 1972. Post-infection development of *Heterodera lespedezae, J. Nematol.*, 4: 104–112.

Bossis, M. 1986. Observations on the population dynamics and control of *Heterodera carotae* in Western France. In *Cyst Nematodes*. Lamberti, F., and Taylor, C. E., eds. Plenum Press, New York, pp. 349–353.

Bossis, M., and Mugniery, D. 1988. *Heterodera carotae* Jones, 1950. 2. Dynamique des populations dan l'ouest de la France. *Rev. Nematol.* 11: 315–320.

Bovien, P. 1953. On havrealen (*Heterodera major*) of resultanterne af nogleforsog pa smittet jord. *Tidsskr. Planteavl*, 56: 581–591.

Brown, E. B. 1958. Pea root eelworm in the eastern countries of England. *Nematologica*, 3: 257–268.

Brown, R. H. 1982a. Cultural practices and their effects on *Heterodera avenae* and grain yields of wheat in Victoria, Australia. *OEPP/EPPO Bull.* 12: 477–484.

Brown, R. H. 1982b. Studies on the Australian pathotype of *Heterodera avenae. OEPP/EPPO Bull.* 12: 413–412.

Brown, R. H. 1984. Cereal cyst nematide and its chemical control in Australia. *Plant Dis.* 68: 922–928.

Brown, G., Callow, R. K., Green, C. D., Jones, F. G. W., Rayner, J. H., Shepherd, A. M., and Williams, T. D. 1971. The structure, composition and origin of the subcrystalline layer in some species of the genus *Heterodera, Nematologica* 17: 591–599.

Brzeski, M. W. 1969. Nematodes associated with cabbage in Poland. *Ekol. Pol. A.* 17: 227–240.

Burrows, P. R. and Perry, R. N. 1988. Two cloned DNA fragments which differentiate *Globodera pallida* from *G. rostochiensis. Rev. Nematol.* 11: 441–445.

Campos-Vela, A. 1967. Taxonomy, life cycle and host range of *Heterodera mexicana*. n. sp. (Nematode: Heteroderidae). Ph.D. thesis, University of Wisconsin.

Canto Saenz, M., and de Scurrah, M. M. 1977. Races of the potato cyst nematode in the Andean region and a new system of classification. *Nematologica* 23: 340–349.

Carlsson, B. and Videgard, G. 1971. Sanering av potatiscystnematoden med hjalp av fangstgroda. *Potatis* 31–32.

Carta, L. K., and Baldwin, J. G. 1990a. Ultrastructure of phasmid development in *Meloidodera floridensis* and *M. charis* (Heteroderinae). *J. Nematol.* 22:250–283.

Carta, L. K., and Baldwin, J. G. 1990b. Phylogenetic implications of phasmid absence in males of three genera of Heteroderinae. *J. Nematol.* 22:284–301.

Cliff, G. M., and Baldwin, J. G. 1985. Fine structure of body wall cuticle of females of eight genera of Heteroderidae. *J. Nematol.* 17: 286–296.

Chitwood, B. G. 1949. Root-knot nematodes. I. A revision of the genus *Meloidogyne* Goeldi, 1887. *Proc. Helm. Soc. Wash.* 16: 90–104.

Chitwood, B. G., Hannon, C. J., and Esser, R. P. 1956. A new nematode genus *Meloidodera*, linking the genera *Heterodera* and *Meloidogyne*. *Phytopathology* 46: 264–266.

Clark, A. J., and Hennessy, J. 1987. Hatching agents as stimulants of movement of *Globodera rostochiensis* juveniles. *Rev. Nematol.* 10: 471–476.

Clarke, A. J., and Perry, R. N. 1977. Hatching of cyst-nematodes. *Nematologica* 23: 350–368.

Clark, A. J., and Perry, R. N. 1985. Egg-shell calcium and the hatching of *Globodera rostochiensis*. *Int. J. Parasitol.* 15: 511–516.

Clark, S. A., Shepherd, A. M., and Kempton, A. 1973. Spicule structure in some *Heterodera* spp. *Nematologica* 19: 242–247.

Cordero, D. A. 1989. Comparative morphology and development of the cone of *Heterodera schachtii* and *Cactodera cacti*. Ph.D. thesis, University of California, Riverside.

Dalmasso, A., Dedryver-Person, F., and Thomas, D. 1982. Polymorphisme génétique chez *Heterodera avenae*. *OEPP/EPPO Bull.* 12: 349–352.

D'Herde, J. 1966. Het erwtencystenaaltje *Heterodera goettingiana* een nieuw parasiet van de erwtenteelt voor Belgie. *Landen Tuinb Jaarb* 1965–1966, 427–430.

De Segurea, C. B. 1952. The golden nematode in Peru. *Plant Dis. Rep.* 36: 253.

Di Vito, M., and Greco, N. 1986. The pea cyst nematode. In *Cyst Nematodes*. Lamberti, F., and Taylor, C. E., eds. Plenum Press, New York, pp. 321–332.

Di Vito, M., and Perrino, P. 1978. Reazione di *Pisum* spp. agli attacchi di *Heterodera goettingiana*. *Nematol. Medit.* 6: 113–118.

Di Vito, M., Greco, N., and Lamberti, F. 1980. Comportamento di poplazioni di *Heterodera goettingiana* su specie diverse di leguminose. *Inf. Fitopaatol.* 30: 7–10.

Dijkstra, J. 1971. Breeding for resistance to *Heterodera* in white clover. *Euphytica* 20: 36–46.

Doncaster, C. C., and Shepherd, A. M. 1967. The behaviour of second-stage *Heterodera rostochiensis* larvae leading to their emergence from the egg. *Nematologica* 13: 476–478.

Doncaster, C. C., and Seymour, M. K. 1973. Exploration and selection of penetration site by Tylenchida. *Nematologia* 19: 137–145.

Ellelnby, C. 1946a. Nature of the cyst wall of the potato-root eelworm *Heterodera rostochiensis*, Wollenweber, and its permeability to water. *Nature* 157: 302–303.

Ellenby, C. 1946b. Ecology of the eelworm cyst. *Nature* 157: 451–452.

Ellenby, C. 1955. The permeability to the hatching factor of the cyst wall of the potato-root eelworm, *Heterodera rostochiensis* Wollenweber. *Ann. Appl. Biol.* 43: 12–18.

Ellenby, C., and Perry, R. N. 1976. The influence of the hatching factor on the water uptake of the second stage larva of the potato cyst nematode *Heterodera rostochiensis*. *J. Exp. Biol.* 64: 141–147.

Elliott, A. P., Phipps, P. M., and Terrill, R. 1986. Effects of continuous cropping of resistant and susceptible cultivars on reproduction potentials of *Heterodera glycines* and *Globodera tabacum solanacearum*. *J. Nematol.* 18: 375–379.

Endo, B. Y. 1980. Ultrastructure of the anterior neurosensory organs of the larvae of the soybean cyst nematode, *Heterodera glycines*. *J. Ultrastruct. Res.* 72: 349–366.

Endo, B. Y. 1984. Ultrastructure of the esophagus of larvae of the soybean cyst nematode *Heterodera glycines*. *Proc. Helm. Soc. Wash.* 51: 1–24.

Endo, B. Y. 1987. Ultrastructure of esophageal gland secretory granules in juveniles of *Heterodera glycines*. *J. Nematol.* 19: 469–483.

Endo, B. Y. 1988. Ultrastructure of the intestine of second and third juvenile stages of the soybean cyst nematode, *Heterodera glycines*. *Proc. Helm. Soc. Wash.* 55: 117–131.

Ennik, G. C., Kort, J., and Luesink, B. 1964. The influence of soil disinfection with DD, certain components of DD and some other compounds with nematicidal activity on the growth of white clover. *Neth. H. Pl. Pathol.* 70: 117–135.

Epps, J. M. 1971. Recovery of soybean cyst nematodes (*Heterodera glycines*) from the digestive tracts of blackbirds. *J. Nematol.* 3:417–419.

Esser, R. P., and Langdon, K. R. 1967. Soybean cyst nematode. *Florida Depart. Agr. Div. Plant Ind.* 8: 1–2.

Evans, K. 1970. Longevity of males and fertilization of *Heterodera rostochiensis. Nematologica* 16: 369–374.

Evans, K., and Brodie, B. B. 1980. The origin and distribution of the golden nematode *Globodera rostochiensis* and its potential in the USA. *Am. Potato J.* 57: 79–89.

Evans, K., and Stone, A. R. 1977. A review of the distribution and biology of the potato cyst nematodes *Globodera rostochiensis* and *G. pallida. Pest Artic News. Summ.* 23:178–189.

Evans, K., Franco, J., and de Scurrah, M. M. 1975. Distribution of species of potato cyst nematodes in South America. *Nematologica* 21: 365–369.

Fagbenle, H. H., Edwards, D. I., and Malek, R. B. 1986. Cyst production by two geographical isolates of *Heterodera lespedezae* on selected lelgumes. *Plant Dis.* 70: 643–644.

Faghihi, J., Ferris, J. M., and Ferris, V. R. 1986a. *Heterodera glycines* in Indiana: I. Reproduction of geographic isolates on soybean differentials. *J. Nematol.* 18: 169–172.

Faghihi, J., Ferris, J. M., and Ferris, V. R. 1986b. *Heterodera glycines* in Indiana: II. Morphology of geographic isolates. *J. Nematol.* 18: 173–177.

Faulkner, L. R., and Bolander, W. J. 1966. Occurrence of large nematode populations in irrigation canals of south central Washington. *Nematologica* 12: 591.

Ferris, V. R. 1979. Cladistic approaches in the study of soil and plant parasitic nematodes. *Am. Zool.* 19: 1195–1215.

Ferris, V. R. 1985. Evolution and biogeography of cyst-forming nematodes. *OEPP/EPPO Bull.* 15:123–129.

Ferris, V. R., Ferris, J. M., and Murdock, L. L. 1985. Two–dimensional protein patterns in *Heterodera glycines. J. Nematol.* 17: 422–427.

Ferris, V. R., Ferris, J. M., Murdock, L. L., and Faghihi, J. 1986. *Heterodera glycines* in Indiana. III. 2-D protein patterns of geographical isolates. *J. Nematol.* 18: 177–182.

Filipjev, I. N. 1934. (Harmful and beneficial nematodes in agriculture) (In Russian) Moscow and Leningrad.

Filipjev, I. N., and Schuurmans Stekhoven, Jr., J. H. 1941. *A Manual of Agricultural Helminthology.* E. J. Brill, Leiden.

Fox, P. C., and Atkinson, H. J. 1985a. Enzyme variation in pathotypes of the potato cyst nematodes *Globodera rostochiensis* and *G. pallida. Parasitology* 91: 499–506.

Fox, P. C., and Atkinson, H. J. 1985b. Immunochemical studies on pathotypes of the potato cyst nematode, *Globodera rostochiensis* and *G. pallida. Parasitology* 90: 471–483.

Franklin, M. T. 1938. On the occurrence of *Heterodera* cysts in various soil and on the roots of *Agrostis stolonifera L. J. Helminthol.* 16: 5–16.

Franklin, M. T. 1939. On the structure of the cyst wall of *Heterodera schachtii* (Schmidt). *J. Helminthol.* 17: 127–134.

Franklin, M. T. 1940. On the specific status of the so-called biological strains of *Heterodera schachtii* Schmidt. *J. Helminthol.* 18: 193–208.

Franklin, M. T. 1945. On *Heterodera cruciferae* n. sp. of brassicas, and on a *Heterodera* strain infecting clover and dock. *J. Helminthol.* 21: 71–84.

Franklin, M. T. 1951. The cyst-forming species of *Heterodera*. Commonwealth Agricultural Bureaux, Bucks, England.

Franklin, M. T. 1957. Note on the nomenclature of the cereal root eelworm. *Nematologica* 2: 149–151.

Franklin, M. T. 1972. *Heterodera schachtii. C. I. H. Descriptions of Plant–Parasitic Nematodes* Set 1, No. 2. St. Albans, Herts, England.

Franklin, M. T., Thorne, G., and Oostenbrink, M. 1959. Proposal to stabilize the scientific name of the cereal-root eelworm (Class Nematoda). *Bull. Zool. Nomencl.* 17: 76–85.

Fujita, K., and Miura, O. 1934. On the parasitism of *Heterodera schachtii* Schmidt on beans. *Trans. Sapporo Nat. Hist. Soc.* 13: 359–364.

Geraert, E., and DeGrisse, A. 1981. The male copulatory system in tylenchid taxonomy (Nematode). *Nematologica* 27: 432–442.

Goffart, H. 1936. *Heterodera schachtii* Schmidt an gemeiner Hanfnessel (*Galesopsis tetrahit* L.) und an Kakteen. *Z. Parasitenkd.* 8: 528–532.

Goffart, H. 1941. Der Göttinger Erbsennematode (*Heterodera göettingiana*), ein Ruckblick auf eine 50 jährige Entwicklung. *Zentralbl. Bact.* 104: 8–87.

Golden, A. M. 1986. Morphology and identification of cyst nematodes. In *Cyst Nematodes*. Lamberti, F., and Taylor, C. E., eds. Plenum Press, New York, pp. 75–99.

Golden, A. M., and Cobb, G. S. 1963. *Heterodera lespedezae* (Heteroderidae), a new species of cyst forming nematode. *Proc. Helm. Soc. Wash.* 30: 281–286.

Golden, A. M., and Ellington, D. S. 1972. Redescription of *Heterodera rostochiensis* (Nematoda: Heteroderidae) with a key and notes on related species. *Proc. Helm. Soc. Wash.* 39: 64–78.

Golden, A. M., and Epps, J. M. 1965. Morphological variations in the soybean-cyst nematode (abstract). *Nematologica* 11: 38.

Golden, A. M., Epps, J. M., Riggs, R. D., Duclos, R. D., Fox, J. A., and Bernard, R. L. 1970. Terminology and identity of infraspecific forms of the soybean cyst nematode (*Heterodera glycines*) *Plant Dis. Rep.* 54: 544–546.

Golden, A. M., and Mulvey, R. H. 1983. Redescription of *Heterodera zeae*, the corn cyst nematode, with SEM observations. *J. Nematol.* 15: 60–70.

Golden, A. M., and Raski, D. J. 1977. *Heterodera thornei* n. sp. (Nematoda: Heteroderidae) and a review of related species. *J. Nematol.* 9: 93–112.

Golden, A. M., Franco, J., Jatala, P., and Astocaza, E. 1983. Description of *Thecavermiculatus andinus* n. sp. (Meloidoderidae), a round cystoid nematode from the Andes mountains of Peru. *J. Nematol.* 15: 357–362.

Goodey, J. B., Franklin, M. T., and Hooper, D. J. 1965. T. Goodey's *The Nematode Parasites of Plants Catalogued Under Their Hosts*, (3rd ed.). Farnh. Roy. Commonw. Agr. Bur.

Grainger, J. 1964. Factors affecting control of the eelworm diseases. *Nematologica* 10: 5–24.

Grandison, G. S. 1982. The status of *Heterodera avenae* on cereals in New Zealand. *OEPP/EPPO Bull.* 12: 333–334.

Granek, I. 1955. Additional morphological differences between the cysts of *Heterodera rostochiensis* and *Heterodera tabacum. Plant Dis. Rep.* 39: 716–718.

Graney, L. S. O. 1985. Observations on the morphology of *Heterodera carotae* and *Heterodera avenae* in Michigan USA (abstract). *J. Nematol.* 17: 519.

Graney, L. S. O., and Miller, L. I. 1982. Comparative morphological studies of *Heterodera schachtii* and *H. glycines*. In *Nematology in the Southern United States*. Riggs, R. D., ed. *Southern Coop. Ser. Bull.* 276, pp. 96–107.

Greco, N. 1981. Hatching of *Heterodera carotae* and *H. avenae. Nematologica* 27: 366–371.

Greco, N. 1986. The carrot cyst nematode. In *Cyst Nematodes*. Lamberti, F., and Taylor, C. E., eds. Plenum Press, New York, pp. 333–346.

Greco, N., Brandonisio, A., and Elia, F. 1985. Control of *Ditylenchus dipsaci, Heterodera carotae* and *Meloidogyne javanica* by solarization. *Nematol. Medit.* 13: 191–197.

Greco, N., Di Vito, M., and Brandonisio, A. 1982. Effect on temperature and plant age on the hatching of *Heterodera carotae* and *H. fici* (abstract). *J. Nematol.* 14: 443.

Greco, N., Di Vito, M., and Lamberti, F. 1986a. Studies on the biology of *Heterodera goettingiana* in Southern Italy. *Nematol. Medit.* 14: 23–39.

Greco, N., Elia, F., and Brandonisio, A., 1986b. Control of *Heterodera carotae*, *Ditylenchus dipsaci*, and *Meloidogyne javanica* with fumigant and nonfumigant nematicides. *J. Nematol.* 18: 359–364.

Green, C. D. 1971a. Mating and finding behaviour of plant nematodes. In *Plant Parasitic Nematodes*, Vol. 2. Zuckerman, B. M., Mai., W. F., and Rohde, R. A., eds. Academic Press, New York, pp. 247–266.

Green, C. D. 1971b. The morphology of the terminal area of the round–cyst nematodes, S. G. *Heterodera rostochiensis* and allied species. *Nematologica* 17: 34–46.

Green, C. D., and Greet, D. N. 1972. The location of the secretions that attract male *Heterodera schachtii* and *H. rostochiensis* to their females. *Nematologica* 18: 347–352.

Green, C. D., Greet, D. N., and Jones, F. G. W. 1970. The influence of multiple mating on the reproduction and genetics of *Heterodera rostochiensis* and *H. schachtii*. *Nematologica* 16: 309–326.

Greet, D. N., Green, C. D., and Poulton, M. E. 1968. Extraction, standardization and assessment of the volatility of the sex attractants of *Heterodera rostochiensis* Woll. and *H. schachtii* Schm. *Ann. Appl. Biol.* 61: 511–519.

Griffin, G. D. 1982. The quantification of sugarbeet cyst nematode soil populations and their relationship to sugarbeet yields. In *Proceedings of the 5th Cyst Nematode Workshop*. Schimdt, D. P., ed. North Carolina State Univ., Raleigh.

Griffin, G. D., and Waite, W. W. 1982. Pathological interaction of a combination of *Heterodera schachtii* and *Meloidogyne hapla* on tomato. *J. Nematol.* 14: 182–187.

Griffin, G. D. 1988. Factors affecting the biology and pathogenicity of *Heterodera schachtii* on sugarbeet. *J. Nematol.* 20: 396–404.

Guevera–Benitez, D. C., Tobar–Jamenez, A., and Guevara Pozo, D. 1970. Estudio cuantitativo del ciclo vital de *Heterodera goettingiana* Liebscher, 1892 y de la posibilidad de su control mediante cultivos cebo. *Revta. Iber. Parasitol.* 30: 113–256.

Gupta, P., and Edward, J. C. 1973. A new record of a cyst-forming nematode (*Heterodera chaubattia* n. sp.) from the hills of Uttar Pradesh. *Curr. Sci.* 42: 618–620.

Gurner, P. S. 1982. Jectarow – A commercial unit for applying low volumes of toxicants to the seed furrow of row crops. *OEPP/EPPO Bull.* 12: 513–516.

Hackney, R. W. 1981. Oat cyst nematode found in San Mateo county. Plant Pathology Reports. *California Dept. Food Agr.* 5: 31.

Hansen, K. 1904. Nogle lagttagelser over Havreaalens Optraeden. *Tidsskr. Landbr. Planteavl.* 11: 279–299.

Hansen, L. R. 1986. Incidence of *Heterodera avenae* in spring barley fields in Denmark. In *Cyst Nematodes*. Lamberti, F., and Taylor, C. E., eds. Plenum Press, New York, pp. 75–99.

Hartwig, E. G. 1981. Breeding productive soybean cultivars resistant to soybean cyst-nematode for the southern United States. *Plant Dis.* 65: 303–307.

Heald, C. M. 1984. Histolopathology of okra and ridgeseed spurge infected with *Meloidodera charis*. *J. Nematol.* 16: 105–108.

Heald, C. M., and Golden, A. M. 1969. *Meloidodera charis*, a cystoid nematode infecting St. Augustine grass. *Plant Dis. Reptr.* 53: 527.

Hernandez, A. L. 1965. Comportamiento de tres variedades de maiz al ataque de *Heterodera punctata* Thorne. Tesis de maestria, Escuela Nacional de Agricultura Chapingo, Estado de Mexico.

Hesling, J. J. 1973. The estimation of Granek's ratio in round-cyst *Heterodera*. *Nematologica* 19: 119–120.

Hesling, J. J. 1978. Cyst nematodes: Morphology and identification of *Heterodera*, *Globodera* and *Punctodera*. In *Plant Nematology*. Southey, J. F., ed. Her Majesty's Stationary Office, London, pp. 125–155.

Hill, N. S., and Schmidt, D. P. 1989. Influence of temperature and soybean phenology on dormancy induction of *Heterodera glycines*. *J. Nematol.* 21: 361–369.

Hirschmann, H. 1956. Comparative morphological studies on the soybean cyst nematode, *Heterodera glycines* and the clover cyst nematode, *H. trifolii* (Nematoda: Heteroderidae). *Proc. Helm. Soc. Wash.* 23: 140–151.

Hirschmann, H., and Triantaphyllou, A. C. 1973. Postembryogenesis of *Meloidodera floridensis* with emphasis on the development of the male. *J. Nematol.* 5: 185–195.

Hirschman, H., and Triantaphyllou, A. C. 1979. Morphological comparison of members of the *Heterodera trifolii* species complex. *Nematologica* 25: 458–481.

Hori, S. 1915. [Phytopathological notes. Sick soil of soybean caused by a nematode] (In Japanese). *Byotyugai-Zasshi* 2: 927–930.

Horne, C. W. 1965. The taxonomic status, morphology and biology of a cyst nematode (Nematoda: Heteroderidae). Ph.D. Dissertation, Texas A & M University.

Hussey, R. S. 1988. Production of monoclonal antibodies specific for secretory granules formed in the esophageal glands of *Meloidogyne* species (abstract). *J. Nematol.* 20: 641.

Hussey, R. S., and Boerma, H. R. 1983. Influence of planting date on damage to soybean caused by *Heterodera glycines*. *J. Nematol.* 15: 253–258.

Hutzell, P. A. 1984. Descriptions of males of *Heterodera zeae*. *J. Nematol.* 16: 83–86.

Ichinohe, M. 1952. On the soybean nematode *Heterodera glycines* n. sp. from Japan. *Oyo-Do-but-sogaku-Zasshi (Magazine of Applied Zoology)* 17: 1–4.

Ichinohe, M. 1961. Studies on the soybean cyst nematode, *Heterodera glycines*. *Hokkaido Natl. Agr. Exp. Sta. Rep.* 55: 1–77.

Inagaki, H. 1979. Race status of five Japanese populations of *Heterodera glycines*. *Jap. J. Nematol.* 9: 1–4.

Ito, S. 1921. Studies on "Yellow dwarf" disease of soybean. (In Japanese). *Hokkaido Agr. Exp. Sta. Rep.* 11: 47–59.

Jatala, P., and Garzon, C. 1987. Detection of the potato cyst nematode *Globodera pallida* in pre-colombian agricultural terraces of Peru (abstract). *J. Nematol.* 19: 532.

Jeronimo, J. 1988. El Nematodo enquistado del maiz *Punctodera* sp y su expresion en 17 municipios del estado de Michoacan. Thesis, Universidad Michoacana de San Nicolas de Hidalgo, Uruapan, Mich.

Johnson, A. W. 1968. Cactus cyst nematode found in Georgia. *Plant Dis. Rep.* 52: 114.

Johnson, C. S., Komm, D. A., and Jones, J. L. 1989. Control of *Globodera tabacum solanacearum* by alternating host resistance and nematicide. *J. Nematol.* 21: 16–23.

Johnson, P. W., and Fushtry, S. G. 1966. The biology of the oat cyst nematode *Heterodera avenae* in Canada II. Nematode development and related anatomical changes in roots of oats and corn. *Nematologica* 12: 630–636.

Jones, F. G. W. 1950a. A new species of root eelworm attacking carrots. *Nature* 165: 81.

Jones, F. G. W. 1950b. Observations on the beet eelworm and other cyst forming species of *Heterodera*. *Ann. Appl. Biol.* 37: 407–440.

Jones, F. G. W. 1965. Pea, cabbage and carrot root eelworms. In *Plant Nematology*. Southey, J. F., ed. Tech. Bull. No. 7. Min. Agr. Fish. Fd., London.

Jones, F. G. W., Parrott, D. M., and Perry, R. N. 1981. The gene for gene relationship and its significance for potato cyst nematodes and their solanaceous hosts. In *Plant Parasitic Nematodes*, Vol. 3. Zuckerman, B. M., and Rohde, R. A., eds. Academic Press, New York.

Jones, M. G. K. 1981a. Host cell responses to endoparasitic nematode attack: structure and function of giant cells and syncytia. *Ann. Rev. Appl. Biol.* 97: 353–372.

Jones, M. G. K. 1981b. The development and function of plant cells modified by endoparasitic nematodes. In *Plant Parasitic Nematodes*, Vol. 3. Zuckerman, B. M., and Rohde, R. A., eds. Academic Press, New York, pp. 255–279.

Jones, M. G. K., and Rodríguez-Kábana, R. 1985. Phytonematode pathology: fungal modes of action. a perspective. *Nematropica* 15: 107–14.

Kalinski, A., and Huettel, R. N. 1988. Use of DNA restriction fragment length polymorphism in detection of races of the soybean cyst nematode (abstract). *J. Nematol.* 20: 643.

Kerry, B. R. 1975. Fungi and the decrease of cereal cyst nematode populations in cereal monoculture. *OEPP/EPPO Bull.* 5: 353–361.

Kerry, B. R., and Crump, D. H. 1977. Observations on fungal parasites of females and eggs of the cereal cyst-nematode, *Heterodera avenae*, and other cyst nematodes. *Nematologica* 23: 193–201.

Kirjanova, E. S. 1969. [On the structure of the subcrystalline layer of the nematode genus *Heterodera* (Nematoda: Heteroderidae) with a description of two new species.] (In Russian) *Parazitologiya* 3: 81–91.

Koenning, S. R., Anand, S. C., and Wrather, J. A. 1988. Effects of within-field variation in soil texture on *Heterodera glycines* and soybean yield. *J. Nematol.* 20: 373–380.

Koliopanos, C. N., and Triantaphyllou, A. C. 1971. Host specificity and morphometrics of four populations of *Heterodera glycines* (Nematoda: Heteroderidae). *J. Nematol.* 3: 364–368.

Koliopanos, C. N., and Triantaphyllou, A. C. 1972. Effect of infection density on sex-ratio of *Heterodera glycines*. *Nematologica* 18: 131–137.

Kontaxis, D. G., Lofgreen, G. P., Thomason, I. J., and McKinney, H. E. 1971. Survival of the sugar beet cyst nematode in the alimentary canal of cattle. *Calif. Agr.* 30: 15.

Kort, J., Ross, H., Rumpenhorst, H. J., and Stone, A. R. 1977. An international scheme for identifying and classifying pathotypes of *Globodera rostochiensis* and *G. pallida*. *Nematologica* 23: 333–339.

Koshy, P. K., Swarup, G., and Sethi, C. L. 1970a. Further notes on the pigeon-pea cyst nematode, *Heterodera cajani*. *Nematologica* 16: 477–482.

Koshy, P. K., Sawrup, G., and Sethi, C. L. 1970b. *Heterodera zeae* n. sp. (Nematoda: Heteroderidae), a cyst forming nematode on *Zea mays*. *Nematologica* 16: 511–516.

Krall, E. L., and Krall, H. A. 1978. [The revision of plant nematodes of the family Heteroderidae (Nematoda: Tylenchida) by using a comparative ecological method of studying their phylogeny] in *Printsipy i methody izucheniya vzaimootnoshenii mezhdu parazitichenskimi nematodami i rasteniyami. Phytohelm. Proc. Akad. Nauk Estonoskoi SSR. Tartu*, pp. 39–56.

Krusberg, L. R., and Sardanelli, S. 1982. Corn cyst nematode (abstract). *J. Nematol.* 14: 453.

Krusberg, L. R., and Sardanelli, S. 1989. Survival of *Heterodera zeae* in soil in the field and in the laboratory. *J. Nematol.* 21: 347–355.

Kühn, J. 1874. über das Vorkommen von Rübennematoden an den Wurzeln den Halmfruchte. *Z. Ver Rübenzucker Ind. Zollver* 24: 149–153.

Kuiper, K. 1960. Resistance of white clover varieties of the clover cyst-eelworm *Heterodera trifolii* Goffart. *Nematologica (Suppl. II):* 95–100.

Kumar, A. C. 1964. A note on the occurrence of *Heterodera cacti* from Mysore. *Curr. Sci.* 33: 534.

La Mondia, J. A. 1988. Tobacco resistance to *Globodera tabacum*. *Ann. Appl. Nematol.* 2: 77–80.

Langdon, K. R., and Esser, R. P. 1969. Cactus cyst nematode, *Heterodera cacti*, in Florida, with host list. *Plant Dis. Rep.* 53: 123–125.

Lauritis, J. A., Rebois, R. V., and Graney, L. S. 1983. Life cycle of *Heterodera zeae* Koshy, Swarup, and Sethi on *Zea mays* L. Axenic root explants. *J. Nematol.* 15: 115–119.

Lear, B. 1971. Reproduction of the sugarbeet nematode and the cabbage root nematode on several cultivars of brussels sprouts. *Plant Dis. Rep.* 55: 1005–1006.

Lear, B., Johnson, D. E., Miyagawa, S. T., and Sciaroni, R. H. 1965a. Response of Brussels sprouts to applications of nematicides in transplant water for control of sugarbeet nematode and cabbage root nematode. *Plant Dis. Rep.* 49: 900–902.

Lear, B., Sciaroni, R. H., Johnson, D. E., and Miyagawa, S. T. 1965b. Response of Brussels sprouts to soil fumigation for control of cabbage root nematode, *Heterodera cruciferae*. *Plant Dis. Reptr.* 49: 903–904.

Lear, B., Johnson, D. E., Miyagawa, S. T., and Sciaroni, R. H. 1966. Yield response of Brussels sprouts associated with control of sugarbeet nematode and cabbage root nematode in combination with the club root organism, *Plasmodiophora brassicae*. *Plant Dis. Rep.* 50: 133–135.

Lehman, P. S., Barker, K. R., and Huisingh, D. 1971. The influence of races of *Heterodera glycines* on nodulation and nitrogen–fixing capacity of soybean. *Phytopathology* 61: 1239–1244.

Liebscher, G. 1890. Eine Nematode als Ursache der Erbsenmudigkeit des Bodens. *Deuts. landw. Pr.* 17: 436–437.

Lownsbery, B. F. 1953. Host preferences of the tobacco cyst nematode, *Heterodera* sp. *Phytopathology* 43: 106–107.

Lownsbery, B. F., and Lownsbery, J. W. 1954. *Heterodera tabacum* new species, a parasite of solanaceous plants in Connecticut. *Proc. Helm. Soc. Wash.* 21: 42–47.

Luc, M., Weischer, B., Stone, A. R., and Baldwin, J. G. 1986. On the definition of heteroderid cysts. *Rev. Nematol.* 9: 418–421.

Luc, M., Maggenti, A. R., and Fortuner, R. 1988. A reappraisal of Tylenchina (Nemata). 9. The family Heteroderidae Filip'ev & Schuurmans Stekhoven, 1941. *Rev. Nematol.* 11: 159–176.

Maas, P. W. Th., Du Bois, E., and Dede, J. 1982. Morphological and host range variation in the *Heterodera trifolii* complex. *Nematologica* 28: 263–270.

Mackintosh, G. M. 1960. The morphology of the brassica root eelworm *Heterodera cruciferae* Franklin, 1945. *Nematologica* 5: 158–165.

Macara, A. M. 1963. Contribuição para o estudo morfólogico e biológico dos nematodes *Heterodera göttingiana* Liebscher, 1892 e *H. rostochiensis* Wollenweber, 1923 encontrados em Portugal. *Broteria.* 32: 25–119.

Mägi, E. A. 1971. [Factors influencing development of potato cyst eelworm.] *Sots. Põllumajandus* 26: 16–18.

Mai, W. F. 1952. Temperature in relation to retention of viability of encysted larvae of the golden nematode, *Heterodera rostochiensis* Wr. (abstract). *Phytopathology*, 42: 113.

Mai, W. F. 1952. Susceptibility of *Lycopersicon* species to the golden nematode. *Phytopathology* 42: 461–462.

Mai, W. F. 1977. Worldwide distribution of potato cyst nematodes and their importance in crop production. *J. Nematol.* 9: 30–34.

Mankau, R., and Linford, M. B. 1960. Host-parasite relationships of the clover cyst nematode, *Heterodera trifolii* Goffart. *Ill. Agr. Exper. Sta. Bull.* 667: 1–50.

Maqbool, M. A. 1981. Occurrence of root-knot and cyst nematodes in Pakistan (abstract). *J. Nematol.* 13: 448–449.

Marshall, J. W., and Crawford, A. M. 1987. A cloned DNA fragment that can be used as a sensitive probe to distinguish *Globodera pallida* from *Globodera rostochiensis* and other common cyst forming nematodes (abstract). *J. Nematol.* 19: 541.

Mathews, H. J. P. 1975. *Heterodera carotae* C. I. H. Descriptions of Plant-parasitic Nematode Set 5, No. 61, Williams Clowes & Sons Ltd., London.

McLeod, R. W. 1968. Cereal cyst nematode: A nematode new to New South Wales. *Agr. Gaz. N.S.W.* 79: 293–295.

Meagher, J. W. 1982a. The effect of environment on survival and hatching of *Heterodera avenae*. *OEPP/EPPO Bull.* 12: 361–369.

Meagher, J. W. 1982b. Yield loss caused by *Heterodera avenae* in cereal crops grown in a Mediterranean climate. *OEPP/EPPO Bull.* 12: 325–331.

Meredith-C, J. A. 1976. Estudio de la reproducion de *Heterodera rostochiensis* Wollenweber, 1923, en diferentes condiciones ambientales sobre varias solanaceas. PhD Thesis, Universidad Central de Venezuela.

Miles, M. 1951. Cyst-forming eelworms on cruciferous plants. *Nature* 4248: 533.

Miller, P. M. 1969. Suppression by benomyl and thiabendazole of root invasion by *Heterodera tabacum. Plant Dis. Rep.* 53: 963–966.

Miller, P. M. 1970. Failure of several non-volatile and contact nematicides to kill eggs in cysts of *Heterodera tabacum. Plant Dis. Rep.* 54: 781–783.

Miller, L. I. 1983. Diversity of selected taxa of *Globodera* and *Heterodera* and their interspecific and intergeneric hybrids. In: *Concepts in Nematode Systematics.* Stone, A. R., Pratt, H. M., and

Khall, L. F., eds. The Systematics Association Special Volume No. 22, Academic Press, New York, pp. 207–220.

Miller, L. I. 1986. Economic importance of cyst nematodes in North America. In *Cyst Nematodes*. Lamberti, F., and Taylor, C. E., eds. Plenum Press, New York, pp. 373–385.

Miller, L. I., and Duke, P. L. 1967. Morphological variation of eleven isolates of *Heterodera glycines* in the United States (abstract). *Nematologica* 13: 145–146.

Miller, L. I., and Gray, B. J. 1968. Horsenettle cyst nematode, *Heterodera virginiae* n. sp., a parasite of solanaceous plants. *Nematologica* 14: 535–543.

Miller, L. I., and Gray, B. J. 1972. *Heterodera solanacearum* n. sp., a parasite of solanaceous plants. *Nematologica* 18: 404–413.

Miller, L. I., Fox, J. A., and Spasoff, L. 1972. Genetic relationship of resistance to *Heterodera solanacearum* in dark-fired and burley tobacco (abstract). *Phytopathology* 62: 778.

Miller, L. I., Stone, A. R., Spasoff, L., and Evans, D. M. 1974. Resistance of *Solanum tuberosum* to *Heterodera solanacearum*. *Proc. Am. Phytopathol. Soc.* 1: 153.

Minton, N. A., and Golden, A. M. 1971. A cystoid nematode, *Meloidodera charis*, in Nebraska (abstract). *Plant Dis. Reptr.* 55: 115.

Moore, W. F., ed. 1984. *Soybean Cyst Nematode*. Mississippi Coop. Ext. Service.

Moortensen, M. L., Rostrup, S., and Kolpin Kavn, R. 1908. Oversight over Landbrugsplanternes Sygdomme i 1907. *Tidsskr. Landbr. Planteavl* 15: 145–158.

Mugniery, D., and Balandras, C. 1984. Examen des possibilities d'eradication du nematode a kystes, *Globodera pallida* Stone. *Agronomie* 4: 773–778.

Mugniery, D., and Bossis, M. 1988. *Heterodera carotae* Jonees, 1950. 1. Gamme d'hotes, vitesse de de'veloppement, cycle. *Rev. Nematol.* 11: 307–313.

Müller, J. 1986. Integrated control of the sugar beet cyst nematode. In *Cyst Nematodes*. Lamberti, F., and Taylor, C. E., eds. Plenum Press, New York, pp. 235–250.

Mulvey, R. H. 1959. Giant eggs of the clover cyst nematode, *Heterodera trifolii* Goffart, 1932. *Nature* 184: 1662–1663.

Mulvey, R. H. 1960a. Oogenesis in some species of *Heterodera* and *Meloidogyne* (Nematoda: Heteroderidae). In *Nematology*. Sasser, J. N., and Jenkins, W. R., eds. University of North Carolina Press, Chapel Hill, pp. 212–215.

Mulvey, R. H. 1960b. The value of cone top and underbridge structures in the separation of some cyst-forming nematodes. In *Nematology*. Sasser, J. N., and Jenkins, W. R., eds. University of North Carolina Press, Chapel Hill, pp. 212–215.

Mulvey, R. H. 1972. Identification of *Heterodera* cysts by terminal and cone top structures. *Can. J. Zool.* 50: 1277–1292.

Mulvey, R. A., and Anderson, R. V. 1974. *Heterodera trifolii*, C. I. H. Descriptions of Plant–Parasitic Nematodes Set 4, No. 46, St. Albans, Herts, England.

Mulvey, R. H., and Golden, A. M. 1983. An illustrated key to the cyst-forming genera and species of Heteroderidae in the western hemisphere with species morphometrics and distribution. *J. Nematol.* 15: 1–59.

Mulvey, R. H., and Stone, A. R. 1976. Description of *Punctodera matadorensis* n. gen., n. sp., (Nematoda: Heteroderidae) from Saskatchewan with lists of species and generic diagnosis of *Globodera* n. rank), *Heterodera* and *Sarisodera*. *Can. J. Zool.* 54: 772–785.

Mundo-Ocampo, M., and Baldwin, J. G. 1983. Host-parasite relationships of *Atalodera* spp. (Heteroderidae). *J. Nematol.* 15: 234–243.

Mundo-Ocampo, M., Baldwin, J. G., and Jeronimo, R. J. 1987. Distribution, morphological variation, and host range of *Punctodera chalcoensis* (abstract). *J. Nematol.* 19: 545–546.

Muñoz, R. C. 1983. Ciclo de vida de *Punctodera chalcoensis* Stone, Sosa Moss y Mulvey, 1976 (Nemat. Heteroder). Tesis Intituto Politecnico Nacional. Escuela Nacional de Ciencias Biologicas, Mexico.

Nakata, K., and Asuyama, H. 1938. Survey of the principal diseases of crops in Manchuria. (In Japanese). *Manchuria Indus. Rep.* 32: 166.

Nausbaum, C. J. 1969. In *Proceedings of the Symposium on Tropical Nematology*. Ramos, J. A., ed. University of Puerto Rico, Mayaguez Campus, Agr. Exp. Sta. Rio Piedras, pp. 58–67.

Norton, D. C. 1967. Relationships of *Heterodera trifolii* to some forage legumes. *Phytopathology* 57: 1305–1308.

Norton, D. C. 1984. Nematode parasites of corn. In *Plant and Insect Nematodes*. Nickle, W. R., ed. Marcel Dekker, New York, pp. 61–94.

Norton, D. C., and Isely, D. 1976. Cyst production of *Heterodera trifolii* on some Leguminosae. *Plant Dis. Rep.* 12: 1017–1020.

Oostenbrink, M. 1951. Het erwtencystenaaltje, *Heterodera gottingiana* Liebscher, in Nederland. *Tijds. Pl. Ziekten.* 57: 52–64.

Oostenbrink, M. 1967. Studies on the emergence of encysted *Heterodera* larvae. *Meded. Rijksfac. LandbWet. Gent.* 32: 503–539.

Othman, A. A., and Baldwin, J. G. 1985. Comparative morphology of *Meloidodera* spp. and *Verutus* sp. (Heteroderidae) with scanning electron microscopy. *J. Nematol.* 17: 297–309.

Othman, A. A., and Baldwin, J. G. 1986. Comparative morphology of *Sarisodera hydrophila*, *Rhizonema sequoiae*, and *Afenestrata africana* (Heteroderidae) with scanning electron microscopy. *Proc. Helm. Soc. Wash.* 53: 69–79.

Othman, A. A., Baldwin, J. G., and Bell, A. H. 1986. Comparative morphology of *Ataladera* spp. and *Thecavermiculatus* spp. (Heteroderidae) with scanning electron microscopy. *J. Nematol.* 18: 275–287.

Othman, A. A., Baldwin, J. G., and Mundo-Ocampo, M. 1988. Comparative morphology of *Globodera*, *Cactodera*, and *Punctodera* spp. (Heteroderidae) with scanning electron microscopy. *Rev. Nematol.* 11: 53–63.

Parrot, D. M. and Miller, L. I. 1977. NC-95 flue cured tobacco as a host of *Globodera pallida*. (abstract) *Virginia J. Sci.* 28: 54.

Perry, R. N. 1986. Physiology of hatching. In *Cyst Nematodes*. Lamberti, F., and Taylor, C. E., eds. Plenum Press, New York, pp. 119–131.

Perry, R. N., and Beane, J. 1988. Effects of activated charcoal on hatching and infectivity of *Globodera rostochiensis* in pot tests. *Rev. Nematol.* 11: 229–233.

Perry, R. N., and Clarke, A. J. 1981. Hatching mechanisms of nematodes. *Parasitology* 83: 435–449.

Perry, R. N., and Feil, J. J. 1986. Observations on a novel hatching bioassay for *Globodera rostochiensis* using fluorescence microscopy. *Rev. Nematol.* 9: 280–282.

Perry, R. N., Wharton, D. A., and Clarke, A. J. 1982. The structure of the egg-shell of *Globodera rostochiensis* (Nematoda: Tylenchida). *Int. J. Parasitol.* 12: 481–485.

Perry, R. N., Zunke, U., and Wyss, U. 1989. Observations on the response of the dorsal and subventral esophageal glands of *Globodera rostochiensis* to hatching stimulation. *Rev. Nematol.* 12: 91–96.

Radice, A. D., Powers, T. O., Sandall, L. J., and Riggs, R. D. 1988. Comparisons of mitochondrial DNA from the sibling species *Heterodera glycines* and *H. schachtii*. *J. Nematol.* 20: 443–450.

Raski, D. J. 1950. The life history and morphology of the sugar beet nematode *Heterodera schachtii* Schmidt. *Phytopathology* 40: 135–152.

Raski, D. J. 1952. The first record of the brassica—root nematode in the United States. *Plant Dis. Rep.* 36: 438–439.

Rice, S. L., Leadbeater, B. S. A., and Stone, A. R. 1986. Changes in cell structure in roots of resistant potatoes parasitized by potato cyst-nematodes. I. Potatoes with resistance gene H1 derived from *Solanum tuberosum* spp. *andigena. Physiol. Pl. Pathol.* 27: 219–234.

Riggs, R. D. 1977. Worldwide distribution of soybean-cyst nematode and its economic importance. *J. Nematol.* 9: 34–39.

Riggs, R. D. 1982. Cyst nematodes in the southern U.S.A. In *Nematology in the Southern United States*. Riggs, R. D., ed. *Southern Cooperative Series Bulletin* 276: 77–93.

Riggs, R. D., Hamblen, M. L., and Rakes, L. 1981. Infra-species variation in reactions to host in *Heterodera glycines* populations. *J. Nematol.* 13: 171–179.

Riggs, R. D., Rakes, L., and Hamblen, M. L. 1982. Morphometric and sereologic comparisons of a number of populations of cyst nematodes. *J. Nematol.* 14: 188–199.

Riggs, R. D., and Schmitt, D. P. 1988. Complete characterization of the race scheme for *Heterodera glycines*. *J. Nematol.* 20: 392–395.

Riggs, R. D., Schmitt, D. P., and Noel, G. R. 1988. Variability in race tests with *Heterodera glycines*. *J. Nematol.* 20: 565–572.

Ringer, C. E., Sardanelli, S., and Krusberg, L. R. 1987. Investigation of the host range of the corn cyst nematode, *Heterodera zeae*, from Maryland. *Ann. Appl. Nematol.* 1:97–106.

Robbins, R. T. 1978. A new Ataloderinae (Nematoda: Heteroderidae), *Thecavermiculatus gracililancea* n. gen., n. sp. *J. Nematol.* 10: 250–254.

Robbins, R. T. 1986. Description of *Thecavermiculatus carolynae* n. sp. (Nematoda: Ataloderinae). *J. Nematol.* 18: 548–555.

Roberts, P. A., and Stone, A. R. 1981. Host ranges of *Globodera* species within *Solaum* subgenus *Leptostemonum*. *Nematologica* 27: 172–189.

Robinson, M. P., Atkinson, H. H., and Perry, R. N. 1987. The influence of temperature on the hatching activity and lipid utilization of second stage juveniles of the potato cyst nematodes *Globodera rostochiensis* and *G. pallida*. *Rev. Nematol.* 10: 349–354.

Robinson, M. P., Atkinson, H. J., and Perry, R. N. 1988. The association and partial characterization of a fluorescent hypersensitive response of potato roots to the potato cyst nematode *Globodera rostochiensis* and *G. pallida*. *Rev. Nematol.* 11: 99–108.

Roessner, J. 1986. Parasitism of *Globodera rostochiensis* by nematophagous fungi. *Rev. Nematol.* 9: 307–308.

Ross, J. P. 1962. Physiological strains of *Heterodera glycines*. *Plant. Dis. Rep.* 46: 766–769.

Rostrup, E. 1896. Oversigt over Sygdommenes Optraeden hos Landrugets Avlsplanter i Aaret 1894. *Tidsskr. Landbr. Planteavl.* 2: 40–71.

Sandall, L. J., and Powers, T. O. 1988. Estimates of nucleotide sequence divergence among mitochondrial genomes of the soybean cyst nematode (abstract). *J. Nematol.* 20: 658.

Santacruz, U. H. 1982. Identificacion de generos de nematodos y evaluacion de sus poblaciones en quince Municipios del Estado de Michoacan. Tesis professional. Dpto. de Parasitologia agricola. Fac. de Agrobiologia, Presidente Juarez Uruapan Mich., Mexico.

Sardanelli, S., Krusberg, L. R. and Golden, A. M. 1981. Corn cyst nematode, *Heterodera zeae*, in the United States. *Plant Dis.* 65: 662.

Schmidt, A. 1871. Über den Rüben-Nematoden (*Heterodera schachtii A.S.*) Z. Ver Ruben Zucker Industr. Zolliver 21: 1–19.

Schmidt, O. 1930. Sind Rüben. und Hafernematoden identisch? *Arch. Pfl. Bau.* 3: 420–464.

Schneider, J. 1961. Les nematodes parasites des cactees. *Phytoma* 107: 17–20.

Seinhorst, J. W. 1986. Effects of nematode attack on growth and yield of crop plants. In *Cyst Nematodes*. Lamberti, F., and Taylor, C. E., eds. Plenum Press, New York, pp. 119: 75–131.

Seinhorst, J. W., and Sen, A. K. 1966. The population density of *Heterodera trifolii* in pastures in the Netherlands and its importance for the growth of white clovers. *Neth. J. Plant Pathol.* 72: 169–183.

Shepherd, A. D. 1963. The emergence of larvae of *Heterodera goettingiana* Liebs, in vitro and a comparison between field populations of *H. goettingiana* and *H. rostochiensis* Woll. *Nematologica* 9: 143–151.

Shepherd, A. M. 1965. *Heterodera: Biology..*In *Plant Nematology*. Southey, J. F., ed. Tech. Bull. No. 7. Her Majesty's Stationary Office, London, pp. 89–102.

Shepherd, A. M., Clark, S. A., and Dart, P. J. 1972. Cuticle structure in the genus *Heterodera*. *Nematologica* 18: 1–17.

Shahina, F., and Maqbool, M. A. 1989. Embryonic and postembryonic developmental stages of *Heterodera zeae* Koshy et al., 1971 of *Zea mays*. *Pak. J. Nematol.* 7: 3–16.

Shmal'co, V. F. 1959. The cactus nematode, *Heterodera cacti* Filipjev et Schuurmans-Stekhoven, 1941. (In Russian) *Tr. Gel' min. Lab. Akaad. Nauk. SSSR* 9: 389–390.

Sikora, R. A., and Maas, P. W. Th. 1986. An analysis of the *Heterodera trifolii* complex and other species in the schachtii group attacking legumes. In *Cyst Nematodes*. Lamberti, F., and Taylor, C. E., eds. Plenum Press, New York, pp. 293–313.

Slack, D. A., Riggs, R. D., and Hamblen, M. L. 1972. The effect of temperature and moisture on the survival of *Heterodera glycines* in the absence of a host. *J. Nematol.* 4: 263–266.

Smart, G. C., Jr. 1964. Physiological strains and one additional host of the soybean cyst nematode, *Heterodera glycines*. *Plant Dis. Rep.* 48: 542–543.

Sosa Moss, C. 1965. Recherches sur *Heterodera punctata* Thorne, parasite du maiss au Mexique. Comptes rendus du huiteme Symposium International de Nematologie, Antibes, 1965. E. J. Brill, Leiden, p. 54.

Sosa Moss, C. 1966. Contribution a l'etude d'un nematode: *Heterodera avenae* Woll. These, Faculte des Sciences, Univ. Paris.

Sosa Moss, C., and Gonzales, C. P. 1973. Respuesta de maiz chalqueño fertilizado y no fertilizado a 4 diferentes niveles de *Heterodera punctata* raza Mexicana (Nematoda: Heteroderidae). *Nematropica* 3: 13–14.

Southey, J. F. 1965. Physical methods of control. In *Plant Nematology*. Southey, J. F., ed. *Tech. Bull. No. 7, M. A. F. F.* Her Majesty's Stationery Office, London.

Spears, J. F. 1968. The golden nematode handbook, survey, laboratory, control and quarantine procedures. U.S. Dept. Agr. Handbook, 353.

Spiegel, Y., Cohn, E., and Chet, I. 1989. Use of chitin for controlling *Heterodera avenae* and *Tylenchulus semipenetrans*. *J. Nematol.* 21: 419–422.

Srivastava, A. M., and Swarup, G. 1975. Preliminary studies on some graminaceous plants for their susceptibility to the maize cyst nematode, *Heterodera zeae* Kosky et al. 1970. *Ind. J. Nematol.* 5: 257–259.

Steele, A. E. 1965. The host range of the sugar beet nematode, *Heterodera schachtii* Schmidt. *J. Am. Soc. Sugar Beet Techn.* 13: 573–603.

Steele, A. E. 1984. Nematode parasties of sugar beet. In *Plant and Insect Nematodes*. Nickle, W. R., ed., Marcel Dekker, New York, pp. 507–569.

Stelter, H. 1969. Zur Biologie der Mohrennematoden, *Heterodera carotae* Jones, 1950. *Biol. Zbl.* 68: 365.

Stirling, G. R., and Wicks, T. J. 1975. *Heterodera cruciferae* on cabbages in South Australia and its chemical control. *Plant Dis. Rep.* 59: 43–45.

Stone, A. R. 1972. The round cyst species of *Heterodera* as a group. *Ann. Appl. Biol.* 71: 280–283.

Stone, A. R. 1973. *Heterodera pallida* n. sp. (Nematoda: Heteroderidae), a second species of potato cyst nematode. *Nematologica*, 18: 591–606.

Stone, A. R. 1974. Application for a ruling on the availability of five specific names proposed as new for the genus *Heterodera* A. Schmidt, 1871 (Nematoda) in "A preliminary key to British species of *Heterodera* for use in soil examination" by B. A. Cooper, 1955. *Bull. Zool. Nom.* 31: 225–227.

Stone, A. R. 1983. Three approaches to the status of a species complex, with a revision of some species of *Globodera* (Nematoda: Heteroderidae) in *Concepts in Nematode Systematics*, Systematics Association Special Volume No. 22. Stone, A. R., Platt, H. M., and Khalil, L. F., eds. Academic Press, London.

Stone, A. R. 1985. Co-evolution of potato cyst nematodes and their hosts: implications for pathotypes and resistance. *OEPP/EPPO Bull.* 15: 131–138.

Stone, A. R. 1986. Taxonomy and phylogeny of cyst nematodes. In *Cyst Nematodes*. Lamberti, F., and Taylor, C. E., eds. Plenum Press, New York, pp. 119–131.

Stone, A. R., and Course, J. A. 1986. *Heterodera goettingiana* C. I. H. Descriptions of Plant-Parasitic Nematodes Set 4, No. 47, St. Albans, Herts, Egland.

Stone, A. R., ad Hill, A. J. 1982. Some problems posed by the *Heterodera avenae* complex. *OEPP/EPPO Bull.* 12: 317–320.

Stone, A. R., and Miller, L. I. 1974. Multiplication of *Heterodera solanacearum* on potatoes. Nematology Department, Rothamsted Report for 1973: 155–156.

Stone, A. R., and Rowe, J. A. 1976. *Heterodera cruciferae*. C. I. H. Descriptions of Plant-Parasitic Nematodes Set 6, No. 90, St. Albans, Herts, England.

Stone, A. R., and Williams, T. D. 1974. The morphology and soluble protein electrophoresis of *Heterodera avenae* pathotypes. *Ann. Appl. Biol.* 76: 231–236.

Stone, A. R., Sosa–Moss, C., and Mulvey, R. H. 1976. *Punctodera chalcoensis* n. sp. (Nematoda: Heteroderidae), a cyst nematode from Mexico parasitizing *Zea mays*. *Nematologica* 22: 381–398.

Stoyanov, D. 1982. Cyst-forming nematodes on cereals in Bulgaria. *OEPP/EPPO Bull.* 12: 341–344.

Tanaka, T. 1921. On the soybean nematode (in Japanese) *Byôyūgai-Zasshi* 8: 551–553.

Taylor, A. L. 1957. *Heterodera* taxonomy, Proceedings Workshop in Phytonematology, Univ. Tennessee, 2: 5–19.

Thorne, G. 1928. *Heterodera punctata* n. sp. a nematode parasite on wheat roots from Saskatchewan. *Sci. Agr.* 8: 707–710.

Thorne, G. 1961. *Principles of Nematology*. McGraw-Hill, New York.

Todd, T. C., and Pearson, C. A. S. 1988. Establishment of *Heterodera glycines* in three soil types. *Ann. Appl. Nematol.* 2: 57–60.

Triantaphyllou, A. C. 1970. Cytogenic aspects of evolution of the family Heteroderidae. *J. Nematol.* 2: 26–32.

Triantaphyllou, A. C. 1975a. Oogensis and the chromosomes of twelve bisexual species of *Heterodera* (Nematoda: Heteroderidae). *J. Nematol.* 7: 34–40.

Triantaphyllou, A. C. 1975b. Genetic structure of races of *Heterodera glycines* and inheritance of ability to reproduce on resistant soybeans. *J. Nematol.* 7: 356–364.

Triantaphyllou, A. C. 1983. Chromosomes in evolution of nematodes. In *Chromosomes in Evolution of Eukaryotic Groups*. Sharma, A. K., and Sharma, A., eds. Chapter 4, Vol. II. CRC Press, Boca Raton, pp. 77–101.

Triantaphyllou, A. C., and Hirschmann, H. 1962. Oogenesis and mode of reproduction in the soybean cyst nematode, *Heterodera glycines*. *Nematologica* 7: 235–241.

Triantaphyllou, A. C., and Hirschmann, H. 1973. Environmentally controlled sex expression in *Meloidodera floridensis*. *J. Nematol.* 5: 181–185.

Triantaphyllou, A. C., and Hirschmann, H. 1978. Cytology of the *Heterodera trifolii* parthenogenetic species complex. *Nematologica*, 24: 418–424.

Triffit, M. J. 1931. On the occurrence of *Heterodera radicicola* associated with *Heterodera schachtii* as a field parasite in Britain. *J. Helminthol.* 9: 205–208.

Trudgill, D. L., Webster, J. M., and Parrott, D. M. 1967. The effect of resistant solanaceous plants on the sex ratio of *Heterodera rostochiensis* and the use of the sex ratio to assess the frequency and genetic constitution of pathotypes. *Ann. Appl. Biol.* 60: 421–428.

Tsatsum, M., and Sakurai, K. 1966. [Influence of root diffusates of several host and non-host plants on the hatching of the soybean cyst nematode *Heterodera glyciunes* Ichinohe, 1952.] *Jap. J. Asppl. Ent. Zool.* 10: 129–137.

Valdeolivas, A., and Romero, M. D. 1985. The biology of *Heterodera avenae* in Spain. In *Cyst Nematodes*. Lamberti, F., and Taylor, C. E., eds. Plenum Press, New York, pp. 187–290.

Vallotton, R. 1980. Lé nematode a kyste *Heterodera carotae*, un dangereux ravageur de la carotte en Suisse romande. *Le maraicher* 43: 259.

Varo Alcala, J., Tobar-Jimenez, A, and Muñoz Medina, H. M. 1970. Lesiones causadas y reacciones provocadas por algunos nematodes en las raices de ciertas plantas. *Revta. ibér. Parasit.* 30: 547–566.

Vazquez, J. T. 1976. Infestaciones de nematodos fitoparasitos como factor limitante en la produccion de maiz en el altiplano mexicano. Produccion del Departamento Mexicano. CODAGEM, pp. 79.

Verma, A. C., and Yadav, B. S. 1975. Occurrence of *Heterodera cajani* in Rajasthan and susceptibility to certain sesame varieties. *Ind. J. Nematol.* 5: 235–237.

Verma, A. C., and Yadav, B. S. 1978. New hosts of maize cyst nematode *Heterodera zeae* (abstract). *Ind. J. Mycol. Plant Pathol.* 8: 72.

Villnueva, R. M. de J. 974. Comparación morfometrica entre una población inglesa y una mexicana de *Heterodera punctata* Thorne (nematodo Heteroderidae), influencia de exudados radiculares en la población Mexicana. Tesis profesional, I. P. N. Mexico.

Volkmar, K. M. 1989. Cereal cyst nematode (*Heterodera avenae*) on oats. II. Early development and nematode tolerance. *J. Nematol.* 21: 384–391.

Vovlas, N. 1978. Studio istopatologico di radici di carota infestate da *Heterodera carotae* e *Meloidogyne javanica. Inf. Fitopatol.* 28: 25–29.

Vovlas, N., and Inserra, R. N. 1978. Osservazioni comparative delle alteraziani istologiche indotte da *Heterodera goettingiana* e *Meloidogyne javanica* in radici di pisello. *Nematol. Medit.* 6: 187–195.

Whitehead, A. G. 1986. Chemical and integrated control of cyst nematodes. In *Cyst Nematodes.* Lamberti, F., and Taylor, C. E., eds. Plenum Press, New York, pp. 413–432.

Wieser, W. 1953. On the structure of the cyst wall in four species of *Heterodera* Schmidt. Statens Vaxtskyddsanst Meddel. 65: 3–15.

Williams, T. D., and Siddiqi, M. R. 1972. *Heterodera avenae*, C. I. H. Descriptions of Plant-Parasitic Nematodes Set 1, No. 2., St. Albans, Herts, England.

Winslow, R. D. 1953. Hatching responses to some *Heterodera* species. *Ann. Appl. Biol.* 40: 225–226.

Winslow, R. D. 1954. Provisional lists of host plants of some root eelworm (*Heterodera* spp.). *Ann. Appl. Biol.* 41: 591–605.

Winslow, R. D. 1955. The hatching responses of some root eelworms of the genus *Heterodera. Ann. Appl. Biol.* 43: 19–36.

Winstead, N. N., Skotland, C. B., and Sasser, J. N. 1955. Soybean-cyst nematode in North Carolina. *Plant Dis. Rep.* 39: 9–11.

Wollenweber, H. W. 1923. Krankheiten und Beschädigungen der Kartoffel. *Arb. d. Forschungs Inst. Kartoff. Berlin* 7: 1–56.

Wollenweber, H. W. 1924. Zur Kenntnis der Kartoffel-Heteroderen. *Illustrierte Landwirtschaftliche Zeitung* 44: 100–101.

Wouts, W. M. 1973a. A revision of the family Heteroderidae (Nematoda: Tylenchoidea). I. The family Heteroderidae and its subfamilies. *Nematologica* 18: 439–446.

Wouts, W. M. 1973b. A revision of the family Heteroderidae (Nematoda: Tylenchoidea). III. The subfamily Ataloderinae. *Nematologica* 19: 279–284.

Wouts, W. M. 1985. Phylogenetic classification of the family Heteroderidae (Nematoda: Tylenchida). *Syst. Parasitol.* 7: 295–328.

Wouts, W. M., and Sher, S. A. 1971. The genera of the subfamily Heteroderinae (Nematoda: Tylenchoidea) with a description of two new genera. *J. Nematol.* 3: 129–144.

Wouts, W. M., and Sturhan, D. 1978. The identity of *Heterodera trifolii* Goffart, 1932 and the description of *H. daverti* n. sp. (Nematoda: Tylenchida). *Nematologica* 24: 121–128.

Wouts, W. M., and Weischer, B. 1977. Eine-Klassifizierung von Fünfzehn in West Europa häufigen Arten der Heteroderinae auf grund von Larvenmerkmalen. *Nematologica* 23: 289–310.

Wrather, J. A., Anand, S. C., and Dropkin, V. H. 1984. Soybean cyst nematode control. *Plant Dis.* 68: 829–833.

Wyss, U. and Zunke, U. 1986. Observations on the behaviour of second stage juveniles of *Heterodera schachtii* inside host roots. *Rev. Nematol.* 9: 153–165.

Yeates, G. W. 1973. Annual cycle of root nematodes on white clover in pasture. I. *Heterodera trifolii* in a yellow grey earth. *New Zealand J. Agr. Res.* 17: 569–574.

Yeates, G. W., and Risk, W. H. 1976. Annual cycle of root nematodes on white clover in pasture. III. *Heterodera trifolii* in a yellow-brown earth. *New Zealand J. Agr. Res.* 19: 393–396.

Yeates, G. W., Ross, D. J., Bridger, B. A., and Visser, T. A. 1977. Influence of the nematodes *Heterodera trifolii* and *Meloidogyne hapla* on nitrogen fixation by white clover under glasshouse condition. *New Zealand J. Agr. Res.* 20: 401–413.

Yokoo, T. 1936. Host plants of *Heterodera schachtii* Schmidt and some instructions. (In Japanese) *Korea Agr. Exp. Sta. Bull.* 8: 167–174.

Zunke, U. 1986. Zur Bildung der subkriustallinen Schicht bei *Heterodera schachtii* unter aseptischen Bedingungen. *Nematologica* 31: 117–120.

8

The Family Pratylenchidae Thorne, 1949

PIETER A. A. LOOF   *Wageningen Agricultural University, Wageningen, The Netherlands*

The family Pratylenchidae contains at present eight genera with about 160 species, many of which are insufficiently described. Several genera contain species of great importance as pests of crops and ornamentals. All species parasitize higher plants; they are predominantly endoparasites of roots, but the adult females remain motile (except in *Nacobbus*).

The family is not very homogeneous. Generally spoken, it contains species of small body size (under 1 mm) with low, flat lip region, heavily sclerotized cephalic framework, short (under 25 µm) and stout stylet with large basal knobs; overlapping pharyngeal glands, fine transverse body striation; distinctly postanal phasmids; and a bursa which usually envelops the tail terminus. Deirids are absent except in one genus. Pharyngointestinal valve not well developed (except in *Pratylenchoides*), the junction being indicated only by sudden widening of the lumen and cessation of the sclerotized lining. However, in *Hoplotylus* and some species of *Hirschmanniella* the lip region is rather high. *Hirschmanniella* species are long (1.5–4 mm) and slender, some species have a long stylet (up to 50 µm), and the bursa does not extend till the tail terminus; the latter also holds for most species of the genus *Radopholus*.

The species belonging to this family are distinguished from Hoplolaimidae by the body posture (straight to curved, but never spiral), longer tail, postanal position of phasmids, usually flattened lip region, short stylet, and much finer transverse striation. The shape of lip region, the heavy cephalic sclerotization, and the overlapping pharyngeal glands distinguish Pratylenchidae from the majority of Belonolaimidae; however, the borderline between the genera *Pratylenchoides* (Pratylenchidae) and *Amplimerlinius* (Belonolaimidae) is somewhat vague.

Type genus: *Pratylenchus* Filipjev, 1936.

The family was revised by Luc (1987). An extensive discussion of its morphology and biology, with keys and descriptions of all species, was given by Ryss (1988).

KEY TO GENERA OF PRATYLENCHIDAE

1. Females monoprodelph 2
    Females didelph 7
2. Young adult females free in soil, with undifferentiated, short genital tubes; later they
    penetrate into roots with the anterior part of the body; they become sessile, immobile,
    swollen, and provoke root galls. Pharyngeal overlap long, dorsal ....
    .... *Nacobbus* Thorne and Allen, 1944
    (Dealt with in Chap. 11)
    Adult females, with fully grown out and differentiated genital tubes, vermiform,
    motile, at most slightly swollen 3
3. Pharyngeal glands forming a short ventral lobe; lip region flattened; female tail
    usually rounded .... *Pratylenchus* Filipjev, 1936*
    Pharyngeal glands forming a long dorsal lobe 4
4. Lateral field with three lines; female tail subacute; lip region conoid; stylet knobs
    narrow and high .... *Hoplotylus* s'Jacob, 1960
    Lateral field with four (rarely 3) lines; spear knobs not conspicuously narrow and high
    5
5. Lip region flattened in both sexes; male stylet not reduced, male pharynx slightly
    reduced .... *Apratylenchoides* Sher, 1973
    Lip region in females flattened or conoid, in males knob-shaped; male stylet and
    pharynx strongly reduced 6
6. Female lip region flattened; adult female vermiform; $c' = >3$ ....
    .... *Radopholus* Thorne, 1949 partim
    (former genus *Radopholoides* de Guiran, 1967)
    Female lip region conoid; adult female slightly swollen; $c' = <2$ ....
    .... *Acontylus* Meagher, 1968
    (Considered belonging to Hoplolaimidae by Fortuner, 1987, but might be mistaken
    for a pratylenchid)
7. Lip region flattened in female, knob-shaped in male; male stylet and pharynx strongly
    reduced .... *Radopholus* Thorne, 1949*
    No or slight sexual dimorphism in lip region shape; at most a very slight one in
    pharynx 8
8. Body length well over 1 mm; tail long, terminus often mucronate; body
    conspicuously slender; bursa not extending to terminus ....
    .... *Hirschmanniella* Luc and Goodey, 1964*
    Body length 0.9 mm or less; terminus not mucronate; body not conspicuously slender;
    bursa extending to terminus 9
9. Deirids conspicuous; pharyngeal glands surround intestine on all sides, the greatest
    overlap dorsally; slight sexual dimorphism in median pharyngeal bulb ....
    .... *Pratylenchoides* Winslow, 1958*
    Deirids absent; pharyngeal glands overlap ventrally; no sexual dimorphism in median
    bulb .... *Zygotylenchus* Siddiqi, 1963*

NOTE: The genera marked * are dealt with in this chapter; the main diagnostic characters
are shown in Fig. 1.

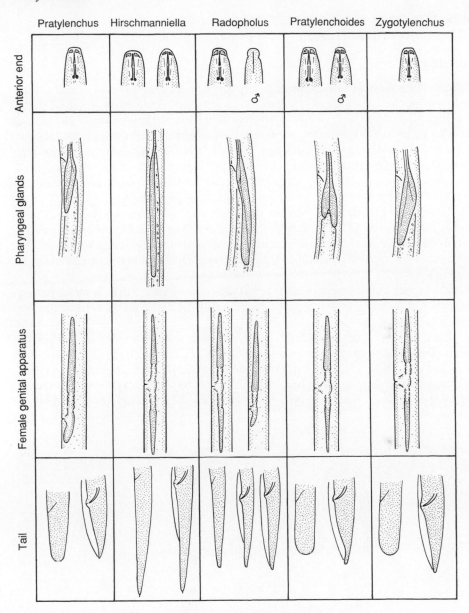

FIGURE 1   Main diagnostic characters of the genera treated.

GENUS *PRATYLENCHUS* FILIPJEV, 1936 (ROOT LESION NEMATODES)

Morphology

The species of *Pratylenchus* are morphologically very similar. This makes the genus as such easily recognizable, but the specific identification is very difficult. Body length of adults ranges from 0.3 to 0.9 mm. The body is rather stout, the index *a* being usually 20–30, in a few species up to 40. Gravid females are stouter than nongravid ones owing to the increase of uterus volume and the presence of eggs. The cuticle is thin and shows fine transverse striation. The lateral field basically is marked by four longitudinal lines, but additional ones may be present in the central zone. In gravid females the lateral field is indistinct due to stretching of the cuticle.

The head is low, distinctly broader than high, and built from two (Fig. 2A, D, E), three (Fig. 2B, C), or four (Fig. 2F) annules (one, two or three striae). It is not offset from the neck by a constriction, but often there is a small discontinuity in body outline at its base. The cephalic framework is heavily sclerotized, the outer margins of the basal plate extend one to two annules posteriad.

Most species have the edges of the apical annule rounded, but in *P. brachyurus* they are angular, so different from all other species that this by itself is sufficient to recognize this species (Fig. 2A).

The anterior surface of the apical annule may be raised, low-conical (*P. coffeae*, *P. neglectus*, Fig. 2E), or more flat (*P. penetrans*, Fig. 2B, *P. scribneri*, Fig. 2D, *P. thornei*, Fig 2C). In *P. scribneri* (Fig. 2D), *P. hexincisus*, and *P. jordanensis* the second annule is conspicuously wider than the apical one; by this character these species are clearly different from *P. neglectus*. In *P. thornei* the lip region is conspicuously high and wholly continuous with the neck.

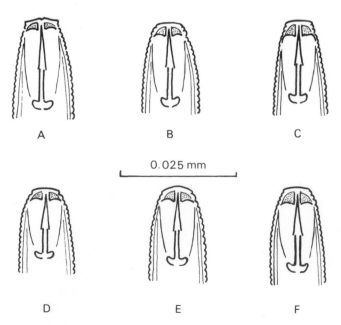

FIGURE 2   Head ends of (A) *Pratylenchus brachyurus*; (B) *P. penetrans*; (C) *P. thornei*; (D) *P. scribneri*; (E) *P. neglectus*; (F) *P. goodeyi*.

Corbett and Clark (1983) found that under SEM three types of head structure can be distinguished. This might be helpful in determining the status of some species, but it is not yet suitable for routine identification.

The stylet is short (11–25 μm, usually 14–17 μm) and stout, with well-developed basal knobs. The pharynx consists of a tapering procorpus, usually roundish median bulb, short isthmus, and a glandular region which over a distance of 16–70 μm overlaps the anterior end of the midintestine on the ventral side. This lobe contains three unicellular glands. The ventrosublateral ones are of unequal length, both are longer than the dorsal gland (Seinhorst, 1971). The orifice of the dorsal pharyngeal gland duct is located 2–4 μm behind the base of the stylet. Deirids are absent. The phasmids lie about midway between anal resp. cloacal aperture and tail terminus. In males they are elongate, running partly in the bursal flaps.

Females are monoprodelph. The posterior genital branch is present as a short sac, often undifferentiated, though in some species some cellular tissue is present at its distal end. The uterus is tricolumellar. The tail usually is two to three anal body diameters long. Males are generally smaller and more slender than females; they possess typical tylenchid spicules and gubernaculum; the caudal alae extend to the terminus of the tail. The genus contains at present some 70 species, but about one-third are insufficiently described. Type species: *P. pratensis* (de Man, 1880).

Biology

All *Pratylenchus* species are mobile endoparasites. They live usually in roots, rhizomes, or tubers, but occasionally they are found in above-ground parts like stems or fruits. Having penetrated into a root, they may multiply to very large numbers (10,000–35,000 specimens per 10 g of roots). All stages from the J-2 on may enter a root, but they can also wander out, live for some time in the soil (it is unknown how); and make for a new host root. The root parts where these nematodes multiply often assume a dark red or brown color, caused by necrosis of the invaded cells and invasion of secondary pathogens like fungi or bacteria. These discolored patches (lesions) have earned *Pratylenchus* species the vernacular name of root lesion nematodes and are a fairly good diagnostic.

Eggs are laid in the root and the whole life cycle may be passed there. The generation time is mostly 6–8 weeks, so that several generations may develop during one growing season. The functioning of infested roots may be heavily impaired. When, due to activity of fungi and bacteria, root rot has set in, the root becomes unfavorable for the nematodes and they wander out. Hence, in the beginning of the growing season there is often a good correlation between numbers of *Pratylenchus* specimens per quantity of roots and damage symptoms, but later this correlation disappears.

Several "sickness symptoms" (slowly expanding patches where plants show poor growth and reduced yield) are due to *Pratylenchus* species. Beside poor growth and yellowing there are mostly no clear above-ground symptoms. Often infestation by *Pratylenchus* predisposes plants to other diseases.

Several interactions between *Pratylenchus* species and fungi are known.

Most of the better known species have a wide distribution which is zonal in character. Some species occur throughout the northern temperate zone but occur also in mountainous regions of the tropics (Loof, 1964). Other species occur throughout the tropics but may maintain themselves in temperate zones in greenhouses. Gotoh (1974) found the species assortment in Japan gradually changing from north to south, from predominantly "temperate" species to predominantly "tropical" species.

Undoubtedly there has been much transport of *Pratylenchus* species with plants and this may be why several species are now almost cosmopolitan.

The species are very polyphagous.

It is not unusual to find two or three species together in a soil sample. For diagnostic work, therefore, it is necessary to identify at least 25 adult female specimens.

Identification Characters: Intraspecific Variation

Characters commonly used to diagnose and to distinguish species are the demanian indexes, body length, number of head annules, shape of head, length of stylet, shape of stylet knobs, structure of lateral field, shape of spermatheca, presence/absence of males, length and structure of posterior uterine sac (PUS), shape of female tail and terminus. Isolated males are generally not identifiable, so that only female characters will be dealt with.

In the following section the various characters will be discussed with particular emphasis on their intraspecific variation, which determines their reliability.

During recent years several studies on intraspecific variation were published, e.g., Taylor and Jenkins (1957), Roman and Hirschmann (1969), Tarte and Mai (1976a,b), Singh and Khan (1981), Olowe and Corbett (1984a,b). The general outcome of these studies is that the majority of commonly used diagnostic characters are highly variable. Tarte and Mai (1976a) even found in populations of *P. penetrans*, reared from single females, specimens possessing all characters of *P. fallax*, *P. convallariae*, and *P. pseudopratensis*, and most characters of *P. subpenetrans*, *P. vulnus*, and *P. sudanensis*. They consider it highly probable that some of these species are actually identical with *P. penetrans*.

Body length. In general specimens extracted from roots are longer (and stouter) than specimens extracted from soil. The descriptions of the five new specimens by Das and Sultana (1979) were based on specimens from soil; body length ranged from 0.41 to 0.56 mm, whereas for example, root specimens of *P. convallariae*, *P. goodeyi*, *P. kasari*, *P. loofi*, and *P. coffeae* have body lengths from 0.50 to 0.77 mm. Unfortunately, most of the literature does not give separate data of soil and root specimens, and often it is not even mentioned explicitly where described specimens came from. Nevertheless there are indications that body length has some importance. Soil specimens of *P. cruciferus* were 0.65–0.79 mm, of *P. macrostylus* 0.51–0.68 mm (Wu, 1971), 0.43–0.70 mm (Minagawa, 1982). Root specimens of *P. estoniensis* measured 0.33– 0.50 mm, of *P. hexincisus* 0.34–0.54 mm, of *P. sefaensis* 0.40–0.53 mm, of *P. subpenetrans* 0.33–0.48 mm.

However, host plant may influence length. On favorable hosts body length is greater than on unfavorable ones (Olowe and Corbett, 1984a). Fischer (1894) found *Tylenchus gulosus* (= *P. penetrans*) to reach great lengths on *Clematis jackmanii*, which I could verify experimentally (1960): a range of 0.54–0.81 mm was found. Roman and Hirschmann (1969) found in this species a mean length of 0.56 mm (0.51–0.66), Taylor and Jenkins (1957) 0.53 mm (0.43–0.62). Roman and Hirschmann found in *P. scribneri*, reared under standard conditions a mean body length of 0.50 mm (0.44– 0.55), whereas I found in a population on *Hippeastrum* in the Netherlands a mean length of 0.58 mm (0.50–0.66).

In general, therefore, body length should be used with great care as a diagnostic character, and in descriptions it should always be mentioned whether the specimens came from soil or from roots.

Vulva position. This appears to be a rather good character and it may be helpful in distinguishing between e.g. *P. penetrans* and *P. convallariae*, but again some 20 specimens should be measured because of overlap. Even this character shows a greater variation in *Pratylenchus* than in most other genera.

The other demanian indexes are of much less value.

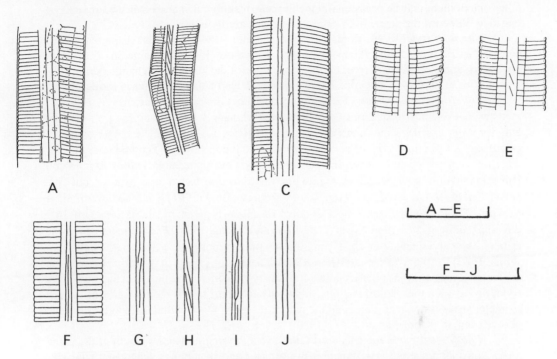

FIGURE 3   Lateral field of (A) *Pratylenchus penetrans* (four lines only); (B) *P. neglectus* (four lines plus oblique striae); (C) *P. crenatus* (six lines, inner ones interrupted); (D,E) *P. bolivianus* (partial areolation); (F–J) *P. sefaensis* (great variation). [(A–C) from Loof, 1960; courtesy of Nederlandse Plantenziektenkundige Vereniging, Wageningen; (D–E) from Corbett, 1984; courtesy of E. J. Brill, Leiden; (F–J) from Fortuner, 1973; courtesy of ORSTOM, Paris.] The scale lines correspond to 25 μm.

Lateral field. In general this consists of four longitudinal lines (Fig. 3A), occasionally with a fifth in the prevulvar region (e.g., in *P. penetrans*). The three bands between these lines are often of equal width, or the central one is wider; in *P. vulnus*, however, the central band is usually narrower. *P. hexincisus* was described as having six lines. In *P. crenatus* there are six lines too, but the two inner ones are interrupted at regular distances (Fig. 3C). In *P. neglectus* the central band is obliquely striated in the prevulvar region (Fig. 3B). These fine details can be observed, however, only on specimens from soil, since in the stouter root specimens the cuticle is stretched and often only the four main lines are visible. Fortuner (1973) described great variation in *P. sefaensis*: a fifth line, often irregular interrupted or branching, or *neglectus*-like oblique striae (Fig. 3F–J). In some species areolation has been observed (Fig. 3D,E), but the reliability of this as a diagnostic character is dubious.

Shape of lip region. Roman and Hirschmann found this a reliable character; also mean width may be diagnostic, but accurate measuring is very difficult. Among the six species studied by them *P. coffeae* had the narrowest lip region (females 6.4μm, males 7.3μm), *P. brachyurus* the widest (females 8.4 μm).

The number of head annules is constant in most species, but variable in *P. pseudopratensis*, *P. vulnus*, and *P. wescolagricus* (three or four). In several species with three annules one of them may be occasionally incomplete, resulting in a head with two annules on one side, three on the other. In *P. zeae* the annules are indistinct.

Shape of stylet knobs. Roman and Hirschmann (1969) found the shape reasonably constant in *P. brachyurus*, *P. scribneri*, and *P. vulnus*, highly variable in *P. zeae*, *P. penetrans*, and *P. coffeae*. Moreover, the appearance of the knobs may change during storage or after mounting in glycerin. The diagnostic value of this character is limited.

Stylet length. Seinhorst (1968) pronounced that the range in samples of 10 specimens was never more than 2 µm. Roman and Hirschmann (1969) found in samples of 50 females ranges of the same order: 1.2 µm in *P. vulnus*; 1.8 µm in *P. brachyurus* and *P. penetrans*; 2.4 µm in *P. coffeae* and *P. scribneri*, 3.0 µm in *P. zeae*. Yet the literature contains rather more strongly diverging statements. For example, in *P. vulnus* Roman and Hirschmann found stylet length 14.4–15.6 µm (this agrees with own observations), but Corbett (1974) gives a range of 14–19 µm, whereas van den Berg (1971) gives the lower limit as 12.5 µm. For *P. brachyurus* mostly a length of 18–20 µm is given, but the extremes are 16.9 µm (van den Berg, 1971) and 25 µm (Café-Filho and Huang, 1989). *Pratylenchus penetrans* usually has a stylet length of 15–17µm , but the range reported is 14 µm (Gotoh and Ohshima, 1963) to 19 µm (Sher and Allen, 1953). These wide ranges may partly be due to the difficulty of measuring a stylet accurately (it is very difficult to find the exact location of the stylet tip between the cheilorhabdions, and with such short stylets the thickness of the micrometer striae is not negligible) and the practical consequence is that the diagnostic value of stylet length is rather small.

Pharyngeal gland lobe. Corbett (1969) was the first to use the length of the gland lobe (from pharyngointestinal junction to posterior end of lobe) as diagnostic character when he described *P. pinguicaudatus*. Loof (1978) gave some data showing that this character shows high intraspecific variation (CV = 11–17%), but nevertheless range and mean are fairly constant and reliable (cf. Table 1). The character can be used only in specimens which have been killed, fixed, processed, and mounted carefully; otherwise pharynx and gland

TABLE 1   Length of Pharyngeal Overlap in Females of Some Species of *Pratylenchus*

Species	Source	n	Range (µm)	Mean (µm)
P. crenatus	Loof, 1978	50	17–38	25
P. neglectus	Loof, 1978	51	19–35	27
P. fallax	own obs.	30	20–44	31
P. zeae	own obs.	25	18–40	31
P. scribneri	Loof, 1978	40	26–41	35
P. bolivianus	own obs.	25	18–49	35
P. vulnus	own obs.	19	26–53	38
P. jordanensis	Hashim, 1983	15	29–53	39
P. thornei	Loof, 1978	47	30–55	40
P. convallariae	own obs.	26	31–55	41
P. sensillatus	Anderson and Townshend, 1985	21	39–66	46
P. goodeyi	own obs.	28	40–63	49
P. penetrans	Loof, 1978	48	32–62	50
P. penetrans	own obs.	24	34–65	50
P. brachyurus	Loof, 1978	51	31–68	51
P. pinguicaudatus	Corbett, 1969	5	54–66	61

FIGURE 4   (A–D) spermathecae: (A) Round (*P. penetrans*); (B) oval (*P. coffeae*); (C,D) elongate to rectangular (C: *P. goodeyi*; D: *P. loosi*). (E–G) posterior uterine branch: (E) short, undifferentiated (*P. neglectus*); (F,G) long, differentiated (F: *P. vulnus*; G: *P. coffeae*). [(D) from Seinhorst, 1977a; (F) from Corbett, 1974; both courtesy of C. I. P., St. Albans; (G) from Loof, 1960; courtesy Nederlandse Plantenziektenkundige Vereniging, Wageningen.] The scale lines correspond to 25 μm.

lobe may be distorted and shrunken, and the exact location of the pharyngointestinal junction can no longer be determined.

Shape of spermatheca (in amphimictic species). This is reasonably useful. Three main types can be distinguished:

Round to squarish, no longer than wide, e.g., *P. penetrans* (Fig. 4A), *P. convallariae*
Oval, e.g. *P. coffeae* (Fig. 4B)
Rectangular, e.g., *P. pratensis*, *P. goodeyi* (Fig 4C), *P. loosi* (Fig 4D) yet *P. coffeae*
    occasionally has a round spermatheca.

Length and structure of posterior uterine sac (PUS). Often it is difficult to determine the exact length of this structure, and also whether some cellular tissue is present at the distal end or not. Many species formerly reported to have an undifferentiated PUS are now known to show differentiation, at lease in part of the specimens (*P. crenatus*, *P. zeae*, and others). Roman and Hirschmann (1969) found the length of the PUS very variable within species, though the range differed for various species. *Pratylenchus vulnus* has a long PUS (30–56 μm), always with cellular differentiation. In *P. zeae* the PUS usually is short (11– 31 μm) but Roman and Hirschmann recorded one case where it measured 58 μm and carried a rudimentary ovary. In *P. coffeae* its length is highly variable (Loof, 1960).

　　Number of tail annules. Seinhorst (1968) considered this of some value though admitting that the number may change with level of focus (two annules may merge into one) and that it is hardly possible to determine objectively which annules belong to the tail proper and which to the terminus in species like *P. crenatus*. The number shows a wide range within species.

　　Shape of tail. The tail always tapers from the anus backward. Tails tapering strongly, with the terminus narrow (Fig. 5A–C), are generally called "conoid"; those tapering but little, with broad terminus (Fig. 5D, E) are called "subcylindrical." Actually, however, these two terms are inadequate, suggesting the existence of two well-separated classes of tail shape, whereas there is a continuous grading from markedly conoid (*P. zeae*) to nearly cylindrical (*P. thornei*). Moreover, females from roots tend to show, owing to their fatter bodies, a more conical tail shape than females of the same species from soil. Accurate drawings or photographs are essential, they say more than words.

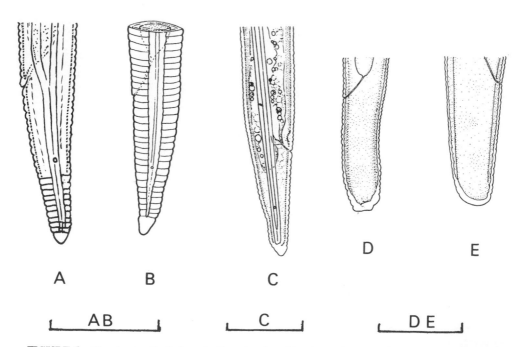

FIGURE 5　Two types of tail shape in *Pratylenchus*. (A–C) conoid (A: *P. loosi*; B: *P. vulnus*; C: *P. zeae*); (D,E) subcylindrical (D: *P. mediterraneus*; E: *P. andinus*). [(A) from Seinhorst, 1977a; (B) from Corbett, 1974; (C) from Fortuner, 1976; all courtesy of C. I. P., St. Albans; (D,E) from Corbett, 1984　(courtesy of E. J. Brill, Leiden).]

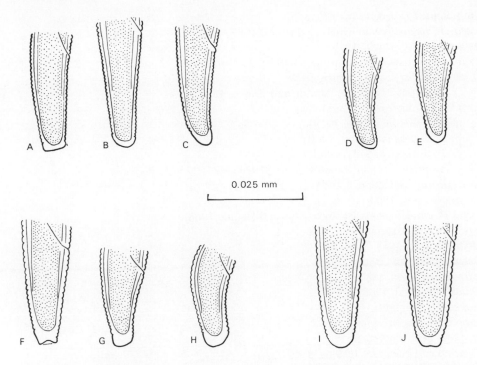

0.025 mm

FIGURE 6   Variation in female tail shape of (A–C) *P. thornei*; (D,E) *P. neglectus*; (F–H) *P. brachyurus*; (I,J) *P. coffeae*.

Shape of terminus. This is fairly constant in females, e.g., of *P. goodeyi* (dorsally sinuate), *P. thornei* (broadly rounded to truncate), (Fig. 6A–C), *P. brachyurus* (Fig. 6 F–H) and *P. coffeae* (Fig. 6 I,J) (terminus broadly rounded, truncate or slightly indented), *P. pratensis* (rounded to oblique, with fine striae) and *P. neglectus* (rounded to oblique, smooth). In soil specimens of *P. crenatus* the terminus often is clavate. In *P. vulnus* and *P. loosi* the terminus is narrowly rounded to subacute. *Pratylenchus zeae* has a narrow terminus too, but Taylor and Jenkins (1957) found its shape very variable. *Pratylenchus penetrans* typically has an evenly rounded, smooth terminus, but specimens with some annulation around it are far from rare (Tarte and Mai, 1976b). Provided that sufficient specimens are available, however, the shape of the terminus has a fair diagnostic value. When the cuticular striation extends around the terminus, the latter is called crenate; when not, it is called smooth.

Occasionally, other characters have been used, e.g., shape and backward extension of labial framework and distance between excretory pore and hemizonid, but more studies are needed to assess their value.

List of Species

P. acuticaudatus Braasch and Decker, 1989
P. agilis Thorne and Malek, 1968
P. alleni Ferris, 1961
P. andinus Lordello, Zamith, and Boock, 1961

P. bolivianus Corbett, 1984
   syn. *P. australis* Valenzuela and Raski, 1985
P. brachyurus (Godfrey, 1929)
   syn. *P. pratensis* apud Thorne, 1940
      P. leiocephalus Steiner, 1949
      P. steineri Lordello, Zamith, and Boock, 1954
P. coffeae (Zimmermann, 1898)
   syn. *P. musicola* (Cobb, 1919)
      P. mahogani (Cobb, 1920)
      P. pratensis apud Yokoo, 1956
   nec *Tylenchus musicola* apud Goodey, 1928 and 1932 = *P. goodeyi*
P. convallariae Seinhorst, 1959
P. crenatus Loof, 1960
   syn. *P. clavicaudatus* Baranovskaya and Haque, 1968
P. cruciferus Bajaj and Bhatti, 1984
P. delattrei Luc, 1958
P. ekrami Bajaj and Bhatti, 1984
P. estoniensis Ryss, 1982
P. fallax Seinhorst, 1968
P. flakkensis Seinhorst, 1968
P. gibbicaudatus Minagawa, 1982
P. goodeyi Sher and Allen, 1953
P. hexincisus Taylor and Jenkins, 1957
P. jordanensis Hashim, 1984
P. kasari Ryss, 1982
P. kralli Ryss, 1982
P. loosi Loof, 1960
P. macrostylus Wu, 1971
P. mediterraneus Corbett, 1984
P. microstylus Bajaj and Bhatti, 1984
P. morettoi Luc, Baldwin, and Bell, 1985
P. mulchandi Nandakumar and Khera, 1970
P. neglectus (Rensch, 1924)
   syn. *P. minyus* Sher and Allen, 1953
      P. capitatus Ivanova, 1968
      P. neocapitatus Khan and Singh, 1975
      P. similis Khan and Singh, 1975 (Jadid population)
P. penetrans (Cobb, 1917)
   syn. *Tylenchus gulosus* Kühn, 1890 (nomen oblitum)
      P. globulicola Romaniko, 1960
P. pinguicaudatus Corbett, 1969
P. pratensis (de Man, 1880)
   syn. *P. helophilus* Seinhorst, 1959
      P. irregularis Loof, 1960
P. pratensisobrinus Bernard, 1984
P. pseudofallax Café-Filho and Huang, 1989
P. pseudopratensis Seinhorst, 1968
P. scribneri Steiner in Sherbakoff and Stanley, 1943
P. sefaensis Fortuner, 1973
P. sensillatus Anderson and Townshend, 1985

P. subpenetrans Taylor and Jenkins, 1957
P. sudanensis Loof and Yassin, 1971
P. teres Khan and Singh, 1975
P. thornei Sher and Allen, 1953
  ?syn. *P. ranjani* Khan and Singh, 1975
P. typicus Rashid, 1974
P. ventroprojectus Bernard, 1984
P. vulnus Allen and Jensen, 1951
P. wescolagricus Corbett, 1984
P. zeae Graham, 1951
  syn. *P. cubensis* Razzhivin and O'Relly, 1975
   ?*P. impar* Khan and Singh, 975

Species *inquirendae*

P. barkati Das and Sultana, 1979
P. bicaudatus Meyl, 1954
P. brevicercus Das, 1960
P. cerealis Haque, 1966
P. chrysanthus Edward, Misra, Rai, and Peter, 1969
P. codiaei Singh and Jain, 1984
P. coffeae brasiliensis Lordello, 1956
P. crassi Das and Sultana, 1979
P. dasi Fortuner, 1985
  nom. nov. pro *P. capitatus* Das and Sultana, 1979
  syn. *P. hyderabadensis* Singh and Gill, 1986
P. emarginatus Eroshenko, 1978
P. exilis Das and Sultana, 1979
P. heterocercus (Kreis, 1930)
P. indicus Das, 1960
P. kolourus Fortuner, 1985
  nom. nov. pro *Tylenchus coffeae brevicauda* Rahm, 1925
P. loofi Singh and Jain, 1984
P. manohari Quraishi, 1983
P. montanus Zyubin, 1966
P. nizamabadensis Mahraju and Das, 1981
P. obtusicaudatus Romaniko, 1977
P. pratensis tenuistriatus Meyl, 1953
P. sacchari (Soltwedel, 1889)
P. singhi Das and Sultana, 1979
P. stupidus Romaniko, 1977
P. tenuis Thorne and Malek, 1968
P. tulaganovi Samibaeva, 1966
P. tumidiceps Merzeevskaya, 1951
P. uralensis Romaniko, 1966
P. variacaudatus Romaniko, 1977

Taxonomy

It is extremely difficult to construct a satisfactory key. Recently, three keys were published: Ryss (1988), Café-Filho and Huang (1989), and Handoo and Golden (1989). All three included all species known to authors, except the notorious *species inquirendae*. The key of Ryss covered 45 species; 11 species were sunk into synonymy. He tried to use few key characters. Café-Filho and Huang dealt with 55 species, Handoo and Golden with 63. The latter key gives much additional morphological information but it contains some errors, e.g., *P. brachyurus* was included under species with filled spermatheca and common males; *P. mulchandi* under species with a reflexed ovary (which in my opinion is an individual aberration rather than a sound taxonomic character). In addition, this paper gives a tabular compendium of the species treated. All three could not avoid making use of body length, stylet length, areolation of lateral field, and number of tail annules.

The uncertain state of the taxonomy of this genus is well illustrated by the widely diverging synonymies that have been proposed by e.g. Ryss (1988), Tarjan and Frederick (1988), and Frederick and Tarjan (1989).*

I have tried to make a shorter key by:

Leaving out all species which are insufficiently described and/or illustrated, or which are not clearly separated from several other species

Grouping some very similar species together.

Use of less reliable characters could not be wholly avoided.

Key to Species (Based on Females)

1. Lip region composed of two annules (one stria) 2
   Lip region composed of three annules (two striae) 13
   Lip region composed of four annules (three striae) 38
2. Spermatheca filled with sperm (males generally common) 3
   Spermatheca empty (males rare or absent) 6
3. Terminus crenate .... *P. flakkensis*
   (V = 73–77; stylet = 17 µm; apical lip annule flattened, distinctly narrower than second annule; spermatheca round to quadrangular; PUS undifferentiated)
   Terminus smooth, unstriated 4
4. Terminus narrowly rounded to subacute .... *P. loosi*
   (V = 79–85; body slender, a = 28–36; apical lip annule slightly rounded; spermatheca oval)
   Terminus broadly rounded, truncate or indented 5
5. Spermatheca spherical; stylet length under 15 µm .... *P. alleni*
   (V = 78–83; stylet = 13.5–15 µm; PUS short (just over body width), undifferentiated; tail bluntly rounded, terminus often with one or two indentations)
   Spermatheca oval; spear length over 15 µm .... *P. coffeae*
   (V = 76–84; stylet = 15–18 µm; PUS very variable in length, usually with rudimentary ovary; terminus broadly rounded, truncate, or indented)
6. Terminus crenate 7
   Terminus smooth, or with single terminal indentation 8

---

*The paper of Frederick and Tarjan (1989) came out after the text of this chapter was sent in, so that it could not be taken into account.

7. Vulva 70–77 .... *P. gibbicaudatus*
   (Stylet = 14–16 μm; lip region flattened, second annule distinctly wider than first; pharyngeal overlap about 40 μm; tail with 24–39 annules, terminus often irregular; PUS with rudimentary ovary, 2 BW long; lateral field with four lines)
   Vulva 79–86 .... *P. estoniensis*
   (Stylet 15.5–17 μm; lip region low domeshaped; tail with 22–30 annules, terminus regularly rounded; lateral field with six lines)

8. Lateral field with six lines .... *P. hexincisus*
   (V = 75–82; stylet 14.5–15.4 μm; lip region flattened; pharyngeal overlap short (24 μm in illustration); PUS short, undifferentiated; terminus round to truncate)
   Lateral field with four (occasionally five) lines, sometimes with oblique striation in central band 9

9. Stylet length over 19 μm; V = 82–89 10
   Stylet length 18 μm or less 11

10. Lip region low, angular; tail distally often subcylindrical .... *P. brachyurus*
    (Pharyngeal overlap about 50 μm (31–68); PUS short, undifferentiated; terminus smooth, broadly rounded to truncate)
    Lip region higher rounded; tail conoid .... *P. macrostylus*
    (Pharyngeal overlap 21–48 μm; PUS 1.0–1.7 BW long, slightly differentiated; tail tapering to narrowly rounded or irregular terminus)

11. Vulva 80–87 .... *P. neglectus*
    (Pharyngeal overlap 27 μm (19–35); apical head annule low dome-shaped; stylet usually 15–17 μm; PUS short, undifferentiated; lateral field in prevulvar region with oblique striation in central band; tail tapering to rounded, often oblique terminus)
    Vulva 73–80 12

12. Orifice of dorsal pharyngeal gland 3 μm or more from stylet base; c' = 2.1–3.0; cuticle of tail indented terminally .... *P. jordanensis*
    (Stylet 14.5–15.0 μm; pharyngeal overlap 39 μm (29–53); PUS short, undifferentiated)
    Orifice of dorsal pharyngeal gland less than 2.5 μm from stylet base; c' = 2.0–2.3; terminal cuticle not indented .... *P. scribneri*
    (Stylet 12–16 μm; pharyngeal overlap 35 μm (20–45); apical head annule flattened; PUS 1.1–1.4 BW long, undifferentiated; terminus usually broadly rounded). Very similar are *P. agilis* Thorne and Malek, 1968 (cuticular striation more distinct) and *P. acuticaudatus* Braasch and Decker, 1989 (cuticular striation more distinct, tail narrowly rounded)

13. Spermatheca filled with sperm (males generally common) 14
    Spermatheca empty (males rare or absent) 27

14. Terminus smooth 15
    Terminus annulated 20

15. Terminus oblique-truncate with ventral projection .... *P. ventroprojectus*
    (V = 78–80; stylet = 14–16 μm; spermatheca oval, rectangular; or elongate; PUS 0.9–1.8 BW long, usually with one terminal cell; terminus smooth or with 1-2 indentations). Terminus regularly rounded 16

16. Terminus bluntly pointed to narrowly rounded. Body long (0.46–0.91 mm) and slender (a = 25–40); PUS long, differentiated .... *P. vulnus*
    (Head with three or four annules; stylet = 13–19 μm; central zone of lateral field usually narrower than marginal ones; median bulb narrower, more oblong than in most other species; pharyngeal overlap 38 μm (26–53). Very similar is *P. ekrami* Bajaj and Bhatti, 1984, apparently distinguished only by shorter stylet (11–13 μm and constantly three head annules)
    Terminus broadly rounded 17

17. Vulva 70–76; spermatheca oval .... *P. sudanensis*
     (Stylet = 14–16 μm; PUS 1–1.5 BW, undifferentiated; tail subcylindrical, with
     broadly rounded, smooth terminus)
     Vulva 76–84; if <76 then spermatheca round 18
18. Lip region high; tail almost truncate; spermatheca round to broadly oval ....
     .... *P. mediterraneus*
     (*V* = 77–80; stylet = 13–15 μm; central zone of lateral field in midbody with oblique
     striae which sometimes give the impression of six lateral lines; PUS about one body
     width long)
     Lip region normal 19
19. Spermatheca oval; *V* = 76–80 .... *P. pseudopratensis*
     (Stylet = 15 μm; body often narrowed ventrally behind vulva; lip region flattened
     anteriorly; terminus round to truncate; PUS short, undifferentiated)
     Spermatheca round; *V* = 77–84 .... *P. penetrans*
     (*V* = usually 78–84; stylet = 15–17 μm; pharyngeal overlap 50 μm (32– 65); PUS
     1.0–1.5 BW long; spermatheca round to squarish; terminus evenly rounded, terminal
     cuticle slightly thickened. Very similar are *P. kralli* Ryss, 1982, which apparently
     differs only by the more angular terminus, and *P. subpenetrans* Taylor and Jenkins,
     1957, differing by longer, differentiated PUS. Ryss (1988) considered both species
     identical with *penetrans*)
20. Terminus pointed; tail annules > 30 21
     Terminus round; tail annules < 30 22
21. Spermatheca round; posterior uterine sac more than 2.5 body widths long ......
     .... *P. morettoi*
     (Body long, 0.56–0.93 mm, and slender, *a* = 26–40, stylet = 14–19 μm, *V* = 73–80, tail
     46–62 μm long, usually with mucro)
     Spermatheca oblong; posterior uterine sac less than 1.8 body widths long ....
     .... *P. kasari*
     (Body rather long, 0.56–0.77 mm; *V* = 75–81; stylet = 16–18 μm; tail length under 40
     μm, terminus without mucro. Ryss (1988) considered *kasari* and *morettoi* identical)
22. Spermatheca oblong 23
     Spermatheca not clearly longer than wide 24
23. *c* = 12–15; stylet 15–17 μm .... *P. pratensisobrinus*
     (*V* = 75–80; spermatheca round to elongate. Considered possibly an extreme variant
     of *P. pratensis* by Bernard (1984); Ryss (1988) synonymized it with *P. kasari*)
     c = 15–24; stylet 13–16 μm .... *P. pratensis*
     (*V* = 75–79; spermatheca large, elongate, farther anterior to vulva than usual in this
     genus; terminus rounded to oblique, sometimes irregularly knobbed)
24. Terminus truncate, tail annules < 20 .... *P. convallariae*
     (*V* = 76–81; stylet 16–18 μm; PUS 1.4–2 BW, undifferentiated; terminus truncate to
     indented)
     Terminus not truncate or oblique-truncate 25
25. Terminus oblique-truncate with ventral projection .... *P. ventroprojectus*
     (see under no. 15)
     Terminus round 26
26. Posterior uterine sac 1.6–2.7 body widths long .... *P. pratensisobrinus*
     (see under no. 23)
     Posterior uterine sac no longer than 1.4 body widths .... *P. fallax*
     (*V* = 77–81; stylet = 16–17 μm; pharyngeal overlap 31 μm (20–44); PUS often with
     some terminal differentiation; terminus rounded or slightly irregular. Very similar is
     P. pseudofallax Café-Filho and Huang, 1989, differing from *P. fallax* by shorter stylet
     (14–16 μm), deeper body striation and undifferentiated PUS)

27. Terminus smooth 28

Terminus annulated 37

28. Stylet length 11–12 μm .... *P. microstylus*

(*V* = 75–77; pharyngeal overlap 28–46 μm; PUS short, undifferentiated; terminus broadly rounded)

Stylet length 14–20 μm 29

29. Head high, not offset; tail truncate; body narrowed ventrally behind vulva; *V* = 73–80 .... *P. thornei*

(Stylet = 15–19 μm; pharyngeal overlap 40 μm (30–55); PUS about 1.5 BW, undifferentiated)

Head normal, usually slightly offset; body not narrowed behind vulva 30

30. Vulva 65–76 .... *P. zeae*

(Stylet = 15–18 μm; pharyngeal overlap = 31 μm (18–40); head annules often indistinct; PUS 1–2 BW long, often with some ovarial tissue; intestine overlaps rectum)

Vulva 73–83 31

31. Tail subcylindrical, broadly rounded 32

Tail tapering, more narrowly rounded 36

32. Posterior uterine sac shorter than body width .... *P. pinguicaudatus*

(*V* = 78–84; stylet = 15–20 μm; lateral field areolated; tail very plump, with broadly rounded terminus. Very similar is *P. cruciferus* Bajaj and Bhatti, 1984, distinguished by greater length (0.65–0.79 vs. 0.42–0.63 mm) and shorter pharyngeal overlap (31–51 μm vs. 52–70). However, Brzeski and Szczygieł (1977) illustrated an overlap of 37 μm in *P. pinguicaudatus*, so the two species cannot be satisfactorily distinguished at present)

Posterior uterine sac longer than body width 33

33. Posterior uterine sac undifferentiated .... *P. andinus*

(*V* = 78–81 (Corbett, 1984; in original description 81–85), stylet = 15–18 μm; PUS 1.5 BW (19–34 μm); pharyngeal overlap 28–53 μm; tail very plump, terminus broadly rounded; lateral field partly areolated)

Posterior uterine sac differentiated 34

34. Vulva 75–78, stylet length 16–20 μm .... *P. mulchandi*

(PUS more than 1.5 BW; terminus truncate, oblique, broadly or occasionally narrowly rounded)

Vulva 77–81, stylet length 14–18 μm 35

35. Body slender (*a* = 28–42); pharynx short (*b* > 7) .... *P. sensillatus*

(*L* = 0.57–0.69 mm; stylet = 15–17 μm ; pharyngeal overlap 46 μm (39–66); PUS 23 μm (14–32); tail very plump)

Body stouter (*a* = 17–31); pharynx longer (*b* = <7) .... *P. sefaensis*

(*L* = 0.40–0.53 mm; stylet = 14–16 μm; tail very plump; PUS 15–30 μm; pharyngeal overlap 21–41 μm; lateral field very variable)

36. Vulva 73–81 .... *P. delattrei*

(Stylet = 14–18 μm; PUS 1.2–1.4 BW long, undifferentiated; a small (0.39–0.55) and stout (*a* = 20–28) species)

Vulva 77–83 .... *P. bolivianus*

(*V* = 80–82; stylet = 17–20 μm; pharyngeal overlap = 18–49 μm; PUS 22–31 μm or 1.25 BW long, undifferentiated. In accordance with Frederick and Tarjan (1989) I consider *P. australis* Valenzuela and Raski, 1985 identical with *P. bolivianus*)

37. Vulva 80–86; posterior uterine sac longer than body width .... *P. crenatus*

(Stylet = 14–18 μm; pharyngeal overlap short, 25 μm (17–38); terminus coarsely striated, often clavate in younger females; cuticular sculpture rather deep and conspicuous)

Vulva 70–77; posterior uterine sac no longer than body width .... *P. teres*
(Stylet = 17–18 μm; pharyngeal overlap = 54 μm (illustration); tail tapering, terminus
round or oblique)

38. Tail dorsally sinuate just before terminus; $V = 73$–75 .... *P. goodeyi*
(Stylet = 15–17 μm; pharyngeal overlap = 48 μm (40–63); spermatheca large,
rectangular)
Tail regular, not sinuate 39

39. Tail annules 34–53; tail pointed; spermatheca round .... *P. morettoi*
(see under no. 21)
Tail annules < 30; tail rounded; spermatheca oval or empty 40

40. Spermatheca empty 41
Spermatheca oval, filled with sperm 42

41. Stylet length 15–17 μm .... *P. typicus*
($V = 79$–84; PUS long, differentiated; tail conoid, terminus generally narrowly
rounded. Males unknown and it is uncertain if the filled spermatheca is not rather a
developing oocyte with granular contents)
Stylet length 17–20 μm .... *P. wescolagricus*
($V = 78$–82; pharyngeal overlap = 21–49 μm; PUS 1 BW long, undifferentiated; tail
subcylindrical, terminus smooth, broadly rounded)

42. Tail narrowly rounded to subacute; posterior uterine sac two body widths long,
differentiated; body slender ($a = 25$–39) .... *P. vulnus*
(see under no. 16)
Tail broadly rounded to truncate; posterior uterine sac one body width long,
undifferentiated; body stout ($a = 21$–25) .... *P. pseudopratensis*
(see under no. 19)

Description of Some Important Species

Pratylenchus coffeae (Zimmerman, 1898) (Fig. 7)
Syn. *Pratylenchus pratensis* apud Yokoo, 1956

Dimensions: $L = 0.37$–0.83 mm; $a = 18$–35; $b = 5.0$–8.7; $b' = 3.8$–6.5; $c = 14$–28; $c' = 1.5$–2.5; $V = 74$–84; stylet = 14–18 μm; pharyngeal overlap = 40 μm (24–54).

Lip region composed of two annules, the anterior one somewhat convex and not
conspicuously narrower than second. Spermatheca round to broadly oval, filled with sperm;
males are common. Terminus broadly rounded, truncate, or indented. Posterior uterine
branch very variable in length: 17–50 μm; when long. It usually carries some ovarial tissue.
More than one egg in the uterus is not uncommon in this species (Siddiqi, 1972); even en-
dotakia matricida does occur (Loof, 1959).

Geographic distribution: Pantropic; also in Japan, Australia, South Africa, Brazil,
and the southern parts of the USA.

Hosts: coffee, banana, mahogany, abacà, citrus, apple, potato, and many other crops
and weeds. Heavy damage often is inflicted to coffee, banana, and citrus. Cobb (1920)
found it inhabiting the bark of the mahogany tree.

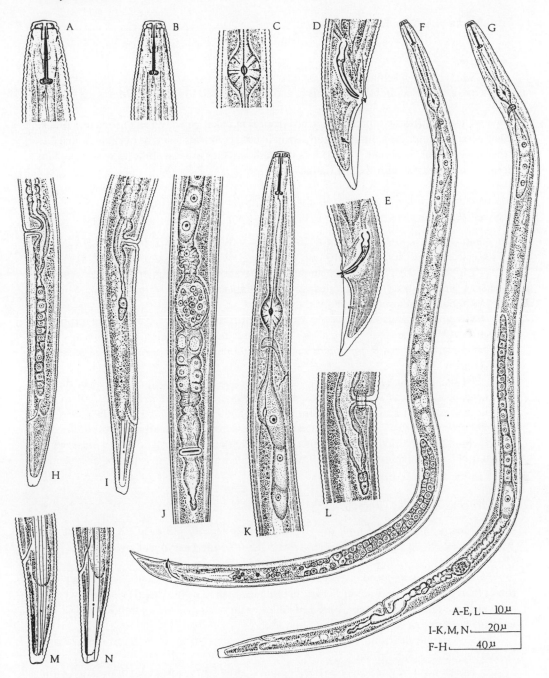

FIGURE 7 *Pratylenchus coffeae* (A) female head end; (B) male, head; (C) male, median bulb; (D,E) male tail; (F) whole male; (G) whole female; (H, I, M, N) female tails; (J) reproductive region; (K) female, neck region; (L) vulva and posterior uterine sac. (From Siddiqi, 1972; courtesy of C. I. P., St. Albans.)

Pratylenchus loosi Loof, 1960 (Fig. 8)

Dimensions: L = 0.48–0.64 mm; a = 28–36; b = 5.7–7.1; c = 18–25; V = 79–85; stylet = 14–18 μm.

This species resembles *P.coffeae* to a large degree, it is more slender, the vulva position is more posterior, and the terminus is narrowly rounded to subacute. Males are numerous.

Geographic distribution: Described from Sri Lanka, where it is a serious pest of tea. It has been reported also from India and Japan, in both countries from tea but also from a few other hosts. After Seinhorst (1977a) it is not sure whether at least some Japanese populations are conspecific with the Sri Lanka ones.

Pratylenchus scribneri Steiner in Sherbakoff and Stanley, 1943 (Fig. 9)
   Syn. *Tylenchus penetrans* Cobb, 1917 (illustration)

Dimensions: L = 0.40–0.66 mm; a = 20–33; b = 4.6–7.6; b' = 3.2–6.0; c = 13–22; c' = 2.1–3.3; V = 72–80; stylet = 13–16 μm; pharyngeal overlap = 35 μm (26–41).

Lip region composed of two annules, the basal one being higher and wider than the apical one; spermatheca empty (males very rare); vulva anterior; terminus smooth, without terminal indentation, usually broadly rounded.

Geographic distribution: USA, Mexico, Japan, India, Israel, Turkey, Egypt, Nigeria, South Africa. The Near East and African records need confirmation; the species may have been confused with *P. jordanensis*. In the Netherlands it has been found imported in *Hippeastrum* spp.

Hosts: The most important hosts are corn, tomato, sugar beet, onion, soybean, potato, snap bean, *Amaryllis* and *Hippeastrum* species, and *Cymbidium* spp. On potato this species causes pimples on the tubers. In *Amaryllis* and *Hippeastrum* this species causes a growth decline, the root system is reduced, shows lesions, and is finally destroyed; the disease is common in Florida but occurs also in the Netherlands. The orchid *Cymbidium*, when infested, shows poor growth, reduced flowering and yellowing of leaves; roots and pseudobulbs show lesions or rottings.

Pratylenchus neglectus (Rensch, 1924) (Fig. 10)

Dimensions: L = 0.31–0.59 mm; a = 17–32; b = 4.0–9.8; b' = 3.6–5.3; c = 14–26; c' = 1.5–2.7; V = 76–88 (usually 81–86); stylet = 15–17 μm; pharyngeal overlap = 27 μm (19–35).

Lip region composed of two annules, the anterior one convex and not clearly narrower than posterior one. Lateral field with oblique striae in central band (visible in young, not yet reproducing females only). Spermatheca empty (males are extremely rare). Tail conical, terminus smooth, rounded to oblique.

Geographic distribution: Northern temperate zone (North America, Europe, Japan), Australia, India, South Africa.

Hosts: cereals, grasses, crucifers, legumes, strawberry, tobacco, peppermint, corn, bean. Damage has been reported in cereals (especially barley), tobacco, and peppermint.

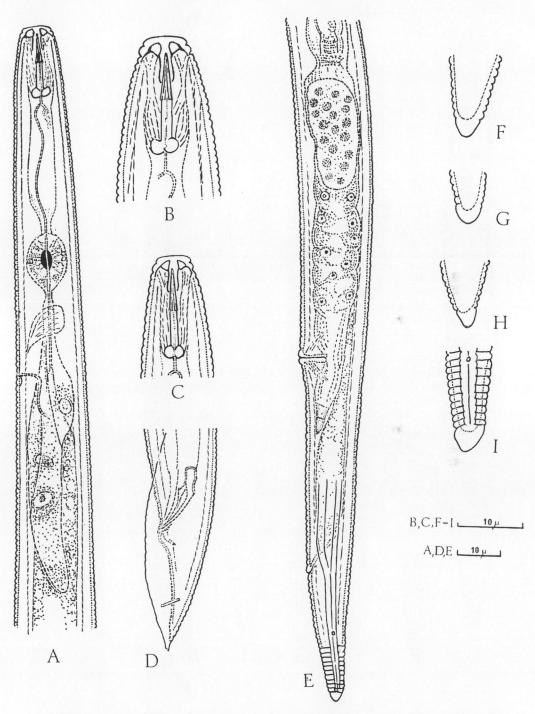

FIGURE 8 *Pratylenchus loosi*. (A) female, anterior end; (B) female head; (C) male head; (D) male tail; (E) posterior part of female; (F–I) variations in terminus of female. (From Seinhorst, 1977a; courtesy of C. I. P., St. Albans.)

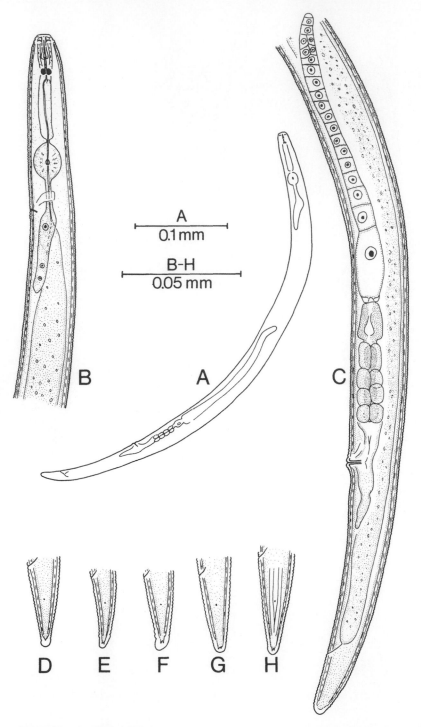

FIGURE 9   *Pratylenchus scribneri*, female. (A) Whole specimen; (B) anterior part; (C) posterior part; (D–H) variation in tail shape. (From Loof, 1985; courtesy of C. I. P. St. Albans.)

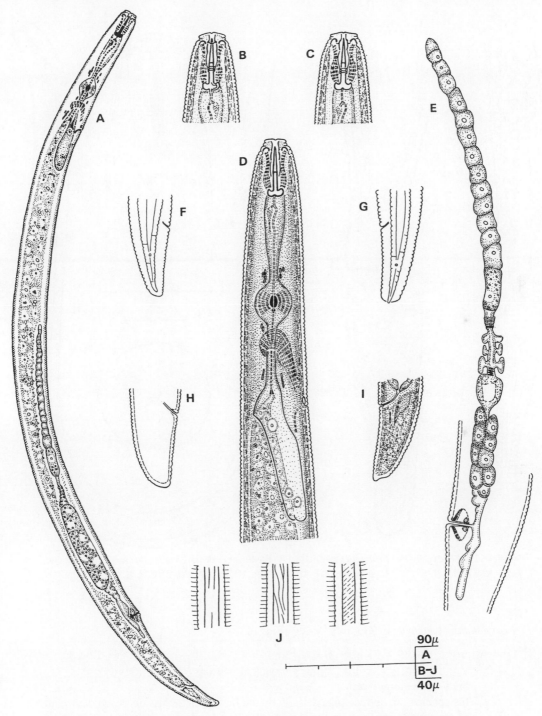

FIGURE 10 *Pratylenchus neglectus*, female (A) whole specimen; (B,C) head end; (D) neck region; (E) reproductive apparatus; (F–I) variation in tail shape; (J) variation in lateral field in midbody. (From Anderson, 1976; courtesy of C. I. P., St. Albans.)

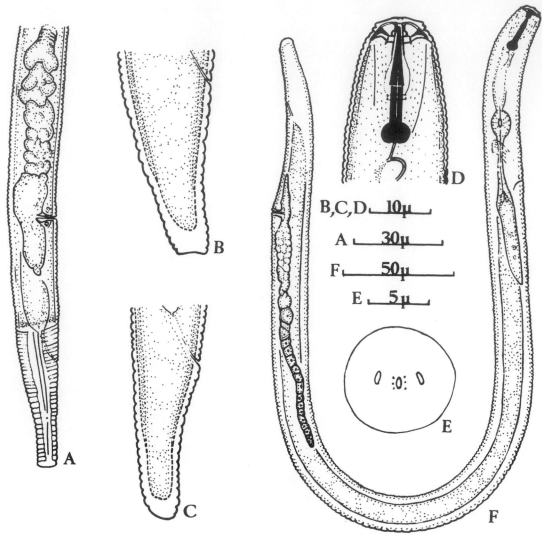

Figure 11   *Pratylenchus brachyurus*, female. (A) Posterior end; (B,C) tails; (D) head end; (E) end-on view of head; (F) whole specimen. (From Corbett, 1976; courtesy of C. I. P., St. Albans.)

Pratylenchus brachyurus (Godfrey, 1929) (Fig. 11)

Dimensions: L = 0.39–0.75 mm; a = 15–30; b = 4.6–9.3; b' = 3.0–4.4; c = 12–28; c' = 1.6–2.8; V = 82–89; stylet = 17–22 μm; pharyngeal overlap = 51 μm (31–68).

Males are extremely rare.

This species can be recognized easily by the lip region being composed of two annules, the anterior one showing an angular contour; long stylet; empty spermatheca; very posterior vulva position; smooth, rounded, often truncate terminus.

Geographic distribution: Pantropic; also in USA, South Africa, Australia, Japan, and Turkey.

Hosts: Severe damage is done to peanut in the USA and Australia. The nematode occurs chiefly in pegs and pod shells; often the pegs are weakened and the infested pods drop off, reinfecting the soil. On the other hand, carefully removing pegs and stalks at harvest has the effect of cleaning the soil. Other important hosts are potato (tuber attack in South Africa), pineapple (Hawaii, Brazil, western Africa), citrus, cotton, peach, soybean, tobacco, coffee, rubber, several Poaceae; *Eucalyptus* in Brazil.

Pratylenchus zeae Graham, 1951 (Fig. 12)

Dimensions: L = 0.34–0.64 mm; a = 17–32; b = 4.9–9.6; b' = 3.2–4.6; c = 11–24; c' = 2.3–3.7; V = 65–76; stylet = 15–18 μm; pharyngeal overlap = 31 μm (18–40).

Males are extremely rare.

This species can be recognized easily by the three, often indistinct head annules, empty spermatheca, anterior vulva position, tapering tail with narrowly rounded to subacute terminus, and high number of tail annules (20–35). The posterior uterine sac often shows some ovarial cell tissue.

Geographic distribution: Pantropic, also in USA, Egypt, Zimbabwe, South Africa, Iraq, Japan, and Australia.

Host plants: Many Poaceae: corn, sorghum, rice, sugar cane, wheat, barley, and several grasses; furthermore tobacco and peanut. Corn, rice, sugar cane, and especially tobacco often suffer heavy damage.

Pratylenchus thornei Sher and Allen, 1953 (Fig. 13)

Dimensions: L = 0.45–0.77 mm; a = 26–38; b = 4.8–8.3; b' = 3.9–5.8; c = 17–28; c' = 2.2–3.2; V = 73–80; stylet = 15–19 μm; pharyngeal overlap = 40 μm (30–55).

This species is recognizable by the high, not offset lip region composed of three annules; cephalic sclerotization extending farther backward than in most other species; empty spermatheca (males are extremely rare), long distance vulva–anus, tail terminus unstriated, broadly rounded to truncate. Often the body is narrowed on the ventral side posterior to the vulva. Several workers from the Mediterranean area have reported populations of *P. thornei* with numerous males, but these were considered a distinct species *P. mediterraneus* by Corbett (1984).

A detailed discussion of variation in numerical characters was given by Singh and Khan (1981).

Geographic distribution: USA, Mexico, India, Australia, Egypt, Canary Islands, South Africa, India, Iran, Japan, and several European countries (the Netherlands, Belgium, Italy, Germany, Yugoslavia).

Hosts: *P. thornei* is a serious pest of wheat in the USA (Utah), Mexico, India, Australia, and Yugoslavia; it also damages corn, barley, and apple, but has many other host plants among which are fruit trees, roses, and other ornamentals. The data were summarized by Fortuner (1977).

Pratylenchus crenatus Loof, 1960 (Fig. 14)

Syn. *P. pratensis* auctt. (prior to 1960) nec (de Man, 1880)

Dimensions: L = 0.32–0.60 mm; a = 20–32; b = 4.9–7.9; b' = 4.2–5.5; c = 16–27; c' = 1.9–2.7; V = 78–86 (usually 81–84); stylet = 14–18 μm; pharyngeal overlap = 25 μm (17–38).

This species is easy to recognize by lip region consisting of three annules (occasionally four?), posterior vulva position, distinct cuticular striation, lateral field with six lines

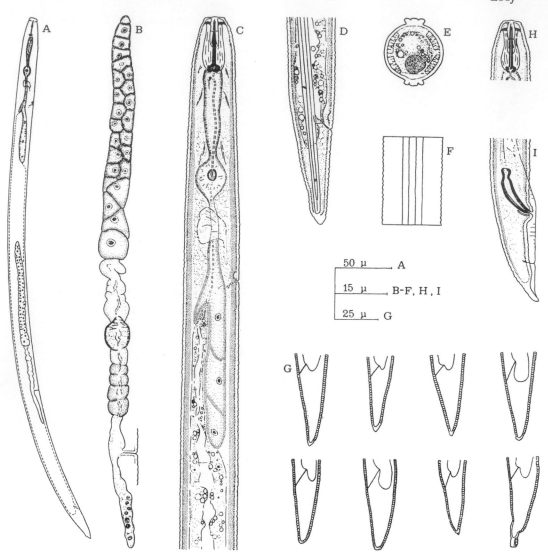

FIGURE 12 *Pratylenchus zeae.* (A–G) Female. (A) Whole specimen; (B) reproductive apparatus; (C) neck region; (D) tail; (E) cross section through midbody; (F) lateral field; (G) variation in tail shape. (H,I) male. (H) head end; (I) tail. (From Fortuner, 1976; courtesy of C. I. P., St. Albans.)

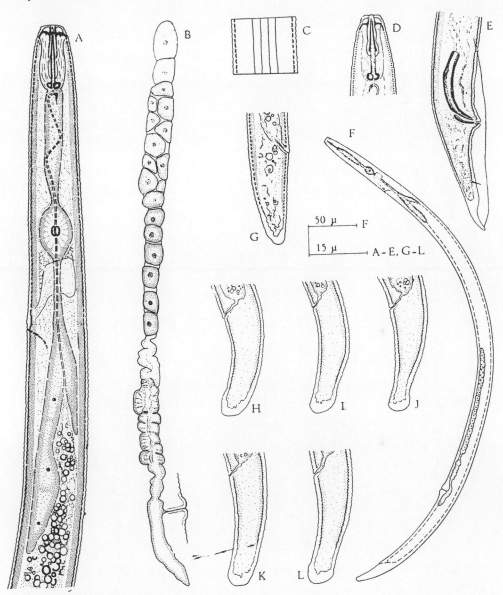

FIGURE 13 *Pratylenchus thornei.* (A–C, F–I) Female. (A) Neck region; (B) reproductive apparatus; (C) lateral field; (F) whole specimen; (G–L) variation in tail shape. (D,E) Male. (D) Head end; (E) tail. (From Fortuner, 1977; courtesy of C. I. P., St. Albans.)

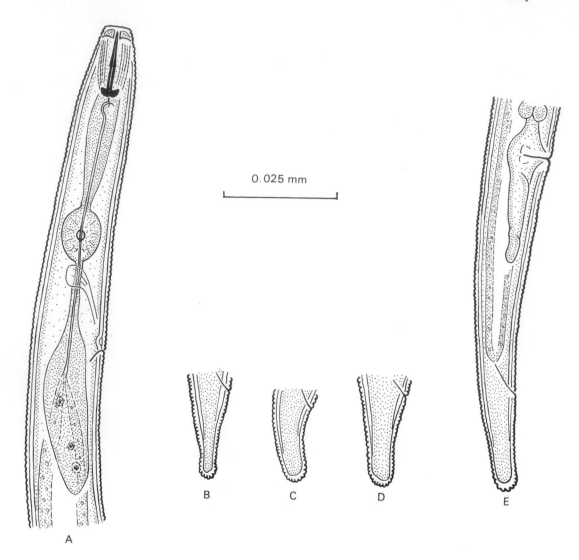

0.025 mm

FIGURE 14   *Pratylenchus crenatus*, female. (A) Neck region; (B–D) variation in tail shape; (E) posterior part.

(the inner ones interrupted), empty spermatheca, long posterior uterine sac (up to two body widths, or (after Seinhorst, 1968) 40–50% of vulva–anus distance, rounded, often clavate tail terminus with coarse crenations running all along it. Males have never been found with certainty.

Geographic distribution: Northern temperate zone (North America, Europe, Japan), South Africa, Venezuela (Andes mountains, 3500 m above sea level).

It occurs predominantly in light sandy soils and in Europe is often found together with *P. penetrans*. Hosts include cereals (which suffer heavy damage), grasses (this species is suspected of being the main cause of difficulties often encountered when breaking up old meadows and sowing new grasses immediately); carrots (causing a "carrot sickness": the main root is small and often branched, the other roots are short, with lesions and dead tips; above-ground symptoms are patches of poorly growing, thin, pale plants).

Pratylenchus penetrans (Cobb, 1917) (Fig. 15)

Dimensions: L = 0.34–0.81 mm (usually 0.43–0.65 mm); a = 17–34; b = 4.1–8.1; b' = 2.6–4.7; c = 14–26; c' = 1.6–2.7; V = 78–84 (occasionally 75–77); stylet = 15–17 μm; pharyngeal overlap 50 μm (32–65). This species can be recognized by the lip region composed of three annules, the apical one flattened with rounded edges; relatively long pharyngeal overlap; round spermatheca, filled with sperm; PUS generally short, undifferentiated; tail tapering to broad, evenly rounded unstriated terminus; and common occurrence of males. However, females with one or two striae around the terminus are not rare (Tarte and Mai, 1976b), so that differentiation from *P. fallax* may be difficult; however, *P. fallax* has exclusively crenate-tailed females and the pharyngeal overlap is shorter (20–44μm). *Pratylenchus convallariae*, which also may be difficult to distinguish from crenate-tailed *P. penetrans*, has a more anterior vulva position (75–81) and the terminus is mostly truncate, sometimes indented. After Roman and Hirschmann (1969) the PUS may be as long as 36 μm which casts doubt on the validity of *P. subpenetrans* Taylor and Jenkins, 1957.

Distribution: Almost cosmopolitan, but chiefly in the temperate zones. In Europe it is probably imported (Seinhorst, 1968). *Pratylenchus penetrans* is a well-known and important pest of many crops and ornamentals. Corbett (1973) summarized the information. Notable is that the host plant list includes a species of fern, i.e., *Rumohra adiantiformis* (Forst.) Ching (Rhoades, 1968). There are indications that several pathotypes exist within this species: Slootweg, 1956, Olthof, 1968; Townshend, Tarte, and Mai (1978) suggested that populations with many crenate-tailed females could represent a pathotype different from such with few crenate-tailed females.

Several interactions with fungi are known, e.g., *Verticillium alboatrum* (Muller, 1974). *Trichoderma* species suppress *P. penetrans* (Miller and Anagnostakis, 1977). In slowly dehydrated soils *P. penetrans* may survive in a coiled, anhydrous condition for more than 2 years (Townshend, 1984). On alfalfa *P. penetrans* was found to inhibit *Ditylenchus dipsaci* (Vrain, 1987).

Pratylenchus fallax Seinhorst, 1968 (Fig. 16)

Dimensions: L = 0.42–0.58 mm; a = 22–33; b = 4.8–6.7; b' = 3.5–4.9; c = 15–26; c' = 2.0–2.9; V = 77–82; stylet = 15–17 μm; pharyngeal overlap = 31 μm (20–44).

Lip region composed of three annules. Spermatheca round, filled with sperm; males are fairly frequent (one per five females). Tail terminus broadly rounded or slightly truncate, annulated to nearly smooth. Posterior uterine sac 0.9–1.6 body widths long, often slightly differentiated.

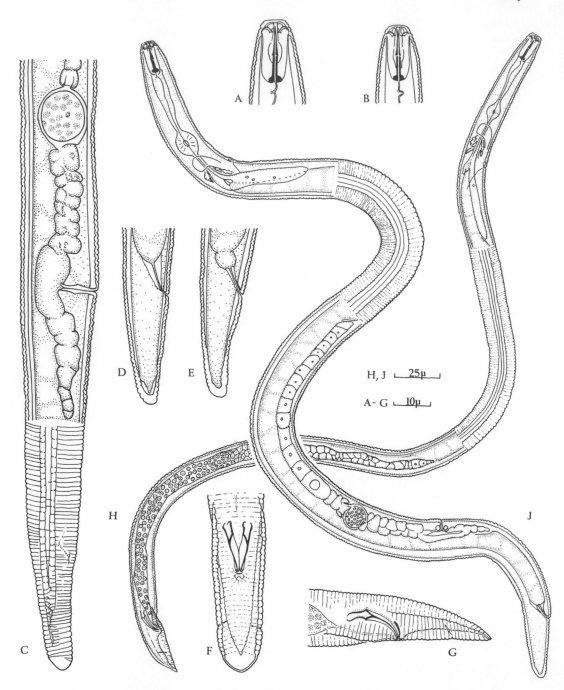

FIGURE 15 *Pratylenchus penetrans.* (A) Female head end; (B) male head end; (C) female, posterior part; (D,E) female tails; (F) male tail, ventral view; (G) male tail, lateral view; (H) whole male; (J) whole female. (From Corbett, 1973; courtesy of C. I. P., St. Albans.)

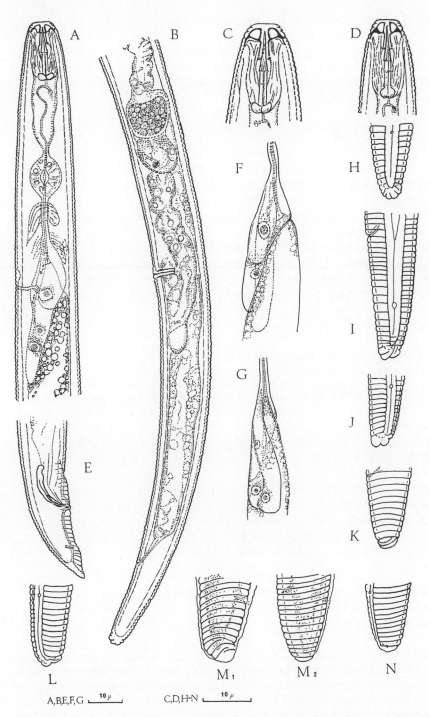

A,B,E,F,G ⊢ 10 μ ⊣   C,D,H-N ⊢ 10 μ ⊣

FIGURE 16 *Pratylenchus fallax*. (A) Female, neck region; (B) female, posterior part; (C) female, head end; (D) male, head end; (E) male tail; (F,G) pharyngeal glands; (H–J) female tail, lateral view; (K–N) female tail, median view. (From Seinhorst, 1977b; courtesy of C. I. P., St. Albans.)

Geographical distribution: Europe (the Netherlands, Belgium, France, Italy, England, Poland) and Japan.

Hosts; cereals [barley shows patchy growth when infested (Corbett, 1970, 1972)], fruit trees (cherry), ornamentals (roses), strawberry; in Japan the species was found on *Chrysanthemum*. In the Netherlands it is common in meadows. It occurs in sandy, loamy, and peaty soils.

Pratylenchus vulnus Allen and Jensen, 1951 (Fig. 17)

Dimensions: L = 0.46–0.91 mm; a = 25–39; b = 5.3–8.7; b' = 3.6–5.0; c = 13–26; c' = 1.8–2.9; V = 77–82; stylet = 13–19 μm; pharyngeal overlap = 26–53 μm. This species can be easily recognized by the slender body, head composed of three or four annules, oval median bulb, oblong spermatheca filled with sperm (males are common), long and differentiated posterior uterine sac (21–64 μm long), and narrowly rounded tail tip. The central band of the lateral field is narrower than the lateral ones.

Geographic distribution: USA, Europe (free in the southern part, in greenhouses in the northern part), Australia, Cuba, Egypt, Japan, India, Central America, Philippine Islands, USSR, and South Africa.

Hosts: About 80 plant species are known to be hosts, mostly woody ones. In the USA this species inflicts severe damage to deciduous fruit and nut crops; in Europe and the USA it is a pest of roses and peach.

Pratylenchus goodeyi Sher and Allen, 1953 (Fig. 18)
Syn. *Tylenchus musicola* apud Goodey, 1928 nec Cobb, 1919.

Dimensions: L = 0.40–0.68 mm; a = 24–37; b = 5.4–7.3; b' = 3.3–4.8; c = 14–18; c' = 2.5–3.2; V = 73–76; stylet = 14–17 μm; pharyngeal overlap = 48 μm (40–63).

Males are common.

This species is easily recognizable by high lip region, composed of four annules; large oblong spermatheca filled with large sperm; dorsally sinuate, smooth tail terminus; and anterior vulva position.

Geographic distribution: East Africa, Canary Islands. Imported to greenhouses in England and Russia. The main host is banana, which suffers heavy damage, but there are more host plants since the species has also been reported from Greece, Crete, and Australia. de Guiran and Vilardebo (1962) found it on the Canary Islands in pure stands of citrus, kenaf, and tomato and considered it probable that these are hosts.

GENUS *HIRSCHMANNIELLA* LUC AND GOODEY, 1964
SYN. *HIRSCHMANNIA* LUC AND GOODEY, 1962 (NOM. PREOCC.)

Long (1–4 mm) and slender nematodes. No sexual dimorphism in head and pharynx region. Cuticle with distinct transverse striae. Lateral field with four lines, often with partial, occasionally with complete, areolation. Head either low, flattened with rounded edges, or higher, hemispheric, composed of three to seven fine annules, not offset from neck. Stylet strong, with large basal knobs, its length three to five times body width at base of head. Median bulb well developed. Pharyngeal glands in tandem, forming a long, ventrally overlapping lobe. Female with two equally developed genital branches. Male with bursa extending over about two-thirds of tail. Spicules tylenchoid, gubernaculum variable in shape. Tail in both sexes elongate, tapering to terminus which often bears a mucro (seldom two) and in

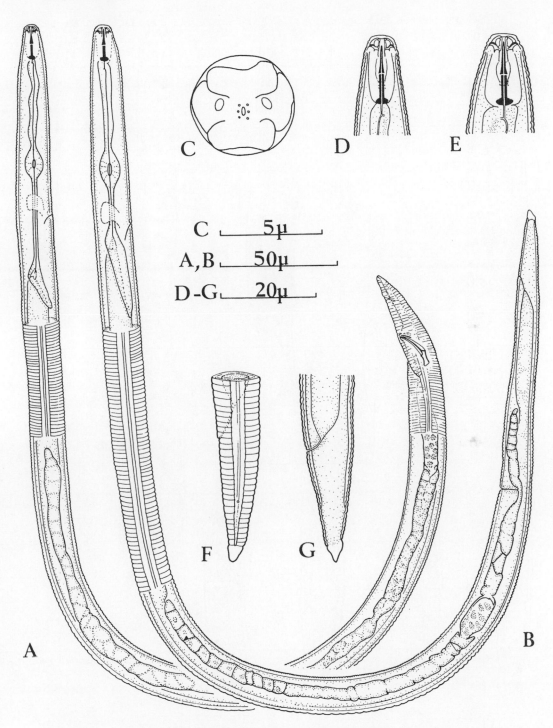

FIGURE 17 *Pratylenchus vulnus.* (A) Whole male; (B) whole female; (C) end-on view of head, drawn from SEM photo; (D) male head end; (E) female head end; (F,G) female tails. (From Corbett, 1974; courtesy of C. I. P., St. Albans.)

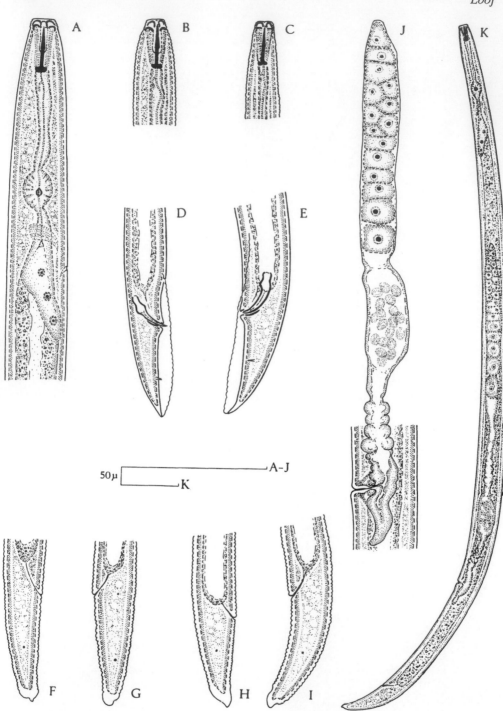

FIGURE 18   *Pratylenchus goodeyi*. (A) Female, neck region; (B) female head end; (C) male head end; (D,E) male tails; (F,I) female tails; (J) female reproductive apparatus; (K) whole female. (From Machon and Hunt, 1985; courtesy of C. I. P., St. Albans.)

some species shows a ventral notch just anterior to terminus (Fig. 19 C–H). Phasmids in posterior third of tail. The midintestine does not fill the whole body cavity; between it and the body wall transverse strands are present giving the impression of a row of cells on either side ("thorneian cells").

Type species: *H. spinicaudata* (Schuurmans Stekhoven, 1944) Luc and Goodey, 1964.

The genus was revised by Sher (1968). Keys were also given by Razzhivin, Fernandez, Ortega, and Quincosa (1981), Ebsary and Anderson (1982), and Sivakumar and Khan (1983).

The species are motile endoparasites of roots and occur predominantly in moist habitats; they can even be found in roots of aquatic plants. Species mixtures are common, hence records older than Sher (1968) may refer to more than one species (Siddiqi, 1973). Babatola and Bridge (1980) gave an account of the feeding behavior and histopathology of three species.

Geographic Distribution

Over half of the species have been recorded as parasites of rice; these occur in the warmer regions of the earth up to the southern USA, China, and Japan. Five species were described from the same regions but not on rice, whereas six species occur in the northern temperate zone.

Occasionally species have been reported outside their normal area. The temperate species *H. gracilis* has been recorded on rice in China (Yin, 1986; Zhang, 1987) and India (Mathur and Prasad, 1971). *Hirschmanniella asteromucronata*, described from rice in Cuba, cannot be distinguished from the European *H. loofi*. Decker and Dowe (1974) reported the tropical *H. diversa* from the German Democratic Republic.

Identification Characters

Head shape and stylet length are prime characters; other important characters are body length, occurrence of males (spermatheca full or empty), number of annules between phasmids and terminus, and presence or absence of a ventrosubterminal notch on the tail. Other characters commonly used are of much less value owing to great intraspecific variation; in the first place areolation of the lateral field. Sher (1968) noted variations in nine of the 15 species he dealt with. Also the structure of the terminus may be quite variable (see, e.g., Luc and Fortuner, 1975), as may be the shape of the gubernaculum.

List of Species

H. anchoryzae Ebsary and Anderson, 1982 (Canada)
H. areolata Ebsary and Anderson, 1982 (Hong Kong)
H. behningi (Micoletzky, 1925) (USSR, Sweden)
H. belli Sher, 1968 (California, China)
H. caudacrena Sher, 1968 (Florida, Louisiana, China)
H. diversa Sher, 1968 (Hawaii, Vietnam, DDR)
H. furcata Razzhivin, Fernandez, Ortega, and Quincosa, 1981 (Cuba)
H. gracilis (de Man, 1880) (Europe, North America, India, Vietnam, China)
    syn. *Radopholus gigas* Andrássy, 1954
H. imamuri Sher, 1968 (Japan, China)
H. loofi Sher, 1968 (the Netherlands, Italy, Poland)

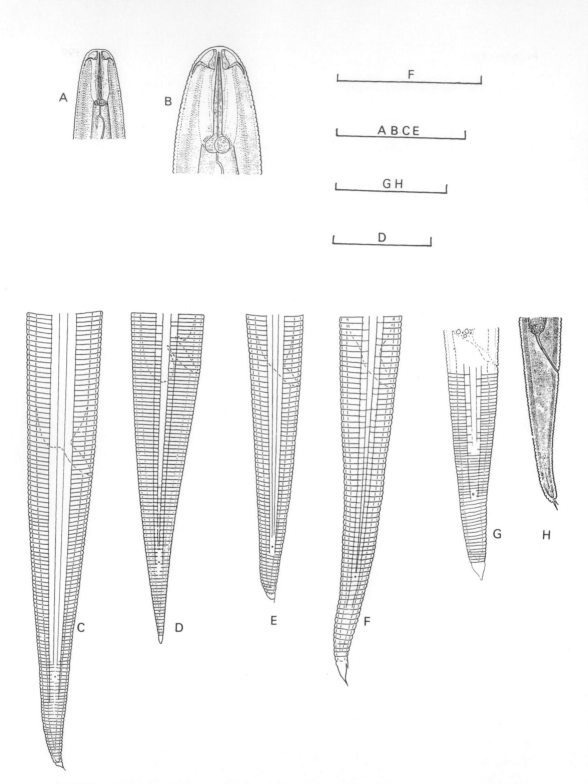

FIGURE 19   *Hirschmanniella*. (A,B) Two types of head shape; (A) flattened, stylet knobs low (*H. gracilis*); (B) high, conoid, stylet knobs high (*H. spinicaudata*). (C–H) female tails: (C) *H. gracilis*, with axial mucro; (D) *H. spinicaudata*, no mucro; (E) *H. oryzae*, with ventral mucro; (F) *H. caudacrena*, axial mucro plus ventral notch; (G) *H. miticausa*, tip narrowed but no real mucro; (H) *H. shamimi*, two mucros. [(A–F) from Sher, 1968a; courtesy of E. J. Brill, Leiden; (G) from Bridge et al., 1984; courtesy of ORSTOM, Paris; (H) from Ahmad, 1974; courtesy of Nematological Society of India, New Delhi.] The scale lines correspond to 50 μm.

syn. *H. asteromucronata* Razzhivin, Fernandez, Ortega, and Quincosa, 1981 (Cuba)

H. marina Sher, 1968 (Florida)

H. mexicana (Chitwood, 1951) (Texas)

H. microtyla Sher, 1968 (Florida, China)

H. miticausa Bridge, Mortimer, and Jackson, 1984 (Solomon Islands)

H. mucronata (Das, 1960) (India, Thailand, Philippines, China)

    syn. *H. dubia* Khan, 1972; *H. indica* Ahmad, 1974; *H. kaverii* Sivakumar and Khan, 1983; *H. magna* Siddiqi, 1966; *H. mangaloriensis* Mathur and Prasad, 1971; *H. ornata* Eroshenko and Nguen Ngoc Chau, 1985; possibly also *H. orycrena* Sultana, 1979 (description incomplete)

H. ngetinhiensis Eroshenko and Nguen Ngoc Chau, 1985 (Vietnam)

H. obesa Razzhivin, Fernandez, Ortega, and Quincosa, 1981 (Cuba)

H. oryzae (van Breda de Haan, 1902) (Pantropic)

    syn. *Tylenchus apapillatus* Imamura, 1931

    syn. *H. exigua* Khan, 1972

    syn. *H. nana* Siddiqi, 1966

H. pisquidensis Ebsary and Pharoah, 1982 (Canada)

H. pomponiensis Abdel-Rahman and Maggenti, 1987 (California)

H. shamimi Ahmad, 1974 (India)

H. spinicaudata (Schuurmans Stekhoven, 1944) (tropical Africa and America)

    syn. *Radopholus lavabri* Luc, 1957

H. thornei Sher, 1968 (Indonesia, China)

H. truncata Razzhivin, Fernandez, Ortega, and Quincosa, 1981 (Cuba)

H. zostericola (Allgén, 1934) (Sweden)

Key to Species

1. Spermatheca empty, males rare or unknown 2
   Spermatheca filled with sperm, males generally common 3
2. Stylet = 20–22 µm; phasmids about 20 annules from terminus; tail with a ventral, often needle-shaped mucro .... *H. belli*
   Stylet = 23–25 µm; phasmids 30–42 annules from the terminus; tail with axial mucro (occasionally with two mucros or a notch) .... *H. pisquidensis*
3. Head hemispheric (it is not possible to distinguish the anterior surface from the sides); the anterior narrowing of the body increases in the stylet region (Fig. 19B). Stylet knobs large, high. Stylet = 22–50 µm 4
   Head low, mostly flattened, with distinct anterior surface and sides; the body is not narrowed more strongly in the stylet region (Fig. 19A). Stylet knobs more clearly offset, not conspicuously high. Stylet = 15–25 µm 14
4. Stylet = 34 µm or more 5
   Stylet = 32 µm or less 9
5. Stylet = 40 µm or more; body length often over 3 mm 6
   Stylet = 39 µm or less; body length always under 3 mm 7
6. Intestine overlaps rectum; lateral field with at least partial areolation ....
   ....*H. spinicaudata*
   Intestine does not overlap rectum; lateral field without areolation ..... *H. obesa*
7. c' = 2.5 or less; tail truncate .... *H. truncata*
   c' = over 3; tail not truncate 8
8. Tail with ventral notch; phasmids 18 annules from terminus .... *H. furcata*
   Tail without ventral notch; phasmids 20–25 annules from terminus ..... *H. loofi*

9. Tail with ventral mucro 10
   Tail with axial mucro, or without mucro 12
10. Stylet = 29–32 μm; phasmids 17–18 annules from terminus .... *H. imamuri*
    Stylet = 23–25 μm; phasmids 9–14 annules from terminus 11
11. c' = over 4.7 .... *H. diversa*
    c' = under 4.3 .... *H. areolata*
12. Phasmids 24 annules from terminus .... *H. behningi*
    Phasmids 19 or less annules from terminus 13
13. Phasmids 4–12 annules from terminus; stylet = 27–30 μm .... *H. thornei*
    Phasmids 12–19 annules from terminus; stylet = 22–29 μm ...... *H. mucronata*
14. c' = 9; stylet = 15 μm .... *H. zostericola*
    c' = 7.5 or less; stylet = 16–25 μm 15
15. Tail with two needle-shaped mucros .... *H. shamimi*
    Tail with 0–1 mucro 16
16. Tail with ventral subterminal notch 17
    Tail without notch 19
17. Phasmids 10 annules from terminus .... *H. mexicana*
    Phasmids 18 or more annules from terminus 18
18. Stylet = 22–25 μm c' = <4; marine species .... *H. marina*
    Stylet = 18–22 μm c' = >4; freshwater species .... *H. caudacrena*
19. Tail with ventral, needle shaped, sharply offset mucro; stylet = 16–20 μm ......
    ....*H. oryzae*
    Tail with axial mucro 20
    Tail without mucro 21
20. Stylet = 16–19 μm .... *H. anchoryzae*
    Stylet = 20–24 μm .... *H. gracilis*
21. Head composed of five to six annules .... *H. miticausa*
    Head composed of three annules 22
22. L = <1.5 mm; c' = <4.7 .... *H. microtyla*
    L = >1.7 mm; c' = >5 .... *H. pomponiensis*

Not included:

H. minor (Goffart, 1933) Siddiqi, 1986: described from a potato tuber in Germany, from a
    J–4 of 0.42 mm body length. Probably this species does not belong in
    Hirschmanniella.
H. ngetinhiensis: text and illustrations do not concord with respect to structure of terminus.

H. oryzae (Fig. 20)

Dimensions: *Females*: L = 1.03–1.63 mm; a = 47–67; b = 7.9–12.1; b' = 4.2–7.2; c = 15–20;
c' 3.9–5.5; V = 50–55; stylet = 16–20 μm. *Males*: L = 1.01–1.40 mm; a = 52–61; b =
8.3–11.3; b' = 4.1–5.7; c = 16–18; c' = 3.9–5.4; stylet = 16–19 μm; spicules = 18–26 μm;
gubernaculum = 7–10 μm.

Specific characters: Head composed of four annules (occasionally three or five). Female terminus rounded, with sharply offset, needleshaped, ventrally placed mucro. Phasmids 12–17 annules from terminus. The spicules are shorter than in other species (18–26 vs.
25–54 μm) except in *H. shamimi*, which differs from *H. oryzae* only in the presence of two
mucros on the terminus.

Geographic distribution: Asia: India, Indonesia, Japan, China, Malaysia, Sri Lanka,
Thailand, Taiwan; Africa: Egypt, Ghana, Madagascar, Nigeria, Senegal; Americas: USA,
San Salvador, Sierra Leone, Venezuela, Brazil.

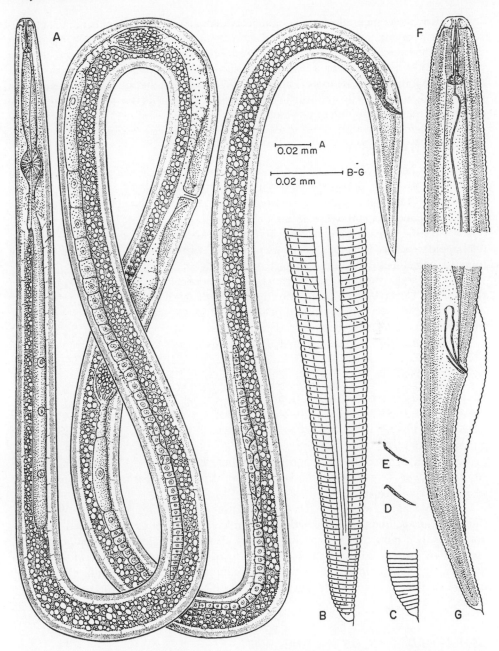

FIGURE 20 *Hirschmanniella oryzae.* (A) Whole female; (B) female tail; (C) female terminus; (D,E) variation in shape of gubernaculum; (F) male, head end; (G) male tail. (From Sher, 1968a; courtesy of E. J. Brill, Leiden.)

Host plants are predominantly monocotyledons (rice, corn, sugar cane) but also dicotyledons (cotton). Several weeds are also hosts (Mohandas et al., 1980). Of particular interest is the leguminous plant *Sesbania rostrata* Brem., which acts as a trap crop (Germani et al., 1983). *Hirschmanniella oryzae* can feed and multiply on this plant, but after 8 weeks they can no longer leave its roots (Pariselle, 1987) until these decay. This plant enhances rice yield, though Pariselle and Rinaudo (1988) state that this is due to the green manure effect rather than to trapping of nematodes.

H. spinicaudata (Fig. 21)

Dimensions: *Females*: L = 1.86–4.02 mm (usually over 2.5 mm); a = 45–74; b = 10–19; b' = 4.0–9.5; c = 18–28; c' = 3.0–4.7; V = 50–58; stylet = 40–50 μm. *Males*: L = 2.12–3.15 mm; a = 49–79; b = 11–17; b' = 4.2–9.2; c = 17–24; c' = 3.0–5.8; stylet = 38–47 μm; spicules = 41–54 μm; gubernaculum = 13–20 μm.

Specific characters: Head with five annules (occasionally four or six). Intestine overlapping rectum. Phasmids 17–24 annules from terminus. Tail tapering to narrowly rounded, sometimes drawn out terminus, without mucro.

Geographic distribution: Africa: Cameroon, Senegambia, Ivory Coast, Zaire, Nigeria, Zambia, Upper Volta. Known also from California and Venezuela.

Hosts: rice (mostly in flooded fields) and various Cyperaceae.

Often occurs together with *H. oryzae*.

Sesbania rostrata has the same effect on this species as on *H. oryzae*.

H. gracilis (Fig. 22)

Dimensions: *Females*: L = 1.48–2.22 mm; a = 50–65; b = 11–17; b' = 5.2–8.7; c = 14–21; c' = 4.0–6.1; V = 48–55; stylet = 20–24 μm. Males: L = 1.38 – 2.02 mm; a = 45–66; b = 12–15; b' = 4.5–7.3; c = 15–22; c' = 3.8–6.1; stylet = 20–23 μm; spicules = 27–38 μm; gubernaculum = 9–15 μm.

Specific characters: Head composed of three to five annules. Tail tapering to acute terminus, often with needle-shaped ventral or axial mucro. Gubernaculum sinuate, often proximally hook-shaped.

Geographic distribution: USA (Texas, California, Ohio, Wyoming, Wisconsin, Kansas); Canada; Europe (the Netherlands, Germany, Poland; other records need confirmation). Usually in moist meadows and ditch banks, also aquatic; tolerates a slight salt content. Prejs (1986) records it from many lakes in northern Poland, chiefly in *Potamogeton pectinatus* L. (sometimes in high densities and possibly causing rhizome damage), less in *P. lucens* L. and *P. pectinatus* L. As mentioned above, it has also been recorded from India and China.

GENUS *RADOPHOLUS* THORNE, 1949
SYN. *RADOPHOLOIDES* DE GUIRAN, 1967, TYPE
SPECIES *R. LITORALIS* DE GUIRAN, 1967;
NEORADOPHOLUS KHAN AND SHAKIL, 1973,
TYPE SPECIES *N. INAEQUALIS* (SAUER, 1958).

Diagnosis: *Females*: Slender, worm-shaped, body length 0.4–0.9 mm. Head hardly offset, low, with strong internal sclerotization. Stylet short (14–23μm) and stout, with well-developed basal knobs. Median bulb of pharynx well developed with distinct valves. The

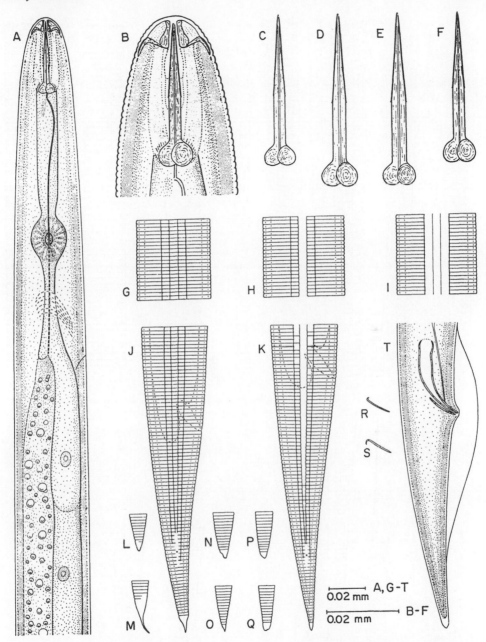

FIGURE 21 *Hirschmanniella spinicaudata.* (A) Female, neck region; (B) female, anterior end; (C–F) female stylets; (G–I) female, lateral field near midbody; (J–K) female, tails; (L–Q) female, variation in shape of terminus; (R,S) variation in shape of gubernaculum; (T) male tail. (From Sher, 1968a; courtesy of E. J. Brill, Leiden.)

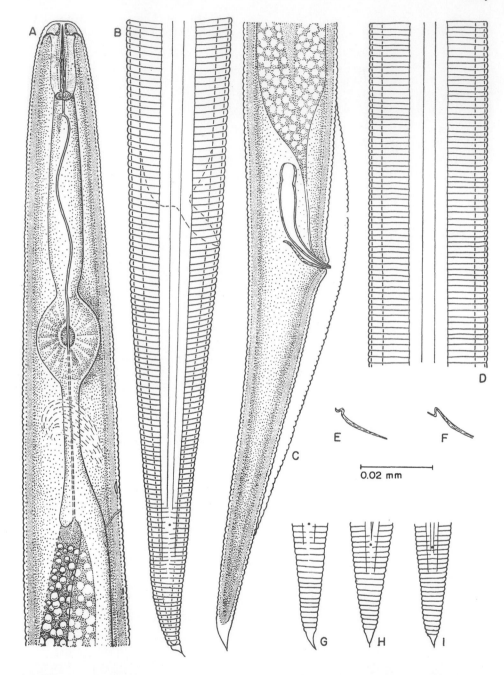

FIGURE 22 *Hirschmanniella gracilis*. (A) Female, anterior part; (B) female tail; (C) male tail;
(D) female, lateral field in midbody; (E,F) variation in shape of gubernaculum; (G–I) female, variation in shape of terminus. (From Sher, 1968a; courtesy of E. J. Brill, Leiden.)

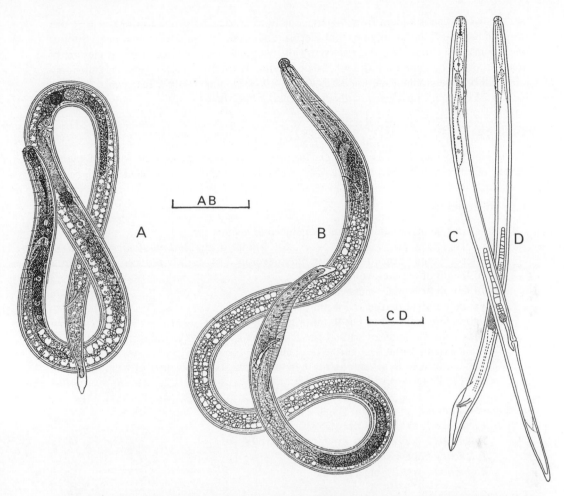

FIGURE 23   General view of *Radopholus*. (A) *R. similis*, female (didelphic); (B) *R. similis*, male (bursa not extending to terminus); (C) *R. litoralis*, female (monoprodelphic); (D) *R. litoralis*, male (bursa extending to terminus). [(A, B) from Cobb, 1915; (C,D) from de Guiran, 1967 (courtesy of E. J. Brill, Leiden).]

glands form a long lobe along the dorsal side of the midintestine. Deirids absent. Vulva postequatorial (54–71%). Anterior genital branch fully developed and functional; the posterior one is either also fully developed (Fig. 23A), or reduced to a short sac (Fig 23C). Tail elongate-conoid with narrowly rounded terminus. Lateral field with three to six longitudinal lines. *Males*: Head offset, knob-shaped, unsclerotized. Stylet thin, knobs rudimentary or absent. Median pharyngeal bulb and valves and gland lobe poorly developed. Bursa in most species not quite extending till tail terminus (Fig. 23B, D). Gubernaculum slightly protrusible. Lateral field as in females.

Type species: *R. similis* (Cobb, 1983) Thorne, 1949.

The genus comprises 28 species, the great majority of which are known only from natural habitats in Australia (19 species) or New Zealand (two species). Of the remaining seven species one is known from Nigeria, one from Madagascar, one from Mauritius, one from India, one from Japan, one from Florida; one occurs nowadays almost in all tropical and subtropical countries and occasionally even in the temperate zone. Two species are of major economic importance and will be dealt with. Taxonomic literature to the other species will be given in the References (Egunjobi, 1968; Kumar, 1980; Minagawa, 1984a). A key to 22 didelphic species was given by Colbran (1970).

Biology

Radopholus species are, like *Pratylenchus*, migratory endoparasites of roots of higher plants, often provoking development of brownish lesions and often associated with fungi.

Key to Species

1. Posterior genital branch of female reduced to a short sac ....
   .... former genus *Radopholoides* (3 species)
   Posterior genital branch of female fully developed or slightly reduced 2
2. Tail length of female over 52 µm 3
   Tail length of female under 52 µm .... 22 species
3. Hyaline terminal part of tail 4 µm or less .... 1 species
   Hyaline terminal part of tail 9 µm or more 4
4. Males with 0–2 anterior genital papillae (hypoptygmata); lateral sectors of head in female extend to the base of the third annule .... *R. similis*
   Males with three to seven anterior hypoptygmata; lateral sectors of head in female extend to beyond third annule .... *R. citrophilus*

Radopholus similis (Cobb, 1893) (Figs. 24; 25C,D; 26C,D)
  Syn. *Tylenchus granulosus* Cobb, 1893;
    Tylenchus acutocaudatus Zimmermann, 1898;
    Tylenchus biformis Cobb, 1909;
    Radopholus similis, banana race.

Dimensions: *Females*: $L = 0.52–0.88$ mm; $a = 22–30$; $b = 4.7–7.4$; $b' = 3.5–5.2$; $c = 8–13$; $c' = 2.9–4.0$; $V = 55–61$; stylet = 17–20 µm. *Males*: $L = 0.54–0.67$ mm; $a = 31–44$; $b = 6.1–6.6$; $b' = 4.1–4.9$; $c = 8–10$; $c' = 5.1–6.7$; stylet = 12–17 µm; spicules = 18–22 µm; gubernaculum = 8–12 µm.

Specific characters: Female: Head composed of three to four annules. Lateral field with four longitudinal lines. Spermathecae round, with small, rod-shaped sperm. Hyaline part of tail 9–17µm long; terminus striated. Phasmids in anterior third of tail. Male: Head four-lobed, lateral sectors strongly reduced. Gubernaculum with small titillae. Bursa extends over about two-thirds of tail.

Geographic distribution: Originally described from the Fiji Islands, but occurs also in Australia, Florida, Central and South America, several Caribbean islands, tropical Africa. In the 1960s it was imported into several European countries (France, Belgium, the Netherlands, Germany) with ornamental plants.

Hosts: The most important hosts are banana ("toppling-over disease") and pepper ("yellows"; during the years after World War II, pepper culture on the Islands of Banka, Indonesia, was almost completely destroyed). More than 250 other plant species are known to be hosts, among them many ornamentals such as *Philodendron* and *Maranta*. To what extent it is a major pest of these has not yet been completely determined.

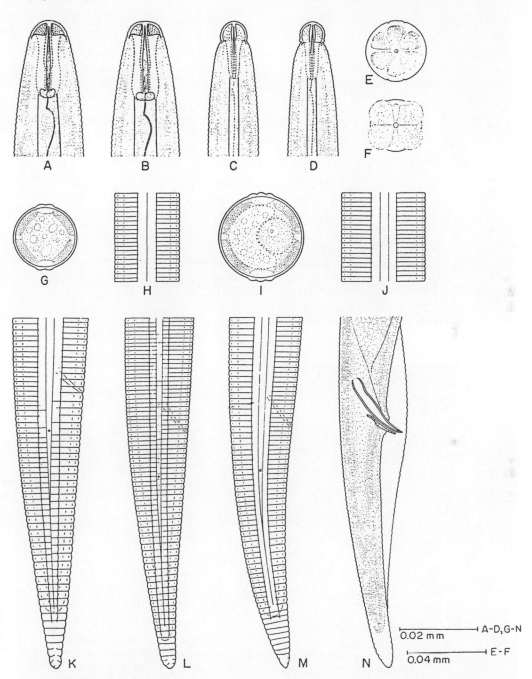

FIGURE 24 *Radopholus similis.* (A,B) Female, anterior end; (C,D) male, anterior end; (E) female, end-on view of head; (F) male, end-on view of head; (G) juvenile, cross-section near midbody; (H) juvenile, lateral field near midbody; (I) female, cross-section near midbody; (J) female, lateral field near midbody; (K–M) female tails; (N) male tail. (From Sher, 1968b; courtesy of Allen Press, Lawrence.)

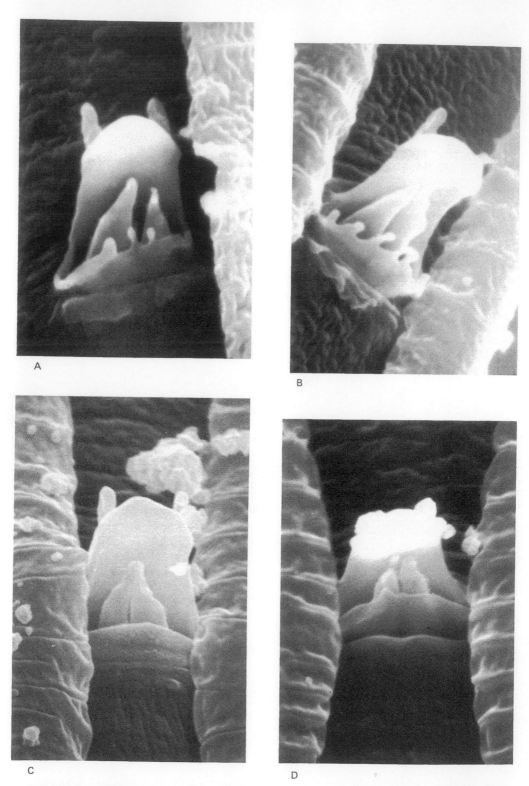

FIGURE 25   SEM photos of males of (A,B) *Radopholus citrophilus*; (C,D) *R. similis*, showing anterior hypoptygmata. (From Huettel and Yaegashi, 1988; courtesy of Society of Nematology.)

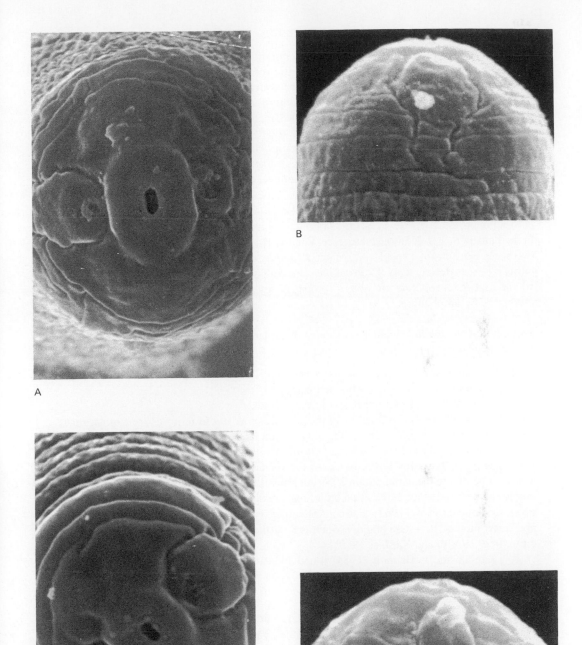

FIGURE 26   SEM photos of females of (A,B) *Radopholus citrophilus*; (C,D) *R. similis*. A and C: end-on view of head; B and D: head, lateral view.(From Huettel and Yaegashi, 1988; courtesy of Society of Nematologists.)

It has long been suspected that there exist different pathotypes within this species. DuCharme and Birchfield (1956) found indications for the occurrence of three pathotypes in Florida: one parasitizing citrus only, one banana and citrus, and one banana but not citrus. The first was not examined further, but the existence of the two others was proven and these have since been known as the citrus race and the banana race, respectively. The citrus race is known from Florida only; the banana race is widely distributed (see above). Later there came indications that there exists more than one "banana race" (Edwards and Wehunt, 1971).

During the 1980s several differences between the citrus race and the banana race were discovered. Huettel and Dickson (1981b) found differences in oocyte maturation; Huettel et al. (1983a) in sex pheromones; Huettel et al. (1983b) in enzymes; Huettel et al. (1983c) in proteins. Huettel et al. (1984a) found the haploid chromosome number to be four in the banana race, five in the citrus race. Finally, Huettel et al. (1984b) concluded that the two races were distinct species; the name *R. similis* was restricted to the banana race; the citrus race was described as *R. citrophilus*. Huettel and Dickson (1981a) found that both *R. similis* and *R. citrophilus* are able to propagate by parthenogenesis, though the normal method is by amphimixis.

Radopholus citrophilus Huettel, Dickson and Kaplan, 1984 (Fig. 25A,B; 26A,B)
    Syn. *R. similis*, citrus race

Dimensions and morphological characters are identical to those of *R. similis*, except those mentioned above in the key (Huettel and Yaegashi, 1988).

    Geographic distribution: Known from Florida only.

    Hosts: The recent splitting up of the old species *R. similis* requires reexamination of the host plant lists. At this moment the only certain hosts of *R. citrophilus* are apparently various species of the genera *Citrus* and *Musa*, though Orton Williams and Siddiqi (1973) say it has many noncitrus hosts. In *Citrus* the species causes "spreading decline," determined to be due to the nematode in 1953 but known much longer. Probably this disease is due to the nematode in combination with fungi. A detailed description of the disease was given by Poucher et al. (1967). For control emphasis is being shifted from the old "push-and-treat" method to culture practices such as P fertilization, to avoid stress situations of the citrus trees (Smith and Kaplan, 1988).

    Kaplan and O'Bannon (1985) found indications of the existence of pathotypes within *R. citrophilus*.

GENUS *PRATYLENCHOIDES* WINSLOW, 1958
SYN. *HOPLORHYNCHUS* ANDRÁSSY, 1985, TYPE SPECIES
H. RIPARIUS ANDRÁSSY, 1985

Diagnosis: Lip region flattened anteriorly in females, often more conoid in males. Median pharyngeal bulb usually more strongly developed in females. Pharyngointestinal valve well developed. Pharyngeal glands overlapping hardly to strongly on dorsal side. Lateral field with four or six lines in middle of body. Deirids present. Female genital system double. Female tail cylindrical or tapering, terminus round (rarely subacute), mostly annulated. Male with bursa enveloping tail. Male tail ventrally contracted just anterior to terminus.

    Type species *P. crenicauda* Winslow, 1958.

    The genus comprises 19 species and is distributed mainly in the temperate zones. The species are endoparasites of plant roots. The genus is distinguished from all other

Pratylenchidae by the presence of a well-developed pharyngointestinal valve and deirids. Through these characters it resembles the belonolaimid genus *Amplimerlinius* Siddiqi, 1976, from which it differs by the ventral contraction of the male tail; the position of several species in which males are unknown is therefore somewhat unsettled.

The genus is morphologically uniform. On the basis of the shape of the pharyngeal gland region the species can be separated into three groups (Baldwin et al., 1983)

List of Species

P. alkani Yüksel, 1977 (Turkey)
P. bacilisemenus Sher, 1970 (California)
P. crenicauda Winslow, 1958 (USSR, Europe, India)
    syn. *Anguillulina obtusa* (Bastian, 1865) apud Goodey, 1932 and 1940;
    Scutellonema sexlineatum Razzhivin, 1971 (after Ryss, 1988)
P. epacris Eroshenko, 1978 (Sakhalin Island)
P. erzurumensis Yüksel, 1977 (Turkey)
P. heathi Baldwin et al., 1983 (Utah)
P. ivanovae Ryss, 1980 (Tadzhikistan)
P. laticauda Braun and Loof, 1966 (Europe)
P. leiocauda Sher, 1970 (France)
    syn. *P. orientalis* Eroshenko and Kazachenko, 1984 (USSR)
P. magnicauda (Thorne, 1935) (North America, north and central Europe, Spitzbergen Island)
P. magnicaudoides Minagawa, 1984 (Japan)
P. maqsoodi Maqbool and Shahina, 1989 (Pakistan)
P. maritimus Bor and s'Jacob, 1966 (the Netherlands)
P. megalobatus Bernard, 1984 (Aleutian Islands)
P. riparius (Andrássy, 1985) (Hungary)
P. ritteri Sher, 1970 (Europe)
P. sheri Robbins, 1985 (California)
P. utahensis Baldwin et al., 1983 (Utah)
P. variabilis Sher, 1970 (California)

Key to Species

1. Ventrosublateral pharyngeal glands not elongate; both nuclei usually anterior to pharyngointestinal valve (Fig. 27D) 2
   One ventrosublateral pharyngeal gland elongate, nucleus posterior to pharyngointestinal junction; the other not elongate, nucleus close to junction (Fig 27E) 9
   Both ventrosublateral glands elongate, nuclei posterior to junction (Fig. 27F) . 14
2. Six lateral lines on distal half of female tail 3
   Four lateral lines on distal half of female tail 4
3. Stylet length 21–22 μm .... *P. riparius*
   Stylet length over 27 μm .... *P. epacris*
4. Spermatheca empty, males unknown 5
   Spermatheca filled with sperm 6
5. Lateral field with four lines, sometimes 5 in middle of body .... *P. ivanovae*
   Lateral field with six lines in middle of body .... *P. magnicauda*

6.   Female tail with 29–42 annules . . . . *P. magnicaudoides* *
      Female tail with 18–28 annules 7
7.   Female tail with 27–28 annules . . . . *P. heathi*
      Female tail with 18–26 annules 8
8.   Body length under 0.85 mm; $c = 14$–18 . . . . *P. laticauda*
      Body length over 0.90 mm; $c = 17$–26 . . . . *P. sheri*
9.   Stylet under 18 μm, tail tapering to subacute terminus . . . . *P. maritimus*
      Stylet over 19 μm, tail more broadly rounded 10
10. Female terminus smooth, or with one to two striae . . . . *P. leiocauda*
      Female terminus regularly annulated 11
11. Males rare or absent, spermatheca usually empty 12
      Males common, spermatheca filled 13
12. Fasciculi absent; female terminus flattened . . . . *P. erzurumensis*
      Fasciculi present; female terminus rounded . . . . *P. crenicauda*
13. Lateral field with four lines . . . . *P. variabilis*
      Lateral field with six lines . . . . *P. maqsoodi*
14. Sperm rod-shaped, tail tapering 15
      Sperm round, tail cylindrical 16
15. Body length under 0.63 mm; stylet = 18–21 μm pharyngeal
      overlap = 3–6 body widths, $b' = 2.4$–3.3 . . . . *P. megalobatus*
      Body length over 0.65 mm; stylet = 21–23 μm; pharyngeal overlap
      = < 2 body widths, $b' = 3.5$–4.1 . . . . *P. bacilisemenus*
16. Female terminus smooth . . . . *P. utahensis*
      Female terminus annulated 17
17. Lateral field punctate, usually with six lines . . . . *P. alkani*
      Lateral field not punctate, four lines . . . . *P. ritteri*

P. crenicauda (Fig. 27A, B, C, E)

Dimensions: *Females*: $L = 0.55$–0.91 mm; $a = 19$–32; $b = 3.3$–6.4; $b' = 3.5$–5.2; $c = 13$–18; $c' = 2.2$–2.7; $V = 55$–63; stylet = 20–23 μm. *Males*: $L = 0.60$–0.74 mm; $a = 24$–33; $b = 4.5$–6.2; $c = 12$–15; stylet = 20–21 μm; spicules = 20–28 μm; gubernaculum = 4–7 μm.

Specific characters: Pharyngointestinal junction surrounded by the glands on all sides, anterior part of intestine projecting knoblike between the glands. Female tail tapering, terminus very coarsely annulated, almost lobate; number of annules to center of terminus 28–36. Male head higher and narrower than female head; median bulb narrow, with small valves. Males are rather rare.

Geographic distribution: Europe. Also recorded from Tadzhikistan (Ryss, 1988) and India (Jairajpuri, 1964).

Hosts: Mainly grasses. The species is common in meadows and turf grasses (Siddiqi, 1974).

P. laticauda (Fig. 27 D, G, H)

Dimensions: *Females*: $L = 0.76$–0.94 mm; $a = 24$–31; $b = 4.1$–5.6; $b' = 3.8$–4.9; $c = 14$–21; $c' = 2.3$–2.5; $V = 51$–61; stylet = 22–24 μm. *Males*: $L = 0.70$–0.85 mm; $a = 24$–35; $b =$

---

*In his diagnosis of this species, Minagawa (1984b) says that it differs from *P. magnicauda* by shorter stylet (56.6–58.8 μm vs. 56–64 μm). These values actually refer to vulva position. Stylet length is 27–34 μm in *magnicauda*, 22–24 μm in *P. magnicaudoides* females.

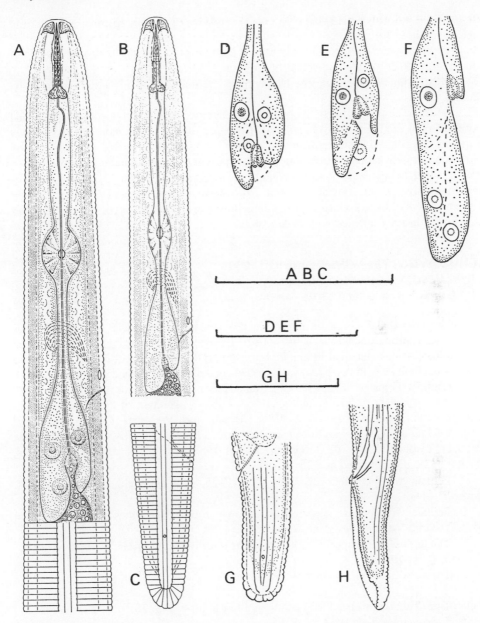

FIGURE 27   Genus *Pratylenchoides*. (A–C) *P. crenicauda*. (A) Female neck region; (B) male neck region; (C) female tail. (D–F) the three types of pharyngeal gland region; (D) type 1 (*P. laticauda*); (E) type 2 (*P. crenicauda*); (F) type 3 (*P. alkani*). (G,H) *P. laticauda*: (G) female tail; (H) male tail. [(A–C) from Sher, 1970; courtesy of Allen Press, Lawerence; (D–F) from Baldwin, Luc, and Bell, 1983; courtesy of ORSTOM, Paris; (G–H) from Braun and Loof, 1966; courtesy Neder-landse Plantenziektenkundige Vereniging, Wageningen.] The scale lines correspond to 50 μm.

5.0–6.7; b' = 4.6–6.0; c = 13–18; c' = 2.6–2.8; stylet = 18–22 μm spicules = 24–28 μm; gubernaculum = 7–10 μm.

Specific characters: Female tail almost cylindrical, with very broadly rounded terminus; with 18–26 annules (to center of terminus).

Geographic distribution: the Netherlands, Germany (both FRG and GDR), Italy, Greece.

Hosts: The species lives endoparasitically in roots, causing brown discoloration and root rot. Braun and Loof (1966) found that *Monarda fistulosa* L. var. *mollis* L. and *Mentha piperita* L. were good hosts. In *Monarda didyma* L. Hijink and van Rossen (1968) found also strong growth suppression. Sprau (1969) reported that *P. laticauda* probably caused growth reduction in *Mentha piperita* and *Petroselinum sativum* Hoffm. in Bavaria. The species might have an antagonistic effect on *Pratylenchus penetrans* and *Longidorus elongatus*: numbers of these species were distinctly lower in (*P. penetrans*) or close to (*L. elongatus*) roots heavily infested with *P. laticauda*.

GENUS *ZYGOTYLENCHUS* SIDDIQI, 1963
SYN. *MESOTYLUS* DE GUIRAN, 1964,
TYPE SPECIES *M. GALLICUS* DE GUIRAN, 1964.

Diagnosis: No sexual dimorphism in head and pharynx. Lateral field with four lines. Deirids absent. Head strongly sclerotized, not offset, flattened. Pharyngeal glands in line, the ventrosublateral ones unequal in length, overlapping the anterior part of intestine on the ventral side. Female with two equally developed genital branches. Male with bursa extending to terminus. Female tail with smooth, rounded terminus. Phasmids near middle of tail.

Type species: *Z. browni* Siddiqi, 1963, a junior synonym of *Z. guevarai* (Tobar Jiménez, 1963) Braun and Loof, 1966. Tarjan and Weischer (1965) considered *Zygotylenchus* as identical with *Pratylenchoides*, but it is now generally accepted that these genera are distinct (see e.g., de Guiran and Siddiqi, 1967). The genus contains two species, which may be distinguished as follows:

Female tail subcylindrical, short (c = 17–25, c' = 2–3) with 13–21 annules *Z. guevarai*.

Female tail elongate, tapering to narrowly rounded terminus (c = 12–16, c' = 3.0–4.5) with 32–45 annules *Z. taomasinae*

The third nominal species, *Z. biterminalis* Razzhivin and Milan, 1978, does not belong here (Luc, 1987).

The species are endoparasites of plant roots.

Z. guevarai (Tobar Jiménez, 1963) Braun and Loof, 1966 (Fig. 28)
   syn. *Z. browni* Siddiqi, 1963
   Mesotylus gallicus de Guiran, 1964

Dimensions: *Females*: L = 0.39–0.77 mm; a = 23–31; b = 3.6–5.7; c = 17–25; c' = 1.4–3.0; V = 53–67; stylet = 15–20 μm. *Males*: L = 0.47–0.66 mm; a = 24–37; b = 3.7–5.5; c = 16–22; stylet = 14–18 μm; spicules = 16–21 μm; gubernaculum = 5–7 μm. Head composed of three to four annules, flattened anteriorly, with rounded edges, Spermatheca round, filled with sperm. Males numerous. Bursa rather narrow, with crenate edge. Spicules cephalate anteriorly, pointed posteriorly. Gubernaculum trough-shaped.

Geographic distribution: Europe, North Africa, and the Asian part of the USSR (Uzbekistan, Tadzhikistan).

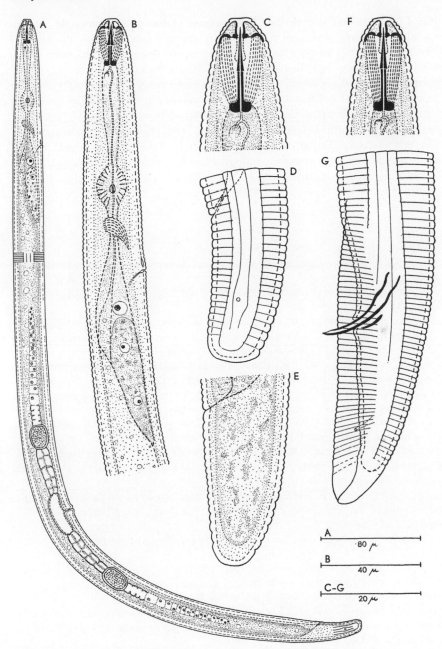

FIGURE 28 *Zygotylenchus guevarai*. (A) Whole female; (B) female neck region; (C) female anterior end; (D,E) female tails; (F) male anterior end; (G) male tail. (From de Guiran, 1964; courtesy of E. J. Brill, Leiden.)

Hosts: Predominantly dicotyledons, but also some grasses (Vovlas et al., 1976). The species is a migratory endoparasite of plant roots. Infestation results in the formation of large cavities in the roots of *Cicer arietinum* L., *Vicia* spp. and *Avena sativa* L., more superficial ones in *Lathyrus cicera* (Varo Alcala et al., 1970). It does considerable damage to *Cicer arietinum* (combined presence of the nematode and a fungus, possibly *Phyllosticta rabiei*) (Tobar Jiménez, 1973), *Viola odorata* and *Cupressus sempervirens* (Tobar Jiménez, 1963). *Zygotylenchus guevarai* has been repeatedly reported from vineyards: Cyprus (Antoniou, 1981); Spain, Rioja area (Pinochet and Cisneros, 1986), Hungary (Decker and Manninger, 1976) and Germany (Tarjan and Weischer, 1965). However, Pinochet and Cisneros did not find it in vine roots and considered it possible that it fed instead on weeds.

The second species, *Z. taomasinae* (de Guiran, 1964), is known only from the type locality (Madagascar) where it parasitizes *Musa acuminata* gr. *sinensis* var. Poyo.

REFERENCES

Anderson, R. V. 1976. *Pratylenchus neglectus*. C. I. H. Descriptions of plant-parasitic nematodes, *6, No. 82.*

Andrássy, I. 1985. A dozen new nematode species from Hungary. *Opusc. Zool. Budapest*, 19–20:3–39.

Antoniou, M. 1981. A nematological survey of vineyards in Cyprus. *Nem. Medit.*, 9:133–137.

Babatola, J. O., and Bridge, J. 1980. Feeding behavior and histopathology of *Hirschmanniella oryzae*, *H. imamuri* and *H. spinicaudata* on rice. *J. Nematol.* 12:48–53.

Baldwin, J. G., Luc, M., and Bell, A. H. 1983. Contribution to the study of the genus *Pratylenchoides* Winslow (Nematoda: Tylenchida). *Rev. Nématol.*, 6:111–125.

Berg, E.van den. 1971. The root-lesion nematodes of South Africa (Genus *Pratylenchus*, Family Hoplolaimidae). *Dept. Agr. Tech. Services Tech. Commun.*; 99.

Braun, A. L., and Loof, P. A. A. 1966. *Pratylenchoides laticauda* n. sp., a new endoparasitic phytonematode. *Netherlands J. Plant Path.*, 72:241–245.

Brzeski, M. W., and Szczygiet, A. 1977. Materialy do poznania krajowych nicieni (Nematoda)—pasozytow roślin 1. Rodzaj *Pratylenchus* Fil. (Tylenchida: Pratylenchidae). *Fragmenta Faunistica*, 23:1–11.

Café-Filho, A. C., and Huang, C. S. 1989. Description of *Pratylenchyus pseudofallax* n. sp. with a key to species of the genus *Pratylenchus* Filipjev, 1936 (Nematoda: Pratylenchidae). *Rev. Nématol.*, 12:7–15.

Cobb, N. A. 1915. *Tylenchus similis,* the cause of a root disease of sugar cane and banana. *J. Agr. Res.* 4:561–568.

Cobb, N. A. 1917. A new parasitic nema found infesting cotton and potatoes. *J. Agr. Res.* 11:27–33.

Cobb, N. A. 1920. A newly discovered parasitic nema (*Tylenchus mahogani* n. sp.) connected with a disease of the mahogany tree. *J. Parasitol.* 6:188–191.

Colbran, R. C. 1970. Studies of plant and soil nematodes. 15. Eleven new species of *Radopholus* Thorne and a new species of *Radopholoides* de Guiran (Nematoda: Tylenchoidea) from Australia. *Qd. J. Agr. Anim. Sci.*, 27:437–460.

Corbett, D. C. M. 1969. *Pratylenchus pinguicaudatus* n. sp. (Pratylenchinae: Nematoda) with a key to the genus *Pratylenchus. Nematologica* 15:550–559.

Corbett, D. C. M. 1973. *Pratylenchus penetrans*. C. I. H. Descriptions of plant-parasitic nematodes, *2, No. 25.*

Corbett, D. C. M. 1974. *Pratylenchus vulnus*. C. I. H. Descriptions of plant-parasitic nematodes, *3, No. 37.*

Corbett, D. C. M. 1976. *Pratylenchus brachyurus*. C. I. H. Descriptions of plant-parasitic nematodes, *6, No. 89.*

Corbett, D. C. M. 1984. Three new species of *Pratylenchus* with a redescription of *P. andinus* Lordello, Zamith and Boock, 1961 (Nematoda: Pratylenchidae). *Nematologica* 29 (1983): 390–403.

Corbett, D. C. M. and Clark, S. A. 1983. Surface features in the taxonomy of *Pratylenchus* species. *Rev. Nématol.* 6:85–98.

Das, V. M., and Sultana, S. 1979. Five new species of the genus *Pratylenchus* from vegetable crops of Hyderabad (Andhra Pradesh). *Ind. J. Nematol.* 9:5–14.

Decker, H., and Dowe, A. 1974. Ueber das Aufreten von Arten der Gattungen *Pratylenchus*, *Pratylenchoides* und *Hirschmanniella* (Nematoda, Pratylenchidae) in der DDR. *Helminthologia* (Bratislava), 15: 829–834.

Decker, H., and Manninger, G. A. 1976. Zum Auftreten von *Zygotylenchus guevarai* (Tobar Jiménez, 1963) Braun & Loof, 1966 in der VR Ungarn. *2. Vortragstagung zu aktuellen Problemen der Phytonematologie am 27.5.1976 in Rostock*:117–125.

DuCharme, E. P., and Birchfield, W. 1956. Physiological races of the burrowing nematode. *Phytopathology* 46:615–616.

Edwards, D. I. and Wehunt, E. J. 1971. Host range of *Radopholus similis* from banana areas of Central America with indications of additional races. *Plant Dis. Reptr.* 55:415–418.

Egunjobi, O. A. 1968. Three new species of nematodes from New Zealand. *New Zealand. J. Sci.*, 11:488–497.

Eroshenko, A. S., Nguen Ngoc Chau, Nguen Vu Tkhan' and Doan Kan' 1985. *Paraziticheskie fitonematody severnoi chasti Vietnama*. Nauka, Leningrad.

Fischer, M. 1894. Ueber eine Clematis-Krankheit. *Ber. Phys. Lab. Vers. landw. Inst. Univ. Halle*, 3:1–11

Fortuner, R. 1973. Description de *Pratylenchus sefaensis* n. sp. et de *Hoplolaimus clarissimus* n. sp. (Nematoda: Tylenchida). *Cah. ORSTOM Sér. Biol.* 21: 25–34.

Fortuner, R. 1976. *Pratylenchus zeae*. C. I. H. Descriptions of plant-parasitic nematodes, *6, No. 77.*

Fortuner, R. 1977. *Pratylenchus thornei*. C. I. H. Descriptions of plant-parasitic nematodes, *7, No. 93.*

Fortuner, R. 1987. A reappraisal of Tylenchina (Nemata). 8. The family Hoplolaimidae Filip' ev, 1934. *Rev. Nématol.* 10:219–232.

Frederick, J. J., and Tarjan, A. C. 1989. A compendium of the genus *Pratylenchus* Filipjev, 1936 (Nemata: Pratylenchidae). *Rev. Nématol.* 12: 243–256.

Germani, G., Reversat, G., and Luc, M. 1983. Effect of *Sesbania rostrata* on *Hirschmanniella oryzae* in flooded rice. *J. Nematol.* 15:269–271.

Goodey, T. 1928. Observations on *Tylenchus musicola* Cobb, 1919 from diseased banana roots. *J. Helminthol.* 6: 193–198.

Gotoh, A. 1974. Geographic distribution of *Pratylenchus* spp. (Nematoda: Tylenchida) in Japan. (Japanese text, English summary). *Bull Kyushu Agr. Exp. Sta.* 17:139–224.

Gotoh, A. and Ohshima, Y. 1963. *Pratylenchus*-Arten und ihre geographische Verbreitung in Japan. (Japanese text, German summary). *Jap. J. Appl. Ent. Zool.* 7:187–199.

de Guiran, G. 1964. *Mesotylus*: nouveau genre de Pratylenchinae (Nematoda: Tylenchoidea). *Nematologica* 9 (1963):567–575.

de Guiran, G. 1967. Description de *Radopholoides litoralis* n.g., n. sp. (Nematoda: Pratylenchinae). *Nematologica* 13:231–234.

de Guiran, G., and Siddiqi, M. R. 1967. Characters differentiating the genera *Zygotylenchus* Siddiqi, 1963 (syn. *Mesotylus* de Guiran, 1964) and *Pratylenchoides* Winslow, 1958 (Nematoda: Pratylenchinae). *Nematologica* 13:235–240.

de Guiran, G. and Vilardebo, A. 1962. Le bananier aux Iles Canaries IV: Les nématodes parasites du bananier. *Fruits* 17:263–277.

Handoo, Z. A. and Golden, A. M. 1989. A key and diagnostic compendium to the species of the genus *Pratylenchus* Filipjev, 1936 (lesion nematodes). *J. Nematol.* 21:202–218.

Hashim, Z. 1983. Description of *Pratylenchus jordanensis* n. sp. (Nematoda: Tylenchida) and notes on other Tylenchida from Jordan. *Rev. Nématol.* 6:187–192.

Hijink, M. J., and van Rossen, H. 1968. Inoculaties met *Pratylenchoides laticauda* Braun & Loof bij *Monarda* hybride "Croftway Pink" en *Mentha piperita* L. *Meded. Rijksfac. Landbouwwet. Gent,* 33:729–737.

Huettel, R. N. and Dickson, D. W. 1981a. Parthenogenesis in the two races of *Radopholus similis* from Florida. *J. Nematol.* 13:13–15.

Huettel, R. N., and Dickson, D. W. 1981b. Karyology and oogenesis of *Radopholus similis* (Cobb) Thorne. *J. Nematol.* 13:16–20.

Huettel, R. N., Dickson, D. W. and Kaplan, D. T. 1983a. Sex attractants and behavior in the two races of *Radopholus similis. Nematologica* 28 (1982):360–369.

Huettel, R. N., Dickson, D. W., and Kaplan, D. T. 1983b. Biochemical identification of the two races of *Radopholus similis* by starch gel electrophoresis. *J. Nematol.* 15:338–344.

Huettel, R. N., Dickson, D. W., and Kaplan, D. T. 1983c. Biochemical identification of the two races of *Radopholus similis* by polyacrylamide gel electrophoresis. *J. Nematol.* 15:345–348. *Proc. Helm. Soc. Wash.* 51:32–35.

Huettel, R. N., Dickson, D. W., and Kaplan, D. T. 1984a. Chromosome number of populations of *Radopholus similis* from North, Central, and South America, Hawaii and Indonesia. *Rev. Nématol.* 7:113–116.

Huettel, R. N. Dickson, D. W., and Kaplan, D. T. 1984b. *Radopholus citrophilus* n.sp., a sibling species of *Radopholus similis. Proc. Helm. Soc. Wash.* 51:32–35.

Huettel, R.N., and Yaegashi, T. 1988. Morphological differences between *Radopholus citrophilus* and *R. similis. J. Nematol.* 20:150–157.

Jairajpuri, M. S. 1964. On *Pratylenchoides crenicauda* Winslow, 1958 (Nematoda: Pratylenchidae) from Srinagar (Kashmir), India. *Curr. Sci.,* 33:339.

Kaplan, D. T., and O'Bannon, J. H. 1985. Occurrence of biotypes in *Radopholus citrophilus. J. Nematol.* 17:158–162.

Katalan-Gateva, S. D., and Gudurova, L. B. 1979. [The endoparasitic species *Zygotylenchus guevarai* (Tobar Jiménez, 1963) Braun and Loof, 1966 (Nematoda: Tylenchida) found in Bulgaria]. *Acta Zool. Bulg.* 12:44–49.

Khan, S. H., and Shakil, M. 1973. On the proposal for *Neoradopholus* n. gen. (Nematoda: Radopholinae). *Proc Natl. Acad. Sci. India. 43rd A. Session. Sect. Biol. Sci.*:16–17.

Kumar, A. C. 1980. Studies on nematodes in coffee soils of South India. 3. A report on *Radopholus similis* and description of *R. colbrani* n. sp. *J. Coffee Res.* 10:43–46.

Kumar, A. C. and Kasiviswanathan, P. R. 1972. Studies on physiological races of *Pratylenchus coffeae. J. Coffee Res.* 2:10–15.

Loof, P. A. A. 1959. Ueber das Vorkommen von Endotokia matricida bei Tylenchida. *Nematologica* 4:238–240.

Loof, P. A. A. 1960. Taxonomic studies on the genus *Pratylenchus* (Nematoda). *LEB Fonds Publ.* 39.

Loof, P. A. A. 1964. Free-living and plant-parasitic nematodes from Venezuela. *Nematologica* 10:201–300.

Loof, P. A. A. 1978. The genus *Pratylenchus* Filipjev, 1936 (Nematoda: Pratylenchidae): A review of its anatomy, morphology, distribution, systematics and identification. *Sveriges Lantbruksuniversitet Växtskyddsrapporter 5.*

Loof, P. A. A. 1985. *Pratylenchus scribneri.* C. I. H. Descriptions of plant-parasitic nematodes, *8, No. 110.*

Luc, M. 1987. A reappraisal of Tylenchina (Nemata). 7. The family Pratylenchidae Thorne, 1949. *Rev. Nématol.* 10:203–218.

Luc, M., and Fortuner, R. 1975. *Hirschmanniella spinicaudata.* C. I. H. Descriptions of plant-parasitic nematodes, *5, No. 68.*

Luc, M., and Goodey, J. B. 1962. *Hirschmannia* n. g. differentiated from *Radopholus* Thorne, 1949 (Nematoda: Tylenchoidea). *Nematologica* 7:197–202.

Luc, M., and Goodey, J. B. 1964 *Hirschmanniella* nom. nov. for *Hirschmannia. Nematologica* 9 (1963), 471.

Machon, J. E., and Hunt, D. J. 1985. *Pratylenchus goodeyi*. C. I. H. Descriptions of plant-parasitic nematodes, *8, No. 120*.

Maqbool, M. A., Ghazala, P., and Qasim, M. 1985. *Zygotylenchus guevarai* (Tobar Jiménez, 1963) Braun and Loof, 1966—a new record from Pakistan. *Pakistan J. Nematol.* 3:15–16.

Mathur, V. K., and Prasad, S. K. 1971. Occurrence and distribution of *Hirschmanniella oryzae* in the Indian Union with description of *H. mangaloriensis* sp. n. *Ind. J. Nematol.* 1:220–226.

Miller, P. M., and Anagnostakis, S. 1977. Suppression of *Pratylenchus penetrans* and *Tylenchorhynchus dubius* by *Trichoderma viride*. *J. Nematol.* 9:182–183.

Minagawa, N. 1982. Descriptions of *Pratylenchus gibbicaudatus* n. sp. and *P macrostylus* Wu, 1971 (Tylenchida: Pratylenchidae) from Kyushu. *Jap. J. Appl. Ent. Zool.* 17:418–423.

Minagawa, N. 1984a. Description of *Radopholoides triversus* n. sp. from Japan with a reference to the classification of the family Pratylenchidae (Nematoda: Tylenchida). *Appl. Ent. Zool.* 19:21–26.

Minagawa, N. 1984b. New species of *Hoplotylus* and *Pratylenchoides* (Tylenchida: Pratylenchidae) from Japan. *Jap. J. Nematol.* 14:15–19.

Mohandas, C., Pattanaik, N. K. C., and Prasad, J. S. 1980. Host range of the rice root nematode, *Hirschmanniella oryzae*. *Ind. J. Nematol.* 9 (1979): 177–178.

Mueller, J. 1974. Vergleichende Versuche über den *Verticillium-Pratylenchus*-Krankheitskomplex unter sterilen und unsterilen Bedingungen. *Nematologica* 19 (1973):249–258.

Olowe, T., and Corbett, D. C. M. 1984a. Morphology and morphometrics of *Pratylenchus brachyurus* and *P. zeae* II. Influence of environmental factors. *Ind. J. Nematol.* 14:6–17.

Olowe, T., and Corbett, D. C. M. 1984b. Morphology and morphometrics of *Pratylenchus brachyurus* and *P. zeae* III. Influence of geographical location. *Ind. J. Nematol.* 14:30–35.

Olthof, Th. H. A. 1968. Races of *Pratylenchus penetrans*, and their effect on black root rot resistance of tobacco. *Nematologica* 14:482–488.

Orton Williams, K., and Siddiqi, M. R. 1973. *Radopholus similis*. C. I. H. Descriptions of plant-parasitic nematodes, *2, No. 27*.

Pariselle, A. 1987. Comportement du nématode *Hirschmanniella oryzae* (Van Breda de Haan) dans les racines de la légumineuse *Sesbania rostrata* Brem. *Rev. Nématol.*, 10: 333–336.

Pariselle, A., and Rinaudo, G., 1988. Etude des interactions entre *Sesbania rostrata*, *Hirschmanniella oryzae* et les rendiments du riz. *Rev. Nématol.* 11:83–87.

Pinochet, J., and Cisneros, T. 1986. Seasonal fluctuations of nematode populations in three Spanish vineyards. *Rev. Nématol.* 9:391–398.

Poucher, C., Ford, H. W., Suit, R. F., and DuCharme, E. P. 1967. Burrowing nematode in citrus. *Fla. Dept. Agr. Div. Plant Industry Bull.* 7.

Prejs, K. 1986. Occurrence of stylet-bearing nematodes associated with aquatic vascular plants. *Ekologia Polska* 34:185–192.

Rashid, A., and Khan, A. M. 1978. Morphometric studies on *Pratylenchus coffeae* with description of *Pratylenchus typicus* Rashid, 1974. *Ind. J. Nematol.* 6 (1976): 63–72.

Rhoades, H. L. 1968. Pathogenicity and control of *Pratylenchus penetrans* on leatherleaf fern. *Plant Dis. Reptr.* 52:383–385.

Roman, J. and Hirschmann, H. 1969. Morphology and morphometrics of six species of *Pratylenchus*. *J. Nematol.* 1:363–386.

Ryss, A. Yu. 1988. *Kornevye paraziticheskie nematody semeistva Pratylenchidae (Tylenchida) mirovoi fauny*. (Russian text). Nauka, Leningrad.

Seinhorst, J. W. 1968. Three new *Pratylenchus* species with a discussion of the structure of the cephalic framework and of the spermatheca in this genus. *Nematologica* 14:497–511.

Seinhorst, J. W. 1971. The structure of the glandular part of the esophagus of Tylenchidae. *Nematologica* 17:431–443.

Seinhorst, J. W. 1977a. *Pratylenchus loosi*. C. I. H. Descriptions of plant-parasitic nematodes, *7, No. 98*.

Seinhorst, J. W. 1977b. *Pratylenchus fallax*. C. I. H. Descriptions of plant-parasitic nematodes, *7, No. 100*.

Sher, S. A. 1968. Revision of the genus *Hirschmanniella* Luc and Goodey, 1963 (*sic*) (Nematoda: Tylenchoidea). *Nematologica* 14: 243– 275.

Sher, S. A. 1968. Revision of the genus *Radopholus* Thorne, 1949 (Nematoda: Tylenchoidea). *Proc. Helm. Soc. Wash.* 35:219–237.

Sher, S. A. 1970. Revision of the genus *Pratylenchoides* Winslow, 1958 (Nematoda: Tylenchoidea). *Proc. Helm. Soc. Wash.* 37:154–166.

Sher, S. A. and Allen, M. W. 1953. Revision of the genus *Pratylenchus* (Nematoda: Tylenchidae). *Univ. Calif. Publ. Zool.* 57:441–470.

Siddiqi, M. R. 1963. On the classification of the Pratylenchidae (Thorne, 1949) nov. grad. (Nematoda: Tylenchida), with a description of *Zygotylenchus browni* nov. gen. et nov. sp. *Z. Parasitenkunde* 23:390–396.

Siddiqi, M. R. 1972. *Pratylenchus coffeae*. C. I. H. Descriptions of plant-parasitic nematodes, *1, No. 6.*

Siddiqi, M. R. 1973. *Hirschmanniella oryzae*. C. I. H. Descriptions of plant-parasitic nematodes, *2, No. 26.*

Siddiqi, M. R. 1974. *Pratylenchoides crenicauda*. C. I. H. Descriptions of plant-parasitic nematodes, *3, No. 38.*

Siddiqi, M. R. 1975. *Zygotylenchus guevarai*. C. I. H. Descriptions of plant-parasitic nematodes *5, No. 65.*

Singh, D. B., and Khan, E. 1981. Morphological variations in populations of *Pratylenchus thornei* Sher and Allen, 1953. *Ind. J. Nematol.* 11:53–60.

Slootweg, A. F. G. 1956. Rootrot of bulbs caused by *Pratylenchus* and *Hoplolaimus* spp. *Nematologica* 1:192–201.

Smith, G. S., and Kaplan, D. T. 1988. Influence of mycorrhizal fungus, phosphorus, and burrowing nematode interactions on growth of rough lemon citrus seedlings. *J. Nematol.* 20:539–544.

Sprau, F. 1969. Schwere Schäden an Pfefferminze (*Mentha peperita* L.) und Peterselie (*Petroselinum sativum* Hoffm.) durch einige freilebende Nematoden. *Mitt. biol. BdAnstalt Land–u. Forstw. Berlin-Dahlem* 136:65–76.

Tarjan, A. C., and Frederick, J. J. 1988. Proposed synonymies within the genus *Pratylenchus* Filipjev, 1936. *Nematropica* 18:21–22.

Tarjan, A. C., and Weischer, B. 1965. Observations on some Pratylenchinae (Nemata), with additional data on *Pratylenchoides guevarai* Tobar Jiménez, 1963 (syn. *Zygotylenchus browni* Siddiqi, 1963 and *Mesotylus gallicus* de Guiran, 1964). *Nematologica* 11:432– 440.

Tarte, R., and Mai, W. F. 1976a. Morphological variation in *Pratylenchus penetrans*. *J. Nematol.* 8:185–195.

Tarte, R. and Mai, W. F. 1976b. Sex expression and tail morphology of female progenies of smooth-tail and crenate-tail females of *Pratylenchus penetrans*. *J. Nematol.* 8:196–200.

Taylor, D. P., and Jenkins, W. R. 1957. Variation within the nematode genus *Pratylenchus*, with the descriptions of *P. hexincisus* n. sp. and *P. subpenetrans* n. sp. *Nematologica* 2:159–174.

Tobar Jiménez, A. 1963. *Pratylenchoides guevarai* n. sp., nuevo nematode tylénchido, relacionado con el ciprés (*Cupressus sempervirens* L.). *Revta Ibér. Parasitologia* 23:27–36.

Tobar Jiménez, A. 1973. Nematodes de los "secanos" de la Comarca de Alhama. I. Niveles de poblacion y cultivos hospedadores. *Revta Ibér. Parasitologia* 33:525–556.

Townshend, J. L. 1984. Anhydrobiosis in *Pratylenchus penetrans*. *J. Nematol.* 16:282–289.

Townshend, J. L., Tarte, R., and Mai, W. F. 1978. Growth response of three vegetables to smooth- and crenate-tailed females of three species of *Pratylenchus*. *J. Nematol.* 10:259–263.

Varo Alcala, J., Tobar Jiménez, A., and Munoz Medina, J. M. 1970. Lesiones causadas y reacciones provocadas por algunos nematodos en las raices de ciertas plantas. *Revta Ibér. Parasitologia* 30:547– 566.

Volvas, N., Inserra, R. N., and Lamberti, F. 1976. Osservazioni sull'epidemologia e sulla patogenicità di *Zygotylenchus guevarai* (Tobar) Braun et Loof. *Nem. Medit.*, 4:183–193.

Vrain, T. C. 1987. Effect of *Ditylenchus dipsaci* and *Pratylenchus penetrans* on *Verticillium* wilt of alfalfa. *J. Nematol.* 19: 379–383.

Winslow, R. D. 1958. The taxonomic position of *Anguillulina obtusa* Goodey, 1932 and 1940. *Nematologica* 3:136–139.

Wu, L. Y. 1971. *Pratylenchus macrostylus* n. sp. (Pratylenchinae: Nematoda). *Can. J. Zool.* 49:487–489.

Yin, K. C. 1986. The identification of rice root nematodes in a Guangzhou suburb. *Plant Protection* 1:14–17.

Zhang, S. S. 1987. [Identification of seven species of *Hirschmanniella* in Fujian]. *J. Fujian Agr. College* 16:155–159.

9

Stem and Bulb Nematodes, *Ditylenchus* spp.

DIETER STURHAN   *Federal Biological Research Center for Agriculture and Forestry, Münster, Germany*

MICHAŁ W. BRZESKI   *Instytut Warzywnictwa, Skierniewice, Poland*

I. INTRODUCTION

Ditylenchus is one of the most "difficult" genera of plant parasitic nematodes. Its systematic position within the Tylenchida changed several times, and a great number of former *Ditylenchus* species were transferred to other genera within Tylenchina as well as Hexatylina. There are 81 species presently recognized in the genus, but 82 species were moved to other genera, considered as *species inquirendae* and *species incertae sedis* or are *nomina nuda* or synonyms of other *Ditylenchus* species. Species differentiation is not easy because of many species that are morphologically similar, and there are few characters available for successful species discrimination. The majority of species in the genus have not been well studied and inclusion of several of the species in *Ditylenchus* appears (still) doubtful. Some of the species are known from a single collection only; others are reported from several continents. The wide geographic distribution of species of this genus may be interpreted as evidence of an ancient (in an evolutionary sense) origin of the genus.

Species of *Ditylenchus* were successful in gaining access to many ecological niches, and *Ditylenchus* is one of the genera among phytonematodes with the greatest variety of feeding habits and other bionomics, in this respect comparable only to the genus *Aphelenchoides*. Only three species (*D. dipsaci, D. destructor, D. angustus*) are of great economic importance as parasites of cultivated higher plants (agricultural crops, ornamentals, etc.); *D. myceliophagus* plays a role as a pest in mushroom growing, and *D. phyllobius* is of interest as an organism which may be used for biological control of weeds. A few other species only are proven or suspected plant parasites, some of them highly host-specialized, such as *D. dryadis* and *D. drepanocercus*. The majority of *Ditylenchus* species are soil inhabiting and probably feed on mycelia; also several of the plant parasites thrive on fungi. Occasionally some of the soil inhabiting species may be encountered in large numbers in the tissue of higher plants, although they have not been observed associated with plant diseases. Several species are found in the frass of bark beetles (*D. dipsacoideus, D. drymocolus, D. parasimilis, D. petilus*).

Most of the phytoparasitic species live as endoparasites in above-ground parts of plants or in roots, stolones, tubers, and rhizomes; *D. angustus* feeds ectoparasitically on plant tissue. More than 500 plant species of dicotyledonous and monocotyledonous angiosperms are known as *Ditylenchus* hosts. There are no published records of parasitism in gymnosperms. The reports on bryophytes as hosts refer to species now in *Subanguina (S. askenasyi, S. brenani)*, or even the genus identity is questionable (Goodey et al. 1965). A wide variety of genera and species of fungi, including nematode-trapping fungi, are known as food sources of *Ditylenchus* species.

With *D. dipsaci*, which is among the nematodes of greatest economic impact worldwide, the genus *Ditylenchus* contains the most "problematical" species in Tylenchida. There is no other nematode known with a comparable excessive intraspecific variation, mainly in host preference, genetic incompatibility, and even partial or complete reproductive isolation of many biological races and "forms" presently combined in this collective species, and with such a wide range in chromosome number. *Ditylenchus dipsaci* is also one of the first plant parasitic nematodes recorded. Only three species, now in *Anguina (A. tritici, A. agrostis, A. graminis)*, had been described earlier.

A thorough treatment of the genus and of all aspects of morphology, biology, ecology, control, etc., of the economically most important *Ditylenchus* species could not be accomplished in this chapter. The number of literature citations has been drastically restricted. For further information, see, for example, reviews by Brzeski (1991), Decker (1969), Gubina (1982), Hesling (1974), Hooper (1972, 1973), Kirjanova and Krall (1971), Russell (1982), and Seshadri and Dasgupta (1975).

II. THE GENUS *DITYLENCHUS* FILIPJEV, 1936

A. Position Within Tylenchida

The genus *Ditylenchus* has been treated differently by recent authors. Traditionally, the generic diagnosis was narrow and included species with thickenings of the esophageal lumen walls within median bulb only, whereas species without these thickenings were collected in the genus *Nothotylenchus* Thorne, 1941. Andrássy (1976), e.g., placed *Ditylenchus* within the family Anguinidae, superfamily Tylenchoidea, and accommodated the genus *Nothotylenchus* in the family Nothotylenchidae, superfamily Neotylenchoidea, both in the suborder Tylenchina. Siddiqi (1980) combined Anguinidae and Nothotylenchidae together with Sychnotylenchidae into the superfamily Anguinoidea, still within Tylenchina. Later he transferred the superfamily to the suborder Hexatylina and placed *Ditylenchus, Nothotylenchus, Diptenchus, Safianema, Orrina*, and other genera in the same family, Anguinidae (Siddiqi, 1986).

Siddiqi (1986) considers the suborders Tylenchina and Hexatylina as two main evolutionary lines—the former evolving to root parasites through algal feeders, the latter developed as fungal feeders or insect parasites and seldom to higher plant parasitism. This view was substantiated by the presence of a musculous metacorporeal bulb in Tylenchina, while this character is generally lacking in Hexatylina. However, there are arguments opposing this point of view. An atrophy of the metacorporeal bulb probably appeared more than once in evolution within the order Tylenchida and the general structure of the nematodes in Tylenchina in Siddiqi's system and Anguinoidea in Hexatylina (according to Siddiqi) is so similar that it would be difficult to imagine they evolved convergently. Maggenti et al. (1987) did not accept Siddiqi's (1986) concept of the suborder Hexatylina mainly because the reproductive system, the head, esophagus, and intestine of *Hexatylus* had been shown to

be different from all other Tylenchida, and they consequently brought back the Anguinoidea to Tylenchina, where they are considered as family Anguinidae.

Fortuner and Maggenti (1987) reviewed Anguinidae and synonymized the genera
Nothotylenchus and *Orrina* with *Ditylenchus*; the genera *Anguillulina*, *Boleodoroides*,
Diptenchus, and *Safianema* had already been synonymized with this genus before (*Orrina*
is considered as a separate genus by Krall, this volume). We concur with this approach and
the present systematic position of *Ditylenchus*, which is as follows:

Order: Tylenchida Thorne, 1949
Suborder: Tylenchina Chitwood, 1950
Superfamily: Tylenchoidea Örley, 1880
Family: Anguinidae Nicoll, 1935 (1926)
Genus: *Ditylenchus* Filipjev, 1936

B. Genus Characteristics

Ditylenchus Filipjev, 1936
  Anguillulina (*Ditylenchus*) Filipjev, 1936
  Nothotylenchus Thorne, 1941
  Boleodoroides Mathur, Khan, and Prasad, 1966
  Diptenchus Khan, Chawla, and Seshadri, 1969
  Safianema Siddiqi, 1981
  Orrina Brzeski, 1981

Diagnosis. Body more or less attenuated. Cuticle thin, striated. Lateral field usually
with three or five bands separated by shallow grooves. Head generally continuous, sometimes slightly expanded or narrower than adjacent body, usually low and flattened. Amphids in form of lateral slits on labial plate. Deirids near excretory pore level. Phasmids not
visible. Stylet delicate, its length mostly within 7–11 µm limit; cone about a third to half of
stylet length; knobs small, except for few species where they are medium-sized. Cephalic
framework only slightly refractive, extends posteriorly from basal plate for about two annuli. Dorsal esophageal gland orifice 1–3 µm posterior to knobs. Esophageal lumen slightly
more refractive in procorpus than in isthmus and glandular part of esophagus. Metacorporeal bulb either musculous and then with thickenings of lumen walls, or devoid of muscles and lumen walls thickenings, or occasionally no esophageal expansion present. Glandular part of esophagus in form of an offset bulb or lobe overlapping intestine. No cardiac
cells, but most anterior intestine cells hyaline.

Female reproductive system prodelphic; postvulval uterine sac present with few exceptions. Vulva posterior, *V* mostly 73–83. Ovary with oocytes arranged in one or two rows,
seldom more; spermatheca axial as a long stretchable sac; crustaformeria composed of four
rows of cells, each row of four cells (in the descriptions of a few species more cells are
drawn). Most anterior part of uterine sac usually with few projections.

Males described for most of the known species, similar to females, although usually
shorter. Spicules slightly ventrally bent, capitulum not very distinct, gubernaculum simple
and not protrusible. Sperm large, usually 3–5 µm in diameter, nucleus surrounded by some
protoplasm. Caudal alae from adcloacal to almost reaching tail end. Tail of both sexes similar, conical to almost filiform, mostly about four to seven body widths long, seldom of different shape.

(The descriptions of some species contain certain characters which are not covered by the diagnosis, e.g., on the shape of the amphidial openings, presence of phasmids, and structure of the female genital tract. A reconfirmation appears necessary.)

Typus generis: *D. dipsaci* (Kühn, 1857) Filipjev, 1936.

C. Variability and Diversity Within Genus

The genus *Ditylenchus* shows little intrageneric diversity in most morphological characters, but considerable intraspecific variation is known to occur for certain features, and identification of species is thus not easy. As shown by Fortuner (1982) already a few characters are of diagnostic significance.

The body length as well as other dimensions can be strongly influenced by the food supply and by environmental conditions. The head may be either smooth or faintly striated, but the lip annulation is not easily seen in bright-field microscopy and can seldom be used as differentiating character. The width of the cuticular annules may be variable within a species. The number of lateral lines is constant and a valuable diagnostic character. However, these lines may be difficult to observe on older specimens as the lateral fields appear to be stretchable zones of the integument. As far as it is known, the number of lateral lines is the same in juvenile and adult specimens of *Ditylenchus* species, and it may often be easier to observe the lateral field on immature specimens.

The cephalic framework is similar in most species and generally does not show any diagnostic character, although the outer margins of the basal plate of *D. longimatricalis* appear slightly more refractive, and the crescentic blades in *D. myceliophagus* are important for identification of this species. The stylet of most species known or suspected to be mycetophagous has a cone shorter than the shaft, while in parasites of higher plants cone and shaft are generally of equal length. The stylet length is rather constant, though variation of 1 μm either shorter or longer than the mean is commonly observed; it is sometimes difficult to see the tip of the stylet. The shape of the stylet knobs shows little intraspecific variation, but often the knobs are so small that it is difficult to draw them properly. Moreover, knobs are similar in many species.

The metacorporeal bulb is variable in shape even among specimens of the same population; its shape depends on the degree of muscle contraction. The thickenings of the esophageal lumen walls in the metacorporeal bulb, which are attachments for bulb muscles, are a constant feature of a species. The observability of this structure depends on the way of killing of the specimens; properly killed specimens, i.e., when the fixative and not the water first penetrates the nematode body, promise further muscle transparence and then the wall thickenings are clearly visible. These thickenings are either small (1–3 μm in length), which is probably a primitive state of the character, or completely absent and then no bulb musculature can be seen; or they are well developed (3–5 μm in length) mainly in parasites of higher plants. The size of these thickenings is a specific character. The esophageal glands may be enclosed in a basal bulb offset from the intestine, or there may be a small overlap of the glands over the intestine, but some species show a long glandular lobe. The shape of the glandular part of the esophagus is variable at least within some species. Generally, the esophageal lobe, if present, is shorter in males than in females. The remaining parts of the alimentary tract do not show diagnostic features. The position of the excretory pore in relation to the esophagus exhibits a certain degree of variation within species, although some species demonstrate distinct interspecific differences: *D. khani* has the excretory pore very close to the esophagointestinal junction, whereas in *D. longimatricalis* it is opposite from the middle of the isthmus.

The female reproductive system is longer in some species than in others, but it changes much during the lifetime of specimens and apparently is influenced by environmental conditions. Consequently, this character can be used with great caution only. The spermatheca can be longer or shorter, but as this organ is stretchable and the degree of stretching depends on the number of sperms it contains, it cannot be used to differentiate species. The position of the vulva, expressed in percentage of body length, is very constant. A difference of about 10% may be expected between specimens with most anterior and most posterior vulva if several populations are examined. The variability within a single population is usually smaller, but occasionally specimens with aberrant vulva position are found. In most cases this is related to an unusually long or short part of the body between vulva and anus. The post vulval uterine sac shows large intraspecific variability, although in most species it can be either longer, shorter, or occasionally absent. Fortuner (1982) correctly concluded that this character could be used in differential diagnoses only if the differences are large.

In males, spicule and bursa length offer diagnostic characteristics. The spicule length is fairly constant, although some variation among populations is observed, and shorter specimens have shorter spicules. The caudal alae are either shorter and adcloacal or they envelop some part of the tail, sometimes even reaching almost to the tail tip. However, because of variability within species and the difficulty to measure exactly the posterior end of the bursa, it could be used for species differentiation providing the differences are large enough.

The tail shape is fairly constant within species, but it is rather uniform among many known species. The length of female tail is often expressed as the ratio of vulva–anus distance to tail length. Generally, this ratio is stable within a population, but fairly large differences could be observed between populations of the same species. However, if enough populations are examined, often populations bridging the extreme values are found. The shape of the tail tip of some species is very constant, as, for example, rounded in *D. destructor* and *D. myceliophagus* or sharply pointed in *D. dipsaci* and *D. convallariae*. However, many species show large variation with intermediate forms. Various authors described these forms as minutely rounded, sometimes as subacute. Therefore, three shapes of the tail extremity have been differentiated in the tabular key (Table 1): pointed, dull, rounded. Some specimens of certain species show a tail tip with a small, more or less pronounced mucro. This cuticular projection may be tracelessly broken. Therefore, it should be used as a specific character with caution only. *Ditylenchus angustus*, for example, is described as having a mucronate tail tip.

D. Nominal *Ditylenchus* Species

Lists of *Ditylenchus* species were recently published by Fortuner (1982), Sumenkova (1982), Siddiqi (1986), and Fortuner and Maggenti (1987). Because of many recent additions and changes (cf. Brzeski, 1991), an up-to-date compilation of nominal species of the genus *Ditylenchus* is given in the inventory below. The species presently considered as valid are given in one list; species which were synonymized with other species, transferred to other genera, considered as *species inquirendae* or as *species incertae sedis* and *nomina nuda* are compiled in a separate list. For a full listing of synonyms of cited species, the review by Brzeski (1991) should be consulted.

Inventory of valid *Ditylenchus* species

D. acuminatus Fortuner and Maggenti, 1987

D. acutatus Brzeski, 1991

D. acutus (Khan, 1965) Fortuner and Maggenti, 1987

D. adasi (Sykes, 1980) Fortuner and Maggenti, 1987

D. affinis (Thorne, 1941) Fortuner and Maggenti, 1987

D. anchilisposomus (Tarjan, 1958) Fortuner, 1982

D. angustus (Butler, 1913) Filipjev, 1936

D. antricolus (Andrássy, 1961) Fortuner and Maggenti, 1987

D. apus Brzeski, 1991

D. attenuatus (Mulvey, 1969) Fortuner and Maggenti, 1987

D. ausafi Husain and Khan, 1967

D. australiae Brzeski, 1984

D. basiri (Khan, 1965) Fortuner and Maggenti, 1987

D. bhatnagari (Tikyani and Khera, 1969) Fortuner and Maggenti, 1987

D. buckleyi (Das, 1960) Fortuner and Maggenti, 1987

D. caudatus Thorne and Malek, 1968

D. citri (Varaprasad, Khan, and Lal, 1980) Fortuner and Maggenti 1987

D. clarus Thorne and Malek, 1968

D. convallariae Sturhan and Friedman, 1965

D. cylindricollis (Thorne, 1941) Fortuner and Maggenti, 1987

D. cylindricus (Khan and Siddiqi, 1968) Fortuner and Maggenti, 1987

D. cyperi Husain and Khan, 1967

D. danubialis (Andrássy, 1960) Fortuner and Maggenti, 1987

D. deiridus Thorne and Malek, 1968

D. destructor Thorne, 1945

D. dipsaci (Kühn, 1857) Filipjev, 1936

D. dipsacoideus (Andrássy, 1952) Andrássy, 1956

D. drepanocercus Goodey, 1953

D. dryadis Anderson and Mulvey, 1980

D. drymocolus (Rühm, 1956) Brzeski, 1991

D. elegans Zell, 1988

D. emus Khan, Chawla, and Prasad, 1969

D. equalis Heyns, 1964

D. exilis Brzeski, 1984

D. ferepolitor (Kazachenko, 1980) Fortuner and Maggenti, 1987

D. filenchulus Brzeski, 1991

D. filimus Anderson, 1983

D. fotedari (Mahajan, 1977) Fortuner and Maggenti, 1987

D. geraerti (Paramonov, 1970) Bello and Geraert, 1972

D. goldeni (Maqbool, 1982) Fortuner and Maggenti, 1987

D. hexaglyphus (Khan and Siddiqi, 1968) Fortuner and Maggenti, 1987

D. indicus (Sethi and Swarup, 1967) Fortuner, 1982

D. khani Fortuner, 1982

D. kheirii Fortuner and Maggenti, 1987

D. loksai (Andrássy, 1959) Fortuner and Maggenti, 1987

D. longicauda Choi and Geraert, 1988

D. longimatricalis (Kazachenko, 1975) Brzeski, 1984

D. lutonensis (Siddiqi, 1980) Fortuner, 1982

D. major (Thorne and Malek, 1968) Brzeski, 1991

D. medians (Thorne and Malek, 1968) Fortuner and Maggenti, 1987

D. medicaginis Wasilewska, 1965

(Inventory continued)

D. mirus Siddiqi, 1963
D. montanus (Kiknadze and Eliashvili, 1988) Brzeski, 1991
D. myceliophagus Goodey, 1958
D. nanus Siddiqi, 1963
D. nortoni (Elmiligy, 1971) Bello and Geraert, 1972
D. obesus Thorne and Malek, 1968
D. oryzae (Mathur, Khan, and Prasad, 1966) Fortuner and Maggenti, 1987
D. parasimilis (Massey, 1974) Fortuner and Maggenti, 1987
D. parvus Zell, 1988
D. petilus (Massey, 1974) Fortuner and Maggenti, 1987
D. phyllobius (Thorne, 1934) Filipjev, 1936
D. silvaticus Brzeski, 1991
D. similis (Thorne and Malek, 1968) Fortuner and Maggenti, 1987
D. singhi (Das and Shivaswamy, 1980) Fortuner and Maggenti, 1987
D. solani Husain and Khan, 1976
D. sorghii Verma, 1966
D. taylori (Husain and Khan, 1974) Fortuner and Maggenti, 1987
D. tenuidens Gritzenko, 1971
D. terricolus Brzeski, 1991
D. thornei (Andrássy, 1958) Fortuner and Maggenti, 1987
D. triformis Hirschmann and Sasser, 1955
D. truncatus (Eliashvili and Vacheishvili, 1980) Fortuner and Maggenti, 1987
D. tuberosus (Kheiri, 1971) Fortuner and Maggenti, 1987
D. turdus Brzeski, 1991
D. turfus (Yokoo, 1968) Fortuner and Maggenti, 1987
D. uniformis (Truskova and Eroshenko, 1977) Fortuner and Maggenti, 1987
D. utschini (Gagarin, 1974) Fortuner and Maggenti, 1987
D. valveus Thorne and Malek, 1968
D. varaprasadi Fortuner and Maggenti, 1987
D. virtudesae Tobar-Jimenez, 1987
D. websteri (Kumar, 1983) Brzeski, 1991

Inventory of departures from the list of presently recognized *Ditylenchus* species (abbreviations used: *nn = nomen nudum, si = species inquirenda, sis = species incertae sedis*):

D. abieticola Rühm, 1956, to *Sychnotylenchus*
D. acris (Thorne, 1941) Fortuner and Maggenti, 1987, *si*
D. acutus (Khan and Nanjappa, 1972) Fortuner, 1982 = *D. acuminatus* Fortuner and Maggenti, 1987
D. allii (Beijerinck, 1883) Filipjev and Schuurmans Stekhoven, 1941, syn. of *D. dipsaci*
D. alliphilus Fortuner and Maggenti, 1987, syn. of *D. acutus*
D. allocotus (Steiner, 1934) Filipjev and Schuurmans Stekhoven, 1941, *si*
D. amsinckiae (Steiner and Scott, 1935) Filipjev and Schuurmans Stekhoven, 1941, to *Subanguina*
D. arboricolus (Cobb, 1922) Goodey and Franklin in Goodey, 1956, to *Deladenus*
D. askenasyi (Bütschli, 1873) Goodey, 1951, to *Subanguina*
D. atypicus (Khera and Chaturvedi, 1977) Fortuner and Maggenti, 1987, to *Boleodorus*
D. autographi Rühm, 1956, to *Sychnotylenchus*
D. bacillifer (Micoletzky, 1922) Filipjev, 1936, *si*
D. balsamophilus (Thorne, 1926) Filipjev and Schuurmans Stekhoven, 1941, to *Subanguina*
D. beljaevae Karimova, 1967, *si*
D. brassicae Husain and Khan, 1976, *sis* (probably to *Filenchus*)

(Inventory of Departures continued)

D. brenani (Goodey, 1945) Goodey, 1951, to *Subanguina*
D. brevicauda (Micoletzky, 1925) Filipjev, 1936, *si*
D. cafeicola (Schuurmans Stekhoven, 1951) Andrássy, 1954, *si*
D. callidus (Izatullaeva, 1967) Fortuner and Maggenti, 1987, *nn*
D. communis (Steiner and Scott, 1935) Filipjev and Schuurmans Stekhoven, 1941, syn. of *D. dipsaci*
D. compactus (Massey, 1974) Fortuner and Maggenti, 1987, *si*
D. damnatus (Massey, 1966) Fortuner, 1982, *si*
D. darbouxi (Cotte, 1912) Filipjev, 1936, *si*
D. dendrophilus (Marcinowski, 1909) Filipjev and Schuurmans Stekhoven, 1941, to *Sychnotylenchus*
D. devastatrix (Kühn, 1869) Filipjev and Schuurmans Stekhoven, 1941, syn. of *D. dipsaci*
D. durus (Cobb, 1922) Filipjev, 1936, to *Deladenus*
D. elongatus (Husain and Khan, 1974) Fortuner and Maggenti, 1987, syn. of *D. cylindricus*
D. eremus Rühm, 1956, to *Sychnotylenchus*
D. eurycephalus (de Man, 1921) Filipjev, 1936, *si*
D. exiguus (Andrássy, 1958) Fortuner and Maggenti, 1987, *si*
D. falcariae Pogosjan, 1967, syn. of *D. dipsaci*
D. fragariae Kirjanova, 1951, syn. of *D. dipsaci*
D. galeopsidis Teploukhova, 1968, *nn*
D. galeopsidis Paramonov, 1970, syn. of *D. dipsaci*
D. gallicus (Steiner, 1935) Filipjev, 1936, to *Sychnotylenchus*
D. glischrus Rühm, 1956, to *Sychnotylenchus*
D. graminophilus (Goodey, 1933) Filipjev, 1936, to *Subanguina*
D. havensteinii (Kühn, 1881) Siddiqi, 1986, syn. of *D. dipsaci*
D. humuli Skarbilovich, 1972, *si*
D. hyacinthi (Prillieux, 1881) Filipjev and Schuurmans Stekhoven, 1941, syn. of *D. dipsaci*
D. innuptus (Andrássy, 1961) Fortuner and Maggenti, 1987, to *Boleodorus*
D. inobservabilis (Kirjanova, 1938) Kirjanova, 1951, *si*
D. intermedius (de Man, 1880) Filipjev, 1936, *si*
D. istatae Samibaeva, 1966, *si*
D. karakalpakensis Erzhanova, 1964, *si*
D. kischkeae (Meyl, 1961) Loof, 1986, *si*
D. longistylus (Khera and Chaturvedi, 1977) Fortuner and Maggenti, 1987, *si*
D. maleki Fortuner and Maggenti, 1987, syn. of *D. major* (Thorne and Malek, 1968) Brzeski, 1991
D. major (Fuchs, 1915) Filipjev, 1936, syn. of *D. maleki*
D. melongena Bhatnagar and Kadyan, 1969, *si*
D. microdens Thorne and Malek, 1968, *si*
D. minutus Husain and Khan, 1967, *si*
D. misellus Andrássy, 1958, to *Filenchus*
D. mycophagus Tarjan and Hopper, 1954, *nn*
D. ortus (Fuchs, 1938) Filipjev and Schuurmans Stekhoven, 1941, to *Sychnotylenchus*
D. panurgus Rühm, 1956, to *Sychnotylenchus*
D. paragracilis (Micoletzky, 1922) Sher, 1970, *si*
D. paramonovi (Gagarin, 1974) Fortuner and Maggenti, 1987, syn. of *D. acutus*
D. petithi (Fuchs, 1938) Rühm, 1956, to *Sychnotylenchus*
D. phloxidis Kirjanova, 1951, syn. of *D. dipsaci*
D. pinophilus (Thorne, 1935) Filipjev, 1936, to *Sychnotylenchus*
D. pityokteinophilus Rühm, 1956, to *Sychnotylenchus*
D. procerus (Bally and Reydon, 1931) Filipjev, 1936, *si*
D. protensus Brzeski, 1984, syn. of *D. ferepolitor*

(Inventory of Departures continued)

D. pumilus Karimova, 1957, *si*

D. pustulicola (Thorne, 1934) Filipjev and Schuurmans Stekhoven, 1941, *sis*

D. putrefaciens (Kühn, 1879) Filipjev and Schuurmans Stekhoven, 1941, syn. of *D. dipsaci*

D. radicicolus (Greeff, 1872) Filipjev and Schuurmans Stekhoven, 1941, to *Subanguina*

D. rarus Meyl, 1954, to *Sychnotylenchus*

D. sapari Atakhanov, 1958, *si*

D. saxenai Fortuner and Maggenti, 1987, syn. of *D. acutus*

D. secalis (Nitschke, 1868) Siddiqi, 1986, syn. of *D. dipsaci*

D. sibiricus German, 1969, syn. of *D. equalis*

D. silvestris (Kazachenko, 1980) Fortuner and Maggenti, 1987, syn. of *D. ferepolitor*

D. sonchophilus Kirjanova, 1958, *nn*

D. sonchophilus Paramonov, 1970, syn. of *D. dipsaci*

D. srinagarensis (Fotedar and Mahajan, 1974) Fortuner and Maggenti, 1987, syn. of *D. acutus*

D. striatus (Fuchs, 1938) Rühm, 1954, to *Sychnotylenchus*

D. sycobius (Cotte, 1920) Filipjev, 1936, *si*

D. taleolus (Kirjanova, 1938) Kirjanova, 1961, *si*

D. tausaghysatus (Kirjanova, 1938) Kirjanova, 1961, *si*

D. tobaensis (Schneider, 1937) Kirjanova, 1951, *si*

D. trifolii Skarbilovich, 1958, syn. of *D. dipsaci*

D. tulaganovi Karimova, 1957, *si*

D. valkanovi (Andrássy, 1958) Zell, 1988, *sensu* Zell, 1988 nec Andrássy, 1958, *sis*

E.   Key to *Ditylenchus* Species

A tabular key to the presently recognized *Ditylenchus* species (with the exception of *D. sorghii*), which was first published by Brzeski (1991), is presented in Table 1. It comprises a few morphological characters that are of diagnostic significance. Data such as stylet length, c and c' values concern females.

III.   SPECIES OF AGRICULTURAL IMPORTANCE

A.   *Ditylenchus dipsaci* (Kühn, 1857) Filipjev, 1936

Stem nematode, stem and bulb nematode

1.   Taxonomic Problems

A total of 13 nominal species have been synonymized with *D. dipsaci*. Whereas synonymization of *D. communis*, *D. devastatrix*, *D. devastatrix narcissi*, *D. havensteinii*, *D. hyacinthi*, *D. putrefaciens*, and *D. secalis* has long been commonly accepted, the names *D. allii*, *D. fragariae*, *D. phloxidis*, and *D. trifolii* have still been used in the recent literature. *Ditylenchus falcariae* (resp. *D. dipsaci falcariae*), *D. galeopsidis*, and *D. sonchophilus* are mostly considered as valid and separate species.

Most species were generally characterized and distinguished from other nominal stem nematode species by certain host preferences. Morphological characters given for these are generally within the range of variation known for *D. dipsaci* s. str. and species already synonymized with *D. dipsaci*. Moreover, there are remarkable differences in pub-

TABLE 1 Tabular key to *Ditylenchus* species

Species	MB	Lateral lines	Stylet (μm)	V	Spicule (μm)	PUS		Tail tip	c	c'	Bursa (% of tail)
D. longimatricalis	+	4	5.5–7.5	60–73	12–15	1.8–5.1	23–73	P	5.3–10.3	5.9–10.2	9–33
D. acuminatus	+	4	5–7	79–84	?	0.7–1		P,?D	9–11.9	5	?
D. exilis	+	4	6–8	77–80	15–17	0.6–1.0		R	8.9–10.7	6.5–9	29–35
D. terricolus	+	4	7–7.5	71–77	13–14	2.1–3.2	35–39	R,D,m	9.1–12.1	4.5–7.1	26–39
D. parvus	+	4	6.5–9	71–77	12–18	2–4	28–51	P	8.6–15	3.6–8.1	17–50
D. equalis	+	4	7–8	77–82	14–15	0.9–1.7	22–50	P,D	9–13	4–6	29–35
D. emus	+	4	7–9	79–81	15–16	2–2.2	37–39	R	12–15	5–5.3	24–33
D. nortoni	+	4	7–8.5	79–86	21–25	about 1		P	9–11.6	5.4–7.6	up to 60
D. filimus	+	4	7–9	81–85	29	0.5–1.1		P	9–11	4.3–6.5	61
D. deiridus	+	4	8	87	?	absent		P,D	18	(3.9)	?
D. solani	+	4	8–11	80–83	18–20	>2	>50	D	10–16	6–7	<20
D. ausafi	+	4	10–11	72–75	12–15	>1		D	9–10	(5.7)	<33
D. angustus	+	4	10–11	78–80	16–21	2–2.5	33–67	P,m	18–24	5.2–5.4	almost 100
D. dipsaci	+	4	10–12	76–86	23–28		40–70	P	11–20	3–6	40–70
D. indicus	+	4	11–12	80–84	14	<2		R	10–14	<5	<33
D. cyperi	+	5	10–11	75–83	15–18		50	D	17–18	(3.1)	almost 100
D. geraerti	+	6	6	80	?		>50	R	10–13.8	(3.5)	?
D. ferepolitor	+	6	6–7.8	68–75	9–14	1.1–3.5	13–45	P	7–10.6	6–9	14–35
D. filenchulus	+	6	7–8	68–72	13–16	2.1–4.5	30–45	R,P	6.1–9.8	7–14	11–15
D. elegans	+	6	7–9	71–77	16–18	(3.2)		P	7.6–10.9	7.2–12.3	(14)
D. apus	+	6	8	75–76	?	0.2–0.4	4–9	R	9.6–10.9	4.1–5.3	?
D. silvaticus	+	6	7–8	78–81	?	0.8–1.4	27–45	P,m	9.2–14.2	3.6–6.3	?
D. acutatus	+	6	7–8	77–81	19	2.8–4.1	45–59	P	14.7–16.2	4.1–5.7	23–57
D. tenuidens	+	6	7–9	76–82	15–18	0.9–1.8	18–37	P	9–12.9	4.1–6.8	24–50
D. medicaginis	+	6	7–9	75–84	15–22	1.2–2.2	30–40	B,R,P	8.4–13.6	4–8.6	20–44
D. valveus	+	6	7–9	76–82	16–23	0.8–2.9	21–50	R,m	8.4–14.1	5–8.8	23–47
D. myceliophagus	+	6	6.5–8.5	77–86	15–20	1.5–2.8	30–69	R	8.2–17	3–7	20–55
D. virtudesae	+	6	7.3–8.2	80–82	11		40–60	R,D	17.3–20.9	(2.8)	100
D. nanus	+	6	7–8	81–85	13–15		52–71	R	15.1–19	2.9–3.6	55–almost 100
D. triformis	+	6	7.5–9.5	75–82	13–15		25–33	R,D	9.6–11.9	(4.4–5.1)	33–50
D. lutonensis	+	6	8–9.4	70–74	15–17	2.6–3.3		P,D	7.2–8.5	8–11	17–20

Species											
D. longicauda	+	6	8–10	75–81	16–23		1.2–2.1	R,P	6.5–11.7	6.7–12	17–23
D. anchilisposomus	+	6	7.6–10.8	78–83	17–20		1.5–2.5	R,D	11.1–13.3	(3.9–5.1)	50–67
D. khani	+	6	8–9	82–90	22		absent	P,D,R	12–18.5	2.9–5	<67
D. caudatus	+	6	10	75	(22)		1	R	10	(4.6)	33
D. australiae	+	6	9–10	80–82	18–19		0.5–1.2	D	10.2–12.7	4–6	23–36
D. clarus	+	6–8	9–10	78–83	19		1.3–2.3	R	12.8–15.4	4–5	49–54
D. dryadis	+	6	10–11	80–83	23–31	61–86		P	17–24	3.4–4.8	64–76
D. destructor	+	6	10–13	77–84	24–27	53–90		R	14–20	3–5	50–70
D. convallariae	+	6	11–13.5	74–79	20–26	25–47		P	12–15	4.5–6	50–66
D. dipsacoideus	+	?	8.3–9	72–79	16		(1.5)	R	12.3–17	(2.8)	75
D. drepanocercus	+	?	8–9	75–80	10		1		16.2–18	(3.1)	50
D. mirus	+	?	8–9	83–85	17	50		sickle, m	17–20	(2.4)	almost 100
D. obesus	+	?	10	79	?		>1	R	12.5	(4.3)	?
D. petilus	–	2	7.5	76	(14)		>1	P, ?D	18	(3.1)	88
D. basiri	–	4	6–7	73–75	13–15		2.3	R	8.8–10.4	5	<50
D. thornei	–	4	6.4	76	?		<1	R, m	10	9	?
D. antricolus	–	4	6.6–7	71–72	?		<1	P, D	6.7–6.9	9.6–10	?
D. loksai	–	4	7.8	67	14		>0.5	R	6.3	8	(12)
D. danubialis	–	4	7.7	72–73	15–16		3.4–5.7	P	9.3–10.4	6–8	17
D. turfus	–	4	7.5	71–85	19		1.5	P	8.8–16	(3.8–5.8)	almost 100
D. parasimilis	–	4	7.5	81	(16)		>1	P	16	(3)	67
D. cylindricollis	–	4	7	90	20		1	P	17	(4.1)	50
D. acutus	–	4	7.5–9	72–78	14–20	25–50	1.3–3	P, D	5.6–12	5–9	12–40
D. cylindricus	–	4	7–8	75–82	14	30–50	1.5–2	R	7–11.4	4.8–6	29–30
D. oryzae	–	4	8–9	74–76	19–21		1.5	R	9	5.5	(40)
D. attenuatus	–	4	9–10	74–77	?		>2	P	13–16	(7.5)	?
D. singhi	–	4	9	77	13–15	50	>1.5	R	8.3	6–7	33
D. websteri	–	4	8–9	80–82	11–13		1.5	P	10.5–13.7	4	20
D. bhatnagari	–	4	9–10	80–82	17–19	20	<1	R	11.2–17.5	(4.2)	>50
D. phyllobius	–	4	9–10	78–84	18–22	46–77	1.6–2.4	R	11.4–17.6	2.9–4.5	52–77
D. truncatus	–	4	11.2–12.5	57–66	?	14–20	1.2	D	6.3–7.4	7–7.5	?
D. adasi	–	4	11–13	68–76	22–24	16–34	1.2–2.3	R, D	8.7–10.1	5.1–7.5	31–43
D. utschini	–	4	12–13.5	75–79	13–15		2	P, ?D	9.8–11.2	6.7	up to 50
D. drymocolus	–	5	11–12	81–85	18–21		(<1)	P	9.7–11.8	(4.9)	(31)
D. varaprasadi	–	6	6–8	79–80	18–21		>1	R	13	4–5	50
D. medians	–	6	6.5–8	79–84	15–18	39–65	1.3–2.6	R	10.9–14	3.7–4.6	27–84

TABLE 1 (Continued)

Species	MB	Lateral lines	Stylet (μm)	V	Spicule (μm)	PUS		Tail tip	c	c'	Bursa (% of tail)
D. hexaglyphus	–	6	7–8.5	82–84	21–22	0.5–1	18–21	R	11.5–14.5	3.6–5.1	38–46
D. taylori	–	6	8–9	75–77	20	2–3		R	10–11	6–7	>50
D. affinis	–	6	8–9	76–80	15–17	1.1–1.3		R	9.2–11.7	4.7–7.3	50
D. uniformis	–	6	8.4	80	?	1.2–1.4		P	12.5	(4.6)	?
D. citri	–	6	8–10	78–80	20–22	2		P, D	12–15	4–6	75
D. kheirii	–	6	8–9	78–83	16	1.5–2.6		R	10–14.3	3.4–6	33
D. tuberosus	–	6	9	81–82	?	2–2.4	50	R	13–15	4–5.2	?
D. fotedari	–	6	9	84–86	17	>1		P, ?D,?R	14–16	4	?
D. major	–	6	10	82	(17)		33–50	R, D	14	(3.2)	50
D. similis	–	6	10	84	22	1		D	17	(2.9)	<75
D. buckleyi	–	6	11	71	15		>50	P	11.2	(4.8)	33
D. goldeni	–	6	11–12	82–83	20–21	0.5		P,D,R	12–14	3.3–3.8	>50
D. montanus	–	6	12–14	80–83	?		<50	P	12–13	5	?

MB = median bulb with (+) or without (–) thickening of lumen walls

PUS = length of postvulval part of uterine sac expressed in relation to vulval body width and/or as percentage of vulva–anus distance

Tail tip (in female) = P, pointed; R, rounded; D, dull; m, mucronate

> = slightly more than , < = slightly less than

( ) = figure obtained from measured drawing and not given in description

? = no information available

lished data characterizing some of these species (e.g., on morphology of *D. sonchophilus* and *D. galeopsidis*). Since many biological races and "strains" differing in host range are known to occur within *D. dipsaci* (see below) and morphological characters, such as measurements, were shown to vary considerably due to environmental factors (host plant, temperature, etc.), the mentioned species appear not sufficiently defined to consider them as separate taxa. More detailed studies are required to evaluate their taxonomic status. At the time being, they are considered here under the collective species *D. dipsaci*.

Knowledge of karyotypes can contribute much in the solving of taxonomic problems. Chromosome numbers were studied (mainly) in the gonial cells of males and females from many biological races and populations of different origin of stem nematodes. The results available so far are compiled on Table 2.

Most populations from cultivated plants proved to have $2n = 24$ chromosomes. Populations from wild plants and from field beans of various origin differ in chromosome number, which mostly varies between $2n = 36$ and $2n = 60$. Two populations, from *Vicia faba* and *Medicago sativa*, were reported to have $2n = 12$ chromosomes. Aneuploidy and other chromosomal aberrations were often observed; reported variation in the high chromosome numbers may be partly due to difficulties in reliable counting.

The basic haploid chromosome number of *D. dipsaci* is assumed to be $n = 12$, eventually $n = 6$. The populations with high chromosome numbers are considered as polyploid (Barabashova, 1978; Sturhan, 1970; Triantaphyllou and Hirschmann, 1980). The volumes of nuclei in populations originating from five species of weeds, with numbers ranging from $2n = 36–38$ to $2n = 56$, were greater than those of populations from onion, phlox, red clover, and parsley with a chromosome number of $2n = 24$. This supports the assumption that the karyotype of the first group of stem nematodes is of polyploid origin (Barabashova, 1984).

All taxa, races, strains, and populations presently combined in the collective species *D. dipsaci* appear to reproduce by obligate amphimixis. Hybridization experiments can thus be used as another important means to study the species entity or possible reproductive isolation.

Crossings have been conducted with many biological races and stem nematode populations of different origin. Some races, e.g., from sugar beet, rye, oat, onion, teasel, and spinach, interbred freely and the progeny appeared to be fully fertile. With certain races fertile offspring was produced in one combination but not in the reciprocal one, as shown in experiments of Eriksson (1974), where the red clover race ♀♀ x lucerne race ♂♂ combination failed as well as other combinations with lucerne race ♂♂, red clover race ♀♀ and white clover race ♀♀. Webster (1967) obtained sterile hybrids in the combination tulip race ♀♀ x red clover race ♂♂ and oat race ♀♀ x tulip race ♂♂. Sturhan (1964) and Ladygina (1974) reported on negative results of crosses with the phlox race. Also crossings of a population from *Plantago maritima* with races from rye and spinach failed as did crossings of the "giant race" from *Vicia faba* with races from tobacco, spinach, and *Plantago maritima* (Sturhan, unpublished). Genetic incompatibility was also recorded for combinations of stem nematodes from *Cirsium setosum* and *Taraxacum officinale* with the onion race and the red clover race and in experiments with races from parsley and parsnip (Ladygina, 1973, 1978). In certain cases and combinations only eggs were produced or a few viable hybrids only developed or the hybrids proved to be infertile. Eggs and juveniles were often abnormal and died in the early stages of development. In a few combinations rarely a vital hybrid progeny was observed, which gradually lost its vitality and fertility and disappeared after a few generations (Ladygina, 1974).

According to our present state of knowledge the collective species *D. dipsaci* has to be considered a species complex or a superspecies, composed of a great number of races and

TABLE 2 Chromosome Numbers in Stem Nematodes

Host plant	Origin	2n	Ref.
Vicia faba	Portugal	12	D'Addabbo et al. (1982)
Medicago sativa	Chile	12	Lamberti et al. (1988)
Fragaria sp.	USSR	16	Paramonov (1962)
Beta vulgaris	FRG	24	Sturhan (unpubl.)
Secale cereale	FRG	24	Sturhan (unpubl.)
Vicia faba	FRG	24	Sturhan (unpubl.)
Spinacia oleracea	FRG	24	Sturhan (unpubl.)
Trifolium pratense	FRG	24	Sturhan (unpubl.)
Medicago sativa	FRG	24	Sturhan (unpubl.)
Valeriana officinalis	FRG	24	Sturhan (unpubl.)
Nicotiana tabacum	FRG	24	Sturhan (unpubl.)
Fragaria sp.	FRG	24	Sturhan (unpubl.)
Rheum rhabarbarum	FRG	24	Sturhan (unpubl.)
Plantago lanceolata	Azores	24	Sturhan (unpubl.)
Allium cepa	USSR	24	Barabashova (1972,1974)
Allium sativum	USSR	24	Barabashova (1972,1978)
Petroselinum crispum	USSR	24	Barabashova (1974,1978)
Pastinaca sativa	USSR	24	Barabashova (1978)
Phlox paniculata	USSR	24	Barabashova (1974)
Narcissus sp.	USSR	24	Barabashova (1972)
Trifolium pratense	USSR	24	Barabashova (1972,1978)
Fragaria sp.	USSR	24	Barabashova (1972,1974)
Allium cepa	Italy	24	D'Addabbo et al. (1982)
Fragaria sp.	Italy	24	D'Addabbo et al. (1982)
Vicia faba	Italy	24	D'Addabbo et al. (1982)
Vicia faba	Malta	24	D'Addabbo et al. (1982)
Allium cepa	Holland	24	D'Addabbo et al. (1982)
Allium cepa	Chile	24	Lamberti et al. (1988)
Picris sp.	USSR	36, 38	Barabashova (1976, 1978, 1979)
Taraxacum officinale	USSR	44,48	Barabashova (1974, 1976, 1978, 1979)
Hieracium pratense	USSR	46	Barabashova (1974, 1976, 1978, 1979)
Hieracium pilosella	USSR	46	Barabashova (1976, 1978, 1979)
Cirsium setosum	USSR	52	Barabashova (1976, 1978, 1979)
Sonchus oleraceus	USSR	52–56	Barabashova (1976, 1978)
Falcaria vulgaris	USSR	56	Barabashova (1976, 1978, 1979)
Plantago maritima	FRG	48–54	Sturhan (1969,1970, unpubl.)
Vicia faba	FRG	48–60	Sturhan (1969,1970, unpubl.)
Vicia faba	Morocco	ca.50	Sturhan (unpubl.)
Vicia faba	Syria	ca.50	Sturhan (unpubl.)
Vicia faba	Malta	54	D'Addabbo et al. (1982)
Vicia faba	Malta	60	D'Addabbo et al. (1982)

populations differing mainly in host preference and many of them at different stages of speciation and partially or completely reproductively isolated from others (e.g., lucerne, red clover, white clover, and phlox races and races from parsley and parsnip). Possible recognition of certain biological races as subspecies should be evaluated (cf. Sturhan, 1983). Differences in karyotype and genetic incompatibility indicate that "forms"are also presently grouped under the nominal species *D. dipsaci*, which probably deserve species status, such as the giant race from *Vicia faba* and populations from *Plantago maritima, Taraxacum officinale*, and *Cirsium setosum*, which had already been suggested by several authors (Ladygina and Barabashova, 1976; Barabashova, 1976, 1978; Ladygina, 1982; Sturhan, 1983; and others). This may hold also for races from *Plantago lanceolata, Hypochoeris radicata, Cirsium arvense, Picris*, and *Hieracium* as well as for *D. falcariae* described from *Falcaria vulgaris* and *D. sonchophilus* from *Sonchus oleraceus*. Some of these "races" appear to be closely related and are probably members of the same species.

More detailed investigations are required to elucidate the complex taxonomic situation of the collective species *D. dipsaci,* in particular, to clarify the taxonomic status of the forms with high chromosome numbers. Besides morphological and karyological studies and further crossing experiments, electrophoretic studies and DNA analyses will probably support the attempts to understand the true nature of *D. dipsaci* and to resolve the taxonomical problems in stem nematodes. Some additional information on races, populations, and nominal species differing in host preference is given below (under "Host Races").

2. Morphological Characters (Fig. 1)

Dimensions (from various sources):

Females: L = 1.0–2.2 mm, a = 36–64, b = 6.5–12, c = 11–20, c' = 3–6, V = 76–86, stylet = 10–13 μm.

Males: L = 1.0–1.9 mm, a = 37–74, b = 6–15, c = 12–19, stylet = 10–12 μm.

Description: Body straight or almost so when relaxed. Lateral field with four incisures. Head unstriated, continuous with adjacent body part. Stylet cone about half of stylet length, knobs rounded. Median esophageal bulb muscular, with thickenings of lumen walls about 4–5 μm long. Basal bulb offset or overlapping intestine for a few micrometers. Excretory pore opposite posterior part of isthmus or glandular bulb. Postvulval part of uterine sac about half of vulva–anus distance long or slightly more. Male cloacal alae envelop about three-quarters of tail length. Spicules 23–28 μm long. Tail of both sexes conical, always pointed.

3. Distribution and Economic Importance

Ditylenchus dipsaci is one of the most devastating plant parasitic nematodes on a wide range of crops. In heavy infestation crop losses of 60–80% are not unusual; e.g., in Italy up to 60% of onion seedlings died before reaching the transplanting stage and for garlic crop losses of about 50% were recorded from Italy and more than 90% from France and Poland. In Morocco *D. dipsaci* was found in 79% of seed stocks of *Vicia faba* examined (Schreiber, 1977).

Stem nematodes are distributed worldwide, especially in temperate regions. They have been recorded from most European countries, including Russia and other parts of the Soviet Union, where *D. dipsaci* is a serious pest in sugar beet, onion, rye, oat, maize, strawberry, lucerne, red clover, and many other agricultural and vegetable crops as well as many horticultural plants, especially flower bulbs, such as narcissus, hyacinth, and tulip. In countries of the Mediterranean region, including the north African countries, it is known mainly

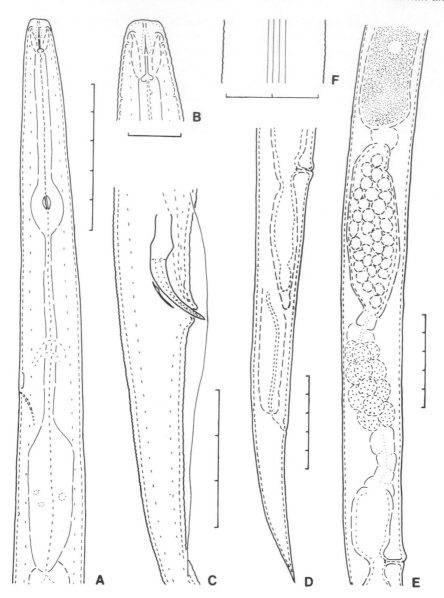

FIGURE 1   *Ditylenchus dipsaci*. (A) Female, esophageal region. (B) Head of female. (C) Male, spicule region. (D) Posterior portion of female. (E) Part of female reproductive system. (F) Lateral field at midbody. Each unit on bars = 10 μm.

as a pest of field beans and onions. In North America lucerne and red clover are the preferred hosts. In warmer regions of the world, e.g., Central and South America, southern Africa, Australia, Iran, and India, mainly lucerne, garlic, and onion are attacked. Stem nematodes are also known to occur in New Zealand, Japan, Hawaii, and many other countries.

The present cosmopolitan distribution of *D. dipsaci* is partly a result of man's activities, and dissemination had been much favored through the ability of the nematode to survive desiccation. Flower bulbs, phlox, strawberry, and other perennials and nursery stock have been important distributing agents. Also seeds of lucerne, clover, field beans, onions, and other plants as well as garlic cloves have been among the common carriers. *Ditylenchus dipsaci* is also readily dispersed through wind, water, and implements.

The majority of stem nematode records in wild plants (except weeds) and from non-agricultural biotopes is from northern Europe, USSR, and USA, where *D. dipsaci* may be endemic.

Stem nematodes in, for example, flower bulbs and seeds of lucerne, field beans, onions, and other *Allium* species are on the list of quarantine organisms of many countries all over the world.

4. Hosts

Ditylenchus dipsaci is an obligate plant parasitic nematode which feeds on the tissues of higher plants. Viglierchio (1971) reported that a Californian population from garlic could reproduce to some extent on soil fungi (*Verticillium, Cladosporium*) under laboratory conditions.

Goodey et al. (1965) recorded a total of more than 400 plant species as hosts of *D. dipsaci* (*sensu lato*), and the number of known hosts may have increased to some 500 species since. Stem nematode hosts are among a wide range of plants from more than 40 plant families of about 30 orders. Around one-third of the known hosts are monocotyledonous species from the subclasses Lilidae and Commelinidae; the majority of the dicotyledonous plants belong to the Asteridae, Dilleniidae, and Rosidae subclasses.

Ditylenchus dipsaci exhibits an extensive intraspecific variation in host range. Very polyphagous populations as well as highly specific populations with a limited host range are known. A great number of "biological races," which are characterized by their host ranges but (generally) not distinguishable by morphological or other characteristics, were differentiated (see below). It is a remarkable fact that even specialized host races are often capable of reproducing on plant species of systematically very distant families, whereas closely related plant species—though known as hosts of other races—are not attacked. *Ditylenchus dipsaci* is readily cultured on, for example, lucerne callus tissue, even biological races whose host range does not include lucerne; their respective host specificities are retained (Bingefors and Bingefors, 1976).

5. Host Races

The existence of variation in host range among stem nematode populations of different origins was first observed more than 100 years ago (Ritzema Bos, 1888). Since then many biological or physiological races in *Ditylenchus dipsaci* have been differentiated, mostly named after their principal host or the plant species on which they had been found for the first time. Eleven distinctive host races were mentioned by Seinhorst (1957) and 12 by Winslow (1960). Subsequently their number grew to 20 and 21 (Hesling, 1966; Decker, 1969; Sturhan, 1969), while Kirjanova and Krall (1971) recognized "only" 15 and Ladygina (1982) at least 30 races (in addition to some species, which are considered as synonyms of *D. dipsaci* in this chapter).

The host ranges of most races are still incompletely known and the tables with differential hosts presented by Seinhorst (1957) and Jones and Jones (1964) do not allow distinction of all cited races, mainly due to lack of information on the host status of the plant species. Most host species and even such used as race differentials are attacked by several races, e.g., onion by the onion, rye, oat, beet, potato, white clover, red clover, strawberry, teasel, tulip, hyacinth, narcissus, ribwort plantain, and probably some more races; strawberry by the strawberry, onion, rye, oat, wild oat, potato, teasel, red clover, tulip, narcissus, and hyacinth races; field beans even by at least 15 races, from rye, oat, onion, beet, potato, strawberry, tobacco, field bean, lucerne, teasel, tulip, narcissus, *Digitalis*, *Valeriana*; and by the "giant race."

It is also recognized that local populations of a race can vary greatly in their host ranges and pathogenicity, including their reaction toward resistant cultivars of a plant (e.g., Sturhan, 1969; Whitehead et al., 1987). Also the existence of intrapopulation variation in host preference was shown. These observations indicate that the host races are not genetically uniform with regard to host specificity. Published information on hosts and nonhosts of biological races is often inconsistent, and the most complete table by Metlitskij (1972; see also Ladygina, 1982) to differentiate 15 races contains several data diverging from those given elsewhere. According to Whitehead et al. (1987), it is unlikely that a workable host differential test for distinguishing host races will be attained.

Moreover, the complexity of racial characterization is increased by the fact that the biological races can interbreed (see above), which may take place in cases of simultaneous infection of common hosts, and that hybrid populations may develop differing in host range from both parental races or populations. There is evidence that generally no linkage of virulence genes exists.

The problem of biological races in stem nematodes has been treated and discussed in more detail by several authors (Hesling, 1966; Sturhan, 1969; Viglierchio, 1971; Ladygina, 1982; and others).

Though the usefulness of designating biological races in stem nematodes is questionable, a brief review on races recorded so far is given below. Populations or strains which are sometimes treated as separate taxa and those which may deserve taxonomical recognition (see above) are also considered. Since the literature on hosts and distribution of biological races is very extensive (the race identity of populations, however, is often unknown), a complete survey can be hardly attained. Some unpublished data of the authors were added. For further information and references see, for example, Decker, 1969; Kirjanova and Krall, 1971; Gubina, 1982.

Teasel race. *Ditylenchus dipsaci* was originally described from the type host *Dipsacus fullonum*, collected at Bonn, Germany. The teasel race, characterized by a rather wide host range which includes strawberries, cucumber, field and *Phaseolus* beans, is also recorded from England, Holland, France, Italy, Soviet Union, Algeria, and the USA.

Rye race. The host specialization of this biological race is low: cultivated and wild plants from numerous families are attacked. The preferred host is rye, but also oat, maize, beet, sunflower, bush and field beans, pea, cucumber, onion, tobacco, etc., along with many common weed species are infected. The rye race is known from central Europe, United Kingdom, Scandinavia, and the Soviet Union.

Oat Race. A special oat race was mainly differentiated in Britain, where it was also designated as oat-onion race or oat-onion-rye race; it is common throughout Europe. The race is very polyphagous, its host range covering *Allium cepa, Vicia faba, Phaseolus vulgaris, Pisum sativum, Beta vulgaris, Fragaria* sp., *Avena fatua*, and many weed species. As with the rye race, wheat and barley are not among the host plants nor are, say, red clover and lucerne. The stem nematodes infesting hydrangeas in England were identified as the oat-onion-race.

Beet race. This biological race has a wide host range, which includes sugar and fodder beet, red beet, mangel, rye, oat, maize, sunflower, onion, field bean, bush bean, pea, cucumber, phlox, and many other crops as well as numerous weeds. Its occurrence has been reported from Belgium, Holland, Federal Republic of Germany, German Democratic Republic, Switzerland, Czechoslovakia, Poland, Yugoslavia, Soviet Union, and the USA.

Potato race. Among the hosts of the polyphagous potato race are important crops such as onion, pea, broad bean, rye, and oat. Because of differences in host range observed among certain populations, two and even three potato races had been distinguished by certain authors. This biological race is known from Britain, the Netherlands, FRG, GDR, Switzerland, and the Soviet Union.

Onion race (*Ditylenchus allii*). The onion race of *D. dipsaci* was originally described as a separate species from the Soviet Union, where it is widely distributed. Besides onion, garlic, and other *Allium* species, a great number of agricultural and horticultural plants as well as many wild plant species are hosts of this polyphagous race, among others, *Phaseolus vulgaris, Glycine max, Pisum sativum, Vicia faba,* and *Beta vulgaris*. Infections of onions and garlic have been reported worldwide, from Britain, Ireland, France, Spain, Italy, Malta, Holland, Belgium, FRG, GDR, Denmark, Sweden, Poland, Czechoslovakia, Hungary, Yugoslavia, Romania, Bulgaria, Canada, the USA, Mexico, Dominican Republic, Cuba, Colombia, Ecuador, Argentina, Brazil, Israel, India, Japan, New Zealand, and other countries. Stem nematodes from garlic were sometimes named "garlic race"; it is unknown if they represent a race different from the onion race.

Tobacco race. This biological race appears to be highly specific though not much is known about its host range. *Faba* and *Phaseolus* beans, cucumber and *Phlox* are hosts, whereas rye, clover, lucerne, potato, and beet are nonhosts. The race has been observed on tobacco in the Netherlands, France, FRG, Italy, and the Soviet Union.

Strawberry race (*Ditylenchus fragariae*). According to host range studies in the USSR, the strawberry race is polyphagous, the host range comprising also pea, onion, garlic, celery, cucumber, red clover, lucerne, and many other cultivated plants, weeds, and wild plant species. Differences in host range are recorded for other countries, but it is known that strawberry is a host of many biological races of *D. dipsaci*. Damage in strawberries through stem nematodes is known from Britain, France, Netherlands, FRG, GDR, Denmark, Sweden, Poland, the USSR, Iran, the USA, and Canada. Stem nematode infestations were also recorded from wild strawberries, for example, in the USA.

Red clover race (*Ditylenchus trifolii*). This biological race belongs to the group of narrowly specialized races. Besides many varieties of red clover, *Trifolium hybridum, Phaseolus vulgaris, Cucumis sativus, Phlox* sp., *Fragaria* sp., *Phacelia heterophylla, Agrostemma githago,* and a few other plants are known as hosts, while white clover, lucerne, wheat, rye, barley, oat, sugar beet, field bean, and other plants proved to be nonhosts. The race is widely distributed in the northern hemisphere and preferentially inhabits countries with a cool, damp climate (Britain, France, Norway, Sweden, Finland, Denmark, Netherlands, FRG, GDR, Switzerland, Czechoslovakia, Poland, USSR, USA).

White clover race. This race is highly host-specific as well. Onions and tulips are also recorded as hosts. Red clover, lucerne, beans, sugar beet, and other common *D. dipsaci* hosts are generally not infested. The white clover race is known from Britain, Denmark, USSR, and other European countries, and from Canada and New Zealand.

Lucerne race. The lucerne or alfalfa race also belongs to the narrowly specialized stem nematode races. Hosts include *Medicago* and *Melilotus* species, *Trifolium hybridum, Onobrychis viciaefolia, Phaseolus* beans, *Phlox*, and several weeds and wild plants, but not onion, beet, and other common stem nematode hosts. This biological race is distributed worldwide in lucerne-growing regions: United Kingdom, France, Holland, FRG, GDR, Denmark, Sweden, Czechoslovakia, Hungary, Yugoslavia, Bulgaria, USSR, Canada, USA,

Mexico, Brazil, Bolivia, Peru, Argentina, Chile, South Africa, Iraq, Iran, Australia, New Zealand, and other countries.

Field bean race. A biological race that is specialized on field and broad beans (*Vicia faba*) and reproducing well on the weed *Galeopsis tetrahit* has been recorded from southern Germany. *Vicia faba* is also known as an excellent host of many other biological races and of the "giant race" of *D. dipsaci*.

Phlox race (*Ditylenchus phloxidis*). This race, which had been described as a separate species and was often recorded as such from the Soviet Union, infests several *Phlox* species and other ornamental plants such as species of *Dianthus, Campanula, Schizanthus, Primula, Oenothera, Solidago*, and *Gilia*. Peas and other crops are also among the hosts of this race, which has been found in Britain, Netherlands, Denmark, FRG, GDR, Czechoslovakia, Switzerland, USSR, Canada, USA, and other countries.

Hyacinth race. The narrowly specialized race mainly infests several hyacinth species, but it reproduces to a lesser extent also in onions and strawberries, but not in, for example, daffodils and tulips. It has been recorded from the United Kingdom, France, Netherlands, Denmark, Switzerland, Yugoslavia, Greece, USSR, Algeria, USA, Canada, and other countries.

Narcissus race. This biological race has a wide host range, which also includes *Galanthus, Begonia, Tigridia, Gladiolus, Scilla, Campanula*, and *Primula* species as well as onion, field bean, *Phaseolus* bean, pea, maize, and other cultivated plants and weeds, whereas hyacinths and *Phlox* are nonhosts. The race is known from Britain, Holland, Denmark, Germany, Switzerland, Yugoslavia, USSR, Canada, USA, and other countries.

Tulip race. Besides tulips *Narcissus, Hyacinthus, Scilla, Primula*, and *Phlox* species are hosts of this rather polyphagous biological race, which also infests onions, *Faba* and *Phaseolus* beans, oats, and strawberries. The tulip race is common in Holland and the United Kingdom but is also known from, for example, Ireland, Germany, Switzerland, Yugoslavia, and Canada.

Wild oat race. From southern Germany a race from *Avena fatua* was described, which is commonly found as a pathogen also in strawberries and multiplies well in peas, but not in barley, wheat, and beet. *Avena sativa* is known as a host of several other races.

Carrot race. A race from carrot (*Daucus carota*) was differentiated in England, which reproduced also on celery, broad bean, pea, oat, and potato.

Flax-hemp race. Flax (*Linum usitatissimum*), hemp (*Cannabis sativa*), and rye (*Secale cereale*) are hosts of this race, which may be identical with the rye race. It is known from USSR, Czechoslovakia, GDR, FRG, Holland, Italy, and Spain. Considerable damage to hemp by stem nematodes is reported also for Turkey.

Giant race. The giant race of *D. dipsaci* was first recorded from *Vicia faba* in Algeria. It is a serious pest of faba beans in many countries in the Mediterranean region including France, but has also been recorded from Portugal, England, and Germany. The giant race seems to have a limited host range; wild plants such as *Ranunculus arvensis, Convolvulus arvensis, Lamium* spp., and *Avena sterilis* are among the hosts. It is generally more damaging to field beans than other races and produces more infested seed. Differences in host range of populations of different origin were observed, isolates from Germany, Morocco, and Italy being more polyphagous. Hooper (1984) even assumes the existence of more than one giant race.

Ribwort plantain race. Infestation of *Plantago lanceolata* by stem nematodes is commonly found in grassland in southern Germany. It has also been recorded from Britain, Holland, France, USSR, and USA. Isolates from England did not reproduce in *Narcissus, Hypochoeris radicata, Allium vineale*, and oat. A German isolate had onion, *Phlox drum-*

mondii, cucumber, and pea as hosts, but not field and bush bean, red and white clover, lucerne, and oat. A population from the Azores invaded field beans and cucumber, but did not reproduce.

Race from *Plantago maritima*. Stem nematodes causing leaf galls and other foliar symptoms in seaside plantain at several foreshore localities in Germany proved to be extremely host-specific. Among more than 40 cultivated and wild plants tested, only in *Phlox drummondii* sometimes slight reproduction occurred and in cotyledons of cucumber deposition of eggs and slow development was observed. A population collected from *P. maritima* in Britain severely attacked *Plantago lanceolata* and *Narcissus*. There are also records from other European countries and from the USA.

Race from *Hypochoeris radicata*. Stem nematodes were found in leaf galls of the false dandelion *H. radicata* and other species of the genus in several European countries and in the western USA, where they are dispersed by seeds of the weed. British strains did not reproduce in *Narcissus, Allium vineale*, clover, and *Plantago lanceolata*.

D. falcariae. Stem nematodes from the umbelliferous host *Falcaria vulgaris* collected in wheat fields at several localities in the Armenian SSR were originally described as *D. dipsaci falcariae*, which was later raised to species rank. An infection of *Falcaria rivini* by stem nematodes (?) had already been recorded from Austria in 1872.

D. galeopsidis. This stem nematode was described from *Galeopsis tetrahit* in the eastern Carpathians and also found in Latvia, the Ukraine, and other regions of the Soviet Union. The same plant species has also been recorded as host of the field bean race and other biological races of *D. dipsaci*; the symptoms of attack were similar to those described and figured by Kirjanova and Krall (1971).

D. sonchophilus. The stem nematode of thistle (*Sonchus oleraceus*) was also described from the Soviet Union. According to Teploukhova (see Ladygina, 1982), Ukrainian *Ditylenchus* populations from *Cirsium arvense. Plantago lanceolata*, and *Taraxacum vulgare* and the Estonian race from *Hieracium* belong to *D. sonchophilus*. Sumenkova (1982) considers *S. oleraceus* to be the only reliable host known today.

Several additional races or populations of stem nematodes with certain host preferences (and/or peculiarities in karyotype or existing genetic incompatibility; see above) were mentioned in the literature. Most of them are not well studied so far and not much information is available on their hosts. Such races, strains, or isolates came from the following hosts: *Valeriana officinalis* (Germany), *Digitalis* spp. (Holland, Germany, USA), *Petroselinum crispum* (USSR, Germany, Britain, USA), *Pastinaca sativa* (USSR, Britain), *Taraxacum officinale* (France, Holland, Italy, Germany, USSR, USA, Canada), *Hieracium pratense* and *H. pilosella* (several European countries), *Picris* spp. (Germany, Czechoslovakia, USSR), *Cirsium arvense* and *C. setosum* (Holland, USSR, Canada).

6. Bionomics

Ditylenchus dipsaci is a migratory endoparasite. All stages of development outside the egg are capable of infecting plants, but the fourth-stage juvenile is the most important infective stage due to its outstanding ability to withstand desiccation and to undergo anabiosis. Mating, which is necessary for reproduction, deposition of eggs, and development take place within plant tissues.

Development and duration of the life cycle are influenced by the temperature and appear to differ among isolates of different origin. According to Ladygina (1957), deposition of eggs starts at temperatures between 1 and 5°C, has its optimum at 13–18°C, and ceases at 36°C. On onion and red clover the life cycle was completed in 17–23 days at temperatures from 13 to 22°C (Yuksel, 1960; Stoyanov, 1964; Zakrzewski, 1977). Goodey

(1922) observed that development from egg to adult required 24–39 days on clover. Between 200 and 500 eggs were laid per female. Temperatures of 15, 19, and 20–25°C have been reported as the optimum temperatures for development and reproduction. Maximum activity and highest invasive ability is generally between 10 and 20°C.

Several consecutive generations develop in the host tissue; development may be arrested in the fourth stage. In dying plant tissues the nematodes tend to aggregate and when dried form "eelworm wool." Preferably preadult juveniles often leave the plants, in particular when these are wilting or dying, and can survive in soil for months or even years in the absence of a host. If dried slowly, mainly fourth-stage juveniles in the stage of anhydrobiosis can survive for a long time. The longest period of dormancy known is from the onion race: specimens which were slowly dried on filter paper and kept for most of the time in a refrigerator revived after 26 years and were even able to raise an infection in pea and to reproduce (Sturhan, unpublished). Stem nematodes are very resistant to low temperatures. In freezing experiments some juveniles even survived an exposure of -150°C for 18 months (Sayre and Hwang, 1975).

The soil type appears of major importance in the persistence of stem nematodes. Populations survived best in clay soils but rapidly declined in sandy soils (Seinhorst, 1956). Populations of "oat race" and "giant race" survived at least 8–10 years in the soil without weed or crop host plants (Hooper, 1984).

7. Host–Parasite Relationships

The stem nematode attacks various parts of higher plants. In most hosts it feeds in the parenchymatous tissue of the stem, but it is also found in the foliage, in inflorescences, buds, rhizomes, stolons, and may rarely also invade roots. Usually young growing tissues are invaded, especially seedlings while below the soil surface, or the nematodes migrate to the apical region of developing or developed shoots. They enter plant tissues through stomates or penetrate directly at the base of stems and leaf axils. Invasion of plants is favored by cool, moist conditions; under dry conditions plants may outgrow stem nematode invasion.

Ditylenchus dipsaci mostly causes breakdown of the middle lamellae of cells so that cells become detached, the tissue swells and may become spongy, and cavities may be formed; in particular, in resistant plants cells are destroyed and extended lesions may develop.

Infested plants are generally stunted, malformed, and may eventually die. Stems are mostly swollen, internodes may be shortened, and distortions, gall-like swellings, discoloration, and local lesions can develop. Leaves are often distorted, discolored and smaller, with reduced and twisted leaf stems or they may obtain a shape morphologically quite different from the normal (in *Phlox*); in many plant species galls or necrotic spots may develop. In cereals extra tillers at the swollen stem base are produced. Inflorescences are malformed in certain hosts (e.g., teasel) and infested seed, where the nematodes are mostly under the testa or in any debris or persistent tissues surrounding the seed, is produced in field beans, vetches, peas, clover, lucerne, spurrey, onion and other *Allium* species, teasel, red beet, carrot, buckwheat, and some composites. Severe crown rot develops, e.g., in mature beet, parsley, and carrot plants; in celeriac decay starts from the lower part of the tap roots and proceeds to the crown. Rotting and decaying of host tissue in association with fungi and bacteria is also common in onions, flower bulbs, rhubarb, and potatoes.

Stem nematodes often invade nonhosts; feeding, further development, and even oviposition may occur but no reproduction takes place. Symptoms produced in nonhosts or resistant plants are mostly necroses and swelling of the stem is often lacking. On heavy invasion even mortality of nonhosts can occur (e.g., Griffin, 1975).

Associations of *D. dipsaci* with some pathogenic bacteria and fungi are known. It was shown, for instance, that stem nematodes transmitted *Corynebacterium insidiosum* to lucerne plants and predisposed wilt-resistant varieties to infection (Hawn, 1963; Hawn and Hanna, 1967).

The economic thresholds of *D. dipsaci* are very low. Already at population densities of 10 nematodes/500 g of soil serious damage to onion, sugar beet, carrot, rye, and other crops can be expected (Seinhorst, 1956; and others).

8. Control

Dissemination of *D. dipsaci* through seed, bulbs, rhizomes, garlic cloves, or any other infested planting material or with straw, hay, or any debris of host plants as well as through irrigation water, farm machinery, etc., should be avoided. Hot water treatment, sometimes combined with soaking in nematicides, is commonly used to disinfest dormant flower bulbs, onions, garlic, and strawberry runners and methyl bromide to fumigate infested seed (e.g., Powell, 1974; Caubel et al., 1985). Such control measures can decrease drastically the population density or the infestation rate, but complete elimination of stem nematodes is mostly not achieved.

For the most part rotation with nonhost crops for 3 or 4 years will drastically decrease soil populations of *D. dipsaci*, but special attention should be given to weeds as possible reservoir hosts and a source of infection. Under nonhost crops *D. dipsaci* races with a wide host range including many weeds appear to persist "almost indefinitely" in heavy soils (Hooper, 1984). Damaging attacks of stem nematodes can be avoided when the biological race is identified or—better—the host range of the local field population is known.

The use of resistant varieties is usually the most efficient method of controlling *D. dipsaci* and in certain crops probably the only economic method. Stem nematode-resistant varieties have been selected or bred in lucerne, red clover, white clover, oat, rye, and other cultivated plants; differences in susceptibility were also recorded for maize, field beans, onions, and garlic (e.g., Bingefors, 1970; Cotten, 1969; Smith, 1958).

Soil fumigation to control stem nematodes is usually not economical. Seed furrow applications and row treatments to protect seedlings and application of systemic nematicides were effective (Hooper, 1984; Whitehead and Tite, 1987; and others). Nematicides incorporated in pelleted seeds showed promising control effects (Schiffers et al., 1984).

Significant control of *D. dipsaci* has been obtained by soil solarization (Siti et al., 1982; Greco et al., 1985). Flooding of the soil for 1 month was also effective. The fungus *Hirsutella rhossiliensis* appears to be a promising agent for biological control of stem nematodes (Cayrol and Frankowski, 1986). Several other fungi were found attacking *D. dipsaci*.

B. *Ditylenchus destructor* Thorne, 1945

Potato rot nematode, tuber rot nematode

1. Morphological Characters (Fig. 2)

Dimensions (after Brzeski, 1991):

Females: $L = 0.69–1.89$ mm, $a = 18–49$, $b = 4–12$, $c = 14–20$, $c' = 3–5$, $V = 77–84$, stylet = 10–13 μm.

Males: $L = 0.63–1.35$ mm, $a = 24–50$, $b = 4–11$, $c = 11–21$, stylet = 10–12 μm.

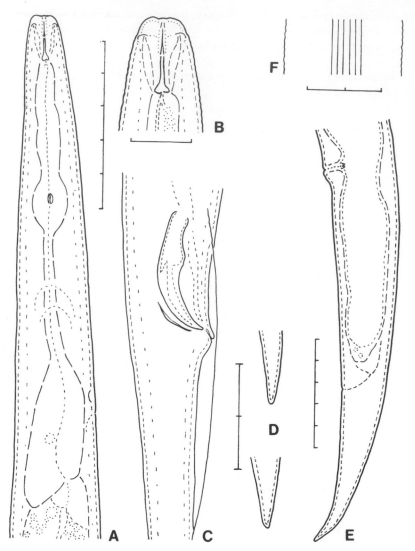

FIGURE 2   *Ditylenchus destructor*. (A) Female, esophageal region. (B) Head of female. (C) Male, spicule region. (D) Ends of female tail. (F) Posterior portion of female. (F) Lateral field at midbody. Each unit on bars = 10 μm.

Description: Body of relaxed specimens slightly ventrally arcuate. Cuticular an-nulation fine, lateral field with six incisures. Head often narrower than adjacent body, stria-tion seldom seen. Posterior blades of cephalic skeleton short. Stylet length mostly 10–12 μm, occasionally specimens with stylet up to 14 μm were described. Stylet cone 45–50% of stylet length, knobs distinct, rounded, with anterior surfaces sloping backward. Median esophageal bulb muscular, with thickenings of lumen walls about 3 μm long. Posterior bulb overlaps intestine for a short distance on dorsal body side, although specimens with offset glandular bulb were occasionally seen. Excretory pore opposite esophageal glands. Vulva position usually within limits $V = 78$–83. Postvulval part of uterine sac 40–98% of vulva–anus distance. Male bursa surrounds 50–90% of tail length. Spicules 24–27 µm long. Tail of both sexes conical, usually ventrally curved, terminus rounded.

2. Distribution and Economic Importance

After *D. destructor* was differentiated from *D. dipsaci* as a separate species in 1945, this nematode was recorded from many countries, mainly from temperate regions: USA, Canada, Peru, most European countries, especially the European part of the USSR, Iran, Pakistan, Bangladesh, China, Japan, Hawaii, New Zealand, South Africa, and the Canary Islands. In some reports the species identity may be questioned.

Kühn (1888) was the first to observe and describe the progressive dry rot of potato tubers produced by this nematode. Today, *D. destructor* is of great economic importance and causing great losses in potatoes, mainly in the European part of the Soviet Union, whereas its importance declined in other European countries, possibly due to intensified weed control and sanitary measures. The tuber rot nematode plays also a role as a pathogen in flower bulbs and tubers. Recently, it was recorded as a parasite of peanuts in South Africa (Jones and de Waele, 1988), and from New Zealand and Hungary it was reported to cause severe damage to the roots of hop plants (Foot and Wood, 1982; Benedek, 1962).

Ditylenchus destructor in seed potatoes and flower bulbs is on the list of quarantine organisms of many countries and organizations (e.g., EPPO).

3. Hosts

Ditylenchus destructor is a polyphagous nematode. About 90 plant species of a wide variety of families are known as hosts (see Esser, 1985; and others); according to Gubina (1982), the known host range comprises as many as 120 species of plants. Potato is the principal host; other common hosts are bulbous iris, tulip, hyacinth, gladiolus, dahlia, vetch, sugar beet, mangold, carrot, parsley, and red, white, and alsike clover. Among the other cultivated plants attacked by *D. destructor* are onion, garlic, celery, cucumber, squash, soybean, chick pea, peanut, lucerne, bush bean, sunflower, rhubarb, hop, sweet potato, tomato, tobacco, sugarcane, barley, and wheat. Hosts among the weeds and wild plants are *Agropyron repens, Artemisia vulgaris, Bellis perennis, Capsella bursa-pastoris, Festuca pratensis, Mentha arvensis, Plantago major, Potentilla anserina, Rumex* spp., *Sonchus* spp., and *Taraxacum officinale*. Not all hosts are heavily attacked; in some, such as paprika, tomato, pumpkin, cucumber, and garlic, infestation is mostly low; other plant species are only rarely attacked.

In the absence of higher plants *D. destructor* is capable of reproducing readily on the mycelium of about 70 species of fungi belonging to 40 genera (*Agaricus, Alternaria, Armillaria, Aspergillus, Botrytis, Cephalosporium, Cylindrocarpon, Fusarium, Heterosporium, Penicillium, Phoma, Pyrenochaeta, Thielaviopsis, Trichoderma, Verticillium*, and others). The potato rot nematode can be successfully cultured on fungus cultures. Knowledge on fungi as food source is based on laboratory results; its actual importance in field conditions is unknown.

Ditylenchus destructor has many host plants in common with *D. dipsaci* and fungus hosts in common with *D. myceliophagus*.

4. Host Races

There are a number of records on differences in host range and virulence among *D. destructor* populations from different hosts which suggests that host races exist in this nematode species, but thus far no biological races have been characterized and designated (as in *D. dipsaci*). Isolates from narcissus, iris, and hyacinth showed considerable differences in pathogenicity in potatoes, the population from hyacinth not attacking potatoes at all. Also *D. destructor* isolates from hop, iris, and mint differed in host range; certain populations infested beet and strawberry, other populations did not. Although *D. destructor* is widespread in South Africa, no damage to potatoes or other crops has been reported, which may be an indication that the populations found on peanut could form a distinct race with a limited host range (Goodey, 1952; Smart and Darling, 1963; de Waele et al., 1989; and others).

Isolates of different host and geographic origin interbred in crossing experiments, but *D. destructor* and *D. myceliophagus* did not (Smart and Darling, 1963; Wu, 1960), which confirms that both have to be considered as separate species. This is substantiated by, for example, differences in morphology and serological response.

5. Bionomics

The potato rot nematode is a migratory endoparasite mainly of underground parts of plants; invasion in aerial parts is rarely observed. In the absence of higher host plants it apparently survives is soil by feeding on terrestrial fungi. *Ditylenchus destructor* does not produce "nematode wool" on infested plants as *D. dipsaci* does.

Development and reproduction are possible from 5 to 34°C, with an optimum temperature at 20–27°C. At 27–28°C the development of one generation takes 18 days, at 20–24°C 20–26 days, at 6–10°C 68 days (Ladygina, 1957; Ustinov and Tereshchenko, 1959). In the Alma-Ata region (USSR) six to nine generations developed in potatoes during the vegetation period (according to Safyanov; see Decker, 1969). The most serious damage in potatoes was observed at temperatures between 15 and 20°C and 90–100% relative humidity, the nematodes not surviving relative humidities below 40%.

There are contradictory data on the ability of *D. destructor* to withstand desiccation and low temperatures. According to Ustinov and Tereshchenko (1959), Thorne (1961), Decker (1969), Hooper (1973), and others, *D. destructor* does not form a resistant resting stage and is unable to withstand desiccation and appears to overwinter as eggs. Kirjanova and Krall (1971) report that juveniles of all stages and adults are capable of anabiosis and that experiments conducted in the Ukraine revealed that the nematodes were easily revived from a state of anabiosis after 1 month. Survival of air-dried specimens for up to 5 months is recorded (see Gubina, 1982). Whereas Makarevskaya (1983) observed that *D. destructor* survived in plant tissue at temperatures up to –2°C, but was killed at –4.5°C, Ladygina (1956) found that the nematodes survived at –28°C.

6. Host–Parasite Relationships

Ditylenchus destructor most frequently infests the underground parts of plants (tubers and stolons of potato, bulbs of lilies, rhizomes of mint, and roots of hop and lilac), but the nematode may also invade above-ground parts and cause dwarfing, thickening, and branching of the stem and dwarfing, curling, and discoloration of leaves (e.g., in potatoes). Eventually, heavily infested potatoes may die, but more often no symptoms of infestation are found in the above-ground parts of plants. The nematode penetrates the tubers from the stem through the stolons or through lenticels and eyes of the tubers. Early symptoms are small white spots just below the skin, which later increase, gradually darken through greyish to dark brown or black, and become spongy in texture. Badly affected tubers have slightly sunken areas with cracked and papery skin. There generally occurs secondary invasion by bacteria, fungi, mites, other nematodes, etc., whereas the potato rot nematodes are accumulated on the boundary between distinctly diseased parts and healthy sections. Diseased tubers in storage may shrink or show various types of dry or wet rot, which mostly spreads to neighboring tubers. Also in carrots and parsnip disease progresses during winter storage.

Infested dahlia tubers develop similar symptoms; in sugar and fodder beet and in carrots symptoms resemble those of *D. dipsaci* attack. On roots and rhizomes in general brown to dark necrotic lesions are caused. Infestations on iris and tulip usually begin at the base of the bulb and extend up to the fleshy scales with yellow to dark brown lesions; secondary rotting may occur and the bulbs are eventually destroyed.

Ditylenchus destructor is easily cultured on plant tissues and many fungi. This nematode is also known to destroy the hyphae of cultivated mushroom, but the closely related *D. myceliophagus* is more usually responsible for crop losses (see Hesling, 1972).

7. Control

The use of healthy seed is an essential step in control and prevention of further spread of *D. destructor*. Other phytosanitary measures include destruction and removal of infested tubers and other plant parts left in the field. Crop rotation, with, say, cereals and maize and potatoes once in 3–4 years, together with careful weed control has proved exceptionally effective.

Potato varieties vary in degree of susceptibility to rotting caused by *D. destructor*. Some varieties as well as many wild *Solanum* species are recorded to be resistant (Bukasov et al., 1983; Ivanova, 1983; Kostina and Zholudeva, 1974), but Stefan (1980) found infestation by potato rot nematodes in all of the several hundred potato varieties and lines tested. In general, early and medium varieties are more severely infested than late varieties.

Soil fumigation can be successful but is usually uneconomical and no longer recommended because of ecotoxicological reasons. Control of dormant iris and other flower bulbs by hot-water treatment is possible, but damage may occur, especially in tulips. Dry heat treatment of potato tubers was also effective against *D. destructor*.

C. *Ditylenchus angustus* (Butler, 1913) Filipjev, 1936

Rice stem nematode

1. Morphological Characters (after Seshadri and Dasgupta, 1975) (Fig. 3)

Dimensions:

Females: $L = 0.8–1.2$ mm, $a = 50–62$, $b = 6–9$, $c = 18–24$, $c' = 5.2–5.4$, $V = 78–80$, stylet = 10–11 μm.

Males: $L = 0.7–1.18$ mm, $a = 40–55$, $b = 6–8$, $c = 19–26$, stylet = 10 μm.

Description: Body slender, almost straight. Lateral field with four incisures. Lip region unstriated. Stylet moderately developed, cone attenuated, about 45% of stylet length; knobs small but distinct, with posteriorly sloping anterior surfaces, about 2 μm across. Median esophageal bulb oval, muscular, with distinct thickenings of lumen walls. Isthmus gradually expanding into basal bulb. The latter slightly overlapping intestine. Excretory pore opposite posterior part of isthmus. Spermatheca very long, vagina somewhat oblique, and more than half of vulval body width long. Postvulval uterine sac 2–2.5 vulval body widths or 50–67% of vulva–anus distance long. Caudal alae of male extending almost to tail tip. Spicules 16–21 μm long. Tail of both sexes conoid, tapering to a sharply pointed terminus resembling a mucro.

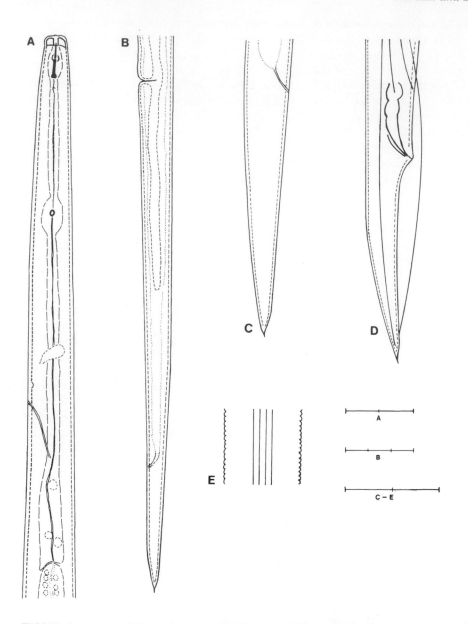

FIGURE 3   *Ditylenchus angustus.* (A) Female, esophageal region. (B) Posterior portion of female. (C) Female tail. (D) Male tail. (E) Lateral field at midbody. Each unit on bars = 10 μm. (After Seshadri and Dasgupta, 1975.)

2. Distribution and Economic Importance

The rice stem nematode is known from rice-growing regions in Asia (India, Pakistan, Bangladesh, Burma, Thailand, Malaysia, Indonesia, Vietnam, Philippines, USSR-Uzbekistan) and Africa (Egypt, Sudan, Malagasy Republic, South Africa).

The disease caused by *D. angustus* is commonly known as "ufra," a name originally used in Bangladesh from where the disease was first recorded in 1912. Other local names are "dak pora" in Bangladesh, "akhet pet" in Burma, and "yad-ngo" or twisting disease in southern Thailand.

Ditylenchus angustus is principally a problem in deep-water rice, but reports of ufra disease occurrence in shallow paddy are also known. Rice grown during the dry season usually escapes severe infestation because the nematodes are unable to travel up the stems.

The economic losses caused by rice stem nematodes have been reported as high as 50% in Uttar Pradesh (India) and from 20 to 90% in Thailand. The area infested by *D. angustus* worldwide was estimated to be 10 million acres with an average crop loss of 30% (Hollis and Keonboonrueng, 1984). The high losses the rice stem nematode can cause pose a deadly threat to small farmers, and in many instances so little grain was produced that harvest was not attempted.

3. Hosts

Ditylenchus angustus is a specialized plant parasite, rice being the principal host. Besides *Oryza sativa*, the host range includes *O. alta*, *O. cubensis*, *O. eichingeri*, *O. globerrima*, *O. latifolia*, *O. meyriana*, *O. minuta*, *O. nivara*, *O. officinalis*, *O. perennis*, *O. rufipogon*, and *O. spontanea*. The wild rice *Leersia hexandra* has been recorded as a host in Madagascar and Burma, and as additional gramineous hosts *Echinochloa colona* and *Sacciolepis interrupta* were reported from rice fields in Vietnam (Nguyen-Thi, 1982). *Zea mays, Panicum repens, Eleusine, Brachiaria, Digitaria, Echinochloa* species, and other grasses proved to be nonhosts (Hashioka, 1963).

According to Vuong (1969), propagation of the rice stem nematode is also possible on various fungi.

4. Bionomics

Rice stem nematodes feed ectoparasitically on young and tender tissues of rice and other host plants. At no time are they found living as endoparasites in plant tissue. During humid periods the nematodes migrate from the soil along the stem of young plants and invade the growing point, where they can be found in the terminal buds a few days after the seedlings have been transplanted. Later they are mainly found in the leaf sheaths and stems, on young stem portions immediately above the uppermost node, in peduncles, panicles, and around young seed.

Under dry conditions and upon ripening of the plants, the nematodes coil up and become inactive. They rapidly regain full motility when moistened. Toward the end of the growing season cottony masses are formed by the nematodes in the state of anabiosis on the ripening and drying plants. Quiescent nematodes can survive desiccation for more than 15 months.

The optimum temperature for infection of rice plants is between 20 and 30°C. The most severe infestations occur in the rainy season crop, whereas rice grown during the dry seasons with lower temperatures suffers least because migration and reproduction of the nematodes are arrested or reduced (see Seshadri and Dasgupta, 1975).

5.   Host–Parasite Relationships

The first symptom of *D. angustus* attack on rice seedlings is a chlorosis or striping of the uppermost leaves, which later often turn a whitish green color, with a yellowish midrib which is sometimes thickened. The leaves can be twisted and malformed and frequently have a wavy leaf margin. In severe attack, the panicle may remain enclosed inside the leaf sheath, while the stem shows a tendency to branch in the affected portions ("swollen ufra"), or the panicles are distorted, have a dark brown discolored peduncle, and only form normal grain near the tip, while the rest of the seed is distorted, shriveled, and sterile or the spikelets remain empty ("ripe ufra"). Heavily infested plants may show retarded growth, severe stunting, and wilted leaves (see, for example, Butler, 1919).

Nematodes are found mainly at the base of the peduncle, the darkly discolored stem just above the upper nodes or inside the glumes of the panicle, which mostly contain enormous numbers of nematodes in all stages from eggs to adults.

No definite associations between the rice stem nematode and other pathogens have been recorded.

6.   Control

Most effective control measures are burning of stubbles and other plant residues, removal of ratoons, stubble ploughing and fallowing during the dry season. The nematodes also usually die within 1 or 2 months in inundated fields. Lengthening the critical overwinter period of *D. angustus* by sowing deepwater rice later than normal or transplanting much later was suggested as a control measure against ufra (McGeachie and Rahman, 1983). Spread through irrigation water, rice straw, plant debris mixed with seed and contaminated soil must be prevented. *Ditylenchus angustus* can be found in freshly harvested rice seeds but is killed by sun drying, so in normal circumstances it is not seed-transmitted. Because of the extreme host specificity of the nematode, crop rotation for 1 year or fallow may eliminate soil infestation.

Rice varieties and wild *Oryza* species showed different degrees of infection and susceptibility, and some appear to be resistant or tolerant to *D. angustus* (Miah and Bakr, 1977; Rahman, 1987).

D. *Ditylenchus myceliophagus*, Goodey, 1958

Mushroom spawn nematode

1.   Morphological Characters (after Brzeski, 1991) (Fig. 4)

Dimensions:

Females: $L = 0.54$–0.92 mm, $a = 29$–50, $b = 4.8$–8.9, $c = 9.8$–17.0, $c' = 3.1$–6.7, $V = 78$–86, $V' = 85$–91, stylet $= 6.5$–8.5 μm.

Males: $L = 0.50$–$.75$ mm, $a = 33$–46, $b = 4.5$–7.2, $c = 8.3$–14.3, $c' = 3.9$–6.2, stylet $= 7$–9 μm.

Description: Body length variable depending on environmental conditions. Relaxed specimens straight or ventrally arcuate. Lateral field with six incisures. Head striation fine, often indistinct. Posterior blades of cephalic skeleton crescentic, short, refractive. Stylet thin, delicate, cone about one-third of stylet length, knobs small and rounded. Metacorporeal bulb oval, muscular, with small thickenings of lumen walls. Esophageal glands usually overlapping intestine, although 17% of examined females and 28% of males from various populations showed glands in a basal bulb offset from intestine. Postvulval uterine sac

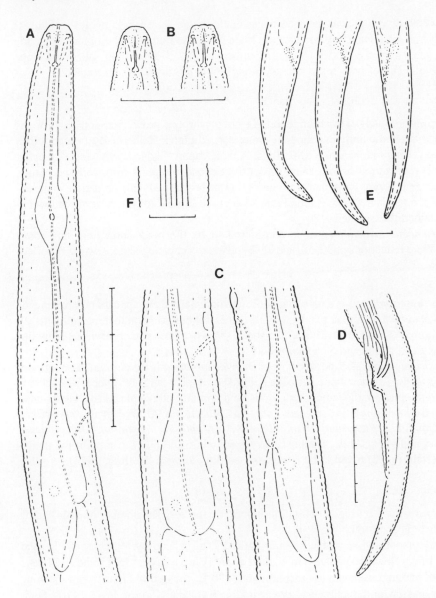

FIGURE 4   *Ditylenchus myceliophagus.* (A) Female, esophageal region. (B) Head of females. (C) Female, posterior part of esophagus. (D) Male tail. (E) Female tails. (F) Lateral field at midbody. Each unit on bars = 10 μm.

usually 45–55% of vulva–anus distance, occasionally either shorter or longer. Post–vulval part of body 8.1–13.3 anal body widths long. Male similar to female, but esophageal glands less developed, shorter. Spicules 15–20 μm, gubernaculum 5–7 μm long. Bursa envelops 20–55% of tail. Tail of both sexes conoid, rather thick; terminus always rounded.

2.   Distribution and Economic Importance

Ditylenchus myceliophagus has been reported from mushroom-growing countries through-out the world, predominantly in temperate areas, e.g., England, Netherlands, France, FRG, GDR, Poland, Bulgaria, Malta, USSR, India, China, Japan, Sudan, USA, and Australia. Most records refer to mushroom compost but there are also reports from soil and plants, including several records from native biotopes. In mushroom beds this nematode is gener-ally common, e.g., Cayrol (1967) found that 85% of the samples of mushroom compost in France contained this species.

    Ditylenchus myceliophagus is considered to be the most damaging mushroom nematode. Yield reductions of 30, 50, and 75% and even complete yield loss were reported.

3.   Hosts

Ditylenchus myceliophagus is a mycophagous nematode, and there appears no proven evi-dence that it can feed on higher plants, though successful culturing on lucerne callus tissue has been reported (Khera et al., 1968). The species was occasionally observed in plant tis-sues such as rice panicle and *Sorghum* root.

    The principal and also the type host of *D. myceliophagus* is the cultivated mushroom *Agaricus bisporus*. The nematode has been cultured on a wide range of other fungi, includ-ing saprophytic, plant pathogenic, predaceous, and animal pathogenic forms. Hosts have been reported in the fungal genera *Acrostalagmus, Alternaria, Arthrobotrys, Aspergillus, Botrytis, Candida, Fusarium, Penicillium, Phytophthora, Pythium, Rhizoctonia, Trichoderma*, and *Verticillium*. The hosts vary greatly in their ability to support the nema-tode, and certain fungi proved to be nonhosts (see, for example, Hesling, 1974).

4.   Bionomics

Ditylenchus myceliophagus is an obligate amphimictic species. The number of eggs laid by a single female is about 60.

    Development and reproduction are markedly influenced by temperature. In agar cultures the life cycle was completed in 40 days at 13°C, in 26 days at 18°C, and in 11 days at 23°C, with maximum population increase at 18–20°C, the optimal mushroom-growing temperature, and the reproduction was almost inhibited at 26°C (Cayrol, 1962, 1970). Opti-mum temperature for development of a population studied in Australia was near 25°C (Evans and Fisher, 1969). In mushroom farms in France maintained at 14°C and 95% hu-midity the life cycle varied from 20 to 25 days (Cayrol, 1962). High humidity seems to favor the development of *D. myceliophagus*.

    If dried slowly at low temperatures and high humidities, all stages of *D. myceliophagus*, but especially the advanced juvenile and young adult stages, are capable of anabiosis, some individuals reviving after up to 3 1/2 years of dormancy. Such desiccated and spirally curled up nematodes are considerably more heat resistant, hence more likely to survive in fermenting or heat-treated compost.

    When *D. myceliophagus* has "overcolonized" and destroyed the mushroom myce-lium, the nematodes migrate from the substrate, aggregate in swarms, and form sticky clus-ters on the surface or hang in stalactite-like columns, formed by up to almost 1 million individuals, beneath boxes of compost in mushroom houses. Mushroom boxes beneath are easily infested, and nematodes in such a state become easily attached to and dispersed by insects and are spread also in draining water and on contaminated implements, etc. (Hesling, 1974; and others). Haglund and Milne (1973) reported Diptera as the main means of dissemination in a mushroom house.

5.   Host–Parasite Relationships

Ditylenchus myceliophagus pierces fungal cells with its stylet and sucks out the contents. Feeding on *Botrytis cinerea* cultures was studied in detail by Doncaster (1966); conidia, young and older cells were chosen as food, but not senile cells. Pierced cells generally die and sometimes also adjacent cells. Mushroom or other fungus mycelium may appear unaffected until a critical population level is reached. The growth then stops and thereafter the mycelium degenerates rapidly.

Populations as low as three individuals/100 g compost at spawning caused a yield loss of about 30% and 20 or more specimens/100 g compost at spawning prevented cropping; 20, 100, and 300 nematodes/100 g at casing caused yield reductions of 50, 68, and 75%, respectively (Goodey, 1960; Arrold and Blake, 1968). Mushroom production ceases when the nematodes number about 17,000/g compost (Cayrol, 1970).

Compost in affected areas of mushroom beds becomes sunken, soggy, and foul-smelling; the surface may be covered by a greyish mold formed by the fruiting bodies of *Arthrobotrys* fungi predaceous on nematodes.

Of 16 commercial strains of *Agaricus bisporus*, seven were shown to be good or very good hosts of *D. myceliophagus*, four were fair hosts, and five were poor hosts. The differences were due to the diameter of the hyphae. In seven other hyphomycetous fungi tested, similar differences in their suitability as hosts were found, also associated with hyphae diameter (Cayrol, 1970). Consecutive experiments with *Agaricus bisporus* and *A. sylvicola* have shown that differences in host suitability are due to the chemical composition of the cell sap rather than to the diameter of the hyphae (Cayrol and Combettes, 1973).

6.   Control

Control of *D. myceliophagus* is essentially preventive. When mushroom beds have become infested by the nematode, there is no feasible cure known once the mushrooms are growing. Good standards of hygiene are essential, which include prevention of dispersion by implements, draining water, and insects. Chemical pretreatment of compost and heating compost and the casing materials to 60°C give good control and protection (for further information, see review by Hesling, 1974).

Among several commercial strains of mushroom tested, strain 252 (Somycel) which allowed feeding up to stage J3 of *D. myceliophagus* only, was considered as resistant (Cayrol, 1972).

E.   *Ditylenchus phyllobius* (Thorne, 1934) Filipjev, 1936

Nightshade gall nematode, silverleaf nightshade nematode

1.   Morphological Characters (after Brzeski, 1991) (Fig. 5)

Dimensions:

Female: $L = 0.59–0.84$ mm, $a = 20–32$, $b = 7.4–10.5$, $b' = 4.1–6.5$, $c = 11.4–17.6$, $c' = 2.9–4.5$, $V = 78–84$, $V' = 85–89$, stylet = 9–10 μm.

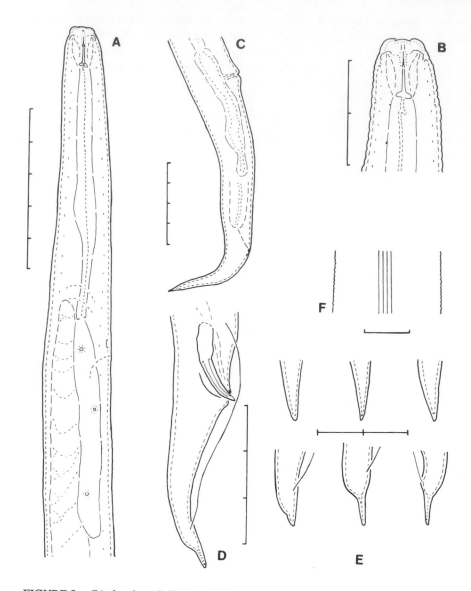

FIGURE 5   *Ditylenchus phyllobius*. (A) Female, esophageal region. (B) Head of female. (C) Posterior portion of female. (D) Mail tail. (E) Tail end of females (above) and males (below). (F) Lateral field at midbody. Each unit on bars = 10 μm.

Males: L = 0.67–0.80 mm, a = 27–34, b = 7.0–9.7, b' = 4.7–7.5, c = 16.3–19.2, c' = 2.5–3.1, stylet = 9–10 μm.

Description: Cuticle very finely and indistinctly striated. Lateral field about 2 μm wide, with four incisures. Head unstriated, 7–9 μm wide at basal plate level. Stylet cone about 40% of stylet length, knobs relatively large. Esophagus usually devoid of metacorporeal bulb, although small swelling near orifice of subventral glands may be present. Glandular part of esophagus forms a long lobe overlapping the intestine. Excretory pore near level of esophagointestinal junction. Gonad may extend up to stylet knobs. Postvulval uterine sac

1.6–2.4 vulva body widths or 45–77% of vulva–anus distance long. Postvulval part of body 8.4–12.3 anal body widths long. Male bursa envelops 52–77% of tail. Spicules 18–22 µm long. Tail of both sexes conical, tip rounded, distinct, 5–7-µm-long projection on male tail end usually present.

2. Distribution and Economic Importance

Ditylenchus phyllobius is known from the southwest USA, in Arizona and Texas (Thorne, 1934; Orr et al., 1975). It also occurs in southern India (Coimbatore district, Tamil Nadu), where it probably gained entry with the introduced weed *Solanum elaeagnifolium* (Sivakumar and Mohanasundaram, 1983). Because of differences in certain morphological characters, Brzeski (1991) suggests that the Indian population may represent a different species.

The nematode is of economic interest because of its potential as a biological control agent of the cosmopolitan silverleaf nightshade, which is an economically important perennial weed species and, in particular, troublesome in cotton production and difficult to control. In addition to the importance of this weed in cropland, it also causes some losses by poison to livestock.

3. Hosts

Ditylenchus phyllobius is a highly host-specific phytonematode. The only known host is silverleaf nightshade (*Solanum elaeagnifolium*). Field observations in areas with high populations of this nematode as well as host range studies, which included also solanaceous species such as *Solanum tuberosum*, *S. rostratum*, and *Lycopersicon esculentum*, confirmed the pronounced host specificity of *D. phyllobius*.

4. Bionomics

Ditylenchus phyllobius is an endoparasite which feeds and reproduces in above-ground parts of its host plant. The very motile infective juveniles (J4) ascend stems in moisture films and accumulate in the apical region, where they generally invade actively growing young tissues, but also preemergent shoots can be infected through the soil (Robinson et al., 1978). The nematodes develop and multiply rapidly in the host tissue; they become dormant when the galled leaves dry. In particular, desiccated fourth-stage juveniles remain viable for years, whereas only short survival of active J4 has been noted in soil. Survival is greatly prolonged at low temperatures. The thermal optimum for motility is about 24°C (Robinson et al., 1981).

5. Host–Parasite Relationships

This foliar gall-forming nematode invades mostly young leaf tissue, although floral and stem parts are occasionally infected. Visual indications of infection are usually evident 10–14 days later. Galls arise on the leaf surfaces, increase in size, and often engulf the entire leaf surface. Hundreds of thousands of nematodes can be found in a single gall. Heavily infected plants are severely stunted or even killed; flowering and seed production are decreased. Galled leaves abscise after a few weeks and defoliated plants die. Infective juveniles overwinter both in the soil and in galled leaves, which are especially subject to dispersal by wind and water.

6. Biological Control Potential

Its extreme host specificity, the capability to survive desiccation, and the detrimental effects it causes on its host make *D. phyllobius* a promising agent for biological weed control. Experiments in which the nematodes were disseminated by irrigation water, by broadcasting nematode-infected leaves or dried plant debris, or by applying infective juveniles in a foliar spray to *S. elaeagnifolium* populations gave promising results (Robinson et al., 1978; Northam and Orr, 1982; Parker, 1986).

F. Other Plant Parasitic Species

1. *Ditylenchus dryadis* Anderson and Mulvey, 1980

This leaf-galling parasite was found in the Canadian High Arctic. It is likely host specific to *Dryas integrifolia*, belonging to the Rosaceae family, where it produces red leaf galls of various configurations and sizes. Among the over 100 species of vascular plants occurring at the type locality no other hosts were found (Anderson and Mulvey, 1980).

2. *Ditylenchus drepanocercus* Goodey, 1953

This nematode is easily differentiated from other species of the genus by the presence of a ventral falciform appendage at the tail tip in both sexes. It was found associated with yellow blotches in leaves of *Evodia roxburghiana*, an evergreen tree belonging to the Rutaceae, growing naturally in the forests of Madras State, India. The nematode apparently does not survive desiccation (Goodey, 1953).

3. *Ditylenchus convallariae* Sturhan and Friedman, 1965

The stem nematode of lily of the valley (*Convallaria majalis*) is known from Denmark, Federal Republic of Germany, German Democratic Republic, and the Netherlands. All stages of the nematode, including eggs, were found in dormant pips of this plant species, where the nematode lived between and also within the bud scales; in developed plants, it occurred mainly in the lower part of the stem and in other aerial parts, but it was not seen in the roots. Distinct symptoms of infestation have not been reported. The nematodes survived freezing in lily of the valley buds for at least 4 months (Sturhan and Friedman, 1965; Decker, 1969).

4. *Ditylenchus medicaginis* Wasilewska, 1965

Ditylenchus medicaginis is a common soil nematode. It was originally described from Poland and is also known from Bulgaria, USSR, Syria, Mexico, and other countries. Occasionally it may occur in large numbers in plant tissues, as found in lucerne (Wasilewska, 1965) and wheat (Brzeski, unpublished). Though this species is considered as a plant parasite by some authors, there is no evidence that the nematodes affected the plants.

5. *Ditylenchus acris* (Thorne, 1941) Fortuner and Maggenti, 1987

Nematodes identified as the species *Nothotylenchus acris* (which is now considered as *species inquirenda*) have been reported from Japan causing disease symptoms in strawberries resembling those of *Aphelenchoides fragariae*: twisting of young petioles, crimping of young leaves, and dwarfing of the whole plants. The "strawberry bud nematodes" were found in large numbers (up to almost 10,000 specimens per plant) mainly in the buds, where they apparently fed ectoparasitically on the tissue (Nishizawa and Iyatomi, 1955; Kondo and Ishibashi, 1977). The nematode was also recorded living in sugar beet and in the roots and the rhizosphere of grasses, red clover, poppy, sage, and other plants from USSR, Czechoslovakia, USA, and Canada (see Kirjanova and Krall, 1971; Thorne, 1961; and others).

6.   Other Parasites of Uncultivated Plants and Mushrooms

Some additional *Ditylenchus* species have been described from roots or were found in above-ground parts of vascular plants. It is unknown whether they fed on plant tissue or on associated fungi. A new species parasitizing leaves of *Carpinus laxiflora* is presently being described from Korea; infested leaves become slightly swollen, turn to yellow and dark brown, and finally drop (Geraert and Choi, 1990). High numbers of an apparently undescribed *Ditylenchus* species were recently isolated from discolored leaves of the fern *Blechnum spicant*, which were collected in the German Alps at an altitude of 1300 m; this appears to be the only record of stem nematode parasitism on higher plants outside the angiosperms (Sturhan, unpublished). The same *Ditylenchus* species was isolated from the moss *Polytrichum* collected in the Tatras, Poland (Brzeski, unpublished). *Ditylenchus valveus* Thorne and Malek, 1968 and *D. filimus* Anderson, 1983 were found in mushroom compost in Canada (Anderson, 1983).

REFERENCES

d'Addabbo Gallo, M., Morone de Lucia, M. R., Grimaldi de Zio, S., and Lamberti, F. 1982. Caryo-phenotype relationships in *Ditylenchus dipsaci*. *Nematologia Mediterranea* 10: 39–47.

Anderson, R. V. 1983. An emended description of *Ditylenchus valveus* Thorne and Malek, 1968 and description of *D. filimus* n. sp. (Nematoda: Tylenchidae) from mushroom compost in Canada. *J. Zool.* 61: 2319–2323.

Anderson, R. V., and Mulvey, R. H. 1980. Description, relationships, and host symptoms of *Ditylenchus dryadis* n. sp. (Nematoda: Tylenchidae) from the Canadian High Arctic, a transitional species of gall-forming parasite attacking *Dryas integrifolia* M. Vahl. *Can. J. Zool.* 58: 363–368.

Andrássy, I. 1976. *Evolution as a Basis for the Systematization of Nematodes*. Pitman, London.

Arrold, N. P., and Blake, C. D. 1968. Some effects of the nematodes *Ditylenchus myceliophagus* and *Aphelenchoides composticola* on the yield of the cultivated mushroom. *Ann. Appl. Biol.* 61: 161–166.

Barabashova, V. N. 1972. [Karyotypes of stem nematodes.] In *Problemy parazitologii. Trudy VII Nauchnoj Konferentsii Parasitologov USSR*. Part 1. Izdatel'stvo "Naukova Dumka," Kiev, pp. 72–74.

Barabashova, V. N. 1974. [Karyotypical peculiarities of some forms of stem nematodes of the collective species *Ditylenchus dipsaci*.] *Parazitologiya* 8: 408–412.

Barabashova, V. N. 1976. [Features of the karyotype of *Ditylenchus dipsaci* from wild plants.] In *VIII Vsesoyuznoe soveshchanie po nematodnym boleznyam sel'skokhozyajstvennykh kul'tur*. Tezisy dokladov i soobshchenij. Izdatel'stvo "Shtiintsa", Kishinev, pp. 70–71.

Barabashova, V. N. 1978. [Karyological investigations on stem nematodes from the *Ditylenchus dipsaci* complex.] Nauchnye Doklady Vysshej; Shkoly, Biologicheskie Nauki, No. 5: 109–114.

Barabashova, V. N. 1979. [The karyotypes of stem nematodes of wild plants.] *Parazitologiya* 13: 257–261.

Barabashova, V. N. 1984. [Polyploidy in stem nematodes of the *Ditylenchus dipsaci* complex (Nematoda, Tylenchida)]. *Vestnik Khar'kovskogo Universiteta* 262: 76–78.

Benedek, I. M. 1962. [Nematode pest of hops (*Ditylenchus destructor* Thorne)]. *Magyar. mezögazd.* 17: 12.

Bingefors, S. 1970. Resistance against stem nematodes, *Ditylenchus dipsaci* (Kühn) Filipjev. *EPPO Publ., Ser. A.* 54: 63–75.

Bingefors, S., and Bingefors, S. 1976. Rearing stem nematode inoculum for plant breeding purposes. *Swed. J. Agr. Res.* 6: 13–17.

Brzeski, M. W. 1991. Review of the genus *Ditylenchus* Filipjev, 1936 (Nematoda: Anguinidae). *Rev. Nematol.* 14:9–59.

Bukasov, S. M., Olefir, V. V., and Turuleva, L. M. 1983. [Resistance of some species of wild potato to the stem nematode.] In *Steblevye nematody sel'skokhozyajstvennykh kul'tur i mery bor'by s nimi.* Vserossijskij NII Zashchity Rastenij, pp. 68–78.

Butler, E. J. 1919. The rice worm (*Tylenchus angustus*) and its control. *Memoirs of the Department of Agriculture in India* 10: 1–37.

Caubel, G., Ducom, P., and Marre, R. 1985. La fumigation au bromure de méthyle dirigée contre le nématode des tiges, *Ditylenchus dipsaci,* contenu dans des lots de semences et de bulbes. *Bull. OEPP* 15: 17–22.

Cayrol, J.-C. 1962. Importance des maladies vermiculaires dans les champignonnieres francaises. *Mushroom* 5: 480–496.

Cayrol, J.-C. 1967. Études préliminaires des relations entre quelques champignons pathogénes des végétaux et deux espéces de nématodes mycophages: *Ditylenchus myceliophagus* J.-B. Goodey 1958 et *Aphelenchoides composticola* M.-T. Franklin 1957. *Ann. Épiphyties* 18: 317–329.

Cayrol, J.-C. 1970. Contribution à l'étude de la biologie de *Ditylenchus myceliophagus* Goodey, 1958, nématode édaphique mycétophage. *Rev. Ecol. Biol. Sol* 7: 311–350.

Cayrol, J.-C. 1972. Possibilités d'utilisation de souches résistantes d'*Agaricus bisporus* dans la lutte contre le nématode mycophage *Ditylenchus myceliophagus* Goodey 1958. *Mushroom Sci.* 8: 631–640.

Cayrol, J.-C., and Combettes, S. 1973. Différence de tolérance de deux espèces d'agaric (*A. bisporus* Lange et *A. sylvicola* Vitt. et Sacc.) a l'égard du nématode mycophage *Ditylenchus myceliophagus* J. B. Goodey, 1958. *Comptes Rendus Hebdomadaires des Séances de l'Académie d'Agriculture de France.* 59: 1022–1030.

Cayrol, J.-C., and Frankowski, J. P. 1986. Influence of the number of parasitizing conidia of *Hirsutella rhossiliensis* on the mortality of *Ditylenchus dipsaci. Rev. Nematol.* 9: 411–412.

Cotton, J. 1969. Cereal varieties resistant to *Heterodera avenae* and *Ditylenchus dipsaci. Proceedings of the 5th British Insecticide Fungicide Conference* 1: 164–168.

Decker, H. 1969. *Phytonematologie.* VEB Deutscher Landwirtschaftsverlag, Berlin.

Doncaster, C. C. 1966. Nematode feeding mechanisms. 2. Observations on *Ditylenchus destructor* and *D. myceliophagus* feeding on *Botrytis cinerea. Nematologica* 12: 417–427.

Eriksson, K. B. 1965. Crossing experiments with races of *Ditylenchus dipsaci* on callus tissue cultures. *Nematologica* 11: 244–248.

Eriksson, K. B. 1974. Intraspecific variation in *Ditylenchus dipsaci.* 1. Compatibility tests with races. *Nematologica* 20: 147–162.

Esser, R. P. 1985. Characterization of potato rot nematode, *Ditylenchus destructor* Thorne, 1945 (Tylenchidae) for regulatory purposes. Florida Department of Agriculture and Consumer Service, Nematology Circular No. 124.

Evans, A. A. F., and Fisher, J. M. 1969. Development and structure of populations of *Ditylenchus myceliophagus* as affected by temperature. *Nematologica* 15: 395–402.

Foot, M. A., and Wood, F. H. 1982. Potato rot nematode, *Ditylenchus destructor* (Nematoda: Tylenchidae), infecting hops in New Zealand. *New Zealand J. Exp. Agr.* 10: 443–446.

Fortuner, R. 1982. On the genus *Ditylenchus* Filipjev, 1936 (Nematoda: Tylenchida). *Rev. Nematol.* 5: 17–38.

Fortuner, R., and Maggenti, A. R. 1987. A reappraisal of Tylenchina (Nemata). 4. The family Anguinidae Nicoll, 1935 (1926). *Rev. Nematol.* 10: 163–176.

Geraert, E., and Choi, Y. E. 1990. *Ditylenchus leptosoma* sp. n. (Nematoda: Tylenchida), a parasite of *Carpinus* leaves in Korea. *Nematologia Mediterranea* 18:27–31.

Goodey, J. B. 1952a. Investigations into the host ranges of *Ditylenchus destructor* Thorne and *D. dipsaci. Ann. Appl. Biol.* 39: 221–229.

Goodey, J. B. 1952b. The influence of the host on the dimensions of the plant parasitic nematode, *Ditylenchus destructor. Ann. Appl. Biol.* 39: 468–474.

Goodey, J. B. 1960. Observations on the effect of the parasitic nematodes *Ditylenchus myceliophagus, Aphelenchoides composticola* and *Paraphelenchus myceliophthorus* on the growth and cropping of mushrooms. *Ann. Appl. Biol.* 48: 655–664.

Goodey, J. B., Franklin, M. T., and Hooper, D. J. 1965. T. Goodey's *The Nematode Parasites of Plants Catalogued Under Their Hosts.* 3rd ed. Commonwealth Agricultural Bureaux, Farnham Royal.

Goodey, T. 1922. On the susceptibility of clover and some other legumes to stem-disease caused by the eelworm, *Tylenchus dipsaci*, syn. *devastatrix* Kühn. *J. Agr. Sci.* 12:20–30.

Goodey, T. 1953. On two new species of nematodes associated with leaf-blotch in *Evodia roxburghiana* an Indian evergreen tree. Thapar Commemoration Volume, pp. 95–103.

Greco, N., Brandonisio, A., and Elia, F. 1985. Control of *Ditylenchus dipsaci, Heterodera carotae* and *Meloidogyne javanica* by solarization. *Nematologia Mediterranea* 13:191–197.

Griffin, G. D. 1975. Parasitism of nonhost cultivars by *Ditylenchus dipsaci. J. Nematol.* 7: 236–238.

Gubina, V. G. (ed.). 1982. *Plant and soil nematodes. Genus Ditylenchus.* Izdatel'stvo "Nauka," Moscow. [See also English translation for Agricultural Research Service, U.S. Department of Agriculture, Washington, D.C.; Saad Publ., Karachi, Pakistan, 1988.]

Haglund, W. A., and Milne, D. R. 1973. Nematode dissemination in commercial mushroom houses. *Phytopathology* 63: 1455–1458.

Hashioka, Y. 1963. The rice stem nematode *Ditylenchus angustus* in Thailand. FAO Plant Protection Bulletin No. 11, pp. 97–102.

Hawn, E. J., 1963. Transmission of bacterial wilt of alfalfa by *Ditylenchus dipsaci* (Kühn). *Nematologica* 9: 65–68.

Hawn, E. J., and Hanna, M. R. 1967. Influence of stem nematode infestation on bacterial wilt reaction and forage yield of alfalfa varieties. *Can. J. Plant Sci.* 47: 203–208.

Hesling, J. J. 1966. Biological races of stem eelworm. *Report Glasshouse Crops Research Institute 1965*: 132–141.

Hesling, J. J. 1972. Nematode pests of mushroom. In *Economic Nematology.* Webster, J. M. ed. Academic Press, London, pp. 435–468.

Hesling, J. J. 1974. *Ditylenchus myceliophagus.* C.I.H. descriptions of plant-parasitic nematodes. Set 3, No. 36. Commonwealth Institute of Helminthology, St. Albans, Herts., England.

Hollis, J. P., and Keoboonrueng, S. 1984. Nematode parasites of rice. In *Plant and Insect Nematodes.* Nickle, W. R. ed. Marcel Dekker, New York, pp. 95–146.

Hooper, D. J. 1972. *Ditylenchus dipsaci.* C.I.H. descriptions of plant-parasitic nematodes. Set 1, No. 14. Commonwealth Institute of Helminthology, St. Albans, Herts., England.

Hooper, D. J. 1973. *Ditylenchus destructor.* C.I.H. descriptions of plant-parasitic nematodes. Set 2, No. 21. Commonwealth Institute of Helminthology, St. Albans, Herts., England.

Hooper, D. J. 1984. Observations on stem nematode, *Ditylenchus dipsaci*, attacking field beans, Vicia faba. *Report Rothamsted Experimental Station for 1983*: 239–260.

Ivanova, B. P. 1983. [The potato stem nematode in Belorussia.] In *Steblevye nematody sel'skokhozyajstvennykh kul'tur i mery bor'by s nimi.* Vserossitskij NII Zashchity Rastenij, pp. 27–33.

Jones, B. L., and Waele, D. de. 1988. First report of *Ditylenchus destructor* in pods and seeds of peanut. *Plant Dis.* 72: 453.

Jones, F. G. W., and Jones, M. G. 1964. *Pests of field crops.* Edward Arnold, London.

Khera, S., Bhatnagar, B. C., Kumar, N., and Tikyani, M. G. 1968. Studies on the culturing of *Ditylenchus myceliophagus* Goodey, 1958. *Ind. Phytopathol.* 21: 103–106.

Kirjanova, E. S., and Krall, E. L. 1971. *Plant-Parasitic Nematodes and Their Control*, Vol. 2. Izdatel'stvo "Nauka" Leningrad. (See also English translation for Agricultural Research Service, U.S. Department of Agriculture and National Science Foundation, Washington, D.C. and New Delhi: Amerind Publishing, Ltd., 1980.)

Kondo, E., and Ishibashi, N. 1977. Seasonal occurrence, dissemination, and survival of the strawberry bud nematode, *Nothotylenchus acris*. *Jap. J. Nematol*. 7: 39–44.

Kostina, K., and Zholudeva, Z. 1974. [Promising potato varieties]. *Kartofel' i Ovishchi* 3: 20–21.

Kühn, J. 1888. Die Wurmfaule, eine neue Erkrankungsform der Kartoffel. *Z. Spiritusind*. 22: 335.

Ladygina, N. M. 1957. [Effects of temperature and humidity on the stem nematodes of potato and onion.] *Trudy NI Inst. Biol. pri Biol. Fak. Khar'kovskogo Gos. Univ*. 27: 101–114.

Ladygina, N. M. 1973. [On physiological compatibility of different forms of stem nematodes. III. Crossing of ditylenchs from parsley, parsnip, onion and strawberry.] *Parazitologiya* 7: 67–71.

Ladygina, N. M. 1974. [On genetic-physiological compatibility of various forms of stem nematodes. IV. Crossing of the phlox nematode with other ditylenchs.] *Parazitologiya* 8: 63–69.

Ladygina, N. M. 1976. [The genetic and physiological compatibility of different forms of the stem nematode. V. Crossing of the red clover race with other stem nematodes.] *Parazitologiya* 10: 40–47.

Ladygina, N. M. 1978. [The genetic and physiological compatibility of different forms of stem nematodes. VI. Cross-breeding of *Ditylenchus* from cultivated plants and from weeds.] *Parazitologiya* 12: 349–353.

Ladygina, N. M. 1982. [Biological races, karyotypes and hybridization.] In Gubina, V.G. ed. [*Plant and Soil Nematodes. Genus* Ditylenchus.] Izdatel'stvo "Nauka", Moscow, pp. 69–86.

Ladygina, N. M., and Barabashova, V. N. 1976. [The genetic and physiological compatibility and the karyotypes of stem nematodes.] *Parazitologiya* 10: 449–456.

Lamberti, F., Grimaldi de Zio, S. and Agostinelli, A. 1988. Caryophenotypes of *Ditylenchus dipsaci* in Chile. *Nematologia Mediterranea* 16: 147.

Maggenti, A. R., Luc, M., Raski, D. J., Fortuner, R., and Geraert, E. 1987. A reappraisal of Tylenchina (Nemata). 2. Classification of the suborder Tylenchina (Nemata: Diplogasteria). *Rev. Nematol*. 10: 135–142.

Makarevskaya, Z. S. 1983. [Laboratory culture of the potato tuber nematode.] In *Steblevye nematody sel'skokhozyajstvennykh kul'tur i mery bor'by s nimi*. Vserossijskij NII Zashchity Rastenij, pp. 131–133.

McGeachie, I., and Rahman, L. 1983. Ufra disease: A review and a new approach to control. *Trop. Pest Management* 29: 325–332.

Metlitski, O. Z. 1972. [Stem nematode races parasitic on cultivated strawberries.] In *Kul'tura zemlyaniki v SSSR*. Izdatel'stvo "Kolos," Moscow, pp. 422–426.

Miah, S. A., and Bakr, M. A. 1977. Sources of resistance to ufra disease of rice in Bangladesh. *Int. Rice Res. Newslett*. 2, No. 5: 8.

Nguyen-Thi, T. C. 1982. New weed host of rice stem nematode identified in Vietnam. *Int. Rice Res. Newslett*. 7: 15.

Nishizawa, T., and Iyatomi, K. 1955. [*Nothotylenchus acris* Thorne, as a parasitic nematode of strawberry plant.] *Jap. J. Appl. Zool*. 20: 47–55.

Northam, F. E., and Orr, C. C. 1982. Effects of a nematode on biomass and density of silverleaf nightshade. *J. Range Management* 35: 536–537.

Orr, C. C., Abernathy, J. R., and Hudspeth, E. B. 1975. *Nothanguina phyllobia*, a nematode parasite of silverleaf nightshade. *Plant Dis. Reptr*. 59: 416–418.

Paramonov, A. A. 1962. [*Principles of Phytonematology*], Vol. 1. Izdatel'stvo "Nauka," Moscow.

Parker, P. E. 1986. Nematode control of silverleaf nightshade (*Solanum elaeagnifolium*); a biological control project. *Weed Sci*. 34 (Suppl. 1): 33–34.

Powell, D. F. 1974. Fumigation of field beans against *Ditylenchus dipsaci*. *Plant Pathol*. 23: 110–113.

Rahman, M. L. 1987. Source of ufra-resistant deep water rice. *Int. Rice Res. Newslett*. 12: 8.

Ritzema Bos, J. 1888. L'anguillule de la tige (*Tylenchus devastatrix* Kühn) et les maladies des plantes dues à ce nématode. *Arch. Mus. Teyler* 2: 161–348.

Robinson, A. F., Orr, C. C., and Abernathy, J. R. 1978. Distribution of *Nothanguina phyllobia* and its potential as a biological control agent for silver-leaf nightshade. *J. Nematol.* 10: 362–366.

Robinson, A. F., Orr, C. C., and Heintz, C. E. 1981. Effects of oxygen and temperature on the activity and survival of *Nothanguina phyllobia. J. Nematol.* 13: 528–535.

Russell, C. C. 1982. The stem nematode in the Southeastern United States. In *Nematology in the Southern Region of the United States*. Riggs, R. D., and Editorial Committee, eds. Southern Cooperative Series Bulletin 276: 157–173.

Sayre, R. M., and Wang, S.-W. 1975. Freezing and storing *Ditylenchus dipsaci* in liquid nitrogen. *J. Nematol.* 7: 199–202.

Schiffers, B. C., Fraselle, J., Hubrecht, F., and Jaumin, L. 1984. Efficacité contre le *Ditylenchus dipsaci* (Kühn) Fil. de nématicides incorporés dans l'enrobage de graines de féverole (*Vicia faba* L.). *Meded. Fac. Landbouwwetensch. Rijksuniv. Gent.* 49: 635–641.

Schreiber, E.-R. 1977. Lebensweise, Bedeutung und Bekämpfungsmöglichkeiten von *Ditylenchus dipsaci* (Kühn) Filipjev an Ackerbohnen *Vicia faba* L. in Marokko. Diss. Technical University Berlin.

Seinhorst, J. W. 1956. Population studies on stem nematodes (*Ditylenchus dipsaci*). Nematologica 1: 159–164.

Seinhorst, J. W. 1957. Some aspects of the biology and ecology of stem eelworms. *Nematologica* 2 (Suppl.): 355–361.

Seshadri, A. R., and Dasgupta, D. R. 1975. *Ditylenchus angustus*. C.I.H. descriptions of plant-parasitic nematodes. Set 5, No. 64. Commonwealth Institute of Helminthology, St. Albans, Herts., England.

Siddiqi, M. R. 1980. The origin and phylogeny of the nematode orders Tylenchida Thorne, 1949 and Aphelenchida n. ord. *Helminth. Abstr., Ser. B* 49: 143–170.

Siddiqi, M. R. 1986. Tylenchida. Parasites of plants and insects. Commonwealth Institute of Parasitology, St. Albans, England.

Siti, E., Cohn, E., Katan, J., and Mordechai, M. 1982. Control of *Ditylenchus dipsaci* in garlic by bulb and soil treatments. *Phytoparasitica* 10: 93–100.

Sivakumar, C. V., and Mohanasundaram, M. 1983. Occurrence of *Orrina phyllobia* (Thorne, 1934) Brzeski 1981 (Anguinidae: Nematoda) in Tamil Nadu, India. *Ind. J. Nematol.* 12 (1982): 416–418.

Smart, G. C., and Darling, H. M. 1963. Pathogenic variation and nutritional requirements of *Ditylenchus destructor. Phytopathology* 53: 374–381.

Smith, O. F. 1958. Reactions of some alfalfa varieties to the stem nematode. *Phytopathology* 48:107.

Stefan, K. 1980. (Problem of potato resistance against *Ditylenchus destructor*). *Zeszyty problemowe Postepów Nauk Rolniczych* 232: 55–61.

Stoyanov, D. 1964. Bioekologicheski prouchvanija na st'blenata nematoda po luka—*Ditylenchus allii* (Beijerinck). *Gradinarska i Lozarska Nauka* 1: 63–71.

Sturhan, D. 1964. Kreuzungsversuche mit biologischen Rassen des Stengelälchens (*Ditylenchus dipsaci*). *Nematologica* 10: 328–334.

Sturhan, D. 1969. Das Rassenproblem bei *Ditylenchus dipsaci. Mitteilungen aus der Biologischen Bundesanstalt für Land- und Forstwirtschaft* 136: 87–98.

Sturhan, D. 1970. *Ditylenchus dipsaci*: doch ein Artenkomplex? *Nematologica* 16: 327–328.

Sturhan, D. 1971. Biological races. In *Plant Parasitic Nematodes*. Vol. 2. Zuckerman, B. M., Mai, W. F., and Rohde, R. A. eds. Academic Press, New York, pp. 51–71.

Sturhan, D. 1983. The use of the subspecies and the superspecies categories in nematode taxonomy. In *Concepts in Nematode Systematics*. Stone, A. R., Platt, H. M., and Khalil, L. F. eds. Academic Press, London, pp. 41–53.

Sturhan, D., and Friedman, W. 1965. *Ditylenchus convallariae* n. sp. (Nematoda: Tylenchida). *Nematologica* 11: 219–223.

Sumenkova, N. I. 1982. [Taxonomic review of the nematode genus *Ditylenchus*. In *Plant and Soil Nematodes. Genus Ditylenchus.*] Gubina, V. G. ed. Izdatel'stvo "Nauka," Moscow, pp. 5–69.

Thorne, G. 1934. Some plant-parasitic nemas, with descriptions of three new species. *J. Agr. Res.* 49: 755–763.

Thorne, G. 1945. *Ditylenchus destructor*, n. sp., the potato rot nematode, and *Ditylenchus dipsaci* (Kühn, 1857) Filipjev, 1936, the teasel nematode (Nematoda: Tylenchidae). *Proc. Helm. Soc. Wash.* 12: 27–34.

Thorne, G. 1961. *Principles of Nematology.* McGraw-Hill, New York.

Triantaphyllou, A. C., and Hirschmann, H. 1980. Cytogenetics and morphology in relation to evolution and speciation of plant–parasitic nematodes. *Ann. Rev. Phytopathol.* 18: 333–359.

Ustinov, A. A., and Tereshchenko, E. F. 1959. [The stem nematode of potato.] *Zashchita Rastenii ot Vreditelei i Boleznei* 6: 29–31.

Viglierchio, D. R. 1971. Race genesis in *Ditylenchus dipsaci. Nematologica* 17: 386–392.

Vuong, H.-H. 1969. The occurrence in Madagascar of the rice nematodes *Aphelenchoides besseyi* and *Ditylenchus angustus*. In *Nematodes of Tropical Crops.* Peachey, J. E. ed. Commonwealth Agricultural Bureaux, Farnham Royal: 274–288.

de Waele, D., Jones, B. L., Bolton, C., and van den Berg, E. 1989. *Ditylenchus destructor* in hulls and seeds of peanut. *J. Nematol.* 21: 10–15.

Wasilewska, L. 1965. *Ditylenchus medicaginis* sp. n., a new parasitic nematode from Poland (Nematoda, Tylenchidae). *Bulletin de l'Académie Polonaise des Sciences Cl. II, Sér. Scie. Biol.* 13: 167–170.

Webster, J. M. 1967. The significance of biological races of *Ditylenchus dipsaci* and their hybrids. *Ann. Appl. Biol.* 59: 77–83.

Whitehead, A. G., Fraser, J. E., and Nichols, A. J. F. 1987. Variation in the development of stem nematodes, *Ditylenchus dipsaci*, in susceptible and resistant crop plants. *Ann. Appl. Biol.* 111: 373–383.

Whitehead, A. G., and Tite, D. J. 1987. Chemical control of stem nematode, *Ditylenchus dipsaci*, in field beans (*Vicia faba*). *Ann. Appl. Biol.* 110: 341–349.

Winslow, R. D. 1960. Some aspects of the ecology of free-living and plant-parasitic nematodes. In *Nematology.* Sasser, J. N., and Jenkins, W. R. eds. University of North Carolina Press, Chapel Hill, pp. 341–415.

Wu, L.-Y. 1960. Comparative study of *Ditylenchus destructor* Thorne, 1945 (Nematoda: Tylenchidae), from potato, bulbous iris, and dahlia, with discussion of the de Man's ratios. *Can. J. Zool.* 38: 1175–1187.

Yuksel, H. S. 1960. Observations on the life cycle of *Ditylenchus dipsaci* on onion seedlings. *Nematologica* 5: 289–296.

Zakrzewski, J. 1977. Studies on the occurrence, biology and pathogenicity of stem nematode (*Ditylenchus dipsaci* Kühn) on red clover in Poland. 1. Biology of stem nematode. *Hodowla Roślin Aklimatyzacja i Nasiennictwo* 21: 313–320.

10

The Aphelenchina: Bud, Leaf, and Insect Nematodes

WILLIAM R. NICKLE   *Agricultural Research Service, United States Department of Agriculture, Beltsville, Maryland*

DAVID J. HOOPER   *Institute of Arable Crops Research, Rothamsted Experimental Station, Harpenden, Hertfordshire, England*

The members of the nematode suborder Aphelenchina Geraert, 1966 have a worldwide distribution and have adapted to a wide range of ecological niches including, plant parasitism, insect parasitism, fungus feeding, and even predation on other nematodes. In contrast to the root feeding found in the Tylenchina, most of the Aphelenchina have evolved, perhaps because of the organization of the esophageal glands, to colonize leaves, buds, stems, tree trunks, and insects. One usually associates aphelenchid nematodes with the species *Aphelenchoides fragariae* and *A. besseyi* which feed on and damage strawberry plants (Raski and Krusberg, 1984). *Aphelenchoides ritzemabosi* causes typical interveinal necrosis on leaves of chrysanthemum plants. Two aphelenchs have been implicated in serious diseases of trees, such as the pine wood nematode, *Bursaphelenchus xylophilus*, on Japanese pine in Japan and the red ring disease of coconut palms in the West Indies caused by *Rhadinaphelenchus cocophilus*. However, the majority of the aphelenchid genera feed on fungi, both in the soil and in association with insects in a wide assortment of habitats. Some have been described from the galleries and frass of bark beetles. *Aphelenchoides composticola* is a common pest of mushroom culture. Four genera are obligate insect parasites in the body cavities of the host insects and two genera are ectoparasites of insects. Members of the Seinuridae are known (Hechler, 1963) to be predaceous on other nematodes. This is a relationship not known to occur in the Tylenchina.

Schistonchus caprifici* was the first aphelenchoid nematode described in the literature (Gasparrini, 1864). Thorne (1961) redescribed this nematode, and his contribution along with the description of a recently discovered species by Reddy and Rao (1986) laid a good foundation for this genus showing its unique association with fig-pollinating wasps. Bastian (1865) described two additional species, *Aphelenchus avenae* and *Aphelenchoides parietinus*. These are common nematodes known to feed on soil fungi. The first plant parasitic member of this group, *Aphelenchoides fragariae*, was described by Ritzema Bos (1890). Cobb (1923) observed that both the dorsal and the subventral esophageal gland orifices were contained within the median esophageal bulb (metacorpus) in specimens of the genus *Aphelenchus*, whereas in the Tylenchoid forms only the subventral glands emptied into the metacorporeal esophageal bulb, posterior to the valve, and the dorsal gland orifice was located anteriorly behind the stylet knobs. This observation provided a sound basis for separating the aphelenchoid and the tylenchoid nematodes. The German scientists Fuchs (1929, 1930, 1931, 1937) and Rühm (1954, 1955, 1956) described several aphelenchoid nematodes from galleries of bark beetles. Skarbilovich (1947) and Paramonov (1964) contributed to the placement of these nematodes into higher taxa and gave some evolutionary ideas with respect to this group. Goodey (1969a, b) Goodey and Hooper (1965), and Franklin (1955) made significant contributions to our knowledge of the aphelenchoid nematode taxonomy. Allen (1952), Massey (1974), Thorne (1961), Nickle (1970a, b), and Sanwal (1961, 1965) provided leadership in North America on the systematics and biology of the aphelenchoid nematodes. More recently, Hunt (1980) described an interesting new nematode ectoparasite of the American cockroach from St. Lucia. From France come the works of Baujard (1984, 1985) from bark beetles and a remarkable new aphelench genus ectoparasitic on adult moths of the agriculturally important pest insect genus *Spodoptera* by Remillet and Silvain (1988) from French Guiana.

The "taxonomic inflation" noted by Fortuner et al. (1987–88) for the Tylenchina also applies to the Aphelenchina. Bearing in mind their observations, we have retained the Aphelenchina at suborder level, as proposed by Geraert (1966). The classification presented here is a compromise between the levels proposed by Andrássy (1976) and Siddiqi (1980).

At present the Aphelenchina contains five families and 32 genera. The families are separated on the basis of both morphological and biological characteristics. Some of the genera, e.g., *Aphelenchoides*, are replete with poorly described species. Only one-third of the 180 described *Aphelenchoides* species have any chance of being identified again. Other genera are monotypic. The following classification is concerned with only those genera which appear to be valid and identifiable.

CLASSIFICATION OF THE APHELENCHINA

Suborder Aphelenchina Geraert, 1966

Family Aphelenchidae Fuchs, 1937 (Steiner, 1949)
    Genus *Aphelenchus* Bastian, 1865
Family Paraphelenchidae T. Goodey, 1951 (J. B. Goodey, 1960)
    Genus *Paraphelenchus* Micoletzky, 1922 (Micoletzky, 1925)
Family Aphelenchoididae Skarbilovich, 1947 (Paramonov, 1953)
    Subfamily Aphelenchoidinae Skarbilovich, 1947
        Genus *Aphelenchoides* Fischer, 1894
        Genus *Anomyctus* Allen, 1940
        Genus *Laimaphelenchus* Fuchs, 1937
        Genus *Megadorus* J. B. Goodey, 1960
        Genus *Ruehmaphelenchus* J. B. Goodey, 1963
        Genus *Schistonchus* (Cobb, 1927) Fuchs, 1937
        Genus *Sheraphelenchus* Nickle, 1970
        Genus *Tylaphelenchus* Rühm, 1956
    Subfamily Ektaphelenchinae Paramonov, 1964
        Genus *Ektaphelenchus* (Fuchs, 1937) Skrjabin et al., 1954
        Genus *Berntsenus* Massey, 1974
        Genus *Ektaphelenchoides* Baujard, 1984
        Genus *Ipsaphelenchus* Lieutier and Laumond, 1978
        Genus *Cryptaphelenchus* (Fuchs, 1937) Rühm, 1954

Genus *Cryptaphelenchoides* J. B. Goodey, 1960
Subfamily Bursaphelenchinae Paramonov, 1964
   Genus *Bursaphelenchus* Fuchs, 1937
   Genus *Huntaphelenchoides* Nickle, 1970
   Genus *Omemeea* Massey, 1971
   Genus *Parasitaphelenchus* Fuchs, 1929
   Genus *Teragramia* Massey, 1974
Subfamily Rhadinaphelenchinae Paramonov, 1962
   Genus *Rhadinaphelenchus* J. B. Goodey, 1960
Family Seinuridae Husain and Khan, 1967 (Baranovskaya, 1981)
   Subfamily Seinurinae Husain and Khan, 1967
      Genus *Seinura* Fuchs, 1931
      Genus *Paraseinura* Timm, 1960
      Genus *Papuaphelenchus* Andrássy, 1973
      Genus *Aprutides* Scognamiglio, Talame and s'Jacob, 1970
Family Entaphelenchidae Nickle, 1970
   Subfamily Entaphelenchinae Nickle, 1970
      Genus *Entaphelenchus* Wachek, 1955
      Genus *Peraphelenchus* Wachek, 1955
      Genus *Praecocilenchus* Poinar, 1969
      Genus *Roveaphelenchus* Nickle, 1970
   Subfamily Acugutturinae Hunt, 1980
      Genus *Acugutturus* Hunt, 1980
      Genus *Noctuidonema* Remillet and Silvain, 1988
Genera Inquirendae:
      Genus *Caballeroides* Chaturvedi and Khera, 1977
      Genus *Devibursaphelenchus* Kakulia, 1967
      Genus *Paraphelenchoides* Haque, 1967

Suborder Aphelenchina Geraert, 1966

Tylenchida. Stylet with or without basal knobs or thickenings. Dorsal esophageal gland emptying into lumen of esophagus within the large, easily seen median muscular bulb, anterior to the crescentic valve plates. Female genital organ always unpaired, prevulvar. Male with caudal papillae. Bursa sometimes with papillary ribs; sometimes short, enveloping only the tip of the tail, often absent. Spicules sometimes narrow, but usually rose-thorn shaped. Living either free in the soil, predaceous on other nematodes, mycophagous, parasitic in leaves, roots, stems, and bulbs or associated with or parasitic in or on insects. The suborder includes five families.

Family Aphelenchidae Fuchs, 1937

Aphelenchina. With overlapping esophageal glands. Male with bursa, supported by four pairs of papillae.

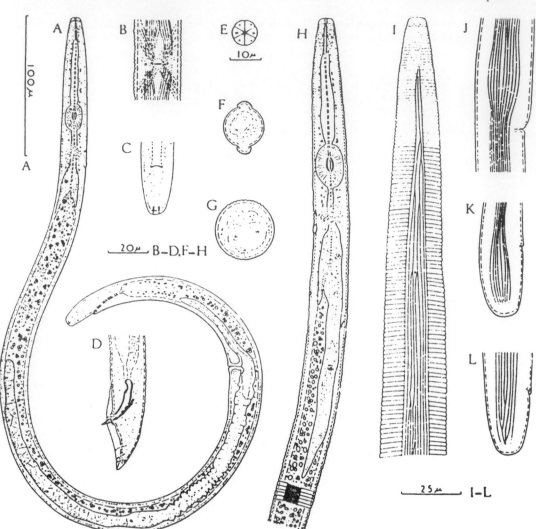

FIGURE 1   *Aphelenchus avenae.* (A) Female. (B) Vulva, ventral view. (C) Female tail, ventral view. (D) Male tail, lateral view. (E) Head, end-on view. (F) T. S. through young female. (G) T. S. through mature female. (H,I) Anterior parts of female body, lateral view. (J) Vulval region, lateral view. (K,L) Female tails, lateral view. (After Goodey and Hooper, 1965.)

Aphelenchus Bastian, 1865 (Fig. 1)
syn. *Isonchus* Cobb, 1913
**   *Metaphelenchus* Steiner, 1943**

Body 0.5–1.2 mm long, tapering anteriorly. Lips slightly offset. Stylet with slight thicken-ings at base. Lateral field with numerous incisures (about 10 or more). Esophageal glands usually with a lobe overlapping the intestine dorsolaterally. Hemizonid usually visible, three to eight annuli posterior to excretory pore. Body often narrowing abruptly behind vulva. Vulva posterior, ovary outstretched, amphidelphic; postvulval sac present, reaching about halfway from vulva to anus. Tail between one and four anal body widths long, cylindrical, bluntly rounded. Phasmids subterminal. Male rare, with rhabditid-like bursa usually supported by one preanal and about three postanal subterminal pairs of ribs. Spicules paired, slender, curved ventrally with only a minute rostrum. Gubernaculum V-shaped, about a third as long as the spicule.

Bionomics: Mycophagous, soil inhabiting; also often found in diseased plant tissue where they feed on associated fungi.

Type Species: *Aphelenchus avenae* Bastian, 1865. Of some 30 species described 12 currently have valid status. Anderson and Hooper (1980) noted that *A. avenae* has a vagina of unequal proportions, "anisomorphic," whereas in other populations, mainly from warmer climates, the vagina has equal proportions termed "isomorphic," for which they proposed the new subgenus *Anaphelenchus*. Although, *A. avenae* is a common cosmopolitan species, males are rare in some populations (CIH Description Set 4, No. 50).

Family Paraphelenchidae Goodey, 1951

Aphelenchina. With a nonoverlapping posterior esophageal bulb. Male without caudal alae, four or five pairs of caudal papillae present.

Paraphelenchus (Micoletzky, 1922) Micoletzky, 1925 (Fig. 2)
syn. *Aphelenchus (Paraphelenchus)* Micoletzky, 1922

Body 0.5–1.0 mm long tapering slightly anteriorly. Lips plump, forming a flat cap, not offset. Stylet with or without small basal thickenings. Esophageal glands contained within a flask-shaped posterior bulb. Excretory pore posterior to level of median bulb. Hemizonid adjacent and posterior to excretory pore. Vulva posterior; ovary directed anteriorly and outstretched; postuterine sac well developed. Female tail short, conical, sometimes with terminal mucro. Male tail with characteristic pattern of caudal papillae; pair 1 (which may be absent or single median papilla) preanal, pairs 2, 3, and 4 subventral, 2 adanal, 3 midtail-length, 4 subterminal, pair 5 also subterminal but subdorsal. Spicules paired slender; apex enlarged, rostrum short, with acute posteriorly directed terminus. Gubernaculum linear, V-shaped.

Bionomics: Mycophagus, soil-inhibiting; *P. myceliophthorus* Goodey, 1958 and *P. pseudoparietinus* (Micoletzky, 1922) Micoletzky, 1925 are occasionally recorded feeding on mushroom hyphae and causing crop decline.

Type Species: *Paraphelenchus pseudoparietinus* (Micoletzky, 1922) Micoletzky, 1925.

Fifteen species have been described of which *P. pseudoparietinus* seems to be the most common. Baranovskaya (1958) gave an excellent description of *P. tritici* from the Soviet Union; for *P. myceliophthorus*, see CIH Description Set 8, No. 115.

Family Aphelenchoididae Skarbilovich, 1947 (Paramonov, 1953)

Aphelenchina. Spear with or without knobs. Esophagus without a terminal bulb, isthmus lacking, glands lying free in the body cavity and forming long lobes. Female tail pointed or rounded, often with a mucron. Bursa, if present at all, a very small lobe on tip of tail. Spicules with a sharply pointed ventral process directed backward; gubernaculum generally absent. Four subfamilies.

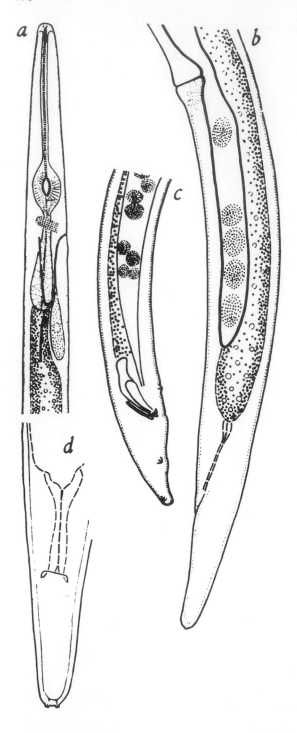

FIGURE 2   *Paraphelenchus myceliophthorus.* (a) Fore part of body. (b) Hind part of female. (c) Hind part of male. (d) Ventral view of female tail end. (After J. B. Goodey, 1958.)

Subfamily Aphelenchoidinae Skarbilovich, 1947

Aphelenchoididae. Spear thin with narrow lumen, knobs usually present but small. Post uterine sac only rarely absent. Spicules usually separate. Tail tip of males without bursal rudiment. Free-living plant parasites or fungus feeders. Four genera.

Aphelenchoides Fischer, 1894 (Fig. 3)
syn. *Pseudaphelenchoides* Drozdovsky, 1967
Asteroaphelenchoides Drozdovsky, 1967

Body slender, length variable. Lips six, fused, similar, appear nonannulated under light microscope, slightly offset from body. Stylet usually with basal knobs. Oocytes in one or more rows; postuterine sac usually well developed, length variable. Spicules paired, rose-thorn-shaped, not fused, rostrum usually prominent. Male tail without bursa or gubernaculum; with three pairs of ventrosubmedian papillae. Tails of both sexes never elongate-filiform, but usually more or less tapering, conical, and frequently ending in one or more mucrones.

Bionomics: Phytophagous, mycophagous. Soil and leaf-inhabiting. *Aphelenchoides fragariae* (Ritzema Bos, 1891) Christie, 1932 is particularly damaging on ferns and succulent plants; *A. ritzemabosi* (Schwartz, 1911) Steiner and Buhrer, 1932 attacks many plants, especially Compositae, including florists chrysanthemum; both species are common in temperate climates. *Aphelenchoides besseyi* Christie, 1942 tends to occur in warmer regions causing white tip disease of rice (Hollis et al. 1984). All three species damage strawberry plants. *Aphelenchoides fragariae* and *A. besseyi* readily reproduce on fungi; *A. composticola* Franklin, 1957 is a common pest of mushroom culture (CIH Descriptions Set 1, No. 4; Set 3, No. 32; Set 5, No. 74; Set 7, No. 92).

Type Species: *Aphelenchoides kuehni* Fischer, 1894.

Some 180 species have been described of which the above mentioned are frequent crop pests.

Anomyctus Allen, 1940 (Fig. 4)

Body 0.7–1.0 mm long. Cuticle strongly annulated. Lips offset; shallow, cuticularized, saucer-like frontal disk present. Six small projections surround oral opening. Stylet linear, without basal knobs, 35 μm long. Postuterine sac 1 1/2 body widths long. Female tail conical to short terminal process. Male similar to female. Testis with anterior flexure. Spicules very large, paired, not fused, similar to those of *Aphelenchoides*. Tail conical, with short terminal process similar to that of female, with three pairs of caudal papillae; bursa and gubernaculum absent.

Bionomics: Unknown, soil-inhabiting; rarely recorded.

Type Species: *Anomyctus xenurus* Allen, 1940.

No other species.

Laimaphelenchus Fuchs, 1937 (Fig. 5)
syn. *Ruidosaphelenchus* Laumond and Carle, 1971

Body 0.5–1.0 mm long. Lips rounded, offset. Stylet with or without basal knobs. Excretory pore usually about two body widths behind median bulb. Vulva with or without flap; vagina sloping forward from vulva; postuterine sac variable, one to nine vulval body widths long. Tails of both sexes conical; suddenly narrowing dorsally to short, narrow, cylindrical isthmus before ending in four stalked, fringed tubercles. Males rare. Testis tip may

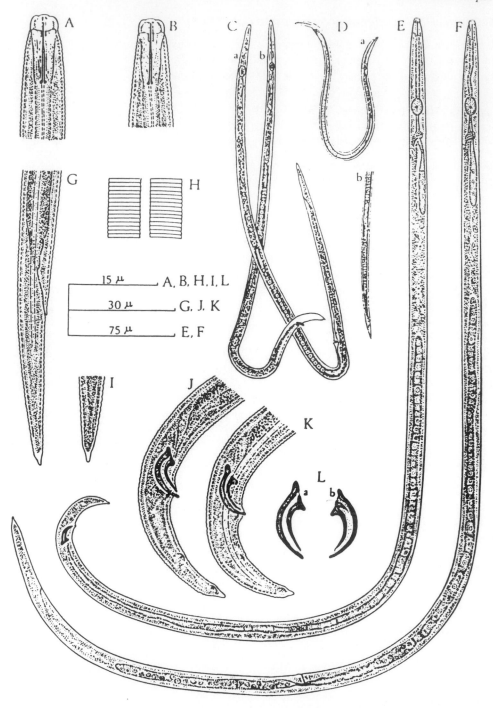

FIGURE 3   *Aphelenchoides fragariae*. (A) Female head end. (B) Male head end. (C) a, female; b, male of *A. olesistus* Ritzema Bos, 1893 [= *A. fragariae*]. (D) a, male; b, posterior portion of female, of *A. fragariae* Ritzema Bos, 1891; (E) Male. (F) Female. (G) Female tail. (H) Lateral field. (I) Female tail tip. (J,K) Male tails. (L) Spicules. (From CIH Series.)

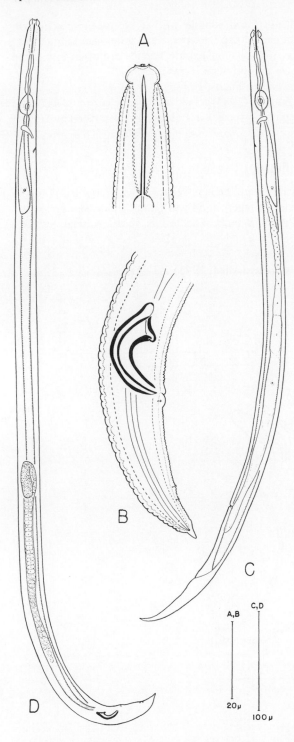

FIGURE 4   *Anomyctus xenurus*. (A) Anterior end of male. (B) Male tail, lateral view. (C) Full body of female, lateral view. (D) Full body of male, lateral view. (After Allen, 1940.)

be reflexed slightly. Spicules paired, not fused, similar to that of *Aphelenchoides ritzemabosi*. Male tail with three pairs of caudal papillae; bursa and gubernaculum absent.

Bionomics: Predaceous and/or mycophagous, some species are often associated with bark beetles but also occur in moss and on plant stems.

Type Species: *Laimaphelenchus moro* Fuchs, 1937.

Of some 19 species described, Baujard (1985) considers six to be currently valid; we concur. *Laimaphelenchus penardi*, a common species in Europe, is redescribed by Hirling (1986).

Megadorus J. B. Goodey, 1960 (Fig. 6)

Body about 0.5 mm long. The chief features are the massive spear with large basal knobs; the anterior part of the esophagus is wide and tapers rapidly to a thin cylinder before joining the large, prominent, oval-rounded, median bulb. The tail is short, conical, without mucrones. Only the female is known.

Bionomics: Unknown; rarely recorded.

Type Species: *Megadorus megadorus* (Allen, 1941) J. B. Goodey, 1960.

No other species.

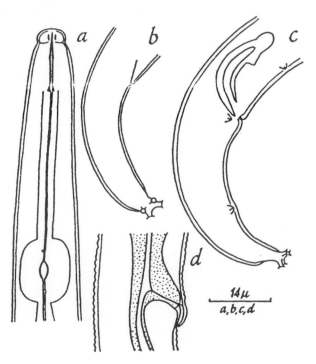

FIGURE 5 *Laimaphelenchus penardi* (Steiner, 1914) Filipjev and Sch. Stek., 1941. (a) Fore part of body. (b) Female tail. (c) Male tail. (d) Vulva detail, lateral. (After Steiner, 1914.)

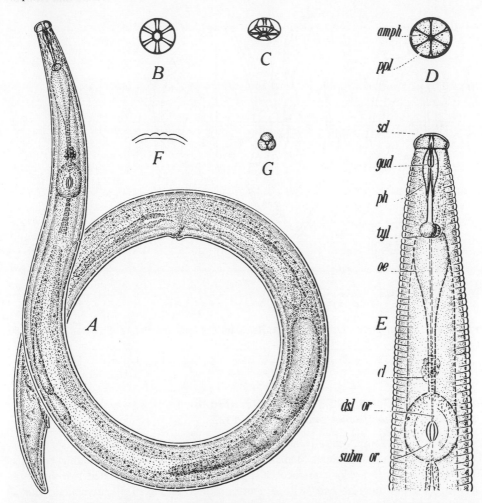

FIGURE 6 *Megadorus megadorus*. (A) Female; ×730. (B) Basal plate of cephalic framework; ×1350. (C) Cephalic framework; ×1350. (D) En face view of head; *amph*, amphid opening; *ppl*, one of the four submedian papillae; ×1350; (E) Neck region; *scl*, cephalic framework; *gud*, guiding sheath of spear; *ph*, heavily sclerotized pharyngeal walls; *tyl*, knobs of spear; *oe*, esophagus; *cl*, group of cells surrounding esophagus; *dsl or*, dorsal gland orifice; *subm or,* submedian gland orifice; ×1350. (F) Wing area in cross-section; ×1350. (G) Knobs of spear in cross-section; ×1350. (After Allen, 1941.)

Ruehmaphelenchus J. B. Goodey, 1963 (Fig. 7)

Body 0.6–1.1 mm long. Lips high, rounded, offset. Stylet with small basal knobs. Vulval lips protruding; postuterine sac long. Female tail short, dome-shaped, with terminal spike. Spicules paired, not fused, each with a triangular ventral flange. Male tail with three pairs of caudal papillae; with finer terminal spike, bursa, and gubernaculum absent.

Bionomics: Unknown, insect associate.

Type Species: *Ruehmaphelenchus martinii* (Rühm, 1955) J. B. Goodey, 1963.

No other species.

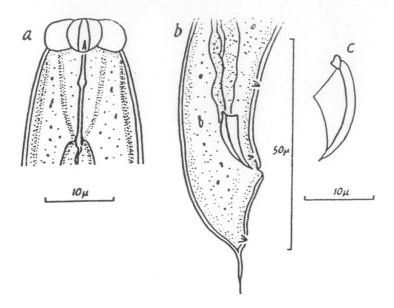

FIGURE 7 *Ruehmaphelenchus martinii.* (a) Head end. (b) Male tail end. (c) Spicules lateral. (After Rühm, 1955.)

Schistonchus (Cobb, 1927) Fuchs, 1937 (Fig. 8)
syn. *Aphelenchus (Schistonchus)* Cobb, 1927

Body 0.6–0.8 mm long, ventrally arcuate C-shaped, finely striated, head offset. Stylet 21–24 µm long with strong knobs. Median esophageal bulb oval to rounded with valves just posterior to center. Ovary outstretched or slightly reflexed. Vulva located two-thirds of body length from anterior end, with a short postuterine sac. Tail attenuated with a mucronated tip. Male tail strongly arcuate, with three pairs of submedian copulatory papillae present. Spicules with wide, elongated apex; distal tip of dorsal limb hooked ventrally. Gubernaculum present. Bursa absent. Tail with a mucronated tip.

Bionomics: *Schistonchus caprifici* is an associate of *Blastophaga psenes* (L.), an insect which pollinates the Smyrna fig, *Ficus carica* L.

Type Species: *Schistonchus caprifici* (Gasparrini, 1864) Cobb, 1927.

One other species *S. racemosa* Reddy and Rao (1986), also associated with a fig-pollinating wasp in India.

Sheraphelenchus Nickle, 1970 (Fig. 9)

Body 0.5–1.2 mm long. Lips slightly offset. Stylet without knobs. Esophageal glands short. Anterior part of gonad in both sexes usually with three cells across. Uterus often containing two or more eggs with embryos; postuterine sac absent. Female tail elongate, attenuates conically posterior to vulva, sharply pointed at tip. Spicules paired, apparently partially fused, slender with elongated apex and narrow rostrum, ventral element continuing less strongly along curve of shaft. Male tail spikelike, narrowing abruptly just posterior to spicule; three pairs of caudal papillae; bursa and gubernaculum absent.

Bionomics: Mycophagous, insect associate.

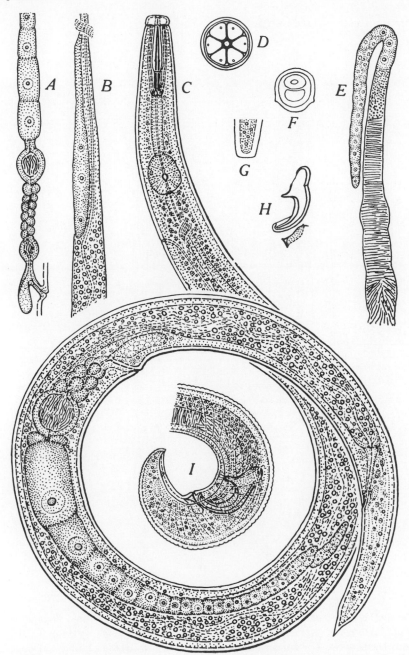

FIGURE 8 *Schistonchus caprifici*. (A) Reproductive tract of recently matured and fertilized female. (B) Esophageal gland region. (C) Female with double flexure in ovary. (D) Face view. (E) Testis with flexure. (F) Cross-section just anterior to spicula. (G) Ventral view of male terminus. (H) Spiculum and gubernaculum. (I) Male tail. (After Thorne, 1961.)

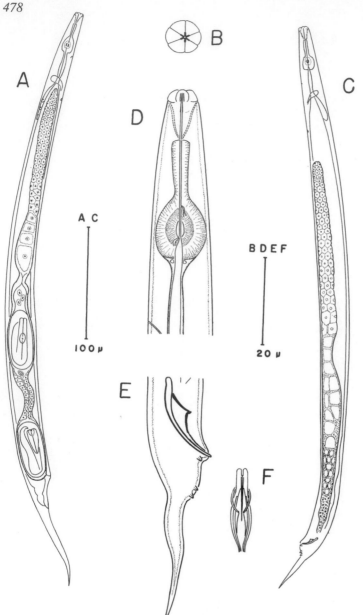

FIGURE 9   *Sheraphelenchus entomophagus*. Drawing of the Kellen fig-cultured population phoretic on *Haptonchus luteolus*. (A) Female, full body, lateral view. (B) An en face view of female. (C) Male, full body lateral view. (D) Female, anterior end, lateral view. (E) Male tail, lateral view, showing spicule. (F) Spicule, ventral view. (After Nickle, 1970.)

Type Species: *Sheraphelenchus entomophagus* Nickle, 1970.
No other species.

Tylaphelenchus Rühm, 1956 (Fig. 10)

Body 0.4–0.8 mm long. Lips slightly offset, high. Stylet robust, with strongly developed tip and shaft, with well-developed basal knobs. Vulval lips not protruding; postuterine sac long. Spicules paired, not fused, similar to those of *Aphelenchoides*. Gubernaculum and bursa absent. Tails of both sexes conical, with ragged wedge-shaped terminus, sometimes with three pointed tips. Male with two to four pairs of caudal papillae.

Bionomics: Unknown, found in moss, lichens, and bark beetle frass.

Type Species: *Tylaphelenchus leichenicola* Rühm, 1956.

Six other species described.

Subfamily Ektaphelenchinae Paramonov, 1964

Aphelenchoididae. Lumen of spear relatively wide, knobs generally present. Rectum and anus lacking in females, gut ending in a blind sac in tail. Postvulval sac short or absent. Bark beetle associates. Six genera.

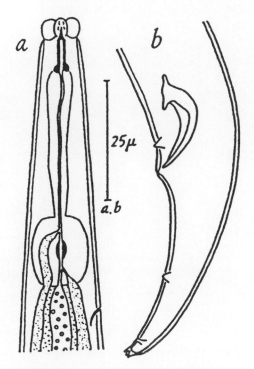

FIGURE 10 *Tylaphelenchus leichenicola*. (a) Fore part of body. (b) Male tail. (After Rühm, 1956.)

Ektaphelenchus (Fuchs, 1937) Skrjabin et al., 1954 (Fig. 11)
syn. *Parasitaphelenchus (Ektaphelenchus)* Fuchs, 1937
Ectaphelenchus Rühm, 1956

Body 0.5–1.2 mm long. Lips offset, with six distinct lips, the laterals narrower and project-ing slightly in front of the others. Stylet not less than 15 μm, with or without basal knobs. Median bulb appears rectangular. Vulval lips not protruding, without a vulval flap; pos-tuterine sac present, length variable. Tail conical to sharp or rounded point. Male tail with one preanal and one or two postanal pairs of subventral papillae and occasionally a median ventral preanal papilla. Spicules paired, not fused, scooped, troughlike with a delicate ros-trum, ventral element continuing line of shafts, bursa, and gubernaculum absent.

Bionomics: Unknown, insect associate.

Type species: *Ektaphelenchus hylastophilus* (Fuchs, 1930) Skrjabin et al., 1954. Twenty other species.

Berntsenus Massey, 1974 (Fig. 12)

Body about 0.9 mm long. Cuticle with faint transverse striae, with or without lateral in-cisures. Lips expanded, three times wider than deep, petiolate. Stylet, under phase illumina-

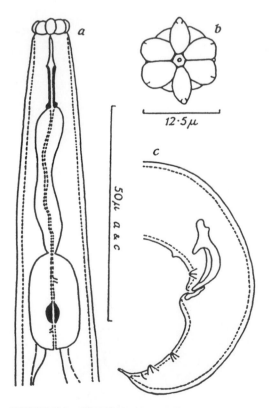

FIGURE 11   *Ektaphelenchus goffarti*. (a) Fore part of body. (b) End-on view of head. (c) Male tail end. (After Rühm, 1956.)

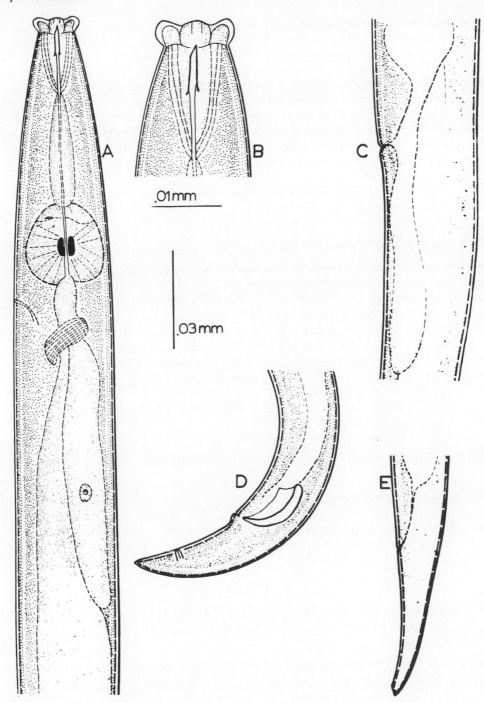

FIGURE 12 *Berntsenus brachycephalus.* (A) Head and neck. (B) Head. (C) Female, vulval region. (D) Male, tail. (E) Female, tail. (After Massey, 1974.)

tion, with heavily cuticularized conus, lightly cuticularized ventral aperture visible. Esophagus short. Metacorpus massive, filling body cavity in some specimens, round to oblong. Dorsal esophageal gland stout. Lips of vulva continuous with body wall. Vagina short. Posterior uterine branch two to three body widths in length. Ovary single. Anus and rectum conspicuous. Terminus subacute. Spicules paired, dorsal shaft lightly sclerotized, ventral segment less sclerotized than dorsal shaft. Apex low. One pair of postanal and ventrosubmedian papillae. Tail ventrally arcuate, conoid; subacute to acute terminus. Related to *Ektaphelenchus* but easily distinguished by its narrower lip region, shape of metacorpus, its longer postuterine sac, and the presence of a conspicuous anus and rectum. Male without bursa, spicules much less refractive with a low apex and delicate rostrum.

Bionomics: Bark beetle associate.

Type Species: *Berntsenus brachycephalus* (Thorne, 1935) Massey, 1974.

One other species.

Ektaphelenchoides Baujard, 1984 (Fig. 13)

Ektaphelenchinae. Body 0.5–0.9 mm long. Tails of both sexes conical with elongate terminus. Lips high, rounded, not offset. Stylet slender, 17–26 μm long, anterior part cylindrical, shorter than posterior portion, with basal thickenings; posterior part cylindrical, long, approximately five-eighths of total spear length, without basal knobs or thickenings. Constriction present at the junction between procorpus and metacorpus. Metacorpus rectangular with prominent posteriorly located valve plates. Rectum and anus absent in female. Postuterine sac present. Spicules with elongated rostrum and rounded prominent apex. Male tail arcuate, with two to four pairs of caudal papillae. Bursa absent.

Bionomics: Insect associates.

Type Species: *Ektaphelenchoides pini* (Massey, 1966) Baujard, 1984.

Three other species.

Ipsaphelenchus Lieutier and Laumond, 1978 (Fig. 14)

Body 0.7–1.2 mm long. Lips round. Cephalic framework prominent. Stylet with slight basal swellings. Median bulb ovoid. Spicules large, with large apex in the shape of a hook and pointed rostrum. Gubernaculum absent. With three pairs of caudal papillae; one preanal, two postanal. Bursa absent. Vulva with flap. Postuterine sac long. Female tail elongated.

Bionomics: Bark beetle associate.

Type Species: *Ipsaphelenchus silvestris* Lieutier and Laumond, 1978.

No other species.

Cryptaphelenchus (Fuchs, 1937) Rühm, 1954 (Fig. 15)
syn. *Parasitaphelenchus (Cryptaphelenchus)* Fuchs, 1937
P. (Steineria) Fuchs, 1937.

Body length short, 180–500 μm, C-shaped, not slender (*a* = about 20). Lips rounded, forming cap, slightly offset. Stylet delicate, less than 10 μm, with small rounded basal knobs. Ovary short; vulval lips not protruding; postuterine sac absent. Female tail conical. Spicules paired, not fused, with prominent narrow rostrum. Small gubernaculum-like structure present. Bursa absent. Male tail gradually attenuated to point; with at least two pairs of caudal papillae.

Bionomics: Unknown, insect associate.

Type Species: *Cryptaphelenchus macrogaster* (Fuchs, 1915) Rühm, 1956.

Twenty other species.

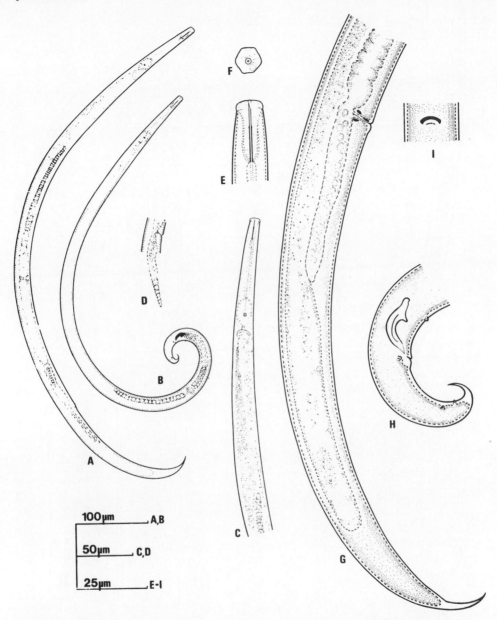

FIGURE 13   *Ektaphelenchoides compsi.* Female. (A) Total view. (C) Esophageal region. (D) Vulval region. (E) Anterior part. (F) En face view anterior part. (G) Posterior part. (I) Ventral view vulval region. Male. (B) Total view. (H) Posterior part. (After Baujard, 1984.)

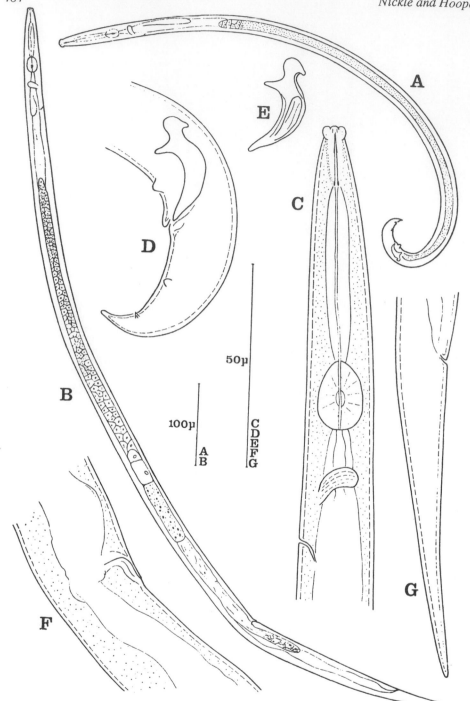

FIGURE 14   *Ipsaphelenchus silvestris*. (A) Male, general view. (B) Female, general view. (C–E) Male: (C) Anterior region. (D) Posterior part. (E) Spicule. (F,G) Female: (F) vulval region. (G) Tail. (After Lieutier and Laumond, 1978.)

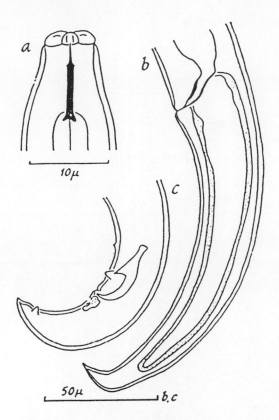

FIGURE 15 *Cryptaphelenchus macrogaster macrogaster.* (a) Head end. (b) Hind part of female. (c) Male tail end. (After Rühm, 1956.)

Cryptaphelenchoides J. B. Goodey, 1960 (Fig. 16)

Head offset; lips equal, rounded, slightly higher than wide. Spear 16 μm, with wide lumen and insignificant basal knobs. Procorpus relatively short, median bulb large, elongate-round, longer than body width at that level. Large valves in second half of the bulb. Anus and rectum visible but gut ends in a blind sac that projects well into the female tail. Female tail conical to a rounded tip. Vulva posterior; vagina at right angles to the body; postuterine sac short. Male tail bluntly rounded with one preanal (head of spicule level), one adanal (postanal) and one postanal (midcaudal) pair of subventral papillae. Bursa absent. Spicules deeply curved ventrally with prominent rostrum and large apex which continues the line of the shafts. An apophysis is connected with the distal ends of the spicules.

Bionomics: Insect associate.

Type Species: *Cryptaphelenchoides macrobulbosus* (Rühm, 1956) J. B. Goodey, 1960.

Two other species.

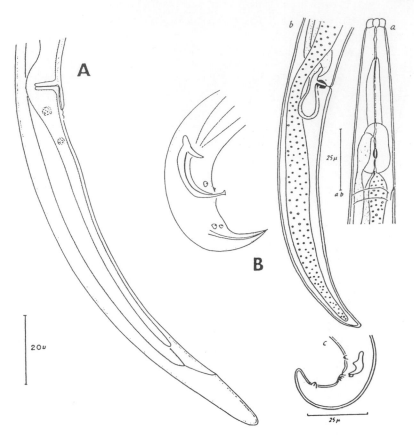

FIGURE 16   *Bursaphelenchus xylophilus* and *Cryptaphelenchoides macrobulbosus*. (A,B) *B. xylophilus*. (A) Female tail showing large vulvar flap and long postuterine sac. (B) Male tail showing typical spicule with expanded tip, caudal alae, and tail papillae. (After Nickle, et al., 1981.) (a–c) *C. macrobulbosus* (Rühm, 1956) J. B. Goodey, 1960. (a) Fore part of body. (b) Hind part of female. (c) Male tail. (After Rühm, 1956.)

Subfamily Bursaphelenchinae Paramonov, 1964

Aphelenchoididae. Spear thin, mostly knobbed. Spicules stout, may be distally fused. Tail tip of males surrounded by a small bursal rudiment. Five genera.

Bursaphelenchus Fuchs, 1937 (Fig. 16)
syn. *Aphelenchoides (Bursaphelenchus)* (Fuchs, 1937)
Rühm, 1956

Body 0.4–1.5 mm long, often slim (a = more than 30). Lips high, offset. Stylet with small rounded knobs. Excretory pore usually behind median bulb. Vulva with cuticular flap or with lips protruding; postuterine sac usually long. Female tail sometimes rounded, conoid sharply pointed or with a mucro. Spicules fairly narrow, usually with elongated apex and prominent rostrum. Male tail strongly arcuate, variously pointed with a short, terminal bursa. Caudal papillae variable, two to three pairs with occasional additional single preanal papilla. Gubernaculum absent.

Bionomics: Mycophagous, insect associates, *B. xylophilus* (Steiner and Buhrer, 1934) Nickle, 1970 is the cause of pine wilt disease, which has devastated pine forests (*Pinus densiflora* Sieb. and Zucc. and *P. thunbergii* Parl.) in Japan and which occurs in North America on various pines. This nematode reproduces readily on fungi and is closely associated with beetles that also feed on pines. In Japan the beetle *Monochamus alternatus* Hope is the main vector; individual beetles may carry thousands of dauerlarvae of *B. xylophilus* which are spread to healthy trees during maturation feeding of the beetles on young healthy pine twigs. The nematodes reproduce very rapidly within the tree tissue causing the death of cells and the blockage of resin canals, resulting in rapid, premature death of the trees. *Bursaphelenchus mucronatus* Mamiya and Enda, 1979 is also often present in the tissue of dead trees; it is similar in appearance to *B. xylophilus* but the female has a longer tail mucro and it is not considered to be the cause of pine wilt (Mamiya, 1984).

Type Species: *Bursaphelenchus piniperdae* Fuchs, 1937.

Yin et al. (1988) give a key to 44 valid species.

Huntaphelenchoides Nickle, 1970 (Fig. 17)

Body length 0.3–1.2 mm. Lips slightly offset. Stylet delicate, 11–14 µm long, with small, almost insignificant basal knobs. Esophageal glands short. Postuterine sac long; may be ovoviviparous. Female tail conical, tapering to a pointed tip. Spicules paired, fused, wide, with prominent rostrum. Tip of heavily cuticularized portion of ventral element appears to terminate some distance posterior to and separated from tip of dorsal element, similar to that of *Parasitaphelenchus* and *Rhadinaphelenchus*. Male tail shorter than that of female, more conical, with two pairs of caudal papillae and with short terminal bursa. Gubernaculum absent.

Bionomics: Unknown for type species but *H. fungivorus* (Franklin and Hooper, 1962) Nickle, 1970 readily feeds on fungi.

Type Species: *Huntaphelenchoides hunti* (Steiner, 1935) Nickle, 1970.

Two other species.

Omemeea Massey, 1971 (Fig. 18)

Body about 0.7 mm long. Lip region heavily cuticularized umbrellalike in lateral view. Cephalic framework refractive, distinct. Stylet long, 23–26 µm, with very prominent basal knobs, subulate cone longer than the shaft. Metacorpus oblong, ovate, the anterior portion glandular. Dorsal esophageal gland robust, relatively short. Excretory pore obscure, anterior to metacorpus. Ovary outstretched, postuterine branch several body widths in length. Female anal opening obscure. Male tail usually arcuate with terminal bursa. Spicules paired with prominent ventral rostra.

Bionomics: Bark beetle associate.

Type Species: *Omemeea maxbassiensis* Massey, 1971.

No other species.

Parasitaphelenchus Fuchs, 1929 (Fig. 19)
syn. *Aphelenchoides (Parasitaphelenchus)* Fuchs, 1937

Body 1.4–3.7 mm long. Lips clearly offset. Stylet relatively long 11–18 µm and slender, with small basal knobs. Vulva posterior 85–90%, body often narrowing abruptly posterior

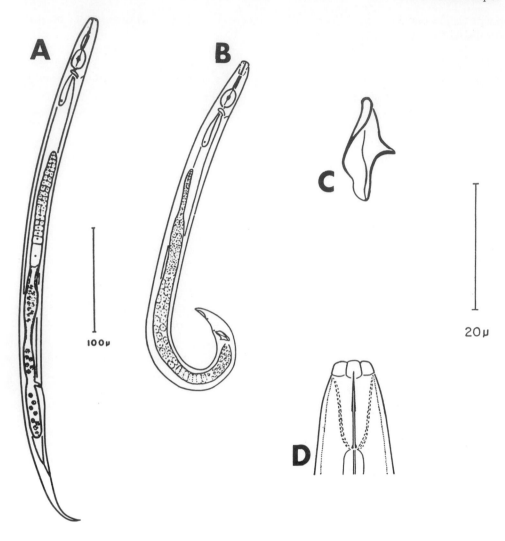

FIGURE 17   *Huntaphelenchoides hunti.* (A) Female. (B) Male. (C) Spicule. (D) Female stylet. (After Nickle, 1970.)

to vulva; postuterine sac long. Female tail short, conical. Spicules paired, apparently fused along shafts, rostrum prominent, ventral element well developed, continuing line of shaft. Tip of cuticularized portion of ventral element appears to terminate some distance from tip of dorsal element, similar to that of *Huntaphelenchoides* and *Rhadinaphelenchus.* Male tail short, conical, sometimes with offset tip and short terminal bursa; with three pairs of caudal papillae. Gubernaculum absent. Juveniles usually with cuticular projections at anterior and posterior ends of body.

Bionomics: Insect parasite associate; juvenile parasitic in hemocele, adults fungivorous.

Type Species: *Parasitaphelenchus uncinatus* (Fuchs, 1929) Fuchs, 1929.

Fifteen other species.

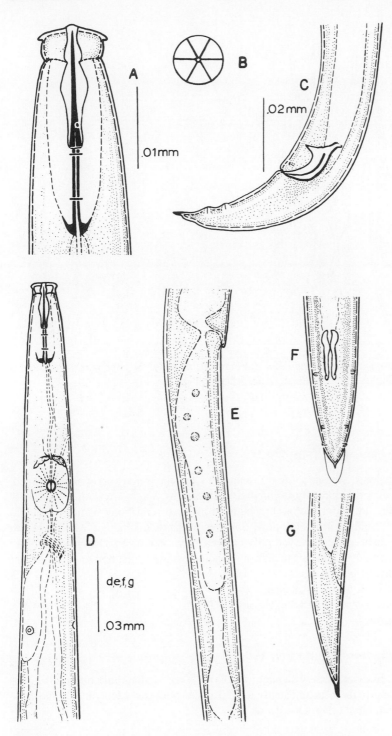

FIGURE 18 *Omemeea maxbassiensis.* (A) Head. (B) Face view. (C) Lateral view, male, tail. (D) Head and neck. (E) Female, midbody. (F) Ventral view, male, tail. (G) Female, tail. (After Massey, 1971.)

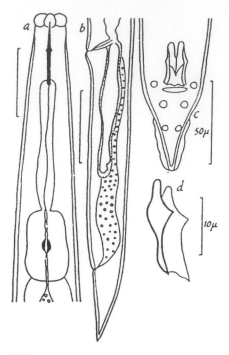

FIGURE 19 *Parasitaphelenchus papillatus.* (a) Fore part of body. (b) Hind part of female. (c) Male tail end ventral. (d) Spicules lateral. (After Rühm, 1956.)

Teragramia Massey, 1974 (Fig. 20)

Body about 0.6 mm long. Lip region set off, rounded. Cephalic framework well developed. Stylet slender, subulate portion only slightly wider than shaft. Vagina extending anteriorly from vulval opening. Posterior uterine branch prominent, several body widths in length. Female tail obtuse, clublike in lateral view. Spicules massive, distinctive. Male tail with terminal bursa, distinctively curved ventrally on preserved specimens. Differs from other members of the subfamily in shape and size of the spicules, in the distinctive vulva, and in the shape of the tails of both sexes.

Bionomics: Bark beetle associate.
Type species: *Teragramia willi* Massey, 1974.
No other species.

Rhadinaphelenchinae Paramonov, 1964

Aphelenchoididae. Body extremely slender (a = about 100). Median bulb oblong. Vagina sloping anteriorly, postvulvar sac unusually long. Tails of both sexes long and cylindrical, tail tip of males with bursal flap. One genus.

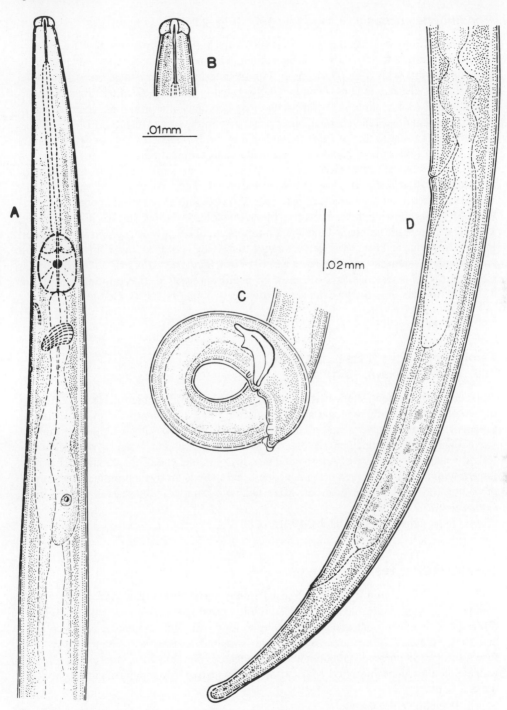

FIGURE 20 *Teragramia willi.* (A) Head and neck. (B) Head. (C) Male, tail. (D) Female, tail. (After Massey, 1974.)

Rhadinaphelenchus J. B. Goodey, 1960 (Fig. 21)

Body about 1 mm long, slender (a = about 100). Lips high, slightly offset. Stylet with well-developed knobs. Median esophageal bulb about twice as long as wide, with valve posterior to center. Vulva with wide overlapping flap; postuterine sac very long, extending about three-fourths distance to anus. Female tail elongated to small rounded terminus. Spicules paired, small (dorsal limb 9–11 µm), with elongated apex and prominent rostrum; tip of ventral element of spicule recurved, but appears to terminate some distance from tip of dorsal element, similar to that of *Parasitaphelenchus* and *Huntaphelenchoides*. Tail strongly arcuate, with four pairs of caudal papillae; with short terminal bursa.

  Bionomics: *Rhadinaphelenchus cocophilus* is the cause of red-ring disease of coconut palms (*Cocos nucifera* L.) in the West Indies and Latin America; oil palms (*Elaeis guineensis* Jacq.) are also attacked, particularly in Colombia. Diseased coconut trees initially show a yellowing then a browning of the lower leaves which spreads to other leaves and infested trees die within 3–4 months. Enormous numbers of *R. cocophilus* are produced in the stem tissue and the characteristic orange to red ring appears about 3 cm wide and 2.5 cm beneath the stem surface. The palm weevil (*Rhynchophorus palmarum*) seems to be the main vector of nematodes from diseased to healthy trees (CIH Description Set 5, No. 72).Type Species: *Rhadinaphelenchus cocophilus* (Cobb, 1919) J. B. Goodey, 1960.

  No other species.

Family Seinuridae Husain and Khan, 1967 (Baranovskaya, 1981)

Aphelenchina. Tail long, elongated to filiform without a terminal mucron. Head continuous or set off. Stylet usually long and slender, anterior pointed part is jointed in *Paraseinura*; lumen of spear relatively wide. Median bulb elliptical, oblong, or long oval, with prominent valve. Ovary single, prodelphic. Postuterine sac present or absent. Spicules somewhat differently shaped from those of *Aphelenchoides*; the proximal end of the transverse bar prolonged with dorsal limb into a prominent apex, and there is always a prominent rostrum at the other end of the transverse bar. Male tail with papillae. Gubernaculum present in *Paraseinura*.

  Type genus: *Seinura* Fuchs, 1931.

Seinura Fuchs, 1931 (Fig. 22)

Body length 0.3–1.1 mm. Tails of both sexes usually variously elongate-filiform, not short-conical. Lips high. Stylet slender, 10–27 µm long, basal knobs absent or with small basal thickenings. Median esophageal bulb oblong to long-oval with prominent posteriorly located valve plates. Postuterine sac present or absent. Spicules rose thorn-shaped, paired, not fused, similar to *Aphelenchoides*, but with an elongated apex and prominent rostrum. Dorsal wall of spicular pouch thickened. Male tail arcuate, with two to four pairs of caudal papillae; bursa absent.

  Bionomics: Predaceous.
  Type Species: *Seinura mali* Fuchs, 1931.
  Thirty-four other species.

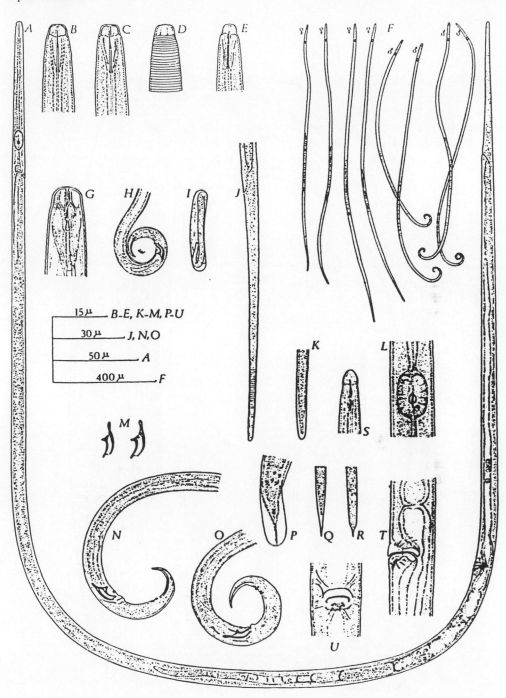

FIGURE 21 *Rhadinaphelenchus cocophilus.* (A–F) Adults. (B–D,G) Female head ends. (E) Male head end. (H,N,O) Male tail ends. (I) Egg. (J) Female tail. (K) Female tail tip. (L) Female median esophageal bulb. (M) Spicules. (P) Bursa in dorsal view. (Q,R) Larval tail tips. (S) Larval head end. (T,U) Vulva in lateral and ventral view, respectively. (From CIH Series.)

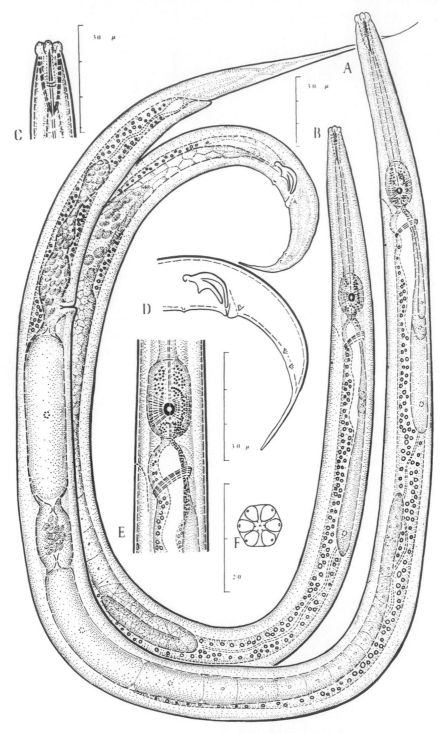

FIGURE 22 *Seinura tenuicaudata.* (A) Female. (B) Male. (C) Female head. (D) Male tail variation. (E) Median bulb. (F) En face view, female. (After Hechler, 1963.)

Paraseinura Timm, 1960 (Fig. 23)

Body length 0.5–0.9 mm. Tails of both sexes long, filiform, with exceedingly fine posterior portion. Lips somewhat rounded, not offset. Stylet 22–25 μm long, anterior pointed part with prominent ventral slant, appearing jointed; shaft with slightly thickened basal knobs. Median esophageal bulb with posteriorly located valves. Postuterine sac short, testis short. Spicules paired, not fused, with prominent rostrum and apex, more robust than that of *Seinura*. Gubernaculum-like structure small, apex rounded, same refractivity as spicules. Male tail coiled once when relaxed; with three pairs of caudal papillae; bursa absent. Otherwise similar to *Seinura*.

Bionomics: Unknown, probably predaceous.

Type Species: *Paraseinura musicola* Timm, 1960.

No other species.

Papuaphelenchus Andrássy, 1973 (Fig. 24)

Body about 0.6 mm long, quite slender (*a* = 55–58). Body cuticle not visibly annulated. Lateral lines plain. Stylet 17–19 μm long and narrow, without knobs. Esophageal bulb long, esophageal glands lie free in the body cavity. Rectum long, weakly defined. Postvulval uterine sac very short. Tail attenuated—in the case of juveniles it is more plump—with a fine rounded end. Male unknown. The genus *Papuaphelenchus* is closely related to the genera *Aphelenchoides* and *Seinura*, but because of the distinct amphids, the long and thin stylet, the long rectum, the relatively short and plump gonad, as well as the tail shape, it differs from these genera.

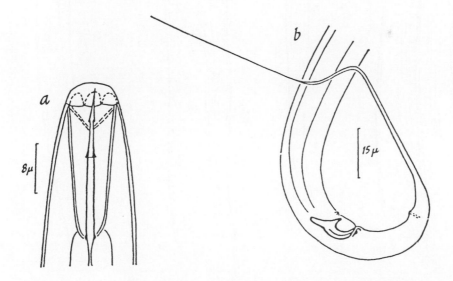

FIGURE 23   *Paraseinura musicola.* (a) Head end. (b) Male tail end. (After Timm, 1961.)

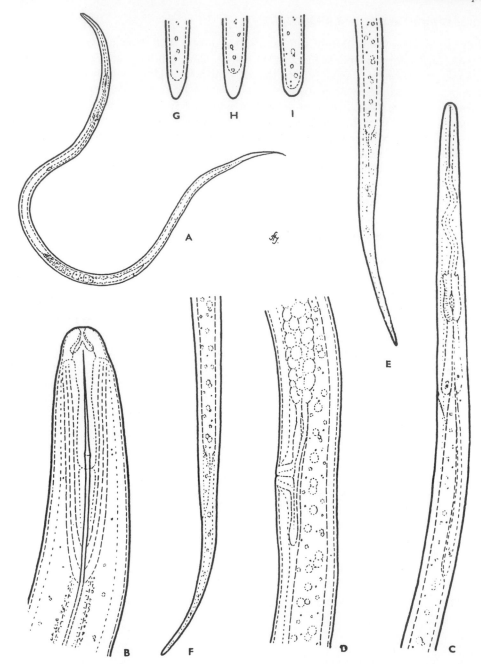

FIGURE 24 *Papuaphelenchus trichodorus.* (A) Full body, 260×. (B) Anterior end, 3800×. (C) Anterior region, 830×. (D) Vulval region, 1650×. (E,F) tail, 1100×. (G–I) Shapes of tail end. (After Andrássy, 1973.)

Bionomics: Wet moss on mountain rocks, New Guinea.
Type Species: *Papuaphelenchus trichodorus* Andrássy, 1973.
No other species.

Aprutides Scognamiglio, Talame, and S'Jacob, 1970 (Fig. 25)

Body length medium. Labial framework lightly cuticularized. Stylet faint without a jointed tip; with small basal swellings. Median bulb well developed, longer than wide with conspicuous valve with crescentic thickenings. Female reproductive system is monovarial; ovary outstretched; postuterine sac present. Female tail as well as the male tail is filiform with a clavate tip but less distinct in the male. Testis of male single outstretched. Spicules typically aphelenchoid; paired with prominent anterior knoblike apex and the distal end of the ventral limb is almost as far posterior as that of the dorsal limb; gubernaculum present, very simple in structure.

Bionomics: Unknown, soil inhabiting.
Type Species: *Aprutides martuccii* Scognamiglio, Talame, and s'Jacob, 1970.
One other species.

Family Entaphelenchidae Nickle, 1970

Aphelenchina. Usually with at least three distinct adult forms, including a vermiform male and female and a swollen endoparasitic female. Stylet present with or without basal flanges. Esophagus with a large median bulb and overlapping glands. Male usually without caudal alae. Spicules rose-thorn-shaped. Gubernaculum absent. Obligate parasites in body cavity of insects.

Type genus: *Entaphelenchus* Wachek, 1955.

Subfamily Entaphelenchinae Nickle, 1970
Entaphelenchus Wachek, 1955 (Fig. 26)

Three adult forms. Small vermiform males and females, 0.5–1.0 mm long, not found in the insect body cavity. Cuticle smooth. Lips offset. Stylet 17–22 μm long, without basal knobs. Excretory pore behind median bulb. Median esophageal bulb with large valves, just posterior to center. Female: Ovary with only a few cells; postuterine sac short. Tail conical with rounded tip. Male: Spicules paired, not fused, with large prominent rostrum; ventral element continuing line of shaft. Male tail short, conical to rounded point, usually with three pairs of caudal papillae; caudal alae and gubernaculum absent. Adult parasitic female found in the insect body cavity. Body large, 1.3–2.5 mm long, convoluted, oviparous; perivaginal glandular cells present around vulva. Tail often dome-shaped, with spikelike tip.

Bionomics: Obligate insect parasites (staphylinids).
Type Species: *Entaphelenchus oxyteli* Wachek, 1955.
Seven other species.

Peraphelenchus Wachek, 1955 (Fig. 27)

Three adult forms, all found in insect body cavity. Lips slightly offset. Stylet 17–18 μm long, with wide lumen, with ventral bend at anterior end. Excretory pore about one body width behind median bulb. Female: C-shaped ventrally when relaxed. Ovary few-celled; uterus filled with sperm (as in infective stage sphaerulariids); postuterine sac long. Tail short, conical. Male: Corkscrew-shaped. Spicules paired, possibly fused at tips, with promi-

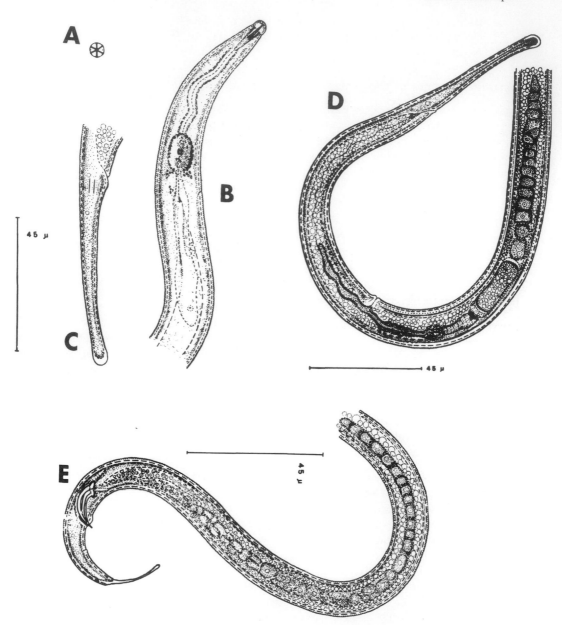

FIGURE 25   *Aprutides martuccii* (A–D) Female. (A) En face view. (B) Esophagus region. (C) Posterior part of the body. (D) Gonad region. (E) Male, posterior half of the body. (After Scognamiglio et al., 1970.)

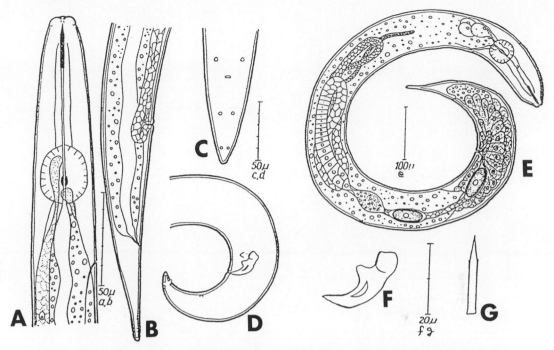

FIGURE 26 *Entaphelenchus oxyteli.* (A) Female anterior end. (B) Female posterior end. (C) Male posterior end ventral view. (D) Male posterior end, lateral view. (E) Mature parasite. (F) Spicule. (G) Stylet. (After Wachek, 1955.)

nent rostrum and apex. Tail bluntly rounded with two pairs of postanal caudal papillae; gubernaculum and caudal alae absent. Adult parasitic female: Ovary short; oviparous.

Bionomics: Obligate insect parasites (*Necrophorus*: Silphidae); rarely recorded.

Type Species: *Peraphelenchus necrophori* Wachek, 1955.

Five other species.

Praecocilenchus Poinar, 1969 (Fig. 28)

Three adult forms, all found in insect body cavity. Female and male found in uterus of adult parasitic female. Cuticle with fine annulation. Lips small, low, not offset. Stylet short (8– 10 μm) with wide lumen, without basal knobs. Female: Very small, less than 0.4 mm long. Esophageal glands long, extending to midbody. Ovary short, postuterine sac short. Tail conical. Male: Small 0.4–0.5 mm long. Spicules paired, not fused, rose-thorn-shaped, with prominent rostrum and apex. Tail curved, conical; bursa and gubernaculum absent. Adult parasitic female: Found in insect body cavity. Body large, 0.9–2.3 mm long swollen, C-shaped ventrally, uterus containing sexually mature males and females. Stylet short, with wide lumen, without basal knobs. Vulval lips protruding; postuterine sac absent; ovovivipa-rous, larval nematodes reaching sexual maturity within uterus of swollen female. Tail bluntly rounded.

Bionomics: Obligate insect parasites (*Rhynchophorus*: Curculionidae).

Type Species: *Praecocilenchus rhaphidophorus* Poinar, 1969.

One other species.

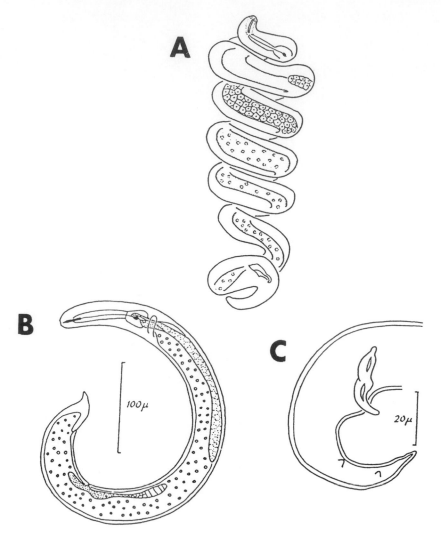

FIGURE 27   *Peraphelenchus necrophori.* (A) Male, corkscrew shape. (After Nickle, 1970.) (B) Female. (C) Male tail end. (After Wachek, 1955.)

Roveaphelenchus Nickle, 1970 (Fig. 29)

Three adult forms, all found in body cavity of insect. Lips slightly offset. Stylet slender, without basal knobs. Esophageal glands short. Female: Ovary with only one large larva in uterus; postuterine sac absent. Tail cylindrical, ending in four mucronate points. Male: When relaxed, tail remains in tight coil. Testis flexed. Spicules paired, not fused, with prominent rostrum. Tail bluntly rounded. Caudal papillae not seen; gubernaculum and bursa absent. Adult parasitic female: Body swollen. Ovary convoluted, extending to neck region; vulva posterior (96%); ovoviviparous, uterus containing pre-adult stage larvae. Tail short, digitate.

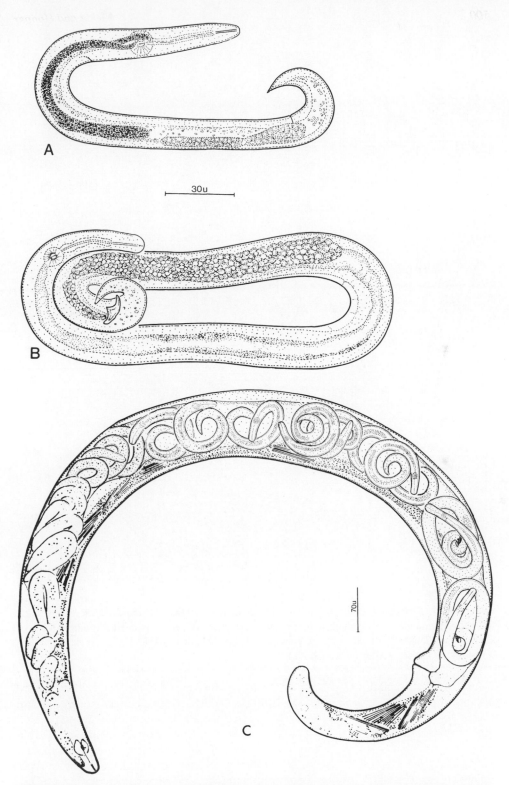

FIGURE 28   *Praecocilenchus rhaphidophorus*. (A) Intrauterine female of *P. rhaphidophorus*. (B) Intrauterine male of *P. rhaphidophorus*. (C) Mature parasitic female of *P. rhaphidophorus* from the hemocele of *Rhynchophorus bilineatus*. (After Poinar, 1969.)

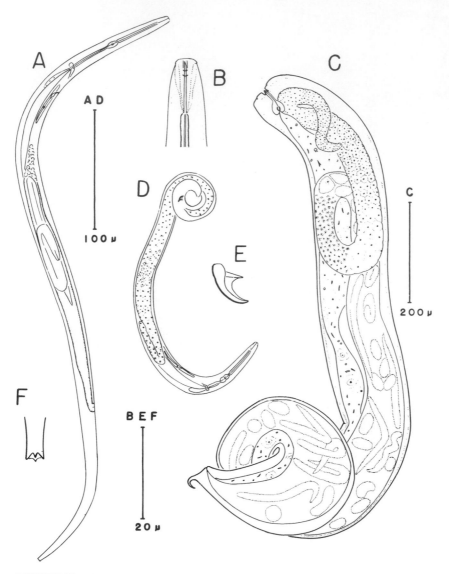

FIGURE 29   *Roveaphelenchus jonesi.* Parasitic in the rove beetle, *Aleochara tristis.* (A) Female,
full body, lateral view. (B) Female, anterior end, lateral view. (C) Adult parasitic female, full body,
lateral view. (D) Male, full body, lateral view, showing coiled tail. (E) Spicule, lateral view. (F) Tail
tip of sexual female. (After Nickle, 1970.)

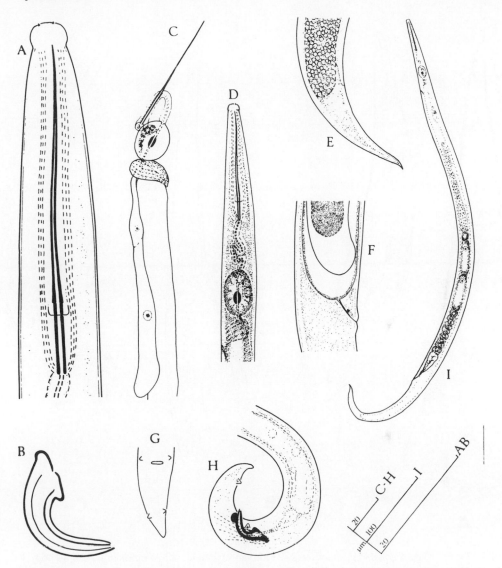

FIGURE 30 *Acugutturus parasiticus*. (A) Stylet. (B) Spicules. (C,D) Esophageal region. (E) Female tail. (F) Vulval region with part of uterine egg. (G) Ventral view of male tail. (H) Male tail region. (I) Female, full body. (After Hunt, 1980.)

Bionomics: Obligate insect parasites (staphylinids), Nebraska.
Type Species: *Roveaphelenchus jonesi* Nickle, 1970.
No other species.

Subfamily Acugutturinae Hunt, 1980
Acugutturus Hunt, 1980 (Fig. 30)

Body 0.6–0.9 mm long, with a single lateral line. Head offset and rather knoblike. Stylet long, 50–60 μm, slender and with a conus three times longer than the shaft which lacks basal thickenings. Procorpus long and reflexed. Female genital tract single, prodelphic; postvul-

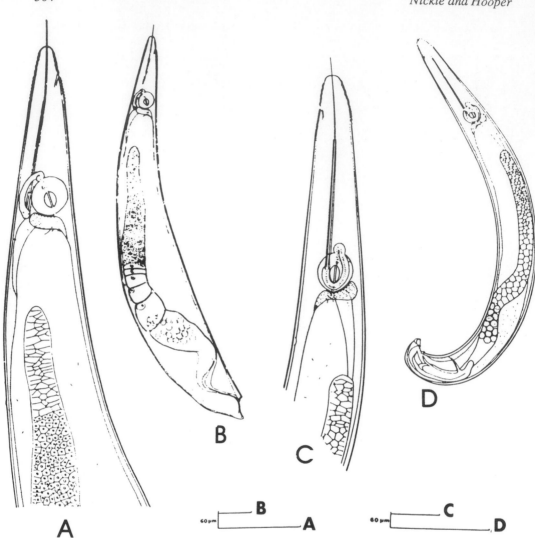

FIGURE 31   *Noctuidonema guyanense.* (A,B) Immature female. (A) Anterior end. (B) Full body. (C,D) Male. (C) Anterior end. (D) Full body. (After Remillet and Silvain, 1988.)

val sac absent. Developing oocytes in multiple tows. Female rectum and anus very much reduced or apparently absent; tail conical. Spicules rose-thorn-shaped with prominent apex and rostrum and gubernaculum-like structure. Two pairs of caudal papillae. Bursa absent.

    Bionomics: Ectoparasite on American cockroach, in St. Lucia, West Indies.

    Type Species: *Acugutturus parasiticus* Hunt, 1980.

    No other species.

Noctuidonema Remillet and Silvain, 1988 (Fig. 31)

Aphelenchoididae; Acugutturinae. Body 0.5–0.8 mm long, females swollen, club-shaped. Head not enlarged but well set off. Stylet very long (over 100 μm), with cone much longer than basal part, the latter with small basal thickenings. Procorpus long and reflexed. Excretory pore very anterior. Genital system with a single branch, not flexed. Oocytes in several rows. Rectum and anus (of the females) more visible before fixing. Tail very short. Large spicules, 90 μm long, with two long basal rostra, gubernaculum absent. One pair of preanal papillae and one pair of caudal papillae. Terminal "bursa" present.

Bionomics: Ectoparasitic on noctuid moths in French Guiana, South America.

Type Species: *Noctuidonema guyanense* Remillet and Silvain, 1988. Related only to *Acugutturus parasiticus* Hunt, 1980, by the length and shape of the stylet, the length of the procorpus, and the type of the gonads. It differs from it by the following characteristics: club shape of the body, much longer stylet length, shape of head and tail, position of the excretory pore, shape and size of the spicules, presence of a bursa.

No other species.

REFERENCES

Allen, M. W. 1940. *Anomyctus xenurus*, a new genus and species of Tylenchoidea (Nematoda). *Proc. Helm. Soc. Wash.* 7:96–98.

Allen, M. W. 1941. *Aphelenchus megadorus*, a new species of Tylenchoidea (Nematoda). *Proc. Helm. Soc. Wash.* 8:21–23.

Allen, M. W. 1952. Taxonomic status of the bud and leaf nematodes related to *Aphelenchoides fragariae* (Ritzema Bos, 1891). *Proc. Helm. Soc. Wash.* 19:108–120.

Anderson, R. V. and Hooper, D. J. 1980. Diagnostic value of the vagina structure in the taxonomy of *Aphelenchus* Bastian, 1865 (Nematoda: Aphelenchida) with a description of *A. (Anaphelenchus) isomermus* n. subgen., n.sp. *Can. J. Zool.* 58:924–928.

Andrássy, I. 1973. Ein Meeresrelikt und einige andere bemerkenswerte Nematodenarten aus Neuguinea. *Opusc. Zool. Bpest.* 12:3–19.

Andrássy, I. 1976. *Evolution as a Basis for the Systematization of Nematodes.* Pitman, London.

Baranovskaya, I. A. 1958. Contribution to the knowledge of the genus *Paraphelenchus* (Micoletzky, 1922) Micoletzky, 1925 (Nematoda: Aphelenchidae) *Zool. Zh.* 37:13–19.

Bastian, H. C. 1865. Monograph on the Anguillidae or free nematoids, marine, land, and freshwater; with descriptions of 100 new species. *Tr. Linn. Soc. London* 25:73–184.

Baujard, P. 1984. Remarques sur la sous-famille des Ektaphelenchinae Paramonov, 1964 et proposition d'*Ektaphelenchoides* n.g. (Nematoda: Aphelenchoididae). *Rev. Nematol.* 7:147–171.

Baujard, P. 1985. Remarks on the genus *Laimaphelenchus* Fuchs, 1937. *Rev. Nematol.* 8:186–187.

C.I.H. *Descriptions of plant-parasitic nematodes.* 1972–1976. 8 sets, 15 descriptions in each set. Issued by Commonwealth Institute of Helminthology, St. Albans, Herts, England.

Cobb, N. A. 1923. Observations on nemas: salivary glands of the nemic genera *Tylenchus* and *Aphelenchus*. *J. Parasitol.* 9:236–239.

Fortuner, R., Geraert, E., Luc, M., Maggenti, A. R., and Raski, D. J. 1987–1988. A reappraisal of Tylenchina (Nemata). Parts 1–11. *Rev. Nematol.*, 10–11.

Franklin, Mary T. 1955. A redescription of *Aphelenchoides parietinus* (Bastian, 1865) Steiner, 1932. *J. Helminthol.* 29:65–76.

Fuchs, A. G. 1929. Die parasiten einiger Russel-und Borkenkafer. *Z. Parasitenk.* 2: 248–285.

Fuchs, A. G. 1930. Neue an Borken-und Russelkafer gebundene Nematoden, halbparasitische und Wohnungseinmieter. Freilebende Nematoden aus Moos und Walderde in Borken-und Russelkafergangen. *Zool. Jb. (Syst)* 59:505–646.

Fuchs, A. G. 1931. *Seinura* gen. nov. *Zool. Anz.* 94:226–228.

Fuchs, A. G. 1937. Neue Parasitische und halbparasitische Nematoden bei Borkenkafern und einige andere Nematoden. I. Teil. Die Parasiten der Waldgartner, *Myelophilus piniperda* L. und *minor* Hartig und die Genera *Rhabditis* Dujardin, 1845, und *Aphelenchus* Bastian, 1865. *Zool. Jb. (Syst.)* 70:291–380.

Gasparrini, G. 1864. Sulla maturazione e la qualita dei fichi dei contorni di Napoli. (Read Nov. 29, 1863). *Atti Accad. Pontaniana* 9:99–118.

Geraert, E. 1966. The systematic position of the families Tylenchulidae and Criconematidae. *Nematologica* 12: 362–368.

Goodey J. B. 1958. *Paraphelenchus myceliophthorus* n. sp. (Nematoda: Aphelenchidae). *Nematologica* 3:1–5.

Goodey, J. B. 1960a. The classification of the Aphelenchoidea Fuchs, 1937. *Nematologica*, 5:111–126.

Goodey, J. B. 1960b. *Rhadinaphelenchus cocophilus* (Cobb, 1919) n. comb., the nematode associated with "red-ring" disease of coconut. *Nematologica* 5:98–102.

Goodey, J. B. 1963. *Soil and Freshwater Nematodes*. John Wiley and Sons, New York.

Goodey, J. B. and Hooper, D. J. 1965. A neotype of *Aphelenchus avenae* Bastian, 1865 and the rejection of *Metaphelenchus* Steiner, 1943. *Nematologica* 11:55–65.

Hechler, H. C. 1963. Description, developmental biology, and feeding habits of *Seinura tenuicaudata* (de Man) J. B. Goodey, 1960 (Nematoda: Aphelenchoididae), a new nematode predator. *Proc. Helm. Soc. Wash.* 30:182–195.

Hirling, W. 1986. *Laimaphelenchus penardi* (Nematoda: Tylenchida) von Typenfundort und Beitrage zur Eidonomie, Biologie und Verbreitung dieses Nematoden und dreier nahe verwandter Arten. *L. sylvaticus* und *L. praepenardi* n. sp. *Zool. Beitr.* (1984/1985) 29:349–375.

Hollis, J., and Keoboonrueng, S. 1984. Nematode parasites of rice. In *Plant and Insect Nematodes* Nickle, W. R. ed., Marcel Dekker, New York, pp. 95–146.

Hunt, D. J. 1980. *Acugutturus parasiticus* n. g., n. sp., A remarkable Ectoparasitic Aphelenchoid Nematode from *Periplaneta americana* (L.), with proposal of Acugutturinae n. subf. *Sys. Parasitol.*, 1(3/4):167–170.

Lieutier, F., and Laumond, C. 1978. Nematodes parasites et associes a *Ips sexdentatus* et *Ips typographus* (Coleoptera: Scolytidae). *Nematologica* 24:184–200.

Mamiya, Y. 1984. The pine wood nematode. In *Plant and Insect Nematodes*. Nickle, W. R. ed., Marcel Dekker, New York, pp. 589–626.

Massey, C. L. 1971. *Omemeea maxbassiensis* n. g., n. sp. (Nematoda: Aphelenchoididae) from galleries of the bark beetle *Lepersinus californicus* Sw. (Coleoptera: Scolytidae) in North Dakota. *J. Nematol.* 3:289–291.

Massey, C. L. 1974. *Biology and Taxonomy of Nematode Parasites and Associates of Bark Beetles in the United States*. USDA, Forest Service, Agricultural Handbook No. 446.

Nickle, W. R. 1970a. A taxonomic review of the genera of the Aphelenchoidea (Fuchs, 1937) Thorne, 1949 (Nematoda: Tylenchida). *J. Nematol.* 2:375–392.

Nickle, W. R. 1970b. Description of Entaphelenchidae n. fam., *Roveaphelenchus jonesi* n. g., n. sp., (Nematoda:Aphelenchoidea). *Proc. Helm. Soc. Wash.* 37:105–109.

Nickle, W. R., Golden, A. M., Mamiya, Y., and Wergin, W. P. 1981. On the taxonomy and morphology of the pine wood nematode, *Bursaphelenchus xylophilus* (Steiner and Buhrer, 1934) Nickle, 1970. *J. Nematol.* 13:385–392.

Paramonov, A. A. 1953. Revision of the superfamily Aphelenchoidea Fuchs, 1937 (Nematoda:Tylenchata). Papers of Helminthology presented to Academician K. I. Skrjabin on his 75th birthday. Contributions to Helminthology. *Dokl. Acad. Nauk SSSR, Moscow*, pp. 488–496.

Paramonov, A. A. 1954. Suborder Tylenchata Chitwood 1950. In: Skrjabin, K. I. et al. Camallanata, Rhabditata, Tylenchata, Trichocephalata and Dioctophymata and the distribution of parasitic nematodes by hosts. *Dokl. Acad Nauk. SSSR, Moscow*, pp. 245–322.

Paramonov, A. A. 1962. Plant-parasitic nematodes. Vol. 1. Origin of nematodes. Ecological and morphological characteristics of plant nematodes. Principles of taxonomy. *Dokl. Acad. Nauk SSSR, Moscow*.

Paramonov, A. A. 1964. *Fundamentals of Phytohelminthology*. Vol. II. Taxonomy of Phytonematodes. *Dokl. Acad. Nauk SSSR., Helminthol. Lab., Moscow*.

Poinar, G. O. 1969. *Praecocilenchus rhaphidophorus* n.gen., n.sp. (Nematoda:Aphelenchoidea) parasitizing *Rhynchophorus bilineatus* (Montrouzier) Coleoptera: Curculionidae) in New Britain. *J. Nematol.* 1:227–231.

Raski, D. J., and Krusberg, L. 1984. Nematode parasites of grapes and other small fruits. In *Plant and Insect Nematodes*. Nickle, W. R., ed., Marcel Dekker, New York, pp. 457–506.

Reddy, Y. N., and Rao, P. N. 1986. *Schistonchus racemosa* sp. n. a nematode parasite of wasp (*Ceratosolen* sp.) associated with the fig, *Ficus racemosa* L. *Ind. J. Nematol.* 16(1): 135–137.

Remillet, M., and Silvain, J. F. 1988. *Noctuidonema guyanese* n. g., n. sp., (Nematoda: Aphelenchoididae) ectoparasite de noctuelles de genre *Spodoptera* (Lepidoptera:Noctuidae). *Rev. Nematol.* 11:21–24.

Ritzema Bos, J. 1890. De bloemkoolziekte der aardbeien, veroorzaakt door *Aphelenchus fragariae* nov. spec. (Voorloopige mededeeling) *Maandbl. Natuurwetensch.* 16:107–117.

Ruhm, W. 1954. Einige neue, ipidenspezifische Nematodenarten. *Zool. Anz.* 153:221–242.

Ruhm, W. 1955. Uber einige an holzbrutende Ipiden gebundene Nematodenarten. *Zool. Anz.* 155:70–83.

Ruhm, W. 1956. Die nematoden der Ipiden. *Parasitolog. Schrift.* 6:1– 437.

Sanwal, K. C. 1961. A key to the species of the nematode genus *Aphelenchoides* Fischer, 1894. *Can. J. Zool.* 39:143–148.

Sanwal, K. C. 1965. Appraisal of taxonomic characters of "parietinus group" of species of the genus *Aphelenchoides* Fischer, 1894 (Nematoda: Aphelenchoididae). *Can. J. Zool.* 43:987–995.

Scognamiglio, A., Talame, M., and s'Jacob, J. J. 1970. *Aprutides martuccii* (Nematoda: Aphelenchoididae) n.sp., n. g. *Boll. Lab. Ent. Agr.* 28:3–11.

Siddiqi, M. R. 1980. The origin and phylogeny of the nematode orders Tylenchida Thorne, 1949 and Aphelenchida N. Ord. *Helm. Abstr. Ser. B.* 49:143–170.

Skarbilovich, T. S. 1947. Revision of the systematics of the nematode family Angillulinidae Baylis and Daubney, 1926. *Dokl. Akad. Nauk SSSR* 57:307–308.

Steiner, G. 1914. Freilebende Nematodes aus der Schweiz. *Arch. Hydrobiol.* 9:259–276, 420–438.

Thorne, G. 1961. *Principles of Nematology*. McGraw-Hill, New York.

Timm, R. W. 1960. *Paraseinura* (Nematoda:Aphelenchoididae) a new genus from East Pakistan. *Nematologica* 5:171–174.

Wachek, F. 1955. Die entoparasitichen Tylenchiden. *Parasitol. Schrift.* 3.

Yin, K., Fang, Y., and Tarjan, A. C. 1988. A key to species in the genus *Bursaphelenchus* with a description of *Bursaphelenchus hunanensis* sp. n. (Nematoda:Aphelenchoididae) found in pine wood in Hunan Province, China. *Proc. Helm. Soc. Wash.* 55:1–11.

11

Reniform and False Root-Knot Nematodes, *Rotylenchulus* and *Nacobbus* spp.

PARVIZ JATALA   *International Potato Center, Lima, Peru*

I. INTRODUCTION

Reniform nematodes *Rotylenchulus* spp. and false root-knot nematodes *Nacobbus* spp. are important parasites of a diverse group of food and fiber crops. Reniform nematodes are rather cosmopolitan in distribution, occurring in a wide spectrum of habitats in most continents, attacking a rather wide variety of plants. *Rotylenchulus reniformis* is the only species of the genus *Rotylenchulus* known to be of major economic importance to agriculture. False root-knot nematodes are rather restricted in their distribution, being found principally in South and North America. However, their incidental occurrence is primarily reported in greenhouses in the Netherlands, England, and from the fields in Russia, and India (Jatala, 1978). They attack a diverse group of food crops and are of major importance in the production of potatoes and sugar beets in South and North America, respectively.

II. TAXONOMY OF *ROTYLENCHULUS*

A.   Family Hoplolaimidae Filipjev, 1934
Nemonchidae Skarbilovich, 1959
Aphasmatylenchidae Sher, 1965
Rotylenchulidae Husain and Khan, 1967

Diagnosis

Tylenchoidea. Juveniles vermiform with distinct or indistinct hyaline terminal portion of tail. Cuticle annulated. Lateral field with four incisures, not areolated. Cephalic region strongly sclerotized. Excretory pore in esophageal region.

Male with secondary sexual dimorphism present. Reduced cephalic sclerotization, sometimes degenerated and nonfunctional. Caudal alae present or absent. If present, generally enveloping the tail. Testis single, outstretched. Spicula cephalated, ventrally arcuate, with narrow distal tip having dorsally subterminal pore and lacking distinct flanges. Gubernaculum large, trough-shaped, fixed, and with titillae.

Female vermiform to kidney-shaped, often spiral when vermiform. Lip region higher than half the diameter of the basal lip annulus, with rounded or trapezoidal lateral view, continuous or rarely offset lacking longitudinal indentation. Stylet strong, 2.5–3 times longer than the diameter of basal lip annulus. Stylet knobs strong, rounded to indented, sometimes anchor-shaped. Dorsal esophageal gland orifice (DGO) at least 4 μm and sometimes more than 20 μm from the stylet base. Stylet in female and juveniles well developed, two to three cephalic region widths long, conus as long or shorter than shaft, knobs prominent and rounded. Median bulb strong and rounded. Esophageal glands generally overlap intestine mostly ventrally, but also laterally and dorsally. Females with two genital branches, posterior branch often reduced to postuterine sac (PUS). Epiptygma and vulval flaps generally present, but sometimes inconspicuous. Cuticle annulation and lateral field obliterated in mature female. Deirids absent. Phasmid porelike, typically near anus, rarely on tail, sometimes migrated far anteriorly (*Hoplolaimus*), sometimes enlarged to scutella and rarely absent. Tail typically short and less than two tail diameters long, rarely longer, and generally more curved dorsally, sometimes regularly rounded and rarely conoid. Eggs laid singly in soil, but in some genera they are laid in a gelatinous matrix, not retained in body in large numbers. Adult female typically ecto- or semiendoparasite of higher plants.

B. Subfamily Rotylenchulinae Husain and Khan, 1967
Acontylinae Fotedar and Handoo, 1978

Diagnosis

Hoplolaimidae. Small-sized (usually 0.5 mm or less in length). Body of mature female swollen or kidney-shaped. Cephalic region not as high as in Hoplolaiminae, continuous, with or without distinct annulation. Cephalic sclerotization, stylet and median esophageal bulb well developed in juveniles and females and regressed in males. Male stylet weaker than that of female. DGO far to very far from stylet. Esophageal glands enlarged into long lobe and overlapping intestine mostly ventrally or laterally. Ovaries in mature female didelphic, reflexed or coiled, posterior branch sometimes reduced to PUS. Tail of young females and males elongated to conoid and with a long hyaline terminal portion. In mature females tails persist. Tail in juvenile tapers to a round tip with two to three anal body widths in length with hyaline terminal portion smaller than in young female and male. Phasmid always porelike, near anus or on tail. Male sexual dimorphism well marked with anterior end smaller than female, sometimes esophagus degenerated and nonfunctional. Caudal alae enveloping tail and sometimes not reaching tail end. Members of Rotylenchulinae are sessile, semiendoparasites, and lay their eggs in a gelatinous matrix.

Key to the Females of the Genera in Rotylenchulinae

1. Esophageal glands overlapping intestine dorsally. Outstretched anterior genital branch. Posterior genital branch degenerated to PUS .... *Acontylus*
   Esophageal glands overlapping intestine laterally or ventrally. Two convoluted or reflexed genital branches 2
2. Two genital branches outstretched in immature female and convoluted in mature female. DGO 5–7 μm from stylet base .... *Senegalonema*

Two genital branches opposed with double flexure in immature female and convoluted in mature female. DGO 13–33 µm from stylet base . . . .
. . . . *Rotylenchulus*

Key to the Species of Rotylenchulus* Dasgupta et al. (1968) revised

1. Stylet 29 µm or more, broadly rounded tail in both the female and male . . . .
   . . . . *sacchari*
   Stylet less than 29 µm, tail of female and male (when present) not broadly rounded
   2
2. Stylet knobs anchor-shaped, stylet 22 µm or more . . . . *macrodoratus*
   Stylet knobs spheroid, stylet usually less than 22 µm 3
3. h = 14 or more; 15 or less in male, L of male 0.5 mm or more 4
   h = 13 or less; 15 or less in male; L of male when present 0.49 mm or less . . 5
4. O = <100 . . . . *clavicaudatus*
   O = >100 . . . . *macrosomus*
5. v = 66 or less 6
   v = 67 or more 9
6. h = >8; >9 in male . . . . *borealis*
   h = <7; <7 in males when present 7
7. c' = 3.3 or more, lip region conoid without visible annulation, males unknown *leptus*
   c' = 3.2 or less, lip region hemispheric or rounded, annulations present, males
   common or rare 8
8. h = <3; lip annulation very fine, not distinct; swollen female tail without spikelike
   process . . . . *parvus*
   h = >3; lip annulation distinct; swollen female tail with spikelike process . . . . . .
   . . . . *veriabilis*
9. Lip region high; stylet 15 µm or more; L = 0.34 mm or more; O = <106 . . . . . .
   . . . . *reniformis*
   Lip region low; stylet 14 µm or less; L = 0.33 mm or less; O usually >110 . . . . .
   . . . . *anamictus*

*Characters mentioned above refer to immature females unless indictated otherwise.

C. Genus *Rotylenchulus* Linford and Oliveira, 1940
Spyrotylenchus Lordello and Cesnik, 1958
Leiperotylenchus Das, 1960

Diagnosis

Rotylenchulinae. Juveniles, males, and young females vermiform, arcuate to spiral upon relaxation. Stylet in juveniles and female two to three times cephalic region width. Tail more rounded terminally and with shorter hyaline terminal end than that of female.
    Immature female. Vermiform, spiral to C-shaped. Labial region from low rounded to high flattened, continuous, annulated or not annulated. Anterior lip annulus divided into six similar-sized sectors. Lateral fields with four incisures, nonareolated. Labial framework and stylet strong. Stylet length ranges from 10 to 26 µm. The stylet knobs slope backward except in *R. macrodoratus* where the knobs are anchor shaped in immature female and juveniles. Orifice of dorsal esophageal gland about 0.56–1.9 times the stylet length behind the

stylet base (13–33 μm from stylet base). Median bulb oval with strong valve. Glandular overlap very long and mostly lateral. The excretory pore located posterior to median bulb. Hemizonid two and three annules long and immediately anterior to the excretory pore. Didelphic, amphidelphic ovaries with double flexures near the distal end. Tail elongate to conoid with prominent hyaline terminal portion.

Mature female. Body swollen, kidney-shaped, with an irregular, less swollen neck, a postmedian vulva, and a short, pointed tail. Cuticle thick and annulated and lateral fields obliterated. Tail conical, pointed, with or without hyaline terminal portion. Ovaries very long and convoluted. Eggs deposited in a gelatinous matrix.

Male. Vermiform. Stylet and esophagus regressed. Tail similar to that of young female. Caudal alae difficult to see, but subterminal, low, not quite reaching tail end. Phasmid in anterior region of tail. Reproductive system monorchic and outstretched. Spicule slender, lacking distal flanges. Gubernaculum fixed, lack titallae and telamon. Cloacal lip pointed, but not forming a tube. Hypoptygma absent.

Type species: *Rotylenchulus reniformis* Linford and Oliveira, 1940
        syn. *Tetylenchus nicotiana* Yokoo and Tanaka, 1954
           Rotylenchulus nicotiana (Yokoo and Tanaka, 1954) Baker, 1962
             Dasgupta, Raski, and Sher (1968) also proposed this combination
           Rotylenchus elisensis Carvalho, 1957
           Helicotylenchus elisensis (Carvalho, 1957) Carvalho, 1959
           Spirotylenchus queirozi Lordello and Cesnik, 1958
           Rotylenchulus queirozi (Lordello and Cesnik, 1958) Sher, 1961
           Leiperotylenchus leiperi Das, 1960
           Rotylenchulus leiperi (Das, 1960) Loof and Oostenbrink, 1962
           Rotylenchulus stakmani Husain and Khan, 1965

Other species: *Rotylenchulus parvus* (Williams, 1960) Sher, 1961
        syn. *Helicotylenchus parvus* Williams, 1960
           R. borealis Loof and Oostenbrink, 1962
           R. anamictus Dasgupta, Raski, and Sher, 1968
           R. clavicaudatus Dasgupta, Raski, and Sher, 1968
           R. leptus Dasgupta, Raski, and Sher, 1968
           R. macrodoratus Dasgupta, Raski, and Sher, 1968
           R. macrosoma Dasgupta, Raski, and Sher, 1968
           R. variabilis Dasgupta, Raski, and Sher, 1968
           R. sacchari Van der Berg and Spaull, 1981

The type species *R. reniformis* was found on cowpea roots *Vigna sinensis* Endl. on the island of Oahu, Hawaii. *Rotylenchulus reniformis* and *R. parvus* are the most widely distributed species and are of major global economic importance to agriculture.

1.   *Rotylenchulus reniformis* Linford and Oliveira, 1940

Measurements (after Dasgupta et al., 1968), Fig. 1 (after Siddiqi, 1986).
    (26 immature female topotypes): $L = 0.34$–0.42 mm; $a = 22$–27; $b = 3.6$–4.3; $b' = 2.4$–3.5; $c = 14$–17; $c' = 2.6$–3.4; $v = 68$–73; spear $= 16$–18 μm; $O = 81$–108.
    (Immature female neotype): $L = 0.4$ mm; $a = 24$; $b = 3.8$; $b' = 2.8$; $c = 16$; $c' = 2.9$; $v = 72$; spear $= 16$ μm; $O = 82$.

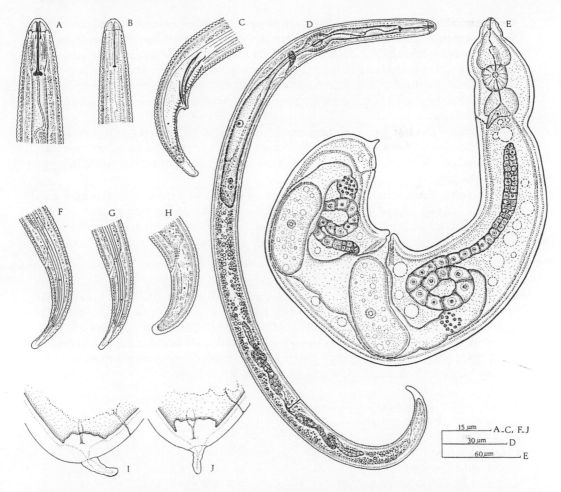

FIGURE 1 *Rotylenchulus reniformis* Linford and Oliveira. (A,F,G) Head end and tail ends of immature females, respectively. (B,C) Head end and tail end of male, respectively. (D) Entire immature female. (E) Entire mature female. (H) Tail end of juvenile. (I,J) Tail ends of mature female. [(A–D) and (F–J) after M. R. Siddiqi, 1986; (E) after D. R. Dasgupta, D. J. Raski, and S. A. Sher, 1968.]

(16 mature females): $L = 0.38–0.52$ mm; $a = 4–5$; $v = 68–73$; body width at vulva = 100–140 μm.

(10 males): $L = 0.38–0.43$ mm; $a = 24–29$; $b' = 2.8–4.8$; $c = 12–17$; $T = 35–45$; spear = 12–15 μm; spicules = 19–23 μm; gubernaculum = 7– 9 μm.

(10 juveniles): $L = 0.35–0.41$ mm; $a = 20–24$; $b' = 3.5–4.1$; $c = 12– 16$; spear = 13–15 μm.

Juveniles. Resemble immature female, but more rounded and stouter tail. Posterior part of esophagus asymmetrical. Overlapping of the esophageal glands laterally and ventrally. DGO about one length of stylet behind the stylet knob. Genital primordium four-celled at the same position as vulva.

Immature females. Vermiform, migratory, body slender assuming an open spiral or C shape. Lip region high, conoid, continuous with four to six (usually five) annules. Heavily sclerotized labial framework. Lateral field nonareolated and one-fifth to one-quarter body width. Deirids absent. Spear knobs round and slope posteriorly. DGO distinct and about one spear length behind spear base. Median bulb oval, with prominent valve. Esophageal glands overlap intestine laterally and mainly ventrally. Excretory pore near base of isthmus behind hemizonid. Vulva not prominent. Reproductive organs didelphic amphidelphic with double flexures near the distal end. Tail tapers to a narrow, rounded terminus with about 20–24 annules. Hyaline portion of the tail 4–8 μm long. Phasmids porelike, about a body width or less behind anus.

   Mature females. Body ventrally arcuate, swollen, kidney-shaped with an irregular less swollen neck. Vulva raised. Spermatheca rounded to irregular, usually with sperms. Body hemispheric beyond anus and with 5–9 μm-long terminal portion. Eggs laid in a gelatinous matrix.

   Males. Vermiform. Stylet and esophagus regressed with reduced median bulb and indistinct valvular apparatus. Labial sclerotization and stylet weaker than in immature female. Elongated to slender, ventrally arcuate spicules. Gubernaculum linear and nonprotruding. Caudal alae reduced and subterminal.

2.   Hosts of *Rotylenchulus reniformis*

Rotylenchulus reniformis was first found on cowpea roots *Vigna sinensis* Endl. on the island of Oahu, Hawaii. It has since been found on a large number of cultivated plants and fruit trees in many tropical and subtropical countries. It has been found attacking over 140 species of more than 115 plant genera belonging to 46 families. Of this large list of host range, 57 plant species belonging to over 40 genera and 28 families are considered crops of agricultural importance. Some of these important crops are as follows:

Anacardiaceae: *Manquipera indica* L.
Annonaceae: *Annona squamosa* L.
Caricaceae: *Carica papaya* L.
Chenopodiaceae: *Chenopodium album* L., *C. murale* L.
Compositae: *Lactuca sativa* L.
Convolvulaceae: *Ipomoea batatas* (L.) Poir, *Convolvulus arvensis* L.
Cruciferae: *Brassica juncea* (L.) Czernjak, *B. oleracea*
Cucurbitaceae: *Cucurbita maxima* Duchesne, *C. moschata* Duchesne, *C. pepo* L., *Cucumis melo* L.
Euphorbiaceae: *Ricinus communis* L.
Gramineae: *Hordeum vulgare* L., *Oryza sativa* L., *Saccharum officinarum* L., *Triticum aestivum* L., *Zea mays* L.
Labiatae: *Mentha* sp.
Lauraceae: *Persea americana* Mill., *P. borbonia* (L.) K. Spreng.
Leguminosae: *Glycine max* (L.) Merr., *Phaseolus vulgaris* L., *Vicia hirsuta* (L.) St. Gray, *Vigna aconitifolius* (Jacq.) Marechal, *V. sinensis* Endl.
Liliaceae: *Allium cepa* L.
Malvaceae: *Gossypium arboreum* L., *G. hirsutum* L.
Moraceae: *Ficus carica* L., *Morus alba* L.
Musaceae: *Musa paradisiaca* L.
Palmae: *Cocus mucifera* L.
Passifloraceae: *Passiflora edulis* Sims.
Punicaceae: *Punica gramatum* L.
Rosaceae: *Prunus ameniaca* L., *P. persica* (L.) Batsch., *Pyrus communis* L., *p. malus* L.
Rubiaceae: *Coffea arabica* L.

Rutaceae: *Citrus aurantifolia* (Christm.) Swingle, *C. decumana* L. [*C. maxima* (Burm.)], *C. limon* (L.) Burm., *C. medica* L., *C. paradisi* Macf.

Solanaceae: *Lycopersicon esculentum* Mill., *Nicotiana tabacum* L., *Solanum melongena* L., *S. tuberosum, S. tuberosum* ssp. *andigena* L.

Sterculiaceae: *Theobroma cacao* L.

Theaceae: *Thea sinesis* [*Camellia sinensis* (L.) Kuntze].

Umbelliferae: *Daucus carota* L.

Vitaceae: *Vitis vinifera* L., *V. rotundifolia* L.

3.   Distribution of *Rotylenchulus reniformis*

Rotylenchulus reniformis is widely distributed throughout tropical and subtropical countries, attacking a large number of cultivated plants and fruit trees. Ayala and Ramirez (1964) compiled a biography on the host range and distribution of this nematode. Dasgupta et al. (1968) reported the presence of this nematode in 17 countries. Siddiqi (1972) reported its presence in five additional countries. Additional distribution data have been reported by other authors (in Siddiqi, 1972). Its wide distribution and capability in attacking a rather large number of agriculturally important crops should place this nematode among the most injurious nematodes of plants and subject to quarantine regulations.

4.   Bionomics of *Rotylenchulus reniformis*

The adult female of *R. reniformis* is an obligate, sedentary, semiendoparasite of roots while the male is nonparasitic. It is bisexual and reproduces by amphimixis (Triantaphyllou and Hirschmann, 1964). However, it has also been reported to rarely reproduce parthenogenetically (Dasgupta and Seshardi, 1971). Apparently there are some behavioral differences of this nematode on various hosts. Dasgupta and Seshardi (1971) reported the occurrence of races A and B on three differential hosts in India. This nematode is capable of surviving in air-dried soil for an extended period of time (Birchfield and Martin, 1967). *Rotylenchulus reniformis* feeds on cortical tissue, phloem, and pericycle and its infection may cause the formation of necrosis on the roots of certain crops. Symptoms such as root discoloration, shedding of the leaves, and formation of malformed fruit or seeds are also associated with the infection by *R. reniformis*. The life cycle of *R. reniformis* was discussed initially by Linford and Oliveira (1940). The first molt occurs within eggs and the eggs are hatched in water without the influence of root exudates. Juveniles develop to preadult stage without feeding and growing through at least three superimposed molts. After infecting the roots of young females become orientated perpendicularly to the longitudinal axis of the root with the posterior portion of the body outside the root. The posterior portion of females becomes swollen and they start laying eggs in 7–10 days. Eggs are laid in a gelatinous matrix and their number may vary from 40–100 eggs per egg mass.

  R. reniformis, in addition to causing direct damage to plant roots, interacts with some important plant pathogens such as *Fusarium* and *Verticillium* species and *Rhizoctonia solani* in the development of disease complexes. It has also been reported to parasitize the bacterial nodules (Ayala, 1962) and interact with other plant parasitic nematodes such as *Meloidogyne* and *Pratylenchus* species.

  Although there are some crops resistant to *R. reniformis*, their number is rather limited for each crop species. Although rotation with the nonhost crops for two or more years reduces the nematode population, its extensive host range limits the number of rotation crops available. Nematicides and soil fumigants are effectively used in controlling *R.*

reniformis populations. Because of its extensive host range and importance as an injurious plant parasitic nematode, regulatory measures should be applied to limit its spread to uninfested areas.

D. *Rotylenchulus parvus* (Williams, 1960) Sher, 1961
   syn. *Helicotylenchus parvus* Williams, 1960

Measurements (after Dasgupta et al., 1968), Fig. 2.

Ten immature female paratypes (Mauritius): $L = 0.23–0.27$ mm; $a = 18–25$; $b = 3.5–3.8$; $b' = 2.4–3.1$; $c = 16–19$; $c' = 2.1–2.7$; $h = $ less than 3 μm; $v = 61–65$; stylet $= 12–13$ μm; $O = ?$.

Forty immature females (Imperial Valley, California): $L = 0.25–0.34$ mm; $a = 20–26$; $b = 3.1–3.7$; $b' = 2.1–3.0$; $c = 16–20$; $c' = 2.0–2.7$; $h = $ less than 3 μm; $v = 60–66$; stylet $= 12–14$ μm; $O = 90–107$.

Fifteen mature females (Greenhouse culture, Davis; originally from Imperial Valley, California): $L = 0.25–0.36$ mm; width at vulva $= 0.04–0.08$ mm; $a = 4.7$; $v = 61–66$; swollen portion of body plus tail $= 0.15–0.25$ mm; stylet $= 12–15$ μm; median esophageal bulb diameter $= 12–15$ μm; egg $= 58–69$ μm × 30–38 μm.

Two males (Imperial Valley, California): $L = 0.38–0.46$ mm; $a = 28–32$; $b' = 36–40$; $c = 18–23$; $h = $ less than 3; $T = 32–34$; stylet $= 10$ μm; gubernaculum $= 5$ μm; spicule $= 16$ μm.

One male (Greenhouse culture, Sacramento; originally from Imperial Valley, California): $L = 0.39$ mm; $a = 29$; $b' = 4.0$; $c = 14$; $h = 3$; $T = 41$; stylet $= ?$; gubernaculum $= 7$ μm; spicules $= 16$ μm.

Ten juveniles (Greenhouse culture, Davis; originally from Imperial Valley, California): $L = 0.25–0.34$ mm; $a = 21–25$; $b' = 2.5–3.5$; $c = 14–18$; stylet 12–13 μm.

Juveniles. Resembling immature females, except with more bluntly rounded tail.

Immature females. Relaxed specimens assume an open C to loose spiral shape. Lip region low and rounded with very fine indistinct annules. Cuticle annulated and annules about 0.8 μm wide. Stylet slender with knobs sloping backward. Excretory pore 60–75 μm from anterior end. Median bulb spheroid with valvular apparatus about 3 μm long. Esophageal glands overlap intestine laterally and ventrally, but more frequently laterally. Tail 13–18 μm or less than three body anal widths long, conoid, ventrally arcuate, and the terminus frequently with a short ventral projection. Hyaline terminal portion of tail less than 3 μm long. Lateral field with four incisures. The outer incisures extend almost to the tail terminus and the inner ones terminate near phasmid that is located about halfway between the anus and tail tip. DGO about halfway between the stylet base and the metacorpus. In the specimens from California and South Africa, DGO about one stylet length from the base of length from the base of stylet.

Mature females. Body swollen and twisted such that the posterior part often crosses the neck region. Anterior part of the body irregularly swollen and widest at vulva. Body tapers abruptly beyond vulva and with conoid postanal part shaped as in immature female but wider. Vulva with prominent raised lips. Vagina funnel-shaped, extending half across the body. Tail without spikelike process or projections. Anus often indistinct; amphids distinct. Annules distinct at neck and tail regions, varying from 1.0 to 1.5 μm in width. Metacorpus rounded and more conspicuous than in immature female.

Males. Rare; lip region hemispheric and higher than in female. Inconspicuous cephalic framework. Esophagus reduced. Tail rounded and roughly similar to that of imma-

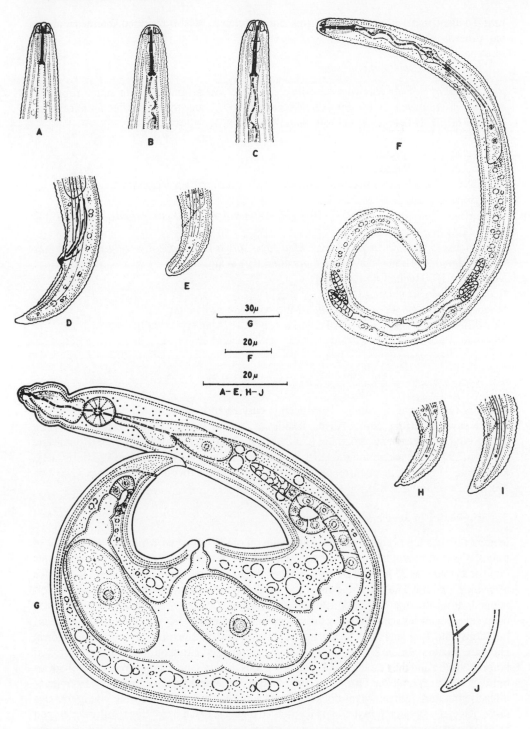

FIGURE 2 *Roylenchulus parvus.* (A) Male, anterior end. (B) Larva, anterior end. (C) Immature female, anterior end. (D) Male, tail. (E) Larva, tail. (F) Immature female. (G) Mature female. (H,I) Immature female, tail. (J) Mature female, tail. (After Dasgupta et al., 1968.)

ture female. Caudal alae reduced, ad-anal, and annulated. Well-developed spicules with linear gubernaculum.

1.   Hosts of *Rotylenchulus parvus*

Rotylenchulus parvus was first described from the soil around the sugarcane root *Saccharum officinarum* L. in Mauritius by Williams (1960). Since then it has been reported on a number of plants belonging to more than 10 families as follows:

Caricacae: *Carica papaya* L.
Chenopodiaceae: *Beta vulgaris* L.
Cruciferae: *Brassica oleracea* var. *boteytis* L., *B. oleracea* var. *capitata* L.
Cucurbitaceae: *Cucumis sativus* L.
Gramineae: *Cynadon dactylon* (L.) Pers., *Hordeum vulgare* L., *Pennisetum glaucum* (L.) R.
       Br., *Zea mays* L.
Leguminosae: *Crotalaria juncea* L., *Medicago sativa* L., *Phaseolus vulgaris* L., *Vigna*
       sinensis (Torner) Savi, *V. unguiculata* (L.) Walp.
Malvaceae: *Gossypium hirsutum* L.
Rosaceae: *Prumus persica* L.
Solanaceae: *Lycopersicon esculentum* Mill; *Nicotiana tabacum* L.
Umbelliferae: *Apium graveolens* L. var. *dulce* P.C.; *Daucus carota* L.
Vitaceae: *Vitis vinifera* L.

2.   Distribution of *Rotylenchulus parvus*

Rotylenchulus parvus is widely distributed in eastern and southern Africa, being reported from Kenya, Mauritius, South Africa, Rhodesia, Zambia, and Zimbabwe (Heyns, 1976). It also occurs in Queensland, Australia (Colbran, 1964) and Dominican Republic (Roman and Grullon, 1975). In the continental USA, *R. parvus* is known to occur in the Imperial Valley, California. However, its occurrence is also recorded in St. Croix and St. Thomas, U.S. Virgin Islands (Garcia, 1982).

3.   Bionomics of *Rotylenchulus parvus*

Mature females are obligate sedentary semiendoparasites of roots while the occurrence of males is very rare and the life cycle can be complete parthenogenetically. Mature females deposit as many as 37 individual eggs in a gelatinous matrix that usually hatch 16–22 days after they are laid. The life cycle from egg to egg completes in 27–36 days. The first molt occurs within the egg 5–8 days after the first cleavage. From 9 to 12 days after the first cleaveage, juveniles with well-developed cephalic framework, stylet knobs, transverse annulations, and four incisures in the lateral field could be observed in second-stage form. Hatching occurs within 2 days thereafter. Apparently second-stage juveniles do not feed before molting and they develop into young females through three superimposed molts. Cuticle of the fourth molt can be observed 9–14 days after hatching. Change of size from juveniles to immature female stages does not occur. Once the penetration by the young females takes place, the swollen females can be observed on the roots in 3 to 4 days. Fully developed females start depositing eggs in a gelatinous matrix after a further 5–7 days. Temperatures of 20–35°C support *R. parvus* reproduction. The highest population in the soil and the highest number of egg masses per root system are produced at 30°C.

   Rotylenchulus parvus infection causes severe root distortion and occasional necrosis on the roots of sugarcane. There is no information on the association of this nematode with other plan pathogens. Although there are no experimental data on the control of this nematode, chemical treatments effective against *R. reniformis* should also be effective against *R. parvus* (Siddiqi, 1972).

III. TAXONOMY OF *NACOBBUS*

A. Family Pratylenchidae Thorne, 1949
Nacobbidae Chitwood, 1950
Radopholldae Allen and Sher, 1967

Diagnosis (after Luc, 1987)

Tylenchoidea. Marked sexual dimorphism in respect to males and females. Mature female of *Nacobbinae* saccate, spindle-shaped or batatiform. Regression of esophagus and stylet in males. Short and slitlike amphidial aperture. Absence of deirids (except in *Pratylenchoides*). Porelike phasmids always on tail. Labial framework heavily sclerotized and lip region low, flattened or rounded anteriorly with no longitudinal striations. Lips either are not set off or are weakly set off from body, and are one-half to three-fifths as wide as lengths of spear. Spear strong with well-developed knobs, rounded or flat anteriorly (except for some species of *Hoplotylus*). Strong, rounded or ovate median esophageal bulb, well distinct from procorpus and with prominent valve. Esophageal glands lobelike and overlapping anterior end of intestine (some species of *Pratylenchoides* with esophageal glands abutting and almost no overlapping of intestine). Except for *Pratylenchoides* and *Apratylenchoides* esophageal-intestinal valve not well developed. Female genital tract with one or two branches. When two branches are present they are equally developed or nearly so. In some cases, posterior branch is reduced to a PUS of variable length, with or without degenerated cellular remnants of the posterior ovary. Uterus tricolumellar and vulva plain without protruding or overlapping vulval lip, flaps, or epiptygmata. Sperms globular or more rarely rodlike. Spicules plain. Gubernaculum plain and, except for *Radopholus*, not protruding. Terminal or, more rarely, subterminal caudal alae. Except for short tail of Nacobbinae, tail of both sexes of *Pratylenchinae* at least twice as long as anal body diameter. Members of the two subfamilies of *Pratylenchinae* and *Nacobbinae* are obligate plant parasites.

B. Subfamily *Nacobbinae* Chitwood, 1950

Diagnosis

Pratylenchidae. Marked sexual dimorphism in body shape. Juveniles, young females, and male vermiform. Mature female saccate, spindle-shaped or batatiform. Anterior region of young female and male similar. Absence of deirids and postanal position of phasmids. Cephalic region low, continuous, broadly rounded, annulated, and with strongly sclerotized framework. Indistinct round to oval labial disk with six pits on surface around oral opening. Rounded dorsal and ventral lip sectors are clearly separated from each other and distinct from the oral disk. Lateral lip sectors are inconspicuous. Well-developed stylet with round knobs about twice the maximum width of cephalic region. Dorsal esophageal gland orifice close to stylet base. Strongly muscular metacorpus. Elongated esophageal glands dorsally overlap intestine. Asymmetrical subventral glands extend beyond the dorsal gland. Vulva near anus. Females monodelphic, prodelphic with absent postvulval uterine sac. Tubular uterus and long serpentine ovary in early mature females. Young females with short, rounded tail (one to two anal body width long). Juveniles with small, indistinct hyaline sub-cylindrical tail terminal. Male tail conical, arcuate, and completely enveloped by a caudal alae. Well-developed, cephalated, and arcuated spicules. Small, simple, and fixed gubernaculum. Eggs are retained in body and/or deposited in a gelatinous material. Juveniles, young females, and males are migratory endoparasites of roots. Mature females are obligate, sedentary endoparasites inciting root galls.

Type and only genus: *Nacobbus* Thorne and Allen, 1944

C. *Nacobbus* Thorne and Allen, 1944

Diagnosis

Nacobbinae. Strong sexual dimorphism.

Juveniles. Vermiform, migratory, similar to immature females.

Immature females. Vermiform, elongated to slender, about 1 mm long. Distinctly annulated cuticle. Lateral field with four incisures, irregularly areolated. Deirids not present. Phasmids in anterior region of tail. Lips are rounded, not set off (except in mature females), with three to four annules in the cephalic region. Rounded, conspicuous labial disk. Submedial sectors rounded and separated. Lateral lip sectors are reduced or absent. Strong stylet with rounded knobs about 21–25 µm long. Esophageal glands dorsally overlapping intestine for more than two body widths. Asymmetrical subventral glands extend past dorsal gland; nuclei of three glands lie in tandem behind the junction of esophagus and intestine. Esophageal-intestinal valve undeveloped. Transverse vulval slit located within two anal body widths of anus. Ovary immature and female genital tractus with very well-developed anterior branch and completely regressed posterior branch with no trace of post-uterine sac. Tail tapering to a broadly rounded terminus, one to two anal body widths long. Young females and juveniles migratory in soil and roots.

Mature females. Body saccate, spindle-shaped, or batatiform tapering anteriorly from median bulb and posteriorly from uterine region. In early stages mature females are usually batatiform with subterminal vulva and anus; very long vulva and tubular uterus together with long serpentine ovary reaching esophagus. Mature females sedentary endoparasites inciting root galls.

Males. Vermiform, similar to immature female except for sexual dimorphism, with well-developed cephalic sclerotization. Stylet about 23–27 µm long. Esophagus structurally similar to that of immature female. Elongated esophageal glands overlapping intestine mostly dorsally. Single outstretched testis. Cephalated, ventrally arcuate, 20–35 µm long spicules. Simple, nonprotruding gubernaculum linear to trough-shaped, fixed cloacal lips. Tail completely enveloped by a caudal alae. Although male are present in the two species, no evidence of fertilization has been demonstrated.

Type species: *Nacobbus dorsalis* Thorne and Allen, 1944

Other species: *Nacobbus aberrans* (Thorne 1935) Thorne and Allen, 1944

        syn. *Anguillulina aberrans* Thorne, 1935

            Pratylenchus aberrans (Thorne) Filipjev, 1936

            N. batatiformis Thorne and Schuster, 1956

            N. serendipiticus Franklin, 1959

            N. serendipiticus bolivianus Lordello, Zamith, and Boock, 1961

Key to Species of *Nacobbus*

1. Number of annules between vulva and the anus in immature female 8–14. Mature females almost round with elongated posterior region and the entire body filled with eggs .... *N. dorsalis*
2. Number of annules between vulva and the anus in immature female 15–24. Mature females spindle-shaped. Only posterior portion of the body contains eggs and rest deposited in a gelatinous material outside the body .... *N. aberrans*

1. *Nacobbus dorsalis* Thorne and Allen 1944

Measurements (after Sher, 1970), Fig. 3.

Six mature female topotypes: L = 1.4 mm (1.3–1.6); stylet = 22 µm (20–24).

Ten immature female topotypes: L = 0.78 mm (0.59–1.06); a = 30 (26–34); b = 6.1 (5.1–7.3); b' = 2.8 (2.0–3.9); c = 39 (30–52); c' = 1.1 (0.8–1.4); v = 95 (94–97); VA = 10 (8–14); stylet = 22 µm (19– 24).

Ten male topotypes: L = 0.89 mm (0.72–1.16); a = 30 (25–41); b = 6.6 (5.2–8.7); b' = 3.6 (2.7–4.7); c = 34 (25–38); c' = 1.6 (1.4– 1.8); stylet = 23.5 µm (20–27); gubernaculum = 9 µm (8–11); spicules = 30 µm (27–35).

Juveniles. The first-stage juvenile can only be observed in egg. At this stage the juvenile tail is about 0.45 mm long and rounded (Fig. 3D). The cuticle is strongly striated and lateral field is marked by four bright lines. The spear is well developed and is about 25 µm long. The second-stage juvenile is most commonly found in egg and develops to near 0.5 mm in length. Tail tip is blunt and round, similar to that of immature female (Fig. 3E). Labial framework, stylet, esophagus, and body annulation are well developed. Esophageal glands are well developed and extend far back along the intestine.

Immature females. Body vermiform and migratory in soil and roots. Lip region with three to four annules, hemispheric and not set off. Round stylet knobs and well-developed median bulb with large conspicuous valve. Elongated esophageal glands dorsally overlap the intestine by more than three time body width. Lateral field with four incisures and incompletely areolated. Rounded tail with 10–18 annules. Number of annules between vulva and the anus 8–14. Phasmid below the level of anus and in anterior portion of tail. Distal annulations are generally wider than rest.

Mature females. Body saccate, oval to spherical shape, short anterior projection contains lip region to median bulb and the elongated posterior portion contains tail, vulva, and portion of uterus with usually one to four eggs in tandem. Spindle-shaped female with single, convoluted ovary becomes wider and rounder with more eggs to almost spherical as it is completely filled with eggs. The body of female may contain up to 1500 eggs.

Males. Except for sexual dimorphism, they are similar to immature females. Male with single, outstretched testis. Large and strongly arcuate spicules rest upon a thin and troughlike gubernaculum. Caudal alae inconspicuous, and enveloping tail. Phasmid conspicuous with prominent connections leading to the lateral lines, but with only minutes fiber extending to the margin of the caudal alae.

Hosts of Nacobbus dorsalis

Nacobbus dorsalis was originally collected from the type host, *Erodium cicutarium* (L.) L'Hér. It is also known to attack the roots of *Beta vulgaris* L., and has been found in the roots of a *Salvia* sp. and in soils around the roots of *Hordeum vulgare* L., and a *Prunus* sp. However, its association with the roots of these two plants has not been demonstrated.

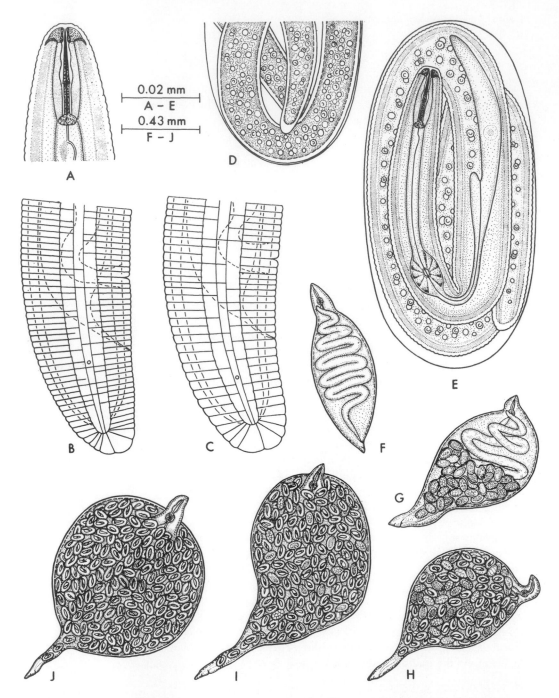

FIGURE 3 *Nacobbus dorsalis*. (A) Immature female, anterior region. (B,C) Immature female, posterior region. (D) Portion of egg with first-stage juvenile. (E) Egg with second-stage juvenile. (F,G) Early stages of mature female. (H–J) Later stages of mature female. (After Sher, 1970.)

Distribution of Nacobbus dorsalis

Nacobbus dorsalis is a native of North America and is only found in the state of California. Its occurrence in other localities is not known.

Bionomics of Nacobbus dorsalis

Although *N. dorsalis* has been found on a limited number of plants, it appears that it is not of much economic importance. However, additional surveys may reveal the importance of this species to agriculture. The similarity of the symptoms caused by *N. dorsalis* to those caused by *N. aberrans* and *Meloidogyne* species and the lack of knowledge of its host range and geographic distribution makes this nematode and interesting specie to observe.

Nacobbus abberans Thorne and Allen, 1944
  syn. *Anguillulina aberrans* (Thorne, 1935
    Pratylenchus aberrans (Thorne) Filipjev, 1936
    N. batatiformis Thorne and Schuster, 1956
    N. serendipiticus Franklin, 1959
    N. serendipticus bolivianus Lordello, Zamith, and Boock, 1961

Sher (1970) in a revision of the genus *Nacobbus* Thorne and Allen, 1944, found no morphological differences between type material of *N. aberrans, N. batatiformis, N. serendipiticus,* and *N. serendipiticus bolivianus.* Jatala and Golden (1977), on the basis of morphometric differences of 26 South American populations, considered *N. aberrans* as a species complex involving two or more forms rather than a single homogeneous species. They suggested that additional morphological and parallel host range studies are needed to clarify classification of this group. Thus, synonymization of *N. serendipiticus, N. serendipiticus bolivianus,* and *N. batatiformis* to *N. aberrans* by Sher (1970) requires consideration. For simplicity, morphological data presented here are after Sher (1970). However, *N. aberrans* is considered as species complex.

Measurements (after Sher, 1970), Fig. 4.

Six mature female topotypes: $L = 1.0$ mm (0.8–1.2); stylet = 22 μm (20–24).

Twelve immature female topotypes: $L = 0.84$ mm (0.71–0.93); $a = 27$ (23–40); $b = ?$; $b' = 3.8$ (2.8–4.1); $c = 37$ (24–40); $c' = 1.2$ (0.9–1.5); $v = 93$ (92–94); $VA = 20$ (15–24); stylet = 23 μm (21–25).

Seven male topotypes: $L = 0.86$ mm (0.71–0.92); $a = 29$ (24–31); $b = 7.0$ (6.4–7.2); $b' = 3.6$ (3.4–4.0); $c = 38$ (32–42); $c' = 1.3$ (1.2–1.4); stylet = 25 μm (23–27); gubernaculum = 7 μm (6–8); spicule = 27 μm (21–30).

Nebraska six mature females: $L = 1.1$ mm (1.0–1.4); stylet = 22 μm (20–24).

Six immature females: $L = 0.78$ mm (0.74–0.84); $a = 30$ (27–32); $b = 7.6$ (7.0–9.3); $b' = 4.2$ (4.1–4.5); $c = 34$ (30–38); $c' = 1.5$ (1.3–2.0); $v = 93$ (92–94); $VA = 22$ (18–25); stylet = 21 μm (20–23).

Five males: $L = 0.80$ mm (0.68–0.88); $a = 33$ (28–39); $b = 6.7$ (6.5–8.3); $b' = 4.3$ (4.2–4.4); $c = 35$ (33–36); $c' = 1.5$ (1.3–1.7); stylet = 24 μm (23–27); gubernaculum = 9 μm (7–11); spicules = 28 μm (24–34).

England seven mature females: $L = 1.08$ mm (0.91–1.45); stylet = 22 μm (20–23).

Ten immature females: $L = 0.71$ mm (0.60–0.78); $a = 28$ (23–31); $b = 5.4$ (5.1–5.8); $b' = 3.3$ (2.8–3.9); $c = 29$ (23–32); $c' = 1.5$ (1.3–1.6); $v = 92$ (91–93); $VA = 20$ (16–24); stylet = 20 μm (19–21).

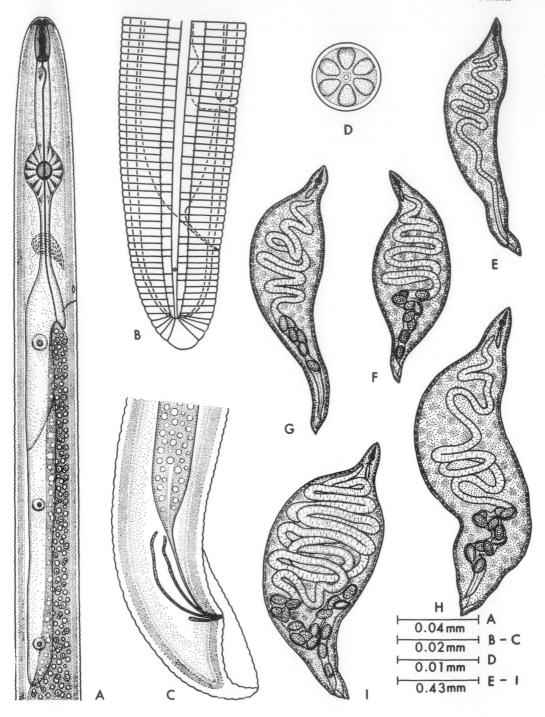

FIGURE 4 *Nacobbus aberrans*. (A) Immature female, anterior region. (B) Immature female, posterior region. (C) Male, posterior region. (D) Male, face view. (E) Early stage of mature female. (F–I) Later stages of mature female. (After Sher, 1970.)

Six males: *L* = 0.78 mm (0.67–0.88); *a* = 30 (24–32); *b* = 6.1 (5.6– 6.6); *b′* = 4.5 (3.5–5.4); *c* = 34 (30–38); *c′* = 1.2 (1.0–1.5); stylet = 25 μm (23–27); gubernaculum = 8 μm (7–10); spicules = 27 μm (26– 32).

Juveniles. Similar to *N. dorsalis*. Body vermiform. The cuticle is annulated and the lateral field is marked by four bright incisures occupying about one-third of the body width. Head skeleton hexaradiate and heavily sclerotized. Head offset with three annules. Stylet robust with round basal knobs. Median bulb well developed. Hemizonid located ventrally, posterior to median bulb. Phasmid small and located at the middle of tail. Tail tip blunt and rounded, similar to that of immature female.

Immature females. Vermiform, lateral field with four incisures incompletely areolated. Head not offset with three to four annules. Esophageal gland both overlap intestine dorsally. Hemizonid located adjacent to excretory pore. Number of annules between vulva and the anus 15–24 and slightly higher position of the vulva in *N. aberrans*. Tail rounded with 10–17 annules; irregular and wider distal annules. Phasmid below level of the anus in anterior portion of tail. Young females migratory in soil and roots.

Mature females. Body white-cream enlarged, spindle-shaped, usually tapering anteriorly and posteriorly. Head skeleton well developed with three to four annules. Stylet slender with small, rounded basal knobs. Matures females sedentary in roots. Vulva and anus posteriorly subterminal. Unsegmented eggs are seen only in posterior portion of body. Most eggs are deposited in a gelatinous material outside the body. The gelatinous matrix is often located outside and in the lower axis of the gall. Mature females closely resemble the early maturing females of *N. dorsalis*. Tail terminus of the mature *N. aberrans* female is often nipple-shaped.

Males. Except for sexual dimorphism, males closely resemble immature females. Head skeleton hexaradiate with heavily sclerotized cephalic framework. Stylet with strong basal knobs projecting anteriorly. As in *N. dorsalis* males with single outstretched testis; large, ventrally arcuated, and slightly cephalated spicules; simple troughlike gubernaculum about one-third to one-fourth as long as the spicules. Tail short, arcuated, enveloped by caudal alae. Phasmid prominent and slightly anterior to the middle of tail.

Hosts of Nacobbus aberrans

Nacobbus aberrans was originally described from a native plant, *Atriplex confertifolia* (Torr. and Frem.) S. Wats. collected in the desert foothills of western Utah, at Lake Utah, on May 25, 1927. Since then it has been found that *N. aberrans* has a rather wide host range attacking a wide variety of crops. The host range includes:

Amaranthaceae: *Amaranthus* sp.
Basellaceae: *Ullucus tuberosus* Loz.
Cactaceae: *Mamillaria vivipara* (Nutt.) Haw., *Opuntia fragilis* (Nutt.) Haw., *O. tortispina* Nutt.
Caryophyllaceae: *Spergula arvensis* L., *Stellaria media* (L.) Cyz.
Chenopodiacea: *Atriplex confertifolia* (Torr. and Frem.) S. Wats.,*Beta vulgaris* L. (sugar beets, red beets, and turnips), *B. vulgaris cicla* L., *B. vulgaris macrorhiza* L., *Chenopodium album, C. quinoa* Willd., *Kochia scoparia* (L.) Shrad., *Solsola kali* var. *tenuifolia* Tausch., *spinacia oleracea* L.
Compositae: *Gallardia pulchella* L., *Lactuca sativa* L., *Tragopogon porrifolius* L., *Tagete mandonii*.
Cruciferae: *Brassica campestris* L., *B. napbrassica* Mill., *B. nigra* (L.) Koch., *B. oleracea viridis* L., *B. oleracea botrytis* L., *B. oleracea gongylodes* L., *B. oleracea gem-

mifera L., *B. oleracea* var. *capitata, B. pekinensis* (Lour.) Rupr., *B. rapa* L., *Matthiola* sp., *Raphanus sativus* L.

Cucurbitaceae: *Cucurbita pepo* L., *Cucumis sativus* L.

Euphorbiaceae: *Calandrinia alba.*

Leguminosae: *Pisum sativum* L., *Phaseolus vulgaris* L.

Solanaceae: *Capsicum annuum* L., *C. baccatum* L., *Lycopersicon esculentum* Mill., *L. peruvianum* (L.) Mill., *Physalis* sp. *Solanum melongena* L., *S. tuberosum* L., *S. tuberosum* ssp. *andigena* L., *S. nigrum* L.

Tropaeolaceae: *Tropaeolum tuberosum* Ruiz. et Pav.

Umbelliferae: *Daucus carota* L.

Zygophyllaceae: *Tribulus terrestis* L.

Distribution of Nacobbus aberrans

Nacobbus aberrans is native of western parts of North America and South America. In the United States it has been found in California, Colorado, Nebraska, Utah, and Wyoming. It also occurs in Mexico. Its distribution in South America is restricted to the central part of Ecuador; north, central, and southern highlands of Peru; all the altiplano potato-growing areas of Bolivia; northern highlands of Chile; and northern and central parts of Argentina. Because of its great morphological and behavioral diversity on different crops that occur around Lake Titicaca, both on Peruvian and Bolivian sides, it is postulated that its center of origin may have been in this area. *N. aberrans* has been reported in the Netherlands (Bruijn and Stemerding, 1968) and England (Franklin, 1959) under greenhouse conditions. Its presence in USSR was reported by Kiryanova and Lobanova (1975). Although the presence of this nematode has been reported in India (Prasad et al., 1975), its presence could not be confirmed.

Bionomics of Nacobbus aberrans

As this species attack a large number of important agricultural crops causing galls on the roots similar to those produced by the genus *Meloidogyne,* it is possible that it is more widely spread than its present distribution data indicate. Juveniles and young females are infective stages of this nematode. Apparently the information on its biology is somewhat confusing. It is apparent, however, that there seems to be some behavioral differences among the various populations, indicating the possibility of the existence of races. Juveniles attack roots and generally do not cause galls. After penetration takes place, the presence of necrosis in the cortical tissue becomes evident. Nematodes molt to J3 and J4 which may or may not go through a quiescent period. This is apparently dependent on the host and the temperature. Once the immature females are formed they may come out of the root tissue, attacking adjacent roots, or migrate within the root tissue and become established and cause root galling. Although the process of internal migration may occur, it appears to be occasional. Young immature females penetrate roots causing initial necrosis followed by gall formation. The life cycle from juvenile to juvenile may take from 35 to 90 days and is dependent on the temperature and the host. Although there is a great similarity in the galls formed by *N. aberrans* and those caused by *Meloidogyne* species, there is a basic difference in that the galls caused by *N. aberrans* are always formed laterally on the root. In the larger sized galls, this difference can be detected by following the root axis through the gall and noting the lateral formation of the gall. Apparently, *N. aberrans* can withstand adverse soil conditions including extremely low soil humidity (1%). This together with its rather wide host range and temperature adaptability makes this nematode an important pest and subject to strict regulatory conditions.

Juveniles and immature females may enter potatoes without causing visible symptoms. Thus cautions should be taken to prevent the spread of potato tubers from the infested area. Similarly, care should be taken to avoid the removal of soil and farm machinery from an infested area.

Information on the availability of resistant cultivar is lacking. The best method of control is the use of nematicides and soil fumigants as well as long-term (6 or more years) rotation with nonhost crops. Chemical dip treatment of the potato tubers seems to control the nematodes rather effectively.

REFERENCES

Ayala, A. 1962. Parasitism of bacterial nodules by reniform nematodes. *J. Agr. Univ. Puerto Rico* 46: 67–69.

Ayala, A., and Ramirez, C. T. 1964. Host-range, distribution, and bibliography of the reniform nematode, *Rotylenchulus reniformis,* with special reference to Puerto Rico. *J. Agr. Univ. Puerto Rico* 48: 140–161.

Birchfield, W., and Martin, W. J. 1967. Reniform nematode survival in air-dried soil. *Phytopathology* 57: 804.

Bruijn, N. De, and Stemerding, S. 1968. *Nacobbus serendipiticus,* a plant parasitic nematode new to the Netherlands. *Neth. J. Plant Pathol.* 74: 227–228.

Colbran, R. C. 1964. Studies on plant and soil nematodes. I. Queenland records of the order *Tylenchida* and the genera *Trichodorus* and *Xiphinema. Queensland J. Agr. Sci.* 21: 77–123.

Dasgupta, D. R., and Raski, D.J. 1960. The biology of *Rotylenchulus parvus. Nematologica* 14: 429–440.

Dasgupta, D. R., Raski, D. J., and Sher, S. A. 1968. A revision of the genus *Rotylenchulus* Linford and Oliveira, 1940. (Nematoda: Tylenchidae). *Proc. Helm. Soc. Wash.* 35: 169–192.

Dasgupta, D. R., and Seshardi, A. R. 1971. Races of the reniform nematode *Rotylenchulus reniformis* Linford and Oliveira, 1940. *Ind. J. Nematol.* 1: 21–24.

Franklin, M. T. 1959. *Nacobbus serendipiticus* n. sp., a root-galling nematode from tomatoes in England. *Nematologica* 4: 286– 293.

Garcia, M. R. 1982. Plant parasitic nematodes of the U.S. Virgin Islands. *J. Nematol.* 14: 441.

Heyns, J. 1976. *Rotylenchulus parvus. J. Commonwealth Inst. Helminthol.* 6: 1–3.

Jatala, P. 1978. Review of the false root-knot nematode (*Nacobbus* spp.) research progress in developments in the control of nematode pests of potato. Report of the 2nd nematode planning conference. International Potato Center, Lima, Peru, pp. 66–69.

Jatala, P., and Golden, A. M. 1977. Taxonomic status of *Nacabbus* species attacking potatoes in South America. *Nematropica* 7: 9–10.

Kiryanova, E.S., and Lobanova, N.A. 1975. A potato parasitic nematode. *ZSRSB/ZASHCH RAST* (*MOSC*) 9: 49.

Linford, M. B., and Oliveira, J.M. 1940. *Rotylenchulus reniformis,* nov. gen., n. sp., an nematode parasite of roots. *Proc. Helm. Soc. Wash.* 7: 35–42.

Luc, M. 1987. A reappraisal of Tylenchina (Nemata). 7. The family *Pratylenchidae* Thorne, 1949. *Rev. Nematol.* 10: 203–218.

Prasad, S. K., Khan, K.E., and Chawla, M.L. 1965. New records of nine nematode genera from the Indian Union. *Ind. J. Entomol.* 27: 360–361.

Roman, J., and Grullon, L. 1975. Nematodes associated with sugarcane in the Dominican Republic. *J. Agr. Univ. Puerto Rico* 59: 138–140.

Sher, S. A. 1970. Revision of the genus *Nacobbus* Thorne and Allen, 1944 (Nematoda: Tylenchoidea). *J. Nematol.* 2: 228–235.

Siddiqi, M. R. 1972. *Rotylenchulus reniformis.* Commonwealth Institute of Helminthology. Plant-Parasitic Nematodes. Set 1, No. 5.

Siddiqi, M. R. 1986. *Tylenchida* Parasites of Plants and Insects. *Slough,* U.K. Commonwealth Institute of Parasitology.

Triantaphyllou, A. C., and Hirschmann, H. 1964. Reproduction in plant and soil nematodes. *Ann. Rev. Phytopathol.* 2: 57–80.

Williams, J. R. 1960. Studies on the nematode soil fauna of sugarcane fields in Mauritius. Tylenchoidea (Partim) *Maruitius Sugar Ind. Res. Inst.* 4: 1–30.

12

Stunt Nematodes: *Tylenchorhynchus, Merlinius*, and Related Genera

R. V. Anderson *Biosystematics Research Centre, Agriculture Canada, Ottawa, Ontario, Canada*

John W. Potter *Research Station, Agriculture Canada, Vineland Station, Ontario, Canada*

INTRODUCTION

The stunt nematodes are a broadly distributed group of genera and species having global existence in agricultural soils and uncultivated land. Of the 267 now valid species, few have been proven as pathogens in the strict sense, but about 8% are known ectoparasites. Many are part of a complex of plant parasitic nematodes in soils around crop plants; generally they can be thought of as a stress factor to their hosts. Most species of stunt nematodes are known only from the type locality, are distinguishable by a few minor characters only, and with few exceptions, are poorly known as to their limits of variation and biology.

The classification of stunt nematodes presently is unstable and one of the most controversial problems in plant nematode taxonomy. Undoubtedly many species will prove to be conspecific with later described species. Consequently, those distributional and host records which cannot be verified by restudy of voucher specimens must be generally regarded as unreliable.

HISTORICAL REVIEW OF STUNT NEMATODE TAXONOMY

The genus *Tylenchorhynchus* was established by Cobb (1913) to accommodate a single species, *T. cylindricus* Cobb, 1913. It remained monotypic until 1935 when Thorne, following the classification of Goodey (1932), synonymized it with the genus *Anguillulina* Gervais and van Beneden, 1859. At this time, Thorne also synonymized *T. cylindricus* with *A. dubia* (= *Tylenchus dubius* Bütschli, 1873), which remained its synonym until 1955, when revalidated by Allen. In 1936, Filipjev reinstated the genus *Tylenchorhynchus*.

Filipjev and Schuurmans-Stekhoven (1941) in a major treatise on plant nematodes, *A Manual of Agricultural Helminthology*, accepted 11 species in the genus *Tylenchorhynchus*. These included *T. dubius* (Bütschli, 1873), *T. macrurus* (Goodey, 1932), *T. magnicauda* (Thorne, 1935), and *T. claytoni* (Steiner, 1937), none of which was considered to be of agricultural importance. By 1954, however, *T. claytoni* was recognized as a serious pest of tobacco in South Carolina (Graham, 1954), and, based on host symptoms, gained the common name "tobacco stunt nematode." As more species became associated with similar symptoms of other crops, the common name "stunt nematode" became broadly applied to all members of the genus *Tylenchorhynchus*.

By 1970, the number of stunt nematode species was 96, which various taxonomists subsequently divided into more manageable groupings. By 1986, two subfamilies and nine genera had been established containing former members of *Tylenchorhynchus*. While these groupings facilitated the order and identification of these morphologically diverse species, their genotypic relationships became less evident. To arrange this inflated assembly of nematode genera and subgenera into a more natural classification based on phylogenetic relationships, Fortuner and Luc (1987) more broadly defined the taxa. By this action, many genera and subfamilies were reduced to synonyms. For this review of stunt nematodes of agricultural importance, we follow the conservatively objective classification scheme of Fortuner and Luc (1987), who recognize a single subfamily and eight genera. Stunt nematodes then, by historical inference, are members of the genera *Tylenchorhynchus* Cobb,1913; *Trichotylenchus* Whitehead,1960; *Nagelus* Thorne and Malek,1968; *Paratrophurus* Arias,1970; *Merlinius* Siddiqi, 1976, which are included, together with *Trophurus* Loof,1956, in the subfamily Telotylenchinae Siddiqi,1960 as defined by Fortuner and Luc.

Species of agricultural importance considered in this review are members of the genera *Tylenchorhynchus*, *Merlinius,* and *Amplimerlinius*. Some species of little agricultural importance, such as *T. cylindricus*, are included to more completely illustrate certain diagnostic features. Illustrations are based solely on museum specimens obtained from a number of sources, as cited in the figure legends. When possible, new descriptive data and measurements with standard deviations are offered to broaden the base for identification of the species. To facilitate recognition of the species groups within the genera we refer the reader to Siddiqi (1986).

Scanning electron microscopy (Powers, 1983; Powers, Baldwin, and Bell,1983; Siddiqi,1986; Fortuner and Luc,1987) has produced a major advance in the taxonomy of stunt nematodes by permitting species to be arranged in well-defined groups, based on external cuticular features, particularly head morphology. These features include: presence and number of longitudinal grooves, or their absence; distinctiveness and shape of the perioral disc or its degree of fusion with the first head annule; presence, number, and arrangement of submedian and lateral lobes, or their absence; and placement of the amphidial apertures. This information, however, is derived from a paucity of specimens. It is evident that some species, like *T. dubius* (Fig. 14E,F,G), exhibit a range of variation in form. Therefore, we consider it premature for the purpose of this review to place too much reliance on these features at the species level.

DESCRIPTIONS OF SPECIES OF AGRICULTURAL IMPORTANCE

Tylenchorhynchus claytoni Steiner, 1937
Syn. *Tessellus claytoni* (Steiner, 1937) Jairajpuri and Hunt, 1984

MEASUREMENTS (after Golden, Maqbool, and Handoo, 1987)

Female

(lectotype; after Loof, 1974): L=650 µm; a=25; b=6; c=16.8; c'=2.15; V=55.5. Stylet length=19.5 µm. Head annules=3. Tail annules=15. Egg 66 × 21 µm.

(n=4; paralectotypes): L=525–664 (604 ± 58.1)µm; a=22–28 (24.3 ± 2.9); b=5.3–5.5 (5.4 ± 0.1); c=15.2–17.0 (16.5 ± 1.0);V=57–58 (57 ± 0.4). Stylet length=18.4–18.9 (18.5 ± 0.2)µm.

(n=15; topotypes): L=589–753 (667 ± 47,8)µm;a=22–28 (25.4 ± 1.1); b=4.3–5.7 (4.8 ± 0.4); c=15–20 (16.8 ± 1.1). V=54–60 (58.6 ± 1.3). Stylet length=18.4–21.0 (19.3 ± 0.5)µm.

(n=20; Boskoop, Netherlands, after Loof, 1974): L=520–620 µm; a=23–28; b=4.5–5.6; c=17–20; c'=2.0–2.7; V=54–59. Stylet length=17–19 µm Tail annules=11–16.

(n=41; Muscatine County, Iowa): L=516–734 (610 ± 57.5)µm; a=21–27 (23.6 ± 1.7); b=4.0–5.7 (4.7 ± 0.4); c=16.0–22.3 (18 ± 1.4); V=53–60 (57 ± 1.3). Stylet length=18.1–20.2 (19.1 ± 0.4)µm.

Male

(n=4; paralectotypes): L=518–654 (586 ± 55.6)µm; a=22–28 (24.3 ± 2.2); c=16–19 (16.8 ± 1.5). Stylet length=18.0–18.5 (18.2 ± 0.2)µm. Spicule length=22–23 (21.9 ± 0.4)µm. Gubernaculum length=13–14 (13.1–0.4)µm.

(n=8; topotypes): L=576–662 (619 ± 31.3)µm; a=28–33 (29.8 ± 1.6); b=4.6–5.5 (5.0 ± 0.3); c=15–17 (15.9 ± 0.5). Stylet length=18.0–19.3 (18.7 ± 0.6)µm. Spicule length=22–24 (23 ± 0.7)µm. Gubernaculum length=13–14(13.6 ± 0.4)µm.

(n=15; Boskoop, Netherlands; after Loof, 1974); L=470– 590 µm; a=26–34; b=4.7–6.2; c=13–17; c'=2.3–2.8; T=58–66. Stylet length=17–19 µm. Spicule length=19–24 µm. Gubernaculum length=11–13 µm.

(n=21; Muscatine County, Iowa): L=511–654 (580 ± 37.8)µm; a=22.2–27.5 (25.1 ± 1.4); b=4.2–5.4 (4.6 ± 0.2); c=12.2–18.5 (16.3 ± 1.6). Stylet length = 17.6–19.7 (19.1 ± 0.5) µm. Spicule length=22.0–25.8 (24.1 ± 0.9) µm. Gubernaculum length = 12.9 –16.0 (14.2 ± 0.8)µm.

DESCRIPTION

Female (after Golden et al. 1987; Loof, 1974): Body linear to slightly arcuate when relaxed, width near midbody (n=44) 21.5–30.9 (25.8 ± 1.9)µm; annule width 1.7–2.5 (2.1 ± 0.2)µm, divided into distinct blocks by 23–29 longitudinal striae. Lateral field 6.4–8.6(7.2 ± 0.8)µm wide, without areolation, outer of four incisures crenate. Deirids absent.

Head 7.0–9.8 (8.6)µm wide, set off from body by slight constriction (Fig. 1B,C), nearly continuous, marked by three or four coarse annules, rarely five. First neck annule posterior to head often reduced (Fig. 1B). Stylet knobs rounded, anterior surfaces sometimes flattened, dorsal esophageal gland orifice about 2.1–2.5µm posterior. Esophagus from head end to center of metacorpus 58–75 (63 ± 3.8)µm long, metacorpus 13 × 9–10 µm,

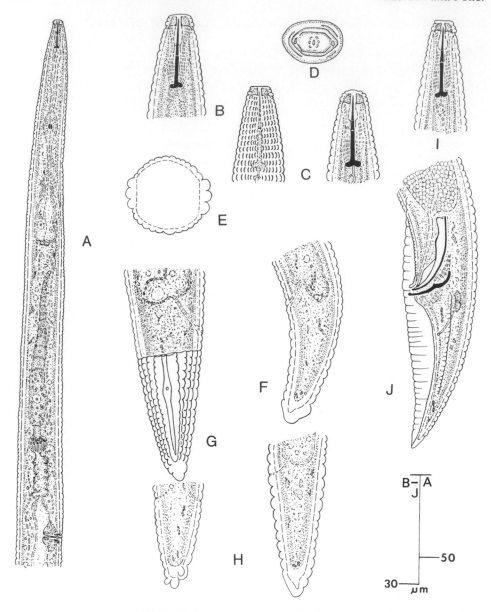

FIGURE 1 *Tylenchorhynchus claytoni*. Female. (A) Anterior half showing characteristic features of the esophagus and reproductive systems (anterior branch). (B,C) Head ends. (D) Surface view of head, redrawn from SEM micrographs, Agricultural Canada. The head annules may be discontinuous, accounting for differences in numbers observed from a side view. (E) Cross section of body cuticle having 22 longitudinal striae in addition to the eight incisures of the lateral fields. (F–H) Tail showing variation in shapes of annules. Typical shape of the tail terminus is represented by (F) male. (I) Head end. (J) Tail; note distinctive shape of the gubernaculum. Other illustrations: Allen (1955); Golden, Maqbool, and Handoo (1987); Kleynhans and Heyns (1984); Steiner (1937); Wang (1971). Specimens, in part, courtesy of A. M. Golden; others Canadian National Collection of Nematodes.

basal bulb elongate-pyriform, containing a distinct dorsal gland nucleus. Excretory pore 94–115 (102 ± 5.5)μm posterior to head end, located at or near the junction of the isthmus and basal bulb. Intestine containing fasciculi through entire length. Ovaries outstretched, anterior one may extend to esophagus, oocytes in a single row, spermatheca rounded, usually with sperms, vagina extending to less than half a body width.

Tail dorsally convex-conoid to uniformly conoid, 27–41 (32)μm long, tapering to a smooth, large, typically heart-shaped terminus. Tail with 9–20 (14) coarse annules, which occasionally extend to tail tip (Fig. 1F–H). Postanal intestinal (rectal) sac absent or with short overlap to about level of anus (Fig. 1F,H). Phasmids centered or off center in lateral field, located from level of anus to 11μm posterior, or 23–38 (29.9 ± 2.9)μm from tail tip.

Male: Similar to female in gross morphology and body dimensions. Bursa marked by coarse annules, gubernaculum protrusible, distinctive by an arcuate distal portion and thickened, knoblike distal end (Fig. 1J).

Type host and locality: Soil around roots of tobacco; Florence, South Carolina.

Tylenchorhynchus clarus Allen, 1955
Syn. *T. tener* Erzhanova, 1964

MEASUREMENTS

Female

(holotype): L=560 μm; a=28; b=5; c=16; V=58.

(n=10; after Allen, 1955): L=490–610 μm; a=28–33; b=4–5; c=16–20; V=57–61. Stylet length=15–17 μm.

(n=5; after Ray and Das, 1983): L=560–595 μm; a=30–33; b=4.5–5.5; c=17–21; c'=2.5–3.0; V=59–60.

(n=31; after Kleynhans and Heyns, 1984): L=512–686 (587 ± 43.0)μm; a=24.7–35.0 (30.5); b=4.6–5.8 (5.0); c=15.5 ± 19.4 (17.7); V=55.5–69.9 (58.5). Stylet length=14.4–17.8 (16.8 ± 0.8)μm. Head width=6.7–8.1 (7.5 ± 0.3)μm. Head end to excretory pore=84.5–109.0 (95.6 ± 5.5)μm. Tail length=26.5–40.9 (33.3 ± 3.9)μm. Tail annules=10–18.

(n=10; intercepted in Canada with roses from California): L=499–654 (572.1 ± 53.1)μm; a=25–34 (29.1 ± 2.8); b=3.9–5.6 (4.7 ± 0.5); c=14–20 (16.9 ± 1.6); c'=2.1–3.0 (2.5 ± 0.3); V=56–61 (59 ± 1.6). Stylet length=16.5–18.1 (17.1 ± 0.4)μm; M=45–55 (50.2).

(n=17; associated with citrus crops, Casablanca, Morocco): L=530–608 (576.8 ± 22.5)μm; a=30–38 (32.9 ± 1.9); b=4.5–5.2 (4.8 ± 0.2); c=14.5–18.8 (17.7 ± 1.3); c'=2.1–3.3 (2.7 ± 0.3); V=51–61 (57.2 ± 2.1). Stylet length=16.2–18.5 (17.3 ± 0.8)μm; M=48–55 (51.4).

(n=10; after Elmiligy, 1969; turnip populations from Giza, Egypt): L=500–680 (560)μm; a=21–34 (28); b=4.3–6.0 (5.0); c=14.5–22.0 (19.0); c'=2.5–2.9 (2.7); V=55–59 (57). Stylet length=16–18 (17)μm. Tail annules=14–18 (16). See Elmiligy, 1969 for other measurements.

Male

(n=3; after Elmiligy, 1969; associated with *Trifolium alexandrinum* L. (n=2) and *Brassica rapa* L. (n=1), respectively; Egypt): L=530, 580, 510 μm; a=24, 26, 22; b=4.6, 4.7, 4.2; c=13.7, 15.0, 15.7; c'=2.5, 2.6, 2.6. Stylet length=16, 17,17 μm. Spicule length=17, 19, 20 μm. Gubernaculum length=9, 10, 11 μm.

DESCRIPTION (emended)

Female: Body ventrally arcuate when relaxed, width 16–19 (17.6 ± 1.1)μm, annules coarse, particularly on neck and tail, 1.5–2.5 μm wide at midbody. Lateral field 6–8 μm wide, rarely transversed by striae, outer of four incisures crenate. Excretory pore 86–99 (90.0 ± 3.2)μm posterior to head end; hemizonid immediately, to 3 μm anterior.

Head typically truncated, 3.1–3.9 μm wide, 6.9–7.7 μm wide, continuous with body, slightly setoff in some specimens, marked by five, sometimes four, distinct annules. Stylet knobs rounded, anterior surfaces flattened or slightly concave, dorsal esophageal gland duct 1.5–2.5 μm posterior. Esophagus 111–127 (119.3 ± 5.2) μm long, from head end to base of metacorpus 58–62 (60.4 ± 1.5) μm. Metacorpus 12–14 ± 9–12 μm, isthmus 25–32 (27.7 ± 2.2) μm long, basal bulb 24–35 (31) ± 11–14 (12)μm, elongate pyriform to saccate, overlapping intestine laterally 2–12 (7.6 ± 2.6)μm. Esophagointestinal valve very large (Fig. 2C, D). Intestinal fasciculi present.

Ovaries outstretched, oogonia usually in a single row, spermatheca present, but underdeveloped, being comprised of a solid sphere of multiple cells (Fig. 2F). Vagina short, no more than half body diameter deep, vulval lips, if present, minute (Fig. 2F).

Tail conoid, usually with slight ventral curvature, tapering to a smooth, conoid tip 5–10 (7)μm long. Tail 29–41 (32.8 ± 3.4)μm long, ventral surface with 7–17 (13.1 ± 2.9) annules. Phasmids 9–17 (13.1)μm posterior to level of anus, at anterior 29– 50 (40)% of tail.

Male (after Elmiligy, 1969): "A single male examined from Giza, Egypt (Fig. 2I) had a linear gubernaculum, which was slightly thickened distally. Testis single, outstretched; bursa large, with crenate margins, arising about one spicule length anterior to cloaca opening, enveloping tail completely. Spicules strong, ventrally arcuate; gubernaculum appearing rod-shaped in lateral view, with proximal half slightly curved upwards. Phasmids slightly posterior to mid-tail. The other characters are the same as in the female."

Type host and locality: Soil around citrus roots; Bates lemongrove, near Ventura, California.

Remarks: Males apparently are rare and not functional in reproduction. A particularly distinctive feature of this species is the underdeveloped, multicellular spermatheca, which lacks the characteristic structure of a functional receptacle. Maqbool et al., 1983, reported finding 10 males with 15 females of *T. clarus* associated with *Allium cepa* and *Oryzae sativa* in Islamabad, Punjab, Pakistan. However, the specimens were atypical of this species in having 3–4 head annules and a more anterior vulva, V=54–56 (55). More study is needed to establish the identity of this species in Pakistan.

Tylenchorhychus agri Ferris, 1963

MEASUREMENTS

Female

(holotype, after Ferris, 1963): L=670 μm; a=29; b=4.9; c=16; V=57. Stylet length=21 μm.
(n=10; after Ferris, 1963): L=700 (660–770)μm; a=30 (28–33); b=5.1 (4.7–5.5); c=18 (15–21); V=56 (55–58). Stylet length=21 (20–23)μm.
(n=6; Alabama population): L=603 (566–652)μm; a=29 (27– 33); b=5.2 (4.8–5.6); c=17 (16–19); c'=2.5 (2.1–2.6); V=55 (53–58). Stylet length=19 (17–20)μm.

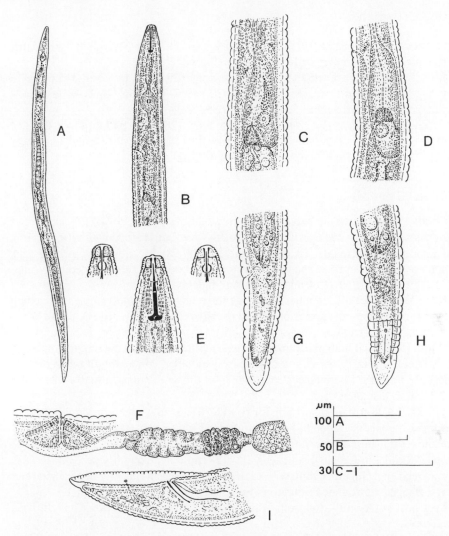

FIGURE 2 *Tylenchorhynchus clarus*. (A) Adult female. (B) Esophageal region. (C,D) Basal esophageal bulb. Note the distinctive overlapping basal esophageal bulb and large esophagointestinal valve which characterize the species. (E) Head end. (F) Vagina and posterior part of gonoduct. (G,H) Female tail. (I) Male tail. Specimen from turnip, Giza, Egypt, Hashim (1983); other illustrations: Allen (1955); Elmiligy (1969); Kleynhans and Heyns (1984); Sauer (1985). Specimens Canadian National Collection of Nematodes, others courtesy of R. Voisin, Antibes, France; male courtesy of E. Geraert, Ghent, Belgium.

Male

(allotype; after Ferris, 1963):L=540 µm; a=29; b=4.3; c=16; T=56. Stylet length=19.5 µm.
    Spicule=22 µm. Gubernaculum length=13.2 µm.
(n=10; after Ferris, 1963): L=660 (540–720)µm; a=33 (29– 36); b=5.1 (4.3–5.6); c=16
    (15–18); T=52 (48–56). Stylet length=20 (19.5–21.0)µm. Spicule length=24
    (22–25)µm. Gubernaculum length=13.4(12.5–14.4)µm.
(n=4; Alabama population): L=620 (510–719)µm; a=27 (24– 32); b=5.2 (4.6–6.3); c=16
    (15–19); c'=2.4 (2.2–2.7). Stylet length=19 (18–19)µm. Spicule length=22
    (21–22)µm. Gubernaculum length=14 (13–15)µm.

DESCRIPTION (emended)

Female: Body ventrally arcuate, curvature usually slight, width 21 (19–23) µm, annules of
neck region coarse, 2.5 µm wide, on body 1.5–2.0 µm wide. Lateral field without areolation,
outer of four incisures crenate. Excretory pore 92 (88–98) µm posterior to head end.
Hemizonid conspicuous in fixed specimens, located one body annule or immediately anterior excretory pore, from level of mid–isthmus to mid–basal bulb.

Head slightly set off from neck, bearing three or four distinct annules (including
labial disk). First neck annule posterior to head usually reduced (Fig. 3B–D). Stylet knobs
with concave anterior surfaces, dorsal gland orifice 2–4 µm posterior. Esophagus 116
(108–125)µm long, basal bulb pyriform to elongate pyriform, usually overlapping intestine
dorsally or laterally as in *T. clarus*. Intestine densely globular, containing fasciculi extending posteriorly into tail. Spermatheca rounded, 8–13 µm in diameter, usually filled with
sperms 2.3 µm in diameter.

Tail 35 (32–39)µm long, subcylindrical, with 20 (16–26) annules on ventral surface,
terminus broadly rounded, smooth. Postanal intestinal sac present, occupying 45–69% of
tail. Phasmids 9–11 µm posterior to level of anus, 23–31% of tail.

Male: Similar to female. Esophagus 111–128 µm long, excretory pore 88–105 µm
posterior to head end. Gubernaculum with a raised ventromedian ridge (cuneus), distal end
with lateral flanges, tip rounded (knoblike)(Fig. 3G).

Type host and locality: Soil from field cropped with corn for 85 years; Univ. Illinois,
Urbana, Illinois.

Remarks: Recognition of *T. agri* is complicated by its similarity to *T. ewingi* Hopper, 1959 in general morphology and body dimensions. Both species have three or four head
annules, four being more common to *T. agri*, an overlapping basal esophageal bulb, a postanal intestinal sac, which extends into about half of the tail, and the same distinctive shape
of the gubernaculum. Based on studies of relatively few specimens, *T. agri* and *T. ewingi*
differ, respectively, only in shape of the female tail, subcylindrical versus conoid, in number
(and size) of tail annules, 16–22 versus 12–15 (four paratypes), and in spicule length, 21–25
µm versus 18.5–20.0 µm (three paratypes).

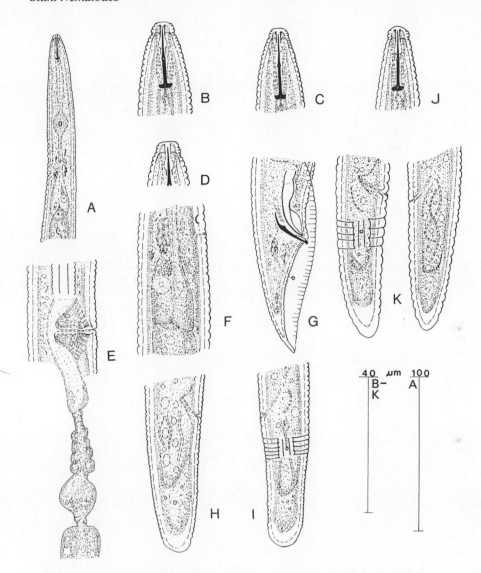

FIGURE 3   *Tylenchorhynchus agri* (Burkville, Alabama population). (A) Esophageal region, female. The basal esophageal bulb typically overlaps the intestine in this species. (B–D) Head end, female. Note variation in annule patterns on the head and neck. (E) Vaginal region (midbody), showing character of the lateral field and a portion of the posterior reproductive branch with spermatheca. (F) Basal esophageal bulb. (G) Tail, male. (H,I) Tail, female. *Tylenchorhynchus ewingi* (paratypcs). (J) Head end. (K) Tails. As in *T. agri*, the postanal intestinal sac does not occupy the entire tail. Other illustrations: Ferris (1963). Specimens Canadian National Collection of Nematodes.

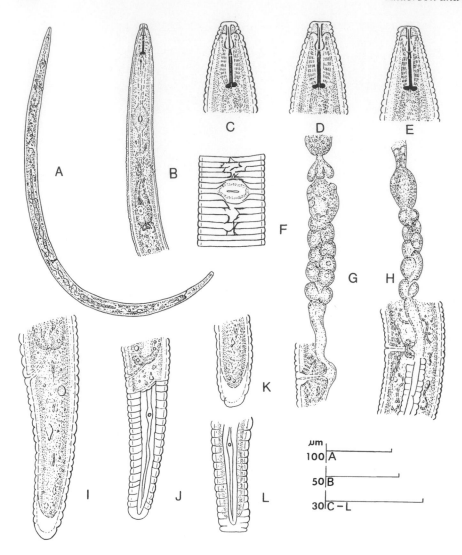

FIGURE 4 *Tylenchorhynchus annulatus*. (Cassidy, 1930) Golden, 1971. (A) Adult. (B) Esophageal region. (C–E) Head end. (F) Ventral view at vagina showing irregularly broken and thickened
striae characteristic of older females. (G,H) Reproductive branch, anterior, showing undeveloped
spermathecae. (I,J) Tail. (K,L) Tail termini. Specimens of (B,D,E,I) from Tifton, Georgia. (A,C,F)
from Baton Rouge, Louisiana. (H,J,K) Paratypes from Kailua, Hawaii. (G,L) Colombia, S. America, possibly representing another species. Other illustrations: Cassidy (1930); Fielding (1956);
Kleynhans and Heyns (1984); Lopez and Salazar (1986); Siddiqi (1976); Timm (1963). Specimens
courtesy of A. M. Golden.

Tylenchorhynchus annulatus (Cassidy, 1930) Golden, 1971
Syn. *Tylopharynx annulatus* Cassidy, 1930
Anguillulina annulatus (Cassidy, 1930) Goodey, 1932
Chitinotylenchus annulatus (Cassidy, 1930) Filipjev, 1936
Ditylenchus annulatus (Cassidy, 1930) Sher, 1970
Tylenchorhynchus martini Fielding, 1956

MEASUREMENTS

Female

(lectotype; after Siddiqi, 1976a): L=660 µm; a=31; b=5.4; c=15; c'=3.6; V=54. Stylet
  length = 17 µm; M=50. (n=3; paralectotypes; after Siddiqi, 1976a): L=660–720 µm;
  a=34–35; b=5.0–5.6; c=15–16; c'=3.1–3.4; V=54–56. Stylet length=17–18 µm; M=
  49–50.
(n=20; Klondike, Louisiana; after Siddiqi, 1976a): L=640–810 (690)µm; a=29–35 (31.4
  um); b=4.6–5.7 (5.0); c=13.5–16.0 (14.8); c'=2.9–3.7 (3.3); V=53–58 (55). Stylet
  length=17–19 (18)µm; M=49–51 (50).
(n=15; from glasshouse rice, Louisiana; after Timm,1963): L=710–830 (770)µm;
  a=28.0–36.8 (31.0); b=4.7–5.1 (4.9); c=13.4–15.9 (14.7); V=53–57 (54.6). Stylet
  length=19–20 µm.
(n=25; after Maqbool, Fatima, and Hashmi, 1983): L=640–720 (680)µm; a=29.2–33.4
  (31); b=4.5–5.5 (4.9); c=13.3–15.5 (14.2); c'=3.0–3.7 (3.3); V=52.0–55.8 (54).
  Stylet length=17.0– 17.8 (17.5)µm. Tail annules=22–28.
(n=13; after Kleynhans and Heyns, 1984): L=615–856 (754 ± 72.9)µm; a=28.2–34.3 (31.8);
  b=6.1–6.8 (6.4); c=15.0–16.4 (15.7); V=54.8–58.8 (56.3). Stylet length=18.5–20.8
  (19.3 ± 0.6)µm.

Male

Unknown

DESCRIPTION

Female (after Siddiqi, 1976a; Kleynhans and Heyns, 1984): Body ventrally arcuate, annules
coarse, 1.7–2.0 µm wide at midbody. Lateral field 30–35% of body width, irregularly areolated, outer of four incisures crenate. Excretory pore (n=11) 102–122 (116 ± 4.2)µm (55–62
body annules) posterior to head end, hemizonid immediately to two body annules anterior,
generally conspicuous. Deirids not observed.

   Head (n=12) 7.2–8.8 (7.9 ± 0.52)µm wide, rounded or somewhat truncated, continuous or slightly set off from body, marked by three, sometimes two, distinct annules. Stylet
knobs often irregular in shape with flat or concave anterior surfaces, sometimes sloping
posteriorly; dorsal esophageal gland orifice 1.5–2.5 µm posterior. Metacorpus 12.5–14.5 ×
8.5–9.5 µm, at 44–49% (47.2) of total esophagus length, basal bulb elongate-pyriform,
23–25 × 10–12 µm, esophagointestinal valve large. Intestine with fasciculi, terminating at
or slightly posterior to the rectum. Ovaries outstretched, oogonia and oocytes in a single
row, spermathaca rarely discernible.

   Tail subcylindrical, about 42.4–52.0 (48.8 ± 2.9)µm long, with 18–27 (22) annules,
terminus broadly rounded, rarely flattened, without annules. Phasmids in anterior half of
tail, 13–17 µm posterior to level of anus.

   Type host and locality: Soil around roots of sugarcane; Kailua, Hawaii.

Tylenchorhynchus nudus Allen, 1955
nec. *T. nudus* of Loof (1959): Timm (1963): Choi and Geraert (1971); Baqri and Ahmad (1981).

MEASUREMENTS

Female

(holotype): L=710 µm; a=30; b=4.1; c=14; V=56. Stylet length=23 µm.

(n=2;after Allen, 1955): L=710 µm; a=30; b=4.1,4.8; c=14,15; V=54,56. Stylet length=19,23.

(n=7;from Ceylon, after Loof, 1959): L=563–660 µm; a=28.1–30.5; b=4.7–5.3; c=12.2–15.3; V=53.4=56.2. Stylet length=17– 21 µm.

(n=9;from Dacca, West Pakistan;after Timm, 1963): L=680–810 (730 µm; a=27.9–32.1 (30.2); b=4.7–6.1 (5.2); c=13–15 (14.2); V=52–58 (54.7). Stylet length = 18.0–19.5 (19.0)µm.

(after Baqri and Ahmad, 1981): L = 602–863 (688 ± 45.0)µm; a = 5.1–6.4 (29 ± 2.1); b = 5.1–6.4 (5.6 ± 0.3); c = 12.5–16.4 (12.9 ± 1.3); V = 53–57 (55 ± 1.2). Stylet length = 18–21 (19 ± 0.6) µm. Annule width = 2.0–2.5 (2.0 ± 0.08)µm. Head width = 6–8 (7 ± 0.3)µm; head height = 3.5–4.0 (3.41 ± 0.09)µm. Esophagus length = 103–144 (123 ± 8.1) µm. Excretory pore = 87–124 (101 ± 7.0)µm. Tail length = 40–59 (47 ± 3.7)µm.

(n=21;Virgin prairie, Matador Ranch, Saskatchewan, Canada): L=699–905 (804.9 ± 48.9)µm; a=31–39 (33.9 ± 2.4); b=4.4–6.6 (5.3 ± 0.4); c=15.0–21.7 (17.5 ± 1.7); c'=2.2–3.6 (2.9 ± 0.3); V=52–56 (54.5 ± 1.2). Stylet length=21.5–23.8 (22.5 ± 0.6)µm; M=46–54(50).

Male

(n=2; after Allen, 1955): L=770,780 µm; a=35; b=5.2,5.8; c=14,16. Stylet length=21 µm.

(n=3;after Loof, 1959): L=518–572 µm; a=24.8–32.9; b=4.3–4.9; c=14.3–15.7; T= 45.8–62.4 Stylet length=18.20 µm.

(after Baqri and Ahmad, 1981): L = 516–726 (645 ± 46.0)µm; a = 25–35 (30 ± 2.6); b = 4.3–6.0 (5.3 ± 0.4); c = 11–17 (14.6 ± 1.4); T = 43–59 (51 ± 5.5). Stylet length = 18–20 (19 ± 0.4)µm. Spicule length = 22–27 (25 ± 1.4)µm. Gubernaculum length = 11–14 (12.8 ± 0.99)µm. Esophagus length = 111–142 (124 ± 8.0)µm. Excretory pore from head end = 85–115 (102 ± 6.6)µm. Tail length = 36–51 (44 ± 4.1)µm.

(n=3; Saskatchewan, Canada,same data as for female): L=759,790,732 µm; a=35,33,35; b=5.5,5.3,5.0; c=15.8,13.6,15.2; c'=2.7,3.6,3.1. Stylet length = 22.3,21.5,21.5 µm; M=48,50,54. Spicule length = 26.2,27.0,25.4 µm. Gubernaculum length = 15,15,15.4 µm.

REDESCRIPTION

Female: Posture variable, from nearly linear to crescentic when relaxed, width 22–25 (23.8 ± 1.3)µm wide, body annules coarse, 1.9–2.3 µm wide, on neck 2.3–3.1 µm. Lateral field 6.2–8.5 (7.2 ± 0.5) µm wide, occupying 26–32 (30)% of body diameter, outer of four incisures coarsely crenate. Excretory pore 109–130 (118.7 ± 5.1)µm from head end, hemizonid immediately or within a body annule anterior.

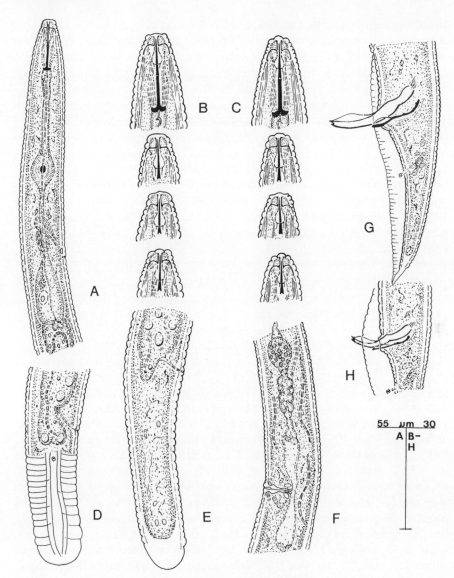

FIGURE 5 *Tylenchorhynchus nudus*. (A) Esophageal region, female. (B,C) Anterior ends of female showing variation in head annulation. (D,E) Female tail. A slight or moderate postrectal intestinal sac if characteristic for this species. (F) Midbody region of female showing anterior portion (to spermatheca) of the reproductive branch. (G) Male tail. *Tylenchorhynchus elegans*. (H) Male tail. Note on (G) and (H) that the spicule blade tips are notched (bifurcate). Other illustrations: Allen (1955); Thorne and Malek (1968). Specimens Canadian National Collection of Nematodes, others of Canadian origin courtesy of J. Baldwin.

Head rounded, continuous with neck, 6.2–9.0 (7.6) μm wide, 2.7–3.9 (3.3) μm high, with two, occasionally three large annules and conspicuous labial disk. Basal ring of head framework thin, extending well into neck. Stylet knobs strongly developed, about 5 μm in diameter, anterior surfaces indented or flattened; dorsal gland orifice 2.0–2.5 μm posterior.

Esophagus 137–163 (151.5 ± 6.7) μm long, from head end to base of metacorpus 83–89 (85.3 ± 2.1)μm. Metacorpus 12–18 (15.4)μm long, 10–13 (11.4)μm wide, isthmus 31–43 (35.7 ± 3.8)μm long, basal bulb pyriform, 24–39 (30.2 ± 3.6)μm long, 12–17 (14.4 ± 1.1)μm wide, slightly overlapping intestine. Esophagointestinal valve large, lobate, intestine densely globular, containing a dense network of thick, intestinal fasciculi. Posteriorly intestine dorsally overlays rectum to level of anus or slightly posterior into tail. Ovaries outstretched, oogonia in a single or double row, spermatheca spherical or oval, containing sperms with a diameter of 2.5–3.0 μm, uteri more than a body length long.

Tail subcylindrical, linear or with ventral curvature, 39–59 (46.3 ± 4.5)μm long, terminating in a large, broadly rounded smooth tip. Ventral surface with 13–26 (18.2 ± 2.6) coarse annules ranging in widths from 1.3–3.5 μm. Cuticle of tail tip markedly thickened, comprising 15–20 (17)% of tail, being 6.2–8.5 (7.8 ± 0.6)μm thick. Phasmids 8–21 (11.9 ± 3.0)μm posterior to level of anus, at anterior 17–40 (25)% of tail.

Male: Similar to female in general morphology and dimensions. Body width 21–24 μm, esophagus 137–147 μm, from head end to base of metacorpus 76–85 μm, excretory pore 101–118 μm posterior to head end. Tail 48–58 μm long, phasmids at 31–36% of tail. Spicules distinctive by having bifurcate tips. Gubernaculum more than half spicule length, strongly sclerotized with thickened distal end, variably arcuate depending on degree of protraction.

Type host and locality: Soil at roots of alfalfa; Experimental Farm, Ottawa, Canada.

Remarks: *Tylenchorhynchus nudus* is a member of a complex of closely similar bisexual species which are widely distributed in the tropics and subtropics. These include: *T. badliensis* Saha and Khan, 1982; *T. bohrrenis* Gupta and Uma, 1980; *T. coffeae* Siddiqi and Basir, 1959; *T. crassicaudatus* Williams 1960; *T. elegans* Siddiqi, 1961; *T. goldeni* Rashid and Singh, 1982; *T. penniseti* Gupta and Uma, 1980; *T. punensis* Khan and Darekar, 1978. These species all have in common: a continuous head with two or three annules; large, coarse body annules about 2–3 μm wide; four incisures; stylet lengths with overlapping ranges from 15–20 μm; stylet knobs with concave anterior surfaces; a large, well–developed esophagointestinal valve; a subcylindrical tail (to slightly clavate) with few annules, usually 17–27, ending in a large, smoothly rounded or conoid tip; similarly structured gubernaculum; and, with the exception of *T. punensis*, absence of a postrectal intestinal sac. Some species undoubtedly are conspecific; others may prove to be more appropriately classified as subspecies or sibling species. An unknown, but important differential character for these species is the form of the spicule blade tips, which are distinctly bifurcate (notched) in *T. nudus*.

T. nudus can be tentatively separated from the above species by its longer mean stylet (23 μm) and, with the exception of *T. punensis*, presence of a postrectal intestinal sac. Specimens of *T. nudus* described from Ceylon by Loof (1959), from Bangladesh by Timm (1963), from Korea by Choi and Geraert (1971) and from West Bengal, India by Baqri and Ahmad (1981) lack a postrectal intestinal sac. They are therefore more representative of other species in the *T. nudus* complex. The species in West Bengal is most similar or identical to *T. musae* Kumar, 1981 and *T. leviterminalis* Siddiqi, Mukherjee, and Dasqupta, 1982, which tend to have a smooth head, lacking annules.

Tylenchorhynchus mashhoodi Siddiqi and Basir, 1959

MEASUREMENTS

Female

(holotype): L=750 μm; a=28; b=5.4; c=19; V=57.8

(n=8; after Siddiqi,1961): L=615–760 (690)μm; a=25.8–30.0 (27.8); b=4.9–5.5 (5.2); c=16.0–19.4 (17.0); c′=2.3; V=55–58 (56.3). Stylet length=17–18 μm. Egg(n=1) 67 × 20 μm.

(n=2 paratypes; respectively): L=608,626 μm; a=35,31; b=5.1,5.4; c=18,17.4; c′=2.8,2.5; V=58,57. Stylet length=16.0,15.5 μm; M=50. Length from head end to base of metacorpus 61,60 μm; to excretory pore 94,90 μm. Tail annules 13,12. Phasmids 8,9 μm posterior to level of anus. Annule width at midbody (n=36)1.8–2.8 (2.28 ± 0.25)μm.

Male

(n=4;after Siddiqi, 1961): L=590=710 (650)μm; a=28.4–33.5 (31.3); b=4.7–5.5 (5.0); c=14–17 (15.6); T=43–52. Stylet length=16.0–17.5 μm. Spicule length=22–24 μm. Gubernaculum length=12–13 μm.

DESCRIPTION

Female (after Siddiqi, 1961): Body slightly ventrally arcuate to varying degrees when relaxed; annules particularly coarse, about 3 μm wide. Lateral field about 30% of body diameter, outer of four incisures strongly crenate. Excretory pore opposite anterior end of basal esophageal bulb; hemizonid immediately, or a body annule anterior, distinct.

Head round, continuous with body or slightly set off, 6.5 μm wide, 3.6 μm high, bearing three large annules. Stylet knobs round with sloping or concave anterior surfaces, diameter about 3.5 μm; dorsal gland orifice 2.5 μm posterior. Metacorpus set off from procorpus, basal bulb elongate-pyriform, slightly overlapping intestine. Esophagointestinal valve large. Intestine containing large globules and intestinal fasciculi, terminating at rectum, without overlapping. Ovaries outstretched, oocytes in single rows; spermathecae spherical, containing sperms with diameters of 1.8–2.0 μm; vagina depth no more than half body diameter.

Tail uniformly conoid, slightly arcuate, ventral surface with 13–16 irregularly coarse annules. Tail tip large, smooth, bluntly conoid. Phasmids well anterior to midtail.

Male: Similar in gross morphology to female, but tending to be more arcuate when relaxed. Spicules arcuate, cephalated, tips of blades narrowly rounded. Gubernaculum strongly sclerotized, rod shaped in lateral aspect, proximal half slightly arcuate.

Type host and locality: Soil around roots of sugarcane, *Saccharum officinarum L.*; Coimbatore, Madras State, South India.

Remarks: Distinctive of the characters established by Siddiqi (1959,1961) for *T. mashhoodi* are: continuous head (or slightly set off) with three (sometimes four) annules; large, prominent body annules about 3 μm wide (1.8–3.0 (2.34 ± 0.33)μm; n =58 annules in two paratypes); and a relatively short, slightly arcuate conoid tail with 13–16 coarse annules, which ends in a large, smooth, bluntly conoid tip. A similar species, *T. elegans* Siddiqi, 1961 was differentiated from *T. mashhoodi* by having smaller and less prominent body annules (1.5–2.5 (1.9 ± 0.22)μm; n =47 annules in two topotypes), and by its longer and

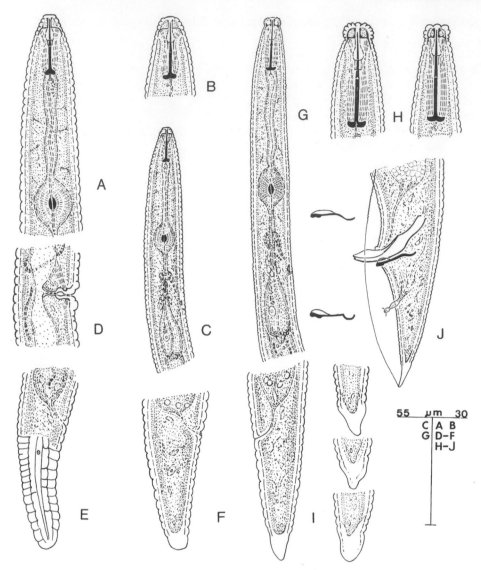

FIGURE 6   *Tylenchorhynchus mashhoodi.* Paratypes. (A) Esophageal region. Note the distinctively large, coarse body annules. (B) Head end. (C) Esophageal region. (D) Body at vagina. (E) Female tail. (F) Female tail of specimen from Malawi, courtesy of M. R. Siddiqi. *Tylenchorhynchus cylindricus.* Specimens from California. (G) Esophageal region. (H) Head ends, the left being the more typical. (I) Female tail, and tail tips showing variation in shape. (J) Male tail with gubernaculum to left. The distinctly hook-shaped proximal end of the lower left gubernaculum is an artifact. Other illustrations: *T. mashhoodi*: Siddiqi (1961); *T. cylindricus*: Allen (1955); Lewis and Golden (1981); Sher and Bell (1975); Siddiqi (1972b); Thorne (1961). Specimens of *T. mashhoodi* courtesy of M. R. Siddiqi (paratypes) and A. M. Golden; of *T. cylindricus*, courtesy of J. Baldwin.

subcylindrical tail with more annules. In addition, based on male topotypes provided by M.R. Siddiqi, the spicule tips are bifurcate, whereas they are rounded in *T. mashhoodi*, thus confirming the distinctiveness of these species. Recently, Siddiqi (1986) synonymized *T.*

goldeni Rashid and Singh, 1982 with *T. elegans*. However, specimens examined from Senegal, which appear identical to those of *T. goldeni*, have spicules with rounded tips. Therefore, we provisionally recognize *T. goldeni* Rashid and Singh, 1982 as a valid species. [Note: Measurements and morphology of 20 females and 20 males from Senegal (courtesy of M. Luc) identified as *T. mashhoodi* proved to be similar or identical to *T. goldeni* which Siddiqi (1986) considers a synonym of *T. elegans,* but is herein regarded as a valid species. See also remarks for biology of the species.]

Undoubtedly, *T. mashhoodi* has been confused with *T. elegans* or other similar species occurring in the tropics or subtropics. Based on studies of nine isolated populations in India, Baqri and Jairijpuri (1970) concluded that *T. mashhoodi* was more variable than diagnosed in the shape of the head, size of body annules, length and shape of the tail and number of tail annules, and in length and shape of the gubernaculum. Accordingly, they redescribed the species and proposed *T. dactylurus* Das, 1960; *T. digitatus* Das, 1960; *T. crassicaudatus* William, 1960, 1960; *T. elegans* Siddiqi, 1961; and *T. zeae* Sethi and Swarup, 1968 as synonyms. These synonymies, however, have not been accepted by Siddiqi (1986) and Fortuner and Luc (1987) and their status and relationship remain uncertain. Subsequent redefinitions of *T. mashhoodi* by Singh and Khera (1978), Ray and Das 1983), Maqbool, Fatima, and Hashmi (1983), Kleynhans and Heyns (1984) and Gupta and Uma (1985) similarly are more representative of *T. elegans* or other close species. Consequently, records of distribution and pathogenicity reported for *T. mashhoodi* may be attributed to other species.

Tylenchorhynchus cylindricus Cobb, 1913
Syn. *Tylenchus cylindricus* Cobb, 1913
Tylenchus (Tylenchorhynchus) cylindricus Cobb, 1913 (Filipjev, 1934)
Anguillulina cylindrica (Cobb, 1913) Thorne, 1935
Tylenchorhynchus dubius of Thorne, 1949; *nec T. dubius* (Bütschli, 1873) Filipjev, 1936

MEASUREMENTS

Female

(lectotype; after Lewis and Golden, 1981): L=1029 μm; a=39.3; b=6; c=21.2; c'=2.35; V=57. Stylet length=27.5 μm.

(n=29 paralectotypes; after Lewis and Golden, 1981): L=721.3–1179.9 (1009.6 ± 110.3)μm; a=30.6–40.8 (36.3 ± 2.76); b=4.5–6.9 (6.1 ± 0.52); c=15.3–26.7 (22.4 ± 2.28); c'=1.64–2.35 (2.1 ± 0.26); V=51–61 (56.9 ± 2.01). Stylet length=25.1–29.5 (27.4 ± 1.34)μm.

(n=13; after Allen, 1955): L=650–990 μm; a=28–35; b=4.2– 6.0; c=13–20; V=54–64. Stylet length=24–27 μm.

Male

(allolectotype; after Lewis and Golden, 1981): L=975 μm; a=43.5; b=5.8; c=22.7; c'=2.2. Stylet length=26.7 μm.

(n=15 paralectotypes; after Lewis and Golden, 1981): L=627.2–1140.7 (937.8 ± 135.5)μm; a=34.7–43.7 (38.2 ± 3.64); b=5.0–7.0 (5.97 ± 0.70); c=20.5–27.3 (22.2 ± 2.46); c'=1.7–2.3 (2.0 ± 0.18). Stylet length=25.1–28.2 (26.3 ± 1.25)μm.

(n = 7; after Allen, 1955): L = 670–1000 μm; a = 30–40; b = 4.4–5.8; c = 14–16. T = 64. Stylet length = 22–28 μm.

DESCRIPTION

Female (after Lewis and Golden, 1981, with minor changes): Body slightly arcuate ventrally when relaxed, annules coarse, 1.5–2.5 (1.94 ± 0.27)μm wide (n=36 annules). Lateral field about one–fourth body width, outer of four incisures crenate. Excretory pore from level of isthmus base to middle of basal esophageal bulb; hemizonid, when observed, immediately anterior. Deirids absent.

Head bulbous, bearing four or five, rarely three or six annules; septum (basal plate) of head framework conspicuously thickened at periphery and central tube (cheilostom). Stylet knobs with concave anterior surfaces; dorsal gland orifice 1.6–2.2 (1.8)μm posterior.

Esophagus 123–157 μm long (n=6); to center of metacorpus 73.9–91.8 (84.7 ± 5.87)μm; to base of metacorpus (n=6) 70–94 (81)μm. Basal esophageal bulb elongate-pyriform, slightly overlapping intestine; esophagointestinal valve large, hemispherical. Intestine not overlapping rectum, containing intestinal fasciculi.

Ovaries outstretched, oogonia usually in single row; spermatheca spherical, sperms with diameters of 2.5–4.0 μm.

Tail 37.4–51.5 (45.1 ± 4.66)μm long, typically conoid, subcylindrical in some specimens, tapering to a smooth, elongated conoid tip, Tail annules 15–22 (19.3 ± 2.3), as few as 11–13 in some specimens. Phasmids in anterior half of tail, 26.6–39.2 (32.3 ± 3.79)μm anterior to terminus.

Male (after Lewis and Golden, 1981): Similar to female in general morphology and dimensions. Spicule tips indented (notched). Gubernaculum linear, distal end thickened, flanged, proximal end variably arcuate. Bursa weakly annulated, margin smooth or indistinctly crenate.

Type host and locality (lectotype): Brackish soil near a marine estuary; Los Patos, California.

Tylenchorhynchus brassicae Siddiqi, 1961

MEASUREMENTS

Female

(holotype;after Siddiqi, 1961;from *Brassica oleracea* L., Aligarh, India): L=705 μm; a=29; b=5; c=14; V=52.4.

(n=10;same data as for holotype): L=580–720; a=26–35; b=5–6; c=14–17; V=52–58. Stylet length=16–17 μm.

(n=25;after Maqbool, Fatima, and Hashmi, 1983; collected from around roots of Citrus sp., *Oryza sativa*, and *Zea mays*, at Sind and Punjab, Pakistan): L=590–673 (643)μm; a=27–33 (30.5); b=5.0–5.7 (5.4); c=15.5–27.0 (20.1); c′=1.9–2.5 (2.2); V=56–57 (56.8). Stylet length 16–17 (16.5)μm.

(n=11;collected from coriander, India and onions, Rampus, U.P., India): L=499–663 (592.3 ± 50.6)μm; a=25–39 (30.5 ± 3.9); b=4.8–5.9 (5.2 ± 0.3); c=16–22 (18.2 ± 1.6); c′ =1.7–3.7 (2.6 ± 0.5); V=56–59 (57). Stylet length=16.9–18.5 (17.8 ± 0.7)μm; M=48–58 (51).

Male

(n=6;same data as for holotype): L=530–670 μm; a=29–35; b=4.8–5.8; c=15–17; T=47–60. Stylet length 15.0–16.5 μm. Spicule length=18–21 μm. Gubernaculum length 9–11 μm.

(n=17;after Maqbool et al., 1983;same data as for female): L=540–668 (590)μm; a=27–31 (29); b=4.8–5.6 (5.2); c=15.0–18.5 (16.7); T=47–61 (54). Spicule length=20.0–20.5 (20.2)μm. Gubernaculum length 10.5–11.0 (10.7)μm.

DESCRIPTION (emended)

Female: Body ventrally arcuate when relaxed, width 17–22 (19.5 ± 1.7)μm, annules (n=60) coarse, 1.5–3.5 (2.1 ± 0.6)μm wide. Lateral field 5–6 μm wide, outer of four incisures weakly crenate. Excretory pore 79–108 (92.5 ± 10.9)μm posterior to head end, hemizonid one or two annules anterior.

Head bulbous, bearing four distinct annules, capped by conspicuous labial disk. Basal ring of head framework extending well into body. Stylet knobs rounded, anterior surfaces concave; dorsal gland orifice 2 or 3 μm posterior.

Esophagus 97–133 (114.5 ± 12.3)μm long, from head end to base of metacorpus 54–71 (62.1 ± 6.4)μm. Metacorpus 12–15 × 9–12 μm, isthmus 21–30 (26.7 ± 3.3)μm long, basal bulb elongate-pyriform, 19–40 (26.7 ± 3.3) × 9–13 (11.3 ± 1.2)μm, its base slightly overlapping intestine, and surrounding a large, lobate esophagointestinal valve. Intestine containing network of intestinal fasciculi, not extending posterior to rectum.

Ovaries generally outstretched, rarely reflexed, may extend to near esophagus and anus in gravid females; oogonia in single row. Spermatheca spherical, containing sperms with diameters of about 2.3 μm (3.1 μm in males). Vagina short, thick-walled, depth less than 50% of body width.

Tail 23–40 (32.7 ± 4.5)μm long, conoid, linear or slightly arcuate, tapering to a smoothly rounded, conoid tip. Ventral surface of tail with 10–21 (16.1 ± 3.1) coarse annules. Phasmids 3–14 (9.9 ± 2.8)μm posterior to level of anus, at anterior 13–40 (29)% of tail.

Male: Similar to female in general morphology and dimensions. Measurements of three males: body width 18–20 μm; length of esophagus 107–127 μm, from head end to base of metacorpus 59–72 μm; excretory pore 88–98 μm posterior to head end. Gubernaculum strongly sclerotized linear in lateral aspect, distally flanged. Spicules distinctive by their bifurcate tips. Bursa margins coarsely annulated.

Type host and locality: Soil around roots of cauliflower, *Brassica oleracea* L.; Aligarh (U.P.), India.

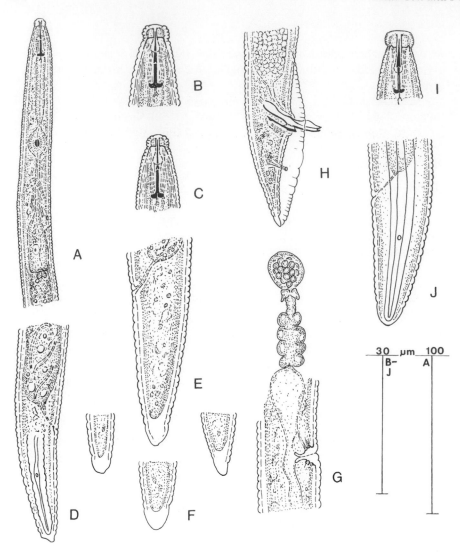

FIGURE 7   *Tylenchorhynchus brassicae.* (A) Esophageal region. (B,C) Head end. (D,E) Female
tail. (F) Variation in shape of tail terminus. (G) Midbody region showing anterior portion (to spermatheca) of female reproductive branch. (H) Male tail. Note the bifurcate (notched) tips of the
spicule blades. As shown, the margin of the bursa may be irregularly crenate. *Tylenchorhynchus
latus.* [For comparison, redrawn to scale from Allen, (1955)]. (I) Head end. (J) Tail. Other illustrations: *T. brassicae*: Siddiqi (1961). Specimens courtesy of M. R. Siddiqi, A. M. Golden, and P. A. A.
Loof.

Tylenchorhynchus capitatus Allen, 1955

> Syn. *Quinisulcius capitatus* (Allen, 1955) Siddiqi, 1971
> > *Tylenchorhynchus acti* Hopper, 1959
> > *Quinisulcius acti* (Hopper, 1959) Siddiqi, 1971
> > *Tylenchorhynchus nilgiriensis* Seshadri, Muthukrishnan, and Shun-
> > mugam, 1967
> > *Quinisulcius himalayae* Mahajan, 1974
> > *Tylenchorhynchus himalayae* (Mahajan, 1974) Fortuner and Luc, 1987
> > *Quinisulcius solani* Maqbool, 1982

MEASUREMENTS

Female

(n = 13 paratypes; after Allen, 1955); L = 630–850 µm; a = 30–38; b = 5.0–5.8; c = 12–17; V = 51–58. Stylet length = 17 µm.

(paratypes; after Jairajpuri, 1985): L=680–800 (720)µm; a=28–34 (31); b=4.8–5.7 (5.3); c=15–19 (17); c'=2.1–3.1 (2.5); V=54–58 (56). Stylet length=17–19 (18)µm.

(n=9 paratypes; after Saltukoglu and Coomans, 1975): L=670–800 µm; a=32–35; b=4.7–5.4; c=13–16; V=52–57. Stylet length=16–18 µm. Tail length=46–53 µm; tail annules 35–47.

(n=14; after Seshadri, Muthukrishnan, and Shunmugan,1967): L=640–790 µm; a=28.5–36.8; b=4.6–5.4; c=14.2–17.0; V=53–57. Stylet length=16.5 µm.

(n=5; after Mahajan, 1974): L=620–780 (720)µm; a=27–31 (29); b=3.8–4.5 (4.0); c=12–16 (14); V=51–56 (53). Stylet length=17–24 (21)µm.

(n=15; specimens collected from tobacco, Harrow, Ontario, Canada): L=711–832 (790 ± 42)µm; a=29–38 (34 ± 2.8); b=4.9–5.7 (5.3 ± 0.2); c=15–20 (17 ± 1.1); c'=2.0–3.1 (2.7 ± 0.3); V=52–58 (56 ± 1.5). Stylet length=15.4–18.0 (17.0 ± 0.8)µm.

Male

(n=4; after Allen, 1955): L=570–700 µm; a=31–38; b=4.3–5.0; c=13–14; T=50–65.

DESCRIPTION (emended, based on Canadian specimens)

Female: Body strongly arcuate to C- or spiral-shape, being typically depressed (concave) midventrally for about 30 µm anterior and posterior to vulva (Fig. 8G). Body width 21–27 (23 ± 1.8)µm, annules fine, particularly along ventral side, width usually 1 µm or less (0.6–1.5 µm). Lateral field 8–10 (8.8)µm wide, 33–42 (37)% of body width, marked by five incisures and scattered areolations (Fig. 8B). Outer incisures crenulate, inner ones irregularly linear. Excretory pore 108–137 (125 ± 7.2)µm posterior to head end, usually opposite anterior half of esophageal basal bulb. Hemizonid indistinct, immediately, or 1–8 annules anterior to excretory pore.

Head rounded, with 6–8 annules, set off from body by a constriction well posterior to septum of head framework (fig. 8D–F). Stylet conus 45–50 (48)% of total length, knobs rounded, sloping posteriorly, dorsal gland orifice 2–4 µm posterior. Esophagus 138–158 (148 ± 5.7)µm long, 79–90 (84 ± 3.1) µm to base of metacorpus. Metacorpus 15–20 × 10–12 µm, isthmus narrow, 23–35 (31) µm long, basal bulb elongate-pyriform, 30–35 × 10–14 µm. Esophagointestinal valve well developed, spherical. Intestine densely globular, with thick fasciculi throughout its entire length. Postrectal intestinal sac absent. Ovaries

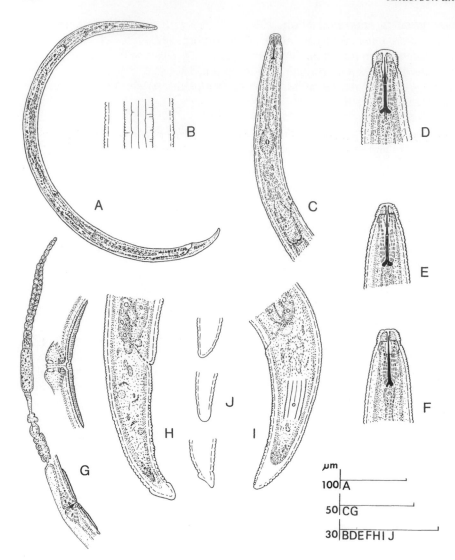

FIGURE 8   *Tylenchorhynchus capitatus*. Female. (A) Adult female of a typical relaxed posture.
(B) Lateral field. (C) Esophageal region. (D–F) Head ends. (G) Anterior reproductive branch with
inset enlargement of vaginal region showing the characteristic midventral depression of the body
anterior and posterior to the vagina, bordered by paired subventral ridges (one shown). (H,I) Tail.
The annules along the ventral surface may be more regular than shown. (J) Variation in tail termini,
the posterior-most being typical for the species. Other illustrations: Hopper (1959); Jairajpuri
(1985); Mahajan (1974); Maqbool (1982); Saltukoglu and Coomans (1975); Seshadri, Muthukrishnan, and Shunmugam (1967); Siddiqi (1986). Specimens Canadian National Collection of Nematodes.

paired, outstretched, oogonia in single or double row, spermatheca small, axial, never observed with sperms. Vulva depressed, always observed centered within a relatively long, trough-shaped depression (Fig. 8G). Tail conoid, arcuate, tapering to a narrowly rounded tip, 37–53 (47 ± 4.1)μm long, marked by 34–50 (43 ± 4.5) fine, regular or irregular annules. Tail terminus without annules, of variable shape, but usually heart shaped and distinctly set off from annulated part. Phasmid 11–20 (13 ± 4.4)μm posterior to level of anus, at 34–50 (43 ± 4.5)% of tail.

Male (after Allen, 1955): Males are rare and have been found only by Allen (1955). The paired spicules and gubernaculum appear similar to those of the *T. dubius* male (Fig. 14J).

Type host and locality: Soil around roots of pear; near Watsonville, California.

Tylenchorhynchus maximus Allen, 1955
Syn. *Bitylenchus maximus* (Allen, 1955) Siddiqi, 1986

MEASUREMENTS

Female

(n=12; after Allen, 1955): L=980–1400 μm; a=37–47; b=5.4– 8.1; c=16–20; V=47–54. Stylet length=21.3–24.0 μm.

(after Thorne and Malck, 1968): L=1000 μm; a=28; b=8.3; c=20; V=50. Stylet length=15–18 μm.

(n=20; Col de la Cayolle, France; from prairie grass): L=1039–1406 (1280 ± 98.3)μm; a=39–58 (45 ± 5.2); b=6.6–8.5 (7.1 ± 0.5); c=17–20 (18.6 ± 0.9); c'=2.6–3.8 (3.3 ± 0.3); V=51–54 (52.3 ± 1.0). Stylet length=21.5–24.6 (22.8 ± 1.0)μm.

(n=20; after Maqbool and Shahina,1987): L=1270–1490 (1350 ± 86.0)μm; a=34–41 (38.2 ± 2.29); b=7.5–8.6 (8.0 ± 0.48); c=21–22 (21 ± 0.70); c'=2.3–2.6 (2.4 ± 0.129); V=49–53 (51 ± 1.58). Stylet length=24 μm; M=53–56 (53).

Male

(n=8; after Anderson, 1977): L=941–1247 (1107)μm; a=37–52 (44); b=6.0–7.4 (6.9); c=17–21 (19); c'=2.6–3.3 (3.0). Stylet length=20–22 (21)μm; M=50–56 (52). Spicule length=30–34 (32)μm. Gubernaculum length 15–18 μm.

DESCRIPTION (emended)

Female: Body strongly arcuate when relaxed, often spiral, width 23–32 (28 ± 2.6)μm, annules coarse, width 1.9–2.5 μm on neck, narrowing abruptly to 1.2–1.5 μm about a body width posterior to esophagus. Lateral field 8–11(10)μm wide, or 31–42% (36) of body width, regularly areolated, the outer of four incisures strongly crenate. Excretory pore 122–148 (139.1 ± 6.9)μm posterior to head end, hemizonid 1.5–6.0 μm (4) anterior, generally very large. Deirids absent.

Head high, rounded, usually set off from neck by a slight constriction, sometimes continuous. Head cuticle thick, marked by 5–7 distinct annules. Basal ring of head framework deep, extending 4–5 μm into body. Stylet delicate, knobs small, about 3 μm in diameter, sloping posteriorly, dorsal gland opening 2–4 μm posterior. Esophagus 153–180 (171.4 ± 7.2)μm long, 91–110 (101.4 ± 4.1)μm to base of metacorpus. Metacorpus 19–22 (20.3 ± 1.1)μm × 12–15 (13.9 ± 0.9)μm, basal bulb pyriform, 27–29 (27.7 ± 0.7)μm × 14–16 (15 ±

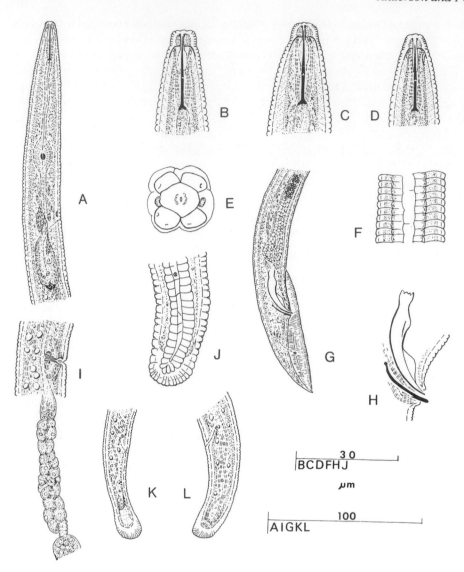

FIGURE 9   *Tylenchorhynchus maximus.* Female. (A) Esophageal region. (B,C) Head end. (E) Head surface, redrawn from SEM micrographs, Agriculture Canada. (F) Lateral field. (I) Vagina and proximal portion of posterior reproductive tract, showing the tubular columella. (J) Tail end. (K,L) Tail variation. Male. (D) Head end. (G) Posterior end of body. (H) Spicule and gubernaculum. The crochet hook like shape of the gubernaculum is a diagnostic feature of this species. Other illustrations: Allen (1955); Anderson (1977); Thorne and Malek (1968). Specimens Canadian National Collection of Nematodes; others courtesy of R. Voisin, Antibes, France.

0.5)μm. Esophagointestinal valve conspicuous, conoid. Intestine with dense network of fasciculi and large globules, postanal intestinal sac present, occupying 59–86 (71.8)% of tail.

Spermatheca and spermatozoa never observed, columella tubular, with cells in three columns.

Tail cylindrical or subcylindrical, occasionally clavate, 59–83 (69 ± 5.4)μm long, with 35–52 (40.7 ± 4.7) annules to mid-terminus; terminus broadly rounded, marked by numerous fine annules. Phasmids 19–30 (23.6 ± 3.7)μm posterior to level of anus, at 25–40(34.4)% of tail.

Male: Rare. Similar to female in general body features, but less extreme in body curvature. Esophagus 150–173 (161)μm long; length of metacorpus 16–19 μm, isthmus 40–50 μm, and basal bulb 20–26 μm. Excretory pore 114–136 (129)μm posterior to headend, ranging from near midisthmus to middle of basal bulb. Testis single, outstretched, reproductive system 485–653 (574)μm long. Spermatozoa spherical, small, 0.7–1.0 μm in diameter. Gubernaculum linear, distal end thickened, proximal end often rounded (knoblike).

Type locality: Lake Placid, New York.

Tylenchorhynchus dubius (Bütschli, 1873) Filipjev, 1936

Syn. *Tylenchus dubius* Bütschli, 1873

Tylenchus browni Kreis, 1929

Tylenchorhynchus robustus typicus var. *cylindricus* of Rahm, 1929

Anguillulina dubia (Bütschli, 1873) Goodey, 1932

Anguillulina (Tylenchorhynchus) dubius Bütschli, 1873 (Goodey, 1932)

Tylenchus (Bitylenchus) dubius Bütschli, 1873 (Filipjev, 1934)

Tylenchorhynchus cylindricus of Thorne, 1949

Tylenchorhynchus (Bitylenchus) dubius Bütschli, 1873 (Filipjev, 1934)

Bitylenchus dubius (Bütschli, 1873) Siddiqi, 1986

MEASUREMENTS

Female

(neotype): L=670 μm; a=30; b=5; c=14; V=56. Stylet length=19 μm.

(n=10; after Allen, 1955): L=620–780 μm; a=30–35; b=5.6; c=13–16; v=54–57. Stylet length=18–19 μm.

(n=20; Ottawa population originating from the Netherlands): L=675–870 (778 ± 50.64) μm; a=27–33 (30 ± 1.7); b=5.1–6.9 (5.7 ± 0.48); c=10–14 (13 ± 1.2); c′=2.5–35 (2.9 ± 0.32); V=52–57(55). Stylet length=17.5–19.5 (18.5 ± 0.55) μm. Conus length=8.5–11.0 (9.5 ± 0.87) μm.

(n=20; population from Vineland, Ontario): L=720–820 (759 ± 25.43)μm; a=26–31 (28 ± 1.10); b=5.4–6.5 (5.8 ± 0.27); c=13–17 (14 ± 1.2); c′=2.2–3.4 (2.8 ± 0.30); V=52–57 (55). Stylet length=16.0–19.5 (18.4 ± 0.86)μm. Conus length = 8.5–10.0 (8.9 ± 0.8)μm.

Male

(n=10; after Allen, 1955): L=670–710; a=33–37; b=5.0–5.6; c=12–15. Stylet length=16–17 μm.

(n=10; Netherlands population): L=620–760 (696 ± 40.8)μm; a=26–31 (29 ± 1.9);
    b=4.8–6.3 (5.5 ± 0.38); c=13–16 (15 ± 1.1); c'=2.8–3.4 (3.0 ± 0.19). Stylet
    length=16–20 (18 ± 1.19)μm. Spicule length=25–30 (27 ± 1.4)μm. Gubernaculum
    length=10–15(13 ± 1.69)μm.

DESCRIPTION (emended)

Female (based on Netherlands and Canadian specimens): Body ventrally arcuate, width
25–29 (27 ± 1.1)μm, annules 0.8–1.3 (11.0)μm wide. Body of older or gravid females may
be depressed at vulva (Fig.10A), and midventral body annules anterior and posterior to
vulva irregularly broken or corrugated (Fig. 10K). Lateral field with outer bands irregularly
areolated, the outer of four incisures crenate. Excretory pore 103–121 (111 ± 5.1)μm from
head end, hemizonid 1.0–3.5 μm anterior. Deirids absent.

    Head usually sharply set off from body (Fig. 10B), sometimes to varying lesser de-
grees (Fig. 10C,D), marked by 6–8 fine, distinct annules. First head annule (SEM) typically
quadrangular, sometimes partitioned into 4–6 lobes (Fig. 10E,F,G). Stylet knobs relatively
small, about 3–4 μm wide, sloping posteriorly; dorsal gland orifice 2–3 μm posterior.

    Esophagus 125–141 (131 ± 5.87)μm; from metacorpus to base of basal bulb 56–73
(65.47 ± 4.56)μm long; from head end to base of metacorpus 60–76 (67.77 ± 3.85)μm, basal
bulb elongate-pyriform sometimes slightly overlapping intestine. Intestine extending to
near tail terminus, densely globular, with conspicuous network of intestinal fasciculi (sinuous canals) through its length.

    Ovaries outstretched, oogonia in single or double rows. Spermatheca spherical,
often filled with sperms.

    Tail subcyclindrical or cylindrical, the proportions varying between populations,
ventral surface marked by 39–60 (53 ± 4.33) annules. Tail terminus subhemispherical (often
dorsally convex) or hemispherical, marked by fine, distinct annules. Phasmids 16–25 (19 ±
3.1)μm posterior to level of anus, in anterior 28–55 (36.3 ± 5.9)% of tail.

    Male (based on Netherlands and Ontario specimens): Similar to female in general
morphology and body proportions. Body width 22–26 (24 ± 1.3)μm, annules about 1.0 μm
wide. Esophagus 112–138 (127 ± 7.7)μm long. Excretory pore 98–120 (109 ± 7.0)μm from
head end. Tail region ventrally arcuate. Tail 43–51 (47 ± 3.5)μm long, phasmids 14–20 (17
± 2.9)μm posterior to cloacal opening at anterior 29–47 (37)% of tail. Spicules arcuate, usually fully extended in fixed specimens, proximal end cephalated, distal end with velum.
Gubernaculum protrusive, flexible, linear or proximally arcuate to varying degrees in lateral profile, distal end clavate, distal flanges weakly developed.

    Type host and locality: Soil around roots of *Centaurea cyanus* L.; Frankfort-am-
Main, West Germany.

    Remarks: The intestine of members of the genus *Tylenchorhynchus* is ramified by a
connecting network of fibrillar bundles, termed intestinal fasciculi (= lateral or sinuous canals). In *T. dubius*, fasciculi range in diameters from 0.3 to over 2 μm and are comprised of
thin (14 nm) and thick (70–90 nm) elements arranged in closely packed, linear arrays (Byers
and Anderson, 1973). Their function is unknown, but they are believed to be most likely
contractile or strengthening.

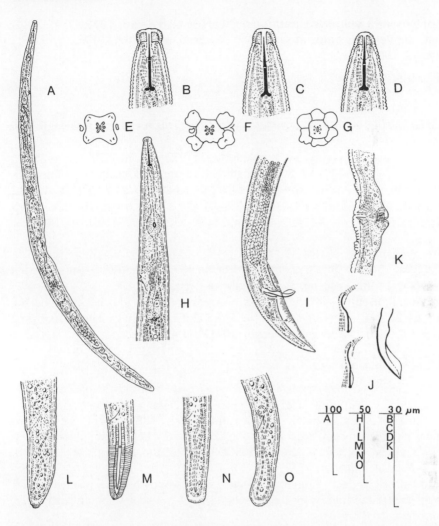

FIGURE 10   *Tylenchorhynchus dubius*. (A) Gravid female. As shown, the vulvar region may be depressed in actively reproducing females. (B–D) Anterior end showing variation in shape of heads and varying degrees to which it is set off from the body (neck). (E–G) Head surface, redrawn from SEM micrographs (Agriculture Canada), showing marked variation in configuration of the first head annule. (H) Esophageal region (neck). (I) Male tail. (J) Spicule and variable forms of the gubernaculum. (K) Midventral region of body at vagina showing irregularities of the body annules characteristic of mature adults. (L–O) Female tail; variation in shape within a Dutch population. Proportions of these forms vary between populations. Other illustrations: Allen (1955); Anderson and Byers (1975); Bridge (1974a); Byers and Anderson (1973); Goodey (1963); Siddiqi (1986). Specimens Canadian National Collection of Nematodes.

Tylenchorhynchus vulgaris Upadhyay, Swarup, and Sethi, 1972
Syn. *Bitylenchus vulgaris* (Upadhyay, Swarup, and Sethi, 1972) Siddiqi, 1986

MEASUREMENTS

Female

(holotype): L=550 µm; a=30; b=5.1; c=14.1; V=56. Stylet length =14 µm. Tail length = 40 µm.

(n=15 paratypes; after Upadhyay, Swarup, and Sethi, 1972): L=560–670 µm; a=25–30; b=4–6; c=14–20; V=52–57. Stylet length=14–16 µm. Tail length=35–45 µm.

(n=7; from Anand, India): L=460–563 (529.6 ± 34.3)µm; a=24–30 (27.7 ± 1.8); b=4.2–5.2 (4.6 ± 0.3); c=13–15 (13.7 ± 0.7); c′=2.2–3.1 (2.7 ± 0.3); V=54–57 (55.4 ± 1.1). Stylet length=13.1–13.8 (13.3 ± 0.3)µm; M=47–53 (50 ± 1.7).

Male

(n=15 paratypes; after Upadhyay, Swarup, and Sethi, 1972): L=530–750 µm; a=25–36; b=4–6; c=14–18; T=43–60. Stylet length=14–17 µm.

(n=8; from Anand, India): L=484–563 (537 ± 29.4)µm; a=29–37 (31.4 ± 2.4); b=4.2–4.8 (4.5 ± 0.2); c=12–15 (13.1 ± 0.9); c′=2.6–3.5 (3.2 ± 0.3). Stylet length=10.8–14.2 (12.8 ± 1.0)µm. Gubernaculum length=10–12 (11.4 ± 0.8)µm.

DESCRIPTION (emended)

Female: Body linear to ventrally arcuate when relaxed. Body width 17–20 (19.1 ± 1.1)µm, annules very fine, less than 1 µm wide, being wider, 1.0–1.5 µm, on neck and tail, without longitudinal striae or ridges. Lateral field about a third of body width, outer of four incisures crenate, inner incisures smooth, generally fusing from anterior to posterior phasmid (Fig. 11D). Deirids absent. Excretory pore 88–103 (93.1 ± 6.5)µm posterior to head end, hemizonid 2–5 annules anterior.

Head bulbous, about 3–5 µm high, 6–8 µm wide, weakly bilobed, marked by 5–7 transverse annules. Head framework lightly sclerotized, basal knobs sloping posteriorly; dorsal gland duct about 1.5–2.0 posterior.

Esophagus 105–128 (116 ± 7.8)µm long, to base of metacorpus 59–65 (62.6 ± 3.4)µm. Metacorpus oval, 13–15 (14)µm long, 9–10 µm wide. Basal bulb pyriform, 23–29 (25.4 ± 1.9)µm long, usually overlapping intestine laterally (Fig. 11C). Esophagointestinal valve conoid. Intestinal fasciculi present, extending through intestine to tail terminus.

Tail conoid, linear, 34–43 (38.9 ± 3.2)µm long, tapering to a smoothly rounded terminus. Ventral surface of tail marked by 32– 43 (36 ± 4.2) annules; cuticle of tail tip markedly thickened. Postanal intestinal sac extending to tail terminus, reported to anterior third of tail in paratypes (see legend for Fig. 11E). Phasmids 11–15 (12.9 ± 1.3)µm posterior to level of anus, at 32–36% of tail.

Male: Similar to female in general morphology. Esophagus length 114–123 (118.8 ± 3.3)µm, to base of metacorpus 59–69 (63.8 ± 3.4)µm. Excretory pore 89–101 (93.1 ± 6.5)µm from head end. Tail 35–46 (40.7 ± 3.7)µm long, bursa continuous. Gubernaculum slightly arcuate, extendable, distal third with lateral flanges (Fig. 11F). Phasmids 14–20 (15.7 + 1.9)µm posterior to level of anus, at 34–43% of tail.

Type host and locality: Soil around roots of *Zea mays* L.; Indian Agricultural Research Institute Farm, New Delhi, India.

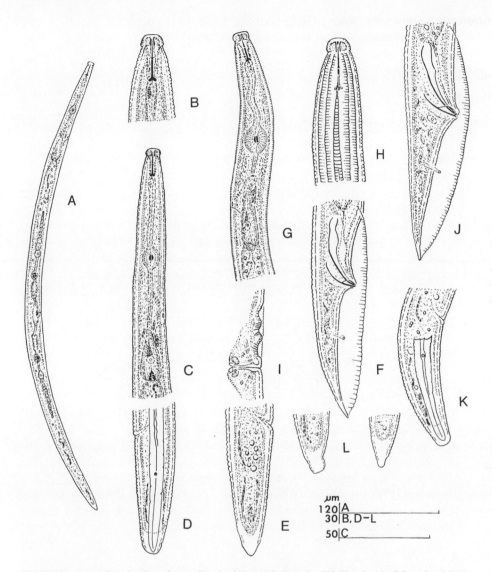

FIGURE 11 *Tylenchorhynchus vulgaris*. (A) Adult female. (B) Head end of female. (C) Female esophagus. (D) Female tail showing lateral field. (E) Female tail. Note that while the intestinal globules are confined within the anterior half of the tail, the post-anal intestinal sac extends to the tail terminus. The presence of intestinal fasciculi in the posterior half of the tail is a reliable indicator for delimiting the intestine. (F) Male tail. *Tylenchorhynchus brevilineatus*. (G) Esophagus of female. (H) Anterior end showing the longitudinal body ridges in the esophageal region which characterize this species. (I) Vaginal region showing irregularities of the midventral body cuticle of some mature individuals. This also is characteristic for some females of *T. vulgaris*, and several other species in Telotylenchinae. (J) Male tail. (K) Female tail. (L) Variation in tail terminus of female. Other illustrations: *T. vulgaris*: Upadhyay, Swarup, and Sethi (1972). *T. brevilineatus*: Gupta and Uma (1985); Kleynhans and Heyns (1984); Siddiqi (1961); Williams (1960). Specimens of *T. vulgaris* courtesy of M. R. Siddiqi, of *T. brevilineatus* courtesy of M. R. Siddiqi, St. Albans and D. J. Hooper, Rothamsted.

Merlinius brevidens (Allen, 1955) Siddiqi, 1970
Syn. *Tylenchorhynchus brevidens* Allen, 1955

MEASUREMENTS

Female

(holotype, after Siddiqi, 1972a): L=690 μm; a=26; b=4.9; c=12; V=56. Stylet length = 16 μm.

(n=11; after Allen,1955): L=540–690 μm; a=23–27; b=4.2– 5.2; c=11–13; V=52–58. Stylet length = 14–16 μm.

(n=15; after Siddiqi, 1961): L=550–850 μm; a=22–29; b=4.6–6.0; c=12–17; V=54–61. Stylet length = 13–15 μm.

(n=9; after Zepp, 1974): L=550–680 (640)μm; a=26–38 (31); b=4.6–5.5 (5.1); c=10.4–13.7 (12.0); c'=2.9–4.2 (3.5); V=54–58 (56). Stylet length = 14–15 (15)μm; M=47–60 (55).

(n=18; after Maqbool, Fatima, and Hashmi, 1983): L=600–700 (640)μm; a=24.7–29.7 (27.2); b=4.9–5.8 (5.3); c=12.6=15.2 (13.8); V=50.8–59.0 (54.8). Stylet length =15.0–16.3 (15.4)μm.

(n=52; after Kleynhans and Heyns, 1984): L=503–696 (586 ± 54.0); a=25.9–40.2 (30.4); b=4.3–6.4 (5.0); c=10.4–16.0 (12.8); V=52.3–59.2 (56.0). Stylet length (n=45)= 13.6–16.4 (14.8 ± 0.60)μm.

Male

(n=6; after Siddiqi, 1961): L=560–640 μm; a=25–31; b=4.6–5.0; c=11.5–13.0; T=40–60. Spicule length=22 μm. Gubernaculum length = 7–8 μm.

(n=4; after Maqbool, Fatima, and Hashmi, 1983): L=560–650 (600)μm; a=25.7–28.8 (27.2); b=4.6–5.2 (4.9); c=10.8–12.1 (11.5); T=40.4–42.6 (41.5). Stylet length = 15.7 μm. Spicule length = 21 μm. Gubernaculum length = 7.35 μm.

(n=9; after Zepp, 1974): L=530–670 (620)μm; a=32–37 (34); b=4.2–5.1 (4.8); c=10.5–12.0 (11.5); c'=3.1–4.5 (3.8). Stylet length = 14–15 (15)μm; M=50–60 (54). Spicule length = 21–24(23). Gubernaculum length =6–7 (7)μm.

DESCRIPTION

Female (after Siddiqi, 1972a; Kleynhans and Heyns, 1984): Body linear to strongly arcuate when relaxed, width 19–24 μm; annules fine, width generally 1 μm or less. Cuticle containing refractive inclusions of varying density. Lateral field about a third of body diameter, with six incisures the outermost of which may be irregularly areolated throughout much of body. Excretory pore (n=45) 73.7–108.0 (89.9 ± 7.3)μm posterior to headend, located anterior to level of basal esophageal bulb. Hemizonid 1–3 μm anterior to excretory pore, hemizonion 9–13 μm posterior. Deirids present, but often obscured by cuticular inclusions, located at or near level of excretory pore.

   Head 6.5–8.5 (7.4 ± 0.50)μm wide, typically low, broadly rounded, continuous or slightly set off from body, marked by five or six fine annules. Basal ring of head skeleton shallow, distinctively arched (Fig. 12, B, C, I). Stylet knobs rounded, sloping posteriorly; dorsal gland duct about 2.5–3.0 μm posterior. Esophagus 110–118 μm long, 55–65 μm to base of metacorpus. Metacorpus 13–16 × 9–12 μm, isthmus 31–34 μm long, basal bulb

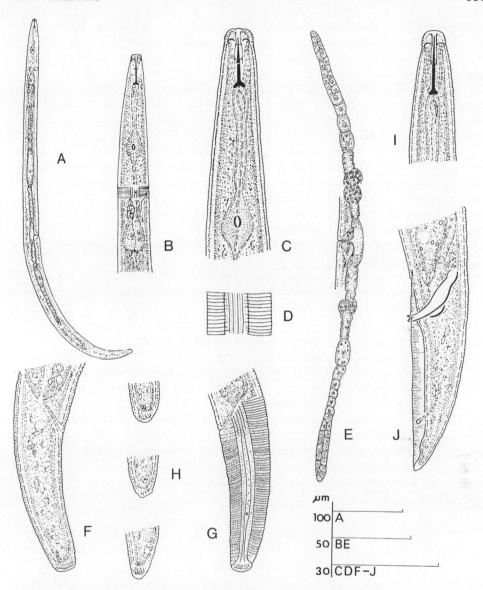

FIGURE 12 *Merlinius brevidens*. Female. (A) Adult. (B) Esophageal region. (C) Head end including metacorpus. (D) Lateral field at midbody. The incisures often are difficult to resolve because of density of cuticular inclusions. (E) Reproductive system. (F,G) Tail. (H) Tail tips showing variation in shape. Male. (I) Head end. (J) Tail. Other illustrations: Siddiqi (1961); Siddiqi (1979); Kleynhans and Heyns (1984); Zepp (1974). Specimens Canadian National Collection of Nematodes; others courtesy of R. Voisin, Antibes, France.

pyriform, 23–25 × 12–15 μm. Esophagointestinal valve rounded, intestine densely globular, without fasciculi. Ovaries outstretched, the anterior often extending to or near the base of the esophagus in gravid females, oogonia in double row. Spermatheca irregularly saccate, typically bi-or tri-lobed, often filled with sperm. Intrauterine egg 61–65 × 22–23 μm. Tail conoid, curved, 34.4–57.8 (45.9 ± 5.1)μm long (n=45), gradually tapering to a smooth broadly rounded or flattened tip. Ventral surface of tail marked by 39–62 fine annules, cuticular inclusions being particularly dense at tail tip. Phasmids near or slightly posterior to midtail.

Male: Uncommon, when present, often in small numbers; similar to female, but head tends to be higher and narrower. Bursa narrow, with dense inclusions, margins finely crenate. Cloacal opening bordered by a pair of subventral tubercles. Gubernaculum small, arcuate, of uniform thickness; not protrusible.

Zepp (1974) reported an abnormal male, which had the tail of a female and lacked a bursa. In other characters and dimensions it was normal.

Type host and locality: Soil around roots of grass; Strawberry Canyon; University of California Campus, Berkeley, California.

Remarks: The arched basal ring of the head framework, distinctive for this species, is rare for members of the Telotylenchidae. With few exceptions, the basal ring has a straight rim, which extends well into the body (neck).

Amplimerlinius macrurus (Goodey, 1932) Siddiqi, 1976

Syn. *Anguillulina macrura* Goodey, 1932, in part

Anguillulina (Tylenchorhynchus) macrura Goodey, 1932 (W. Schneider, 1939)

Aphelenchus dubius Steiner, 1914

Anguillulina (Tylenchorhynchus) dubia Steiner, 1914 (W. Schneider, 1939)

Tylenchorhynchus macrurus (Goodey, 1932) Filipjev, 1936

Tylenchorhynchus macrurus of Allen, 1955, in part

Amplimerlinius dubius (Steiner, 1914) Siddiqi and Klinger, 1980

MEASUREMENTS

Female[i]

(n=20; after Wallace and Greet, 1964): L=830–1190 μm; a=19–25; b=4.8–6.3; c=14–20; V=54–59. Stylet length = 25–34 μm.

(n=10; after Saltukoglu, Geraert, and Coomans, 1976): L=940–1020 μm; a=27–35; b=4.9–5.6; c=17–19; c′=2.2–2.6; V=55–56. Stylet length= 23–23 μm. Esophagus length = 182–190 μm. Tail length=54–57.

(n=12; from Dampvitoux, France; associated with winter barley): L=802–938 (862 ± 51.8)μm; a=23–29 (26 ± 1.8); b=4.8–5.4 (5.0 ± 0.1); c=14–16 (15 ±0.7); c′=2.2–2.7 (2.4 ± 0.2); V=55–60 (57 ± 1.6). Stylet length = 26.9–30.0 (28.5 ± 0.7)μm; M=43–51 (46 ± 2.5).

Male

(n=20; after Wallace and Greet, 1964): L=840–1150 μm; a=21– 26; b=4.7–5.4; c=10–13. Stylet length = 30–39 μm.

(n=8; France, same data as for female): L=796–971 (877 ± 68.0)μm; a=26–30 (27 ± 1.9); b=4.6–5.4 (5.1 ± –0.2); c=10–13 (12 ± 0.8); c′=2.7–3.3 (3.1 ± 0.1). Stylet length = 26.9–30.0 (28.6 ± 1.1)μm. Spicule length = 32–40 (27 ± 2.9)μm. Gubernaculum length = 11–14 (13 ± 1.0)μm.

(n=10; after Saltukoglu, Geraert, and Coomans, 1976): L=940–1050 μm; a=29–35; b=5.2–5.9; c=15–18; c′=2.4–2.8; T=28.5–40.0. Stylet length = 24 μm. Gubernaculum length = 9–11 μm.

DESCRIPTION (emended)

Female: Body robust, C-shaped to spiral when relaxed, width 30– 36 (34 ± 1.9)μm, annule width 1.2–1.5 μm. Cuticle containing fine inclusions particularly evident in lateral field. Lateral field 9–10 μm wide, 26–37% of body width, outer of six incisures cenate, sometimes appearing smooth. Deirids present, 17–32 (24 ± 5.5)μm posterior to level of base of metacorpus, at 44–80% of isthmus. Excretory pore 114–139 (128 ± 6.2)μm posterior to head end, at level of posterior half of isthmus. Hemizonid narrow, immediately, to 4 μm interior to excretory pore.

Head conoid, apex rounded or somewhat flattened, continuous with body or slightly set off, bearing six or seven annules. Head framework (lamina) conspicuously sclerotized, basal ring less so as compared to *A. icarus* (Fig. 13, A, F). Stylet robust, basal knobs rounded, anterior surfaces sloping posteriorly, indented in some specimens; dorsal gland orifice 4–5 μm posterior.

Esophagus 157–186 (179 ± 25.7)μm long, from head end to base of metacorpus 91–111 (101 ± 6.2)μm. Metacorpus ovoid, 21–26 (23 ± 1.4) × 12–15 (14 ± 0.9)μm, isthmus uniformly narrow, 35–45 (40 ± 3.3)μm long, basal bulb ellipsoidal, 24–34 (29 ± 2.8) × 12–22 (17 ± 2.6)μm. Esophagointestinal valve conoid, well developed.

Ovaries outstretched, oogonia typically in a double row, spermatheca small, spherical, axial, usually filled with sperms having a diameter of about 2 μm.

Tail cylindrical, 52–62 (57 ± 4.0)μm long, with 31–43 (38 ± 3.4) annules along ventral surface to midtip of terminus; terminus broadly rounded, marked by numerous fine-coarse annules. Hyaline part of tail end 11–15 (12 ± 3.6)μm thick, occupying 15–25% of tail. Lateral field open at termination on tail end. Phasmids 15–32 (22 ± 4.2)μm posterior to level of anus, at 15–32% of tail.

Male: Similar to female in general morphology. Body width 29–39 (33 ± 3.1)μm, esophagus 155–183 (171 ± 9.6)μm long, from head end to base of metacorpus 92–109 (101 ± 5.3)μm. Excretory pore 123–139 (131 ± 5.7)μm posterior to head end. Tail 67–81 (74 ± 4.2)μm long, postanal lip bearing a pair of large, subventral tubercles. Spicules slight arcuate distally, tip indented. Gubernaculum arcuate, rod shaped in lateral profile.

Type host and locality: Soil around roots of grass; Winches Farm, St. Albans, England.

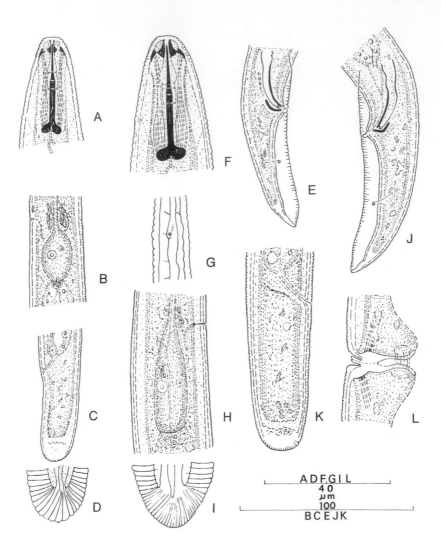

FIGURE 13   *Amplimerlinius macrurus*. (A) Head end. (B) Basal esophageal bulb. (C) Female tail.
(D) Tail terminus. (E) Male tail. *Amplimerlinius icarus*. (F) Head end. (G) Deirid in lateral field,
which is a feature of the species in this genus. (H) Basal esophageal bulb. (I) Tail terminus. (J) Male
tail. (K) Female tail. (L) Vagina showing prominent vulval lips (epiptygma), which are less well
developed in *A. macrurus*. Note that the cuticle of *A. icarus* is particularly thick and distinctly dou-
ble-layered. Other illustrations: *A. macrurus*: Siddiqi (1976b); Siddiqi and Klinger (1980); Wallace
and Greet (1964). *A. icarus*: Siddiqi (1976b); Wallace and Greet (1964). Specimens of *T. macrurus*
courtesy of R. Voisin, Antibes, France; of *T. icarus* Canadian National Collection of Nematodes.

Amplimerlinius icarus (Wallace and Greet, 1964) Siddiqi, 1976
    Syn. *Tylenchorhynchus icarus* Wallace and Greet, 1964
        Anguillulina macrura Goodey, 1932, in part
        Tylenchorhynchus macrurus of Filipjev, 1936, in part
        Tylenchorhynchus macrurus of Allen, 1955, in part
        Merlinius icarus (Wallace and Greet, 1964) Siddiqi, 1970

MEASUREMENTS

Female

(n=20; after Wallace and Greet, 1964): L=1450–1960 μm; a=29– 34; b=5.9–6.9; c=19–25; V=50–57. Stylet length=34–42 μm; M=50–59.

(n=16; intercepted in Canada with *Prunus* sp. from France): L=1489–2008 (1700 ± 156.4) μm; a=30–39 (34 ± 2.6); b=6.0–7.9 (6.9 ± 0.6); c=18–24 (22 ± 2.1); c′=1.8–2.7 (2.3 ± 0.3); V=50–58 (54 ± 2.6). Stylet length=34.6–41.2 (37.8 ± 1.6)μm. M=43–48(45 ± 1.6).

Male

(n=20; after Wallace and Greet, 1964): L=1310–1680 μm; a=28–37; b=5.1–6.7; c=14–18. Stylet length = 36–48 μm.

(n=6; intercepted, as for female): L=1352–1637 (1512 ± 96.1)μm; a=34–39 (38 ± 2.3); b=5.7–7.0 (6.6 ± 0.4); c=14–16 (15 ± 0.8); c′=2.5–3.5 (3.3 ± 0.4). Stylet length=35.4–37.7 (36.7 ± 0.9)μm. Spicule length=41–46 (43 ± 1.7)μm. Gubernaculum length =11–14 (13 ± 1.4)μm.

DESCRIPTION (emended)

Female (based on intercepted specimens from France): One of the largest, most robust species in the Telotylenchidae. Body C-shaped to spiral when relaxed, width 45–58 (50 ± 4.0)μm, annules fine, 0.9–1.5 (1.1)μm wide. Cuticle with two distinct layers. Lateral field 13–22 (16 ± 2.3)μm wide, 29–34% of body width, irregularly areolated in neck and tail region, outer of six incisures smooth or, regularly or irregularly crenate. Deirids present, 21–51 (34 ± 9.2)μm posterior to level of metacorpus base. Excretory pore 136–209 (181 ± 16.0)μm posterior to head end, located from midlevel to base of isthmus, to midlevel of basal bulb in some specimens. Hemizonid narrow, sometimes indistinct, located opposite excretory pore, to 5 μm anterior.

    Head conoid, rounded, continuous with body, marked by four to six transverse annules. Head framework (lamina and basal ring) strongly sclerotized. Stylet robust, basal knobs rounded, anterior surfaces sloping posteriorly; dorsal gland orifice 4–5 μm posterior.

    Esophagus 223–299 (249 ± 17.6)μm long, from head end to base of metacorpus 138–177 (146 ± 9.6)μm. Metacorpus ovoid, 33–42 (36 ± 2.4) × 16–23 (21 ± 2.6)μm, isthmus 33–63 (50 ± 7.3)μm long, basal bulb ovoid, 45–57 (53 ± 3.6) × 15–27 (20 ± 2.6)μm. Esophagointestinal valve large, conoid.

    Ovaries outstretched, oogonia in single or often double row; spermatheca small, spherical, axial, usually filled with sperms having a diameter of about 3 μm. Vulva depressed, vulval lips present, vagina walls very thick.

Tail cylindrical or subcylindrical, 65–93 (78 ± 8.9)μm long, with 34–53 (44 ± 5.2) annules on ventral surface; terminus broadly rounded, marked by numerous fine annules. Hyaline part of tail end 14–18 (16 ± 1.4)μm thick occupying 16–24% of tail. Incisures of lateral field open at termination on tail end. Phasmids 24–44 (31 ± 5.9)μm posterior to level of anus, at 32–53 (40)% of tail.

Male: Similar in general morphology to female, but of smaller dimensions. Body width 39–43 (40)μm, esophagus 221–237 (230)μm long, from head end to base of metacorpus 132–139 (137)μm. Excretory pore 170–184 (177)μm from head end. Tail 91–111 (99)μm long, postanal lip bearing a pair of large, subventral tubercles. Spicules slightly arculate distally, spicule tip broad, deeply indented. Gubernaculum arcuate, rod shaped in lateral profile.

Type host and locality: Soil around roots of grass; Winches Farm, St. Albans, England.

BIOLOGY OF SPECIES OF AGRICULTURAL IMPORTANCE

Tylenchorhynchus claytoni

Distribution

North temperate zone, N. America, Europe, N. Asia; common in eastern and treed areas of the United States; its distribution in North America and Europe is catalogued by Loof (1974), Mathias (1978b) and Norton (1984); the latter presents a map indicating the easterly distribution in North America. *Tylenchorhynchus claytoni* was also found by Kiryanova (1951) in Uzbekistan SSR; by Katalan-Gateva and Baicheva (1976) in Bulgaria; by Mamiya (1969) and by Yamazaki et al. (1981) in Japan; by Choi (1975) in Korea; by Prasad and Rao (1977) and Krishnaprasad and Krishnappa (1982) in India; and by Braasch (1976) in Germany (DDR).

Ecology

Reproduction took place in temperatures favorable to the host (Krusberg,1959); on tobacco, the highest numbers were obtained at 29–32°C (85–90°F), and on wheat 21–27°C (70–80°F). No reproduction occurred between 2–13°C (35–55°F) but the nematode lived in moist soil without host plants for 10 months at 2–24°C (35–75°F). Population increased on Kentucky 31 tall fescue at 21–34°C (McGlohon et al., 1961); in greenhouse studies, soil type had little effect on *T. claytoni*. Exposure for 24 hr to –15°C killed 75–90% of *T. claytoni*, and all were killed by –43°C (Miller,1968). *Tylenchorhynchus claytoni* showed positive thermotaxis in a thermal gradient from 24–48°C and migrated until death occurred at 44–48°C (Hitcho and Thorson,1972). Maize roots in carbofuran-treated soil were not attractive to *T. claytoni* (DiSanzo,1973).

Development

On lucerne seedlings in monoxenic nutrient medium culture at 24°C, eggs were laid close to the roots, in particular the root caps and root hairs (Krusberg,1959). Eggs hatched after 6 days and development to the adult took 33 days. Males were necessary for reproduction. Wang (1971), in similar culture at 28°C, found the life cycle took 31–38 days, egg-to-egg, with egg laying beginning 8–10 days after the last moult. The first moult was inside the egg, and the J2 larvae hatched in 5–6 days at 22–25°C. The sexes were recognizable by the gonad primordium in the third stage; embryology was described.

Feeding

On lucerne, *T. claytoni* fed on the epidermal cells of the root cap and among but not actually on the root hairs (Krusberg,1959). The nematode glided over the root surface seeking a feeding area, and after puncturing the epidermis, fed in the same spot for about one half hour; no apparent injury to root cells occurred, similarly as for annual ryegrass, creeping red fescue and Kentucky bluegrass (Troll and Rohde,1966). Attacked tobacco roots did not show definite lesions (Graham,1954), although they were shrivelled, sparsely developed and much retarded in growth. *Tylenchorhynchus claytoni* did not reach the xylem in feeding on tobacco (Lucas and Krusberg,1956).

Host-Parasite Relationships

Ectoparasitic, pathogenic. Nuclei in cells of the outer cortex and epidermis of maize roots parasitized by *T. claytoni* were spheroidal instead of ellipsoidal, even in cells not directly attacked, although meristem cells were not affected (Deubert et al., 1967). Reduced growth and off-color symptoms have been reported on various plants (Loof,1974). *Tylenchorhynchus claytoni* has been associated in particular with azaleas (Jenkins,1956; Sher,1958; Oostenbrink,1959; Heungens, 1965; Vogel and Bernet,1958; Lindhardt and Bagger,1967; Townshend,1967; Yokoo and Koga, 1967; Yamazaki et al., 1961; Barker and Worf, 1964,1966; Streu,1975); holly (Haasis et al.,1961; Springer,1964; Barker et al.,1974; Aycock et al.,1976; Barker et al.,1979; Benson and Barker,1977); and conifers, chiefly pines (Hopper,1958; Ruehle and Sasser,1962; Ruehle,1966; Sutherland and Adams,1965; Barham et al.,1974).

Tylenchorhynchus clarus

Distribution

Subtropical North America, Europe, and Mediterranea. The species is described from California (Allen,1955) and known from six other states (Norton,1984); Mexico (Knobloch, 1975); Holland (Dao, 1970); Egypt (Elmiligy,1969); and Jordan (Hashim,1983).

Ecology

The optimum temperatures, 24 and 27°C, for population increase (Noel and Lownsbery,1978) were similar to those determined by Malek (1980) namely, range of 20–35°C, within optimum at 30°C, but were contrary to those found by Edongali and Lownsbery (1980). Noel and Lownsbery (1978) found greater multiplication of *T. clarus* at 20°C than at 25 or 30°C on lucerne. They also found more rapid reproduction under elevated light intensities. The cultivar of lucerne could also have been a factor in the above contradictory results. The nematode reproduced to a greater extent in clay loam and least in sandy loam (Kheir et al.,1977), and the female:male ratio was 2:1 at low population levels (ca. 250–500/plant) but approached 1:1 as population levels increased (ca. 4000–8000).

Host-Parasite Relationships

Ectoparasitic, occasional endoparasitic. *Tylenchorhynchus clarus* penetrated lucerne roots in the zone of differentiation, with up to half of the body length inside the root (Noel and

Lownsbery,1978); about 10% of the population were completely within cortical tissue. Nematodes were also abundant in the meristematic and elongated zones. Although lesions from feeding activity were not found, host feeder roots were reduced in number.

Tylenchorhynchus agri

Distribution

North America Midwest. Further geographic locations within the United States are presented by Norton (1984).

Ecology

Between 5–20°C storage temperatures, nematode numbers generally remained constant or increased in storage (Elmiligy,1971). Reproduction occurred over 20–35°C (Malek,1980) on red clover, with a maximum population increase at 30°C; the species is in a group (including *T. agri, T. nudus, T. clarus, and T.annulatus)* which requires relatively high temperatures for maximum population increase. Species in this group multiply very little during winter in greenhouse culture, but will increase rapidly to high numbers in the summer (Malek,1980).

Development

The life cycle was 25 days at 27°C, with a 7 day period between oviposition and egg hatch (Coates-Beckford,1982a); population increases were more rapid at higher temperatures up to 36°C. Embryonic development was followed (Coates,1977) from one-celled eggs deposited by the female through to J2's hatching on day 7 after oviposition. Juvenile stages were seen at 7, 13, 19 days, respectively, for J2's, J3's, and J4's. Female:male ratio was 2:0 (Malek,1980).

Feeding

Feeding was ectoparasitically on root-hair zone epidermal cells, usually for less than 5 min duration (Coates-Beckford,1982a). Amosu and Taylor (1974b) observed feeding on root epidermal cells in the elongation zone.

Host-Parasite Relationships

Ectoparasite. *Tylenchorhynchus agri* alone slightly stimulated *Trifolium pratense* plants, which increased their top growth and weight progressively with increasing nematode inoculum levels (Amosu and Taylor,1974a). In contrast, with clipping, red clover plant weights were reduced by the nematode (Coates-Beckford,1982b), and the nematode had no effect on a first crop of winter wheat but plant stature and dry matter of subsequent wheat crops were reduced (Coates-Beckford and Malek,1982).

Tylenchorhynchus annulatus

Distribution

Subtropical and tropical in all continents except Europe; heavily researched in North America. The species *annulatus* was first described from Hawaii (Cassidy,1930). Known from southern and central United States in N. America; Nigeria, Senegal, Sierra Leone in Africa; Trinidad and Venezuela in Central and S. America; India and Bangladesh in Asia (Siddiqi,1976a). Subsequent reports from Cuba (Fernandez and Ortega,1985), Costa Rica (Lopez and Salazar,1986), and Guyana (Singh,1972); Pakistan (Saeed,1974), Taiwan (Lin, 1970), Philippines (Gargantiel and Davide,1973), and Okinawa, Japan (Fukudome, 1978); and Australia (Stirling and Vawdrey,1985). Norton (1984) shows the distribution in the Mississippi Valley and the southern United States.

Ecology

*Tylenchorhynchus annulatus (*syn. *T. martini)* is thermophilic (Malek,1980), along with *T. agri*, *T. nudus*, and *T. clarus*. *Tylenchorhynchus annulatus* multiplied best on oat and wheat seedlings at 30°C (Patel and MacDonald,1966) and less well at 18 and 24°C. The preferred temperature of 30°C and optimum range of 20–35°C was confirmed by Malek (1980) on red clover and high reproductive rate, an increase of 625× (from 200 to 125,000) in 90 days. *Tylenchorhynchus annulatus* numbers increased on sorghum, were static on soybeans, and remained low on corn and wheat, at unspecified temperatures (Dickerson et al.,1978); this disagrees with Patel and MacDonald (1966). Adults and J4 juveniles survived freezing soil temperatures of –3°C and –16°C, with J2's and J3's being less resistant (Patel and Mac-Donald,1968). The optimum soil moisture content for survival of *T. annulatus* (Johnston, 1958) was 40–60% of field capacity, compared with 75–100% for other nematodes tested. Populations in rice fields were inversely correlated with rainfall or irrigation (Hollis and Fielding,1958). In Trinidad (Singh,1978), numbers of *T. annulatus* were low from January to May and peaked in August.

Host-Parasite Relationships

Migratory ectoparasite, pathogenic. Inoculated sugarcane plants had more sparse, irregular and stubby roots than the check plants (Birchfield and Martin,1956). Ectoparasitic feeding was seen on young succulent roots of bermuda grass (Johnson,1970); damage resulted from cell destruction, presumed injection of digestive secretions and removal of cell contents; lateral root feeding resulted in root pruning, decreased water and mineral uptake and depressed plant growth; root-tip feeding caused severe stunting and necrosis. Alhassan (1968) found radioactive ^{32}P was removed from rice by *T. annulatus* and traced in the intestine of the nematode; the greatest quantity was in the soil, next in rice tops and roots, and least in the nematodes. *Tylenchorhynchus annulatus* reproduced equally well on both rice root tip and embryo callus and remained infective when re-inoculated onto rice roots (Plowright,1988). "Swarming" nematodes caused greater reduction in root length and dry weight of two rice cultivars, with severe stunting and chlorosis, than "non-swarmers" (Joshi and Hollis,1976).

"Swarming"

"Swarming" of *T. annulatus* adults and all juvenile stages was observed first by Hollis (1958). Swarming was initiated only with abundant and rapid growth of the host plant, permitting rapid nematode multiplication, and resulted from a stickiness of the nematode cuticle (Hollis,1962; Hollis and McBride,1962). Endopectidases, especially trypsin, masked the stickiness and inhibited swarming, but removal of the enzymes by washing reversed the inhibition. Reversion to the original non-swarming state also occurred when the nematodes were reared with poor nutrition (McBride, 1964). Swarming nematodes showed swelling of the external cuticular cortex, separation of the cortical layers, and light-colored spots in the matrix and fiber layers (Ibrahim,1967); the stickiness on the cuticle probably resulted from partial dissolution of the cortex and matrix. Swarming nematodes had cytoplasmic virus particles in the hypodermis and somatic muscle layer, and on the cuticle surface; non-swarmers did not have such particles (Ibrahim et al.,1973,1978). The authors concluded that swarming was symptomatic of a cytoplasmic polyhedrosis virus disease, suggested that swarming localized and separated diseased nematodes from the healthy population, and speculated that swarmers would die of starvation, if not of disease.

Tylenchorhynchus nudus

Distribution

Central N. American, Orient, India. The nematode has been reported mainly from the U.S. midwest and Mississippi River Valley (Norton, 1984), from Taiwan (Hu et al., 1969), from Okinawa and the Japanese Ryukyu Islands (Gotoh, 1968; Gotoh, 1976), from Korea (Choi, 1975) and from India (Misra and Das, 1979; Baqri and Ahmad,1981).

Ecology

In a series of studies, Smolik and Malek (1972a) determined the prevalence of *T. nudus* in almost 50% of samples from Kentucky bluegrass (*Poa pratensis*) in South Dakota; found (Smolik and Malek,1973) an eight-fold nematode increase in 4 months; and established an optimum temperature for reproduction of 30°C on wheat (Smolik and Malek,1972b), on sorghum (Smolik,1977) and on red clover (Malek,1980); the latter also found a 20–35°C range for reproduction on *Poa pratensis*. This species is thermophilic (Malek,1980). Thomas (1978) found that *T. nudus* numbers were unaffected by various ploughing, disking, and "no-till" tillage regimes. Under soybeans, the fewest *T. nudus* were found in the "toeslope" of a "toposequence" (Nyhan et al.,1972). Schmitt and Norton (1972) found *T. nudus* on high dry locations of prairies, but not in low poorly drained potholes, and Norton and Schmitt (1978) found *T. nudus* dominating a nematode community of 21 species only on a flat upland soil drift. *Tylenchorhynchus nudus* was widely distributed on sugarcane in Taiwan, in loam and clay soils; sugarcane yield increased by 10–20% with nematodes controlled by fumigation (Hu et al.,1969).

Host-Parasitic Relationships

Ectoparasitic. *Poa pratensis* crowns and roots were reduced significantly by *T. nudus*, especially in sandy loam and under low moisture and nutrient regimes (Smolik and Malek,1973). Spring wheat showed growth reduction by *T. nudus* and mutual suppression of *T. nudus* and *Helicotylenchus leiocephalus* when co-inoculated (Smolik,1972);sorghum, especially unfertilized, was reduced by *T. nudus* at 15, 20, 30, 35°C (Smolik,1977). When *Solanum melongena* was co-inoculated with *T. nudus* and *Meloidogyne incognita*, *Hoplolaimus indicus* or *Criconemoides ornatus*, or with all three species, *T. nudus* did not affect root growth, but reproduction was reduced in the four species co-inoculation (Misra and Das,1979).

Tylenchorhynchus mashhoodi

Distribution

Africa, Asia. Known primarily within India and Pakistan and within the Gambia-Senegal-Mauritania region (Fortuner and Merny,1973) and in Cameroon (Samsoen and Geraert,1975). Singh and Khera (1978) reported *T. mashhoodi* from Calcutta; Gupta and Uma (1985) found it on the roots of *Saccharum officinarum*; Pakistan (Qasim et al.,1988).

Ecology

The species was frequent and numerous in "medium" elevation paddyfields containing no saltwater, but were absent from lowlying "mangrove" fields with high salt content (Fortuner and Merny,1973). Fortuner (1975) found *T. mashhoodi* in a topographical gradient, with the species gradually disappearing as rice field elevation increased from lowlying flooded fields toward nonflooded upland fields with water tables near, but below surface. *Tylenchorhynchus mashhoodi* population levels increased in wet aerated soils but were apparently not adapted to constant flooding (Fortuner,1976). In sandy loam, the species declined sharply after heavy rain (Chaturvedi and Khera,1979). Dissolved hydrogen sulphide was toxic to *T. mashhoodi*, due to pH lowering (Fortuner and Jacq,1976); hydrogen sulphide released by sulphate-reducing bacteria reached lethal concentrations within 3 days, a possible mechanism for biological control of *T. mashhoodi* between crops. In Asia, peas were the preferred host and 23°C the optimum temperature for population increase (Chaturvedi and Khera,1979). Gaur and Haque (1986) found weak interspecific competition among *T. mashhoodi*, *T. vulgaris*, *Hoplolaimus indicus*, and *Rotylenchulus reniformis* because of their different niches on the host. On rice, Nadakumar and Rao (1974) found *T. mashhoodi* on the boot leaves and sheaths in high air humidity conditions. The species moved rapidly toward grass roots on water agar in petri plates (Azmi and Jairajpuri,1977b); showed declining activity on agar with increasing water-film thickness or increasing agar concentration (Azmi and Jairajpuri,1976); and tended to move deeper into agar when exposed to light (Azmi and Jairajpuri,1977a).

Host-Parasite Relationships

Ectoparasite? *Tylenchorhynchus mashhoodi* produced significant stunting, weight reductions, and discoloration of maize in pots at inoculum levels of 100/pot or more (Mahapatra and Das,1984).

Tylenchorhynchus cylindricus

Distribution

North America. Known from Mediterranea and Orient. This type species for the genus *Tylenchorhynchus* is known from 15 U.S. states (Norton, 1984) as well as from Egypt (Oteifa and Tarjan, 1965; Nesterov and Hamed, 1972), Iran (Abivardi, 1973) and Taiwan (Tu et al., 1972).

Ecology

The species is indigenous to the arid soils of the western United States, associated with roots of desert plants (Thorne,1961). Abivardi (1973) reported up to 2000 *T. cylindricus* per 100 cc of soil around roots of guava in Iran. The species is also very numerous in vineyards in Egypt (Nesterov and Hamed,1972).

Host-Parasite Relationships

Ectoparasite. *Tylenchorhynchus cylindricus* caused moderate stunting of Hopi M cotton and tepary bean (Reynolds and Evans,1953); feeding at the root tips resulted in stunted root systems and tops. Riffle (1970) did not find reduction of growth of ponderosa pine over 13 months in a greenhouse, although the nematode parasitized and multiplied well on this plant; *Pinus edulis* was not a host for the nematode (Riffle,1972). Populations of *T. cylindricus* were greater on sites where cotton showed poor growth (Tu et al.,1972).

Tylenchorhynchus brassicae

Distribution

Mediterranea, India, Orient. The species was described (Siddiqi,1961) from North India, and is now also known from Egypt (Oteifa and El–Sharkawi,1965); Spain (Romero and Arias, 1969); Israel (Cohn et al.,1973); China (Li et al.,1986); and the Georgian SSR (Bagaturiya,1971). Norton (1984) quotes Siddiqi as reporting an occurrence in Florida.

Ecology

High temperature and low moisture are unfavorable for survival of the nematode in the absence of hosts (Siddiqi and Khan,1974); at moderate temperature and moisture, the nematode can survive up to 240 days. *Tylenchorhynchus brassicae* populations around *Citrus sinensis* (best host) peaked in the rainy season at 0–10 cm depth and were minimal in winter at 10–20 cm (Dwivedi et al.,1985); around *Psidium guajava*, maximum numbers were at 20–40 cm in spring; around *Zizyphus jujuba* populations were greatest in autumn at 0–10 cm. Nematode increase seemed positively correlated with soil potassium and phosphorus, but not with soil organic carbon. Cauliflower growth in pots of nematode-infested soil increased with soil quantity (Khan et al.,1985) although nematodes also increased more rapidly in the greatest soil volume.

Feeding

Ectoparasite. Feeding of *T. brassicae* on cabbage (Khan et al.,1986) was ectoparasitic on the meristematic zone, elongation zone, and differentiation zone but rarely on the root cap or cotyledons; feeding lasted 30–60 sec. Long starvation periods resulted in longer initiation of feeding and also longer duration of feeding.

Tylenchorhynchus capitatus

Distribution

Widespread on all continents; broad host range (Jairajpuri,1985). Norton (1984) shows the distribution in North America.

Ecology

Tylenchorhynchus capitatus on red clover reproduced over 10–20°C with an optimum at 20°C but did not reproduce at 30°C (Malek). Siddiqi (1961), Swarup et al. (1964), Seshadri et al. (1967), Mahajan (1974), and Maqbool (1982) observed that the species in India and Pakistan occurs in the cooler climates of hilly regions (Jairajpuri,1985). Numbers were highest 20–25 cm vertically in soil and 20 cm horizontally from host potato plants (Maqbool and Hashmi,1985), and population increase was greatest in a 90% sand/10% clay soil (Maqbool and Hashmi,1986). Nematode numbers gradually declined in soils without a potato host. Intersexes were described as essentially female but with poorly developed spicules and gubernaculum (Wouts,1966; Seshadri et al.,1967).

Host-Parasite Relationships

Ectoparasite. Whether or not pathogenic in the strictest sense, *T. capitatus* is clearly parasitic on some hosts; hollyhock plants showed 70% stunting, and apple, plum, strawberry, and citrus also showed poor growth (Swarup et al.,1964). The external cell walls of the hypodermis of potato and wheat showed cellulose thickening due to *T. capitatus* (Varo-Alcala,1970). Bridge (1976) considered the nematode a root ectoparasite in association with various plants in Ecuador, and Katalan-Gateva and Baicheva (1978) regarded it as pathogenic to tobacco in the Bulgarian Mesta River basin. Vovlas (1983) found *T. capitatus* (syn. *Q. acti*) as semi-endoparasitic on corn feeder roots, causing lesions and cavities in epidermal, subepidermal, and cortical tissue.

Tylenchorhynchus maximus

Distribution

Northern Hemisphere, Europe, North America. The species was described (Allen,1955) for several U.S. locations and noted from Canada; Mathias (1978) quotes others (Sturhan, 1966; Doa,1970; Brzeski,1977) for its distribution in Germany, Holland, and Poland. Waseem and Craig (1962) reported the species in Nova Scotia, Canada, and Powell (1964) found it in Georgia, United States. Norton et al. (1984) indicated a general distribution in the United States and Canada.

Ecology

Malek (1980) cites the temperatures for *T. maximus* population increase as a range, 15–25°C, with an optimum at 25°C, with no survival at 30°C on *Poa pratensis* (Kentucky bluegrass). Thus *T. maximus*, along with *T. dubius*, *T. canalis*, *T. capitatus*, and *M. brevidens*, fits a group of species preferring cooler soil temperatures.

Development

Bridge (1974b) found that, similar to *A. icarus*, the J1 of *T. maximus* inside the eggshell moved in figure-eights for 3 days, the eggshell softened and increased in size about 12 hr before hatching and the J2 emerged from an endwall protrusion after piercing it with a stylet-thrust. On *Lolium perenne* seedlings cultured on agar, 15 *T. maximus* females laid 95 eggs over a 19-day period, hatching occurring 17–19 days after deposition. Bridge (1974b) called *T. maximus* parthenogenetic, although Anderson (1977) has described the male, noting that no sperm were observed in any female from populations with males.

Feeding

Ectoparasite. Three feeding phases of *T. maximus* were observed (Bridge and Hague,1974); first,stylet thrusting until penetration of the host cell wall was achieved; then a delay or salivation period when a spherical mass could be seen increasing around the stylet tip; finally, pulsation of the median esophageal bulb during ingestion of cell contents. *Tylenchorhynchus maximus* browsed while on epidermal cells of the root hair zone and elongation zone, moving continuously from cell to cell, feeding for short periods of 3–16 min. However, at root tips, nematodes remained 2–3 days at one site, feeding up to 3 hr. Penetration of epidermal cells of the root tip by one nematode attracted others which aggregated at the site, usually within 6 days. This aggregation did not occur at sites other than the root tip. With intensive feeding, the root usually ceased growth within 48 hr, after bending abnormally.

Mechanical breakdown of epidermal cells resulted in cavities, where nematodes partially or wholly entered to feed on cortical cells and undifferentiated vascular cells. When primary root tips ceased growth because of attack by large numbers of *T. maximus*, the nematodes often dispersed back along the roots and re-aggregated at young lateral root tips.

Tylenchorhynchus dubius

Distribution

Northern Hemisphere; common in Europe, probably also in Asia (USSR, India, Korea); frequent in northcentral and northeastern United States and Canada; Norton (1984) maps the North American distribution as including 13 states of the United States and Ontario, Canada. Reports from the USSR include Skuodite-Rudzyavichene (1970) in Lithuania; Dement'eva (1971) in Moldavia; Glotova (1972) from Amur, far eastern USSR; Baranovskaya et al. (1975) near Moscow; Shesteperov (1975) at Moscow; Mavlyanov (1972) in Uzbekistan; Sabova et al. (1979) in Czechoslovakia; and Stoyanov (1979) in Bulgaria. Prasad et al. (1964) reported *T. dubius* from India, and Choi (1975) from Korea.

Ecology

Tylenchorhynchus dubius is characteristically found in sandy soils; stress processes typical of such soils such as desiccation, flooding, and high and low temperatures all affect the species. Desiccation is detrimental to all life stages including eggs, which are the most susceptible, followed by J2's, males, and J3's, with females and J4's the most resistant; the species was almost eradicated over 19 weeks in dry soil (Simons,1973). Adverse conditions of temperature (whether high or low), flooding, and starvation are also deleterious, especially to J2's (Sharma, 1971). Optimal reproduction temperature for *T. dubius* was considered as 20°C (Brzeski, 1970), 19–22°C (Sharma, 1971), or 10–25°C (Malek,1980), although the minimum for reproduction was 13°C (Sharma,1971) or 10°C (Malek, 1980). Saynor (1972) found survival of the nematode in soil without a host for 12 months at 5 and 10°C, but no survival after 4 months at 20 or 30°C, while Malek (1980) found no survival at 35°C. Short-term survival (7–14 days) at –11°C was determined by Sharma (1971). The variation in acceptable temperature range or thermal deathpoints may relate to acclimation to local conditions, which in the U.S. Midwest may be warmer than in the United Kingdom or the Netherlands. Brzeski and Dowe (1969) determined that, although egg hatch was greatest at pH 7.0 and was inhibited at pH 3.0, the nematode was more prevalent in acidic soils and reproduction increased with decreasing pH.

Development

At 20°C in tap water, the *T. dubius* life cycle from egg to egg required 40 to 48 days (Sharma,1971). The species is bisexual, reproducing by amphimixis; 8 days after egg laying, the first molt took place inside the shell, followed within hours by hatching of the J2 juvenile. The J3 stage was reached after 12 days; the J4 after 16 days; and the adult after 32 to 40 days. All the active stages, J2 to adult, fed parasitically on roots of hosts.

Feeding of *T. dubius* on root hairs and epidermal cells just behind the apical meristems never lasted more than 10 min in one spot (Klinkenberg,1963); epidermal cells were punctured after 20 to 50 rapid stylet thrusts, after which a substance was seen to flow from the stylet into the cell forming a spherical mass. Following a quiet period of about a half-minute, the esophageal bulb pulsated rapidly while the spherical mass shrank and disappeared, and the stylet was withdrawn leaving no visible trace of damage. Feeding was in the root elongation zone (Sharma,1971); no feeding on root hairs was seen; the head did not penetrate the root; and cells used for feeding showed a yellow-brown discoloration visible only at high magnification. The nematodes moved actively and seemed not to interfere with one another although in close contact (Wyss,1973); the flow of salivary granules from synthesis in the dorsal gland was to the gland reservoir. Saliva was not injected into plant cells as granules but a clear zone was seen at the stylet tip in the granular plant cytoplasm. During the ingestion process, liquid food only was withdrawn but not to the point of emptying the cell; accumulation of cytoplasm at the clear zone around the stylet was seen. After feeding, when the stylet was withdrawn, no cell content exuded from the site, but the accumulated cytoplasm ceased streaming and gradually coagulated; root hairs ceased growth and were killed but root growth was not seriously affected. Anderson and Byers (1975), and Byers and Anderson (1973) have studied extensively the ultrastructure of the stylet, esophagus, and related musculature in *T. dubius*. Their observations probably apply as well to all stunt nematodes; they provide a thorough and reasonable explanation of stylet movement and the controlled release of substances from secretion granules.

Host-Parasite Relationships

Browsing ectoparasite. Many of the reports of detrimental *T. dubius* association with plants have indicated stunted growth and weight reduction as the main effects (Bridge, 1974). Varo-Alcala et al. (1970) reported destruction of root epidermal cells and necrosis of the whole root cortical parenchyma of carnations in Spain. Laughlin and Vargas (1972a) found reduction of foliar and root weights of "Toronto" creeping bentgrass (*Agrostis palustris*) and also found suppression of secondary stolon formation, shortening of internodes and premature initiation of flowering; the symptoms suggest a nematode-induced early maturity.

Tylenchorhynchus vulgaris

Distribution

Common in India, one report from Poland (Kornobis,1981). The species was described (Upadhyay et al.,1972) in India.

Ecology

Upadhyay and Swarup (1972) cultured the species monoxenically on excised corn roots and callus; determined the species is polyphagous but prefers Gramineae, and multiplied to the greatest extent in a 3:1 sandy loam:sand mixture at pH 5.5–6.5. Multiplication of up to 40 × in 90 days depended on the corn host cultivar (Upadhyay and Swarup,1976). Infested wheat showed severe yellowing, leaf blight, and stunting and dwarfing of plants (Patel et al.,1987).

Development

Ectoparasite. Siyanand et al. (1982) report four molts, the first being within the egg. Males were necessary for reproduction (Upadhyay et al.,1972); the life cycle was completed in 25–27 days at 25–30°C; all stages were attracted to the host by diffusible substances released by maize roots, and fed on root hairs. Combined inoculation with *Pratylenchus thornei* or *Hoplolaimus indicus* shortened the life cycle (egg-to-egg) of *T. vulgaris* by 2–3 days, down to 13–15 days, at 30°C (Siyanand et al.,1982).

Merlinius brevidens

Distribution

Known from all continents (Siddiqi,1972a); from Korea (Choi,1975); in Hungary (Decker and Manninger,1977); and in the far eastern USSR near Vladivostok (Eroshenko,1981). A distribution map for the species in North America is in Norton (1984).

Ecology

In a California alfalfa field, *M. brevidens* populations clustered in zones of fine-textured soil (23% sand,43% silt,34% clay)(Goodell and Ferris,1980). A population increase after rainfall was due to increased hatching (Norton,1959). The species survived for only short periods in wet soils, but persisted for up to 9 months in dry soils (Meagher,1970). Over a 90-day period, the species reproduced at 10–20°C, with an optimum increase at 20°C, and did not survive at 30°C (Malek,1980). A 5-month reproductive period occurred in spring and early summer, with a population peak towards autumn (Bergé et al., 1973).

Development

A feeding female was observed to lay 11 eggs over 120 hr while browsing along a root. Other details of the life cycle are not known.

Feeding

The nematode fed ectoparasitically on epidermal cells and grass root hairs, moving rapidly from one feeding site to another (Bridge and Hague,1974). Stylet penetration lasted from 33 sec to about 4 min; the complete feeding process on average was 11 min to a maximum 27 min. No apparent mechanical damage to root hair or epidermal cells was noted nor were secretions from the stylet tip seen during feeding.

Host-Parasitic Relationships

Ectoparasitic. Stunting, yellowing, reduced tillering and reduced yields of wheat and barley were severe in the presence of *M. brevidens* and *Olpidium* sp. together (Langdon et al.,1961),but only after wheat was vernalized. However, Mayol (1981) found that inoculation with nematodes only at planting of wheat or at the start of vernalizational caused foliage and weight reductions. Association of *M. brevidens* with fungi, usually *Olpidium brassicae*, has seriously depressed wheat yields (Schlehuber et al., 1965; Sturhan,1966; (Mathias,1978b). *Merlinius brevidens* caused cellulose thickenings of the external cell walls of the root hypodermis of wheat and barley (Varo-Alcala et al.,1970).

Amplimerlinius macrurus

Distribution

Europe, Africa, Asia, Mediterranea; localized in the United Kingdom. The species was separated (Wallace and Greet,1964) from *A. icarus* which is longer. Filipjev and Schuurmans Stekhoven (1941) indicated the species was found in England, Switzerland, Holland, Japan, and the Belgian Congo. Saltukoglu et al. (1976) reported the species from the Istanbul area of Turkey. Norton (1984) cites one published occurrence (as *Tylenchorhynchus macrurus*) in North Dakota.

Ecology

Wallace and Greet (1974) determined that *A. macrurus* could survive several days in dry soil, tolerating high osmotic pressures rarely achieved in most soils.

Feeding

Ectoparasite and semi-endoparasite. Goodey (1943) observed *A. macrurus* feeding on oat and ryegrass roots, with the anterior third of the body embedded. *Amplimerlinius macrurus* and *A. icarus* feed as fairly sedentary ectoparasites, (Bridge and Hague,1974) but also penetrate roots and feed on internal tissues, unlike *M. brevidens* or *T. dubius*. Ectoparasitic feeding on epidermal cells of the root hair region, elongation region, root tip, and root cap cells took place by penetrating with the head at right angles to the cell, lips pressed firmly against the wall, and direct thrusting of the stylet. During the following delay and median bulb pulsation period, *A. macrurus* fed continuously for 13.5 hr. As semi-endoparasites, they fed on cortical tissues with heads embedded in the root, using multiple thrusts at varying angles to breach the cell wall. Similar multiple thrusts and slow side-to-side head movements achieved penetration of subepidermal cells. With the head embedded, median bulb pulsation continued for hours or days. Nematodes were firmly attached and not easily dislodged. Little observable damage, except for darkening of the cells surrounding the embedded head was evident; root growth appeared normal. The authors suggest that *A. macrurus* and *A. icarus*, with strong cephalic framework and robust stylets, represent advanced ectoparasites tending toward "passing over into endoparasitism."

Amplimerlinius icarus

Distribution

Known from Europe. The species was described from England (Wallace and Greet,1964) and reported from Sicily (Vinciguerra and La Fauci,1978). Wyss (1969) found *A. icarus* in lower Saxony and Sturhan (1966), Dao (1970), and Brzeski (1977) report the species from Germany, Holland, and Poland, respectively.

Ecology

Wallace and Greet (1964) found the species associated with a grass plot of timothy and meadow fescue. In the field, the greatest populations of *A. icarus* occurred at about 5 cm and few were found below 24 cm, the topsoil depth. Numbers decreased during dry periods and increased after a rain; greater numbers were found at deeper depths during a wet period. Since the nematode migrates to the wet end of a moisture gradient, it may move in response to rain and thus avoid desiccation. It moved most readily in soil at field capacity and could survive 4 days, at approximately the permanent wilting point of most plants. Oxygen consumption studies also confirmed that brief periods of drought could be survived. In soil with plants, 35% of a population survived 32 weeks, while in damp sand without a host, at 10°C, nematodes were alive after 32 weeks, but dead after 48 weeks.

Development

The fully formed larvae of *A. icarus* moved continually in the eggshell in forward and backward figure-eight movements, 3 days before hatching (Bridge,1974). Rigid eggshells softened and increased in size 12 hr before hatching. Nematode movement distorted the elastic walls and the head pressing against the end walls caused protrusions. Before hatch, with the head pressed against an eggshell end wall, the larva thrust its stylet until the wall was punctured, unlike the purposeful accurate thrusting employed by cyst nematodes.

Feeding

Ectoparasitic and semi-endoparasitic. Feeding by *A. icarus* on perennial ryegrass (Bridge and Hague,1974) is essentially similar to that of *A. macrurus*, except duration of feeding by *A. icarus* was longer, up to 31.75 hr on epidermal cells and, when embedded in cortical cells, up to 5 days. Stylet penetration time for *A. icarus* was 73 sec–2 min 18 sec and salivation period was 5 min 55 sec–15 min. These times were much longer than for *M. brevidens* or *T. dubius*. Unlike *T. maximus*, *A. icarus* did not aggregate.

ACKNOWLEDGMENTS

We are indebted to the following contributors of specimens on which many of the illustrations and descriptions are based: G. Germani, Antenne ORSTOM d'Antibes, Antibes, France; E. Geraert, Museum voor Dierkunde, Universiteit Gent, Gent, Belgium; D. J. Hooper, Rothamsted Experimental Station, Harpenden, Herts, England; J. Baldwin and M. Mundo-Ocampo, University of California, Riverside, California; A. M. Golden, U. S. Department of Agriculture, Beltsville, Maryland; P. A. A. Loof, Vakgroep Nematologie, Wageningen, the Netherlands; and R. Voisin, IRNA, Station de Recherches de Nématologie et de Génétique Moléculaire des Invertésbrés, Antibes, France.

REFERENCES

Abivardi, C. 1973. A stylet nematode, *Tylenchorhynchus cylindricus* Cobb, 1913, infesting the common guava, *Psidium guajava* L., in Iran. Nematol. Mediterranea 1:139–140.

Alhassan, S. A. 1968. Removal of phosphorus–32 from rice roots by plant-parasitic nematode. Dissn. Abs. Int. 29:9–10.

Allen, M. W. 1955. A review of the nematode genus *Tylenchorhynchus*. Univ. Calif. Publ. Zool. 61:129–166.

Amosu, J. O., and Taylor, D. P. 1974a. Interaction of *Meloidogyne hapla, Pratylenchus penetrans* and *Tylenchorhynchus agri* on kenland red clover, *Trifolium pratense*. Indian J. Nematol. 4:124–131.

Amosu, J. O., and Taylor, D. P. 1974b. Stimulation of growth of red clover by *Tylenchorhynchus agri*. Indian J. Nematol. 4:132–137.

Anderson, R. V. 1977. Description of the male of *Tylenchorhynchus maximus* Allen, 1955 (Nematoda: Tylenchorhynchinae). Can. J. Zool. 55:1921–1923.

Anderson, R. V., and Byers, J. R. (1975). Ultrastructure of the esophageal procorpus in the plant parasitic nematode, *Tylenchorhynchus dubius*, and functional aspects in relation to feeding. Can. J. Zool. 53:1581–1595.

Andrássy, I. 1954. Drei neue arten aus superfamilie Tylenchoidea. Nematologische Notizen 3. Ann. Biol. Univ. Hung. 2:9–15.

Andrássy, I. 1966. Erd und süsswasser-nematoden aus Ghana klasse Secernentea (Phasmidia). Ann. Univ. Sci. Budapest, Rolando Eötvös 8:5–24.

Andrássy, I. 1985. *Scutylenchus quadrifer*. C. I. H. Descriptions of plant-parasitic nematodes. Set 8, No. 108. Comm Inst. Parasitol. Spottiswoode Ballantyne Printers Ltd., Great Britain.

Aycock, R., Barker, K. R., and Benson, D. M. 1976. Susceptibility of Japanese holly to *Criconemoides xenoplax, Tylenchorhynchus claytoni*, and certain other plant-parasitic nematodes. J. Nematol. 8:26–31.

Azmi, M. I., and Jairajpuri, M. S. 1976. Studies on nematode behaviour. IV. Comparative activity of *Tylenchorhynchus mashhoodi* and *Rhabditis* sp. Indian J. Nematol. 6:96–98.

Azmi, M. I., and Jairajpuri, M. S. 1977a. Studies on nematode behaviour. IX. Effects of light on the movement patterns of *Hemicriconemoides mangiferae*, *Tylenchorhynchus mashhoodi* and juveniles of *Acrobeloides* sp. Indian J. Nematol. 7:87–91.

Azmi, M. I., and Jairajpuri, M. S. 1977b. Attraction of plant-parasitic nematodes to host roots. Nematologica 23:119–121.

Bagaturiya, N. L. 1971. Nematode fauna of sugar beet in eastern Georgian SSR (In Russian.) Sobscheniya Akad. Nauk Gruzinskoi SSR 63:217–220.

Baqri, Q. H., and Ahmad, N. 1981. Nematodes from West Bengal (India) VII. Morphometric and allometric variations in *Tylenchorhynchus Nudus* Allen, 1955 (Tylenchorhynchidae: Tylenchida).Bull. Zool. Surv. India 3:239–247.

Baqri, Q. H., and Jairajpuri, M. S. 1970. On the Interspecific variations of *Tylenchorhynchus mashhoodi* Siddiqi and Basir, 1959 and an amended key to species of *Tylenchorhynchus* Cobb, 1913 (Nematoda). Rev. Brasil. Biol. 30:61–68.

Baranovskaya, I. A., Abdel-Khadi, M. A., and Petrova, Z. I. 1975. The effect of some agronomic and ecological factors on the nematode fauna of winter rye. (In Russian.) Doklady Moskovkoi Sel'skokhozyaistvenoi Akademii im. K. A. Timiryazeva. No. 209:127–133.

Barham, R. O., Marx, D. H., and Ruehle, J. L. 1974. Infection of ectomycorrhizal and nonmycorrhizal roots of Shortleaf Pine by nematodes and *Phytophthora cinnamomi*. Phytopathology 64:1260–1264.

Barker, K. R., Aycock, R., and Daughtry, B. I. 1974. Host suitability of four woody ornamentals to certain nematodes. Plant Dis. Rep. 63:113–116.

Barker, K. R., Benson, D. M., and Jones, R. K. 1979. Interaction of Burfordi, Rotunda, and Dwarf Yaupon hollies and Aucuba with selected plant parasitic nematodes. Plant Dis. Rep. 63:113–116.

Barker, K. R., and Worf, G. L. 1964. Parasitism of 'Southern Stock' azaleas in Wisconsin by *Tylenchorhynchus claytoni*, *Trichodorus christiei*, and *Meloidogyne incognita*. Phytopathology 54:887.

Barker, K. R., and Worf, G. L. 1966. Effect of nutrients on nematode activity on azalea. Phytopathology 56:1024–1027.

Benson, D. M., and Barker, K. R. 1977. Efficacy of 1, 2-Dibromo-3-chloropropane and Aldicarb for control of parasitic nematodes in the root zone of American boxwood, and Japanese and Chinese hollies. J. Nematol. 9:263–264.

Bergé, J. B., Dalmasso, A., and Kermarrec, A. 1973. Studies of the fluctuations in population numbers of the nematode fauna of a pasture. Revue D'Ecol. Biol. du Sol 10:271–285.

Birchfield, W., and Martin, W. J. 1956. Pathogenicity on sugarcane and host plant studies of a species of *Tylenchorhynchus*. Phytopathology 46:277–280.

Braasch, H. 1976. Nematodes of plants grown in peat soils. Nachr. (In German.) Pflanzenschutz DDR 30:228–231.

Bridge, J. 1974a. *Tylenchorhynchus dubius*. C. I. H. Descriptions of plant-parasitic nematodes. Set 4, No. 51. Comm. Inst. Parasitol. Spottiswoode Ballantyne Printers Ltd., Great Britain.

Bridge, J. 1974b. Hatching of *Tylenchorhynchus maximus* and *Merlinius icarus*. J. Nematol. 6:101–102.

Bridge, J. 1976. Plant parasitic nematodes from the lowlands and highlands of Ecuador. Nematropica 6:18–23.

Bridge, J., and Hague, N. G. M. 1974. The feeding behaviour of *Tylenchorhynchus* and *Merlinius* species and their effect on growth of perennial ryegrass. Nematologica 20:119–130.

Brzeski, M. W. 1970. Wplyw temperatury na rozwój *Heterodera schachtii* Schm. *Tylenchorhynchus dubius* Büt. (Nematoda, Tylenchida). Roczn. Nauk Roln., Ser. E, 1:203–211.

Brzeski, M. W. 1977. Plant parasitic nematodes, suborder Tylenchina. (In Polish.) Panstwowe Wydawnictwo Naukowe, p. 87.

Brzeski, M. W., and Dowe, A. 1969. Effect of pH on *Tylenchorhynchus dubius* (Nematoda, Tylenchidae). Nematologica 15:403–407.

Bütschli, O. 1873. Beiträge zur Kenntnis der freilebenden Nematoden. Nova Acta Acad. Nat. Curios. 36:1–124. pls. 17–24.

Byers, J. R., and Anderson, R. V. 1972. Ultrastructural morphology of the body wall, stoma, and stomatostyle of the nematode, *Tylenchorhynchus dubius* (Bütschli, 1873) Filipjev, 1936. Can. J. Zool. 50:457–465.

Byers, J. R., and Anderson, R. V. 1973. Morphology and ultrastructure of the intestine in a plant-parasitic nematode, *Tylenchorhynchus dubius*. J. Nematol. 5:28–37.

Cassidy, G. H. 1930. Nematodes associated with sugarcane in Hawaii. Hawaiian Planters Rec. 34:379–387.

Chaturvedi, Y., and Khera, S. 1979. Studies on taxonomy, biology and ecology of nematodes associated with jute crop. Tech. Monograph, Zool. Surv. India 2:1–105.

Choi, Y. E. 1975. A taxonomical and morphological study of plant parasitic nematodes (Tylenchida) in Korea. Korean J. Plant Prot. 14:1–19.

Choi, Y. E., and Geraert, E. 1971. Two new species of Tylenchida from Korea with a list of other nematodes new for this country. Nematologica 17:93–106.

Coates, P. L. 1977. Studies of the developmental biology and host-parasite relationships of *Tylenchorhynchus agri* Ferris (Nematoda: Tylenchidae). Dissn. Abs. Int. 37b:4783–4784.

Coates-Beckford, P. L. 1982a. Developmental biology and feeding behaviour of *Tylenchorhynchus agri* on two hosts, *Trifolium pratense* and *Poa pratensis*. Nematropica 12:1–5.

Coates-Beckford, P. L. 1982b. Influence of temperature and initial population density on population development and pathogenicity of *Tylenchorhynchus agri* on *Trifolium pratense* and *Poa pratensis*. Nematropica 12:15–20.

Coates-Beckford, P. L., and Malek, R. B. 1982. Influence of time on population development and pathogenicity of *Tylenchorhynchus agri* on *Trifolium pratense* and *Triticum aestivum*. Nematropica 12:7– 14.

Cobb, N. A. 1913. New nematode genera found inhabiting fresh water and non-brackish soils. J. Wash. Acad. Sci. 3:432–444.

Cohn, E., Sher, S. A., Bell, A. H., and Minz, G. 1973. Soil nematodes occurring in Israel. Special Pub., Agr. Res. Organization, Bet Dagan 22:1–12.

Dao-D., F. 1970. Climatic influence on the distribution pattern of plant parasitic and soil inhabiting nematodes. Meded. Landbhoogesch, Wageningen 70:1–181.

Decker, H., and Manninger, G. A. 1977. Zur Nematodenfauna von *Prunus amygdalus* Batch. Biol. Gesellschaft DDR. Sekt. Phytopath. und Wilhem-Pieck-Universität, Rostock, pp. 140–150.

Dement'eva, S. P. 1971. Parasitic species of nematodes on solanaceous crops in Moldavia. (In Russian.) Parazity Zhivotnykh i Rastenii. Kishinev:Izdatel'stvo "Shtiintsa" 7:139–142.

Deubert, K. H., Norgren, R. L., Paracer, S. M., and Zuckerman, B. M. 1967. The influence of *Tylenchus agricola* and *Tylenchorhynchus claytoni* on corn roots under gnotobiotic conditions. Nematologica 13:56–62.

DeGuiran, G. 1967. Description de deux espèces nouvelles du genre *Tylenchorhynchus* Cobb, 1913 (Nematoda:Tylenchinae) Accompagnée d' une clé des femelles, et précisions sur *T.mamillatus* Tobar-Jimenez, 1966. Nematologica 13:217–230.

Dickerson, O. J., Franz, T. J., and Lash, L. D. 1978. Influence of crop rotation on nematode populations in Kansas. J. Nematol. 10:284.

Disanzo, C. P. 1973. Nematode response to carbofuran. J. Nematol. 5:22–27.

Doucet, M. E., and Luc, M. 1981. Description de *Nagelus alpensis* n. sp. et observations sur *Scutylenchus tessellates* et *S. quadrifer* (Nematoda: Tylenchida). Rev. Nématol. 4:47–58.

Dwivedi, B. K., Malhotra, S. K., and Misra, S. L. 1985. Effect of certain organic compounds in soil on the population dynamics of *Tylenchorhynchus brassicae* at Allahabad. J. Soil Biol. Ecol. 5:11– 115.

Edongali, E. A., and Lownsbery, B. G. 1980. Reproduction of mixed populations of *Tylenchorhynchus clarus* and *Pratylenchus* spp. on 10 host plants. Plant Disease 64:458–459.

Elmiligy, I. A. 1969. Redescription of *Tylenchorhynchus clarus* Allen, 1955. Nematologica 15:288–290.

Elmiligy, I. A. 1971. Recovery and survival of some plant-parasitic nematodes as influenced by temperature, storage time and extraction technique. Meded. Fac. Landbouw. Rijks. Gent 36:1333–1339.

Eroshenko, A. S. 1981. Plant parasitic nematodes of underwood; families Tylenchorhynchidae and Hoplolaimidae (Nematoda). In Svobodnozhivuschie i fitopatogennye nematody fauny Dal'nego Vostoka Biol. Pedolog. Inst. of Far-Eastern Res. Cent. of USSR Acad. Sci. Vladivostok, 22–27;85–92.

Fernandez, M., and Ortega, J. 1985. Distribution of rice-parasitic nematodes in Cuba. III. Province of Havana. Ciencias de la Agricultura 23:25–30.

Ferris, V. R. 1963. *Tylenchorhynchus silvaticus* N. sp. and *Tylenchorhynchus agri* n. sp. (Nematoda:Tylenchida). Proc. Helm. Soc. Wash. 30:165–168.

Fielding, M. J. 1956. *Tylenchorhynchus martini*, A new nematode species found in the sugarcane and rice fields of Louisiana and Texas. Proc. Helm. Soc. Wash. 23:47–48.

Filipjev, I. N. 1934. The classification of the free-living nematodes and their relation to the parasitic nematodes. Smithson. Misc. Collect. 89:110–112.

Filipjev, I. N. 1936. On the classification of the Tylenchinae. Proc. Helm. Soc. Wash. 3:80–82.

Filipjev, I. N., and Schuurmans Stekhoven Jr., J. H. 1941. A manual of agricultural helminthology. E. J. Brill, Leiden.

Fortuner, R. 1975. Nematode root-parasites associated with rice in Senegal (High Casamance and central and northern regions)and in Mauritania. Cahiers ORSTOM., Serie Biologie, Nematologie. 10:147–159.

Fortuner, R. 1976. Ecological study of nematodes of rice paddies in Senegal. Cahiers ORSTOM, Serie Biologie, Nematologie 11:179–191.

Fortuner, R., and Jacq, V. A. 1976. In vitro study of toxicity of soluble sulphides to three nematodes parasitic on rice in Senegal. Nematologica 22:343–351.

Fortuner, R., and Luc, M. 1987. A reappraisal of Tylenchina (Nemata). 6. The family Belonolaimidae Whitehead, 1960. Rev. Nématol. 10:183–202.

Fortuner, R., and Merny, G. 1973. Nematode root-parasites associated with rice in Lower Casamance (Senegal) and Gambia. Cahiers ORSTOM, Serie Biologie 21:3–20.

Fukudome, N. 1978. Plant-parasitic nematodes found in the tobacco growing areas in Okinawa (Japan). Bull. Kagoshima Tob. Exp. Sta. 21:43–62.

Gargantiel, F. T., and Davide, R. G. 1973. Survey and pathogenicity tests of plant parasitic nematodes associated with sugarcane in Negros Occidental. Philippine Phytopathol. 9:59–66.

Gaur, H. S., and Haque, M. M. 1986. Interspecific relations between concomitant populations of *Tylenchorhynchus* spp., *Rotylenchulus reniformis* and *Hoplolaimus indicus* under different crops. Indian J. Nematol. 16:241–246.

Geraert, E. 1966. On some tylenchidae and Neotylenchidae from Belgium with the description of a new species, *Tylenchorhynchus microdorus*. Nematologica 12:409–416.

Glotova, L. E. 1972. On the nematode fauna of oats in the Amur region (USSR). (In Russian.) In Zashchita sel'skokhozyaĭstvennykh rasteniĭ to vrediteleĭ i bolezneĭ v priamur'e. (Sbornik Nauchnykh Rabot.) Blagoveshehenskiĭ Sel'skokhozyaĭstvennyi Inst. 33–36.

Golden, A. M. 1971. Classification of the genera and higher categories of the order Tylenchida (Nematoda). In Plant parasitic nematodes. Volume I. Morphology, Anatomy, Taxonomy and Ecology. (B. M. Zuckerman, W. F. Mai,and R. A. Rohde, eds.) Academic Press, New York, pp. 191–232.

Golden, A. M., Maqbool, M. A., and Handoo, Z. A. 1987. Descriptions of two new species of *Tylenchorhynchus* Cobb, 1913 (Nematoda: Tylenchida), with details on morphology and variation of *T. claytoni*. J. Nematol. 19:58–68.

Goodell, P., and Ferris, H. 1980. Plant-parasitic nematode distributions in an alfalfa field. J. Nematol 12:136–141.

Goodey, T. 1932. The genus *Anguillulina* Gerv. and v. Ben., 1859, *Tylenchus* Bastian, 1865. J. Helminthol. 10:75–180.

Goodey, T. 1943. A note on the feeding of the nematode *Anguillulina macrura*. J. Helminthol. 21:17–19.

Goodey, T. 1963. Soil and freshwater nematodes (Rewritten by J. B. Goodey). Methuen and Co. Ltd., London.

Gotoh, A. 1968. The plant-parasitic nematodes found associated with major crops in Okinawa, the Ryukyu Islands. Proc. Assoc. Plant Prot. Kyushu 14:77–82.

Gotoh, A. 1976. A review of the plant-parasitic nematodes in warm and subtropical regions in Japan. Misc. Bull. Kyushu Agric. Exp. Sta. 54:1–61.

Graham, T. W. 1954. The tobacco stunt nematode in South Carolina. Phytopathology 44:332.

Gupta, N. K., and Uma. 1985. On two species of the genus *Tylenchorhynchus* Cobb, 1913. Res Bull. Punjab Univ. 36:19–22.

Haasis, F. A., Wells, J. C., and Nusbaum, C. J. 1961. Plant parasitic nematodes associated with decline of woody ornamentals in North Carolina and their control by soil treatment. Plant Dis. Rep. 45:491–496.

Hashim, Z. 1983. Description of *Pratylenchus jordanensis* n. sp. (Nematoda: Tylenchida) and notes on other Tylenchida from Jordan, Rev. Nematol. 6:187–192.

Heungens, A. 1965. Chemische bestrijding van bodemmoeheid in de azaleateelt. Meded. Landbhoogesch. Opzoekstns. Gent 30:1444–1453.

Hitcho, P. J., and Thorson, R. E. 1972. Behavior of free-living and plant-parasitic nematodes in a thermal gradient. J. Parasitol. 58:599.

Hollis, J. P. 1958. Induced swarming of a nematode as a means of isolation. Nature (London) 182:956–957.

Hollis, J. P. 1962. Nature of swarming in nematodes. Nature (London) 193:798–799.

Hollis, J. P., and Fielding, M. J. 1958. Population behavior of plant parasitic nematodes in soil fumigation experiments. Louisiana Agric. Exp. Stn. Bull. No. 515 p. 30.

Hollis, J. P., and McBride, J. M. 1962. Induction of swarming in *Tylenchorhynchus martini* (Nematoda, Tylenchida). Phytopathology 52:14.

Hopper, B. E. 1958. Plant-parasitic nematodes in the soils of southern forest nurseries. Plant Dis. Rep. 42:308–314.

Hopper, B. E. 1959. Three new species of the genus *Tylenchorhynchus* (Nematoda: Tylenchida). Nematologica 4:23–30.

Hu, C. H., Tsai, T. K., and Chu, H. T. 1969. The nematode investigation in sugarcane fields of Taiwan and effects of soil fumigation. Proc. 13th Congr. Int. Soc. Sugarcane technol., Taiwan, pp. 1262–1269.

Ibrahim, I. K. A. 1967. Morphological differences between the cuticle of swarming and nonswarming *Tylenchorhynchus martini*. Proc. Helm. Soc. Wash. 34:18–20.

Ibrahim, I. K. A., and Hollis, J. P. 1973. Electron microscope studies on the cuticle of swarming and nonswarming *Tylenchorhynchus martini*. J. Nematol. 5:275–281.

Ibrahim, I. K. A., Joshi, M. M., and Hollis, J. P. 1978. Swarming disease of nematodes: Host range and evidence for a cytoplasmic polyhedral virus in *Tylenchorhynchus martini*. Proc. Helm. Soc. Wash. 45:233–238.

Jairajpuri, M. S. 1985. *Quinisulcius capitatus*. C. I. H. descriptions of plant-parasitic nematodes. Set 8, No. 111, Comm. Inst. Parasitol. Spottiswoode Ballantyne Printers Ltd., Great Britain.

Jenkins, W. F. 1956. Decline of azaleas, a possible new nematode disease. Maryland Florist 35:1–2.

Johnson, A. W. 1970. Pathogenicity and interaction of three nematode species on six bermuda grasses. J. Nematol. 2:36–41.

Johnston, T. M. 1958. The effect of soil moisture on *Tylenchorhynchus martini* and other nematodes. Proc. Louisiana Acad. Sci. 20:52–55.

Joshi, M. M., and Hollis, J. P. 1976. Pathogenicity of *Tylenchorhynchus martini* swarmers to rice. Nematologica 22:123–124.

Katalan-Gateva, S. H., and Baicheva, O. 1976. The nematode fauna on two tobacco varieties, rila 9 and kulsko in the Blagoevgrad region. Khelminthologiya, Sofia 1:49–61.

Katalan-Gateva, S. H., and Baicheva, O. 1978. The nematode fauna of tobacco oriental-261 in the Blagoevgrad region II. Khelminthologiya, Sofia 5:47–59.

Khan, M. A., Hasan, A., and Mahmood, I. 1986. Feeding behavior of *Tylenchorhynchus brassicae* Siddiqi on cabbage seedlings. Bangladesh J. Botany 15:175–179.

Khan, A. T., Mahmood, I., and Saxena, S. K. 1985. Effect of space on the multiplication of *Tylenchorhynchus brassicae* on cauliflower and *Meloidogyne incognita* on eggplant. Pakistan J. Nematol. 3:23– 30.

Kheir, A. M., Shohla, G. S., and Elgindi, D. M. 1977. Population behaviour of *Tylenchorhynchus clarus* infecting egyptian cotton, *Gossypium barbadense*, in relation to soil type. Zeits. Pflanzenkr. Pflanzensch. 84:663–665.

Kiryanova, E. S. 1951. Soil nematodes of Golodnoi Steppe in Uzbekistan. Trudi Zool. Inst. Akad. Nauk SSSR 9:625–657.

Kleynhans, K. P. N., and Heyns, J. 1984. Nematodes of the families Dolichorhynchidae, Tylenchorhynchidae, and Belonolaimidae in South Africa (Tylenchida:Hoplolaimoidea). Phytophylactica 16:143–153.

Klinkenberg, C. H. 1963. Observations on the feeding habits of *Rotylenchus uniformis*, *Pratylenchus crenatus*, *P. penetrans*, *Tylenchorhynchus dubius* and *Hemicycliophora similis*. Nematologica 9:502–506.

Knobloch, N. A. 1975. *Quinisulcius tarjani* sp.n. (Nematoda:Tylenchorhynchinae) with key to *Quinisulcius* species and notes on other plant parasitic nematodes from Mexico. Proc. Helm. Soc. Wash. 42:52–56.

Kornobis, S. 1981. Nematodes occurring on maize. In Malerialy XXI Sesji Naukowej Inst. Ochrony Róslin, Poznan, Poland, pp. 389–394.

Krishnaprasad, K. S., and Krishnappa, K. 1982. Reactions of a few rice varieties to the build up of stunt nematode *Tylenchorhynchus claytoni*. Indian J. Nematol. 12:173–174.

Krusberg, L. R. 1959. Investigations on the life cycle, reproduction, feeding habits and host range of *Tylenchorhynchus claytoni* Steiner. Nematologica 4:187–197.

Langdon, K. R., Struble, F. B., and Young, H. C.,Jr. (1961). Stunt of small grains, a new disease caused by the nematode *Tylenchorhynchus brevidens*. Plant Dis. Rep. 45:248–252.

Laughlin, C. W., and Vargas, J. M. 1972a. Pathogenic potential of *Tylenchorhynchus dubius* on selected turfgrass. J. Nematol. 4:277– 280.

Lewis, S. A., and Golden, A. M. 1981. Redescription and lectotype designation of *Tylenchorhynchus cylindricus* Cobb, 1913. J. Nematol. 13:521–528.

Li, S. M., et al. 1986. Description of the nematodes of crops in Henan province. Acta Agr. Univ. Henan. 20:349–357.

Lin, Y.Y. 1970. Studies on the rice root parasitic nematodes in Taiwan. J. Agric. and For., Chung-Hsing Univ. (Taiwan). 19:13–27.

Lindhart, K., and Bagger, O. 1967. Plantesygdomme i Danmark, 1966. 83. Aarsoveright samlet ved Statens Plante pathologiske Forsøg. Tidsskr. PlAvl., 71: 285–337.

Loof, P. A. A. 1959. Miscellaneous notes on the genus *Tylenchorhynchus* (Tylenchinae:Nematoda). Nematologica 4:294–306.

Loof, P. A. A. 1974. *Tylenchorhynchus claytoni*. C. I. H. descriptions of plant parasitic nematodes. Set 3, No. 39. Comm. Inst. Parasitol. Spottiswoode Ballyntyne Printers Ltd., Great Britain.

Lopez-C., R., and Salazar-F., L. 1986. Nematodes associated with rice (Oryza sativa L.) in Costa Rica. II. Intraspecific variations of *Tylenchorhynchus annulatus* (Cassidy, 1930) Golden, 1971. (In Spanish.) Turrialba 36:355–362.

Lucas, G. B., and Krusberg, L. R. 1956. The relationship of the stunt nematode to granville wilt resistance in tobacco. Plant Dis. Rep. 40:150–152.

Maggenti, A. R., Luc, M., Raski, D. J., Fortuner, R., and Geraert, E. 1987. Reappraisal of Tylenchina (Nemata). 2. Classification of the suborder Tylenchina (Nemata:Diplogasteria). Rev. Nématol. 10:135–142.

Mahajan, R. 1974. *Tylenchorhynchus kashmirensis* sp.n. and *Quinisulcius himalayae* sp.n. (Nematoda:Tylenchorhynchinae) from India. Proc. Helm. Soc. Wash. 41:13–16.

Mahapatra, S. N., and Das, S. N. 1984. Pathogenicity of *Tylenchorhynchus mashhoodi* on maize, (*Zea mays*). J. Research Assam Agric. Univ. 5:119–121.

Malek, R. B. 1980. Population response to temperature in the subfamily Tylenchorhynchinae. J. Nematol. 12:1–6.

Mamiya, Y. 1969. Plant parasitic nematodes associated with coniferous seedlings in forest nurseries in eastern Japan. Bull. Govt. Forest Exp. Sta. 219:95–119.

Maqbool, M. A. 1982. Description of *Quinisulcius solani* n. sp. (Nematoda: Tylenchorhynchidae) with a key to the species and data on *Scutylenchus koreanus* from Pakistan. J. Nematol. 14:221–225.

Maqbool, M. A., Fatima, N., and Hashmi, S. 1983. *Merlinius niazae* n. sp. (Nematoda: Merlininae) and the occurrence of some members of Merlininae and Tylenchorhynchinae in Pakistan. Pakistan J. Nematol. 1:111–121.

Maqbool, M. A., and Hashmi, S. 1985. Studies on horizontal and vertical distribution of *Quinisulcius solani* in a potato field. Pakistan J. Nematol. 3:95–98.

Maqbool, M. A., and Hashmi, S. (1986). Effect of soil type on the population of *Quinisulcius solani* on potato. Nematropica 16:92–98.

Maqbool, M. A., and Shahina, F. 1987. Nematodes of the northern areas of Pakistan. Description of *Nagelus saifulmulukensis* n.sp. and *Merlinius montanus* n.sp. (Nematoda: Merlininae) with notes on three species of *Tylenchorhynchus* Cobb, 1913. Rev. Ném10:289– 294.

Mathias, P. L. 1978a. Report on the European cereal and maize nematode symposium, Amsterdam, December, 1977. MAFF/ADAS Report FR 4339:1–10.

Mathias, P. L. 1978b. III. Tylenchorhynchidae. The biology, ecology and economic importance of stunt nematodes (Tylenchorhynchidae) especially those in western Europe. In Spiral and stunt nematodes. A workshop manual, nematology group of association of applied biologists. Rothamsted Exp. Sta. Harpenden,England, pp. 1–15.

Mavlyanov, O. M. 1972. Parasitic nematodes on sugar-cane in Uzbekistan. Gel'minty Pishchevykh Produktov. Tezisy Dokladov Mezhrespublikanskoi Nauchoi Koferentsii. 22–25 Dekabrya 1972 GODA:70–71.

Mayol, P. S. 1981. Pathogenicity of *Merlinius brevidens* as related to host development. Plant Disease 65:248–250.

McBride, J. M. 1964. Studies on the induction of swarming in *Tylenchorhynchus martini* Fielding, 1956 (Nematoda: Tylenchida). Dissn. Abs. Int. 25: 2190–2190.

McGlohon, N. W., Sasser, J. N., and Sherwood, R. T. 1961. Investigations of plant-parasitic nematodes associated with forage crops in North Carolina. N. C. Agric. Exp. Stn. Tech. Bull, Raleigh, No. 148.

Meagher, J. W. 1970. Seasonal fluctuations in numbers of larvae of the cereal cyst nematode (*Heterodera avenae*) and of *Pratylenchus minyus* and *Tylenchorhynchus brevidens* in soil. Nematologica 16:333– 347.

Miller, P. M. 1968. The susceptibility of parasitic nematodes to sub-freezing temperatures. Plant Dis. Rep. 52:768–772.

Misra, C., and Das, S. N. 1979. Interaction of some plant parasitic nematodes on the root-knot development in Brinjal. Indian J. Nematol. 7:46–53.

Mulk, M. M. 1978. On the validity of *Merlinius rugosus* (Siddiqi, 1963) Siddiqi, 1970 and description of *M. siddiqi* n.sp. Med. Fac. Landbouww. Rijksuniv. Gent 43/2.

Nandakumar, C., and Rao, Y. S. 1974. On the migratory behaviour of some subterranean parasitic nematodes to aerial parts of rice plants. Nematologica 20:106.

Nesterov, P. I., and Hamed, M. 1972. Study of the nematode fauna of vineyards in the Moldavian SSR and Egypt. Problemy Parazitologii. Trudy VII Nauchnoi Konf. Parazitol. USSR. Part II. Kiev, USSR. Izdatel'stro "Nauk Dunka." 68–69.

Noel, G. R., and Lownsbery, B. F. 1978. Effects of temperature on the pathogenicity of *Tylenchorhynchus clarus* to alfalfa and observations on feeding. J. Nematol. 10: 195–198.

Norton, D. C. 1959. Relationship of nematodes to small grains and native grasses in north and central Texas. Plant Dis. Rep. 43:227– 235.

Norton, D. C., and Schmitt, D. P. 1978. Community analyses of plant-parasitic nematodes in the Kalsow Prairie, Iowa. J. Nematol. 10:171–176.

Norton, D. C., Donald, P., Kimpinski, J., Meyers, R.F., Noel, G. R., Noffsinger, E. Mae., Robbins, R. T., Schmitt, D. P., Sosa-Mose, C., and Vrain, T. C. 1984. Distribution of plant-parasitic nematode species in North America. Society of Nematologists.

Nyhan, J. W., Frederick, L. R., and Norton, D. C. 1972. Ecology of nematodes in Clarion-Webster toposequences associated with *Glycine max* (L.) Merrill. Proc. Soil Sci. Soc. Amer. 36:74–78.

Oostenbrink, M., 1959. Enige bijzondere aaltjesaantastingen in 1958. (Some special nematode attacks in 1958). T. Pl. Ziekt. 65:64.

Oteifa, B. A., and El-Sharkawi, S. E. 1965. Species identity of some Egyptian parasitic nematodes of onion. Bull. Zool. Soc. Egypt 20:55–62.

Oteifa, B. A., and Tarjan, A. C. 1965. Potentially important plant-parasitic nematodes present in established orchards of newly-reclaimed sandy areas of the United Arab Republic. Plant. Dis. Rep. 49:596–597.

Patel, D. J., Thakar, N. A., Patel, H. R., and Patel, C. C. 1987. Outbreak of stunt nematodes on wheat in Gujarat. Curr. Sci. India 56:731.

Patel, K. P., and MacDonald, D. H. 1966. Effect of temperature on the rate of buildup of *Tylenchorhynchus martini*. Phytopathology 56:894.

Patel, K. P., and MacDonald, D. H. 1968. Winter survival of *Tylenchorhynchus martini* in Minnesota. Phytopathology 58:266–267.

Plowright, R. A. 1988. Rice callus for the monoxenic culture of root ectoparasitic nematodes. Nematologica 34:121–123.

Powell, W. M. 1964. The occurrence of *Tylenchorhynchus maximus* in Georgia. Plant. Dis. Rep. 48:70.

Powers, T. O. 1983. Systematic analysis of Merlininae Siddiqi, 1971 (Nematoda: Dolichodoridae). Ph.D. Thesis. University of California, Riverside, Ca.

Powers, T. O., Baldwin, J. G., and Bell, A. H. 1983. Taxonomic limits of the genus *Nagelus* (Thorne and Malek, 1968) Siddiqi, 1979 with a description of *Nagelus borealis* n. sp. from Alaska. J. Nematol. 15:582–593.

Prasad, S. K., and Rao, Y. S. 1977. Interaction between *Tylenchorhynchus claytoni* and *Helicotylenchus crenatus* in rice. Indian J. Nematol. 7:170–172.

Prasad, S. K., Dasgupta, D. R., and Mukhopadhyaya, M. C. 1964. Nematodes associated with commercial crops in northern India and host range of *Meloidogyne javanica* (Treub, 1885) Chitwood, 1949. Indian J. Entom. 26:438–446.

Qasim, M., Hashmi, S., and Maqbool, M. A. 1988. Distribution of parasitic nematodes and their importance in fruit production of Baluchistan. Pakistan J. Nematol. 6:17–22.

Ray, S., and Das, S. N. 1983. Three new species in the family Tylenchorhynchidae (Tylenchoidea: Nematoda) from Orissa, India. Indian J. Nematol. 13:16–25.

Reynolds, H. W., and Evans, M. M. 1953. The stylet nematode, *Tylenchorhynchus dubius*, a root parasite of economic importance in the southwest. Plant Dis. Rep. 37:540–544.

Riffle, J. W. 1970. Nematodes parasitic on *Pinus ponderosa*. Plant. Dis. Rep. 54:752–754.

Riffle, J. W. 1972. Effect of certain nematodes on the growth of *Pinus edulis* and *Juniperus monosperma* seedlings. J. Nematol. 4:91– 94.

Romero, M. D., and Arias, M., 1969. Nematodes of Solanaceous plants cultivated in the Southern Spanish Mediterranean area.1. Tylenchida. Bol. Real. Soc. Esp. Hist. Nat. 67:121–142.

Ruehle, J. L. 1966. Nematodes parasitic on forest trees. I. Reproduction of ectoparasites on pines. Nematologica 12:443–447.

Ruehle, J. L., and Sasser, J. N. 1962. The role of plant-parasitic nematodes in stunting of pine in southern plantations. Phytopathology 52:56–68.

Sabova, M., Valocka, B., and Liskova, M. 1979. Species of corn crop nematodes and their seasonal dynamics. Helminthologia, Bratislava 16:35–44.

Saeed, M. 1974. Studies on some stylet-bearing nematodes associated with Sapodilla (*Achras zapota* L.) with special reference to *Hemicriconemoides mangiferae* Siddiqi, 1961. Ph.D. Thesis. Univ. of Karachi, Karachi.

Saltukoglu, M. E., and Coomans, A. 1975. The identity of *Quinisulcius acti* with *Q. capitatus* (Nematoda: Dolichodoridae). Meded. Fac. LandWet. Gent. 40:497–500.

Saltukoglu, M. E., Geraert, E., and Coomans, A. 1976. Some Tylenchida from the Istanbul Area (Turkey). Nematologia Mediterranea 4:139–153.

Samsoen, L., and Geraert, E. 1975. Nematode fauna of rice paddies in the Cameroon. I. Tylenchida. Rev. Zool. Africaine 89:536–554.

Sauer, M. R. 1985. A scanning electron microscope study of plant and soil nematodes. Comm. Sci. Ind. Res. Org. (CSIRO), Australia.

Saynor, M. 1972. The influence of host plants and soil cultivation of *Pratylenchus*, *Tylenchorhynchus*, and *Helicotylenchus* species in a natural habitat. Ph.D. Thesis, Fac. Sci., Univ. London, Imp. Coll., London.

Schlehuber, A. M., Pass, H., and Young, H. C.,Jr. 1965. Wheat grain losses caused by nematodes. Plant Dis. Rep. 49:806–809.

Schneider, W. 1939. Würmer oder Vermes. 11. Fadenwürmer oder Nematodes. 1. Freilebende und pflanzenparasitische Nematoden. Tierwelt Deutschlands (Dahl). 36:1–260.

Schmitt, D. P., and Norton, D. C. 1972. Relationships of plant parasitic nematodes to sites in native Iowa prairies. J. Nematol. 4:200–206.

Scognamiglio, A., and Talame, M., 1974. Su alcune variazioni morfologiche nella popolazione Italiana del nematode *Merlinius quadrifer* (Andrássy, 1954) Siddiqi, 1970. Boll. Lab. Ent. Agr. Portici 31:1–7.

Seshadri, A. R., Muthukrishnan, T. S., and Shunmugan, S. 1967. A new species of *Tylenchorhynchus* (Tylenchidae:Nematoda) from Madras State, India. Curr. Sci. 36:551–553.

Sharma, R. D. 1971. Studies on the plant parasitic nematode *Tylenchorhynchus dubius*. Meded. Landbhogesch. Wageningen 71–1, pp. 1– 154.

Sher, S. A. 1958. The effect of nematodes on azaleas. Plant. Dis. Rep. 42:84–85.

Sher, S. A., and Bell, A. H. 1975. Scanning electron micrographs on the anterior region of some species of Tylenchoidea (Tylenchida: Nematoda). J. Nematol. 7:69–83.

Shesteperov, A. A. 1975. The susceptibility of red clover varieties and some leguminous species to the ectoparasitic nematodes, *Paratylenchus projectus* and *Tylenchorhynchus dubius*. Byulleten' Vsesoyuznogo Instituta Gel'Mintologii Im. K. I. Skryabina 15:125– 130.

Siddiqi, M. R. 1961. Studies on *Tylenchorhynchus* spp. (Nematoda:Tylenchida) from India. Zeits. Parasitkde. 21:46–64.

Siddiqi, M. R. 1972a. *Merlinius brevidens*. C. I. H. Descriptions of plant-parasitic nematodes. Set 1., No. 8. Comm. Inst. Parasitol. Spottiswoode Ballantyne Printers, Ltd., Great Britain.

Siddiqi, M. R. 1972b. *Tylenchorhynchus cylindricus*. C. I. H. Descriptions of plant-parasitic nematodes. Set 1, No. 7. Comm. Inst. Parasitol. Spottiswoode Ballantyne Printers, Ltd., Great Britain.

Siddiqi, M. R. 1976a. *Tylenchorhynchus annulatus* (= *T. martini*). C. I. H. Descriptions of plant-parasitic nematodes. Set 6, No. 85. Comm. Inst. Parasitol. Spottiswoode Ballantyne Printers, Ltd., Great Britain.

Siddiqi, M. R. 1976b. New plant nematode genera *Pleisiodorus* (Dolichodorinae), *Meiodorus* (Meiodorinae sub-fam. n.), *Amplimerlinius* (Merlininae) and *Gracilacea* (Tylodoridae grad. n.). Nematologica 22:390–416.

Siddiqi, M. R. 1979. Taxonomy of the plant nematode subfamily Merliniinae Siddiqi, 1970, with descriptions of *Merlinius processus* n. sp., *M. loofi* n.sp. and *Amplimerlinius globigerus* n.sp. from Europe. Syst. Parasitol. 1:43–59.

Siddiqi, M. R. 1986. Tylenchida parasites of plants and insects. Commonwealth Agr. Bureaux, Slough, Great Britain.

Siddiqi, M. R., and Basir, M. A. 1959. On some plant-parasitic nematodes occurring in South India, with the description of two new species of the genus *Tylenchorhynchus* Cobb, 1913. Proc. 46th Meet. Indian Sci. Cong., Pt. IV (Abs.):35.

Siddiqi, M. R., and Klinger, J. 1980. *Amplimerlinius dubius* comb. n. for *Aphelenchus dubius* Steiner, 1914. Nematologica 26:376–379.

Siddiqui, Z. A., and Khan, A. M. 1974. Studies on the longevity of *Tylenchorhynchus brassicae* Siddiqi in absence of host. Indian J. Nematol. 4:97–99.

Simons, W. R. 1973. Nematode survival in relation to soil moisture. Meded. Landbhogesch. Wageningen 73–3: 1–85.

Singh, N. D. 1972. Plant-parasitic nematodes associated with some economic crops in Guyana. Plant Dis. Rep. 56:1059–1062.

Singh, N. D. 1978. Observations on seasonal population changes of selected plant parasitic nematodes on sugarcane. Nematropica 9:78.

Singh, R. V., and Khera, S. 1978. Plant parasitic nematodes from the rhizosphere of vegetable crops around Calcutta.2. Family Tylenchorhynchidae. Bull. Zool. Surv. India 1:25–28.

Siyanand, Seshadri, A. R., and Dasgupta, D. R. 1982. Investigation of the lifecycles of *Tylenchorhynchus vulgaris*, *Pratylenchus thornei* and *Hoplolaimus indicus* individually and in combined infestations in maize. Indian J. Nematol. 12:272–276.

Skuodite-Rudyzyavichen, Z. 1970. Distribution of nematodes in hothouses of the botanical gardens of the Academy of Sciences of the Lithuanian SSR. Acta Parasitol. Lithuanica 10:137–141.

Smolik, J. D. 1972. Reproduction of *Tylenchorhynchus nudus* and *Helicotylenchus leiocephalus* on spring wheat and effect of *T. nudus* on growth of spring wheat. Proc. S. Dakota Acad. Sci. 51:153–159.

Smolik, J. D. 1977. Effects of *Trichodorus allius* and *Tylenchorhynchus nudus* on growth of sorghum. Plant Dis. Rep. 61:855–858.

Smolik, J. D., and Malek, R. B. 1972a. *Tylenchorhynchus nudus* and other nematodes associated with Kentucky bluegrass turf in South Dakota. Plant. Dis. Rep. 56:898–900.

Smolik, J. D., and Malek, R. B. 1972b. Temperature and host suitability studies on *Tylenchorhynchus nudus*. Proc. S. Dakota Acad. Sci. 51:142–145.

Smolik, J. D., and Malek, R. B. 1973. Effect of *Tylenchorhynchus nudus* on growth of Kentucky bluegrass. J. Nematol. 5:272–274.

Springer, J. K. 1964. Nematodes associated with plants in cultivated woody plant nurseries and uncultivated woodland areas in New Jersey. New Jersey Dept. Agr. Circ. 429:1–40.

Steiner, G. 1914. Freilebenden Nematoden aus der Schweiz. Arch. Hydrobiol. Planktonk. 9:259–276.

Steiner, G., 1937. Opuscula miscellanea nematologica. V. Proc. Helm. Soc. Wash. 4:33–38.

Stirling, G. R., and Vawdrey, L. L. 1985. Distribution of a needle nematode, *Paralongidorus australis*, in rice fields and areas of natural vegetation in North Queensland. Australasian Plant Pathol. 14:71–72.

Stoyanov, D., 1979. Endoparasites of plant nematodes found in Bulgaria. Rastitelna Zashchita 26:29–31.

Streu, H. T. 1975. Insect, mite, and nematode control on azaleas. In Growing azaleas commercially (A. M. Kofranek and R. A. Larson,eds.). Univ. California, Davis, pp. 89–96.

Sturhan, D. 1966. Über Verbreitung, Pathogenität, und Taxonomie der Nematoden Gattung *Tylenchorhynchus*. Mitt. biol. Bundanst. Ld.-u. Forstw. 118:82–99.

Sutherland, J. R., and Adams, R. E. 1965. The parasitism of red pine and other forest nursery crops by *Tylenchorhynchus claytoni* Steiner. Nematologia 10:637–643.

Swarup, G., Sethi, C. L., and Gill, J. S. 1964. Some records of plant parasitic nematodes in India. Curr. Sci. Bangalore 33:593.

Thomas, S. H., 1978. Populations densities of nematodes under seven tillage regimes. J. Nematol. 10:24–27.

Thorne, G. 1935. Notes on free-living and plant-parasitic nematodes. I. Proc. Helm. Soc. Wash. 2:46–47.

Thorne, G. 1961. Principles of nematology. McGraw-Hill Book Co., Inc., New York, Toronto.

Thorne, G., and Malek, R. B. 1968. Nematodes of the Northern Great Plains. Part I. Tylenchida (Nematoda: Secernentea). S. Dakota Agr. Exp. Sta. Tech. Bull. 31:1–111.

Timm, R. W., 1963. *Tylenchorhynchus trilineatus* n.sp. from West Pakistan, with notes on *T. nudus* and *T. martini.* Nematologica 9:262–266.

Townshend, J. L. 1967. Economically important nematodes in Ontario—1966. Proc. Ent. Soc. Ont. 97:5–6.

Troll, J. and Rohde, R. A. 1966. Pathogenicity of *Pratylenchus penetrans* and *Tylenchorhynchus claytoni* on turfgrasses. Phytopathology 56:995–998.

Tu, C. C., Cheng, Y. S., and Kuo, G. F. L. 1972. An investigation on cotton nematodes of Taiwan and a preliminary study on the effects of reniform nematode, root-knot nematode and stubby-root nematode on cotton. Plant Protection Bull., Taiwan 14:95–109.

Upadhyay, K. D., and Swarup, G. 1972. Culturing, host range and factors affecting multiplication of *Tylenchorhynchus vulgaris* on maize. Indian J. Nematol. 2:139–145.

Upadhyay, K. D. and Swarup, G. 1976. Reaction of some maize varieties against *Tylenchorhynchus vulgaris.* Indian J. Nematol. 6:105–106.

Upadhyay, K. D., Swarup, G., and Sethi, C. L. 1972. *Tylenchorhynchus vulgaris* sp.n. associated with maize roots in India, with notes on its embryology and life history. Indian J. Nematol. 2:129–138.

Varo-Alcala, J., Tobar-Jimenez, A., and Munoz-Medina, J. M. 1970. Lesiones causadas y reacciones provocadas por algunos nematodes en las raices de ciertas plantas. Revta. Iber. Parasit. 30:547–566.

Vinciguerra, M. T., and L A Fauci, G. 1978. Nematodes of moss on the island of Lampedusa, Sicily. Animalia 5:13–37.

Vogel, W., and Bernet, R. 1958. Root die-back in azaleas, a disease caused by nematodes. (In German.) Schweiz. GartenBl., No. 13. p.3.

Vovlas, N. 1983. Observations on the morphology and histopathology of *Quinisulcius acti* on corn. Rev. Nématol. 6:79–83.

Wallace, H. R. and Greet, D. N. 1964. Observations on the taxonomy and biology of *Tylenchorhynchus macrurus* (Goodey, 1932) Filipjev, 1936 and *Tylenchorhynchus icarus* sp.nov. Parasitology 54:129–144.

Wang, L. H. 1971. Embryology and life cycle of *Tylenchorhynchus claytoni* Steiner, 1937 (Nematoda:Tylenchoidea). J. Nematol. 3:101– 107.

Waseem, M., and Craig, D. L. 1962. A survey of plant-parasitic nematodes in strawberry fields of Nova Scotia. Plant Dis. Rep. 46:586–590.

Williams, J. R. 1960. Studies on the nematode soil fauna of sugar cane fields in Mauritius. 4. Tylenchoidea (Partim). Occas. Paper. Mauritius Sugar Industry Res. Inst. 4:1–30.

Wouts, W. M. 1966. The identity of New Zealand populations of *Tylenchorhynchus capitatus* Allen, 1955, with a description of an intersex. New Zealand J. Sci. 9:878–881.

Wyss, U. 1969. Untersuchungen über das Schadauftreten wandernder Würzelnematoden an Erdbeerkulturen in Niedersachsen. Mitt. Biol. Bundanst. u. Forstw. 136:110–126.

Wyss, U. 1973. Feeding of *Tylenchorhynchus dubius.* Nematologica 19:125–136.

Yamazaki, K., Okabe, M., Aihara, T., and Yuhara, I. 1981. Studies on injury of azalea in continuous cropping field. I. Distribution of *Tylenchorhynchus claytoni* in azalea fields in Kanagawa Prefecture and its control. Bull. Kanagawa Hort. Exp. Sta. 28:73– 83.

Yokoo, T., and Koga, S., 1967. Nematode species detected from the rhizosphere in an azalea nursery. Proc. Assoc. Plant. Prot. Kyushu 13:145.

Zepp, A. L. 1974. Abnormal male of *Merlinius brevidens* (Allen,1955) Siddiqi, 1970. Nematologia Mediterranea 2:81–82.

13

Stubby Root and Virus Vector Nematodes: *Trichodorus, Paratrichodorus, Allotrichodorus,* and *Monotrichodorus*

WILFRIDA DECRAEMER *Koninklijk Belgisch Instituut voor Natuurwetenschappen, Brussels, Belgium*

I. INTRODUCTION

Trichodorid species first received wide attention in 1951 by the discovery of their plant pathogenic role. Christie and Perry (1951) demonstrated that a species now known as *Paratrichodorus minor* (syn. *P. christiei*) caused damage to Florida crops. The symptoms it produced on host plants were abnormally stunted roots, a condition descriptively called "stubby root." Since then, the trichodorid nematodes in general have been referred to as stubby root nematodes.

Interest in the group has increased since the early 1960s, when trichodorid species were discovered to be vectors of the plant viruses: tobacco rattle virus and pea early browning virus.

The economic consequence of crop losses due to trichodorids, both as causal agents of direct damage to plants and as vectors of plant virus diseases, is a worldwide problem.

Even though Allen (1957) considered establishing two subgenera within the genus, for a long time the family Trichodoridae contained only the genus *Trichodorus* Cobb, 1913. Then Siddiqi (1974) erected a new genus *Paratrichodorus* for a number of species in *Trichodorus*, and divided the new genus *Paratrichodorus* into three subgenera: *Paratrichodorus, Atlantadorus,* and *Nanidorus*. Later, the three subgenera were raised to genus level by Siddiqi (1980). Decraemer (1980), Decraemer and De Waele (1981), and Sturhan (1985) rejected *Atlantadorus* and *Nanidorus* as valid subgenera or genera, and considered them synonyms of the genus *Paratrichodorus*. More recently, the monodelphic species were placed in two separate genera *Monotrichodorus* Andrássy, 1976 and *Allotrichodorus* Rodriguez-M, Sher, and Siddiqi, 1978, respectively, for species closely related to *Trichodorus* and to *Paratrichodorus*. Both monodelphic genera show a very close relationship to each other.

A. Agricultural Importance

1. Direct Damage

Trichodorid species in general are root ectoparasites, usually aggregating at the root tip. They feed by repeated thrusts of the onchiostyle on epidermal cells, root hairs, with a preference for the meristemic tissue near the tip of the growing root. They cause a cessation of root elongation and an overall reduction in root development.

Many species of Trichodoridae have been recorded as plant pathogens, but some species are considered more economically important than others because of distribution, type of hosts, and as virus vectors. The species *P. minor* (syn. *P. christiei*) appears to be fairly cosmopolitan as a pathogen and is responsible for stubby root injury, root weight loss, and decreased seedling height and weight of more than 100 plant species (economic crops, broad-leaf plants, weeds, grasses, trees, etc.). It is considered to be the most economically important species of stubby root nematodes in the USA. In the Netherlands, Kuiper (1977) found that *P. teres* was responsible for stunting of sugar beet in a new polder soil. In Great Britain *T. cylindricus* and *P. anemones*, for example, caused "docking disorder" of sugar beet, a condition resulting from damage to seedlings by the feeding of these ectoparasites and characterized by patches of stunted plants in the crop in the early stage growth (Whitehead and Hooper, 1970). *Trichodorus viruliferus* caused severe necrosis, arrested growth, and swelling of extending roots of apple (Pitcher, 1967). *Paratrichodorus porosus* was responsible for extensive damage to the root system of *Camellia* (Barriga, 1965) and was also plant pathogenic to maize and sorghum (Chèvres-Roman et al., 1971).

2. Role as Virus Vectors

Although nematodes had long been suspected as possible vectors of plant viruses, the first records of trichodorid species shown to be vectors of tobacco rattle virus are as recent as 1961: *P. Pachydermus* in the Netherlands (Sol and Seinhorst, 1961), *P. minor* (syn. *P. christiei*) in the USA (Walkinshaw et al., 1961), *T. primitivus* in England (Harrison, 1961) and in Germany (Sänger, 1961).

Trichodroids only transmit three tobraviruses: tobacco rattle virus (TRV), pea early browning (PEBV), and pepper ringspot (PVR). Tobacco rattle virus infects a wide range of weed and crop species and is responsible for spraing disease in potato and color break in bulbous ornamentals, two of the most economically important diseases caused by tobraviruses in crops in Europe (Table 1). Specific association between nematode species and serological variants of TRV have been identified. One given species of (*Para*)*trichodorus* can transmit several isolates of TRV, but several species only transmit one isolate.

Pea early browning virus occurs naturally in pea and lucerne, and is only recorded from Western Europe, particularly the Netherlands and Great Britain (Gibbs and Harrison, 1964; van Hoof, 1968). Forms of PEBV known in the Netherlands and in Great Britain generally have similar properties, but differ in the details of their behavior (Gibbs and Harrison, 1964) and are transmitted by different vectors (see Table 1). Experimental evidence of specificity is shown by *P. anemones*, which transmitted an English but not a Dutch strain of PEBV (Harrison, 1967). Pepper ringspot virus has only been reported from South America with *P. minor* as vector.

TABLE 1   Nematode-Transmitted Tobacco Rattle Virus and Pea Early Browing Vectors, Infected Crops, and Main Geographic Occurrence of Disease

Virus	Vector	References	Example of crops infected	Main geographic occurrence of disease
Pea early browning				
Dutch isolates	*P. pachydermus*	van Hoof, 1962		
	P. teres	van Hoof, 1962		Western Europe (the Netherlands)
English isolates	*P. anemones*	Harrison, 1967	Pea, lucerne	
	T. primitivus	Harrison, 1966		Western Europe (particularly eastern England)
	T. viruliferus	Gibbs and Harrison, 1964		
Tobacco rattle				
American isolates	*P. allius*	Jensen and Allen, 1964		
	P. minor (= *P. christiei*)	Walkinshaw et al., 1961	Potato, tomato, onion, tobacco	USA
	P. porosus	Ayala and Allen, 1968		
	P. teres	Jensen et al., 1974		
European isolates	*P. anemones*	van Hoof, 1968		
	P. cylindricus	van Hoof, 1968		
	P. nanus	van Hoof, 1968		
	P. pachydermus	Sol et al., 1960	Potato, tobacco,	
	P. primitivus	Harrison, 1961; Sänger, 1961	sugar beet, bulbous	Europe
	P. similis	Cremer and Kooistra, 1964	ornamentals	
	P. teres	van Hoof, 1964		
	P. viruliferus	van Hoof, 1964		
Japanese isolate	*P. minor*	Komuro et al., 1970	Aster	Japan

Source: Adapted from Taylor (1978).

B.   Ecological Association

1.   Distribution and Biology

Trichodoridae are known from all over the world, and are widespread in Europe and North America. They have a preference for coarser textured soils, such as free-draining sandy soils (Winfield and Cooke, 1975); occasionally they are found in sandy-loam and peat soils. Since it was also recorded from clay soil (Seinhorst, 1963; Cooper, 1971), *T. primitivus* seems to be an exception to this soil type distribution pattern.

Distribution of Trichodoridae in the soil may be irregular both horizontally and vertically. They are rarely numerous in the surface layers, except after periods of prolonged rainfall (Cooke and Draycott, 1971), and usually inhabit areas below the cultivation furrow or in deeper layers (Cooke, 1971). Trichodorids appear to be very susceptible to desiccation. They are able to compensate for this sensitivity to water loss by living in deeper, moister soil layers in summer and in the upper soil layers in autumn (Rössner, 1972). This group of ectoparasites is also susceptible to mechanical injury (Bor and Kuiper, 1966). Below the tilth layers, where nematodes are not subject to moisture or mechanical stress, a relatively large and more stable population is maintained. When the soil remains at or above field capacity for long periods of time, large populations can be maintained, especially where host roots are abundant (Winfield and Cooke, 1971). Kuiper and Loof (1962) observed the greatest number of *P. teres* 5–10 cm below the soil surface in a field where they were pathogenic on sugar beet, fewer numbers at 15–30 cm, but numbers increased again below 30 cm. De Pelsmaeker et al. (1985) found that in a sandy soil *T. primitivus* had a vertical distribution between 0 and 70 cm but preferred the zone between 20 and 40 cm. According to Rössner (1969), there is a difference in the vertical distribution between males of *P. pachydermus* and *T. viruliferus*, with the greatest numbers between 30 and 60 cm and 10 and 40 cm, respectively. He also found *Trichodorus* specimens reaching depths between 220 and 250 cm.

Rohde and Jenkins (1957) found that *P. minor* (syn. *P. christiei*) completed its life cycle in 21–27 days at 22°C on *Zea mays*. On apples, *T. viruliferus* can complete a life cycle in about 45 days at 15–20°C (Pitcher and McNamara, 1970). Ayala et al. (1970) observed that the optimum temperature for reproduction of *P. minor* (syn. *P. christiei*) was 16–24°C and for *P. porosus* 24°C.

2.   Plant Virus Transmission and Retention

Transmission of a virus by a nematode consists of a complex series of events: uptake of virus particles from infected plants, retention of the infected particles within the vector nematode, inoculation by the vector to another plant cell, and infection of the receptor plant by the virus (Harrison et al., 1973).

Nematodes acquire viruses during their feeding on infected plants, and transmission efficiency increases as the access time increases. In trichodorids retained virus particles are associated with the esophageal wall throughout its length, but not with the onchiostyle (Harrison et al., 1973). They are probably selectively and specifically adsorbed on the cuticular lining of the esophagus as infected plant sap is taken up. Specificity of transmission seems related to the specificity of the association between the protein surface of the virus particles and the cuticular surface in the vector where the particles are retained (Taylor, 1978). Dissociation of the virus particles in the vector seems most likely to occur during the initial phases of feeding, when the nematode's saliva passes into the plant cell. This process is followed by inoculation of the dissociated infective virus particle with the saliva of the nematode into a plant cell. In order for the process to be a success, the cell cannot be damaged by the feeding (Harrison et al., 1973). Soil moisture can influence the successful transmission of tabroviruses by trichodorids.

Adult and juvenile stages of vector species appear to be equally efficient in transmitting the associated viruses. However, the virus is not retained during the molt (Harrison et al., 1973). Yet, the viruses can persist for long periods within the vector and still remain infective (Ayala and Allen, 1966; van Hoof, 1970).

C. Control

Usually the first indication of a nematode-transmitted virus infection in a crop is the appearance of patches of diseased plants. An additional characteristic is the persistence of an infection in the soil for long periods of time.

Probably one of the most important and common ways to spread an infestation is by the distribution and use of infected planting material. Ornamental bulbs, for example, are effective disseminators of TRV (Brunt, 1966). Soil adhering to farm machinery, to transplants, and to the feet of man and animals as well as nematodes transported by the wind may be responsible for the spread of virus vectors (Boag, 1985). Infected weeds and root pieces remaining after the harvest of an infected crop constitute an overwintering reservoir of viruses (Cooper and Harrison, 1973). Also viruses in trichodorids may easily survive winter fallow between susceptible crops.

Trichodorids are polyphagous; and have a wide host range. Crop rotation in general offers little prospect of practical prevention of damage. Since these ectoparasites are sensitive to soil disturbance, tillage prior to planting may give good results (Kuiper, 1977).

Control of virus vector nematodes by chemical nematicides gives satisfactory results. However, some of the chemicals included in control measures may have use banned in some countries. Large amounts of methylbromide, D-D, or chloropicrin kill a high percentage of a trichodorid population in the soil (Taylor, 1978). Systemic nematicides, such as aldicarb, fenamiphos, and oxamyl, do not kill the nematodes but will control virus spread.

The problem of nematode-transmitted viruses can be avoided by not planting susceptible crops on infected land. Potentially infected nematode vectors can be detected through soil sampling and the extent of virus infection by bait testing. Low soil moisture contents and lower soil temperatures reduce the vector mobility and give a substantial reduction in, e.g., spraing disease in potatoes or color break in tulip bulbs.

II. TAXONOMY

A. Trichodoridae (Thorne, 1935) Siddiqi, 1961

Order Dorylaimida (de Man, 1876) Pearse, 1942
    Suborder Diphtherophorina Coomans and Loof, 1970
        Superfamily Trichodoroidea (Thorne, 1935) Siddiqi, 1974
            Family Trichodoridae (Thorne, 1935) Siddiqi, 1961
                Type genus: *Trichodorus* Cobb, 1913

1. Diagnosis of Superfamily Trichodoroidea

Body cigar-shaped. Onchiostyle relatively long, dorsally curved without basal knobs; guiding ring simple, surrounds anterior end of onchiostyle. Nerve ring about midesophagus, narrowed region. Esophagus with swollen posterior glandular bulb; five gland nuclei present: one posterior ventrosublateral pair, one small anterior ventrosublateral pair, and a single large dorsal nucleus; dorsal nucleus position variable, varies within same species, varies between anterior and posterior ventrosublateral pairs. Male with one testis, outstretched; precloacal supplements well developed; caudal alae present or absent; spicules straight to ventrally curved, with or without ornamentation. Female reproductive system didelphic-amphidelphic or monodelphic-prodelphic, ovary (ovaries) reflexed; oviduct consisting of

two finely granular cells; spermatheca(e) present or absent; rectum almost parallel to longitudinal body axis; anus subterminal. Tail short, maximum length one anal body width.

Diagnosis of family Trichodoridae: same as superfamily.

2.   Differentiating Characters Between the Genera of the Trichodoridae

See Table 2.

B.   Differential Diagnosis of Genera

1.   Genus *Trichodorus,* Cobb, 1913 (Fig. 1A–C)

a.   Diagnosis

Trichodoridae. Cuticle usually not strongly swollen when fixed; basal layer with pseudoan-
nulation. Esophagus with offset bulb; rarely species with ventral overlap of esophagus
glands or dorsal intestinal overlap. Excretory pore opening usually at level of narrowed
midesophageal region or anterior region of the esophagus bulb. Female reproductive system
didelphic-amphidelphic; spermathecae present; vagina well developed, length about one-
half body width; vaginal constrictor muscles well developed; vaginal sclerotization conspicuous, small to large, variable in shape. Ventrally vulva a pore, a transverse slit, or,
rarely, a longitudinal slit. One to four lateral body pores present, rarely absent; one advulvar
pair within one body width posterior to the vulva. Males common. One to three
ventromedian cervical papillae; rarely none or four. Lateral cervical pores usually present,
rarely absent; usually one pair near onchiostyle base or just behind nerve ring. Sperm with
large sausage-shaped nucleus or rounded nucleus. Spicules ventrally arcuate, rarely
straight; smooth or with ornamentation of bristles, striations, flanges. Capsule of spicule
suspensor muscles well developed. Diagonal copulatory muscles well developed, extending far anterior to spicular region, causes ventral curvature of male tail in death. Caudal alae
absent, except in *T. cylindricus.* Three precloacal ventromedian supplements, rarely four,
more or less evenly spaced, usually at least one within range of retracted spicules. Tail with
one pair postcloacal subventral papillae; one pair of caudal pores.

Type species: *Trichodorus primitivus* (de Man, 1880) Micoletzky, 1922
syn. *Dorylaimus primitivus,* de Man, 1880
T. castellanensis, Arias, Jiminez, and Lopez, 1965

b.   List of Trichodorus *species*

T. aequalis, Allen, 1957
T. aquitanensis, Baujard, 1980
T. azorensis, Almeida, De Waele, Santos, and Sturhan, 1989
T. beirensis, Almeida, De Waele, Santos, and Sturhan, 1989
T. borai, Rahman, Jairajpuri, and Akmad, 1985
T. borneoensis, Hooper, 1962
T. californicus, Allen, 1957
T. cedarus, Yokoo, 1964; syn. *T. kurumeensis,* Yokoo, 1966; syn. *T. longistylus,* Yokoo,
1964
T. complexus, Rahman, Jairajpuri, and Ahmad, 1985
T. coomansi, De Waele and Carbonell, 1983
T. cottieri, Clark, 1963
T. cylindricus, Hooper, 1962

TABLE 2 Differentiating Characters Between the Genera of the Trichodoridae

♂ characters	Trichodorus Fig. 1A–C	Paratrichodorus Fig. 1G–I	Monotrichodorus Fig. 1D–F	Allotrichodorus Fig. 1J–K
Caudal alae	Absent, except T. cylindricus	Present, obscure or distinct	Absent or obscure in M. sacchari	Present, obscure to distinct
Tail in death	Curved ventrad (1 exc.)	Straight	Curved ventrad	Curved ventrad or straight
Capsule of spicule suspensor muscles	Well developed	Poorly developed	Well developed	Moderately developed
Diagonal copulatory muscles	Well developed, extending far anterior to spicular region	Poorly developed, usually restricted to caudal alae region	Moderately developed, extending just anterior to retracted spicules	Weak to well developed, restricted to caudal alae region or just anterior to it
Number precloacal supplements	3	1–3	3	3, exceptionally 4
SP within range retracted spicules	1–2 (mainly 1)	1–2	1–2	2–3
Number VP	1–3 exceptionally 0 or 4	0 or 1, exceptionally 2 P. mirzai	1	Usually 0 or 1
Spicule capitular extension	Absent	Absent	Absent	Present or absent
Shape spicules	Usually ventrally curved	Usually straight	Long, slender, ventrally curved	Usually long, slender, ventrally curved
Ornamentation spicules	Smooth, striated, with or without bristles, flanges	With striation	With striations, bristles	With or without striations, bristles
LP	1 pair, exceptionally 0	Usually absent or one pair	One pair	One pair
♀ characters				
Genital system	Didelphic-amphidelphic	Didelphic-amphidelphic	Monodelphic-prodelphic	Monodelphic-prodelphic
Spermatheca(e)	Present	Absent or present	Absent	Absent
Vagina	Well developed, one-half body width long	Short	Well developed, more than one-half body width	Usually well developed, about one-half body width long, except short in A. westindicus
Vaginal sclerotization	Distinct, small to large	Obscure, small	Distinct, small to large	Distinct, small to large
Vaginal constrictor muscles	Well developed	Poorly developed	Well developed	Well developed
V	Pore, transverse slit exceptionally a longitudinal slit	Pore, or transverse or longitudinal slit	Transverse slit, except longitudinal slit in M. sacchari	Transverse slit
♂ and ♀ characters				
Lateral advulvar pores	One pair postadvulvar	Present in 50% of species	One pair	Absent
Lateral body pores	1–2 pairs, rarely absent	Present in 25% of species	Usually absent	Absent
Number of caudal pores	One pair	One pair	One pair	One pair
Body cuticle when fixed	Usually not swollen	Usually well swollen	Usually not swollen	Variable

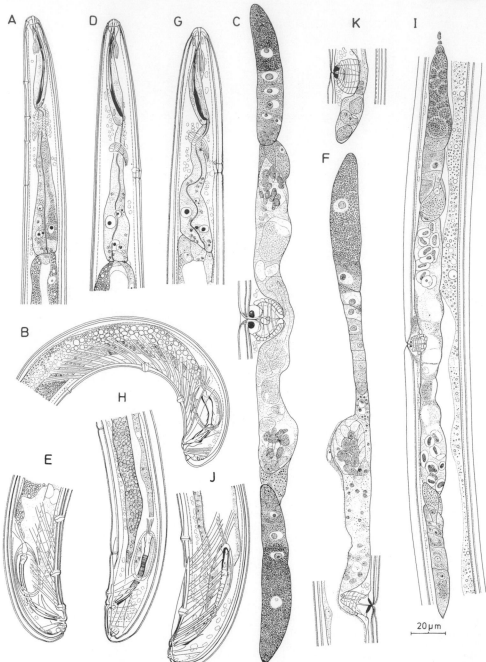

FIGURE 1   The genera of the Trichodoridae. Anterior body region in male: (A) *Trichodorus velatus* (paratype), (D) *Monotrichodorus vangundyi*, (G) *Paratrichodorus weischeri* (paratype). Male posterior body region: (B) *T. velatus* (paratype), (E) *M. vangundyi*, (H) *P. pachydermus*, (J) *Allotrichodorus campanulatus*. Female reproductive system: (C) *T. taylori* (holotype), (F) *M. vangundyi*, (K) *A. campanulatus*, (I) *P. weischeri* (paratype).

T. dilatatus, Rodriguez-M and Bell, 1978

T. eburneus, De Waele and Carbonell, 1983

T. elegans, Allen, 1957

T. hooperi, Loof, 1973

T. intermedius, Rodriguez-M and Bell, 1978

T. lusitanicus, Siddiqi, 1974

T. obscurus, Allen, 1957; syn. *T. primitivus* apud, Thorne, 1939 and Goodey, 1951

T. obtusus, Cobb, 1913

T. orientalis, De Waele and Hashim, 1984

T. pakistanensis, Siddiqi, 1962

T. persicus, De Waele and Sturhan, 1987

T. petrusalberti, De Waele, 1988

T. philipi, De Waele, Meyer, and Van Mieghem, 1990

T. primitivus (de Man, 1880) Micoletzky, 1922; syn. *Dorylaimus primitivus*, de Man, 1880; syn. *T. castellanensis*, Arias Delgado, Jiminez Millan, and Lopez Pedregal, 1965

T. proximus, Allen, 1957

T. rinae, Vermeulen and Heyns, 1984

T. sanniae, Vermeulen and Heyns, 1984

T. similis, Seinhorst, 1963

T. sparsus, Szczygiel, 1968

T. taylori, De Waele, Mancini, Roca, and Lamberti, 1982

T. tricaulatus, Shishida, 1979

T. variopapillatus, Hooper, 1972

T. velatus, Hooper, 1972

T. viruliferus, Hooper, 1963

T. yokooi, Eroshenko and Teplyakov, 1975

2.   Genus *Paratrichodorus*, Siddiqi, 1974 (Fig. 1G–I)
        syn. *Atlantadorus*, Siddiqi, 1974
            Nanidorus, Siddiqi, 1974

a.   *Diagnosis*

Trichodoridae. Cuticle usually strongly swollen when heat-killed or treated with acid fixative. Usually ventral esophageal and/or dorsal intestinal overlap present, rarely absent. Female reproductive system didelphic-amphidelphic; spermathecae present or absent and sperm throughout uterus. Lateral body pores present in almost 50% of species; lateral body pores rarely located advulvar. Vagina short, constrictor muscles inconspicuous, vaginal sclerotization small to inconspicuous. Ventrally vulva a pore, or a longitudinal or transverse slit. Males rare or unknown in about 40% of species. Males ventromedian cervical papillae absent or, if present, usually only one papillae near excretory pore, except *P. mirzai* with two. Lateral cervical pores absent or, if present, usually one pair near onchiostyle base or excretory pore. Male tail straight in death; caudal alae present, obscure to distinct. Sperm cells large subcylindrical with large sausage-shaped nucleus, or small oval to rounded, or threadlike. Diagonal copulatory muscles poorly developed, usually restricted to caudal alae region; capsule of spicule suspensor muscles inconspicuous. Spicules usually straight; corpus striated. One to three ventromedian precloacal supplements, exceptionally four; usually two supplements present within caudal alae region, third supplement less developed, anterior to caudal alae; or some species with only one supplement present. Usually one pair, rarely two, ventromedian postcloacal papillae. Caudal pores usually present, rarely absent.

Type species: *Paratrichodorus tunisiensis* (Siddiqi, 1963) Siddiqi, 1974
   syn. *Trichodorus tunisiensis*, Siddiqi, 1963

b.   List of Paratrichodorus *Species*

P. acaudatus (Siddiqi, 1960) Siddiqi, 1974; syn. *T. acaudatus*, Siddiqi, 1960

P. acutus (Bird, 1967) Siddiqi, 1974; syn. *T. acutus*, Bird, 1967

P. alleni (Andrássy, 1968) Siddiqi, 1974; syn. *T. alleni*, Andrássy, 1968

P. allius (Jensen, 1963) Siddiqi, 1974; syn. *T. allius*, Jensen, 1963; syn. *P. tansaniensis*,
   Siddiqi, 1974

P. anemones (Loof, 1965) Siddiqi, 1974; syn. *T. anemones*, Loof, 1965

P. anthurii, Baujard and Germani, 1985

P. atlanticus (Allen, 1957) Siddiqi, 1974; syn. *T. atlanticus*, Allen, 1957

P. catharinae, Vermeulen and Heyns, 1983

P. grandis, Rodriguez-M and Bell, 1978

P. hispanus, Roca and Arias, 1986

P. lobatus (Colbran, 1965) Siddiqi, 1974; syn. *T. lobatus*, Colbran, 1965; syn. *T. clarki*,
   Yeates, 1967

P. macrostylus, Popovici, 1989

P. minor (Colbran, 1965) Siddiqi, 1974; syn. *T. minor*, Colbran, 1965; syn. *P. christiei*
   (Allen, 1957) Siddiqi, 1974; syn. *P. obesus* (Rasjivin and Penton, 1975)
   Rodriguez-M and Bell, 1978

P. mirzai (Siddiqi, 1960) Siddiqi, 1974; syn. *T. mirzai*, Siddiqi, 1960; syn. *T. musambi*,
   Edward and Misra, 1970

P. nanus (Allen, 1957) Siddiqi, 1974; syn. *T. nanus*, Allen, 1957

P. pachydermus (Seinhorst, 1954) Siddiqi, 1974; syn. *T. pachydermus*, Seinhorst, 1954

P. porosus (Allen, 1957) Siddiqi, 1974; syn. *T. porosus*, Allen, 1957; syn. *T. bucrius*,
   Lordello and Zamith, 1958

P. renifer, Siddiqi, 1974

P. rhodesiensis (Siddiqi and Brown, 1965) Siddiqi, 1974; syn. *T. rhodesiensis*, Siddiqi and
   Brown, 1965

P. sacchari, Vermeulen and Heyns, 1983

P. teres (Hooper, 1962) Siddiqi, 1974; syn. *T. teres*, Hooper, 1962; syn. *T. flevensis*, Kuiper
   and Loof, 1962

P. tunisiensis (Siddiqi, 1963) Siddiqi, 1974; syn. *T. tunisiensis*, Siddiqi, 1963

P. weicheri, Sturhan, 1985

3.   Genus *Monotrichodorus,* Andrássy, 1976 (Fig. 1D–F)

a.   Diagnosis

Trichodoridae. Cuticle usually not swollen when fixed. No esophageal or intestinal overlaps. One pair of caudal pores. Female reproductive system monodelphic-prodelphic; spermatheca present; postvulvar uterine sac present, minute to large. Vagina well developed,
length one-half body width or more. Vaginal sclerotization small to large. One pair of lateral
advulvar body pores present. Ventrally vulva a transverse slit, except in *M. sacchari* where
longitudinal. Males with one ventromedian cervical papilla near excretory pore. One pair
lateral cervical pores near the nerve ring.

Spicules long, slender; without a spicule capitular extension. Caudal alae usually
absent, except obscure in *M. sacchari*. Diagonal copulatory muscles extending just anterior
to retracted spicules; capsule of spicule suspensor muscles well developed. Three

medioventral precloacal supplements present. One pair large subventral postcloacal papillae.

Type species: *Monotrichodorus monohystera* (Allen, 1957) Andrássy, 1976
syn. *Trichodorus monohystera*, Allen, 1957

b.   List of Monotrichodorus *species*

M. monohystera (Allen, 1957) Andrássy, 1976; syn. *T. monohystera*, Allen, 1957
M. sacchari, Baujard and Germani, 1985
M. vangundyi, Rodriguez-M and Siddiqi, 1978

4.   Genus *Allotrichodorus,* Rodriguez-M, and Bell, 1978 (Fig. 1J,K)

a.   Diagnosis

Trichodoridae. Cuticle variable when fixed: not swollen, or moderately to distinctly swollen. Anteriorly directed dorsal intestinal overlap present or absent. One pair caudal pores, except absent in *A. westindicus*. Female reproductive system monodelphic-prodelphic; spermatheca present; postvulvar uterine sac present, minute to large. Vagina usually well developed, length about one-half body width, except smaller in *A. westindicus*. Vaginal sclerotization small to conspicuous. Lateral advulvar body pores absent. Ventrally vulva a transverse slit. Males usually without ventral cervical papillae; or one papilla near excretory pore. One pair of lateral cervical pores near nerve ring. Spicules long, slender; with or without spicule capitular extension. Caudal alae present, minute to well developed. Diagonal copulatory muscles restricted to caudal alae region or just anterior to it. Capsule of spicule suspensor muscles moderately to poorly developed. Three to four medioventral precloacal supplements present. One pair of large subventral to lateroventral postcloacal papillae.

Type species: *Allotrichodorus campanullatus*, Rodriguez-M, Sher, and Siddiqi, 1978

b.   List of Allotrichodorus *species*

A. brasiliensis, Rashid, De Waele, and Coomans, 1986
A. campanullatus, Rodriguez-M, Sher, and Siddiqi, 1978
A. guttatus, Rodriguez-M, Sher, and Siddiqi, 1978
A. longispiculis, Rashid, De Waele, and Coomans, 1986
A. loofi, Rashid, De Waele, and Coomans, 1986
A. sharmae, Rashid, De Waele, and Coomans, 1986
A. westindicus (Rodriguez-M, Sher, and Siddiqi, 1978) Rashid, De Waele, and Coomans, 1986; syn. *Paratrichodorus westindicus*, Rodriguez-M, Sher, and Siddiqi, 1978

C.   Description and Figures of Economically Important Species

1.   *Trichodorus primitivus* (de Man, 1880) Micoletzky, 1922 (Fig. 2A–J)

a.   Measurements

Males: L = 500–1000, onch = 47–58.5, spic = 34–54, a = 18.6–29.5, b = 4.0–6.5, c = 40–65. Females: L = 500–950, onch = 40–57, a = 18–27, b = 4.4–5.8, V = 54–58%; juveniles 2nd stage: L = 330–550, onch = 34–42, gen. syst. = 10–22; juveniles 3rd stage: L = 390–660, onch = 39–50, gen. syst. = 22–60; juveniles 4th stage: L = 550–660, onch = 43–47, gen. syst. 60–110 (from de Man, 1880; Allen, 1957; Coomans, 1962; Seinhorst, 1963; Loof, 1963; Decraemer, 1979; Arias and Roca, 1986).

b. Description

Males with three VP anterior to excretory pore; VP1 about mid-onchiostyle region. VP2 at base onchiostyle. One LP opposite posterior third onchiostyle. Sperm cells large, striated; nucleus sausage-shaped, length 9 μm. Spicules, proximal end 4 μm wide, corpus gradually tapering to narrow distal half; a few fine bristles about midspicule, when spicules retracted obscure, position indicated by serrated outline of spicules. Gubernaculum convex anteriorly, usually positioned between the spicules, sometimes positioned ventrally. Three SP present, SP1 opposite head of retracted spicules. Female with one pair LAP just posterior to vulva; two pairs lateral pores anterior to vulva, one pair three body widths anterior to vulva, one pair five to seven body widths anterior to vulva. Vaginal sclerotization rod-shaped, parallel to vaginal lumen. Type host and locality: Clay soil covered with grass (see Seinhorst, 1963), near Leiden, the Netherlands.

c. Differential Diagnosis

Trichodorus primitivus is close to *T. similis, T. variopapillatus,* and *T. aquitanensis* for most characters. It is distinguished from them by shape of spicules, shape and position of gubernaculum, shape of vulval sclerotization. *Trichodorus primitivus* also differs from *T. similis* and *T. variopapillatus* by the number of VP in onchiostyle region and from *T. similis* and *T. aquitanensis* by sperm structure.

d. Distribution and Hosts

Trichodorus primitivus has been reported mainly from temperate regions, mostly from Europe. It is widely distributed in Great Britain, Belgium, West Germany, Poland, southern Sweden, and occurs at various sites in France, the Netherlands, Norway, Denmark, Bulgaria, Portugal, and Rumania. It has not been found in Italy or Yugoslavia. Occasionally it has been reported from the USA (Oregon, California, Arkansas, Virginia, Maryland) and New Zealand. Known hosts are *Beta vulgaris, Brassica oleracca* var. *acephala, Trifolium pratense* (Whitehead and Hooper, 1970); *Albizia julibrissin, Apium graveolens* var. *dulce, Zea mays, Medicago sativa, Cucumis sativus,* tobacco (Christie et al., 1950; Gibbs and Harrison, 1964). Other plant associations are *Avena sativa, Buxus* sp., *Camellia japonica, Gossypium hirsutum, Juglans cinera, Lespedeza* sp., *Minosa sp., Pinus* sp., *Triticum aestivum* (Noffsinger, 1984); *Pisum sativum, Solanum tuberosum, Stellaria media, Anemone coronaria, Fragaria ananassa, Brassica oleracea* var. *capitata, Malus sylvestris, Prunus domestica, Prunus* sp., *Ribes nigrum* var. *europeum,* tobacco, in vineyards, carrots (Hooper and Siddiqi, 1972); *Picea sitchensis, P. abies, Pinus sylvestris, P. concorta, Larynx* sp., and other coniferous and deciduous trees in Scotland (Boag, 1978); *Poa trivialis, Trifolium pratense, Menthor* sp., *Rumese* sp. in Portugal (Almeida et al., 1989).

e. Biology

Trichodorus primitivus prefers sand and sandy-loam soils but occasionally is found in heavy soils (Sturhan, 1967; Cooper, 1971). It can occur in a wide range of depths: from 0 to 70 cm with the highest populations occurring at 20–40 cm (De Pelsmaeker et al., 1985) and from 60 to 120 cm under clover (Pietler, 1977). *T. primitivus* has a higher tolerance for copper in the soil than the other *Trichodorus* sp. and shows a relatively high tolerance for soil acidity (pH 5.1–6.5 in sand; 6.3–6.9 in sandy loam) (Cooper, 1971; De Pelsmaeker et al., 1985). *Trichodorus primitivus* reproduces bisexually. Its life cycle is similar to that of *T. viruliferus,* about 45 days at 15–20°C (Pitcher and McNamara, 1970).

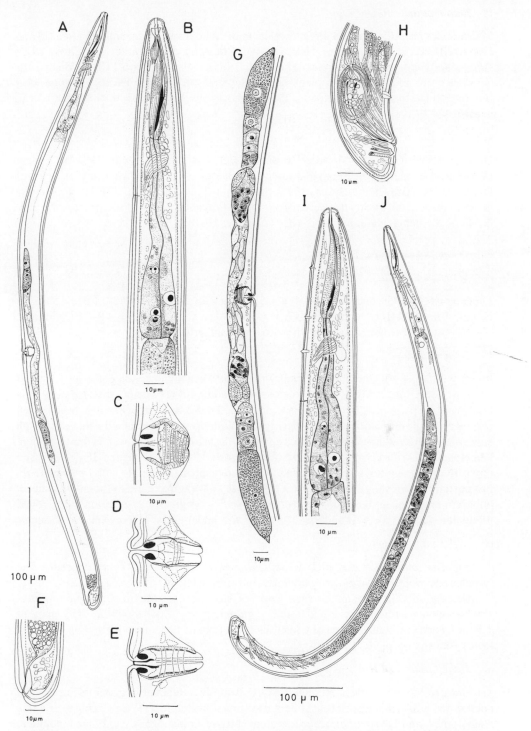

FIGURE 2 *Trichodorus primitivus*. Female (A) total view, (B) anterior body region, (C–E) vaginal region, (F) tail, (G) genital system. Male: (H) copulatory apparatus and tail, (I) anterior body region, (J) total view.

f. Host–Parasite Relationship

Trichodorus primitivus causes direct damage to the plant roots by feeding (Whitehead and Hooper, 1970). It is a vector of TRV, causing spraing in potato tubers (McElroy, 1972; Hooper and Siddiqi, 1972; De Pelsmaeker et al., 1985). It also transmits PEBV, which damages pea crops and infects other plants such as lucerne, cucumber, tobacco, and sugar beet (Gibbs and Harrison, 1964; Harrison, 1964). It is also responsible for docking disorder in sugar beet.

g. Control

Good control of docking disorder in sugar beet was obtained with D-D or Telone (Whitehead et al., 1970). Spraing disease in potatoes was controlled using aldicarb or DuPont 1410 (Brown and Sykes, 1973) or axamyl (Alphey, 1978). The latter systemic nematicide does not greatly decrease the numbers of trichodorids but does decrease the incidence of the nematode-transmitted TRV (Allphey, 1978).

2. *Trichodorus similis,* Seinhorst, 1963 (Fig. 3)

a. Measurements

Males: L = 550–1059, onch = 35–49, spic = 35–44, a = 23.4–32.7, b = 5.2–11.3, c = 54–120. Females: L = 600–1000, onch = 36–47, V = 51–59%, a = 19.8–30, b = 4.9–11 (Seinhorst, 1963; Wyss, 1974; Decraemer, 1979; Roca and Lamberti, 1981).

b. Description

Males with three VP anterior to excretory pore; VP1 usually opposite posterior fourth of onchiostyle region, rarely at base onchiostyle or some just posterior. One LP near base onchiostyle. Sperm cells large; nucleus elongated, 2×5.5–6.5 μm. Spicules, head offset, corpus midregion widened, tapered distally to a blunt open end; retracted spicules smooth appearance, protruding spicules with a few visible bristles in narrower calomus region. Gubernaculum with a characteristic proximal hook and distal knob. Three SP present, SP1 opposite or just anterior to proximal end of retracted spicules. Female with one pair LAP, just posterior to vulva. Ventrally vulva a transverse slit. Vagina round to rhomboid shaped, only proximal end surrounded by constrictor muscles. Vaginal sclerotization large, round triangular shape. Type host and locality: peaty soil under grass, Vinkeveen, Netherlands.

c. Differential Diagnosis

Trichodorus similis is closest to *T. variopapillatus, T. obtusus,* and *T. aquitanensis.* The spicules resemble those of *T. variopapillatus*, but are stouter, and those of *T. obtusus*, but are shorter. Sperm cells are smaller than, and not striated as, those in *T. variopapillatus.* *Trichodorus similis* is distinguished from *T. aquitanensis* and *T. variopapillatus* by shape of spicules, shape of vagina, shape of vulval sclerotization, by lack of lateral body pores other than LAP, and by shorter onchiostyle.

d. Distribution and Hosts

Trichodorus similis has been recorded mainly from the temperate regions in Europe and once in the USA (Michigan). It is widely distributed in Belgium, West Germany, and the Netherlands, and is known to occur less frequently in England, France, Poland, southern Norway, southern Sweden, southern Italy, and Sicily. It is found in various habitats, such as woodlands, meadows, and arable land (Wyss, 1974; De Waele and Coomans, 1983; Roca and Lamberti, 1984; Knobloch and Bird, 1981). Plant associations are *Prunus persica,*

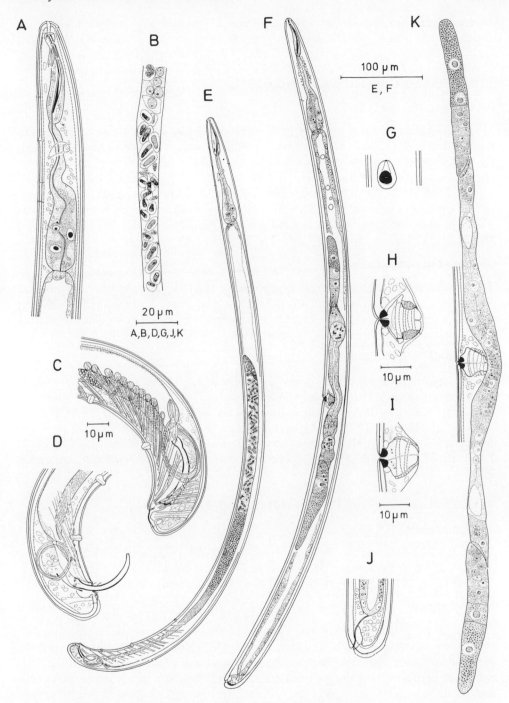

FIGURE 3 *Trichodorus similis*. Male: (A) anterior body region, (B) spermatozoa, (C, D) posterior body region (C after Decraemer, 1980), (E) total view. Female: (F) total view, (G) ventral view of vulva, (H, I) vaginal region (after Decraemer, 1980), (J) tail, (K) reproductive system.

Juglans regia, Cypressus sempervirens, Sorghum sp., *Pyrus domestica, Prunus domestica, Gladiolus* sp., *Picea* sp., *Hordeum vulgare, Brassica rapa* var. *silvestris, Nicotiana tabacum, Fragaria vesca* var. *semperflorens, Chenopodium quinoa, Stellaria media, Daucus carota, Medicago sativa, Poa annua, Lolium perenne, Trifolium repens,* onion, sugar beet, pepper, pea, and potato (Wyss, 1974).

e. Biology

Trichodorus similis prefers sandy or light sandy-loam soils, but is occasionally found in medium loams (Rau, 1975; Wyss, 1974). It commonly occurs between 0 and 40 cm, with the highest populations between 20 and 30 cm; rarely found below 40 cm (i.e., 40–60 cm) (De Pelsmaeker et al., 1985; Steudel, 1969; Wyss, 1974). It prefers a slightly acid soil (pH 5.1–6.1 in sand, pH 6.8–6.9 in sandy loam) (De Pelsmaeker et al., 1985). *Trichodorus similis* reproduces bisexually. Under optimum conditions the life cycle is probably completed in about 45 days, which is similar to *T. viruliferus* (Wyss, 1974; Pitcher and McNamara, 1970).

f. Host–Parasite Relationships

Trichodorus similis causes damage from feeding, resulting in stubby root symptoms. *Trichodorus similis* is a vector of TRV, infecting several plants. It can retain TRV for long periods of time (van Hoof, 1970). In potato it causes spraing, in *Gladiolus* notched leaf disease (Cremer and Kooistra, 1964; van Hoff, 1968, 1970; Erikson, 1974).

g. Control

Satisfactory control on sugar beets growing in infested sandy soils was obtained by fumigating with small doses of D-D or Telone (Whitehead et al., 1970). Best control of potato spraing disease caused by TRV in sandy soils was achieved by injecting D-D before planting (Cooper and Thomas, 1971). Methomyl, Nemacur P, and oxamyl controlled the virus spread more effectively than they controlled *Trichodorus* (Reepmeyer, 1974; Alphey, 1978). However, aldicarb greatly reduced spraing and numbers of *Trichodorus* (Steudel, 1969).

3. *Trichodorus viruliferus,* Hooper, 1963 (Fig. 4A–J)

a Measurements

Males: L = 620–890, onch = 37–53, spic = 26–37, a – 20.4–33, b = 4.2–11.5, c = 47–92. Females: L = 578–919, onch = 38–56, a = 18.7–28, b = 4.7–12.4, V = 51–62% (Hooper, 1963; Hooper, 1976; Mancini et al., 1979; Roca and Lamberti, 1984).

b. Description

Esophageal ventrolateral overlap over the intestine present or absent. Males with three VP anterior to excretory pore; VP1 within onchiostyle region. One LP just behind base onchiostyle. Sperm cells large; nucleus large, sausage-shaped 2 × 8–9.5 µm. Spicules, head not offset, corpus wide at the proximal end, gradually tapers to small constriction about midspicule, then widening, tapers at distal end; narrow midspicule with a few bristles. Gubernaculum usually positioned between spicules, posteriorly with distinct dorsal keel. Three SP present, SP1 just anterior to retracted spicules. Female with one pair LAP in vulva region. Ventrally vulva porelike. Vaginal sclerotization oval-shaped, oblique, or almost parallel to vaginal lumen. Vagina more or less rhomboid-shaped, internal differentiation present, well-developed constrictor muscles, divided in two groups. Type host and locality:

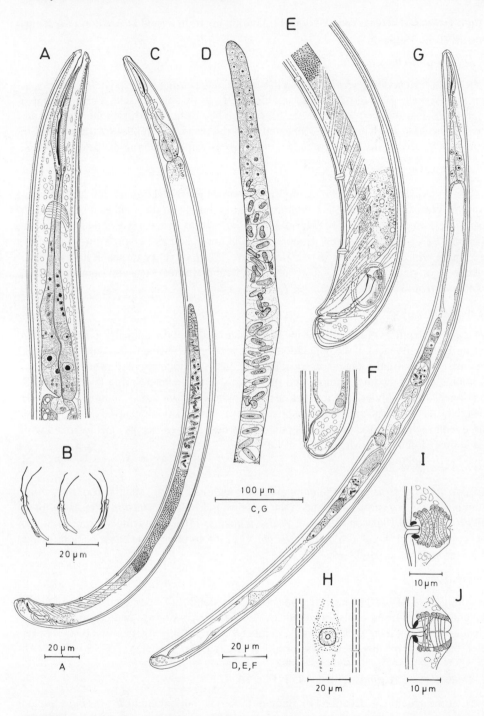

FIGURE 4 *Trichodorus viruliferus.* Male: (A) anterior body region, (B) copulatory apparatus (after Decraemer, 1980), (C) total view, (D) testis, (E) posterior body region. Female: (F) tail, (G) total view, (H) ventral view vulva (redrawn after Hooper, 1964), (I,J) vaginal region (after Decraemer, 1980).

In light sandy soil around roots of wheat following grass, in a field at Roydon near Kings Lym, Norfolk, England.

c. Differential Diagnosis

In the male, *Trichodorus viruliferus* is close to the *T. lusitanicus* group (*T. lusitanicus, T. azorensis, T. beirensis*) and *T. velatus*, all characterized by spicules with a constriction at about midcorpus and the presence of bristles. It can be distinguished from them by the degree of narrowing and length of the indented zone and by a more slender spicule corpus. In the female, *T. viruliferus* differs mainly in the shape of vagina and vaginal sclerotizations.

d. Distribution and Hosts

Trichodorus viruliferus is known mainly from Western Europe (England, the Netherlands, Belgium, West Germany) and is common in Poland, Italy, and the southeast of France. It was also reported from the USA (Florida). Hosts are wheat, rye, barley, potato, apple, pea (Symalla, 1972; van Hoof, 1964; Hooper, 1976). Other plant associations are sugar beet, grass, grape vine, maize (Riter in van Hoof et al., 1966); olive, pear, peach, tomato, artichoke, pepper, poplar, hazelnut, oak, orange, fig, lemon, walnut (Roca and Lamberti, 1984) and *Pinus strobus* (Mancini et al., 1979).

e. Biology

Trichodorus viruliferus prefers sand and sandy-loam soils but occasionally can be found in loam soils. Usually most of the population is found at depths between 10 and 40 cm, and the remainder at 40–60 cm. General seasonal fluctuations in the total population are reflected in minimums during summer/autumn and maximums during winter/spring (Seddon, 1973; Noffsinger, 1984). Males and females usually found in about equal numbers. On apple roots, the life cycle was completed in about 45 days; reproduction occurred throughout the year, even in winter at 5°C at 20 cm depth. Oocyte development seems closely linked with food source (Pitcher and McNamara, 1970).

f. Host–Parasite Relationship

Trichodorus viruliferus causes direct damage through its feeding, such as root browning, stunting, and the occasional swelling of the root tips. In England *T. viruliferus* is a vector of PEBV (Gibbs and Harrison, 1964). It is also a vector of TRV, causing spraing in potato tubers (van Hoof, 1964, 1968). In fallow soil a viruliferous population can remain infective for 3 years (Symalla, 1972).

g. Control

The fumigants D-D or Telone will give satisfactory control of docking disorder of sugar beet and spraing disease in potato tubers (Cooke, 1975; Cooper and Thomas, 1971). Nemacur P did not reduce the nematode number, but apparently it either inactivated or repelled the TRV-carrying nematodes for a time (Reepmeyer, 1973).

4. Trichodorus proximus, Allen, 1957 (Fig. 5A–I)

Note that the specimens described by Baujard (1983) do not belong to *T. proximus* but are now known as *Trichodorus burneus* De Waele and Carbonell, 1983 (De Waele, 1986).

a. Measurements

Males: L = 850–1320, onch = 46–65, spic = 43–65, gub = 17–26, a = 20.4–33.7, b = 6.6–7.9, c = 60–89.2. Females: L = 1100–1500, onch = 49–70, a = 19–30.9, b = 6.7–9.3, V = 49–60% (Allen, 1957; Thorne, 1974).

b. Description

Male with excretory pore opening about midesophageal bulb. One VP at less than one body width anterior to excretory pore. One pair LP at level VP. Sperm cells small, 3.6 × 6.4–6.8 µm, nucleus oval-shaped, 1.8 × 3.5–5.0 µm. Spicules, proximal end wide, a slight depression about midcorpus. Gubernaculum usually positioned between spicules, posterior with dorsal keel. Three SP present, SP1 opposite head retracted spicules or just anterior to them. Females with excretory pore near posterior end of esophagus. Two pairs lateral body pores; one pair of pores about three body widths anterior to vulva; one pair LAP one-half body width posterior to vulva. Vagina barrel-shaped (when relaxed, Fig. 5G) or elongated. Vaginal sclerotization small, triangular-shaped. Type host and locality: Soil around roots of St. Augustine grass, Tampa, Florida.

c. Differential Diagnosis

Trichodorus proximus resembles *T. eburneus* in many respects. Males of *T. proximus* differ from those of *T. eburneus* by smaller sperm (sperm cells 5.7 × 10.7–11.8 µm and nucleus 2.1–2.9 × 5.4–6.4 µm in *T. eburneus*); by shape spicules, total corpus ventrally curved instead of proximal portion only in *T. eburneus*. Females of both species are difficult to distinguish, a difference in vaginal sclerotization and shape vagina is obscure; only sperm when present in spermathecae can separate both species clearly. The geographic distribution of *T. proximus* is limited to the USA and that of *T. eburneus* to Africa.

d. Distribution and Hosts

In its distribution *T. proximus* is restricted to the USA (Florida, Iowa, Kansas, Michigan, New York, South Dakota) (Allen, 1957; Thorne, 1974; Noffsinger, 1984). The only record outside of the USA was from the south of France (Scotto la Massese, 1985), but the report appeared to be an error (pers. commun., Scotto la Massese). Known hosts are *Lycopersicon esculentum, Stenotaphrum secundatum* (Harrison, 1976; Harrison and Smart, 1975; Rhodes, 1965). Other plant associations are *Andropogon gerardi, Bouteloua curtipendula, Cynodon dactylon, Eucalyptus* sp., *Poa pratensis, Rhododendron* sp., *Sabal palmetto, Solanum tuberosum, Tilia cordata, Magnolia virginiana* (Noffsinger, 1984).

e. Biology

In a potato field with a soil moisture of 20–29%, the highest population of *T. proximus* occurred between the depths of 20 and 50 cm. Under controlled conditions on tomato, survival and reproduction of *T. proximus* was 20% (Harrison and Smart, 1975). Males and females apparently are in equal numbers.

f. Host–Parasite Relationship

Trichodorus proximus causes direct damage to the plant, through feeding. On grass, it was directly responsible for a distinct chlorotic condition as well as a reduction in plant growth. Associations with other plant pathogens are unknown.

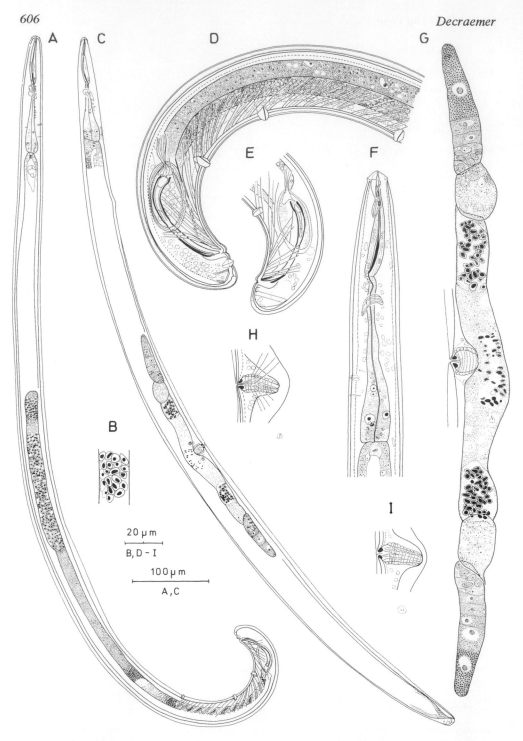

FIGURE 5 *Trichodorus proximus*. Male: (A) total view (paratype), (B) spermatozoa (paratype), (D) posterior body region (paratype) (after Decraemer, 1980), (E) copulatory apparatus and tail. Female: (C) total view (paratype), (F) anterior body region, (G) reproductive system, (H, I) vaginal region.

g. Control

No specific data are available for *T. proximus*, but probably similar control methods for other trichodorid species could be used for this species.

5. *Trichodorus cedarus*, Yokoo, 1964 (Fig. 6A–J)
syn. *Trichodorus longistylus*, Yokoo, 1964
Trichodorus kurumeensis, Yokoo, 1966

a. Measurements

Males: L = 510–850, onch = 51–63, spic = 36–53, a = 11.9–23, b = 3.5–6.1, c = 51–83. Females: L = 494–870, onch = 51–63, a = 13.1–20, b = 3.6–6.2, V = 53–61% (Yokoo, 1964; Mamiya, 1967; Shishida, 1979).

b. Description

Males with three VP, rarely with one or two VP; one specimen with four VP, all posterior to onchiostyle and anterior to excretory pore. One LP at level excretory pore or VP3. Sperm cells large, clearly fibrillar-shaped; nucleus large, oval-shaped, 3 × 6.5 μm. Spicules slightly ventrally curved; cephalated; spicule corpus with transverse striae except at extremities; some with delicate bristles at midcorpus (Shishida, 1979). Gubernaculum almost straight, dorsal keel on posterior third. Three SP present, rarely four, SP1 about midway of retracted spicules. Tail terminal cuticle more or less thickened, contour not evenly rounded. Females with one pair advulvar ventrosubmedian to ventrosublateral body pores, one-third to one-half a body width posterior to vulva. Ventrally vulva a narrow transverse slit. Vaginal sclerotization small, more or less triangular-shaped. Tail terminal cuticle not thickened.Type host and locality: *Cryptomeria japonica* D. Don (Cedar), Asakura-Machi, Asakura-Gun, Fukuoka Prefecture, Japan.

c. Differential diagnosis

Trichodorus cedarus is close to *T. tricaulatus*. It differs from it in males by shape of spicules, position of SP1, and in females by shape of ventral vulva, shape and position of vaginal sclerotization, and lack of lateral body pores other than a single pair of LAP.

d. Distribution and Hosts

Trichodorus cedarus has only been reported from the Far East, Japan, and South Korea (Yokoo, 1964; Mamiya, 1969; Lee, 1976). It has a wide distribution on a wide variety of plants. The widespread occurrence in Japan has made it one of the most important plant parasitic nematodes, in Japanese forest nurseries (Mamiya, 1969). Plant associations are *Cryptomeria japonica, Chamaecyparis obtusa, Larix leptolepis, Abies sachalinensis, A. homolepis, Picea jezoensis v. hondoensis, Pinus densiflora, P. sylvestris, P. strobus, P. resinosa* (Mamiya, 1969). *Torreya nucifera, Pinus thunbergii, Carpinus tschonoskii, Quercus acutissima, Castanea crenata, Castaneopsis cuspidata, Zelkova serrata, Celtis sinensis, Cercidiphyllum japonicum, Magnolia obovata, Cinnamomum camphora, Neolitsea sericea, Kerria japonica, Prunus yedoensis, Rhus succedanea, Daphniphyllum macropodum, Camelia sinensis, C. japonica, Eurya japonica, Fatsia japonica, Cornus controversa, Rhododendron indicum, Callicarpa japonica* (Shishida, 1979), soybean, cabbage, apple, barley (Lee, 1976).

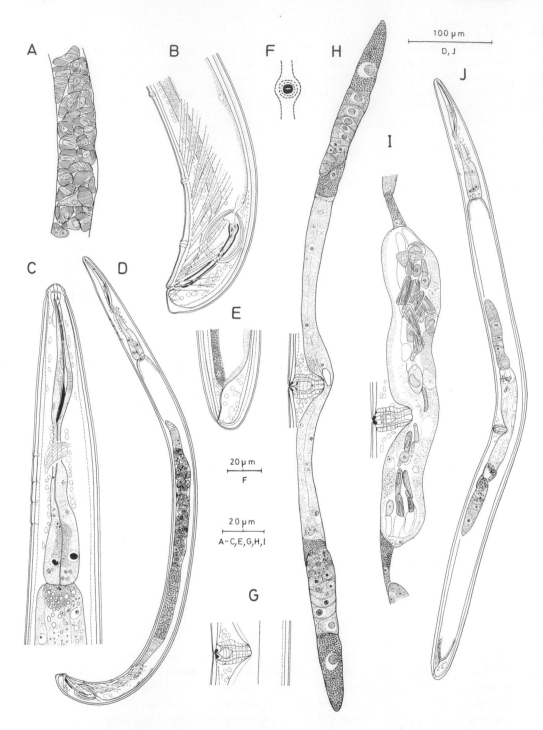

FIGURE 6 *Trichodorus cedarus.* Male: (A) spermatozoa, (B) posterior body region, (C) anterior body region, (D) total view. Female: (E) tail, (F) ventral view vulva (after Mamiya, 1967), (G) vaginal region (H) reproductive system, (I) distal part of reproductive system, (J) total view.

e. Biology

According to Yokoo (1964), *T. cedarus* appeared to be the cause of stubby root in nurseries of black pine and cedar. No data are available on its life cycle.

f. Control

In a loamy soil moderately infested with *T. cedarus* on *Cryptomeria* seedlings, the nematode was effectively controlled by applying a D-D mixture (Mamiya, 1969).

6. *Paratrichodorus minor* (Colburn, 1956) Siddiqi, 1974 (Fig. 7A–I)
   syn. *Trichodorus minor,* Colbran, 1956
   Trichodorus christiei, Allen, 1957
   Paratrichodorus christiei (Allen, 1957) Siddiqi, 1974
   Trichodorus obesus, Razjivin and Penton, 1975

a. Measurements

Females: L = 440–1530, onch = 26–47, a = 15–33, b = 3–6.8, V = 50–64%. Males: L = 540–678, onch = 32–39, spic = 48–73, a = 16–25, b = 3.6–6.5, c = 39–48 (from Colbran, 1956; Allen, 1957; Siddiqi, 1962, 1963; Mamiya, 1967; Bird and Mai, 1968; Hooper, 1962; Heyns, 1975; Shishida, 1978; Vermeulen and Heyns, 1983; Roca and Lamberti, 1984).

b. Description

Onchiostyle widest portion with insertion of smaller inner spear, obscure. Excretory pore opening near esophagus base. Esophageal glands usually overlap intestine ventrally and subventrally, rarely absent, no anterior directed intestinal overlap. Females without lateral body pores. No spermatheca. Sperm present or absent; when present, throughout uterus. Sperm cells very small. Vaginal sclerotization rod-shaped. Ventrally vulva short (2–2.5 µm), transverse slit. Caudal pores obscure subterminal. Males rare. No VP or LP present. Spicule capitulum marked slightly, corpus transversely striated, except at ventrally curved distal end. Gubernaculum with small dorsal keel. Single SP present, 9–14 µm anterior to cloacal opening. Caudal alae extending from just anterior to SP1, posteriorly to postcloacal papillae. Type host and locality: No type host was indicated in the original description; Moggill, Queensland, Australia.

c. Differential Diagnosis

Paratrichodorus minor is close to *P. nanus* having like it males with a single SP, lacking LP and VP. It differs from it in male and female by position of excretory pore opening, by a longer onchiostyle, and usually by esophageal overlap; in female by shape of vagina and shape of vaginal sclerotization and in male by sperm structure.

d. Distribution and Hosts

Paratrichodorus minor is cosmopolitan but may have partially been distributed by man. It is widespread in the USA, Australia, South Africa, and occurs in Japan, Java, Fiji, New Zealand, Philippines, Venezuela, Nicaragua, Brazil, Cuba, Argentina, Puerto Rico, Israel, Egypt, Senegal, Mauritania, Upper Volta, Ivory Coast, Canary Islands, India, Afghanistan, Taiwan, and USSR. According to Sturhan in Wyss (1970), the species is not endemic in Europe. Occasionally it has been found in Italy (Sicily), Belgium, the Netherlands, Sweden, Switzerland, Portugal, and West Germany. Over 100 species of plants (economic crops, grasses, broad-leaf plants, weeds) are known hosts of *P. minor*. Some economically important crops that are common hosts are *Lycopersicon esculentum, Persea americana,*

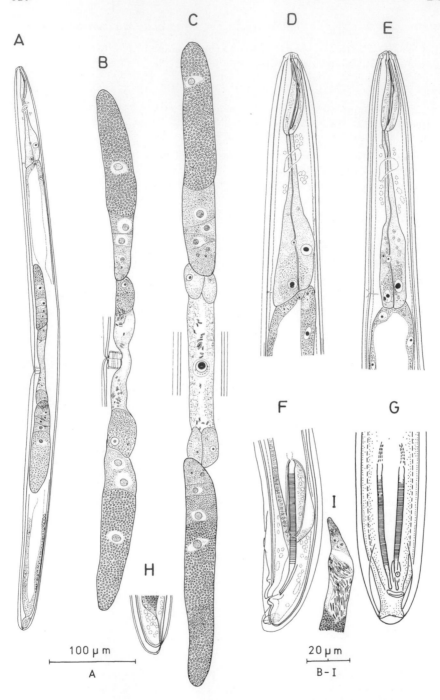

FIGURE 7 *Paratrichodorus minor*. Female: (A) total view (paratype *P. christiei*), (B) reproductive system, (C) ventral view of reproductive system, (D) anterior body region (paratype *P. christiei*), (H) total view. Male: (E) anterior body region, (F) copulatory apparatus and tail, (G) ventral view of copulatory apparatus (after Heyns, 1975), (I) testis and part of vas deferens.

Medicago sativa, Saccharum officinarum, Vitis vinifera, Solanum tuberosum, Phoenix canariensis, red beet, sugar beet, endive, lettuce, cabbage, cauliflower, brussels sprouts, broccoli, mustard, radish, muskmelon, barley, millet, sweet corn, peanut, soybean, onion, okra, eggplant, celery, red clover, sweet pepper, boysenberry, peach, persimmon, walnut, wheat, blueberry, cranberry, chayote, lima bean, grapefruit, cowpea, carrot, castor bean, cotton, and azalea (Colbran, 1956, 1964; Mamiya, 1967; Hooper, 1977; Heyns, 1975; Allen, 1957; Rohde and Jenkins, 1957; McGowan, 1983). The nematode has also been associated with many other species of plants.

e. Biology

Paratrichodorus minor prefers sandy soils but has also been found in peat and muck soils (Christie, 1959; Noffsinger, 1984). Brodie (1976) found the highest population densities of *P. minor* at 30 cm depth in soil with an 83% sand content. Above and below this depth (to 60 cm), significant reduction in density occurred. Soil temperature, texture, and moisture exert a marked influence on the vertical distribution of this species. Highest population densities occurred from December to March, when soil temperature at a 30-cm depth was 11–17°C and soil moisture was 18–23% by volume (Brodie, 1976). In cranberry bogs in the northern United States, highest populations of *P. minor* occurred in November and December, while in the southeastern USA (Georgia) the highest was in June (Noffsinger, 1984).

In field populations males are usually rare. Bird and Mai (1968) found only 12 males among thousands of females; only Chaves (1984) found a population from Santa Fe, Mexico with a male-to-female ratio of 1:3. *Paratrichodorus minor* is parthenogenetic. On tomato, the life cycle is completed in 21–22 days at 22°C and in 16–17 days at 30°C (Rohde and Jenkins, 1957); in 60 days, will have a 10-fold or more increase in population. Under controlled conditions, survival and reproduction on tomato seedlings was greatest at 10% moisture.

f. Host–Parasite Relationship

Paratrichodorus minor causes direct damage to the meristemic tissues by feeding and results in an overall reduction in root development. Above-ground symptoms of infected plants are retarded growth, wilted foliage, and sensitivity to drought (McGowan, 1983; Christie, 1959; Heyns, 1975). Chimaera symptoms involve several kinds of partial chlorophyll defects which were found by Hafez et al. (1981). When maintained on different plant species, Bird and Mai (1965) reported morphometric and allometric variations among specimens from a single population of *P. minor.* Physiological races have also been demonstrated.

Paratrichodorus minor is a vector of the Wisconsin isolate of TRV, which is the cause of stem mottle and tuber spraing, or corky ringspot, in potato (Walkinshow et al., 1961; Ayala and Allen, 1966) and in tests it is found to transmit TRV from roots of aster plants (Komuro et al., 1970). Liu and Ayala (1970) found a positive interaction between *Fusarium moniliforme* and *P. minor* on root growth but not on top growth of five sugar cane varieties. It is a vector of pepper ringspot virus in South America.

g. Control

Although *P. minor* is polyphagous, in some circumstances crop rotation may be advantageous, since populations increase only slightly in cropping systems involving *Crotalaria* sp. or fallow. Brodie et al. (1969) found that "coastal" Bermuda grass suppressed populations of *P. minor.* On beggarweed or marigold populations of *P. minor* did not increase (Brodie et al., 1980). McGowan et al. (1961) found that on cock's foot, damage caused by *P. minor* is reduced 30–40% by applying dry foliage to the soil. Christie (1959) observed that fallow and dry tillage may be a fairly effective control, whereas flooding is not. As higher amounts of fertilizer are applied, populations of *P. minor* will decrease. *Asparagus officinalis* is known to be antagonistic to many nematode species, including *P. minor* (Rohde and Jenkins, 1958). *Paratrichodorus minor* increased rapidly on corn and cotton, but the population was suppressed by peanut and soybean. The peanut sequence was the most effective monocrop system for suppressing most nematode species, including *P. minor* (Johnson et al., 1975).

After fumigation, Perry (1953) found that *P. minor* reestablished itself much more quickly than any of the other numerous plant parasitic species. Rhoades (1968) also found higher populations of *P. minor* in cabbage plots 4 months after fumigation with D-D or EDB than in unfumigated plots. Benson and Barker (1977) stated that control of *P. minor* on Japanese and Chinese hollies with aldicarb or dibromochloropropane was effective for 8 months, but after 12 months the nematode densities were increasing. Nonvolatile nematicides such as aldicarb and carbofuran controlled the parasite better than D-D, EDB, and DBCP (Johnson and Chalfaut, 1972). Rhoades (1985) found in field trials on Myakka fine sand in Florida that fenamiphos applied at 2.24 kg ai/ha increased early growth and subsequent grain yield of field corn (*Zea mays*). Population reestablishment was slower and the onion yields higher after treatment with fensulfothion, carbofuran, prophos, aldicarb, thionazin, and phorate than with D-D (Rhoades, 1969).

7. *Paratrichodorus pachydermus* (Seinhorst, 1954) Siddiqi, 1974 (Fig. 8A–I)
    syn. *Trichodorus pachydermus,* Seinhorst, 1954
        Paratrichodorus (Atlantadorus) pachydermus (Seinhorst, 1954) Siddiqi, 1974

a. Measurements

Males: L = 610–990, onch = 45–60, spic = 40–54, a = 17.2–29, b = 4.1–8.1, c = 25–45. Females: L = 600–1020, onch = 44–58, V = 51–63%, a = 13.9–32, b = 4.1–7.7 (from Seinhorst, 1954; Allen, 1957; Baujard, 1980).

b. Description

Excretory pore opening at anterior end of esophageal bulb, or just anterior. Usually a distinct anteriorly directed intestinal overlap of esophagus dorsally and dorsolaterally, some specimens without distinct overlap; or posterior ventrosublateral esophageal glands overlap intestine; or both intestinal and esophageal overlaps present. Males with a single VP, just anterior to excretory pore. One pair LP about level of VP or just anterior. Sperm cells more or less oval-shaped, nucleus 8.5–9.5 μm. Spicules distal part slightly ventrally curved, corpus transversely striated in midregion. Gubernaculum parallel to spicules, slightly thickened at distal end. Caudal alae present. Three SP present: SP1 just anterior to cloacal opening, SP2 near anterior end of caudal alae, SP3 about two body widths anterior to SP2. Two pair ventrosubmedian postcloacal papillae present; anterior pair minute, just posterior to anus, usually obscure; posterior pair large, subterminal near caudal pores. Females with one pair LP usually within onchiostyle region. Lateral body pores posterior to vulva; number variable between two and five on each side. One or two pairs of caudal pores; if one pair, positioned subterminally; if two pairs, positioned lateroterminally and subterminally (Weischer, 1985). Ventrally vulva porelike. Vaginal sclerotization small, shape elongate-rounded. No spermathecea; uterus functions as sperm reservoir, sperms throughout uterus.

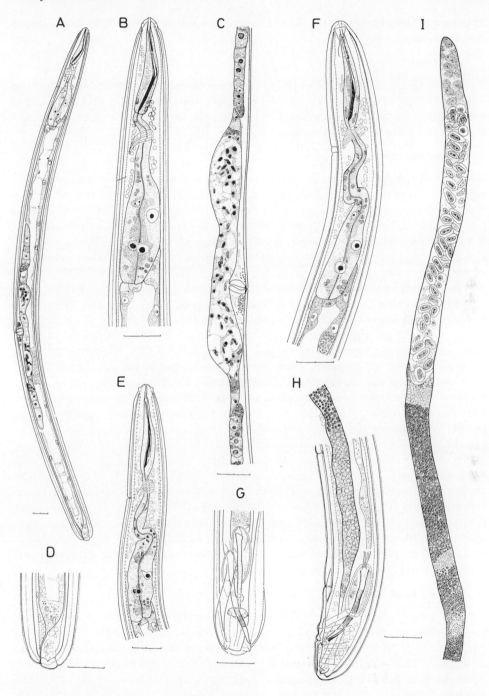

FIGURE 8 *Paratrichodorus pachydermus*. Female: (A) total view, (B) anterior body region, (C) reproductive system (after Geraert et al., 1981), (D) tail. Juvenile fourth stage: (E) anterior body region (after Decraemer, 1979). Male: (F) anterior body region, (G) ventral view of copulatory apparatus and tail, (H, I) reproductive system and tail.

Type host and type locality: soil around the roots of *Prunus serotina* Ehrh, and other shrubs, Ede, the Netherlands.

c. Differential Diagnosis

Paratrichodorus pachydermus is close to *P. anemones* in most respects. It differs from it in males by shape of spicules, arrangement of supplements, tail shape, and two pairs of ventrosubmedian postcloacal papillae; in females by number of lateral body pores posterior to vulva and absence of spermathecae.

d. Distribution and Hosts

Paratrichodorus pachydermus is known only from temperate regions, mainly Europe, where it is one of the most common Trichodoridae species. It has also been reported from Portugal. Occasionally it has been found in Ontario, Canada, and the USA (Florida, Maryland, Michigan, New Jersey, Virginia) where it was probably introduced (Noffsinger, 1984). *Paratrichodorus pachydermus* is polyphagous and occurs in a wide range of habitats: arable land, grassland, and woodland. Known hosts are potato, tobacco, grape vine, mulberry, strawberry, sugar beet, red clover, onion, beech, tulip (van Hoof, 1973); *Gladiolus* sp. (Cremer and Schenk, 1967); oak, *Prunus serotina* (Seinhorst, 1954); *Crataegus* sp. m, *Clematis vitalba, Ligustrum sinense, Solarium nigrum, Syringa vulgaris, Tropaeolum majus* (in De Waele et al., 1985; Alphey and Barbez, 1985). Other plant associations are barley, rye, rye grass, lettuce, maize, fruit plantations, meadows, *Humulus lupulus, Acer* sp., *Alnus* sp., *Betula* sp., *Sambucus* sp., *Castanea* sp., *Liquidambar* sp., *Prunus* sp., *Salix* sp., *Ulnus* sp., *Populus* sp., *Fagus* sp., *Quercus* sp., *Geranium molle, Mercurialis annua, Stellaria media, Capsella bursapastoris, Pinus pinaster, P. sylvestris, P. contorta, Picea sitchensis, P. abies, Larix* sp., *Ammophila arenaria, Vaccinium corymbosum, Dahlia pinnata, Juniperus, Pinus strobus, Rosa* sp., *Aesculus* sp., *Chamaecyparus* sp., *Cotinus coggygria, Cupressus* sp. (De Waele et al., 1985).

e. Biology

Paratrichodorus pachydermus is common in sandy or light sandy-loam soils, but occasionally occurs in loamy or peat soil. It prefers slightly acid soils and usually occurs in depths between 15 and 70 cm (De Waele et al., 1985). Boag (1981) recorded no correlation between the vertical distribution of *P. pachydermus* and root distribution of the host plant, *Sitka* spruce. Also *P. pachydermus* can withstand both drought and waterlogged conditions. Males and females occur in equal numbers. Under controlled conditions, populations will increase 30-fold in 4 months (Reepmeyer, 1973).

f. Host–Parasite Relationship

Through feeding effects alone, *P. pachydermus* may cause serious damage and decrease yield on pine and onion (Whitehead and Hooper, 1970; Wasilewska, 1971; Banck, 1977). *Paratrichodorus pachydermus* is a vector of TRV (Sol and Seinhorst, 1961) and PEBV (van Hoff, 1962). TRV with *P. pachydermus* as vector can infect a wide range of wild and cultivated plants; in potato it produces spraing and stem mottle (De Waele et al., 1985). This species can retain TRV for long periods (van Hoof, 1964) and it shows some specificity in the transmission of TRV (van Hoof, 1968).

g. Control

Paratrichodorus pachydermus has such a wide host range, that crop rotation is not an effective means of control. The best control was achieved by fumigation of soil with D-D or chloropicrin at 371 liters/ha (Whitehead et al., 1970). More recently, systemic nematicides have been used to control virus vector nematodes. Alphey et al. (1975) showed that D-D decreased spraing by killing the nematode vectors, whereas the oxime carbamates oxamyl and aldicarb controlled spraing not by decreasing the numbers of nematodes but by altering their behavior and ability to transmit the virus (De Waele et al., 1985).

8. *Paratrichodorus teres* (Hooper, 1962) Siddiqi, 1974 (Fig. 9A–L)
    syn. *Trichodorus teres*, Hooper, 1962
    Trichodorus flevensis, Kuiper and Loof, 1962

a. Measurements

Males: L = 537–860, onch = 46–60, spic = 40–54, a = 20–22.2, b = 5.1–5.5, c = 43–54. Females: L = 711–1230, onch = 41–61, V = 51–68%, a = 20.7–33.1, b = 4.5–9.3; juveniles 2nd stage: L = 475–503, gen. syst. = 14.0–14.8; juveniles 3rd stage: L = 563–571, gen. syst. = 22.4–23.6; juveniles 4th stage: L = 674–684, gen. syst. = 55.6–55.8 (from Hooper, 1962; Kuiper, 1977, Decrarmer and De Waele, 1981; Vermeulen and Heyns, 1983).

b. Description

Excretory pore opening about opposite anterior end of esophageal bulb. Anteriorly directed intestinal overlap of esophagus varies from slight to pronounced, sometimes without overlap; esophageal gland overlap of intestine varies from minute to long, between different populations. Ventrally vulva a short longitudinal slit. No spermathecae; uterus uniform sac-shaped sperm reservoir, sperm present or absent; sperm cells small. Vaginal sclerotization minute, oval-shaped. Usually one pair LP, not at same level on each side of body. Sometimes numbers variable from one to two, similar numbers or differing by one pore between both sides. Males rare. No VP nor LP present. Sperm cells small, fibrillar-shaped. Spicule corpus slightly tapering toward extremities, and transversely striated, except both ends. Gubernaculum wider at distal end; end slightly bifurcated. Three SP present; SP1 and SP2 within caudal alae region; SP3 less developed, about 2.5 body widths anterior to SP2. Caudal alae present. One large pair of subventral postcloacal papillae. Type host and locality: light sandy soil, following lettuce; Hellesdon, Norwich, Norfolk, England.

c. Differential Diagnosis

Paratrichodorus teres resembles *P. minor* by absence of VP, minute sperm cells and the rarity of males. It differs from it in position of excretory pore, in males by shape of spicules, number of SP, in female by shape of vagina and vaginal sclerotizations. It is close to *P. lobatus*, but differs from it by rarely males, few sperm cells, 3 SP and geographic distribution.

d. Distribution and Hosts

Paratrichodorus teres is known mainly from temperate regions in Europe (Great Britain, the Netherlands, Belgium, West Germany, Poland, Italy, and France). It also occurs in South Africa, and throughout the intermountain region in Oregon (USA). It has been found in arable land, grasslands, and is polyphagous. Known hosts are sugar beet, potato, and onion. Other plant associations are wheat, carrots, lettuce, *Gladiolus*, rye grass, cole seed, barley, rye, red clover, *Dahlia, Tagetes*, cock's foot, *Trollius* sp., *Glycine max*, and ferns (Kuiper, 1977; Noffsinger, 1984).

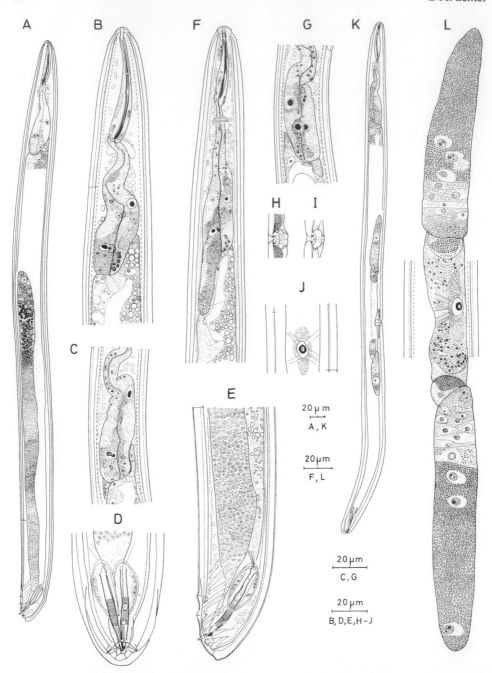

FIGURE 9 *Paratrichodorus teres*. Male: (A) total view, (B) anterior body region, (C) esophageal region (after Decraemer and De Waele, 1981), (D) ventral view of copulatory apparatus and tail, (E) posterior body region. Female: (F) anterior body region, (G) esophageal region (F, G after Decraemer and De Waele, 1981), (H, I) vaginal region (paratype), (J) ventral view of vulva, (K) total view (paratype), (L) reproductive system.

e. Biology

Paratrichodorus teres is common in marine sandy or sandy-loam soils with a preference for soils with a low amount of silt or organic matter. It mainly occurs in the upper soil layers and aggregates around the roots of host plants at depths of 5–10 cm, and between 30–50 cm (Kuiper, 1977). It is highly sensitive to soil disturbance. Males are rare, but do occur in some populations, and not in others (Kuiper and Loof, 1964; Loof, 1965). According to Wyss (1970), reproduction is mainly parthenogenetic. Under controlled conditions, this nematode will increase 75-fold in 4 months (Reepmeyer, 1973).

f. Host–Parasite Relationship

Paratrichodorus teres may cause serious damage through feeding effects alone, and acts as a vector of TRV. A relationship was found between the degree of virus infection and type of preceeding crop (Kuiper, 1977). After growing a number of agricultural crops (spring wheat, potatoes, sugar beet, winter alseed) and following with fallow, a general transmission of rattle virus could be demonstrated from soil samples.

g. Control

Crop rotation offers little prospect for a practical prevention of damage by *P. teres*. Good reduction of these nematode populations was obtained with relatively low doses of nematicides (D-D: 1–3 liters/acre); but a larger quantity of nematicide is necessary to prevent virus transmission. When farmyard manure is used, symptoms of plant infection with rattle virus can be masked; but populations were not reduced by application of large amounts of farmyard manure. *Paratrichodorus teres* is sensitive to soil disturbance, so tillage prior to planting can give good results (Kuiper, 1977).

9. *Paratrichodorus porosus* (Allen, 1957) Siddiqi, 1974 (Fig. 10A–K)
syn. *Trichodorus porosus*, Allen, 1957
Trichodorus bucrius, Lordello and Zamith, 1958

a. Measurements

Males: L = 530–770, onch = 43–48, spic = 36–38, gub = 12–13, a = 15–25, b = 4–6, c = 60–90. Females: L = 420–800, onch = 39–58, V = 51–59%, a = 13.6–25.4, b = 3.1–6.2 (from Allen, 1957; Lordello and Zamith, 1958; Siddiqi, 1962; Mamiya, 1967; Shishida, 1979; Vermeulen and Heyns, 1983).

b. Description

Excretory pore opening between nerve ring and anterior end of esophageal bulb. Usually with a short anteriorly directed intestinal overlap of esophagus. Males rare, only known from type locality. A single VP just anterior to excretory pore opening. One LP at level of VP. Sperm cells 3.5 × 10 μm, nucleus sausage-shaped, length 8.5 μm. Two SP present, within range of retracted spicules. One pair postcloacal subventral papillae present. Spicule corpus transversely striated. Gubernaculum parallel to spicules, distal end slightly enlarged. Caudal alae inconspicuous. Female without spermathecae; sperm in uterus. Ventrally, vulva porelike. Vagina short, barrel-shaped. Vaginal sclerotization small, inconspicuous. Usually one pair of ventromedian advulvar body pores, positioned anterior and posterior to vulva, anterior pore nearest to vulva. Frequently Japanese specimens with three advulvar pores (two anterior, one posterior; or one anterior, two posterior); one Californian specimen with five pores (three anterior, two posterior). Type host and locality: soil around the roots of *Musa* sp.; Corona, California.

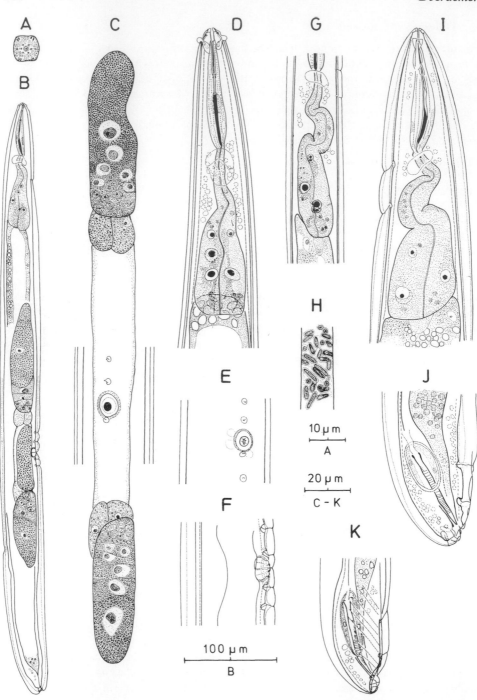

FIGURE 10 *Paratrichodorus porosus.* Female: (A) en face view of head (after Siddiqi, 1962), (B) total view, (C) ventral view of reproductive system, (D) ventral view of anterior body region, (E) ventral view of vulva, (F) vaginal region. Male: (G) esophageal region (paratype), (H) spermatozoa (paratype), (I) anterior body region (paratype), (J, K) copulatory apparatus and tail (paratypes).

c. Differential Diagnosis

Paratrichodorus porosus can be differentiated from all other species of the genus, in female by the ventromedian position of the advulvar body pores. Males of *P. porous* are rare as in *P. teres, P. minor, P. allius, P. nanus,* and *P. lobatus,* but have well-developed sperms as in the last three named species. They differ from all former species by shape and length of spicules, presence of one VP and number of SP (except *P. allius*).

d. Distribution and Hosts

Paratrichodorus porosus is widely spread in Australia, Japan, South Africa, and the USA (Alabama, Arkansas, California, Florida, Michigan, New Jersey, Tennessee, North Carolina, and South Carolina). It also occurs in Hawaii, Brazil, India, and Korea. It was recorded from Azores, Madeira. In the USA known hosts of *P. porosus* are *Brassica oleracea* var. *capitata, Capsicum frutescens, Daucus carota* var. *sativa, Gossypium hirsutum, Juglans californica, Lactuca ativa, Lycopersicon esculentum, L. peruvianum, Malus sylvestris, Medicago sativa, Nicotiana glutinosa, Pisum sativum, Prunus persica, P. americana, P. mahaleb, Pyrus communis, Rosa multiflora, Spinacea oleracea, Vigna sinensis, Vitis vinifera, Zea mays* var. *saccharata* (Ayala and Allen, 1968; Ayala et al., 1970). *Paratrichodorus porosus* has also been found associated with many other plants such as *Camellia* sp., *Citrus* sp., *C. limon, C. sinensis, Ficus carica, Gardenia* sp., *Glycine max, Ilex* sp., *I. crenata rotundifolia, Juglans* sp., *J. regia, Ligusrum ovalifolium, Lupinus* sp., *Malus coronaria, Morus* sp., *Musa* sp., *M. nana, Olea europaea, Persea americana, Philodendron* sp., *Prunus amygdalus, P. domestica, Rhododendron* sp., *Rosa* sp., *Rubus ursinus* var., *loganbaccus, Vigna sinensis* (Noffsinger, 1984); *Saccharum officinarum* in Madeira, Australia, and Hawaii; *Carica papaya, Cypressus sempervirens* var. *stricta, Persea americana* and *Vitis vinifera* in Australia; *Camellia sinensis* in India and South Africa; pear orchards and pumpkin plants in South Africa and cabbage in Hawaii. In Japan *P. porosus* has been found associated with five species of conifers (*Ginko biloba, Taxus cuspidata, Pinus thunbergii, P. densiflora, P. pentaphylla*) and 26 species of broad–leaf trees: *Juniperus* sp., *Carpinus tschonoskii, Quercus acutissima, Q. serrata, Q. salicina, Castanea crenata, Castanopsis cuspidata, Celtis sinensis, Aphanathe aspera, Cercidiphyllum japonicum, Magnolia obovata, Cinnamomum camphora, Neolitsea sericea, Daphniphyllum macropodum, Ilex integra, Acer plamatum, Ternstroemia japonica, Cleyera japonica, Eurya japonica, Idesia polycarpa, Aucuba japonica, Fatsia japonica, Rhododendron indicum, Callicarpa japonica, Viburnum awabuki, Cryptomeria japonica* (Mamiya, 1967; Shishida, 1979).

e. Biology

Paratrichodorus porosus occurs in sandy and sandy-loam soils. Under sugarcane plantings, populations were concentrated in the upper 60 cm with highest numbers at 45 cm. From 60 to 105 cm depth, populations were low and remained practically constant (Gomes Carneiro et al., 1982). Under controlled conditions, reproduction on maize occurred at temperatures between 18 and 35°C, with the optimum at 24°C. At 29°C populations decreased suddenly, and the lowest reproduction occurred at 12 and 35°C (Ayala et al., 1970). In North Carolina *P. porosus* reproduced at high rates; populations increased from 750 to 17,730 specimens in 7 months on *Camellia* (Barriga, 1965).

f. Host–Parasite Relationships

Paratrichodorus porosus can cause extensive damage to the root system of *Camellia* (Barriga, 1965). Under controlled conditions, *P. porosus* damaged maize and sorghum (*Sorghoum vulgare*) (Chèvres-Roman et al., 1971). Nishizawa (1973) suspected *P. porosus* as the causal agent for black-rot disease of Chinese yam. Ayala and Allen (1968) found that under controlled conditions *P. porosus* transmitted the Californian isolate of TRV after feeding on virus-infected Glurk tobacco and *N. glutinosa*.

g. Control

No detailed studies have been reported on the control of *P. porosus* but soil treatments with fumigants, fallow (Wallace, 1971), or oxime carbamates should give control. Since these treatments are known to be effective against other trichodorids, they should provide satisfactory control for *P. porosus*.

D. Key to Economically Important Species Described

1. Males

1. Caudal alae absent; body cuticle usually not swollen with fixation 2
    Caudal alae present; body cuticle usually swollen with fixation 6
2. 1 VP . . . . *T. proximus*
    3 VP 3
3. No VP within onchiostyle region . . . . *T. cedarus*
    1–2 VP within onchiostyle region 4
4. Spicules distinctly constricted at midregion . . . . *T. viruliferus*
    Spicules not or hardly constricted at midregion 5
5. Spicules wider at proximal end, gradually tapers to very narrow distal half; gubernaculum positioned usually ventral to spicules; or between spicules; sperm cells large, striated . . . . *T. primitivus*
    Spicules midregion wider, capitulum offset; gubernaculum not ventral to spicules; sperm cells with smaller nucleus, not striated . . . . *T. similis*
6. 1 VP, 1 LP, sperm cells large, oval-shaped 7
    Without VP and LP, sperm cells minute 8
7. 3 SP; 2 pairs of postcloacal papillae; spicules slightly ventrally curved, caudal alae conspicuous; male and female in equal numbers . . . . *P. pachydermus*
    2 SP; 1 pair of postcloacal papillae; spicules straight, caudal alae inconspicuous; males rare . . . . *P. porosus*
8. 1 SP; caudal pores rarely observed . . . . *P. minor*
    3 SP; caudal pores distinct . . . . *P. teres*

2. Females

1. Body cuticle usually not swollen with fixation; vagina long, usually with well-developed constrictor muscles; vaginal sclerotizations usually well developed . . . 2
    Body cuticle well swollen with fixation; vagina short, weakly developed constrictor muscles, vaginal sclerotizations poorly developed 6
2. Vaginal sclerotization as elongated oval pieces, well separated, about parallel to vaginal lumen 3
    Vaginal sclerotizations different 4
3. 2 prevulvar LAP; vagina elongated to rhomboid, constrictor muscles not differentiated; vaginal sclerotizations long oval (Fig. 2C–E) . . . . *T. primitivus*
    No prevulvar LAP; vagina rhomboid, constrictor muscles in two groups; vaginal sclerotizations short oval . . . . *T. viruliferus*

4. Vaginal sclerotizations large triangular, close to one another; vagina rhomboid, constrictor muscles in proximal half (Fig. 3H, I) .... *T. similis*
   Vaginal sclerotizations small (rounded) triangular, vaginal constrictor muscles with different arrangement 5
5. Vagina slightly pear-shaped; vaginal sclerotizations close to one another; long elongated sperm cells .... *T. cedarus*
   Vagina barrel-shaped, vaginal sclerotizations more separated; sperm cells smaller .
   .... *T. proximus*
6. Usually 1 pair ventromedian advulvar body pores, sometimes 3–5 pairs ....
   .... *P. porosus*
   No ventromedian body pores present 7
7. Sperm cells large oval-shaped, dispersed throughout uterus; ventrally vulva a pore .
   .... *P. pachydermus*
   Sperm cells minute; vulva not a pore ventrally 8
8. No lateral body pores present; vaginal sclerotization minute, oval, parallel to longitudinal body axis; ventrally vulva a transverse slit .... *P. minor*
   1 pair of LAP, sometimes 0–2 on each side; vaginal sclerotization minute, oval–shaped; ventrally vulva a longitudinal slit .... *P. teres*

III. ABBREVIATIONS

exc	position excretory pore opening
gen. syst.	length genital system
gub	length of gubernaculum
L	body length
LAP	lateral advulvar body pore
LP	lateral cervical pore
onch	length of onchiostyle
spic	length of spicules measured along the median line
SP	precloacal ventromedian supplement
SP1	precloacal ventromedian supplement nearest to the anus
V	position of vulva from anterior, expressed as a percentage of the total body length
VP	ventromedian cervical pore
VP1	anterior-most ventromedian cervical papilla

ACKNOWLEDGMENTS

I am very much obliged to the following individuals for all their specimens: J. B. Baldwin, P. Baujard, A. Bell, E. C. Bernard, A. C. Colbran, M. De Pelsmaeker, A. M. Golden, J. Heyns, J. Hooper, P. A. A. Loof, M. Luc, Y. Mamiya, E. M. Noffsinger, D. C. Norton, Y. Shishida, D. Sturhan, and W. Wouts. I also thank E. Geraert and E. M. Noffsinger for their comments on the manuscript.

BIBLIOGRAPHY

For individual references not listed, please see the following:

Decreamer, W. 1980. Systematics of the Trichodoridae (Nematoda) with keys to their species. *Rev. Nematol.* 3(1): 81–99

Esser, R. P. 1971. A compendium of the genus *Trichodorus* (Dorylaimoidea: Diphtherophoridae). *Proc. Soil Crop Sci. Soc. Fla.* 31: 244–253.

Taylor, C. E. 1978. Plant-parasitic Dorylaimida: Biology and virus transmission. In *Plant Nematology.* J. E. Southey, ed. London, pp. 232–243.

REFERENCES

Almeida, T., De Waele, D., Santos, S., and Sturhan, D. 1989. Species of *Trichodorus* (Nematoda: Trichodoridae) from Portugal. *Rev. Nematol.* 12(3): 219–233.

Alphey, T. J. W. 1978. Oxamyl sprays for the control of potato spraing disease caused by nematode-transmitted tobacco rattle virus. *Ann. Appl. Biol.* 88: 75–80.

Alphey, T. J. W., Cooper, J. I., and Harrison, B. D. 1975. Systematic nematicides for the control of trichodorid nematodes and of potato spraing disease caused by tobacco rattle virus. *Plant Patho.* 24: 117–121.

Arias, M., and Roca, F. 1986. *Trichodorus castellanensis* Arias Delgado, Jiminez Millan, and Lopez Pedregal, a junior synonym of *Trichodorus primitivus* (de Man). *Nematologia Mediterranea* 14(2): 279–281.

Banck, A. 1977. Survey of root nematodes in onions. *Vaxtskyddsnotiser* 41(4): 111–114.

Barriga, R. 1965. Parasitism of camellias by some nematode species. Abstract. *Nematologica* 11(1): 34.

Baujard, P. 1980. *Trichodorus aquitanensis* n. sp. et donnés nouvelles sur *Paratrichodorus (Atlantadorus) pachydermus* (Seinhorst, 1954) Siddiqi, 1974 (Nematoda: Trichodorus). *Rev. Nematol.* 3(1): 21–27.

Baujard, P. 1983. Observations sur les Trichodoridae Thorne, 1935 (Nematoda) de l'Afrique de l'Ouest. *Rev. Nematol.* 6(2): 223–228.

Baujard, P., and Germani, G. 1985. Description de *Monotrichodorus sacchari* n. sp., *Paratrichodorus anthurii* n. sp. et d'une population de *Paratrichodorus westindicus* Rodriguez-M, Sher and Siddiqi, 1978 (Nematoda: Trichodoridae). *Rev. Nematol.* 8(1): 35–39.

Benson, D. M., and Barker, K. R. 1977. Efficacy of 1,2–dibromo–3–chloropropane and aldicarb for control of parasitic nematodes in the root zone of American boxwood, and Japanese and Chinese hollies. *Nematologica* 9(4): 263–264.

Boag, B. 1978. Nematodes in Scottish forest nurseries. *Ann. Appl. Biol.* 88: 279–286.

Boag, B. 1981. Observations on the population dynamics and vertical distribution of trichodorid nematodes in a Scottish forest nursery. *Ann. Appl. Biol.* 98(3): 463–469.

Boag, B. 1985. The localized spread of virus-vector nematodes adhering to farm machinery. *Nematologica* 31(2): 234–235.

Brodie, B. B. 1976. Vertical distribution of three nematode species in relation to certain soil properties. *J. Nematol.* 8(3): 243–247.

Brodie, B. B., Good, J. M., and Adams, W. E. 1969. Population dynamics of plant nematodes in cultivated soil: effect of sod-based rotations in Cecil sandy loam. *J. Nematol.* 1: 309–312.

Chèvres-Roman, R., Gross, H. D., and Sasser, J. N. 1971. The influence of selected nematode species and number of consecutive plantings of corn and sorghum on forage production, chemical composition of plant and soil, and water use efficiency. *Nematropica* 1(2): 4041–4046.

Christie, J. R. 1959. *Plant Nematodes: Their Bionomics and Control.* University of Florida, Gainesville.

Christie, J. R., Perry, V. G., and Wilson, J. W. 1950. Control of nematodes injurious to vegetable crops. *Report Florida Agr. Exp. Stat.* 4950: 144–145.

Decraemer, W. 1988. Morphometric variability and value of the characters used for species identification in *Trichodorus* Cobb, 1913. *Bulletin van het Koninklijk Belgisch Instituut voor Natuurwetenschappen, Biologie* 58: 29–44.

Decraemer, W. 1989. Morphologic variablility and value of the characters used for species identification in *Paratrichodorus* Siddiqi, 1974 (Nematoda: Trichodoridae). *Nematologica 35*: 37–61.

Decreamer, W., and De Waele, D. 1981. Taxonomic value of the position of oesophageal gland nuclei and of oesophageal gland overlap in the Trichodoridae (Diphtherophorina). *Nematologica* 27(1): 82–94.

De Waele, D., 1988. *Trichodorus petrusalberti* n. sp. (Nematoda: Trichodoridae) from rice with additional notes on the morphology of *T. sanniae* and *T. rinae*. *J. Nematol.* 20(1): 85–90.

De Waele, D., Alphey, T. J. W., and Barbez, D. 1985. *Paratrichodorus pachydermus*. *C. I. II. Descriptions of plant-parasitic Nematodes.*, Set 8, no. 112.

De Waele, D., and Carbonell, E. 1982. Two new species of *Trichodorus* (Nematoda: Diphtherophorina) from Africa. *Nematologia* 28(4): 387–397.

De Waele, D., and Coomans, A. 1983. Distribution of Longidoridae and Trichodoridae. *Atlas of Plant Parasitic Nematodes of Belgium*, 1–42.

De Waele, D., and Hashim, Z. 1984. *Trichodorus orientalis* n. sp. (Nematoda: Trichodoridae) from Jordan and Iran. *Sys. Parasitol.* 6(1): 63–67.

De Waele, D., Mancini, G., Roca, F., and Lamberti, F. 1982. *Trichodorus taylori* n. sp. (Nematoda: Dorylaimida) from Italy. *Nematologica Mediterranea* 10(1): 27–37.

Eriksson, K. B. 1974. Virus-transmitting nematodes. *Vaxtskyddsnotiser* 38(3): 43–51.

Eroshenko, A. S., and Teplyakov, A. A. 1975. A new species of ectoparasitic nematode in *Trichodorus* from coniferous forest in the Far East. (In Russian). *Parazitologiya* 9(6): 545–547.

Esser, R. P. 1967. Foliar and other plant-parasitic nematodes associated with azalea in Florida. *Plant Dis. Reptr.* 51:46–49.

Gomes Carneiro, R. M. D., Carneiro, R. G., and Monteiro, A. R. 1982. Vertical distribution of four sugarcane nematodes in relation to soil factors. *Nematologica* 6: 117–132.

Hafez, S. L., Raski, D. J., and Lownsbery, B. F. 1981. Effect of *Paratrichodorus christiei* on Thompson seedless grape. *Rev. Nematol.* 4(1): 115–120.

Harrison, B. D. 1967. Pea early-browning virus (PEBV). *Rep. Rothamsted Exp. Stn for 1966*: 115.

Harrison, B. D., Robertson, W. M., and Taylor, C. E. 1974. Specificity of retention and transmission of viruses by nematodes. *J. Nematol.* 6(4): 155–164.

Harrison, R. E., and Smart, G. C. Q. Jr. 1975. Vertical distribution of *Trichodorus christiei* and *Trichodorus proximus* relative to soil moisture. *J. Nematol.* 7(4): 324.

Heyns, J. 1975. *Paratrichodorus christiei*. *C. I. H. Descriptions of Plant–Parasitic Nematodes* 5.

van Hoof, H. A. 1964. Het tijdstip van infectie en veranderingen in de concentratie van ratelvirus (kringerigheid) in de aardappelknol. *Bull. OpzoekStns Gent* 29: 944–955.

van Hoof, H. A. 1966. Nematode populations active and inactive with regard to transmission of nepoviruses. *Nematologica* 12: 615–618.

van Hoof, H. A. 1970. Some observations on retention of tobacco rattle virus in nematodes. *Neth. J. Plant Patho.* 76(6): 329–330.

van Hoff, H. A., Maat, D. Z., and Seinhorst, J. W. 1966. Viruses of the tobacco rattle virus group in northern Italy: their vectors and serological relationships. *Neth. J. Plant Pathol.* 72: 253–258.

Hooper, D. J. 1976. *Trichodorus viruliferus*. *C. I. H. Descriptions of Plant-Parasitic Nematodes*, Set 6, no. 86.

Hooper, D. J. 1977. *Paratrichodorus (Nanidorus) minor*. *C. I. H. Descriptions of Plant-Parasitic Nematodes*, Set 7, no. 103.

Hooper, D. J., and Siddiqi, M. R. 1972. *Trichodorus primitivus*. *C. I. H. Descriptions of Plant-Parasitic Nematodes*, Set 1, no. 15.

Johnson, A. W., and Chalfaut, R. B. 1972. Control of nematodes and corn earworm on sweet corn. *J. Nematol.* 4: 227–228.

Johnson, A. W., Dowler, C. C., and Hauser, E. W. 1975. Crop rotation and herbicide effects on population densities of plant-parasitic nematodes. *J. Nematol.* 7(2): 158–168.

Komuro, Y., Yoshino, M., and Ichinohe, M. 1970. Tobacco rattle virus isolated from aster showing ringspot syndrome and its transmission by *Trichodorus minor* Colbran. (In Japanese). *Ann. Phytopathol.. Soc. Jap.* 36: 17–16.

Kuiper, K. 1977. Introductie en vestiging van plantenparasitaire aaltjes in nieuwe polders, in het bijzonder van *Trichodorus teres. Meded. Landb. Hogesch.* 77-4, separate series no. 55: 1–140.

Kuiper, K., and Loof, P. A. A. 1964. Observations on some *Trichodorus* species. Abstract. *Nematologica* 10(1): 77.

Lee, Y. B. 1976. Two genera of Trichodoridae (Trichodoroidea: Nematoda) new to Korea. *Korean J. Plant Protect.* 15(2): 75–78.

Liu, L. J., and Ayala, A. 1970. Pathogenicity of *Fusarium moniliforme* and *F. roseum* and their interaction with *Trichodorus christiei* on sugarcane in Puerto Rico. Abstract. *Phytopathology* 60(11): 15.

McGlohon, N. E., Sasser, J. N., and Sherwood, R. T. 1961. Investigations of plant-parasitic nematodes associated with forage crops in North Carolina. *Tec. Bull. N. Carol. Agric. Exp. Stn* 148.

MacGowan, J. B. 1983. The stubby-root nematode, *Paratrichodorus christiei* (Allen, 1957) Siddiqi, 1974. Nematology Circular, Division of Plant Industry, Florida Dept. of Agriculture and Consumer Services, no. 97.

Mamiya, Y. 1969. Plant parasitic nematodes associated with coniferous seedlings in forest nurseries in eastern Japan. *Bull. Gov. Forest Exp. Stat.* 219: 95–119.

Mamiya, M. 1969. Effects of nematicides on nematode populations and growth of *Cryptomeria* seedlings at a forest nursery. *Bull. Gov. Forest Exp. Stat.* 220: 121–132.

Mancini, G., Moretti, M., Cotroneo, A., Palenzona, M., and Ferrara, A. M. 1979. *Trichodorus aequalis* e *T. viruliferus* (Nematoda, Trichodoridae) su semine di *Pinus strobus. Informatore Fitopatologico* (29)10: 3–7.

Noffsinger, E. M. 1984. Adenophorea. In *Distribution of plant-parasitic nematode species in North America.* A project of the Nematode Geographical Distribution Committee of the Society of Nematologists, pp. 20–29.

Pelsmaeker, M. de, Coomans, A., Coolen, W., De Meyer, P., and Saverwyns, A. 1985. Trichodoridae in het pootaardappelareaal van West- en Oost-Vlaanderen (1978–1984). *Landbouwtijdschrift* 38(4): 631–636.

Perry, V. G. 1954. Return of nematodes following fumigation of Florida soils. *Proc. Florida State Horticultural Society*, Vol. 66, pp. 112–114.

Pietler, K. 1977. Zur populationsdynamik virusübertragender Nematoden der Gattungen *Longidorus, Xiphinema* und *Trichodorus* unter Berücksichtigung der Standortverhältnisse und des Witterungverlaufes. *Biol. Gesellsch. Deutsch. Demokr. Rep. Rostock*: 70–80.

Pitcher, R. S. 1967. The host-parasite relations and ecology of *Trichodorus viruliferus* on apple roots, as observed from an underground laboratory. *Nematologica* 13(4): 547–557.

Rahman, M. F., Jairajpuri, M. S., and Ahmad, I. 1985. Two new species of the genus *Trichodorus* Cobb, 1913 (Nematoda: Dorylaimida) from India. *Rev. Nematol.* 8(2): 103–107.

Rashid, F., De Waele, D., and Coomans, A. 1986. Trichodoridae (Nematoda) from Brazil. *Nematologica* 31(3): 289–320.

Rau, J. 1975. Das Vorkommen virusübertragender Nematoden in ungestörten Biotopen Niedersachsens. Dissertation.

Rebois, R. V., and Cairns, E. J. 1968. Nematodes associated with soybeans in Alabama, Florida and Georgia. *Plant Dis. Reptr.* 52(1): 40–44.

Reepmeyer, H. 1973. Untersuchungen zur Biologie einzelner *Trichodorus*-Arten und zu ihrer bekampfung. Abstract. *Mitt. biol. BundAnst. Ld-u. Forstw.* 151: 301.

Rhoades, H. L. 1969. Nematicide efficacy in controlling sting and stubby-root nematodes attacking onions in Central Florida. *Plant Dis. Reptr.* 53: 728–730.

Rhoades, H. L. 1985. Comparison of fenamiphos and *Arthrobotrys amerospora* for controlling plant nematodes in central Florida. *Nematropica* 15(1): 1–7.

Roca, F., and Arias, M. 1986. A new *Paratrichodorus* species (Nematoda: Trichodoridae) from Spain. *Nematologia Mediterrranea* 14(2): 181–185.

Roca, F., and Lamberti, F. 1984. Trichodorids (Nematoda) from Italy. *Nematologica Mediterranea* 12(1): 95–118.

Rössner, J. 1969. Populationsentwicklung pflanzenparasitarer Nematoden unter Koniferen. *Mitt. biol. BundAnst. Ld- u. Forstw.* 136: 50–58.

Rössner, J. 1971. Einfluss der austrocknung des bodens auf wandernde wurzelnematoden. *Nematologica* 17(1): 127–144.

Scotto la Massese, C. 1985. Atlas of plant parasitic nematodes of France. Distribution of Longidoridae, Xiphinemidae and Trichodoridae. (gen. ed. Alphey, T. J. W.). European Plant Parasitic Nematode Survey.

Shishida, Y. 1979. Studies on nematodes parasitic on woody plants. 1. Family Trichodoridae (Thorne, 1933) Clark, 1961. *Jap. J. Nematol.* 9: 28–44.

Siddiqi, M. R. 1980. On the generic status of *Atlantadorus* Siddiqi, 1974 and *Nanidorus* Siddiqi, 1974 (Nematoda: Trichodoridae). *Syst. Parasitol.* 1(2): 151–152.

Siddiqui, I. A., Sher, S. A., and French, A. M. 1973. Distribution of plant parasitic nematodes in California. Department of Food and Agriculture, Division of Plant Industry, Sacramento, California.

Sturhan, D. 1967. Vorkommem von *Trichodorus*-Arten in Westdeutschland. *Mitt. biol. BundAnst. Ld-u. Forstw.* 121: 146–151.

Sturhan, D. 1985. Ein neuer Phytonematode aus Deutschland: *Paratrichodorus weischeri* spec. nov. (Nematoda, Trichodoridae). *Mitt. biol. BundAnst. Ld-u. Forstw.* 226: 31–45.

Symalla, H. J. 1972. Influence of different crops on TRV-infectivity and population density of *Trichodorus*. Abstract. *Sum. Inter. Nem. Sym., Eur. Soc. Nem., Reading*: 73–74.

Thorne, G. 1939. A monograph of the nematodes of the superfamily Dorylaimoidea. *Capitata Zool.* 8(5): 1–261.

Thorne, G. 1974. Nematodes of the Northern Great Plaina. Part II Dorylaimoidea in part (Nemata: Adenophorea). Tech. Bull. 41.

Vermeulen, W. J. J., and Heyns, J. 1983. Studies on Trichodoridae (Nematoda: Dorylaimida) from South Africa. *Phytophylactica* 15(1): 17–34.

Vermeulen, W. J. J., and Heyns, J. 1984. Further studies on southern African Trichodoridae (Nematoda: Dorylaimida). *Phytophylactica* 16: 301–305.

Wallace, H. R. 1971. The influence of the density of nematode populations on plants. *Nematologica* 17(1): 154–166.

Wasilewska, L. 1971. Nematodes in a young pine plantation in the Laski Forest Administration District of the Kampinos Forest. *Zesz. probl. Postop. Nauk. rohp.* 121: 159–167.

Whitehead, A. G., and Hooper, D. J. 1970. Needle nematodes (*Longidorus* spp.) and stubby-root nematodes (*Trichodorus* spp.) harmful to sugar beet and other field crops in England. *Ann. Appl. Biol.* 65: 339–350.

Winfield, A. L., and Cooke, D. A. 1975. The ecology of *Trichodorus*. In *Nematode Vectors of Plant Viruses*. Lamberti, F., Taylor, C. E., and Seinhorst, J. W. eds. Plenum Press, London, pp. 309–341.

Wyss, U. 1970. Zur Bestimmung von *Trichodorus* Arten. *PflKrankh.* 9: 506–514.

Wyss, U. 1974. *Trichodorus similis. C. I. H. Descriptions of Plant-Parasitic Nematodes*. Set 4, no. 59.

14

Sting and Awl Nematodes: *Belonolaimus* spp. and *Dolichodorus* spp.

GROVER C. SMART, JR., and KHUONG B. NGUYEN   *University of Florida, Gainesville, Florida*

STING NEMATODES, *BELONOLAIMUS* SPP.

I. AGRICULTURAL IMPORTANCE

A. Geographic Distribution

In the United States, sting nematodes have been found in deep sand soils in the Atlantic coastal plain regions of Florida (Steiner, 1949), Georgia (Holdeman, 1955), South Carolina (Graham, 1952), North Carolina (Holdeman, 1955), and Virginia (Owens, 1950, 1951), and in the gulf coast states of Alabama (Christie, 1959; Minton and Hopper, 1959), Mississippi (Robbins, pers. commun.), Louisiana (Holdeman, 1955); and Texas (Christie, 1959; Norton, 1959). Additionally, it has been found in sandy soils in Arkansas (Riggs, 1961), Kansas (Dickerson et al., 1972), Missouri (Perry in Perry and Rhoades, 1982), Oklahoma (Russell and Sturgeon, 1969), New Jersey (Hutchinson and Reed, 1956; Hutchinson et al., 1961; Myers, 1979), Nebraska (Kerr and Wysong, 1979), and in a greenhouse in Connecticut (Holdeman, 1955). All of these sting nematodes, except in Nebraska, were almost certainly *B. longicaudatus* even though some of the earlier reports listed them as *B. gracilis*; those in Nebraska were reported as near *B. nortoni*.

Outside the United States, sting nematodes have been reported from the Bahamas and Bermuda (*B. longicaudatus*) (Perry and Rhoades, 1982), Brazil, Sao Paulo State (*B. anama* and *B. jara*) (Monteiro and Lordello, 1977), Costa Rica (*B. longicaudatus*) (Lopez, 1978, 1979), Mexico (*B. longicaudatus*) (R. Robbins, pers. commun.), New South Wales, Australia (*B. lolii*) (Siviour, 1978; Siviour and McLeod, 1979), and Puerto Rico (*B. lineatus* and *B. longicaudatus*) (Perry and Rhoades, 1982; Roman, 1964). In the Bahamas, Bermuda, and Puerto Rico, *B. longicaudatus* was found only in golf courses which had imported turf grass sod from Florida or Georgia (Perry and Rhoades, 1982).

B. Host Range of *Belonolaimus longicaudatus*

The sting nematode *B. longicaudatus* has been recognized as an important pathogen since about 1950. A large number of plants, including most vegetable crops, agronomic crops, turf grasses and forage grasses, have been shown to be hosts (Christie et al., 1952; Holdeman and Graham, 1953; Johnson, 1970; Khuong and Smart, 1975; Owens, 1951; Rau, 1958, Robbins and Barker, 1973). Several researchers have shown that there are physiological races of the nematode (Abu-Gharbieh and Perry, 1970; Good, 1968; Owens, 1951; Perry and Norden, 1963; Perry and Smart, 1967, 1969; Robbins and Barker, 1973). For example *B. longicaudatus* in Virginia is pathogenic to peanut (Owens, 1951), but in Georgia (Good, 1968) and Florida (Perry and Norden, 1963) it causes no injury to peanut and does not reproduce well on it (Good, 1968). The plants listed below were reported by Robbins and Barker (1973) as poor hosts or nonhosts for three populations of *B. longicaudatus* from North Carolina, but good to excellent hosts for one population from Georgia:

> bermudagrass, *Cynodon dactylon* (L.) Pers.
> bushbean, *Phaseolus vulgaris* L. "Wade Stringless"
> curled dock, *Rumex crispus* L.
> eggplant, *Solanum melongena* L. "Black Beauty"
> fescue, *Festuca arundinacea* "Kentucky 31"
> highbush blueberry, *Vaccinium corymbosum* L. "Atlantic"
> lettuce, *Lactuca sativa* L. "Great Lakes"
> turnip, *Brassica rapa* L. "Seven Top"

They reported also that loblolly pine, *Pinus taeda* L., was a good host for the North Carolina populations but a poor host for the Georgia population. The above data indicate that hosts of the sting nematode in one geographic area may not be hosts in other areas.

Although Robbins and Barker (1973) found other differences in host preferences between the three North Carolina populations, they found that all four populations had the same preference or nonpreference for the plants in the categories listed below:

Excellent Host Plants

> Chinese elm, *Ulmus parvifolia* Jacq.
> Johnson grass, *Sorghum halepense* (L.) Pers.
> muscadine grape, *Vitis rotundifolia* Michx.
> pecan, *Carya illinoensis* (Wang.) K. Koch
> strawberry, *Fragaria virginiana* Duchesne "Earlibelle"
> white clover, *Trifolium repens* L. "Regal Ladino"
> corn, *Zea mays* L. "Pioneer 309 AMF"
> hairy crabgrass, *Digitaria sanguinalis* (L.) Scop.
> potato, *Solanum tuberosum* L.
> pearl millet, *Pennisetum glaucum* (L.) R. Br. "Starr"
> soybean, *Glycine max* (L.) Merr. "Lee"

Good Host Plants

> barley, *Hordeum vulgare* L. "Wade"
> centipedegrass, *Eremochloa ophiuroides* (Munro.) Hack.
> crimson clover, *Trifolium incarnatum* L.
> rye (*Secale cereale* L. "Abruzzi")

wheat, *Triticum aestivum* L. "Blueboy"
sweet corn, *Zea mays* L. "Golden Midget"
annual morning glory, *Ipomoea purpurea* (L.) Roth

Poor Host Plants

camellia, *Camellia sasanqua* Thunb.
cocklebur, *Xanthium pennsylvanicum* Wallr.
cotton, *Gossypium hirsutum* L. "Stoneville 7A" (excellent host in other experiments)
gladiolus, *Gladiolus hortulanus* Bailey "Beverly Ann"
Japanese holly, *Ilex crenata* Thunb. "Convexa"
jimson weed, *Datura stramonium* L.
lambsquarter, *Chenopodium album* L.
sweet potato, *Ipomoea batatas* (L.) Lam. "Century"

Nonhost Plants

asparagus, *Asparagus officinalis* L.
buckhorn plantain, *Plantago lanceolata* L.
okra, *Hibiscus esculentus* L. "Clemson Spineless"
pokeweed, *Phytolacca americana* L.
tobacco, *Nicotiana tabacum* L. "Hicks"
watermelon, *Citrulus lanatus* (Thunb.) Matsum. and Nakai "Garrisonian"

Since Robbins and Barker worked with a limited number of populations of *B. longicaudatus* and found differences as well as similarities among reactions to host plants, we suggest that plants be tested against local populations of the sting nematode to determine host status.

C. Pathogenicity and Symptoms

Since 1949 when Steiner (1949) described the sting nematode *Belonolaimus gracilis* as a parasite injuring the roots of slash- and long-leaf pine seedlings in forest nurseries near Ocala, Brooksville, and Valparaiso, Florida, much research has been done regarding its effects on crop plants. [Researchers in the southern states of the USA are convinced that Steiner described an extremely rare species of *Belonolaimus* and that the widespread nematode causing damage in the USA is not *B. gracilis* but *B. longicaudatus* described by Rau (1958). We make that assumption in this chapter.] Brooks and Christie (1950) and Owens (1950, 1951) reported the pathogenicity of the sting nematode to strawberry and peanut, respectively. Christie et al. (1952) published on its pathogenicity to strawberry, sweet corn, and celery. Graham and Holdeman (1953) showed its pathogenicity to cotton, corn, soybean, and cowpea; Tomerlin and Perry to soybean (1967); Schenck et al. (1962) to cuttings of grapevine; Rhoades (1967) to onion; and Khuong and Smart (1975) to collard, kale, and cauliflower. The latter researchers showed that populations as low as three nematodes per 100 g of soil at the time of transplanting susceptible plants can result in significant loss of yield. The sting nematode *B. longicaudatus* is now known to be an important pathogen of many crop plants (Boyd and Perry, 1971; Christie, 1959; Christie et al., 1954; Kelsheimer and Overman, 1953; Nutter, 1955; Perry et al., 1971; Robbins and Barker, 1973; Smart, 1969).

FIGURE 1   Sweet corn damaged by *Belonolaimus longicaudatus*. The center two rows and the rows which look similar to the left and to the right did not receive a nematicide while all rows in which corn is growing well did. Sting nematode populations were rather uniformly distributed throughout the field which is not the usual situation.

While symptoms caused by sting nematodes vary somewhat depending on such factors as inoculum level, host plant, and age of plant, in general the root system is greatly reduced and exhibits a combination of stubby roots and coarse roots with dark lesions along the root and at root tips (Perry and Rhoades, 1982) (Fig. 1). Sometimes roots are girdled completely. Shoot symptoms consist of stunting, premature wilting, and leaf chlorosis. Infested areas in fields vary in size and shape but the boundary between diseased and healthy plants is fairly well defined (Perry and Rhoades, 1982). Histological symptoms on infected roots consist of external lesions (feeding areas) at the apices or along the margins of roots, cavities overlapped by external root cells, rupture of cell walls, coagulation of the protoplasm of cells bordering the cavities, and a maturation of the meristem near lesions (Khuong, 1974; Perry and Rhoades, 1982; Standifer, 1959; Standifer and Perry, 1960). Apparently, salivary secretions from the feeding nematodes affect cells beyond those fed on (Standifer, 1959). Above-ground plants are stunted (Fig. 1) and chlorotic, and some plants may die. In the case of corn, even if the plants survive, rarely are usable ears of corn produced.

Interactions

Belonolaimus longicaudatus in association with fusarium wilt of cotton (Holdeman and Graham, 1952) makes symptom expression more severe (Holdeman and Graham, 1954) and breaks the resistance in cotton resistant to fusarium wilt (Holdeman and Graham, 1953). This work was confirmed by Cooper and Brodie (1962). The sting nematode also has been reported to increase the severity of *Pythium* root rot on chrysanthemum (Johnson and Littrell, 1970), and it, together with *Verticillium*, suppressed growth of tomato by 69% (Overman and Jones, 1977).

D. Crop Losses

The sting nematode causes serious damage to agricultural crops such as bean, beet, cabbage and other crucifers, celery, corn, cucumber, okra, onion, pea, pepper, potato, soybean, and turf grasses. In Florida, where the nematode is distributed widely in the sandy soils (it does not occur naturally in peat or clay soils), it may be responsible for more crop losses than any other single plant pest of any type (Perry and Rhoades, 1982). In other states in which it is found its distribution is less extensive, but it causes severe damage wherever it occurs.

In a cotton field heavily infested with the sting nematode, fumigated plots averaged 340 cotton bolls per plot compared to 16.5 bolls in nonfumigated plots (Graham and Holdeman, 1953). Preplant control of this nematode with a nematicide in North Carolina increased the yield of soybean up to 500%, of peanut up to 400%, and of corn up to 100% (Cooper et al., 1959). An at-planting application of a nematicide in North Carolina increased the yield of peanut up to 2200 kg/ha (Sasser et al., 1967). In Florida, yield of corn grain was increased up to 121% by the use of a nematicide either before or at planting (Johnson and Dickson, 1973), and the yield of forage grasses was increased an average of 70% when using a resistant variety compared to a susceptible variety (Boyd and Perry, 1969). Even trees do not escape damage, but an estimate of losses has not been given (Ruehle, 1964, 1967, 1973; Ruehle and Sasser, 1962).

E. Control

The sting nematode can be controlled with nematicides (Bistline et al., 1967; Christie, 1953, 1959; Christie et al., 1952; Cooper et al., 1959; Heald and Burton, 1968; Holdeman, 1956; Johnson and Dickson, 1972; Miller, 1952; Overman, 1959, 1964a,b; Rhoades, 1964, 1965; Robbins and Barker, 1973; Sasser and Cooper, 1961; Weingartner et al., 1978), with resistant cover crops and alternate crops (Brodie et al., 1970; Overman, 1959; Rhoades, 1964), by rotating susceptible crops with resistant varieties (Hunt et al., 1973; Perry and Norden, 1963), and by using organic soil amendments (Heald and Burton, 1968; Hunt et al., 1973; Tomerlin and Smart, 1969).

II. BIOLOGY AND TAXONOMY

A. Biology

The distribution of *Belonolaimus longicaudatus* is limited to sandy soil (Brodie, 1976; Miller, 1960, 1972; Robbins and Barker, 1974; Thames, 1959). Fine-textured soils may inhibit the movement and reproduction of the nematode (Thames, 1959). Four hundred and thirty-one farms in Virginia were surveyed for the sting nematode; it was found only in the "A" horizon of the soil profile with a sand content of 84–94% (Miller, 1972). Miller (1972) postulated that soil texture and soil moisture are two factors affecting survival of the sting nematode. In controlled research, the sting nematode increased only in soils with a minimum of 80% sand and a maximum of 10% clay. The optimum size of soil particles for reproduction was from 120 to 370 μm (Robbins and Barker, 1974). *Belonolaimus longicaudatus* is widespread in the sandy soils of Florida but has never been found in muck or marl soils (Christie, 1959). Experimentally, the nematode will not reproduce in untreated muck soil, but will reproduce very well in muck soil if the soil is steam-sterilized, or treated with the nematicide DD or with methyl bromide (Rhoades, 1980b).

The rather large-bodied sting nematode is not as active as smaller nematodes, and recovery is generally better with a centrifugation technique than with a Baermann funnel technique. Nonetheless, in laboratory tests Khuong (1974) recovered 55–60% of the sting nematodes with a Baermann funnel set up for 24 hr.

Temperature is important in the life cycle of this nematode. In Florida, reproduction was greater at 29.4°C (85°F) than at 26.7°C (80°F), and was greatly reduced at 35°C (95°F) (Perry, 1964). In North Carolina, reproduction was greatest between 25 and 30°C (Robbins and Barker, 1974). In Florida, no sting nematodes were present in the top 2.5 cm of bare soil when the temperature at that depth reached 39.5°C (103°F); the nematodes either migrated downward or died (Boyd and Perry, 1969; Boyd et al., 1972).

Soil moisture also is an important factor in the reproduction and distribution of the nematode. Populations of the sting nematode were highest when soil moisture was from 15 to 20% (Brodie and Quattlebaum, 1970), and reproduction was greater at a moisture level of 7% than at 30% or at 2% (Robbins and Barker, 1974).

Temperature and moisture also are important in vertical distribution and overwintering of the nematode. In Georgia, sting nematode populations were greatest at 15–30 cm depths between November and April. Some were found at 30–45 cm and very few at 0–15 cm. The nematode may move deeper into the soil for overwintering. Also, at 30–45 cm they might escape fumigation or other control measures and could reestablish infestation of the root zone of the subsequent crops (Potter, 1967). In North Carolina, populations of the sting nematodes were highest at 7.5–15 cm depth from October to January. In the fall the populations were highest at 0–7.5 cm and at 15–30 cm, with very few specimens at the 30- to 35-cm depth at any time. In March and April the nematode almost disappeared at all depths, but populations increased slightly in May and June (Barker, 1968).

Light intensity may increase the rate of reproduction of the sting nematode. When cotton was grown in a greenhouse with 12 hr of supplemental light, reproduction of the nematode increased compared to natural light alone (Barker et al., 1975).

The life cycle and feeding habits of the nematode have not been studied thoroughly. The nematode is classified as an ectoparasite, but occasionally specimens are found inside roots (Christie et al., 1952). The survival stage or stages of the nematode may not be the same in all areas where it occurs, but in Florida, all stages are found in the soil throughout the year. The nematode reproduces bisexually with males comprising about 40% of the population. Adult females usually have sperm in the spermatheca. Details of the life cycle are not known, but under optimum conditions the life cycle is completed in about 28 days.

B. Taxonomy

Historical: The genus *Belonolaimus* was first reported by Steiner in 1942 at a meeting of the Soil Science Society of Florida in a paper entitled, "Plant Nematodes the Grower Should Know" (Steiner, 1949). Because of the war years, this paper was not published until 1949. *Belonolaimus gracilis* was considered to be the prevalent sting nematode until Rau (1958) described *Belonolaimus longicaudatus* and showed that it is the commonly occurring sting nematode in the southern USA while *B. gracilis* is found rarely. The sting nematode referred to by most authors as *B. gracilis* prior to Rau's 1958 paper almost certainly was *B. longicaudatus*.

Rau later (1963) described three new species of *Belonolaimus—B. euthychilus, B. maritimus,* and *B. nortoni*—and gave a key to the five nominal species. Three other species of *Belonolaimus* were described later: *B. hastulatus* Colbran, 1960, which was transferred to *Tylenchorhynchus* by Fortuner and Luc (1987); *B. lineatus* Roman, 1964 from Puerto Rico, which was transferred to *Ibipora* by Monteiro and Lordello (1977) and back to *Belonolaimus* by Fortuner and Luc (1987) when they proposed *Ibipora* as a junior synonym of *Belonolaimus*; and *B. lolii* Siviour, 1978 from Australia, which was transferred to *Ibipora* by Siviour and McLeod (1979) and back to *Belonolaimus* by Fortuner and Luc (1987). Monteiro and Lordello (1977) described *Ibipora anama* and *I. jara*, which were transferred to *Belonolaimus* by Fortuner and Luc (1987). According to Fortuner and Luc (1987), *Belonolaimus* contains nine species: *B. gracilis* (type species), *B. anama, B. euthychilus, B. jara, B. lineatus, B. lolii, B. longicaudatus, B. maritimus,* and *B. nortoni.*

The position of *Belonolaimus* in the order Tylenchida has been a source of conjecture almost since the genus was first described. Steiner (1949) placed it in the family Tylenchidae; Chitwood and Chitwood (1950) placed it in the subfamily Dolichodorinae, family Criconematidae; Loof (1958) stated that it should be transferred to the subfamily Hoplolaiminae, family Tylenchidae; Whitehead (1960) proposed the subfamily Belonolaiminae for *Belonolaimus* and *Trichotylenchus*; Golden (1971) proposed that the family Belonolaimidae be erected for the subfamilies Belonolaiminae and Telotylenchinae; Andrássy (1976) accepted that proposal; Siddiqi (1986) suggested that the subfamily Belonolaiminae be placed in the family Dolichodoridae; and, finally, Fortuner and Luc (1987) accepted the family Belonolaimidae to contain the subfamilies Belonolaiminae and Telotylenchinae.

Family Belonolaimidae Whitehead, 1960

syn. Telotylenchidae Siddiqi, 1960
Tylenchorhynchidae Eliava, 1974

Diagnosis (after Fortuner and Luc, 1987)

Tylenchoidea. Medium- to large-sized nematodes, with tail cylindroid to conoid, more than twice as long as wide but never elongate filiform (typically $c' = 2$–5). Phasmids always on posterior half of tail, never enlarged into scutella. Deirids present or absent. Face view as seen with SEM either ancestral (first lip annulus six-sectored) or with lateral sectors regressed and face view evolving toward either a grossly quadrangular shape or a four-leaf clover shape. Females typically with two genital branches (except *Trophurus*). Columned uterus with three rows of cells. Males with peloderan caudal alae, rarely lobed or stopping just short of the tail tip. Spicules with or without pronounced velum.

Belonolaimidae are ectoparasites of plant roots, but a few species appear to be migratory endoparasites as well.

Type subfamily: Belonolaiminae Whitehead, 1960
Other subfamily: Telotylenchinae Siddiqi, 1960

Key to subfamilies:

Labial region bulbous; stylet slender, elongate with cone longer than shaft; labial disk
        rounded .... Belonolaiminae
Labial region never bulbous; stylet shorter, not slender, cone about as long as shaft; labial
        disk lemon-shaped or variously fused with lip sectors .... Telotylenchinae

Subfamily Belonolaiminae Whitehead, 1960

Diagnosis (after Fortuner and Luc, 1987)

Belonolaimidae. Cephalic framework often very weak, sometimes heavily sclerotized. Stylet slender, elongate, usually 60–150 μm long, with cone longer than shaft ($m = 60–80$). In forms with elongate stylets, procorpus enlarged and separated from the median bulb by a constriction. Median bulb strong, muscular, with large valve. Labial region often offset, bulbous in lateral view, sometimes continuous with body contour. Face view, as seen with the SEM, generally with a well-marked, round, labial disk and a first-lip annulus with submedian sectors well marked and lateral sectors regressed, almost absent. Rarely, lateral sectors only slightly regressed. In one genus, *Morulaimus*, labial disk and lateral sectors are fused into a lemon-shaped structure. Female tail long, generally cylindroid to broadly rounded end, sometimes more conoid. Deirids always absent.

Belonolaiminae differs from Telotylenchinae by its biology and by a tendency toward an elongation of the stylet to reach inside roots. SEM face views with well-marked round labial disk are characteristic for most genera.

Type genus: *Belonolaimus* Steiner, 1949
    syn. *Ibipora* Monteiro and Lordello, 1977

Other genera: *Carphodorus* Colbran, 1965
                    Geocenamus Thorne and Malek, 1968
                        syn. *Hexadorus* Ivanova and Shagalina, 1983
                        Morulaimus Sauer, 1966
                        Sauertylenchus Sher, 1974

Key to genera:

1. Lateral field with six incisures .... *Geocenamus*
    Lateral field with four incisures or less 2
2. Stylet short, 37 μm long .... *Sauertylenchus*
    Stylet long, 60 μm or longer 3
3. Body more than 2 mm long .... *Belonolaimus*
    Body 1.5 mm or less in length 4

4. Labial framework weak, perioral disk lemon-shaped, epiptygma present ....
.... *Morulaimus*
Labial framework massive, perioral disk rounded, epiptygma absent ....
.... *Carphodorus*

Genus *Belonolaimus* Steiner, 1949 (Figs. 2, 3)
syn. *Ibipora* Monteiro and Lordello, 1977

Diagnosis (after Fortuner and Luc, 1987)

Belonolaiminae. SEM face view shows a well-marked rounded labial disk, and first lip annulus divided into six sectors; lateral sectors almost completely regressed, seen only as

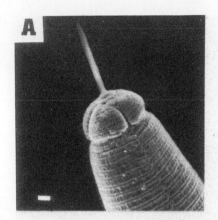
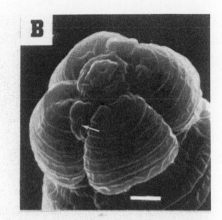
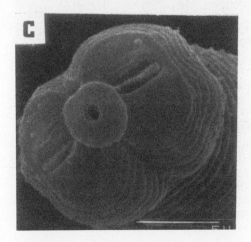
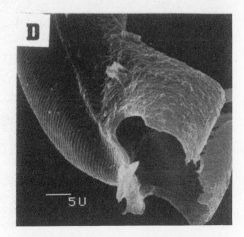

FIGURE 2   SEM photographs of *Belonolaimus* and *Dolichodorus*. (A, B) *Belonolaimus longicaudatus.* (A) Anterior region with protruding stylet. (B) Labial region. (C, D) *Dolichodorus miradvulvus.* (A) Head showing perioral disk, four lips, two amphids and labial annules. (D) Male tail with large bursa. (After Smart and Khuong, 1985.)

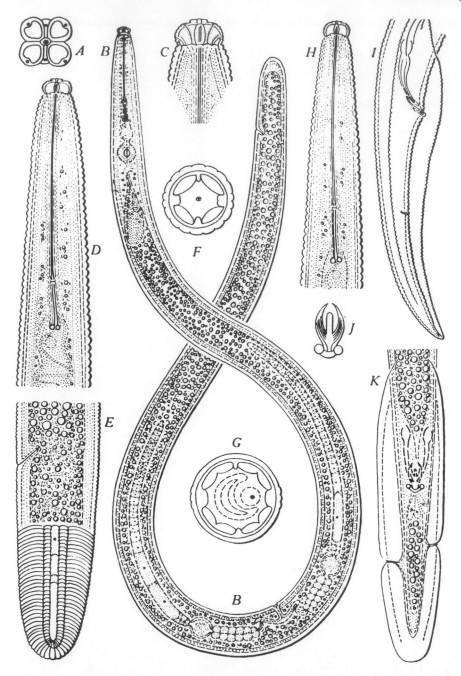

FIGURE 3 *Belonolaimus lineatus*: (A–G) Female. (A) Face view. (B) Entire body. (C) Head, showing papillae. (D) Anterior end. (E) Tail, showing lateral field and phasmid. (F) Cross-section through middle of stylet. (G) Cross-section through anterior ovary. (H–K) Male. (H) Anterior end. (I) Tail, lateral view. (J) Gubernaculum, ventral view. (K) Tail, ventral view. (After Roman, 1964.)

small interruptions of the first one or two labial annuli. Labial region marked by deep longitudinal grooves. Stylet very long, 60–150 μm, its cone 70–80% of total stylet length. Procorpus thickened to accommodate the long stylet and separated from the median bulb by a constriction; median bulb with strong valve. Esophageal glands overlapping beginning of intestine. Female tail cylindroid with a broadly rounded terminus. Lateral field with four lines or less.

Type species: *Belonolaimus gracilis* Steiner, 1949

Other species: *B. anama* (Monteiro and Lordello, 1977) Fortuner and Luc, 1987
        syn. *Ibipora anama* Monteiro and Lordello, 1977
    B. euthychilus Rau, 1963
    B. jara (Monteiro and Lordello, 1977), Fortuner and Luc, 1987
        syn. *Ibipora jara* Monteiro and Lordello, 1977
    B. lineatus Roman, 1964
        syn. *Ibipora lineatus* (Roman, 1964) Monteiro and Lordello, 1977
    B. lolii Siviour, 1978
        syn. *Ibipora lolii* (Siviour, 1978) Siviour and McLeod, 1979
    B. longicaudatus Rau, 1958
    B. maritimus Rau, 1963
    B. nortoni, Rau, 1963

Key to species:

1. Lateral field with one incisure 2
   Lateral field with four incisures 6
2. Tail longer than stylet 3
   Tail not longer than stylet 4
3. Stylet less than 100 μm; vulva with protruding lips . . . . *B. nortoni*
   Stylet more than 100 μm; vulva without protruding lips. . . . . *B. longicaudatus*
4. Labial region of female continuous with the body; stylet and esophagus of male degenerate (sexual dimorphism) . . . . *B. euthychilus*
   Labial region of female offset by constriction; stylet and esophagus of male normal (no sexual dimorphism) 5
5. Female tail hemispheric; median bulb elongate; paired opposing pieces in vagina large . . . . *B. maritimus*
   Female tail convex-conoid; median bulb nearly spherical; paired opposing pieces in vagina absent . . . . *B. gracilis*
6. Female body 2–2.6 mm; vulval lips and sclerotized pieces in vagina protruding . . . . *B. lolii*
   Female body 1.9 mm or less; vulval lips and sclerotized pieces in vagina not protruding 7
7. Body with longitudinal striae in region anterior to the esophageal glands; stylet 98–100 μm long . . . . *B. lineatus*
   Body without longitudinal striae in region anterior to the esophageal glands; stylet 92 μm long or less 8
8. Female stylet 66 μm long; *c* ratio about 18 . . . . *B. jara*
   Female stylet 77–92 μm; *c* ratio about 24 . . . . *B. anama*

Species *Belonolaimus gracilis* Steiner, 1949

Measurements (in μm) (after Steiner, 1949 and Rau, 1961)

Female (after Steiner, 1949): Length = 2150; a = 52; b = 6.1; c = 19.2; V = 52; stylet = 157; tail = 120; tail/ABW = 3; stylet/tail = 1.31.

Females (61) (after Rau, 1961): Length = 1900 (1400–2300); a = 49 (39–63); b = 6.7 (5.1–9.8); c =23 (16–28); V = 53 (50–57); stylet = 152 (130–168); tail = 78 (53–134); tail/ ABW = 2.6 (1.8–3.6); stylet/tail = 1.76 (1.33–2.31).

Male (after Steiner, 1949): Length = 1700; a = 52; b = ?; c = 14.7; stylet = ?; tail = 116; spicules = 51; gubernaculum = 20.

Male (58) (after Rau, 1961): Length = 1900 (1400–2500); a = 52 (44–61); b = 6.3 (5.1–7.2); c = 17 (13–29); stylet = 137 (99=154); tail = 99 (60–140); stylet/tail = 1.37 (1.07=1.96); spicule = 45 (35–50); gubernaculum = 16 (14–18).

Description (after Steiner, 1949 and Rau, 1961)

Body cylindrical, coarsely annulated; annules laterally interrupted by single groove. Head distinctly offset, four lobed, annulated. Stylet about 157 μm long with rounded basal knobs. Procorpus short, containing large ampulla of dorsal esophageal gland, its outlet near base of stylet; esophageal canal much folded when stylet is retracted. Median bulb spherical, with large valves. Isthmus short; end portion of esophagus consisting of three much enlarged gland cells overlapping anterior end of intestine. Intestine granular, rather opaque, its cavity obscure; rectum and anus also rather inconspicuous. Gonads amphidelphic. Spicula in side view slightly arched, of about the same width their entire length, distally pointed. Gubernaculum about two-fifths length of spicula; bursa with one short rib representing the phasmid on each wing near middle of tail.

Type habitat: Soil around the roots of Cuban pine *Pinus caribaea* Morelet, Ocala, Florida.

Genus *Sauertylenchus* Sher, 1974 (Fig. 4)

Diagnosis (after Sher, 1974 and Fortuner and Luc, 1987)

Belonolaiminae. Body large sized (1.7 mm in type species). Labial region annulated, offset by constriction. Face view with six sectors, lateral sectors slightly smaller than submedians. Labial framework weakly developed. Stylet thin, long (37 μm in type species). Esophagus with basal bulb elongate, rounded posteriorly. Deirids absent. Lateral field with four incisures. Epiptygma present. Tail length more than two times anal body width. Male spicules with flanges. Gubernaculum slightly protruding from cloaca.

Type and only species: *Sauertylenchus labiodiscus* Sher, 1974.

Measurements (μm) (after Sher, 1974)

Holotype (female): Length = 1670; a = 46; b = 8; c = 21; V = 51; stylet = 39 μm.

Paratypes (female): Length = 1710 (1460–2060); a = 48 (42–53); b = 8.2 (6.3–10.0); c = 23 (20–26); V = 53 (49–55); stylet = 37 (35–40).

Allotype (male): Length = 1580; a = 48; b = 7.6; c = 21; stylet = 37; spicules = 36; gubernaculum = 15.

Paratypes (male): Length = 1420 (1230–1650); a = 46 (36–54); b = 7.4 (6.3–8.0); c = 22 (20–26); stylet = 36 (33–39); spicules = 36 (33–38); gubernaculum = 15 (12–16).

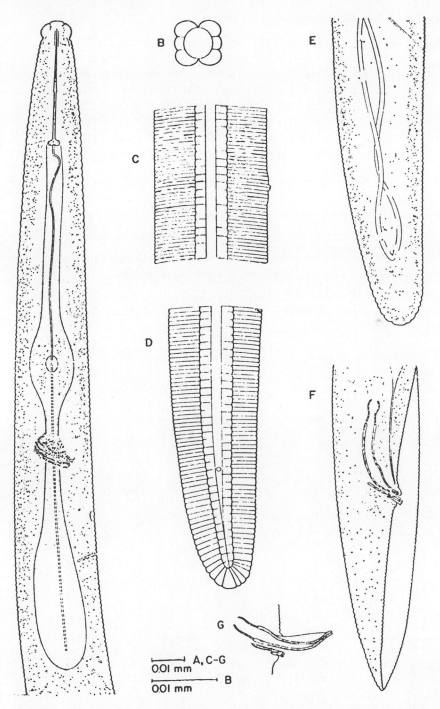

FIGURE 4 *Sauertylenchus labiodiscus*: (A–E) Female. (A) Anterior region. (B) Face view. (C) Center of body, surface view. (D) Tail region, surface view. (E) Tail region, longitudinal section. (F, G) Male. (F) Tail region. (G) Spicules and gubernaculum. (After Sher, 1974.)

Description (after Sher, 1974)

Female: Lip region rounded, distinctly set off from body with six to eight fine annules, labial disk conspicuous. Face view with six lip sectors; the lateral sectors considerably smaller than the other sectors. Amphid apertures not seen with light microscope and obscure with SEM. Stylet knobs rounded with slightly sloping anterior surface. Dorsal esophageal gland opening 4 μm posterior to stylet knobs. Median bulb oval. Basal bulb elongate, rounded posteriorly with conspicuous esophagointestinal valve. Epiptygma double. Lateral canal conspicuous. Intestine overlapping rectum into tail. Outer field of lateral incisures incompletely areolated. Phasmid on tail. Tail terminus rounded, annulated with the annules often irregular.

Male: Similar to female except for sexual differences. Spicules notched distally; gubernaculum recurved distally, titillae present.

Type habitat: Soil around the roots of *Rhagodia* sp., New South Wales, Australia.

Genus *Geocenamus* Thorne and Malek, 1968 (Fig. 5)

Diagnosis (after Thorne and Malek, 1968 and Fortuner and Luc, 1987)

Belonolaiminae. Body medium-sized, tapering near extremities. Labial region offset by constriction. Face view with six sectors, lateral sectors smaller. Perioral disk present. Cephalic framework weakly developed. Lateral field with six incisures. Deirids absent. Stylet

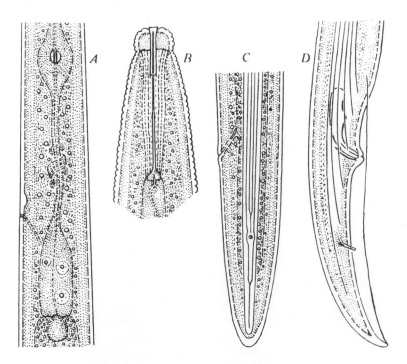

FIGURE 5   *Geocenamus tenuidens*: (A) Esophageal region. (B) Head region. (C) Female tail, showing six incisures and phasmid in lateral field. (D) Male tail, showing six incisures and phasmid in lateral field, spicules, gubernaculum, and bursa. (After Thorne and Malek, 1968.)

slender, 25–130 μm long, cone longer than shaft. Tail conoid to almost cylindroid with conoid to bluntly rounded terminus. Male tail arcuate, enveloped by a broad bursa. Spicules without velum, bluntly ended. Gubernaculum not protruding from cloaca.

Type species: *Geocenamus tenuidens* Thorne and Malek, 1968
  syn. *Tylenchorhynchus polonicus* Szczygiel, 1970
    Geocenamus polonicus (Szczygiel, 1970) Sturhan, 1981

Species *Geocenamus tenuidens* Thorne and Malek, 1968

Measurements (in μm) (after Thorne and Malek, 1968)
    Female: Length = 833; a = 30; b = 6.7; c = 12; V = 50.
    Male: Length = 800; a = 37; b = 6.1; c = 12; testis = 60.

Description (after Thorne and Malek, 1968)

Body cylindroid, tapering near extremities. Lip region with five annules, set off by deep constriction. Cephalic framework weakly developed. A refractive, perioral disk present. From this disk a slender, tubular stylet guide extends back about 8 μm. Stylet exceedingly slender, 27 μm long with strong sloping knobs, its muscles attached to cephalic framework. Lateral fields marked by six incisures which decrease in number on tail as illustrated. Hemizonid occupying about four annules. Nuclei in basal bulb unusually conspicuous. Cardia elongate spheroid. Intestinal cells filled with variable-sized granules. Ovaries outstretched, spermatheca spheroid, about half of body width, oocytes arranged in single file. Postrectal blind sac extending two-thirds distance into tail. Female terminus rounded, male tail slightly arcuate, enveloped by broad bursa.

Type habitat: Prairie sod near Rugby, North Dakota.
Other species: *G. arealoferus* (Razzhivin, 1971) Fortuner and Luc, 1987
        syn. *Morulaimus arealoferus* Razzhivin, 1971
      G. arcticus (Mulvey, 1969) Tarjan, 1973
      G. deserticola (Ivanova and Shagalina, 1983) Fortuner and Luc, 1987
        syn. *Hexadorus deserticola* Ivanova and Shagalina, 1983
      G. kirjanovae (Sagitov, 1973) Fortuner and Luc, 1987
        syn. *Dolichodorus kirjanovae* Sagitov, 1973
          H. kirjanovae (Sagitov, 1973) Siddiqi, 1986
      G. longus (Wu, 1969) Tarjan, 1973
        syn. *Tylenchorhynchus longus* Wu, 1969
      G. tokobaevi (Sultanalieva, 1983) Fortuner and Luc, 1987
        syn. *Morulaimus tokobaevi* Sultanalieva, 1983
          H. tokobaevi (Sultanalieva, 1983) Siddiqi, 1986
      G. uralensis Baidulova, 1983

Key to species:

1. Stylet length 100 μm or more 2
    Stylet length 70 μm or less 3
2. Stylet length 100 μm, body length 1800–2150 μm .... *G. tokobaevi*
    Stylet length 120–131 μm, body length 1295–1489 μm .... *G. arealoferus*
3. Body length more than 1500 μm (1531–1813) .... *G. kirjanovae*
    Body length 1500 μm or less 4

4. Body length less than 900 μm, *c* ratio less than 15 5
    Body length more than 900 μm, *c* ratio more than 15 6
5. Stylet length 27 μm, body length about 830 μm .... *G. tenuidens*
    Stylet length 20 μm, body length 621–751 μm .... *G. uralensis*
6. Stylet length less than 40 μm .... *G. arcticus*
    Stylet length more than 40 μm 7
7. Tail terminus smooth, *c* ratio 19.8 or more .... *G. deserticola*
    Tail terminus annulated, *c* ratio 17 or less .... *G. longus*

Genus *Morulaimus* Sauer, 1966 (Fig. 6)

Diagnosis (after Fortuner and Luc, 1987)

Belonolaiminae. SEM face view with labial disk lemon-shaped; first labial annulus divided into six sectors, lateral sectors somewhat flattened. Labial region not marked by deep longitudinal indentation. Stylet elongate, 60–100 μm. Cone 60–80% of total stylet length. Labial framework always weak. Procorpus thickened to accommodate the long stylet and separated from the median bulb by a constriction; median bulb with strong valve. Esophageal glands overlapping beginning of intestine. Tail sometimes short for Belonolaimidae (c' = 2–3). Tail shape varies from cylindroid with a broadly rounded terminus to almost conoid.

Type species: *Morulaimus arenicolus* Sauer, 1966
Other species: *Morulaimus geniculatus* Sauer, 1966
        syn. *Scutellonema magnum* Yeates, 1967
     M. sclerus Sauer, 1966
     M. simplex Sauer and Annells, 1981
     M. soldus Colbran, 1969
     M. whitei (Fischer, 1965) Sauer, 1966
        syn. *Telotylenchus whitei* Fisher, 1965

Key to species:

1. Labial region not offset, sclerotization behind labial plate prominent ....
    .... *M. simplex*
    Labial region offset by constriction, sclerotization behind labial plate absent 2
2. Conspicuous sclerotized pieces in dorsal and ventral sectors of cephalic framework
    present 3
    Conspicuous sclerotized pieces in dorsal and ventral sectors of cephalic framework
    absent 4
3. Tail cylindrical, more than two anal body widths long .... *M. sclerus*
    Tail tapers, less than two anal body widths long .... *M. whitei*
4. Head with platelets on lateral side, esophagointestinal junction about midesophageal
    lobe .... *M. geniculatus*
    Head without platelets on lateral side, esophagointestinal junction in anterior part of
    basal bulb 5
5. Stylet length 94 μm (84–101) .... *M. arenicolus*
    Stylet length 75 μm (69–81) .... *M. soldus*

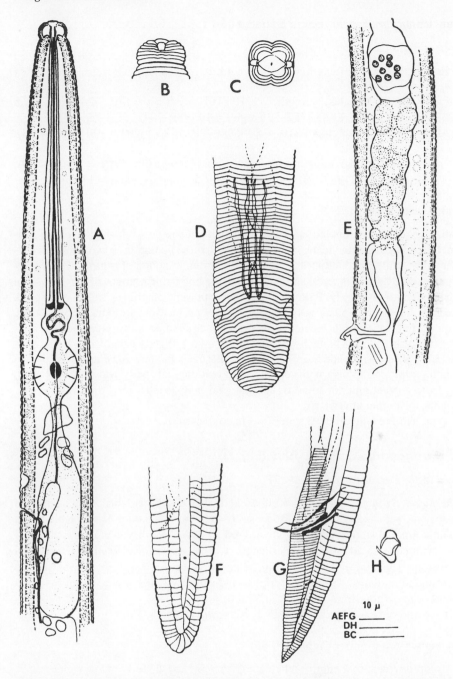

FIGURE 6 *Morulaimus arenicolus*: (A) Anterior end of female. (B) Female head, lateral view. (C) Face view, female. (D) Male tail, ventral view. (E) Vulval region. (F) Female tail with lateral field and phasmid. (G) Male tail. (H) End-on view of spicules showing overlapping flanges. (After Sauer, 1966.)

Species *Morulaimus arenicolus* Sauer, 1966

Measurements (in μm) (after Sauer, 1966)

Holotype (female): Length = 1210; a = 43; b = 6.9; c = 31; V = 58; stylet = 104 (conus = 76).

Paratypes (36 females): Length = 1210 (1070–1340); a = 40 (32–47); b = 7.6 (6.5–8.7); c = 30 (26–36); V = 54 (53–58); stylet = 103 (89–108).

Allotype (male): Length = 1210; a = 48; b = 7.0; c = 22; stylet = 100 (conus = 73); spicules = 32; gubernaculum = 15.

Paratypes (19 males): Length = 1060 (920–1250); a = 43 (35–49); b = 7.1 (5.9–8.9); c = 19 (17–22); stylet = 94 (84–101); spicules = 30–35; gubernaculum = 13–16.

Description (after Sauer, 1966)

Body tapering anteriorly from latitude of median bulb to lip region, which is set off by constriction. Lateral field commencing about 20 annules behind lip region as two lines, increasing to three lines anterior to the spear junction, and to four in region of isthmus. Outer lines crenate, crossed by striae in neck and tail region, with variable amount of often obscure striation in other areas of body. Prominent labial disk present, amphid apertures close to its lateral edges. Cephalic papillae not observed in any en face preparations, though there would appear to be one papilla located near edge of each submedian lip, since the innervating nerve can be seen lower in the head. Labial framework slightly sclerotized. Stylet very long. Junction of esophagus and intestine close to base of short isthmus. Female with prominent double epiptygma. Vulva between one-third and one-half body width. Tail conical, tapering from a point anterior to anus, 1.7–2.6 anal body widths long. Phasmids small, slightly anterior to midpoint of tail.

Type habitat: Uncultivated sandy soil in southeastern Australia.

Genus *Carphodorus* Colbran, 1965 (Fig. 7)

Diagnosis (after Fortuner and Luc, 1987)

Belonolaiminae. SEM face view with well-marked labial disk; first labial annulus divided into six sectors, with lateral sectors a little smaller than the submedian. Labial region with deep indentation. Labial framework massive, strongly developed. Stylet elongate, about 95 μm long, with cone about 68% of stylet length. Corpus as in *Belonolaimus*. Esophageal glands overlap beginning of intestine. Tail relatively short (c' = 1.7), cylindroid with a broadly rounded terminus. Original description indicates two lines in the lateral field, but SEM pictures show four lines (Sauer et al., 1980).

Type and only species: *Carphodorus bilineatus* Colbran, 1965

Measurements (in μm) (after Colbran, 1965)

Holotype (female): Length = 1111; a = 37; b = 6.3; c = 23.8; V = 55.4; stylet = 96.7.

Paratypes (10 females): Length = 970–1111; a = 33.2–37.0; b = 6.0–6.6; c = 22.0–24.1; V = 55.4–57.3; stylet = 90.3–96.7.

Allotype (male): Length = 875; a = 34.6; b = 6.8; c = 21.2; stylet = 83.1; spicules = 28.2; gubernaculum = 12.3.

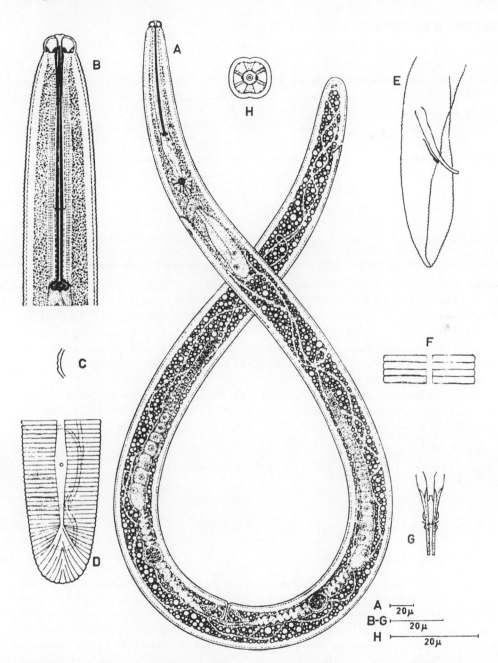

FIGURE 7 *Carphodorus bilineatus*: (A) Female, entire body. (B) Female head. (C) Cross-section of lateral field. (D) Female tail. (E) Male tail. (F) Female, lateral field. (G) Spicules and gubernaculum. (H) En face view of female. (After Colbran, 1965.)

Description (after Colbran, 1965)

Female: Body cylindrical, forming an open C when relaxed. Transverse striae fine, 1.0–1.2 μm apart in midbody. Lateral field with two fine incisures (four with SEM, Sauer et al., 1980) reduced to one near isthmus and terminating about 20 μm from anterior end; field anterior to incisure marked by occasional breaks in striae. Lip region subspherical, bearing 10–12 annules, set off from body by constriction. Anterior end of lip region marked by dorsal and ventral grooves. Lip cap present. Amphids opening through small pores on lateral lips at edge of lip cap. Hexaradiate framework well developed in lower part of lip region; dorsal and ventral blades wider than subdorsal and subventral blades. Single papilla near outer margin of each subdorsal and subventral lip. Dorsal esophageal gland orifice 3.2 μm behind stylet. Esophagus with ovate metacorpus and elongate, swollen terminal portion with posterior third overlapping intestine laterally. Excretory pore opposite isthmus; tubes of excretory system coiled in pseudocoel. Hemizonid 10 (5.6–10.5) μm behind excretory pore. Vulva equatorial; ovaries paired, opposed, outstretched; spemathecae subspherical; oocytes for most part in double row. Tail 1.7 (1.8–2.0) anal body widths long, subcylindrical; terminus hemispheric, striated. Phasmids one-fourth to one-fifth tail length behind anus.

Male: Shorter and slenderer than female. Transverse striae 1.2 μm apart. Lateral field with two incisures (four incisures?). Dorsal esophagal gland orifice 3.6 (3.5–4.3) μm behind stylet. Hemizonid 10.6 (9.0–10.6) μm behind excretory pore. Caudal alae extending to terminus. Spicules slightly curved, tip blunt. Gubernaculum half as long as spicules, proximal end with dorsally directed hook, distal end with titillae. Phasmids about one-fourth tail length behind cloaca.

Type habitat: Coarse sandy soil about the roots of *Eucalyptus andrewsi* Maiden (messmate) in open forest, roadside at Pozieres (Portion 104, parish of Marsh), Queensland.

AWL NEMATODES, *DOLICHODORUS* SPP.

I. AGRICULTURAL IMPORTANCE

A. Geographic Distribution

In the United States, at least one species of awl nematode has been reported from eight rather widely separated states. Four species have been reported from Florida, two from Arkansas, and one each from California, Michigan, Maryland, New York, North Carolina, and West Virginia. It is probable that the nematode is located in other states in isolated areas. The locations and species are Arkansas (*D. cobbi* from woodland by Golden et al., 1986, and *D. grandispicatus* from *Acer rubrum* L. by Robbins, 1982); California (*D. similis* from *Sparganium greenei* Morong by Golden, 1958); Florida (*D. aestuarius* from *Juncus roemerianus* Scheele by Chow and Taylor, 1978, *D. heterocephalus* from Silver Springs by Cobb, 1914, *D. miradvulvus* from *Anubias nana* Engler by Smart and Khuong, 1985, and *D. profundus* from *Strelitzia reginae*, Tri-ology, Fla. Dept. Agr. and Consum. Ser., 1977); Michigan (*D. heterocephalus* from Douglas Lake by Cobb, 1914); Maryland (*D. marylandicus* from *Zoysia japonica* L. by Lewis and Golden, 1981); New York (*D. heterocephalus* from Lake Champlain by Fisher and Hugo, 1969); North Carolina (*D. marylandicus* from

Pinus sp. by Lewis and Golden, 1981); and West Virginia (*D. silvestris* from *Pinus strobus* L. by Gillespie and Adams, 1962).

The genus has been reported from seven other rather widely separated countries. Two species have been reported from Argentina and India and one each from Brazil, Italy, Nigeria, Togo, and the Philippines. The locations and species are Argentina, Cordoba (*D. aquaticus* from *Bromus* sp. by Doucet, 1985 and *D. longicaudatus* from *Poa* sp. and other Graminae by Doucet, 1981); Brazil, Bahi (*D. minor* from *Cocos nucifera* L., *Psidium guajava* L., *Ananas comosus* (L.) Merrill, and *Hevea brasiliensis* Muell.-Arg. by Loof and Sharma, 1975); India, Gujarat (*D. kishansinghi* from *Mangifera indica* L. by Jairajpuri and Rahmani, 1979), and India, Kerala (*D. pulvinus* from *Cocos nucifera* L. by Kahn et al., 1971); Italy (*D. heterocephalus* from *Olea europaea* L. by D'Errico et al., 1977); Nigeria, Abidjan (*D. nigeriensis* from grasses by Luc and Caveness, 1963); Togo, Akodessewa (*D. profundus* from *Cocos nucifera* L. by Luc, 1960); and the Philippines, Mindanao (*Dolichodorus* sp. from *Boehmeria nivea* by Epstein et al., 1971).

B. Host Range of *Dolichodorus heterocephalus*

The host range of *D. heterocephalus* is quite extensive and includes agronomic crops, vegetable crops, fruit trees, ornamental trees, shrubs, grasses, and weeds. The following listing is alphabetical by Latin name.

Acer rubrum L., red maple
Acoelorrhaphe wrightii (Griseb. and H. Wendl.) H. Wendl, ex Becc., Everglades palm
Andropogon virginicus L., broomsedge
Andropogon sp.
Apium graveolens L. var. dulce (Mill.) Pers., celery
Arecastrum romanzoffianum (Cham.) Becc., queen palm
Brassica oleracea L., Capitata group, cabbage
Camellia japonica L., camellia
Capsicum annuum L., Grossum group, bell pepper
Carya floridana Sarg., scrub hickory
Casuarina sp., Australian pine
Celtis laevigata Willd., sugarberry, hackberry
Chamedorea elegans Mart., parlor palm
Citrus aurantium L., sour orange
Citrus sinensis (L.) Osbeck, sweet orange
Colocasia esculenta (L.) Schott, taro, dasheen
Cynodon dactylon (L.) Pers., Bermuda grass
Cynodon sp.
Cyperus alternifolius L., umbrella sedge
Cyperus rotundus L., nutgrass
Cyperus sp.
Dianthus plumarius L., cottage pink
Diodia virginiana L., buttonweed
Eleocharis dulcis (Burm. f.) Trin. ex Henschel, water chestnut
Eremochloa ophiuroides (Munro) Hack., centipedegrass
Fragaria sp.
Fraxinus caroliniana Mill., pop ash
Gladiolus sp.
Hydrocotyle sp., dollar weed

Ilex glabra (L.) A. Gray, gallberry
Impatiens balsamina L., garden balsam
Juncus sp., rush
Juniperus silicicola (Small) L. H. Bailey, southern red cedar
Liquidambar styraciflua L., sweetgum
Livistona chinensis (Jacq.) R. Br. ex Mart., Chinese fan palm
Lycopersicon esculentum Mill., tomato
Mentha spicata L., spearmint
Mikania scandens (L. F.) Willd., climbing hempweed
Myrica cerifera L., waxmyrtle
Nerium oleander L., oleander
Nymphaea sp., water lily
Pallavicinia leyllii (W. J. Hook.) Gray ex Trev.
Panicum abscissum Swallen, cutthroat grass
Panicum hemitomum Schult., maidencane
Paspalum urvillei Steud., vasey grass
Phaseolus vulgaris L., bean
Pinus caribaea Morelet, Cuban pine
Pinus elliottii Engelm., slash pine
Pinus palustris Mill., longleaf pine
Pinus taeda L., loblolly pine
Podocarpus macrophyllus (Thunb.) D. Don, Japanese yew
Polygonum persicaria L., smartweed
Pontederia sp.
Quercus virginiana Mill., southern live oak
Rumohra adiantiforme (G. Forst.) Ching, leatherleaf fern
Roystonea regia (HBK) O. F. Cook, Cuban royal palm
Sabal palmetto (Walt.) Lodd. ex. Schult. and Schult. f., cabbage palmetto
Saccharum officinarum L., sugarcane
Salix sp., willow
Scirpus validus Vahl, soft-stem bulrush
Scoparia dulcis L., sweet broom
Sesbania exaltata (Raf.) V. L. Cory, sesbania
Sesbania grandiflora (L.) Poir., scarlet wistaria
Solidago sp., goldenrod
Solanum tuberosum L., potato
Sorghum bicolor (L.) x *S. sudanense* (Piper) Stapf, sorghum-Sudan grass hybrid
Stenotaphrum secundatum (Walt.) O. Kuntze, St. Augustine grass
Tagetes patula L., French marigold
Taxodium sp., cypress
Tradescantia fluminensis Vell., Wandering Jew
Vaccinium corymbosum L., highbush blueberry
Vaccinium elliottii Chapm., Elliott's blueberry
Vaccinium fuscatum Ait., blueberry
Vaccinium macrocarpon Ait., cranberry
Zea mays L., corn
Zoysia japonica Steud., zoysia grass

Many of the plants in the above list of hosts of *D. heterocephalus* came from unpublished records in the files of the Nematology Bureau, Division of Plant Industry, Florida Department of Agriculture and Consumer Services, Gainesville. Other listings are from Esser and Rhoades, 1976; Gerber and Smart, 1987; Gerber et al., 1987; Perry, 1953; Rhoades, 1980a, 1983; Steiner, 1949; and Tarjan, 1952.

C. Pathogenicity and Symptoms

The pathogenicity of *D. heterocephalus* and the symptoms expressed by its host plants are well known. Those of some other species of *Dolichodorus* are not so well known but from the information available appear to be very similar to those of *D. heterocephalus*. The information given here is for the latter species.

The nematode feeds from the surface of roots by injecting its long stylet into cells mostly at root tips. If feeding occurs in the piliferous area for 1–4 hr, the root becomes constricted in that area. Normal growth of the root tip is inhibited in the vicinity of a feeding site and the elongation of cells on the nonparasitized areas causes the root to curve. Root tips may appear enlarged because the tissues mature right to the tip, and sometimes on some plants galls appear on root tips. Root tips fed on for more than 2 days cease to grow (Paracer et al., 1967, Fig. 8). When an infested plant produces new root initials, the nematode feeds

FIGURE 8   The aquatic plant *Anubias nana*. As new root initials were formed, they were fed on by *Dolichodorus miradvulvus*. Hence the plant was totally devoid of a root system.

on them stopping their growth also. The result is a skeleton root system with virtually no feeder roots and few if any secondary roots. This is often referred to as stubby root or coarse root or both. Root tissue surrounding the feeding site is destroyed and discolored. Sometimes the nematode feeds on the hypocotyl as well as the roots (Perry, 1953). Above-ground the plants are stunted and chlorotic due to the lack of an adequate root system.

Using a mixture of nematodes containing 77% *D. heterocephalus*, Tarjan et al. (1952) reported that celery plants that received 1000 *D. heterocephalus* (77% of nematode mixture) showed symptoms as early as 3 weeks after inoculation. Plants were noticeably stunted and chlorotic with chlorosis and necrosis especially evident in the older leaves. Based on observations in the field and on controlled experiments using pure populations of *D. heterocephalus*, Perry (1953) reported that seedlings of celery set in the field made little or no growth, and new root production was almost nonexistent. Seeds of snap beans planted in infested soil failed to germinate, were swollen but did not decay. Nematodes were present both inside and outside the seed coat and apparently fed on the embryo, stopping growth and germination. Seeds of corn germinated and grew but the roots were injured giving a stubby appearance. Tomato roots were almost totally destroyed. Roots of bell pepper were damaged but less severely than those of tomato, corn, and beans.

D. Crop Losses Due to *Dolichodorus heterocephalus*

Data for crop losses due to *D. heterocephalus* are scarce, which is somewhat surprising since its damage is similar to that caused by the sting nematode. The reason may be that the awl nematode is not as widely distributed as the sting nematode. Perry (1953) reported a yield loss of more than 50% of celery in one infested field. Paracer et al. (1967) showed an 81% loss in dry weight of corn roots and a 58% loss in dry weight of garden balsam roots after 4 months with an initial inoculum level of 50 awl nematodes (calculated from data). Nematode populations increased 17-fold on corn and 14-fold on garden balsam. Rhoades (1983) reported that spearmint harvested at intervals over a 6-month period from February through August showed a loss of 24.5%; however, by July yield was about half that from control plants (calculated from data). For four successive plantings of corn with each planting growing for 6 weeks, Rhoades (1985) reported a loss of only 4% fresh plant weight (calculated from data) for the awl nematode when starting with 500 or 1000 nematodes as inoculum. The awl nematode populations had increased 22 times when starting with 500 and 26 times when starting with 1000. The only plausible explanation for the small loss reported by Rhoades, which does not coincide with the losses reported by others, is that his plants received such good care in the greenhouse that they were able to compensate for the damage caused by the nematode.

E. Control

Nematicides may be used to control the awl nematodes, and most of the materials that have been tried appear to be satisfactory. However, since few materials are cleared for use, one should check with the county agent, farm advisor, or extension specialist to determine what materials are available and effective.

II. BIOLOGY AND TAXONOMY

A. Biology

The awl nematode was first described from lakes as a freshwater nematode. Its natural habitat appears to be either aquatic or wet soil. In Florida, where the nematode is found most often, it is not widespread but is usually found in the wet soil around the edges of lakes, streams, and ponds or in the water. Water hazards on golf courses are usually excellent places to find the nematode. When in other locations, it is often in soil that is flooded or kept moist through irrigation as occurred in the area of Sanford, Florida where it was recorded damaging celery. According to Christie (1959), speaking of the Sanford area:

> Several years ago some of the growers of this area adopted the practice of spreading over their fields a silt-like soil obtained from the banks of the St. Johns River. It seems probable that this was the source of the infestation.

Also, the nematode has been pumped onto seedling nurseries from ponds used as a source of irrigation water (Esser, 1979). However, it is found sometimes in areas remote from bodies of water, so would not appear to be dependent on wet conditions. Perhaps this nematode is in a state of evolution from aquatic to terrestrial.

The life span of the nematode is not known but it did not survive 3 months of fallow in greenhouse pots (Paracer et al., 1967). Populations of the nematode normally are maintained on plants in a greenhouse, but Paracer and Zuckerman (1967) cultured it successfully on corn root callus tissue.

The nematode is classified as an ectoparasite and feeds mostly near root tips but some feeding occurs in the piliferous region. One or more nematodes may feed at one site for several days. The stylet penetrates the root 20–50 μm, and once it is inserted muscular contractions of the median esophageal bulb begin. Otherwise the nematode is quiescent during feeding (Paracer et al., 1967).

B. Taxonomy

Historical: The genus *Dolichodorus* was established by Cobb in 1914 when he described the genus and the type species *Dolichodorus heterocephalus* from Douglas Lake in Michigan and Silver Springs in Florida (Cobb, 1914). Steiner (1949) was the first to refer to the nematode as the awl nematode alluding to its "awl-shaped, long, buccal stylet." He reported it feeding on celery roots near Sanford, Florida. For 43 years, 1914–1957, the genus contained only the one species until Allen (1957) described *D. obtusus*. (This species was later transferred to *Neodolichodorus*). The next year, Golden (1958) described *D. similis* and currently the genus contains a total of 15 species.

Andrássy (1976) erected a new genus, *Neodolichodorus*, to contain those dolichodorids in which the female has a bluntly rounded tail terminus and four incisures in the lateral field. Thus, *D. obtusus* becomes the type species of *Neodolichodorus*.

The taxonomic relationship of *Dolichodorus* has been discussed by several authors. Filipjev and Schuurmans Stekhoven (1941) and Thorne (1949) placed the genus in the family Tylenchidae. Chitwood (1950) erected the subfamily Dolichodorinae in the family Criconematidae with the single character, a very long stylet, to contain the two genera *Dolichodorus* and *Belonolaimus*. Loof (1958) discussed the taxonomic status of the subfamily Dolichodorinae and suggested that the subfamily be rejected and that *Dolichodorus* be transferred back to Tylenchidae in the subfamily Tylenchinae, and *Belonolaimus* to

Hoplolaiminae. Skarbilovich (1959) raised the subfamily Dolichodorinae to family rank. Goodey (1963) placed the subfamily Dolichodorinae in the family Hoplolaimidae; Allen and Sher (1967) included it in the family Tylenchidae.

Siddiqi (1970) redefined the family Dolichodoridae and included in it the subfamilies Dolichodorinae, Trophurinae, Tylenchorhynchinae, and Tylodorinae. Golden (1971) and Andrássy (1976) considered Dolichodoridae as a small family containing only *Dolichodorus* and *Brachydorus*. Fotedar and Handoo (1978) followed Golden's concept and added a new subfamily, Dolichorhynchinae, to contain the new genus *Dolichorhynchus,* the male of which has a bursa similar to that of *Dolichodorus.*

Siddiqi (1986) again considered the family Dolichodoridae and included in it the subfamilies Belonolaiminae, Dolichodorinae, Macrotrophurinae, Meiodorinae, Merliniinae, Telotylenchinae, Trophurinae, and Tylenchorhynchinae. He also raised the family to the superfamily level, Dolichodoroidea.

Luc and Fortuner (1987) considered Dolichodoridae as a family containing *Dolichodorus* Cobb, 1914, *Neodolichodorus* Andrásssy, 1976, and *Brachydorus* De Guiran and Germani, 1968 (which they considered a genus dubium).

Family Dolichodoridae Chitwood, 1950

Diagnosis (after Luc and Fortuner, 1987)

Tylenchoidea. Large, slender nematodes with cylindroid bodies. No secondary sexual dimorphism. Lateral field with three or four lines (incisures). Deirids absent. Labial region distinctly offset, annulated (rarely smooth). Labial sclerotization strong, with a very thick basal plate and thick arches. Amphid apertures seen as small slits. Stylet generally well developed (up to 150 μm); cone markedly longer than shaft. Esophagus with procorpus fused with median bulb, strong valve, short isthmus, and pyriform glandular region not overlapping the intestine. Female tail rounded to hemispheric, or with spikelike extension, rarely elongate-conoid. Female with two genital branches; columned uterus with four rows of cells; vagina vera heavily sclerotized. Male caudal alae terminal, winglike, trilobed. Amphimictic reproduction. Obligate migratory ectoparasites of plant roots.

Type and only subfamily: Dolichodorinae Chitwood, 1950

Subfamily Dolichodorinae
Diagnosis. Dolichodoridae (as in the family)

Type genus: *Dolichodorus* Cobb, 1914
Other genera: *Neodolichodorus* Andrássy, 1976
                      syn. *Plesiodorus* Siddiqi, 1976
Brachydorus de Guiran and Germani, 1968

Key to genera:

1. Stylet less than 40 μm, tail attenuated .... *Brachydorus*
    Stylet more than 50 μm, tail conoid, spicate, rarely attenuated 2
2. Female tail conically pointed, lateral field with three incisures ..... *Dolichodorus*
    Female tail bluntly rounded, lateral field with four incisures .... *Neodolichodorus*

Genus *Dolichodorus Cobb,* 1914 (Figs. 2, 9)

Diagnosis

Dolichodorinae. Female: Body long, slender. Labial region offset by constriction, striated, four-lobed in face view. Labial disk present. Amphids laterally directed slits. Stylet long (50–160 µm). Esophagus not overlapping intestine generally (slightly overlapping intestine in some cases). Lateral field with three incisures, areolated. Vulva at midbody; gonads two. Female tail symmetrical, usually tapering suddenly to form spicate end. Phasmid on tail. Male: Body smaller than female, bursa large, trilobed. Phasmids on tail.

Type species: *Dolichodorus heterocephalus* Cobb, 1914
Other species: *D. aestuarius* Chow and Taylor, 1978
                D. aquaticus Doucet, 1985
                D. cobbi Golden, Handoo, and Wehunt, 1986
                D. grandaspicatus Robbins, 1982
                D. kishansinghi Jairajpuri and Rahmani, 1979
                D. longicaudatus Doucet, 1981
                D. marylandicus Lewis and Golden, 1981
                D. minor Loof and Sharma, 1975
                D. miradvulvus Smart and Khuong, 1985
                D. nigeriensis Luc and Caveness, 1963
                D. profundus Luc, 1960
                D. pulvinus Khan, Seshadri, Weischer, and Mathen, 1971
                D. silvestris Gillespie and Adams, 1962
                D. similis Golden, 1958

Key to species:

1.   Excretory pore anterior to median bulb, usually opposite base of stylet . . . .
    . . . . *D. pulvinus*
    Excretory pore more posteriorly 2
2.   Cuticle with deep grooves anterior and posterior to vulva . . . . *D. miradvulvus*
    Cuticle without deep grooves anterior and posterior to vulva 3
3.   Female tail tapering gradually from anus to tip; male tail with large hypoptygma . .
    . . . . *D. longicaudtus*
    Female tail narrowing suddenly between anus and tip; male tail without hypotygma
    4
4.   Excretory pore opposite median bulb 5
    Excretory pore posterior to median bulb 7
5.   Female T/ABW ratio 3; body less than 2 mm long . . . . *D. nigeriensis*
    Female T/ABW ratio 2 or less; body more than 2 mm long 6
6.   Lip region cushion-shaped; anterior cuticle tessellate; stylet length less than 120 µm
    . . . . *D. profundus*
    Lip region rounded; anterior cuticle not tessellate; stylet length more than 130 µm
    . . . . *D. silvestris*
7.   Excretory pore usually opposite middle of basal bulb . . . . *D. similis*
    Excretory pore opposite isthmus 8
8.   Female tail 88–140 µm 9
    Female tail less than 75 µm 11

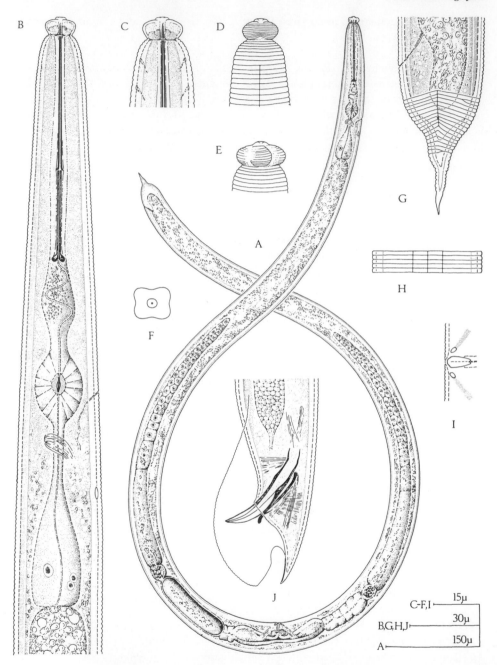

FIGURE 9   *Dolichodorus heterocephalus*: (A) Female, entire. (B) Female, anterior region. (C, D) Female heads, lateral views. (E) Female head, not quite lateral. (F) Outline of female head and lip region en face. (G) Female tail. (H) Lateral field. (I) Lateral view of vulval region showing sclerotization of vagina and opposing pieces to which muscles are attached (somewhat diagrammatic). (J) Male tail. (After Orton Williams, 1974.)

9.   Female tail cylindrical to the first one-sixth, then narrowing suddenly to form a spike
     .... *D. grandaspicatus*
     Female tail narrowing gradually the first one-third, then suddenly to form a spike .
     10
10. Bursa extending one tail length beyond tail tip .... *D. kishansinghi*
     Bursa not extending quite so far beyond tail tip .... *D. aquaticus*
11. Body length less than 2 mm .... *D. minor*
     Body length more than 2 mm 12
12. Stylet length 62–76 µm .... *D. aestuarius*
     Stylet length 83 µm or more 13
13. Stylet length 83–92 µm; tail short, abruptly reduced in diameter, acuminate terminus
     often spicate .... *D. marylandicus*
     Stylet length 99 µm or more; tail smoothly tapering or irregularly conoid, acute
     terminus .... *D. heterocephalus*

Note: The species *Dolichodorus kirjanovae* Sagitov, 1973 was transferred to the genus *Geocenamus* by Fortuner and Luc (1987).

Species *Dolichodorus heterocephalus* Cobb, 1914

Measurements (in µm) (after Golden et al., 1986)

Specimens from Sanford, Florida. Lectotype (female): Length = 3573; width = 81.7; a = 43.7; b = 11.7; c = 38.8; V = 51.5; stylet = 95; stylet conus = 55; stylet shaft = 40; anterior end to center of median bulb = 192; DGO = ?; anterior end to excretory pore = 224; tail terminus to phasmid = 92; ABW = 46; tail = 92. Female (16): Length = 2366; width = 54.2; a = 43.6; b = 9.6; c = 39.4; V = 54.2; stylet = 104.2; stylet conus = 64.2; stylet shaft = 39.9; DGO = 4.5; anterior end to center of median bulb = 154; tail terminus to phasmid = 41.7; ABW = 37.7; tail = 60.9; hemizonid to excretory pore = 39.8. Paralectotype (male, Cobb original, from Florida): Length = 2878; a = 41.7; b = 9.7; stylet = 100; stylet conus = 60; stylet shaft = 40; DGO = ?; anterior end to center of median bulb = 185; spicules = 56; gubernaculum = 26; hemizonid to excretory pore = 27.9. Male (8): Length = 1987; width = 45.1; a = 44; b = 8.1; c = 63.9; stylet = 100.6; stylet conus = 62.3; stylet shaft = 37.9; DGO = 4.7; anterior end to center of median bulb = 152.3; excretory pore = 165.3; spicules = 44.1; gubernaculum = 25.6; ABW = 20.6; tail = 26.4.

Description

Female: Body long, cylindrical; cuticle annulated. From the face view the head is seen to be somewhat flattened dorsoventrally. Labial region offset by constriction with four lobes, the dorsal and ventral divisions are distinct but the lateral one is not. Labial disk circular. The amphid opening elongate laterally. Stylet long, usually more than 100 µm with strong backward-sloping knobs. Esophagus with the procorpus somewhat swollen at the center containing a wavy lumen. Metacorpus muscular, well developed containing an ellipsoid valvular apparatus. Isthmus narrow, as long as the corresponding body width, with nerve ring located from middle to its base. Postcorpus elongate-pyriform, glandular. Esophago-intestinal valve present. Excretory pore opposite isthmus, usually near the nerve ring. Hemizonid present, posterior to the excretory pore. Vulva a transverse slit with a pair of sclerotized pieces in lateral view anterior and posterior to vagina. Vagina sclerotized. Gonads paired, outstretched; spermatheca prominent. Tail irregular conoid tapering at about midtail to form a spike with acute terminus. Lateral field with three incisures, areolated; according to Williams (1974), anteriorly the middle line first appears a few annules behind the head, the outer lines arising some annules farther back. Posteriorly, the outer lines disappear leaving the central incisure on the tail. Phasmid on tail, almost at the end of the central incisure.

Male: Slightly smaller than female but similar except for reproductive structures. Tail conoid with thick cuticle, as long as the anal body width. Bursa trilobed, well developed with bifurcate tail tip. Spicules slightly arcuate. Gubernaculum with distal end reflexed.

Genus *Neodolichodorus* Andrássy, 1976 (Fig. 10)

Diagnosis (after Luc and Fortuner, 1987)

Dolichodorinae. Female: Labial region rounded, striated (rarely smooth), weakly offset, rounded to roughly quadrangular in en face view; labial disk generally not prominent; amphid aperture small slit dorsally, ventrally directed. Lateral field with four lines. Stylet long (50–140 μm), strong. Tail short, hemispheric, rarely conical. Phasmids adanal or slightly anterior to anus.

Male: Spicules not flanged or weakly flanged. Gubernaculum apparently not protruding.

Type species: *Neodolichodorus obtusus* (Allen, 1957) Andrássy, 1976
                 syn. *Dolichodorus obtusus* Allen, 1957
                      Plesiodorus obtusus (Allen, 1957) Siddiqi, 1976
Other species: *N. adelaidensis* (Fisher, 1964) Siddiqi, 1977
                 syn. *Dolichodorus adelaidensis* Fisher, 1964
                      Plesiodorus adelaidensis (Fisher, 1964) Siddiqi, 1976
                 N. arenarius (Clark, 1963) Siddiqi, 1977
                 syn. *Dolichodorus arenarius* Clark, 1963
                      Plesiodorus arenarius (Clark, 1963) Siddiqi, 1976
                 N. brevistilus (Heyns and Harris, 1973), Siddiqi, 1977
                 syn. *Dolichodorus brevistilus* Heyns and Harris, 1973
                      Plesiodorus brevistilus (Heyns and Harris, 1973) Siddiqi, 1976
                 N. cassati (Luc and Dalmasso, 1971) Siddiqi, 1977
                 syn. *Dolichodorus cassati* Luc and Dalmasso, 1971
                      Plesiodorus cassati (Luc and Dalmasso, 1971) Siddiqi, 1976
                 N. leiocephalus Doucet, 1981
                 N. rostrulatus (Siddiqi, 1976) Siddiqi, 1977
                 syn. *Plesiodorus rostrulatus* Siddiqi, 1976

Key to species:

1. Labial region smooth, tail conoid, narrowing suddenly in posterior half of the tail to
   form a spike .... *N. leiocephalus*
   Labial region striated, tail hemispheric 2
2. Stylet 47–56 μm long .... *N. brevistilus*
   Stylet 91 μm or more 3
3. Labial region with two annules, but these annules not complete as seen under the SEM
   .... *N. rostrulatus*
   Labial region with more than five annules 4
4. Female stylet 132–136 μm long, phasmid on tail .... *N. obtusus*
   Female stylet 116 μm or less, phasmids preanal 5

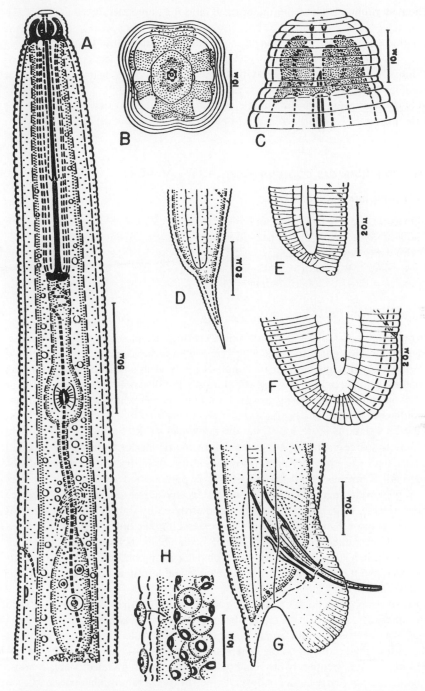

FIGURE 10 *Neodolichodorus obtusus*: (A) Anterior end of female. (B) Face view. (C) Lip region. (D) Larval tail. (E) Preadult female tail. (F) Female tail. (G) Male tail. (H) Parasite, internal and external. (After Allen, 1957.)

5. Labial region slightly offset from the body, *b* ratio 5.3 and *c* ratio 73–81 ....
    .... *N. cassati*
    Labial region well offset from the body by constriction, *b* ratio 7–9 and *c* ratio 71 or less
    6

6. Middle bursal lobe rounded at the end with spinule-like processes on its margin; lateral
    field regularly areolated .... *N. arenarius*
    Middle bursal lobe slightly bifurcate at end, no spinule–like processes on its margin;
    lateral field irregularly areolated (central field blank except on tail) ....
    .... *N. adelaidensis*

Species *Neodolichodorus obtusus* (Allen, 1957) Andrássy, 1976

Measurements (μm) (after Allen, 1957)

Holotype (female): Length = 2500; a = 35; b = 7; c = 58; V = 57; stylet = 133. Female (14): Length = 1900–2700; a = 35–45; b = 6.2–8.7; c = 45–65; V = 51–60; stylet = 132–136. Allotype (male): Length = 2250; a = 42; b = 7.5; c = 48; stylet = 138; spicules = 73; gubernaculum = 24. Male (8): Length = 1800–2260; a = 36–45; b = 7; c = 47–51; stylet = 120–138; spicules = 62–73; Gubernaculum = 21–26.

Description (after Allen, 1957)

Female: Lip region globular, set off by constriction, bearing seven to eight transverse striae. Lip cap conspicuous, hexagonal in shape when viewed en face. Amphid openings oval in shape, located at lateral margins of lip cap. Amphids easily visible in face views, and seen without difficulty in dorsal or ventral views of lip region. Papillae not visible from lateral or face view. Oral opening surrounded by six small refractive pieces which resemble the inner circlet of papillae but located in a position which precludes their being papillae. The most anterior of the lip striations is six-lobed, but the remainder of the lip region is obscurely four-lobed. Cuticle coarsely striated, striae about 4 μm apart anterior to excretory pore and 2–3 μm apart posterior to excretory pore. Body cylindrical from base of esophagus to obtusely rounded tail. Terminus of tail striated. Excretory pore located at beginning of posterior bulb of esophagus; hemizonid located a short distance posterior to opening of excretory duct. Lateral incisures, four, with the outer lines interrupted by transverse body striae at regular intervals. In most instances alternate striae extend into the outer lateral fields. Central lateral line not interrupted by striae. In region of median bulb, incisures reduced to three which extend to within a short distance of lip region. Deirids not observed. Phasmid openings small, located about midway between anal opening and terminus of tail and ventral to center of middle lateral line.

Corpus of esophagus thick expanding gradually into median bulb. Median bulb roughly oval anteriorly, but reduced rather abruptly posteriorly. Isthmus a slender tube slightly longer than one body width, expanding rather abruptly into the elongate posterior bulb. Esophago-intestinal valve or cardia conspicuous. Intestine overlaps posterior end of esophageal bulb, which is frequently obscured by dense intestinal granules. Vulva a transverse slit; vagina extends nearly half of body width interiorly. Uteri muscular; spermathecae not seen. Oocytes arranged in a single row.

Male: Similar to female. Phasmids opening between middle lateral lines near cleft of caudal alae. Caudal alae conspicuously lobed, bearing striations and enveloping terminus of tail. Viewed from ventral position, caudal alae smoothly rounded at terminus. Spicules massive, heavily sclerotized. Gubernaculum linear with distal end slightly enlarged. Viewed from dorsal side gubernaculum consists of three pieces at its proximal end; these pieces are fused at about one-third the distance from the proximal end.

Genus *Brachydorus* de Guiran and Germani, 1968 (Fig. 11)

Diagnosis (after de Guiran and Germani)

Dolichodorinae. Labial region rounded, subcircular in cross-section. Cephalic sclerotization prominent. Stylet strong, three times as long as width of labial region; knobs rounded. Procorpus long, cylindrical, enlarged at its base; median bulb pyriform, well developed; basal bulb oval to pyriform. Female with two gonads, vulva near middle of body. Lateral field with four incisures, not areolated. Tail long, attenuated. Male tail trilobed. Phasmid in both sexes post-anal. No sexual dimorphism in anterior part of the bodies.

Type species: *Brachydorus tenuis* de Guiran and Germani, 1968
Other species: *B. swarupi* Koshy, Raski, and Sosamma, 1981

Key to species:

1. Stylet 20–23 μm long, female body 1030–1320 μm .... *B. tenuis*
2. Stylet 28–32 μm long, female body 1870–2340 μm .... *B. swarupi*

Species *Brachydorus tenuis* de Guiran and Germani, 1968

Measurements (in μm) (after de Guiran and Germani, 1968)

Holotype (female): Length = 1150; $a = 41$; $b = 6.9$; $c = 9.5$; $V = 52$. Female (23): Length = 1180; $a = 42.3$; $b = 7$; $c = 9.7$; $V = 50.5$; stylet = 21.6. Allotype (male): Length = 1060; $a = 41.5$; $b = 6.2$; $c = 48$. Male (23): Length = 990; $a = 41.2$; $b = 6.1$; $c = 42.3$; stylet = 21.7; spicules = 22–39; gubernaculum = 9–12.

Description (after de Guiran and Germani, 1968)

Female: Body slightly curved ventrally; cylindrical; annulated. Annules averaging 2 μm wide at midbody. Cuticle with two layers, exterior layer thin and hyaline, interior layer thicker, also annulated. Lateral field with four incisures not areolated forming three smooth bands, expanding two-fifths of body width at midbody.

Labial region rounded, flat anteriorly, slightly offset, with a slight incisure at the apex to form one obscure annule. Cephalic sclerotization well developed; basal plate thick and strongly sclerotized. Vestibule wall also sclerotized, especially in anterior part. Cross-section of labial region circular at base and subhexagonal at apex. Stylet robust, 20–23 μm long, about three times as long as width of labial region; stylet knobs subspherical. Procorpus of esophagus long and cylindrical, enlarged at base. Dorsoesophageal gland opening about 4 μm posterior to base of stylet. Median bulb pyriform. Isthmus narrow, cylindrical, about half as long as procorpus. Basal bulb oval to pyriform. Cardia in basal bulb. Nerve ring located in the anterior third of the isthmus. Excretory pore, 124–152 μm from anterior end; hemizonid, two annules in length, positioned anterior to excretory pore. Gonads paired, amphidelphic, outstretched; oocytes in one row, sometimes two rows; spermatheca oval, 15 × 30 μm, filled with spherical sperms about 2 μm in diameter.

Tubular structure convoluted, accompanying intestine throughout its length; postanal sac present. Tail long, 104–133 μm, attenuated, narrows gradually to posterior end.

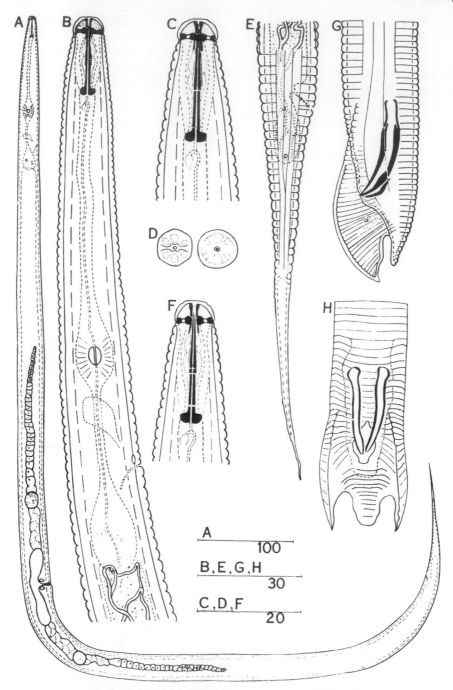

FIGURE 11 *Brachydorus tenuis*: (A–E) Female. (A) Entire female. (B) Anterior region. (C) Head. (D) Cross-section at top and base of lip region. (E) Posterior region. (F–H) Male. (F) Head. (G) Posterior region, lateral view. (H) Posterior region, ventral view. (After De Guiran and Germani, 1968.)

Anal body width 5.5–7.8 μm . Phasmids dotlike situated at 0.3–1.5 anal body widths posterior to anus.

Male: Similar to female. Stylet 19.5–23 μm long. Excretory pore situated at 112–142 μm from anterior end. Testis single, outstretched; spicule curved 22–39 μm long. Gubernaculum 9–12 μm long. Bursa trilobed. Lateral field with longitudinal incisures.

ACKNOWLEDGMENT

We thank Dr. K. R. Langdon, Botanist, Division of Plant Industry, Florida Department of Agriculture and Consumer Services, Gainesville, for assistance with botanical names.

REFERENCES

Abu–Gharbieh, W. I., and Perry, V. G. 1970. Host differences among Florida populations of *Belonolaimus longicaudatus* Rau. *J. Nematol.* 2: 209–216.

Allen, M. W. 1957. A new species of the genus *Dolichodorus* from California (Nematoda: Tylenchida). *Proc. Helm. Soc. Wash.* 24: 95–98.

Allen, M. W., and Sher, S. A. 1967. Taxonomic problems concerning the phytoparasitic nematodes. *Ann. Rev. Phytopathol.* 5: 246–247.

Andrássy, I. 1976. *Evolution as a Basis for the Systematization of Nematodes.* Pitman, London.

Barker, K. R. 1968. Seasonal population dynamics of *Belonolaimus longicaudatus, Meloidogyne incognita, Pratylenchus zeae, Trichodorus christiei* and *Tylenchorhynchus claytoni. Nematologica* 14: 2–3.

Barker, K. R., Hussey, R. S., and Yang, H. 1975. Effects of light intensity and quality on reproduction of plant parasitic nematodes. *J. Nematol.* 7: 364–368.

Bistline, F. W., Collier, B. L., and Dieter, C. E. 1967. Tree and yield response to control of a nematode complex including *Belonolaimus longicaudatus* in replanted citrus. *Nematologica* 13: 137–138.

Boyd, F. T., and Perry, V. G. 1969. The effect of sting nematode on establishment, yield, and growth of forage grasses on Florida sandy soils. *Proc. Soil Crop Sci. Soc. Fla.* 29: 288–300.

Boyd, F. T., and Perry, V. G. 1971. Effect of seasonal temperatures and certain cultural treatments on sting nematodes in forage grass. *Proc. Soil Crop Sci. Soc. Fla.* 30: 360–365.

Boyd, F. T., Schroder, V. N., and Perry, V. G. 1972. Interaction of nematodes and soil temperature on growth of three tropical grasses. *Agron. J.* 64: 497–500.

Brodie, B. B. 1976. Vertical distribution of three nematode species in relation to certain soil properties. *J. Nematol.* 8: 243–247.

Brodie, B. B., and Quattlebaum, B. H. 1970. Vertical distribution and population fluctuations of three nematode species as correlated with soil temperature, moisture and texture. *Phytopathology* 60: 1286 (Abstr.)

Brodie, B. B., Good, J. M., and Jaworski, C. A. 1970. Population dynamics of plant nematodes in cultivated soil: Effect of summer cover crops in old agricultural land. *J. Nematol.* 2: 147–151.

Brodie, B. B., Good, J. M., and Marchant, W. H. 1970. Population dynamics of plant nematodes in cultivated soil: Effect of sod-based rotations in Tifton sandy loam. *J. Nematol.* 2:135–138.

Brooks, A. N. 1954. Sting nematode, *Belonolaimus gracilis* Steiner. *Proc. Soil Crop Sci. Fla.* 14:157–158.

Brooks, A. N., and Christie, J. R. 1950. A nematode attacking strawberry roots. *Proc. Fla. State Hort. Soc.* 63:123–125.

Chitwood, B. G. 1950. An outline classification of the Nematoda. In *An Introduction to Nematology. I. Anatomy.* Chitwood, B. G., and Chitwood, M. B., eds. Monumental Printing, Baltimore, pp. 12–27.

Chitwood, B. G., and Chitwood, M. B. 1950. *An Introduction to Nematology. I. Anatomy.* Monumental Printing, Baltimore.

Chow, F. H., and Taylor, A. L. 1978. *Dolichodorus aestuarius* n. sp. (Nematoda: Dolichodoridae). *J. Nematol.* 10: 201–204.

Christie, J. R. 1953. The sting nematode can be controlled by soil fumigation. *Down to Earth* 9: 8–9.

Christie, J. R. 1959. *Plant Nematodes, Their Bionomics and Control.* Univ. Fla., Agr. Exp. Stat., Gainesville.

Christie, J. R., Brooks, A. N., and Perry, V. G. 1952. The sting nematode, *Belonolaimus gracilis,* a parasite of major importance on strawberries, celery, and sweet corn in Florida. *Phytopathology* 42: 173–176.

Christie, J. R., Good, J. M., and Nutter, G. C. 1954. Nematodes associated with injury to turf (*Belonolaimus gracilis*). *Proc. Soil Crop Sci. Soc. Fla.* 14: 167–169.

Clark, W. C. 1963. A new species of *Dolichodorus* (Nematoda: Tylenchida) from coastal dune sands. *N. Zealand J. Sci.* 6: 531–534.

Cobb, N. A. 1914. The North American freeliving freshwater nematodes. *Trans. Am. Microscop. Soc.* 33: 69–134.

Colbran, R. C. 1960. Studies of plant and soil nematodes. 3. *Belonolaimus hastulatus, Psilenchus tumidus,* and *Hemicycliophora labiata,* three new species from Queensland. *Qd. J. Agr. Anim. Sci.* 17: 175–181.

Cooper, W. E., and Brodie, B. B. 1963. Correlation between *Fusarium* wilt indices of cotton varieties with root-knot and sting nematodes as predisposing agents. *Phytopathology* 52: 6 (Abstr.).

Cooper, W. E., Wells, J. C., and Sasser, J. N. 1959. Sting nematode control on four crops with pre- and postplant applications of Nemagon. *Phytopathology* 49: 316 (Abstr.).

D'Errico, F. P., Lamberti, F., and Fiume, F. 1977. Ritrovamento di *Dolichodorus heterocephalus* Cobb nell'Italia Meriodanale. *Nematologia Mediterranea* 5: 99–101.

Dickerson, O. J., Willis, W. G., Dainello, F. J., and Pain, J. C. 1972. The sting nematode, *Belonolaimus longicaudatus,* in Kansas. *Plant Dis. Reptr.* 56: 957.

Doucet, M. E. 1981. Description de *Dolichodorus longicaudatus* n. sp. et *Neodolichodorus leiocephalus* n. sp. (Nematoda: Tylenchida). *Rev. Nematol.* 4: 191–197.

Doucet, M. E. 1985. Description de *Dolichodorus aquaticus* n. sp. (Nematoda: Dolichodoridae). *Nematologica* 31: 143–150.

Epstein, J., Walawala, J. J., Davide, R. G., and Palis, R. K. 1971. The extent of nematode infestation on ramie in Mindanao and its cellular damage. *Univ. Philippines, Coll. Agr., Laguna, Philippines* 54(9/10): 448–473.

Esser, R. P. 1979. Nematode entry and dispersion by water in Florida nurseries. Nematol. Circ. No. 54. Fla. Dept. Agr. Consum. Ser., Div. Plant Indust., Gainesville.

Esser, R. P., and Rhoades, H. L. 1976. *Dolichodorus heterocephalus* Cobb, 1914 (awl nematode). *Nematol. Circ. No. 14,* Fla. Dept. Agr. Consum. Ser., Div. Plant Indust., Gainesville.

Filipjev, I. N., and Schuurmans Stekhoven, J. H., Jr. 1941. *A Manual of Agricultural Helminthology.* E. J. Brill, Leiden.

Fisher, J. M. 1964. *Dolichodorus adelaidensis* n. sp. and *Paralongidorus eucalypti* n. sp. from S. Australia. *Nematologica* 10: 464–470.

Fisher, K. D., and Hugo, G. 1969. Distribution of *Dolichodorus heterocephalus* Nematoda in the littoral zone of Lake Champlain. *Limnol. Oceanogr.* 14: 617–620.

Fortuner, R., and Luc, M. 1987. A reappraisal of Tylenchina (Nemata). 6. The family Belonolaimidae Whitehead, 1960. *Rev. Nematol.* 10: 183–202.

Fotedar, D. N., and Handoo, Z. A. 1978. A revised scheme of classification to the order Tylenchida Thorne, 1949 (Nematoda). *J. Sci. Univ. Kashmir* 3: 55–82.

Gerber, K., and Smart, G. C., Jr. 1987. Plant-parasitic nematodes associated with aquatic vascular plants. In *Vistas on Nematology.* Veech, J. A., and Dickson, D. W., eds. Soc. Nematol., Hyattsville, Maryland.

Gerber, K., Smart, G. C., Jr., and Esser, R. P. 1987. *A Comprehensive Catalogue of Plant Parasitic Nematodes Associated with Aquatic and Wetland Plants. Bulletin 871 (Technical).* Agr. Exp. Stat., Inst. Food Agr. Sci., Univ. Florida, Gainesville.

Gillespie, W. H., and Adams, R. E. 1962. An awl nematode, *Dolichodorus silvestris* n. sp., from West Virginia. *Nematologica* 8: 93–98.

Golden, A. M. 1958. *Dolichodorus similis* (Dolichodorinae) a new species of plant nematode. *Proc. Helm. Soc. Wash.* 25: 17–20.

Golden, A. M. 1971. Classification of the genera and higher categories of the order Tylenchida (Nematoda). In *Plant Parasitic Nematodes,* Vol. 1. Zuckerman, B. M., Mai, W. F., and Rohde, R. A., eds. Academic Press, New York, pp. 191–232.

Golden, A. M., Handoo, Z. A., and Wehunt, E. J. 1986. Description of *Dolichodorus cobbi* n. sp. (Nematoda: Dolichodoridae) with morphometrics and lectotype designation of *D. heterocephalus* Cobb, 1914. *J. Nematol.* 18: 556–562.

Good, J. M. 1968. Relation of plant parasitic nematodes to soil management practices. In *Tropical Nematology.* Smart, G. C., Jr., and Perry, V. G. eds. Univ. Fla. Press, Gainesville, pp. 113–138.

Goodey, J. B. 1963. *Soil and Fresh Water Nematodes.* Methuen, London.

Graham, T. W. 1952. Nematodes as ectoparasites on tobacco, cotton and other plants. *Phytopathology* 42: 9 (Abstr.).

Graham, T. W., and Holdeman, Q. J. 1953. The sting nematode, *Belonolaimus gracilis* Steiner: A parasite on cotton and other crops in South Carolina. *Phytopathology* 43: 434–439.

Guiran, G. de, and Germani, G. 1968. *Brachydorus tenuis* n. g., n. sp. (Nematoda: Dolichodorinae) associee a *Ravelana madagascariensis* sur la Cote Est Malgache. *Nematologica* 14: 447–452.

Heald, C. M., and Burton, G. W. 1968. Effect of organic and inorganic nitrogen on nematode populations feeding on turf. *Nematologica* 14: 8 (Abstr.).

Heyns, J., and Harris, R. H. G. 1973. *Dolichodorus brevistilus* n. sp. (Tylenchida) a plant parasitic nematode found in sugarcane fields in Natal. *Phytophylactica* 5: 123–126.

Holdeman, Q. L. 1955. The present known distribution of the sting nematode, *Belonolaimus gracilis,* in the coastal plain of the southeastern United States. *Plant Dis. Reptr.* 39: 5–8.

Holdeman, Q. L. 1956. Effectiveness of EDB, DD, and Nemagon in controlling the sting nematode on sandy soil in South Carolina. *Phytopathology* 46: 15 (Abstr.).

Holdeman, Q. L., and Graham, T. W. 1952. The association of the sting nematode with some persistent cotton wilt spots in northeastern South Carolina. *Phytopathology* 42: 283–284 (Abstr.).

Holdeman, Q. L., and Graham, T. W. 1953. The sting nematode breaks resistance to cotton wilt. *Phytopathology* 43: 475 (Abstr.).

Holdeman, Q. L., and Graham, T. W. 1954. Effect of sting nematode on expression of fusarium wilt in cotton. *Phytopathology* 44: 683–685.

Hunt, P. G., Smart, G. C., Jr., and Eno, C. F. 1973. Sting nematode, *Belonolaimus longicaudatus,* immotility induced by extracts of composted municipal refuse. *J. Nematol.* 5: 60–63.

Hutchinson, M. T., and Reed, J. B. 1956. The sting nematode *Belonolaimus gracilis,* found in New Jersey. *Plant Dis. Reptr.* 40: 1049.

Hutchinson, M. T., Reed, J. P., Streu, H. T., DiEdwardo, A. A., and Schroeder, P. H. 1961. Plant parasitic nematodes of New Jersey. *N. Jersey Agr. Exp. Stat.* 796.

Jairajpuri, M. S., and Rahmani, S. A. 1979. *Dolichodorus kishansinghi* n. sp. (Nematoda: Dolichodoridae). *Ind. J. Nematol.* 7: 183–185.

Johnson, A. W. 1970. Pathogenicity and interaction of three nematode species on six bermuda grasses. *J. Nematol.* 2: 36–41.

Johnson, A. W., and Littrell, R. H. 1970. Pathogenicity of *Pythium aphanidermatum* to chrysanthemum in combined inoculations with *Belonolaimus longicaudatus* or *Meloidogyne incognita. J. Nematol.* 2: 255–259.

Johnson, J. T., and Dickson, D. W. 1973. Evaluation of methods and rates of application of three nematicide-insecticides for control of the sting nematode on corn. *Proc. Soil Crop Sci. Soc. Fla.* 32: 171–173.

Kelsheimer, E. G., and Overman, A. J. 1953. Nematodes in lawns. *Fla. Agr. Exp. Stat. Ann. Rep.* 1953: 249.

Kerr, E. D., and Wysong, D. S. 1979. Sting nematode, *Belonolaimus* sp., in Nebraska. *Plant Dis. Reptr.* 63: 506–507.

Khan, E., Seshadri, A. R., Weischer, B., and Mathen, K. 1971. Five new nematode species associated with coconut in Kerala. *Ind. J. Nematol.* 1: 116–127.

Khuong, N. B. 1974. Some nematodes associated with vegetables in North Florida, and pathogenicity of *Belonolaimus longicaudatus* to collard, kale and cauliflower. M.S. Thesis, Dept. Entomol. and Nematol., IFAS, Univ. Fla., Gainesville.

Khuong, N. B., and Smart, G. C., Jr., 1975. The effects of *Belonolaimus longicaudatus* on growth of collard, kale and cauliflower. *Plant Dis. Reptr.* 59: 819–822.

Koshy, P. K., Raski, D. J., and Sosamma, V. K. 1981. *Brachydorus swarupi* sp. n. (Nematoda: Dolichodorinae) from soil about roots of arecanut palm in Kerala State, India. *J. Nematol.* 13: 401–404.

Lewis, S. A., and Golden, A. M. 1981. Description and SEM observation of *Dolichodorus marylandicus* n. sp. with a key to species of *Dolichodorus. J. Nematol.* 13: 128–135.

Loof, P. A. A. 1958. Some remarks on the status of the subfamily Dolichodorinae, with description of *Macrotrophurus profundus* n. g., n. sp. (Nematoda: Tylenchidae). *Nematologica* 3: 301–307.

Loof, P. A. A., and Sharma, R. D. 1975. *Dolichodorus minor* n. sp. (Nematoda: Dolichodoridae) with a key to the genus *Dolichodorus. Rev. Theobroma CEPEC, Itabuna, Brazil* 5: 35–41.

Lopez, R. 1978. *Belonolaimus*, un nuevo integrante de la nematofauna de Costa Rica. *Agronomia Costarricense* 2: 83–85.

Lopez, R. 1979. *Belonolaimus longicaudatus* en la costa pacifica de Costa Rica. *Agronomia Costarricense* 3: 61.

Luc, M. 1960. *Dolichodorus profundus* n. sp. (Nematoda: Tylenchida). *Nematologica* 5: 1–6.

Luc, M., and Caveness, F. E. 1963. *Dolichodorus nigeriensis* n. sp. (Nematoda: Dolichodoridae). *Proc. Helm. Soc. Wash.* 30: 297–299.

Luc, M., and Dalmasso, A. 1971. *Dolichodorus cassati* n. sp. (Nematoda: Tylenchida). *Ann. Zool. Ecol. Anim.* 3: 97–101.

Luc, M., and Fortuner, R. 1987. A reappraisal of Tylenchina (Nemata). 5. The family Dolichodoridae Chitwood, 1950. *Rev. Nematol.* 10: 177–181.

Miller, L. I. 1952. Control of the sting nematode on peanuts in Virginia. *Phytopathology* 42: 470.

Miller, L. I. 1960. The influence of soil components on the survival and the development of sting nematode *Belonolaimus longicaudatus. Va. J. Sci.* 11: 160 (Abstr.).

Miller, L. I. 1972. The influence of soil texture on the survival of *Belonolaimus longicaudatus. Phytopathology* 62: 670–671.

Minton, N. A., and Hopper, B. E. 1959. The reniform and sting nematodes in Alabama. *Plant Dis. Reptr.* 43: 47.

Monteiro, A. R., and Lordello, L. G. E. 1977. Dois novos nematoides encontrados associados a canade-acucar. *Rev. Agr., Piracicaba, Brazil* 52: 5–11.

Myers, R. F. 1979. The sting nematode, *Belonolaimus longicaudatus*, from New Jersey. *Plant Dis. Reptr.* 63: 756–757.

Norton, D. C. 1959. *Plant Parasitic Nematodes in Texas.* Texas Agr. Exp. Stat., College Station, Texas.

Nutter, G. C. 1955. Nematode investigation in turf. *Fla. Agr. Exp. Stat. Ann. Rep.* 1955: 58.

Overman, A. J. 1959. Cover crops as a means of nematode control. *Fla. Agr. Exp. Stat. Ann. Rep.* 1959: 316.

Overman, A. J. 1964a. Nematodes: Their effects and control on vegetable and ornamental crops. *Fla. Agr. Exp. Stat. Ann. Rep.* 1964: 308.

Overman, A. J. 1964b. Identification, pathogenicity and control of stylet bearing nematodes associated with chrysanthemum production in Florida. *Fla. Agr. Exp. Stat. Ann. Rep.* 1964: 312.

Overman, A. J., and Jones, J. P. 1977. Effects of *Belonolaimus longicaudatus, Criconemoides* sp., and *Meloidogyne incognita* on *Verticillium* wilt of tomato. *J. Nematol.* 9: 279–280.

Owens, J. V. 1950. Sting nematode found hostile toward Virginia peanuts. *Peanut J. Nut World* 30: 31.

Owens, J. V. 1951. The pathological effects of *Belonolaimus gracilis* on peanuts in Virginia. *Phytopathology* 41: 29 (Abstr.).

Paracer, S. M., and Zuckerman, B. M. 1967. Monoxenic culturing of *Dolichodorus heterocephalus* on corn root callus. *Nematologica* 13: 478–479.

Paracer, S. M., Waseem, M., and Zuckerman, B. M. 1967. The biology and pathogenicity of the awl nematode, *Dolichodorus heterocephalus. Nematologica* 13: 517–524.

Perry, V. G. 1953. The awl nematode, *Dolichodorus heterocephalus*, a devastating plant parasite. *Proc. Helm. Soc. Wash.* 20: 21–27.

Perry, V. G. 1964. Nematode host-parasite relationships. *Fla. Agr. Exp. Stat. Ann. Rep.* 1964: 115–116.

Perry, V. G., and Norden, A. J. 1963. Some effects of cropping sequence on populations of certain plant nematodes. *Proc. Soil Crop Sci. Soc. Fla.* 23: 116–120.

Perry, V. G., and Rhoades, H. L. 1982. The genus *Belonolaimus*. In *Nematology in the Southern Region of the United States*, Southern Coop. Ser. Bull. 276. Riggs, R. D., and Editorial Committee, Southern Regional Research Committees, S-76, S-154, eds. pp. 144–149.

Perry, V. G., and Smart, G. C., Jr., 1967. Factors influencing survival and pathogenicity of plant parasitic nematodes. *Fla. Agr. Exp. Stat. Ann. Rep.* 1967: 116.

Perry, V. G., and Smart, G. C., Jr., 1969. Factors influencing survival and pathogenicity of nematodes. *Fla. Agr. Exp. Stat. Ann. Rep.* 1969: 93.

Perry, V. G., Smart, G. C., Jr., and Horn, G. C. 1971. Nematode problems in turfgrasses in Florida and their control. *Proc. Fla. State Hort. Soc.* 83: 489–492.

Potter, J. W. 1967. Vertical distribution and overwintering of sting, stunt and ring nematodes in Norfolk sandy loam soil following peanuts. *Nematologica* 13: 150 (Abstr.).

Rau, G. J. 1958. A new species of sting nematode. *Proc. Helm. Soc. Wash.* 25: 95–98.

Rau, G. J. 1961. Amended descriptions of *Belonolaimus gracilis* Steiner, 1949 and *B. longicaudatus* Rau, 1958 (Nematoda: Tylenchida). *Proc. Helm. Soc. Wash.* 28: 198–200.

Rau, G. J. 1963. Three new species of *Belonolaimus* (Nematoda: Tylenchida) with additional data on *B. longicaudatus* and *B. gracilis. Proc. Helm. Soc. Wash.* 30: 119–128.

Rhoades, H. L. 1964. Nut grass as a host of plant nematodes. *Fla. Agr. Exp. Stat. Ann. Rep.* 1964: 204.

Rhoades, H. L. 1965. Nematodes, their effects and control on vegetable and ornamental crops. *Fla. Agr. Exp. Stat. Ann. Rep.* 1965: 212.

Rhoades, H. L. 1967. The effects and control of sting and stubby-root nematodes on onions. *Proc. Fla. State Hort. Soc.* 79: 175–180.

Rhoades, H. L. 1980a. Relative susceptibility of *Tagetes patula* and *Aeschynomene americana* to plant nematodes in Florida. *Nematropica* 10: 116–120.

Rhoades, H. L. 1980b. Reproduction of *Belonolaimus longicaudatus* in treated and untreated muck soil. *Nematropica* 10: 139–140.

Rhoades, H. L. 1983. Effect of *Belonolaimus longicaudatus, Dolichodorus heterocephalus*, and *Pratylenchus scribneri* on growth of spearmint, *Mentha spicata*, in Florida. *Nematropica* 13: 145–151.

Rhoades, H. L. 1985. Effects of separate and concomitant populations of *Belonolaimus longicaudatus* and *Dolichodorus heterocephalus* on *Zea mays. Nematropica* 15: 171–174.

Riggs, R. D. 1961. Sting nematode in Arkansas. *Plant Dis. Reptr.* 45: 392.

Robbins, R. T. 1982. Description of *Dolichodorus grandaspicatus* n. sp. (Nematoda: Dolichodoridae). *J. Nematol.* 14: 507–511.

Robbins, R. T., and Barker, K. R. 1973. Comparison of host range and reproduction among populations of *Belonolaimus longicaudatus* from North Carolina and Georgia. *Plant Dis. Reptr.* 57: 750–754.

Robbins, R. T., and Barker, K. R. 1974. The effects of soil type, particle size, temperature and moisture on reproduction of *Belonolaimus longicaudatus*. *J. Nematol.* 6: 1–6.

Robbins, R. T., and Hirschmann, H. 1974. Variation among populations of *Belonolaimus longicaudatus*. *J. Nematol.* 6: 87–94.

Roman, J. 1964. *Belonolaimus lineatus* n. sp. (Nematoda: Tylenchida). *J. Agr. Univ. Puerto Rico* 48: 131–134.

Ruehle, J. L. 1964. Plant-parasitic nematodes associated with pine species in southern forests. *Plant Dis. Reptr.* 48: 60–61.

Ruehle, J. L. 1967. *Distribution of Plant-Parasitic Nematodes Associated with Forest Trees of the World.* Forest Service, USDA, Southeastern Forest Exp. Stat., Asheville, North Carolina.

Ruehle, J. L. 1973. Nematodes and forest trees—types of damage to tree roots. *Ann. Rev. Phytopathol.* 11: 99–118.

Ruehle, J. L., and Sasser, J. N. 1962. The role of plant-parasitic nematodes in stunting of pines in southern plantations. *Phytopathology* 52: 56–68.

Russell, C. C., and Sturgeon, R. V. 1969. Occurrence of *Belonolaimus longicaudatus* and *Ditylenchus dipsaci* in Oklahoma. *Phytopathology* 59: 118 (Abstr.).

Sasser, J. N., and Cooper, W. E. 1961. Influence of sting nematode control with O-diethyl O-2-pyrazinyl phosphorothioate on yield and quality of peanuts. *Plant Dis. Reptr.* 45: 173–175.

Sasser, J. N., Wells, J. C., and Nelson, A. L. 1967. Correlations between sting nematode populations at three sampling dates following nematocide treatments and the growth and yield of peanuts. *Nematologica* 13: 152 (Abstr.).

Sauer, M. R., Brzeski, M. W., and Chapman, R. N. 1980. Observations on the morphology of Belonolaiminae. *Nematol. Medit.* 8: 121–129.

Schenck, N. C., Mortensen, J. A., and Stover, L. H. 1962. Sting nematode on grape cuttings in Florida. *Plant Dis. Reptr.* 46: 446–447.

Siddiqi, M. R. 1970. On the plant-parasitic nematode genera *Merlinius* gen. n. and *Tylenchorhynchus* Cobb and the classification of the families Dolichodoridae and Belonolaimidae n. rank. *Proc. Helm. Soc. Wash.* 37: 68–77.

Siddiqi, M. R. 1976. New plant nematode genera *Plesiodorus* (Dolichodorinae), *Meiodorus* (Meiodorinae subfam. n.) *Amplimerlinius* (Merliniinae) and *Gracilancea* (Tylodorinae grad. n.). *Nematologica* 22: 390–416.

Siddiqi, M. R. 1977. *Plesiodorus* Siddiqi, 1976 (Nematoda: Dolichodoridae), a junior objective synonym of *Neodolichodorus* Andrássy, 1976. *Nematologica* 23: 265.

Siddiqi, M. R., 1986. *Tylenchida, Parasites of Plants and Insects.* Commonwealth Inst. Parasitol., St. Albans, England.

Siviour, T. R. 1978. Biology and control of *Belonolaimus lolii* n. sp. in turf. *Aust. Plant Pathol. Soc. Newslett.* 7: 37.

Siviour, T. R., and McLeod, R. W. 1979. Redescription of *Ibipora lolii* (Siviour, 1978) comb. n. (Nematoda: Belonolaimidae) with observations on its host range and pathogenicity. *Nematologica* 25: 487–493.

Skarbilovich, T. S. 1959. On the structure of nematodes in the order Tylenchida Thorne, 1949. *Acta Parasit. Pol.* 7: 117–132.

Smart, G. C., Jr. 1969. Plant nematode problems on turf grasses. *Fla. Agr. Exp. Stat. Ann. Rep.* 1969: 91.

Smart, G.. C., Jr., and Khuong, N. B. 1985. *Dolichodorus miradvulvus* n. sp. (Nematoda: Tylenchida) with a key to species. *J. Nematol.* 17: 29–37.

Standifer, M. S. 1959. The pathologic histology of bean roots injured by sting nematodes. *Plant Dis. Reptr.* 43: 983–986.

Standifer, M. S., and Perry, V. G. 1960. Some effects of sting and stubby root nematodes on grapefruit roots. *Phytopathology* 50: 152–156.

Steiner, G. 1949. Plant nematodes the grower should know. *Soil Sci. Soc. Fla., Proc. 1942* IV-B: 72–117.

Tarjan, A. C. 1952. Awl nematode injury on Chinese water chestnuts. *Phytopathology* 42: 114.

Tarjan, A. C., Lownsberry, B. F., Jr., and Hawley, W. O. 1952. Pathogenicity of some plant-parasitic nematodes from Florida soils. I. The effect of *Dolichodorus heterocephalus* Cobb on celery. *Phytopathology* 42: 131–132.

Thames, W. H., Jr. 1959. Plant parasitic nematode populations of some Florida soils under cultivated and natural conditions. *Diss. Abstr.* 20: 1109–1110.

Thorne, G. 1949. On the classification of the Tylenchida, new order (Nematoda: Phasmidia). *Proc. Helm. Soc. Wash.* 16: 37–73.

Tomerlin, A. H., Jr., and Perry, V. G. 1967. Pathogenicity of *Belonolaimus longicaudatus* to three varieties of soybean. *Nematologica* 13: 154 (Abstr.).

Tomerlin, A. H., Jr., and Smart, G. C., Jr. 1969. The influence of organic soil amendments on nematodes and other soil organisms. *J. Nematol.* 1: 29–31 (Abstr.).

Weingartner, D. P., Shumaker, J. R., Dickson, D. W., and Littell, R. C. 1978. Nematode control on Irish potatoes in northeast Florida using a soil fumigant and a nonvolatile nematicide both alone and in combination. *J. Nematol.* 10: 301.

Whitehead, A. G. 1960. *Trichotylenchus falciformis* n. g., n. sp. (Belonolaiminae n. subfam.: Tylenchida Thorne, 1949) an associate of grass roots (*Hyparrhenia* sp.) in Southern Tanganyika. *Nematologica* 4: 279–285.

Williams, R. J. O. 1974. *Dolichodorus heterocephalus*. C. I. H. descriptions of plant-parasitic nematodes. Set 4, No. 56.

15

The Hoplolaiminae

RENAUD FORTUNER *California Department of Food and Agriculture, Sacramento, California*

I. INTRODUCTION

The Hoplolaiminae *sensu* Fortuner (1987a) include eight genera: *Pararotylenchus, Rotylenchus, Scutellonema, Aorolaimus, Hoplolaimus, Helicotylenchus, Antarctylus*, and *Aphasmatylenchus*. They belong to the family Hoplolaimidae together with the forms in the related subfamily Rotylenchulinae. Some Hoplolaiminae, and particularly some species in the genera *Scutellonema* and *Helicotylenchus*, have a worldwide distribution and they are very common on many cultivated plants. Other hoplolaimids are found in more restricted areas: *Aphasmatylenchus* in a few sites in West Africa, *Pararotylenchus* in cool localities in the western United States and the Far East, *Antarctylus* on an island in Antarctica, etc.

Host specificity varies from species to species. Most Hoplolaiminae are found in many different plants: the list of hosts of *Helicotylenchus dihystera* includes grasses, rice, corn, soybean, sorghum, wheat, rye, sugarcane, Cyperaceae, potato, strawberry, peanut, cotton, vegetables, trees and bushes, tea, various ornamentals, and more.

Hoplolaimids accomplish part or all of their cycle in the soil. They are obligate plant parasites, and they are closely associated with plant roots. Some are migratory ectoparasites browsing the surface of roots, sometimes attacking the outer layers. Some are migratory semiendoparasites. They penetrate the cortex and remain there for long periods as semiendoparasites, but they can move out of the roots. A few species are true endoparasites, e.g., several *Hoplolaimus* species, *Scutellonema bradys*, etc.

Members of this family are often considered as second-class parasites. In California, all *Helicotylenchus, Hoplolaimus, Rotylenchus*, and *Scutellonema* are classified as parasites and organisms of little or no economic significance by the California Department of Food and Agriculture. However, some hoplolaimids have been shown to cause heavy damage. *Hoplolaimus columbus* is a major parasite of soybean and cotton in southern United States, *Scutellonema bradys* can cause severe damage on yams, *Helicotylenchus multicinctus* is an important parasite of banana, etc.

II. TAXONOMY OF HOPLOLAIMIDAE

A. Diagnosis of the Family

Hoplolaimidae Filip'ev,* 1934
    syn. Nemonchidae Skarbilovich, 1959
        Aphasmatylenchidae Sher, 1965
        Rotylenchulidae Husain and Khan, 1967

Tylenchoidea. Females vermiform to kidney-shaped; when vermiform, habitus often spiral. Lip region high, typically higher than half the diameter of the basal lip annulus; anterior end with rounded or trapezoidal outline in lateral view, annulated, sometimes with longitudinal striae on basal lip annulus, rarely striae on other lip annuli. Lateral field typically with four lines, sometimes regressed (some *Hoplolaimus* and *Aorolaimus* spp.). Phasmids typically near anus level, rarely on tail, sometimes migrated far anteriorly (*Aorolaimus, Hoplolaimus*), generally small porelike structures, sometimes enlarged into scutella, rarely absent (*Aphasmatylenchus*). Tail typically short, less than two tail diameters long, rarely longer; generally more curved dorsally, sometimes regularly rounded, rarely conical. Caudalids and cephalids generally present; deirids absent.

Labial framework strong, with high arches. Stylet strong, its length typically equals to 2–1/2 to 3 times the diameter of the basal lip annulus. Stylet knobs strong, rounded to indented, sometimes anchor-shaped. Dorsal esophageal gland opening (DGO) at least 4 μm, sometimes more than 20 μm, from the stylet base. Median bulb strong, rounded. Esophageal glands arrangement variable but mostly overlapping the intestine. Esophagointestinal junction a small, triangular structure.

Two genital branches, opposed, outstretched or rarely flexed (*Rotylenchulus*); posterior branch may be degenerated or reduced to a PUS. Columned uterus with three rows of four cells. Epiptygmata and vulval flaps generally present but sometimes inconspicuous.

Males with secondary sexual dimorphism present, with anterior end less developed than in females, sometimes degenerated and nonfunctional. Caudal alae generally enveloping the tail end, rarely stopping short of it (*Rotylenchulus*). Gubernaculum often with titillae.

Type subfamily: Hoplolaiminae Filip'ev, 1934.

Members of the family Hoplolaimidae are recognized by the strong and long stylet, and by the high lip region.

Under the dissecting microscope, typical hoplolaimids are seen (Fig. 1) as medium to large nematodes, with a body elegant to robust, regularly cylindroid. Stylet is well visible with knobs appearing as clearly seen dots. Esophagointestinal junction is definitely not straight, but often it is not clearly defined, and it looks like a half-round bulge of the dark intestine. When clearly visible, the junction often shows a dorsal overlap of the intestine by the glands. Vulva at midbody, easy to see as a straight line in a clear area. Body remains very regularly cylindroid down to the short tail, often broadly rounded or quarter rounded.

When present, males are similar to females, but a little smaller and with a stylet less well defined. The tail has the typical conoid ventrally bent shape seen in many tylenchs,

---

*The Russian author Filip'ev used to transliterate his name as " Filipjev," and he is known in the West under this spelling. However, the Russian letter "ь" should be transliterated as ' according to the International Standard ISO/R9 1968 (E)–Table 2. The correct transliteration Filip'ev is used in this chapter.

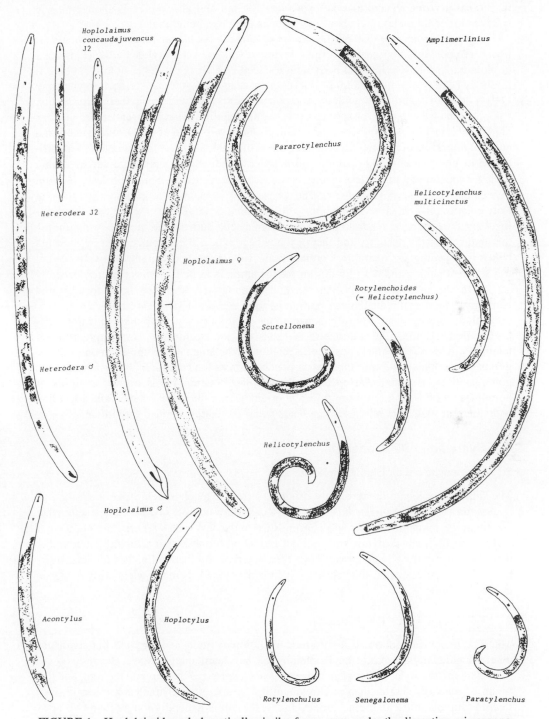

FIGURE 1 Hoplolaimids and phenetically similar forms, seen under the dissecting microscope.

with clearly visible caudal alae and spicules. The tail and alae region is shorter than in tylenchorhynchid males, and about the same as in pratylenchid males.

Some species in the genera *Pararotylenchus, Amplimerlinius, Pratylenchoides* (syn. *Hoplorihynchus*) have forms intermediate between tylenchorhynchids and hoplolaimids: robust cylindroid body with rounded end, strong stylet easy to see, vulva at midbody (like typical hoplolaimids) but junction straight (like typical tylenchorhynchids). Only a detailed examination with the compound microscope will reveal their true identity.

Some taxa, genera, or species of the family exhibit remarkable tendencies that can help in their identification: shortening of tail, posterior migration of DGO, enlargement and anterior migration of phasmids, etc. Enlargement of phasmids into scutella involves modification of both the phasmid opening and the phasmid duct. The opening enlarges from a typical diameter of 1 μm or less in, e.g., *Rotylenchus* or *Helicotylenchus* species to 3–5 μm in *Scutellonema* or *Hoplolaimus*. The phasmidial duct, which is a simple tube in *Rotylenchus*, enlarges in a wide chamber, the ampulla, generally even wider than the opening. As a consequence, reported measurement of a scutellum should include two values— the diameter of the opening, and that of the widest part of the ampulla, seen by focusing below the opening. For example, Germani et al. (1986) report measurements of 2.8 (2–3.5) μm for the opening and 4.1 (3.5–5) μm for the ampulla in *Scutellonema brachyurus*.

In the male genital system the gubernaculum is thickened both dorsally and ventrally. The dorsal cuticularized portion is often described as if it were the whole gubernaculum, and the ventral portion is named the capitulum. The capitulum sometimes has been called telamon, after a structure that exists in the strongylids. Both capitulum and telamon are seen in a ventral position in relation to the spicules but the telamon of strongylids is a cuticularization of the cloacal pouch whereas the capitulum, being a part of the gubernaculum, is a cuticularization of the spicular pouch. This is a good example of homoplasy with two structures of different origin having the same function and appearing equivalent at first. The gubernaculum bears distal projections named the titillae.

B. Systematic Relationships

1. Relationships with Belonolaimidae

The family Hoplolaimidae probably originated from tylenchorhychid-like ancestors, as demonstrated by *Pararotylenchus*. This genus exhibits a number of derived characters that are already hoplolaimid (short tail, phasmids near anus level, DGO more than 4 μm from stylet base, high labial region, strong stylet and labial framework, and spiral habitus), but the criterion traditionally given as the most characteristic for the family, the glandular overlap over the intestine, does not appear, or is little developed, as in the tylenchorhynchids in Belonolaimidae.

2. Relationships with Pratylenchidae

The Pratylenchidae followed a different evolutionary path and became endoparasites whereas generally speaking the Hoplolaimidae are semiendoparasites. Because of endoparasitism, the stylet of Pratylenchidae does not have to reach deep into the root, and this organ is generally shorter than in Hoplolaimidae. Some other characters are less derived than in Hoplolaimidae, e.g., in Pratylenchidae, the tail is generally longer than two body diameters at anus, and the phasmids are situated on the tail, as in Belonolaimidae.

Most pratylenchids are easy to differentiate from hoplolaimids by the lower lip region and the shorter, squat stylet, but some intermediate forms may be quite puzzling. For example, Andrássy recently described a new genus, *Hoplorhynchus*, with characteristics intermediate between hoplolaimids and tylenchorhynchids (hence the genus name). It was eventually shown by Luc (1986) to be a synonym of the genus *Pratylenchoides*, a pratylenchid.

3. Relationships with Heteroderidae

The Heteroderidae probably originated from Hoplolaimidae. They evolved into sedentary endoparasites by establishing very close relationships with their hosts. The host response resulting in specialized feeding structures allowed the female body to enlarge for increased egg production. In some aspects of their morphology, the two families are very close. The anterior end of an heteroderid male has a definite hoplolaimid appearance, and it may be difficult to differentiate under the dissecting microscope second-stage juveniles of *Heterodera* from, e.g., *Hoplolaimus concaudajuvencus* juveniles (Fig. 1).

Adults of heteroderids have unique characteristics that clearly differentiate the family, both for systematics and identification. Males have extremely short tails; females are typically enlarged.

C. Subfamilies in Hoplolaimidae

Hoplolaiminac Filip'cv, 1934
      syn. Nemonchinae Skarbilovich, 1959
           Rotylenchoidinae Whitehead, 1958
           Aphasmatylenchinae Sher, 1965
           Rotylenchinae Golden, 1971
           Pararotylenchinae Baldwin and Bell, 1981

Hoplolaimidae. Adult females remain vermiform. Lateral field with four lines or less. Phasmids near anus or erratic, anteriorly migrated on body, rarely on tail, small or enlarged, rarely absent. Tail short, more curved dorsally with or without a ventral projection, or regularly rounded, rarely conoid. Caudalids and cephalids present, labial framework and stylet strong. DGO more or less far from stylet. Esophageal glands either of the same length and not overlapping the intestine, or variously enlarged and overlapping the intestine. Genital branches always outstretched, two of equal length or posterior branch smaller or reduced to a postuterine sac (PUS).

Males with anterior end smaller than females but still functional. Caudal alae enveloping tail end.

Migratory ecto- or semiendoparasites of higher plants. Eggs laid free in the soil and not deposited in a gelatinous matrix.

Type genus: *Hoplolaimus* von Daday, 1905
        syn. *Nemonchus* Cobb, 1913
             Hoplolaimoides Shakil, 1973
             Basirolaimus Shamsi, 1979
Other genera: *Rotylenchus* Filip'ev, 1936
        syn. *Anguillulina* (*Rotylenchus*) Filip'ev, 1936
             Gottholdsteineria Andrássy, 1958
             Orientylus Jairajpuri and Siddiqi, 1977
             Calvatylus Jairajpuri and Siddiqi, 1977
             Interrotylenchus Eroshenko, 1984
             Scutellonemoides Eroshenko, 1984

> *Varotylus* Siddiqi, 1986
> *Helicotylenchus* Steiner, 1945
> syn. *Rotylenchoides* Whitehead, 1958
>   *Zimmermannia* Shamsi, 1973
> *Scutellonema* Andrássy, 1958
> *Aorolaimus* Sher, 1963
> syn. *Peltamigratus* Sher, 1964
>   *Nectopelta* Siddiqi, 1986
> *Aphasmatylenchus* Sher, 1965
> *Antarctylus* Sher, 1973
> *Pararotylenchus* Baldwin and Bell, 1981

2.   Rotylenchulinae Husain and Khan, 1967

Rotylenchulinae, including *Acontylus, Senegalonema,* and *Rotylenchulus* are studied in Chap. 11.

Rotylenchulinae differs from Hoplolaiminae in having body of mature females swollen or kidney-shaped, lip region not as high as in Hoplolaiminae, glandular overlap long and mostly lateral. Rotylenchulinae are sessile semiendoparasites and they lay their eggs in a gelatinous matrix.

III.   *Pararotylenchus* **Baldwin and Bell, 1981**

A.   Diagnosis

Hoplolaiminae. Females: Body spiral to C-shaped. Labial region annulated; oral opening slitlike; round labial disk present; amphid apertures elongate at lateral edge of disk; first lip annulus divided into six lip sectors; lateral sectors smaller than the others (SEM). Lateral field with four lines. Labial framework, stylet, and stylet knobs average-sized for the family; knobs round to indented. DGO about 4–7 μm from stylet base. Esophageal glands not overlapping intestine, symmetrically arranged around esophageal lumen. Two genital branches outstretched, equally developed. Two epiptygmata present. Tail short, of variable shape, usually more curved on dorsal side; phasmids small, porelike, near level of anus.

Males. Caudal alae enveloping tail. Gubernaculum flexed distally and with small titillae but no capitulum. Spicules trough-shaped. No secondary sexual dimorphism.

Type species: *Pararotylenchus hopperi* Baldwin and Bell, 1984, syn. *P. brevicaudatus* (Hopper, 1959) Baldwin and Bell, 1981.

Except for *P. pini,* known from Japan and Korea, all the species in this genus have been found in the western United States, in Utah and California. All occur in cool regions, either at high elevations or along the cool Pacific coast. They are associated with a wide range of plants, including various conifers, rose, aspen, *Veratrum* spp., and barley.

B.   Systematic Relationships

Pararotylenchus belongs to Hoplolaimidae because of the short tail, phasmids near anus level, DGO more than 4 μm from the stylet base, high labial region, strong stylet and cephalic framework, and spiral habitus. The esophageal glands not overlapping the intestine are similar to the glands in *Tylenchorhynchus* and some other Belonolaimidae. In the hypothesis that hoplolaimids originated from tylenchorhynchid-like ancestral forms,

Pararotylenchus can be seen as a relic of the intermediary forms where some but not all of the characters of Hoplolaimidae were present.

   Pararotylenchus is very close to *Rotylenchus*, except for the arrangement of the glands and the presence of longitudinal striae on the lip annuli in some *Rotylenchus* species. Baldwin and Bell (1981) noted that a morphological continuum exists from *Pararotylenchus*, with glands of equal length, not overlapping the intestine, and esophageal lumen symmetrically arranged between the three glands, to *Rotylenchus fallorobustus* and *R. breviglans* that also have glands of equal length not overlapping the intestine but where the esophageal lumen is shifted ventrally between the subventral glands, and finally to the typical *Rotylenchus* species with dorsal overlap.

C. Representative Species

1. *Pararotylenchus hopperi* Baldwin and Bell, 1984 (Fig. 2)

Measurements (mean ± standard deviation, and range; after Baldwin and Bell, 1981): Females (n = 22): L = 1.413 mm ± 1.134 (1.24–1.61); a = 38.5 ± 2.954 (32.8–44); b' = 7.38 ±0.406 (6.7–8.3); c = 44.22 ± 4.938 (36.5–53.3); c' = 1.15 ± 0.180 (0.8–1.6); m = 51.1% ± 1.353 (49–54); V = 58.5% ± 2.029 (54–62); stylet = 38.3 μm ± 4.239 (36.5–41). Males (n = 21): L = 1.239 mm ± 0.767 (1.12–1.39); a = 42.06 ± 0.154 (38–46.3); b = 6.81 ± 0.989 (5.8–10.7); c = 33.95 ± 2.351 (28.9–39.4); c' = 1.81 ± 0.176 (1.5–2.1); stylet = 35.62 μm ± 0.857 (33–37); spicules = 36 μm ± 2.021 (32.5–39), gubernaculum = 15.9 μm ± 0.088 (14.5–18.5).

   Description. Females: Body generally in loose spiral shape. Lip region slightly offset, flattened to slightly rounded anteriorly, usually with eight annuli. SEM face view shows round oral disk, and first annulus separated in six low sectors, the two lateral sectors smaller than the submedian ones. Labial framework sclerotized. Stylet massive with knobs anteriorly flattened. DGO 5.36 μm ± 1.082 (3.5–7.5) posterior to stylet base. Excretory pore at level of the middle portion of the esophageal glands. Hemizonid one to five annuli anterior to excretory pore. Two large epiptygmata, one inwardly, the other outwardly folded. Spermatheca oval, with round sperms. Tail dorsally convex conoid, 32.32 μm ± 4.239 (24.5–38) long, 28.3 μm ± 2.583 (20.5–32) wide at anus level, tail end rounded and coarsely annulated. Phasmids prominent, at anus level.

   Males similar to females. Spicules arcuate; gubernaculum with titillae.

IV. *ROTYLENCHUS* FILIP'EV, 1936

A. Diagnosis

Hoplolaiminae. Females: Body spiral to C-shaped. Labial region offset or continuous with body contours, anteriorly rounded or flattened, generally annulated, with or without longitudinal striae on basal lip annulus. Lateral field with four lines, with or without scattered transverse striae. Labial framework, stylet, and stylet knobs average-sized for the family; knobs with rounded to indented anterior surface. DGO often close to stylet (6 μm) but with a tendency to posteriorly directed migration (up to 16 μm). Esophageal glands overlap intestine dorsally and laterally; dorsal gland more developed than subventral glands; intestine symmetrically arranged between the subventral glands. Two genital branches outstretched, equally developed; posterior branch rarely degenerated. One or two epiptygmata present.

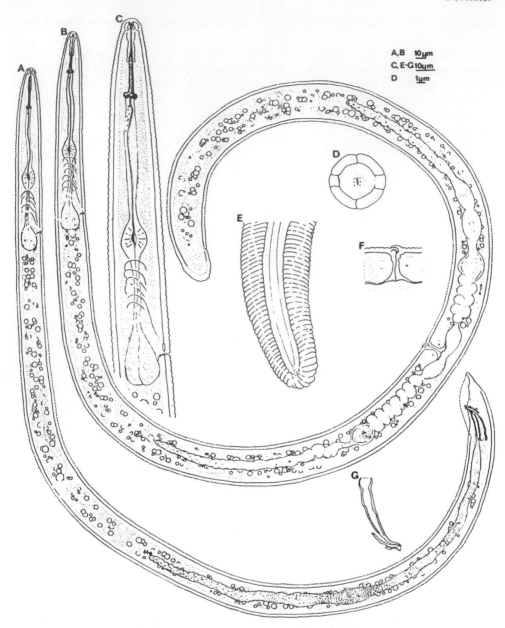

FIGURE 2 *Pararotylenchus hopperi* Baldwin and Bell, 1984. (A) Entire male. (B) Entire female (curvature of specimen slightly exaggerated for convenience of illustration). (C) Cephalic and esophageal region of female. (D) Face view as observed with SEM. (E) Female tail. (F) Vulva. (G) Spicule and gubernaculum.(Baldwin and Bell, 1981, courtesy *J. Nematol.)*

Tail short, hemispheric, rarely with small ventral projection; phasmids porelike, small, near anus level.

Males: Caudal alae enveloping tail, not lobed. Secondary sexual dimorphism not marked, sometimes anterior part of male body slightly smaller than that of female.

Type species: *Rotylenchus robustus* (de Man, 1876) Filip'ev, 1936.

B. Systematic Relationships

From a hypothetical ancestor resembling present-day *Pararotylenchus*, the most obvious synapomorphy in this genus is the development of the dorsal gland and its overlapping of the intestine, and the reduction in size of the ventral glands. However, at the difference of the genus *Helicotylenchus*, the esophageal lumen remains symmetrically situated between the two subventral glands.

The characters used by various authors to differentiate the genera *Orientylus* Jairajpuri and Siddiqi, 1977, *Calvatylus* Jairajpuri and Siddiqi, 1977, *Interrotylenchus* Eroshenko, 1984, *Scutellonemoides* Eroshenko, 1984, *Varotylus* Siddiqi, 1986 have been shown (Fortuner, 1987a; Maggenti et al., 1988) to be irrelevant at the genus level.

C. Representative Species

1. *Rotylenchus robustus* (de Man, 1876) Filip'ev, 1936 (Fig. 3)

Measurements (after Sher, 1965). Females ($n = 20$): $L = 1.22–1.87$ mm; $a = 32–40$; $b = 8.1–10.1$; $b' = 6.0–8.1$; $c = 48–70$; $V = 54–58$; stylet $= 44–50$ μm; $o = 8–16$. Males ($n = 10$): $L = 1.06–1.30$ mm; $a = 35–42$; $b = 6.6–10.0$; $b' = 5.2–6.9$; $c = 33–41$; stylet $= 41–44$ μm; $o = 10–15$; spicules $= 34–40$ μm; gubernaculum $= 17–21$ μm; capitulum $= 9–12$ μm.

Description. Females: Relaxed body usually forming a single spiral, sometimes C-shaped; annuli prominent, about 1.7 μm wide near middle. Lateral fields irregularly areolated on body, with four lines, about one-fifth wide as body. Lip region hemispheric, offset by a constriction, with six to seven distinct annuli; labial disk slightly raised above the lip outline; lip annuli longitudinally indented to give a tiled surface appearance; 24–30 such indentations on basal annulus. In SEM head view, round labial disk with first labial annulus in six sectors about equal in size; other labial annuli with characteristic tiled appearance; amphid apertures wide and kidney-shaped (De Grisse et al., 1974). Anterior and posterior cephalids distinct, at 2–3 and 9–10 annuli behind lip region. Spear well developed, cone 50–56% of its total length; basal knobs large, about 7 μm across, rounded but sometimes with flat or indented anterior surfaces. Median esophageal bulb ovate, very muscular, and with a prominent valvular apparatus in center. Esophageal glands extending over intestine dorsally and dorsolaterally as the two subventral glands are shifted from their normal to a subdorsal position; nuclei of subventral glands varying in position from slightly posterior to slightly anterior to that of dorsal gland. Excretory pore usually close to esophagointestinal valve. Hemizonid distinct, three annuli long, just anterior to the excretory pore. Hemizonion indistinct. Vulva a depressed, transverse slit; two short epiptygmata present, sometimes appearing as a single epiptygma in lateral view. Uterine sac formed by a few flattened cells; at the transition to the uterus not more than four larger cells are found; the columnar uterus consists of no more than 12 cells in three rows. Spermathecae formed by 12 cells in various spatial arrangements, rounded, usually packed with sperms; oviduct as two rows of four small, flattened cells forming a constriction between spermatheca and ovary. At the transition to the oviduct, the ovary sometimes contains eight nuclei lying at about the same level, sometimes slightly overlapping the oviduct (Geraert, 1981). Oocytes in a single row except

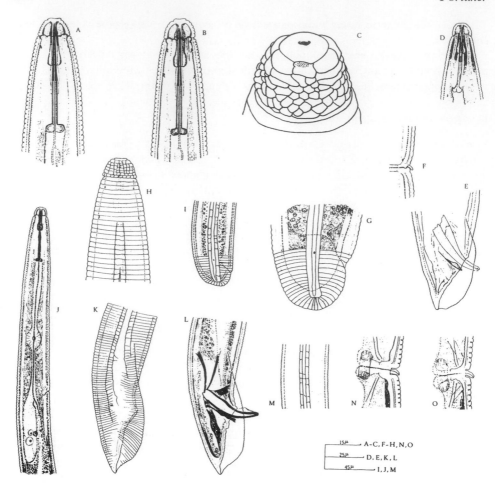

FIGURE 3   *Rotylenchus robustus* (de Man, 1876) Filip'ev, 1936. (A,C,H) Female, anterior ends.
(B,D) Male, anterior ends. (F,N,O) Vulval region with epiptygmata. (G,I) Female, posterior ends.
(E,K,L) Male, posterior ends. (J) Male, esophageal region. (M) Lateral fields. (H,K) Surface views.
(Siddiqi, 1972, courtesy Commonwealth Agricultural Bureaux.)

for the multiplication area. Intestine partially overlapping rectum. Tail hemispheric, regularly annulated, with 8–17 annuli from anus to center of terminus. Phasmids distinct, porelike, usually just preanal but varying from three annuli posterior to seven annuli anterior to anal level.

Males: Body in open C-shape when relaxed. Lip region more distinctly offset and elevated than in female. Bursa crenate, enveloping tail. Spicules slightly cephalated and ventrally arcuate, with well-developed ventral flanges on distal third. Gubernaculum protruding, with prominent titillae.

Rotylenchus robustus is widely distributed in Europe, USSR, Egypt, Zaire, India, Canada, USA, and Brazil. It prefers light sandy soils and it parasitizes many grasses, ornamentals, vegetable crops, and forest trees. It is an ectoparasite of roots causing reduced growth, yellowing, and yield reduction.

Rotylenchus robustus is representative of a group of species within the genus *Rotylenchus* with phasmid not enlarged into scutella, and situated opposite each other, near the anus; lateral fields composed of four lateral lines, sometimes areolated near the phasmids; lip region including longitudinal striae in addition to the usual transverse lip annuli; and opening of the dorsal esophageal gland less than one-quarter stylet length behind stylet base. The females have two equally developed genital branches. The caudal alae of the males are not indented.

2. Species Formerly in the Genus "*Calvatylus*"

The species that were transferred to this genus are quite similar to those in typical *Rotylenchus* except that distinct lip annulation is lacking. Other minor differences include lip region slightly offset (offset in *Rotylenchus*), tail shorter than body width (often a little longer in *Rotylenchus*), and in the males, gubernaculum not protruding, lacking titillae (titillae present in *Rotylenchus*). The genus *Calvatylus* was accepted as valid by Siddiqi (1986), but it was considered a synonym of *Rotylenchus* by Ferraz (1980) and by Fortuner (1987a). It included only three species: *R. calvus, R. heredicus,* and *R. nexus*. None are economically important parasites, and they will not be described here.

3. *Rotylenchus orientalis* Siddiqi and Husain, 1964 (Fig. 4)

Measurements (after Siddiqi and Husain, 1964). Females ($n = 5$): Length = 0.72 mm (0.68–0.76); $a = 29$ (27–32); $b = 6.4$ (6.2–6.7); $c = 42$ (36–56); $V = 69.5\%$ (66–72); stylet = 25–28 μm. After Choi and Geraert, 1972. Females ($n = 19$): $L = 570$–690; $a = 27$–29; $b = 5.8$–7; $c = 37$–52; $c' = 0.8$–1.3; $V = 67$–73%; stylet = 23–26 μm; $m = 43$–47; $o = 50$–61.

Description. Females: Body spirally coiled. Cuticle with coarse striae, 1.7 μm apart near midbody. Lateral fields one-fourth as wide as body. Phasmids porelike, from four annuli anterior to three annuli posterior to anus level. Head conoid-rounded, with four or five annuli, continuous with body. Spear of medium built, with prominent, anteriorly cupped to flattened basal knobs. DGO 12–15.5 μm behind spear base. Corpus cylindrical, narrowed at its junction with median esophageal bulb which is oval in shape. Excretory pore a little behind level of nerve ring. Esophageal glands forming an elongate lobe, extending over dorsal and dorsolateral sides of the anterior end of the intestine; the subventral glands usually are elongated as in *Helicotylenchus*, but in some specimens the subventral glands were not as long giving the esophagus the appearance of a *Rotylenchus* esophagus (Geraert, 1976). Vulva a transverse slit. Reproductive system paired, opposed; posterior genital branch reduced, 50–60% of anterior branch; this difference in length is essentially caused by the lesser development of the ovary. Spermatheca empty, sometimes inconspicuous. Uterus at its distal end with a prominent swelling. Tail 11–18 μm long, dorsally convex, rounded, with 9–11 ventral annuli, about one anal body width long.

Males unknown.

The species is known from India to Korea, but its biology has not been studied. Its systematic position is very unsettled.

Rotylenchus orientalis is representative of the species formerly in the genus *Orientylus*. These species are quite similar to typical *Rotylenchus* except that the posterior genital branch is nonfunctional or absent, and the lip region does not have longitudinal striae. Other minor differences include DGO more posterior, and lip region smaller, narrower than in other *Rotylenchus*. Siddiqi (1986) accepted *Orientylus* as a valid genus in Rotylenchoidinae, a subfamily differentiated from Rotylenchinae by smaller lip region, smooth ba-

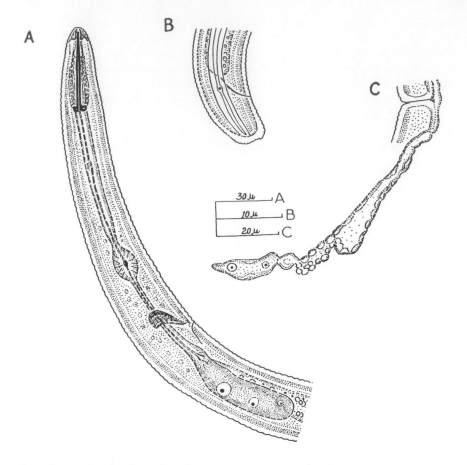

FIGURE 4 *Rotylenchus orientalis* Siddiqi and Husain, 1964. (A) Female, esophageal region. (B) Female, posterior end. (C) Female, regressed posterior genital branch. (Siddiqi and Husain, 1964, courtesy *Proc. Helm. Soc. Wash.*)

sal lip annulus, and DGO more than one-fourth of stylet length behind stylet base. Zancada and Lima (1986) and Fortuner (1987a) considered it a synonym of *Rotylenchus*.

4. *Rotylenchus varus* (Jairajpuri and Siddiqi, 1979) Zancada and Lima, 1986 (Fig. 5).

Measurements (after Jairajpuri and Siddiqi, 1979). Females (n = 20): L = 0.55 mm (0.47–0.61); a = 31 (29–33); b = 6.5 (6–7.5); b' = 4.3 (4–4.7); c = 45 (36–64); c' = 1.2 (0.9–1.5); V = 64% (62–68); stylet = 21 μm (20–23); m = 46 (44–50); o = 51 (49–55).

Description. Females: Heat relaxed body forming a tight double spiral of 1.5–2.5 turns. Cuticle annuli distinct, 1.2 μm wide at midbody. Lateral field with four lines, non-areolated, one-fourth to one-third of body width wide. Labial region narrow, almost hemispherical, continuous with body contours; three or four indistinct lip annuli; cephalic framework strongly sclerotized. Stylet cone 9–11 μm long; stylet knobs anteriorly rounded to indented. DGO 9–12 μm from stylet base. Excretory pore 76–89 μm from anterior end, opposite base of isthmus. Hemizonid zero to two annuli anterior to excretory pore, three annuli wide. Esophagointestinal junction ventral just posterior to the level of the excretory pore. Glands overlapping the intestine over two body widths; dorsal gland in a dorsal posi-

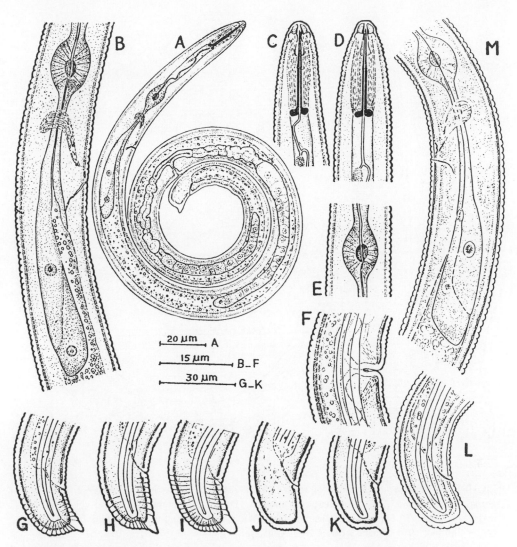

FIGURE 5 *Rotylenchus varus* (Jairajpuri and Siddiqi, 1979) Zancada Lima, 1986. (A) Female, entire body. (B,M) Median esophageal bulb and glandular region. (C,D) Anterior ends. (E) Median bulb. (F) Vulvar region. (G–L) Tails. (Jairajpuri and Siddiqi, 1979 and Siddiqi, 1986, courtesy *Ind. J. Nematol.* and CAB.)

tion, its nucleus about one body width behind intestinal junction; subventral glands in a dorsolateral position, extending past the dorsal gland; subventral gland nuclei posterior to the dorsal one. Vulva depressed, with two indistinct epiptygmata deep at the end of vagina vera. Two genital branches opposed. Spermatheca small, dorsally offset, empty; oocytes in one row except for multiplication area. Tail more rounded dorsally, with distinct terminal ventral projection; 6–10 ventral tail annuli, the terminal annuli are narrower than the other tail annuli. Phasmids dotlike, 3–10 annuli anterior to anus, on or near ventral inner lateral field line.

Males unknown.

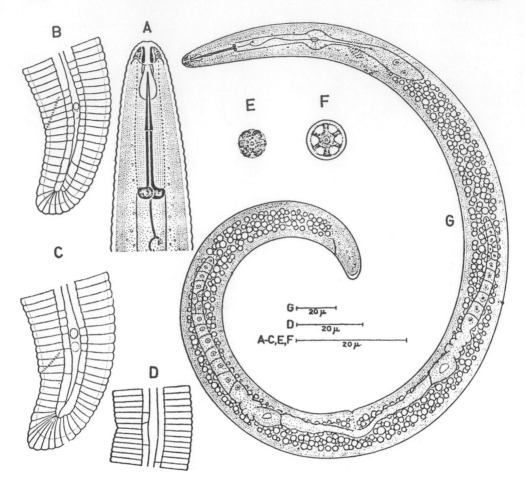

FIGURE 6   *Rotylenchus insularis* (Phillips, 1971) Germani, Baldwin, Bell, and Wu, 1986. (A) Female, anterior end. (B,C) Female, posterior ends. (D) Female, lateral fields near vulva. (E) Female, face view. (F) Female, cross-section through basal labial annulus. (G) Female, entire body. (After Phillips, 1971, courtesy *Qd. J. Agr. Anim. Sci.*)

      Rotylenchus calvus has been found on grass in Malawi, and nothing is known of its biology. It is representative of the species formerly in the genus *Varotylus*. The subventral glands extend past the dorsal gland, whereas they are less developed in typical *Rotylenchus*. However, they remain in a symmetrical position, quite different from the asymmetrical arrangement found in *Helicotylenchus*. As in *Orientylus*, the lip region is smaller, narrower than in typical *Rotylenchus*, and it does not have longitudinal striae. The DGO is more posterior to stylet base. The posterior genital branch is well developed.

5.   *Rotylenchus insularis* (Phillips, 1971) Germani, Baldwin, Bell, and Wu, 1986 (Fig. 6)

Measurements (after Phillips, 1971). Females (n = 7): L = 558 μm (515–596); a = 22 (22–23); b = 6.0 (5.3–6.7); b' = 4.6 (4.1–5.1); c = 37 (30–37); c' = 1.2 (0.9–1.5); V = 61% (57–65); stylet = 23 μm (22–24).

Description. Females: Body in spiral shape. Lip region hemispheric, not set off, with four annuli; basal lip annulus without longitudinal striae. Stylet knobs rounded, with flattened anterior surfaces. DGO 7 μm behind stylet. Hemizonid one annulus anterior to excretory pore. Spermathecae without sperms. Epiptygma not seen. Scutellum opening 1.9 μm (1.8–2.1) in diameter, three annuli anterior to anus. Lateral fields areolated in esophageal region and near scutella, occasional areolation of outer bands in midbody. Tail broadly rounded, slightly more curved dorsally, with 12 ventral annuli; terminal striations similar to other tail annuli.

Males unknown.

Rotylenchus minutus and four of the five *Scutellonema* species described by Phillips (1971) from Queensland have phasmid openings 1.2–2.1 μm in diameter. This is smaller than the diameter of the opening in typical *Scutellonema* (at least 3 μm, up to 9 μm), but larger than the phasmid opening in *Rotylenchus*, often less than 1 μm in diameter. The two phasmids are a little above anus level, and not quite opposed to each other, which is another difference from typical *Scutellonema*. The lateral field with four lines is areolated near the phasmids in at least three of the species (*R. minutus, R. insulare, R. impar*), and striae are seen near the phasmids in a few specimens of *R. laeviflexum*, a character at variance with typical *Rotylenchus*, but more frequent in *Scutellonema*. The basal lip annulus has no longitudinal striae. These four species seems somewhat intermediate between *Rotylenchus* and *Scutellonema*. They were placed in *Rotylenchus* by Germani et al. (1986) and Fortuner (1987a), but left in *Scutellonema* by Siddiqi (1986).

V. *SCUTELLONEMA* ANDRÁSSY, 1958

A. Diagnosis

Hoplolaiminae. Females: Body spiral to C-shaped or almost straight. Labial region narrow truncate to offset rounded; annulated, with or without longitudinal striae. First labial annulus divided into six sectors, lateral sectors smaller than the others (SEM). Amphid apertures oval between labial disk and lateral sectors. Lateral field with four lines usually areolated near phasmids and anteriorly, sometimes transverse striae scattered over whole field. Labial framework, stylet and stylet knobs average-sized for the family; knobs rounded to indented. DGO 4–8 μm from stylet base. Esophageal gland overlap dorsal and lateral. Two genital branches outstretched, equally developed. Epiptygmata present. Tail short, rounded. Phasmids enlarged (scutella) situated opposite each other, near anus level.

Males. Caudal alae enveloping tail tip, regular or rarely deeply lobed. No secondary sexual dimorphism.

Type species: *Scutellonema bradys* (Steiner and Le Hew, 1933) Andrássy, 1958.

B. Systematic Relationships

The species in *Scutellonema* are identified by the phasmids, enlarged into scutella, and opposite each other near the anus level. The development of scutella seems to be a valid apomorphy because it probably has some evolutionary advantage (a wide open scutellum may be a better chemoreceptor than a regular phasmid with a small opening). *Rotylenchus*, with a small phasmid opening and no phasmidial ampulla, is accepted and differentiated from *Scutellonema* and the genera with scutella.

C. Representative Species

1. *Scutellonema brachyurus** (Steiner, 1938) Andrássy, 1958 (Fig. 7)

Measurements: Females: Composite description from Steiner (1938), Williams (1960), Sher (1964a), van den Berg and Heyns (1973), Sharafat-Ali et al. (1973), Geraert et al. (1975), Ratanaprapa and Boonduang (1975), van den Berg and Kirby (1979), Khan and Nanjappa (1970: *S. bangalorensis*), and Sivakumar and Selvasekaran (1982: *S. conicaudatum*). For each measurement is given the average and standard deviation calculated from the mean values, and the range of individual values. Females ($n = 11$): $L = 0.715$ mm ± 0.0758 (0.5–1); $a = 26 ± 2.770$ (16.7–36); b ($n = 9$) = 6.6 ± 0.780 (4.5–10.3); b' ($n = 6$) = 5.7 ± 0.451 (4–9.1); $c = 70.6 ± 15.444$ (37.2–127.7); c' ($n = 4$) = 0.55 ± 0.083 (0.4–0.9); $V = 58.75\% ± 1.636$ (51.9–67); stylet = 27 μm ± 1.684 (21.7–31). Males after Sher, 1964a. ($n = 8$): $L = 0.63–0.85$ mm; $a = 25–33$; $b = 6.1–7.5$; $b' = 4.5–6.6$; $c = 45–58$; stylet = 24–27 μm; gubernaculum = 12–14 μm; capitulum = 7–9 μm. After van den Berg and Heyns, 1973 ($n = 7$): $L = 0.7$ mm (0.6–0.8); $a = 34.3$ (31.6–36.9); $b = 7.3$ (6.9–8.1); $b' = 5.4$ (5.3–5.6); $c = 41.1$ (37–39.8); stylet = 22.9 μm (22.4–24.3); spicules = 26.1 μm (25–27.2); gubernaculum = 12 μm (10.7–13.2); capitulum = 7.5 μm (7– 8.1).

Description. Females: Body forming a single spiral occasionally arcuate; annuli distinct, average annulus width 1.4 μm near midbody. Lateral fields about one-fourth as wide as body, areolated from the anterior end to the excretory pore and at phasmids, marked by four incisures. Lip region hemispheric, slightly set off by a constriction, with three to four (sometimes five) annuli of irregular width. In SEM face view, labial disk round to oval; first lip annulus divided into six sectors; lateral sectors slightly smaller than submedian ones (Germani et al., 1986). Basal annulus marked by six longitudinal striae. Labial framework strongly sclerotized, its basal ring extending posteriorly over one body annulus. Stylet well developed, cone almost as long as shaft ($m = 48.1 ± 1.737$; 43–52.3); basal knobs prominent, rounded with slightly flattened anterior surfaces, 4.5 μm across. DGO rather obscure, about 5 μm (2–7) behind spear knobs. Cephalids, hemizonid, and hemizonion indistinct. Esophagus with a slight constriction at base of isthmus. Median esophageal bulb ovoid, extending over 8 (8–10) body annuli. Excretory pore at level of esophageal glands, at 132 μm (115–150) or 88 (80–97) body annuli from the anterior end. Esophageal glands overlapping intestine dorsally and dorsolaterally. Hemizonid at level of, or up to eight annuli anterior to, excretory pore; hemizonion 5–10 body annuli posterior to hemizonid. Oocytes in a single row except in multiplication zone. No spermatheca or sperms in uterus. One or two epiptygmata indistinct, inwardly folded. Intestine slightly overlapping rectum. Tail broadly rounded; annuli following tail contour; annuli slightly irregular and coarser than adjacent body annuli; 12 (7–19) annuli between anus and tail tip; inner cytoplasmic core terminally flattened or with shallow depression. Phasmid opening 2–4 μm in diameter from two annuli behind to two annuli in front of anus; phasmid ampulla 3.5–5 μm in diameter, but according to Wang and Chen (1985), the volume of the ampulla is affected by the osmolarity of the environment.

Males: Absent from most populations. Rarely present in small numbers (one for several hundred females). Similar to female in most details. Spicules arcuate, with a swollen head, a crescentic shaft, and a blade composed of a central trunk with two winglike lateral vela. Gubernaculum completely embedded in the cuticle of the spicular pouch; the gub-

*The species name is often spelled *"brachyurum"*; however *urus* (= tail) is a substantive in apposition, and its ending does not change whatever the gender of the generic name with which it is combined [ICZN, Art. 31 (b) (ii)].

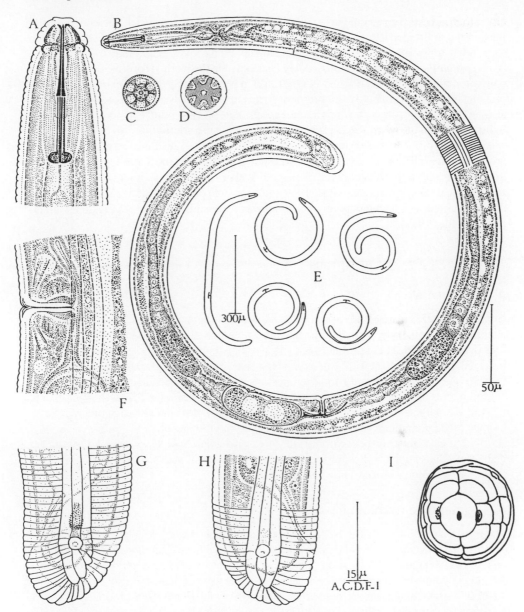

FIGURE 7 *Scutellonema brachyurus* (Steiner, 1938) Andrássy, 1958. (A) Head view. (B) Female, entire body. (C,I) Face view. (D) Cross-section through basal labial annulus. (E) Thermal death body posture. (F) Vulval region. (G,H) Posterior ends. (Siddiqi, 1974, courtesy Commonwealth Agricultural Bureaux.)

ernaculum proper is connected to a capitulum lying in the spicular pouch between the spicules; in the midregion the capitulum separates from the dorsal wall of the spicular pouch to form a bar tongue that bears the proximal titillae (Wang and Chen, 1985b).

Scutellonema brachyurus is widely distributed in the world. It occurs on a number of cultivated plants. It has been found in virgin soil in Natal, and Sher (1964a) infers from this fact that this species is native from southern Africa. It is primarily an ectoparasite, but it may also be found in the root cortex. It is a semiendoparasite on sycamore. High populations may build up on suitable hosts. Reports of injury to the plant are conflicting. No apparent injury was caused by *S. brachyurus* to red clover (Chapman, 1963); it caused a reduction in leaf growth of amaryllis but no concomitant destruction of bulb and roots although it was found in the epidermis and cortical tissues (Nong and Weber, 1964); it was found associated with stunted sugarcane in Malagasy (Luc, 1959). Root growth of pine seedlings increased rather than decreased with high inoculum of *S. brachyurus*, possibly because of parasitic stimulation (Ruehle, 1973).

VI. *AOROLAIMUS* SHER, 1963

A. Diagnosis

Hoplolaiminae. Females: Body spiral to C-shaped, medium-sized. Lip region slightly offset or continuous with body, with or without annuli and/or longitudinal striae. Lateral field with four or less incisures. Labial framework and stylet medium-sized; stylet knobs flattened to indented anteriorly. DGO 3–10 μm from stylet base. Esophageal glands with three nuclei, overlap intestine dorsally and laterally; intestine symmetrically arranged between the subventral glands. Two genital branches outstretched, equally developed. Tail short, rounded. Phasmids enlarged to scutella erratically situated on body, not opposite each other, anterior to anus level; sometimes one scutellum is anterior to vulva level.

Males. Caudal alae enveloping tail, lobed or regular. Secondary sexual dimorphism visible in labial region and esophageal structures smaller in males.

Type species: *Aorolaimus helicus* Sher, 1963.

B. Systematic Relationships

The position of the scutella (on tail vs. on body) may have some biological significance and *Aorolaimus* is accepted as a valid genus and differentiated from *Scutellonema*, with scutella on tail. However, the exact location of the scutella on the body (e.g., anterior or posterior to the vulva) cannot be accepted as a generic character in view of the variability of this character in this group and in *Hoplolaimus*. The synonymy of *Peltamigratus* and *Nectopelta* to *Aorolaimus* proposed by Fortuner (1987a) is upheld here.

The lateral lip sectors of the face view, as seen with SEM in *Rotylenchus* and *Scutellonema*, have either the ancestral arrangement (six sectors of about equal size) or the lateral sectors are smaller than the submedian sectors. Some species in *Aorolaimus* (syn. *Peltamigratus*) have one or the other of these two arrangements, or the lateral sectors are longer than the submedian ones, or the lateral sectors merge with the first labial annulus. The variability of this character makes it difficult to accept it as a valid criterion for evolutionary relationships. It is known only in a few species (particularly no *Aorolaimus stricto sensu* have been studied with SEM), and future studies may shed more light on its value and significance.

C. Representative Species

1. *Aorolaimus christiei* (Golden and Taylor, 1956) Fortuner, 1987 (Fig. 8)

Measurements (after Sher, 1964b). Females (n = 20): L = 0.67–0.87 mm; a = 25–31; b = 6.4–8.3; b' = 5.3–7.3; c = 45–80; V = 53–58%; stylet = 30–34 µm; o = 11–18; anterior phasmid = 76–85%; posterior phasmid = 82–90%. Males (n = 10): L = 0.66–0.81 mm; a = 26–33; b = 6.7–8.6; b' = 5.3–6.7; c = 39–58; stylet = 29–33 µm; o = 11–18; anterior phasmid = 76–84; posterior phasmid = 85–89; gubernaculum = 11–15 µm; spicules = 27–31 µm;

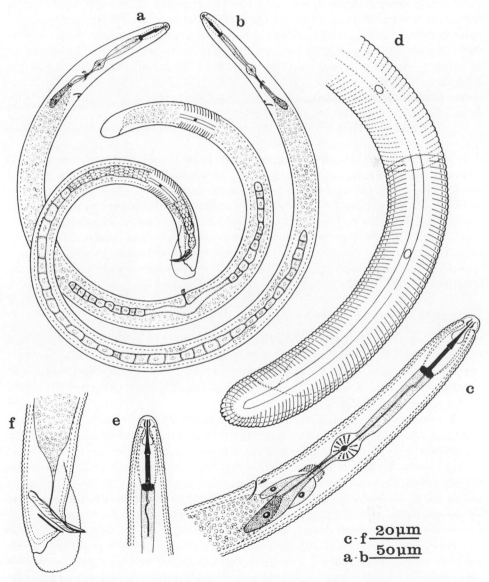

FIGURE 8 *Aorolaimus christiei* (Golden and Taylor, 1956) Fortuner, 1987. (a) Female, entire body. (b) Male, entire body. (c) Female, esophageal region. (d) Female, posterior end. (e) Male, anterior end. (f) Male, posterior end. (Bittencourt and Huang, 1986, courtesy *Rev. Nématol.*)

capitulum = 6–8 μm. After Bittencourt and Huang, 1986; Females (n = 5): L = 0.88 mm ± 0.08; a = 31.5 ± 2.7; b = 8.7 ± 0.9; b' = 6.7 ± 0.9; c = 53.8 ± 2.6; o = 19.9 ± 3.4; V = 57.3% ± 2.3; stylet = 29.6 μm ± 1; anterior phasmid = 80.4% ± 1.4; posterior phasmid = 88.1% ±1. Males (n = 2): L = 0.83–0.91; a = 34.7–39; b = 9.1; b' = 7.1–7.9; c = 56.2–61.3; stylet = 28 μm; spicules = 26–29 μm; gubernaculum = 11–12 μm.

Description. Females: Body usually in an open C-shape. Body annuli 1.3–1.4 μm wide at midbody. Lip region slightly offset or continuous with body, with three indistinct annuli. Stylet knobs often slightly flattened anteriorly. DGO about 5 μm posterior to stylet base. Excretory pore usually opposite anterior part of the esophageal glands. Hemizonid zero to two annuli anterior to excretory pore. Hemizonion about eight annuli below hemizonid. Spermatheca round, usually partially filled with sperms. Two epiptygmata, well developed, projecting outward. Lateral field usually with two lines, sometimes four lines on part of the body. Intestine slightly overlapping rectum. Tail rounded to slightly asymmetrical dorsally, shorter than body anal diameter, with 6–10 annuli; terminal annuli deeply marked.

Males: Similar to females except for sexual dimorphism. Caudal alae deeply indented. Gubernaculum with titillae. Capitulum present.

This species is known from Florida and Brazil. Its biology and pathogenicity have not been studied.

A. christiei and some of the species formerly in the genus *Peltamigratus* were placed by Eerens and Loof (1985) in a phenetic group, "Peltamigratus Group I." In this group of species, the scutella are not opposite each other, but they are both situated between vulva and anus. The lateral fields have two lines, or at least two lines in posterior part of the body even though there may be four lines anteriorly. There are no areolations near the scutella. The lip region is without longitudinal striae, and generally without lip annuli (rarely present). The male caudal alae generally are not indented (rarely indented). This group is represented by *P. christiei* and other species from Central and South America, Florida, and western Africa.

2. *Aorolaimus annulatus* (Mulk and Siddiqi, 1982) Fortuner, 1987

Measurements (after Mulk and Siddiqi, 1982). Females (n = 4): L = 0.75 (0.69–0.81) mm; a = 28 (26–29); b = 8 (6.4–8.6); c = 44 (32–63); c' = 0.8 (0.6–1.2); V = 55 (52–59); stylet = 26 μm (25–27). Males (n = 4): L = 0.69 mm (0.67–0.71); a = 30 (26–33); b = 7.5 (7–8); c = 35 (34–36); c' = 1.2–1.3; T = 53% (49–56); stylet = 26 μm (25–26.5); spicules = 30–31 μm; gubernaculum = 15–15.5 μm.

Description. Females: Body arcuate to C-shaped, maximum width 25–29 μm, annuli 1.6–1.7 μm wide at midbody. Lateral field with four smooth lines extending to near tail tip, areolated only anteriorly at phasmids. Scutella postvulval, anterior one at 64–70% and posterior one at 84–92% of body length. Lip region hemispheric, offset, with four to five distinct annuli. Stylet well developed, cone 11–12.5 μm long or 45% (43–47) of stylet length; knobs rounded, compact 4.5–5 μm across. DGO 7–8 μm from stylet base. Median esophageal bulb oval, about 11 × 9 μm. Esophageal gland lobe about one body width long. Excretory pore near base of gland lobe. Hemizonid indistinct, one to two annuli anterior to excretory pore. Vulva with two epiptygmata which may or may not be projecting beyond the body surface. Spermathecae with sperm. Rectum shorter than anal body width. Tail hemispheric to conoid-rounded, with 7–15 ventral annuli, terminal annuli roughly about the same width as adjacent annuli but slightly offset from them.

Males: Generally similar to females. Body C-shaped. Anterior and posterior phasmids at 60–70% and 84–88% of body from anterior end, respectively.

Siddiqi (1986) proposed a new genus, *Nectopelta* (syn. *Aorolaimus*), for species like *A. annulatus* with both scutella are posterior to vulva as above, but with lateral fields always with four lateral lines, and always areolated near scutella. The lip region always has annuli; labial longitudinal striae may be present or absent. The males have caudal alae indented or not indented. *Nectopelta* was rejected by Fortuner (1987a).

The species that were placed in the genus *Nectopelta* have the same geographic distribution as those in *Peltamigratus* (*sensu* Siddiqi). *A. indicus* has been reported in Iraq and India.

3. *Aorolaimus helicus* Sher, 1963 (Fig. 9)

Measurements (after Sher, 1963). Females ($n = 20$): $L = 0.78$–0.98 mm; $a = 24$–33; $b = 7.9$–9.9; $b' = 6.2$–7.8; $c = 43$–71; $V = 54$–60%; stylet $= 24$–29 μm; $o = 16$–24. Males ($n = 10$): $L = 0.72$–0.84 mm; $a = 26$–36; $b = 6.2$–9.7; $b' = 4.7$–6.5; $c = 28$–36; stylet $= 23$–28 μm; spicules $= 27$–32 μm; gubernaculum $= 11$–15 μm; capitulum $= 10$–14 μm.

Description. Females: Body in spiral shape. Lip region rounded, slightly set off from body, with four to five annuli; basal annulus with longitudinal striae. Stylet knobs rounded to oval-shaped. Excretory pore at level of esophagointestinal valve. Hemizonid just anterior to excretory pore, two annuli long. Hemizonion and caudalid not seen. Anterior phasmid 30–40%, and posterior phasmid 80–86% of body length from anterior end. One or two epiptygmata. Spermatheca round, partially filled with sperms. Intestine slightly overlapping rectum. Tail rounded to broadly rounded, more curved dorsally, with 9–13 ventral annuli.

Males: Generally similar to females. Lip region more broadly rounded, more set off from body. Anterior phasmid at 30– 39%, and posterior phasmid at 78–84% from anterior end. Gubernaculum with inconspicuous titillae.

Aorolaimus helicus represents a group of species that constituted the genus *Aorolaimus* when it was first described by Sher (1963). *Aorolaimus s. str.* has lateral fields always with four lateral lines, and always areolated near scutella; lip region always with annuli; labial longitudinal striae may be present or absent. The only difference with the taxa traditionally placed in *Peltamigratus* is that one phasmid is anterior to the vulva. These species differ from the genus *Hoplolaimus*, also with one phasmid anterior to vulva, in that they have stylet and cephalic framework not massive, and stylet knobs not tulip-shaped. The males have caudal alae not indented.

Species in *Aorolaimus s. str.* have been found in the USA (Great Plains, Maryland, South Carolina), the Mediterranean (Spain, Morocco, Israel), and India (Maharashtra). Their biology and pathogenicity is unknown.

VII. *HOPLOLAIMUS* VON DADAY, 1905

A. Diagnosis

Females: Body straight, large (1–2 mm long). Lip region offset from body, wide, anteriorly flattened, with clearly marked annuli, and with longitudinal striae. Lateral field with four lines or less, generally areolated at level of phasmids and anteriorly, sometimes with striae irregularly scattered over entire field, rarely not areolated. Labial framework and stylet massive; stylet knobs anchor or tulip-shaped. DGO 3–10 μm from stylet base. Esophageal glands overlap intestine dorsally and laterally; sometimes gland nuclei duplicated to a total of six nuclei; intestine symmetrically arranged between the subventral glands. Two genital

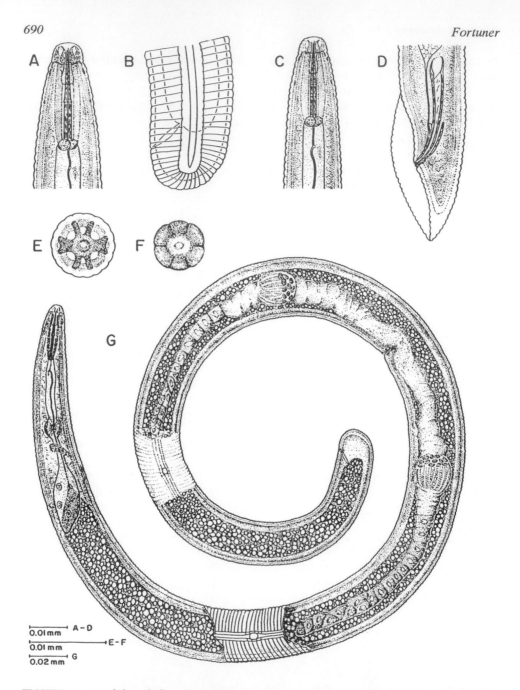

FIGURE 9   *Aorolaimus helicus* Sher, 1963. (A) Female, anterior end. (B) Female, posterior end.
(C) Male, anterior end. (D) Male, posterior end. (E) Female, cross-section through basal labial annulus. (F) Female, face view. (G) Female, entire body. (Sher, 1963, courtesy *Nematologica*.)

branches outstretched, equally developed. Tail short, rounded, phasmids enlarged to scutella erratically situated on body, anteriorly to anus level, and sometimes anterior to vulva level, not opposite each other.

Males: Caudal alae enveloping tail, regular. Secondary sexual dimorphism visible in labial region and esophageal structures smaller in males.

Type species: *Hoplolaimus tylenchiformis* von Daday, 1905.

Members of the genus *Hoplolaimus* are easy to recognize under the dissecting microscope. Under the compound microscope they are seen to have scutella present, erratic on body (not on tail or near anus), and esophageal glandular overlap mostly dorsal. There is a possible confusion with *Aorolaimus* (syn. *Peltamigratus*) that also have erratic scutella. Differences exist in the shape of anterior end, cephalic framework, and stylet knobs. *Hoplolaimus* is more robust, and larger than *Aorolaimus*.

B. Systematic Relationships

All the species in the genus *Hoplolaimus* share a number of apomorphies as identified from an hypothetical *Scutellonema*-like ancestor: body always large, robust, straight, stylet large, strong, with tulip-shaped knobs, each knob with toothlike projection, massive cephalic framework, labial area clearly demarcated from body, with well-marked annuli, basal labial annulus wide, giving an almost trapezoidal outline to the lip region.

Four groups of species can be defined within the genus (Fortuner, 1974) depending on number of gland nuclei (ancestral state: three nuclei; derived state: six nuclei); the number of lateral lines (ancestral four lines, derived less than four lines); the position of the excretory pore below (ancestral) or above (derived) the hemizonid; and the presence of either regular or irregular striae on the basal lip annulus.

A separate genus name, *Basirolaimus*, was proposed by Shamsi (1979) for the species that have extra gland nuclei, regression of the lateral field, migration of the hemizonid above the excretory pore, and diminution of the number of striae on the basal lip annulus. However, *H. clarissimus* has all the characters of *Hoplolaimus sensu stricto*, except that it has six gland nuclei, as in *Basirolaimus*. The group of species including *H. pararobustus* has all the characters of *Basirolaimus* except that it has only three gland nuclei. The migration of the excretory pore above the hemizonid has been observed in several taxa, widely separated from hoplolaimids, such as *Sychnotylenchus* (Anguinidae) and *Meloidogyne* (Heteroderidae).

While the characters used to define *Basirolaimus* are useful for practical identification, they do not exhibit a clear evolutionary pattern. The genus was rejected by Luc (1981) and Fortuner (1987a).

C. Representative Species

1. *Hoplolaimus galeatus* (Cobb, 1913) Thorne, 1935 (Fig. 10)

Measurements (after Sher, 1963). Females (n = 20): L = 1.24–1.94 mm; a = 25–34; b = 7.6–10.8; b' = 6–8.8; c = 42–82; V = 52–60%; stylet = 43–52 μm; o = 10–17; anterior phasmid at 30–46% on body length; posterior phasmid at 75–88% of body length. Males (n = 10): L = 1.05–1.56 mm; a = 25–32; b = 8.3–10.3; b' = 6–8.4; c = 28–40; stylet = 40–48 μm; o = 10–17; anterior phasmid at 29–45% of body; posterior phasmid at 75–89% of body length; gubernaculum = 20–28 μm; spicules = 40–52 μm; capitulum = 16–20 μm. After Thorne and Malek (1968): Females: L = 1.1–1.5 mm; a = 22–26; b = 7.4–8.6; c = 48–54; V = 55%. Males: L = 0.9–1.3 mm; a = 27–30; b = 6.5–7; c = 28–32; T = 42–50%. After Doucet (1980).

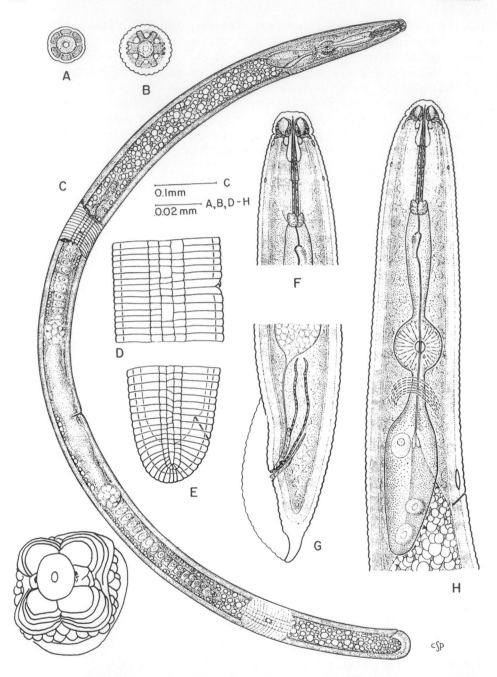

FIGURE 10   *Hoplolaimus galeatus* (Cobb, 1913) Thorne, 1935. (A) Female, face view. (B) Female, cross-section through the basal labial annulus. (C) Female, entire body. (D) Female, surface view at vulva. (E) Female, posterior end. (F) Male, anterior end. (G) Male, posterior end. (H) Female, anterior end. (Sher, 1961, courtesy *Nematologica*.)

Females (n = 10): L=1.4 mm, 0.07; a = 29.1, 1.42; b = 9.7, 0.59; b' = 6.9, 0.47; c = 55.1, 8.24; V = 0.84, stylet = 45.4 μm, 1.67; tail = 27.3 μm, 4.1. Males (n = 10): L = 1.2 mm, 0.07; a = 29.5, 2.53; b = 8.6, 0.89; b' = 6.5, 0.63; c = 33, 1.71; stylet = 43.2 μm, 0.57; tail = 38.2 μm, 3.03; spicules = 49 μm, 2.15; gubernaculum = 24.5 μm, 1.02; capitulum = 15.5 μm, 2.33.

Description. Females: Body slightly ventrally curved when relaxed. Cuticle consisting of six layers (Wen and Chen, 1972). Lateral field with four lines, usually areolated but occasionally only partly so. Cephalic region offset usually with five annuli, forming a low cone generally with flattened sides. All labial annuli divided into tilelike elements, and basal annulus with 32–36 regular longitudinal striations. In SEM face view, labial disk oval, well marked, first labial annulus divided into six sectors, the submedian sectors well developed, rounded, each with a papilla, the two lateral sectors visibly smaller. Amphidial apertures a long slit along the labial disk. Four labial annuli with a general quadrilobed shape, with two longitudinal lines in dorsal and ventral position. Basal annulus somewhat squarish, divided in many tilelike rounded parts, all more or less of the same size, most arranged regularly in a row, but occasionally in two rows (Hirschmann, 1983). Tail rounded, with 10–16 annuli. Cephalic framework massive. Spear knobs with anteriorly directed processes. Cephalic framework massive. Median bulb spheroid. Esophageal glands with three nuclei. Dorsal gland opening near spear base. Intestine overlapping rectum and usually extending into tail. Excretory pore near level of esophagointestinal valve. Hemizonid about two annuli long situated just anterior to excretory pore. Hemizonion about five annuli posterior to excretory pore. Caudalid about eight annuli anterior to anus. Ovaries outstretched, spermathecae round to oval. One or two epiptygmata, usually conspicuous.

Males: Body smaller and more slender than female. Cephalic region higher and less conoid than female, hemispheric with convex sides, without tiling or with striae only on basal annulus. Spicules slightly arcuate; the outer one with a distinct velum seen with SEM (Högger and Bird, 1974). Gubernaculum with two lateral titillae, bent distally in some specimens. The gubernaculum stays protruded out of the cloacal orifice, even when all the other male appendages are retracted. When this occurs, only the gubernaculum and its two lateral titillae are visible in a ventral SEM view (Högger and Bird, 1974). Capitulum present. Caudal alae broad, striated, enveloping tail.

The population described by Doucet (1980) has scutellum diameter 8–9 μm, always two epiptygmata, and intestine overlapping rectum but not continuing into tail.

Hoplolaimus galeatus is widely distributed in the USA. It has also been reported from Canada, Central and South America, and India. It has a large variety of hosts, particularly cotton, trees (pine, oak, sycamore, etc.), turf grasses, other graminaceous plants, etc. It lives as an endoparasite on cotton, causing considerable damage to cortex and vascular tissue. On pine, most of the cortex of infested roots was destroyed. In sycamore, *H. galeatus* causes extensive root necrosis but it is unable to penetrate completely within the roots and its body partly protrudes out of the root.

Hoplolaimus galeatus is representative of a group of *Hoplolaimus* species characterized by lateral field with four lines, and esophageal glands with three nuclei. This group includes *H. tylenchiformis*, *H. californicus*, *H. concaudajuvencus*, *H. galeatus*, and more.

2. *Hoplolaimus pararobustus* (Schuurmans Stekhoven and Teunissen, 1938) Coomans, 1963 (Fig. 11)

Measurements: Composite measurements from descriptions published in Andrássy (1961, *H. kittenbergeri*), Coomans (1963), Elmiligy (1980), Goodey (1957, *H. proporicus*), Maqbool and Ghazala (1988), Sher (1963), Suryawanshi (1971), van den Berg and Heyns (1970), Vovlas and Lamberti (1985), and Whitehead (1959, *H. angustalatus*).

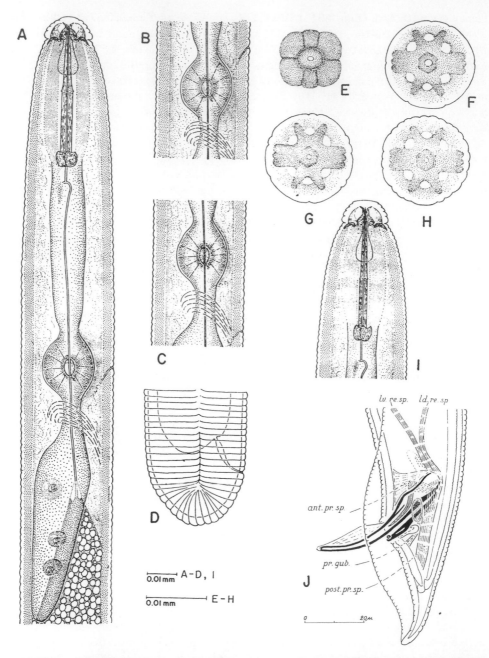

FIGURE 11 *Hoplolaimus pararobustus* (Schuurmans Stekhoven and Teunissen, 1938) Coomans, 1963. (A) Female, esophageal region. (B,C) Female, median esophageal bulb. (D) Female, posterior end. (E) Female, face view. (F–H) Female, cross-section through basal labial annulus. (I) Male, anterior end. (J) Male, posterior end; ant.pr.sp. = *anterior protractor spiculi*; post.pr.sp. = *posterior protractor spiculi*; lv. and ld. re.sp. = *lateroventral* and *laterodorsal retractor spiculi*; pr.gub. = *protractor gubernaculi*. (Sher, 1963; Coomans, 1963, courtesy *Nematologica.*)

Females ($n = 17$): $L = 1.314$ mm ± 0.127 (0.94–1.8); $a = 27.3 \pm 2.689$ (20–39); b ($n = 14$) $= 8.98 \pm 1.361$ (6–14.1); b' ($n = 7$) $= 7.1 \pm 0.682$ (5.1–10); $c = 60.8 \pm 15.185$ (40–164); c' ($n = 7$) $= 0.67 \pm 0.135$ (0.4–0.9); $V = 56.28\% \pm 2.6096$ (51–62); stylet $= 42.3$ μm ± 2.378 (37.5–49). Males ($n = 14$): $L = 1.158$ mm ± 0.102 (0.93–1.5); $a = 29.2 \pm 2.227$ (21–37.2); $b = 8.5 \pm 1.328$ (6.2–13.8); b' ($n = 7$) $= 6.6 \pm 0.570$ (5–8.7); $c = 36.9 \pm 6.605$ (22.2–51.9); c' ($n = 4$) $= 1.6 \pm 0.206$ (1.4–2.1); stylet ($n = 11$) $= 40.8$ μm ± 2.676 (35–46); spicules $= 46.9$ μm ± 3.318 (40–57), gubernaculum $= 21.9$ μm ± 2.4296 (15.4–31); capitulum ($n = 10$) $= 16.1$ μm ± 2.728 (10.3–20).

Description. Females: Body cylindrical, straight to slightly ventrally arcuate when heat-relaxed; cuticle annuli distinct, about 2 μm wide; subcuticular annuli about 1 μm wide. Lip region hemispheric, set off from body by a constriction, usually with four distinct annuli; sometimes with three on one side and four on the other; as many as six annuli may be present. Longitudinal striae on lip annuli variable in number, on basal annulus 7–25 striae may be seen in face views. SEM face views show a round/oval labial disk and six labial sectors; the lateral sectors are smaller than the submedian ones; longitudinal striae divide the basal annulus in irregular sections, triangular or rectangular in shape (Vovlas and Lamberti, 1985). Labial framework hexaradiate, heavily sclerotized; lateral sectors smaller than submedian ones. Lateral fields represented by an interruption of body annuli or by a single incisure toward the extremities or along the entire body length; or by two to three incomplete incisures especially in badly fixed specimens. Phasmids as enlarged scutella, scutellum opening diameter 5.3 μm (4.1–6), anterior one at $34\% \pm 5.725$ (22–52) and posterior one at $79.5\% \pm 2.305$ (58–89) of body length from anterior end; anterior one may be on right side and posterior one on the left, or vice versa. Excretory pore distinct, usually opposite the median esophageal bulb, but may be anterior or posterior to it. Hemizonid always posterior (sometimes by as much as 20 annuli) to the excretory pore, about two body annuli wide. Stylet cone about as long as shaft; basal knobs tulip-shaped, with two or more anterior projections. DGO 5.6 μm ± 1.377 (4–8) behind spear base. Esophagus typical of the genus; median bulb rounded, very muscular and with a distinct valvular apparatus in center; three uninucleate esophageal glands extending dorsally or dorsolaterally over the intestine; esophagointestinal junction indistinct. Intestine overlaps rectum. Vulva a transverse slit; one epiptygma anterior or posterior. Spermathecae spherical, axial, with sperms. Ovaries with oocytes in tandem. Eggs 78–96 $\pm$ 26–28 μm. Tail short, rounded, usually hemispheric and with 7–15 annuli.

Males: Common. Body ventrally arcuate. Lip region rounded, higher than in females, with three to five (usually four) annuli. Excretory pore near median esophageal bulb, 8–23 annuli in front of hemizonid. Stylet, esophagus, lateral field, and phasmids as in female. Spicules ventrally arcuate, slightly cephalated, with large distal flanges. Gubernaculum troughlike, with titillae. The gubernaculum is said to be protrusible through cloaca, and Coomans (1963) did draw *protractor gubernaculi* and *retractor gubernaculi* muscles, but actual outward movement of the gubernaculum has not been documented. In SEM observation of *H. galeatus* (Högger and Bird, 1974), the gubernaculum protrudes, but it is not protrusible. Bursa large, with crenate margins, enclosing the conoid tail.

Hoplolaimus pararobustus is found in Africa, mostly within the roots of banana, but it can also parasitize coffee, tea, sugarcane, palm trees, various tropical fruit trees, rice, yam, and other plants. It has been described from grass in Maharashtra, India (Suryawanshi, 1971) and various plants in Pakistan (Maqbool and Ghazala, 1988).

On banana, it feeds mostly endoparasitically, but it has occasionally been observed only partially embedded within the roots. On coffee, it feeds semiendoparasitically. Cortex penetration results in cavities and ruptured cells. Numerous irregular brown necrotic lesions develop on the roots of infested coffee plants (Vovlas and Lamberti, 1985).

In the group represented by *H. pararobustus*, the lateral field is degenerate and never shows the regular complement of four lines. The esophageal glands have three nuclei, as in typical *Hoplolaimus*. The group includes *H. pararobustus* and a few other species that have been synonymized by various authors to *H. pararobustus*. The species *H. capensis* that was listed by Siddiqi (1986) as a synonym of *H. pararobustus* is in fact a valid species with longer stylet and spicules.

3. *Hoplolaimus columbus* Sher, 1963 (Fig. 12)

Measurements (after Sher, 1963). Females ($n = 20$): $L = 1.26–1.80$ mm; $a = 30–38$; $b = 9.1–12.4$; $b' = 6.3–9.7$; $c = 39–57$; $V = 51–60\%$; stylet $= 40–48$ μm; $o = 9–13$; anterior phasmid $= 34–47\%$; posterior phasmid $= 80–90\%$. After Fassuliotis (1974). Males ($n = 8$): $L = 1.15–1.40$ mm; $a = 31.9$ (25.9–39.2); $b = 10.9$ (9.58–12.18); $c = 29.9$ (26.8–33.1); stylet $= 42$ μm (40.2–43.7); $o = 5.2$ (4.8–5.2); anterior phasmid $= 38\%$ (35.4–42.2); posterior phasmid $= 82\%$ (79.7–83.2); spicules $= 46.8$ μm (36.6–52.5); gubernaculum $= 21.3$ μm (19.5–23.2). *H. chambus* (after Jairajpuri and Baqri, 1973). Females ($n = 14$): $L = 1.40$ (1.24–1.62) mm; $a = 31$ (28–35); $b = 9.8$ (8.4–11); $b' = 8.1$ (7–9.1); $c = 58$ (52–67); $V = 55\%$ (52–56) $o = 12–14$. Males unknown.

Description. Females: Cephalic region offset, usually with three annuli but a fourth annulus is often seen on one side of the cephalic region due to the division of one of the regular annuli. The basal annulus of the lip region has 10–15 irregular longitudinal striae (six striae in *H. chambus*). Basal plate with six arms, the dorsal and ventral ones being tripartite. Spear knobs with two anterior projecting processes. Esophageal glands with six nuclei, one or two often indistinct. Excretory pore behind the level of the esophagointestinal valve. Hemizonid two to five annuli posterior to excretory pore, two annuli long (9–11 annuli in *H. chambus*). Hemizonion usually not seen. Anterior phasmid 29–47%, posterior phasmid at 79–90% of body length from anterior end. Sher (1963) reported one specimen with both phasmids posterior to vulva. Ovaries outstretched; spermatheca absent (indistinct?). Two epiptygmata. Intestine overlaps rectum, extending partly into tail. Tail rounded with 16–22 annuli from anus to tail end (13 annuli in *H. chambus*). In a population from Georgia, Bird and Högger (1974) observed a peglike projection in the anal region of half of the 22 specimens studied under SEM. The peg was 4 μm long and 1–2 μm in diameter. An orifice of about 0.2 μm in diameter was present at the tip of the peg. Each specimen with anal peg lacked a normal anal opening, and the vulva was covered with cuticle and had no definite orifice. Caudalid usually not seen. Lateral field represented by one indistinct incisure.

The cuticle consists of six layers in four zones: cortical, with three sublayers, medial, striated, and granulated. The granular layer is unique to this nematode and it may explain its resistance to osmotic stress and desiccation (Lewis and Huff, 1976).

Males: Extremely rare. Reported only once from a soybean field in South Carolina present in the ratio of one male for 60 females (Fassuliotis, 1974). Body similar to female except for secondary sexual dimorphism. Cephalic region with three to four annuli with seven to eight irregular longitudinal striae on basal lip annulus. Basal plate with six arms, the lateral ones are tripartite. Excretory pore anterior or posterior to esophagointestinal valve. Hemizonid two to eight annuli posterior to excretory pore and hemizonion 10 annuli posterior to hemizonid. No longitudinal lines present. Gubernaculum troughlike, with distinct titillae. Spicules slightly arcuate with a very thin velum observed only when spicules

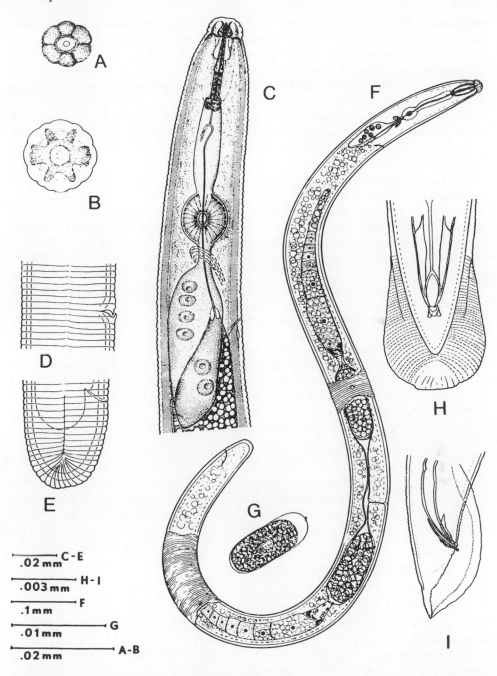

FIGURE 12 *Hoplolaimus columbus* Sher, 1963. (A) Female, face view. (B) Female, cross-section through basal labial annulus. (C) Female, esophageal region. (D) Female, surface view at vulval region. (E) Female, posterior end. (F) Female, entire body. (G) Egg in one-cell stage showing distal stalk. (H) Male, posterior end, ventral view. (I) Male, posterior end, lateral view. (Fassuliotis, 1976, courtesy Commonwealth Agricultural Bureaux.)

are extended. Capitulum distinct, lying between the spicules. Caudal alae begin at about the level of the anterior end of the spicules and extends around the tip of the tail.

Hoplolaimus columbus is known in Georgia, South Carolina, North Carolina. It has been reported in Pakistan. *H. chambus* has been described in India, and it has been reported in Vietnam (Eroshenko and Thanh, 1981). *H. columbus* is reported to be spreading in the USA, which would point to a recent introduction.

In the USA, *H. columbus* is an important parasite of soybean and cotton. It is an endoparasite on soybean roots, penetrating endodermis, pericycle, and phloem. On cotton, it is a semiendoparasite, and it penetrates the cortex but not the endodermis (Lewis et al., 1974).

Life cycle was investigated on alfalfa root and callus cultures by Fassuliotis (1975). Feeding was ectoparasitic on cortical cells in the maturation area of the root. High populations (more than 4000 per 100 cm^3 of soil) were found associated with the roots of dwarfed, chlorotic soybean plants. Few pods were produced. In infested cotton fields, the plants were dwarfed and with purple leaves, and the bolls were small or shedding. Very few other nematodes are found when medium to high populations of *H. columbus* are present (Fassuliotis et al., 1968).

Other crops, wheat, corn, millet (*Panicum miliaceum L.*), lima bean, common bean, watermelon, okra, and numerous weeds are excellent hosts of *H. columbus*. This makes carefully planned crop rotations and good control of weeds important factors for a successful control of the nematode (Stokes, 1977).

A number of species similar to *H. columbus* were grouped in the genus *Basirolaimus* because of degenerate lateral field with less than four lines and six esophageal gland nuclei. Siddiqi (1986) claims that this results from the presence of four nuclei in the dorsal gland instead on the usual single one, but he does not state the origin of this observation. Included in the genus *Basirolaimus* were *H. seinhorsti, H. columbus, H. indicus, H. puertoricensis,* and others, mostly from India. *Basirolaimus* is a synonym of *Hoplolaimus*.

4. *Hoplolaimus clarissimus* Fortuner, 1974

Hoplolaimus clarissimus is the only species in *Hoplolaimus s. l.* with six nuclei, as in *Basirolaimus*, but with a nondegenerated lateral field, with four well-marked lines as in *Hoplolaimus s. str.* This species has been found only in a limited area in Senegal. It has no agricultural importance and will not be described here.

VIII. *HELICOTYLENCHUS* STEINER, 1945

A. Diagnosis

Hoplolaiminae. Females: Body vermiform, spiral to straight. Labial region continuous to slightly offset, rounded or anteriorly flattened, generally annulated but never longitudinally striated; anterior lip annulus generally not divided into sectors, with elongate amphid apertures (SEM); rarely faint or well-marked lip sectors are present. Lateral field with four lines. Phasmids small, near anus; cephalids and caudalid present. Tail 1–2.5 body diameters long, typically more curved dorsally, with or without a terminal ventral process, sometimes rounded. Stylet and labial framework average-sized. DGO from 6 to 16 μm from stylet base. Median bulb rounded with average-sized valve. Glands overlap intestine dorsally and ventrally, all three glands of about the same length. Two genital branches, the posterior one sometimes degenerated or reduced to a PUS. Epiptygmata present but folded inward, into the vagina. Vulval flaps present, inconspicuous.

Males. Slight secondary sexual dimorphism seen in smaller anterior end. Caudal alae enveloping tail end.

Type species: *Helicotylenchus dihystera* (Cobb, 1893) Sher, 1961.

B. Systematic Relationships

Helicotylenchus is distinctive among hoplolaimids by the arrangement of its esophageal glands, overlapping the beginning of the intestine on all sides, and the position of the esophageal lumen, asymmetrically situated between the dorsal gland and one of the subventral glands. Most other hoplolaimids have a symmetrical arrangement, with the lumen between the two subventral glands (Seinhorst, 1971). The esophageal glands are sometimes described as fused together and forming a single structure around the intestine (Siddiqi, 1986), but this affirmation is not backed by any published observation of this region in cross-section.

Helicotylenchus most likely originated from ancestral forms close to *Pararotylenchus*, but the difference in glandular structure between *Helicotylenchus* (asymmetrical) and *Rotylenchus, Scutellonema, Aorolaimus,* and *Hoplolaimus* (symmetrical) is unexplained. It is not known whether *Helicotylenchus* and the other Hoplolaiminae are monophyletic.

The position of the dorsal esophageal gland opening is extremely variable in the subfamily. Its use, either for systematics or for identification, is very delicate. There is a definitive tendency among hoplolaimids for a posterior migration of the DGO. The opening, which is only 1 or 2 µm from the stylet in many tylenchs, is 3–10 µm in many hoplolaimids, up to 16 µm in *Helicotylenchus*, and up to 33 µm in *Rotylenchulus*.

The species in *Helicotylenchus* are identified within hoplolaimids by their peculiar glandular arrangement and by the fact that the DGO is often farther away from the stylet base than in other hoplolaimids. In addition, they are often smaller and slimmer, tail outline is often a quarter-round (instead of a half-round as in, e.g., *Rotylenchus*), and the body posture is often spiral. There are many exceptions to these latter characters and they can be best used to define some phenetic groupings within *Helicotylenchus*.

C. Representative Species

1. *Helicotylenchus dihystera* (Cobb, 1893) Sher, 1961 (Fig. 13)

Measurements (average of the mean values in 20 different populations, standard deviation, after Fortuner, 1987b). Females: $L = 652$ mm ± 44.4; stylet = 25 µm ± 0.7; esophagus length = 114 µm ± 7; esophageal glands length = 139 µm ± 7; DGO = 11.3 µm ± 1.5; distance to excretory pore = 107 µm ± 5; body diameter = 24.3 µm ± 1.9; tail length = 16.5 µm ± 1.8; tail diameter at anus = 14 µm ± 1; tail annuli = 9.6 annuli ± 1.4; ratios $a = 26.9 ± 1.8$; $b = 5.7 ± 0.3$; $b' = 4.7 ± 0.2$; $c = 40.8 ± 3.7$; $c' = 1.2 ± 0.1$; $m = 46 ± 2.5$; $V = 63.4\% ± 0.6$.

Description. Females: Body spiral-shaped, spiral very variable, from loose one-turn spiral in the posterior half with anterior half almost straight to tight spiral of almost three turns. Lip region hemispheroid, outline hemispheric, or with anterior end very slightly flattened four to five labial annuli, more or less distinct. In SEM face view, labial disk oval, first lip annulus not divided into sectors, amphid apertures two longitudinal slits clearly marked at the limit of the labial disk. Stylet knobs anteriorly flattened to indented. Hemizonid always anterior to excretory pore, pore always anterior to esophagointestinal junction. Intestinal fasciculi (canals) absent. Lateral field areolated at the esophagus level; rarely some transverse striae scattered on rest of body. Inner lines of lateral field fusion at posterior end

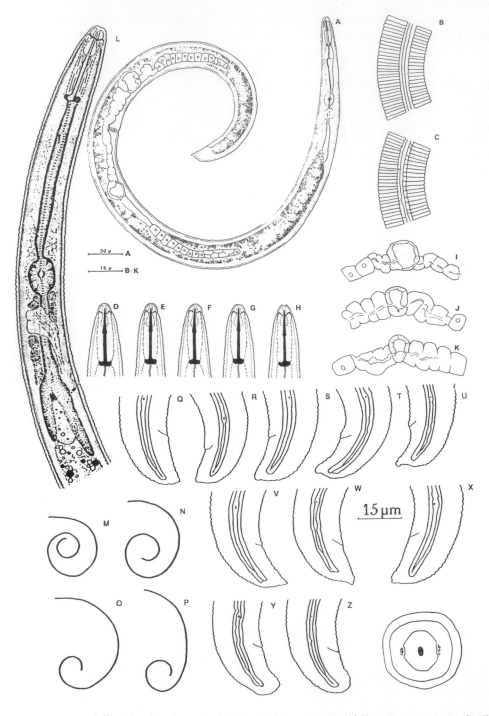

FIGURE 13 *Helicotylenchus dihystera* (Cobb, 1893) Sher, 1961. (A) Female, entire body. (B,C) Lateral fields. (D–H) Anterior ends. (I–K) Position of spermatheca in genital tract. (L) Esophageal region. (M–P) Body thermal death posture. (Q–Z) Tail ends. (Fortuner et al., 1984 and Fortuner, 1987b, courtesy *Rev. Nématol.* and Editions ORSTOM.)

in a Y or V pattern, the length of the leg of the Y varies from 0 to 5 μm. Phasmids 3–13 annuli anterior to anus, situated in the center of the lateral field or closer to one or the other of the inner lines. Tail more curved dorsally, with or without a short ventral terminal projection. Genital system with two branches, both functional, but posterior branch often slightly smaller than anterior one. Spermatheca offset, empty.

Males absent.

Helicotylenchus dihystera is a cosmopolitan species with a very large host list. It is an ecto- or semiendoparasite on the roots of many plants. Feeding behavior has been studied on wheat by Jones (1978a). Typically several cells are penetrated before feeding begins. The nematodes are partially or fully embedded in the root where they feed from a single cell during several days. Feeding resulted in cortical lesions of the roots (Jones, 1978b). Damage to crops was observed in the greenhouse on olive seedlings in Egypt where *H. dihystera* proved to be more damaging than *Xiphinema elongatum* and *Meloidogyne javanica*, with 78% reduction in top weight, and greatly reduced root system (Diab and El-Eraki, 1968). *H. dihystera* was found associated with unhealthy turf bowling greens in Australia (Wallace, 1971). In Nebraska, *H. dihystera* was one of the predominant nematode species found in unhealthy lawns, but attempts to demonstrate pathogenicity in vitro were unsuccessful (Sumner, 1967).

There are many species similar to *H. dihystera* in the genus *Helicotylenchus*, with body spiral, tail more curved dorsally (outline shaped as a quarter-round) with small or large ventral projection. In some species the tail is almost conoid pointed (e.g., *H. caudatus, H. craigi, H. issykkulensis, H. lissocaudatus, H. spicaudatus, H. thornei,* etc.). In *H. martini,* the tail is quite long for the genus, over two diameters long, and it looks like the tail of some *Tylenchorhynchus* species. In other species, the tail is indented dorsally (e.g., *H. crenacauda, H. curvicaudatus, H. digitatus, H. digitiformis, H. pteracercus,* etc.).

2. *Helicotylenchus africanus* Micoletzky, 1916 (Fig. 14)

Measurements (after Fortuner, Maggenti, and Whittaker, 1984). Females ($n = 6$): $L = 0.856 \pm 70$); stylet = 29.9 μm ± 1.2; esophagus = 127 μm ± 7; esophageal glands = 174 μm ± 14; dorsal gland opening = 9.5 μm ± 1.9; excretory pore = 116 μm ± 4; body diameter = 24.3 μm ± 3.5; tail length = 30.9 μm ± 4.3; anal body diameter = 15.6 μm ± 1.3. Ratios: $a = 35.8 \pm 4.3$; $c = 28.2 \pm 3.3$; $c' = 1.98 \pm 2.5$; $m = 44 \pm 0.6$; $V = 59.0\% \pm 1.42$. Males (after Sher, 1966): ($n = 4$): $L = 0.78$–0.84 mm; $a = 34$–38; $b = 4.8$–5.6; $c = 34$–40; stylet = 25–28 μm; $m = 46$–49; $o = 37$–41; spicules = 21–25 μm; gubernaculum = 8–9 μm. After van den Berg and Heyns, 1975. Females ($n = 43$): $L = 0.8$ mm (0.7–1); $a = 36.5$ (29.5–42); $b = 7.4$ (6.2–8); $b' = 5.2$ (4.6–7.3); $c = 38.6$ (32.6–43.5); $c' = 1.6$ (1.4–1.8); $o = 36.6$ (31–37.7); $V = 60\%$ (56–64); stylet = 27.6 μm (26.1–28.7). Males ($n = 3$): $L = 0.8$ mm (0.8–0.9); $a = 35.4$ (31.4–38.1); $b = 6.7$ (6.2–8.5); $b' = 4.7$ (4.4–5.3); $c = 40$ (38.8–41); stylet = 23.6 μm (23.5–23.9); spicules = 25 μm; gubernaculum = 9.2 μm (8.5–9.9).

Description. Females: Body posture C-shaped, annuli 1.5–2 μm wide at midbody. Lip region hemispheric not or slightly set off, with four or five well-marked annuli; labial disk not visible in lateral view. Basal ring of cephalic framework 2 μm deep. Anterior cephalid not seen, posterior cephalid 11–16 μm from the anterior end. Stylet knobs variable in shape, anteriorly indented, flattened or rounded, 5.9 μm (4.8–6.6) across and 2.4 μm (2.2–2.9) wide. Excretory pore anterior to the esophagointestinal junction. Hemizonid just anterior to the excretory pore; hemizonion indistinct, 9–10 annuli posterior to hemizonid. Fasciculi absent. Spermatheca apparently in line with the genital tract, full of rounded sperms. Epiptygmata not seen in most specimens, but two epiptygmata rarely present. Lat-

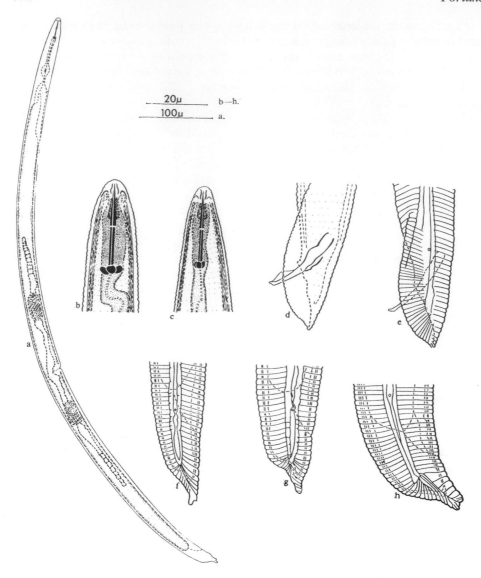

FIGURE 14 *Helicotylenchus africanus* Micoletzky, 1916. (a) Female, entire body. (b) Female, anterior end. (c) Male, anterior end. (d, e) Male, posterior end. (f–h) Female, posterior end. (van den Berg and Heyns, 1975, courtesy *Phytophylactica*.)

eral field 4.5–6.5 μm wide, with scattered transverse striae on body and tail; longitudinal inner lines join together on the tail in a U-shaped pattern, in a V-shaped pattern in one specimen. Phasmids from one annulus posterior to eight anterior to anus level, distinct, and in the center of the lateral field. Caudalid three to seven annuli anterior to anus. Tail with 9–18 ventral annuli, about two body diameters long, with a short nonannulated ventral section, and with dorsal terminal annuli smaller than other tail annuli. Tail dorsally curved with a rounded terminal projection, 2–4 μm long, annulated, rarely the projection is pointed.

Males present, similar to females except for sexual dimorphism.

H. africanus is known from southern Africa. This and some other species in the genus are similar to *H. dihystera*, mostly by the tail shape, but they have body weakly curved, C-shaped.

3.   *Helicotylenchus multicinctus* (Cobb, 1893) Golden, 1956 (Fig. 15)

Composite description from Elmiligy (1970), Goodey (1940), Das (1960), Ratanaprapa and Boonduang (1975), Sher (1966), Sauer and Winoto (1975), van den Berg and Heyns (1975), van den Berg and Kirby (1979), and Vovlas (1984).

Measurements. Females ($n = 12$): $L = 0.546$ mm ± 0.054 (0.393–0.710); $a = 26.9 \pm 2.077$ (18.5–35); $b = 4.95 \pm 0.650$ (3.7–6.4); $b' (n = 9) = 4.2 \pm 0.293$ (3.4–5); $c = 48 \pm 6.273$

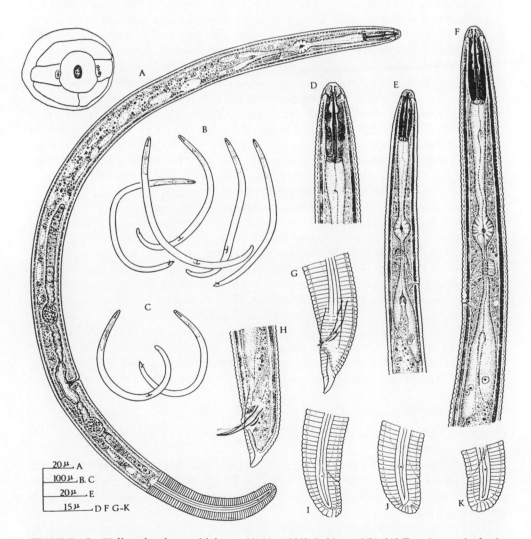

FIGURE 15   *Helicotylenchus multicinctus* (Cobb, 1893) Golden, 1956. (A) Females, entire body. (B,C) Males and females, thermal death posture. (D) Female, anterior end. (E) Female, esophageal region. (F) Male, esophageal region, (G,H) Male, tail ends. (I–K) Females posterior ends. (Siddiqi, 1973 and Vovlas, 1984, courtesy Commonwealth Agricultural Bureaux and *Rev. Nématol.*)

(31–70); c' (n = 9) = 1 ± 0.128 (0.7–1.5); V = 67.6% ± 1.622 (61–75.6); stylet = 23.6 μm ± 1.263 (19.8–28). Males (n = 11): L = 0.5 mm ± 0.065 (370–736); a = 30 ± 2.939 (22.5–40); b = 4.8 ± 0.543 (3.6–6.3); b' (n = 8) = 4.1 ± 0.423 (3.1–3.8); c = 33 ± 3.044 (25–41); stylet (n = 9) = 21.6 μm ± 1.277 (18–24); spicules = 18.4 μm ± 1.975 (15–26); gubernaculum = 5.8 μm ± 0.944 (4–8.5).

Description. Females: Body arcuate to C-shaped when relaxed; annuli distinct, about 1.5 μm wide at midbody; lateral fields not areolated, with four incisures, about one-fourth of the body width. Lip region hemispheric, slightly offset, with three to five (usually four) annuli; labial disk slightly raised above the first labial annulus. In SEM face view, labial disk rounded; first labial annulus with six sectors; submedian sectors low; lateral sectors rectangular, smaller than submedian sectors; amphid openings ovoid (Vovlas, 1985). Labial framework heavily sclerotized, with basal ring conspicuous extending posteriorly along three to four body annuli, which are much narrower at that region than on the rest of the body. Anterior and posterior cephalids usually indistinct, zero to one and four to six annuli posterior to cephalic framework respectively. Stylet well developed, with prominent basal knobs measuring 5–6 μm across, anteriorly flattened or concave. Procorpus usually expanded anteriorly; median esophageal bulb round to oval with small valvular apparatus in center; about six body annuli long; esophageal glands compact, wrapped round front end of intestine, dorsal gland anterior to subventral glands. Excretory pore level with or close to esophagointestinal junction. Hemizonid usually distinct, two to three annuli long, zero to three annuli anterior to excretory pore; hemizonion minute, six to eight annuli behind excretory pore. Posterior genital branch sometimes reduced. Spermathecae slightly offset, rounded, usually filled with sperms. Vulva prominent, a depressed transverse slit. Epiptygma not observed. Intestine not overlapping rectum. Tail slightly tapering, with a hemispheric annulated terminus, usually more curved dorsally than ventrally, devoid of any ventral projection or mucro (but ventral process present in second-stage larvae; Zuckerman and Strich-Harari, 1964), with 6–13 ventral annuli; inner lateral fields lines fusion U-or V-shaped on tail. Phasmids porelike, one to six annuli anterior to anus level.

Males: Similar to females except for sexual dimorphism. Body less curved than in females. Sperms small, rounded. Bursa short, not conspicuously projecting beyond body contour in lateral view, crenate and enclosing tail. Spicules cephalated with narrowed distal half bearing small ventral flanges; gubernaculum simple.

Helicotylenchus multicinctus is an important parasite of banana in all the banana-growing regions of the world. It is an endoparasite in the cortex of the roots where it feeds and produces small superficial lesions. It represents a group of species with body C-shaped, tail conoid/rounded, and vulva more posterior than usual in the genus *Helicotylenchus*.

4. *Helicotylenchus vulgaris* Yuen, 1964 (Fig. 16)

Composite description after Yuen (1964, 1966), d'Errico (1970), Mancini and Moretti (1977), and Ivan (1978).

Measurements. Females (n = 5): L = 0.935 mm ± 0.083 (0.706–1.180); a = 28.7 ± 1.662 (23.8–34); b = 7.3 ± 0.712 (5.8– 9.1); b' = 5.2 ± 0.189 (4.1–6.6); c = 79.4 ± 7.028 (52.5–115.5); c' (n = 2) = 0.7 (0.5–0.9); V = 59.8% ± 0.924 (55–65); stylet = 32.3 μm ± 0.609 (30–36).

Description. Females: Body slender C-shaped or coiled in 1–1/2 spirals. Anterior end continuous with body contour, anteriorly flattened, with four or five annuli. Cuticle annuli approximately 1.7 μm wide. Lateral field occupies about one-quarter body width at region of vulva, anteriorly areolated. Stylet strongly developed with cone and shaft approxi-

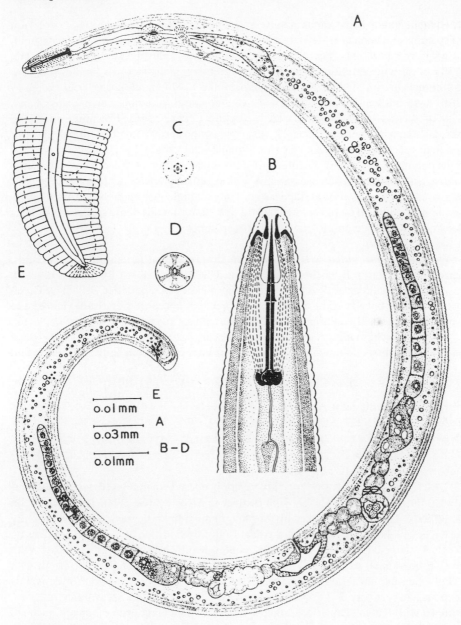

FIGURE 16 *Helicotylenchus vulgaris* Yuen, 1964. (A) Female, entire body. (B) Anterior end. (C) Face view. (D) Cross-section through basal lip annulus. (E) Tail. (Yuen, 1964, courtesy *Nematologica*.)

mately equal in length. Basal knobs massive and slightly concave anteriorly. DGO 9–12 μm behind spear base. Esophageal-intestinal junction close behind nerve ring and always anterior to excretory pore. The nerve ring and excretory pore are located at 104–120 and 120–138 μm from the anterior end, respectively. Hemizonid distinct two annuli wide, zero to one annulus anterior to excretory pore. Hemizonion 11–17 annuli posterior to excretory pore. Tail 8–15 μm long, rounded, with 4–10 ventral annuli. Annuli narrower at distal end. Tail end of second-stage juveniles with projection (Yuen, 1966). Phasmids small and porelike, 6–18 annuli anterior to anus. Tip of anterior or posterior ovary occasionally flexed. Vulva a transverse slit opening into the short, tubular, thick-walled vagina. Uterus consisting of the columned uterus and the thin-walled muscular portion between columned uterus and vagina. Spermatheca offset, occasionally with spherical granules. Oocytes arranged in a single row except at region of multiplication. Only a single egg matures at a time and is retained in the columned uterus before it is laid. Intestine contains dense spherical granules and does not overlap rectum. Rectal glands conspicuous.

    Males unknown.

    Helicotylenchus vulgaris is known from Europe (Great Britain, the Netherlands, Germany, France, Italy). It has been observed in USSR (Moldavia), Romania, Bulgaria, South Africa, and California. Possible damage to pea has been reported in Britain by Green and Dennis (1981), but according to Spaull (1982), it does not cause significant yield loss to sugar beet under normal conditions in the same country.

    Helicotylenchus vulgaris and related species are easily confused with *Scutellonema* species under the dissecting microscope because of their body spiral and hemispheric tail.

5. *Helicotylenchus coomansi* Sharafat-Ali and Loof, 1975 (Fig. 17)

Measurements (after Sharafat-Ali and Loof, 1975). Females ($n = 4$): $L = 1.17–1.30$ mm; $a = 35–40$; $b = 6.5–7.3$; $b' = 5.4–5.5$; $c = 43–67$; $c' = 0.7–1.1$; $V = 58–60$; stylet = $39–42$ μm; $m = 48–50$; $o = 16–19$. Males ($n = 2$): $L = 1.19–1.26$ mm; $a = 39–41$; $b = 7.0–8.0$; $b' = 5.5–6.4$; $c = 33–35$; $c' = 2.0$; $T = 46–51$; stylet = $35–36$ μm; $m = 51–52$; $o = 16–20$; spicules = $34–36$ μm; gubernaculum = $10–12$ μm.

    Description. Females: Body weakly ventrally curved when relaxed; tapering gradually toward both extremities. Cuticle with distinct transverse striae about 1.8 μm apart on middle of body. Lateral field marked by four longitudinal lines, the two outer ones slightly crenate except on the tail of the female; areolated anteriorly. Fusion of the two inner lines on tail mostly V-shaped. Lip region continuous, conoid-hemispheric to slightly truncate, with four to five fine annuli sometimes indistinct. Labial framework strongly sclerotized, with basal ring extending posteriorly over five to six annuli. Anterior and posterior cephalids conspicuous, respectively 5–7 and 12–15 annuli behind lip region. Spear guide massive. Spear knobs with flattened anterior surfaces. Excretory pore anterior to esophagointestinal junction. Hemizonid conspicuous, two to three annuli long, two to three annuli anterior to excretory pore. Vaginal walls thickened. Spermatheca axial, offset, oval to elongate, filled with sperm. Tail straight, tapering ventrally, almost straight dorsally, terminus irregularly hemispheric with a subventral unstriated area more than three annuli wide; the distal subdorsal annuli generally narrower than other annuli. Tail with hyaline distal area 7–13 μm wide. Ten to thirteen tail annuli. Phasmids prominent, five to eight annuli anterior to anus.

    Males: Body almost straight when relaxed. Lip region broadly hemispheric, with five annuli. Tail tapering to a subacute terminus, somewhat ventrally offset. Spicules almost straight. Gubernaculum simple, slightly sinuate. Phasmids anterior to cloacal aperture.

    Helicotylenchus coomansi was described from the Netherlands.

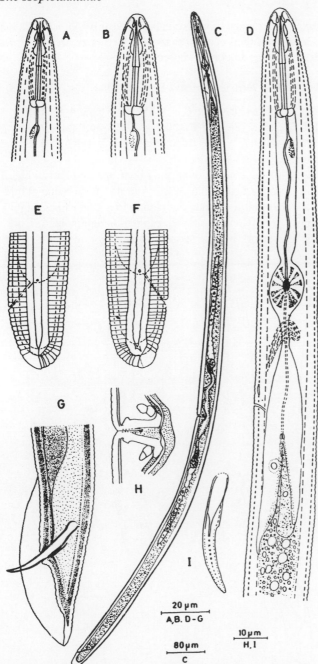

FIGURE 17   *Helicotylenchus coomansi* Sharafat-Ali and Loof, 1975. (A) Male, anterior end. (B) Female, anterior end. (C) Female, entire body. (D) Female, esophageal region. (E,F) Female, tail end. (G) Male, tail end. (H) Female, vaginal structure. (I) Male, spicule. (Sharafat-Ali and Loof, 1975, courtesy *Nematologica*.)

Helicotylenchus coomansi and related taxa have body weakly curved, almost straight, and hemispheric tail. Because of their general shape and large size, these species may be thought to be *Hoplolaimus* species under the dissecting microscope.

6.   *Helicotylenchus intermedius* (Luc, 1960) Siddiqi and Husain, 1964 (Fig. 18)

Measurements (after Luc, 1960). Females (n = 6): L = 0.454 mm (0.394–0.523); a = 22.4–235.5; b = 3.9–4.7; c = 41.4–59.7; V = 78–83.2%; stylet = 26–27 μm. Males (n = 3): L = 0.372 mm (0.343–0.387); a = 25.4–29.7; b = 3.7–4.56; c = 25.7–32.2; stylet = 20.21 μm; spicules = 14.5–15 μm; gubernaculum = 4.5–5 μm.

Description. Females: Body almost straight or slightly bent, marked narrowing of the body at vulva level, narrowing less marked at anus level. Annuli 1.6 μm wide at mid-body. Lateral field with four lines, one-fourth of body diameter wide. Outer lines slightly crenated, anteriorly areolated. Inner lines fusion U-shaped. Lip region hemispheroid, non-offset, with four annuli. Cephalic framework well developed, basal ring three annuli deep.

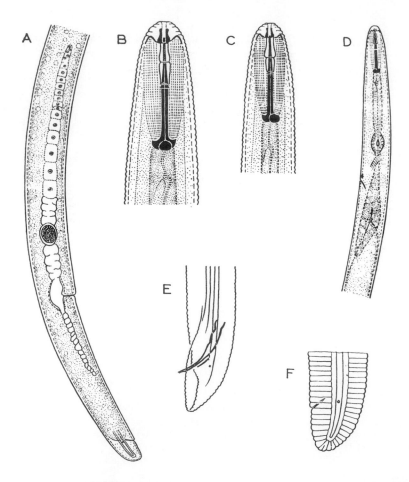

FIGURE 18   *Helicotylenchus intermedius* (Luc, 1960) Siddiqi and Husain, 1964. (A) Female, tractus genital and posterior end. (B) Female, anterior end. (C) Male, anterior end. (D) Male, esophageal region. (E) Male, posterior end. (F) Female, posterior end. (Luc, 1960, courtesy *Nematologica*.)

Stylet cone slightly shorted than shaft, basal knobs anteriorly indented. DGO at 7–9 μm from stylet base. Median bulb oval. Esophageal glands with three nuclei, overlapping the intestine ventrally or lateroventrally. Nerve ring immediately posterior to median bulb. Excretory pore at 78–98 μm from anterior end. Hemizonid just anterior to excretory pore, two annuli wide. Vulva posterior. Anterior genital branch 130 μm (111–162) long; oocytes in one row except in the multiplication area, spermatheca 10 μm in diameter, with thick wall, full of spherical sperms about 1 μm in diameter. Posterior genital branch regressed, 41–53 μm long, appearing as a row of small, degenerated cells. Tail 8–12 μm long, 0.8–1 μm wide at anus level, more rounded dorsally, with rounded end. Eight to ten tail annuli. Phasmids very small, at anus level.

Males: Similar to females. Stylet shorter than in females. Excretory pore at 67–69 μm from anterior end. Spicules 14.5–15 μm long, very slightly curved, cephalated. Small gubernaculum, slightly curved. Caudal alae enveloping tail end. Phasmids anterior to anus level.

Helicotylenchus intermedius is found in virgin forests in Ivory Coast. It was observed in the virgin forest of Taï. After the land was deforested for cultivation, this nematode disappeared from subsequent sampling (Fortuner and Couturier, 1983).

Helicotylenchus intermedius has vulva very posterior and posterior genital branch degenerate but still present. It is considered to belong to the genus *Rotylenchoides* by some authors (Luc, 1960; Sher, 1966, Siddiqi, 1986) and to the genus *Helicotylenchus* by others (Siddiqi and Husain, 1964; Fortuner, 1984).

7. *Helicotylenchus brevis* (Whitehead, 1958) Fortuner, 1984 (Fig. 19)

Measurements (after Whitehead, 1958). Females (n = 6): L = 0.43–0.53 mm; a = 18–22; b = 3.6–4.4; c = 36–55; V = 89.7–92.1%; stylet = 26–29 μm. Males (n = 11): L = 0.37–0.44 mm; a = 24–30; b = 3.4–4.3; c = 23–36; spicule = 21–25 μm; gubernaculum = 4–8 μm; capitulum = 11 μm.

Description. Females: Anterior end flattened, continuous with body, with three to four annuli. Lateral field with four lines, anteriorly areolated. Cephalic framework strongly developed; basal ring extending posteriorly over three body annuli. Stylet stout and about three head widths long. Stylet shaft slightly shorter than cone. Stylet knobs anteriorly flattened. DGO slightly less than one-third of stylet length behind stylet base. Precorpus long; valvular median bulb ovate; isthmus short; esophageal glands in a lobe of varying shape overlapping the anterior portion of the intestine. Esophagointestinal junction posterior to nerve ring. Nuclei of esophageal glands not observed. Excretory pore opposite middle of glandular lobe, hemizonid immediately anterior to excretory pore. Ovary short with oocytes in two rows. Spermatheca rounded. Postvulval uterine sac present, about one vulval body width long. Phasmids small at about anal level. Tail short, less than one anal body diameter long, rounded.

Males: Testis with double row of spermatocytes passing into a longer vas deferens, full of spermatozoa. Walls of vas deferens are composed of large squamous cells. Narrow ejaculatory duct passes between spicules to cloaca. Caudal alae arise just anterior to spicule heads and surround tail tip. Spicules are slightly cephalated, with strongly developed blades. Gubernaculum very thin. A capitulum was observed between the spicules in a single specimen.

Helicotylenchus brevis was found on cultivated plants (banana, mango) and uncultivated bush, ferns, and bulbous plants in southern Africa (Tanzania and South Africa).

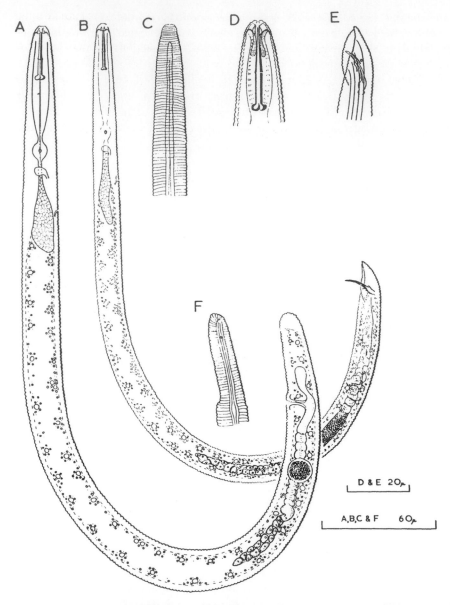

FIGURE 19 *Helicotylenchus brevis* (Whitehead, 1958) Fortuner, 1984. (A) Female, entire body.
(B) Male, entire body. (C) Female, anterior end showing origin of lateral fields. (D) Female, anterior
end. (E) Male, tail end. (F) Female, posterior end. (Whitehead, 1958, courtesy *Nematologica*.)

This species was the type species of the genus *Rotylenchoides*. It is similar to the species in the group *H. intermedius*, but here the regression of the posterior genital branch is complete, and it is represented only by a short postuterine sac. Fortuner (1984; 1987a) synonymized *Rotylenchoides* to *Helicotylenchus*.

IX. *ANTARCTYLUS* SHER, 1973

Hoplolaiminae. Females: Body vermiform, spiral to C-shaped. Labial region rounded, continuous, annulated; anterior lip annulus not divided into sectors (SEM, Sher and Bell, 1975). Lateral field with four lines. Phasmids small, near anus. Caudalids not described. Tail rather long (two to three body diameters long), conoid, pointed. Stylet and cephalic framework average-sized. DGO about 10 μm from stylet base in the only species known in this genus. Median bulb oval/rounded with average sized valve. Glands overlapping the intestine; the dorsal gland and one subventral gland overlap dorsally, the second subventral gland overlaps ventrally for a short distance. Both subventral glands are shorter than the dorsal one. Esophagointestinal junction a small triangular structure. Two equally developed genital branches. One or two epiptygmata present, inconspicuous. Vulval flap not described.

Males: Slight secondary sexual dimorphism seen in smaller anterior end. Tail with long hyaline end. Caudal alae said to envelop tail end, but seen stopping just short of tail tip in original figure. Gubernaculum not described; titillae not figured.

The type and only species, *A. humus* Sher, 1973 (Fig. 20), was found in forest peat soil in Auckland Island, Antarctica. Its pointed tail, rather long for a hoplolaimid ($c' = 2.5$), similar to some taxa in Telotylenchinae such as *Triversus*, and the arrangement of the esophageal glands, somewhat similar to that in *Helicotylenchus*, may indicate that this genus is a relic of the forms that evolved into present day *Helicotylenchus*.

X. *APHASMATYLENCHUS* SHER, 1965

A. Diagnosis

Hoplolaiminae. Females: Body vermiform, circle to C-shaped. Labial region slightly offset from body, annulated, but without longitudinal striae. First labial annulus divided into six equal sectors, elongate amphid apertures (SEM). Lateral field with four lines. Phasmids absent. Cephalids and caudalids not described. Tail 1.5–2 body diameters long, more curved dorsally, with rounded end. Stylet and labial framework well developed. DGO about 8 μm from stylet base. Median bulb rounded. Gland overlapping the intestine ventrally and laterally. Two genital branches equally developed. Epiptygmata and vulval flap not described.

Males: Slight secondary sexual dimorphism seen in the smaller anterior end. Tail conoid, elongate, with a hyaline end. Caudal alae enveloping tail end. Gubernaculum with titillae, but no capitulum.

Type species: *A. nigeriensis* Sher, 1965

B. Systematic Relationships

In this genus, the phasmids have disappeared, which is peculiar in a family where many species have over enlarged phasmids developed into scutella. The glands overlap is lateral, not like any other in Hoplolaiminae. The tail is often conoid rather than cylindroid, and it is rather long for the group. The male tail of *A. straturatus* has a definite tylenchorhynchid

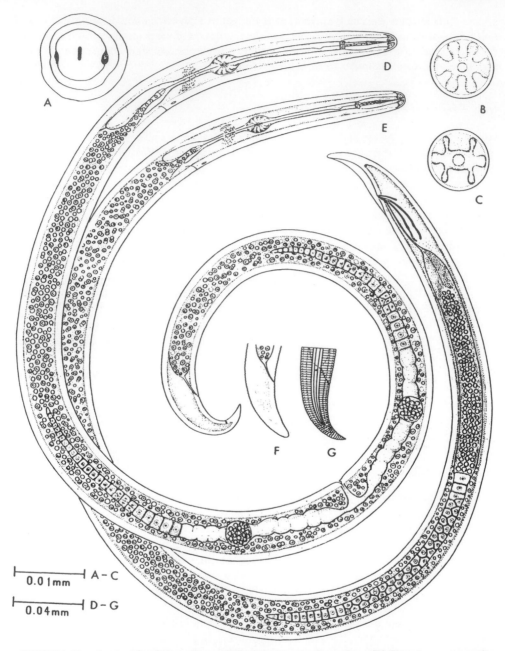

0.01mm ⊢ A–C

0.04mm ⊢ D–G

FIGURE 20 *Antarctylus humus* Sher, 1973. (A) Female, face view. (B) Female, cross-section through basal annulus of lip region. (C) Male, cross section through basal annulus of lip region. (D) Female. (E) Male. (F,G) Female, posterior ends. (Sher, 1973, courtesy *J. Nematol.*)

allure. The face pattern is ancestral and offers no clue to the systematic position of this genus. It is placed in Hoplolaiminae because of posterior DGO; junction esophagointestinal faint, without definite valve; sexual dimorphism (smaller anterior end in males); high lip region with well-developed cephalic framework.

This genus is known only from West Africa. *A. nigeriensis* was found in two localities in Nigeria, and a rain forest in Ivory Coast (Fortuner and Couturier, 1983), *A. variabilis* from Senegal and Mali, and *A. straturatus* is known only from southwest Burkina Faso.

C. Representative Species

Aphasmatylenchus straturatus Germani, 1970 (Fig. 21)

Measurements after Germani (1970). Females ($n = 36$): $L = 1.43$ mm (1.18–1.75); $a = 31.1$ (26–35.6); $b' = 7.8$ (6–9.1); $c = 22.4$ (18–27.5); $V = 52.9\%$ (50–56); stylet = 33 μm (30–36). Males ($n = 6$): $L = 1.13$ mm (0.99–1.26); $a = 34.8$ (29.1–38); $b' = 6.8$ (6.1–7.2); $c = 16.4$ (14.6–17.4); $T = 28.7\%$ (28–29.4); stylet = 29 μm (28–30).

Other population in Germani (1970): Females ($n = 36$): $L = 1.6$ mm (1.2–1.9); $a = 33.9$ (27.3–44); $b = 8.1$ (6.5–10); $c = 24$ (18.4–32.5); $c' = 1.7$(1.3–2.3); $V = 54\%$ (50.7–58.7); stylet = 35 μm (32–39). Males ($n = 12$): $L = 1.3$ mm (1.1–1.5); $a = 37.9$ (34–47.2); $b' = 6.7$ (5.7–7.6); $c = 19.8$ (16.5–30); $T = 24.9\%$ (19–32); stylet = 30 μm (25–33).

Description. Females: Body curved ventrally in an open C, sometimes a closed circle shape. Annuli 1.4–1.8 μm wide at center of body. Cuticle annuli divided by fine longitudinal striae giving it a corncoblike aspect. Lateral field occupying about one-fourth body width with four lines (three longitudinal bands), outer bands regularly areolated, central band irregularly areolated. Lip region slightly offset from body contour, generally rounded, with slight flattening on top; 8–11 labial annuli; the first annulus with the labial disk is conspicuously larger. In face view, with SEM, large oval oral disk present, but amphidial openings not observed; the first annulus is divided into six sectors, the two lateral sectors smaller than the submedian sectors. Cephalic framework massive (reminiscent of *Dolichodorus*). Esophagus: procorpus sometimes narrowing slightly at posterior end; dorsal esophageal gland duct opening 5–9 μm behind spear base; esophageal glands overlapping the intestine ventrally or lateroventrally. Postanal intestinal sac present. Fasciculi (canals) present over the entire length of the intestine including the postanal sac. Excretory pore at 138–208 μm from anterior end. Hemizonid one to three annuli anterior to excretory pore, two to three annuli wide. Spermatheca almost rectangular, containing spermlike bodies. Tail cylindroconical, tapering to a round or bluntly rounded terminus, 41–06 μm long or 1.3–2.5 times anal body width.

Males: Lip region more rounded than in female, distinctly set off. In SEM face view, well-developed, subhexagonal oral disk present. Amphidial openings narrow slits. Stylet and esophagus less well developed than in female. Only one esophageal gland nucleus observed. Excretory pore located 130–206 μm from anterior end. Testis single, outstretched. Spicules curved, 37–57 μm long; gubernaculum 14–21 μm long, with titillae. Caudal alae enveloping tail, with annulated edges. Tail 3–3.3 times body width at cloaca.

This species is known only from Niangoloko, Burkina Faso, where it is a major parasite of peanut and pigeon pea. The nematode seems to destroy the nitrogen-fixing nodules associated with leguminous plants. The normal host of the nematode is the karite tree (*Butyrospermum parkii*). The nematode spends the dry season in the soil, 40–50 cm deep, where it feeds on the tree roots. During the rainy season, the nematode is attracted toward the peanuts, and it moves near the surface (Germani and Luc, 1982).

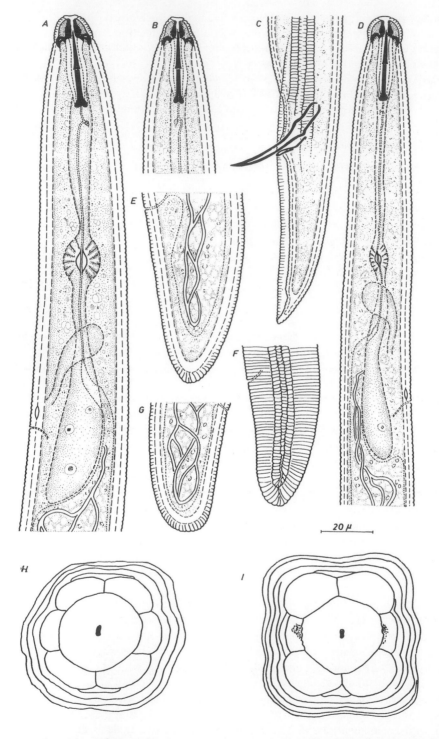

FIGURE 21 *Aphasmatylenchus straturatus* Germani, 1970. (A,B) Female, anterior end. (C)
Male, posterior end. (D) Male, anterior end. (E–G) Female, posterior end. Bottom left, female,
SEM face view. Bottom right, male, SEM face view. (Germany, 1977, courtesy Commonwealth
Agricultural Bureaux.)

REFERENCES

Andrássy, I. 1961. Wissenschaftliche Ergbnisse der ersten ungarischen zoologischen Expedition in Ostafrika. 2. Nematoda. *Ann. Hist-Nat. Mus. Nat. Hung.* 53: 281–297.

Baldwin, J. G., and Bell, A. H. 1981. *Pararotylenchus* n.gen. (Pararotylenchinae n.subfam., Hoplolaimidae) with six new species and two new combinations. *J. Nematol.* 13: 111–128.

Bird, G. W., and Högger, C. 1974. Anal peg of *Hoplolaimus columbus. Nematologica* 20: 103 + plate XI.

Bittencourt, C., and Huang, C.S. 1986. Brazilian *Peltamigratus* Sher, 1964 (Nematoda: Hoplolaimidae) with description of six new species. *Rev. Nématol.* 9: 3–24.

Chapman, R. A. 1963. Population development of the plant parasitic nematode *Scutellonema brachyurum* on red clover. *Proc. Helm. Soc. Wash.* 30: 169–173.

Choi, Y. E., and Geraert, E. 1971. Some remarkable Tylenchida from Korea. *Nematologica* 18: 66–73.

Coomans, A. 1963. Observations on the variability of morphological structures in *Hoplolaimus pararobustus. Nematologica* 9: 241–254.

Das, V. M. 1960. Studies on the nematodes parasites of plants in Hyderabad (Andhra Pradesh, India). *Z. Parazitenkunde* 19: 553–605.

De Grisse, A. T., Lippens, P. L., and Coomans, A. 1974. The cephalic sensory system of *Rotylenchus robustus* and a comparison with some other tylenchids. *Nematologica* 20: 88–95.

D'Errico, F. P. 1970. Su alcuni nematodi fitoparassiti trovati in Italia. *Boll. Lab. Ent. Agr. Filippo Silvestri* 28: 183–189.

Diab, K. A., and El-Eraki, S. 1968. Plant-parasitic nematodes associated with olive decline in the United Arab Republic. *Plant Dis. Reptr.* 52: 150–154.

Doucet, M. E. 1980. Description de deux nouveaux *Peltamigratus* et d'une population d'*Hoplolaimus galeatus* (Nematoda: Tylenchida) de la Province de Cordoba, Argentine. *Nematologica* 26: 34–36.

Eerens, J. P. J., and Loof, P. A. A. 1985. Systematic observations on the genera *Scutellonema* Andrássy, 1958 and *Peltamigratus* Sher, 1964 (Nematoda, Hoplolaimidae). *Meded. Fac. LandbWettens. Gent* 50: 857–860.

Elmiligy, I. A. 1970. On some Hoplolaiminae from Congo and Egypt. *Meded. Fac. LandbWettens. Gent* 35: 1141–1153.

Eroshenko, A. S. 1984. [Plant nematodes of coniferous plants in the Primorsk Territory.] *Parazity zhivotnykh i rastenii*: 87–97.

Eroshenko, A. S., and N. V. Thanh 1981. [Ectoparasitic nematodes of pineapple plantations in the northern and central provinces of Vietnam.] In [*Freeliving and Plant Parasitic Nematodes in the Far East.*] Eroshenko, A. C., and O. I. Belogurov, eds., Dal'nev Nauch. Tsentr Akad. Nauk SSSR, Bio.-Poch. Inst., Vladivostok, pp. 28–34 and 93–98.

Fassuliotis, G. 1974. A description of males of *Hoplolaimus columbus. J. Nematol.* 6: 116–118.

Fassuliotis, G. 1975. Feeding, egg-laying, and embryology of the Columbia lance nematode, *Hoplolaimus columbus. J. Nematol.* 7: 152– 158.

Fassuliotis, G. 1976. *Hoplolaimus columbus.* C.I.H. descriptions of plant-parasitic nematodes, Set 6, No. 81.

Fassuliotis, G., Rau, G. J., and Smith, F. H. 1968. *Hoplolaimus columbus,* a nematode parasite associated with cotton and soybeans in South Carolina. *Plant Dis. Reptr.* 52: 571–572.

Ferraz, S. 1980. Description of *Rotylenchus nexus* n. sp. (Nematoda: Hoplolaiminae) from Brazil, with some observations on the nematode genus *Calvatylus. Syst. Parasitol.* 2: 21–24.

Fortuner, R. 1974. Description de *Pratylenchus sefaensis* n.sp. et de *Hoplolaimus clarissimus* n.sp. (Nematoda: Tylenchida). *Cah. ORSTOM, sér. Biol. No. 21* (1973): 25–34.

Fortuner, R. 1984. Morphometrical variability in *Helicotylenchus* Steiner, 1945. 6: Value of the characters used for specific identification. *Rev. Nématol.* 7: 245–264.

Fortuner, R. 1987a. A reappraisal of Tylenchina (Nemata). 8. The family Hoplolaimidae Filip'ev, 1934. *Rev. Nématol.* 10: 219–232.

Fortuner, R. 1987b. Variabilité et identification des espèces chez les nématodes du genre *Helicotylenchus. Etudes et Thèses*, ORSTOM, Paris.

Fortuner, R. 1989. A new description of the process of identification of plant-parasitic nematode genera. In *Nematode Identification and Expert System Technology*. Fortuner, R., ed., Plenum Press, New York, pp. 35–44.

Fortuner, R., and Couturier, G. 1983. Les nématodes parasites de plantes de la foret de Taï (Côte d'Ivoire). *Rev. Nématol.* 6: 3–10.

Fortuner, R., Maggenti, A. R., and Whittaker, L. M. 1984. Morphometrical variability in *Helicotylenchus* Steiner, 1945. 4: Study of field populations of *H. pseudorobustus* and related species. *Rev. Nématol.* 7: 121–135.

Geraert, E. 1976. Problems concerning the genera *Helicotylenchus* Steiner, 1945 and *Rotylenchus* Filipjev, 1936. *Nematologica* 22: 284– 288.

Geraert, E. 1981. The female reproductive system in nematode systematics. *Ann. Soc. R. Zool. Belg.* 110: 73–86.

Geraert, E., Zepp, A., and Borazanci, N. 1975. Some plant nematodes from Turkey. *Meded. Fac. LandbWettens. Gent* 40: 511–515.

Germani, G. 1970. *Aphasmatylenchus straturatus* sp. n. (Nematoda: Hoplolaimidae) from West Africa. *Proc. Helm. Soc. Wash.* 37: 48–51.

Germani, G. 1977. *Aphasmatylenchus straturatus*. C.I.H. descriptions of plant-parasitic nematodes, Set 7, No. 104.

Germani, G., Baldwin, J. G., Bell, A. H., and Wu, X. Y. 1986. Revision of the genus *Scutellonema* Andrássy, 1958 (Nematoda: Tylenchida). *Rev. Nématol.* 8(1985): 289–320.

Germani, G., and Luc, M. 1982. Etudes sur la "chlorose voltaïque" des légumineuses due au nématode *Aphasmatylenchus straturatus* Germani. II. *Rev. Nématol.* 5: 195–199.

Goodey, J. B. 1957. *Hoplolaimus proporicus* n. sp. (Hoplolaiminae: Tylenchida). *Nematologica* 2: 108–113.

Goodey, T. 1940. On *Anguillulina multicincta* (Cobb) and other species of *Anguillulina* associated with the roots of plants. *J. Helminthol.* 18: 21–38.

Green, C. D., and Dennis, E. B. 1981. An analysis of the variability in yield of pea crops attacked by *Heterodera goettingiana, Helicotylenchus vulgaris* and *Pratylenchus thornei. Plant Path.* 30: 65–71.

Hirschmann, H. 1983. Scanning electron microscopy as a tool in nematode taxonomy. In *Concepts in Nematode Systematics*. Stone, A. R., Platt, H. M., and Khalil, L. F., eds., Academic Press, London, pp. 95–111.

Högger, C., and Bird, G. W. 1974. Secondary male sex characteristics of *Hoplolaimus galeatus. J. Nematol.* 6: 12–16.

Ivan, M. 1978. Trei specii de namatode identificate in culturile de coacaz, noi pentru fauna Romaniei. *Studii si Cercetari de Biologie, Biologie Animala* 30: 13–15.

Jairajpuri, M. S., and Baqri, Q. H. 1973. Nematodes of high altitudes in India. I. Four new species of Tylenchida. *Nematologica* 19: 19–30.

Jairajpuri, M. S., and Siddiqi, M. R. 1979. Observations on the nematode genera *Orientylus* and *Calvatylus* (Rotylenchoidinae: Hoplolaimidae) with descriptions of three new species. *Ind. J. Nematol.* 7: 101–111.

Jones, R. K. 1978a. The feeding behavior of *Helicotylenchus* spp. on wheat roots. *Nematologica* 24: 88–94.

Jones, R. K. 1978b. Histological and ultrastructural changes in cereal roots caused by feeding of *Helicotylenchus* spp. *Nematologica* 24: 393–397.

Khan, E., and Nanjappa, C. K. 1970. Four new species in the superfamily Hoplolaimoidea (Tylenchida: Nematoda) from India. *Bull. Ent.* 11: 143–149.

Lewis, S. A., and Huff, T. F. 1976. Cuticle anatomy of *Hoplolaimus columbus. J. Nematol.* 8: 293.

Lewis, S. A., Smith, F. H., and Powell, W. M. 1974. Histopathology of infection by *Hoplolaimus columbus* on cotton and soybean and aspects of its pathogenicity. *J. Nematol.* 6: 145.

Luc, M. 1959. Nématodes parasites ou soupçonnés de parasitisme envers les plantes de Madagascar. *Bull. Inst. Res. Agron. Madagascar* 3: 89–101.

Luc, M. 1960. Trois nouvelles espèces du genre *Rotylenchoides* Whitehead, 1958 (Nematoda: Tylenchida). *Nematologica* 5: 7–17.

Luc, M. 1981. *Basirolaimus* Shamsi, 1979, a junior synonym of *Hoplolaimus* von Daday, 1905 (Nematoda: Tylenchida). *Nematol. Mediterranea* 9: 197–199.

Luc, M. 1986. *Hoplorhynchus* Andrássy, 1985, a junior synonym of *Pratylenchoides* Winslow, 1958 (Nemata: Pratylenchidae). *Rev. Nématol.* 9: 198.

Maggenti, A. R., Luc, M., Raski, D. J., Fortuner, R., and Geraert, E. 1988. A reappraisal of Tylenchina (Nemata). 11. List of generic and supra-generic taxa, with their junior synonyms. *Rev. Nématol.* 11: 177–188.

Mancini, G., and Moretti, F. 1977. Il genere *Helicotylenchus* Steiner, 1945 in piemonte e Valle d'Aosta, Nota 1. *Redia* 59: 225– 228.

Maqbool, M. A., and Ghazala, P. 1988. Observation of some known species of *Hoplolaimus* von Daday, 1905 (Nemata: Hoplolaimidae) from Pakistan. *Pak. J. Nematol.* 6: 1–7.

Mulk, M. M., and Siddiqi, M. R. 1982. Three new species of hoplolaimid nematodes from South America. *Ind. J. Nematol.* 12: 124– 131.

Nong, L., and Weber, G. F. 1964. Amaryllis diseases caused by two nematodes. *Phytopathology* 54: 902–903.

Phillips, S. P. 1971. Studies of plant and soil nematodes. 16. Eight new species of spiral nematodes (Nematoda: Tylenchoidea) from Queensland. *Qd. J. Agr. Anim. Sci.* 28: 227–242.

Ratanaprapa, D., and Boonduang, A. 1975. Identification of plant parasitic nematodes of Thailand. A second systematic study of Hoplolaimidae in Thailand. *Plant Prot. Serv. Tech. Bull.* (Bangkok) No. 27.

Ruehle, J. L. 1973. Influence of plant-parasitic nematodes on longleaf pine seedlings. *J. Nematol.* 5: 7–9.

Sauer, M. R., and Winoto, R. 1975. The genus *Helicotylenchus* Steiner, 1945 in West Malaysia. *Nematologica* 21: 341–350.

Seinhorst, J. W. 1971. The structure of the glandular part of the esophagus of Tylenchidae. *Nematologica* 17: 431–443.

Shamsi, M. A. 1979. *Basirolaimus* gen.n. (Nematoda: Hoplolaimidae) with the description of *Basirolaimus sacchari* n.sp. from India. *Nematol. Mediterranea* 7: 15–19.

Sharafat-Ali, S., Geraert, E., and Coomans, A. 1973. Some spiral nematodes from Africa. *Biol. Jaarb.* 41: 53–70.

Sharafat-Ali, S., and Loof, P. A. A. 1975. Two new species of *Helicotylenchus* Steiner, 1945 (Nematoda: Hoplolaiminae). *Nematologica* 21: 207–212.

Sher, S. A. 1961. Revision of the Hoplolaiminae (Nematoda). I. Classification of nominal genera and nominal species. *Nematologica* 6: 155–169.

Sher, S. A. 1963. Revision of the Hoplolaiminae (Nematoda). II. *Hoplolaimus* Daday, 1905 and *Aorolaimus* n.gen. *Nematologica* 9: 267– 295.

Sher, S. A. 1964a. Revision of the Hoplolaiminae (Nematoda) III. *Scutellonema* Andrássy, 1958. *Nematologica* 9(1963): 421–443.

Sher, S. A. 1964b. Revision of the Hoplolaiminae (Nematoda) IV. *Peltamigratus* n.gen. *Nematologica* 9(1963): 455–467.

Sher, S. A. 1965. *Aphasmatylenchus nigeriensis* n.gen., n.sp. (Aphasmatylenchinae n. subfam.: Tylenchoidea, Nematoda) from Nigerian soil. *Proc. Helm. Soc. Wash.* 32: 172–176.

Sher, S. A. 1965. Revision of the Hoplolaiminae (Nematoda). V. *Rotylenchus* Filipjev, 1936. *Nematologica* 11: 173–198.

Sher, S. A. 1966. Revision of the Hoplolaiminae (Nematoda). VI. *Helicotylenchus* (Steiner, 1945). *Nematologica* 12: 1–56.

Sher, S. A. 1973. *Antarctylus humus* n.gen., n.sp. from the subantarctic (Nematoda: Tylenchoidea). *J. Nematol.* 5: 19–21.

Sher, S. A., and Bell, A. H., 1975. Scanning electron micrographs of the anterior region of some species of Tylenchoidea (Tylenchida: Nematoda). *J. Nematol.* 7: 69–83.

Siddiqi, M. R. 1972. *Rotylenchus robustus*. C.I.H. descriptions of plant-parasitic nematodes, Set 1, No. 11.

Siddiqi, M. R. 1972. *Helicotylenchus multicinctus*. C.I.H. descriptions of plant-parasitic nematodes, Set 2, No. 23.

Siddiqi, M. R. 1972. *Scutellonema brachyurum*. C.I.H. descriptions of plant-parasitic nematodes, Set 4, No. 54.

Siddiqi, M. R. 1986. Tylenchida parasites of plant and insects. Commonwealth Agricultural Bureaux, Slough, UK.

Siddiqi, M. R., and Husain, Z. 1964. Three new species of nematodes in the family Hoplolaimidae found attacking citrus trees in India. *Proc. Helm. Soc. Wash.* 31: 211–215.

Sivakumar, C. V., and Selvasekaran, E. 1982. Description of two new species of *Scutellonema* Andrássy, 1958 (Hoplolaimoidea: Nematoda). *Ind. J. Nematol.* 12: 118–123.

Spaull, A. M. 1982. *Helicotylenchus vulgaris* and its association with damage to sugar beet. *Ann. App. Biol.* 100: 501–510.

Steiner, G. 1938. Nematodes infesting red spiderlilies. *J. Agr. Res.* 56: 1–8.

Stokes, D. E. 1977. The Columbia lance nematode, *Hoplolaimus columbus* Sher, 1963. *Nematology Circular*, 34, Florida Depart. Agr. and Consumer Serv.

Sumner, D. R. 1967. Nematodes in bluegrass. *Plant Dis. Reptr.* 51: 457–460.

Suryawanshi, M. V. 1971. Studies on Tylenchida (Nematoda) from Marathwada, India, with descriptions of four new species. *Nematologica* 17: 393–406.

Thorne, G., and Malek, R. B. 1968. Nematodes of the Northern Great Plains. Part I. Tylenchida (Nemata: Secernentea). *South Dakota Agr. Exp. Stat. Tech. Bull.* No. 31.

Van den Berg, E., and Heyns, J. 1970. South African Hoplolaiminae. I. The genus *Hoplolaimus* Daday, 1905. *Phytophylactica* 2: 221–226.

Van den Berg, E., and Heyns, J. 1973. South African Hoplolaiminae. 2. The genus *Scutellonema* Andrássy, 1958. *Phytophylactica* 5: 23–39.

Van den Berg, E., and Heyns, J. 1975. South African Hoplolaiminae 4. The genus *Helicotylenchus* Steiner, 1945. *Phytophylactica* 7: 35– 52.

Van den Berg, E., and Kirby, M. F. 1979. Some spiral nematodes from the Fiji Islands (Hoplolaimidae: Nematoda). *Phytopathology* 11: 99– 109.

Vovlas, N. 1984. Morphology of a local population of *Helicotylenchus multicinctus* from southern Italy. *Rev. Ném3atol.* 6(1983): 327–329.

Vovlas, N. 1985. Head structure of five species in the subfamily *Hoplolaiminae* (Nematoda). *Nematol. Mediterranea* 12(1984): 163–168.

Vovlas, N., and Lamberti, F. 1985. Observations on the morphology and histopathology of *Hoplolaimus pararobustus* attacking coffee in Sao Tome. *Nematol. Mediterranea* 13: 73–80.

Wallace, H. R. 1971. The influence of the density of nematode populations on plants. *Nematologica* 17: 154–166.

Wang, K. C., and Chen, T. A. 1985a. Ultrastructure of the phasmids of *Scutellonema brachyurum*. *J. Nematol.* 17: 175–186.

Wang, K. C., and Chen, T. A. 1985b. Ultrastructure of male sexual apparatus of *Scutellonema brachyurum*. *J. Nematol.* 17: 435–444.

Wen, G. Y., and Chen, T. A. 1972. Fine structures of the cuticle of *Hoplolaimus galeatus*. *J. Nematol.* 4: 236–237.

Whitehead, A. G. 1958. *Rotylenchoides brevis* n.g., n.sp. (Rotylenchoidinae n.subfam.: Tylenchida). *Nematologica* 3: 327–331.

Whitehead, A. G. 1959. *Hoplolaimus angustalatus* n.sp. (Hoplolaiminae: Tylenchida). *Nematologica* 4: 99–105.

Williams, J. R. 1960. Studies on the nematode fauna of sugar cane fields in Mauritius. 4. Tylenchoidea (partim). *Mauritius Sug. Ind. Res. Inst. Occ. Pap.* No. 4.

Yuen, P. H. 1964. Four new species of *Helicotylenchus* Steiner (Hoplolaiminae: Tylenchida) and a redescription of *H. canadensis* Waseem, 1961. *Nematologica* 10: 373–387.

Yuen, P. H. 1966. Further observations on *Helicotylenchus vulgaris* Yuen. *Nematologica* 11(1965): 623–637.

Zancada, M. C., and Lima, M. B. 1986. Numerical taxonomy of the genera *Rotylenchus* Filipjev, 1936 and *Orientylus* Jairajpuri and Siddiqi, 1977 (Nematoda: Tylenchida). *Nematologica* 31: 44–61.

Zuckerman, B. M., and Strich-Harari, D. 1964. The life stages of *Helicotylenchus multicinctus* (Cobb) in banana roots. *Nematologica* 9(1963): 347–353.

16

Wheat and Grass Nematodes: *Anguina*, *Subanguina*, and Related Genera

EINO L. KRALL   *Institute of Zoology and Botany, Estonian Academy of Sciences, Tartu, Estonia, USSR*

I. INTRODUCTION

Nematodes of the *Anguina* group were known in the eighteenth century. In the past, the wheat nematode *Anguina tritici* caused significant economic losses in many countries. Due to their high specialization, several species have potential value as biological control agents of weeds. They induce specific leaf, stem, or seed gall formation; only rarely are such galls incited on roots. Few changes have been made in the taxonomy of anguinid nematodes for a long period, since Brzeski (1981), Chizhov and Subbotin (1985), Siddiqi (1986), and Fortuner and Maggenti (1987) proposed their original revisions of the group.

In dividing anguinids into subfamilies, the system by Siddiqi (1986) is accepted here. However, the generic taxonomy of the most important subfamily, Anguininae, is followed according to the proposal by Chizhov and Subbotin (1985). We feel that the latter system reflects some evolutionary trends in the group and if so it would be regarded as the most promising for further discussion.

Up to now more than 40 nominal gall–forming anguinid species have been described. Several of them have been synonymized by various authors. Of 27 species accepted here, 15 species are parasites of monocotyledons, mostly of grasses. The remaining species parasitize dicotyledons mostly of the family Asteraceae (Compositae), and only occasionally single specialized species attack plants of some other families.

II. TAXONOMY OF ANGUINIDAE

Family Anguinidae Nicoll, 1935 (1926)
   syn. **Anguillulinidae Baylis and Daubney, 1926**
       Anguinidae Paramonov, 1962 (Siddiqi, 1971)
       Ditylenchidae Golden, 1971
       Nothotylenchidae Thorne, 1941 (Jairajpuri and Siddiqi, 1969)

Diagnosis (after Siddiqi, 1986, abridged)

Anguinoidae. Adults to 3 (5) mm long, slender or obese. Cephalic region low, smooth. Vulva generally at less than 85% of body length. Postvulval uterine sac mostly present. Tails similar in both sexes, female tail rarely subcylindrical, never cylindrical or hooked. Bursa not enclosing tail tip. Fungus feeders or plant parasites.

> Type subfamily: Anguininae Nicoll, 1935 (1926)
> Other subfamilies: Nothanguininae Fotedar and Handoo, 1978
>   Nothotylenchinae Thorne, 1941

Key to subfamilies of Anguinidae (adopted after Siddiqi, 1986)

1. Median esophageal bulb present, with valves .... Anguininae
    Median esophageal bulb absent, esophageal lumen sometimes with refractive
    thickenings 2
2. Females spiral, obese, males without gubernaculum .... Nothanguininae
    Females not spiral, less or not obese; males with gubernaculum ....
    .... Nothotylenchinae

Subfamily Anguininae Nicoll, 1935 (1926)
   syn. **Anguillulininae Baylis and Daubney, 1926**
       Anguininae Paramonov, 1962
       Cynipanguininae Fotedar and Handoo, 1978
       Ditylenchinae Golden, 1971
       Pseudhalenchinae Siddiqi, 1971

General Characteristics (after Chizhov and Subbotin, 1985)

Anguinidae. Body length from about 1.0 to 5.0 mm. The variation in the body length is remarkable within one and the same species if two generations per year develop. Body length is also dependent on the localization in various parts (leaves, stems) of the same host plant as well as on the degree of susceptibility of various host plant species. Body vermiform or obese, after heat relaxation spirally coiled or "C"–shaped. The obese females of some species almost immobile. Cuticle finely annulated, the annules often being visible only on the forepart of the body. Some species also with fine longitudinal striae on the cuticle. Lateral field not visible on older specimens, the number of incisures varying from 2 to 20 may be different in various parts of the body or between sexes. Head six radial, annulation absent or with two to six annules. Tails conoid, their tips often varying within the same species, sometimes with mucro. Stylet short and thin, mostly about 10 μm in length. Procorpus cylindrical, medium bulb ovoid or almost spherical with valves. Isthmus short. Three esophageal glands well developed, forming the basal bulb. Dimensions and location of the latter are greatly dependent on the age of the specimen as well as on the generation observed. Thus in young specimens of some species belonging to the second generation the basal bulb may be

Ditylenchus–like. In the specimens of the first generation, however, the bulb is clearly trapezoid and overlapping the intestine. The older the specimens the more the esophageal glands extend on the intestine. Ovary in the germinative zone with oocytes in one row or in several rows arranged about a rachis. Preuteral gland as a multicellular tube containing several eggs. Uterus short. Morphology of the genital tract of anguinids is also greatly variable and is dependent on the generation to which specimens belong as well as on the specimen's age, but also on the host plant species. Thus, the ovary may be outstretched or once reflexed in the second generation, whereas oocytes are laying in single row. In the first generation of the same species, however, the ovary may be two to three times reflexed and the oocytes in the germinative zone are laying in two to five rows. However, as a rule, the position of the oocytes in the ovary is certainly of diagnostic value for anguinids. The preuteral gland in the specimens of the first generation may consist of up to 400 cells and contain up to 18 synchronous eggs. The corresponding numbers for the second generation do not exceed 200 and 2–4, respectively. The number of cell rows in the preuteral gland is varying from 4 to 10 in different species. In males, spicules are often complex and thus may be of value for taxonomy. Unfortunately, they are not yet adequately described for most species.

As highly specialized plant parasites, the gall–forming anguinids have original ontogenetic cycles of development. Except juveniles, they cannot exist outside of their host plants. Without doubt, the latter may be useful also for taxonomy of these nematodes. In anguinins, invasive stages are different between species (second–, third–, and fourth–stage juveniles). The degree of development of the genital tract in anguinins is negatively correlated with increasing the stages in which infestation takes place. Thus, species having second–stage juveniles infective possess more complex genital organs than species attacking plants only on the fourth juvenile stage. In *Heteroanguina graminophila* (fourth stage infective), *Mesoanguina millefolii* (third–stage infective) and *Anguina agrostis* (second–stage infective), maximal egg production per female does not exceed 600, 1500, and 2500, respectively.

Type genus: *Anguina* Scopoli, 1777
Other genera: *Heteroanguina* Chizhov, 1980
              Mesoanguina Chizhov and Subbotin, 1985
              Subanguina Paramonov, 1967

The genera *Ditylenchus* Filipjev, 1936 and *Pseudhalenchus* Tarjan, 1958 certainly also belong to anguinins together with some other groups not accepted generally on the generic level. As they are not gall–inducing, they will not be discussed further.

Key to the gall–inducing genera of Anguininae (after Chizhov and Subbotin, 1990)

1. Parasites of monocotyledonous plants 2
   Parasites of dicotyledonous plants 4
2. Incite galls on the roots of grasses. Female preuteral gland consisting of 48–80 cells in four to five rows.Juveniles of the second–stage infective, their length only 390–530 µm . . . . *Subanguina*
   Incite galls on aerial parts, rarely on rhizomes, but not on roots 3
3. Female preuteral gland consisting of 180–600 cells in 6–12 rows; juveniles of the second–stage infective, their length less than 1.0 mm. Incite leaf, stem, or seed galls on grasses and cereals . . . . *Anguina*
   Female preuteral gland consisting of 60–90 (72) cells; juveniles of the fourth–stage infective, their average length more than 1.0 (up to 1.5) mm. Incite leaf or stem galls on grasses or on Cyperaceae (see also below) . . . . *Heteroanguina*

4. Female preuteral gland consisting of 32–52 (48) cells; juveniles of the third–stage infective, with average length less than 1.0 mm. Incite leaf or stem galls with an expressed internal cavity mostly Asteracea, but also on Boraginaceae and Plantaginaceae. Usually two morphologically different generations developing in one gall .... *Mesoanguina*
Female preuteral gland consisting of 60–90 (72) cells; juveniles of the fourth–stage infective; their average length exceeds 1.0 mm. Incite galls without a pronounced cavity on plants belonging to families Umbelliferae and Polygonaceae (see also below). Only one generation developing in the gall .... *Heteroanguina*

Anguina Scopoli, 1777
　　syn. *Paranguina* Kirjanova, 1985
　　　Afrina Brzeski, 1981
　　　Cynipanguina Maggenti, Hart, and Paxmann, 1974

Diagnosis (after Chizhov and Subbotin, 1990)

Anguininae. Females after heat relaxation crescentic or spirally coiled. Ovary reflexed two or three times. Oocytes in the zone of maturation usually in two or more rows (in the second–generation females of *A. agropyri* in a single row). Preuteral gland irregular, consisting of 180–600 cells in 6–12 rows (20–60 cells in each of it). Up to 18 synchronous eggs present in preuteral gland. Infective juveniles of the second stage. Incite usually pigmented galls with an internal cavity on above–ground parts or sometimes on the rhizomes (underground stems) of monocotyledons of the family Poaceae (grasses). Only one (depending on host, one or two in *A. agropyri*) generation in the same gall.
　　Type species: *A. tritici* (Steinbuch, 1799) Filipjev, 1936.

Key to species of *Anguina* Scopoli, 1777 (after Chizhov and Subbotin, 1990)

1. Incite seed galls 2
　 Incite galls on leaves, stems, rhizomes (= underground stems) or spikelets 5
2. Female average length under 1.8 mm, stylet length 6–8 μm. Parasites of tropical grasses 3
　 Female length more than 1.8 mm, stylet length 8–12 μm. Parasites of nontropical grasses 4
3. Postvulval uterine sac length 39–49 μm, c in males 22–24. Incite seed galls in *Saccharum* .... *spermophaga*
　 Postvulval uterine sac length 46–53 μm, c in males 26–35. Incite seed galls on *Hyparrhenia* .... *hyparrheniae*
4. Tail terminus bluntly rounded. Spicules sharply enlarged in the middle, with capitulum possessing a ventral fold. Parasites of wheat and rye .... *tritici*
　 Tail terminus pointed. Spicules only slightly enlarged in the middle, their capitulum without or with a weakly expressed ventral fold. Parasites of fodder grasses ......
　 .... *agrostis*
5. Postvulval uterine sac considerable shorter than half the vulva–anus distance. Gubernaculum length 18–21 μm. Incite galls on *Cynodon* .... *tumefaciens*
　 Postvulval uterine sac either slightly shorter or slightly longer than half the vulva–anus distance. Gubernaculum length 9–18 μm 6

6. *a* in females 18–36 7
    a in females 12–18 9
7. *a* in males 19–20. Incite galls on *Ehrharta* .... *australis*
    a in males 21–44 8
8. Female tail terminus rounded. Incite small pigmented galls on leaves, stems, and spikelets of *Festuca* .... *graminis*
    Female tail terminus pointed. Incite great unpigmented galls on stem base or on rhizomes of *Elytrigia, Poa, Triticum, Secale, Hordeum,* and *Elymus* .... *agropyri*
9. *V* = 90–93. Spicules with distinct suture between calamus and lamina. Incite galls on *Danthonia* .... *danthoniae*
    V = 87–88. Spicules without suture between calamus and lamina. Incite galls on *Microlaena* .... *microlaenae*

Note: *A. pustulicola* has not been included in the key because of lack of sufficient information.

Anguina tritici (Steinbuch, 1799) Chitwood, 1935 (Figs. 1–3)
> syn. *Vibrio tritici* Steinbuch, 1799
> > *Rhabditis tritici* (Steinbuch, 1799) Dujardin, 1845
> > *Anguillula tritici* (Steinbuch, 1799) Grube, 1849
> > *Anguillulina tritici* (Steinbuch, 1799) Gervais and Van Beneden, 1859
> > *Tylenchus tritici* (Steinbuch 1799) Bastian, 1865
> > *Anguillula scandens* Schneider, 1866
> > *Tylenchus scandens* (Schneider, 1866) Cobb, 1890
> > *Anguillulina scandens* (Schneider, 1866) Goodey, 1932

Measurements (compiled after Kirjanova and Krall, 1971 and Southey, 1972)

Females: L = 3.0–5.2 mm; a = 13–30; b = 9.8–25.0; c = 24–63; V = 70–95; stylet = 8–11 μm. Males: L = 1.9–2.5 mm; a = 21–30; b = 6.3–13.0; c = 17–28; stylet 8–11 μm; spicules 35–40 μm; gubernaculum = 10 μm. Eggs: on the average 85 × 38 mm, but may also be larger (130 × 63 μm). Second–stage juveniles: L = 0.75–0.95 μm.

Description (after Southey, 1972): Females: Body obese, spirally coiled ventrally. Lip region low and flattened, slightly offset. Cuticle very finely annulated. Procorpus of esophagus swollen but constricted at junction with median bulb. Isthmus of esophagus sometimes posteriorly swollen, then offset from glandular basal bulb by a deep constriction. Basal bulb pyriform, sometimes with irregular lobes visible not overlapping the intestine. Ovary with two or more flexures with many oocytes arranged around a rachis. Spermatheca pyriform, separated from oviduct by a sphincter. Postvulval uterine sac present. Tail conoid, tapering to an obtuse or rounded tip, not mucronate.

Males: More slender than females. After heat relaxation may be slightly curved both ventrally or dorsally. Testis with one or two flexures. Spicules stout, arcuate, with two ventral ridges running from tip to widest part. The head of the spicules is rolled of folded ventrally. Gubernaculum simple, trough–like. Bursa does not reach the tail tip.

Hosts: Common wheat (*Triticum aestivum* L.), emmer (*T. dicoccum* Shrank), durum wheat (*T. durum* Desf.), small spelt (*T. monococcum* L.), spelt (*T. spelta* L.), and *T. ventricosum* Ces. Rye (*Secale cereals* L.) is also a good host, but oats and barley are poor hosts or residents.

FIGURE 1   The wheat nematode *(Anguina tritici)*. Female. (After Marcinowski, 1909 from Kirjanova and Krall, 1971.)

FIGURE 2   Wheat seedling severely attacked by *Anguina tritici*. (From Kirjanova and Krall, 1969.)

Bionomics (after Kirjanova and Krall, 1971; Southey, 1972): Incite seed galls (ear-cockles) in cereals. The invasive juveniles attack young seedlings, move upward along the plant to the meristem or concentrate in the leaf axis of shoots where they remain until ears develop. The attacked plants become stunted and bear shorter and deformed stems and leaves. Severely infested plants do not form ears or form only stunted ears on stunted stems. A diseased ear is much wider and shorted than the normal one and has short deformed awns. The appearance of the diseased ear varies widely in different wheat varieties. The galls vary from light to dark brown to almost black shades in different wheat varieties. They may be mistaken for weed seeds or be confused with grain attacked by bunt (*Tilletia tritici*). The galls are hard to the touch and filled with a white powdery mass consisting of invasive juveniles. They may remain quiescent (anhydrobiotic) for decades when desiccated. Up to 40 adult nematodes could be found inside one seed gall. Each female is often reported to be

FIGURE 3   Wheat grains (left) and earcockle (middle and right) filled by juveniles of *Anguina tritici*. [Original figures by E. Aomets; (material collected in 1930s from Georgia, USSR).]

associated with bacterial yellow ear rot disease caused by *Corynebacterium tritici* ("tundu" disease in India).

Distribution: Formerly widespread in wheat–growing areas of Asia, Australia, Europe, North Africa, and North America. The earcockle disease has now been almost extented in most countries. Outbreaks of the disease were reported more recently in India. It also probably occurs in some other Asian (or Mediterranean) countries. The disease may be effectively controlled by mechanical seed cleaning and crop rotation.

Anguina agrostis (Steinbuch, 1799) Filipjev, 1936 (Figs. 4 and 5)
    syn. *agropyronifloris* Norton, 1965 (Chizhov and Subbotin, 1990)
        agrostidis Bastian, 1865
        funesta Price, Fisher, and Kerr, 1979
        lolii Price, 1973 (*nomen nudum*)
        phalaridis Steinbuch, 1799
        phlei Horn, 1888
        poophila Kirjanova, 1952
        Afrina wevelli Van den Berg, 1985 (Chizhov and Subbotin, 1990)

Measurements from *Agrostis* spp (after Chizhov, 1980)

Females: $L = 1.39$–2.60 mm; $a = 13.8$–25.4; $b = 12.6$–28.7; $c = 25.2$–43.0; $V = 87$–92; stylet $= 10$–12 µm. Males: $L = 1.05$–1.45 mm; $a = 23.8$–30.0; $b = 6.5$–8.9; $c = 21.5$–28.4; spicules $= 25$–32 µm; gubernaculum $= 10$–13 µm, stylet $= 10$–12 µm. Eggs: 67–92 (79) $\times$ 33–38 (35) µm. Second–stage infective juveniles: $L = 0.55$–0.82 mm; $a = 47.2$–65.0; $b = 3.2$–4.5; $c = 11.7$–20.0; stylet $= 10$ µm. Third–stage juveniles: $L = 0.80$–1.05 mm; $a = 27.1$–30.4; $b = 4.7$–6.5; $c = 14.6$–16.7; stylet $= 10$–12 µm. Fourth–stage juveniles: $L = 0.97$–1.70 mm; $a = 22.4$–26.6; $b = 6.2$–11.5; $c = 16.4$–31.0; stylet $= 10$–12 µm.

Measurements from *Poa* sp. and *Phleum phleoides* (after Chizhov, 1980)

Females: $L = 2.54$–4.14 mm; $a = 21.8$–30.7; $b = 10.3$–26.5; $c = 28.8$–48.9; $V = 90$–92; stylet $= 10$–12 µm. Males: $L = 1.54$–2.31 mm; $a = 21.6$–34.4; $b = 6.8$–11.6; $c = 20.4$–32.7; stylet $= 10$–12 µm; spicules $= 30$–37 µm; gubernaculum $= 13$–15 µm. Eggs: 74–110 (92) $\times$ 38–46

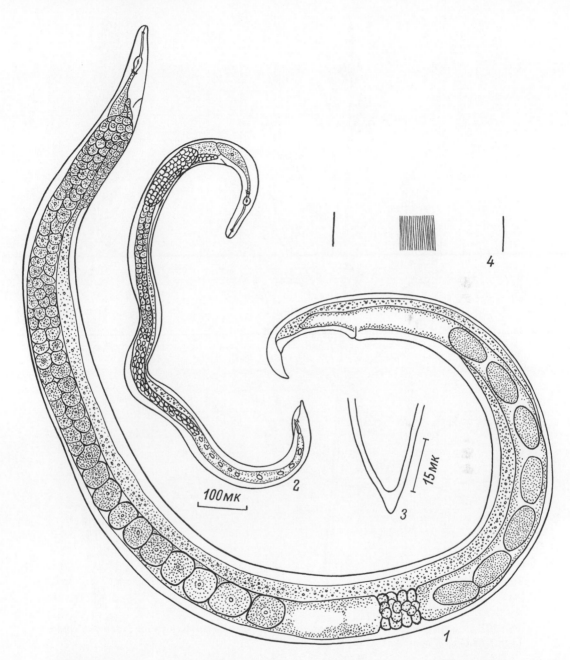

FIGURE 4 *Anguina agrostis.* 1, female; 2, male; 3, female tail tip; 4, lateral field. (From Kirjanova and Krall, 1971.)

FIGURE 5   Seed galls induced by *Anguina agrostis* on *Agrostis tenuis* (left) and *Phleum phleoides* (right). (From Kirjanova and Krall, 1971.)

(42) μm. Second–stage infective juveniles: $L = 0.81–1.25$ mm; $a = 43.5–71.0$; $b = 4.0–6.1$; $c = 12.1–15.4$; stylet = 10 μm. Third–stage juveniles: $L = 1.09–1.28$ mm; $a = 24.5–37.1$; $b = 4.0–6.4$; $c = 14.2–17.6$; stylet = 10–12 μm. Fourth–stage juveniles: $L = 1.20–2.18$ mm; $a = 17.8–34.7$; $b = 4.6–8.7$; $c = 14.8–21.6$; stylet = 10–12 μm.

Description (after Chizhov, 1980): Females: Body spirally coiled or crescentic (C–shaped). Cuticle with fine annulation. Lip region set off, 3–4 μm high. Procorpus cylindrical, somewhat enlarged in the middle part with a constriction before joining median bulb. The latter ovoid. Isthmus short and narrow. Basal bulb well developed, trapezoid, although not overlapping the intestine. The maximum width of the body in specimens parasitizing *Poa* ssp. or *Phleum phleoides* is reached at the level immediately behind the esophagus. In more slender specimens from *Agrostis* (bent grass) the maximum width of the body is reached only at the level of spermatheca. Ovary reflexed two or three times. Oocytes in the zone of multiplication in two or three rows. Spermatheca elongate, sometimes with up to 10–12 synchronous oocytes, separated from the preuteral gland by a short constriction. Preuteral gland long, with up to 20 synchronous egg, separated from the uterus by a constriction (oviduct). Postvulval uterine branch reaching half the distance from the vulva to anus, filled with sperms. Tail short, conoid, with acute terminus.

Males: Body after heat relaxation almost outstretched. Testis usually with one or two flexures. Bursa subterminal, does not reach the tail tip.

Hosts: *Agrostis tenuis* Sibth. (type host). Besides various bent grass species, this nematode is reported from other grass genera, including *Apera, Arctagrostis, Calamagrostis, Dactylis, Eragrostis, Festuca, Hordeum, Koeleria, Lolium, Phalaris, Phleum, Poa, Puccinellia, Sporobolus,* and *Trisetum* (Kirjanova and Krall, 1971; Southey, 1973). Their host status will be discussed below.

Bionomics: Second–stage juveniles hatch and invade young grasses early in May (Europe) or at the end of August (Australia). The juveniles migrate to the inflorescence and incite seed galls. Heavily infested grasses have been repeatedly mistakenly described by botanists as new species or forms (*f. vivipara* a.0.). Maturation and oviposition began at mid–June in European countries. In Australia, new juveniles began to hatch in the middle of October. Only one generation develops per year. Chizhov (1980) demonstrated that the bent grass nematode from *Agrostis tenuis* and *A. stolonifera* did not infest either *Poa* ssp. or *Phleum phleoides*. On the other hand, *Anguina agrostis* from *Poa angustifolia, P. pratensis,* or *Phleum phleoides* did not incite seed galls on two species of the bent grass mentioned above. Using also the morphological data (see above),he stated (Chizhov, 1980) that two different species incite seed galls on the grasses in the European part of the USSR. However, in the next revision (Chizhov and Subbotin, 1990) the same author considered *Anguina agrostis* to be a complex species of a very high morphological variability depending on host range. Similar data on host specialization of *A. agrostis* are reviewed by Southey (1973).

Anguina agrostis is considered to be a serious or potentially important nematode pest of bent grass, especially in the Pacific Northwest USA and New Zealand (Southey, 1973). Fresh galls of *A. agrostis* from chewing's fescue have been found to be poisonous to cattle and sheep in the USA and this species (= *A. funesta*) is known also as the vector of annual rye grass toxicity in Australia (Kirjanova and Krall, 1971; Southey, 1973; McKay a.o., 1981).

Distribution: Europe, Asia, (the far east of the USSR), North America (USA and Canada), Australia, New Zealand, South Africa.

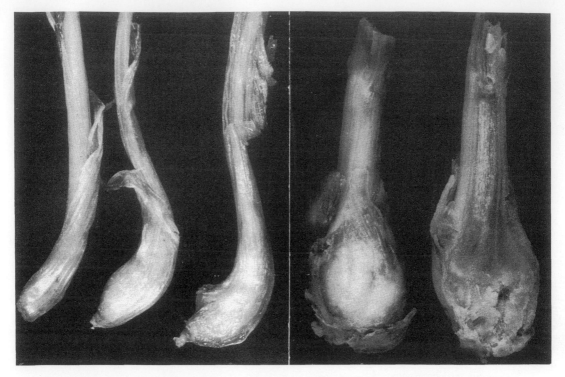

FIGURE 6   Stem galls induced by *Anguina agropyri* on rye (left) and on couch grass (right). Estonia, USSR. (Original photograph by E. Krall.)

Anguina agropyri Kirjanova, 1955 (couch grass nematode) (Figs. 6 and 7)
    syn. *Paranguina agropyri* Kirjanova, 1955
        A. pacifica Cid del Prado Vera and Maggenti, 1984

Measurements (after Chizhov and Berezina, 1986)

From *Agropyron repens*: Females (first generation): L = 2.23–4.78 mm; a= 16.6–25.7; b = 14.5–24.4; c = 21.2–47.7; V = 84–90; stylet = 10–12 μm. Males: L = 1.91–3.11 mm; a = 22.7–29.5; b = 10.6–21.3; c = 18.6–34.0; stylet = 10–12 μm; spicules = 33–40 μm; gubernaculum = 13–15 μm. Juveniles (second–stage): L = 0.81–1.13 mm; a = 51.5–75.5; b = 3.8–5.2; c = 9.7–11.8; stylet = 10–12 μm. Females (second generation): L = 1.94–3.23 mm; a = 26.0–36.5; b = 9.3–14.6; c = 16.9–23.2; V = 77–84; stylet = 10–12 μm. Males: L = 1.53–2.27 mm; a = 28.6–44.0; b = 7.2–11.4; c = 14.0–21.0; stylet = 10–12 μm; spicules = 31–37 μm; gubernaculum = 13–15 μm. Juveniles (second–stage): L = 0.83–1.15 mm; a = 52.3–67.4; b = 4.0–5.2; c = 10.0–12.6; stylet = 10–12 μm.

    From *Poa annua*: Females (second generation): L = 1.86–2.56 mm; a = 26.2–35.4; b = 9.0–14.6; c = 16.6–21.5; V = 74–84; stylet = 10.2–11.6 μm. Males: L = 1.47–2.32 mm; a = 22.2–36.3; b = 7.4–13.1; c = 14.1–22.9; stylet = 10–12 μm; spicules = 26–33 μm; gubernaculum = 8.7–14.5 μm. Juveniles (second–stage): L = 0.78–1.24 mm; a = 49.0–68.0; b = 4.5–5.6; c = 10.4–12.2; stylet = 9.0–12.2 μm.

FIGURE 7 One healthy and two severely stunted by *Anguina agropyri* wheat plants. Estonia, USSR. (Original photograph by E. Krall.)

Description (after Kirjanova and Krall, 1971, emended): Females of the first generation considerably more obese than vermiform adults of the second generation. Lip region flattened. Cuticle with fine striations which intercepts the folds of the lateral fields. Procorpus of the esophagus swollen. Median bulb with 8 typical valves. Esophageal glands overlapping intestine only slightly. Three nuclei present in the esophageal glands. The fourth gland was wrongly misidentified in the first description of this species, leading to establishment of the genus *Paranguina*. Excretory pore a little above the level of the esophageal-intestinal junction. Ovaries reflexed. Postvulval uterine branch long, extending up to halfway or more from the vulva to anus. Tail conoid tapering to a finely rounded tip.

Males: Considerably more slender than females. Testis reflexed. Sometimes the testical reflex occurs at the commencement of the reproductive tube but sometimes the loop occurs at some distance below the beginning of the intestine. Spicules elongate, almost straight. Capitulum appears close–ended. Each spicule is laterally reinforced by two longitudinal rays which commence at one level in the last third of proximal part of the spicule. Gubernaculum 15 μm in length, uneven in thickness throughout its length, and in an optical section looks proximally bifurcate. Bursa narrow and seems practically to extend to the tail tip.

Hosts: Type host, couch (quack) grass, *Agropyron repens* P.B. with several varieties. Other host plants are rye (*Secale cereale* L.), wheat (*Triticum vulgare* L.), barley (*Hordeum sativum* L.), annual bluegrass (*Poa annua* L.), and *Elymus arenarius* L. Galls in the base of the stem of the annual bluegrass in California and galls on the same host plant in the Moscow region have been proved to be similar and the nematodes inducing them identical to each other. Thus, *A. pacificae* has been synonymized with *A. agropyri* (Chizhov and Subbotin, 1990).

Bionomics (after Kirjanova and Krall, 1971, emended): *A. agropyri* induces bulbous galls at the base of the stems of grasses (sometimes also on rhizomes). In Estonia, USSR, the main host is couch grass in which the nematodes reach maturity at the beginning of June. Egg laying continues to the end of June. At the beginning of July the first generation dies and the generation is developing within the same gall, reaching maturity by the second half of July. Egg laying occurs at the end of July and in August. Juveniles of the latter generation hatch and emerge from the galls in August and September. From this natural host infective juveniles accumulating in field soils, apparently in the autumn, readily attack wheat and rye. By the beginning of June next year, infested cereals become severely stunted. Instead of bulbous thickenings (galls) characteristic of couch grass, lower parts of stems of wheat and rye become stouted and swelled. In rye, small bulbous thickenings at the base of the stem may also be present. In the phase of development when the grains emerge in the tubule, lower internodes of infested plants remain abnormally short and thickened. They lose their pliancy but never show a typical bulb–shaped gall. As a result of the severe reduction of the lower internodes, the leaves of infested cereals take on the form characteristic of the disease—rosettes or sheaves. Depending on the degree of infestation, the plants either do not form ears or give an extremely low yield, the number and weight of grains being much smaller than from uninfested plants. As a result of their stoutness, the infested shoots are easily broken.

Winter wheat and rye are infested early in spring, but some invasive juveniles may penetrate the plant in autumn. Nematodes localize in the lateral portion of the stem or in the newly developing tissues directly above the nodes. In the lower internodes they incite slit-like necrotic areas about 3–5 mm in length. Sometimes juveniles are able to penetrate upward and infest the young tissues above the following node. They have occasionally been found up to a height of 10 cm in the third and fourth internodes. In about half of the observed cases, two to five juveniles per plant may cause severe stunting of cereals. Even a single juvenile of *A. agropyri* without success in the plant may cause typical symptoms of the disease. In July and August, severely infested wheat and rye plants may be observed in the field but they never contain any developmental stages of the nematode. Fragments of the adults of the first generation decompose rather rapidly in the young developing tissues.

In the case of 40 to 70% infested plants of rye, the grain yield has been estimated as 46 and 12% of the average harvest in slightly infested areas of the yield, respectively.

Distribution: European part, Caucasus and Siberia of the USSR, Poland, and North America (California). Damage to cereals has been hitherto documented only in Estonia, Latvia, and Lithuania and most probably also in Poland.

Control measures include control of the main host plant, couch grass, using herbicides. Sowing susceptible cereals immediately after perennial cultivation of grasses is not recommended in infested areas.

Anguina graminis (Hardy, 1850) Filipjev, 1936 (Fig. 8)
   syn. *Vibrio graminis* Hardy, 1850
         Tylenchus graminis (Hardy, 1850) Marcinowski, 1909
         Anguillulina graminis (Hardy, 1850) Goodey, 1932

Measurements (after T. Goodey from Kirjanova and Krall, 1971)

Females: L = 1.87–2.70 mm; a = 18.7–20.0; b = 9–10; c = 27–30; V = 84–86; stylet = 10 μm. Males: L = 1.12–1.58 mm; a = 21.0–22.4; b = 5–6; c = 16–17; stylet = 10 μm; spicules = 40–42 μm; gubernaculum = 13 μm. Second–stage juveniles: L = 0.67–0.79 μm. Eggs 71–83 × 33–37 μm.

After Southey, 1974: Females: L = 1.72–2.48 mm; a = 18.3–30.9; b = 9.4–12.1; c = 24.0–37.2; V = 84.5–89.2. Males: L = 1.25–1.54 mm; a = 25.3–26.9; b = 7.0–8.4; c = 18.9–23.5

Description (after Southey, 1974): Females: Body obese, spirally coiled, frequently having the near–circular spiral form after death. Lip region flattened, barely offset from body by a faint indentation. Cuticle marked by fine annulation. Lateral fields not observed. Esophageal procorpus and isthmus more or less constricted at their junctions with median and basal glandular bulb, respectively. The latter more or less pyriform. Excretory pore at the level of anterior part of glandular bulb, further forward than in *A. agrostis* or *A. tritici*. Ovary reflexed one or two times. Postvulval uterine branch extending up to halfway from vulva to tail tip and terminating in a small knob. Tail conoid, tapering to a finely rounded terminus with or without a minute lobe at its tip.

Males: More slender than females. After heat relaxation, may be curved ventrally but also dorsally. Testis reflexed one time to about half of its body length. Spicules arcuate with capitulum which appears to be open–ended anteriorly with dorsal and ventral edges curving toward one another. Gubernaculum simple. Bursa does not reach the tail tip.

Hosts: Sheep fescue, *Festuca ovina* L. (type host), also other species of *Festuca* (*F. rubra* L., *F. dumetorum* Hack a.o.). Some records on other grasses need critical re–examination (Kirjanova and Krall, 1971).

Bionomics: Incites small leaf galls. Older galls are of deep purple color. Second–stage juveniles are infective.

Distribution: Common in Europe, also in northwestern European part of the USSR.

FIGURE 8   Stem galls incited by *Anguina graminis* on the red fescue, *Festuca rubra* L. Estonia, USSR. (Original photograph by E. Krall.)

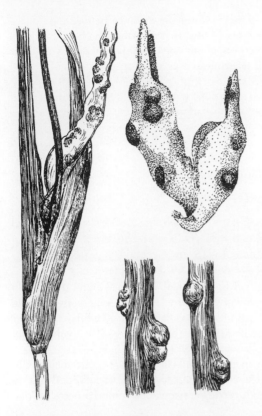

FIGURE 9　Galls incited by *Anguina australis* on the leaves and stems of the grass, *Ehrharta longifolia* S. M. (After Steiner, 1940 from Kirjanova and Krall, 1971.)

Anguina australis Steiner, 1940 (Fig. 9)

Measurements (after Steiner from Kirjanova and Krall, 1971)

Females: L = 2.1–2.5 mm; a = 19–20; b = 17–20; c = 28–31; V = 88–93; stylet = 8–9 µm. Males: L = 1.38–1.56 mm; a = 19–20; b = 11–12; c = 14–20; stylet = 8–9 µm; spicules = 31 µm; gubernaculum = 10 µm.

After Van den Berg, 1986: Females (paralectotypes, n = 3): L = 1.79–2.11 (1.90) mm; a = 11.0–12.2; b = ?; c = 29.9–37.0 (31.1); V = 89–91 (90); stylet = 9.9–11.1 (10.5) µm. Males (paralectotypes, n = 5): L = 1.12–1.42 (1.23) mm; a = 15.8–19.3 (18.0); b = 7.2–9.7 (8.2); c = 19.8–22.9 (21.2); stylet = 10.0–11.0 (10.4) µm; spicules = 26.5–35.3 (31.0) µm; gubernaculum = 9.6–12.1 (10.7) µm.

The differences in measurements made by Van den Berg (1986) from those in the original description (see above) can probably be attributed to remounting and flattening of the specimens.

Description (after Van den Berg, 1986): Females: body curved into a tight circle. Lip region flattened, set off, with faint annules. Lateral field with four incisures. Procorpus cylindrical, not narrowing markedly before ovate median bulb. Basal bulb large and irregular. Postvulval uterine branch reaching 55–60% (58%) of the distance between the vulva and anus. Tail tapering gradually to a pointed tip. Males: Body posture an open C. Lip region with four faint annules. Bursa extending from about two cloacal body widths anterior to the cloaca to just anterior to tail tip. Tail tapering to a pointed tip.

Host: A grass, *Ehrharta longifolia* S.M.

Bionomics: Incite small oval to round elevations with flattened bases on the leaves, stems, and inflorescence. Seed development may be prevented by the nematode.

Distribution: Australia.

Anguina hyparrheniae Corbett, 1966
    syn. *Afrina hyparrheniae* (Corbett, 1966) Brzeski, 1981

Measurements and Description (after Corbett, 1966)

Females (n = 10): L = 1.51–1.84 mm; a = 18–29; b = 7.0–11.3; c = 25–34; V = 90–92; stylet = 7–8 µm. Males (n = 10); L = 1.26–1.39 mm; a = 31–39; b = 6.8–9.6; c = 27–35; stylet = 6–7 µm; spicules = 32–40 µm; gubernaculum = 13–18 µm. Second–stage infective juveniles: L = 0.66–0.80 mm; a = 43–56; b = 4.9–6.5; c = 12.9–16.6; stylet = 6–8 µm.

Females: Body tightly coiled ventrally. Lips flattened, smooth. Lateral field rarely seen, with four incisures. Procorpus wide, median bulb ovate, isthmus long, convoluted in front of basal bulb. The latter large and spathulate. Cardia distinct. Genital tract outstretched, rarely with one small flexure. Postvulval uterine branch large, 1–1.5 vulval body width long. Tail terminus acutely pointed.

Males: Body coiled ventrally into an open to closed circle. Testis outstretched without flexures. Spicules massive and curved. Gubernaculum roughlike and curved. Bursa rejoins tail tip after forming two lobes which may overlap conspicuous ventral process beyond terminus.

Hosts: Tropical grasses of the genus *Hyparrhenia*. Type host *H. collina* (Pilger) Stapf. Alternative hosts include *H. rufa* (Nees) Stapf, *H. newtoni* (Hack.) Stapf, *H. gazensis* (Rendle) Stapf, *H. Variabilis* Stapf, *H.* ? *nyassae* and *Hyparrhenia* sp.

Bionomics: Associated with severe clumping of the inflorescence into "witches broom." Initiate seed gall formation. Galls are shriveled, light greenish brown to a dark olive, and much smaller than the light brown seeds. The galls contain eggs, second–stage juveniles, and adults. Third– and fourth–stage juveniles may be found only at the growing point of infected plants.

Distribution: Southeastern part of central Africa: Malawi, Mozambique, Zambia.

Anguina danthoniae Maggenti, Hart, and Paxman, 1974

Measurements and Description (after Maggenti a.o., 1974)

Females (n = 20): L = 1.82–3.09 mm; a = 12.1–16.2; b = 7.9–13.7; c = 25.3–42.4; V = 90–93; stylet = 8.0–12.5 µm. Males (n = 20): L = 1.59–2.35 mm; a = 16.2–21.4; b = 8.5–11.7; c = 20.4–31.3; stylet = 10.5–14.0 µm; spicules = 27.5–34.5 µm; gubernaculum = 11.0–17.5 µm. Eggs 100–135 × 37.5–47.5 µm. Second–stage infective juveniles: L = 0.74–0.82 mm; a = 49.6–66.0; b = 4.2–4.4; c = 7.7–9.5; stylet = 12.0–13.0 µm.

Females: Body coiled, very stout, tapering rapidly to both extremities. Head region set off. Cuticle finely annulated. Lateral field with longitudinal, transverse, and oblique striae. Procorpus cylindroid, narrow, median bulb ovate. Isthmus cylindroid, set off from glandular basal bulb by constriction. Basal bulb with posterior digitlike extension, 45–69 µm in length. Ovary outstretched, generally with one flexure in impregnated females. Oocytes arranged more than half the distance from vulva to anus.

Males: Body crescentic, less stout than in females. Testis outstretched. Spicules with distinct suture between calamus and lamina.

Host: California oat grass, *Danthonia californica* Boland.

Bionomics: Incite spherical galls (0.5–3.0 mm) on leaves and occasionally also on the seed head stems. They may be arranged on the leaves singly or in groups of 50 or more. Galls unpigmented, from green to straw yellow in color. Adults are the resistant stage which withstand the dry period in cryptobiotic conditions. Then they revive, their gonads mature and copulation and egg laying takes place. The second–stage juveniles are infective, they leave old galls and attack leaves.

Distribution: Widespread along the western coast of the United States (California, Oregon, Texas).

Anguina spermophaga Steiner, 1937
   syn. *Afrina spermophaga* (Steiner, 1937) Van den Brande, 1985
       Subanguina spermophaga (Steiner, 1937) Siddiqi, 1986

Measurements and Description (after G. Steiner from Kirjanova and Krall, 1971 with remarks by Van den Brande, 1985)

Females: L = 1.7–1.8 mm; a = 35.8–37.6; b = 8.3–10.0; c = 33–35; V = 87–91. Males: L = 1.3–1.4 mm; a = 31–35; b = 9; c = 21.6–24.0. Eggs 64.5–68.8 × 30.0–32.3 µm.

This type of material has been re–examined by Van den Brande (1985) as follows: Females (n = 5): L = 1.28–1.7 mm; a = 18.4–29.9; b = 9.8–11.9; c = 18.0–34.5; V = 88–92; stylet = 6.6–7.4 µm. Males (n = 9): L = 0.96–1.4 mm; a = 25.8–40.2; b = 4.9–7.3; c = 18.2–24.3; stylet = 5.5–7.0; spicules = 36.8–42.3 µm; gubernaculum = 14.0–18.4 µm. Juveniles (n = 6): L = 0.61–0.68 mm; a = 39.6–44.4; b = 5.3–7.4; c = 11.1–15.4; stylet = 7.4–8.5 µm.

Females: Body spindle–shaped, annulated. Lip region flattened. Stylet with very weakly developed basal knobs. Procorpus not exceptionally wide. Median bulb ovate. Basal bulb large and irregular. Ovary with two flexures. Postvulval uterine branch length about half the distance between vulva and anus. Tail with narrow, rounded, hyaline tip, in some specimens with slight indentations just anterior to tip.

Males: Spicules anguinoid with a long gubernaculum. Tail with a digitate projection (mucro). Bursa wide, extending to the base of this projection and expanded in two lobes in ventral view.

Hosts: A grass, *Saccharum spontaneum* L.

Bionomics: Described from the inflorescence (seed galls).

Distribution: Infested plants had been grown in Virginia, originating from the seeds collected in Turkestan, USSR. No other records from elsewhere.

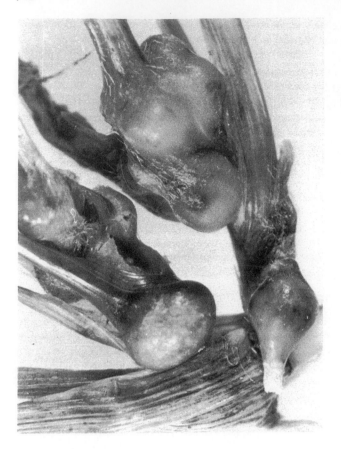

FIGURE 10   Galls incited by *Anguina tumefaciens* on the bradley grass in South Africa. (Courtesy of J. Heyns; original photograph by E. Krall.)

Anguina tumefaciens (Cobb, 1932) Filipjev and Schuurmans Stekhoven, 1941 (Fig. 10).
    syn. *Tylenchus tumefaciens* Cobb, 1932
        Anguillulina tumefaciens (Cobb, 1932) Goodey, 1933
        Pratylenchus tumefaciens (Cobb, 1932) Filipjev, 1936
        Afrina tumefaciens (Cobb, 1932) Brzeski, 1981

Measurements and Description (after N. A. Cobb from Kirjanova and Krall, 1971 with the notes by Van den Brande, 1985

Females: $L = 1.4$ mm; $a = 22.2$; $b = ?$; $c = 31.4$; $V = 88$. Males: $L = 1.2$ mm; $a = 31.2$; $b = ?$; $c = 19.6$; spicules = 39 μm; gubernaculum = 13 μm.
    After Van den Brande (1985): Females ($n = 7$): $L = 1.38–2.06$ mm; $a = 20.1–39.3$; $b = 6.7–10.9$; $c = 17.0$; $V = 86–92$; stylet = 8.8–10.7 μm. Males ($n = 10$): $L = 1.35–2.03$ mm; $a = 23.7–34.6$; $b = 7.1–8.0$; $c = 15.0–24.6$; stylet = 8.0–11.1 μm. Juveniles ($n = 2$): $L = 0.74–0.78$ mm; $a = 38.9$; $b = 5.9–6.0$; $c = 13.4$; stylet = 9.2–9.6 μm.

Females: Body posture ranging from an open C to a complete circle. Body narrowing rapidly behind vulva. Lip region flattened with indistinct annulation. Procorpus enlarged, median bulb rounded. Basal bulb large and irregular slightly overlapping intestine. Ovary single or double reflexed. Short postvulval uterine branch present. Tail tapering to an acute terminus.

Males: Bradley grass, *Cynodon transvaalensis*.

Bionomics: Incites ovoid galls 2–8 mm long on stems, leaves, and in flower heads. Infested plants are stunted and may die out on lawns.

Distribution: South Africa: Johannesburg, Pretoria, and Heidelberg in Transvaal.

Anguina microlaenae (Fawcett, 1938) Steiner, 1940
    syn. *Anguillulina microlaenae* Fawcett, 1938

Measurements and Description (after Fawcett from Kirjanova and Krall, 1971 with notes by Van den Brande, 1986)

Females: L = 1.3–3.0 mm; a = 12–19; b = 9–19; c = 15–32; V = 87–88; stylet = 9–9 µm. Males: L = 1.1–2.0 mm; a = 17–27; b = 8–14; c = 16–26; stylet = 8–9 µm; spicules = 27–35 µm; gubernaculum = 9–10 µm. Eggs 85–120 × 37–52 µm. Second–stage (originally described as first–stage) juveniles: L = 0.5–1.0 mm.

Females: Body obese, spirally coiled. Median esophageal bulb indistinct, elongated. Basal bulb overlaps the intestine. Ovary outstretched, six or eight oocytes around a rachis. Postvulval uterine branch extending about halfway of the distance from vulva to anus. Tail tip with an offset peglike process.

Males: Body curved dorsally. Testis outstretched with apparently about six spermatocytes around a rachis. Bursa not extending to tail terminus.

Hosts: A grass, *Microlaenae stipoides* R. Br. (type host). Another host plant is *Astrebla pectinata* (Lindl.) F. Muell.

Bionomics: Incites pediculate galls on shoots, leaves, and inflorescence. Up to 2,000 eggs may be found in each gall with dimensions 0.5–1.5 to 1.0–3.0 mm.

Distribution: Victoria, Australia.

Anguina pustulicola (Thorne, 1934) Goodey, 1951
    syn. *Anguillulina pustulicola* Thorne, 1934
        Ditylenchus pustulicola (Thorne, 1934) Filipjev and Schuurmans Stekhoven, 1941
        Subanguina pustulicola (Thorne, 1934) Siddiqi, 1986

Measurements (after G. Thorne from Kirjanova and Krall, 1971)

Females: L = 1.3–1.8 mm; V = 96; stylet = 10–12 µm. Males unknown.

This scarcely described species is characterized by the nonoffset head region, very short tail which is conoid and with bluntly rounded terminus as well as by a postvulval uterine branch, length of which exceeds the length of the tail. As no information is available on the median esophageal bulb, this species may belong to other subfamilies of Anguinidae as well.

Host: Unknown grass (Poaceae).
Bionomics: Incites leaf and stem galls of irregular form.
Distribution: Mexico.

Mesoanguina Chizhov and Subbotin, 1985 (after Chizhov and Subbotin)

Anguininae. During each growing season, usually two generations of nematodes develop in each gall. Adults of the first generation greater, after heat relaxation crescentic or spirally coiled. Adults of second generation smaller and not curved remarkably. Ovary in the females of first generation with two or three flexures. Oocytes in several rows. Preuteral gland consists of 32–60 cells in four to five rows, 8–12 eggs in each, and may contain up to 16 synchronous eggs. Ovary in females of the generation outstretched or with only one flexure. Oocytes in the zone of multiplication generally in single file. Preuteral gland shorter than in the females of first generation and may contain up to four synchronous eggs. Infective juvenile of the third stage. Incite unpigmented galls with well–expressed internal cavity on above–ground parts of dicotyledon, mostly on species of the family Asteraceae (Compositae), but also on Boraginaceae and Plantaginaceae.

Type species: *M. millefolii* (Löw, 1874) Chizhov and Subbotin, 1985

Key to species of *Mesoanguina* (after Chizhov and Subbotin, 1990)

1. Bursa encloses half of the tail 2
   Bursa encloses much more than half the tail 3
2. Postvulval uterine branch shorter than half of vulva–anus distance, its length equal to the corresponding body width. Stylet length more than 12 μm. Incite galls on *Balsamorhiza, Wyethia*, and *Helianthus* .... *balsamaphila*
   Postvulval uterine branch longer than half of vulva–anus distance, its length almost twice the corresponding body width. Stylet length less than 12 μm. Incite galls on *Guizotia* .... *guizotiae*
3. Postvulval uterine branch considerably shorter than half of vulva–anus distance. Average length of eggs equal to or less than 60 μm 4
   Postvulval uterine branch either slightly shorter or slightly longer than half of vulva–anus distance. Average length of eggs more than 60 μm 7
4. Postvulval uterine branch length 39–48 μm. Incite galls on *Achillea* .... *millefolii*
   Postvulval uterine branch length 17–25 μm. Incite galls on *Artemisia* ..... *moxae*
5. Spicule length under 27 μm, incite galls on *Arctotheca* .... *mobilis*
   Spicule length more than 27 μm 6
6. Stylet length under 11 μm. Orifice of dorsal esophageal gland less than 11.5 μm behind stylet base. Incite galls on plants of the families Plantaginaceae or Boraginaceae ..
   7
   Stylet length 11 μm or more. Orifice of dorsal esophageal gland more than 1.5 μm behind stylet base. Incite galls on plants of the tribe Cardueae of Asteraceae (Compositae) family .... *picridis*
7. Postvulval uterine branch shorter than half of vulva–anus distance. Incite galls on *Plantago* (family Plantaginaceae) .... *plantaginis*
   Postvulval uterine branch longer than half of vulva–anus distance. Incite galls on *Amsinckia* (family Boraginaceae) .... *amsinckiae*

Mesoanguina millefolii (Löw, 1874) Chizhov and Subbotin, 1985
  syn. *Tylenchus millefolii* Löw, 1874
      Anguillulina millefolii (Löw, 1874) Goodey, 1932
      Anguina millefolii (Löw, 1874) Filipjev, 1936
      Subanguina millefolii (Löw, 1874) Brzeski, 1981

Measurements (after Filipjev and Schuurmans Stekhoven, 1941)

Females of the first generation: L = 1.58–2.28 mm; a = 20–28; b = 8.8–12; c = 22.4–31.1; V = 88–92; stylet = 10–11 μm. Second generation: L = 0.93–1.38 mm; a = 28–46; b = 10; c = 15–20; V = 81.9–86.8. Males of the first generation: L = 1.33–1.55 mm; a = 27–37; b = 7.8–9.1; c = 16–20; stylet = 10–11 μm; spicules = 33–36 μm; gubernaculum = 15–17 μm. Second generation: L = 0.85–1.26 mm; a = 30–48; b = 5.7–8.9; c = 13–18; spicules = 30–32 μm; gubernaculum = 14–16 μm.

    Description (after Kirjanova and Krall, 1971): Females: Body usually coiled in a circle or in the form of C. Lateral field obscure. Valve of the median esophageal bulb shifted forward. Posterior glandular part of the esophagus overlaps intestine ventrally. Ovaries twice reflexed. Length of the spermatheca and the postvulval uterine branch equal to the corresponding body width. Males: Bursa narrow, extends almost to the tail tip. The latter slightly dorsally curved in some specimens. Eggs small, more rounded than cylindrical in shape, 42–61 × 28–35 μm in size. Invasive juveniles of the third stage are 0.65–0.76 mm in length.

    Hosts: Several species of the milfoil genus (*Achillea*) as well as *Tanacetum* (= *Pyrethrum*) *chiliophyllum* (F. and M.) Sch. Bip.

    Bionomics: Incite galls on the leaves, stems, and peduncles. Individual galls range from 3 to 6 mm in length, while sungalls are very much larger. The leaves may become twisted. Two generations develop in one gall.

    Distribution: Widely distributed in Europe. Also in the Caucasus (Armenia, USSR).

Mesoanguina picridis (Kirjanova, 1944) Chizhov and Subbotin, 1985 (Fig. 11)
    syn. *Anguillulina picridis* Kirjanova, 1944
      Anguina picridis (Kirjanova, 1944) Kirjanova, 1957
      Paranguina picridis (Kirjanova, 1944) Kirjanova and Ivanova, 1968
      Subanguina picridis (Kirjanova, 1944) Brzeski, 1981
      Paranguina centaureae Kirjanova and Ivanova, 1968
      Subanguina centaureae (Kirjanova and Ivanova, 1968) Brzeski, 1981
      Anguina chartolepidis Pogossian, 1966
      Subanguina chartolepidis (Pogossian, 1966) Brzeski, 1981
      Paranguina cousininae Kirjanova and Ivanova, 1968
      Subanguina cousininae (Kirjanova and Ivanova, 1968) Brzeski, 1981
      Anguina kopetdaghica Kirjanova and Shagalina, 1969
      Subanguina kopetdaghica (Kirjanova and Shagalina, 1969) Brzeski, 1981
      Paranguina montana Kirjanova and Ivanova, 1968
      Subanguina montana (Kirjanova and Ivanova, 1968) Brzeski, 1981
      Anguina pharangii Chizhov, 1984
      Subanguina pharangii (Chizhov, 1984) Siddiqi, 1981
      Paranguina varsobica Kirjanova and Ivanova, 1968
      Subanguina varsobica (Kirjanova and Ivanova, 1968) Brzeski, 1981

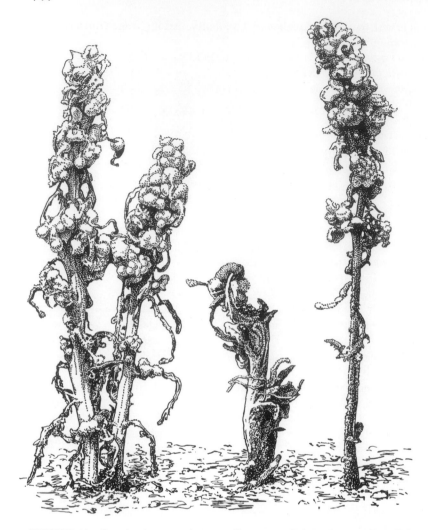

FIGURE 11   Russian knapweed, *Acroptilon repens* (L.) D. C. severely galled and stunted by the knapweed nematode, *Mesoanguina picridis* in Tadjikistan, USSR. (After Kirjanova and Ivanova, 1968 from Kirjanova and Krall, 1971.)

Measurements and Description (after Kirjanova and Ivanova, 1968)

Females: L = 1.5–2.1 mm; a = 22–28; b = 6.0–12.5; c = 14.0–25.0; V = 81–84; stylet = 12 μm. Males: L = 1.3–1.8 mm; a = 36.0–40.0; b = 6.7–7.0; c = 14.0–18.7; stylet = 12 μm; spicules = 40–43 μm; gubernaculum = 10–15 μm. Eggs 63–80 × 31–34 μm. Invasive third–stage juveniles: L = 0.75–0.85 mm.

Females: Cuticle annulated at esophageal region only, with longitudinal striae. Lateral field variable within a single specimen, in the middle–body region smooth or with six or more (to 20) incisures. Head region with three annules. Procorpus slightly swollen posteriad with a constriction before spherical median bulb. Ovary with two flexures. Spermatheca ovoid, 76 × 60.8 μm. Postvulval uterine branch long, with reduced ovary consisting of one to two cells. Tail conoid with pointed terminus.

Males: Testis with two flexures. Spicules with transverse suture and with a longitudinal ventral membrane.

The length of spicules has been thought to be one of the reliable diagnostic features to distinguish between species. But in *M. picridis* from the type host, Russian knapweed to several other host plants, Watson (1986c) demonstrated the spicules length being as follows:

28.5–45.0 (34.5) μm from *Cynara scolymus*
36.0–48.0 (40.5) μm from *Acroptilon repens* (type host)
36.0–54.0 (42.8) μm from *Onopordum acanthium*
40.5–48.0 (43.5) μm from *Centaurea diffusa*

From various records in the USSR the data on spicules length of different populations attacking closely related plant species but having been described as new species are as follows; 40.0–43.0 and 40.0–45.0 μm from *Acroptilon repens* in Tajikistan and Armenia, respectively (Kirjanova and Ivanova, 1968; Pogossian, 1966); 40.0–43.0 μm for *Paranguina centaurea* from *Centaurea squarrosa* Willd. In Tajikistan (Kirjanova and Ivanova, 1968); 39.0–42.0 μm for *P. cousiniae* from *Cousinia* spp. In Tajikistan (Kirjanova and Ivanova, 1968), 39.0–40.0 μm for *P. varsobica* from *Rhaponticum integrifolia* Winkl. In Tajikistan (Kirjanova and Ivanova, 1968); 36.0–43.0 (38.6) μm for *Anguina chartolepidis* from *Chartolepis biebersteinii* J. and Sp. in Armenia at altitude 2300 m (Pogossian, 1966); 33.4–42.0 μm for *A. kopetdaghica* from *Cousinia onopordioides* Ledeb. In Turkmenia (Kirjanova and Shagalina, 1969); 28.0–34.0 μm for *A. pharangii* from *Centaurea leucophylla* Bied. In foothills of northern Caucasus (Chizhov, 1984); 27.0–30.0 μm for *Paranguina montana* from *Cousinia* spp. in Tadjikistan at an altitude of 1500 to 2500 m (Kirjanova and Ivanova, 1969). Here a not very clear tendency appears, the spicules being smaller in populations from higher altitudes.

Hosts: *Mesoanguina picridis* has a wide range of host plants within the tribe Cynareae of the family Asteracea. All experimentally infested host plants belong to this tribe. Russian knapweed, *Acroptilon repens* (L.) D. C. (type host) is the only highly susceptible host plant in the experiments. Other experimental hosts are *Centaurea diffusa* Lam., *C. maculosa* Lam., *C. x pratensis* Thuill., *Carduus nutans* L., *Cirsium flodmanii* (Rydb.) Arthur, *Cynara scolymus* L., *Onopordum acanthum* L. (Watson, 1986a–c) as well as *Centaurea atropurpurea* Waldst. and Kit., *C. jacea* L., *C. phrygia* L., and *C. scabiosa* L. (Krall and Krall, 1976). There are seven further species of *Mesoanguina*, parasitizing plants of the same tribe Cynareae in southern regions of the USSR. They are morphologically closely related to *M. picridis*, which is a greatly variable species (Watson, 1986c). Thus two and even three different species have since been described from plants of the genera *Centaurea* and *Cousinia*, respectively. All they have been synonymized with *M. picridis* by Chizhov and Subbotin (1990). Some of the host plants (*Cousinia*) are highly susceptible to specialized populations (*A. kopetdaghica* in Turkmenia), so that pathotypes of *M. picridis* possibly exist.

Bionomics: Incites heavy gall formation on leaves and stems of Russian knapweed. This is a noxious weed in the southern part of the USSR as well as in the USA and Canada. Many authors have discussed the potential of this nematode as a biological control agent. It was introduced from Soviet mid–Asia and released in Kazakhstan, USSR, and Canada for this purpose (Ivanova, 1966; Watson, 1986a a.o.).

Distribution: Tadjik SSR (type locality) as well as other mid–Asian regions of the USSR and Caucasus (Armenia).

Mesoanguina moxae (Yokoo and Choi, 1968) Chizhov and Subbotin, 1985
    syn. *Anguina moxae* Yokoo and Choi, 1968
        Subanguina moxae (Yokoo and Choi, 1968) Brzeski, 1981 (after Choi and
           Loof, 1973)

Measurements and Description

Females (n = 31): L = 1.13–1.75 mm; a = 17–29; b = 88.9–14.2; c = 20–33; stylet = 9–10 μm. Males (n = 14): L = 1.11–1.45 mm; a = 25–35; b = 8.1–11.2; c = 18–23; Stylet = 9–10 μm; spicules = 30–34 μm; gubernaculum = 12–14 μm. Eggs 48–68 × 22–30 μm.
    Females: Body stout, usually coiled. Cuticle finely annulated. Lateral field with 9–10 incisures. Lip region almost continuous. Procorpus enlarged. Median bulb broadly oval, separated by a constriction from both procorpus and isthmus. Isthmus slightly swollen. Glandular basal bulb sharply offset from the isthmus. Dorsal gland overlapping the intestines. Ovary generally with two flexures. Oocytes arranged in numerous rows. Spermatheca oblong. Preuteral gland much longer than the spermatheca. Up to 11 synchronous eggs may be present in the genital tract. Postvulval uterine sac collapsed, shorter than one–half of the corresponding body width. Tail terminus sharply pointed.
    Males: Body straight or only slightly curved. Testis outstretched, without or with one or two flexures. Spicules with stout, parallel–sided head and curved shaft. Gubernaculum slightly curved. Bursa faintly crenate and does not reach the tail terminus.
    Host: *Artemisia asiatica* Nakai (family Asteraceae).
    Bionomics: Incite leaf galls which reach a diameter of 3–4 mm. Infested plants are generally stunted and are less branched than healthy ones. The juveniles may withstand desiccation for at least 3 years.
    Distribution: Korea and far eastern USSR.

Mesoanguina mobilis (Chit and Fisher, 1975) Chizhov and Subbotin, 1985
    syn. *Anguina mobilis* Chit and Fisher, 1975
        Subanguina mobilis (Chit and Fisher, 1975) Brzeski, 1981

Measurements and Description (after Chit and Fisher, 1975)

Females: L = 1.3–2.5 mm; a = 20.7–31.8; b = 6.4–9.4; c = 16.9–30.7; V = 71.0–86.8; stylet = 9–11 μm. Males: L = 1.2–1.5 mm; a = 21.2–39.1; b = 5.9–6.9; c = 16.3–25.4; stylet = 8–9 μm; spicules = 23–25 μm; gubernaculum = 7.2–7.8 μm. Eggs 82–114 × 50–57 μm. Second–stage juveniles: L = 0.26–0.27 mm.
    Females: Body slightly curved. Lateral field with eight incisures. Procorpus wide. Median bulb rounded. Isthmus slender. Basal bulb massive, spatulate, sometimes oblique and overlapping the intestine slightly. Ovary with one, two, or three flexures. Oocytes arranged around a rachis. Spermatheca ovate or slightly more elongate. Preuteral gland a long tube and separated from the spermatheca by a constriction. Up to four eggs in uterus and up to six eggs in preuteral gland at any one time. Tail conically pointed. Postvulval uterine branch occupies 65–72% of vulva–anus distance.
    Males: Body straight. Testis outstretched or with one or two flexures. Spicules massive and curved, the manubrium separated from the shaft by a sharp bulge. Gubernaculum tough–like. Bursa nearly enveloping tail terminus. Terminus sharply pointed.
    Host: Cape weed, *Arctotheca calendula* (L.) Levyns (family Asteraceae).
    Bionomics: Incite leaf and stem galls. In some cases twisting of leaves and stems occur. Third–stage juveniles withstand desiccation and attack plants when the first true leaves appear at second half of April. Small galls containing adults appear at the end of May.

Egg laying is observed from mid–May. Third–stage juveniles of the second generation may be found at the end of May, and young adults in the same galls in mid–June. Third–stage juveniles may also migrate through the tissues to incite new galls.

Distribution: South Australia.

Mesoanguina guizotiae (Van den Berg, 1986) Chizhov and Subbotin (1990)
   syn. *Anguina guizotiae* Van den Berg, 1986

Measurements and Description (after Van den Berg, 1986)

Females ($n = 19$): $L = 1.28–2.30$; $a = 24.6–38.7$; $b = 6.4–9.5$; $c = 14.1–22.3$; $V = 79–85$; stylet = 8.5–11.1 μm. Males ($n = 11$): $L = 1.05–1.48$; $a = 29.0–46.7$; $b = 6.2–8.4$; $c = 13.5–16.3$; stylet = 9.6–11.1 μm; spicules = 27.2–31.6 μm; gubernaculum = 8.1–10.3 μm. Second–stage juveniles ($n = 7$): $L = 0.62–0.82$ mm; $a = 29.9–39.1$; $b = 4.3–5.5$; $c = 9.6–12.1$; stylet = 8.8–10.6 μm. Third–stage (?) juveniles: $L = 0.94$ μm; $a = 42.7$; $b = 4.7$; $c = 10.4$; stylet = 9.9 μm. Fourth–stage (?) juveniles: $L = 1.22$ mm; $a = 53.4$; $b = 4.8$; $c = 12.0$; stylet = 9.6 μm.

    Females: Body S– or U–shaped. Lip region flattened, not offset, with three or four annules. Lateral field with four incisures. Procorpus cylindrical, narrowing slightly before the median bulb. Isthmus narrow, basal bulb overlapping the intestine dorsally. Ovary outstretched or with one to four flexures. In younger females, the ovary consists of a single row of cells. In older females there are many cells around the rachis. Postvulval uterine branch reaching 59% (45–74)% of the distance between the vulva and the anus. Tail tapering to an acute tip.

    Males: Testis outstretched or with one flexure. Bursa extending to about middle of tail.

    Host and bionomics: Incite gall formation on niggerwood (*Guizotia abyssinica*) of Asteraceae family. Data on the localization of galls are lacking. According to the figure of genital tract of this nematode given by Van den Berg (1986), it is a typical *Mesoanguina* and so apparently third–stage juveniles should be infective.

    Distribution: Africa (Ethiopia).

Mesoanguina balsamophilus (Thorne, 1926) Chizhov and Subbotin, 1985
   syn. Tylenchus balsamophilus Thorne, 1926
      Anguillulina balsamophila (Thorne, 1926) Goodey, 1932
      Anguina balsamophila (Thorne, 1926) Filipjev, 1936
      Ditylenchus balsamophilus (Thorne, 1926) Filipjev and Schuurmans Stekhoven, 1941

Measurements and Description (after G. Thorne from Kirjanova and Krall, 1971)

Females: $L = 2.6–3.4$ mm; $a = 17$; $b = 7$; $c = 30$; $V = 88$; stylet = 16 μm. Males: $L = 1.5–2.0$ mm; $a = 30$; $b = 7$; $c = 20$; eggs 60 × 37 μm. Second–stage juveniles: $L = 0.35–0.40$ mm. Third–stage (?) juveniles: $L = 0.9–0.11$ mm.

    Females: Body obese. Lateral field with four incisures or smooth on fully developed specimens. Ovary with two flexures. Postvulval uterine branch about as long as the corresponding body width. From 3 to 10 synchronous eggs in uterus and preuteral gland.

    Males: More slender than females. Spicules massive, curved, and with a weakly developed round capitulum. Bursa extending slightly past middle of tail.

    Hosts: Cutleaf balsamroot. *Balsamorhiza macrophylla* Nutt., also *B. sagitatta annuus* L. of the family Asteraceae.

Bionomics: Apparently the overwintering third–stage juveniles (? preadults by G. Thorne) attack young plants in spring. The incite hallow galls varying from 1 to 4 mm in length and 1 to 2 mm in height on the undersurface of the leaves.

Distribution: USA: Utah, Colorado, Idaho, and Washington.

Mesoanguina amsinckiae (Steiner and Scott, 1935) (Chizhov and Subbotin, 1985)
    syn. *Anguillulina dipsaci* var. *amsinckiae* Steiner and Scott, 1935
        Ditylenchus amsinckiae (Steiner and Scott, 1935) Thorne, 1961

Measurements and Description (after Thorne, 1961 from Kirjanova and Krall, 1971)

Females: L = 1.0–1.4 mm; a = 31; b = 8; c = 22; V = 83–91; stylet = 10 µm. Males: L = 1.0–1.3 mm; a = 30; b = 7; c = 15; stylet = 10 µm. Eggs 90 × 30 µm.

Females: Lateral field with eight incisures. Ovary usually reflexed two times. Oocytes arranged in four to six rows. Postvulval uterine sac occupies two–thirds of the distance between vulva and anus, a few rudimentary cells attached to its posterior end.

Males: Ratio of length of spicules to length of tail 1:2.38. Bursa extends almost to the tip of the tail.

Host: Type and single host *Amsinckia intermedia* Fish and Mey (family Boraginaceae).

Bionomics: Juveniles penetrate the young plants, migrate to the seed head, and initiate galls 5–10 mm long. Two generations develop inside each gall. One gall may contain 40,000–180,000 juveniles. Second–stage juveniles may withstand desiccation for about 4 years.

Distribution: California.

Mesoanguina plantaginis (Hirschmann, 1977) Chizhov and Subbotin, 1985
    syn. *Anguina plantaginis* Hirschmann, 1977
        Subanguina plantaginis (Hirschmann, 1977) Brzeski, 1981

Measurements and Description (from Hirschmann, 1977)

Females of the first generation: L = 1.19–2.06 mm; a = 31.0–47.0; b = 5.8–7.8; c = 14.7–24.4; stylet = 9.6–10.2 µm; spicules = 27.5–34.2 µm; gubernaculum = 10.2–13.9 µm. Second generation: L = 1.14–1.51 mm; a = 32.8–48.8; b = 5.2–7.5; c = 13.9–20.2l; stylet = 9.2–10.2 µm; spicules = 26.5–31.1 µm; gubernaculum = 8.8–13.6 µm. Eggs 66.8–96.4 × 30.6–43.8 µm. Third–stage infective female juveniles: L = 0.63–0.83 mm; a = 26.5–32.6; b = 5.0–6.3; c = 10.9–13.3; stylet = 9.2–10.0 µm. Third–stage infective male juveniles: L = 0.69–0.87 mm; a = 26.6–34.2; b = 4.4–6.4; c = 10.1–13.2; stylet = 9.1–9.9 µm.

Females: Body slightly curved ventrally. Cuticle finely annulated. Lateral field with mostly four to nine incisures. Lip region slightly set off, flattened, with two to three annules. Procorpus wide, median bulb oval, isthmus slightly swollen posteriad. Basal bulb overlaps intestine dorsally and laterally. Ovary usually twice reflexed in females of first and outstretched of second generation. Oogonia in germinal zone in several rows. Oocytes arranged around a rachis. Posterior growth zone with large oocytes in single file. Oviduct consisting of 12 cells. Spermatheca elongate, shorter than preuteral gland (crustaformeria), consisting of 32 cells (8 cells in 4 rows). Up to 12 or 3–4 synchronous eggs in uterus (preuteral gland) in first– and second–generation females, respectively. Postvulval uterine sac length about 43% of vulva–anus distance. Tail conical with peglike nonannulated tip.

Males: Spicules not fused, curved and massive with two sclerotized thickenings protruding into cytoplasmic core near junction of shaft and blade. Blades with sclerotized wings in ventral aspect. Gubernaculum toughlike, curved. Bursa crenate, not enveloping tail tip.

Host: Type and single host *Plantago aristata* Michx. (Plantaginaceae).

Bionomics: Incite hard globular or oval galls up to 3 – 5 mm in diameter on leaves, peduncles, and inflorescence (bracts, sepals, petals). Two generations often develop in one gall.

Distribution: North Carolina, U.S.A.

Heteroanguina Chizhov, 1980

Diagnosis (after Chizhov and Subbotin, 1980)

Anguininae. Adult specimens not curved remarkably after heat relaxation. Ovary outstretched or with one to three flexures. Oocytes in the zone of multiplication in single file. Preuteral gland consists of 60–90 cells in 4–6 rows, 12–18 cells in each row, and many contain up to 6 synchronous eggs. Infective juveniles of fourth stage, their average length exceeds 1.0 mm (up to 1.5 mm). Incite often pigmented galls without a pronounced internal cavity on above–ground parts of either mono– or dicotyledonous plants. Only one generation developing in each gall.

Type species: *H. graminophila* (Goodey, 1933) Chizhov, 1980

Key to species of *Heteroanguina* Chizhov, 1980 (after Chizhov and Subbotin, 1990)

1. Postvulval uterine branch collapsed, less than 50 μm in length; spicule length over 33 μm 2
   Postvulval uterine branch well developed, more than 50 μm in length; spicule length under 33 μm 3
2. Cuticle with well–developed annulation, lateral field with four incisures. Tail without thorn–like procession. Gubernaculum 11–12 μm. Incite leaf galls on dicotyledons (*Polygonum*) . . . . *polygoni*
   Cuticle with very faint annulation, lateral field with more than six incisures. Tail terminus often with a thorn–like procession. Gubernaculum 13–19 μm. Incite leaf and stem galls on monocotyledons (grasses) . . . . *graminophila*
3. Stylet length 9–12 μm. Orifice of dorsal esophageal gland 1.2–1.5 μm behind stylet base. Lateral fields with six incisures. Incite leaf galls on umbelliferous dicotyledons (*Ferula*) . . . . *ferulae*
   Stylet length 12–14 μm. Orifice of dorsal esophageal gland 8–10 incisures. Incite extremely elongated galls on leaves and stems of cyperaceous monocotyledons (*Carex*) . . . . *caricis*

Heteroanguina graminophila (T. Goodey, 1933) Chizhov, 1980
    syn. *Anguillulina graminophila* T. Goodey, 1933 (Fig. 12)
        Ditylenchus graminophilus (T. Goodey, 1933) Filipjev, 1936
        Anguina graminophila (T. Goodey, 1933) Thorne, 1961
        Subanguina graminophila (T. Goodey, 1933) Brzeski, 1981
        Anguina tridomina Kirjanova, 1958
        Anguina calamagrostis Wu, 1967
        Subanguina calamagrostis (Wu, 1967) Brzeski, 1981

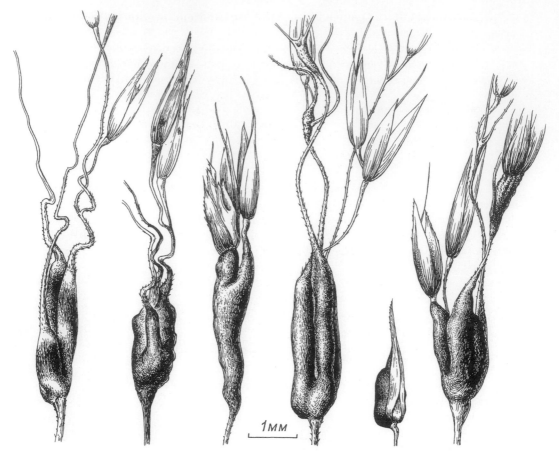

FIGURE 12   Stem galls incited by *Heteroanguina graminophila* in the inflorescence of *Agrostis tenuis* Sibth. Leningrad region, USSR. (From Kirjanova and Krall, 1971.)

Measurements and Description (after Chizhov, 1980)

Females: L = 1.56–2.60 mm; a = 25.5–49.1; b = 7.9–14.1; c = 15.7–38.4; V = 88–92; stylet = 13–15 μm. Males: L = 1.40–1.95 mm; a = 26.9–50.0; b = 7.0–13.1; c = 19.4–26.7; stylet = 13–15 μm; spicules = 34–44; gubernaculum = 15–18 μm. Second–stage juveniles: L = 0.25–0.68 mm; a = 22.2–35.0; b = 2.7–4.7; c = 9.4–19.2; stylet = 10–12 μm. Third–stage juveniles: L = 0.64–1.26 mm; a = 31.6–60.0; b = 3.4–6.4; c = 13.0–22.4; stylet = 10–12 μm. Fourth–stage infective juveniles: L = 0.88–1.55 mm; a = 24.8–84.6; b = 4.3–8.3; c = 12.5–26.8; stylet = 12–15 μm.

Females: Almost outstretched. Head offset. Ovary mostly outstretched and does not reach the esophagus. Oocytes in the zone of multiplication in single file. Spermatheca prolonged ovoid containing up to five synchronous oocytes. Preuteral gland long of irregular type with three to six synchronous eggs. Postvulval uterine sac greatly reduced. Tail sharply conoid.

Males: Body outstretched. Testis outstretched not reaching the esophagus. Spicules more resemble ditylenchoid than those of anguinoid type. Bursa narrow, not reaching the tail tip.

Bionomics: Infective fourth–stage juveniles overwinter in the soil and attack leafs of young grasses early in spring. They incite elongate leaf galls. Within 2 or 3 weeks after infestation mature specimens appear in the galls and after 5–7 days the oviposition begin. From 100 to 600 eggs may be laid by one female. One and one half or two weeks later, the embryonal development of juveniles within eggs has been finished and second–stage juveniles hatch in the galls. About another week is needed for the development of third– and fourth–stage juveniles in the same gall. Thereafter the galls split from the lower surface of the leaf and invasive juveniles move out. They may survive in the soil as well as attack new leaves or even stems immediately. Depending on the host species and humidity conditions, this nematode may develop in two or three generations in the mid–European part of the USSR.

Hosts: *Agrostis alba* L. (type host). Also several other species of bent–grass (*Agrostis* sp.) and *Calamagrostis* spp. Rarely oats (*Avena sativa* L.), timothy (*Phleum pratense* L.), and *Bromus* sp.

Distribution: Europe (including USSR), northern Asia (Siberia and far eastern USSR), North America (USA and Canada).

Heteroanguina caricis (Solovyova and Krall, 1982) Chizhov and Subbotin, 1985
    syn. *Anguina caricis* Solovyova and Krall, 1982

Measurements and Description (after Solovyova and Krall, 1983)

Females: L = 1.96–2.65 mm; a = 25–45; b = 13–22; c = 17.3–26.5; V = 83–88; stylet = 11.7–14.2 µm. Males: L = 2.02–2.96 mm; a = 34–42; b = 9–27; c = 19–25; stylet = 11.7–14.2 µm; spicules = 26.7–33.5 µm; gubernaculum = 10.0–11.7 µm. Eggs 40–86 × 27–69 µm. Fourth–stage infective juveniles: L = 1.47–1.50 mm; a = 42.9–44.0; b = 11.3; c = 14.7; stylet = 13.4 µm.

Females: Body outstretched or crescent. Head region offset. Lateral field with 8–10 incisures. Procorpus strongly developed, median bulb ovoid. In young females the glandular part of the esophagus well separated from the intestine. In impregnated females esophageal glands overlapping intestine ventrally or laterally and ventrally (Ryss and Krall, 1981). Ovary outstretched or with one to three flexures, with oocytes in several files. Spermatheca ovoid. Preuteral gland consisting of about 60–70 cells. Uterus 75–142 µm in length. Postvulval uterine branch either degenerated or may be developed as well as the true uterus being 50–134 µm in length. Tail terminus from roundly conoid to sharply pointed, sometimes with mucro.

Males: Body slender, almost straight. Head low, rounded. Testis outstretched or with one flexure. Tail terminus fingerlike, sharply pointed. Bursa narrow and exceeds 15–42 µm from tail tip.

Hosts: Type host *Carex nigra* (L.) Reichard of the family Cyperaceae. Also five other species of this genus: *C. acuta* L., *C. caespitosa* L., *C. heleonastes* Ehrb., *C. loliacea* L., *C. cinerea* Poll. Several other species of sedges proved not to be hosts in infested localities, however.

Bionomics: Infective juveniles may survive in tap water during 3–4 months (whereas those of *H. graminophila* died mostly within 3 days). On the other hand, they cannot withstand desiccation. Young shoots of sedges become infested in early May. At the beginning of June infested leaves become wrinkled and in the middle of June elongated galls appear on them. Gall length varies greatly from 4.5 to 87.0 mm and width from 0.5 to 5.0 mm. Up to 20 galls may develop on one leaf. Severely infested plants are stunted, their height being two or three times less than in healthy ones.

Distribution: Northwest USSR (Europe): Karelia, Estonia, Latvia, Lithuania, Leningrad region.

Heteroanguina ferulae (Ivanova, 1977) Chizhov and Subbotin, 1985
    syn. *Anguina ferulae* (Ivanova, 1977) Brzeski, 1981

Measurements and Description (after Ivanova, 1977 and 1981)

Females: L = 1.5–2.4 mm; a = 30–50; b = 8.2–12; c = 18–30; V = 85–90; stylet = 8.4–11.2 μm. Males: L = 1.2–2.0 mm; a = 40–60; b = 6–11; c = 18–21 μm; stylet = 8.4–12; spicules = 27–30 μm; gubernaculum = 9.6–11 μm. Eggs 70–78 × 40–50 μm. Second–stage juveniles: L = 0.19–0.65 μm; a = 20.6–33.0; b = 3.4–5.1; c = 7.4–14.0; stylet = 8.4–9.8 μm.Third-stage (cryptobiotic) juveniles: L = 0.67–1.1 mm; a = 34–43; b = 5.5–7.8; c = 12.6–17.2; stylet = 9.5–10.8 μm. Fourth–stage (infective) juveniles: L = 1.09–1.3 mm; a = 48.5–54.0; b = 5.3–7.8; c = 16.0–17.5; stylet = 9.5–10.8 μm.

Females: Lip region not offset, smooth. Cuticle finely annulated. Lateral field with six incisures. Procorpus expanded, median bulb ovoid. Isthmus long. Basal bulb dorsally overlapping intestine. Ovary with two flexures. Spermatheca ovoid. Postvulval uterine branch about a half or more the distance between vulva and anus. Tail conoid, terminus sharp or bluntly rounded.

Males: Body very slender. Testis flexured one or two times. Gubernaculum baton-like. Bursa extends only to the middle part of the tail.

Hosts: *Ferula Jaeschkeana* Vatke (type host), also *F. Eugenii* R. Kam. of the family Umbelliferae.

Bionomics: Incite leaf gall formation. Single galls are 2–4, sometimes to 10 mm in diameter. Galls are smooth, parenchymous, without an expressed cavity. Third–stage juveniles are cryptobiotic and withstand desiccation. Apparently the fourth–stage juveniles are infective. They attack plants at the end of March and at the beginning of April are already mature.

Distribution: Asia: Tajikistan, USSR.

Heteroanguina polygoni (Pogossian, 1966) Chizhov and Subbotin, 1985
    syn. *Anguina polygoni* Pogossian, 1966
        Subanguina polygoni (Pogossian, 1966) Brzeski, 1981

Measurements and Description (after Pogossian, 1966)

Females: L = 1.22–1.84 mm; a = 30.0–51.0; b = 6.0–8.8; c = 18.4–28.6; V = 74.5–94.5; stylet = 12–13 μm. Males: L = 1.26–1.52 mm; a = 30.0–58.5; b = 5.0–8.9; c = 16.4–28.5; stylet = 12–13 μm; spicules = 35–40 μm; gubernaculum = 11–12 μm. Eggs 60.0–100.8 × 29.8–43.2 μm. Fourth–stage infective juveniles: L = 1.02–1.22 mm; a = 32.6–41.9; b = 5.4–8.9; c = 17.0–33.0; stylet = 12 μm.

Females: Cuticle well annulated with four incisures on the lateral field. Body sharply narrowing behind the degenerated postvulval uterine branch. Basal esophagus bulb overlapping the intestine. Ovary with two or three flexures. Tail conical, terminus not extremely sharp.

Males: Spicules long and slender, almost straight, not curved. Bursa does not extend to the tail terminus.

Host: *Polygonum alpestre* C. A. M. (Polygonaceae).

Bionomics: Incite red gall–like thickenings on the leaves and petoles. Nematodes develop and reproduce in high numbers in the parenchymatous tissue. No internal cavity could be established within these thickenings.

Distribution: Asia: Armenia, USSR.

Subanguina Paramonov, 1967

Diagnosis (after Chizhov and Subbotin, 1990)

Anguininae. Adult specimens almost straight after heat relaxation. Ovary outstretched or with one to two flexures. Oocytes in the zone of maturation in a single row. Preuteral gland consists of 48–80 cells in four to five rows, 12–20 cells in each, and may contain up to four synchronous eggs. Infective juveniles of the second stage. Incite hooklike galls with well–developed internal cavity on roots of grasses. Only one generation develops in the gall.

Type and single species: *S. radicicola* (Greeff, 1872) Paramonov, 1967.

Subanguina radicicola (Greeff, 1872) Paramonov, 1967 (Figs. 13 and 14)

syn. *Anguina radicicola* Greeff, 1872
Tylenchus radicicola (Greeff, 1872) Oerley, 1880
Heterodera radicicola (Greeff 1872) Müller, 1884
Heterobolbus radicicola (Greeff, 1872) Julien, 1989
Caconema radicicola (Greeff, 1872) Cobb, 1924
Anguillulina radicicola (Greeff, 1872) Goodey, 1932
Ditylenchus radicicola (Greeff, 1872) Filipjev, 1936
Anguina radicicola (Greeff, 1872) Teploukhova, 1967
Tylenchus hordei (Greeff, 1872) Schøyen, 1885
Subanguina hordei (Schøyen, 1865) Siddiqi, 1986

Measurements and Description (after Chizhov and Marjenko, 1984 from specimens collected on annual meadow grass, *Poa annua* L.)

Females: L = 1.11–2.24 mm; a = 26.2–37.6; b = 6.8–13.8; c = 13.5–21.5; V = 75–81; stylet = 12–16 μm. Males: L = 0.86–1.98 mm; a = 26.0–36.9; b = 7.3–10.1; c = 13.6–20.8; stylet = 12–16; spicules = 25–31 μm; gubernaculum = 6–11 μm. Second–stage infective juveniles: L = 0.39–0.53 mm; a = 24.7–40.0; b = 3.4–4.1; c = 6.4–7.7; stylet = 10–11 μm. Eggs 79–125 × 31–40 μm.

Females: Labial region offset. Cuticle finely annulated with 10–12 incisures on the lateral field. Procorpus cylindrical, median bulb ovoid. Isthmus short and narrow. Esophagus glands as a wide bulb slightly overlapping the intestine dorsally. Ovary simple or once or twice reflexed. In the maturation zone the oozytes arranged in one row. Postvulval uterine branch long, more than half the vulva–anus distance. Tail tip sharply pointed.

Males: Testis outstretched, sometimes with one flexure. Bursa not extending the sharply pointed tail tip.

Bionomics: Females, males, eggs, and second–stage juveniles hibernate within the root galls. Juveniles may also leave galls by autumn. At the end of April they attack root tips of grasses and induce hypertrophy of parenchymatous tissue with the subsequent formation of a cavity between parenchyma and the root central cylinder. Only one generation develops in each gall but three subsequent generations have been observed in the Moscow region per growing season.

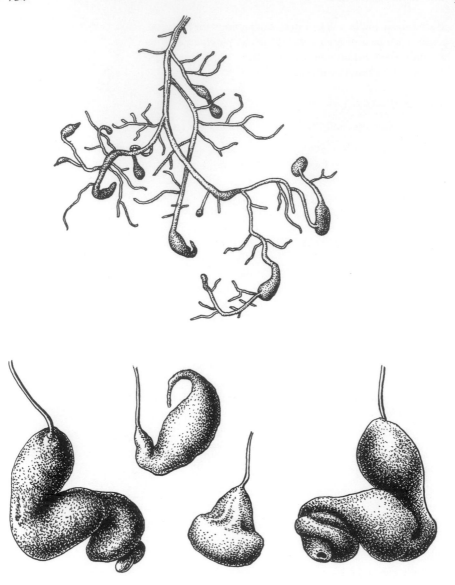

FIGURE 13   Hook–like root galls incited by *Subanguina radicicola* on *Phalaris arundinaceae* in Estonia (above) and on meadow grass, *Poa* sp. in the Ukraine (below). [(Original figure by E. Aomets (top) and from Kirjanova and Krall, 1969 (bottom).]

Hosts (after Kirjanova and Krall, 1971): Rye, wheat, barley, and many food grasses belong to the hosts of *S. radicicola*. Apparently not fewer than six races or pathotypes of this nematode exist. The Elymus race is attacking *Elymus arenarius* and may be transferred to barley, rye, wheat, and oats. The barley or Scandinavian race has been reported as a pest of cereals in northern Europe. It may also parasitize roots of foxtail, timothy, meadow grass,

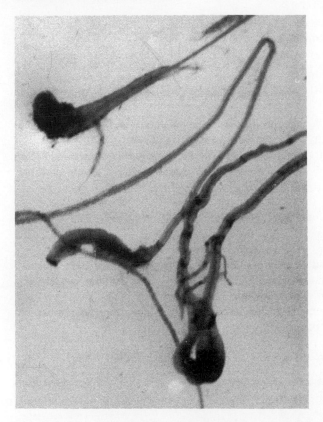

FIGURE 14   Roots of barley, experimentally infested with the meadow grass race of *Subanguina radicicola*. Estonia, USSR. (Original photograph by E. Krall.)

and hair grass. The Saskatchewan race parasitizes Smith's couch grass (*Agropyron smithii* Rydb.) and can migrate to wheat in Canada. Barley and rye are also susceptible. The Poa race is the most widespread in Europe. It severely infests Kentucky bluegrass, meadow grass a.o. The Phalaris race severely infests *Phalaris arundinaceae* L. in Estonia. The Rhode Island or Ammophila race is distributed in Rhode Island on *A. breviliguata* fern.

Distribution: Widely distributed in Europe. Also in North America (USA and Canada).

Subfamily Nothanguininae Fotedar and Handoo, 1978, after Siddiqi, 1986

Anguinidae. Adults 1–2 mm long, obese, stout, female spirally curved. Corpus of the esophagus nonmuscular, nonvulvate. Muscular median bulb absent. Basal bulb large, dorsal gland base may extend over intestine. Vulva at about 90%, without lateral membranes. Crustaformeria with more than four cells in each row. Several synchronous eggs in uterus. Ovary and testis reflexed. Gubernaculum in males absent. Bursa short. Gall inducing in grasses.

Type and single genus: *Nothanguina* Whitehead, 1959
Type and single species: *N. cecidoplastes* (Goodey, 1934) Whitehead, 1959

Nothanguina cecidoplastes (Goodey, 1934) Whitehead 1959
   syn. *Anguillulina cecidoplastes* Goodey, 1934
          Anguina cecidoplastes (Goodey, 1934) Filipjev, 1936, after Whitehead from
          Kirjanova and Krall, 1971

Females: Ovary reflexed twice. Postvulval uterine branch well expressed. Tail tip more or
less digitiform.
          Males: Bursa extends to half the distance between the cloaca and tail tip. Spicules
wide and flat with longitudinal rays, spicule head somewhat dilated.
          Bionomics, host and distribution: Induces gall formation on leaves, stems, and inflo-
rescence of *Andropogon pertusus* (wild.), a common fodder grass in India. Juveniles of the
second stage are invasive.

Subfamily Nothotylenchinae Thorne, 1941, after Siddiqi, 1986, abridged

Diagnosis:

Anguinidae. Adults slender, females rarely partially obese. Corpus nonmuscular, nonval-
vate, median bulb absent, or as a nonmuscular fusiform swelling, Monodelphic, prodelphic.
Crustaformeria with four rows of cells. Gubernaculum present.
Type genus: *Nothotylenchus* Thorne, 1941 (not gall inducing). Three other genera, two of
the (*Pterotylenchus* and *Orrina*) presented by gall–inciting species.

Key to gall–inducing genera of Nothotylenchinae (abridged after Siddiqi, 1986)

1.   Females with conspicuous lateral vulval flaps .... *Pterotylenchus*
     Lateral vulval flaps inconspicuous or absent .... *Orrina*

Pterotylenchus Siddiqi and Lenné, 1984, after Siddiqi and Lenné, 1984

Diagnosis

Nothotylenchinae. Females straight to slightly arcuate. Lateral field with four incisures.
Corpus nearly cylindrical, not differentiated into pre– and postcorpus; posterior region
slightly swollen lacking musculature and valva plates. Basal bulb elongate–saccate, with
dorsal esophageal gland extending over intestines for about one body width. Vulva a long
transverse slit flanked and partly covered by large prominent cuticular flaps. Postvulval
uterine sac present. Preuteral gland (crustaformeria) comprising 32–36 cells in four rows of
eight to nine cells each. Spermatheca elongate, empty. Ovary outstretched, with oocytes in
one or two rows. Tail elongate–conoid with a pointed tip.
          Type and single species: *P. cecidogenus* Siddiqi and Lenné, 1984.

Pterotylenchus cecidogenus Siddiqi and Lenné, 1984

Measurements (after Siddiqi and Lenné, 1984)

Females: L = 0.59–0.80 mm; a = 2–35; b = 4.4–5.8; c = 9.6–12.5; V = 80–84; stylet = 8–11
µm. Males not known. Second–stage juveniles: L = 0.21–0.31 mm; a = 18–26; c = 6.1–7.5;
stylet = 6–7.5 µm.
          Description (see above).
          Host: A tropical pasture legume *Desmodium ovalfolium* Wall. (Leguminosae).

Bionomics: Incite stem galls ranging from 0.5 to 2.0 cm on nodes and stem divisions. It may cause considerable disruption of the vascular systems and eventual death of the plant.

Distribution: South America: Colombia.

Orrina Brzeski, 1981

Diagnosis (after Brzeski, 1981)

Nothotylenchinae. Mature females slightly or nor swollen. No median bulb and no refractive thickenings in esophageal lumen. Esophageal glands form a lobe overlapping intestine. Isthmus short and narrow. Crustaformeria in form of quadricolumella. Ovary with few oogonia in circumference. Gall forming on leaves of solanaceous plants.

Type and single species: *Orrina phyllobia* (Thorne, 1934) Brzeski, 1981 (Fig. 15)
syn. *Anguillulina phyllobia* Thorne, 1934
Ditylenchus phyllobius (Thorne,
1934) Filipjev, 1936
Nothanguina phyllobia (Thorne,
1934) Thorne 1961

Measurements (after Thorne, 1961)

Females: $L = 0.75$–0.95 mm; $a = 23$; $b = 8.3$; $c = 18$; $V = 82$. Males: $L = 0.7$–0.85 mm; $a = 30$; $b = 7.4$; $c = 18$.

Females: Cuticle finely striated. Lateral field with four incisures. Lip region set off. Stylet short, slightly longer than width of lip region. Ovary a single line of oocytes except for a short section of multiplication. Postvulval uterine branch extending two–thirds the distance from vulva to anus. Tail conoid, acute.

Males: Spicula slightly arcuate, somewhat cephalated. Gubernaculum thin, toughlike. Bursa not extending to tail terminus.

Juveniles: The fourth–stage preadult juveniles ($L = 0.62$ mm) have conical tail with slightly rounded tip. This stage is the quiescent form.

Host: Silver–leaf nightshade (*Solanum elaeagnifolium* Cav.).

Bionomics (after Esser and Orr, 1979): In early stages of infection, small leaves are crinkled and thickened, then distorted abnormally and necrotized within about 6 weeks. Severely infested plants are stunted, their flowers and fruit sets reduced. According to Thorne (1961), in a single leaf, 842,000 specimens have been found. The only known host plant of *O. phyllobia* is considered one of the most important weed species in Africa, Australia, and southern parts of America. In field conditions of the USA in heavily infested soil 50% of the infected plants have been killed each year whereas growth and reproduction of the surviving plants were reduced 50%. The nematode is considered to be very promising in a biological weed control

Distribution: Arizona and Texas.

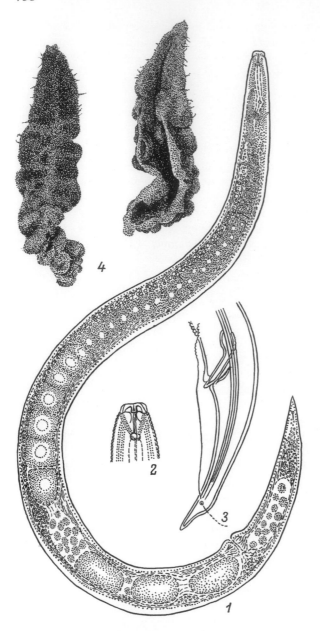

FIGURE 15   *Orrina phyllobia.* 1, general view; 2, labial region; 3, male tail; 4, galls on leaves of
Solanum eleaegnifolium Cav. (After Thorne, 1961 from Kirjanova and Krall, 1971.)

REFERENCES

Brzeski, M. W. 1981. The genera of Anguinidae (nematoda: Tylenchida). *Rev. Nematol.* 4: 23–24.

Chit, W., and Fisher, J.M. 1975. *Anguina mobilis* n.sp., a parasite of cape weed (*Arctotheca calendula*). *Nematologica* 21: 53–61.

Chizhov, V.N. 1980. On the taxonomic status of some species of the genus *Anguina* Scopoli, 1777. *Byull. Vsesoyuznogo Inst. gel'mintol.* 26: 83–95. In Russian.

Chizhov, V.N., and Berezina, N.V. 1986. *Poa annua* L.—a new host of *Paranguina agropyri* (Nematoda: Tylenchida). *Byull. Vsesoyznogo Inst. gel'mintol.* 45: 68–73. In Russian.

Chizhov, V.N., and Marjenko, A.Y. 1984. The morphology and biology of *Subanguina radicicola* (Nematoda: Tylenchida). *Zoologichesky Zh.* 63: 767–769. In Russian.

Chizhov, V.N., and Subbotin, S.A. 1985. Revision of the nematode subfamily Anguininae (Nematoda: Tylenchida) on the basis of their biological characteristics. *Zoologichesky Zh.* 64: 1476–1486. In Russian.

Chizhov, V.N., and Subbotin, S.A. 1990. Phytoparasitic nematodes of the subfamily Anguininae (Nematoda: Tylenchida). Morphology, trophic specialization, system. *Zoologichesky Zh.* 69.

Choi, Y.E., and Loof, P.A.A. 1973. Redescription of *Anguina moxae* Yokoo and Choi, 1968. *Nematologica* 19: 285–292.

Corbett, D.C.M. 1966. Central African nematodes. III. *Anguina hyparrheniae* n.sp. associated with witches broom of *Hyparrhenia* spp. *Nematologica* 12: 280–286.

Esser, R.P., and Orr, C.C. 1979. *Nothanguina phyllobia*: A nematode pest of a noxious weed *Solanum elaeagnifolium*. Florida Dept. Agr. and Consumer Service, Div. Plant Industry. Nematology Circular 51.

Filipjev, I.N., and Schuurmans Stekhoven, J.H., Jr. 1941. *A Manual of Agricultural Helminthology*. Brill, Leiden.

Fortuner, R., and Maggenti, A.R. 1987. A reappraisal of Tylenchina (Nemata). 4. The family Anguininae Nicoll, 1935 (1926). *Rev. Nematol.* 4: 163–176.

Hirschmann, H. 1977. *Anguina plantaginis* n.sp. parasitic on *Plantago aristata* with a description of its developmental stages. *J. Nematol.* 9: 229–243.

Ivanova, T.S. 1977. A new nematode species, *Anguina ferula* sp.n. (Anguininae: Tylenchida) from Ferula in Tadjikistan. *Doklady Akademii nauk Tadzhikskoi SSR* 20: 65–69. In Russian.

Kirjanova, E.S., and Ivanova, T.S. 1968. New species of the genus *Paranguina* Kirjanova, 1955 (Nematoda: Tylenchida) from Tadjikistan. Ushtchel'ye Kondara 2. Narzikulov. M.N. ed. Donish, Dushanbe: 200–217. In Russian.

Kirjanova, E.S., and Krall, E.L. 1971. *Plant Parasitic Nematodes and Their Control*. Leningrad, Nauka, Vol.2. In Russian. (English translation: New Delhi, 1980).

Krall, E., and Krall, H. 1976. Nematodes and biological weed control. European Society of Nematologists. *XIIIth Int. Nematol. Symposium, Abstracts.* Dublin, Ireland, pp. 50–51.

Maggenti, A.R., Hart, W.H., and Paxmann, G.A. 1974. A new genus and species of gall forming nematode from *Danthonia californica*, with a discussion of its life history. *Nematologica* 19(1973): 491–497.

Mckay, A.C., Fisher, J.M., and Dubé, A.J. 1981. Ecological field studies on *Anguina funesta*, the vector in annual ryegrass toxicity. *Aust. J. Agr. Res.* 32: 917–926.

Paramonov, A.A. 1967. A critical review of the suborder Tylenchina (Filipjev, 1934)(Nematoda: Secernentea). *Trudy gel'mintologicheskoi laboratorii Akad. nauk SSSR* 18: 78–101. In Russian.

Pogossian, E.E. 1966. New findings of the parasitic nematodes of the genera *Anguina* Scopoli, 1777 and *Paranguina* Kirjanova, 1955 in the Armenian SSR. *Doklady Akademii nauk Armjanskoi SSR* 42: 177–184. In Russian.

Ryss, A., and Krall, E. 1981. Classification of the superfamilies Tylenchoidea and Hoplolaimoidea with notes on the phylogeny of the suborder Tylenchida (Nematoda). *Eesti NSV Teaduste Akadeemia Toimetised (Izvestiya Akad, nauk Estonskoi SSR). Biol.* 30: 288–298.

Siddiqi, M.R. 1986. *Tylenchida: Parasites of Plants and Insects*. Commonwealth Institute of Parasitology. Commonwealth Agr. Bureaux, Farnham Royal, Slough, England.

Siddiqi, M.R., and Lenné, J.M. 1984. *Pterotylenchus cecidogenus* n.gen., n.sp., a new stem–gall nematode parasitizing *Desmodium ovalifolium* in Columbia. *J. Nematol.* 16: 62–64.

Solovyova, G.I., and Krall, E.L. 1983. *Osokovaya anguina* (The sedge leaf gall nematode). Leningrad, Nauka. In Russian.

Southey, J.F. 1972. *Anguina tritici*. C.I.H. descriptions of plant–parasitic nematodes. Set 1, No. 13. Farnham Royal, UK: Commonwealth Agricultural Bureaux.

Southey, J.F. 1974. *Anguina graminis*. C.I.H. descriptions of plant–parasitic nematodes. Set 4, No. 53. Farnham Royal, UK: Commonwealth Agricultural Bureaux.

Thorne, G. 1961. *Principles of Nematology*. McGraw–Hill, New York.

Van den Berg, E. 1985. Notes on the genus *Afrina* Brzeski, 1981 (Anguinidae: Nematoda) with descriptions of new and known species. *Phytophylactica* 17: 67–79.

Van den Berg, E. 1986. *Anguina guizotiae* sp.n. from Ethiopia with a redescription of *A. australis* Steiner, 1940 (Nematoda: Anguinidae). *Phytophylactica* 18: 11–16.

Watson, A.K. 1986a. Biology of *Subanguina picridis*, a potential biological control agent of Russian knapweed. *J. Nematol.* 18: 149–154.

Watson, A.K. 1986b. Host range of, and plant reaction to, *Subanguina picridis*. *J. Nematol. 18: 112–120.*

Watson, A.K. 1986c. Morphological and biological parameters of the knapweed nematode, *Subanguina picridis*. *J. Nematol. 18: 154–158.*

17

Tylenchulidae in Agricultural Soils

DEWEY J. RASKI   *University of California, Davis, Davis, California*

I. INTRODUCTION

There are at present in this family about 110 species distributed in nine genera. The citrus nematode, *Tylenchulus semipenetrans*, is the only species known to be of major economic importance to agriculture and it is worldwide in distribution.

Though few in species, the tylenchulids have a remarkable diversity in biology and feeding habits. Their hosts are predominantly perennial trees and vines, but some are parasites of vegetable crops, ornamentals, grasses, range land, etc.

All species are obligate parasites of roots except for *Tylenchocriconema alleni*, which feeds on the leaf surface of bromeliads in the crowns below the water line.

II. TAXONOMY OF TYLENCHULIDAE

A. Superfamily Criconematoidea Taylor, 1936

        Syn. Hemicycliophoroidea Skarbilovich, 1959
            Tylenchocriconematoidea Raski and Siddiqui, 1975

Diagnosis (after Raski and Luc, 1987) Tylenchina.

All stages usually under 1 mm long, rarely up to 1.9 mm (Hemicycliophorinae). Marked sexual dimorphism: male slender, female sausage-shaped, cylindrical, or spheroidal.

Female and juvenile with very variable cuticle: Thick with retrose annuli lacking lateral field, provided or not with lobation, crenation, spines, scales; or thick with smooth, coarse, rounded annuli covered or not with an extracuticular layer; or thin cuticle with fine rounded annuli and lateral fields often marked with lines (obliterated in swollen stages). Labial area in female and juvenile with usually one or two often modified annuli; oral aperture dorsoventrally longitudinal on a raised area or labial disk. Amphidial apertures round to oval, close to labial disk area. Basically, there are six pseudolips of which the four submedian ones can bear each a submedian lobe; no sensillae visible on surface of lip area.

Labial framework hexaradiate, with light to strong sclerotization. Deirids reported in thin-cuticled genera *Tylenchulus* and *Paratylenchus*. Phasmids absent. Females and most juveniles with well-developed stylet, often very long, with cone markedly longer than the shaft; basal knobs well developed, either sloping backward or anchor-shaped. Female and juvenile esophagus with median bulb enormously developed, muscular, containing a large often elongated cuticular valvular apparatus and being amalgamated with procorpus which is usually broad and surrounds the basal region of the stylet; isthmus either slender and offset from glandular bulb or short and broad being amalgamated with glandular bulb. Esophageal glandular bulb small, offset from intestine (except in *Sphaeronema whittoni* and *Meloidoderita kirjanovae* in which the glands are free). Orifice of dorsal esophageal gland at a short distance (usually under 4 μm) behind stylet base. Vulva transversely oval or slit-like, located posteriorly, usually at over 75% of body length. Female genital tract: one branch, anterior, outstretched (may be coiled in swollen females). Postvulval uterine sac absent. In juveniles: Female genital primordium showing no element of a posterior branch. Spermatheca usually offset and inclined laterally or ventrally. Uterus with a distinct columned part, but number of rows of cells apparently not constant; in swollen females ovijector can have a thickened wall, transformed into a cyst in *Meloidoderita*. Swollen female may deposit numerous eggs in a gelatinous matrix produced by the excretory system. Intestine syncytial, lacking a definite lumen, often extending beyond anal level. Female anus a small pore, rarely absent.

   Male: Small, slender. Cuticle thin, with narrow annuli; no extracuticular layer; typical lateral field present. Stylet mostly absent, or degenerated and nonfunctional. Esophagus degenerated, nonfunctional. One testis. Spicules often very long and setaceous, with small narrow head, elongate-slender shaft, and finely pointed distal end; variable in shape but often arcuate. Gubernaculum linear or crescent-shaped in lateral view, not protrusible. Caudal alae when present, usually low, rarely peloderan; but well developed, leptoderan in Tylenchocriconematinae and mostly Hemicycliophorinae. Cloacal lips usually narrow and elevated, or drawn out as a penial tube. Hypotygma present or absent.

   Inhabitant of soil: Females an obligate plant root parasite. Juveniles feed on plant roots, with rare exceptions. Male nonfeeding, free-living in the soil.

   Type family: Criconematidae Taylor, 1936.
   Other family: Tylenchulidae Skarbilovich, 1947.

B. Family *Tylenchulidae* Skarbilovich, 1947
Syn. Paratylenchidae Thorne, 1949
   Sphaeronematidae Raski and Sher, 1952
   Meloidoderitidae Kirjanova and Pogosyan, 1973
   Tylenchocriconematidae Raski and Siddiqui, 1975

Diagnosis (after Raski and Luc, 1987) Criconematoidea.
   Female: Small (except *Tylenchocriconema*: up to 0.83 mm). Body slender, swollen, or globose. Cuticle thin, except in some swollen or globose forms; without ornamentation, except some species with fine punctation (*Paratylenchus*) or minute spines (*Meloidoderita*). Typical lateral field present except in some swollen or globose forms. Weakly developed. Stylet delicate, of variable length; basal knobs rounded to sloping backward. Isthmus clearly marked; esophageal glandular bulb slightly reduced to medium size.
   Male: Stylet degenerate or absent. Caudal alae absent except *Tylenchocriconema*. Lateral field identical to females. *Juvenile*: Slender. Cuticle thin, without ornamentation.

Lateral field identical to female. Stylet present, functional (except fourth stage of *Paratylenchus*).

Bionomics: Ectoparasitic mostly on roots of higher plants, in some cases under bark of perennial host roots (some species of *Gracilacus*); exceptionally (*Tylenchocriconema*) on leaves and in crowns (mostly below water line of bromeliads).

Type subfamily: Tylenchulinae Skarbilovich, 1947.

Other subfamilies: Paratylenchinae Thorne, 1949
Tylenchocriconematinae Raski and Siddiqui, 1975

Key to Subfamilies of Tylenchulidae

1. Long (females <0.83 mm), slender species; head truncate (obliquely ventrad in males), squarish; males with long, leptoderan caudal alae .... Tylenchocriconematinae
   Short (females <0.50 mm), females slender to variously swollen or spherical in shape; males usually without caudal alae (rudimentary, if present) 2
2. Female stylet 12–119 μm, with elongate cone usually much longer than shaft plus knobs; mostly elongate-slender, if swollen females are elongate-obese ....
   .... Paratylenchinae
   Female stylet 8–20 μm; cone about equal to shaft plus knobs; mature females mostly swollen dorsally, lemon-shaped, or spherical .... Tylenchulinae

C. Subfamily Tylenchulinae
Syn. Sphaeronematinae Raski and Sher, 1952

Diagnosis (after Raski and Luc, 1987): Tylenchulidae.

All stages: Labial sclerotization discrete; stylet, when present, short, with rounded basal knobs. Excretory pore at very variable distance from anterior end. Strong sexual dimorphism.

Female: Fixed into the root (totally or partially); body obese. Excretory gland very developed producing a gelantinous matrix. Vulva very posterior to terminal. Uterine wall often thickened, transformed into a cyst in *Meloidoderita*. Anus and rectum often obscure, sometimes absent. Eggs embedded in a gelatinous matrix.

Male: Vermiform. Stylet degenerate or absent. Esophagus degenerate. Penial tube, if present, very short. Tail elongated. No caudal alae.

Juvenile: All stages provided with a functional stylet. (Deirids present in males and juveniles of some species.)

Type genus: *Tylenchulus* Cobb, 1913

Other genera: *Sphaeronema* Raski and Sher, 1952
Trophonema Raski, 1957
Trophotylenchulus Raski, 1957
Meloidoderita Pogosyan, 1966

Key to the Genera of Tylenchulinae

1. Female globose, lemon-shaped, or irregularly ovoid; uterine wall thickened; vulva subterminal to terminal, anus dorsal 2
   Female variously swollen; uterine wall not abnormally thickened; vulva and anus ventral 3
2. Uterus forms cystoid body enclosing some eggs .... *Meloidoderita*
   Uterus thickened but does not form cystoid body enclosing eggs ... *Sphaeronema*

3. Female excretory pore near vulva, with cuticular outgrowths .... *Tylenchulus*
    Female excretory pore more anterior, lacking cuticular outgrowths 4
4. Female with circumoral elevation, all stages encapsulated in round, brittle structure
    .... *Trophotylenchulus*
    Female without circumoral elevation, head rounded; various stages covered with
    gelatinous matrix .... *Trophonema*

1. *Tylenchulus* Cobb, 1913

Diagnosis (after Raski and Luc, 1987) Tylenchulinae.

All stages: Excretory pore situated very posteriorly (68–85% of body length); pore surrounded by small, irregularly shaped lobes; excretory duct forward-directed.

Female: Fixed into the root by its anterior part only. Body slightly crescent-shaped. Vulva posterior, but not terminal. Postvulval part short, tapering, ventrally curved and bluntly rounded (*T. semipenetrans*); almost straight, conoid, finely rounded (*T. furcus*). Anus obscure or absent (*T. semipenetrans*). No uterine cyst.

Male: Stylet degenerate. No penial tube. Lateral field with two lines (*T. semipenetrans*), two plus two faint inner lines (*T. furcus*).

Type species: *T. semipenetrans* Cobb, 1913.

Key to Species of *Tylenchulus* Cobb, 1913 (after Inserra et al., 1988)

1. Second-stage juveniles and adult females with distinct rectum and anus. Adult mature females with body swollen posteriorly 60% or more of its total length 2
    Second-stage juveniles and adult females without distinct rectum and anus; adult mature females with body swollen posteriorly for 58% or less of its total length . 3
2. Second-stage juveniles with furcate or bifid tail tips .... *furcus*
    Second-stage juveniles with tail tapering or narrowly conoid, not furcate ....
    .... *graminis*
3. Adult mature females with conoid postvulval section with broad base; males with stylet knobs 1.6 µm wide or more, basal bulb 8.1 µm wide or more, and tail cylindrical with bluntly rounded terminus .... *palustris*
    Adult mature females with digitate postvulval section with round terminus; male stylet knobs 1.2 µm wide or less, basal bulb 8.0 µm wide or less, tail tapering ....
    .... *semipenetrans*

2. *Trophotylenchus* Raski, 1987
    syn. *Ivotylenchulus* Hasim, 1983

Diagnosis (after Raski and Luc, 1987) Tylenchulinae.

All stages: Excretory pore situated posteriorly (33–61% of body length), excretory duct variable, apparently perpendicular to body line (*T. clavicaudatus, T. piperis, T. saltensis*), forward-directed (*T. floridensis, T. obscurus*), or posteriorly directed (*T. mangenoti*). Tail long, slender-conoid, finely rounded tip. Encapsulated in round, brittle structure.

Female: Fixed into root by anterior part only. Body swells on ventral side, curls tightly more than 360°. Circumoral disk protrudes prominently. Anus obscure, lips of anus slightly raised. No uterine cyst.

Juvenile: Circumoral disk present. Lateral field with two lines (four in *T. piperis*).

Type species: *Trophotylenchulus floridensis* Raski, 1957.

Key to Species of *Trophotylenchulus*

1. Female length 440 µm; juvenile length 380–440 µm 2
   Female length <410 µm; juvenile length <371 µm 3
2. *V* = 80; excretory pore of female 123 µm from anterior end; male length 410–580 µm,
   spicules 20 µm .... *floridensis*
   V = 70.5; excretory pore of female 182 µm from anterior end; male length 390–420 µm,
   spicules 14–15 µm .... *saltensis*
3. Juvenile tail slightly clavate .... *clavicaudatus*
   Juvenile tail bluntly rounded or with very small digitate terminus, never clavate ...
   4
4. Juveniles with distinct small, ventral digitate terminus .... *piperis*
   Juveniles with rounded terminus 5
5. Female stylet 13–15 µm .... *mangenoti*
   Female stylet <13 µm 6
6. Female tail tapering conoid; males known; juvenile stylet 14–15 µm .... *obscurus*
   Female tail bluntly conoid, rounded; males unknown; juvenile stylet 11–12 µm ...
   .... *andhraensis*

3. *Sphaeronema* Raski and Sher, 1952
   syn. *Goodeyella* Siddiqi, 1986
   Tumiota Siddiqi, 1986

Diagnosis (after Raski and Luc, 1987) Tylenchulinae.
   All stages: Excretory pore situated close to nerve ring level.
   Female: Globose to lemon-shaped or irregularly swollen. Vulva appearing as terminal, and anus as dorsal, when located. Uterine wall thickened.
   Male: No stylet. Spicular sheath present or absent.
   Type species: *Sphaeronema californicum* Raski and Sher, 1952.

Key to Species of *Sphaeronema*

1. Female head rounded, lip region not set off or bearing circumoral elevation; cuticle
   roughly reticulate .... *californicum*
   Female head with distinctly set off lip region or with circumoral elevation; cuticle
   smooth or finely annulated 2
2. Vulva of female with prominent, sharp ridges on each side of convoluted, coarse vulval
   lips .... *cornubiense*
   Vulva of female with smooth, rounded lips or flush with body contour 3
3. Female length = 0.13–0.21 mm; stylet = 12–13 µm; both females and juveniles with
   circumoral elevations; juvenile length = 0.30–0.34 mm; stylet = 12–13 µm ......
   .... *minutissimum*
   Female length >0.18 mm; stylet >15 µm; juvenile length 0.36 mm, stylet >14 µm; all
   juvenile heads rounded, not set off or with circumoral elevation 4
4. Female shape essentially globose except for neck region; vulva slit flush with body
   contour; juveniles with glands overlapping intestine .... *whittoni*
   Female mostly lemon-shaped to irregularly swollen, vulva lips rounded, protruding
   beyond body contour; juveniles with glands symmetrically arranged, set off from
   intestinal junction 5
5. Female length = 0.38–0.55 µm, head with circumoral elevation .... *camelliae*
   Female length <0.36 mm, head with lips set off by distinct incisure 6

6. Female with irregularly swollen shape, stylet = 15–20 μm; juvenile length = 0.36–0.45 mm, stylet = 14–17 μm, short tail ($c' = 3$) .... *sasseri*
   Female with regular, lemon-shape, stylet = 20–27 μm; juvenile length >0.44 mm, stylet >19 μm, tail longer ($c' > 5.8$) 7
7. Stylet knobs of females and juveniles strongly developed, rounded, cone on juvenile stylet about equal in length to base plus knobs, juvenile tail longer ($c' = 7.1–9.6$) . ... *alni*
   Stylet knobs of females and juveniles less developed, backwardly directed, cone on juveniles distinctly longer than base plus knobs, juvenile tail shorter ($c' = 5.8$) .... *rumicis*

4. *Trophonema* Raski, 1957

Diagnosis (after Raski and Luc, 1987) Tylenchulinae.
   All stages: Excretory pore at level of nerve ring.
   Female: Fixed into the root by their anterior portion. Body coiled, regularly inflated at the median part. Vulva posterior, but not terminal; postvulval part tapering, ventrally curved; anus and rectum present.
   Uterus wall not thickened. No uterine cyst.
   Male: No stylet. Penial tube short.
   Type species: *Trophonema arenarium* (Raski, 1956) Raski, 1957
      syn. *Sphaeronema arenarium* Raski, 1956.

Key to Species of *Trophonema* (after Minagawa, 1983)

1. (2) Female tail end bluntly rounded .... *okamotoi*
2. (1) Female tail end pointed.
3. (4) Female tail long ($c = 8.8–13.8$); male gubernaculum 6.0–6.3 μm; second-stage larva shorter than 320 μm .... *asoense*
4. (3) Female tail short ($c = 12–18$); male gubernaculum 5 μm; second-stage larva longer than 320 μm .... *arenarium*

5. *Meloiderita* Pogosyan, 1966

Diagnosis (after Raski and Luc, 1987) Tylenchulinae.
   All stages: Excretory pore situated around nerve ring level.
   Female: Globose. Vulva appearing as terminal and anus as dorsal. Cuticle bearing numerous small spines. Uterus wall greatly thickened, transformed into a cystoid body which retains part of the eggs.
   Male: No stylet. Penial tube short (*M. polygoni*) or absent (*M. kirjanovae*).
   Type species: *Meloidoderita kirjanovae* Pogosyan, 1966.

Key to Species of *Meloidoderita* (after Minagawa, 1983)

1. Female cystoid bodies average 388 μm long, 356 μm wide, excretory pore averages 112 μm from anterior end, spines on cystoid body average 22 μm long, width at base 5.3 μm .... *polygoni*
   Female cystoid bodies average 256–268 μm long, 218–222 μm wide, excretory pore averages 62–75 μm long from anterior end, spines on cystoid bodies averaging 8.5 μm long, width at base 2.3 μm or spines widely dispersed, swollen or knoblike 2

2. Spines on female cystoid bodies fine, averaging 8.5 µm long, width at base 2.3 µm, more dense; juvenile stylet 12–14 µm . . . . *kirjanovae*
    Spines on female cystoid bodies widely dispersed, coarse, swollen knoblike; juvenile stylet 14–16 µm . . . . *safrica*

D. *Tylenchulus semipenetrans* Cobb, 1913 (after Siddiqi, 1974)

Measurements (after Van Gundy, 1958): 25 females (immature): $L = 0.29$ (0.25–0.36) mm, $a = 17.5$ (15–20), $b = 2.52$ (2.15–3.00), spear = 13.5 (12–15) µm, excretory pore = 80.2 (76.9–84.2)% of body length. 10 females (mature): $L = 0.375$ (0.349–0.406) mm, $a = 4.5$ (3.75–5.07), $b = 2.97$ (2.80–3.12), excretory pore = 82.5 (79.4–86.0)% of body length, eggs = 67 × 33 µm. 25 males: $L = 0.37$ (0.33–0.41) mm, $a = 33.9$ (29–39), $b = 3.57$ (3.28–4.23), $c = 7.89$–10.1, $T =$ (approx.) 36, spear = 11 (10–12) µm, excretory pore = 53.1–58.4% of body length.

Second-stage larvae: Female larvae: $L = 0.323$ (0.295–0.364) mm, spear = 13 (12–14) µm. Male larvae: $L = 0.307$ (0.284–0.344) mm, spear = 12 (11–13) µm.

Description

Immature female: Body vermiform with distinct transverse striae. Lip region conoid-rounded, smooth, continuous with body; labial disk absent; labial framework moderately sclerotized. Spear about 13 µm long, with well-developed, rounded basal knobs. Orifice of dorsal esophageal gland about 4 µm behind spear knobs. Procorpus elongate cylindrical with sclerotized lumen; metacorpus or median bulb strongly muscular, oval, with large oval cuticular thickening in center. Isthmus also elongate-cylindrical; basal bulb saccate containing three esophageal glands, with esophagointestinal junction at base slightly shifted toward ventral side; cardia present; intestinal lumen indistinct; rectum and anus atrophied, nonfunctional. Vulva near posterior end, with thick labia. Excretory duct distinct, opening through a prominent pore close in front of vulva. Ovary single, anteriorly outstretched, immature, with a few oocytes. Postuterine sac absent. Tail bluntly rounded.

Mature female (measurements based on 10 females from lemon roots from Jhansi, India): Body behind neck swollen irregularly, 68–100 µm at its widest, ventrally arcuate; neck region distorted, often broken off; body behind vulva digitate 36 (32–43) µm long; diameter at vulva 23 (20–27) µm. Body cuticle 4 (2.5–5.5) µm thick near middle, not annulated; layer of somatic musculature absent. Ovary coiled; spermatheca with sperms; uterus with single egg (eggs oval 60–70 µm × 35–37 µm). Intestine syncytial, filling most of the body cavity, lacking a lumen; rectum and anus absent. Excretory pore 17 (12–19) µm in front of vulva. Maggenti (1962) discovered that the gelatinous matrix is produced through the excretory pore (and not through the vulva as, for example, in *Meloidogyne*); the excretory cell is enormously developed, is ventral, and has a large central nucleus with conspicuous nucleolus.

Male: Body slender, mostly straight with slightly arcuate tail end when relaxed; striae (0.8–0.9 µm apart; lateral fields inconspicuous. Lip region smooth, conoid; framework lightly sclerotized. Spear and esophagus degenerated; spear knobs minute; median bulb not muscular, spindle-shaped; basal bulb offset from intestine. Hemizonid opposite or just behind nerve ring. Excretory pore behind middle of body. Testis single, outstretched. Bursa absent. Spicules slender, arcuate, 14–18 µm long; gubernaculum straight or crescent-shaped, fixed 3–4 µm long. Cloacal aperture on a prominent conoid protuberance of body. Tail elongate-conoid to a rounded terminus.

Second-stage larva: Body straight or arcuate; striae 0.8 μm apart: lateral fields obscure, with two incisures. Lip region, spear, and esophagus as in immature female; tapering portion of spear less than half spear length with terminal third needlelike; orifice of dorsal esophageal gland 4 μm behind spear base; median bulb with spindle-shaped cuticular thickening in center; esophagointestinal junction pushed into basal bulb so that glands appear to form a short overlap (Fig. 1F). Deirids present, a little behind level of nerve ring; hemizonid distinct. Excretory pore behind middle of body. Genital primordium two-to four-celled, posterior to excretory pore. Rectum and functional anus absent; clear area indicating anus present near tail end of male larvae which are slightly shorter, more slender, and have a more pointed tail end than the female larvae.

Trophotylenchulus floridensis Raski, 1957

One female: 0.44 mm, a = 5.3, b = 2.4, c = 18.6, V = 57$_8$01.5, stylet = 15 μm. Seven males: 0.41–0.58 mm, a = 32–41, b = ?, c = 7.6–9.9, T = 27–41%, stylet = 10–12 μm.

Female (holotype): 0.44 mm, a = 5.3, b = 2.4, c = 18.6, V = 57$_8$01.5. Body swollen near vulva, tapering anteriorly, greatly reduced in diameter posterior to vulva; body annuli 2.5 μm near middle of body; lip region with distinct circumoral elevation, without definite annulation; labial framework lightly sclerotized; stylet 15 μm long, prorhabdion approximately 50% of stylet length, knobs 2 μm × 3 μm, rounded; dorsal esophageal gland orifice 7 μm behind stylet base; lining of corpus unusually thick; bulb elongate, massive; excretory pore posterior to esophagus, 123 μm (33% of body length) from anterior end; spermatheca not observed but found present in other paratype specimens; ovary well developed, extending to base of median bulb and reflexed twice; posterior uterine branch very short or absent; anal opening obscure, marked by slight elevation of cuticle, probably nonfunctional; tail arcuate, terminus bluntly rounded at end; lateral field not observed.

Male (allotype): 0.44 mm, a = 31.6, b = ?, c = 9.2; T = 31%. Body slender, cylindroid, tapering anteriorly and posteriorly; annuli 1.1 μm wide; lip region continuous with body contour, without definite annulation; labial framework very slightly sclerotized; stylet 11.5 μm long, weakly developed knobs represented only by slight swellings; excretory pore 175 μm (39.5% of body length) from anterior end of body; hemizonid approximately two body annuli wide, immediately posterior to nerve ring; esophagus degenerate; spicule slightly curved, approximately 20 μm long, with distinct bend near distal end; gubernaculum simple, slightly curved, 5 μm long; spicule sheath conspicuous; caudal alae absent; tail conoid, ventrally curved, with bluntly rounded terminus.

Larva: 0.41–0.38 mm, a = 27–31, b = 2.6–3.5, c = ?. Body slender, tapering anteriorly and posteriorly; annulation 1.2–1.3 μm wide; lateral field not observed; lip region continuous with body contour, without definite annulation but with distinctive circumoral elevation similar to that of female; labial framework lightly sclerotized; stylet 14–15 μm long, knobs 2 μm × 3 μm, rounded; dorsal gland orifice 4–4.5 μm behind spear base; excretory pore 164–194 μm (38–44% of body length) from anterior end; nerve ring immediately anterior to hemizonid; anus obscure, probably nonfunctional; tail elongate conoid with blunt terminus; genital primordium 140 μm from tip of tail.

Sphaeronema californicum Raski and Sher, 1952

Female: L = 0.134–0.209 mm, a = ?, b = ?, c = ?, V = subterminal.

Male: L = 0.395–0.470 mm, a = 33.2–44.8, b = ?, c = 12.0–14.8, T = 20.3–29.8%, gub. = 3–4.5 μm, spicule = 19–21 μm.

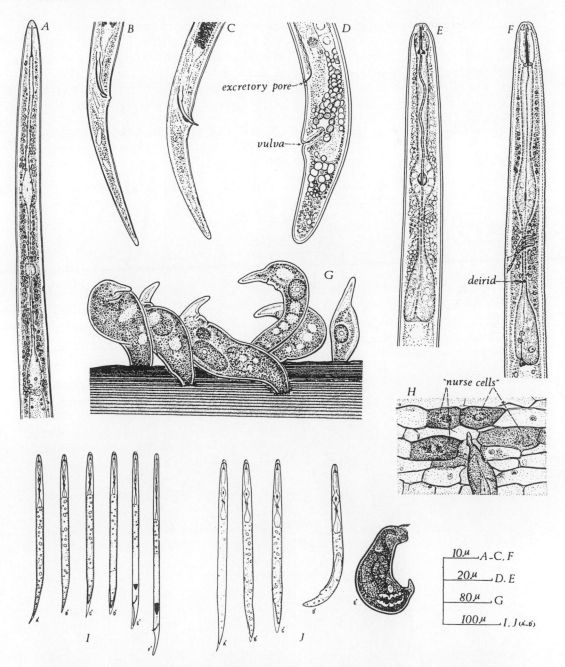

FIGURE 1 *Tylenchulus semipenetrans* Cobb, 1913. (A) Esophageal region. (B,C) Tail regions of males from lemon roots in Jhansi, India. (D,E) Posterior and anterior regions of an immature female from Brookville, Florida. (F) Esophageal region of a female II-stage larva from grapefruit soil in Melinda Forest area, British Honduras. (G) Females on citrus roots from Brookville, Florida. (H) A "feeding site" of adult female in cortex of a citrus root; note dense "nurse cells" around the head of nematode. (I,J) Developmental stages of male and female, respectively. (JE') Adult female from Riverside, California. [(D,E,G and JE') Redrawn from pencil sketches of N. A. Cobb, courtesy of A. M. Golden, USDA, Maryland; (H) after Van Gundy and Kirkpatrick, 1964; I(A'–F') and J(A'–D'), after Van Gundy, 1958, courtesy of *Nematologica*.]

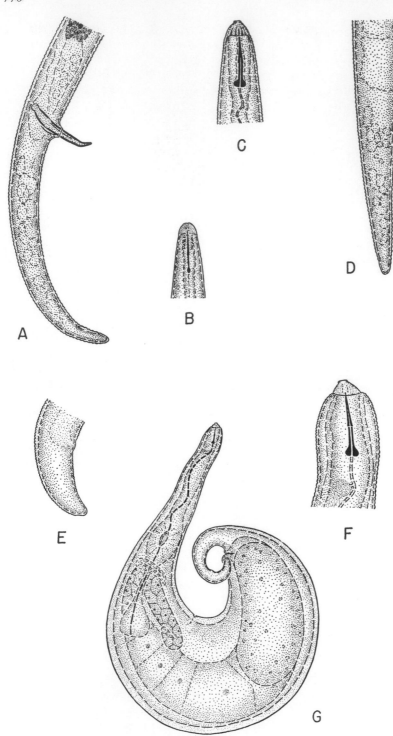

FIGURE 2   *Trophotylenchulus floridensis.* (A) Male tail, 2160×. (B) Head of male, 2160×. (C) Head of larva, 2160×. (D) Tail of larva, 2160×. (E) Tail of female, 2160×. (F) Head of female, 2160×. (G) Female, 710×.

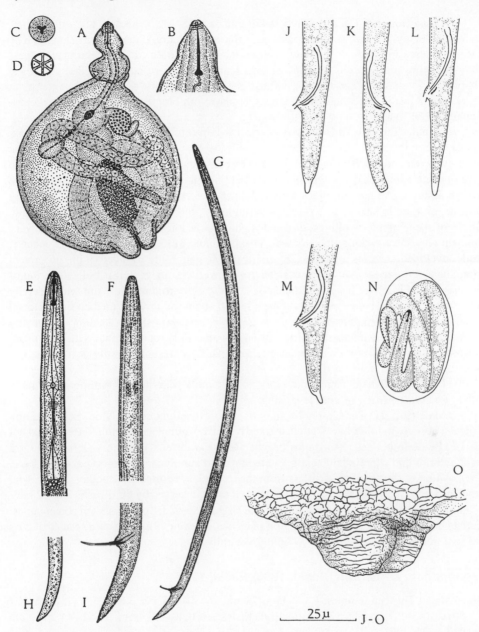

FIGURE 3 *Sphaeronema californicum.* (A) Female, entire, 290×. (B) Female, head, 750×. (C) Female, cross-section of median bulb, 290×. (D) Larva, face view, 750×. (E) Larva, esophageal region, 490×. (F) Male, esophageal region, 750×. (G) Male, entire, 300×. (H) Larval tail, 375×. (I) Male tail, 750×. (J–M) Male tails. (N) Egg, containing larva. (O) Female, vulvar region. [(A–I) After Raski and Sher, 1952. Courtesy of the Helminthological Society of Washington. (J–O) Original, specimens from type locality, courtesy of Rothamsted Experimental Station.]

Larva (Fig. 1A–C): L = 0.390–0.470 mm, a = 25–31, b = 3.3–4.4, c = 10.7–11.5. Body slender, cylindrical. Annulation of body fine and obscure. Lateral field not marked by definite lines. Lip region continuous with neck. Sclerotization of lips hexaradiate. Spear 14.4–16.7 μm, prorhabdion 53–57% of total length. Knobs of spear rounded swellings. Esophagus long, slender with well-defined isthmus. Excretory pore slightly anterior to nerve ring. Anus very obscure. Tail conoid, arcuate with bluntly rounded terminus, occasionally more sharply conoid than illustrated.

Female (Fig. 1G–I): Body subspherical with protruding neck which frequently assumes elaborate shapes due to conformity with the host plant cellular structure. Cuticle strongly developed, up to 9 μm thick, marked by conspicuous reticulate pattern. One or two very obscure annuli near lip region. Lateral field not marked by definite lines. Reticulate pattern assuming vague crosslines near and on lips of vulva. Sclerotization of head very delicate, obscure in face view. Spear 14–20 μm long, prorhabdion 51–60% of total length. Knobs of spear simple swellings, directed slightly posteriorly. Lip region set off by a very faint annulus, the anterior surface of which bears minute, smooth, rounded lips. Esophagus well developed. Corpus elongate, cylindrical. Median bulb rounded with prominent triradiate sclerotized valve. Isthmus slender with well set off posterior bulb. Junction of esophagus and intestine obscure. Lumen of esophagus approximately 1 μm wide, heavily sclerotized. Excretory pore present on neck region about the level of median bulb. Dorsal gland orifice approximately 4–5 μm posterior to spear. Ovary single, leading to a greatly developed uterus with unusually thick, muscular walls. Vulva subterminal with prominent, protruding lips. Small somewhat indefinite cuticular flap present laterally near vulva lips. Anus and phasmids not observed.

Immature female nematodes prior to last molt pear-shaped without protruding vulva, reticulate pattern of cuticle lacking.

Male (Fig. 1D–F): Lip region smooth, set off by slight constriction. Sclerotization delicate, obscure in face view. Cuticle marked by narrow transverse annuli. Lateral field not marked by definite lines. Esophagus degenerate, spear absent. Excretory pore 66–99 μm from anterior end. Hemizonid small, located on third annulus posterior to excretory pore. The position of the hemizonid differs from the examples reported by Goodey (1951), who described it as anterior to the excretory pore in all cases observed by him. Testis 20.3–29.8% of length. Spicules 19–21 μm long, simple and slightly curved. Spicule sheath conspicuous. Cuticle bulges characteristically in region of cloacal opening. Phasmids not seen. Tail conoid with rounded terminus, curved slightly ventrally.

Trophonema arenarium (Raski, 1956) Raski, 1957

Dimensions: Thirteen females: L = 0.45–0.58 mm, a = 10–16, b = 3.3–4.5, c = 12–18, V = 47–60$_{74-80}$1.3–2.7, stylet = 13–15 μm. Eight males: L = 0.43–0.57 mm, a = 30–44, b = ?, c = 8–10, T = 24–39%, stylet = lacking.

Female (holotype): L = 0.58 mm, a = 16, b = 3.8, c = 13; V = 59$_{75}$2.1. Body swollen near vulva, tapering anteriorly and posteriorly. Body annuli 1–1.3 μm wide. Lip region continuous with body contour, without definite annulation. Labial framework lightly sclerotized. Stylet 14 μm, long, prorhabdion approximately 50% of stylet length. Stylet knobs moderate in size, rounded. Dorsal esophageal gland orifice 6 μm behind stylet base. Esophagus very strongly developed with heavily cuticularized lumen. Excretory pore at level of nerve ring, 103 μm from anterior end. Hemizonid approximately two body annuli wide, immediately anterior to excretory pore. Spermatheca present. Ovary well developed, usually extending to base of median bulb, often reflexed once or twice. Posterior uterine

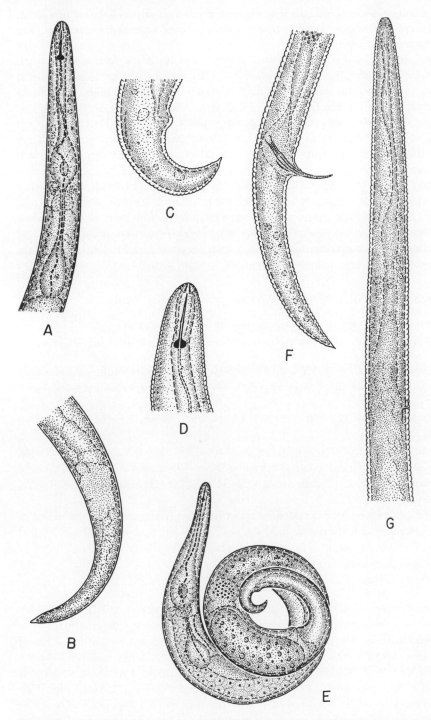

FIGURE 4 *Trophonema arenarium.* (A) Neck region of larva, 905×. (B) Tail of larva, 1385×. (C) Tail of female, 1385×. (D) Head of female, 1385×. (E) Female, 460×. (F) Male tail, 1385×. (G) Neck region of male, 1385×.

branch very short or lacking. Anal opening obscure, located on slight elevation of cuticle. Tail conical, ventrally arcuate, tapering to a minutely rounded terminus. Lateral field not observed. Phasmids not observed.

Male (allotype): L = 0.56 mm, a = 42.2, b = ?, c = 10. Body slender, cylindrical, tapering anteriorly and posteriorly. Body annuli 1–1.3 μm wide. Lip region continuous with body contour, without definite annulation. Labial framework very lightly sclerotized. Stylet lacking. Excretory pore 107 μm from anterior end. Hemizonid two body annuli wide, immediately anterior to excretory pore. Esophagus degenerate. Testis 34.4% of body length. Spicules slightly curved, 22 μm long. Gubernaculum simple, 5 μm long. Spicule sheath conspicuous. Bursa absent. Phasmids not observed. Tail conoid, slightly ventrally curved, with acute terminus.

Larva: L = 0.32–0.40 mm, a = 21–29, b = 3.1–3.7, c = 10.1. Body slender, tapering anteriorly and posteriorly, generally assumes coiled position in fixative. Annulation fine and obscure. Lateral field not observed. Lip region continuous with body contour. Labial framework lightly sclerotized. Spear 12–14 μm long, prorhabdion approximately 50% of stylet length. Knobs of spear with rounded swellings. Excretory pore 71–93 μm from anterior end. Anus very obscure. Tail conoid, ventrally arcuate, with rounded terminus.

Meloidoderita kirjanovae Pogosyan, 1966 (after Siddiqi, 1985)

Measurements (after Pogosyan, 1966 and Kirjanova and Pogosyan, 1973):

Holotype female: L = 0.352 mm, body width = 0.292 mm, neck = 4.8 μm, head annuli 3.

Paratype females (n = 127): L = 0.216–0.450 (0.33) mm, body width = 0.144–0.371 (0.255) mm, stylet = 15–19 μm, excretory pore = 44–88 (62.5) μm, vulva-anus distance = 44–180 (90) μm.

Cystoid bodies (n = 22): 180–355 (268) × 151–345 (218) μm. Eggs (n = 150): 60–92 (76) × 35–50 (42) μm.

Second-stage juveniles (live material, n = 15): L = 0.218– 0.522 (0.46) mm, body width = 17–20 (19) μm (fixed material after Pogosyan, 1966) (n = 33), L = 0.317–0.44 (0.409) μm, body width = 12–17 (13) μm, a = 22–34 (30.7), b = 3.0–3.4, c = 7.2–9.9 (8.2), stylet = 13–14 μm.

Second-stage juveniles from *Mentha* sp. in Armenia (fixed specimens mounted in glycerine—personal observations) (n = 20): L = 0.36–0.41 mm, a = 26–33 (28.5), b = 3.1–4.3 (3.5), c = 7–9.3 (7.6), c' = 5.1–6.3 (5.7), genital primordium = 70–73 (71.7)% or 97–122 (108) μm from anterior end, stylet = 12–14 (13) μm.

Description (after Kirjanova and Pogosyan, 1973 and Pogosyan, 1966)

Mature females: Body swollen, pyriform, without a neck or tail (Fig. 1G,H,L). Vulva at the tip of a small conical elevation of the posterior end of body; longitudinal axis of body from head to vulva with anus shifted dorsally, 44–180 (90) μm from vulva. Ratio of body length to width 1.08–1.68:1–1.32. Cuticle about 3.8 μm thick near head and about 6 μm thick near middle, with numerous spinelike outgrowths about 3 μm long with conus longer than shaft and basal knobs 4.4–5.5 (4.6) μm across and 2.2–3.3 (2.5) μm high. Orifice of dorsal esophageal gland 2.5–6.6 (4.3) μm from anterior end. Anterior and posterior cephalids opposite base of cephalic region and basal knobs of stylet, respectively. Excretory pore opposite median esophageal bulb, 44–88 (62) μm (in holotype 33 μm) from anterior end. Precorpus cylindrical. Median bulb oval, variable in size with age, 15–40 μm in diameter, with large oval valve plates. Vulval region about 50 μm in diameter, lacking a perineal pattern. Vulval slit 24–36 (25) μm long. Monodelphic, prodelphic. Uterus spherical, with

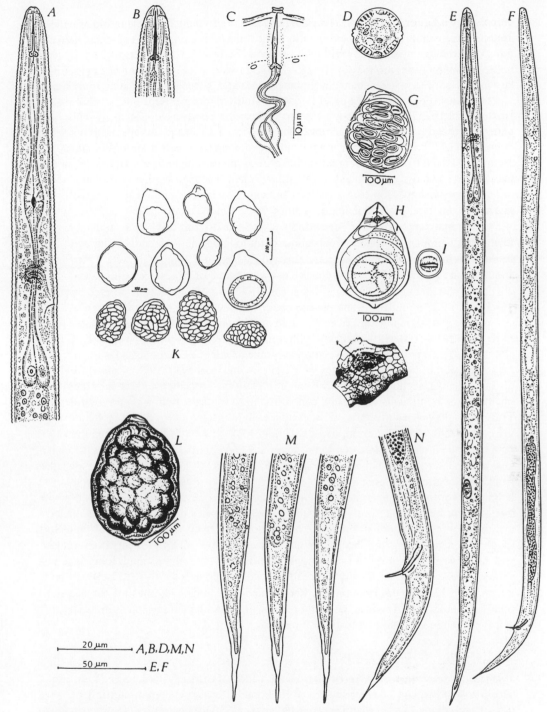

FIGURE 5 *Meloidoderita kirjanovae* (A–E), (G–M): (A, B) second-stage juveniles, anterior end. (C) Female anterior. (D) Second-stage juvenile, cross-section. (E) Second-stage juvenile, entire. (G, L) Mature females. (I) Vulva, ventral view. (J) Female in root. (K) Development of cystoid body by uterine enlargement. (M) Second-stage juvenile, tails. *Meloidoderita* sp. from *Mentha*, Israel: (F, N) males (original). [(A, B, D, E, M) Original from specimens ex. *Mentha* sp., Armenia, USSR (courtesy E. L. Krall). (G) From Poghossian (1966); (H–L) from Kirjanova and Poghossian (1973).]

very thick walls, occupying up to half the body cavity in young females. In old females it assumes a branched, palmate shape and fills most of the body cavity. The egg-filled uterus is transformed into a cystoid body within the maternal body wall which then withers. Cystoid bodies 180–355 (268) μm long, 151– 345 (218) μm wide. A large number of eggs are also laid in a gelatinous matrix which covers the females and, subsequently, the cystoid bodies.

Males: Pogosyan (1966) and Pogosyan and Kirjanova (1973) did not find males. However, in 1975 Pogosyan reported the male as having a body as long as the juvenile, and underdeveloped esophagus, and a bursa. The presence of a bursa in this species is doubtful; Pogosyan (1975) might have confused it with the molting cuticle. In a *Meloidoderita* sp. on *Mentha* in Israel, Cohn and Mordechai (1982) encountered one to four males in 2% of egg masses, several having the molted cuticle still attached. The males are very similar to those of *Sphaeronema* spp. (Fig. 1F,N). A stylet and bursa are lacking and the testis and esophagus are degenerate. The vas deferens is filled with minute sperm.

Second-stage juveniles: Vermiform. Cuticle 0.7–1 μm thick near midbody, with fine distinct annulation. Lateral fields about 4 μm long or one-fifth body width, with four equidistant incisures. Cephalic region 8 μm wide at base and 3 μm high, with three to four annuli. Stylet 14–15 μm long, with conus 8.6–9 μm long. Orifice of dorsal esophageal gland 2.3 (? 2–3) μm behind stylet base. Tail conical, with blunt or, more often, pointed terminus 36–53 μm long; terminal hyaline portion about a stylet length or more long.

Second-stage juveniles from a culture on *Mentha* sp. from Armenia, donated by E. L. Krall (personal observation): Body cylindroid, considerably tapering posteriorly. Cuticle annuli 0.9–1.1 μm wide, lacking longitudinal striae or ridges (somatic muscles usually give the impression of longitudinal ridges; cf. Kirjanova and Pogosyan, 1973). Lateral fields indistinct, in cross-section appear as two small ridges joined together (= three incisures). Cephalic region continuous with body, conoid-round to truncate, with a central oral depression; labial disk absent; framework hexaradiate, lightly sclerotized. Stylet moderately developed; conus well sclerotized, 6.6–7.2 (7) μm long, often a little longer than the shaft; knobs rounded, closely adpressed to the shaft, 2.6–3 μm across. Orifice of dorsal esophageal gland 3–4 (3.4) μm behind stylet knobs. Esophagus 90–123 (108) μm long; distance from anterior end to center of median bulb 49–68 (58) μm, or 49–60 (53.5)% of esophageal length. Precorpus narrows posteriorly, then gradually expands to merge imperceptibly with oval median bulb having prominent valve plates. Isthmus long and narrow. Esophageal glands forming a terminal bulb, slightly extending over anterior end of intestine; esophagointestinal junction inconspicuous. Excretory pore 76–84 (79) μm from anterior end, usually opposite middle of esophagus. Hemizonid two to three annuli long, at or just anterior to excretory pore. Genital primordium two to four cells at 70–73 (71–5)% of body length, 95–122 (109) μm from tail tip. Rectum and anus indistinct; intestine forms a short blind postanal sac. Tail conical, posterior region with irregular outline and terminus usually spicate; hyaline terminal part 9–14 (12) μm long (Fig. 1M).

1.   Hosts of *Tylenchulus semipenetrans*

Tylenchulus semipenetrans is reported from 29 species of *Citrus* (Vilardebo and Luc, 1961), 21 citrus hybrids, and 11 other species of Rutaceae. It is also known to multiply on olive (*Olea*), grapevine (*Vitis*), lilac (*Syringa vulgaris*) (USA), persimmon (*Diospyros lotus*) (USA), *Mikania batatifolia* (USA), loquat (India), pear (Japan), and *Calodendrum capense* (Israel).

Host plants of *Trophotylenchulus* include a wide variety of types: *Pinus* spp., *Quercus* spp., *Diospyros* sp., *Liquidambar styraciflua*, *Magnolia* sp. (*T. floridensis*); *Citrus aurantifolia* (*T. andhraensis*); *Deeringia arborescens* (Australia); *Dorstenia embergeri* (*T. mangenoti*); *Hodgkinsonia frutescens* (Australia); *Piper nigrum* (India); Olive, *Olea europaea* (Jordan).

Sphaeronema spp. have as hosts California laurel (*Umbellularia californica*), *Arctostaphylos* sp. (*S. californicum*); *Camellia japonica* (*S. camelliae*); sugarcane, *Saccharum* hybrid (*S. cornubiensis*); *Citrus* sp. (*S. minutissimum*); *Rumex confertus* (*S. rumicis*); Fraser fir, *Abies fraseri*; and red spruce, *Picea rubens* (*S. rumicis*).

Trophonema spp. have host plants as follows: salt rush, *Juncus leseurii, Centella asiatica*, and Sansouire (*T. arenarium*); *Scirpus wichurai* Böckler. f. *concolor* (*T. asoense*); and *Lespedeza bicolor* Turcz. var. *japonica* and *Indigofera pseudo-tinctoria* (*T. okamotoi*).

Host plants of *Meloidoderita* include *Mentha longifolia* (*M. kirjanovae*); *Polygonum* spp., *Rumex* spp., buckwheat, *Fagopyrum esculentum* and rhubarb, *Rheum rhabarbarum* (*M. polygoni*); sugarcane, *Saccharum* hybrid (*M. safrica*).

2. Distribution of *Tylenchulus semipenetrans*

Tylenchulus semipenetrans is found throughout the world in virtually every region where citrus is grown. The other three species so far are reported only from restricted areas: *T. graminis* and *T. palustris* (Florida) and *T. furcus* (South Africa).

Representatives of *Trophotylenchulus* are found widely spread in the world but all save one of the various species are known only from the location description: Florida (*T. floridensis*); India (*T. andhraensis*); Australia and USSR (*T. clavicaudatus*); Ivory Coast (*T. mangenoti*); Australia (*T. obscurus*); India (*T. piperis*); Jordan (*T. saltensis*).

The same is true for *Sphaeronema*: California (*S. californicum*); Florida (*S. whittoni*); North Carolina (*S. sasseri*); USSR (*S. alni* and *S. rumicis*); Japan (*S. camelliae*); Indonesia (*S. minutissimum*); South Africa (*S. cornubiensis*).

Only three species of *Trophonema* are known: They are from California, Madagascar, and France (*T. arenarium*); and Japan (*T. asoense* and *T. okamotoi*).

Meloidoderita is reported from USSR and Israel (*M. kirjanovae*); Maryland and Virginia (*M. polygoni*); and South Africa (*M. safrica*).

3. Bionomics of *Tylenchulus semipenetrans*

As mentioned earlier, *T. semipenetrans* is the only species of this family known to be of major economic importance to agriculture. It is the cause of "slow decline" of citrus which is accompanied by general reduction of tree growth, lack of vigor, yellowing of foliage, and small size of fruit. The critical level of infestation was reported by Cohn (1969) as 4000/g root above which tree growth and production is significantly lowered. Grapevines show a generally unthrifty condition when infested with this nematode.

The larvae of *T. semipenetrans* feed on the surface layers of roots causing some discoloration and necrosis. Young females penetrate deeper into the root tissues where a feeding site is developed around the head of the female. The posterior part of the female body remains outside the root and swells as eggs are produced. Eggs are laid in a gelatinous matrix outside the plant tissue. Van Gundy and Kirkpatrick (1964) reported the feeding site to be composed of 16 cortical cells with thickened walls and an enlarged nucleus and nucleolus. Secondary infections in the field result in decay and root destruction. Males as larvae or adults do not feed.

The life cycle is 6–8 weeks from egg to egg with optimum temperature for development at 25–31°C. Eggs hatch in 12–14 days and second-stage males develop to adults in 7 days with no feeding. Second-stage female larvae required 14 days to find a suitable rootlet, feed, and molt.

Females of *Sphaeronema, Trophonema,* and *Trophotylenchulus* have a feeding habit similar to *Tylenchulus* in that females penetrate to varying depths of the cortical layers of host roots with the posterior part of the body swollen and external to the root. *Trophotylenchulus* spp. have a unique host-parasite relationship. Soon after feeding commences a brittle, brown-to-black covering is formed, apparently originating from the host. This covering enlarges and surrounds the developing larvae and females. *Meloidoderita* females differ from the other genera in that larvae penetrate host plant roots to feed endoparasitically. They become exposed to the exterior as they mature, swell, and rupture the root surface. Eggs are laid in a matrix but some are also retained in the uterus, which hardens into a rough, cystoid body protecting those eggs.

E. Subfamily *Paratylenchinae* Thorne, 1949

Diagnosis (after Raski and Luc, 1987) Tylenchulidae.

Female: Small, slender nematodes, up to 0.50 mm; annulation fine; lateral field with two to four lines; lip region weakly sclerotized, bluntly rounded or with various shapes from small liplike projections near oral aperture; stylet length variable (12– 110 μm); conus longer than shaft plus knobs; procorpus gradually merges into metacorpus, isthmus long, slender, posterior bulb distinctly set off; spermatheca well developed offset from anterior end of uterus, with or without sperm.

Male: Degenerate; stylet if present weakly developed; esophagus reduced; spicules slightly curved; no caudal alae (except weakly developed in *Cacopaurus*).

Juvenile: Similar to females but mostly with shorter stylet [lacking or very reduced, rudimentary in fourth-stage (dauer) juvenile].

Bionomics: Long-stylet species become swollen as sedentary feeders, some under bark of perennial host roots; most others ectoparasitic on roots.

  Type genus: *Paratylenchus* Micoletzky, 1922
  Other genera: *Cacopaurus* Thorne, 1943
       Gracilacus Raski, 1962

Key to Genera of *Paratylenchinae*

1. Female body cylindroid-obese (a = 5–8), cuticle bearing minute tubercles, lateral field with rows of tubercles, vulva very posterior, postvulvar region short, bluntly conoid
  . . . . : *Cacopaurus*
  Female body slender or variously swollen mostly in prevulvar region, tubercles in cuticle rare, postvulvar region more elongate-conoid rounded 2
2. Female stylet 12–40 μm, not flexible; mostly vermiform, not abnormally swollen .
  . . . . *Paratylenchus*
  Female stylet 41–119 μm, flexible; many species swollen in mature stage . . . .
  . . . . *Gracilacus*

1. *Paratylenchus* Micoletzky, 1922
  syn. *Paratylenchoides* Raski, 1973

Diagnosis (after Raski and Luc, 1987) Paratylenchinae.

Females: Small, under 0.5 mm, vermiform, not abnormally swollen.

Labial framework weakly sclerotized (except *P. israelensis* and *P. sheri* where stronger). Stylet small to medium-sized (12–40 μm), not flexible. Excretory pore from level of nerve ring to level of esophagointestinal junction. Annuli smooth.

Juveniles: Resembling female. Stylet rarely present; if so weak in J4 only.

Type species: *Paratylenchus bukowinensis* Micoletzky, 1922.

Key to Species of *Paratylenchus* Micoletzky, 1922

1. Lateral field with three longitudinal incisures 2
   Lateral field with four longitudinal incisures 9
2. V = 72 (69–73) . . . . *minusculus*
   V = 77–86 3
3. Female stylet = 28–40 μm 4
   Female stylet = 15–26 μm 5
4. Female stylet = 35.6 (33–40) μm; lip region strongly set off by constriction, modified into a flattened disk; male length = 220–235 μm; tail sharply conoid . . . . *tui*
   Female stylet = 31 (28–33) μm; lip region less pronounced, truncate; male length = 248–269 μm; tail bluntly rounded . . . . *vandenbrandei*
5. Average female stylet length >22 μm 6
   Average female stylet length <22 μm 7
6. Female length 0.18 (0.17–0.19)mm; stylet length 18 (16–19) μm; tail finely rounded . . . . *humilis*
   Female length .22 (0.20–0.26)mm; stylet length 22 (20–23) μm; tail dorsally indented near terminus, bluntly digitate . . . . *leptus*
7. Female stylet length 16–20 μm; tail evenly conoid; males present . . . . . . *aquaticus*
   Female stylet length 14–16 μm; tail terminus conoid to variously shaped, males unknown 8
8. Dorsal gland orifice of female 4–6 μm from knobs, tail annulated to tip, conoid or variously shaped . . . . *variatus*
   Dorsal gland orifice of female 1–2 μm from knobs, tail smooth at tip, beaklike . . . . *rostrocaudatus*
9. Tail serrated near terminus . . . . *serricaudatus*
   Tail not serrated near terminus 10
10. Body of female appears smooth, annuli indistinct . . . . *leiodermus*
    Body annuli of female distinct 11
11. Body of female with deep cleft in tail 12
    Female without deep cleft in tail 13
12. Female stylet length 18 μm; vulvar flap missing; V = 86.5 . . . . *uncinatus*
    Female stylet length 23–28 μm; vulvar flap present; V = 81–84 . . . . *pseudouncinatus*
13. Female head with slight but definite constriction posterior to lip region 14
    Female head conoid, lips close to oral aperture projecting in a rectangular outline in front 17
    Female head rounded to conoid-truncate 22
14. Female stylet = 19–28 μm 15
    Female stylet = 29–39 μm 16
15. Female stylet = 20 (19–22) μm . . . . *perlatus*
    Female stylet = 24–28 μm . . . . *arculatus*
16. Vulvar flaps absent . . . . *morius*
    Vulvar flaps present . . . . *coronatus*

*Lack of males makes identification difficult and questionable. Eroshenko (1978) considered this species related to *P. halophilus*, *P. minusculus*, *P. nanus*, and *P. neoamblycephalus*. It also bears resemblance to *P. bukowinensis* [as redescribed by Brzeski (1976)]. *Paratylenchus concavus* can be distinguished from all these by total length, *V– a* distance, tail shape, etc., but characteristics of lip region shape is ambivalent between *a* and *b* in Fig. 3 of Eroshenko. In *a* the anterior end is simple truncate, in *b* it is rounded. This is important in precise use of this key.

35. Female length = 0.21(0.19–0.23) mm .... *breviculus*
    Female length >0.25 mm 36
36. V = 80 (78–81) .... *goldeni*
    V = 81–87 37
37. Female stylet = 14 (13–16) μm; male length = 0.31(0.29– 0.33) mm .... *variabilis*
    Female stylet = 18 (16–20) μm; male length = 0.40 mm .... *alleni*
38. Female length = 0.19–0.25 mm 39
    Female length >0.29 mm 40
39. Female stylet = 32 (28–35) μm; spicules gently curved to tip .... *salubris*
    Female stylet = 23 (21–24) μm; spicules with distinct distal bend ventrad ....
    .... *flectospiculus*
40. *V* = 80 (78–82); male tail long, *c* = 10–12 .... *tenuicaudatus*
    V = 81–87; tail shorter, *c* = 12–15 41
41. Female tail broadly conoid with bluntly rounded, almost hemispheric terminus ...
    .... *mexicanus*
    Female tail more slender conoid with rounded terminus .... *dianthus*
42. Female stylet = 38 μm .... *paramonovi*
    Female stylet = 35 μm 43
43. Males with stylet 44
    Males without stylet 47
44. Female stylet = 13 (12–14) μm .... *veruculatus*
    Female stylet = 17 μm .... *besoekianus*
    Female stylet = 22 (21–23) μm .... *holdemani*
    Female stylet = >23 μm 45
45. Tail of male and female sharply conoid .... *baldaccii*
    Tail of male and female conoid-rounded 46
46. Average length of female stylet = 24–25 μm; total range = 27–34 μm .... *hamatus*
    Average length of female stylet = 24–25 μm; total range = 23–29 μm ....
    .... *bukowinensis*
47. Male tail sharply pointed 48
    Male tail conoid-rounded 50
48. Female stylet = 15–16 μm .... *vexans*
    Female stylet = >21 μm 49
49. Female stylet = 31 (27–34) μm; male tail gently curves ventrad .... *ciccaronei*
    Female stylet = 25 (21–29) μm; male tail strongly curved ventrad 90–180° ....
    .... *fueguensis*
50. Average length of female stylet = <22 μm 51
    Average length of female stylet = >24 μm 54
51. Average *V* = 78 (76–80) .... *longicaudatus*
    Average *V* = >81 (total range = 79–89) 52
52. Average length of female = >0.32 mm .... *microdorus*
    Average length of female = <0.30 mm 53
53. Female stylet averages 16–18 μm (total range = 14–21); tail evenly conoid with
    rounded terminus .... *minutus*
    Female stylet averages 20–22 μm (total range = 19–24); tail slender conoid, terminus
    finely rounded to acute .... *elachistus*
54. Average length of female stylet = <25 μm .... *lepidus*
    Average length of female stylet = >27 μm 55

55.   Female tail subacute .... *nanus*
      Female tail slender conoid to finely rounded terminus .... *neoamblycephalus*
Note: *Paratylenchus platyurus* was described by Eroshenko (1978) and a full translation of
the Russian text has not been available. Therefore, it has not been possible to include it in
this key.

2.   *Gracilacus* Raski, 1962

Diagnosis (after Raski and Luc, 1987) Paratylenchinae.
        Females: Small, under 0.5 mm, vermiform or swollen in prevulval region. Labial
framework weakly sclerotized. Stylet long (41–119 μm), flexible. Excretory pore from
level of base of stylet to level of nerve ring. Annuli smooth, more rarely (three species) with
small tubercles.
        Juveniles: Resembling female. Stylet generally present, well developed.
        Type species: *Gracilacus epacris* (Allen and Jensen, 1950) Raski, 1962
                    syn. *Cacopaurus epacris* Allen and Jensen, 1950
                          Paratylenchus epacris (Allen and Jensen, 1950) Goodey,
                          1963.

Key to Species of *Gracilacus*

1.   Lateral field with two incisures 2
     Lateral field with three incisures 4
     Lateral field with four incisures 10
2.   Female stylet = 110–119 μm .... *elegans*
     Female stylet = 54–76 μm 3
3.   Female stylet = 54–59 μm, tail blunt rounded, lateral vulvar membranes lacking ..
     .... *janai*
     Female stylet = 68–76 μm, tail conoid rounded, lateral vulvar membranes present .
     .... *peperpotti*
4.   Lateral vulvar membranes small 5
     Lateral vulvar membranes lacking 6
5.   Refractive cuticular ornamentations present on females .... *colina*
     Refractive cuticular ornamentations lacking on females .... *idalima*
6.   Female tail bluntly rounded 7
     Female tail subacute to finely rounded or deformed 8
7.   Female stylet = 58(48–68) μm; *b* = 2.7 (2.3–4.1) .... *aculenta*
     Female stylet = 75(63–80) μm; *b* = 2.2 (1.7–2.5) .... *latescens*
8.   Female stylet = 61–69 μm .... *acicula*
     Female stylet 69 μm 9
9.   Female length = 0.24(0.21–0.25) mm; tail mostly deformed, some acute ....
     .... *solivaga*
     Female length = 0.31(0.29–0.34) mm; tail subacute .... *costata*
10. Lateral vulvar membranes present 11
     Lateral vulvar membranes absent 28
11. Female head with submedian lobes or lips prominent and/or protruding 12
     Female head conoid to rounded 16
12. Female stylet = 92 (86–95) μm .... *peratica*
     Female stylet < 82 μm 13

13. Female cuticle ornamented with minute refractive dots .... *mutabilis*
    Female cuticle smooth 14
14. Female head with low, round lips .... *parvula*
    Female head with expanded lips set off by deep constriction 15
15. Female length = 0.22–0.32 mm; males with expanded lips as in female, stylet lacking
    .... *longilabiata*
    Female length = 0.41–0.51 mm; male head slightly expanded, lobed; stylet weak ..
    .... *capitatus*
16. Excretory pore at isthmus or more posterior 17
    Excretory pore near valve of median bulb or more anterior 24
17. Spermatheca absent; spermagonium present, round .... *robusta*
    Spermatheca present 18
18. Spermatheca ovoid to ovoid-elongate 19
    Spermatheca spherical 21
19. Female head conical with rounded apex; stylet = 48–56 μm .... *goodeyi*
    Head rounded or a truncated cone; stylet >55 μm 20
20. Female length = 0.25–0.31 mm; stylet = 55–65 μm; male length = 0.28–0.34 mm .
    .... *aonli*
    Female length = 0.38(0.33–0.42) mm; stylet = 66 (63–70) μm; male length = 0.36–0.42
    mm .... *pandata*
21. Posterior edge of annuli on females crenate .... *crenata*
    Posterior edge of annuli on females smooth 22
22. Female head a truncated cone .... *marylandica*
    Female head rounded 23
23. Female head smooth; corpus swells markedly in posterior half .... *abietis*
    Female head with fine annulations; corpus without marked swelling in posterior half
    .... *straeleni*
24. Female tail bluntly rounded .... *epacris*
    Female tail subacute to finely rounded 25
25. Female stylet = 97(91–102) μm .... *anceps*
    Female stylet <92 μm 26
26. Female tail subacute to rounded; stylet = 89(83–92) μm .... *intermedia*
    Female tail finely rounded, almost acute; stylet = 58–85 μm 27
27. Female length = 0.22–0.25 mm; stylet = 58–62 μm; V = 70 .... *brasiliensis*
    Female length = 0.33(0.28–0.38) mm; stylet = 77(70–85) μm; V = 80– 86 ... *mira*
28. Female stylet <62 μm 29
    Female stylet >65 μm 32
29. Female cuticle ornamented with small refractive dots .... *punctata*
    Female cuticle smooth 30
30. Female stylet = 41–46 μm; V = 82–84 .... *micoletzkyi*
    Female stylet = 55–62 μm; V = 69.7–73.5 31
31. Valve in metacorpus of female elongate (about 12 μm long), excretory pore at level of
    valve .... *acti*
    Valve in metacorpus of female small (about 3–4 μm long), excretory pore anterior to
    valve .... *raskii*
32. Female head rounded to conoid but submedian lobes not set off 33
    Female head with submedian lobes set off or protruding 35
33. Female stylet = 67 (65–69) μm .... *steineri*
    Female stylet >72 μm 34

34.   Female length = 0.25–0.28 mm; excretory pore at isthmus, opposite nerve ring . . . .
      . . . . *oostenbrinki*
      Female length = 0.33–0.39 mm; excretory pore near knobs of stylet . . . . *macrodora*
35.   Female stylet = 74 (69–83) µm; tail bluntly rounded . . . . *teres*
      Female stylet = 94 (82–104) µm; tail acute . . . . *enata*

3.   *Cacopaurus* Thorne, 1943

Diagnosis (after Raski and Luc, 1987) Paratylenchinae.

   Females: Body very small (0.2–0.3 mm), cylindroid-obese (a = 5–8). Cuticle thin, bearing minute tubercles. Lateral field with four lines, ornamented with rows of tubercles. Labial framework weakly developed. Stylet very long (92–102 µm) in regard to body size. Vulva very posteriorly situated; postvulval part very short, conoid. Uterus thick-walled.

   Males: No stylet. Esophagus degenerated. Caudal alae weakly developed, adanal.

   Juveniles: Stylet well developed.

   Bionomics: Females sedentary on roots of woody plants.

   Type and only species: *Cacopaurus pestis* Thorne, 1943.

F.   Developmental Stages of *Paratylenchus bukowinensis* (after Brzeski et al., 1974)

Eggs: (Fig. 1). The measurements of 28 freshly laid eggs are 50 (42–60) × 20 (15–25) µm, and the ratio of length to width = 2.4 (2.1–3.4). The first-stage larva has no stylet, develops and molts within egg to the second stage. This stage hatches from the egg shell.

   Second-stage larvae (n = 15): L = 0.16 (0.12–0.18) mm, a = 13 (11–15), b = 2.4 (2.1–2.7), c = 11.6 (8.2–15.4), stylet = 12 (11–13) µm.

   Cuticle transversely annulated, lateral field with four incisures, cuticular structures are hardly visible. Head trapezoid, truncate, and low. Cephalic skeleton with lateral blades extending posterior from head. Stylet length = 8 (6–10)% of body length. Diameter of stylet knobs equals 2 µm. Excretory pore located 52 (44–65) µm or 33 (28–42)% of body length from anterior body end. Pharynx well developed, procorpus occupies 59 (54–69), isthmus 24 (17–27), and basal bulb 18 (12–22)% of total pharynx length. The length of pharynx = 42(41–48)% of body length. A granular structure was seen on dorsal side of basal bulb of all the examined freshly hatched larvae. Genital primordium length = 8 (5–11) µm. Tail tapers to broadly rounded terminus, tail length = 2.1 (1.7–3.4) of anal body width.

   Third-stage larvae (n = 15): L = 0.26 (0.21–0.32), a = 18 (16–22), b = 3.6 (2.8–4.5), c = 14.0 (12.4–15.6), stylet = 17 (15–19 µm).

   Cuticle delicately annulated, lateral field with four incisures. Cuticular structures well visible. Head trapezoid, truncate, higher than in the second-stage larvae. Cephalic skeleton weakly developed. Stylet length = 6 (5–7)% of total body length. Excretory pore 63 (54–69) µm from the anterior body end, which = 25 (18–30)% of total body length. Pharynx well developed, procorpus larger than in second-stage larvae occupies 63 (56–69), isthmus 20 (17–24), and basal bulb 15 (14–20)% of the pharynx length. The length of pharynx = 29 (23–36)% of total body length. Length of genital primordium = 13 (12–14) µm or 5.5 (4.4–6.5)% of body length. Tail length divided by anal body width = 2.0 (1.5– 2.4).

   Fourth-stage larvae (n = 17): L = 0.36 (0.29–0.43), a = 21 (16–24), b = 4.1 (3.4–5.1), c = 13.1 (10.5–14.9), stylet = 11 (8–12) µm.

   Cuticle transversely annulated, lateral field with four incisures. Stylet very delicate, often hardly visible, with small basal swellings. Stylet length = 2.2–3.3% of total body length. Excretory pore located 85 (75–98), isthmus 22 (18–28), and basal bulb 15 (13–21)%

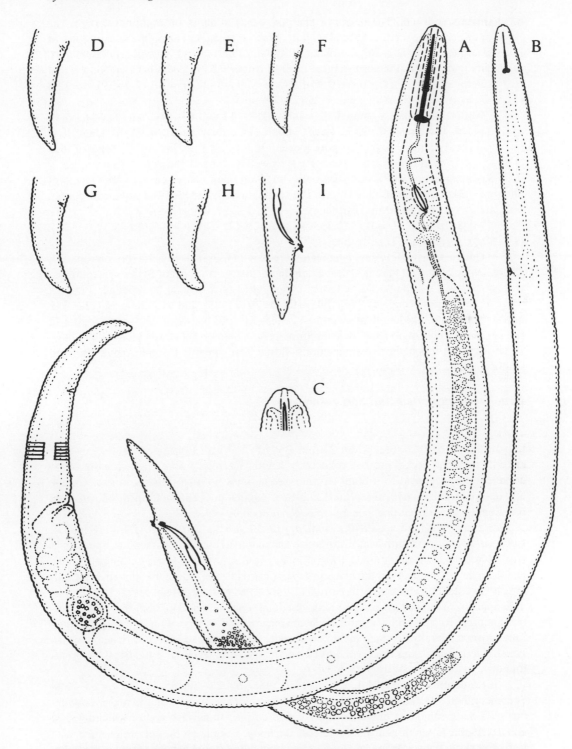

FIGURE 6 *Paratylenchus bukowinensis* (A) Female. (B) Male. (C) Female head. (D–H) Female tails showing variations. (I) Male tail.

of total pharynx length. The length of pharynx = 24 (20–26)% of total body length. The body posterior to the median bulb filled with large granules obscuring the observations of internal organs, especially the reproductive system. Rectum and anus poorly developed, often barely visible. Tail tapers to a blunt tip. Tail length = 2.2 (1.5–2.6) of anal body width.

Females (n = 35): L = 0.40 (0.32–0.54), a = 20 (16–27), b = 4.1 (3.7–5.2), c = 15.3 (11.9–20.6), V = 84 (81–86), stylet = 25 (23–29) μm.

Cuticle transversely annulated, lateral field with four incisures. Lip region more or less rounded, lips rounded. Stylet knobs slightly anteriorly indented. Stylet length = 6 (5–7)% of body length. Excretory pore located 91 (76–106) μm or 24 (20–28)% of total body length from the anterior body end. Pharynx well developed, the procorpus occupies 70 (64–79), isthmus 12 (9–18), and basal bulb 18 (11–21)% of pharynx length. The length of pharynx = 24 (19–29)% of body length. The spermatheca is rounded and filled with sperm. Tail tapers to small round tip. Tail length = 2.4 (2.1–2.9) of anal body width.

Males (n = 15): L = 0.38 (0.33–0.42), a = 28 (25–31), b = 4.9 (4.6–5.3), c = 11.2 (9.8–12.8), stylet = 14 (12–18) μm.

Cuticle transversely annulated, lateral field with four incisures. Lip region rounded. Cephalic skeleton with lateral blades extending posterior from head. Stylet very delicate, with small basal swellings. The length of stylet = 4 (3–5)% of total body length. Pharynx reduced, procorpus occupies 62 (58–64), isthmus 26 (25–28), and basal bulb 12 (11–12)% of pharynx length. The length of pharynx = 20 (19–21)% of body length. Excretory pore 22 (20–24)% of body length from the anterior body end. Gonads short, spicules slightly bent, 23 (21–25) μm in length. Gubernaculum 5–8 μm. Tail tapers to a small rounded tip, tail length = 3.0 (2.5–3.5) of cloacal body width.

Gracilacus epacris Allen and Jensen, 1950

Description

Larvae: L = 0.20–0.26 mm, width 22 μm. Body cylindrical. Lip region continuous with neck. Esophagus occupying one-third body length. Excretory pore slightly posterior to nerve ring. Tail conoid to a blunt terminus, sometimes with an obscure mucro. Cuticle marked by obscure lateral striations. Lines of wing area not observed. Spear 42 μm long, basal knobs well developed, conspicuous. Anal opening obscure.

Young female: L = 0.28–0.29 mm, a = 13–14, b = 2.8–3.0, c = 18–20, V = 83–84%. Lip region continuous with neck. Transverse annulation of cuticle conspicuous. Spear long, well developed, sometimes about 100 μm in length. A sclerotized spear-guiding apparatus visible in the lip region. Position of excretory pore variable in relation to esophagus depending on the extent to which spear is extruded. Valvulated postcorpus not set off from corpus of esophagus. Terminal esophageal bulb glandular, distinctly set off from intestine. Intestine granular. Rectum and anal opening inconspicuous. Vulva large with slightly protruding lips, guarded laterally by small membranous flaps. Ovary single. Uterus with heavy walls. Anterior portion of egg duct a modified spermatheca. tail conoid cylindrical, terminus rounded, frequently slightly digitate.

Adult female: L = 0.24–0.32 mm, a = 7.5–8.0, b = 2.6–3.6, c = 16–20, V = 85–87%. Spear length 82–98 μm. Body obese, straight, curved, or variously bent. Lip region continuous with neck contour. Cuticle marked by well-defined transverse annuli which average about 1.1 μm, in width near the middle of the body. Cuticle plain, not ornamented with refractive dots. Wing area marked by four lines extending from the neck region to the vicinity of the anal opening. Body diameter reduced posterior to vulva. Body posterior to the vulva cylindrical conoid. Tail bluntly rounded. Vulva conspicuous, guarded laterally by

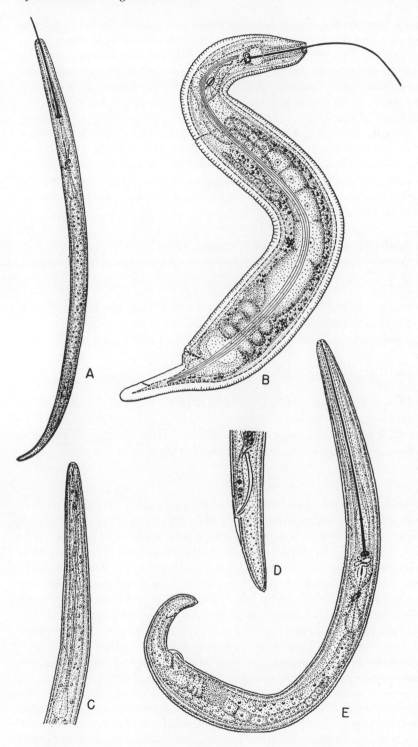

FIGURE 7   *Gracilacus epacris.* (A) Young larva, 805×. (B) Mature female, 805×. (C) Anterior end of male, 1060×. (D) Male tail, 1610×. (E) Young female, 805×.

small membranous flaps. Basal knobs of spear well developed. Corpus of esophagus thick, gradually broadening into the valvulated postcorpus. Isthmus of esophagus distinct, surrounded by the nerve ring. Posterior bulb of esophagus well defined but usually obscured by the ovary and intestine. Excretory duct opening in vicinity of posterior esophageal bulb. Intestine granular. Ovary single, frequently doubly reflexed in older females and extending anteriorly to the region of the postcorpus. Uterus heavily walled. Modified spermatheca present at anterior end of egg duct.

Male: Length 0.24–0.30 mm, $a = 17$–22, $b = 4.0$–4.1, $c = 12$. Lip region continuous with neck contour. Cuticle marked by transverse annuli. Body rather uniformly cylindrical. Tail bluntly rounded, sometimes faintly digitate. Bursa or bursalike structure extremely obscure. Spicules ventrally curved, cephalated, slender, about 16 μm in length. Gubernaculum slightly curved. Testis single. Esophagus degenerate, details obscure. Phasmids not seen. Spear absent in adult males.

Cacopaurus pestis Thorne, 1943 (after Franklin, 1974)

Measurements

Females (after Thorne, 1943): $L = 0.20$–0.26 mm, width = 35–43 μm. Six female (French population): $L = 0.22$–0.25 (0.24) mm, $a = 7.2$–8.4 (7.7).

Lectotype female (after Raski, 1962): $L = 0.21$ mm, $a = 6.9$, $b = 2.2$, $c = ?$, $V = 93\%$, spear = 92 μm.

Males (after Thorne, 1943): $L = 0.25$–0.29 mm, $a = 30$, $b = 3.5$, $c = 11$, $T = 37\%$.

Five males (French population): $L = 0.24$–$0.30(0.27)$ mm, $a = 28$–45 (34). Larvae (seven, ? second stage) (after Raski, 1962): $L = 0.24$–0.30 (0.27) mm, $a = 23.2$–30.0 (25.7), $b = 3.0$–3.5 (3.2), $c = ?$, spear = 39–45 (42) μm.

Larvae (six, ? fourth stage) (after Raski, 1962): $L = 0.21$–0.25 (0.23) mm, $a = 8.0$–16.5 (13), $b = 2.1$–2.5 (2.3), $c = ?$, spear = 55–66 (60) μm.

Description

Female: Body cylindrical, obese, about one-seventh as broad as long, often more or less sharply bent, with broad conical terminus and blunt anterior end. Cuticle with annuli about 1 μm wide at midbody and ornamented with minute refractive tubercles. Head smooth with minute lip region and obscure cephalic framework: excretory pore in region of median esophageal bulb: lateral field divided into three bands by four longitudinal incisures resembling rows of dots: field broadening near vulva to enclose a scutellum-like area, then narrowing and continuing to terminus. Vulva broad, posterior, without flaps, body narrows behind it: anus obscure, subterminal. Spear long, curved, four specimens measured 92–102 (97) μm (Raski, 1962); the anterior part about six times as long as the posterior, which has rounded knobs. Female attached to the host root by the deeply embedded, slender spear which is often broken when the nematode becomes detached. Esophagus with procorpus hardly offset from metacorpus, which has a well-developed valve apparatus: clearly delimited, narrow isthmus encircled by nerve ring: swollen posterior part of esophagus forming a small bulb abutting on intestine, the cells of which are filled with fat globules. Ovary single, with two flexures, reaching anteriorly well beyond the spear base: oviduct made up of a few large cells, one becoming a spermatheca: uterus forming a conspicuous thick-walled chamber: vagina short, directed forward. The uterine egg may be as long as one-quarter of body length.

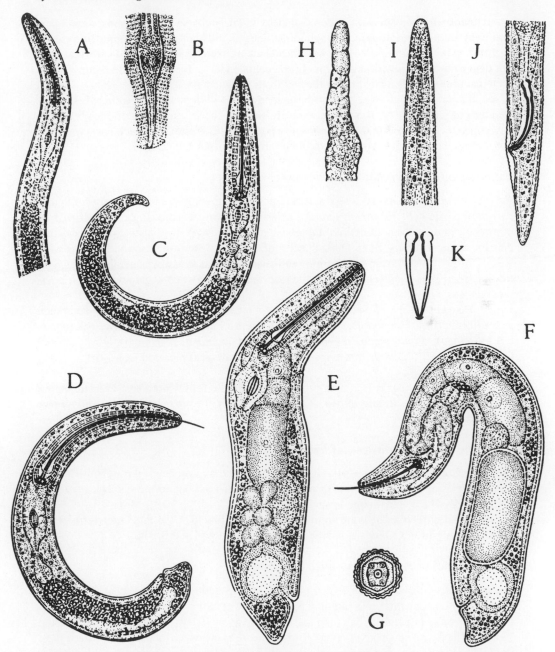

FIGURE 8 *Cacopaurus pestis.* (A) Anterior of young larva. (B) Lateral field of female in vulval region. (C) Female larva. (D) Newly molted female. (E) Adult female showing spermatheca, oviduct, and vagina. (F) Gravid female. (G) Face view of female head. (H) Anterior part of testis. (I) Anterior of male. (J) Posterior of male showing bursa. (K) Spicules and gubernaculum ventral. (C–F) 600×; (A,B,I,J) 2000×; (G,H,K) 2670×. (After Thorne, 1943.)

Male: Body relatively slender. Cuticle with fairly coarse unornamented annuli; lateral field with three or four faint longitudinal lines; deirids conspicuous, opposite anterior end of intestine (Thorne, 1943); rounded head lacking skeletal structure and spear; esophagus degenerate. Testis with small cap cell; spicules slender, curved, cephalated; thin, curved gubernaculum; small, adanal bursa formed from cuticular flaps.

Larvae: Second-stage larva with rounded head with light skeletal framework: spear slender but well developed, with backwardly sloping knobs: esophagus moderately well developed, organized as in female: excretory pore at level of isthmus which is encircled by nerve ring. Tail elongate-conoid with bluntly rounded tip.

1.   Hosts of *Paratylenchus bukowinensis*

Paratylenchus bukowinensis is reported to reproduce in greenhouse tests on carrot (*Daucus carotae*), parsley (*Petrosalinum sativus*), celeriac (*Apium graveoleus*), cabbage (*Brassica oleracea*), and rutabaga (*Brassica napus* var. *napobrassica*). It is also reported causing damage to celery in the field. Other species of *Paratylenchus* are described from a wide variety of host plants or in association with such plants as grasses, vegetable crops, oak trees, cocoa (*Theobroma cacao*), dry land rice, rhubarb, strawberry, grapevines, tea, ornamentals, etc.

Hosts of species of *Gracilacus* are known mostly as trees and perennial plants such as California laurel (*Umbellularia californica*), white birch (*Betula papyrifera*), oak (*Quercus coccinea*), *Pinus taeda*, magnolia, chirimoya (*Annona cherimola*), black walnut (*Juglans hindsii*), *Platanus racemosa*, pear, etc., but are also reported from various grass and weed soils.

Cacopaurus pestis is a monotypic genus with the following hosts: Persian walnut (*Juglans regia*), *Citrus aurantium*, *Rosa indica*, lilac (*Syringa vulgaris*), and poplar (*Populus nigra*).

2.   Distribution of *Paratylenchus bukowinensis*

Paratylenchus bukowinensis is widely spread in Poland and probably throughout Europe. Other species of *Paratylenchus* are variously distributed but one or more species is found almost everywhere nematology is studied.

Species of *Gracilacus* are similarly widely distributed but so far *Cacopaurus pestis* is known only from California, southern France, Iran, Italy, and Spain.

3.   Bionomics of *Paratylenchus bukowinensis*

Paratylenchus bukowinensis is an ectoparasite; second-and third-stage larvae and females feed mostly on epidermal cells but may penetrate one or two layers deeper. Eggs are laid in the soil, up to 23 in a cluster; hatching takes place in about 7 days and the entire cycle about 23 days. The fourth-stage larvae do not feed and serve as the overwintering or resistant stage in the absence of growing host plants. Root diffusates stimulate molting to adult stage. It may injure cabbage but is most damaging to plant species of the Umbellifera.

Some species of *Gracilacus* appear to feed in a very different manner under the bark of perennial host roots (Cid del Prado-Vera and Maggenti, 1988). Most females swell into various irregular shapes, depositing eggs in the same feeding area under the bark.

Cacopaurus pestis feeds as a sessile ectoparasite. Eggs are laid in the soil near the sedentary female and small colonies result as the young feed on nearby host plant tissues. The females turn brown in color after maturation but do not retain eggs. Many empty female cuticles may be found persisting in the soil. Damage appears to be serious on walnut but not on citrus or rose.

G.   Subfamily *Tylenchocriconematinae* Raski and Siddiqui, 1975

Diagnosis (after Raski and Luc, 1987) Tylenchulidae.

Female: Long, slender nematodes, 0.50–0.83 mm; labial area with weak sclerotization, anterior surface flattened bearing squarish plate with H-shaped oral aperture, four small submedian lobes at corners of plate; stylet prominent, conus longer than shaft plus knobs, becomes very slender in anterior half; procorpus enlarges gradually to large muscular metacorpus followed by long, slender isthmus and elongate posterior bulb with enclosed glands; body annulation very fine, lateral field slightly raised from contour of body, appears as two longitudinal lines but SEM photograph shows four equally spaced lines; postvulvar body very long, conoid; spermatheca offset, distinct, at anterior end of uterus; vulva a broad slit with slightly overhanging anterior lip, SEM photograph shows prominent lateral vulval membranes.

Male: Slender, slightly shorter than female, 0.42–0.65 mm; labial area without discernible sclerotization, unusual oblique anterior surface sloping ventrad with small, distinct submedian lobes; stylet absent; esophagus degenerate; lateral field with four equally spaced lines; spicules long, curved; bursa long slender, extending to terminus.

Juvenile: Similar to female in general aspect and stylet structure.

Type and only genus: *Tylenchocriconema* Raski and Siddiqui, 1975.

H.   *Tylenchocriconema* Raski and Siddiqui, 1975

Diagnosis having characters of the subfamily:

Type and only species: *Tylenchocriconema alleni* Raski and Siddiqui, 1975.

Twenty-seven female paratypes: $L = 0.68$ mm (0.50–0.83), $a = 53$ (40–70), $b = 5.6$ (4.8–6.2), $c = 15$ (10–26), $V = 24$ (15–17)$79(78–81)$, stylet $= 23$ μm (20–29), prorhabdion $= 15$ μm (13–19), excretory pore $= 113$ μm (86–142).

Thirteen male paratypes: $L = 0.57$ mm (0.42–0.65), $a = 50$ (42– 63), $b = 5.6$ (4.8–6.2), $c = 12$ (11–13), spicules $= 24.5$ μm (22–28), gubernaculum $= 4.7$ μm (4–6), excretory pore $= 96$ μm (71–103), $T = 24$ (19–44).

Female (holotype): $L = 0.73$ mm, $a = 55$, $b = 6.2$, $c = 12$, $V = 12_{79}$, stylet $= 23$ μm, prorhabdion $= 15$ μm, excretory pore $= 111$ μm. Body slender, narrows only slightly toward anterior end which is truncate. Lips shallow, set off by slight constriction. Amphidial pores ovate, near oral aperture. Four small, separate, submedian, rounded lips with prominent papillary innervations. Cephalic framework weakly developed; tip of stylet in guiding apparatus, which appears as slender, dark rods. Stylet very slender, especially in anterior portion; knobs well developed, posteriorly directed. Dorsal esophageal gland orifice approximately 6 μm posterior to knobs (5 μm in some paratypes). Procorpus not set off, widens gradually to large metacorpus (in some paratypes procorpus longer and of uniform diameter for as much as 30 μm posterior to stylet knobs, but in all paratypes procorpus widens gradually to large metacorpus). Valve in metacorpus large, ovate. Isthmus long, slender, widening gradually to form ovate, posterior bulb with glands enclosed. Lumen of esophagus prominently cuticularized near end of posterior bulb; esophageal-intestinal valve small, rounded. Excretory pore inconspicuous, at same level as the 6-μm-long hemizonid. Deirids not observed. Ovary single, outstretched anteriorly; spermatheca an ovate projection, approxi-

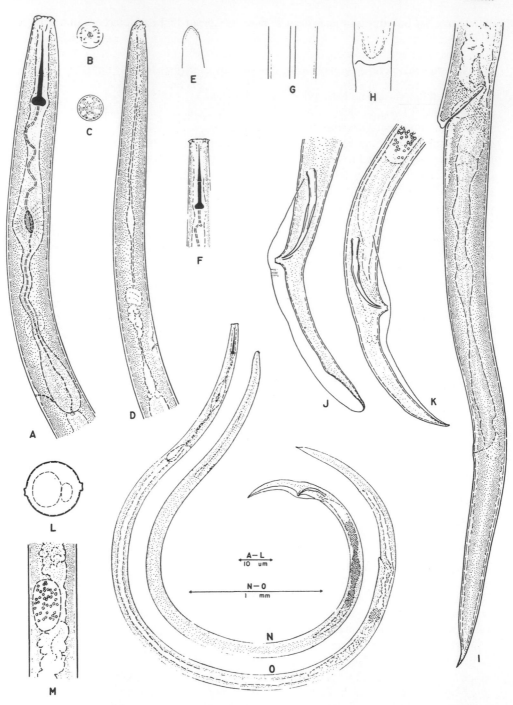

FIGURE 9 (A–O) *Tylenchocriconema alleni* n. g. n. sp. (A) Female, anterior end. (B) Female head, en face view. (C) Female head, transverse section posterior to (B). (D) Male, anterior end. (E) Male, anterior end of head at focus deeper than (D). (F) Juvenile (female?), anterior end. (G) Female, lateral fields. (H) Vulva in ventral view. (I) Female tail. (J,K) Male tails. (L) Female transverse section at midbody; (M) female in region of spermatheca. (N) Male, full length. (O) Female, full length.

mately 7 × 14 μm, on right lateroventral side at anterior end of uterus. Vagina heavily cuticularized, oblique, directed anteriad at about 45°angle from transverse plane. Posterior wall of vagina slightly swollen, but no evidence of posterior ovary or reduced postuterine sac. Several well-developed muscles at right angles to vagina. Body enlarges slightly in region of vulva. Anterior lip of vulva projects posteriorly as overlapping cuticular flap.

Body abruptly narrows immediately posterior to vulva, tapers gradually up to last 10 μm, then more abruptly ending with an acute terminus. Body slightly swollen at anal opening. Rectum weak, anal opening obscure. Phasmids not observed. Body annuli average about 0.6 μm in width, cuticle almost smooth in appearance. Lateral field a narrow, projecting band about 1.3 μm wide, occupying approximately 11% of body width, and appears as two distinct lines in lateral view. In transverse section, band projects beyond contour of body outline. Lateral field extends anteriorly beyond level of metacorpus, posteriorly near terminus.

Male (allotype): $L = 0.47$ mm, $a = 47$, $b = ?$, $c = 12$, spicules = 23 μm, gubernaculum = 4 μm, excretory pore = 74 μm, $T = 25$. Body slender, an open C-shape after fixing; narrows gradually anteriorly to slender, finely rounded, asymmetrical head. Anterior end turned about 30° ventrad with four small rounded lips or submedian lobes on the oblique surface. Cephalic framework indistinct; stylet lacking. Esophagus degenerate, but outlines of metacorpus and posterior bulb partially visible. Excretory pore immediately posterior to hemizonid. Testis single, outstretched. Spicules curved; head (capitulum) and shaft (calomus) composing 36% of spicule length (35–40% in other paratypes). Gubernaculum simple, rod-shaped. Prominently protruding anal sheath. Body narrows abruptly posterior to cloacal opening, then curves ventrad forming slender tail with an acute terminus. One specimen was observed to recurve dorsad posterior to cloacal opening. Caudal alae begin at level of anterior end of retracted spicules, curve out (with rounded outline) at level of anal sheath, then narrows to terminus. Caudal alae with fine transverse striae, crenate margin. Lateral field similar to that of female. Phasmids not seen. Body annuli fine; cuticle appears smooth in outline.

Eight juveniles (probably second-stage): $L = 0.44$ mm (0.36–0.52), $a = 40$ (29–50), $b = 4.5$ (4.1–5.0), $c = ?$, stylet = 18 μm (16–19), prorhabdion = 12 μm (11–12), excretory pore = 89 μm (77–104). Body shape and head end similar to adult female. Lips slightly smaller, shallower, more pointed dorsally and ventrally than in adult female (Fig. 1F). Tail evenly conoid to an acute terminus, but not as slender as adult female at posterior 10–12 μm. Lateral field as in adult female, but narrower. Developing gonad 17 μm (12–23) long, about 109 μm (97–118) from terminus.

1. Hosts of *Tylenchocriconema alleni*

A monotypic genus whose only species is known to attack only one plant so far. That host is a bromeliad, *Tillandsia flabellata*.

2. Distribution of *Tylenchocriconema alleni*

Probably indigenous to Guatemala. It has been intercepted in shipments of an ornamental bromeliad to California from Guatemala and is in nurseries of the Netherlands on plants introduced in shipments from Guatemala.

3. Bionomics of *Tylenchocriconema alleni*

Females and larvae feed entirely ectoparasitically in the crowns of the host just below the water line. Lesions are produced on the top leaf surfaces. Ultimately flowering is inhibited, necrotic leaves die, and severely infected plants may die.

REFERENCES

Brzeski, M. W., Zepp, A. L., and D'Errico, F. P. 1976. Developmental stages of *Paratylenchus bukowinensis* Micoletzky. *Roczniki Nauk Rolniczych Serial E* 5:145–151.

Cid del Prado-Vera, I., and Maggenti, A. R., 1988. Description of *Gracilacus hamicaudata* sp. n. (Nemata: Criconematidae) with biological and histopathological observations. *Rev. Nemat.* 11:29–33.

Cohn, E. 1969. The citrus nematode, *Tylenchulus semipenetrans* Cobb, as a pest of citrus in Israel. *Proc. First Int. Citrus Symp.*, Riverside, CA 2: 1013–1017.

Eroshenko, A. S. 1978. Pathogenic nematodes of pine plantations in the south of Sakhalin Island. *Fitogel'mintologicheskie issledovaniya Moscow*, "Nauka." (1978) 32–39 [Ru] Inst. of Pedology, Far Eastern Sci. Cent., Acad. of Sci. of the USSR, Vladivostok, USSR. Taken from Helmin. Abstr. Ser. B. 1979 Vol. 48(3):106. Item no. 1048.

Franklin, M. T. 1974. *Cacopaurus pestis*. C.I.H. Descriptions of plant-parasitic nematodes. Set 3, No. 44. Commonwealth Institute of Helminthology, St. Albans, Herts., England.

Inserra, R. N., Vovlas, N., O'Bannon, J. H., and Esser, R. P. 1988. *Tylenchulus graminis* n. sp. and *T. palustris* n. sp. (Tylenchulidae), from native flora of Florida, with notes on *T. semipenetrans* and *T. furcus*. *J. Nematol.* 20: 266–287.

Minagawa, N. 1983. Descriptions of two new species of nematode genus *Trophonema* Raski, 1957 (Tylenchida: Tylenchulidae). *Appl. Ent. Zool.* 18: 90–97.

Raski, D. J., and Luc, M. 1987. A reappraisal of Tylenchina (Nemata). 10. The superfamily Criconematoidea Taylor, 1936. *Rev. Nematol.* 10: 409–444.

Siddiqi, M. R. 1974. *Tylenchulus semipenetrans*. C.I.H. descriptions of plant-parasitic nematodes. Set 3, No. 34. Commonwealth Institute of Helminthology, St. Albans, Herts., England.

Siddiqi, M. R. 1985. *Meloidoderita kirjanovae*. C.I.H. descriptions of plant-parasitic nematodes. Set 8, No. 113. Commonwealth Institute of Helminthology, St. Albans, Herts., England.

Van Gundy, S. D. 1958. The life history of the citrus nematode *Tylenchulus semipenetrans* Cobb. *Nematologica* 3: 283–294.

Van Gundy, S. D., and Kirkpatrick, J. D. 1964. Nature of resistance in certain citrus rootstocks to citrus nematode. *Phytopathology* 54: 419–427.

Vilardebo, A., and Luc, M. 1961. Le "slow decline" des citrus du au nématode *Tylenchulus semipenetrans* Cobb. *Fruits Paris* 16: 445–454.

18
Tylenchidae in Agricultural Soils

ETIENNE GERAERT   *Laboratorium voor Morfologie en Systematiek der Dieren,*
Rijksuniversiteit Gent, Ghent, Belgium

I. INTRODUCTION

Typical Tylenchidae are slender; they have a nonsclerotized head framework; a short, delicate stylet; and a short, nonoverlapping esophagus; the intestine contains 20–24 or more
cells arranged in pairs; the female reproductive system is monodelphic, prodelphic with a
quadricolumellar uterus and a short, if any, postvulval uterine sac; the tail is elongated.
Males are abundant and resemble females; they have relatively short spicules and a small, if
any, adanal bursa; sperm is small (less than 2 μm, usually only 1 μm), rounded. It is the only
family where amphidial apertures can be on the lateral side of the head (these apertures are
usually slitlike and oriented in a lateral plane) and where the anterior, usually conical part of
the stylet can be much shorter than the posterior, cylindrical part; both items are considered
to reflect primitivity.

     Also included in the family are (1) a few didelphic genera that show many similari-
ties with some monodelphic genera; (2) an occasional species with an overlapping gland
region; or (3) a sclerotized head framework; or (4) long cephalic setae.

     The assemblage of all species having the characteristics mentioned yield more than
50 nominal genera, 17 subfamilies, and 6 families. In Geraert and Raski (1987) and Maggenti et al. (1988), the five families are considered synonyms of Tylenchidae, the 17 subfamilies are reduced to 5, and 19 genera are considered junior synonyms of existing ones.

     A detailed study of the lip area with SEM, of the female reproductive system, and of
some morphometric relationships helped to diagnose five subfamilies (Geraert and Raski,
1987).

II. DIFFERENTIATING CHARACTERS OF THE SUBFAMILIES

1. Ecphyadophorinae: extremely slender ($a = 60$–180) to slender (*Lelenchus*; $a = 40$–60);
    bursa lobed, directed posteriorly (except *Lelenchus*) head dorsoventrally flattened with

amphidial apertures as longitudinal clefts on lateral side of head (not so in *Ecphyadophora* and *Mitranema*).

2. Boleodorinae: amphidial apertures as open V or oblique slits on lateral side of head; stylet delicate, anterior part one-third stylet length, knobs small, often flangelike or missing; tail elongate, often rounded at the end.

3. Atylenchinae: head with undivided laterally elongated front plate with large, roundish amphidial apertures; stylet well developed, anterior part somewhat smaller than posterior part; vulva covered by lateral or longitudinal flaps; males often with large hypoptygmata.

4. Tylodorinae: head end with an oral disk, slitlike amphidial apertures longitudinally oriented in a lobed or slightly divided labial disk or ring; stylet long to very long; female reproductive system with more than 20 crustaformeria cells in the uterus and an elongated spermatheca in line.

5. Tylenchinae: all species not having the particular characteristics mentioned in the subfamilies 1–4; amphidial openings of two major types: a longitudinally (most species) and a transversely oriented slit; no special lip pattern, usually no distinct oral disk; head quadrangular (D-V flattened in *Malenchus*); stylet medium, small or very small, usually with knobs.

III. GENERA BELONGING TO THE SUBFAMILIES MENTIONED

Subfamily Tylenchinae Oerley, 1880
syn. Dactylotylenchinae Wu, 1969
Duosulciinae Siddiqi, 1979

Type genus: *Tylenchus* Bastian, 1865
            syn. *Aerotylenchus* Fotedar and Handoo, 1979
                 Areotylenchus Fortuner, 1984

Other genera: *Miculenchus* Andrássy, 1951
                  syn. *Ceramotylenchus* Ebsary, 1986
              Filenchus Andrássy, 1954
                  syn. *Dactylotylenchus* Wu, 1968
                       Discotylenchus Siddiqi, 1980
                       Duosulcius Siddiqi, 1979
                       Lambertia Brzeski, 1977
                       Ottolenchus Husain and Khan, 1963
                       Zanenchus Siddiqi, 1979
              Malenchus Andrássy, 1968
                  syn. *Neomalenchus* Siddiqi, 1979
                       Irantylenchus Kheiri, 1972
                       Polenchus Andrássy, 1980
                       Allotylenchus Andrássy, 1984
                       Cucullitylenchus Huang and Raski, 1986
                       Mukazia Siddiqi, 1986

Subfamily Ecphyadophorinae Skarbilovich, 1959
syn. Epicharinematinae Maqbool and Shahina, 1985
Ecphyadophoroidinae Siddiqi, 1986

Type genus: *Ecphyadophora* de Man, 1921
            syn. *Karachinema* Maqbool and Shahina, 1985

Other genera: *Lelenchus* Andrássy, 1954
            Ecphyadophoroides Corbett, 1964
               syn. *Tenunemellus* Siddiqi, 1986
            Epicharinema Raski, Maggenti, Koshy, and Sosamma, 1982
            Mitranema Siddiqi, 1986

Subfamily Tylodorinae Paramonov, 1967
syn. Campbellenchinae Wouts, 1978
Eutylenchinae Siddiqi, 1986

Type genus: *Tylodorus* Meagher, 1963

Other genera: *Eutylenchus* Cobb, 1913
            Cephalenchus Goodey, 1962
               syn. *Imphalenchus* Dhanachand and Jairajpuri, 1980
            Campbellenchus Wouts, 1977

Subfamily Atylenchinae Skarbilovich, 1959
syn. Antarctenchinae Spaull, 1972
Pleurotylenchinae Andrássy, 1976
Aglenchinae Siddiqi and Khan, 1983

Type genus: *Atylenchus* Cobb, 1913

Other genera: *Aglenchus* Andrássy, 1954
            Pleurotylenchus Szczygiel, 1969
            Antarctenchus Spaull, 1972
            Gracilancea Siddiqi, 1976
            Coslenchus Siddiqi, 1978
               syn. *Cosaglenchus* Siddiqui and Khan, 1983
                 Paktylenchus Maqbool, 1983

Subfamily Boleodorinae Khan, 1964
syn. Psilenchinae Paramonov, 1972
Basiriinae Decker, 1972
Leipotylenchinae Sher, 1973

Type genus: *Boleodorus* Thorne, 1941

Other genera: *Psilenchus* de Man, 1921
            Basiria Siddiqi, 1959
               syn. *Clavilenchus* Jairajpuri, 1966
                 Basiroides Thorne and Malek, 1968
                 Neobasiria Javed, 1982
                 Pseudobasiria Jahan, 1986
                 Neopsilenchus Thorne and Malek, 1968

> *Atetylenchus* Khan, 1973
>   syn. *Leipotylenchus* Sher, 1974
> *Neothada* Khan, 1973
> *Duotylenchus* Saha and Khan, 1982
> *Basirienchus* Geraert and Raski, 1986

Genera *incertae sedis*: *Macrotrophurus* Loof, 1958
                    Luella Massey, 1974
                    Sakia Khan, 1964
                    Basiliophora Husain and Khan, 1965

IV. KEY TO THE GENERA

1. Females didelphic 2
    Females monodelphic 5
2. Tail short, subcylindrical, rounded; stylet very long (90– 110 μm) ....
    .... *Macrotrophurus*
    Tail elongated, attenuated; stylet small (less than 20 μm) 3
3. Cephalic framework moderately sclerotized; vulva provided with lateral vulval
    membranes; male cloaca provided with large hypoptygmata .... *Antarctenchus*
    Cephalic framework not sclerotized; vulva without lateral vulval membranes; male
    without hypoptygmata 4
4. Head high with distinct amphidial slit on the lateral side; median bulb usually behind
    middle of esophagus .... *Psilenchus*
    Head low with indistinct amphidial slit; median bulb anterior to middle of esophagus
    .... *Atetylenchus*
5. Stylet = 76–104 μm .... *Tylodorus*
    Stylet = 38–52 μm .... *Epicharinema*
    Stylet = 22–34 μm 6
    Stylet less than 22 μm 7
6. Cuticle with longitudinal ridges .... *Campbellenchus*
    Cuticle without ridges .... *Gracilancea*
7. Head provided with setae 8
    Head without setae 9
8. Vulva covered by longitudinal flap, male without caudal alae; cloaca provided with
    large hypoptygmata .... *Atylenchus*
    Vulva with lateral vulval flaps; male with caudal alae; cloaca raised ....
    .... *Eutylenchus*
9. Head high with distinct, oblique amphidial slit on the lateral side of the head
    (monodelphic Boleodorinae); anterior stylet part one-third of stylet 10
    Head variously shaped, amphidial slit longitudinally oriented or only present on the
    front; anterior stylet part one-third or one-half stylet 13
10. Cuticle provided with longitudinal ridges (or tail hooked or bent near terminus which is
    pointed) .... *Basirienchus*
    Cuticle without longitudinal ridges 11
11. Body ventrally curved, sometimes in a spiral; female reproductive system with offset
    spermatheca filled with refractive sperm and ovary with oocytes in multiple rows .
    .... *Boleodorus*
    Body more or less straight; sperm not refractive and oocytes not in multiple rows .
    12

12. Stylet without knobs, anterior part with wide lumen .... *Neopsilenchus*
     Stylet with or without knobs, anterior part conical with very fine lumen .. *Basiria*
     (and the related genus *Duotylenchus*)
13. Cuticle provided with longitudinal ridges 14
     Cuticle without longitudinal ridges 16
14. Stylet without knobs, anterior part about one-third stylet length, dorsal esophageal
     gland opening 3–5 µm posterior to stylet end .... *Neothada*
     Stylet with knobs, anterior part slightly less to about half stylet length, dorsal
     esophageal gland opening close to stylet knobs 15
15. Vulva covered by longitudinal flap .... *Pleurotylenchus*
     Vulva with lateral vulval flaps or vulva uncovered .... *Coslenchus*
16. Head with disklike structure 17
     Head with smooth contour 19
17. Head with small disk at the front end 18
     Head with large dome-shaped structure .... *Cucullitylenchus*
18. Very slender ($a = 62$–67); caudal alae posteriorly concave .... *Mitranema*
     Not so slender; caudal alae rounded .... *Filenchus*
     (some *Filenchus* species, before considered as a separate genus *Discotylenchus*)
19. Very slender ($a = 60$–180); bursa lobed 20
     Relative body width variable, from thick to slender; caudal alae, if present, rounded
     21
20. Head quadrangular; amphidial pores on front; vulva with overhanging anterior lip
     .... *Ecphyadophora*
     Head D-V flattened; long sinuous amphidial aperture; vulva without overhanging
     anterior lip, sometimes with lateral flaps .... *Ecphyadophoroides*
21. Cuticle deeply incised 22
     Cuticle not deeply incised 24
22. Head quadrangular in front; body annulations zigzag in surface view, the distinct
     annulation continues on the head; lateral field a small ridge 23
     Head flattened; lateral field a small ridge with four, six or many lines (the many lines
     visible only using SEM); males with caudal alae .... *Malenchus*
23. Male without caudal alae .... *Miculenchus*
     Male with caudal alae .... *Mukazia*
24. Very slender species; annulation not so distinct; head D-V flattened with long sinuous
     amphidial aperture .... *Lelenchus*
     Relatively thicker species; annulation usually distinct, head usually quadrangular;
     amphidial aperture when long, not sinuous 25
25. Conus about 1/3 of stylet length 26
     Conus slightly less to about 1/2 of stylet length 27
26. Head high with longitudinal amphidial aperture on lateral side; clavate stylet knobs
     with ventrally situated opening of esophageal lumen; dorsal esophageal gland opening
     one-half to one stylet length posterior to knobs .... *Irantylenchus*
     Head quadrangular; annulation distinct; stylet with rounded knobs; lateral field with
     two, three, or four lines; dorsal gland opening close to knobs .... *Filenchus*
27. Vulva provided with lateral flaps 28
     Vulva without flaps 29
28. Lateral field two lines; vagina thin; postvulval sac short .... *Allotylenchus*
     Lateral field with three lines (central ridge can also appear as two lines close to each
     other); vagina thickened; postvulval sac short .... *Aglenchus*

Lateral field with four or usually six lines; vagina not thickened; postvulval sac well developed .... *Cephalenchus*
29. Lateral field and body annulation inconspicuous; caudal alae very small ....
.... *Polenchus*
Lateral field and body annulation distinct; caudal alae distinct .... *Tylenchus*

V. SELECTED GENERA

Of the 32 genera mentioned in the identification table only *Tylenchus, Filenchus, Miculenchus, Malenchus, Ecphyadophora, Ecphyadophoroides, Lelenchus, Cephalenchus, Aglenchus, Coslenchus, Psilenchus, Boleodorus, Basiria,* and *Neopsilenchus* are frequently represented in agricultural soils, *Filenchus* being the most abundant.

Only some *Cephalenchus* species have economic importance.

A. Genus *Ecphyadophora* de Man, 1921 (Fig. 1)

Diagnosis

Ecphyadophorinae. Moderately sized, extremely slender nematodes. Cuticle with fine transverse annuli; appears often smooth under light microscope but annuli evident under SEM. Lateral field with four equidistant incisures or not observed. Anterior end rounded, with four symmetrical lobes, annulate almost to labial plate, which bears small, ovate amphidial apertures. Stylet short, up to 13 μm, anterior conical part usually shorter than posterior cylindrical part. Esophagus very thin, spindle-shaped median bulb without valve apparatus, anterior to middle of esophagus; terminal bulb elongated, offset from intestine.

Vulva a transverse slit, posteriorly directed because of overhanging anterior lip, lateral dikes at edges of vulval slit; vagina obliquely anterior; spermatheca elongate, sometimes offset, sometimes in line; postvulval uterine sac very short. Spicules straight, needlelike. Gubernaculum present or absent. Caudal alae flaplike, posteriorly directed.

Tail subcylindrical to conical, with finely rounded to pointed terminus.

Type species: *E. tenuissima* de Man, 1921

The six known species can be differentiated by the following key (adapted from Raski et al., 1982).

1. $V = 56–58$ .... *E. vallipuriensis* Husain and Khan, 1968
   $V > 69$ 2
2. Exceptionally thin, $a > 133$ 4
   $a < 122$ 3
3. Beadlike ornamentation of cuticle (only distinct under SEM) ....
   .... *E. caelata* Raski and Geraert, 1986
   Cuticle smooth .... *E. quadralata* Corbett, 1964
4. Stylet fine, knobs moderately developed .... *E. tenuissima*
   Stylet prominent, knobs well developed 5
5. Stylet 9–12 μm, $L = 0.83–1.00$ mm .... *E. teres* Raski et al., 1982
   Stylet 8 μm, $L = 0.68–0.71$ mm ....
   .... *E. elongata* (Maqbool and Shahina, 1985) Geraert and Raski, 1987

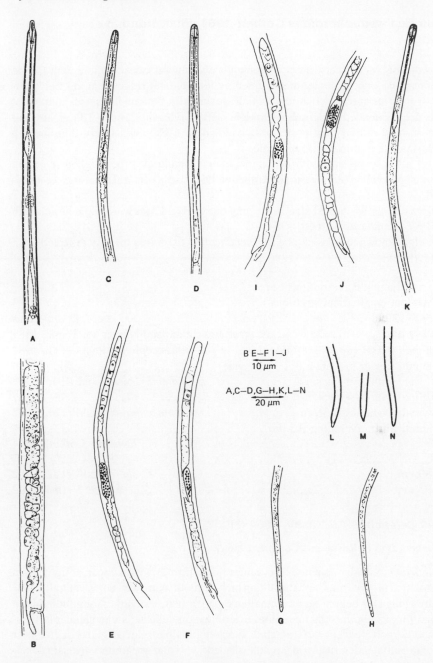

FIGURE 1 *Ecphyadophoroides indicus*. (A) Female, esophageal region. (B) Vulvar region and part of gonad; *Ecphyadophora quadralata*, (C–H) female, Californian specimens, (C,D) esophageal region, (E,F) vulvar region and female reproductive system, (G,H) tail. Female, Texas specimen, (I) vulvar region and female reproductive system, (J–N) *Ecphyadophora caelata*. Female, (J) vulvar region and part of female reproductive system, (K) esophageal region, (L–N), variations in tail. (From Raski and Geraert, 1986a, *Nematologica*).

B.   Genus *Ecphyadophoroides* Corbett, 1964 (Figs. 1 and 2)

Diagnosis

Ecphyadophorinae. Moderately sized, extremely slender nematodes. Cuticle with fine to moderate transverse annuli, sometimes crossed by longitudinal ridges. Lateral fields obscure, two or four incisures. Cephalic region dorsoventrally flattened, smooth, amphidial apertures as deep, extended, longitudinal straight clefts. Stylet short, up to 13 μm, anterior conical part usually shorter than posterior cylindrical part. Esophagus very thin, median bulb spindle-shaped, terminal bulb elongate.

Vulva a transverse slit, with or without lateral membranes; perpendicular vagina with thin walls; postvulval uterine sac short; spermatheca partly offset. Tail elongate, terminus pointed.

Male with flaplike caudal alae, projecting outward and backward. Spicules curved.
Type species: *E. annulatus* Corbett, 1964

The eight known species can be differentiated by the following key (adapted from Raski et al., 1982).

1.   Body with longitudinal striations 2
     Body without longitudinal striations 3
2.   Stylet 9.5–12 μm .... *E. annulatus*
     Stylet 7–9 μm .... *E. theae* Eroshenko and Nguen Vu Thank, 1981
3.   Body annuli coarse (up to 2.0 μm wide) ..... *E. macrocephalus* Raski et al., 1982
     Body annuli fine or moderate (<1.3 μm wide) 4
4.   Lateral field with four lines .... *E. graminis* Husain and Khan, 1968
     Lateral field with two lines or not observed 5
5.   Body annuli averaging 1.3 μm .... *E. leptocephalus* Raski et al., 1982
     Body annuli finer, averaging 0.5 μm 6
6.   *V* > 65 .... *E. tenuis* Corbett, 1964
     V < 62 7
7.   *L* < 0.89 mm .... *E. sheri* Raski et al., 1982
     L > 0.91 mm .... *E. indicus* Verma, 1972

C.   Genus *Lelenchus* Andrássy, 1954 (Fig. 3)

Diagnosis (according to Raski and Geraert, 1986a)

Ecphyadophorinae. Slender nematodes, *a* values range from 37–93; almost straight when fixed and mounted in glycerin. Cephalic region high, smooth, narrowed dorsoventrally; amphid apertures long slits beginning near small oral cap or plate extending longitudinally to posterior margin of smooth cephalic region. Sclerotization delicate; stylet slender but distinct with small knobs (except in *L. elegans* knobs not visible); dorsal gland orifice near knobs. Median bulb slender, fusiform (spindle-shaped); valvular apparatus variously developed; isthmus very slender; posterior bulbar region pyriform to elongate. Spermatheca rounded to oval; uterine cells probably in quadricolumellar arrangement; vagina curved anteriad (*L. leptosoma*) or perpendicular. Vulval opening with lateral membranes; postuterine sac short. Tail long, becoming almost filamentous. Lateral field variable: lacking (*L. leptosoma*), a narrow band forming two lines (*L. filicaudata*) or two distended tube-like bands forming four lines. Males similar to females; caudal alae leptoderan, short.
Type species: Lelenchus leptosoma (de Man, 1880) Meyl, 1961

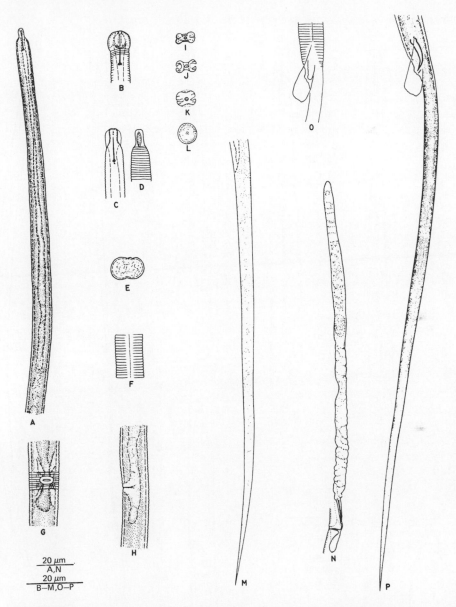

FIGURE 2 *Ecphyadophoroides leptocephalus*. Female, A–N. (A) Esophageal region, (B) cephalic region, dorsoventral view, (C) cephalic region, median level, dorsoventral view, (D) cephalic region, surface level, lateral view, (E) transverse section about midbody, (F) lateral field about midbody, (G) vulva, ventral view, (H) vulvar region, lateral view, (I–L) transverse sections of cephalic region from anterior surface (I), each one successively posterior up to (L), (M) tail, (N) female reproductive system. Male, (O,P). (O) Caudal alae and cloacal opening, surface view, (P) tail, spicules, and gubernaculum. (From Raski et al., 1982, *Rev. Nematol.*)

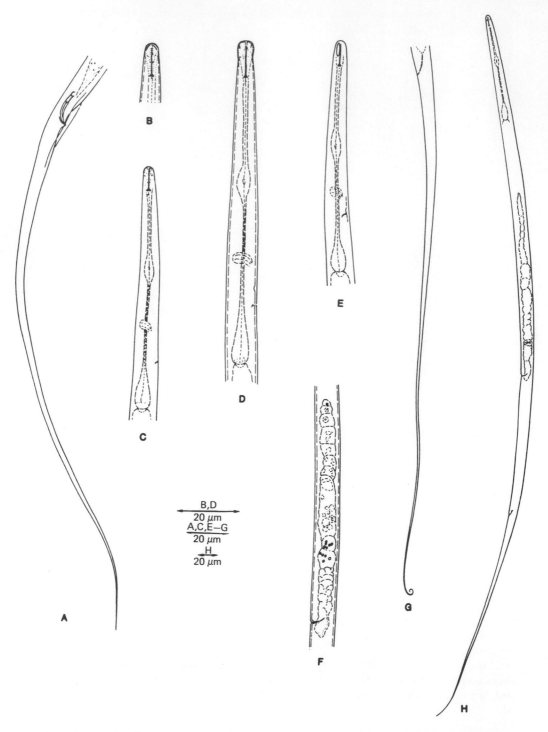

FIGURE 3   *Lelenchus leptosoma*. (A,D, E) Male, (A) tail end, (D,E) esophageal region, (B,C, F–H) female, (B) head, (C) esophageal region, (F) reproductive system, (G) tail, (H) entire female, (From Raski and Geraert, 1986a, *Nematologica*.)

Of the three species described in Raski and Geraert (1986a), only the type species is frequently found. It is a remarkably thin nematode in which the very tail end is often difficult to discern because of its thinness.

D. Genus *Filenchus* Andrássy, 1954 (Figs. 4 and 5)

Diagnosis (according to Raski and Geraert, 1987)

Tylenchinae. Head sclerotization delicate, stylet usually < 15 μm; cone less than half total stylet length. Transverse striae usually extend onto head up to small labial plate which is squarish with rounded corners; four cephalic sensilla present or absent. Amphid apertures usually elongate slits beginning near oral disk or at edge of labial plate extending laterally through three or four head annuli; rarely small elliptical apertures confined to labial plate. Body annuli fine to coarse. Lateral field two lines setting off a single plain band or three or four lines. Tail elongate conoid, curved or straight, to effilate/filiform even hairlike in outline.

The more than 70 described species can be identified with the key in Raski and Geraert (1987).

We selected a few species: the larger *F. vulgaris* and the smaller *F. helenae*, both with a fine tail tip, *F. ditissimus* and *F. facultativus*, both small species with rounded tail tip, *F. facultativus* having coarse body annuli. Little is known about their area or distribution; nevertheless it seems that all four of them are cosmopolitan and common in agricultural soils. They have also been selected because they represent various aspects of the morphology of the genus. Some discussion with closely related species is added.

Filenchus vulgaris (Brzeski, 1963) Lownsbery and Lownsbery, 1985 (Fig. 4)

This very common species can be diagnosed by body length 0.5–0.8 mm; head 5–7 μm wide at base, about half as high, not offset; stylet delicate with tiny knobs, 9–12 μm long; esophagus 89–124 μm long with an oval median bulb, slight central thickenings at 40–48%; posterior glandular region pear-shaped, bulblike; vulva wide, simple slit, vagina straight with thin walls; postuterine sac well developed but usually not exceeding one vulval body diameter; spermatheca offset, oval to elongated; tail elongated, usually straight with very fine but usually not filiform tail end; tail in most specimens 1–1.5 times vulva–anus distance. Annuli usually 1–1.5 μm wide, mostly faint; lateral field four lines, two inner lines sometimes difficult to observe.

A few other large species are very similar to *F. vulgaris* and can be differentiated by the following characteristics: *F. thornei* (Andrássy, 1954) Andrássy, 1963 has a much longer tail (193–233 μm vs. 111–165 μm) and stylet knobs always distinct (often indistinct in *F. vulgaris*); *F. cylindricus* (Thorne and Malek, 1968) Niblack and Bernard, 1985 is a longer animal (0.86–1.15 mm) with a slightly longer stylet (12–13 μm vs. 9–12 μm) and tail similar to vulva–anus distance (in *F. vulgaris* tail usually considerably longer than vulva–anus distance).

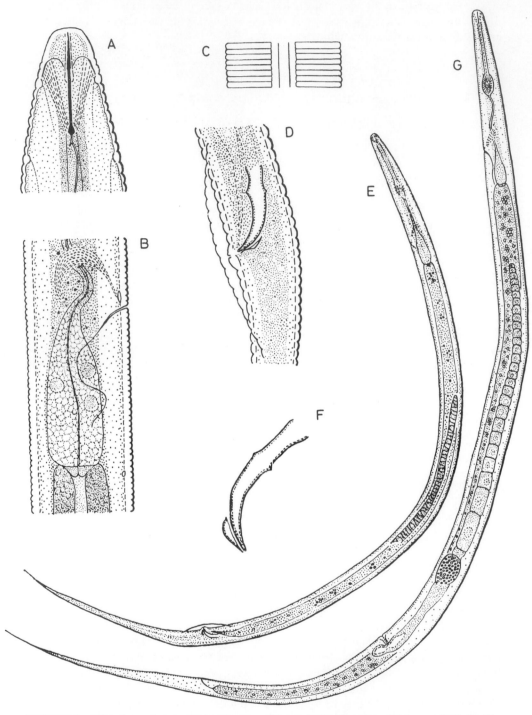

FIGURE 4 *Filenchus vulgaris*, (A) head region, (B) terminal bulb region, (C) lateral field, (D–F) male, (D) cloacal region, (E) entire view, (F) spicule and gubernaculum, (G) female, entire view. (After Brzeski, 1963, *Bull. Acad. Pol. Sci.*)

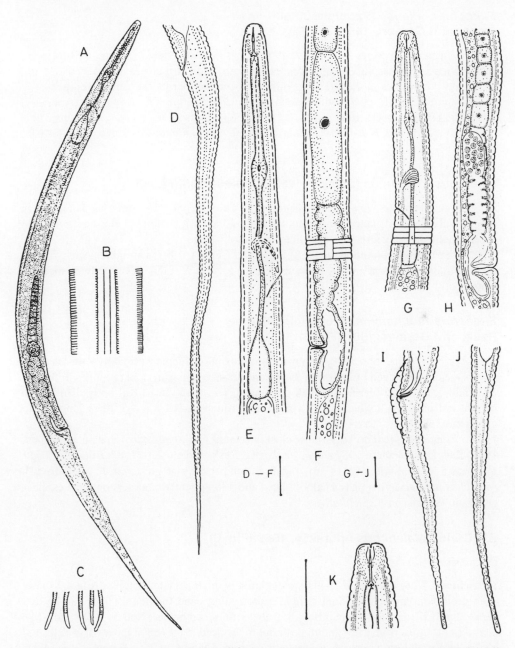

FIGURE 5 *Filenchus* species, (A–C) *F. ditissimus*, (A) female, total view, (B) lateral field, (C) tail tip variation, (D–F) *F. helenae*, (D) tail, (E) esophageal region, (F) female reproductive system, (G–K) *F. facultativus*, (G) esophageal region, (H) female reproductive system, (I) male tail, (J) female tail, (K) head. [(A–C) after Brzeski, 1963b, *Bull. Acad. Pol. Sci.*, (D–F) after Szczygiel, 1969, *Opusc. Zool. Budapest*; (G–K) after Szczygiel, 1970, *Bull. Acad. Pol. Sci.*)]

Filenchus helenae (Szczygiel, 1969) Raski and Geraert, 1987 (Fig. 5)

This much smaller species (L = 0.35–0.49 mm) has a short stylet (7– 8 μm), a filiform tail about twice as long as vulva–anus distance, delicate body striation (annuli less than 1 μm apart), only two lines in the lateral field, and postvulval uterine sac well developed.

Another small species with fine tail and fine body annuli, *F. discrepans* (Andrássy, 1954) Raski and Geraert, 1986 is not easy to differentiate from *F. helenae*: *F. discrepans* is described with males, *F. helenae* without; *F. discrepans* has a vulva position of 61–64%, *F. helenae* 53–61%.

Filenchus ditissimus (Brzeski, 1963) Siddiqi, 1986 (Fig. 5)

Another small species (L = 0.34–0.47 mm) with a small stylet (5–7 μm), has tail about as long as vulva–anus distance, elongated, tail end cylindrical; postvulval uterine sac short to absent; delicate body striation, four lines in lateral field.

In *F. misellus* (Andrássy, 1958) Raski and Geraert, 1987, a similar species, the tail tip is pointed to filiform and the postvulval uterine sac is half the vulval body diameter in length.

Filenchus facultativus (Szczygiel, 1970) Raski and Geraert, 1987 (Fig. 5)

This species with an L of 0.38–0.63 mm and a stylet of 6–8 μm (exceptionally 9 μm) is in the first place characterized by coarse body annuli (1.3–2.6 μm apart) and two lines in the lateral field; postvulval uterine sac well developed, less than one vulval body diameter in length; tail narrowly cylindroid; tail end finely rounded, tail 1.2–2.5 times longer than vulva–anus distance.

Filenchus facultativus is most closely related to *F. equisetus* (Husain and Khan, 1967) Raski and Geraert, 1987 and *F. neoparvus* Raski and Geraert, 1987. *Filenchus equisetus* has a longer stylet (12–14 μm) and a smooth head (striae present in *F. facultativus*). *Filenchus neoparvus* also has a smooth head and a long genital tract reaching the esophageal bulb (shorter in *F. facultativus*).

E.   Genus *Malenchus* Andrássy, 1968 (Fig. 6)

Diagnosis:

Tylenchinae. Very small (= 0.25–0.6) and relatively thick nematodes with distinct, usually even prominent annulation. Lateral field a protruding band (numerous lines can be discerned by SEM; in a few species only four to six lines can be observed by SEM and LM); lateral field begins from a few annuli posterior to the head to the end of the median bulb, and continues for about one-quarter to one-half of the tail length.

Head dorsoventrally compressed; amphidial apertures continuing on the lateral side of the head as longitudinal, sometimes sinuous slits; head bears four to six annuli (exceptionally one to two); cephalic framework weak. Stylet (7–14 μm) usually delicate (one exception), anterior part very thin and about half the posterior part; basal knobs distinct, mostly oblong. Opening of dorsal esophageal gland close to spear base. Esophagus with weakly to moderately developed median bulb.

Female reproductive system prodelphic, straight. Spermatheca shape apparently variable even within a species: from entirely in line to partly in line and partly offset;

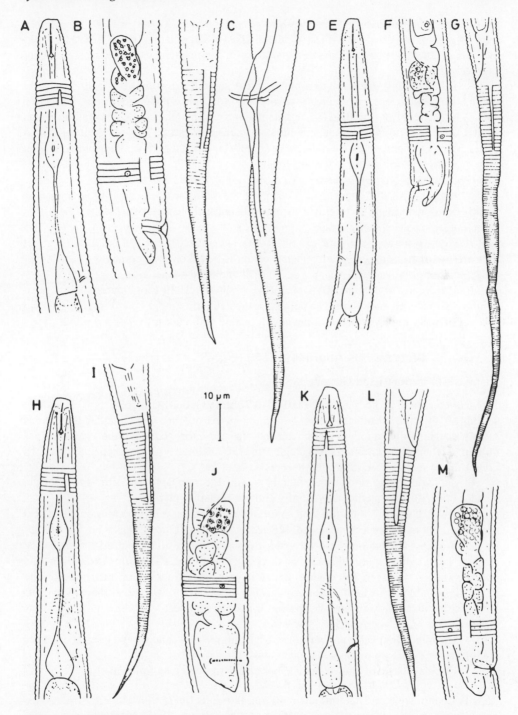

FIGURE 6 *Malenchus* species. (A–D) *M. exiguus*, (A–C) female, (D) male, (E–G) *M. nanellus*, female, (H–J) *M. acarayensis*, female, (K–M) *M. ovalis*, female, (A,E,H,K) anterior region, (B,F,J,M) female reproductive system, (C,G,I,L) female tail, (D) male tail. (From Geraert and Raski, 1986, *Nematologica*.)

postvulval uterine sac short. Vaginal walls more or less thickened; vulva at 55–70%; sunken with epiptygmata; vulval slit with small lateral dikes or larger lateral flaps. Advulval papillae, considered to be phasmids, dorsolateral, from 26 annuli anterior to vulva to a few annuli posterior to vulva.

Tails tapering gradually to a more or less pointed tip, straight or only terminally curved. Tail annuli become suddenly much finer posterior to ending of lateral field.

Males less frequent than females, sometimes absent. Cloacal lips protruding; bursa adanal, small but prominent (can sometimes be considered as an outgrowth of the lateral field). Spicules slightly bent ventrally, very slightly tapering, 12–19 μm long with open tip. Gubernaculum short and thin, 2–6 μm long.

Type species: *Malenchus machadoi* (Andrássy, 1963), Andrássy, 1968.

Andrássy's (1981) excellent revision of the genus was emended by Geraert and Raski (1986), who found undescribed species with unusual characteristics: lateral field with only four or six lines, stylet robust (not delicate).

Figure 6 shows four of the "usual" species: lateral field under LM a plain band, stylet delicate. Apart from measurements identification is based on the beginning of the lateral field (position relative to esophagus) and on tail and head shape.

Malenchus bryophilus (Steiner, 1914) Andrássy, 1980 (not figured) is the most common species, characterized by a tail of less than 100 μm long, a lateral field originating at level of spear knobs or one to three annuli behind and a spear of 7–9.5 μm.

F.   Genus *Miculenchus* Andrássy, 1959 (Fig. 7)

Diagnosis (after Raski and Geraert, 1985)

Tylenchinae. Small to medium species (0.40–0.72 mm). Cuticle with deep, transverse striae setting off thin but prominent annuli bearing complex notches and cuticular projections mostly matching onto another. Annuli become finer anteriorly but continue as discrete annuli up to labial plate. Lateral field a single band raised above body outline; appears as two lines usually scalloped, conforming to transverse striae.

Cephalic region not set off, bluntly rounded to hemispheric. Amphids elongate, slender open slits oriented dorsoventrally entirely on labial plate.

Stylet delicate, 7–11.5 μm long, anterior conical part as long as posterior cylindrical part, small knobs backwardly directed. Metacorpus well developed, muscular, about midway in esophagus, posterior bulb elongate, pyriform. Vulva simple, slitlike, indented V shape in lateral view; spermatheca squarish to rounded, offset. Postuterine sac shorter than one vulval body diameter. Tail conical, with finely rounded to acute terminus. Male similar to female, notable for its total lack of caudal alae; anterior margin of cloacal aperture overlaps opening as a finely pointed lip, posterior margin rounded.

Type species: *M. salvus* Andrássy, 1959.

Miculenchus species are very rare, but cosmopolitan. The wavy appearance of the annuli that continue up to the labial plate make it easily recognizable.

Three species are known, which can be differentiated as follows:

L = 0.33 mm, stylet = 11.5 μm, tail = 46 μm, terminus finely rounded ....
.... *M. tesselatus* (Ebsary, 1986) Maggenti et al., 1988
L = 0.40–0.49 mm, stylet = 7–9 μm, tail = 60–104 μm, terminus pointed ....
.... *M. salvus*
L = 0.71–0.72 mm, stylet = 9 μm, tail = 164–169, terminus finely pointed ....
.... *M. elegans* Raski and Geraert, 1983

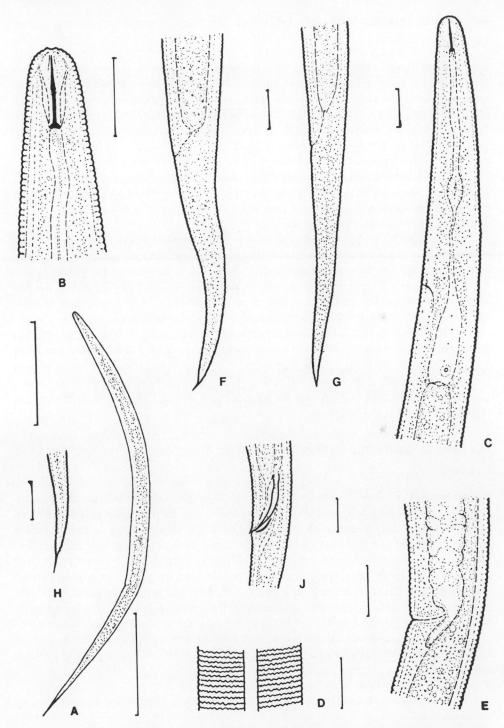

FIGURE 7 *Miculenchus salvus*. (A) Entire female, (B) head region, (C) esophageal region, (D) structure of cuticle, (E) vulval region, (F,G) female tails, (H) tip of female tail, (J) cloacal region of male. Scale in A = 100 μm, B–J: 10 μm. (From Andrássy, 1980.)

G. Genus *Tylenchus* Bastian 1865 (Fig. 8)

Diagnosis

Tylenchinae. Large, robust Tylenchidae, usually ventrally curved. Body length predominantly in the range 0.7–1 mm (extreme values 0.56–1.3 mm). Body annuli distinct, lateral field with four lines. Head not or slightly offset, usually with four to five annuli; amphidial apertures large, pitlike, essentially confined to labial plate and only partly on lateral side of head. Stylet robust, relatively large (10–21 μm); knobs distinct, rounded; anterior, conical part similar in length to posterior, cylindrical part. Dorsal esophageal gland opening close to stylet base. Median bulb well developed, anterior to middle of esophagus.

Vulva behind middle of body; reproductive system anteriorly outstretched; spermatheca in line or offset; postvulval uterine sac shorter than vulval body width. Spicules arcuate; gubernaculum thin. Anterior cloacal lip pointed, posterior one rounded.

Type species: *T. davainei* Bastian, 1865.

Andrássy's (1979) revision contains a key and detailed descriptions. Some of the smaller species in his revision showed a stylet with a smaller anterior conical part; they have been transferred to *Filenchus* (Raski and Geraert, 1987). In Geraert and Raski (1987) *Tylenchus* contains 17 species apart from the type species. Andrássy (1979) pointed out that the most frequent species are *T. davainei*, *T. elegans* de Man, 1876, *T. arcuatus* Siddiqi, 1963, and *T. ritae* Siddiqi, 1963, the type species being the most abundant.

These four species have a similar stylet length of 14–18 μm. *Tylenchus arcuatus* has the shortest tail ($c' = 5$–6 vs. $c' = 7$–10 for the three other species); *T. davainei* has a stylet of 16–18 μm and the tail tip is pointed, *T. ritae* and *T. elegans* have a stylet of 14–15 μm and a rounded tail tip; in *T. ritae* the terminal bulb is as long as the median bulb and the tail measures 120–150 μm while in *T. elegans* the terminal bulb is longer than the median bulb and the tail measures 100–120 μm.

H. Genus *Aglenchus* Andrássy, 1954 (Fig. 9)

Diagnosis

Atylenchinae. Body 0.4–0.7 mm long, straight to slightly curved. Cuticle distinctly annulated, annuli usually 1.5–2 μm wide (extreme values: 1.3–2.4 μm). Lateral field with two ridges that are usually close to each other giving three lateral lines; when the two ridges are slightly separated four lines are visible, the two inner ones closer together. More or less rounded amphidial apertures on head front (subfamily character), with lateral offshoot in *Aglenchus dakotensis* male. Head smooth. Stylet 9–13 μm long, anterior conical part always slightly smaller than posterior cylindrical part. Advulval papillae dorsosublateral, anterior to vulva. Vulva sunken, with smaller inner lips and usually with large lateral flaps. Vagina obliquely forward, walls usually thickened; no postvulval uterine sac. Spermatheca offset, oval to rounded, usually with sperm. Tail filiform. Males with adanal bursa, spicules slightly curved; the cloacal lips form an elongated tube (slightly offset in some *A. agricola* males).

Type species: *A. agricola* (de Man, 1884) Andrássy, 1954

In Geraert and Raski (1989) four species are reported. They can be differentiated by measurements.

1.   Tail 178–276 μm long, $c = 2.5$–3.4 . . . . *A. muktii* Phukan and Sanwal, 1980
     Tail 134–179 μm long, $c = 3.0$–4.4 . . . . *A. agricola*
     Tail 96–136 μm long, $c = 4.2$–6 2

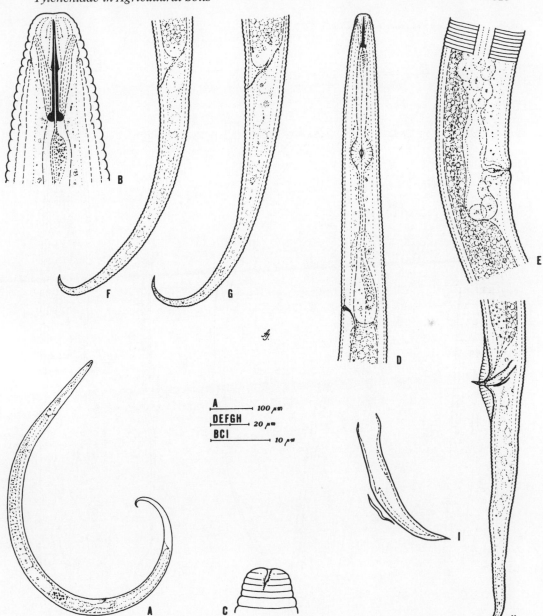

FIGURE 8 *Tylenchus davainei*. (A) Entire female, (B) anterior end, (C) lateral surface view of head, (D) esophageal region, (E) vulval region, (F,G) female tails, (H) male tail, (I) spiculum and gubernaculum. (From Andrássy, 1977, C.I.H. Descriptions of Plant Parasitic Nematodes.)

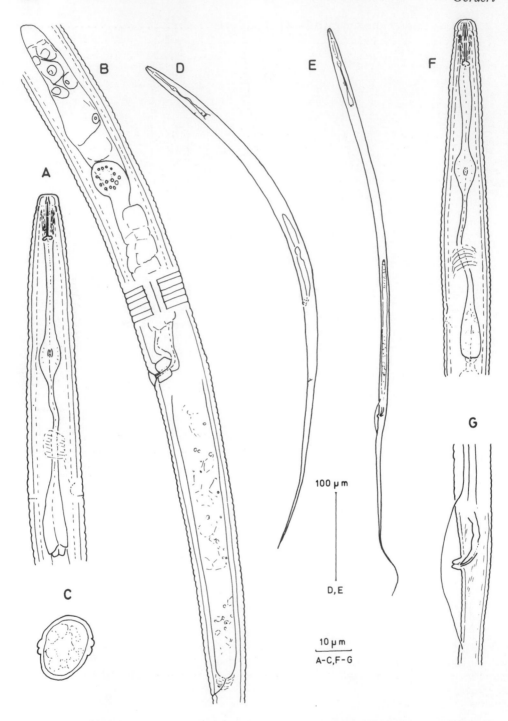

FIGURE 9 *Aglenchus muktii*, (A–D) female, (A) anterior end, (B) vulvar and anal region, (C) transverse section, (D) general view, (E–G), male, (E) general view, (F) anterior end, (G) cloacal region. (From Geraert and Raski, 1989, *Nematologica*.)

2. Annuli 1.2–1.5 µm, stylet 10.5–12 µm ....

    ....*A. mardanensis* Maqbool, Shahina, and Zerina, 1984

    Annuli 1.9–2.6 µm, stylet 9–11 µm .... *A. dakotensis* Geraert and Raski, 1989

Aglenchus agricola is cosmopolitan. Very common and widespread in Europe, it feeds on grasses and clover (Bongers, 1988); it is not only found in arable soil and meadows but in all kinds of noncultivated soils, in mountain regions, and even in freshwater.

Aglenchus muktii is known from India and USA (California), *A. mardanensis* from Pakistan, and *A. dakotensis* from USA (North Dakota). It could be that these species are more widespread but because of their overall resemblance with the type species have been considered as *A. agricola*.

I. Genus *Coslenchus* Siddiqi, 1978 (Fig. 10)

Diagnosis

Atylenchinae. Body 0.33–1.00 mm long, straight to slightly curved. Cuticle distinctly annulated (annuli 1.5–3.6 µm wide), with 10–27 longitudinal ridges, apart from the lateral field which shows three to six lines. Head usually annulated. Stylet 9–15 µm long, well developed, with knobs; anterior conical part always slightly smaller than posterior cylindrical part. Vulva sunken with small inner lips and usually with lateral membranes; vagina variously thickened, straight to slightly obliquely forward-directed; postvulval uterine sac from none to small, exceptionally one vulval body diameter long; spermatheca often empty. Advulval papillae dorsosublateral, usually posterior to vulva. Tail elongate, tail tip from rounded to pointed to effilate. Males often rare, with adanal bursa and protruding cloacal lips (these are thicker and more offset than in *Aglenchus* males).

Type species: *Coslenchus costatus* (de Man, 1921) Siddiqi, 1978

The many *Coslenchus* species are difficult to differentiate; ventral views and sections are needed to know the number and structure of the longitudinal ridges. Mizukubo and Minagawa (1985) published a key for 31 species; Brzeski (1987) considered 23 species as valid; Geraert and Raski (1989) came to 29 species for which a key was presented. Only six species were described before 1981, with the type species as the most cosmopolitan. *Coslenchus costatus* certainly is the most common representative of the genus but all earlier identifications have to be checked. This is particularly so for the host–parasite relationship as *C. costatus* has now and then been reported as a pest. The population studied by Wood (1973) developed well on various grasses, but only as an ectoparasite and not causing root damage.

J. *Basiria* Siddiqi, 1959 (Fig. 11)

Diagnosis (Geraert, 1968a, emended)

Boleodorinae. L from medium to large (0.4–1.1 mm). Head high with a very weakly sclerotized framework; amphidial apertures large slits usually oblique, only on the lateral side of the head; stylet thin-walled usually with flangelike knobs, the anterior part usually half the length of the posterior part. Esophagus with moderately developed median bulb and offset terminal bulb; median bulb situated at 36–58% of the total esophageal length: outlet of dorsal esophageal gland can vary from 1 to 13 µm posterior to stylet end. Vulva postequatorial, one branch anteriorly outstretched, postvulval uterine sac usually not longer than vulval body diameter; spermatheca usually in line. Spicules and gubernaculum slightly curved; bursa adanal, moderately developed. Tail in both sexes elongated; tail tip filiform,

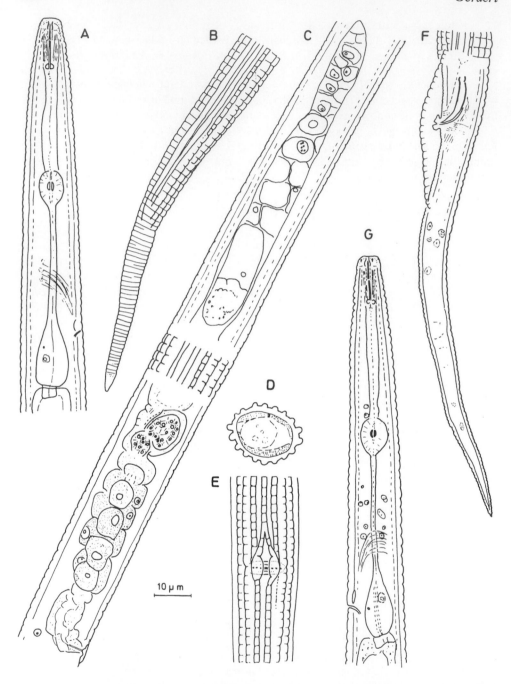

FIGURE 10   *Coslenchus aquaticus* Geraert and Raski, 1989. (A–E) Female, (A) anterior end, (B) tail, (C) reproductive system, (D) transverse section showing the seven ventral and seven dorsal ridges and the lateral field (two variously separated ridges), (E) vulva, ventral view, (F,G) male, (F) tail, (G) anterior end. (From Geraert and Raski, 1989, *Nematologica*.)

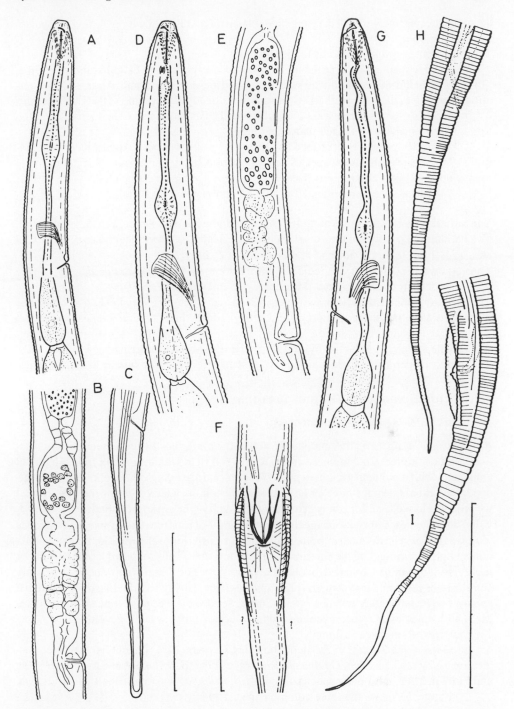

FIGURE 11 (A–C) *Basiria tumida*, (D–F) *B. gracilis*, (G–I) *B. aberrans*, (A,D,G) anterior end, (B,E) female reproductive system, (C,H) female tail, (F) ventral view of cloacal region in male, (I) male tail. All scalcs show 50 μm. (From Geraert, 1968a, *Nematologica*.)

acute, subacute, finely rounded, rounded, or clavate. Lateral field with two, four, or exceptionally six lines.

Type species: *B. graminophila* Siddiqi, 1959.

     Basiria species resemble most *Filenchus* species but at high magnification the large,
oblique amphidial slits on the lateral side of the head are very diagnostic. Another differentiating character is that the spermatheca is usually in line in *Basiria* and offset in *Filenchus*.

     Neopsilenchus species were formerly described as *Basiria* species; their differentia-
tion on stylet shape is explained under that heading.

     Some 40 species are described, but the validity of some can be questioned, e.g., Sid-
diqi (1986): "Several nominal species described from India could be invalid junior synonyms and a revision of the genus, based on the type specimens, is needed."

     The three species represented in Fig. 11 show:

The variation in development of stylet knobs (not always distinct, e.g., in *B. gracilis*).
The variation in position of the dorsal esophageal gland: close to stylet end in *B. tumida*, at
       4–7 µm in *B. gracilis* and at 10–13 µm in *B. aberrans*).
The variation in position of the median bulb: anterior to the middle in *B. tumida*, at about the
       middle in *B. gracilis*, and posterior to the middle in *B. aberrans*.
The variation in the lateral field: two lines in *B. gracilis*, four in *B. tumida* and *B. aberrans*.
The variation in tail shape: posterior third subcylindrical in *B. tumida*, conical in the two
       other species represented.

     The female reproductive system (Fig. 11B, E) shows the spermatheca in line, filled
with rounded sperm. The male reproductive system (Fig. 11F, I) is similar in most species.

K.   Genus *Boleodorus* Thorne, 1941 (Fig. 12)

Diagnosis (after Geraert, 1971, emended)

Boleodorinae. Rather small nematodes (L = 0.3–0.7 mm), mostly curved ventrally, sometimes spirally coiled. Cuticle finely annulated, lateral field with four or six lines. Head not
annulated, with or without a slight depression at the oral aperture, framework weakly developed; amphidial openings more or less oblique slits on the lateral sides of the head. Stylet
weakly developed, anterior conical part half the length of the posterior cylindrical part, provided with flanges. Dorsal esophageal gland opening 1–4 µm from end of stylet. Esophagus
with weakly developed median bulb, situated at or posterior to the middle; posterior bulb
not or slightly surrounding the anterior part of the intestine. Tail elongated, conical, more or
less curved; tip pointed, rounded, or clavate. Vulva posterior, female reproductive tract only
anteriorly developed, outstretched; spermatheca offset, usually filled with small refractive
sperm; oocytes in mature females often in multiple rows; posterior uterine sac very short.
Male with spicules and gubernaculum slightly curved, bursa weakly developed.

Type species: *B. thylactus* Thorne, 1941.

     Geraert and Raski (1987) listed 20 species. Geraert (1971) concluded that three spe-
cies are easily recognized: *B. clavicaudatus* Thorne, 1941, *B. thylactus*, and *B. volutus* Lima
and Siddiqi, 1963; most other species resemble *B. thylactus* but are difficult to distinguish.

     Figure 12 shows the three common species mentioned above:

B. clavicaudatus: head with median involution, tail clavate, *V* anterior (about 60%).
B. thylactus: head with median involution, body arcuate (O to C shape), *V* more posterior (*V*
      = 62–70), tail hooked (described as the typical tail form) to variously curved.

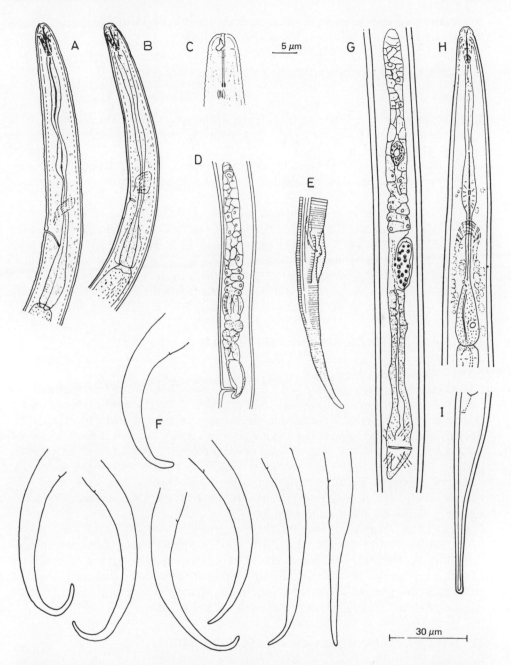

FIGURE 12 *Boleodorus*, (A) *B. thylactus*, female, esophageal region, (B) *B. volutus*, male, esophageal region, (C) *B. volutus*, head, (D) *B. thylactus*, female reproductive system, (E) *B. thylactus*, male tail, (F) *B. thylactus*: variation in tail shape found in populations from the neighborhood of Ghent, (G–I) *B. clavicaudatus*, (G) female reproductive system, (H) esophageal region, (I) female tail. (From Geraert, 1971, *Nematologica*.)

B. volutus: head without median involution; body spirally curved with one, two, or three convolutions; *V* posterior (*V* = 66–72), tail short.

L.   Genus *Neopsilenchus* Thorne and Malek, 1968 (Fig. 13)

Diagnosis

Boleodorinae. Similar to *Basiria*, except for stylet and sperm shape. The delicate stylet is cylindroid and has no knobs the anterior part is slightly curved and has a relatively wide opening. The sperm is rod-shaped.

Thorne and Malek (1968) erected the genus to accommodate *N. magnidens* (Thorne, 1949), type species; the main differentiating character from *Basiria* was the absence of knobs.

Four other species have been described with a cylindroid stylet as in the type species: *N. affinis* Khan and Khan, 1976, *N. minor* (Geraert, 1968) Kheiri, 1972, *N. similis* Khan and Khan, 1976, and *N. varians* Khan and Khan, 1976. Two other species have been transferred because of the absence of knobs: *N. citri* (Jairajpuri, 1968) Bello, 1972 and *N. noctiscriptus* (Andrássy, 1962) Khan, 1973. A restudy of this material is necessary.

Neopsilenchus magnidens and *N. minor* are frequently found; they differ from each other in measurements as the species names suggest.

M.   Genus *Psilenchus* de Man, 1921 (Fig. 14)

Diagnosis

Boleodorinae. Large tylenchids (*L* = 0.7–1.7 mm). Lateral lines two to four, sometimes irregular. Amphidial apertures large, transverse to oblique slits on the lateral side of the head. Head annulated or smooth. Stylet delicate, anterior part about half the cylindrical part; anterior part usually conical, sometimes more cylindrical; no knobs (often difficult to see the exact stylet end). Dorsal esophageal gland opening probably always several micrometers posterior to stylet end (up to two-thirds the stylet length). Median bulb usually posterior to middle of esophagus. Vulva near middle, ovaries paired, opposite, and outstretched; spermatheca elongate, in line. Tail elongate, usually terminally clavate. Bursa adanal: spicules arcuate, gubernaculum thin and troughlike.
Type species: *P. hilarulus* de Man, 1921

Kheiri (1970) presented a key for 11 species. Siddiqi synonymized two and restudied three species. All *Psilenchus* species are very similar to each other and it seems to me that a thorough study will show more similarities than differentiations; some of the doubtful differences now used are annulated vs. smooth head, length of stylet, position of dorsal esophageal gland opening, number of lateral lines, and tail shape.

N.   Genus *Cephalenchus* Goodey, 1962 (Fig. 15)

Diagnosis

Tylodorinae. Small to medium-sized species (*L* = 0.40–0.82 mm). Body with coarse annuli, 1.4–3.0 μm wide; lateral field mostly with six longitudinal incisures, equally spaced (exceptionally four incisures). Cephalic region mostly set off by constriction or change in contour. Stylet slender and long (14–25 μm), anterior conical part about as long as posterior cylindrical part; knobs well developed, rounded, sometimes anteriorly concave. Amphidial apertures squarish/ovoid pits or elongate lateral slips extending from oral plate to edge of labial

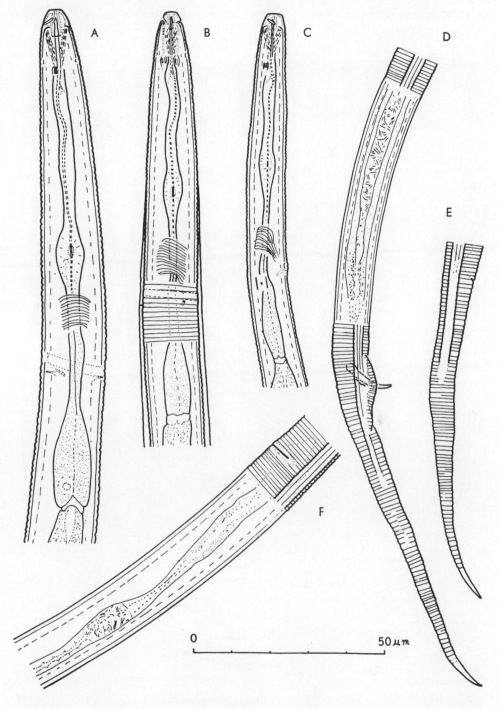

FIGURE 13 *Neopsilenchus magnidens*, (A) female, esophageal region, (B–F) *N. minor*, (B,C) esophageal region of female (B) and male (C), (D) posterior end of male, (E) tail of female, (F) female reproductive system. (From Geraert, 1968, *Nematologica*.)

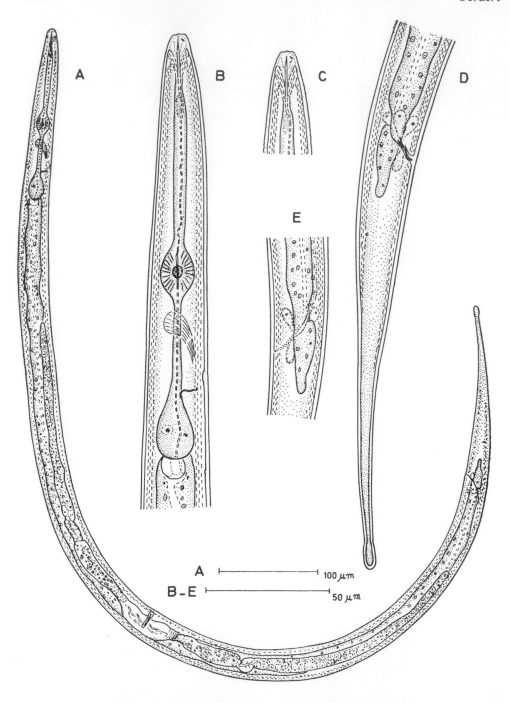

FIGURE 14 *Psilenchus iranicus* Kheiri, 1970. (A) Entire female, (B) esophageal region, (C) female head, (D) female tail, (E) anal region showing postanal intestinal lobe and separate cells, both mentioned as typical for this species. *Psilenchus hilarulus* does not show these cells and has a smooth lip region. (From Kheiri, 1970, *Nematologica*.)

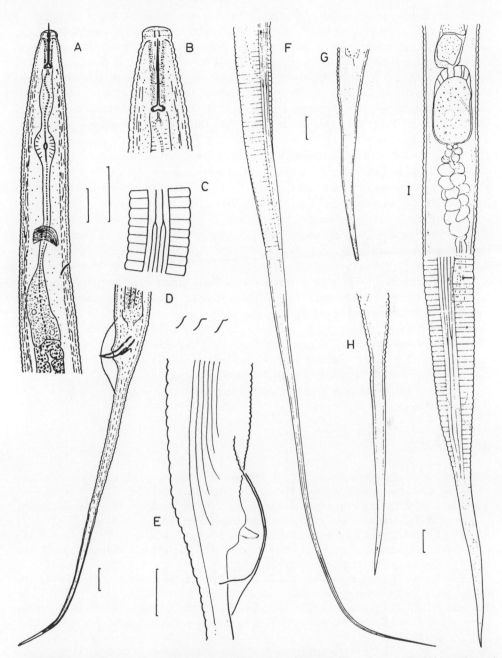

FIGURE 15 *Cephalenchus* species, *C. hexalineatus* (A–E), (G–I), *C. leptus* (F), (A) esophageal region, (B) head region, (C) lateral field at transition esophagus/intestine, (D) male tail with variability in gubernaculum curvature, (E) male bursa, left and right, (F) female, post vulval, (G,H) variability in tail shape, (I) female reproductive system and tail. [(A–D) from Geraert, 1962; (E–I) from Geraert, 1968b, *Meded. Fak. LandbWet. Gent.*]

plate (not continuing on lateral side of head). Procorpus quite short, length to middle of valve 40–44% of esophageal length; posterior glandular region small, pyriform or elongated; glands usually enclosed in bulbar structure but overlapping intestine in two species. Advulval papillae dorsolateral, prevulval (phasmids on tail reported in one species). Tail elongated from conoid with a bluntly rounded tip to very slender with terminus effilate, acute or finely rounded. Vulva usually with small lateral membranes, mostly excentric, displaced laterally usually to right side; long postvulval uterine sac, spermatheca elongated, in line. Caudal alae may be unequal in size; if so, right side smaller and rejoining body posteriorly at an unusual angle.

Type species: *Cephalenchus hexalineatus* (Geraert, 1962) Geraert and Goodey, 1964

In Raski and Geraert (1986) 14 species have been differentiated. Two of them appear to be very common: the type species with a tail of medium length (80–140 µm) and *C. leptus* Siddiqi, 1963 with a very long tail (170–280 µm).

Colbran (1964) considered *C. hexalineatus* as a junior synonym of *C. marginatus* (Cobb, 1893) Geraert, 1968. Andrássy (1984) separated them a.o. by the difference in stylet length *Cephalenchus hexalineatus* has a slightly shorter stylet (14–17 µm vs. 16–20.5 µm). Raski and Geraert (1986) agreed with that action.

Cephalenchus species have several times been reported as migratory root ectoparasites feeding on epidermal cells (review in Hooper, 1974). The species name used was often *C. emarginatus* but at least two other species were involved: *C. hexalineatus* as parasites of roots of various conifer seedings (Sutherland and Keable, 1966; Sutherland, 1967; Geraert, 1968b) and *C. illustris* Andrássy, 1984 as parasites of cabbage and cauliflower roots (Khera and Zuckerman, 1962; 1963, originally identified as *Tylenchus agricola*: Raski and Geraert, 1986).

Severe attack can cause root stunting but no root necrosis or other pathogenic effect (Hooper, 1974).

REFERENCES

Andrássy, I. 1977. *Tylenchus davainei*. C.I.H. descriptions of plant-parasitic nematodes. Set 7, no. 97. Commonwealth Institute of Parasitology. St. Albans, England.

Andrássy, I. 1979. The genera and species of the family Tylenchidae Oerley 1880 (Nematoda). The genus *Tylenchus* Bastian, 1865. *Acta Zool. Hung.* 256: 1–33.

Andrássy, I., 1980. The genera and species of the family Tylenchidae Oerley, 1880 (Nematoda). The genus *Aglenchus* (Andrássy, 1954) Meyl, 1961, *Miculenchus* Andrássy, 1959, and *Polenchus* gen.n. *Acta Zool. Hung.* 26: 1–20.

Andrássy, I., 1981. The genera and species of the family Tylenchidae Oerley, 1880 (Nematoda). The genus *Malenchus* Andrássy, 1968. *Acta Zool. Hung.* 27: 1–47.

Andrássy, I. 1984. The genera and species of the family Tylenchidae Oerley, 1880 (Nematoda). The genera *Cephalenchus* (Goodey, 1962) Golden, 1971 and *Allotylenchus* gen.n. *Acta Zool. Hung.* 30:1–28.

Brzeski, M. W. 1963a. On the taxonomic status of *Tylenchus filiformis* Bütschli, 1873 and the description of *T. vulgaris* sp.n. (Nematoda: Tylenchidae). *Bull Acad. Pol. Sci.* 11: 531–535.

Brzeski, M. W. 1963b. *Tylenchus ditissimus* sp.n. (Nematoda: Tylenchidae). *Bull. Acad. Pol. Sci.* 11: 537–540.

Brzeski, M. W. 1987. Taxonomic notes on *Coslenchus* Siddiqi, 1978 (Nematoda: Tylenchidae). *Ann. Zool.* 40: 417–436.

Colbran, R. E., 1964. Studies of plant and soil nematodes. 7. Queensland records of the order Tylenchida and the genera *Trichodorus* and *Xiphinema*. *Qd. J. Agr.* 21: 77–123.

Geraert, E. 1962. *Bijdragen tot de Kennis der plantenparasitaire en der vrijlevende Nematoden van Kongo*. Ganda-Congo. Instituut voor Dierkunde, Rijksuniversiteit Gent.

Geraert, E., 1968a. The genus *Basiria* (Nematoda: Tylenchina). *Nematologica* 14: 459–481.

Geraert, E. 1968b. Morphology and morphometrics of the subgenus *Cephalenchus* Goodey, 1962—genus *Tylenchus* Bastian, 1865 (Nematoda). *Meded. Rijksfac. LbWet. Gent* 33: 669–678.

Geraert, E. 1971 Observations on the genera *Boleodorus* and *Boleodoroides* (Nematoda: Tylenchida). *Nematologica* 17: 263–276.

Geraert, E., and Raski, D. J. 1986. Unusual *Malenchus* species (Nemata: Tylenchidae). *Nematologica* 31: 27–55.

Geraert, E., and Raski, D. J. 1987. A reappraisal of Tylenchina (Nemata). 3. The family Tylenchidae Oerley, 1880. *Rev. Nematol.* 10: 143–161.

Geraert, E., and Raski, D. J. 1989. Study of some *Aglenchus* and *Coslenchus* species (Nemata: Tylenchidae). *Nematologica* 35: 6–46.

Hooper, D. J. 1974. *Cephalenchus emarginatus*. C.I.H. descriptions of plant-parasitic nematodes. Set 3, no. 33. Commonwealth Institute of Parasitology, St. Albans, England.

Kheiri, A. 1970. Two new species of the family Tylenchidae (Nematoda) from Iran, with a key to *Psilenchus* de Man, 1921. *Nematologica* 16: 359–368.

Khera, S., and Zuckerman, B. M. 1962. Studies on the culturing of certain ectoparasitic nematodes on plant callus tissue. *Nematologica* 8: 272–274.

Khera, S., and Zuckerman, B. M. 1963. In vitro studies of host-parasitic relationships of some plant-parasitic nematodes. *Nematologica* 9: 1–6.

Mizukubo, T., and Minagawa, N. 1985. The genus *Coslenchus* Siddiqi, 1978 (Tylenchidae: Nematoda) from Japan. II. Synonymy of *Coslenchus* over *Cosaglenchus* and *Paktylenchus* based on the Phylogenetic Relationships and a Key to the Species. *Jap. J. Nematol.* 15: 14–25.

Raski, D. J., and Geraert, E. 1985. A new species of *Miculenchus* Andrássy, 1959 and further notes on *M. salvus* (Nemata: Tylenchidae). *Nematologica* 30: 419–428.

Raski, D. J., and Geraert, E. 1986a. New species of *Lelenchus* Andrássy, 1954 and *Ecphyadophora* de Man, 1921 (Ecphyadophorinae: Tylenchidae) from southern Chile. *Nematologica* 31 (1985): 244–265.

Raski, D. J., and Geraert, E. 1986b. Descriptions of two new species and other observations on the genus *Cephalenchus* Goodey, 1962 (Nemata: Tylenchidae). *Nematologica* 32: 56–78.

Raski, D. J., and Geraert, E. 1987. Review of the genus *Filenchus* Andrássy, 1954 and description of six new species (Tylenchidae: Nemata). *Nematologica*, 32(1986): 256–311.

Raski, D. J., Koshy, P. K., and Sosamma, V. K. 1982. A revision of the subfamily Ecphyadophorinae Skarbilovich, 1959 (Tylenchidae: Nematoda). *Rev. Nématol.* 5: 119–138.

Siddiqi, M. R. 1986. *Tylenchida. Parasites of Plants and Insects*. Commonwealth Agricultural Bureaux, Slough, U.K.

Sutherland, J. R. 1967. Parasitism of *Tylenchus emarginatus* on conifer seedling roots and some observations on the biology of the Nematode. *Nematologica* 13: 191–196.

Sutherland, J. R., and Keable, R. 1966. The nematode *Tylenchus (Cephalenchus) hexalineatus* (Geraert, 1962) Geraert and Goodey, 1964 found in three Quebec forest nurseries. *Prog. Rep. Can. Dept. For.* 22: 2–3.

Szczygiel, A. 1969. A new genus and four new species of the subfamily Tylenchinae de Man, 1876 (Nematoda: Tylenchidae) from Poland. *Opusc. Zool. Budapest* 9: 159–170.

Szczygiel, A., 1970. Two new species of the genus *Tylenchus* (Nematoda, Tylenchidae) from Poland. *Bull. Acad. Pol. Sci.* 17: 673–677.

Wood, F. M. 1973. Life cycle and host-parasite relationships of *Aglenchus costatus* (de Man, 1921) Meyl, 1961 (Nematoda: Tylenchidae). *N. Zealand J. Agr. Res.* 16: 373–380.

19
Resistance-Breaking Races of Plant Parasitic Nematodes

ROBERT D. RIGGS   *University of Arkansas, Fayetteville, Arkansas*

I. INTRODUCTION

Ritzema Bos studied host races of the plant parasitic nematode *Ditylenchus dipsaci* as early as 1888 (Ritzema Bos, 1888–1892). The information on races was summarized by Sturhan (Sturhan, 1971); however, he devoted little space to "resistance-breaking" races. Populations of nematodes that invade and parasitize plants thought to be resistant were reported as early as 1944 (Christie and Albin, 1944), but in some cases the races were later described as separate species as in root-knot nematodes (Chitwood, 1949). Since that time many reports of resistance-breaking types or races have been published from all over the world. When a population of a nematode species attacks a cultivar thought to be resistant, the resistance is said to have broken down. Depending on the point of view a change in the plant is indicated in that it is no longer resistant to the parasite or a change in the parasite is indicated in that it can now parasitize the plant that once was resistant. In fact no change has taken place in either plant or nematode but a different population of nematodes has been selected and increased that has the ability to parasitize the particular variety of the plant in question.

Resistance may be defined as the ability of a host to hinder the development or reproduction of a pathogen (Robinson, 1969). It is sometimes divided into vertical resistance and horizontal resistance. Vertical resistance may be defined as differential resistance in which races of the pathogen interact differentially with the host cultivars (Van der Plank, 1975). This type of resistance is generally controlled by one to a few genes or a group of genes acting as a unit. Horizontal resistance is then defined as nondifferential resistance in which all races of a pathogen interact uniformly. This is controlled by a large number of genes and we may assume that the pathogen genes for parasitism cannot overcome all of the genes for resistance.

Infraspecific or infrasubspecific groups of nematodes have been variously called biotypes, biological races, biological species, different isolates, differentially selected populations, host races, host-specific types, pathogenic biotypes, pathotypes, physiological races, physiological strains, physiological variations, and strains. The two major terms used are race and pathotype. Race refers to the pathogen and is a subdivision of a species or sub-

species separated by physiological or pathological criteria while pathotype refers to a population of a pathogen in which all members have common characteristics of pathogenicity
(Robinson, 1969). More commonly pathotype has been used to refer to populations with
differing pathogenic capabilities in the species *Globodera pallida* (Stone) Behrens, 1975;
G. rostochiensis (Wollenweber) Behrens, 1975; and *Heterodera avenae* Wollenweber,
1924. Race has been used to refer to population differences in *H. glycines* Ichinohe, 1952;
Ditylenchus dipsaci (Kuhn) Filipjev, 1936; and *Meloidogyne* spp. However, races of *H.
glycines* are separated by cultivars or lines whereas races of *D. dipsaci* and *Meloidogyne*
spp. are separated by different host species. In this chapter race will be used to refer to all
types of biological or physiological variation in the reproductive capacity of nematodes in
relation to a host. Population will be used for a group of nematodes without distinguishing
isolates, phenotypes, or genotypes.

II. *DITYLENCHUS DIPSACI*

Races of *D. dipsaci* generally have a narrow host range but the races intermate and the progeny at times have broader host ranges than the parents (Eriksson, 1974). Ten races of *D.
dipsaci* were crossed 900 times in 44 combinations. Fertile hybrids were obtained in most
cases but in a few instances (LU × RU, LU × TU, OA × RE, OA × WC, and RX × WC) fertile
hybrids were obtained with certain crosses but not in their reciprocals. The successful hybrids varied greatly in their parasitism and egg production. Some hybrids had wider host
ranges than the parents but resistance-breaking features were not observed (Eriksson,
1974). The incompatibility encountered indicated that different species are involved yet in
some cases hybrids reproduce better than intrarace matings. *Ditylenchus dipsaci* is a good
example of a composite species with broad physiological but very little morphological variation. It does not appear to have resistance-breaking races that cross over the host specificity
barriers.

      Pure cultures of *Ditylenchus dipsaci* derived from single females were inoculated on
eight plant species in pots (Webster, 1967). Onion was a host of all six races, and tulip was a
host of all but the red clover race. Lucerne was a good host only for the lucerne race and a
poor host for the white clover race. The tulip race and the oat race reproduced on more of the
eight hosts than the other races. Hyacinth was not a good host for any race. Hybrids did not
reproduce on any of the hosts as well as did the parent races. Mixtures of races in nature
probably result in hybrids with low reproduction levels and longer life cycles but not new
races.

      Attempts to separate races of *D. dipsaci* by gel diffusion and immunoelectro–
phoresis indicated that differences could be obtained (Gibbins and Grandison, 1968); however, there was variation between runs. This variation may be because of differences in ratios of eggs to juveniles to adults in the antigens.

III. *MELOIDOGYNE* SPP.

Early reports of resistance-breaking types in *Heterodera marioni* syn. *Meloidogyne* sp.
(Christie and Albin, 1944) were explained and dispelled by the description of four species
by Chitwood (1949). However, since that time races have been reported in various species
of *Meloidogyne*.

      Of the four species of *Meloidogyne* most commonly found worldwide (Sasser and
Carter, 1982), physiological variation has been reported frequently in *M. incognita* (Koboid

and White) Chitwood, 1949, less frequently in *M. arenaria* (Neal) Chitwood, 1949, and only occasionally in *M. hapla* Chitwood, 1949, and *M. javanica* (Treub) Chitwood, 1949.

From 1975 to 1985 Sasser, Hirschmann, Taylor, Triantaphyllou, and many others, working under the auspices of the International *Meloidogyne* Project, examined the relationships of populations of *Meloidogyne* species from all over the world. Their variability was assessed on the basis of morphological characters, cytology, and differential host test. Primarily on the basis of the differential host test (Table 1); the total number of populations tested were 52% *M. incognita*, 31% *M. javanica*, 8% *M. hapla*, 7% *M. arenaria*, 1% *M. exigua* Goeldi, 1987, and all others 1% (Sasser and Carter, 1982). Four races of *M. incognita* and two races of *M. arenaria* were described (Taylor and Sasser, 1978) (Table 1). Of the populations of *M. incognita*, 72% were race 1, 13% each races 2 and 3, and 2% race 4 (Sasser and Carter, 1982). The populations of *M. arenaria* were 16% race 1 and 84% race 2. This is interesting in that *M. arenaria* is called the peanut root-knot nematode but only 16% of the populations tested reproduce on peanut.

A. *Meloidogyne incognita*

By far the greatest number of reports of physiological variation in root-knot nematodes has been in *Meloidogyne incognita*, the species most commonly encountered around the world (Sasser and Carter, 1982). Variability in the reproduction and galling of different populations of *M. incognita* were reported as early as 1953 (Martin, 1953). Populations that parasitized and reproduced on "resistant" tomato (*Lycopersicon esculentum* (L.) Mill.) were selected from greenhouse populations not only of *M. incognita* but of *M. arenaria* also (Riggs and Winstead, 1959). Similar races of *M. incognita* also were reported in Hawaii (Gilbert pers. commun.) in field plots used to test resistant tomato breeding lines, in Australian field plantings of tomato (Sauer and Giles, 1959), and in West Virginia (Kish and Adams, 1973). The resistance-breaking races were quite stable (Winstead and Riggs, 1963). There were morphological variations associated with the resistance-breaking type but they were not sufficient to separate them from the parent population. When the resistance-breaking races in West Virginia were tested on normally susceptible cultivars, five were found to be resistant.

Five resistance-breaking races of *M. incognita* were found in Japan (Okamoto, 1979). Three races broke the resistance of tomato cultivars but would not reproduce on resistant tobacco or sweet potato. One race broke the resistance of tobacco but would not reproduce on resistant tomato and sweet potato cultivars. The fifth race broke resistance of sweet potato but would not reproduce on resistant tomato and tobacco cultivars. No morphological differences were observed among the five races. A population (N–2) of *M. incognita* from Japan was pathogenic to the three sweet potato cultivars, Norin No. 2, Norin No. 5, and Taihaku, which were resistant to another *M. incognita* population (K–14) (Nishizawa, 1974).

Seventeen populations of *M. incognita* were tested on tomato (Rutgers), tobacco (*Nicotiana tobacum* L. cv. NC95), cotton (*Gossypium hirsutum* L. cv. McNair 1032), cowpea (*Vigna unguiculata* (L.) Walp. subsp. *unguiculata* cv. M57–13N), watermelon (*Citrullus lanatus* (Thunb.) Matson and Nakai cv. Dixie Queen), and pepper (*Capsicum annum* L. subsp. *annum* cv. California Wonder) (Southards and Priest, 1973). All 17 populations reproduced on tomato and none reproduced on NC95 tobacco. The variation in reproduction on the other four hosts could be divided into six races.

A population of *M. incognita* found on the Wartelle farm in Louisiana reproduced freely on Bragg and other soybean cultivars reported to have resistance to *M. incognita* (Williams et al., 1973). Crosses between cultivars with low levels of resistance to this iso-

TABLE 1 Separation of Species and Races of *Meloidogyne* by Means of Differential Hosts

Meloidogyne species and race	Differential host plants[a]					
	NC95 tobacco	Deltapine 16 cotton	Cal. Wonder pepper	Charleston Grey watermelon	Flo-runner peanut	Rutgers tomato
M. incognita						
Race 1	–	–	+	+	–	+
Race 2	+	–	+	+	–	+
Race 3	–	+	+	+	–	+
Race 4	+	+	+	+	–	+
M. arenaria						
Race 1	+	–	+	+	+	+
Race 2	+	–	–	+	–	+
M. javanica	+	–	–	+	–	+
M. hapla	+	–	+	–	+	+

[a]Tobacco (*Nicotiana tabacum* L.), cotton (*Gossypium hirsutum* L.), pepper (*Capsicum frutescens* L.), watermelon (*Citrullus vulgaris* Schad.), peanut (*Arachis hypogaea* L.), and tomato (*Lycopersicon esculentum* Mill.).
Source: From Taylor and Sasser, 1978.

late resulted in lines with high levels of resistance. This population would not reproduce on Centennial sweet potato, a cultivar that normally is very susceptible to *M. incognita* (Martin and Birchfield, 1973). Four populations of *M. incognita* from Egypt reproduced readily on Bragg soybean (Oteifa and Elgindi, 1982).

Several soybean cultivars with resistance to *M. incognita* have been released without regard to race. Four of these resistant cultivars and two susceptible ones were tested against three populations each of races 1 and 3 and one population each of races 2 and 4 (Windham and Barker, 1986). Coker 156 soybean, reported to be a susceptible cultivar, was resistant to all race 1 and 2 populations tested and two of the race 3 populations. One race 3 and the race 4 population reproduced well on this cultivar. None of the cultivars was resistant to one race 3 population although Centennial allowed less reproduction than the others. Forrest was resistant to two of three race 1 and two of three race 3 populations. In another study (Swanson and Van Gundy, 1984) race 2 reproduced well on Centennial but poorly on Pickett 71, whereas races 1, 3, and 4 reproduced well on Pickett 71 but not well on Centennial.

Two populations of *M. incognita*, one from Louisiana and one from South Carolina, were selected for five generations on resistant tobacco (Triantaphyllou and Sasser, 1960). The original populations caused some galling and produced a few egg masses on resistant tobacco lines (Table 2). However, after five selection generations the resulting population caused considerably more galling and reproduced readily. Three other populations and the South Carolina population were also selected on resistant Hawaii 5229 tomato. Changes to a higher level of galling and reproduction occurred in all four populations as a result of the selection (Table 3). The selected populations from resistant tobacco were tested on resistant tomato and reproduced as poorly as the original population.

The grape cultivars Dogridge, Ramsey, 1613, Harmony, and Freedom are resistant to *M. incognita*, *M. arenaria*, and *M. javanica* (Cain et al., 1984). However, a population of

TABLE 2 Root-Knot Infection Classes of Three Tobacco Cultivars or Lines Inoculated with Four Isolates of *Meloidogyne incognita*

| | Gall rating[a] on tobacco line | | |
| | Susceptible | Resistant | |
Isolate[b]	DB–101	Bel 6–28	8071
3A–2–Original	5	2	2
–Clone B	5	4	4
9A–1–Original	5	2	2
–Clone B	5	4	4

[a]Rating system was: 0 = no evidence of infection; 1 = roots galled, larvae died; 2 = plants resistant but 1–10 egg masses/plant; 3 = slightly suitable host, 10–100 egg masses/plant; 4 = moderately suitable host, 100 to abundant egg masses/plant, roots moderately to severely galled; and 5 = highly suitable host, egg masses abundant, roots moderately to severely galled.
[b]Isolate 3A–2 was from okra and tomato in Louisiana; 9A–1 was from soybean in South Carolina; B isolates were clones that were reproduced on "resistant" tobacco.
Source: From Triantaphyllou and Sasser, 1960.

TABLE 3   Root-Knot Infection Classes on Two Tomato Cultivars
Inoculated with Eight Isolates of *Meloidogyne incognita*a

Isolateb	Infection classesa on tomato	
	Hawaii 5229	Rutgers
1A–1–Original	3	5
–Clone B	3–4	5
1B–3–Original	3	5
–Clone B	3–4	5
7A–6–Original	2	5
–Clone B	4	5
9A–1–Original	3	5
–Clone B	4–5	5

aRating system as in Table 2.
bIsolate 1A–1 is from tomato and 1B–3 from cotton in North Carolina;
7A–6 is from parsnip in Kentucky; and 9A–1 is from soybean in South
Carolina; B clones were reproduced on "resistant" tomato.
Source: From Triantaphyllou and Sasser, 1960.

M. incognita from USDA grape plots at Fresno, California severely galled all of the above cultivars, especially Harmony and Freedom. The host race of this population is not known but it obviously breaks resistance in grapes.

An unusual situation in breaking resistance occurred in *M. incognita* on pepper (Husan, 1985). The cultivars Jowala and Longthin Faizabodi are normally resistant but when they were inoculated with *Rhizoctonia solani* and/or *Pythium aphanidermatum* their resistance to *M. incognita* became ineffective.

B.   *Meloidogyne javanica*

In Florida, two populations of *M. javanica* were tested on a series of nine differential hosts. One population reproduced well on four of the hosts whereas the other population reproduced on six of the hosts (Kirby et al., 1975). In Senegal a population of *M. javanica* caused severe galling of the strawberry cultivar Cambridge when it was planted in an infested soil (Sasser and Carter, 1982). Strawberry is usually susceptible only to *M. hapla*. Host differential and morphological studies indicated that this was a typical *M. javanica* except for its parasitism of strawberry.

Meloidogyne javanica was reported to reproduce on Giza 69 cotton (*G. barbadense* L.) in Egypt but not on Acala 4–42 (*G. hirsutum* L.) (Ibrahim et al., 1979). Root galls and egg masses were numerous on Giza 69. Cotton is not usually a host of *M. javanica*.

Ibrahim (1982) reported a population of *M. javanica* (origin not reported) that reproduced on peanut, not usually a host of *M. javanica*. Studies on susceptibility of commercial crop cultivars to four *M. incognita* and two *M. javanica* populations from Egypt revealed variations. Three barley cultivars were tested and all three were moderately susceptible to one population of *M. javanica* and immune to the other. Three of the four wheat cultivars tested had the same reaction as barley against the two *M. javanica* populations. A fourth wheat cultivar was moderately resistant to one population and immune to the other.

Another population of *M. javanica* was found infecting peanuts in a field in southern Georgia (Minton et al., 1969). The peanuts were showing symptoms similar to those related to parasitism by *M. arenaria*.

Two races of *M. javanica* have been reported from Italy (DeVito and Greco, 1982). One race of *M. javanica* does not reproduce on pepper (more common) and the other race does. Two populations of the second race of *M. javanica* were found. Sasser and Carter (1982) indicated that about 10% of the 311 *M. javanica* populations from 59 countries reproduced on pepper, which is not a host for *M. javanica*, but they did not consider this consistent enough to be called a separate race. About 2.4% of the *M. javanica* populations reproduced on peanut but these too were not considered a separate race. In contrast, only 2% of the *M. incognita* populations reproduced on both cotton and NC95 tobacco, and was separated as race 4 of that species.

C. *Meloidogyne hapla*

Experiments with 33 populations of *M. hapla* from Poland, using lettuce, cucumber, red beet, cabbage, onion, maize, and oats as hosts, indicated that six races could be differentiated (Brzeski and Baksik, 1981). Variations in reproduction were also demonstrated in six populations of *M. hapla* from Bulgaria. A tomato line resistant to *M. hapla* in the United States was susceptible to a *M. hapla* population in Poland (Stoyanov, 1979). Two races of *M. hapla* were reported from Italy (De Vito and Greco, 1982); one that reproduces on NC95 tobacco and Florunner peanut but not on Charleston Grey watermelon (commonly found) and one that did not produce on NC95 tobacco or Florunner peanut but reproduced on Charleston Grey watermelon. When 2000 peanut introduction lines were planted on two fields in Virginia, each infested with *M. hapla*, two lines were resistant to the population in one field but not to the population in the other field (Miller, 1972). This indicates a physiological difference between the two populations of *M. hapla*.

Meloidogyne hapla before selection on cucumber reproduced poorly on cucumber (three egg masses per plant) compared to tomato (74 egg masses per plant) (Stephen, 1982). Egg masses were formed in 35 days on tomato and 51 days on cucumber. When the egg masses formed on cucumber were put back on cucumber for five generations the number of egg masses formed on cucumber increased (45 per plant compared to 140 per plant on tomato) and the time required decreased (33 days on cucumber and 38 days on tomato). Perineal patterns were similar from the two hosts. Apparently there was a mixture of genotypes in this population and a population virulent to cucumber was selected.

Morphological studies of *M. hapla* cytological races A (facultative meiotic parthenogenetic) and B (mitotic parthenogenetic) revealed variation between the two races (Eisenback and Hirschmann, 1979). There were distinct differences in the morphology of the lip region of three populations of race A viewed with scanning electron microscopy. These populations reproduce by amphimixis at least part of the time. In contrast, three populations of race B which reproduce by mitotic parthenogenesis had very similar lip region morphology. This indicates that nematodes that reproduce by amphimixis are more variable than those that reproduce by parthenogenesis, as would be expected.

D. Other *Meloidogyne* Spp.

Race 1 of *M. arenaria*, two populations from Florunner peanut, was compared morphologically—13 female, 22 male, and 21 second-stage juvenile characters—with two populations of race 2, one from G–28 tobacco and one from Davis soybean (Osman et al., 1985). There were no significant differences in any of the 18 characters measured. This supports the placement of these populations as races rather than species or subspecies.

Based on the number of chromosomes, Triantaphyllou (1982) divided *M. incognita*, *M. arenaria*, and *M. hapla* each into two cytological races. Cytological race A of *M. incognita* has 40–46 (3n) chromosomes while race B has 32–36 (2n) chromosomes; race A of *M. arenaria* has 50–56 (3n) chromosomes, race B 34–37 (2n); race A of *M. hapla* has 14–17 (n) chromosomes, race B 30–32 (2n). Race A of *M. hapla* normally reproduces by amphimixis and race B reproduces by parthenogenesis as do both cytological races of *M. incognita* and *M. arenaria*. The cytological races do not correspond in any way to the physiological or host races.

Differences in egg mass production of two population of *M. graminicola* Golden and Birchfield, 1965 on six rice varieties suggest that two races exist (Rao, et al., 1977). Two geographically separated populations of *M. exigua* Goeldi, 1887 exhibit differences in pathogenicity on coffee (Machado, 1974). Five races of *M. naasi* Franklin, 1965 from five different geographic locations were differentiated using 22 plant species (Michell et al., 1973a). Studies on their developmental rate indicated that one race developed much more rapidly than the other four (Michell et al., 1973b).

Meloidogyne chitwoodi does not normally parasitize alfalfa but in 1982 two fields that had been rotated from alfalfa to potato had plants damaged by *M. chitwoodi* (Santos and Pinkerton, 1985). This shift will remove one of the rotation possibilities in an area where other choices are rare. In Mexico *M. chitwoodi* appears to reproduce well on alfalfa with no apparent damage to the plant (L. I. Miller, pers. commun.).

IV. RACES OF *TYLENCHULUS SEMIPENETRANS*

The citrus nematode, *Tylenchulus semipenetrans* Cobb, 1913 is a serious parasite of citrus wherever citrus is grown. Resistance has been found but four races of *T. semipenetrans* were found in California (Table 4) (Baines et al., 1974). They were differentiated on the basis of

TABLE 4   Average Number of Adult Female *Tylenchulus semipenetrans* per Centimeter of Root of Differential Host (Greenhouse Test)

Nematode biotype[a]	Sweet Orange[b]	Differential host				
		Troyer citrange	Pomeroy *P. trif.*	Rubidoux *P. trif.*	Grape	Olive
1	2.93	2.78c	0.00a	0.00a	0.15b	0.01a
2	1.936ab	0.23a	0.01a	0.00a	0.03a	0.15b
3	1.41a	2.25bc	0.49b	1.48b	0.05a	0.00a
4	2.06ab	1.79b	0.00a	0.02a	—	0.43c

[a]Biotypes 1 and 2 were obtained from *Citrus* sp., 3 from *Poncirus trifoliata* and 4 from olive.
[b]"Homosassa" sweet orange, *Citrus sinensis; Poncirus trifoliata* "Pomeroy" and "Rubidoux"; *Vitis vinifera* "Thompson seedless grape," and *Olea europaea* "Manzanillo" olive.
Note: Numbers are means of six replications. Those followed by same letter are not different (*P* = 0.05) according to Duncan's multiple-range test.
Source: From Baines et al., 1974.

relative parasitism of "Homosassa" sweet orange, "Troyer" citrange, "Pomeroy" and "Rubidoux" *Poncirus trifoliata* (L.) Raf., grape (*Vitis vinifere* L.), and olive (*Olea europaea* L.). The differences in number of nematodes per centimeter of root were sometimes quite small but they were consistent and were significant ($P = 0.05$).

V. RACES OF *APHELENCHOIDES BESSEYI*

Spring dwarf and summer dwarf of strawberry were at one time thought to be caused by different races of *Aphelenchoides fragariae* (Ritzema Bos, 1891) Christie, 1932 (Christie, 1938). In later years the cause of summer dwarf was shown to be *A. besseyi* Christie, 1942 (Christie, 1943), the same nematode that causes white tip of rice (Allen, 1952). The populations of *A. besseyi* that cause white tip do not parasitize strawberry and those causing summer dwarf of strawberry do not parasitize rice. However, both nematodes will reproduce on a number of species of fungi.

VI. RACES OF *BURSAPHELENCHUS XYLOPHILUS*

Pine wilt disease caused by *Bursaphelenchus xylophilus* (Steiner and Buhrer) Nickle, 1970 is a serious disease of pines in Japan. A number of species of pines in the United States apparently are resistant to pine wilt or to invasion by *B. xylophilus*. However, a population of *Bursaphelenchus xylophilus* from *Pinus sylvestris* L. in Missouri infected *P. sylvestris* seedlings and to some extent *P. nigra* Arnold seedlings (Bolla et al., 1986). Another population of *B. xylophilus* from *P. strobus* L. in Vermont infected and reproduced only in *P. strobus* seedlings. Apparently, these populations represent two races of *B. xylophilus*.

VII. RACES OF *RADOPHOLUS* SPP.

Burrowing nematode of citrus is a very destructive pest that was originally described as *Radopholus similis* (Cobb, 1893) Thorne, 1949 (Cobb, 1893). In 1956 two races of burrowing nematode were differentiated in greenhouse tests, one reproducing on banana and the other on banana and citrus (DuCharme and Birchfield, 1956). Cytogenetic studies showed that one race of *R. similis* had $n = 4$ chromosomes, the other $n = 5$ (Huettel and Dickson, 1979). No attempt was made to correlate cytological race and host race. Later the citrus-burrowing nematode was described as *R. citrophilus* Huettel, Dickson, and Kaplan (Huettel et al., 1984). Observations indicated that a portion of field populations usually reproduced on resistant citrus rootstocks (Kaplan and O'Bannon, 1985). A population (P1) from rough lemon (*R. citrophilus*-susceptible) was compared in a greenhouse test with a population (P2) from Milam lemon (*R. citrophilus*-resistant). They were tested on Ridge pineapple, Algerian navel orange, Milam lemon, Carrizo citrange, and rough lemon. P1 reproduced well only on rough lemon, whereas P2 reproduced readily on all test hosts although not as much on rough lemon. This is obviously a case of a resistance-breaking type in an important crop pest.

VIII. RACES OF CYST-FORMING NEMATODES

A. *Heterodera trifolii*

Heterodera trifolii Goffart, 1932, is a cyst-forming species of nematode that usually reproduces parthenogenetically. It parasitizes several plants in diverse families, most of which
are not of high economic significance. In southern France two races of *H. trifolii* were found
on *Dianthus caryophyllus* L. (Cuany and Dalmasso, 1975). Morphologically they are similar but one has 27 chromosomes and reproduces by mitotic parthenogenesis like most members of the species. It reproduces on Nice varieties of carnation and on clovers. The other
race has 18 chromosomes and reproduces by amphimixis. It reproduces on American and
Nice varieties of carnation, on *Brassica* spp., clovers, and *Rumex acetosella* L.

A population of *H. trifolii* in the Netherlands was found to be pathogenic to sugar
beet, *Beta vulgaris* L. Previously *H. trifolii* was not known to parasitize sugar beet; in fact,
no cyst nematode other than *H. schachtii* A. Schmidt, 1871 reproduced readily on this host
(Maas and Heijbroek, 1982). It was thought at first to be a variant of *H. schachtii* in which
the female turned yellow before becoming a brown cyst. However, closer examination
showed that it was *H. trifolii* that was causing severe damage to sugar beet in the southern
part of the Netherlands.

Of 40 plants tested, 29 were hosts of *H. trifolii* and seven of those were resistant to *H.
schachtii* (Steele et al., 1983). Four plants that were nonhosts of *H. trifolii* were susceptible
to *H. schachtii*. In this case the resistance-breaking race broke the resistance of a new host
species, not a resistant cultivar of a known host.

Significant differences in reproduction occurred among populations of *H. trifolii*
when they were propagated on various hosts (Table 5) (Singh and Norton, 1970). Of six
populations from Canada, Illinois, Kentucky, Oregon, and Iowa, only the population from
Kentucky reproduced readily on three cultivars of *Trifolium pratense* L. The population
from Kentucky and one from Union County, Iowa reproduced well on one cultivar of
Melilotus officinalis (L.) Lam. but not on two other cultivars; the other four populations had
little or no reproduction. Only the population from Canada reproduced abundantly on *Trifolium fragiferum* L. Four populations had good reproduction on *Trifolium repens* L. cv.
Merit Ladino while two populations had only fair reproduction. The Illinois population had
only fair to poor reproduction on the two white clover cultivars and PI 166371. The Kentucky population had good reproduction only on Merit Ladino of the white clovers tested
and only the Canada and Union County, Iowa populations reproduced on PI 166371. In this
case resistance breaking in the usual sense was not operating but the six populations had
quite different host ranges. Six populations of *H. schachtii*, two each from Idaho, Oregon,
and Utah, were studied for differences in seedling penetration, rate of emergence, and virulence (Griffin, 1981). Those from Idaho and Oregon and Utah were all similar but Utah 2
juvenile penetrated sugar beet seedling in larger numbers, emerged over a shorter period,
and damaged sugar beet significantly more than the others. This is not a resistance-breaking
race but definitely a physiological race.

B. *Heterodera schachtii*

The sugar beet cyst nematode *Heterodera schachtii* causes extensive damage in all sugar
beet-growing areas. Because no commercially acceptable resistant varieties are available,
resistance-breaking types are not a major factor in sugar beet production. Nevertheless races
of *H. schachtii* do occur.

TABLE 5 Cyst Production by Six Isolates of *Heterodera trifolii* on Various Hosts

Host	Relative number of cysts of isolate[a]				Iowa	
	Canada	Illinois	Kentucky	Oregon	Stone Co.	Union Co.
Trifolium pratense						
Dollard	+	+	++++	+	+	+
Kenland	+	+	++++	+	+	++
Lakeland	+	+	++++	+++	+	++
T. repens						
Merit Ladino	++++	++	++++	++++	++++	++
Regal	++++	++	+	++++	++++	++++
PI 166371	++++	+	+	+	++	++++
T. fragiferum	++++	+	+	+	+	+
Melilotus officinalis						
Madrid	+	+	+	+	+	+
Yellow (M204)	+	+	++++	+	+	++++
Yellow Blossom (M38)	+	+	+	+	−	+

a + = 1–50 cysts/10—cm pot; ++ = 51–120; +++ = 121–225; ++++ = over 225 (in most cases over 500).
Source: From Singh and Norton, 1970.

TABLE 6   Relative Reproduction[a] of Five Isolates of *Heterodera schachtii* on Seven Hosts

Host	C1	C2	M1	N1	F1
Cleome	+++	+++	+++	+++	++++
Lespedeza	+++	+++	+++	++	++
Tomato	+++	++	++	++	−
Pokeweed	+++	+	+++	+	+
Purslane	++	++	++	−	+
Soybean	+	+	+	−	+
Sugar beet	+++	+++	+++	++++	+++

[a]+ = very poor host; ++ = poor host; +++ = good host; ++++ = very good host; − = nonhost.
Source: From Graney and Miller, 1980.

Two populations of *H. schachtii* found in Germany—one in Bavaria, one in the Rhineland—reproduced equally well on sugar beet (Anonymous, 1973). The population from the Rhineland formed many more cysts on oil radish and swede rape than the one from Bavaria. Six populations of *H. schachtii*, two each from Idaho, Oregon, and Utah, were studied for differences in seedling penetration, rate of emergence, and virulence (Griffin, 1982). Those from Idaho and Oregon and Utah 1 were all similar but Utah 2 juveniles penetrated sugar beet seedlings in larger numbers, emerged over a shorter period, and damaged sugar beet significantly more than the others. This is not a resistance-breaking race but definitely a physiological race. Five populations of *H. schachtii* from the United States (two from sugar beet in California, and one each from sugar beet in Michigan, cabbage in New York, and cabbage in Florida) were differentiated by their ability to develop egg-bearing females on *Cleome spinosa* Jacq. cv. Ruby Queen (cleome), *Lespedeza striata* (Thunb ex Murr.) Hook and Arn. cv. Kobe (common lespedeza), *Lycopersicon esculentum* Mill. cv. Pearson A–1 (tomato), *Phytolacca decandra* L. (pokeweed), *Portulaca oleracea* L. (common purslane), *Glycine max* (L.) Merr. cv. Lee (soybean), and *Beta vulgaris* L. cv. US 75 (sugar beet) (Table 6) (Graney and Miller, 1980). This is another example of race differentiation by using host species rather than cultivars within a species.

C. *Heterodera cajani*

Heterodera cajani Koshy, 1967 occurs mainly in Africa and Asia as a parasite of *Cajanus cajan* (L.) Millsp. and *Vigna unguiculata* Walp. It had been reported not to reproduce on *Cyamopsis tetragonolobus* (L.) Traub, but a population was found that matured and produced eggs on *C. tetragonolobus* (Walla and Bajaj, 1986). This is another case where the resistance of a different host species was broken.

D. *Heterodera avenae*

The cereal cyst nematode (*Heterodera avenae* was known to occur in various European countries in the early 1900s. Differences in pathogenicity of various populations on grasses and grain crops led to the conclusion that there were races of this species. However, races were not conclusively demonstrated until 1959 (Anderson, 1959).

TABLE 7 Susceptibility of Six Barley Cultivars Grown in Five Fields Infested with *Heterodera avenae* in Denmark

Cultivar	Number of cysts[a]				
	Lynghy	Hoybakke	Svogerslen	Hasle	Odum
Rex II (Check)	19	37	33	71	186
Herta	83	78	60	140	235
Maja	52	57	35	200	345
Drost	68	0	0	2	202
Alfa	15	1	0	2	272
Fero	65	0	0	4	206

[a]Number of cysts on the roots of 30 plants.
Source: From Anderson, 1959, reprinted with the permission of E. J. Brill, Leiden, Netherlands.

Heterodera avenae populations collected in Denmark were tested on different sets of hosts to determine the level of reproduction (Anderson, 1959). When populations from five locations were tested on six barley cultivars, two populations reproduced on all six cultivars and three populations reproduced readily on only three cultivars (Table 7). The differences were confirmed by repeating the test. To determine how widespread the resistance-breaking races were, 99 populations were collected from all over Denmark and were tested on Drost, Alfa, and No. 191 barley cultivars (Table 8). The resistance-breaking races appeared to be quite common; however, there were 36 populations that did not produce any females on either Drost or Alfa, indicating that these are fairly pure nonaggressive types. Parasitism of resistant barley cultivars does not mean that resistant oat cultivars will be parasitized. Four races of *H. avenae* were found in the Netherlands and one was similar to one of the Denmark races (Kort et al., 1964).

In 1972 and 1973 an international variety test was conducted to determine the number of races of *H. avenae* (Nielsen, 1974). Thirteen races were designated and mixtures were studied that possibly contain additional races. The report of the Rothamsted Experiment Station in 1974 indicated that a series of 24–35 oat, wheat, barley, and rye varieties were used to distinguish 15 races (Anonymous, 1974). Tests of *H. avenae* populations from Australia indicated that they are unlike any race reported from Europe but have similarities to a

TABLE 8 Tests of Three Barley Cultivars Against 99 Populations of *Heterodera avenae*

Numbers of cysts	Number of populations		
	Drost	Alfa	No. 191
none	38	43	84
1–5	11	5	10
6–25	8	11	3
26–100 B	22	15	2
over 100	20	23	0

[a]Figures are the number of populations that produced a certain number of cysts on the test cultivars.
Source: From Anderson, 1959, reprinted with the permission of E. J. Brill, Leiden, Netherlands.

race in India (Brown, 1974). Five races were reported from India and they appeared to be different than those in Europe and Australia (Mathur et al., 1974).

In Sweden the *H. avenae*-resistant barley cultivar Ansgar was planted in field plots infested with *H. avenae* for 5 years (Anderson, 1977). The first 2 years the nematode population declined but at the end of 5 years the *H. avenae* population level had increased almost to the same level as plots planted continuously in a susceptible cultivar.

In India Sun II oats, Drost, Herta, 191, and Morocco barley, *Lolium perenne* L. and *Dactylis glomerata* L. were planted in *H. avenae*-infested fields (Mathur et al., 1974). After 14 weeks plants were dug and cyst numbers determined. Eight combinations of host parasitism were found but three were considered to be combinations of pairs of the other five (Table 9). There were five distinct races. Populations 6, 7, and 8 may have been combinations or they could have been distinct races making a total of eight.

E. *Heterodera glycines*

The soybean nematode, *Heterodera glycines* Ichinohe, 1951, has been known to occur in Japan and China since late nineteenth century (pers. commun.). It was reported from Korea in 1936 (Yokoo, 1936) and in the United States in 1954 (Winstead et al., 1955). In 1962 (Ross, 1962) the first report of races was published; a population from North Carolina reproduced much better on PI 88788 than did a population from Tennessee. Another race was reported from Virginia in 1964 (Smart, 1964) and a fourth from Arkansas in 1968 (Riggs et al., 1968). A race also was reported from Japan in 1968 (Sugiyama et al., 1968).

In 1969, a committee of nematologists and soybean breeders met to determine what to call the variants of *H. glycines* and how to separate them. As a result four races were described on the basis of their relative ability to mature on four soybean differentials (Golden et al., 1970). The number of females that matured on a differential was compared to the number on the susceptible cultivar Lee. If the number was 10% or more of the number on Lee it was given a plus rating; if less than 10% it was given a minus rating. A fifth race was described from Japan in 1979 (Inagaki, 1979). In spite of many reports of populations that did not fit in the five races, no other races were described until 1987 when race 6 was described from Arkansas (Robbins, 1987) and race 7 was described from China (Chen et al., 1987). Races 8–16 were described in 1988 (Riggs and Schmitt, 1988) (Table 10). This made possible the identification of the race of any population, even though it did not necessarily include all of the variability in *H. glycines*. Before all of the variation could be categorized the different levels of resistance would need to be delineated and lines with resistance categorized by genotype. One line from each genotype could be selected as a differential. The number of races would be $2n$ where n = number of differentials assuming that a plus/minus rating system is used.

The occurrence of mixtures of genotypes has been demonstrated repeatedly in *H. glycines*. In a field test a resistance-breaking race was selected by planting resistant soybean lines and cultivars 3 consecutive years (Slack et al., 1981; Elliott et al., 1986). The population level in the plots planted to resistant soybean was 42, 47, and 120 second-stage juveniles/500 cm³ soil at the end of the three growing seasons. In greenhouse studies, race 3 of *H. glycines* was subjected to selection pressure by planting Pickett, Peking, PI 88788, or PI 90763 for three sequential inoculations (Triantaphyllou, 1975; Riggs et al., 1977). In each case the resulting population reproduced freely on the formerly resistant cultivar or line. Similar selection has been demonstrated by researchers in other areas (Luedders and Dropkin, 1983; McCann et al., 1980; Young, 1982, 1984).

TABLE 9 Reaction of Indicator Hosts to Cereal Cyst Nematode in India

Soil Source	Reaction[a]							No. populations[b]	Biotype
	Sun II oat	Drost barley	Herta barley	191 barley	L. perenne	D. glomerata	Morocco barley		
Bhagru	−	−	−	−	+	−	−	2	1
Shrimadhopur	−	−	+	+	−	−	−	1	2
Jaitpura	−	−	+	−	−	−	−	50	3
Govindpura	−	−	+	−	−	+	−	4	4
Manda	−	+	+	−	−	−	−	1	5
Kaladera	−	−	+	−	+	−	−	13	1 + 3
Bhabhru	−	−	+	−	+	+	−	4	1 + 4
Bhadwadi	−	+	+	−	+	−	−	1	1 + 5

a + = 200 cysts/plant; − = 0–3 cysts/plant.
b These could be mixtures of the races listed or could be races 6, 7, and 8.
Source: From Mathur et al., 1974, reprinted with the permission of E. J. Brill, Leiden, Netherlands.

F.   *Globodera* Spp.

The golden nematode of potato was described as *Heterodera schachtii* ssp. *rostochiensis* by Wollenweber (1923). From the time resistant cultivars were released, at least as early as 1957 (Jones, 1957), until 1972 variability in pathogenicity was observed and three or five races were recognized depending on whose scheme was followed (Table 11) (Cole and Howard, 1966; Kort, 1974). However, in 1973 part of the variability was separated as *H. pallida* (Stone, 1973). The generic name was changed from *Heterodera* to *Globodera* (Skarbilovich) Behrens, 1975 and *G. rostochiensis* continued to exhibit variability and variations were found between populations of *G. pallida*. New schemes for identification of races were proposed in 1977 by Kort et al. (1977) and Canto Saenz and de Scurrah (1977) simultaneously (Tables 12–14). The scheme developed by Kort et al. (1977) was mainly devised to accommodate the variability known to occur in Europe. The scheme presented by Canto Saenz and de Scurrah (1977) was especially designed to accommodate the variability found in the Andean region of South America but also included that in Europe. Since the potato and the potato cyst are thought to have originated in the Andean region, the variation there could be greater than in other areas. In contrast, no variability has been demonstrated in populations in New York.

Mixes of races occur in fields and may present problems in identification and control (Kort and Baker, 1980). Decker (1981) indicated that repeated planting of resistant potato cultivars must not be practiced because it increases the danger of the development and accumulation of aggressive races. He further stated that one crop of resistant cultivar was equal

TABLE 10   Races of *Heterodera glycines* as Distinguished by Host Differentials

Race	Reaction on differential[a]			
	Pickett	Peking	PI 88788	PI 90763
1	−	−	+	−
2	+	+	+	−
3	−	−	−	−
4	+	+	+	+
5	+	−	+	−
6	+	−	−	−
7	−	−	+	+
8	−	−	−	+
9	+	+	−	−
10	+	−	−	+
11	−	+	+	−
12	−	+	−	+
13	−	+	−	−
14	+	+	−	+
15	+	−	+	+
16	−	+	+	+

[a] + = number of females and cysts on differential is 10% or more of the number on the susceptible cultivar Lee; − = number of females and cysts on differential <10% of the number on Lee.
Source: From Riggs and Schmitt, 1988.

TABLE 11   Pathotype Classification Schemes for *Heterodera rostochiensis* Developed in Great Britain[a] and the Netherlands[b]

Differential host	Pathotype reaction British Scheme		
	A	B	C[c]
Solanum tuberosum ssp. *tuberosum*	+	+	+
S. tuberosum ssp. *andigena* CPC1673	−	+	+
S. multidissectum P55/7	+	−	+
S. tuberosum ssp. *andigena* X *S. multidissectum* K6/34	−	−	+

Differential host	Dutch Scheme					
	A	B	C	D	E	F
S. tuberosum ssp. *tuberosum*	+	+	+	+	+	+
S. tuberosum ssp. *andigena* CPC1673	−	+	+	+	+	−
S. kurtzianum KTJ60.21.19	−	−	+	+	+	+
S. vernei GLKS58.1642.4	−	−	−	+	+	+
S. vernei (VTn)262.33.3	−	−	−	−	+	−

[a]From Cole and Howard, 1966.
[b]From Kort, 1974.
[c]British pathotype C was changed to E to prevent confusion with Dutch pathotype C.

to one crop of a nonhost in reducing the population level. The German Seed Growers Society and the Plant Protection Service were responsible for preventing repeated planting of resistant cultivars in Germany.

IX. *CACTODERA BETULAE*

The birch cyst nematode, *Cactodera betulae* (Hirschmann and Riggs) Krall and Krall, 1978, appears to be a nematode of no economic significance in which two genetic types or races occur. One race reproduces on birch, *Betula lenta* L. (formerly *B. niger*) and other *Betula* spp., and one on black locust, *Robinia pseudoacacia* L., and there appears to be no change in the host ranges that would indicate a break in resistance. Both will reproduce to a limited extent on *Lespedeza striata* and it could possibly bridge between the two hosts.

X. CONCLUSIONS

Races of nematodes may not always be resistance-breaking types. For example, the races of *D. dipsaci* do not appear to be derived one from another by resistant cultivar selection pressure. They may have developed by parallel evolution in situations where only a poor host

TABLE 12 Proposed International Pathotyping Scheme for *Globodera rostochiensis* and *G. pallida* Compared with British, Dutch, and German Schemes

Differential	Plant resistance code	New pathotype reaction[a]							
		Ro1	Ro2	Ro3	Ro4	Ro5	Pa1	Pa2	Pa3
Solanum tuberosum ssp. *tuberosum*		+	+	+	+	+	+	+	+
S. tuberosum ssp. *andigena* CPC1673 hybr.	Ro1, 4	-	+	+	-	+	+	+	+
S. kurtzianum hybr. 60.21.19	Ro1, 2	-	-	+	+	+	+	+	+
S. vernei hybr. 58.1642/4	Ro1, 2, 3	-	-	-	+	+	+	+	+
S. vernei hybr. 62.33.3	Ro1, 2, 3, 4 Pa1, 2	-	-	-	-	+	-	-	+
S. vernei hybr. 65.346/19	Ro1, 2, 3, 4, 5	-	-	-	-	-	+	+	+
S. multidissectum hybr. P55/7	Pa1	+	+	+	+	+	-	+	+
S. vernei hybr. 69.1377/94	Ro1, 2, 3, 4, 5 Pa1, 2, 3	-	-	-	-	-	-	-	-

[a]Ro1 = British A, Dutch A; and Hiltrup; R02 = Dutch B and Obersteinbach; Ro3 = Dutch C; Ro4 = Dutch F; Ro5 = Harmerz; Pa1 = British B; Pa2 = Dutch D; and Pa3 = Frenswegen, Chavorney, Dutch E, and British E.

Source: From Kort et al., 1977, reprinted with the permission of E. J. Brill, Leiden, Netherlands.

TABLE 13 Race Separation in *Globodera rostochiensis*

Differential host	No. designated to differential	Race reactions[a]			
		R1A	R1B	R2A	R3A
Solanum tuberosum spp.					
tuberosum	0	+	+	+	+
S. tuberosum spp.					
andigena (H1)	1	–b	–b	+	+
S. kurtzianum					
KTT/60.21.19	2	–	+	–b	+
S. vernei GLKS58.1642.4	3	–	+	–	–b
S. vernei (VTn)262.33.3	4	–	–	–	–

[a]R1A = British A, Dutch A; R1B = Dutch F, R2A = Dutch B abd R2B = Dutch C.
[b]The most important reaction for the classification of races.
Source: From Canto Saenz and de Scurrah, 1977, reprinted with the permission of E. J. Brill, Leiden, Netherlands.

was available and over a period of isolation the nematode adapted to that host and became specific for it, a form of resistance breaking. They appear to have distinct, narrow host ranges that tend to remain narrow. However, they do interbreed and the progeny have broader host ranges than the parent races.

There are two races of *Aphelenchoides besseyi*: one that causes white tip of rice, *Oryza sativa* L., and one that causes summer dwarf of strawberry, *Fragaria* sp. There appears to be no resistance to either within its host species and no changes in the host range of either have been reported.

In the cases cited above resistance breaking would not be of economic significance because apparently it does not occur. However, the lack of occurrence may be because there have been fewer studies on nematodes of lesser importance. On the other hand, *D. dipsaci* has been studied extensively.

Resistance-breaking races may also become the prevalent populations in situations where cultivar selection pressure is not a factor. Race 2 of *H. glycines* was found in Virginia before resistant soybean cultivars were planted (L. I. Miller, pers.commun.) It could have been brought from the Orient where it developed on resistant soybean selections or it could have developed on a weed host, possibly *Lespedeza striata* as was the case in one Missouri field. Similar situations were observed in Arkansas, where fields that had been in cotton for 15–20 years or pasture for 20–30 years had high population levels of cysts the second year in Forrest soybean. Selection pressure from resistant cultivars is not responsible for all of the races that occur but it is the mechanism for much of the variation.

The races of root-knot nematodes, *Meloidogyne* spp., do not appear to have resistance breaking types as currently constituted. However, there are numerous reports of resistance breaking in *Meloidogyne* spp. Types that break resistance in tomatoes have been reported widely but not in commercial fields. The lack of selection of resistance-breaking types in some areas may be because tomato growers do not plant resistant cultivars on the same field two or more years in succession. If resistance-breaking types do occur on tomatoes they probably will not have much economic impact world wide but could have great

TABLE 14 Race Separation in *Globodera pallida*

Differential host	No. designated to differential	Race reaction[a]						
		P1A	P1B	P2A	P3A	P4A	P5A
Solanum tuberosum ssp. *tuberosum*	0	+	+	+	+	+	+
S. multidissectum (H2)	1	–[b]	–[b]	+	+	+	+
S. kurtzianum KTT.60.21.19	2	+	+	–[b]	+	+	+
S. vernei GLKS58.1642.4	3	+	+	+	–[b]	+	+
S. vernei (VTn)262.33.3	4	–	+	–	–	–[b]	+

[a]P1A = British B; P4A = Dutch D; and P5A = British and Dutch E.
[b]The most important reaction for the classification of races.
Source: From Canto Saenz and de Scurrah, 1977, reprinted with the permission of E. J. Brill, Leiden, Netherlands.

impact on individual tomato production areas. Because of the worldwide distribution of *M. incognita, M. arenaria, M. javanica,* and *M. hapla,* if resistance breaking is a common phenomenon in these species it will have a tremendous impact on agricultural production costs and methods.

The impact of resistance-breaking races of the cyst nematodes *H. avenae, H. glycines, G. pallida,* and *G. rostochiensis* appears to be devastating. The hosts of these four species of cyst nematodes, small grains (wheat, oats, rye, and barley), soybean, and potato are crops that are grown worldwide and provide food substances for a major portion of the world population. Only rice and possibly corn are used for food, directly or indirectly, for as large a number of people.

In the case of each of the major cyst-forming nematode species, resistance has been a major management tactic to provide protection against major damage. The amount of money spent collectively on providing cultivars resistant to any one of them is huge, not only because of the cost of producing a resistant cultivar but because of the necessity to produce one resistant cultivar after another. So far the sources of resistance to each of the cyst nematode species seem to be endless because as soon as a resistant cultivar is parasitized by a resistance-breaking race another source of resistance has been found. Logic would indicate that the limit eventually will be found and other management practices will be necessary. The final economic impact will be determined by the alternative measures that are necessary.

A case study was made of the benefits of a soybean cultivar resistant to *H. glycines,* the soybean cyst nematode (Bradley and Duffy, 1982). Yield loss prevention that resulted from growing Forrest soybean from 1975 through 1980 was estimated to be worth $405 million, measured in terms of the dollar's worth in 1980. The effectiveness of Forrest decreased in succeeding years and, although some benefit is still being derived, the loss to resistance-breaking races of *H. glycines* would have been $67.5 million per year if new sources of resistance had not been available.

In summary, resistance-breaking races or populations of plant parasitic nematodes appear to occur in relatively few species. The species in which they do occur are worldwide in distribution and affect crops of major significance to the world population.

REFERENCES

Allen, M. W. 1952. Taxonomic status of the bud and leaf nematodes related to *Aphelenchoides fragariae* (Ritzema Bos 1891). *Proc. Helm. Soc. Wash.* 19(2): 115–117.

Anderson, S. 1959. Resistance of barley to various populations of the cereal root eelworm (*Heterodera major*). *Nematologica* 4: 91–98.

Andersen, S. 1977. Byg 191 resistance to cereal root eelworm knocked out in Holland trial. *Vaxtskyddsnotises* 41(4): 101–103, 126 (English summary).

Anonymous. 1973. Sugarbeet. 1973 Annual Report of the Federal Biological Institute of Agriculture and Forestry in Berlin and Braunsweig.

Anonymous. 1974. Cereals. Report of the Rothamsted Experiment Station, 1974, Part 1. Her Majesty's Stationery Office, London.

Baines, R. C., Cameron, J. W., and Soost, R. K. 1974. Four biotypes of *Tylenchulus semipenetrans* in California identified and their importance in the development of resistant citrus rootstocks. *J. Nematol.* 6: 63–66.

Bolla, R. I., Winter, R. E. K., Fitzsimmons, K., and Linit, M. J. 1986. Pathotypes of the pinewood nematode *Bursaphelenchus xylophilus. J. Nematol.* 18: 230–238.

Bradley, E. B., and Duffy, M. 1982. The value of plant resistance to soybean cyst nematodes: A case study of Forrest soybeans. N.R.E.D., E.R.S., U.S.D.A. Staff Report No. AGES 820929, Washington, D.C.

Brown, R. H. 1974. Biotype studies of the cereal cyst nematode (*Heterodera avenae*) in Victoria. Simposio Internacional (XII) de Nematologia, Sociedad Europea de Nematologos, 1–7 Septiembre, 1974, Granada, Spain (Abstr.).

Brzeski, M. W., and Baksik, A. 1981. Typy populai *Meloidogyne hapla* w polsce. *Zeszyty Problemowe Postepow Nauk Rolniczych* 249: 73–75 (In Polish) (Helminth. Abstr.)

Cain, D. W., McKenry, M. V., and Tarailo, R. E. 1984. A new pathotype of root-knot nematode on grape rootstocks. *J. Nematol.* 16: 207–208.

Canto Saenz, M., and Mayer de Scurrah, M. 1977. Races of the potato cyst nematode in the Andean region and a new system of classification. *Nematologica* 23: 340–349.

Chen, P. H., Zhang, D. S., and Chen, S. Y. 1987. First report on a new physiological race (race 7) of soybean cyst nematode (*Heterodera glycines*). *J. Chin. Acad. Agr. Sci.* 20: 94 (Abstr.).

Chitwood, B. G. 1949. "Root-knot nematodes"—Part I. A revision of the genus *Meloidogyne* Goeldi, 1887. *Proc. Helm. Soc. Wash.* 16: 90–104.

Christie, J. R. 1938. Two distinct strains of the nematode *Aphelenchoides fragariae* occurring on strawberry plants in the United States. *J. Agr. Res.* 57: 78–80.

Christie, J. R. 1943. Spring dwarf and summer dwarf of strawberry. U.S.D.A. Circular 681: 1–10.

Christie, J. R., and Albin, F. E. 1944. Host-parasite relationships of the root-knot nematode, *Heterodera marioni*, I. The question of races. *Proc. Helm. Soc. Wash.* 11: 31–37.

Cobb, N. A. 1893. Nematodes, mostly Australian and Fijian. *MacLeary Mem. Vol. Linn. Soc. N. South Wales*, pp. 252–308.

Cole, C. S., and Howard, H. W. 1966. The effects on a population of potato-root eelworm (*Heterodera rostochiensis*) of growing potatoes resistant to pathotype B. *Ann. App. Biol.* 58: 487–495.

Cuany, A., and Dalmasso, A. 1975. Characteres et specificite de deux especes biologique d'*Heterodera* se developpant sur *Dianthus caryophyllus*. *Nematologia Mediterranea* 3: 11–21.

Decker, H. 1981. Cyst-forming nematodes on potato. In *Plant Nematodes and their Control*. English translation from 1969 German text. New Delhi Amerind Publishing Co. Pvt. Ltd., pp. 202–215.

DeVito, M., and Greco, N. 1982. Research on root-knot nematodes in Italy. *Proceedings of the Third Research and Planning Conference on Root-knot Nematodes, Meloidogyne* spp. North Carolina State University Graphics, Raleigh, pp. 34–38.

Dropkin, Victor H. 1988. The concept of race in phytonematology. *Ann. Rev. Phytopatho.* 26: 145–161.

DuCharme, E. P., and Birchfield, W. 1956. Physiologic races of burrowing nematode. *Phytopathology* 46: 615–616 (Abstr.).

Eisenback, J. D., and Hirschmann, H. 1979. Morphological comparison of second-stage juveniles of six populations of *Meloidogyne hapla* by SEM. *J. Nematol.* 11: 5–16.

Elliott, A. P., Phipps, P. M., and Terrill, R. 1986. Effects of continuous cropping of resistant and susceptible cultivars on reproduction potentials of *Heterodera glycines* and *Globodera tabacum solanacearum*. *J. Nematol.* 18: 375–379.

Eriksson, K. B. 1974. Intraspecific variation in *Ditylenchus dipsaci*. I. Compatibility tests with races. *Nematologica* 20: 147–162.

Gibbins, L. N., and Grandison, G. S. 1968. An assessment of serological procedures for the differentiation of biological races of *Ditylenchus dipsaci*. *Nematologica* 14: 184–188.

Golden, A. M., Epps, J. M., Riggs, R. D., Duclos, L. A., Fox, J. A., and Bernard, R. L. 1970. Terminology and identity of infraspecific forms of the soybean cyst nematode (*Heterodera glycines*). *Plant Dis. Rep.* 54: 544–546.

Graney, L. S., and Miller, L. I. 1980. Differentiation of five isolates of *Heterodera schachtii* as races of the sugarbeet cyst nematode. *J. Nematol.* 12: 223.

Griffin, G. D. 1981. Pathological differences in *Heterodra schachtii* populations. *J. Nematol.* 13: 191–195.

Huettel, R. N., and Dickson, D. W. 1979. Cytogenetic differences between two Florida races of *Radopholus similis. J. Nematol.* 11: 300–301 (Abstr.).

Huettel, R. N., Dickson, D. W., and Kaplan, D. T. 1984. *Radopholus citrophilus* sp. n. (Nematoda), a sibling species of *Radopholus similis. Proc. Helm. Soc. Wash.* 51: 32–35.

Husan, A. 1985. Breaking resistance in Chili to root-knot nematode by fungal pathogens. *Nematologica* 31: 210–217.

Ibrahim, I. K. A. 1982. Species and races of root-knot nematodes and their relationships to economic host plants in northern Egypt. *Proceedings of the Third Research and Planning Conference on Root-Knot Nematodes, Meloidogyne* spp. North Carolina State University Graphics, Raleigh, pp. 66–84.

Ibrahim, I. K. A., Khalil, H. A. A., and Rezk, M. A. 1979. Root-knot nematodes on cotton in northern Egypt. *J. Nematol.* 11: 301 (Abstr.).

Inagaki, H. 1979. Race status of five Japanese populations of *Heterodera glycines. Jap. J. Nematol.* 9: 1–4.

Jones, F. G. W. 1957. Resistance-breaking biotypes of the potato root eelworm (*Heterodera rostochienses* Woll.) *Nematologica* 2: 185–192.

Kaplan, D. T., and O'Bannon, J. H. 1985. Occurrence of biotypes of *Radopholus citrophilus. J. Nematol.* 17: 158–162.

Kirby, M. F., Dickson, D. W., and Smart, G. C., Jr. 1975. Physiological variation within species of *Meloidogyne* occurring in Florida. *Plant Dis. Rep.* 59: 353–356.

Kish, A. J., and Adams, R. E. 1973. Resistance-breaking biotypes in the root knot nematode. *Phytopathology* 63: 803 (Abstr.).

Kort, J. 1974. Identification of pathotypes of the potato cyst nematode. *EPPO Bull.* 4: 511–518.

Kort, J., and Bakker, J. 1980. The occurrence of mixtures of potato cyst-nematode pathotypes or species. *Nematologica* 26: 272–274.

Kort, J., Ross, H., Rumpenhorst, H., and Stone, A. R. 1977. International scheme for identifying and classifying pathotypes of potato cyst-nematodes *Globodera rostochiensis* and *G. pallida. Nematologica* 23: 333–339.

Kort, J., Dantuma, G., and van Essen, A. 1964. On biotypes of the cereal-root eelworm (*Heterodera avenae*), and resistance in oats and barley. *Tijdschrift fur Pflanzenziekten* 65: 1–4.

Luedders, V. D., and Dropkin, V. H., 1983. Effect of secondary selection of cyst nematode reproduction on soybeans. *Crop. Sci.* 23: 263.

Maas, P. W. Th., and Heijbroek, W. 1982. Biology and pathogenicity of the yellow beet cyst nematode, a host race of *Heterodera trifolii* on sugarbeet in the Netherlands. *Nematologica* 28: 77–93.

Machado Neto, R. 1974. Estudo sobre diferentes patogenicidades de *Meloidogyne exigua*—cafeeiro no estado de Sao Paulo. *Solo* 66(2): 23–27.

Martin, W. J. 1953. Reaction of the Deltapine 15 variety of cotton to different isolates of *Meloidogyne. Phytopathology* 43: 292 (Abstr.).

Martin, W. J., and Birchfield, W. 1973. Further observations of variability in *Meloidogyne incognita* on sweet potatoes. *Plant Dis. Rep.* 57: 199.

Mathur, B. N., Arya, H. C., Mathur, R. L., and Honda, D. K. 1974. The occurrence of biotypes of the cereal cyst nematode (*Heterodera avenae*) in the light soils of Rajasthan and Haryana, India. *Nematologica* 20: 19–26.

McCann, J., Dropkin, V. H., and Luedders, V. D. 1980. The reproduction of differentially selected populations of *Heterodera glycines* on different lines of soybean (*Glycine max*). *J. Nematol.* 12: 230–231 (Abstr.).

Michell, R. E., Malek, R. B., Taylor, D. P., and Edwards, D. I. 1973a. Races of the barley root-knot nematode, *Meloidogyne naasi.* I. Characterization by host preference. *J. Nematol.* 5: 41–44.

Michell, R. E., Malek, R. B., Taylor, D. P., and Edwards, D. I. 1973b. Races of the barley root-knot nematode, *Meloidogyne naasi.* II. Developmental rates. *J. Nematol.* 5: 44–46.

Miller, L. I. 1972. Resistance of plant introductions of *Arachis hypogaea* to *Meloidogyne hapla, Meloidogyne arenaria* and *Belonolaimus longicaudatus. Va. J. Sci.* 23: 101 (Abstr.).

Minton, N. A., McGill, J. Frank, and Golden, A. Morgan. 1969. *Meloidogyne javanica* attacks peanuts in Georgia. *Plant Dis. Reptr.* 53: 668.

Mulvey, R. H., and Stone, A. R. 1976. Description of *Punctodera matadorensis* n. gen., n. sp. (Nematoda: Heteroderidae) from Saskatchewan with lists of species and generic diagnosis of *Globadera* (new rank) *Heterodera*, and *Sarisodera. Can. J. Zool.* 54: 772–785.

Nielsen, C. H. 1974. International test collection for *Heterodera avenae* and possibly other *Heterodera* species. *Nordisk Jordbruksforskning* 56(4): 410 (Abstr.).

Nishizawa, T. 1974. A new pathotype of *Meloidogyne incognita* breaking resistance of sweet potato, and some trials to differentiate pathotypes. *Jap. J. Nematol.* 4: 37–42.

Okamoto, K. 1979. Pathogenicity and larval dimensions of resistance-breaking races of root-knot nematode, *Meloidogyne incognita. Jap. J. Nematol.* 9: 16–19.

Osman, H. A., Dickson, D. W., and Smart, G. C., Jr. 1985. Morphological comparisons of host races 1 and 2 of *Meloidogyne arenaria* from Florida. *J. Nematol.* 17: 279–285.

Oteifa, B. A., and Elgindi, D. M. 1982. Relative susceptibility of certain commercially important cultivars to existing biotypes of *Meloidogyne incognita* and *M. javanica* in Nile Delta, Egypt. *Proceedings of the Third Research and Planning Conference on Root-Knot Nematodes, Meloidogyne* spp. North Carolina State University Graphics, Raleigh, pp. 157–169.

Rao, Y. S., Jena, R. N., and Prasad, K. S. K. 1977. Infectivity of two isolates of *Meloidogyne graminicola* in rice. *Ind. J. Nematol.* 7: 98–99.

Riggs, R. D., and Winstead, N. N. 1959. Studies on resistance in tomato to root-knot nematodes and on the occurrence of pathogenic biotypes. *Phytopathology* 49: 716–724.

Riggs, R. D., and Schmitt, D. P. 1988. Complete characterization of the race scheme for *Heterodera glycines. J. Nematol.* 20: 392–395.

Riggs, R. D., Slack, D. A., and Hamblen, M. L. 1968. New biotypes of soybean cyst nematode. *Ark. Farm Res.* 17(5): 11.

Riggs, R. D., Hamblen, M. L., and Rakes, L. 1977. Development of *Heterodera glycines* pathotypes as affected by soybean cultivars. *J. Nematol.* 9: 312–318.

Ritzema Bos, J. 1888–1892. L'Anguillule de la tige (*Tylenchus devastatrix* Kuhn) et les maladies des plantes dues a ce nematode. *Arch. Mus. Teyler (Ser. 2)* 3: 161–348, 545–588.

Robbins, R. T. 1987. Observed race changes of *Heterodera glycines* in successive plantings of SCN-resistant and susceptible soybeans in Arkansas. *J. Nematol.* 19: 552 (Abstr.).

Robinson, R. A. 1969. Disease resistance terminology. *Rev. Appl. Mycol.* 48: 593–606.

Ross, J. P. 1962. Physiological strains of *Heterodera glycines. Plant Dis. Rep.* 46: 766–769.

Santo, G. S., and Pinkerton, J. N. 1985. A second race of *Meloidogyne chitwoodi* discovered in Washington State. *Plant Dis.* 69: 361.

Sasser, J. N., and Carter, C. C. 1982. Root-knot nematodes (*Meloidogyne* spp.): Identification, morphological and physiological variation, host range, ecology and control. In *Nematology in the Southern Region of the United States.* Riggs, R. D. ed. Southern Cooperative Series Bulletin 276 Ark. Agr. Exp. Stat., Fayetteville.

Sauer, M. R., and Giles, J. E. 1959. A field trial with a root-knot resistant tomato variety. Irrigation Res. Stat. Tech. Paper No. 3, Melbourne, Australia.

Singh, N. D., and Norton, D. C. 1970. Variability in host-parasite relationships of *Heterodera trifoli. Phytopathology* 60: 1834–1837.

Slack, D. A., Riggs, R. D., and Hamblen, M. L. 1981. Nematode control in soybeans: Rotation and population dynamics of soybean cyst and other nematodes. *Ark. Agr. Exp. Stat. Report Ser.* 263.

Smart, G. C., Jr. 1964. Physiological strains and one additional host of the soybean cyst nematode, *Heterodera glycines. Plant Dis. Reptr.* 48: 542–543.

Southards, C. J., and Priest, M. F. 1973. Variation in pathogenicity of seventeen isolates of *Meloidogyne incognita. J. Nematol.* 5: 63–67.

Stephen, Z. A. 1982. Selection of the root-knot nematode, *Meloidogyne hapla* on cucumber. *Nematologica* 28: 174.

Steele, A. E., Toxopeus, H., and Heijbroek, W. 1983. Susceptibility of plant selections to *Heterodera schachtii* and a race of *H. trifolii* parasitic on sugarbeet in the Netherlands. *J. Nematol.* 15: 281–288.

Stone, A. R. 1973. *Heterodera pallida* n. sp. (Nematoda: Heteroderidae): A second species of potato cyst nematode. *Nematologica* 18: 591–606.

Stoyanov, D. 1979. Pathogenicity variations of three species of root knot nematodes. *In Root-Knot Nematodes (Meloidogyne Species): Systematics, Biology and Control*. Lamberti, F., ed. Academic Press, London, pp. 307–310.

Sturhan, Dieter. 1971. Biological races. *In Plant parasitic nematodes. Vol. 2. Cytogenetics, Host-Parasite Interactions and Physiology*. Zuckerman, B. M., Mai, W. F., and Rohde, R. A. eds. Academic Press, New York, pp. 51–72.

Sturhan, D. 1985. Species, subspecies, race, and pathotype problems in nematodes. *Bull. OEPP/EPPO Bull.* 15: 139–144.

Sugiyama, S., Hiruma, K., Miyahara, T., and Kohuhun, K. 1968. Studies on the resistance of soybean varieties to soybean cyst nematode. II. Differences of physiological strains of the nematode from Kerawana and Kikyogahara. *Jap. J. Breed.* 18: 206–212.

Swanson, T. A., and Van Gundy, S. D. 1984. Variability in reproduction of four races of *Meloidogyne incognita* on two cultivars of soybean. *J. Nematol.* 16: 368–371.

Taylor, A. L., and Sasser, J. N. 1978. *Biology, Identification and Control of Root-Knot Nematodes (Meloidogyne Species)*. North Carolina State University Graphics, Raleigh.

Triantaphyllou, A. C. 1975. Genetic structure of races of *Heterodera glycines* and inheritance of ability to reproduce on resistant soybeans. *J. Nematol.* 7: 356–364.

Triantaphyllou, A. C. 1982. Cytogenetics and sexuality of root-knot and cyst nematodes. *In Nematology in the Southern Region of the United States*. Riggs, R. D. ed. Southern Cooperative Series Bulletin 276. Ark. Agr. Expt. Stat., Fayetteville, pp. 71–76.

Triantaphyllou, A. C., and Sasser, J. N. 1960. Variation in perinneal patterns and host specificity of *Meloidogyne incognita*. *Phytopathology* 50: 724–735.

Van der Plank, J. E. 1975. *Principles of Plant Infection*. Academic Press, New York.

Walla, R. K., and Bajaj, H. K. 1986. Existence of host races of pigeon-pea cyst nematode, *Heterodera cajani* Koshy. *Nematologica* 32: 117–119.

Webster, J. M. 1967. The significance of biological races of *Ditylenchus dipsaci* and their hybrids. *Ann. Appl. Biol.* 59: 77–83.

Williams, C., Birchfield, W., and Hartwig, E. E. 1973. Resistance in soybeans to a new race of root-knot nematode. *Crop Sci.* 13: 299–301.

Windham, G. L., and Barker, K. R. 1986. Relative virulence of *Meloidogyne incognita* host races on soybean. *J. Nematol.* 18: 327–331.

Winstead, N. N., and Riggs, R. D. 1963. Stability of pathogenicity of B biotypes of the root-knot nematode *Meloidogyne incognita* on tomato. *Plant Dis. Rep.* 47: 870–872.

Winstead, N. N., Skotland, C. B., and Sasser, J. N. 1955. Soybean-cyst nematode in North Carolina. *Plant Dis. Rep.* 39: 9–11.

Wollenweber, H. W. 1923. Krankheiten und Behadigungen der Kartoffel. *Arb. Forsch. Inst. Kartoff. Berl.* 7: 52.

Yokoo, T. 1936. Host plants of *Heterodera schachtii* Schmidt and some instructions. *Korea Agr. Exp. Sta. Bull.* 8(43): 47–174.

Young, L. D. 1982. Effects of continuous culture of resistant soybeans on soybean cyst nematode development. *J. Nematol.* 14: 475 (Abstr.).

Young, L. D. 1984. Changes in the reproduction of *Heterodera glycines* on different lines of *Glycine max*. *J. Nematol.* 16: 304–306.

PART V
INSECT PARASITIC
NEMATODES

20

Steinernema (Neoaplectana) and *Heterorhabditis* Species

WILHELMUS M. WOUTS   *Mt. Albert Research Centre, Auckland, New Zealand*

I. BIOLOGY

A. Introduction

Steinernema and *Heterorhabditis* species (order Rhabditida) are nematodes parasitic on insects. They transmit bacteria which are lethal to their host, a characteristic which makes them more suitable for biological control of insects than any other nematode group. Over the last two decades steinernematids and heterorhabditids have become increasingly popular in insect control. Among their useful attributes are (a) a wide host range, (b) the infective juveniles can be readily and cheaply cultured, either on hosts or on artificial media, (c) cultures can be stored for extended periods, and (d) they can be easily applied in the field. Provided the humidity of the air is within a suitable range, the nematodes are resistant to a variety of environmental conditions, can actively find and penetrate susceptible hosts, and cause up to 100% mortality within a few days (Figs. 1 and 2). They are harmless to higher organisms.

Twenty years ago probably no more than five or six scientists worldwide were committed to the study of this group of nematodes. At present their number is well over 50 and growing (Poinar, 1985a). This development is largely due to pressure from environmental lobbies for biological means of pest control and bans on persistent insecticides.

The American R. W. Glaser (1932a) was the first person to publish on the biological control potential of steinernematids. In a field in New Jersey, he observed that *Steinernema glaseri* killed large numbers of Japanese beetles, *Popillia japonica* Newn. In his laboratory, in a salve can of sterilized soil, a heavy suspension of this nematode killed the beetles in the manner of a true infectious disease; however, houseflies, silkworms force–fed with nematodes, and plants all remained uninfected (Glaser 1932a,b). Glaser did not understand the mode of action of this disease but did not discount the possible presence of a bacterium. His first field experiments were extremely promising and a program for extensive treatment against Japanese beetles throughout New Jersey was organized. It had to compete, however, with Dutky's milky disease organism (*Bacillus popilliae*), introduced in 1941, and which

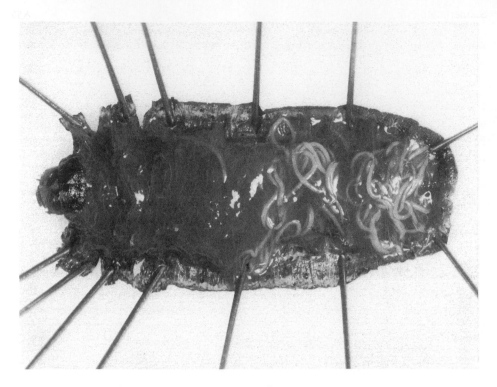

FIGURE 1   Large *Heterorhabditis* females feeding on a dissected host decomposed by the bacterial symbiont of the nematode.

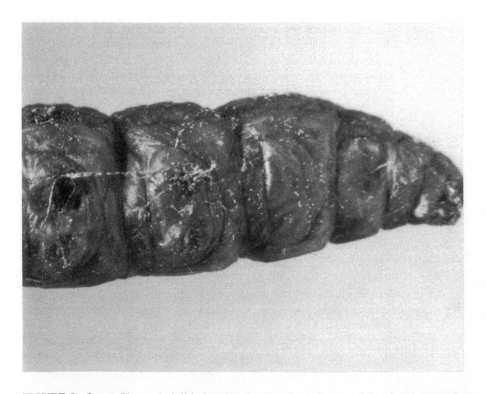

FIGURE 2   Large *Heterorhabditis* females showing through the cuticle of an infected host.

showed astonishing results, and World War II intervened (Briand and Welch, 1963; Stoll, 1973).

The discovery of two other species of this genus, *S. feltiae* Filipjev, 1934 and *S. bibionis* Bovien, 1937, attracted little attention. Filipjev (1934) and Bovien (1937) considered that their species were able to multiply within a host but did not suspect pathogenicity. Bovien assumed that the infective juveniles accumulated in the host's intestine and penetrated the body cavity after the host had died.

Interest in the group revived when Dutky (Anon., 1954) discovered that *S. feltiae* (strain DD–136 from codling moth), after actively seeking out and penetrating a host, releases a symbiont in the hemocele which causes a lethal septicemia (Fig. 3). The symbiont and its breakdown products serve as food for the nematode. The symbiont also produces an antibiotic which prevents putrification of the cadaver. Dutky obtained excellent results with strain DD–136. Unfortunately, only some of his results were published in scientific papers, most becoming public property through press releases, publications of the U.S. Agricultural Research Service, exhibits, talks, and correspondence (Anon., 1956; Dutky, 1959). All these were excellent sources which quickly generated interest at the time, but they are difficult to access today.

Dutky and Hough (1955) developed a practical rearing method using larvae of the greater wax moth, *Galleria mellonella* L. as hosts. *Steinernema feltiae* was made widely available and soon showed promise when used against a large number of insect pests in many parts of the world.

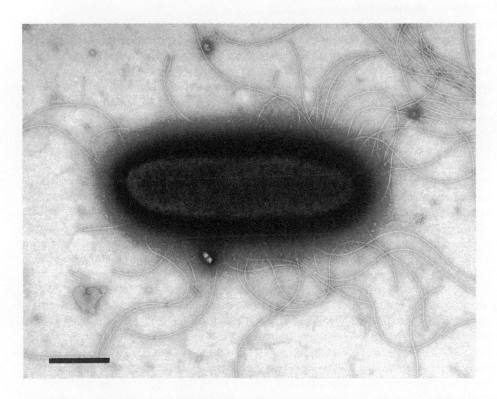

FIGURE 3   Flagellate *Xenorhabdus* cell (bar = 1 µm).

Only Dutky's *S. feltiae* DD–136 strain was known to transmit a lethal symbiont, and most work concentrated on that strain. More literature is available on this strain than on all other *Steinernema* species combined. However, all steinernematid and heterorhabditid infective juveniles carry symbiotic bacteria. The bacterial species belong to one genus, *Xenorhabdus*, but are specific for each nematode species (Akhurst and Boemare, 1988; Poinar and Brooks, 1977; Thomas and Poinar, 1979).

The number of insect species susceptible to steinernematids and heterorhabditids seems unlimited. Few, however, cause plant damage in an environment sufficiently damp to allow the nematodes to reach them. In general, soil insects occupy habitats conducive to nematode survival; however, in undisturbed agricultural soils, nematode mobility is very limited (Reed and Carne, 1967). Steinernematids and heterorhabditids can be very effective agents of control against hosts associated with glasshouse crops, especially those grown in pots, and also against hosts which tunnel in the tissues of live plants.

B. Attraction and Penetration

Early workers assumed that infective steinernematid juveniles could not penetrate the integument of the insect (Hoy, 1954) and believed that they were ingested by chance with food (Briand and Welch, 1963; Reed and Carne, 1967). In fact, the infective juveniles are actively attracted to a host (Bedding and Akhurst, 1975; Schmiege, 1963) through gradients of temperature (Byers and Poinar, 1982) and carbon dioxide (Gaugler et al., 1980), and by chemicals and microorganisms associated with the excretory system. The infectives aggregate near insect larvae, hosts and nonhosts alike (Pye and Burman, 1981; Schmidt and All, 1978, 1979). Sandner (1986a,b) showed that within 7 days, infectives are able to migrate through a 40–cm column of sand to kill *Galleria* larvae. The infectives actively invade the host through natural openings—mouth, anus, and sometimes spiracles. Physical barriers of the insect seem to be the only character that separates a nonhost from a host (Wouts, 1980).

Infective juveniles of *Heterorhabditis* have a sclerotized dorsal labial tooth in the third–stage juvenile (L–3) (Wouts, 1979) and to some extent, therefore, are also able to penetrate the cuticle. For cuticle penetration the infective juveniles explore the cuticle for periods up to several hours, and attempt penetration of crevices, folds, leg joints, and flexible areas of the cuticle generally. At a suitable site, an infective juvenile becomes immobile and, using the labial tooth, abrades or scratches the cuticle of the insect until it ruptures. This may take as long as 30 min. The nematode then enters the insect, progressing 20-100 μm at a time, alternated with pauses of a minute or more. Often more than one nematode enters the same wound. Cuticular penetration may provide a means of entry into insects such as root–feeding scarabaeid larvae, which are protected by spiracles closed by sieve plates, actively pushing nematodes away from the mouth region with their anterior legs, and defecating so frequently that nematodes entering the anus are pushed out again (Bedding and Molyneux, 1982). Infectives which enter the mouth or anus penetrate the wall of the digestive system to invade the body cavity of the insect.

There has been considerable confusion about the pathogenic nature of steinernematids (Jackson and Moore, 1969). Stoll (1959) considered these nematodes to be essentially parasites of live hosts. Welch and Briand (1961a) assumed that they develop in dead hosts. Stammer (1962) accepted their pathogenicity but assumed that each species killed its host in its own unique way. With the discovery that all species harbor a pathogenic bacterial symbiont in the intestines of the infectives, it was realized that all species infect and kill their host in the same way. Infectives in the hemocele of the insect host suck up blood and initially develop on this. In the reactivated nematode gut the symbiont begins to multiply, and passes out from it into the host body cavity to cause a lethal septicemia (Poinar, 1966). A single nematode, introducing from one to three bacterial cells directly into the hemocele of *Galleria* larvae, may cause mortality within 48 hr (Dutky, 1959; Poinar and Thomas, 1967). To kill a host, the symbiont need not originate from the nematode's intestine. If it is injected with or introduced on the surface of the infective juvenile, it will kill the host with equal effectiveness (Poinar and Thomas, 1966). However, when fed to a host, the bacteria have no effect (Khan and Brooks, 1977; Poinar and Thomas, 1967), unless small injuries are present in the digestive system through which the bacteria can reach the body cavity (Sandner et al., 1973). The nematode does not need the bacteria to kill the host; axenic nematodes kill their hosts too. The symbiont serves to break down the host tissues into acceptable food for the nematodes (Fig. 4). In the absence of the symbiont, the nematode dies after about 4 days (Poinar and Thomas, 1966).

Inside the host the symbiont produces wide–spectrum antibacterial substances which in combination with the host's defense system actively suppress contaminating bacteria in the early development of the nematode (Akhurst, 1983b; Anon., 1956; Dutky, 1959).

Poinar (1966), Thomas and Poinar (1979), and Akhurst (1982) maintain that all infective juveniles are monoxenic. This may be true for old infectives. Young infectives, however, develop in an insect cadaver which is heavily contaminated and they themselves therefore contain contaminants (Boemare, 1983; Lysenko and Weiser, 1974; Weiser, 1962a). The use of such freshly harvested, surface–sterilized, contaminated infective juveniles for rapid reinoculation is one reason why cultures collapse. For the infectives to become monoxenic a "starving time" is required between hosts, during which the contaminants can be "expelled or attacked by the digestive secretions of the nematode" (Poinar, 1966).

The presence of contaminating bacteria in cultures of the nematode is usually detrimental to the nematode. However, not all contaminants are equally destructive and some, e.g., *Pseudomonas fluorescens*, may even be beneficial and in combination with the nematode will kill the host and allow the nematode to develop on it as if it were the symbiont (Lysenko and Weiser, 1974).

Insect mortality is determined by the nematode-insect complex and not simply by the virulence of the bacterium. If bacteria are injected or have penetrated through a wound (Sandner et al., 1973), in the absence of the nematode a much higher level of inoculum is required in order to kill the host. Therefore, in the disease development of the bacterium-

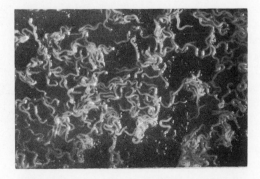

FIGURE 4   Large *Heterorhabditis* females feeding on the bacterial symbiont on a lipid agar plate.

nematode complex the nematode does not only act as a vector or "living syringe" for the bacteria; it also contributes directly to the effective kill of the host (Burman, 1982).

C. *Xenorhabdus* Species: The Bacterial Symbionts

1. History and Isolation

Glaser et al. (1942) give a description of a bacterial contaminant, tolerated by *S. glaseri* in axenic cultures, which fits the description of the symbiont: "a nonputrefactive, non–sporulating, gram–negative motile bacillus."

Bovien (1937) reports: "In the anterior part of the intestine [of the infectives] I found a lemon–shaped body apparently consisting of a mass of rod shaped bacteria held together by some hyaline substance. During the transformations which are leading to the following stage the body disintegrates and the supposed bacteria spread in the gut until they totally disappear." He also accurately illustrates this body.

But it was Dutky who, in his Ph.D. thesis on *S. feltiae*, proved that these "supposed" bacteria are in fact bacteria, and that they are symbiotically associated with the nematode and determine its pathogenicity (Dutky, 1937; Dutky and Hough, 1955) by causing septice-mia in the insect host (Dutky, 1959). The symbiont can be isolated from a drop of blood of an insect host that has been infected with infective juveniles for about 6 hr; of by transfer-ring infectives, surface–sterilized in an aqueous solution of 0.4% Hyamine 10X for 30 min, to a drop of insect blood in which the nematode releases the symbiont within 2 hr (Poinar, 1966); or by crushing the infectives onto nutrient plates (Poinar et al., 1971).

2. The Genus *Xenorhabdus* Thomas and Poinar, 1979 and Its Species

The symbionts were identified as *Pseudomonas septica* (Ritter and Theodorides, 1966; Weiser, 1962a, 1963). *P. aeruginosa* (Weiser, 1962b), and *Achromobacter nematophilus* Poinar and Thomas (1965), and are at present recognized as consisting of species of the genus *Xenorhabdus*, in the family Enterobacteriaceae (Thomas and Poinar, 1979). The ge-nus is exceptional for the Enterobacteriaceae in that it contains species that have polycrys-talline inclusions, species that do not reduce nitrate, one species only which is catalase-positive, and one species having cells with membranous structures which may be photosomes, the source of their bioluminescence (Boemare, 1983; Boemare et al., 1983).

The genus *Xenorhabdus* is characterized by gram–negative, non–spore–forming, rod–shaped cells, measuring 3.0-10.5 μm in length by 0.6-1.7 μm in width, which are closely associated with entomogenous nematodes and are the cause of their pathogenicity in insects. The optimum temperature for bacterial development is about 29°C, optimum pH is 7.0, and the optimum concentration of NaCl is 0.5% (Poinar et al., 1971). *Xenorhabdus* species produce crystals in their cells (Boemare et al., 1983; Grimont et al., 1984).

Originally, two *Xenorhabdus* species were recognized: *X. nematophilus*, the sym-biont of steinernematids, and *X. luminescens*, the symbiont of heterorhabditids.

Xenorhabdus nematophilus colonies on MacConkey agar absorb neutral red, form-ing either red or brown colonies. On nutrient agar plates, at pH 6.9, containing bromothymol blue and triphenyltetrazolium chloride (NBTA), 3– to 5– day–old colonies have a red core overlaid by dark blue (except the strain from *S. glaseri*, *Xenorhabdus nematophilus* which produces red colonies). On these plates the colonies absorb the bromothymol blue and re-duce the triphenyltetrazolium chloride so that they become blue and the agar around them becomes clear. In the infective juveniles *X. nematophilus* is restricted to the anterior portion of the intestinal lumen, in the pouch where Bovien (1937) observed bacteria in *S. bibionis*.

Xenorhabdus nematophilus was recently split up into *X. nematophilus*, the symbiont of *S. feltiae*; *X. bovienii*, the symbiont of *S. bibionis*, *S. affinis*, and *S. kraussei*; *X. poinarii*, the symbiont of *S. glaseri*; and *X. beddingii*, the symbiont of undescribed species from Australia. *Xenorhabdus nematophilus* colonies are white, opaque on nutrient broth agar, and deep blue on NBTA plates; *X. bovienii* colonies are yellow on nutrient broth and deep blue on NBTA; *X. poinarii* colonies are opaque on nutrient broth and red on NBTA; and *X. beddingii* colonies are brown on nutrient broth and blue on NBTA. These species differ further in their temperature tolerance and in their physiological activity (Akhurst and Boemare, 1988).

Xenorhabdus luminescens colonies, after 3-5 days on NBTA plates, are usually a shade of green with red-brown centers and have a clear zone in the agar. They are mucoid and contain a pigment after 5 days. They have lipolytic properties and are catalase–positive (i.e., they react with peroxide) (Thomas and Poinar, 1979). When observed in total darkness for about 10 min, colonies of *X. luminescens* demonstrate a bioluminescence which is sufficiently strong to show through the cuticle of an infected host and to expose photographic film (Figs. 5 and 6); one nonluminescent strain is known (Akhurst and Boemare, 1986). In the infective juvenile nematode, *X. luminescens* is present right through the digestive system from behind the stomal area through to where it reaches the rectum, but bacteria are absent from the rectum itself (Kahn and Brooks, 1977; Poinar et al., 1977).

Different strains of *X. luminescens* differ in their luminescence, pigmentation, protein pattern, and pathogenicity, suggesting that this species too may in fact comprise more than one species (Akhurst and Boemare, 1988; Hotchkin and Kaya, 1984; Han and Wouts, 1990).

3. The Primary Form and Secondary Form *Xenorhabdus*

Xenorhabdus species express themselves on nutrient plates in two different ways. Freshly isolated from native infectives, their colonies are convex; on MacConkey agar they absorb

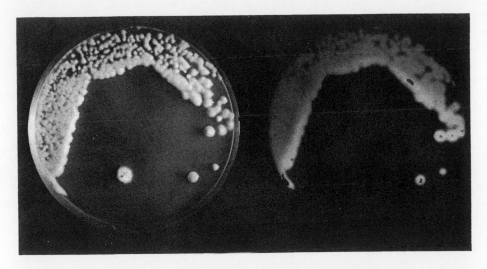

FIGURE 5   *Xenorhabdus luminescens* culture photographed in daylight (left) and in the dark (right). A nonluminescent colony contaminates the lower center of the plate.

FIGURE 6   *Galleria mellonella* larvae infected with *Xenorhabdus luminescens* photographed in daylight (left) and in the dark (right).

neutral red and (except for *X. poinarii*) on NBTA they absorb bromothymol blue and reduce triphenyltetrazolium chloride.

Older cultures may have translucent, rather flat, wide colonies on NBTA plates with no clear zone in the agar and do not absorb neutral red, as if they were contaminants. The fresh blue/green isolates and the old, red isolates are considered two different forms of the same bacterial species. There is no significant difference in the pathogenicity of the two forms. However, the more original isolate seems to be the form preferred by the nematode, and this has become known as the primary form; the somewhat degenerate form, developing from old cultures, is called the secondary form. Protein patterns obtained by electrophoresis suggest that a *Xenorhabdus* species may have more than one secondary form (Hotchkin and Kaya, 1984).

Some isolates of *Xenorhabdus* change more easily into the secondary form than others. Pure cultures of the primary form can be maintained by freeze drying, deep freezing, or frequent subculturing of the original isolate (Akhurst, 1980; Boemare and Akhurst, 1989).

Infective *S. feltiae* juveniles, reared on media with both forms of *X. nematophilus* present, retain only the primary form. Steinernematids grow and reproduce on the symbiont of related steinernematids and on certain contaminants but generally they reproduce most prolifically on the primary form of their own species. *Steinernema glaseri* may grow more rapidly on the symbiont of other steinernematids than on its own symbiont but it retains only its own symbiont. Therefore, in mixed infections, *S. glaseri* can compete favorably with other *Steinernema* species and still maintain its own symbiont in the resulting infective juveniles. *Steinernema* species do not develop on the symbiont of *Heterorhabditis* species (Akhurst, 1983a; Dunphy et al., 1985).

The cause and reason for the presence of the two forms of *Xenorhabdus* species is not yet fully understood but could be related to attack by defective bacteriophages (Poinar et al., 1980) or lysogenic bacteriophage. The secondary form could be immune to (Poinar,

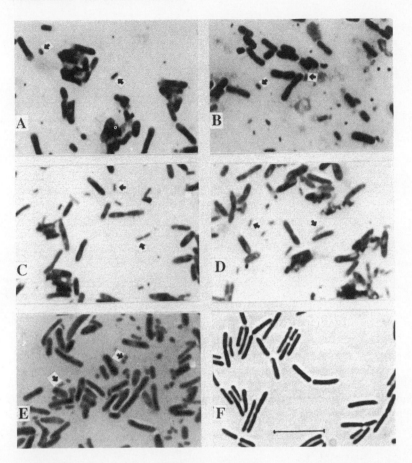

FIGURE 7   (A–E) *Xenorhabdus* species, primary form. (A) *X. luminescens.* (B) *X. poinari.* (C) *X. bovieni.* (D) *X. nematophilus.* (E) *X. bovieni.* (F) *X. bovieni,* secondary form (bar = 20 µm). Note the small cells (arrows) in the primary form cultures.

pers. commun.) such phages and become predominant. In older primary form cultures, especially in *Heterorhabditis* cultures, a cell size is present which is absent in secondary form cultures (Figs. 7 and 8). The small–celled component resembles an independent bacterial species. Subcultured on yeast extract-rich media it forms red colonies on NBTA plates that resemble contaminants but fail to produce a pure culture; among the small cells in these colonies a small percentage of large *Xenorhabdus* cells always remain. The primary form seems to result from a close association between *Xenorhabdus* cells and this small–celled variant. The forces responsible for the two cell sizes seem also to regulate the production of mucoid substances by the primary form. The true nature of these forces has not been elucidated but may be regulated by an as yet unidentified plasmid. When *Xenorhabdus* cells lose the association with the small–celled species they attain the secondary state (Wouts, unpublished).

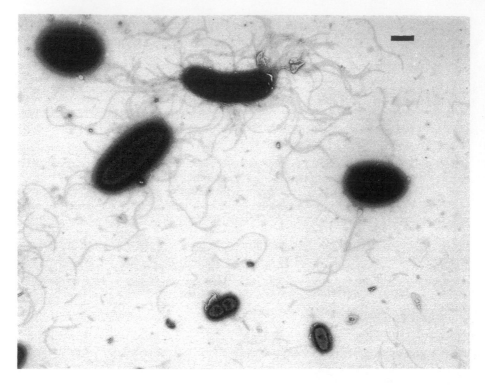

FIGURE 8   *Xenorhabdus luminescens*, primary form, large and small cells (bar = 1 μm). (Courtesy of Dr. I. Hallett, DSIR, Auckland.)

4.   Antimicrobial Activity

Developing *Xenorhabdus* species produce a wide spectrum of antibacterial substances which actively suppress *Bacillus cereus*, *B. subtilis*, *Escherichia coli*, *Enterobacter cloacae*, *Candida albicans*, and *C. krausei* (Akhurst, 1983a). Nutrient agar plates, spot–inoculated with primary form *Xenorhabdus*, incubated for 3 days, killed by exposure to chloroform for 2 hr, allowed to air for 1 hr, covered with sterile soft agar at 45°C, and seeded with a culture of one of these test organisms show inhibition zones after 24 hr. Akhurst reports that only primary form *Xenorhabdus* show distinct antimicrobial activity. This may be true for his test assortment, but generally antimicrobial activity is common to both forms, against unrelated microorganisms as well as against other *Xenorhabdus* species and against the opposite form (Figs. 9,10, and 11). Antibiotic activity of broth cultures is poor (Akhurst, 1983b) when the broth is poor, but in enriched broth cultures it can be very pronounced. It seems to be correlated with the metabolic activity of the bacterium (Han and Wouts, 1990). Antibiotic activity is not noticeably affected by heating at either 60 or 121°C, but is removed by dialysis (Akhurst, 1983b).

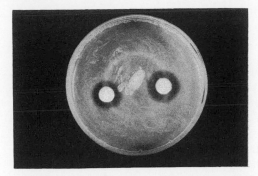

FIGURE 9   Antimicrobial activity of *Xenorhabdus luminescens*, primary form, against the primary form of *Steinernema feltiae.*

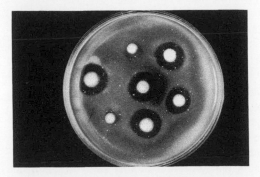

FIGURE 10   Antimicrobial activity of *Xenorhabdus luminescens*, primary form, against *Bacillus thuringiensis.*

FIGURE 11   Antimicrobial activity of *Xenorhabdus luminescens*, secondary form, against *X. luminescens*, primary form, of a different isolate.

D. Host Reaction to Invasion by the Infective Juveniles

1. Susceptible Hosts

Invasion, infection, and death of a susceptible host—*Galleria mellonella* larvae in particular—is very rapid. The nematodes reach the hemocele about 4 hr after being force-fed to *Galleria* larvae. By that time the cell nuclei of the fat body have assumed a variety of shapes. The granules of the karyoplast accumulate near the nuclear membrane. The spaces between the laminae of the nuclear membrane widen and the nucleolus disappears (Sandner and Seryczynska, 1971). The total fat body begins to disintegrate, with fat droplets making the hemolymph opaque (Matha and Mráček, 1984). A defense mechanism seems either absent or very weak (Welch and Bronskill, 1962). However, a cytological defense reaction does take place. It initially fluctuates and does not become appreciable until about 2 hr after the nematodes are force–fed to the *Galleria* larvae. At this time the hemocytes have become damaged and their numbers reduced to about 20% of those of the control. The free hemocytes involved in phagocytosis then increase in number, and reach a maximum of 160-200% of the number of hemocytes of the control some 10-18 hr later. In the period of maximum phagocytic activity one phagocyte can contain up to eight bacteria (Matha and Mráček, 1984 did not observe phagocytosis). By now the hemocytes may be considerably damaged. The cell membrane forms invaginations that develop into long digital cytoplasmic pseudopodia. The channels of the interplasmic reticulum are distinctly enlarged. The karyoplasm has extensive attenuations distributed over the whole area of the nucleus. The Golgi apparatus is well developed. The mitochondria are strongly enlarged, becoming rod–shaped, and numerous lysosomes are present, usually associated with the Golgi apparatus. After the maximum number of hemocytes has been reached their number gradually decreases, lowering the pH of the hemolymph and paralyzing the host, which is killed when all hemocytes are destroyed. It is the presence of the symbiont, not the infective larvae, that causes both the hemocytes to multiply and damage to these hemocytes. The protective lipoidal epicuticle of the nematode protects the nematode, and initially the symbiont inside it, against any direct humoral, sheath, or cellular encapsulation reaction by *Galleria* (Dunphy and Webster, 1985, 1987; Matha and Mráček, 1984; Sandner et al., 1977; Seryczynska, 1972; Seryczynska et al., 1974; Seryczynska and Kamionek, 1972).

2. Partially Resistant Hosts

The reaction is much stronger in somewhat resistant hosts such as the Colorado potato beetle, *Leptinotarsa decemlineata* (Say), especially in imagines emerging from the winter diapause. The number of hemocytes may increase to as much as 800% of normal after 60 hr. Low initial dosages of the bacterium cause slow increases and low maxima. In August, *L. decemlineata* beetles become susceptible. The number of their free hemocytes remains unchanged when the nematode infects. This, unfortunately, has no practical value as the beetles, having laid their eggs, then usually die anyway (Sandner et al., 1977; Seryczynska, 1974, 1975; Seryczynska and Kamionek, 1974).

The infective juveniles are not influenced by, and have no influence on, the defense mechanism of healthy *Galleria* larvae but they can break down an acquired immunity. The LD50 of the symbiont *X. nematophilus* in diapausing pupae of the cecropia moth, *Hyalophora cecropia* (L.), can be increased from 500 to 5×10^5 by injecting the pupae with *Enterobacter cloacae* beta 12 cells, without changing the sensitivity of the insect to the monoxenic (LD50 < 10) or axenic (LD50 = 500) nematode. Axenic and monoxenic infectives, injected into or actively penetrating such immunized pupae, break down this acquired immunity. The symbiont in the sheltered environment of the monoxenic nematode intestine is not released until the immune system has been considerably weakened. The greater the number of nematodes added to hemolymph in vitro, the sooner complete immune suppression is reached. If after 4 hr the nematodes are removed, the inactivation of the immune reaction continues (Goetz et al., 1981). A water–soluble toxin, identified by Burman (1982), produced by axenic *Steinernema* species immediately after the infective starts feeding, may be responsible for this effect. Production of this toxin in a host or culture medium is cyclical: it peaks when L–3 and L–4 and possibly adults are present, and reaches a maximum after 11 days at the end of the third generation. Young juveniles, less than 500 μm long, may inactivate the toxin. The toxin is heat– but not cold–sensitive, and at high dosages will kill *Galleria* larvae (Boemare et al., 1982; Burman, 1982).

3. Hypersensitive Hosts and Vertebrates

The reaction of hypersensitive hosts, such as mosquito larvae of the genera *Aedes* and *Culex*, to invasion by *S. feltiae* or *H. bacteriophora* is much more dramatic than that of *Galleria* or *L. decemlineata* larvae, and is directed against the nematode (Poinar and Kaul, 1982; Poinar and Leutenegger, 1971). Early–instar mosquitoes cannot ingest microspheres larger than 30 and 60 μm in diameter, and are therefore more or less resistant, but the fourth instar can readily ingest infective juveniles. The infective juveniles reach the host's midgut within 2 min and after 8 min penetrate the gut wall at the junction of the esophagus and the anterior midgut. The mortality of the nematodes that become trapped in other parts of the intestine is high. After 12 min, some nematodes reach the body cavity of the thoracic region (Dadd, 1971). Mosquitoes have no free circulating hemocytes. They have sessile hemocytes within the anal papillae, but these play no role in the immune process. The main defense system of the mosquitoes consists of humoral encapsulation of the nematode by coagulation of melanin and other polyphenol-protein complexes in the hemolymph, brought about by the activity of hemolymph–borne phenoloxidases (Andreadis and Hall, 1976). Humeral encapsulation of the nematode begins almost immediately, when a homogeneous nonpigmented layer is deposited on the cuticle of the L–4,followed by a continuous layer of particles of melanin and melanized host blood cells. Forty minutes after penetration, this matrix is 1-4 μm thick. After 5-10 hr two layers are apparent, and the inner layer seems to start melanization of the cuticle of the live nematode. When only one nematode invades, the mosquito larva may survive, but it pupates 1-2 weeks later than noninfected larvae. Adults developing from those larvae display normal flight activity. When mosquito larvae are attacked by two or more nematodes, ingested juveniles become degraded, and the host dies (Bronskill, 1962).

The encapsulation reaction is specific and is activated at the site at which the capsule is formed. This is a common defense mechanism in insects with a low number of blood cells and seems to be a defense typically against parasites. It is not activated by inert material, and not by mermithids. Mermithids, apparently, are experienced by their hosts as inert material (Goetz and Vey, 1974; Hall et al., 1975).

The reaction of higher vertebrates to the presence of *S. feltiae* in the bloodstream resembles that of mosquitoes, but leaves no sign of pathogenicity, toxicity, or nematode–related histopathology. The reaction of a rat to nematodes injected intraperitoneally is basically a "foreign body" response. Peritoneal macrophages encapsulates the nematodes, in a single layer at 8 hr and a multilayer at 48 hr, and gradually kill the nematodes within 3-4 days. After 8 hr the macrophages often adhere so firmly to the nematode cuticle that it requires several days in culture tubes for them to drop off. At that stage the freed infectives are still capable of killing and reproducing in *Galleria* larvae (Gaugler and Boush, 1979; Jackson and Bradbury, 1970). Subcutaneous injection of infective *S. feltiae* and *Heterorhabditis bacteriophora* into 9–day–old White Leghorn chicks and adult Swiss albino (Namru) mice, intracerebral injection into suckling mice, and intraperitoneal injection into monkeys, rats, and rabbits does not cause disease or mortality (Poinar et al., 1982; Wang and Li, 1986). Infective juveniles of *S. feltiae* eaten by rats reappear within 8 hr, virtually all killed by the acidic digestive fluids of the gut. This inability of the nematode to withstand conditions in the intestine, combined with the active defense mechanism in the blood of warm–blooded animals and the fact that the nematodes do not withstand temperatures above 34°C, makes them safe for man to work with (Briand and Welch, 1963; Schmiege, 1963).

Cold–blooded vertebrates, however, may be susceptible and at risk. Complete development of *S. feltiae* in the Antillian toad *Bufa narinus* has been observed, the apodous stage showing mortality 1 day after application of nematodes. Susceptibility decreases with age (Kermarrec and Mauleon, 1985).

E. Environment

In the environment of steinernematids and heterorhabditis, the most critical factors for their successful survival, parasitization, and reproduction are temperature and relative humidity (RH) of the atmosphere (Kamionek et al., 1974; Kamionek and Sandner, 1972).

The range of temperatures for parasitization is greater than that for nematode reproduction, but there are considerable differences between individual nematode species/strains. Heterorhabditids need a somewhat higher temperature, and they occur in nature within a narrower range of temperatures than do steinernematids (Molyneux, 1986).

Infective juveniles are most active between 12 and 32°C. Outside this range penetration and development are very slow. Cold–tolerant strains will infect *Galleria mellonella* larvae at a temperature as low as 8°C (Finney–Crawley, 1985) and infective juveniles injected into the hemocele at 13°C can kill a susceptible host (Creighton et al., 1968; Pye and Burman, 1978; Reed and Carne, 1967).

At low temperatures, infective juveniles can survive for long periods (Briand and Welch, 1963). At 5°C, in 1-1/2 cm of water, *S. glaseri* can survive for more than three years: infectivity and ability to reproduce decline to 41% only in the last 2 years (Jackson, 1973). *Steinernema glaseri* and *S. bibionis* infectives are less active, and for that reason live longer than do *S. feltiae* and *Heterorhabditis* species (Molyneux, 1985). Short periods of extreme low temperatures do not seem to be catastrophic to nematode populations as do high temperatures. Storage at –10°C for 18 hr kills most *S. feltiae* infectives, but those that survive are infective. Infectives that are buried in soil can survive winter temperatures ranging from +9 to –19°C without loss of infectivity (Fedorko, 1971; Schmiege, 1963). But 1 hr at 35°C immobilizes almost all infective *S. feltiae* juveniles, and 16 hr at 37 °C or 1 hr at 41°C causes 100% mortality (Briand and Welch, 1963; Schmiege, 1963). At 20 °C the generation time of *S. feltiae* is about 8 days (Reed and Carne, 1967).

Death of the host is caused by the symbiont and therefore takes place most quickly at the optimum temperature for development of the symbiont, which is higher than the optimum temperature for the nematode. *Steinernema feltiae* kills *Galleria mellonella* larvae at 30°C in 16 hr, at 25°C in 24 hr, at 19°C in 44 hr, at 15°C in 120 hr, and at 9°C in 312 hr (Dutky, 1959).

To enable infection of a host living above ground, the RH of the air should be high enough for the infective juveniles to persist and to form a film of water through which they can move (Anon., 1956). The RH does not need to be 100%; at 22-25°C and 65% RH, survival is still 100% after 12 hr (Moore, 1965). Nematode survival at low RH is determined by the rate of desiccation and rehydration. In field soil, dried slowly at RH 70%, steinernematid and heterorhabditid infective juveniles can survive 20 or more days of drought (Moore, 1965). When RH is reduced rapidly to 80%, at 5°C, *S. feltiae* survives for only 24 hr, but when infectives placed on cellulose nitrate membrane filters are gradually desiccated to this level, more than 90% are still alive after 12 days. Even at 20°C, slowly adjusted to RH 48.4%, 80% of the juveniles survive 4 days. At 20°C and RH only 10%, 45% of such preconditioned infectives survive the first day (Simons and Poinar, 1973).

The efficacy of the infectives against soil insects depends on the mobility of the nematodes, the particle size of the soil, and the presence of a host, and varies between different nematode species. The dispersal rate of infectives in very heavy clay soils is negligible but the rate increases as the proportion of clay and silt decreases. In sandy soils *S. feltiae* disperses up to 7 cm/day (Moyle and Kaya, 1981;Poinar and Hom, 1986). Some infectives of any of the species will move as much as 14 cm up and down a column of pure silica or coarse sandy loam, when attracted by *Galleria* larvae (Georgis and Poinar, 1983a,b). *Steinernema glaseri* infectives generally migrate further than the infectives of the other species (Schroeder and Beavers, 1987). *Steinernema glaseri* shows a greater tendency to move downward than upward; *S. feltiae* juveniles tend to disperse better upward than downward. *Heterorhabditis bacteriophora* migrates faster than *H. heliothidis*. In sandy loam soil, fewer than 30% of the *H. heliothidis* infectives move down more than 2 cm, whereas more than 50% of *H. bacteriophora* infectives travel that distance, and deeper in the first 5 days after application (Georgis and Poinar, 1983c; Moyle and Kaya, 1981).

F. Biological Activity Against Different Hosts

The biological activity of the heterorhabditids and steinernematids against pest insects has been evaluated extensively in laboratory and field tests. Insect species against which promising results were obtained in pot or field tests are listed in Table 1. The most promising results will be discussed here.

Field experiments started with *Steinernema glaseri*, a species which was originally isolated from a soil scarab and therefore most extensively tested against this group of insects. With the discovery that a lethal symbiont kills the host, the emphasis changed to *S. feltiae* and to the more recently discovered *Heterorhabditis* species.

The activity of heterorhabditid and steinernematid species and strains differs from species to species and is influenced by environmental factors such as temperature, humidity, soil type, etc. Therefore, to determine the most suitable candidate for large–scale field applications against an insect species, it is important that the activity of all nematode species/strains available be evaluated under the prevailing field conditions (Bedding et al., 1983).

1. Soil-Inhabiting Beetles and Weevils

Soil–inhabiting beetles are difficult targets for steinernematids and heterorhabditids. They have evolved in the presence of nematodes and have acquired barriers to infection (Bedding and Molyneux, 1982; Sandner and Stanuszek, 1971). *Steinernema glaseri* seems best capable of coping with these obstacles (Carne and Reed, 1963/1964). It may also be more active than the other species because of its greater vertical mobility in soil (Schroeder and Beavers, 1987; Toba et al., 1983; Welch and Briand, 1961b). In recently dug field soil *S. glaseri* controls Japanese beetle grubs successfully resulting in less than 10% of the grubs developing into adults. The nematodes establish and spread over a large area (Glaser et al., 1940). If grub levels are restored in the following spring, the surviving nematode population eliminates the total grub population (Glaser and Farrell, 1935). *Steinernema feltiae* and also *H. heliothidis* may be effective against some beetles provided they are applied at the same time the beetles enter the soil to hibernate or pupate, e.g., just before the first–generation Colorado beetle larvae drop off potato plants or third instar elm leaf beetle descend from their host tree to pupate in the soil (Kaya et al., 1981; Wright et al., 1987).

Weevils, being more migratory than beetles, can physically avoid exposure to nematodes and have not developed the same barriers to infection as beetles. They can be very susceptible and can be excellent potential candidates for biological control with nematodes. *Steinernema glaseri*, *S. feltiae*, and *Heterorhabditis* species are all active against late–instar black vine weevil, *Otiorrhynchus sulcatus* (Fabr.) in strawberries on or near the surface of the soil, or injected against larvae deeper in the soil (Bedding and Miller, 1981a; Georgis and Poinar, 1984a; Scherer, 1987). Applied to the surface against larvae deeper in the soil *S. glaseri* may give 100% control (Evenhuis, 1982; Klingler, 1986; Simons, 1981).

2.   Wood–Tunneling Insects

Steinernema bibionis, the most commonly occurring steinernematid, had no prominence as a biological control agent until Miller and Bedding (1982) proved that it is effective against black currant clearwing, *Synanthedon tipuliformis* (Clerck). In stacked bundles of borer–infested cuttings sprayed with *S. bibionis* and stored damp for 4 days, almost all diapausing and postdiapausing larvae die. Dipping is less effective than spraying. Sprays on 4–year-old, borer–infested black currant bushes in the early grape stage, at 70% RH and 10°C, gives 90% control after 30 days. In old wood the slower drying, fewer exposed entrance holes, and rougher surface provide a better environment for the nematode to control the borer than new wood (Bedding and Miller, 1981b; Miller and Bedding, 1982). In general, however, to control wood–tunneling insects, injection into exposed tunnels is required and sprays are unsuccessful (Clearwater and Wouts, 1981; Deseo and Kovacks, 1982; Simons, 1978; Clearwater and Wildish, unpublished data).

3.   Lepidoptera

Under artificial conditions, e.g., on damp filter paper and in loose soil of flower pots, many Lepidoptera are very susceptible to steinernematids and heterorhabditids. But sprayed on undisturbed soil against soil insects these nematodes are ineffective, except against surface feeders like the underground grass caterpillar, *Oncopera fasciculata* (Walker), which take the surface sprayed nematodes in with food (Creighton et al., 1968; Creighton and Fassuliotis, 1985; Jaques, 1967; Jaques et al., 1968; Reed and Carne, 1967). The nematode's high temperature requirements and the host's cold habitat further reduces the effectiveness against soil–dwelling insects (Carne and Reed, 1963/1964; Dumbleton, 1945; Reed and Carne, 1967).

Prospects for control of lepidopterans developing above ground are good. These insects have evolved in the absence of the nematode and developed no resistance against them. Their environments can be modified and the most suitable conditions for application chosen. To achieve control of leaf–eating lepidopterans the nematodes must be applied on a humid day or wetting agents and antievaporants must be added, especially in the summer months (Chamberlin and Dutky, 1958; Fox and Jaques, 1966; Jaques, 1967; Kaya et al., 1984; Landazabal et al., 1973; Veremchuk, 1974). Applied this way the nematodes may kill as effectively as agricultural chemicals. High mortality, however, does not necessarily mean effective control. Applied against corn earworm, *Heliothis zeae* (Boddie), in various ways, *S. feltiae* causes high mortality (Moore, 1965; Tanada and Reiner, 1962), but eco-

TABLE 1 Insect Species Against Which Control Has Been Obtained with *Steinernema* and *Heterorhabditis* Species in Pot and/or Field Tests.

Insect species	Common name	S.g	S.f	S.b	Het	S.k	Ref.
Beetles							
Agriotes lineatus (L.)	Wireworms		+				Danilov, 1975
Allisonotum impressicola Arrow	Sugarcane beetle	±	+				Wang, J. and Li, L, 1983
Costelytra zealandica (White)	Grass grub			±	±		Hoy, 1954; Kain et al., 1982
Graphognathus peregrinus (Buchanan)	White-fringed beetle		±				Swain, 1943
Leptinotarsa decemlineata Say	Colorado beetle		±	±	±		(1)
Ligyrus subtropicus (Blatchley)	White grub	++	±		±		Sosa and Beavers,1985
Limonius californicus (Mannerheim)	Beet wireworm		±				Toba et al., 1983
Othnonius batesi Oliff.	Dark soil scarab		±				Reed and Carne, 1967
Pyrrhalta luteola (Mueller)	Elm leaf beetle		±		±		Kaya et al., 1981
Selatosomus aeneus (L.)	Wireworms		+				Danilov, 1974
Weevils							
Diaprepes abbreviatus (L.)	Root stalk borer weevil		++				Schroeder, 1987
Hylobius abietis L.	Large pine weevil		++				Pye and Pye, 1985
Hylobius Buchanan	Pine root collar weevil		++				Schmiege, 1963
Nemocestes incomptus (Horn)	Strawberry root weevil		+				Georgis and Poinar, 1984a
Otiorrhynchus sulcatus (Fabr.)	Black vine weevil	++	++		++		(2)
O. ovatus L.	Strawberry root weevil		+		±		Rutherford et al., 1987
Wood-tunneling insects							
Amyelois transitella Walker	Navel orange worm		++				Linegren et al., 1987
Oemona hirta (F)	Lemon tree borer		++				Clearwater and Wouts, 1980
Ostrinia nubilalis (Huebner)	European corn borer		++				Briand and Welch, 1963
Penniseta marginata (Harris)	Raspberry crown borer		++				Capinera et al., 1986
Platypitia carduidactyla (Riley)	Artichoke plume moth		±				Tanada and Reiner, 1960
Prionxystus robinae (Peck)	Carpenter worm		++				Lindegren et al., 1987

TABLE 1 (Continued)

Insect species	Common name	S.g	S.f	S.b	Het	S.k	Ref.
Rhyacionia buoliana (Schiff)	Eur. pine shoot moth		++				Schmiege, 1963
Sciapteron tabaniformis (Rottenburg)	Dusky clearwing		++				Simons, 1978
Synanthedon myopaeformis Brkh.	Apple clearwing		++	++			Deseo and Miller, 1985
S. tipuliformis (Clerck)	Currant clearwing		++				(3)
S. typhiaeformis Brkh.	Clearwing		+	+			Deseo and Miller, 1985
Vitacea polistiformis (Harris)	Grape root borer		++				Saunders and All, 1985; All et al., 1980
Zeuzera pyrina L.	Leopard moth		++				Deseo and Kovacks, 1982
Lepidoptera larvae							
Acrolepia assectella (Zeller)	Leek moth		±				Laumond, 1972
Agrotis ipsilon (Hufn)	Black cutworm	++		++			Lossbroek and Teunissen, 1985
A. segetum (Schiff)	Cutworm	++		++			Lossbroek and Teunissen, 1985
Cirphus compta Mo			+				Isreal et al, 1969
Diabrotica balteata LeConte	Corn root worm		++				Poinar et al, 1983
Heliothis virescens (Fabr.)	Tobacco budworm		++				(4)
H. zeae (Boddie)	Corn earworm		++				Moore, 1965; Tanada and Reiner, 1962
Hyphantria cunea (Drury)	Fall webworm		++				Jacques, 1967; Yamanaka et al., 1986
Laspeyresia pomonella (L.)	Codling moth		++				Kaya et al., 1984
Malacosoma americanum (F)	Eastern tent caterpillar		++				Jacques, 1967
Mythimna separata (Walker)	Cosmopolitan armyworm		+				Isreal et al., 1969
Operophtera brumata (L.)	Winter moth		++				Jacques, 1967; Jacques et al., 1968
Pieris rapae (L.)	Imported cabbage butterfly		++				Jacques, 1967; Jacques et al., 1968
Pseudexentera mali Freeman	Pale apple leafroller		++				Jacques 1967; Jacques et al., 1968
Rhyacionia frustrana (Comstock)	Pine tip moth		±				Nash and Fox, 1969
Spodoptera exigua (Huebner)	Beet armyworm		++				Kaya and Hara, 1980, 1981
S. frugiperda (J. E. Smith)	Fall armyworm		++				Landazabal et al., 1973

Bark beetles

Species	Common name	S.g	S.f	S.b	S.k	Het	Reference
Anoplophora malaciaca L.	In navel orange trees	±					Kashio, 1986
Dendroctonus frontalis Zimmerman	Pine bark beetle	±					Moore, 1970
D. pondorosae Hopk.	Mountain pine beetle	±					(5)

Flies

Species	Common name	S.g	S.f	S.b	S.k	Het	Reference
Cephalcia abietis L.	Saw fly	±				++	Mráček and David, 1986
C. lariciphala (Wachtl)	Larch saw fly	±		±			(6)
C. falleni Dalm	Spruce saw fly	±		±			Sander, 1984
Delia antiqua (Meigen) (= *Hylemya antiqua*)	Onion root fly		++				Coomans and de Grisse, 1985
D. florilega (Fallen) (= *H. florilega*)	Turnip maggot	++					Cheng and Bucher, 1972
D. platura (Meigen) (= *H. platura*)	Seedcorn maggot	++					Cheng and Bucher, 1972
D. radicum (Weideman) (= *H. brassicae*)	Cabbage root fly	±				++	Welch and Briand, 1961a; Georgis and Poinar, 1984b
Heteropeza pygmaea Winnertz	Mushroom cecidomyid fly	±		+			Richardson and Hughes, 1986
Lucilia cuprina (Wiedeman)	Sheep blow fly	+					Molyneux et al., 1983
Mycophilia speyeri (Barnes)	Mushroom cecidomyid fly	±	+				Richardson and Hughes, 1986
M. barnesi Edwards	Mushroom cecidomyid fly	±	+				Richardson and Hughes, 1986
Musca spp.	House flies	+					(7)
Pristiphora erichsonii (Hartig)	Larch saw fly	±	+				Finney and Bennett, 1983
Simulium sp.	Black flies	±					Gaugler and Molloy, 1981

(1) (Scognamiglio et al., 1968; Toba et al., 1983; Welch, 1958; Welch and Briand, 1961b; Wright et al., 1987).(2) (Bedding and Miller, 1981a; Evenhuis, 1982; Georgis and Poinar, 1984b; Klingler, 1986; Scherer, 1987; Simons, 1981; Verbruggen et al., 1985). (3) (Bedding and Miller, 1981b; Miller and Bedding, 1982).(4) (Chamberlin and Dutky, 1958; Samsook and Sikora, 1981). (5) (Finney and Mordue, 1976; Finney and Walker, 1977, 1979; McVean and Brewer, 1981; Poinar and Deschamps, 1981). (6) (Georgis and Hague, 1982; Webster and Bronskill, 1968). (7) (Georgis et al., 1987; Mullens et al., 1987; Ren et al., 1985).

Note: ++ = Commercially acceptable control; + = good control under ideal conditions; ± = some control.

S.g = *Steinernema glaseri*; S.f = *S. feltiae*; S.b = *S. bibionis*; S.k = *S. kraussei*; Het = *Heterorhabditis* species.

nomic damage to the corn ears is not prevented (Bong and Sikorowski, 1983). Chemical insecticides not only kill the insect but also give a long–term foliage protection and therefore generally may result in a better crop (Briand and Welch, 1963).

Not all stages in the life cycle of the insect are equally susceptible. *Spodoptera exigua* prepupae, for instance, are highly susceptible to *S. feltiae* with up to 94% mortality occurring, but the pupae are not very susceptible. The pupae have a pore–free layer of silk within their cocoon which acts as a mechanical barrier (Kaya and Hara, 1980, 1981; Samsook and Sikora, 1981). In contrast to the pupae of *S. exigua*, and also the pupae of *Heliothidis virescens*, the porous cocoons of *Galleria mellonella*, the silk worm *Bombyx mori* (L.), and the winter moth *Operophtora brumata* provide no barrier for the nematodes and in pupae of these species a 100% mortality is achieved as easily with as without cocoon (Hara and Kaya, 1983; Jaques et al., 1968; Kaya and Grieve, 1982; Moyle and Kaya, 1981). The adults are susceptible and may be controlled by nematode application into well–watered ground during pupation (Veremchuk, 1974).

4. Flies

Fruitflies, blowflies, and houseflies are susceptible to *S. feltiae* and *Heterorhabditis* species, but with fruitflies useful results in the field have not yet been obtained (Beavers and Calkins, 1984; Lindegren and Vail, 1986). At temperatures below 20°C, treatment with 4 million *H. heliothidis* infective juveniles per square meter poultry barn or introduced in bait pads with fly sex pheremones and fly–breeding medium or in 5% sucrose on cotton balls *H. heliothidis* reduces the fly population causing a mortality of adult flies of up to 67% (Belton et al., 1987; Geden et al., 1987; Renn et al., 1985, 1986).

5. Social Insects

Bees, wasps (including yellow jackets), and termites on damp filter paper are susceptible to steinernematids and heterorhabditids (Gambino, 1984; Wojcik and Georgis, 1988). *Vespula rufa astropilosa* (L.), *Vespula pensulvanica* (Saussure), and honeybees, *Apis mellifera*, workers may be killed if exposed to nematodes incorporated in a suitable bait. Direct spraying of workers, however, results in less than 10% mortality (Hacket and Poinar, 1973; Poinar and Ennik, 1972). *Glyptotermes dilatatus* (Bugnion and Popoff), a pest of low–grown tea in Sri Lanka, and workers of *Reticuletermes* sp. and *Zootermopsis* sp. are susceptible to *S. feltiae*. Doses of 40 and 30 ml/bush at 4000 and 8000 *S. feltiae* infectives/ml give complete control of *G. dilatatus* within 60-90 days (Danthanarayana, 1983). Within the cadavers juveniles are produced (Danthanarayana and Vitarana, 1987; Georgis et al., 1982).

Social insects remove dead members from their nests. This precludes the nematodes from systemic irradication of nests. However, it also means that nematodes can be safely used against insect pests where bees are present (Kaya et al., 1982).

6. Parasites of Insects

Ideally, nematodes should not affect the natural insect parasites of their target host. Unfortunately they do, by directly parasitizing the susceptible developing parasite in its host, or indirectly by killing the host. Ten- to 11– day–old larvae of *Hyposoter exiguae* (Viereck) and *Glyptapanteles* (=*Apanteles*) *militaris* (Walsh), two slowly developing hymenopterous parasites of the army worm, *Pseudaletia unipuncta* (Haworth), and 2– to 3–day–old larvae of the fast–developing tachinids *Myxexoristops* sp. a parasite of *Cephalcia abietis*, and *Compsilura concinnata* Meigen, a parasite of the army worm, *Pseudaletia unipuncta*, largely escape infection as they complete development or kill the host before the nematode takes over (Georgis and Hague, 1982; Kaya, 1978, 1984; Mráček and Spitzer, 1983). One–day–old tachinids and less than 10– day–old hymenopterous larvae die when the host dies (Georgis and Hague, 1982). Pupae of the hymenopterous parasites *Hyposoter exigua, Glyptapanteles (= Apanteles) militaris, Cotesia (= Apanteles) medicaginis* Muesebeck and *Chelonus* sp., like *Spodoptera exigua* and *Heliothidis virescens* pupae, have a pore–free layer of silk within the cocoon and are resistant to nematodes. Nematodes that are inside already do affect the pupa and will develop, but the resulting infective larvae cannot escape (Kaya and Hotchkin, 1981).

7. Small Arthropods

Many small arthropods, such as larvae of the mosquito *Aedes aegypti* L. and *Culex pipiens* (Roth.); cat flea, *Ctenocephalidis felis* (Bouche), larvae, pupae and prepupae; rat flea, *Xenopsylla cheopsis* (Rothchild), larvae: the collembolan *Sminthurus viridis* (L.); terrestrial isopods, *Porcellio* ssp.; and the garden millipede, *Oxydus gracilis* Koch, are extremely susceptible to steinernematids and heterorhabditids and die within 24-36 hr of exposure. The nematodes destroy the insect's internal tissues, are usually to a certain degree encapsulated, and are rarely able to develop to sexual maturity (Mráček and Weiser, 1983; Poinar and Thomas, 1985; Silverman et al., 1982). At densities of 100-5000 mosquito larvae per m2, rates equivalent to 150,000-500,000 *S. feltiae* infectives per m2 would be required for effective control.

Sminthurus viridis* causes damage to clover in Tasmania and may be controlled by nematodes applied under suitable ecological conditions (Ireson and Miller,1983; Poinar and Kaul, 1982; Poinar and Paff, 1985; Welch, 1960; Welch and Bronskill, 1962).

8. Snails

Steinernema glaseri, S. feltiae, S. bibionis, and *Heterorhabditis heliothidis* can all kill the snail *Oncomelania hupensis*. Applied at 100, 200, and 300 infective juveniles per cm2 soil surface, mortality of the snails in pots is about 45, 90, and 95%, respectively. The nematodes propagate in the snail (Li et al., 1986).

G. Nematode Suspensions For Application in the Field

Steinernematids and heterorhabditids withstand fairly high pressures and resist the toxic action of chemical insecticides. Field application of infectives can be made by adding the infectives, as aqueous suspensions, to irrigation systems (Reed et al., 1986), and by spraying or misting the infectives with traditional or specially designed spray/misting equipment, either on their own or mixed with chemicals in integrated pest control programs (Berg et al., 1987; Briand and Welch, 1963; Lindegren et al., 1987). Once sprayed on the leaves, even on a humid day, the nematodes become vulnerable and may rapidly dry out and die. Longevity of the infectives on the sprayed surface can be improved by adding of wetting agents and antidesiccants to the nematode suspension. Simple compounds such as glycerin, honey, sorbitol, urea, sugar, talcum suspensions, carboxymethyl cellulose, and Tween 20 are all somewhat effective, but commercial antidesiccants are more effective (Moore, 1970; Welch and Briand, 1961). The methylcellulose polymer Methocel J75 (Dow Chemical), Gelgard M, and Norbak have been used successfully, especially when combined with the surfactant Arlaton (ICI Americas). In the glasshouse and in simulated field conditions these substances delay *S. feltiae* desiccation from 20 min to more than 2 hr (MacVean and Brewer, 1981). Commercial desiccants improve control of the larch sawfly, *Pristophora erichsonii* (Hartig), from very low to up to 80%, and control of the potato beetle, *Leptinotarsa decem-*

lineata, from 10% to more than 30% in the laboratory (Webster and Bronskill, 1968) and to more than 60% in the field (MacVean et al., 1982). Unfortunately, and almost without exception, these chemicals have an adverse effect on either the host, the crop, or the nematode (Shapiro et al., 1985). Additives, therefore, are not nearly as important as careful selection of the right climatic conditions at the time of application. Spraying in the evening, for instance, may delay the desiccation of leaf surfaces by up to 14 hr. Alternative ways of protecting the nematodes against desiccation are to apply them embedded in inert hydrogels, paraffin oils, alginate capsules, or as a paste: all these are preparations that may ensure long life of nematodes and improve their shelf life. Partially protected from desiccation in such formulations, the nematodes are able to wait for suitable conditions in the field, and there is more time for the nematodes to find a host or for an insect to ingest the nematodes (Kaya et al., 1987; Kaya and Nelson, 1985; Poinar et al., 1985; Sandner and Pezowick, 1979; Wojcik and Georgis, 1988). When applied in the field, nematodes not only are subjected to desiccation but are also sensitive to short–wavelength UV light. Their reproductive potential is inhibited when they are exposed for 3.5 min, and their pathogenicity is lost when they are exposed for 7 min. When nematodes are to be applied against insects that live exposed to the sun, persistent formulations should contain a UV block if they are to be commercially viable (Gaugler and Boush, 1978).

H.   Mass Production of Infective Juveniles

1.   Mass Production of Infectives on a Host

The easiest and most commonly employed method of maintaining and rearing steinernematids and heterorhabditids for small–scale experimental purposes is mass production on an insect host. The technique was first attempted by Glaser (1931), but Dutky (Anon., 1956) was the first worker to obtain large numbers of infectives in *Galleria* larvae (Dutky, 1959). The number of infectives produced by a host depends on its susceptibility, which in turn correlates with the time taken by the nematode to kill the host. *Galleria* pupae die sooner than *Galleria* larvae, which die more quickly than Colorado beetle larvae, which die before the imagines. Yields decrease accordingly. The rate at which a host dies also depends on the bacterial load introduced by the infectives. Increased numbers of infectives kill the insect more rapidly but do not increase nematode yields.

      For mass production in a host, the host can be exposed to infective juveniles on damp filter paper or can be injected with infectives. The nematodes in the insect complete one or more life cycles, depending on the number of infectives reaching the hemocele. After about 10-12 days, at room temperature, nematode development is completed and infective larvae start leaving the host. To collect the infectives, the host can be placed on a White trap (Fig. 12), which consists of a small, inverted Petri dish, covered with filter paper, in a large Petri dish with water, the edge of the filter paper touching the water in the larger dish. On the wet filter paper the infectives leave the host and migrate into the water of the large dish from where they can be collected for storage (White, 1927). Infective juveniles emerging from a spent infested host move away as far as possible from the cadaver, escaping the low pH caused by accumulation of ammonium (Pye and Burman, 1981). If cadavers are placed in a circle on water agar the emerging infectives will accumulate in the center of the circle and in the outer periphery of the plate, leaving a clear zone under the circle of dead hosts.

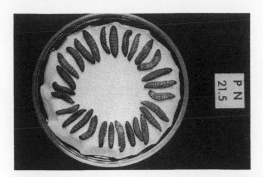

FIGURE 12   White's water trap with *Heterorhabditis*-infected *Galleria* larvae.

2.   Mass Production of Axenic Infectives on Solid Artificial Media

Mass production of axenic *Steinernema* species on artificial media has been attempted ever since the discovery of *S. glaseri* in the early 1930s (Glaser et al., 1942). The first successful media were homogenate animal tissues in dextrose–agar, a potato/sweet potato mixture (3:1) overgrown with baker's yeast, and lean veal pulp mixed with water in layers 4 mm thick. They produced 9000-12,000 nematodes per cm^2 of culture area, and would reach an average production of 140 million infectives per day. *Steinernema glaseri* also develops well on pieces of fresh, sterile rabbit kidney placed on agar and on semisolid gels of autoclaved ground beef kidney or liver in 0.5% sodium chloride–agar (McCoy and Glaser, 1936; Glaser et al., 1940). Reproduction declines with successive transfers. Passage through an insect (Glaser, 1931) or the addition to the culture medium of powdered Japanese beetle grubs, bovine ovaries (Glaser, 1940a) or raw liver extract (Stoll, 1953b) prevents this decline. A starting population of axenic nematodes can be obtained by surface–sterilizing infective juveniles for 1 hr in 0.1% merthiolate (sodium ethylmercurithiosalicylate) or 0.4% Hyamine 10X (methyl benzethonium chloride). The nematodes are washed in sterile water and transferred to the center of a nutrient broth plate, into a well which is surrounded with *Xenorhabdus*–decomposed medium about 2 cm away from the well. The infectives migrate from the well over the agar into the medium, where they develop into females in about 3 days. Females with some eggs inside their uterus are washed in sterile water and transferred to low concentrations of penicillin, streptomycin, and ampicillin, in which they are kept overnight. Some will develop as axenic cultures after being washed and individually transferred to a piece of raw mouse liver on nutrient broth plates or tubes.

3.   Mass Production of Monoxenic Infectives on Solid Artificial Media

With the discovery of the symbiont (Dutky, 1937), cheap substrates could be developed that in the presence of the symbiont would break down into excellent culture media. House et al. (1965), using dog food, obtained yields per square centimeter surface area that were comparable to those obtained on Glaser's expensive agar media. Bedding (1981) further increased yields per unit medium using enlarged surface areas. A yield of 1300 million nematodes per 3 kg medium can be obtained from ground animal tissue, thoroughly mixed with animal fat, squeezed into shredded polyurethane foam–plastic in flasks, or in aerated plastic bags, then autoclaved and inoculated with the symbiont and nematodes; by far the highest yield of any medium known, at a medium cost of less than U.S.$0.01 per million nematodes (Bedding,

FIGURE 13   Bedding's sponge rearing method in Erlenmeyer flasks. From right to left: culture flask with sponge impregnated with medium; culture flask inoculated with *Xenorhabdus luminescens* and incubated for 3 days at 23°C; culture flask 1 month after inoculation with *Heterorhabditis* sp.

1984). As an alternative, nutrient broth, yeast extract, and vegetable oil can be used in place of ground animal tissues and fat (Fig. 13). The resulting medium is somewhat more expensive but is equally effective, easier to prepare, and more pleasant to work with; also, at room temperature the infective larvae can be stored in the spent culture flasks for months (Wouts, 1981).

For long–term mass production of steinernematids, solid media are the most suitable of all media described in the literature. Infectives reared in such media keep their infectivity for over 20 years and probable as long as they live (Jackson and Moore, 1969; Jackson, 1969; Stoll, 1953a).

4.   Mass production of Axenic Infectives in Liquid Artificial Media

Glaser (1940b) reported that he had successfully cultured axenic *S. glaseri* in liquid media in Erlenmeyer flasks, either on a piece of sterile rabbit kidney in 3-4 mm 0.5% NaCl, or in kidney extracts devoid of particulate matter. He provided no detailed recipe, and credit for being the first to develop and describe fully liquid media should go to his successor, Stoll, (1951). A medium consisting of 9 ml veal or heart infusion broth, 1 ml filter–sterilized, heat–sensitive, aqueous raw liver extract, and 25 mg dextrose in 0.5% NaCl at pH 6-6.5 in 22 x 180 mm shaking tubes, incubated at 21-26°C, produces 2500-7500 nematodes in 3 weeks. Yields per unit double by using Erlenmeyer flasks and preparing the raw liver extract from pregnant rabbits (Stoll, 1954). Cultures can be successfully subcultured after 4 weeks, provided the worms are washed once or twice before transfer (Stoll, 1959).

The complexity of the procedure and the need of the nematodes for a rest period between generations (Stoll, 1959) contrast sharply with the relative ease with which consistently large numbers of monoxenic nematodes, with a higher percentage of infective juveniles, can be reared on a live host (Dutky and Hough, 1955) and solid artificial media (Bedding, 1981; House et al., 1965). It was for this reason that only workers interested in the genetics, development, and nutritional requirements of the nematodes continued to rear steinernematids in axenic liquid cultures. Geneticists adapted a medium consisting of soya peptone and yeast extract enriched with liver extract, originally developed by Cryan et al. (1963) for *Panagrellus redivivus*. The medium was used aerated (Buechner and Hansen, 1971) or with chicken embryo extract thinly layered on glass fibers in columns (Tarakanov, 1980). Those interested in the nutrient requirement of the nematode adapted the partially defined medium "S" developed by Trager (1957) for the leptomonad *Leishmania tarentolae*, the chemically defined tissue culture media "199" developed by Morgan et al. (1950), or NCTC developed by McQuilkin, Evans, and Earl (1957), with some of the amino acid concentrations increased.

Steinernema feltiae fails to reproduce in these media (Jackson 1962a,b) but develops in a hemin/protein medium developed by Hansen and Berntzen (1969). However, none of these media shows an activity as great as Stoll's medium with liver growth factor (Beucher et al., 1970).

5. Mass Production of Monoxenic Infectives in Liquid Artificial Media

Bedding's efficient method of mass–producing nematodes on medium–coated sponge is labor–intensive to prepare. Also, sponge is not biodegradable and for commercial use of the nematodes the sponge has to be removed. For commercial purposes, therefore, mass production in monoxenic liquid cultures is more practical and cheaper. Ideally, all the components will decompose and at the end of the production process the nematodes can be used for application in the field without any further processing.

The most important single factor that enables such yields on solid media as are achieved today is the use of the symbiont in the culture medium. Adding the symbiont to liquid media has an equally dramatic effect and has made mass production of steinernematids and heterorhabditids commercially attractive. Unfortunately, little is known about these media. They were developed by or for commercial companies for financial gain and are kept secret or protected by patents. In the patent application "Liquid culture of nematodes," International Publication Number WO 86/01074, by Biotechnology Australia Pty. Ltd. (1986), the following cleaning procedure and medium is listed for *Steinernema feltiae* production in stirring tank fermenter. If infective *S. feltiae* juveniles are washed for 2 hr in 0.1% w/v Thimerasol and transferred to a medium of soya peptone 40 g/liter, yeast extract 30 g/liter, horse hemoglobin 1 g/liter, cholesterol 0.2 g/liter, streptomycin 1 g/liter, and ampicillin 10,000 U/liter, the yield is 95,000 axenic infectives/ml 10 days later. In a homogenate of 100 g/liter ox kidney and 10 g/liter yeast, fermented by *X. nematophylus* at 28°C, pH 6.8, and oxygen saturation 50% for 24 hr, 2000 of these axenic juveniles, at 23°C and 20% oxygen saturation, yield 90,000 infective juveniles per milliliter medium 10 days later. The California–based company Biosys and the English Agricultural Genetics Company Ltd., now supply steinernematid nematodes for field use produced in liquid culture at 10-20 cents per million.

6.   Storage of Infective Juveniles

The infectives last several months when stored in a shallow layer of water or in aerated flasks at room temperature. In this condition, however, they are difficult to handle because of the free water present.

     Several methods have recently been developed to store commercial quantities of nematodes for biological control of insects. In these methods the free water is absorbed into inert material such as a sponge or other layers of porous composition, and hygroscopic gels (Bedding, 1984; Howell, 1979; Popiel et al., 1987). Through or on these carriers the nematodes move freely in a damp, rather than wet, environment. Details of these methods, however, are strictly guarded by the companies using them.

II.   TAXONOMY

A.   Life Cycle And Morphology (Fig. 14)

In nature, steinernematids and heterorhabditids are predominantly present as free–living infective juveniles in soil frequented by insects for at least part of the year. Infective steinernematids are L–3, which, except for *S. bibionis* and *S. rara* (de Doucet 1986), are often still enclosed by the cuticle of the L–2 . Heterorhabditids are infective as early as L–2 and remain infective until they start feeding as parasitic L–3. Infective juveniles can be retrieved from soil via a trap host (Bedding and Akhurst, 1975) or by using any of the standard nematode extraction methods (Saunders and All, 1982). Newly formed infective juveniles will emerge from the trap hosts at the end of the life cycle, about 10 days after burying the host, and can be collected in a White trap. Infective juveniles extracted from soil using standard nematode extraction methods can be distinguished from the other nematodes in the sample by their characteristic movements, their slender bodies, closed stoma, sharp tails, the longitudinal lines on the cuticle of L– 2 heterorhabditids, and the lines in the lateral field of L–3 steinernematids. In nonmoving, live specimens, the tails are slightly bent to one side. Steinernematid infective juveniles have large glands that open out into the excretory duct and dorsally displace the area around the base of the esophagus. Heterorhabditid L–3 possess a characteristic, sclerotized labial tooth above the stoma, which is supported by rays extending between the lips. Ventrally a sclerotized labial plate is present. In parasitic L–3 the stoma gradually opens, the animal increases somewhat in length, doubles in width and molts, usually within a day. The L–4 increases in width and almost doubles in length. The sex organs develop substantially to become functional with the next molt, about 3-4 days after the infectives penetrate the host. The cuticle of the L–4 and female is smooth, without longitudinal lines. The labial region, often retracted in large females, has six lips, each with a papilla. The labial papillae in large heterorhabditid females are formed like flaps and may have a function in food gathering. Except for *S. rara*, steinernematid females have four cephalic papillae near the base of the lips. The amphids are obscure. The stoma is about as wide as it is long and tapers into the narrow esophagus. Relative to the length of the body the esophagus is short, and ends in a valved basal bulb. The ovaries are paired, opposite, and reflexed. The rectum and anus are distinct. Steinernematid males remain small and, except for sexual characteristics, are identical to the females. They appear somewhat earlier than the females. The testis is single and reflexed, the spicules are heavy and paired and a gubernaculum is present. Several pairs of genital papillae (including a pronounced preanal ventral papilla) are located in the caudal region and there is no bursa. Heterorhabditid males are absent in the generation that develops from the infective juveniles; they are present only

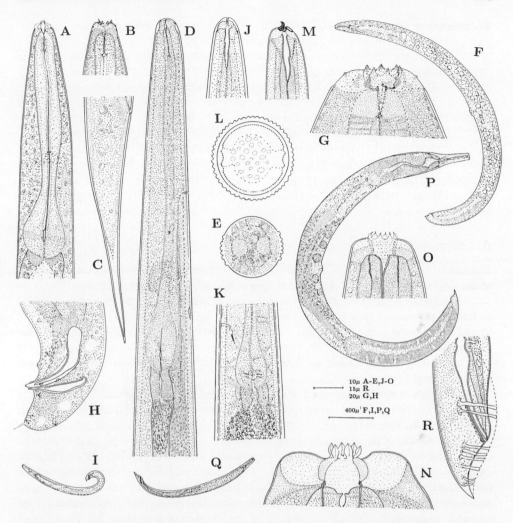

FIGURE 14   (A–I) *Steinernema bibionis*; (J–R) *Heterorhabditis heliothidis*. (A) Esophageal region of the L-1. (B) Cephalic region of the the L-2. (C–E) L-3. (C) Tail lateral aspect. (D) Esophageal region. (E) Cross-section at midbody. (F) Female whole. (G) Cephalic region female. (H) Male tail lateral aspect. (I) Male whole. (J–L) L-2. (J) Cephalic region. (K) Esophageal/intestinal junction. (L) Cross-section at midbody. (M) Cephalic region L-3, showing distinct labial tooth. (N) Anterior region of a large female. (O) Anterior region of a small female. (P) Female whole. (Q) Male whole. (R) Male tail lateral aspect.

when a second generation occurs. They do not seem to feed. The spicules are slender; genital papillae are absent, and the bursa is supported by nine ribs.

The number of generations which steinernematids and heterorhabditids produce in one host is directly related to the quantity of food available (Moore, 1965). When the population density in the host is low the females grow large and lay many of their eggs (Dutky, 1959); the remaining eggs develop inside the female. The lip region of the emerging L–1 is delicate but soon develops its distinct outline with six labial papillae. The L–1 develop rapidly and directly into L–4 and into males and females of the second generation.

When population density in the host is high and food supply becomes insufficient, newly formed females are small, more eggs or all eggs remain inside the female, and the L–1 molt into the L–2 which transform into infectives, which sooner or later leave the host cadaver (Wouts, 1979, 1980). The early stages of heterorhabditids developing inside the female usually remain within the female longer than do the early stages of steinernematids.

B.   Systematics

The family Steinermatidae was proposes by Chitwood and Chitwood (1937) for the species of the genera *Steinernema* Travassos, 1927 and *Neoaplectana* Steiner, 1929.

Steinernema kraussei, the type of the genus *Steinernema* was originally described as *Aplectana kraussei* Steiner, 1923, a new species in the order Oxyurata. Travassos (1927) placed it in the new genus *Steinernema* on the basic of the lack of a valvelike structure in the basal bulb of the esophagus.

In 1929, Steiner described *Neoaplectana glaseri*, the type species of the genus *Neoaplectana*.

Filipjev (1934b) described *Neoaplectana feltiae* and proposed the subfamily Steinernematinae for this group of species. In a different paper published the same year, Filipjev (1934a) states: "Steiner (1923) described a parasite [*S. kraussei*] from a sawfly larva and because of its general similarity he referred it to the oxyurid genus *Aplectana*.... (Now it is referred to a separate genus, *Steinernema*.) The writer [Filipjev 1934b] found a closely related species [*S. feltiae*], probably congeneric with Steiner's, in a cutworm,..." indicating that the *Neoaplectana* species he was most familiar with was congeneric with *Steinernema kraussei* and inferring therewith that *Steinernema* and *Neoaplectana* were probably congeneric. Critical assessment of the characters separating the two species proved that these characters have no diagnostic significance and that the two genera are indeed congeneric (Mráček, 1977; Wouts et al., 1982). Consequently, the genus *Steinernema* was recognized as a valid genus and the genus *Neoaplectana* as its junior synonym. Of the more than 20 nominal species in the genus *Neoaplectana*, three were accepted as valid and transferred to the genus *Steinernema*.

These changes have become accepted generally. The recently described species *N. anomali* Kozodoi, 1984, *N. rara* de Doucet, 1986, and *N. intermedia* Poinar, 1985b and the redescribed species *N. affinis* (Poinar, 1988) transferred to the genus *Steinernema*, together with the type species and the three species earlier transferred from *Neoaplectana*, bring to eight the number of valid *Steinernema* species.

The family Heterorhabditidae was introduced by Poinar (1976) for nematodes with a bacterial association similar to that of *Neoaplectana*, but with a heterogonic life cycle and a different morphology as discussed earlier in this chapter. The type genus *Heterorhabditis* contains the type species *H. bacteriophora* Poinar, 1976 and two species described since.

Morphological characters observed and measured, using light microscopy, have been the main means of identifying steinernematids and heterorhabditids. These characters have lately been supplemented by characteristics of the species–specific symbiont, unique nutritional requirements (Jackson, 1962a, 1962b), serological reactions (Jackson, 1961, 1965), protein and DNA electrophoresis of the species (Akhurst, 1986; Sherman and Jackson, 1963; Southern, 1979), and DNA analysis (Curran et al., 1985; Grimont et al., 1984). Characters that on their own have not yet been used for the characterization of a nematode species.

C. Classification

Phylum: Nematoda Chitwood, 1950
Class: Secernentea von Linstow, 1905
Order: Rhabditida (Oerly, 1880) Chitwood, 1933
Suborder: Rhabditina (Oerly, 1880) Chitwood, 1933
Superfamily: Rhabditoidea (Oerly, 1880) Travassos, 1920
Family: Steinernematidae (Filipjev, 1934) Chitwood and Chitwood, 1937
Type genus: *Steinernema* Travassos, 1927
    syn: *Neoaplectana* Steiner, 1929
    type species: *Neoaplectana glaseri* Steiner, 1929
Type species: *Aplectana kraussei* Steiner, 1923
Other species:
    Steinernema glaseri (Steiner, 1929) Wouts, Mráček, Gerdin, and Bedding, 1982
    Steinernema bibionis (Bovien, 1937) Wouts, Mráček, Gerdin, and Bedding, 1982
    Steinernema feltiae (Filipjev, 1934) Wouts, Mráček, Gerdin, and Bedding, 1982
    Steinernema intermedia (Poinar, 1985, *Neoaplectana*) new combination
    Steinernema rara (de Doucet, 1986, *Neoaplectana*) new combination
    Steinernema affinis (Bovien, 1937, *Neoaplectana*) new combination
    Steinernema anomali (Kozodoi, 1984, *Neoaplectana*) new combination
For the synonymies of the above *Steinernema* species see Wouts et al., (1982).

Family: Heterorhabditidae Poinar, 1976
Type genus: *Heterorhabditis* Poinar, 1976
    syn: *Chromonema* Khan, Brooks, and Hirschmann, 1976
    Type species: *Chromonema heliothidis* Khan, Brooks, and Hirschmann, 1976
Type species: *Heterorhabditis bacteriophora* Poinar, 1976
Other species:
Heterorhabditis heliothidis (Khan, Brooks, and Hirschmann, 1976) Poinar, Thomas, and
    Hess, 1977
Heterorhabditis megidis Poinar, Jackson, and Klein, 1987

D. Diagnosis

Heterorhabditis bacteriophora, the type species of the genus *Heterorhabditis*, is characterized by infective juveniles that are less than 0.6 mm long and have a tail length of about 0.09 mm. *Heterorhabditis heliothidis* can be distinguished from the type species by infective juveniles that are more than 0.6 mm long and have a tail length of about 0.1 mm. *Heterorhabditis megidis* can be distinguished from these two species by infective juveniles that are more than 0.7 mm long (Poinar et al., 1987).

Key to the species of the genus *Steinernema*

1. Average length of the infective juveniles more than 1000 µm long 7
    Average length of the infective juveniles less than 1000 µm long 2
2. Male tail with a projection or spine 3
    Male tail without a projection or spine .... *S. intermedia*
3. Average length of the infective juvenile more than 700 µm long 4
    Average length of the infective juveniles less than 700 µm long 6
4. Male spicules approximately 60-70 µm long 5
    Male spicules approximately 45-55 µm long .... *S. kraussei*
5. Spine absent in tail terminus of the infectives .... *S. bibionis*
    Spine present in tail terminus of the infectives .... *S. affinis*
6. Female with cephalic papillae, L–2 cuticle of infective juvenile tenacious ....
    .... *S. feltiae*
    Females without cephalic papillae L–2 cuticle not tenacious .... *S. rara*
7. Tip spicules with a notch .... *S. glaseri*
    Tip spicules not notched .... *S. anomali*

NOTE ADDED IN PROOF

Since the completion of the text of this chapter Poinar (1989) has suggested that the *Steiner-nema* species with the shortest infective larvae, the species here referred to as *S. feltiae*, should be called *S. carpocapsae* (Weiser, 1955). However, on the basis of the evidence provided by Poinar (1989) *S. carpocapsae* may be identical to *S. menozzi* (Travassos, 1931). Consequently, *S. menozzi*, being the earlier (1931) proposed name, it may be the valid name of the species. As this could not be verified before the deadline of this chapter and to prevent further instability the changes proposed by Poinar (1989) were not adopted in this chapter.

REFERENCES

Akhurst, R. J. 1980. Morphological and functional dimorphism in *Xenorhabdus* spp. bacteria sym-biotically associated with the insect pathogenic nematodes *Neoaplectana and Heterorhab-ditis*. *J. Gen. Microbiol.* 121: 1-7.

Akhurst, R. J. 1982. A *Xenorhabdus* sp. (Eubacteriales: Enterobacteriaciae) symbiotically associ-ated with *Steinernema kraussei* (Nematoda: Steinernematidae). *Rev. Nemat.* 5(2): 277-280.

Akhurst, R. J. 1983a. *Neoaplectana* spp. specificity of association with bacteria of the genus *Xenor-habdus*. *Exp. Parasitol.* 55(2): 258-263.

Akhurst, R. J. 1983b. Antibiotic activity of *Xenorhabdus* spp. bacteria symbiotically associated with insect pathogenic nematodes of the families Heterorhabditidae and Steinernematidae. *J. Gen. Microbiol.* 128(12): 3061-3066.

Akhurst, R. J. 1986. Use of starch gel electrophoresis in the taxonomy of the genus *Heterorhabditis* (Nematoda: Heterorhabditidae). *Nematologica* 33: 1-9.

Akhurst, R. J., and Boemare, N. E. 1986. A non–luminescent strain of *Xenorhabdus luminescens* (Enterobacteriaceae). *J. Gen. Microbiol.* 132(1): 1917-1922.

Akhurst, R. J., and Boemare, N. E. 1988. A numerical taxonomic study of the genus *Xenorhabdus* (Enterobacteriaceae) and proposed elevation of the subspecies of *X. nematophilus* to spe-cies. *J. Gen. Microbiol.* 134(7): 1835-1845.

All, J. N., Saunders, M. C., Dutcher, J. D., and Javid, A. M. 1980. Susceptibility of grape root borer larvae *Vitacea polistiformis* (Lepidoptera: Sesiidae) to *Neoaplectana carpocapsae* (Nematoda: Rhabditida) potential of host kairomones for enhancement of nematode activity in grape vineyards. *Mis. Publ. Entomol. Soc. Am.* 12(2): 9-14.

Andreadis, T. G., and Hall, D. W. 1976. *Neoaplectana carpocapsae*: Encapsulation in *Aedes aegypti* and changes in host hemocytes and hemolymph proteins. *Exp. Parasitol.* 39(2): 252-261.

Anon. 1956. Nematodes on our side. *Agr. Res. Wash.* 4(10): 3-4.

Beavers, J. B., and Calkins, C. O. 1984. Susceptibility of *Anastrepha suspensa* (Diptera: Tephritidae) to steinernematid and heterorhabditid nematodes in laboratory studies. *Env. Entomol.* 13(1): 137-139.

Bedding, R. A. 1981. Low cost in vitro mass production of *Neoaplectana* and *Heterorhabditis* species (Nematoda) for field control of insect pests. *Nematologica* 27: 109-114.

Bedding, R. A. 1984. Large scale production, storage and transport of the insect–parasitic nematodes *Neoaplectana* spp. and *Heterorhabditis* spp. *Ann. App. Biol.* 104: 117-120.

Bedding, R. A., and Akhurst, R. J. 1975. A simple technique for the detection of insect parasitic rhabditid nematodes in soil. *Nematologica* 21: 109-110.

Bedding, R. A., and Miller, L. A. 1981a. Use of a nematode, *Heterorhabditis heliothidis*, to control black vine weevil, *Otiorrhynchus sulcatus*, in potted plants. *Ann. App. Biol.* 99: 211-216.

Bedding, R. A., and Miller, L. A. 1981b. Disinfecting blackcurrant cuttings of *Synanthedon tipuliformis* using the insect parasitic nematode *Neoaplectana bibionis*. *Env. Entomol.* 10: 449-453.

Bedding, R. A., and Molyneux, A. S. 1982. Penetration of insect cuticle by infective juveniles of *Heterorhabditis* spp. (Heterorhabditidae: Nematoda). *Nematologica* 28: 354-359.

Bedding, R. A., Molyneux, A. S., and Akhurst, R. J. 1983. *Heterorhabditis* spp., *Neoaplectana* spp. and *Steinernema kraussei* interspecific and intraspecific differences in infectivity for insects. *Exp. Parasitol.* 55(2): 249-257.

Belton, P, Rutherford, T. A., Trotter, D. B., and Webster, J. M. 1987. *Heterorhabditis heliothidis* a potential biological control agent of house flies in caged layer poultry barns. *J. Nematol.* 19(2): 263-266.

Berg, G. N., Williams, P., Bedding, R. A., and Akhurst, R. J. 1987. A commercial method of application of entomopathogenic nematodes to pasture for controlling subterranean insect pests. *Plant Prot. Quart.* 2(4):174-177.

Boemare, N. 1983. Recherches sur les complexes némato–bacteriens entomophagenes: Etude bactériologique, gnotobiologique et physiopathologique du monde d'action parasitaire de *Steinernema carpocapsae* Weiser (Rhabditida: Steinernematidae). Thèse de Docteur d'Etat Université de Languedoc. Centre de Parasitologie et Pathologie Comparée, Université des Sciences, Montpellier.

Boemare, N. E., and Akhurst, R. J. 1988. Biochemical and physiological characterization of colony form variants in *Xenorhabdus* spp. (Enterobacteriaceae). *J. Gen. Microbiol.* 134: 751-761.

Boemare, N., Laumond, C., and Luciani, J. 1982. Pathologie animale.–Mise en évidence d'une toxicogenese provoquée par le nematode axenique entomophage *Neoaplectana carpocapsae* Weiser chez l'insecte *Galleria mellonella* L. (Note). *Comptes Rendus de l'Academie Scientifique Paris III*, 295: 543-546.

Boemare, N., Louis, C., and Kuhl, E. 1983. Etude ultrastructurale des cristaux chez *Xenorhabdus* spp. bactéries inféodées aux nématodes entomophages Steinernematidae et Heterorhabditidae. *Comptes Rendus des Sciences de la Societe de Biologie* 177(1): 107.

Bong, C. F., and Sikorowski, P. P. 1983. Use of the DD–136 strain of *Neoaplectana carpocapsae* (Rhabditida: Steinernematidae) for control of corn earworm *Heliothidis zea* (Lepidoptera: Noctuidae). *J. Econ. Entomol.* 76(3): 590-593.

Bovien, P. 1937. Some types of association between nematodes and insects. *Videnskabelige Meddelelser fra Dansk Naturhistorisk Forening* 101: 1-114.

Briand, L. J., and Welch, H. E. 1963. Use of entomophilic nematodes for insect pest control. *Phytoprotection* 44(1): 37-41.

Bronskill, J. F. 1962. Encapsulation of rabditoid nematodes in mosquitoes. *Can. J. Zool.* 40(7): 1269-1275.

Buechner, E. J., and Hansen, E. L. 1971. Mass culture of axenic insect parasitic nematodes using continuous aeration. *J. Nematol.* 3(2): 199-200.

Beuchner, E. J., Hansen, E. L., and Yarwood, E. A. 1970. Growth of nematodes in defined medium containing hemin and supplemented with commercially available proteins. *Nematologica* 16(3): 403-409.

Burman, M. 1982. *Neoaplectana carpocapsae*: Toxin production by axenic insect parasitic nematodes. *Nematologica* 28: 62-70.

Byers, J. A., and Poinar, G. O. 1982. Location of insect hosts by the nematode, *Neoaplectana carpocapsae*, in response to temperature. *Behaviour* 79(1): 1-10.

Capinera, J. L, Cranshaw, W. S., and Hughes, H. G. 1986. Suppression of raspberry crown borer *Pennisetia marginata* (Lepidoptera: Sesiidae) with soil applications of *Steinernema feltiae* (Rhabditida: Steinernematidae). *J. Invert. Pathol.* 48(2): 257-258.

Carne, P. B., and Reed, E. M. 1963/1964. Nematode parasites of curl grubs. Annual Report, Division of Entomology, CSIRO, 71.

Chamberlin, F. S., and Dutky, S. R. 1958. Test of pathogens for the control of tobacco insects. *J. Econ. Entomol.* 51(4): 560.

Cheng, H. H., and Bucher, G. E. 1972. Field comparison of the neoaplectanid nematode DD–136 with Diazinon for control of *Hylemya* spp. on tobacco. *J. Econ. Entomol.* 65(6): 1761-1763.

Chitwood, B. G., and Chitwood, M. B. 1937. *An introduction to nematology*. Monumental Printing Co., Baltimore, Maryland.

Clearwater, J. R., and Wouts, W. M. 1980. Preliminary trials on the control of lemon tree borer with nematodes. *Proceedings of the 33rd New Zealand Weed and Pest Control Conference*, Tauranga. Swiftcopy Centre Ltd., Palmerston North, New Zealand, pp. 133–135.

Coomans, P, and de Grisse, A. 1985. Susceptibility of *Delia antiqua* to *Steinernema bibionis*. *Mededelingen van de Faculteit Landbouwwetenshappen Rijksuniversiteit Gent* 50(1): 155-162.

Creighton, C. S., Cuthbert, F. P, and Reid, W. J. 1968. Susceptibility of certain celeopterous larvae to the DD–136 nematode. *J. Invert. Pathol.* 10(2): 368-373.

Creighton, C. S., and Fassuliotis, G. 1985. *Heterorhabditis* sp. (Nematoda: Heterorhabditidae) a nematode parasite isolated from the banded cucumber beetle, *Diabrotica balteata*. *J. Nematol.* 17(2): 150-153.

Cryan, W. S., Hansen, E. L., Martin, F., Sayre, W., and Yarwood, E. A. 1963. Axenic cultivation of the dioecious nematode *Panagrellus redivivus*. *Nematologica* 9: 313-319.

Curran, J., Baillie, D. L., and Webster, J. M. 1985. Use of genomic DNA restriction fragment length differences to identify nematode species. *Parasitology* 90: 137-144.

Dadd, R. H. 1971. Size limitations on the infectibility of mosquito larvae by nematodes during filter–feeding. *J. Invert. Pathol.* 18(2): 246-251.

Danilov, L. G. 1974. Susceptibility of wireworms to the infestation by the nematode *Neoaplectana carpocapsae* Weiser, 1955 str *agriotos*. *Bull. All–Union Res. Inst. for Plant Protection* 30: 54-57.

Danthanarayana, W. 1983. Control of live–wood tea termite *Glyptotermis dilatatus* with the entomogenous nematode *Heterorhabditis* sp. *Pacific Sci. Congr. Proc.* 15 (1/2): 54-55.

Danthanarayana, W., and Vitarana, S. I. 1987. Control of live–wood tea termite *Glyptotermis dilatatus* using *Heterorhabditis* sp. (Nemat.). *Agr. Ecosys. Env.* 19(4): 333-342.

Deseo, K. V., and Kovacs, A. I. 1982. Observations on and experiments with entomopathogenous nematodes in Italy (1980-1982). *Progr. Third International Colloquium on Invertebrate Pathology. Fifteenth Annual Meeting of the Society for Invertebrate Pathology*, 6-10 September 1982, University of Sussex, Brighton, U.K., 115.

Deseo, K. V., and Miller, L. A. 1985. Efficacy of entomogenous nematodes *Steinernema* spp. against clearwing moths, *Synanthedon* spp., in North Italian apple orchards. *Nematologica* 31(1): 100-108.

Doucet de, M. M. A. 1986. A new species of *Neoaplectana* Steiner, 1929 (Nematoda: Steinernematidae) from Cordoba, Argentina. *Rev. Nematol.* 9(4): 317-323.

Dumbleton, L. J. 1945. Bacterial and nematode parasites of soil insects. *New Zealand J. Sci. Technol. A. Agri. Sec.* 27(1): 76-81.

Dunphy, G. B., Rutherford, T. A., and Webster, J. M. 1985. Growth and virulence of *Steinernema glaseri* influenced by different subspecies of *Xenorhabdus nematophilus. J. Nematol.* 17(4): 476-482.

Dunphy, G. B., and Webster, J. M. 1985. Influence of *Steinernema feltiae* (Filipjev) Wouts, Mráček, Gerdin and Bedding DD136 strain on the humoral and haemocytic responses of *Galleria mellonella* (L.) larvae to selected bacteria. *Parasitology* 91:369–380.

Dunphy, G. B., and Webster, J. M. 1987. Partially characterized components of the epicuticle of dauer juvenile *Steinernema feltiae* and their influence on hemocyte activity in *Galleria mellonella. J. Parasitol.* 73(3): 584-588.

Dutky, S. R. 1937. Investigation of the diseases of the immature stages of the Japanese beetle. Thesis, Rutgers University, New Brunswick, N.J.

Dutky, S. R. 1959. *Insect Microbiology*. Advances in Applied Microbiology. Academic Press, New York and London, Vol. 1, pp. 175–200.

Dutky, S. R., and Hough, W. S 1955. Note on a parasitic nematode from codling moth larvae, *Carpocapsa pomonella* (Lepidoptera, Olethreutidae). *Proc. Entomol. Soc. Wash.* 57(5): 244.

Evenhuis, H. H. 1982. Control of black vine weevil *Otiorrhynchus sulcatus* (Coleoptera: Curculionidae). *Mededelingen van de Faculteit Landbouwwetenschappen Rijksuniversiteit Gent* 47(2): 675-678.

Fedorko, A. 1971. Badania nad mozliwoscia przetrewwania w glebie larw inwazyjnych z rodzajow *Pristionchus* i *Neoaplectana. Zeszyty Probleowe Postepow Nauk Rolniczych* 121: 227-233.

Filipjev, I. N. 1934a. Miscellanea Nematologica. I. Eine neue Art der Gattung *Neoaplectana* Steiner, nebst Bemerkungen ueber die systematische Stellung der letzteren. *Magazin de Parasitologie de l'Institut Zoologique de l'Academie de l'URSS* 4: 229-240.

Filipjev, I. N. 1934b. The classification of the free–living nematodes and their relation to the parasitic nematodes. *Smithsonian Misc. Collections* 89: 1-63.

Finney, J. R., and Bennett, G. F. 1983. The susceptibility of some saw flies (Hymenoptera: Tenthredinidae) to *Heterorhabditis heliothidis* (Nematoda: Heterorhabditidae) under laboratory conditions. *Can. J. Zool.* 61(5): 1177-1180.

Finney, J. R., and Mordue, W. 1976. The susceptibility of the elm bark beetle *Scolytus scolytus* to the 136 strain of *Neoaplectana* sp. *Ann. Appl. Biol.* 83: 311-312.

Finney, J. R., and Walker, C. 1977. The DD–136 strain of *Neoaplectana* sp. as a potential control agent for the European bark beetle. *J. Invert. Pathol.* 29: 7-9.

Finney, J. R., and Walker, C. 1979. Assessment of a field trial using the DD–136 strain of *Neoaplectana* sp. for the control of *Scolytus scolytus. J. Invert. Pathol.* 33: 239-241.

Finney–Crawley, J. R. 1985. Isolation of cold–tolerant steinernematid nematodes in Canada. *J. Nematol.* 17(4): 496.

Fox, C. J. S., and Jaques, R. P. 1966. Preliminary observations on biological insecticides against imported cabbageworm. *Can. J. Plant Sci.* 46: 497-499.

Gambino, P. 1984. Susceptibility of the western yellowjacket *Vespula pensylvanica* to three species of entomogenous nematodes. *IRCS Medical Science: Environmental Biology and Medicine; Experimental Animals; Microbiology, Parasitology and Infectious Diseases* 12(3): 264.

Gaugler, R., and Boush, G. M. 1978. Effects of ultraviolet radiation and sunlight on the entomogenous nematode *Neoaplectana carpocapsae. J. Invert. Pathol.* 32: 291-296.

Gaugler, R., and Boush, G. M. 1979. Non–susceptibility of rats to the entomogenous nematode *Neoaplectana carpocapsae. Env. Entomol.* 8: 658-660.

Gaugler, R., Lebeck, L., Nakagaki, B., and Boush, G. M. 1980 Orientation of the entomogenous nematode *Neoaplectana carpocapsae* to carbon dioxide. *Env. Entomol.* 9(5): 649-652.

Gaugler, R., and Molloy, D. 1981. Field evaluation of the entomogenous nematode *Neoaplectana carpocapsae* as a biological control agent of black flies (Diptera: Simuliidae). *Mosquito News* 41(3): 459-469.

Geden, C. J., Arends, J. J., and Axtell, R. C. 1987. Field trials of *Steinernema feltiae* (Nematoda: Steinernematidae) for control of *Alphitobius diaperinus* (Coleoptera: Tenebrionidae) in commercial broiler and turkey houses. *J. Econ. Entomol.* 80(1): 136-141.

Georgis, H., and Hague, N. G. M. 1982. Interactions between *Neoaplectana carpocapsae* (Nematoda) and *Olesicampa monticola*, a parasitoid of the larch sawfly *Cephalcia lariciphila*. *IRCS Medical Science: Environmental Biology and Medicine Experimental Animals; Microbiology, Parasitology and Infectious Diseases* 10(8): 617.

Georgis, R., Mullens, B. A., and Meyer, J. A. 1987. Survival and movement of insect parasitic nematodes in poultry manure and their infectivity against *Musca domestica*. *J. Nematol.* 19(3): 292-295.

Georgis, R., and Poinar, G. O. 1983a. Effect of soil texture on the distribution and infectivity of *Neoaplectana carpocapsae* (Nematoda: Steinernematidae). *J. Nematol.* 15(2): 308-311.

Georgis, R., and Poinar, G. O. 1983b. Effect of soil texture on the distribution and infectivity of *Neoaplectana glaseri*(Nematoda: Steinernematidae). *J. Nematol.* 15(3): 329-332.

Georgis, R., and Poinar, G. O. 1983c. Vertical migration of *Heterorhabditis bacteriophora* and *Heterorhabditis heliothidis* (Nematoda: Heterorhabditidae) in sandy loam soil. *J. Nematol.* 15(4): 652-654.

Georgis, R., and Poinar, G. O. 1984a. Field control of the strawberry root weevil *Nemocestes incomptus* by neoaplectana nematodes (Steinernematidae: Nematoda). *J. Invert. Pathol.* 43(1): 130-131.

Georgis, R., and Poinar, G. O. 1984b. Greenhouse control of the black vine weevil, *Otiorrhynchus sulcatus* (Coleoptera: Curculionidae) by heterorhabditid and steinernematid nematodes. *Env. Entomol.* 13: 1138-1140.

Georgis, R., Poinar, G. O., and Wilson, A. P. 1982. Susceptibility of damp wood termites and soil and wood–dwelling termites to the entomogenous nematode *Neoaplectana carpocapsae*. *IRCS Medical Science: Environmental Biology and Medicine; Experimental Animals; Microbiology, Parasitology and Infectious Diseases* 10(7): 563.

Glaser, R. W. 1931. The cultivation of a nematode parasite of an insect. *Science* 73(1901): 614-615.

Glaser, R. W. 1932a. Studies on *Neoaplectana glaseri*, a nematode parasite of the Japanese beetle (*Popillia japonica*). New Jersey Department of Agriculture Circular No. 211.

Glaser, R. W. 1932b. A pathogenic nematode of the Japanese beetle. *J. Parasitol.* 18(2): 119.

Glaser, R. W. 1940a. Continued culture of a nematode parasitic in the Japanese beetle. *J. Exp. Zool.* 84(1): 1-12

Glaser, R. W. 1940b. The bacteria–free culture of a nematode parasite. *Proc. Soc. Exp. Biol. Med.* 43(3): 512-514.

Glaser, R. W., and Farrell, C. C. 1935. Field experiments with the Japanese beetle and its nematode parasite. *J. NY Entomol. Soc.* 43: 345-371.

Glaser, R. W., McCoy, E. E., and Girth, H. B. 1940. The biology and economic importance of a nematode parasitic in insects. *J. Parasitol.* 26(6): 479-495.

Glaser, R. W., McCoy, E. E., and Girth, H. B. 1942. The biology and culture of *Neoaplectana chresima*, a new nematode parasitic in insects. *J. Parasitol.* 28(2): 123-126.

Goetz, P., Boman, A., and Boman, H. G. 1981. Interactions between insect immunity and an insect– pathogenic nematode with symbiotic bacteria. *Proc. Roy. Soc. London B* 212: 333-350.

Goetz, P., and Vey, A. 1974. Humoral encapsulation in Diptera (Insecta): Defense reactions of *Chironomus* larvae against fungi. *Parasitology* 68: 193-205.

Grimont, P. A. D., Steigerwalt, A. G., Boemare, N., Hickman–Brenner, F. W., Deval, C., Grimont, F., and Brenner, D. 1984. Deoxyribonucleic acid relatedness and phenotypic study of the genus *Xenorhabdus*. *Int. J. Syst. Bacteriol.* 34(4): 378-388.

Hackett, K. J., and Poinar, G. O. 1973. The ability of *Neoaplectana carpocapsae* Weiser (Steiner-nematidae: Rhabditoidea) to infect honey bees (*Apis mellifera*, Apidae: Hymenoptera). *Am. Bee J.* 113(3): 100.

Hall, D. N., Andreadis, T. G., Flanagan, T. R., and Kaczor, E. J. 1975. Melanotic encapsulation of the nematode *Neoaplectana carpocapsae* by *Aedes aegypti* larvae concurrently parasitized by the nematode *Reesimermis nielseni. J. Invert. Pathol.* 26(2): 269-270.

Han, R., Li, L., and Wouts, W. M. 1990. The disparities between morphological and functional dimorphism in *Xenorhabdus luminescens. Natural Enemies of Insects* 11(1):25–32.

Han, R., and Wouts, W.M. 1991. Development of *Heterorhabditis* spp. strains as characters of possible *Xenorhabdus luminescens* isolates. *Rev. Nematol.* (in press).

Hansen, E. L., and Berntzen, A. K. 1969. Development of *Caenorhabditis briggsae* and *Hymenolepsis nana* in interchanged media. *J. Parasitol.* 55: 1012-1017.

Hara, A. H., and Kaya, H. K. 1983. Suitability of *Spodoptera exigua* pupae from different pupation sites to the nematode *Neoaplectana carpocapsae. J. Invert. Pathol.* 42(3): 418-420.

Hotchkin, P. G., and Kaya, H. K. 1984. Electrophoresis of soluble proteins from two species of *Xenorhabdus* bacteria mutualistically associated with the nematodes *Steinernema* spp. and *Heterorhabditis* spp. *J. Gen. Microbiol.* 130(10): 2725-2732.

Howell, J. F. 1979. New storage methods and improved trapping techniques for the parasitic nematode *Neoaplectana carpocapsae. J. Invert. Pathol.* 33: 155-158.

Hoy, J. M. 1954. The biology and host range of *Neoaplectana leucaniae*, a new species of insect parasitic nematode. *Parasitology* 44(4): 392-399.

House, H. L., Welch, H. E, and Cleugh, T. R. 1965. A food medium of prepared dog biscuit for mass production of the nematode DD–136 (Nematoda: Steinernematidae). (Correspondence) *Nature*, 206(4986): 847.

Ireson, J. E., and Miller, L. A. 1983. Susceptibility of the collembolan *Smithurus viridis* to the insect parasitic nematodes *Neoaplectana bibionis* and *Heterorhabditis heliothidis. Entomologia Exp. Appl.* 34(3): 342-343.

Israel, P., Rao, Y. R. V. J., Rao, P. S. P., and Varma, A. 1969. Control of paddy cut worm by DD–136, a parasitic nematode.(Correspondence) *Curr. Sci.* 38(16): 390-391.

Jackson, G. J. 1961. The parasitic nematode, *Neoaplectana glaseri*, in axenic culture. I. Effects of antibodies and anthelminthics. *Exp. Parasitol.* 11(2/3):241-247.

Jackson, G. J. 1962a. On axenic cultures of certain protozoan and worm parasites of insects. *Trans. NY Acad. Sci.* 24(8): 954-965.

Jackson, G. J. 1962b. The parasitic nematode, *Neoaplectana glaseri*, in axenic culture. II. Initial results with defined media. *Exp. Parasitol.* 12(1): 25-32.

Jackson, G. J. 1965. Differentiation of three species of *Neoaplectana* (Nematoda: Rhabditida), grown axenically. *Parasitology* 55(3): 571-578.

Jackson, G. J. 1969. Nutritional control of nematode development. In *Germ–free Biology: Experimental and Clinical Aspects. Mirand, E. A., and Back, N. (eds.).* Advances in Experimental Medicine and Biology, Vol. 3, Plenum Press, New York, 333–341.

Jackson, G. J. 1973. *Neoaplectana glaseri*: Essential amino acids. *Exp. Parasitol.* 34(1): 111-114.

Jackson, G. J., and Bradbury, P. C. 1970. Cuticular fine structure and molting of *Neoaplectana glaseri* (Nematoda), after prolonged contact with rat peritoneal exudate. *J. Parasitol.* 56(1): 108-115.

Jackson, G. J., and Moore, G. E. 1969. Infectivity of nematodes, *Neoaplectana* species for the larvae of the weevil *Hylobius pales*, after rearing in species isolation. *J. Invert. Pathol.* 14(2): 194-198.

Jaques, R. P. 1967. Mortality of five apple insects induced by the nematode DD–136. *J. Econ. Entomol.* 60(3): 741-743.

Jaques, R. P, Stultz, H. T., and Huston, F. 1968. The mortality of the pale apple leafroller and winter moth by fungi and nematodes applied to soil. *Can. Entomologist* 100(8): 813-918.

Kain, W. M., Bedding, R. A., and van der Mespel, C. J. 1982. Preliminary evaluation of parasitic nematodes for grass grub, *Costelytra zealandica*, control in central Hawkes Bay of New Zealand. *N. Zealand J. Exp. Agri.* 10(4): 447-450.

Kamionek, M., Maslana, I., and Sandner, H. 1974. The survival of invasive larvae of *Neoaplectana carpocapsae* Weiser in a waterless environment under various conditions of temperature and humidity. *Zeszyty Probleowe Postepow Nauk Rolniczych* 154: 409-412.

Kamionek, M., and Sandner, H. 1972. Resistance of the invasive larvae of *Neoaplectana carpocapsae* to drying. In *Abstr. International Symposium of Nematology* (11th), European Society of Nematologists, Reading, UK, 3-8 September, 1972, p. 36.

Kashio, T. 1986. Application of bark compost containing entomogenous nematodes, *Steinernema feltiae* DD–136, for the control of white spotted longicorn beetle, *Anoplophora malasiaca*. *Proc. Assoc. Plant Protection of Kyushu* 32: 175-178.

Kaya, H. K. 1978. Interaction between *Neoaplectana carpocapsae* (Nematoda: Steinernematidae) and *Apanteles militaris* (Hymenoptera: Bracondiae), a parasitoid of the armyworm, *Pseudaletia unipunta*. *J. Invert. Pathol.* 31: 358-364.

Kaya, H. K. 1984. Effect of the entomogenous nematode *Neoaplectana carpocapsae* on the tachinid parasite *Compsilura cincinnata* (Diptera: Tachinidae). *J. Nematol.* 16(1): 9-13.

Kaya, H. K., and Grieve, B. J. 1982. The nematode *Neoaplectana carpocapsae* and the beet armyworm *Spodoptera exigua* infectivity of prepupae and pupae in soil and of adults during emergence from soil. *J. Invert. Pathol.* 39(2): 192-197.

Kaya, H. K., and Hara, A. H. 1980. Differential susceptibility of lepidopterous pupae to infection by the nematode *Neoaplectana carpocapsae J. Invert. Pathol.* 36(3): 389-393.

Kaya, H. K., and Hara, A. H. 1981. Susceptibility of various species of lepidopterous pupae to the entomogenous nematode *Neoaplectana carpocapsae*. *J. Nematol.* 13(3): 291-294.

Kaya, H. K., Hara, A. H., and Reardon, R. C. 1981. Laboratory and field evaluation of *Neoaplectana carpocapsae* (Rhabditida: Steinernematidae) against the elm leaf beetle *Pyrrhalta lutolea* (Coleoptera: Chrysomelidae) and the western spruce budworm *Choristoneura occidentalis* (Lepidoptera: Tortricidae). *Can. Entomol.* 113(9): 787-794.

Kaya, H. K., and Hotchkin, P. G. 1981. The nematode *Neoaplectana carpocapsae* Weiser and its effect on selected ichneumonid and braconid parasites. *Env. Entomol.* 10: 474-478.

Kaya, H. K., Joos, J. L., Falcon, L. A., and Berlowitz, A. 1984. Suppression of the codling moth (Lepidoptera: Olethreutidae) with the entomogenous nematode, *Steinernema feltiae* (Rhabditida: Steinernematidae). *J. Econ. Entomol.* 77(5): 1240-1244.

Kaya, H. K., Mannion, C. M., Burlando, T. M., and Nelson, C. E. 1987. Escape of *Steinernema feltiae* from alginate capsules containing tomato seeds. *J. Nematol.* 19(3): 287-291.

Kaya, H. K., Marston, M., Lindegren, J. E., and Pengh, Y. S. 1982. Low susceptibility of the honey bee *Apis mellifera* (Hymenoptera: Apidae) to the entomogenous nematode *Neoaplectana carpocapsae*. *Env. Entomol.* 11(4): 920-924.

Kaya, H. K., Nelson, C. E. 1985. Encapsulation of steinernematid and heterorhabditid nematodes with calcium alginate: A new approach for insect control and other applications. *Env. Entomol.* 14(5): 572-574.

Kermarrec, A., and Mauleon, H. 1985. Potential noxiousness of the entomogenous nematode *Neoaplectana carpocapsae* to the Antillian toad *Bufo marinus*. *Mededelingen van de faculteit Landbouwwetenshappen Rijksuniversiteit Gent* 50(3A): 831-838.

Kahn, A., and Brooks, W. M. 1977. A chromogenic bioluminescent bacterium associated with the entomophilic nematode *Chromonema heliothidis*. *J. Invert. Pathol.* 29: 253-261.

Khan, A., Brooks, W. M., and Hirschmann, H. 1976. *Chromonema heliothidis* n.gen. n.sp. (Steinernematidae: Nematoda), a parasite of *Heliothis zea* (Noctuidae, Lepidoptera), and other insects. *J. Nematol.* 8: 159-168.

Klingler, J. 1986. Einzats und Wirksamkeit insektenparasitischer Nematoden gegen den Gefurchten Dickmaulruessler. *Gaertnermeister* 89(13): 277-279.

Kozodoi, E. M. 1984. A new entomopathogenic nematode, *Neoaplectana anomali* sp.n. (Rhabditida, Steinernematidae) and observations on its biology. *Zool. Zh.* 63: 1605-1609.

Landazabal, A. J., Fernandez, A. F., and Figueroa, P. A. 1973. Biological control of *Spodoptera frugiperda* (J.E. Smith), on maize by the nematode *Neoaplectana carpocapsae*. *Acta Agronomica (Columbia)* 23(3/4): 41-70.

Laumond, C. 1972. Practical use of *Neoaplectana* against various Lepidoptera from vegetable crops. In *Abstr. International Symposium of Nematology* (11th), European Society of Nematologists, Reading, UK, 3-8 September, 1972, p. 41.

Li, P. S., Deng, C. S., Shang, S. G., and Yang, H. W. 1986. Laboratory studies on the infectivity of the nematode *Steinernema glaseri* to *Oncomelania hupensis*, a snail intermediate host of blood fluke, *Schistosoma japonicum*. *Chin. J. Biol. Control* 2(2): 50-53.

Lindegren, J. E., Agudelo–Silva, F., Valero, K. A., and Curtis, C. E. 1987. Comparative small–scale field application of *Steinernema feltiae* for naval orangeworm control. *J. Nematol.* 19(4): 503-504.

Lindegren, J. E., and Vail, P. V. 1986. Susceptibility of Mediterranean fruit fly, *Ceratitis capitata*, melon fly *Dacus cucurbitae*, and oriental fruit fly, *Dacus dorsalis* (Diptera: Tephritidae) to the entomogenous nematode *Steinernema feltiae* in laboratory tests. *Env. Entomol.* 15(3): 465-468.

Lindegren, J. E., Yamashita, T. T., and Barnett, W. W. 1981. Parasitic nematode may control carpenter worm in fig trees. *Cal. Agri.* Jan/Febr.

Lossbroek, T. G., and Theunissen, J. 1985. The entomogenous nematode *Neoaplectana bibionis* as a biological control agent of *Agrotis segetum* in lettuce. *Entomologia Exp. Appl.* 39(3): 261-264.

Lysenko, O., and Weiser, J. 1974. Bacteria associated with the nematode *Neoaplectana carpocapsae* and the pathogenicity of this complex for *Galleria mellonella* larvae. *J. Invert. Pathol.* 24(3): 332-336.

McCoy, E. E., and Glaser, R. W. 1936. Nematode culture for Japanese beetle control. NJ Department of Agriculture Circular 265, pp. 3-9.

McQuilkin, W. T., Evans, V. J., and Earle, W. R. 1957. The adaptation of additional lines of NCTC clone 929 (strain L) cells to chemically defined protein–free medium NCTC 109. *J. Natl. Cancer Institute* 19: 885-907.

McVean, C. M., and Brewer, J. W. 1981. Suitability of *Scolytus multistriatus* and *Dendroctonus ponderosae* as hosts for the entomogenous nematode *Neoaplectana carpocapsae*. *J. Econ. Entomol.* 74(5): 601-607.

McVean, C. M., Brewer, J. W., and Capinera, J. L. 1982. Field tests of antidesiccants to extend the infection period of an entomogenous nematode, *Neoaplectana carpocapsae*, against the Colorado potato beetle, *Lepinotarsa decemlineata*. *J. Econ. Entomol.* 75(1): 97-101.

Matha, V., and Mráček, Z. 1984. Changes in hemocyte counts in *Galleria mellonella* (L) (Lepidoptera: Galleriidae) larvae infected with *Steinernema* sp. (Nematoda: Steinernematidae). *Nematologica* 30: 86-89.

Miller, L. A., and Bedding, R. A. 1982. Field testing of the insect parasitic nematode *Neoaplectana bibionis* (Nematoda: Steinernematidae) against currant borer moth, *Synanthedon tupiliformis* (Lep: Sesiidae) in black currants. *Entomophaga* 27(1): 109-114.

Molyneux, A. S. 1985. Survival of infective juveniles of *Heterorhabditis* spp. and *Steinernema* spp. (Nematoda: Rhabditida) at various temperatures and their subsequent infectivity for insects. *Rev. Nematol.* 8(2): 165-170.

Molyneux, A. S. 1986. *Heterorhabditis* spp. and *Steinernema* spp. temperature and aspects of behavior and infectivity. *Exp. Parasitol.* 62(2): 169-180.

Molyneux, A. M., Bedding, R. A., and Akhurst, R. J. 1983. Susceptibility of larvae of the sheep blow fly, *Lucilia cuprina*, to various *Heterorhabditis* spp. *Neoaplectana* spp. and an undescribed steinernematid (Nematoda). *J. Invert. Pathol.* 42(1): 1-7.

Moore, G. E. 1965. The bionomics of an insect–parasitic nematode. *J. Kansas Entomol. Soc.* 38(2): 101-105.

Moore, G. E. 1970. *Dendroctonus frontalis* infection by the DD–136 strain of *Neoaplectana carpocapsae* and its bacterium complex. *J. Nematol.* 2(4): 341-344.

Morgan, J. F., Morton, H. J., and Parker, R. C. 1950. Nutrition of animal cells in tissue culture. I. Initial studies on a synthetic medium. *Proc. Soc. Exp. Biol. Med.* 73: 1-8.

Moyle, P. L., and Kaya, H. K. 1981. Dispersal and infectivity of the entomogenous nematode *Neoaplectana carpocapsae* (Rhabditida: Steinernematidae) in sand. *J. Nematol.* 13(3): 295-300.

Mráček, Z. 1977. *Steinernema kraussei*, a parasite of the body cavity of the sawfly, *Cephaleia abietis*, in Czechoslovakia. *J. Invert. Pathol.* 30: 87-94.

Mráček, Z., and David, L. 1986. Preliminary field trials of *Cephalcia abietis* (Hymenoptera: Pamphilidae) larvae with steinernematid nematodes in Czechoslovakia. *J. Appl. Entomol.* 102(3): 260-263.

Mráček, Z., and Spitzer, K 1983. Interaction of the predators and parasitoids of the sawfly *Cephalcia abietis* (Pamphilidae: Hymenoptera) with its nematode *Steinernema kraussei*. *J. Invert. Pathol.* 42(3): 397-399.

Mráček, Z., and Weiser, J. 1983. Pathogenicity of *Neoaplectana carpocapsae* (Nematoda) for the flea *Xenopsylla cheopis*. *J. Invert. Pathol.* 42(1): 133-134.

Mullens, B. A., Meyer, J. A., and Cyr, T. L. 1967. Infectivity of insect–parasitic nematodes (Rhabditida: Steinernematidae, Heterorhabditidae) for larvae of some manure–breeding flies (Diptera: Muscidae). *Env. Entomol.* 16(3): 769-773.

Nash, R. F., and Fox, R. C. 1969. Field control of the Nantucket pine tip moth by the nematode DD–136. *J. Econ. Entomol.* 62(3): 660-663; *Archivos do Instituto Biologico Sao Paulo* 8: 215-230.

Poinar, G. O. 1966. The presence of *Achromobacter nematophilus* in the infective stage of a *Neoaplectana* species. (Steinernematidae: Nematoda). *Nematologica* 12(1): 105-108.

Poinar, G. O. 1976. Description and biology of a new insect parasitic rhabditoid *Heterorhabditis bacteriophora* n.gen. n.sp. (Rhabditida; Heterorhabditidae n.fam.). *Nematologica* 21: 463-470.

Poinar, G. O. 1985a. Additional members. Neoaplectana *Newslet.* 6(1): 11.

Poinar, G. O. 1985b. *Neoaplectana intermedia* n.sp. (Steinernematidae: Nematoda) from South Carolina. *Rev. Nematol.* 8(4): 321-327.

Poinar, G. O. 1988. Redescription of *Neoaplectana affinis* Bovien (Rhabditida: Steinernematidae). *Rev. Nematol.* 11(2): 143-147.

Poinar, G. O. 1989. Examination of the neoaplectanid species *feltiae* Filipjev *carpocapsae* Weiser and *bibionis* Bovien (Nematoda: Rhabditida). *Rev. Nematol.* 12(4):375–377.

Poinar, G. O., and Brooks, W. M. 1977. Recovery of the entomogenous nematode, *Neoaplectana glaseri* Steiner, from a native insect in North Carolina. *IRCS Med. Sci.: Environmental Biology and Medicine; Experimental Animals; Microbiology, Parasitology and Infectious Diseases* 5: 473.

Poinar, G. O., and Deschamps, N. 1981. Susceptibility of *Scolytus multistriatus* to neoaplectanid and heterorhabditid nematodes. *Env. Entomol.* 10(1): 85-87.

Poinar, G. O., and Ennik, F. 1972. The use of *Neoaplectana carpocapsae* (Steinernematidae: Rhabditoidea) against adult yellowjackets (Vespula spp., Vespida: Hymenoptera). *J. Invert. Pathol.* 19(3): 331-334.

Poinar, G. O., Evans, J. S., and Schuster, E. 1983. Field test of the entomogenous nematode *Neoaplectana carpocapsae* for control of corn rootworm larvae *Diabrotica* sp. (Coleoptera). *Prot. Ecol.* 5(4): 337-342.

Poinar, G. O., Hess, R., and Thomas, G. 1980. Isolation of defective bacteriophages from *Xenorhabdus* spp. (Enterobacteriaceae). *IRCS Med. Sci.: Cell and Membrane Biology: Environmental Biology and Medicine; Microbiology, Parasitology and Infectious Diseases* 3: 141.

Poinar, G. O., and Hom, A. 1986. Survival and horizontal movement of infective stage *Neoaplectana carpocapsae* in the field. *J. Nematol.* 18(1): 34-36.

Poinar, G. O., Jackson, T., and Klein, M. 1987. *Heterorhabditis megidis* sp.n. (Heterorhabditidae: Rhabditida) parasitic in Japanese beetle, *Popillia japonica* (Scarabaeidae: Coleoptera), in Ohio. *Proc. Helm. Soc. Wash.* 54(1): 53-59.

Poinar, G. O., and Kaul, H. N. 1982. Parasitism of the mosquito *Culex pipiens* by the nematode *Heterorhabditis bacteriophora. J. Invert. Pathol.* 39(3): 382-387.

Poinar, G. O., and Leutenegger, R. 1971. Ultrastructural investigation of the melanization process in *Culex pipiens* (Culicidae) in response to a nematode. *J. Ultrastr. Res.* 36(1/2): 149-158.

Poinar, G. O., and Paff, M. 1985. Laboratory infection of terrestrial isopods (Crustacea: Isopoda) with neoaplectanid and heterorhabditid nematodes (Rhabditida: Nematoda). *J. Invert. Pathol.* 45(1): 24-27.

Poinar, G. O., and Thomas, G. M. 1965. A new bacterium, *Achromobacter nematophilus* sp.nov. (Achromobacteriaceae: Eubacteriales) associated with a nematode. *Int. Bull. Bac. Nomen. Taxon.* 15(4): 249-252.

Poinar, G. O., and Thomas, G. M. 1966. Significance of *Achromobacter nematophilus* Poinar and Thomas (Achromobacteriaceae: Eubacteriales) in the development of the nematode DD–136 (*Neoaplectana* sp.; Steinernematidae). *Parasitology* 56:385-390.

Poinar, G. O., and Thomas, G. M. 1967. The nature of *Achromobacter nematophilus* as an insect pathogen. *J. Invert. Pathol.* 9(4): 510-514.

Poinar, G. O., and Thomas, G. M. 1985. Effect of neoaplectanid and heterorhabditid nematodes (Nematoda: Rhabditoidea) on the millipede (*Oxidus gracilis*). *J. Invert. Pathol.* 45(2): 231-235.

Poinar, G. O., Thomas, G. M., and Hess, R. 1977. Characteristics of the specific bacterium associated with *Heterorhabditis bacteriophora* (Heterorhabditidae: Rhabditina). *Nematology* 23: 97-102.

Poinar, G. O., Thomas, G. M., and Mookerjee, P. 1985. Feasibility of embedding parasitic nematodes in hydrogels for insect control. *IRCS Med. Sci.: Environmental Biology and Medicine; Experimental Animals; Microbiology, Parasitology and Infectious Diseases* 13(8): 754-755.

Poinar, G. O., Thomas, G. M., Presser, S. B., and Hardy, J. L. 1982. Inoculation of entomogenous nematodes *Neoaplectana* and *Heterorhabditis* and their associated bacteria *Xenorhabdus* spp. into chicks and mice. *Env. Entomol.* 11(1): 137-138.

Poinar, G. O., Thomas, G. M., Veremtschuk, G. V., and Pinnock, D. E. 1971. Further characterization of *Achromobacter nematophilus* from American and Soviet populations of the nematode *Neoaplectana carpocapsae* Weiser. *Int. J. Sys. Bateriol.* 21(1): 78-82.

Popiel, I., Itamer, G., and Lindegren, J. E. 1987. Storage and shipment of osmotically desiccated entomogenous nematodes. U.S. Patent No. 031,883.

Pye, A. E., and Burman, M. 1978. *Neoaplectana carpocapsae*: Infection and reproduction in large pine weevil larvae, *Holybius abietis. Exp. parasitol.* 46(1): 1-11.

Pye, A. E., and Burman, M. 1981. *Neoaplectana carpocapsae*: Nematode accumulation on chemical and bacterial gradients. *Exp. Parasitol.* 51(1): 13-20.

Pye, A. E., and Pye, A. 1985. Different applications of the insect parasitic nematode *Neoaplectana carpocapsae* to control the large pine weevil, *Hylobius abietis. Nematologica.* 31(1): 109-116.

Renn, N., Barson, G., and Richardson, P. N. 1985. Preliminary laboratory tests with two species of entomophillic nematodes for control of *Musca domestica* in intensive animal units. *Ann. Appl. Biol.* 106(2): 229-234.

Renn, N., Barson, G., and Richardson, P. N. 1986. Potential housefly control in intensive animal units using entomophilic nematodes. In *British Crop Protection Conference: Pests and Diseases.* Vol. 2. Proceedings of a conference held at Brighton Metropole, England, Nov. 17–20, 1986. Thornton Heath, U.K., British Crop Protection Council, pp. 615-622.

Reed, E. M., and Carne, P. B. 1967. The suitability of a nematode (DD–136) for the control of some pasture insects. *J. Invert. Pathol.* 9(2): 196-204.

Reed, D. K., Reed, G. L., and Creighton, C. S. 1986. Introduction of entomogenous nematodes into trickle irrigation systems to control striped cucumber beetle *Acalymna vittatum* (Coleoptera: Chrysomelidae). *J. Econ Entomol.* 79(5): 1330-1333.

Richardson, P. N., and Hughes, J. S. 1986. Use of the nematode *Heterorhabditis heliothidis* to control mushroom cecidomyiid flies (Diptera: Cecidomyiidae). *Rev. Nematol.* 9(3): 305.

Ritter, M, and Theodorides, J. 1966. Utilisation des nématodes dans la lutte biologique contre les insectes. *Comptes Rendus de 90e Congrés National des Sociétés Savantés.* Nice, 1965, II: 525-531.

Rutherford, T. A., Trotter, D., and Webster, J. M. 1987. The potential of heterorhabditid nematodes as control agents of root weevils. *Can. Entomol.* 119: 67-73.

Rutherford, T. A., and Webster, J. M. 1986. Commercial use of entomophilic nematodes in Canada. *J. Nematol.* 18(4): 630.

Samsook, V., and Sikora, R. A. 1981. Influence of parasite density, host developmental stage and leaf wetness duration on *Neoaplectana carpocapsae* parasitism of *Heliothis virescens*. *Mededelingen van de Faculteit Landbouwwetenschappen Rijksuniversiteit Gent* 46(2): 685-694.

Sandner, H. 1984. On the possibility of utilization of entomophilic nematodes in forest protection, Third symposium on the protection of forest ecosystems. *Ann. Warsaw Agr. Univ. SGGW–AR* 18: 7-10.

Sandner, H. 1986a. The movement of invasive larvae of *Steinernema feltiae* (Filipjev) in soil. *Ann. Warsaw Agr. Univ. SGGW–AR* 20: 41-44.

Sandner, H. 1986b. Movement of invasive larvae of *Heterorhabditis bacteriophora* Poinar in soil. *Ann. Warsaw Agr. Univ. SGGW–AR* 20: 45-47.

Sandner, H., Kamionek, M., and Seryczynska, A. 1973. Ways in which the bacteria *Achromobacter nematophilus* Poinar and Thomas penetrate into the body cavity of test insects. *Bull. l'Academie Polonaise des Sciences Biologique II* 11(3): 245-247.

Sandner, H., and Pezowicz, E. 1979. New possibilities and perspectives in the use of parasitic nematodes in plant protection. *Mat. XIX sesji nauk. IOR, Poznan*: 345-352.

Sandner, H., and Seryczynska, H. 1971. Changes in the ultrastructure of the fatbody of *Galleria mellonella* L. under the influence of invasion by *Neoaplectana carpocapsae* Weiser (Nematoda: Steinernematidae). *Bull. l'Academie Polonaise des Sciences Biologique II* 19(1): 67-69.

Sandner, H., Seryczynska, H., and Kamionek, M. 1977. *Relationship of Nematodes of Genera* Neoaplectana *and* Pristiochus *to Their Hosts*. Warsaw Agriculture University Institute of Environmental Protection, FG–263.

Sandner, H., and Stanuszek, S. 1971. Comparative research of the effectiveness and production of *Neoaplectana carpocapsae s.l. Zeszyty Probleowe Postepow Nauk Rolniczych* 121: 209-226.

Saunders, M. C., and All, J. N. 1982. Laboratory extraction methods and field detection of entomogenous rhabditoid nematodes from soil. *Env. Entomol.* 11(6): 1164-1165

Saunders, M. C., and All, J. N. 1985. Association of entomophilic rhabditoid nematode populations with natural control of first instar larvae of the grape root borer, *Vitacea polistiformis*, in concord grape vineyards. *J. Invert. Pathol.* 45(2): 147-151.

Scherer, W. 1987. Bekaempfung des Gefurchten Dickmaulruesslers in Freiland–Erdbeeren. *Obstbau* 12(5): 225-227.

Schmidt, J., and All, J. N. 1978. Chemical attraction of *Neoaplectana carpocapsae* (Nematoda: Steinernematidae) to insect larvae. *Env. Entomol.* 7(4): 605-607.

Schmidt, J., and All, J. N. 1979. Attraction of *Neoaplectana carpocapsae* (Nematoda: Steinernematidae) to common excretory products of insects. *Env. Entomol.* 8(1): 55-61.

Schmiege, D. C. 1963. The feasibility of using a neoaplectanid nematode for control of forest insect pests. *J. Econ. Entomol.* 56(4): 427-431.

Schroeder, W. J. 1987. Laboratory bioassays and field trials on entomogenous nematodes for control of *Diaprepes abreviates* (Coleoptera: Curculionidae) in citrus. *Env. Entomol.* 16(14): 987-989.

Schroeder, W. J., and Beavers, J. B. 1987. Movement of entomogenous nematodes of the families Heterorhabditidae and Steinernematidae in soil. *J. Nematol.* 19(2): 257-259.

Scognamiglio, A., Giandomenico, N., and Talama, M. 1968. Prova di lotta biologica contra *Leptinotarsa decemlineata* (Say) con l'impiego del nematode DD–136 Dutky e Hough, 1955. *Boll. Lab. Ent. Agr. "Filippo Silvestri" di Portici* 26: 191-204.

Seryczynska, H. 1972. Changes in the ultrastructure of the haemolymph cells of *Galleria mellonella* L. under the influence of the nematode *Neoaplectana carpocapsae* Weiser. *Bull. l'Academie Polonaise des Sciences, Series des Sciences Biologiques II*, 20(1): 45-47.

Seryczynska, H. 1974. Changes in the ultrastructure of haemolymph cells in *Leptimotarsa decemlineata* Say due to the effect of the nematodes *Neoaplectana carpocapsae* Weiser. *Bull. l'Academie Polonaise des Sciences, Series des Sciences, Biologiques* II 22(7/8): 503-505.

Seryczynska, H. 1975. Toxicity of *Achromobacter nematophilus* Poinar et Thomas cell suspension against *Leptinotarsa decemlineata* Say. *Bull. l'Academie Polonaise des Sciences, Series des Sciences Biologiques* II 23(5): 347-350.

Seryczynska, H., and Kamionek, M. 1972. Defense reactions of *Galleria mellonella* L. caterpillars under the influence of the parasitic nematode *Neoaplectana carpocapsae* Weiser. *Bull. l'Academie Polonaise des Sciences, Series des Sciences Biologiques* II 20(10): 739-742.

Seryczynska, H., and Kamionek, M. 1974. Defense reactions of *Leptinotarsa decemlineata* Say in relation to *Neoaplectana carpocapsae* Weiser (Nematoda: Steinernematidae) and *Pristionchus uniformis* Fedorko et Stanuszek (Nematoda: Diplogasteridae). *Bull. l'Academie Polonaise des Sciences, Series des Sciences Biologiques* II 22(2): 95-100.

Seryczynska, H., Kamionek, M., and Sandner, H. 1974. Defense reaction of caterpillars of *Galleria mellonella* L. in relation to bacteria *Achromobacter nematophilus* Poinar and Thomas (Eubacteriales: Achromobacteriaceae) and bacteria–free nematodes *Neoaplectana carpocapsae* Weiser (Nematoda: Steinernematidae). *Bull. l'Academie Polonaise des Sciences, Series des Sciences Biologique* II 22(3): 193-195.

Shapiro, M., McLane, W., and Bell, R. 1985. Laboratory evaluation of selected chemicals as antidesiccants for the protection of the entomogenous nematode *Steinernema feltiae* (Rhabditidae: Steinernematidae) against *Lymantria dispar* (Lepidoptera: Lymantriidae). *J. Econ. Entomol.* 78(6): 1437-1441.

Sherman, I. W., and Jackson, G. J. 1963. Zymograms of the parasitic nematodes, *Neoaplectana glaseri* and *N. carpocapsae*, grown axenically. *J. Parasitol.* 49(3): 392-397.

Silverman, J., Platzer, E. G., and Rust, M. K. 1982. Infection of the cat flea *Ctenocephalides felis* by *Neoaplectana carpocapsae*. *J. Nematol.* 14(3): 394-397.

Simons, W. R. 1978. Preliminary research on entomophagous nematodes in particular on *Neoaplectana* species in the Netherlands. *Mededelingen van de Faculteit Landbouwwetenschappen Rijksuniversiteit Gent* 43: 765-768.

Simons, W. R. 1981. Biological control of *Otiorrhynchus sulcatus* with heterorhabditid nematodes in the glasshouse. *Netherlands J. Plant Pathol.* 87(4): 149-158.

Simons, W. R., and Poinar. G. O. 1973. The ability of *Neoaplectana carpocapsae* (Steinernematidae: Nematodea) to survive extended periods of desiccation. *J. Invert. Pathol.* 22: 228-230.

Sosa, O., and Beavers, J. B. 1985. Entomogenous nematodes as biological control organisms for *Ligyrus subtropicus* (Coleoptera: Scarabaeidae) in sugar cane. *Env. Entomol.* 14(1): 80-82.

Southern, E. M. 1979. Analysis of restriction–fragment patterns from complex deoxyribonucleic acid species. *Biochem. Soc. Symp.* 1979: 44.

Stammer, H. J. 1962. Protozoen und Wuermer als Parasiten in Insekten. *Deutsche Entomologische Zeitschrift*, Neue Folge 9(5): 441-460.

Steiner, G. 1923. *Aplectana kraussei* n.sp., eine in der Blattwespe *Lyda* sp. parasiterende Nematodenform, nebst Bemerkungen ueber das Seitenorgan der parasitischen Nematoden. *Zentralblatt fuer Bakteriologie, Parasitenkunde und Infektionskrankheiten* II 59(1/4): 14-18.

Steiner, G. 1929. *Neoaplectana glaseri* n.gen., n.sp. (Oxyuridae), a new nemic parasite of the Japanese beetle (*Popillia japonica* Newm.). *J. Wash. Acad. Sci.* 19(19): 436-440.

Stoll. N. R. 1951. Axenic *Neoaplectana glaseri* in fluid cultures. *J. Parasitol.* 37(5 Sect. 2): Suppl. p. 18.

Stoll, N. R. 1953a. Infectivity of Japanese beetle grubs retained by *Neoaplectana glaseri* after seven years axenic culture. *J. Parasitol.* 39 (4 Sect. 2): Suppl. p. 33.

Stoll, N. R. 1953b. Axenic cultivation of the parasitic nematode. *Neoaplectana glaseri*, in a fluid medium containing raw liver extract. *J. Parasitol.* 39(4 Sect. 1): 422-444.

Stoll, N. R. 1954. Improved yields in axenic fluid cultures of *Neoaplectana glaseri* (Nematoda). *J. Parasitol.* 40(5 Sect. 2): 14.

Stoll, N. R. 1959. Conditions favoring the axenic culture of *Neoaplectana glaseri*, a nematode parasite of certain insect grubs. *Ann. NY Acad. Sci.* 77(2): 126-136.

Stoll, N. R. 1973. Rudolf William Glaser, and *Neoaplectana*. *Exp. Parasitol.* 33(2): 189-196.

Swain, R. B. 1943. Nematode parasites of the white–fringed beetle. *J. Econ Entomol.* 36(5): 671-673.

Tanada, Y., and Reiner, C. 1960. Microbial control of the artichoke plume moth *Platyptilia carduidactyla* (Riley) (Pterophoridae, Lepidoptera). *J. Insect Pathol.* 2: 230-246.

Tanada, Y., and Reiner, C. 1962. The use of pathogens in the control of the corn earworm, *Heliothis zea* (Boddie). *J. Insect Pathol.* 4(2): 139-154.

Tarakanov, V. I. 1980. Method of permanent axenic cultivation of entomopathogenic nematode *Neoaplectana glaseri*. *Trudy Vsesoyuznogo Instituta Gel'mintologii im, K.I. Skryabina* 25: 106-110.

Thomas, H. A., and Poinar, G. O. 1979. *Xenorhabdus* gen. nov., a genus of entomopathogenic, nematophilic bacteria of the family Enterobacteriaceae. *Int. J. Syst. Bacteriol.* 29(4): 352-360.

Toba, H. H., Lindegren, J. E., Turner, J. E., and Vail, P. V. 1983. Susceptibility of the Colorado beetle *Leptinotarsa decemlineata* and the sugar beet wireworm *Limonius californicus* to *Steinernema feltiae* and *Steinernema glaseri*. *J. Nematol.* 15(4): 597-601.

Trager, W. 1957. Nutrition of a hemoflagellate (*Leishmania tarentolae*) having an interchangeable requirement for choline of pyridoxal. *J. Protozool.* 4: 269-276.

Travassos, L. 1927. Sobre o genera *Oxysomatium*. *Boletim Biologico* (Sao Paulo) 5: 20-21.

Travassos, L. 1931. Uma nova especie do genero *Neoaplectana* Steiner, 1929 (Nematoda). *Boletim Biologico* (Sao Paulo) 19:150–154.

Verbruggen, D., de Grisse, A., and Heungens, A. 1985. Possibilities for integrated control of *Otiorrhynchus sulcatus*. *Mededelingen van de Faculteit Landbouwwetenschappen Rijksuniversiteit Gent* 50(1): 155-162.

Veremchuk, G. V. 1974. Some factors affecting the infection of insects with *Neoaplectana carpocapsae agriotis* (Nematoda: Steinernematidae). *Parazitologiya* 8(5): 402-407.

Wang, J., and Li, L. 1983. Studies on the use of *Steinernema* spp. (Nematoda) to control the sugarcane beetle *Alissonotum impressicolle*. *Natural Enemies of Insects* 5(3):171–176.

Wang, J., and Li, L. 1986. Entomogenic nematode research in China. In *Fundamental and Applied Aspects of Invertebrate Pathology*. Samson, R. A., Vlak, J. M., and Peters, D. (eds). Foundation of the 4th International Colloquium of Invertebrate Pathology, Wageningen, Netherlands.

Webster, J. M., and Bronskill, J. F. 1968. Use of Gelgard M and an evaporation retardant to facilitate control of larch sawfly by a nematode–bacterium complex. *J. Econ. Entomol.* 61(5): 1370-1373.

Weiser, J. 1962a. Protozoonosen und Nematodenbefall bei Insekten. *Colloque International sur la Pathologie des Insectes et la Lutte Microbiologique*, Paris, 1962 (Entomophaga, Mémoire Hors Série, No. 2), 67–75.

Weiser, J. 1962b. Ueber die Benutzung der Nematoden zur biologische Schadlingsbekaempfung. International Congress of Entomology (11th), Vienna, 1960. *Proceedings* II, pp. 880–882.

Weiser, J. 1963. Diseases of insects of medical importance in Europe. *Bull. WHO* 28: 121-127.

Welch, H. E., 1958. Test of a nematode and its associated bacterium for control of Colorado potato beetle *Leptinotarsa decemlineata* (Say). *Rep. Entomol. Soc. Ontario* 88: 53-54.

Welch, H. E. 1960. Potentialities of nematodes in the biological control of insects of medical importance. In Conference on the Biological Control of Insects of Medical Importance, Technical Report, Washington, D.C., 67-75.

Welch, H. E., and Briand, L. J. 1961a. Field experiments on the use of a nematode for the control of vegetable crop insects. *Proc. Entomol. Soc. Ontario* 91: 197-202.

Welch, H. E., and Briand, L. J. 1961b. Tests of the nematode DD–136 and an associated bacterium for the control of the Colorado potato beetle, *Leptinotarsa decemlineata* (Say) *Can. Entomologist* 93(9): 759-763.

Welch, H. E., and Bronskill, J. F. 1962. Parasitism of the mosquito larvae by the nematode DD–136 (Nematoda: Neoaplectanidae). *Can. J. Zoo.* 40(7): 1263-1268.

White, G. F. 1927. A method for obtaining infective nematode larvae from cultures. *Science* 66: 302-303.

Wojcik, W. F., and Georgis, R. 1988. Infection of adult western yellowjackets with desiccated *Steinernema feltiae* (Nematoda). *J. Invert. Pathol.* 52: 183-184.

Wouts, W. M. 1979. The biology and life cycle of a New Zealand population of *Heterorhabditis heliothidis* (Heterorhabditidae). *Nematologica* 25: 191-202.

Wouts, W. M. 1980. The biology, life cycle and redescription of *Neoaplectana bibionis* Bovien, 1937 (Nematoda: Steinernematidae). *J. Nematol.* 12: 62-72.

Wouts, W. M. 1981. Mass production of the entomogenous nematode *Heterorhabditis heliothidis* (Nematoda: Heterorhabditidae) on artificial media. *J. Nematol.* 13: 467-469.

Wouts, W. M., Mráček, Z., Gerdin, S., and Bedding, R. A. 1982. *Neoaplectana* Steiner, 1929 a junior synonym of *Steinernema* Travassos, 1927 (Nematoda: Rhabditida). *Syst. Parasitol.* 4: 147-154.

Wright, R. J., Agudelo, F., and Georgis, R. 1987. Soil applications of steinernematid and heterorhabditid nematodes for control of Colorado potato beetle *Leptinotarsa decemlineata* Say. *J. Nematol.* 19(2): 201-206.

Yamanaka, S., Seta, K., and Yasuda, M. 1986. Evaluation of the use of entomogenous nematode, *Steinernema feltiae* (str Mexican) for the biological control of the fall webworm, *Hyphantria cunea* (Lepidoptera: Arctiidae). *Jap. J. Nematol.* 16: 26-31.

21

Terrestrial and Semiterrestrial Mermithidae

HELMUT KAISER   *Karl-Franzens-Universität Graz, Graz, Austria*

I. INTRODUCTION

Current information on Mermithidae is rather fragmentary; there are a number of scattered individual works that are mainly concerned with the taxonomy and morphology of these nematodes. There are fewer articles that are concerned with the biology, distribution, and host–parasite relationships. Research has been most intensive on the continents of North America and Europe. Our knowledge of the occurrence of Mermithidae in Africa and South America can be said to be very limited, and very little research has been done on the other continents.

There are more recent monographs by Nickle (1972) and Rubtsov (1972b, 1974, 1978). Of Rubtsov's works, the first three are limited to limnic mermithids; the 1978 monograph includes and lists all the genera and species known up to that date, including the terrestrial ones. In all, there are 65 genera and over 500 species. These numbers must be regarded very critically, as shown by Curran and Hominick (1981) very dramatically in a study of the genus *Gastromermis*. The authors checked the description of 95 of the 99 species known at that time. Only 25 of the species were adequately described and the remaining 70 species had to be considered species *inquirenda*.

Reports on infections with mermithids are found for virtually all orders of insects (for a summary, see the host lists by Poinar, 1975). The overwhelming majority of reports are, however, concerned with less closely identified mermithids. This can be explained by the fact that the parasitic stages found in the insects are lacking in distinguishing characteristics and only in a few species is an identification possible. Only the postparasitic, sexually mature animals can be identified positively.

This chapter will tend to disregard the taxonomic problems and is mainly concerned with economically and ecobiologically interesting species whose developmental cycles and host relationships have been better studied.

II. MORPHOLOGICAL AND PHYSIOLOGICAL CHARACTERISTICS

The terrestrial mermithids include species with a size from a few to 405 mm (*Hexamermis lineata* Kaiser, 1977). Most of the species are between 50 and 150 mm long. The bodily proportions are also very different from species to species. The diameter–to–length ratios vary from 1:60 in *Pheromermis vesparum* Kaiser, 1984 to 1:1200 in *Psammomermis filiformis* Kaiser, 1984. The intraspecific variability of the body size of terrestrial mermithids is also extraordinarily large. A population of *H. lineata* studied in Austria showed females to be between 175 and 405 mm and males to be between 35 and 120 mm. These considerable differences in the bodily proportions of a population are due to the regularity of an allometric function shown by the equation:

$$D = K \times L^a$$

where D = diameter and L = length.

The allometric function is species–specific (Couturier, 1960; Kaiser, 1977). The allometric equation can be interpreted as a growth curve. Large animals of a species are always relatively thinner than small animals of the same species. These variable diameter-to-length relationships permit the surface of the body to be kept in a certain relationship to its volume in animals of different sizes (Kaiser, 1977). This is an important prerequisite for the parenteral feeding during the parasitic growth phase, during which food is absorbed through the body surface. Both morphological (Batson, 1979; Kaiser, 1986a; Poinar and Hess, 1977) and physiological (Rutherford and Webster, 1974; Rutherford et al., 1977; Gordon et al., 1974; Gordon and Webster, 1972) findings support the idea of parenteral feeding. The extremely thread–shaped body of the mermithids is perfectly round in cross–section and is covered by a smooth outer cuticle (Fig. 3). The cuticle of most of the terrestrial species is highly developed and reaches a thickness of up to 50 μm. It is made up of different layers (Fig. 1). Most of the terrestrial species characteristically show two overlapping spirally arranged layers of giant fibers in the outer layers of the cuticule (Fig. 2). Some terrestrial genera such as *Aranimermis*, *Orthomermis*, and most of the limnic genera do not have crossed fiber layers. In other genera, such as *Psammomermis* and *Romanomermis*, they are present in postparasitic juveniles but lacking in the adults.

The epidermal nuclei are in longitudinal ridges (cords). The number of cords present is often a criterion for diagnosis of the genus (Fig. 3). There can be six or eight cords in the body middle; in a few cases, two or four cords have been reported. Serial sections of *Psammomermis* have shown that the situation can vary in one species and that this is probably usually the case (Kaiser, 1984). Before the nerve ring, all eight cords have cell nuclei; behind it, only the ventral and two lateral ridges have nuclei while the subdorsal, subventral, and dorsal ridges lack nuclei (Fig. 3). In other species such as *Hexamermis* and *Amphimermis*, the subdorsal ridges join the dorsal ridge and so these genera have only six cords in the middle of the body (Fig. 3).

A. Digestive System

In free-living nematodes, the digestive system is divided into oral cavity, pharynx, intestine, rectum, and anus; in mermithids it has undergone a functional and morphological change. The mouth opening is located terminally or relocated more or less pronouncedly ventrally. Sexually mature animals do not have an oral cavity, but animals in the infectious stage do

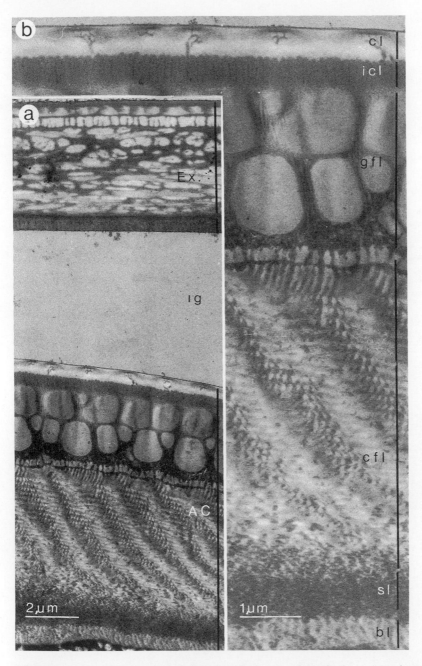

FIGURE 1   Electron micrograph of a transverse section throught the cuticle of a molting stage of *Hexamermis* sp. (a) Total view showing the cuticle of the adult stage (AC) and the already detached larval cuticle, the exuvie (Ex). Intercuticular gap (ig). (b) Adult cuticle of (a) showing six different layers: cortical layer (cl), internal striated cortical layer (icl), giant fiber layer (gfl), concentric fiber layer (cfl), spongy fibrilous layer (sl), basal layer (bl).

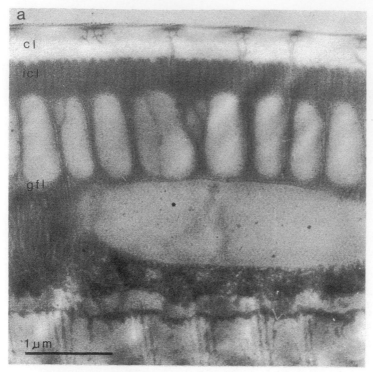

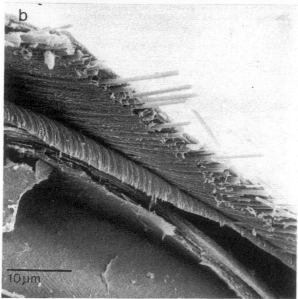

FIGURE 2   Electron microgragh of a cross-section (a) and scanning electron micrograph (b) of the outer layer of the cuticle. (a) *Hexameris* sp. Note the honeycomblike basic structure of the spiral fibers. (b) *Amphimermis elegans*. The fracture of the cuticle clearly shows the multilayered cuticle with protruding spiral fibers.

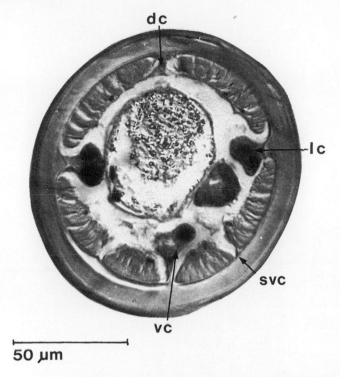

FIGURE 3   Light micrograph of a transverse section through the body of *Amphimermis elegans*. Reserve substances are consumed and crystals as metabolic end products appear in the trophosome. dc, dorsal cord; lc lateral cord; svc, subventral cord; vc, ventral cord.

and it is equipped with a dorsal stylet. In some genera the pharyngeal tube begins directly on the outer cuticular layers (*Agamermis, Hexamermis, Melolonthinimermis, Thaumamermis,* etc.) or else only in the area of the hypodermic tissue of anterior end, often with a thickened ring (*Amphimermis, Mermis, Pheromermis, Oesophagomermis, Psammomermis, Arani-mermis,* etc.) The pharynx itself is modified to a fine capillary lined with cuticule and has a diameter of only 2.5–10 μm. The length of this tube shows only slight intraspecific variability and attains 15% to nearly 100% of the body length. The proximal end is a dead end and is attached to the intestine by fine connective tissue. The covering tissue is thin and usually lacks muscles. There is a remarkable and extensive system of pharyngeal glands that empty serially via fine pores into the capillary (Fig. 4). This "stichosome" is usually made up of 8 or 16 cells arranged in a line or in rows of two. The number of stichocytes is not known for all the terrestrial genera but usually there are 16 (in *Agamermis, Amphimermis,* and *Hexamermis*). *Mermis* and *Pheromermis* have eight stichocytes and *P. villosa* only four. In limnic mermithids, even larger numbers are known (24 and more) as well as varying numbers. After sexual maturity, the stichocytes degenerate continuously.

Intestine: Like the pharynx, the intestine is also very modified. A lumen is lacking in all species as it has lost its functions as a digestive organ. Instead, it has turned into a storage organ, the trophosome, in which intracellular nutrients are saved up for the free-living, nonfeeding phase, e.g., for the formation of sexual products, it is successively emptied and finally only contains metabolic end products in crystal form (Fig. 3). The trophosome is the largest organ within the pseudocoelum, beginning just behind the nerve ring (or, more

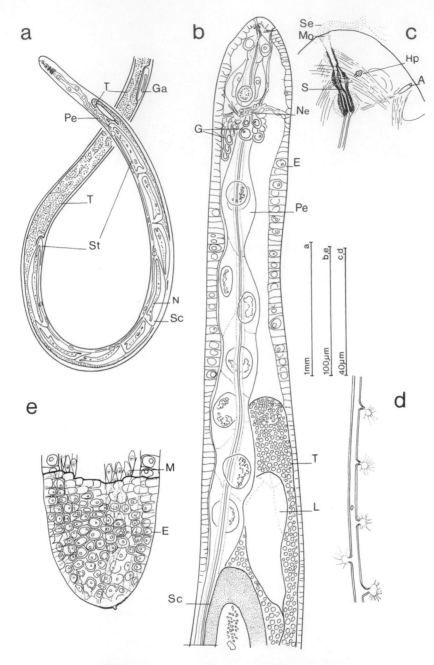

FIGURE 4 Six millimeter long growing stage of *Pheromeris vesparum*. (a) Anterior body with
stichosome and genital anlage. (b) Magnified cut of (a); front end until beginning of the
stichosome. (c) High magnification of mouth part with stylet and its strongly cuticularizing border.
(d) Pharyngeal sheet with fine pores opening into the stichocytes. (e) Tail end. A, amphid; E, epidermis; G, ganglion; Hp, head papillae; Mo, mouth opening; Ne, nerve ring; Pe, pharyngeal epithelium; S, stylet; Sc, stichocyte, Se, secretion; St, stichosome. (After Kaiser, 1987.)

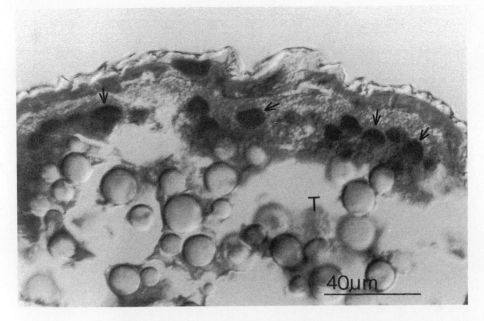

FIGURE 5 Cross section through a growing stage of *Pheromeris vesparum* showing the syncytial trophosome (T), with peripheral nuclei (arrows). (After Kaiser, 1987.)

rarely, just before it) and taking up the entire pseudocoelum except for a space between it and the body wall, leaving only room for the gonads and ending only at the tip of the tail. The cell borders often disappear and then a syncytial sausage is formed with the cell nuclei located peripherally (Fig. 5). In some genera it also remains cellular (*Agamermis, Hexamermis*). The reserved substances in the trophosome give most of the mermithids a white appearance. Only after the reserves have been used up do they appear to be glassy and transparent. Rectum and anus are missing.

B. Ventral Glands

In most mermithids, the ventral glands are also reduced. A ventral gland system made up of three cells was also found in some species of the *Hexamermis brevis* species complex and in an as yet undescribed *Thaumamermis* species (unpublished data). Paired giant glandular cells empty via an unpaired orifice cell into the ventral pore. There is sexual dimorphism, with considerably larger glands in the female than in the corresponding male. Only a ventral pore was found in a number of species.

C. Genital System

Male: The spicula are paired or unpaired; there is no gubernaculum. The form and size of the spicula are species- and genus-specific. There is a great variety of forms, ranging from short, parallel rods to long, corkscrew rods that can reach a length of several millimeters. Other species have spicula of varying length (*Thaumamermis*) or some fused distal ends (*Psammomermis*) or a common tip (*Amphimermis*). Motor muscles include retractors, and protrusors that surround the spicula as a sheath, as well as dorsal fixators attached to the sheath. The position of the hind end of the male (usually rolled up ventrally) is achieved by

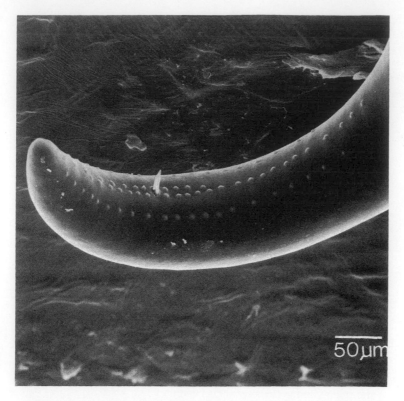

FIGURE 6   Scanning electron micrograph of the tail end of *Amphimermis elegans,* with clearly visible genital papillae. The fused tip of the spicula projects from the cloaca.

extensive bursal muscles running from ventromedial to laterodorsal. In the basic pattern, genital papillae are present in three rows (Fig. 6). The rows can also be double or can virtually turn into fields. The number of papillae is highly variable within certain limits. The vas deferens runs from the cloaca to the middle of the body, where it divides into an anterior and posterior branch. It is surrounded by muscles throughout its entire length. Both branches connect directly with the simple, sac–like testes. The position of the testes can vary; they are located ventrolaterally, left- or right-sided or right- or left-sided.

Female: The vulva is ventral, about in the middle of the body; in larger females of a given species it is more anterior than in smaller females. It is a simple slit or is covered by a cuticular flap directed posteriorly (Fig. 7). The vagina has a cuticularized intima and is surrounded by powerful ring muscles. It varies in form and can be used as a diagnostic characteristic. Vaginas have been described as being barrel-, horn-, S-, and cone-shaped. The paired gonads are divided into uterus, oviduct (also functions as a spermathec), and ovary. Uterus and oviduct have a distinct muscularis. The ovaries are straight or twisted, or sometimes meandering.

D.   Sensory Organs and Nervous System

Sensory organs are concentrated on the anterior and posterior ends of the animal. On the anterior end, we differentiate among head papillae (2–6), mouth papillae (0–2), and am-

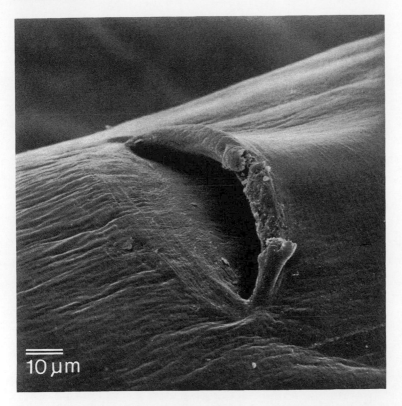

FIGURE 7   Scanning electron micrograph of the vulva of *Amphimermis elegans.*

phids (2). The head papillae are lateral (Fig. 8), subventral, and subdorsal. They are made up of groups of stumps, three for the subventral and subdorsal and two for the lateral. Published data, however, are very divergent and the subject requires further study. The amphids show a cup-shaped cuticularized fovea into which a larger number of sensory stumps (modified cilia) extend (Fig. 9). The aperture can be wide or pore-shaped, and wide openings are often broad ovals (Fig. 10).

Free nerve endings are concentrated between the anterior end and the nerve ring (e.g., canal fibers in *Hexamermis, Agamermis*) and occur laterally over the lateral hypodermal cords on the entire body.

Phasmid–like sensory organs have been found in *Hexamermis* (Kaiser, 1977) and *Amphimermis* (unpublished data). These are paired lateral sensory stumps on the tail end that communicate with the surface via a fine pore, similar to the amphids (Fig. 11).

E.   Sexual Products

Sperm (Figs.12 and 13): Mermithids have characteristic long sperm that show three parts: a long, plasmatic thread; a rod–like, highly refractive nucleus; and a vesicular zone from which pseudopod–like filaments can emerge. The vesicles are lined up in double rows and under electron microscope show the fine structure of tubular mitochondria (Fig. 13). Sperm size and form are definitely species-specific. They vary in length from 10 μm (*Hexamermis brevis* group) to 70 μm (*Amphimermis elegans*).

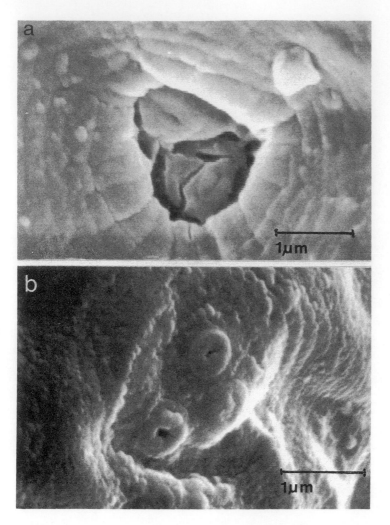

FIGURE 8   Scanning electron micrographs of head papillae of *Amphimermis elegans.* (a) Two lateral head papillae of a female. (b) Three subventral papillae of a male.

Ova: Nickle (1972) provides a survey of sizes and forms. Sizes vary between 35 and 250 μm in diameter; species with an indirect or passive mode of infection such as *Mermis* and *Pheromermis* have small but very numerous eggs, while species with an active mode of infection, i.e., in which the infectious stages themselves seek their hosts, are more likely to have medium to large eggs. Special appendages such as polar byssi or villi and pigmented cases are found in *Mermis,* and sticky hulls in *Pheromermis* (Fig.15).

F.   The Infectious Stages

The free-living infectious stages are most reminiscent in their morphology of free-living nematodes, although they already show parasitic specializations (Fig. 14). Pharynx and intestine are still located one behind the other and a tissue strip recapitulates the position of the anus. The presence of a dorsal stylet is characteristic. In most mermithids it is straight with a

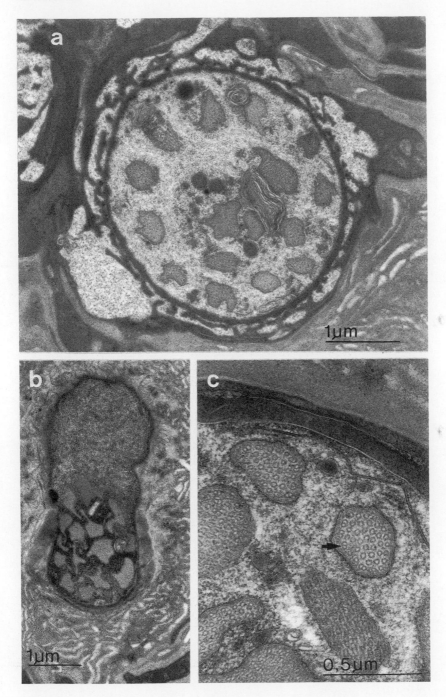

FIGURE 9 Electron micrographs of transverse sections through the amphid of *Pheromermis villosa.* (a) Section through the canalis amphidialis showing six modified cilia. (b) Between canalis and fovea amphidialis. (c) Magnification of (a) showing ultrastructure of cilia with single microtubuli inside and doubled microtubles peripheral. (After Kaiser, 1986a.)

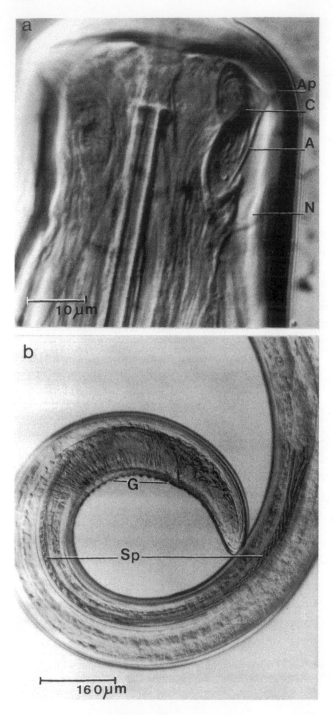

FIGURE 10   Male *Amphimermis elegans.* (a) Head end with porus-like apertura amphidialis. The pharynx ends with a ringlike ridge. (b) Tail end with twisted spicules visible. A, amphid; Ap, apertura amphidialis; C, cilia; G, genital papillae; N, nerve ending in the cuticle; Sp, spicula.

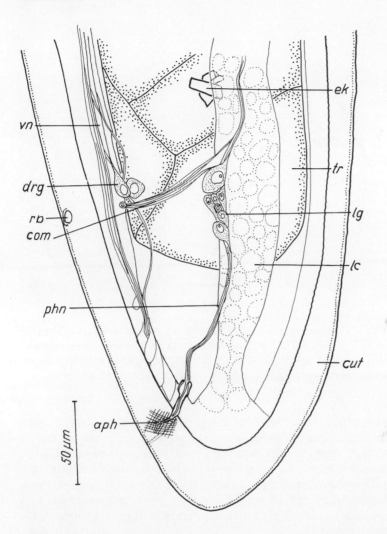

FIGURE 11 Tail end of a female *Hexamermis* sp. showing nervous system and a phasmidlike sense organ. aph, apertura phasmidialis; com, commissure; cut, cuticle; drg, dorsorectal ganglion; ek, excretory crystal; lc, lateral cord; lg, lateral ganglion; phn, phasmidian nerve; ra, rudimentary anus; tr, trophosome. (After Kaiser, 1972.)

sharp tip which can be extended from the oral cavity (Fig. 15c). Only in *Mermis* and *Pheromermis* does the shaft of the stylet have a sinusoid curve and a dorylaimoid guiding ring (Figs. 15a, b; 16). The curved stylet can, when pushed forward, be turned around its long axis (drill principle; Kaiser, 1987), while the straight stylet works as a chisel. The capillary pharyngeal tube is attached to the stylet and has no muscular casing. A muscular bulbus can only be present at the transition to the pharyngeal gland system (stichosome); it regulates the release of penetration gland secretion (Fig. 14). Paired ventral glands have been shown in *Hexamermis* and *Agamermis*, with a pore slightly behind the nerve ring (Fig. 14). The intestine usually has no lumen; it is made up of large cells one behind the other that are

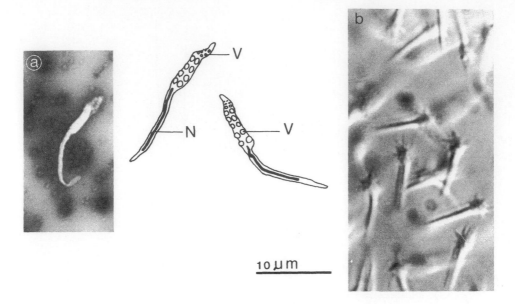

FIGURE 12   Spermatoza of *Pheromermis cornicularis* (a) and *Hexamermis brevis* (b). (a) Light micrograph of an Opal Blue preparation and reconstruction of Opal Blue and Giemsa-stained preparates. N, nucleus; V, specialized vesicles. (After Kaiser, 1984.) (b) Photomicrograph of living spermatozoa of *Hexamermis brevis* showing psuedopodialike filaments at the vesicular region.

filled with globular reserve substances. The genital anlage consists of two to four cells and is located ventrally in midbody. Species that actively find their hosts above the soil surface have very slender infectious stages that are up to 4 mm long, are highly active, and have a negative geotactic orientation. Many species and genera can be recognized from their infectious stages. Species that preferentially infect animals living in the soil, such as *Psammomermis*, have infectious stages that are considerably less motile and sometimes have blunt tail end. Species with oral infection via uptake of embryonated eggs are not capable of survival in an open environment.

G.   Parasitic Stage

After penetration into the abdominal cavity of the host animal, either growth takes place directly in the hemolymph or, more rarely, the parasite causes an internal gall (Fig. 17) to form (Tunicamermis; Couturier, 1953) or develops within membrane-encased ganglia of the central nervous system (*Oesophagomermis leipsandra*; Poinar, 1968)(Fig. 18). It is the parasitic stage that most often confronts the entomologist. Owing to its lack of distinguishing characteristics, it is usually not easy to identify unequivocally. There are three phases of parasitic development as described in the following paragraphs.

Phase 1: After penetration of the host, the mermithid seems to undergo a resting phase in which it grows slightly in diameter but scarcely in length. But in this phase, very important morphological changes take place having to do with food uptake. This is the change of the epidermis to a resorbing epithelium by the formation of microvilli and pinocytes on the outer surface under a delicate cuticle (Fig. 19). As shown by Rutherford and Webster (1974), Rutherford et al. (1977), Craig and Webster (1982), and others, the thin

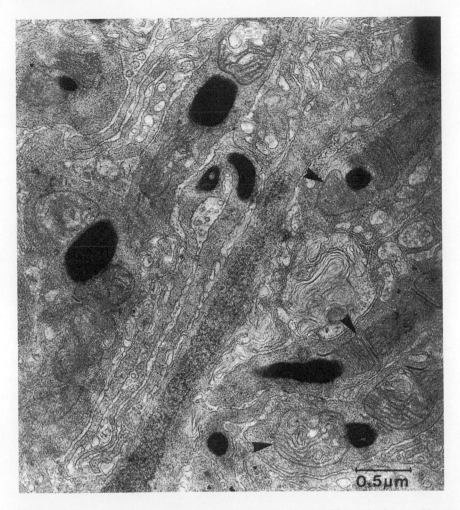

FIGURE 13   Electron micrograph of testal spermatozoa of *Amphimermis elegans.* Electron-dense nuclei, black dots. Specialized vesicles are arranged in rows on each side of the nuclei and show fine structure of tubular mitochondria (arrows).

cuticle is not an obstacle to the uptake of amino acids and glucose. In the pharyngeal gland system, the stichosome grows into an extensive organ (Fig. 4). The function of these glandular cells is unknown, but the pharynx seems to have lost its ability to take up nourishment; its glandular function seems to be of primary importance. The tail end of the infectious stage loses its locomotor function, ceases to grow, and forms the so-called appendage. Only *Agamermis*, when it penetrates its host, drops the part of its tail that is necessary for movement in a water film (Fig. 14). Mermithids with indirect infection and thus less motility do not form an appendage and also have a rounded tail end.

   Phase 2: This is the actual growth phase, in which the animal grows rapidly in length. All reserve substances are deposited; the trophosome turns into a large organ (Fig. 5). It nearly fills the pseudocoel and grows around the stichosome anteriorly to the neural

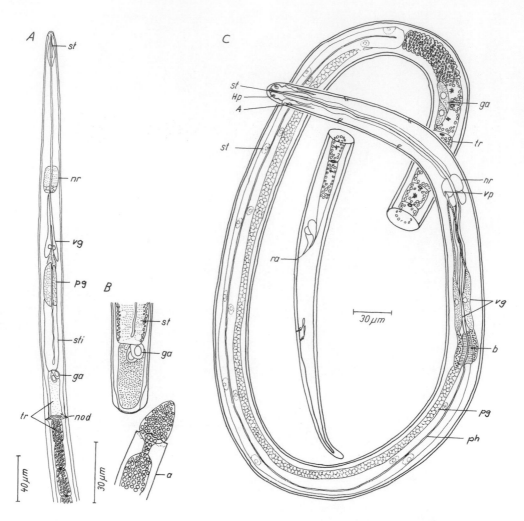

FIGURE 14 Infectious stages of *Agamermis sinuosa* (A, B) and *Hexamermis albicans* (C). (B) shows the automization of the hind end at a special nodus, which happens during the penetration act. A, amphid; a, autotomized body, b, bulbus; ga, genital anlage; Hp, head papillae; nr, nerve ring; nod, nodus; pg, penetration gland; ph, pharynx; ra, rudimentary anus; sti, stichosome; tr, trophosome. (After Kaiser, 1977.)

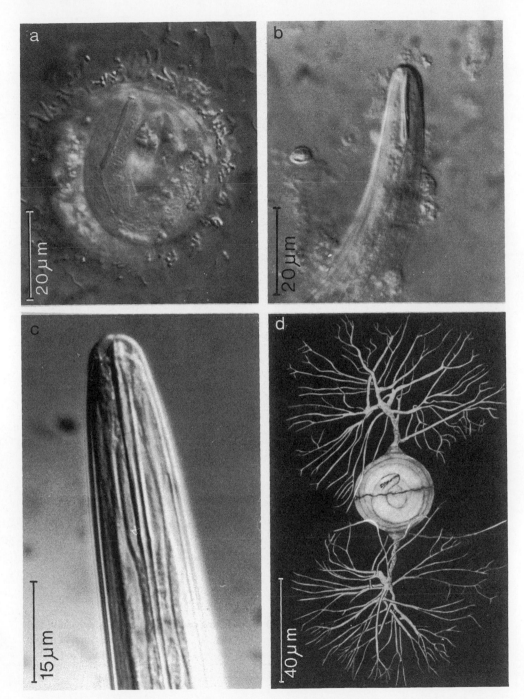

FIGURE 15   Eggs and infectious stages of different species of merithidae. (a) Egg and infectious stage of larva of *Pheromermis villosa* inside. (b) Head end of *Pheromermis villosa* showing stylet and guiding ring. (c) Head end of infectious stage larva of *Hexamermis albicans* with straight stylet. (d) Egg of *Mermis nigrescens*. (After Kaiser and Skofitsch, 1981.)

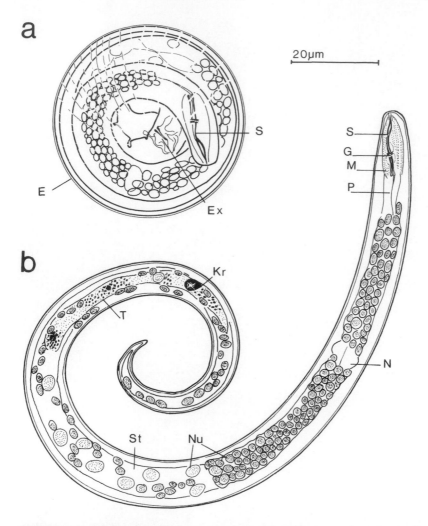

FIGURE 16   Egg (a) and infectious stage larva (b) of *Pheromermis vesparum*. E, egg shell; Ex, exuvie of the first larval stage inside the egg; G, guiding ring; Kr, crystal; M, musculature; N, nerve ring; Nu, nuclei; P, pharynx; S, stylet; St, stichosome; T, trophosome. (After Kaiser, 1987.)

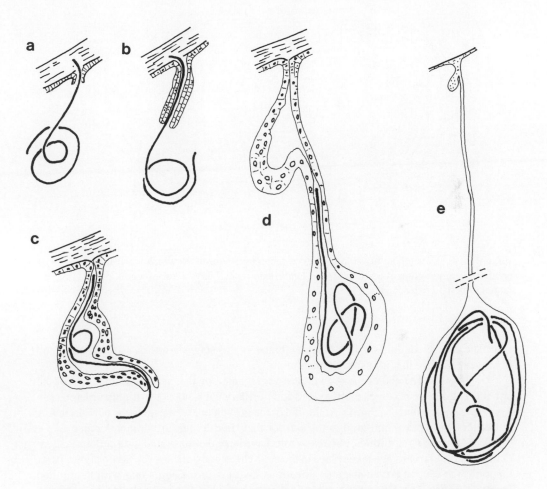

FIGURE 17 *Tunicamermis melolonthae*. Schema of the formation of epidermoidal cyst. (a–c) Under the influence of the infectious larval stage the hypodermis develops actively and envelopes the parasite. (d) The parasitic larva finds itself enclosed in a cellular mass (or occasionally a syncytium) which represents a germinal bud. (e) Finally *T. melolonthae* is isolated in a sac whose walls have become very thin. (After Couturier, 1963.)

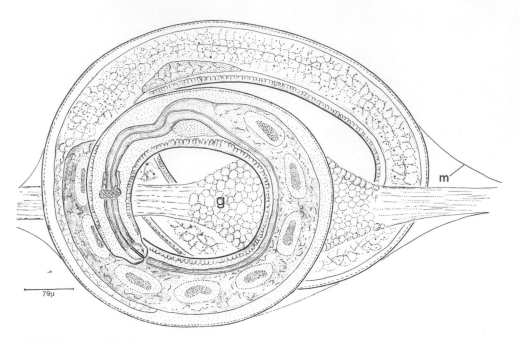

FIGURE 18   Developing juvenile of *Oesophagomermis leipsandra* which has just broken out of an abdominal ganglion of a second instar *Diabrotica* larva 7 days after infection. g, ganglion; m, neural lamella. (After Poinar, 1968.)

ring. The cuticle continues to be a thin membrane. In this stage the animals are very thin and burst very readily in a nonisotonic medium.

Phase 3: The end of the growth phase is signaled by the increasing thickness of the cuticle. The cuticle becomes multilayered and fibrillary but is not yet so highly ordered as in sexually mature animals (Fig. 1). At the end of the growth phase, these stages bore their way out of the host animal and go into the soil for their free-living further development.

Molts: For mermithids, only two molts have been demonstrated with certainty. Generally, the first larval stage takes place in the egg. This stage does not yet have a stylet. After the molt, the second larved stage (= infectious stage) is in the egg along with the molted larval skin (Fig. 16). During the parasitic development, no further molt usually seems to occur. In the course of the parasitic development, the oral cavity can become so highly cuticularized that the stylet is no longer recognizable (see Fig. 4c) (*Pheromermis*), *Hexamermis*) or it is stripped off as a distinctly recognizable organ with the last molt to the sexually mature animal (*Amphimermis*). This last molt appears light microscopically in mermithids with thick cuticle to be double, with a very delicate inner and a tough outer exuvia. This was ground for speculation (Poinar and Grisco, 1962b; Kaiser, 1977) that this inner thin layer might be a rudimentary last molt. Newer studies (Kaiser, 1986) suggest that this is not the case, as electron microscopically in *Pheromermis villosa* the inner layer is to be seen as the basal layer of the larval cuticle, which has separated from the outer layers (Fig. 20). Other ectodermal structures such as the fovea amphidialis and the pharyngeal tube are only molted once. More than two molts are reported by Poinar (1968) for *Oesophagomermis leipsandra* (once in the egg, twice in the postparasitic stage) and by

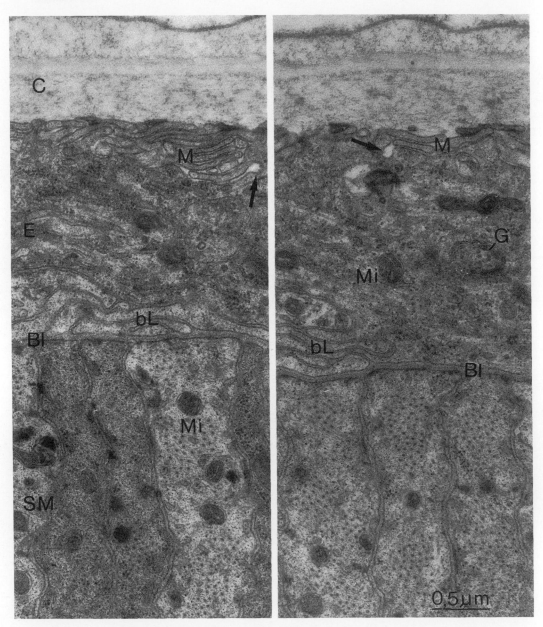

FIGURE 19   Electron micrographs of transverse sections through the cuticle of a young growing stage of *Pheromermis villosa*. bL, basal labyrinth; Bl, basal lamina; C, cuticle; E, epidermis; G, golgi vesicle; M, microvilli; Mi, mitochondria; SM, somatic musculature; the arrows point at micropinocytose vesicles. (After Kaiser 1986a.)

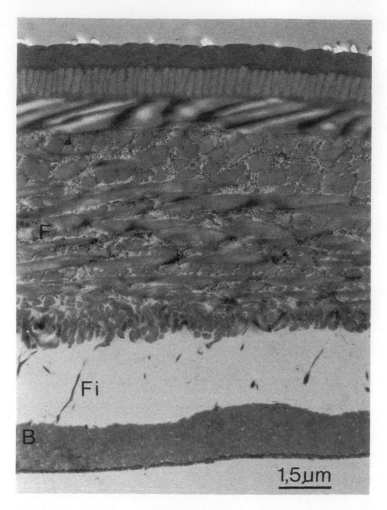

FIGURE 20   Electron micrograph of a section through the exuvia of *Pheromermis villosa* with separating basal layer (B). Separation is not completed, the basal layer is still connected by some fibers (Fi) to the fibrillar layer (F). (After Kaiser, 1986a.)

Poinar and Otieno (1975) in *Reesimermis nielseni* (once in the egg, once in the host, twice in the postparasitic stage).

III.   DIVERSITY, HOSTS, HABITATS, GEOGRAPHIC DISTRIBUTION

The diversity of mermithids in a given place depends first on the diversity of potential host animals and second on the nature and moisture of the soil. If the parameter of a large variety of host animals, permeable soil with the ability to bind water, and sufficient rainfall are fulfilled, a large variety of mermithids can be expected. This can only be confirmed by digging and finely breaking up the soil on the spot, which is a lot of work. The free–living sexually mature and postparasitic juvenile stages are usually found at a depth of 30–40 cm, or considerably deeper in sandy soil. In the humid climate of Central Europe, up to 80 mermithids could be found in just 0.1 m³ of soil, representing seven different species (Kaiser, 1977). Special studies on the species spectrum of mermithids in different biotypes are extremely rare for terrestrial forms of these nematodes, with the exception of such early reports as that of Hagmeier (1912).

As even so-called terrestrial mermithids cannot survive longer dry periods, their occurrence in deserts or semiarid regions is virtually out of the question or limited to a few moist spots. This is most apparent in studies of grasshoppers, the internationally most studied insect order with respect to parasitization by mermithids. Higher infection rates are known only for areas with at least a moderately moist climate or seasonal moist periods. In North America, Glaser and Wilcox (1918) reported up to 75% infection in *Melanoplus atlantis* and *M. bivittatus* in Vermont. Smith (1958) found up to 36% infection of grasshoppers by *Mermis nigrescens* and *Agamermis decaudata* in western Canada. Further reports from temperate zones in North America are summarized by Webster and Thong (1984).

In monsoon regions in Pakistan, the parasitization rate of the grasshoppers *Oxya multidentata*, *Shirakiacris shirakii*, and *Hieroglyphus banian* was greatest after monsoon rains. There are further reports on parasitized grasshoppers from Madagascar (Linstow, 1897) and Argentina for the migratory grasshopper *Schistocera paranensis* (Steiner, 1921).

The works of Baker (1981, 1983, 1984, 1986) in New South Wales in Australia provide the most information on the influence of climatic factors on the distribution of mermithids. Mermithids from grasshoppers were not observed in areas with less than 400 mm annual rainfall. In areas with 400–800 mm annual precipitation, infection of the Acridiidae *Gastrimargus musicus*, *Austroicetes vulgaris*, and *Phaulacridium vittatum* was irregular and slight. In areas with more than 800 mm rainfall, infection plays an important role in the regulation of the grasshopper population. New South Wales is an area with heavy summer rains. In contrast, mermithids were less abundant in areas with high winter precipitation and low summer precipitation, such as southwest Western Australia, South Australia, and Victoria.

Regional differences in infection of grasshopper with the mermithids species complex were as follows: In areas with precipitation between 400 and 800 mm yearly, only one species, *Hexamermis* sp. with special adaptations to the circumstances, was found. In moist tableland with 800–1200 mm precipitation, further unknown *Hexamermis* and *Amphimermis* species were found, as well as *Mermis quirindiensis* Baker and Poinar, 1986. In areas with more than 1200 mm precipitation annually, an *Agamermis* species and *Mermis athysanota* were also found.

As far as is known, mermithid fauna in grasshoppers is similar throughout the world at the generic level. In North America there are species of the genera *Mermis* and *Agamermis*; in Europa, *Mermis*, *Hexamermis*, and *Amphimermis* (author's data); and in Australia, *Hexamermis*, *Agamermis*, *Amphimermis*, and *Mermis*. At the species level, however, the parasite complexes are completely different. Only *Mermis nigrescens* seems to show holarctic distribution.

In general, the worldwide distribution of different (most?) genera is based only on isolated individual reports. The genus *Mermis* is known from North America, Europa, Siberia, Africa, Madagascar, Hawaii, Tahiti, and Australia; *Hexamermis* from Europe, North and South America, Africa, and Australia; *Amphimermis* from Europe, North America, Japan, and Australia.

Except for deserts and polar ice zones, mermithids are known from all the corners of the earth. They can parasitize virtually all orders of insects as well as spiders (Poinar and Benton, 1986; Poinar, 1985; Rubtsov, 1977), harvestmen (Kaiser, 1977; Poinar 1985), crustaceans (Poinar, 1981), myriapods (Hagmeier, 1912; Kaiser, 1977), and snails (Meissner, 1854; Rathke, 1953; Kaiser, 1977).

IV. TAXONOMY OF TERRESTRIAL (AMPHIBIC) MERMITHIDAE

There is quite a lot of uncertainty about the position of the mermithids in the nematode system. This is increasingly apparent in the creation of new ultraclassifications. At present, mermithids are generally placed in the vicinity of the dorylaimida (Adenophorea) owing to the structure of the oral cavity in juvenile animals, the amphids and the preanal papillae in the adult male (Lorenzen, 1981). Goodey (1963) and Rubtsov (1978), in contrast, emphasize the similarity of the pharygeal system with that of the Trichinellidae and Trichuridae and create the order Trichosyringida, or superimpose Trichosyringida with the order Mermithida.

There is more agreement with the formation of the superfamily Mermithoidea, Braun 1883 (Dorylamida) with the two families Mermithidae, Braun 1883 and Tetradonematidae, Cobb 1919.

The Tetradonematidae are diagnosed as follows (after Poinar, 1975): lips rudimentary or absent; juveniles with a minute stylet which disappears in later stages; head papillae reduced; pharynx consists of a simple hollow tube and associated tetrad or four celled structure; anus absent; ovaries and testis paired; vulva in middle of body; spiculum single; bursa absent. Nematodes mature to the adult stage and mate in the body cavity of the insect host.

Chitwood and Chitwood (1950) see the Tetradonematidae as primitive mermithids, and the mermithids with their development to the sexually mature stages outside of the host animal,as more highly evolved. On the other hand, Kaiser (1983) sees the Tetradonematidae as more highly evolved neotenic species, as neoteny is generally a more evolutionary trend in Mermithidae. The Tetradonematidae are therefore considered to be a catch-all for neotenic mermithids which sexual maturity in the body cavity of their hosts. This thesis tends to be confirmed by the recent finding of a mermithid (*Pheromermis villosa*) which, like the Tetradonematidae, only has four stichocytes in the pharyngeal gland system. The single common morphological feature of the Tetradonematidae must be obtained from the evaluation for a relationship analysis, as this occurs in all mermithids as plesiomorphy (Kaiser, 1986a). Artyukhovsky (1971) considers that the two different methods of infection provide evidence for the Mermithidae being a polyphyletic taxon with two subfamilies: Mermithidae with the genus *Mermis* (oral infection by embryonated eggs) and Paramermithinae which includes all other genera (infection by direct penetration). Kaiser (1983) pointed out that the genus *Pheromermis* must then also be placed in the Mermithinae because of oral infection and because of the homologous dorylamid stylet with a guiding ring in *Mermis* and *Pheromermis*; the guiding ring is missing in all other mermithids.

A. Differential Diagnosis of Genera and Key to Terrestrial and Semiterrestrial Mermithids

Mermithidae Braun, Mermithoidea Braun, Doryaimida (De Man)

Diagnosis: Long, slender nematodes, between 5 and 500 mm long, but usually between 10 and 100 mm. Cuticle smooth or containing criss-cross fiber near the outer layers, head containing two, four, or six head papillae and sometimes a pair of lateral mouth papillae. Mouth opening terminal or shifted ventrally. Amphids tubelike or modified pouchlike, pharynx modified into a slender pharyngeal tube which is closed on its posterior end, surrounded in its posterior part by the stichosomal tissue consisting of large cells (stichocytes) numbering 4, 8, 16, or, rarely, more. Intestine modified to a cellular or syncytial sac, a food storage organ called the trophosome, functional anus absent, ovaries and testes paired, muscular vagina straight or curved, males with paired, fused, or single spicules, several rows of genital papillae usually present, preparasitic juveniles with functional stylet, a pair of penetration glands and often one pair of ventral cells. Pharynx and intestine located in a row one behind the other.

Genus *Mermis* Dujardin, 1842

Diagnosis (Figs. 15d and 21): medium to large mermithids, 15–150 mm long. Anterior and posterior ends bluntly rounded, thick cuticle with distinctly criss-crossed fibers. Four head papillae, lateral papillae missing, two mouth papillae. Amphids wide and weakly cuticularized, aperture porelike at the level of the head papillae. Mouth terminal or slightly shifted ventrally. Six hypodermal cords in the body middle. Trophosome syncytial. Vagina muscular and S–shaped, paired spicula, bow-shaped with rounded distal ends, roughly as long as or a bit longer than the diameter of the cloaca. Small but numerous eggs with polar byssi, more rarely with hairs or folds. Stylet in the infectious stages with sinusoid curve and dorilaimoid guiding ring. The host is infected via oral uptake of embryonated eggs. Parasitic and postparasitic stages without appendix.
Distribution: worldwide.

Type species: *Mermis nigrescens* Dujardin, 1842; Host: Orthoptera, Forficula; holarctic.
Syn. *Mermis subnigrescens* Cobb, 1926 = *M. meissneri* Copp, *M. kirgisia*, Kirjanova, Karaveva, and Romanenco, 1959.

Other species: *M.mirabilis* Linstow, 1903 (= *M. tahitiensis* Baylis 1933; Host: not known; Tahiti.
M. changodudus Poinar, Remillet, and Waerebeke 1978; Host: *Heteronychus* spp. (Scarabaeidae); Madagascar.
M. kenyensis Baylis, 1944; Host: not known; Kenya.
M.athysanota Steiner, 1921; Host: *Phaulacridium vittatum* (Acrididae); southeastern Australia.
M. quirindiensis Baker and Poinar, 1986; Host: *Phaulacridium vittatum*; southeastern Australia.
M. savaiiensis Orton Williams, 1984; western Samoa.

Species *inquirenda*: *M. quakensis* Gafurov, 1982; Tadzhikistan

Genus *Pheromermis* Poinar, Lane, and Thomas, 1976
Syn. *Welchimermis* Rubtsov 1978, *Agamomermis* Welch, 1958 (part), *Allomermis* Steiner, 1925 (part)

Diagnosis (Figs. 22, 23, 24) (after Poinar, Lane, and Thomas, 1976 revised): Mermithids lacking lip papillae but containing four cephalic papillae in submedial positions; amphids well developed, anterior in position; vagina S-shaped (not bent in transverse plane to the body); spicules paired, separate, short (less than twice tail diameter); six hypodermal cords at midbody; cuticle with cross–fibers; pharynx with four or eight stichocytes; eggs without processes; eggs very small; infectious stages with styles with sinusoid bend and a guiding

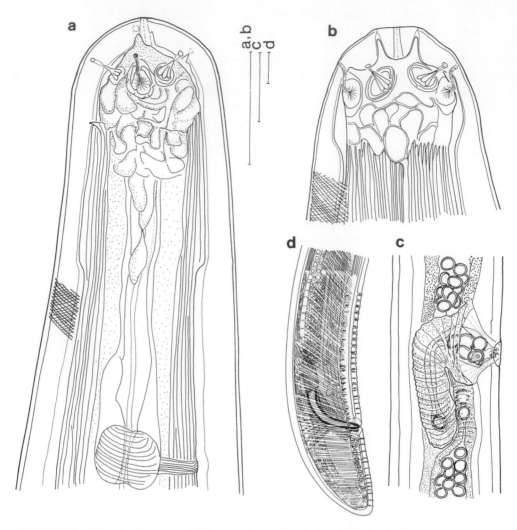

FIGURE 21 *Mermis nigrescens.* (a) Lateral view of female head end. (b) Ventral view of female head end. (c) Lateral view of vulva, vagina, and uterus. (d) Lateral view of male tail end. Scale bars in 100 μm. [(d) after Hagmeier, 1912.]

ring as in *Mermis*; infection via oral uptake of embryonated eggs. Distribution: Europe, North America.

Type species: *Pheromermis pachysoma* Poinar, Lane, and Thomas 1976.

As early as 1905 von Linstow described a *Mermis pachysoma* in English wasps. As only postparasitic stages were available to him, he was unable to provide an exact description. Poinar et al. (1976) thought that they had found the same species in California and described it as *P. pachysoma*. As this is the only description of *P. pachysoma* and the original animal material is no longer available, *P. pachysoma* should be retained for the California species. A European species, *P. vesparum*, was described by Kaiser (1988).

Other species: *P. vesparum* Kaiser, 1987; Host: *Dolichovespula saxonica*, *Paravespula vulgaris* and *P. germanica*, *Polistes* sp.

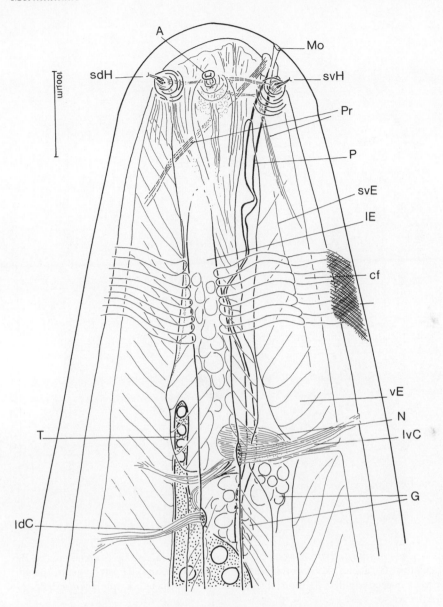

FIGURE 22 *Pheromermis vesparum.* Lateral view of male head end. A, amphid; cf, cuticular fields; G, ganglia; ldc, latero-dorsal commissure; lE, lateral epidermoidal cord; lvC, latero-ventral commissure; Mo, mouth opening; N, nerve ring; P, pharynx; Pr, pharynx retractors; sdH, subdorsal head papillae; svE, subventral epidermoidal cord; svH, subventral head papillae; T, trophosome; vE, ventral epidermoidal cord. (After Kaiser, 1987.)

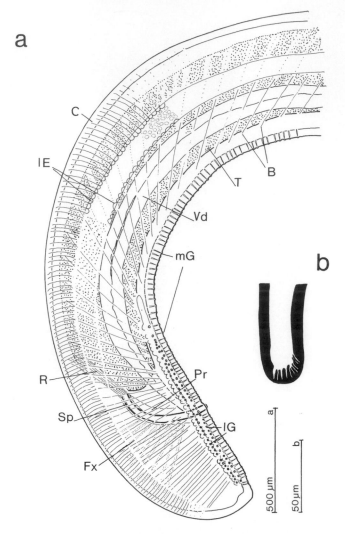

FIGURE 23   *Pheromermis vesparum.* Male tail end, lateral view (a) and tip of spiculum (b). B, bursal musculature; C, cuticle; Fx, fixator; lE, lateral epidermoidal cord; lG, lateral genital papillae; mG, median general papillae; Pr, protrusor; R, retractor; Sp, spicula; T, trophosome; Vd, vas deferens. (After Kaiser, 1987.)

> *P.villosa* Kaiser, 1986; Host: *Lasius flavus* and *L. niger*; hermaphroditic;
> Austria.
> *P. myrmecophila* Crawley and Baylis, 1921; Host: *Lasius alienus*;
> England.

Species *inquirenda*:   *P. (Allomermis) lasiusi* Rubtsov, 1970

Genus *Agamermis* Cobb, Steiner, and Christie, 1923

Diagnosis (Figs. 14A, B and 25) (after Nickle, 1972 and author's data): Mermithidae. Medium to very large nematodes, 10–465 mm in length. Mouth opening terminal. Amphids

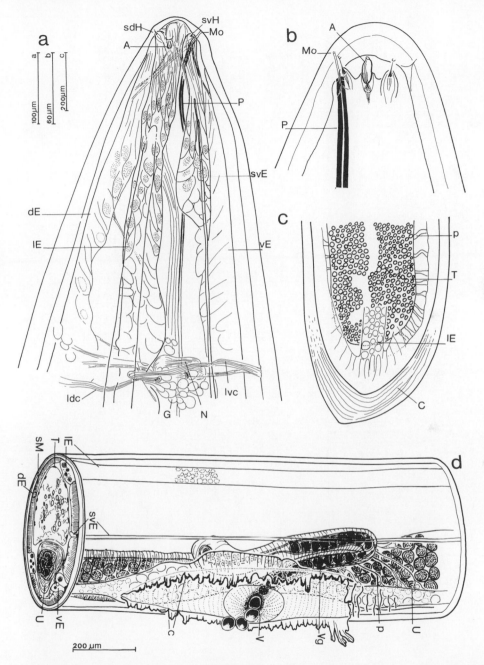

FIGURE 24   Hermaphroditic female of *Pheromermis villosa*. (a,b) Head end, lateral view. (c) Tail end. (d) Midbody area with cuticular bristles around the vulva. A, amphid; C, cuticular bristles; dE, dorsal epidermoidal cord; lE, lateral cord; Mo, mouth opening; P, pharynx; p, parenchymatous tissue; sdH, subdorsal head papillae; sM, somatic musculature, svE, subventral cord; svH, subventral head papillae; T, trophosome; U, uterus; V, vulva; Vg, vagina. (After Kaiser 1986a.)

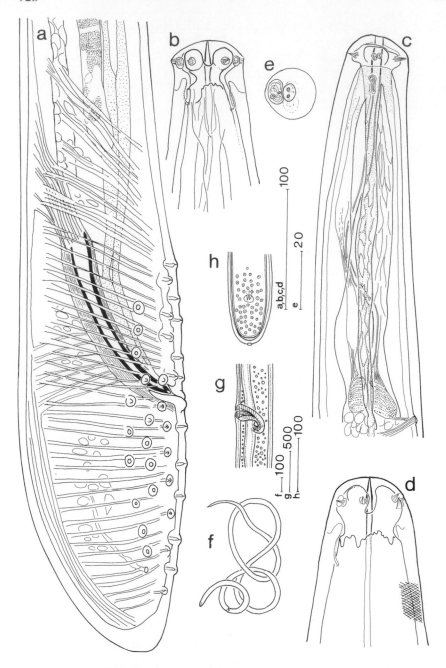

FIGURE 25 *Agamermis decaudata*. (a) Male tail end. (b) Male head end, dorsal view. (c) Male head end, dorsal view. (d) Female head end, ventrolateral view. (e) High magnification of male amphid and lateral head papillae, lateral view. (f) Infectious larval stage. (g) Vagina. (h) Tail end of a male molting stage, ventral view. [(f–h) After Nickle, 1972.]

small. Six head papillae, lateral and submedian in position. Cuticle with criss-cross fibers. In midbody six cords. Tail tip in both sexes bluntly rounded. Male: with two short spicules, genital papillae in four to six irregular rows (as in *Hexamermis*). Female: vagina S-shaped, vulvar cone well developed, cuticularized. Infectious stage: amputates tail, i.e., 75% of the body at a preformed node just before penetrating the host. Postparasitic and parasitic juveniles with craterlike appendage from decaudated tail. Eggs: large without byssi. Distribution: worldwide.

Type species: *Agamermis decaudata* Cobb, Steiner, and Christie, 1923; Host: diverse locusts and grasshoppers but was also found parasitizing chrysomelid and coccinelid beetles; North America.

Other species: *A. cobbi* Schuurmans Stekhoven and Mawson, 1955; hosts unknown; Alsace (France).
A. paradecaudata Steiner, 1925; Host: *Helopeltis antonii* (Miridae); Java.
A. unka Kaburaki and Imamura, 1923; Host: *Calligypona furcifera, Nilaparvata oryzae* (Delphacidae); Japan.
A. sinuosa Kaiser, 1977; Host: *Chlorionidea flava, Typhlocyba* sp., *Liburnia* sp. (Cicadina); Austria.

Species *inquirenda*: *A. hangaica* Rubtsov, 1976
A. saldulae Rubtsov, 1969
A. angusticephala Steiner, 1924
A. dubia Steiner, 1924

Genus *Amphimermis* Kaburaki and Imamura, 1932.
Syn. *Complexomermis* Filipjev, 1934, *Linstowimermis* Rubtsov, 1978 (part.)

Diagnosis (Figs. 10, 26, and 27) (after Nickle, 1972, revised): Mermithidae. Usually medium to long nematodes, 13–260 mm in length. Mouth opening terminal or with slight ventral shift. Amphids large, usually with porelike aperture. Six head papillae lateral and submedial in position. Cuticle thick with clearly visible criss–cross fibers. Longitudinal cords in midbody, six.

Pharyngeal tube long, in some species reaching 60% of body length. Pharyngeal tube does not reach the mouth opening but ends behind it with a ringlike thickening. Tail tip in both sexes bluntly rounded and ventrally curved. Male: with two long, twisted spicula. Genital papillae in three rows, center row bifurcates around genital opening. Female: vagina elongated, S-shaped, strongly muscularized. Often with vulval flap. Preparasites do not amputate tail and have extremely high mobility. Eggs numerous, medium-sized, without byssi. Distribution: seems to be worldwide but until now no records from Africa and South America.

Type species: *Amphimeris zuimushi* Kaburaki and Imamura, 1932; Host: *Chilo simplex* (Lepidoptera); Japan.

Other species: *A. elegans* (Hagmeier) 1912,
Syn. *Mermis elegans* Hagmeier, 1922, *Complexomermis* (Hagmeier, 1912) Filipjev, 1934. Host: *Stenobothrus, Decticus* (Orthoptera), *Leptinotarsa, Gastrophysa. Phyllotreta* (Coleoptera), *Forficula* (Dermaptera); Europe.

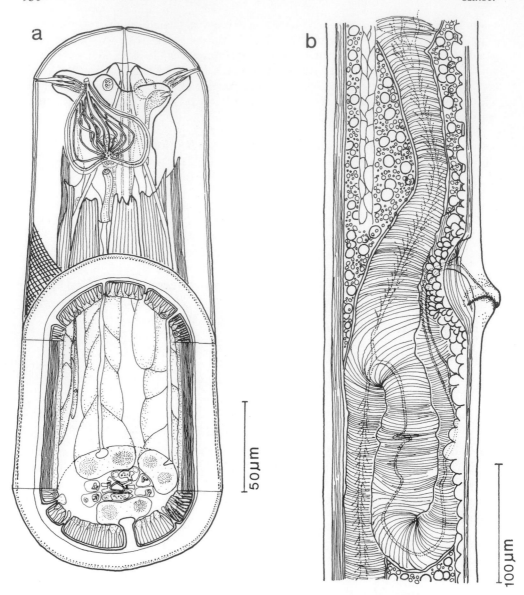

FIGURE 26 *Amphimermis* sp. (from Austria). (a) Male head end, lateral view. Semischematic reconstruction after glycerin preparate and serial cross-sections. (b) Midbody of a female showing vagina and ventral flap of vulva.

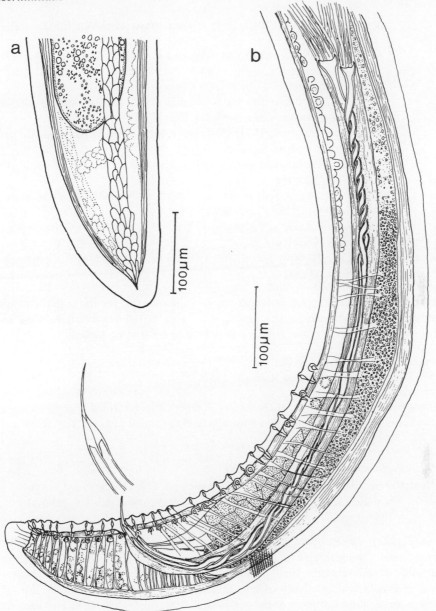

FIGURE 27 *Amphimermis* sp. (from Austria). (a) Female tail end. (b) Male tail end, with for the genus typically twisted spicula and a common tip.

A. bogongae, Welch, 1963; Host: *Agrotis infusa* (Lepidoptera); Australia, capital territory.

A. ghilarovi (Polozhentsev and Artyukhovsky, 1958) Syn. *Complexomemis* Polozhentsev and Artyukhovsky, 1969; Host: unknown; USSR. Voroshilôvgrad.

A. litoralis Artyukhovsky and Karchenko, 1971; Host: unknown; USSR. Tsentralshchno-Chernozemnoy National Park.

A. maritima Rubtsov, 1971; Host: unknown; USSR, Kedrovaya Pad, Subutinsky National Park.

A. avoluta Rubtsov and Koval, 1975; Host: Leptinotarsa (Coleopteria); USSR, Ukraine.

A. tinyi Nickle, 1971; Host: *Ischurna postica*, Anomalagrion hastatum *(Odonata);* USA, Louisiana.

A. avoluta Rubtsov and Koval, 1975; Host: Leptinotarsa (Coleoptera); USSR, Ukraine.

A. volubilis Rubtsov and Koval, 1975; Host: Leptinotarsa (Coleoptera); USSR, Ukraine.

The two species *A. avoluta* and *A. volubilis* are very questionable, as the author (unpublished data) occasionally found animals among *A. elegans* that owing to a developmental defect had spicula that were either completely twister or showed no torsion.

Species *inquirenda*:   *A. lagidzae* Rubtsov, 1975
                        A. mongolica Rubtsov, 1976

The genus *Amphimermis* probably includes any number of undescribed species as Baker (1983) reports undescribed species from Australia, and the author also has such in his collection.

Genus *Oesophagomermis* Artyukhovsky, 1969

Syn. *Mermis*   Hagmeier, 1912 (part.), *Agamermis*, Polozhentsev and Artyukhovsky, 1958 (part.). *Filipjevimermis* Polozhentsev and Artyukhovsky, 1958 (part.)

Diagnosis (Fig. 28) (after Rubtsov, 1978, rev.): Mermithidae of moderate size with rounded head and tail ends. Cuticle thick with slightly developed or missing criss-cross fibers. Longitudinal cords in midbodys six. Head papillae six, in lateral and submedial position. Amphids oval, pouchlike with relatively wide aperture, located close to the lateral head papillae. Mouth opening shifted slightly to the ventral side. Pharyngeal tube does not reach the outer cuticular layers. Vulva oblique slit in ventral body wall. Vagina cylindrical with S-shaped curve. Pharyngeal tube long, may almost run through the whole body in some species. Spicules long and parallel with archlike curve, their length is 1.5–2 times the body diameter at the genital opening. Genital papillae in three rows, medial row bifurcates at genital opening. Parasitic and postparasitic juveniles with small tail appendage.

Type species: *Oesophagomermis terricola* (Hagmeier) 1912; Host: unknown; Europe.

Other species: *O. brevivaginata* Artyukhovsky and Kharchenko, 1971; Host unknown: USSR, Central Chernozem Park.

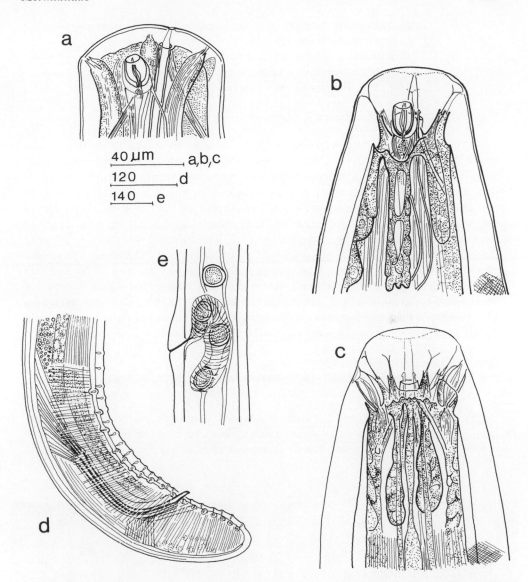

FIGURE 28 *Oesophagomermis terricola.* (a) Male head end, lateral view. (b) Lateral view of male head end after a somewhat wrinkled glycerin preparate. (c) Specimen of (b), ventral view. (d) Male tail end. (e)Vagina. (After Hagmeier, 1912.)

O. hydrophilis Artyukhovsky and Kharchenko, 1971; Host unknown;
Central Chernozem Park.

O. leipsandra (Poinar and Welch) 1968; Host: *Diabrothica* sp.
(Chrysomelidae); USA, South Carolina.
Syn: *Filipjevimermis leipsandra* Poinar and Welch, 1968.

Species *inquirenda*:   *O. coriacea* Rubtsov, 1972

Owing to the precise description by Hagmeier (1912), the genus *Oesophagomermis* can easily be distinguished from *O. terricola*. Rubtsov (1978) places such species as *arvalis* Poinar and Welch, 1962 and *Filipjevimermis leipsandra* Poinar and Welch, 1968 in the genus, although not enough is known about these two species to be able to do this without reservation. Above all, there are no data on the length of the pharynx and how it ends in the area of the mouth opening. The species *leipsandra*, however, corresponds in some criteria such as form and length of the amphids, the ventrally shifted mouth opening, and the morphology of the female genital opening to the situation in *Oesophagomermis;* for this reason, this species should remain in the genus *Oesophagomermis* for the time being. Rubtsov (1978) is correct in abolishing the genus *Filipjevmermis*, as it is arbitrarily constructed with a minimum of characteristics and thus does not permit a relationship analysis. There is no apparent reason for a transfer of the species *arvalis* into the genus *Oesophagomermis* and for this reason it should remain in the genus *Hexamermis* until the necessary studies can be done.

Genus *Hexamermis* Steiner, 1924
Syn. *Ovomermis* Rubtsov, 1976; *Mermis* Meissner, 1854; Siebold, 1848; Hagmeier, 1912

Diagnosis (Figs. 14B, 29, and 30) (after Nickle, 1972, revised): Mermithidae. Medium-sized to very long nematodes (15–450 mm long). Mouth opening terminal. Amphids small or weakly cuticularized, opening usually close to lateral head papillae. Head papillae six, in

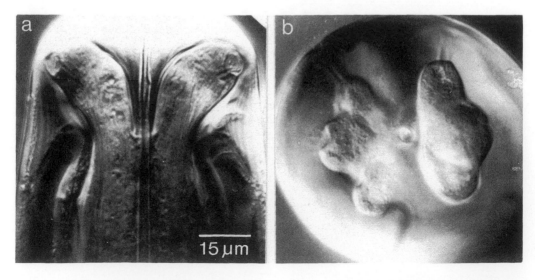

FIGURE 29   Light micrographs of head ends of *Hexamermis albicans*. (a) Lateral view. (b) En face view.

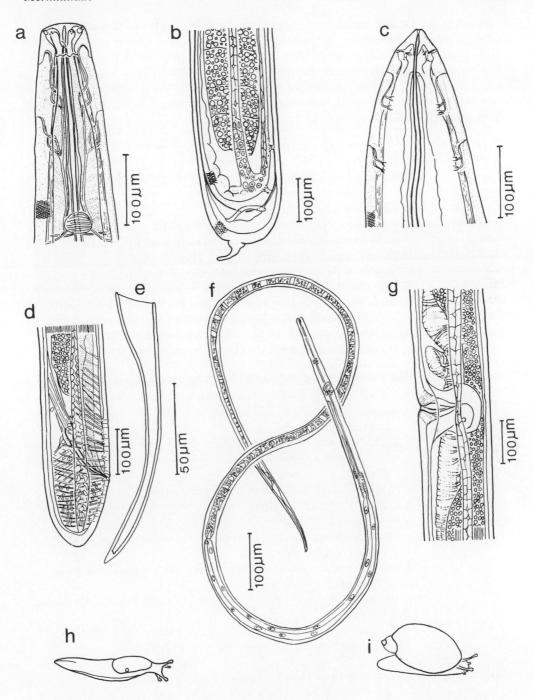

FIGURE 30 *Heximermis albicans*. (a) Male head end, lateral view. (b) Tail end of molting stage. (c) Female head end, lateral view. (d) Male tail end. (e) Spiculum. (f) Infectious stage larva. (g) Vulvar area of female. (h) Host *Deroceras agreste*. (i) Host *Succinea putris*.

submedian and lateral position. Cuticle with clearly visible crisscross fibers. Parasitic in terrestrial insects, harvestman, or mollusks. Tail tip in both sexes bluntly rounded or with small cuticular projection on tail end. Six cords in midbody. Male: with two short spicula. Genital papillae in four to six rows in an irregular pattern. Female: Vagina S-shaped or horn-shaped with well-developed, cuticularized vulval cones. Infective juveniles do not amputate tail. Postparasitic juveniles with short, fingerlike appendage. Eggs medium-sized to large without byssi.

Type species: *Hexamermis albicans* (Siebold, 1948). Polozhentsev and Artyukhovsky, 1959

Syn. *Mermis albicans* Siebold, 1848, Meissner 1854, Hagmeier 1912, *Hexamermis truncata* Wouts, 1981.

H. bussardi Baylis, 1933.

Other species: *H. arsenoidea* (Hagmeier) 1912; Host: unknown: Germany.

H. artjuckovskii Kharchenko 1966; Host: unknown;USSR.

H. brevis (Hagmeier) 1912; Host: Tipulidae; Europe.

H. capitata Rubtsov 1976; Host: unknown; Mongolia.

H. cavicula Welch, 1963; Host: *Agrotis infusa* (Lepidoptera); Australia.

H. ferghanensis Kirjanova, Karaveva, and Romanenko, 1959; Host: *Hypomoneuta malinella* (Lepidoptera); Kirghizia.

H. minutissima Rubtsov, 1976; Host: unknown; Mongolia.

H. pratensis Polozhentsev, Artyukhovsky, and Kharchenko, 1965; Host: unknown; European USSR.

H. pusilla Rubtsov and Koval, 1975; Host: *Leptinotarsa decemlineata*; Ukraine.

H. stepposis Artyukhovsky and Kharchenko, 1975; Host: unknown; USSR central forest steppe.

H. lineata Kaiser, 1977; Host: Saltatoria (Stenobothrus, Mecosthetus, Chorthippus, *Euthystira*. Opiliones: (Oligolophus, *Phalangium*); Austria.

H. elongata Kaiser, 1977; Host unknown; Austria.

H. incisura Kaiser, 1977; Host: unknown; Austria; USSR. Ukraine (unpublished).

H. cathetospicula Poinar and Chang 1985; Host: *Tryporyza incertula* (Lep.); Malaysia.

H. dactylocercus Poinar and Linares, 1985; Host: *Aeneolamia varia* (Homoptera); Venezuela.

H. cornuta ? Rubtsov and Koval, 1975; Host: *Leptinotarsa decemlineata*; Ukraine.

H. glossinae Poinar, Mondet, Gouteux, and Laveissier, 1981; Host: *Glossina palpalis. G. pallicera*; western Africa.

Species *inquirenda*: *H. acuminata* (Leidy), 1875

H. meridionalis Steiner, 1924

H. microamphidis Steiner, 1925

H. alascensis Steiner, 1932

H. polyina Steiner, 1934

H. cornuta Gleiss, 1955

H. angustata Rubtsov and Koval, 1975

H. obtusa Rubtsov and Koval, 1975

H. vaginata Rubtsov, 1976
H. abrevis Rubtsov, 1971
H. paralbicans Rubtsov, 1971

It was not possible to obtain the works containing the descriptions of *H. subaquatilis* Kharchenko, 1966, *H. tumefactis* Kharchenko, 1966, and *H. sujidae* Rubtsov, 1971.

Discussion: The genus *Hexamermis* can be described as being very difficult from the taxonomic point of view for two reasons: first, the genus seems to be very rich in species and second, the individual species are especially similar. This uncertainty is especially apparent in *H. albicans*.

Early students of Mermithidae (Siebold, 1848, 1854; Meissner, 1854; Rauther, 1905) saw all long, white mermithids as *Mermis albicans* when they had a short appendix on the tail end as parasitic or postparasitic juveniles. Thus there are descriptions of *M. albicans* from the most various insect orders as well as from mollusks. Rathke (1953) studied an *H. albicans* from *Succinea putris* and came to the conclusion that there are a variety of mermithid species to be found under the name of *H. albicans* and that only one species is to be viewed as a mollusk parasite. This situation still prevails today and *H. albicans* will have to be viewed as a conglomerate of different but similar mermithids until studies are made of host specificity and the various forms have been isolated reproductively (Kaiser, 1977). Wout's attempt to use an even older name, *Mermis truncata* (Rudolphi, 1809), for not more closely defined mermithids, especially from butterfly caterpillars, is thus not a solution to this problem.

Division of the genus *Hexamermis* into *Hexamermis* and *Ovomermis* (Rubtsov, 1976) based on the position of the amphids directly next to the lateral head papillae or somewhat behind them is impracticable, as there are the most various transitional forms in the genus *Hexamermis* (Kaiser, 1977). Species *inquirenda* include those in which only one sex is known, or those described on the basis of parasitic or postparasitic juveniles.

Genus *Melolonthinimermis* Artyukhovsky, 1963
    Syn. *Pseudomermis* Schuurmans Stekhoven and Mawson, 1955

Diagnosis (Fig. 31) (after Schuurmans Stekhoven and Mawson, 1955): Mermithidae of large size. Cuticle thick, with obvious crossing fibers. Mouth terminal. Head papillae four. Amphids of moderate size, cup-shaped, with small aperture behind the plane of the head papillae. Longitudinal cords six. Vulva straight. Vagina short with V-like curve. Spicules two, of moderate length and slightly curved ventrally. Genital papillae four in three rows, bifurcating around the genital opening. Develop in the larvae of Melolonthinae.

Only one species is known:

M. hagmeieri (Schuurmans Stekhoven and Mawson) 1955. Host: *Melolontha melolontha*; France, Austria (unpublished).

Genus *Psammomermis* Polozhentsev, 1941
    Syn: *Pologenzevimermis* Kirjanova, Karaveva, and Romanenko, 1959

Diagnosis (Figs. 32 and 33) (after Kaiser 1984): Very thin, elongated mermithids, body length depending on species a few millimeters to 35 cm. Cuticle thin, adults without crisscross fibers, or with very fine ones occasionally visible under highest magnification. Postparasitic stages with distinct criss-cross fibers. Mouth opening terminal. Cuticularized pharyngeal lining begins with a thickened collar in the hypodermis. Pharynx maximally

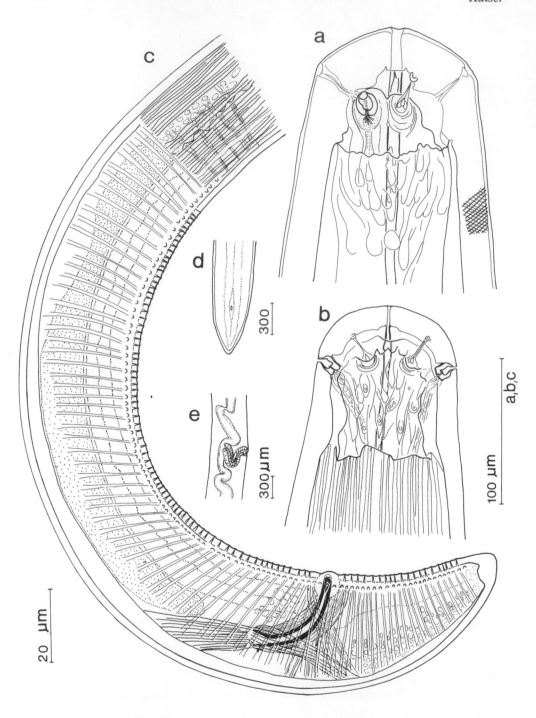

FIGURE 31 *Melolonthinimermis hagmeieri.* (a) Male head end, lateral view. (b) Female head end, ventral view. (c) Male tail end. (d) Male tail end, ventral view. (e) Vagina. [(d, e,) After Schuurmans Stekhoven and Mawson, 1955.]

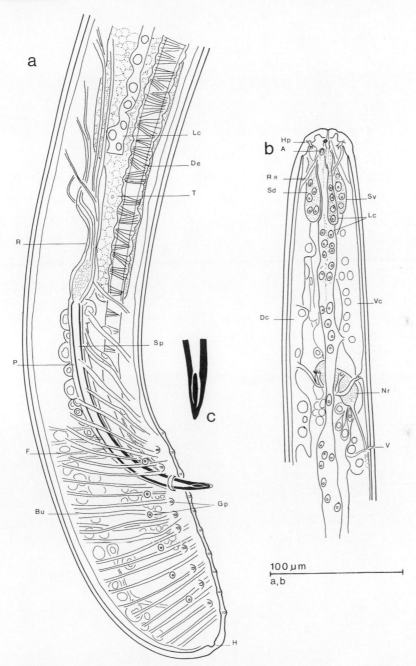

FIGURE 32 *Psammomermis cornicularis*. (a) Mail tail end. (b) Male head end, lateral view. (c) Tip of spiculum. A, Amphid; Bu, bursal musculature; Dc, dorsal cord; De, ductus ejaculatorius; F, fixator; Gp, genital papillae; H, hypodermal tissue; Hp, head papillae; Lc, lateral cord; Nr, nerve ring; P, protrusor; Ra, "retractor amphidialis"; Sd, subdorsal cord; Sp, spicula; Sv, subventral cord; T, trophosome; V, ventral pore. (After Kaiser, 1984.)

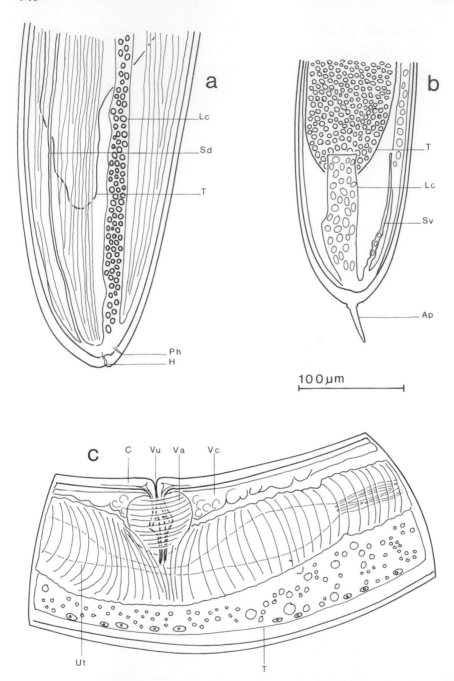

FIGURE 33 *Psammomermis cornicularis.* (a) Female tail end. (b) Tail end of a postparasitic larval stage. (c) Vulva and vagina. Ap, appendage; C, cuticle; H, hypodermal tissue; Lc, lateral cord; Ph, phasmid; Sv, subventral cord; T, trophosome; Ut, uterus; Va, vagina; Vu, vulva; Vc, ventral cord. (After Kaiser, 1984.)

50% of body length. Eight hypodermal cords in all sections of the body. Trophosome syncytial. Six head papillae in one plane. Amphids pore- to cup-shaped, with fine round to wide oval opening. Vagina short and muscular, with straight or slightly bent lumen. Vulva straight or with varying degree of overlap of the genital opening. Spicula paired, thin, bent distal to ventral, and pointed. Proximal with V-spread capitulae, distal calami strictly parallel with pronounced tendency to grow together. Hind end of parasitic and postparasitic stages with small appendage. Adults have a hypodermic cone extending into the cuticle instead of an appendage.

Preparasitic juveniles: With long pharyngeal section—more than 50% of body length, short tail, and relatively slight motility.

Type species: *P. korsakovi* Polozhentsev, 1941, Host: *Melolontha* groups; USSR.
Other species: *P. alechini* Artyukhovsky and Kharchenko, 1965; Host unknown; middle belt of European USSR.
P. kulagini Polozhentsev, 1941; Host: *Melolontha*; USSR.
P. parvula Rubtsov, 1976; Host: unknown; Mongolia.
P. cornicularis Kaiser, 1984; Host: unknown;Austria, Europe.
P. conjuncta Kaiser, 1984; Host: unknown; Austria, Europe.
P. minor Kaiser 1984; Host: unknown; Austria,Yugoslavia (Bosnia).
P. filiformis Kaiser, 1984; Host: *Otiorrhynchus gemmatus* (Curculionidae); Austria, Europe.

Species *inquirenda*: *P. busuluk* Polozhentsev, 1952
P. tiliae Rubtsov, 1972
P. agrotinae Rubtsov and Pukhaev, 1980

Genus *Tunicamermis* Schuurmans Stekhoven, Mawson, and Couturier, 1955

Diagnosis (Fig. 34): Mermithidae of large size, from 80 to 350 mm long and 180 to 300 μm diameter. Cuticle thick, with obvious criss-cross fibers. Mouth terminal. Head papillae six, in one plane. Amphids small, pouchlike, with wide transverse aperture behind the head papillae. Longitudinal cords six. Vulva with thick, swollen lips protruding outward. Vagina short, cylindrical, obliquely directed backward over its greatest part. Tail blunt. Males unknown. Eggs discoid, 60–65 × 45 μm. Juveniles develop in beetle groups of Melolonthidae by forming and internal gall.

Type species: *T. melolonthae* Schuurmans Stekhoven, Mawson, and Couturier, 1955; Host: Melolonthidae; France.

Genus *Thaumamermis* Poinar, 1981

Diagnosis (Fig. 35) (after Poinar, 1981): Medium- to large-sized nematodes with the adult cuticle lacking criss-cross fibers under the light microscope, six head papillae in one plane, no mouth papillae, six hypodermal cords at midbody, vagina elongated, flexed two or three times (modified S-shape), amphids cup-shaped, medium in size, located in the same plane or slightly anterior to the head papillae, spicules paired, showing extreme dimorphism. One is extremely short and thick, equal to or shorter in length than the body width at the cloaca; the other is extremely long and filiform; postparasitic juvenile with a small appendage.

Type species: *T. cosgrovei* Poinar, 1981; Host: terrestrial Isopoda; *Armadillidium vulgare, Porcelloscaber*; USA, California.

Genus *Aranimermis* Poinar and Benton, 1986

Diagnosis (Fig. 36) (after Poinar and Benton, 1981): Medium- to large-sized nematodes, adult cuticle lacking crossed fibers under the light microscope, six cephalic papillae arranged in one plane, oral papillae absent, six hypodermal cords at midbody. Amphids cup shaped, medium in size, located slightly posterior to the ring of cephalic papillae. Vagina elongated, modified S shape, flexed two to three times and running in two different planes (parallel and perpendicular to the body axis). Spicules paired, long, six or more times body width at cloaca, separate, but closely addressed to each other. Postparasitic juvenile cuticle with faint crossed fibers.

Type species: *A. aptispicula* Poinar and Benton, 1986; Host: *Cesonia bilineata, Watodes* sp., *Misumenops* sp., *Wulfila alba, Phidippus* sp., *Verrucosa arenata, Tamarus* sp.; Little River canyon rim, Alabama, USA.

Other, rarely found terrestrial genera include *Amphidomermis* Filipjev, 1934 with type species *A. tenuis* (Hagmeier)1912, *Amphibiomermis* Artyukhovsky, 1969 (this genus is prob-

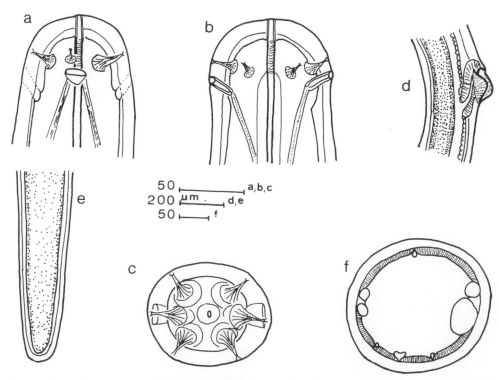

FIGURE 34   *Tunicamermis melolonthae.*(a) Head end, lateral view. (b) Head end, vental view. (c) Head end, en face. (d) Vulval area. (e) Tail end. (f) Cross-section showing dorsal, lateral, subventral, and ventral cords. (After Schuurmans Stekhoven and Mawson, 1955.)

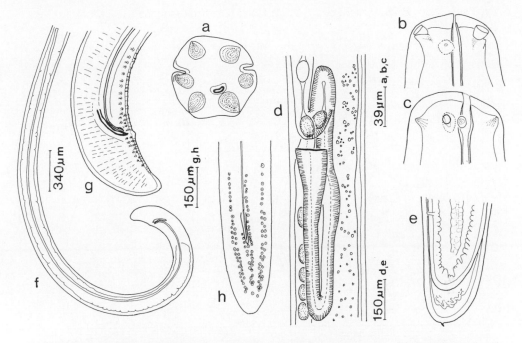

FIGURE 35 *Thaumamermis cosgrovei.* (a) Male head end, en face view. (b) Female head end, ventral view. (c) Female head end, lateral view. (d) Vulvar area, ventrolateral view. (e) Tail end of female molting stage. (f) Posterior portion of male showing filiform spiculum and short spiculum. (g) Male tail end. (h) Male tail, ventral view. (After Poinar, 1981.)

ably identical to *Hexamermis*) and *Skrjabinomermis* Polozhentsev, 1952 (is probably identical to *Psammomermis*).

Genera *inquirenda*: *Onchiomermis* Rubtsov, 1976
  Coccinellimermis Rubtsov, 1978
  Rhynchomermis Rubtsov, 1978
  Arachnomermis Rubtsov, 1978
  Pentatomimermis Rubtsov, 1978
  Dipteromermis Rubtsov, 1976
  Schuurmanimermis Rubtsov, 1978

These genera, with the exception of *Schuurmanimermis*, are based on findings of isolated parasitic and postparasitic juvenile stages and thus cannot be adequately defined or recognized. For the establishment of *Schuurmanimermis*, Rubtsov (1978) refers to the finding of several females described by Schuurmans Stekhoven and Mawson (1955) as *Agamermis couturieri*.

B. Key to the Genera of Terrestrial Mermithidae

1. Mermithidae with four head papillae 2
    Mermithidae with six head papillae 5

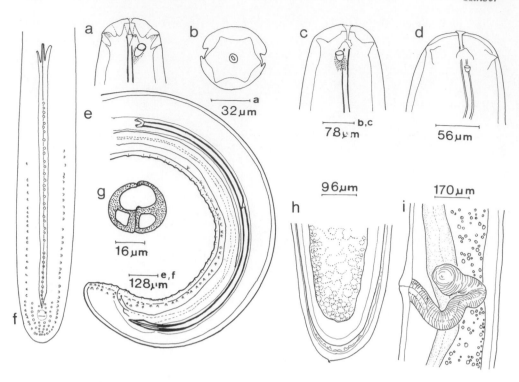

FIGURE 36 *Aranimermis aptispicula*. (a) Male head end, lateral view. (b) Female head end, en face. (c) Female head end, lateral view. (d) Postparasitic juvenile female head end, lateral view. (e) Male tail end. (f) Male tail end, ventral view. (g) Cross-section of male spicula. (h) Tail of a molting postparasitic female. (i) Vagina. (After Poinar and Benton, 1986.)

2. M. with four head papillae and two mouth papillae; mouth opening terminal or slightly shifted to the ventral side; cuticle with criss-cross fibers; parasitic and postparasitic juveniles without tail appendage; vagina muscular, S-shaped; spicules two short and curved; hypodermal cords in midbody six. . . . . *Mermis*
    M. with four head papillae, without mouth papillae 3
3. Mouth opening shifted to the ventral side until level of the head papillae; vagina long and S-curved; spicules two, short and curved; cuticle with criss-cross fibers; hypodermal cords six at midbody; postparasitic juveniles without tail appendage; heterosexual or hermaphroditic; life cycle with intermediate host. . . . .
    . . . . *Pheromermis*
    Mouth opening terminal 4
4. Amphids close to lateral head papillae, strongly cuticularized, cup-shaped, with round aperture; cuticle with criss-cross fibers, hypodermal cords six at midbody; vagina short with V-like curve; spicules short, curved; postparasitic juveniles with short tail appendage. . . . .*Melolonthinimermis*
5. Mouth opening terminal, cuticular sheet of the pharyngeal tube penetrates the body cuticle close to the mouth opening; cuticle with criss-cross fibers; six hypodermal cords in midbody; amphids small, cup-shaped with small pore like aperture near the lateral head papillae; no vulval flap, vagina strongly muscularized with

well-cuticularized lumen, horn- or S-shaped curve; spicules two, relatively short, trophosome cellular. 6

Mouth opening, vagina, amphids different 7

6. Parasitic and postparasitic juveniles with tail appendage .... *Hexamermis*
Parasitic and postparasitic juveniles with excavated ring on tail end; preparasitic juveniles loose their tail end during penetration of host. .... *Agamermis*

7. Mouth opening shifted slightly to the ventral side; cuticular sheet of the pharyngeal tube penetrates into the body cuticle close to the mouth opening; adults without visible criss-cross fibers in the cuticle; six hypodermal cords at midbody; vagina modified S-shape, flexed two or three times; spicules paired, one is short and thick, the other extremely long and filiform; postparasitic juvenile with short tail appendage. .... *Thaumamermis*
Mouth opening terminal or slightly shifted to ventral side; cuticular sheet of pharyngeal tube in adults does not reach the mouth opening but ends abruptly, often with a ring like ridge, in the hypodermal tissue of the head end. 8

8. Mouth opening shifted slightly to the ventral side; adult cuticle with faint or missing criss-cross fibers; hypodermal cords six at midbody; amphids of moderate size, aperture relatively wide, opening at the level of the head papillae; pharyngeal tube very long, can reach nine-tenths of body length; vulva oblique, vagina cylindrical, S-shaped; spicules long, parallel with arch like curve, 1.5–2 times body diameter; parasitic and postparasitic juveniles with small tail appendage. .... *Oesophagomermis*
Mouth opening terminal. 9

9. Vagina short, barrel-shaped; spicules long, thin, parallel, tendency to fuse in the distal part, 2–5.5 times body diameter; trophosome syncytical; eight hypodermal cords at midbody; adults without or with faint criss-cross fibers, postparasitic juveniles with clearly visible criss-cross fibers and short tail appendage; in adults, the position of the juvenile appendage is marked by a hypodermal protuberance into the tail cuticle. .... *Psammomermis*
Vagina short but cylindrical and posteriorly directed; vulva flanked by lateral lips; amphids small, cup-shaped with oval aperture; reproduction without males .... *Tunicamermis*
Vagina S-shaped 10

10. Spicules twisted, parallel only in the midpart and on the distal end, fused on tip. Amphids large, weakly cuticularized, aperture pore like; cuticle with criss-cross fibers; hypodermal cords six; parasitic and postparasitic juveniles with tail appendage. .... *Amphimermis*
Spicules parallel throughout its length 11

11. Vagina elongated, flexed two or three times in two different planes; adults lacking criss-cross fibers; six hypodermal cords, amphids cup-shaped, opening round, slightly behind the lateral head papillae; spicules paired, six or more times times body width; separate but closely addressed to each other; postparasitic juveniles with or without faint tail appendage. .... *Aranimermis*
Vagina short but S-shaped; spicules short 12

12. Adult cuticle without criss-cross fibers; amphids very wide, cup-shaped with wide, transverse oral aperture .... *Amphidomermis*

V. LIFE CYCLES AND BIONOMICS

Mermis nigrescens and related species

The life cycles of *Mermis* species have been studied in Europe (Hagmeier, 1912), in America (Christie, 1937; Cobb, 1926), and in Australia (Baker, 1983). (Fig. 37). The life cycles are very similar and can be summarized as follows: The fully developed parasitic juveniles leave their hosts and migrate into the soil. They penetrate to a depth of 20–40 cm, depending on the degree of moisture needed to survive. The juveniles molt into adults. After copulation, the females begin to produce eggs. The eggs are retained in the uterus, where the embryonal development takes place and where they are provided with an additional case to provide a degree of protection against dehydration after they have been laid. The secondary egg case has polar byssi to hold the eggs onto vegetation after laying. After rainfall, the soil cells containing the females fill with water. This causes them to emerge from the soil and creep slowly over the vegetation, laying their eggs as they do so. Females that are not affected by the moisture remain in the soil until heavier rains cause them to emerge. When the eggs are ingested by the hosts with vegetation, the infectious stages hatch in the intestine, bore through the intestinal wall into the abdominal cavity, and grow to a length of 9–15 cm in 4–10 weeks.

Rainfall is the only factor that determines the activity of the animals, while the parasitization rate depends on a number of additional factors: the number of females that are stimulated to lay eggs during a rainy period, the density of the eggs on the vegetation, the period during which the eggs remain viable (depends on temperature and sunshine), and, finally, the number of eggs that are actually consumed by insects before they fall victim to such influences as dryness, mowing, or consumption by grazing animals

This risky developmental cycle is compensate for by the large number of eggs. A gravid female *Mermis nigrescens* with a body length of 85 mm contains about 14,000 eggs before laying (Christie, 1937).

Another interesting point is the existence of a red pigment spot on the anterior end of the female. *M. nigrescens*, the chromotrope. Christie (1937) and Croll (1966) both found that oviposition was arrested or greatly reduced in the dark and resumed on illumination. Adult female *M. nigrescens* moved from red to green light if given the choice. Ellenby (1964) found distinct α and β absorption bands in the chromatrope, and he concluded that the chromatrope contains oxyhemoglobin. After illumination of the chromatrope with light of different wavelengths, Croll (1966) found a peak of uterine contractions at 540 μm. This is the β absorption band of oxyhemoglobin.

Other *Mermis* species such as *M. changodudus* (a parasite of soil-dwelling Scarabaeidae) do not have a pigment spot or byssi on the eggs. Specializations of this sort would be pointless in the soil).

Pheromermis *pachysoma*, *P. vesparum*, and *P. villosa*

Pheromermis species are parasitoid in social Hymenoptera, wasps, and ants. Poinar et al. (1976) describe the Californian *P. pachysoma* from *Vespula pensylvanica*. The adult nematodes occur in water or saturated soil and the eggs are fully embryonated at oviposition. These mermithids need an intermediate host to complete their life cycle. The eggs hatch in the gut of various insects and infective stage juveniles penetrate the gut wall and enter a quiescent state in the tissues of paratenic hosts. Wasp larvae may be infected when they are fed paratenic hosts captured by worker yellowjackets.

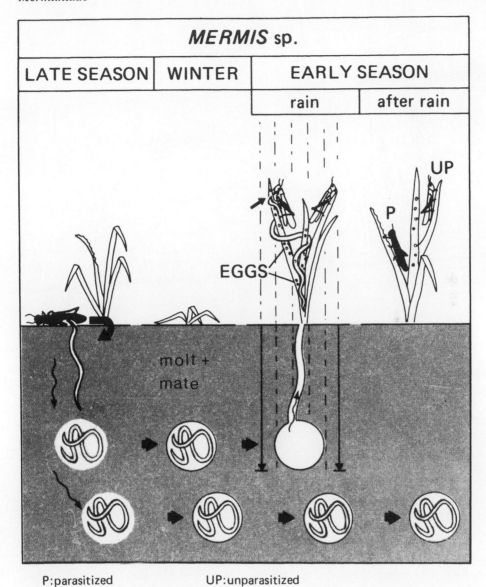

MERMIS sp.

LATE SEASON	WINTER	EARLY SEASON	
		rain	after rain

UP

P

EGGS

molt + mate

P:parasitized   UP:unparasitized

FIGURE 37 Seasonal life cycle of *Mermis* sp. in New South Wales, Australia. (After Baker, 1983.)

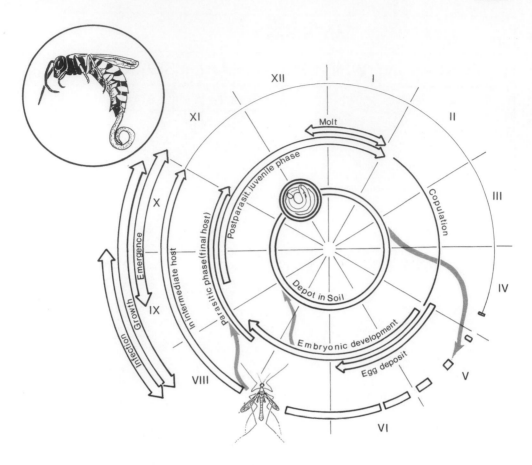

FIGURE 38   Seasonal aspects of the life cycle of *Pheromermis vesparum* in Austria. (After Kaiser, 1987.)

Similar developmental cycles were found by Kaiser (1986b, 1987) in Europe (Figs. 38–40). In Austria, *P. vesparum* parasitizes *Vespula vulgaris, V. germanica,* and *Dolichovespula saxonica. P. vesparum* appears from the end of August in the hemocele of wasp larvae in aerial and soil nests. Owing to the late infection, parasitism is concentrated in the reproductive stages of the wasp. Outwardly the infected wasps show no parasitic pathology. In the inner organs, the fat body is reduced and the ovaries shrunken. Flight capability is retained. For their further development, the emerging parasitoids have to reach a moist subtrate, where they molt after 3 months. After molting, egg laying begins in May. Unfertilized females retain their viability for years. Embryonal development is completed after 2 months. The infectious stage remains in the egg until consumed by a satisfactory intermediate host (Tipulidae, Limnophilidae, Trichoptera). The developmental cycle lasts 1 year or can stretch over a number of years owing to the longevity of the infectious stage in the protective egg case. The numbers of eggs are extraordinarily large. For example, in 2 months, a female with a length of 11 cm lays some 1.2 million eggs.

There is also a high level of synchronization between *P. villosa* and its hosts, the ants *Lasius flavus* and *L. niger* (Kaiser, 1986b). The parasitic growth stage of *P. villosa* occurs

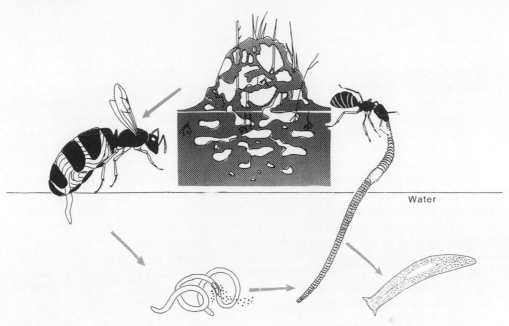

FIGURE 39   Simplified schema of the life cycle of *Pheromermis villosa* in Austria. (After Kaiser, 1986b.)

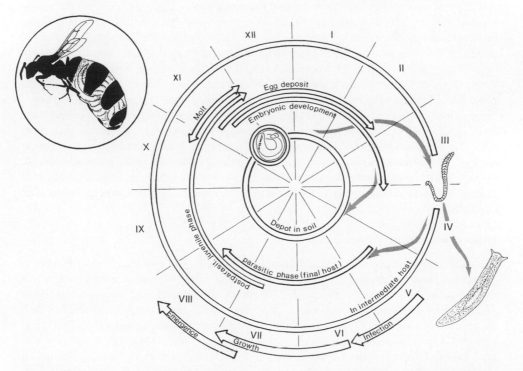

FIGURE 40   Seasonal aspects of the life cycle of *Pheromermis villosa* in Austria. (After Kaiser, 1986b.)

only for a short time, from mid-July to the beginning of August, in pupae and winged reproductive stages of the two *Lasius* species. The infection occurs in the larval stage of the ant via consumption of muscular tissue from Oligochaetae containing the infectious stages. Growth of *P. villosa* is initiated by pupation of the host. In the pupation stage, *Pheromermis* competes for nutrition with the host and prevents the development of flight musculature. Infected females show narrowing of the head and thorax regions and shortened wings. Parasitized males usually show normal wings. The reduction of the flight musculature and the tendency of the ants to seek moist substrate to discharge their parasites are advantageous for the further development of *P. villosa*. At the same time, the occurrence of *P. villosa* is limited to moist biotopes.

With few exceptions, all the animals manage to reproduce, as they are proterandric hermaphrodites. A special fringed border around the vulva that develops from a splitting of the cuticle seems to play a role in the occasional pseudocopulation of two hermaphrodites.

Egg laying goes on until January or February of the following year and embryonal development until March. One animal lays as many as 300,000 eggs. The infectious stages can remain virulent for as long as 3 years in the egg. They can be seen year-round in Oligochaetae. Infectious stages and oligochaetes thus have a depot function in nature. Tricladic turbellaria are also often infected by afflicted oligochaetes. This, however, is a dead end for the development of *P. villosa*.

Agamermis decaudata

Christie's (1936) work on the developmental cycle of *A. decaudata* in North America is a classical study of the biology of mermithids. The free-living stages of these nematodes are found in loose bunches at depths of 5–70 cm in the soil. Each cavity contains one female and a number of males. Copulation is necessary for reproduction. Eggs are laid from the beginning of July on into autumn and collect on the walls of the cavity. It is here that the embryonal development takes place as well as the first molting within the egg. The output of a female can be as many as 6500 eggs. In the following year, the females continue to lay eggs. The eggs that were laid first hatch in July of the following year at the same time that their hosts, the grasshoppers *Melanoplus femur-rubrum* and *Conocephalus brevipennis*, appear. The infectious stages wander to the surface and onto low-lying vegetation, then bore through the body wall into the body cavity. At this point they drop the propagative end piece from a preformed breakage point and begin to grow phenomenally. Males remain in the host for 1–1.5 months, females for 2–3 months. The mermithids emerge head foremost , forcing their way through the body wall between the segments, fall to the ground, and enter the soil. During the first winter, the males and females remain isolated in their cavities and only after the molt in the following spring do the males seek the females. Egg laying begins soon after copulation.

Hexamermis and *Amphimermis*

The developmental cycles of *Hexamermis* and *Amphimermis* species are very similar to those of *Agamermis decaudata*. These genera as well parasitize their hosts by penetrating the cuticle of the host from the outside. Differences are due to the chronological sequence and length of the developmental periods depending on the hosts and the climates in which the animals occur. Baker (1983) gives a short summary of the developmental cycle for *Amphimermis* sp., a parasitoid of grasshoppers in New South Wales (Fig. 41). Late in the season, fully fed *Amphimermis* worms emerge from their hosts and enter the soil. During the winter, the worms remain deep in the soil, tightly coiled in earthen cells. In the spring, fe-

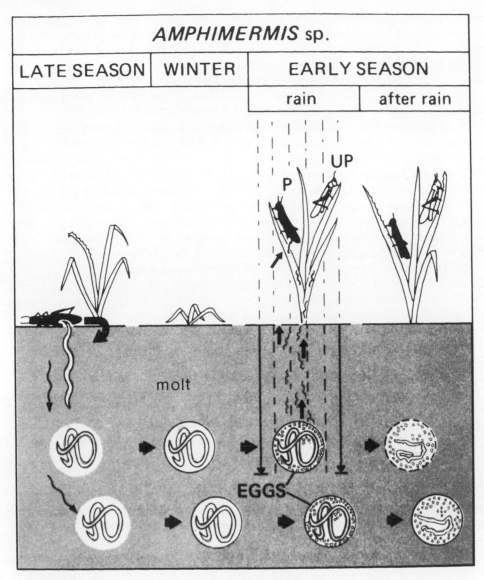

FIGURE 41 Seasonal life cycle of *Amphimermis* sp. in New South Wales, Australia. (After Baker, 1983.)

males become sexually mature and commence to lay thousands of small eggs over the walls of the cell. Eggs at different stages of development are present at any given time due to protracted laying by the female. After rain, water fills the cell and those eggs which are fully developed hatch. The small infective juveniles larvae wriggle through the damp soil to the surface, where they locate and penetrate grasshopper hosts. They remain in the host for a period of 4–5 weeks, during which time they grow from 2–3 mm to 150–200 mm in length. Undeveloped eggs continue to develop and hatch in response to later rain.

Amphimermis elegans was studied by the author in Austria (unpublished). The developmental cycle is very much like that described above. _A. elegans_, however, seems to need a cold period before it can lay eggs. Fertilized females were kept at room temperature for 2 years and were not seen to lay eggs. Only after the temperature was lowered to 8°C for 3 weeks did the females begin to lay eggs after the temperature was raised to 18–20°C.

The developmental cycle of _A. zuimushi_ as parasitoid of the rice border as described by Kaburaki and Imamura (1932) also resembles the cycle described above. It may, however, be assumed that the hatching of the infectious stages is a continuous process in this case, as the rice fields are continuously flooded during the infectious period in August.

Developmental cycles for _Hexamermis_ species were described by Couturier (1950) for leaf-eating insects of the willow _Salix fragilis_, by Rathke (1953) for the snails _Succinea putris_ and _Deroceras agreste_, and by Kaiser (1972) for the Colorado potato beetle _Leptinotarsa decemlineata_. All these cycles follow the scheme described above for _Amphimermis_ sp. Couturier showed that the infectious stages of _Hexamermis_ sp. are capable of climbing several meters up trees to reach their hosts. Regarding the massive propagation of _Hexamermis_ sp. from the Colorado potato beetle, it is important to note that the eggs can be stored under moderate humidity for up to 24 months without an appreciable number of them hatching. Flooding will cause the infectious stages to emerge and they are then available for infection experiments; this had already been done by Peterson (1978) for the artificial propagation of _Romanomermis culicivorax_, a parasitoid of mosquitoes. This _Hexamermis_ species does not need a winter dormancy period (unpublished data). They are capable of penetrating larvae, pupae, and imagines of _Leptinotarsa_.

Oesophagomermis (= Filipjevimermis) leipsandra

Oesophagomermis leipsandra shows a rather unusual developmental cycle. It is a parasitoid of certain chrysomelids (Coleoptera) such as _Diabrotica undecimpunctata_, a sweet-potato pest in South Carolina.

It has a short life cycle; theoretically it could be as short as 47 days and it is capable of parthenogenetic reproduction. Females begin to lay eggs just 2 weeks after hatching. Life expectancy is only 2–3 weeks. During this time they lay several thousand eggs. The infectious stages emerge from the eggs after 12– 15 days and penetrate the _Diabrotica_ larvae, but not the pupae and beetles (Cuthbert, 1968). The extremely motile infectious stages penetrate their host via the body surface, using the stylet and secretion from the pharyngeal gland system (Poinar, 1968). They first enter the hemocele directly and then penetrate the nerve ganglia, especially the protocerebral lobes and the subesophageal ganglion. As they grow, they stretch the neuronal lamella to a wide sack, which then bursts, letting the nematodes into the hemocele (see Fig. 18). In this way it is possible for the young parasitic larvae to avoid such defensive mechanisms of the host as encapsulation, since they do not come into contact with blood cells within the neuronal lamella.

VI. HOST–PARASITE RELATIONSHIPS

The influences of mermithids on their hosts show a wide variety of varying reactions. While the penetration into the host animal, whether via the body surface or the intestinal wall, scarcely causes reactions, the growth stages interfere seriously with the metabolism of their hosts, causing changes in form and behavioral disorders. It should be emphasized that the emergence of the parasitoids always is lethal for the host animal. When mermithids emerge, they damage their hosts so seriously that sometimes body parts such as antennae or legs are torn off, or they bore such a large hole in the body wall that a lot of blood is lost. And when such a relatively large parasite leaves the host, such a large vacuum develops that the blood can no longer circulate properly.

The influence of mermithids on the metabolism of their hosts has been most thoroughly studied in grasshoppers (Gordon and Webster, 1971, 1972; Rutherford and Webster, 1974, 1978; Rutherford et al., 1977). Mermithids grow extremely fast in their insect hosts. *Mermis nigrescens* needs only 3 weeks to grow from 370 μm to 10 cm in length in the migratory grasshopper *Schistocerca gregaria*. Gordon and Webster (1972) showed that parasitic juveniles consume large amounts of amino acids, lipids, and carbohydrates from the host's hemolymph. Under experimental conditions, protein was synthesized most quickly with uptake of [14c] leucine. Dipeptides and polypeptides were not taken up. The parasitic stages of *M. nigrescens* mainly consumed glucose from the hemolymph of *S. gregaria* and very little of the actual blood sugar of insects, trehalose. Rutherford and Webster (1974) assumed that an active transport system in the troposome is responsible for the uptake of glucose through the cuticle, which provides a diffusion gradient from the cuticle to the hypodermis and the pseudocoelom. In contrast, the amino acid uptake seems to be controlled by a stereospecific transport system that is localized in the nematode cuticle (Rutherford et al., 1977).

The influence of parasitizing mermithids on the various organs of infected insects does not seem to be a direct interference, e.g., by hydrolytic enzymes, but instead works indirectly by removing essential substances from the hemolymph; this then indirectly affects to supply of nutrition to the fat body, ovaries, and musculature. Two weeks after infection of *S. gregaria* with *M. nigrescens* there is a significant reduction of glycogen and nonglycogenic carbohydrates in the fat body. This is simultaneously accompanied by a progressive decrease in the level of active and inactive glycogenic phosphorylase (Gordon et al., 1971). The constant uptake of glucose from the hemolymph causes a reduction in glycogenesis in the fat body of the host and so makes more glucose available in the hemolymph for the parasite (Rutherford and Webster, 1978). After 2 weeks of parasitism, vitellogenic and novitellogenic proteins are also reduced in the fat body.

It is very probable that mermithids stimulate the catabolism of fat body proteins in the host and prevent the anabolism of these proteins in order to provide a suitably high amino acid level in the hemolymph. With the prevention of vitellogenesis, the egg cells can only take up small amounts of protein from the hemolymph. As the growing mermithids are constant competition for vitellogenic proteins, their parasitism can lead to sterilization of the female host (Webster and Thong, 1984).

Not infrequently do mermithoses cause deformation of the wings in insects. Deformation in grasshoppers, such as shortened or inextensible wings, depends not so much on the number of mermithids in one host as on the time of infection (Sugiyama, 1958; Webster, 1972). Deformations occur with early infection in nymphs, when the grasshoppers have yet to turn into imagines. The same is true for the chrysomelids *Leptinotarsa decemlineata* and *Melasoma populi* (Kaiser, 1972; Couturier, 1950). When these beetles are infected with

Hexamermis sp. as larvae and are parasitized throughout pupation into the imaginal stage, the outer wings cannot be properly extended. Mermithoses have the most apparent effects in caste-forming Hymenoptera, especially in ants.

In Hymenoptera, only females form castes. The unfertilized ant eggs develop into males that have wings and do not show polymorphism. Fertilized eggs develop into winged queens or unwinged workers. In some species there are also special soldier castes. There is no doubt as to the importance of the diet in the determination of castes in ants. In general, there is a trophogenic determination at certain sensitive moments in the larval phase that with sufficient food includes a predisposition to development to reproductive stages. The sexually differentiating mechanism of growth was studied by Brian (1974) in *Myrmica rubra*. With protein deprivation in certain larval stages, males and females showed different reactions. Male larvae responded with growth stop and accelerated development. The wing buds were not affected and their growth activity was coordinated with the extremities. Females, in contrast, responded to protein deprivation with growth stop and accelerated development as well, but with a preference for the ventral as opposed to the dorsal anlagen. In larvae, this happens irregularly and so prevents the formation of intercastes. The mermithids interfere in these complicated processes as food competitors with the ant larvae and cause a variety of form changes. Ant specialists became aware early on of the form changes caused by mermithoses and created a good deal of literature on this phenomenon (Wheeler, 1907, 1928; Emery, 1904; Vandel, 1927, 1930, 1934; Gösswald, 1930; Kloft, 1950; Passera, 1976) and intercastes are described as mermithogates and mermithostratiotes. Mráček (1908) described short–winged females with a tendency to narrowed head and thorax as mermithogyne. Crawley and Baylis (1921) studied mermithogynes from *Lasius flavus* and found atrophied ovaries and very reduced fat bodies. Kaiser (1986a, b) described the developmental cycle of *Pheromermis villosa*. Parasitic stages of these mermithids could only be found in the short period from mid–July to the beginning of August in the pupae and winged imagines of *L. flavus* and *L. niger*. The ant larvae were infected by consumption of muscle tissue from oligochaetes containing the infectious stages. *P. villosa* only begins to grow when the host pupates. At this stage, when the ants are incapable of feeding, the growing nematodes provide extreme competition for food and so prevent the development of the flight muscles (Fig. 42). In the pupal phase , the caste has long since been determined and so the protein deprivation caused by the mermithids takes place too late to completely prevent the development of wings. In parasitized males, which have a completely different genetic predisposition than females, the wing length remains virtually unchanged. In contrast, Vespidae parasitized by mermithids *(Pheromermis vesparum)* showed no changes in form (Kaiser, 1987) (Fig. 43). Here the parasitic development begins in the larval stage of the wasps and the effects of parasitism are distributed evenly throughout the juvenile development.

Upon emergence, both wasps and ants show distinct behavioral changes that help the mermithids reach the moist substrate they require for their further development. Poinar et al. (1976) and Kaiser (1987) observed in Californian and European Vespidae, respectively, that individuals infected with *Pheromermis* sought a moist milieu to discharge their parasites. This behavior was shown even more pronouncedly by female *L. flavus*, which purposefully sought open water to discharge their *Pheromermis* parasites and even submerged themselves under the surface of the water (Kaiser, 1986b) whereupon the worms promptly departed from their hosts' abdomens.

Another interesting subject in the host–parasite relationship is the phenotypic sexual determination in mermithids. Since Christie (1929) and Cobb et al. (1929), it has been known that high infection rates shift the sexual ratio in favor of the male. In *Mermis nigres-*

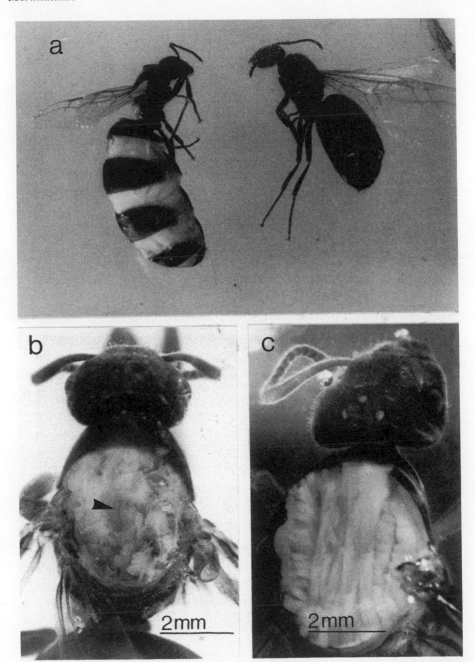

FIGURE 42   Influence of mermithid infections to queens of the ant *Lasius flavus*. (a) Mermithized queen with short wings and swollen abdomen (left) and normal unparasitized queen (right). (b)Parasitized queens have instead of flight musculature just a tissue which resembles the fat body(arrow). (c) Healthy queen with normal flight musculature. (After Kaiser, 1986b.)

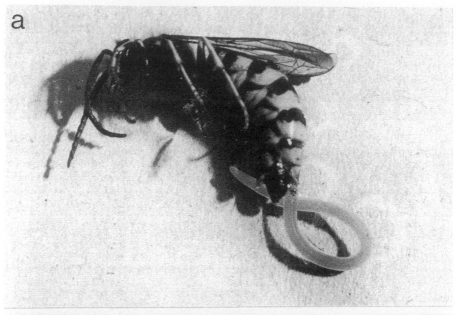

FIGURE 43   (a) Male of *Dolichovespula saxonica* with escaping *Pheromermis vesparum*. (b) Queen of *Vespula vulgaris* with prepared abdomen and escaping parasitic stage larva of *Pheromermis vesparum* (arrow). (After Kaiser, 1987.)

cens, however, under laboratory condition more than 50 parasites per adult locust are necessary to shift the sexual ratio significantly (Craig, 1973). As Webster and Thong (1984) also note, this is due not to the number of parasites in a host but to the food supply within a host. Another factor is certainly whether a juvenile animal or a more developed one is infected, as the latter would offer a far greater food supply. Terrestrial mermithids often have only slight host specificity and in addition to their main hosts can occasionally use other insects as food supply (survival strategy), which offer feeding potential varying with their size.

Sommer (1981) studied the parasite complex of different *Phyllotreta* species (Coleoptera, Halticinae). These very small beetles also sometimes harbored Mermithidae; infection was maximally 10%. The mermithids removed from the flea beetles were determined to be *Hexamermis albicans, H. brevis, Amphimermis elegans, Amphimermis* sp., and *Psammomermis* sp. Without exception, only male animals were removed from these small hosts. The flea beetles are obviously merely occasional hosts which are too small to permit normal growth and reproduction of the nematodes.

Kaburaki and Iyatomi (1933) studied the effects of multiple infections by *Amphimermis zuimushi* in *Chilo simplex*. Of 285 infected rice borers, 80 contained only a single nematode. With the exception of one male and an intersex, all the hatched nematodes in the group were females. In all cases when between two and 36 nematodes parasitized a host, they developed inside the host into males and females. If the number in one host was 39 or more, all the nematodes developed into males.

VII. AGRICULTURAL IMPORTANCE

Mermithids are common parasitoids of agricultural pests. The infection rates are identical to the mortality rates. There are plentiful reports on this subject but most of them only concern individual parasitized insects (see lists in Poinar, 1975; Wouts, 1984; Chatterjee and Singh, 1965). This chapter will discuss only articles that contain more precise information on the nature of the parasitoids, infection rates, biological details, etc.

A. Orthoptera

The extent of mermithid infection in Saltatoria was discussed above. There seems to be no doubt about the importance of mermithids as one of the factors regulating population dynamics in grasshoppers. Mongkolkiti and Hosford (1971) considered the breakdown of a natural population of *Hesperotettix viridis* in North Dakota to be due to parasitization by *Mermis nigrescens*. In his very differentiated analysis of the total parasite complex of grasshopper pests, Baker (1983) showed that in the overlapping area of Diptera (Nemestrinidae, Tachinidae, Sarcophagidae) and mermithids, the control mechanisms is most effective. The significance of mermithids *(Amphimermis, Mermis)* is greatest when punctual rains in early summer permit the infection of nymphs, as this prevents many animals from becoming sexually mature and thus from reproducing. When infection takes place only after egg laying has occurred, the effect of the parasitization is insignificant for the next generation.

B. Hemiptera

There are relatively few reports on mermithoses in hemipteran pests. From Venezuela, Poinar and Linares (1985) reported heavy infestation of *Aenolamia varia*, a sugarcane pest, by *Hexamermis dactylocercus*. In some sugarcane fields, up to 50% of the pests were infected. Small cicadas were often mentioned as mermithid hosts. In the USA, Christie (1963) and Weaver and King (1954) demonstrated *Agamermis decaudata* in small cicadas. In Japan, the Delphacidae *Calligypoda furcifera* and *Nilaparvata oryzae* are infected by *Agamermis unka* (Kaburaki and Imamura, 1932). In Austria, *Agamermis sinuosa* parasitizes *Chlorioidae flava*, *Typhlocyba* sp., and *Liburnia* sp. Other insects in the area studied were not infected (Kaiser, 1977). Finally, *Agamermis paradecaudata* and *Hexamermis microamphidis* were found in Miridae, *Helopeltis antonii*, a pest in tea plantations in Java (Steiner, 1935). More precise data on parasitization rates are not available.

C. Coleoptera

The mermithid fauna of beetles is somewhat better known. The families Chrysomelidae, Curculionidae, and Scarabaeidae are mentioned especially often as mermithid hosts.

In Europe, the mermithids of the field cockchafer *Melolontha melolontha* became known through the classical works of Couturier (1953, 1959, 1963). He described the developmental cycles in France of two biologically very different species, the parthenogenetic *Tunicamermis melolonthae* and the heterosexual *Melolonthinimermis hagmeiri*.

In the USSR, *Psammomermis* species play in important role as opponents of the forest cockchafer. Polozhentsev (1941) and Polozhentsev and Artyukhovsky (1958) found populations of cockchafer larvae in pine woods around Voronezh that showed infection rates of up to 60% by *Psammomermis* sp. In the USA, a *Psammomermis* species has become known as a parasite of the Japanese beetle, *Popilia japonica* (Klein et al., 1976).

The biology of *Hexamermis arvalis*, a parasitoid of the alfalfa weevil, *Hypera postica*, was studied in New York by Poinar and Gyrisco (1962). Although a parasitization rate of 33% was reached in some fields, the authors found that the economic significance of this mermithid is not too great as it does not occur that often.

In France, Couturier (1949) studied the leaf-eating insect fauna on willows on the Garonne in the Gironde from 1932 to 1940. His study included Chrysomelidae (*Phyllodecta vitellinae*, *Plagiodera versicolor*, and *Melasoma populi*) as well as Tenthredinidae, Hymenoptera (*Pteronidae salicis*) and Noctuidae, Leptidoptera (*Apatele leporina*). From year to year, infestation with a *Hexamermis* sp. varied greatly, from 0 to 90%. Preference was shown for the populations of *Melasoma populi*.

The Colorado potato beetle, imported into Europe, showed itself in a number of countries to be a suitable host for native mermithid species. From the USSR, Rubtsov and Koval (1975) reported parasitization rates of 60–100% by the genera *Amphimermis* and *Hexamermis*. Similar parasitization rates are known for Austria (Kaiser, 1972, 1977) and Poland (Stanuszek, 1970) (Fig. 44). Owing to the high parasitization rates, the potato beetle is no longer an economic problem in Austria. An attempt to test this species for introduction into the USA (Nickle and Kaiser, 1984; Kaiser and Nickle, 1986) had to be dropped as this nematode shows too slight host specificity and can also parasitize such useful insects as Coccinellidae.

In South Carolina, Cuthbert (1986) studied a field population of the banded cucumber beetle *(Diabrotica* sp.), a sweet-potato pest. Between August and October, the infestation of beetle larvae with *Oesophagomermis leipsandra* varied between 33% and 100%.

D. Lepidoptera

The lists of lepidoptera parasitized with mermithids are especially long. Usually these are reports of not more closely defined mermithids (see Wouts, 1984), or the parasite is given as *Hexamermis albicans* (in Wouts also *Hexamermis truncata*). In this formulation, *H. albicans* stands for a whole group of not more closely defined *Hexamermis* species. High in-

FIGURE 44   Colorado potato beetle with escaping *Hexamermis* sp. (Courtesy of S. Stanuszek.)

fection rates can also be attained in field populations of lepidopteran larvae. Chatterjee and Singh (1965) report a 73% infection rate of an *Antigastra catalaunis* population on *Sesamum indicum* and 92.5% infection of *Achara janata* on cotton. Kaburaki and Imamura (1932a) found 76% of a population of the rice borer *Chilo simplex* to be infested with *Amphimermis zuimushi*. Polozhentsev and Artyukhovsky (1953) report 60% infestation of gypsy moth caterpillars in the vicinity of Voronezh. *Hexamermis cathetospicula* was found in Malaysia as an opponent of the rice stem borer *Tryporyza incertulae* (Pyralidae) (Poinar and Chang, 1985).

Although mermithid infections have repeatedly been named as an important factor regulating butterfly populations, unequivocal studies on their actual significance are not available.

E. Hymenoptera

Mermithidoses can reach spectacular dimensions in ants, with effects ranging from changes to from to intercastes (see section on host–parasite relationships). Gösswald (1930, 1932) reported from Germany (mid-Main region) on local destruction of ant populations by mermithids. The mermithids themselves were not identified in these studies, although a variety of different species must have been involved.

In the species of the genus *Pheromermis*, parasitoids in ants and Vespidae, parasitism is concentrated in the reproductive stages (Poinar et al., 1976; Kaiser, 1986a, b, 1987). Thus the potency of the population regulatory effect on these hymenopterans is especially great.

On the economic significance of mermithids, one can join Webster (1972, 1980) in saying in summary that they have a notable potency as biological control agents for various insect pests but that our knowledge of diverse aspects of their biology and species composition is still very limited.

REFERENCES

Artyukhovsky, A. K. 1963. A new genus of mermithids, *Melolontinimermis* gen. nov. (Nematoda, Mermithidae). *Meater. nauk. konf. Vsesoyuz. ob–va Gel'mintol.*, Part 1. Moscow, pp. 22–23.

Artyukhovsky, A.K. 1969. New genera of the family Mermithidae (Enoplida, Nematoda). *Zool. Zh.* 48: 1309–1319.

Artyukhovsky, A. K. 1971. On the basis of the subfamilies Mermithinae and Paramermithinae subfam. nov. (Mermithidae, Nematoda). *Sb. Tr., Voronesh. lesotekhnich. Inst.* 33: 56–58.

Artyukhovsky, A. K., and Kharchenko N. A., 1965. Key to mermithids (Mermithidae, Nematoda) of the Streletskaya steppe. *Tr. Tsentralno–Chernosemn. Goszapovednika* 9: 159–185.

Artyukhovsky, A. K., and Karchenko N. A., 1971. New mermithids (Mermithidae, Nematoda) from the central chernozem belt of the Soviet Union. *Tr. Tsentralno–Chernosemn. Goszapovednika.* 11: 109–131.

Baker, G. L. 1981. The seasonal abundance of dipterous and mermithid parasites of wingless grasshopper, *Phaulacridium vittatum* (Sjöst.), in the central tablelands of New South Wales. *Proc. 3rd Austr. Conf. Grassl. Invert. Ecol.* S.A. Govt. Printer, Adelaide, pp. 169–176.

Baker, G. L. 1983. Parasites of locusts and grasshoppers. Agfacts A. E. 2. N. S. W. Dept. of Agriculture, Sydney.

Baker, G. L. 1984. High incidence of mermithid nematode parasitism in grasshopper populations during the 1983/84 season. *Aust. Invert. Pathol. Working Group Newslett.* 5: 15.

Baker, G. L., and Casimir M., 1986. Current locust and grasshopper control technology in the Australasian Region. *Proc. 4th Triennial Meeting, Pan. Am. Acridol. Soc.*, 1985, 191–220.

Baker, G. L., and Poinar G. O., Jr. 1986. *Mermis quirindiensis* n. sp. (Nematoda, Mermithidae), a parasite of locusts and grasshoppers (Orthoptera: Acrididae) in south-eastern Australia. *Rev. Nematol.* 9: 125–134.

Baker, G. L., and Poinar, G. O., Jr. 1988. A description of the male and redescription of female *Mermis athysanota* Steiner, 1921 (Nematoda: Mermithidae). *Rev. Nematol.* 11: 343–350.

Batson, B. S. 1979. Body wall of juvenile and adult *Gastromermis boophthorae* (Nematoda: Mermithidae): Ultrastructure and nutritional role. *Int. J. Parasitol.* 9: 495–503.

Baylis, H. A. 1933. Two new species of the nematode genus *Mermis. Ann. Mag. Natur. Hist.* 105–420.

Baylis, H. A. 1944. Observations on the nematode *Mermis nigrescens* and related species. *Parasitology* 36: 122–132.

Brian, M. V. 1974. Kastendetermination bei *Myrmica rubra* L. In *Sozialpolymorphismus bei Insekten*. Schmidt, G. H. ed. Wiss. Verlagsges., Stuttgart, pp. 565–589.

Chatterjee, P. N., and Singh P., 1965. Mermithid parasites and their roles in natural control of insect pests. *Ind. For.* 91: 714– 721.

Chitwood, B. G., and Chitwood, M. B. 1950. *An Introduction to Nematology* (rev. ed.). Monumental Printing, Baltimore.

Christie, J. R. 1929. Some observations on sex in the Mermithidae. *J. Exp. Zool.* 53: 59–76.

Christie, J. R. 1936. Life history of *Agamermis decaudata*, a nematode parasite of grasshoppers and other insects. *J. Agr. Res.* 52: 161–178.

Christie, J. R. 1937. *Mermis subnigrescens*, a nematode parasite of grasshoppers. *J. Agr. Res.* 55: 353–364.

Cobb, N. A. 1919. *Tetradonema plicans* nov. gen. et spec., representing a new family, Tetredonematidae as now found parasitic in larvae of the midge-insect *Sciara coprophila* Lintner. *J. Parasitol* 5: 176–185.

Cobb, N. A. 1926. The species of *Mermis*, a group of very remarkable nemas infesting insects. *J. Parasitol.* 13: 66–72.

Cobb, N. A., Steiner, G., and Christie, J. R. 1923. *Agamermis decaudata* Cobb, Steiner and Christie; a nema parasite of grasshoppers and other insects. *J. Agr. Res.* 23: 921–926.

Cobb, N. A., Steiner, G., and Christie, J. R. 1927. When and how docs sex arise? *Off. Rec. U.S. Dep. Agr.* 6: 6.

Couturier, A. 1950. Biologie d'un *Hexamermis* (Nematoda Mermithidae) parasite des insectes dèfoliateurs de l'osier. *Ann. Epiphyties* 1: 1–24.

Couturier, A. 1953. Mode de formation d'un kiste simple chez *Tunicamermis melolonthinarum* Cout., nèmatode parasite des larves de Melolonthinae (Colèoptères). *C. R. Acad. Sci.* 236: 1201– 1203.

Couturier, A. 1959. Observation sur le polymorphisme des larves infectieus chez les Mermithidae (Nèmatodes). *C. R. Acad. Sci.* 248: 2123–2125.

Couturier, A. 1960. Observation sur la dysharmonie de taille chez quelques Mermithidae (Nèmatodes). *C. R. Acad. Sci.* 251: 436–438.

Couturier, A. 1963. Recherches sur des Mermithidae, Nèmatodes parasites du hanneton commun (*Melolontha melolontha* L. Colèopt. Scarab.). *Ann. Epiphyties* 14: 203–267.

Craig, S. M. 1973. The host-parasitic relationship between *Mermis nigrescens* Dujardin and the dessert locust, *Schistocerca gregaria* Forskal. M. Sc. Thesis, Simon Fraser University, Burnaby, British Columbia.

Craig, S. M., and Webster, J. R. 1982. Influence of host nutrients on the parasitic development of *Mermis nigrescens* Mermithidae. *J. Nematol.* 14: 398–405.

Crawley, W. R., and Baylis, H. A. 1921. *Mermis* parasitic on ants of the genus *Lasius*. *J. Roy. Microsc. Soc.* 257: 353–372.

Croll, N. A. 1966. *The Behavior of Nematodes: Their Activity, Senses, and Responses*. Edward Arnold, London.

Cuthbert, F. P., Jr., 1986. Bionomics of a mermithid (nematode) parasite of soil-inhabiting larvae of certain chrysomelids (Coleoptera). *J. Invert. Pathol.* 12: 283–287.

Curran, J., and Hominick, W. M. 1981. Description of *Gastromermis metae* sp. n. (Nematoda: Mermithidae) with an assessment of some diagnostic characters and species in *Gastromermis*. *Nematologica* 27: 259–274.

Dujardin, F. 1842. Mèmoire sur la structure anatomique des *Gordious* et d'un autre helminthe, le *Mermis*, qu'on a confondu avec eux. *Ann. Sci. Natur. Zool.*, 2nd series, 18: 129–151.

Ellenby, C. 1964. Haemoglobin in the chromotrope of an insect parasitic nematode. *Nature*, 202: 615–616.

Emery, C. 1904. Zur Kenntnis des Polymorphismus der Ameisen. *Zool. Jb.*, Suppl. VII: 587–610.

Filipjev, I. N. 1934. *Parasites of Insects*. Chapter 5, Harmful and useful nematodes in rural economy. OGIZ–Sel'khozgiz, Moscow.

Gafurov, A. K. 1982. New species of Mermithidae (Nematoda) from Lepidoptera and Orthoptera. *Izvestiya Akademii Nauk Tadzhikskoi SSR, Biologicheskie Nauki* 3: 18–24.

Glaser, R. W., and Wilcox, A. M. 1918. On the occurrence of a *Mermis* epidemic amongst grasshoppers. *Psyche* 25: 12–15.

Gleiss, H. G. W. 1955. Der Knäuelwurm *Hexamermis cornuta* nov. sp. (Mermithidae, Nematoda) als Endoparasit des Kartoffelkäfers. *Zool. Anz.* 155: 139–143.

Goodey, T. 1963. *Soil and Freshwater Nematodes*, 2nd ed., rewritten by J. B. Goodey. Methuen, London; John Wiley and Sons, New York.

Gordon, R., Bailey, C. H., and Barber, J. M. 1974. Parasitic development of the mermithid nematode *Reesimermis nielseni* in the larval mosquito *Aedes aegypti. Can. J. Zool.* 52: 1293–1302.

Gordon, R., and Webster, J. M. 1971. *Mermis nigrescens*: Physiological relationship with its host, the adult desert locust *Schistocerca gregaria. Exp. Parasitol.* 29: 66–79.

Gordon, R., and Webster, J. M. 1972. Nutritional requirements for protein synthesis during parasitic development of the entomophilic nematode *Mermis nigrescens. Parasitology* 64: 161–172.

Gordon, R., Webster, J. M., and Med, D. E. 1971. Some effects of the nematode *Mermis nigrescens* upon carbohydrate metabolism in the fat body of its host, the desert locust *Schistocerca gregaria. Can. J. Zool.* 49: 431–434.

Gösswald, K. 1930. Weitere Beiträge zur Verbreitung der Mermithiden bei Ameisen. *Zool. Anz.* 90: 13–27.

Gösswald, K. 1932. Ökologische studien über die Ameisenfauna des mittleren Maingebietes. *Z. wiss. Zool.* 142: 1–125.

Hagmeier, A. 1912. Beiträge zur Kenntnis der Mermithiden. I. Biologische Notizen und systematisch Beschreibung einiger alter und neuer Arten. *Zool. Jb. Syst.* 32: 521–612.

Kaburaki, T., and Imamura, S. 1932a. A new mermithid-worm parasitic in the rice borer with notes on its life history and habits. *Proc. Imp. Acad. Japan* 8: 109–112.

Kaburaki, T., and Imamura, S. 1932b. Mermithid-worm parasitic in leaf-hoppers, with notes on its life history and habits, *Proc. Imp. Acad. Japan* 8: 136–141.

Kaburaki, T., and Iyatomi, K. 1933. Notes on sex in *Amphimermis zuimushi* Kab. et Im. *Proc. Imp. Acad. Japan* 9: 333–336.

Kaiser, H. 1972. Mermithidae (Nematoda) als Parasiten des Kartoffelkäfers (*Leptinotarsa decemlineata* Say) in der Steuermark. Ph. D. Thesis, Karl-Franzens-Universität, Graz, Austria.

Kaiser, H. 1977. Untersuchungen zur Morphologie, Biometrie, Biologie und Systematik von Mermithiden. Ein Beitrag zum Problem der Trennung morphologisch schwer unterscheidbarer Arten. *Zool. Jb. Syst.* 104: 20–71.

Kaiser, H. 1983. Phylogenetic relationships in the Mermithidae (Nematoda) based on traditional and physiological evidence. In *Systematics Association Special. Vol. 22. Concepts in Nematode Systematics.* Stone, A. R., Platt, H. M., and Khalil, L. F. eds. Academic Press, London, pp. 249–259.

Kaiser, H. 1984. Die Gattung *Psammomermis* nov. copmb. (Mermithidae, Nematoda) – Morphologie, Beschreibung und Verbreitung einiger neuer Arten. *Mitt. Naturwiss. Ver. Steiermark* 114: 303–330.

Kaiser, H. 1986a. Morphologische Analyse des Ameisen-Parasitoiden *Pheromermis villosa* n. sp. (Nematoda, Mermithidae). *Mitt. Naturwiss. Ver. Steiermark* 116: 269–294.

Kaiser, H. 1986b. Über Wechselbeziehungen zwischen Nematoden (Mermithidae) und Ameisen. *Zool. Anz.* 217: 156–177.

Kaiser, H. 1987. Biologie, ökologie und Entwicklung des europäischen Wespen-Parasitoiden *Pheromermis vesparum* n. sp. (Mermithidae, Nematoda). *Zool. Jb. Syst.* 421–449.

Kaiser, H., and Nickle, W. R. 1985. Mermithiden (Mermithidae, Nematoda) parasitieren Marienkäfer (Coccinella septempunctata L.) in der Steiermark. *Mitt. Naturwiss. Ver. Steiermark* 115: 115–118.

Kaiser, H., and Skotitsch, G. 1981. Zur fragedes systematischen Wertes von Larval- und Adultmerkmalen von Mermithiden (Nematoda) unter Berücksichtigung morphologischer, biologischer und immunologischer Merkmale. *Zool. Jb. Syst.* 108:70–83.

Kharchenko, N. A. 1966. A new mermithid, *Hexamermis artyukhovskii* sp. n. (Mermithidae, Nematodes). *Zool. Zh.* 45: 494–499.

Kirjanova, E. S., Karaveva, R. P., and Romanenko, K. E. 1959. Mermithids (Mermithidae, Nematodes)—parasites of *Hypomoneuta malinella* and H. *padella* in southern Kirgiz. *Tr. Kirg. Lecn. Op. Stan.* 2: 195–240.

Klein, M. G., Nickle W. R., Benedict, P. R., and Dunbar, D. E. 1976. *Psammomermis* sp. (Nematoda: Mermithidae): A new nematode parasite of the Japanese Beetle, *Popilia japonica* (Coleoptera: Scarabaeidae). *Proc. Helm. Soc. Wash.* 43: 236–236.

Kloft, W. 1950. Ökologische Untersuchungen zur Verbreitung der Mermithiden bei Ameisen. *Zool. Jb. Syst.* 78: 526–530.

Leidy, L. 1875. Notes on some parasitic worms. *Proc. Acad. Natur. Sci. Phila.* 27: 14–16, 17–18, 400.

Linstow von, O. F. B. 1897. Nemathelminthen grösstenteils in Madagascar gesammelt. *Arch. Naturgesch.* 1: 27–34.

Linstow von, O. F. B. 1905. Helminthologische Beobachtungen. *Arch. Mikroskop. Anat.* 66: 355–366.

Lorenzen, S. 1981. Entwurf eines phylogenetischen System der freilebenden Nematoden. *Veröff. Inst. Meeresforsch. Bremerh.* Suppl. 7: 1–472.

Meissner, G. 1854. Beiträge zur Anatomie und Physiologie von *Mermis albicans*. *Z. Wiss. Zool.* 5: 207–284.

Mongkolkiti, S., and Hosford, R. M., Jr. 1971. Biological control of the grasshopper *Hesperotettix viridis pratensis* by the nematode *Mermis nigrescens*. *J. Nematol.* 3: 256–363.

Mrazek, A. 1908. Myrmekologiske poxnamky. III. Brachypterni mermithogyni u *Lasius alienus*. *Act. Soc. Ent. Bohem.* 5: 1–8.

Nickle, W. R. 1972. A contribution to our knowledge of the Mermithidae (Nematoda). *J. Nematol.* 4: 113–146.

Nickle, W. R., and Kaiser, H. 1984. An Austrian mermithid nematode parasite offers biological control of the Colorado potato beetle, *Leptinotarsa decemlineata* (Say). *Proc. Helm. Soc. Wash.* 5: 340– 341.

Orton William, K. J. 1984. *Mermis savaiiensis* n. sp. (Mermithidae: Nematoda) from western Samoa. *Syst. Parasitol.* 6: 257–260.

Passera, L. 1976. Origine des intercastes dans les sociètès de *Pheidole pallidula* (Ny L.) (Hymenoptera, Formicidae) parasitèes par *Mermis* sp. (Nematoda, Mermithidae). *Insectes Soc.* 23: 559–575.

Petersen, J. J. 1978. Observations on the mass production of *Romanomermis culicivorax*, a nematode parasite of mosquitoes. *Mosq. News* 38: 83–86.

Poinar, G. O., Jr., 1968. Parasitic development of *Filipjevimermis leipsandra* Poinar and Welch (Mermithidae) in *Diabrotica udecimpunctata* (Chrysomelidae). *Proc. Helm. Soc. Wash.* 35: 161–169.

Poinar, G. O., Jr., 1975. *Entomogenous Nematodes: A Manual and Host List of Insect– Nematode Associations.* E. J. Brill, Leiden.

Poinar, G. O., Jr., 1981. *Thaumamermis corgrovei* n. gen., n. sp. (Mermithidae: Nematoda) parasitizing terrestrial isopods (Isopoda: Oniscoidea). *Syst. Parasitol.* 2: 261–266.

Poinar, G. O. Jr., 1985. Nematode parasitism of spiders and harvestman. *J. Arachnol.* 13: 121–128.

Poinar, G. O., Jr., and Benton C. L. B., Jr. 1986. *Aranimermis aptispicula* n. g., n. sp. (Mermithidae: Nematoda), a parasite of spiders (Arachnida: Araneida). *Syst. Parasitol.* 8: 33–38.

Poinar, G. O., Jr., and Chang, P. 1985. *Hexamermis cathetospiculae* n. sp. (Mermithidae: Nematoda), a parasite of the rice stemborer, *Tryporyza incertulas* (Wlk.) (Pyralidae: Lepidoptera) in Malaysia. *J. Nematol.* 17: 360–363.

Poinar, G. O., Jr., and Gyrisco, G. G. 1962a. A new mermithid parasite of the alfalfa weevil, *Hypera postica* (Gyllenhal). *J. Insect Pathol.* 4: 201–206.

Poinar, G. O., Jr., and Gyrisco, G. G. 1962b. Studies on the bionomics of *Hexamermis arvalis* Poinar and Gyrisco, a mermithid parasite of the alfalfa weevil, *Hypera postica* (Gyllenhal). *J. Insect Pathol.* 4: 469–483.

Poinar, G. O., Jr., and Hess, R. 1977. *Romanomermis culicivorax*: morphological evidence of transcuticular uptake. *Exp. Parasitol.* 42: 27–33.

Poinar, G. O., Jr., Lane, R. S., and Thomas, G. M. 1976. Biology and redescription of *Pheromermis pachysoma* (V. Linstow) n. gen., n. comb. (Nematoda: Mermithidae), a parasite of yellowjackets (Hymenoptera: Vespidae). *Nematologica* 22: 360–370.

Poinar, G. O., Jr., and Linares, B. 1985. *Hexamermis dactylocercus* sp. n. (Mermithidae: Nematoda), parasite of *Aeneolamia varia* (Cercopidae: Homoptera) in Venezuela. *Rev. Nematol.* 8: 109–111.

Poinar, G. O., Jr., Mondet, B., Gouteux, J. P., and Laveissièr, C. 1981. *Hexamermis glossinae* sp. nov. (Nematoda: Mermithidae), a parasite of tsetse flies in West Africa. *Can. J. Zool.* 59: 858–861.

Poinar, G. O., Jr., Remillet, M., and Van Waerebeke, D. 1978. *Mermis changodudus* sp. n. (Mermithidae) a nematode parasite of Heteronychus beetles (Scarabaeidae) in Madagascar. *Nematologica* 24: 100–104.

Poinar, Jr., G. O., and Welch, W. E. 1968. A new nematode, *Filipjevimermis leipsandra* sp. n. (Mermithidae), parasitic in Chrysomelid larvae (Coleoptera). *J. Invert. Pathol.* 12: 259–262.

Polozhentsev, A. P. 1941. The mermithid fauna of *Melolontha hippocastani* Fabr. *Tr. Bashk. Nauch.-Issled. Veterin. Stantsii* 3: 301–334.

Polozhentsev, A. P. 1952. New Mermithidae of sandy soil of pine forest. *Tr. Helm. Lab.* 6: 376–382.

Polozhentsev, A. P. 1953. On the nematodes of the family Mermithidae Braun, 1883. *Tr. Gel'mintol. 75th let Skrjabin.* 532– 542.

Polozhentsev, A. P., and Artyukhovsky, A. K. 1953. Beneficial worm–destroyers of the gipsy moth. *Lesnoe Khozyastvo* 4: 46.

Polozhentsev, A. P., and Artyukhovsky, A. K. 1958. New mermithid species and genera: *Complexomermis ghilarovi* n. sp.; *Hexamermis kirjanovae* n. sp.; *Filipjevimermis* n. g.; *Filipjevimermis paramovi* n. sp. *Zool. Zh.* 37: 997–1005.

Polozhentsev, A. P., and Artyukhovsky, A. K. 1959. On the systematics of the family Mermithidae Braun, 1883 (Dorylaimata, Enoplida). *Zool. Zh.* 38: 816–828.

Polozhentsev, A. P., Artyukhovsky, A. K., and Kharachenko, N. A. 1965. A new mermithid *Hexamermis pratensis* n. sp. from a meadow bottom sand soil flooded by the Khoper river. *Materials from Scientific Conference All Union Soviet of Gel'minthologists,* pp. 209–213.

Rathke, B. 1953. Zur Biologie von Hexamermis albicans (Siebold). *Veroff. überseemus. Bremen* 2: 161–210.

Rauther, M. 1906. Beiträge zur Kentnis von *Mermis albicans* v. Sieb., mit besonderer Berücksichtigung des Haut-Nerven-Muskelsystems. *Zool. Jb. Abt. Anat.* 23: 1–76.

Rubtsov, I. A. 1969. On a new species of *Agamermis* (Mermithidae) of a bug in Poland. *Acta Parasitol. Pol.* 16: 97–100.

Rubtsov, I. A. 1970a. A new species of mermithids from ants. *Parasitologij* 4: 338–341.

Rubtsov, I. A. 1970b. A new species of mermithid from a bug. New and little known species of the fauna of Siberia. *Novosibirsk. Sect.* 3: 102–106.

Rubstov, I. A. 1971. New species of mermithids from the south marine territory. *Zool. Zh.* 50: 1143–1153.

Rubtsov, I. A. 1972a. Mermithids from Bashkiria. *Zool. Zh.* 51: 954–962.

Rubtsov, I. A. 1972b. *Aquatic Mermithids.* Part 1. Nauka, Leningrad.

Rubtsov, I. A. 1974. Aquatic mermithids. Part 2. Nauka, Leningrad.

Rubtsov, I. A. 1976a. New species of mermithids. *Zool. Zh.* 55: 1292–1298.

Rubtsov, I. A. 1976b. Mermithids (Nematoda. Mermithidae) – parasites of insects from Mongolia. In *Nasekomye Mongolii IV.* Nauka, Leningrad, pp. 596–614.

Rubtsov, I. A. 1978. *Mermithidae: Classification, Importance, Application.* Nauka, Leningrad.

Rubtsov, I. A., and Koval Yu V. 1975. Mermithids of the Colorado Potato Beetle. *Proceedings of the All-Union Scientific-Research Institute for Plant Protection,* Vol. 44, Nauka, Leningrad, pp. 126–153.

Rudolphi, C. A. 1809. *Entozoorum sive vermium intestinalium historia naturalis,* Vol. 2, Amsterdam.

Rutherford, T. A., and Webster, J. M. 1974. Transcuticular uptake of glucose by the entomophilic nematode, *Mermis nigrecens. J. Parasitol.* 60: 804–808.

Rutherford, T. A. and Webster, J. M. 1978. Some effects of *Mermis nigrescens* on the hemolymph of *Schistocerca gregaria. Can. J. Zool.* 56: 339–347.

Rutherford, T. A., Webster, J. M., and Barlow, J. S. 1977. Physiology of nutrient uptake by the entomophilic nematode *Mermis nigriscens* (Mermithidae). *Can. J. Zool.* 55: 1773–1781.

Schuurmans Stekhoven, J. H., Jr., and Mawson, P. M. 1955. Mermithidès d'Alsace. *Ann. Parasitol. Hum. Comp.* 50: 69–82.

Siebold, C. T. E. von. 1848. Über die Fadenwürmer der Insekten. *Entomol Z.* 9: 290–300.

Siebold, C. T. E. von. 1854. Beitrag zur Naturgeschichte der Mermithiden. *Z. Wiss. Zool.* 5: 201–206.

Smith, R. W. 1958. Parasites of nymphal and adult grasshoppers (Orthoptera: Acrididae) in western Canada. *Can. J. Zool.* 36: 217–262.

Sommer, G. 1981. Biology and parasites of *Phyllotreta* spp. (Coleoptera, Halticinae). Commonwealth Institute of Biological Control, European Station Report, Delemont.

Stanuszek, S. 1970. Pathogenicity of nematodes of the genus *Hexamermis* (Steiner, 1934) in relation to the female Colorado beetle (*Leptinotarsa decemlineata* Say). *Proc. IX. Int. Nematol. Symposium* (Warsaw, 1967), pp. 359–367.

Steiner, G. 1921. Beiträge zur Kenntnis der Mermithiden. T. I. Mermithiden von- Neu-Mecklenburg und Revison einiger v. Linstowscher Arten und Rudolphis, *Filaria truncatula* = *Mermis truncatula. Centralbl. Bakt. Abt.* 1: 57.

Steiner, G. 1924 Beiträge zur Kenntnis der Mermithidae. 2. Teil. Mermithiden aus Paraguay in der Sammlung des Zoolog. Meseums zu Berlin. *Centralbl. Bakt. Parasitenk. Infektionsk.* 62: 90–110.

Steiner, G. 1925. Mermithids parasitic in the tea bug (*Helopeltis antonii* Sign.) *Med. Proefstation Thee, Batavia* 94: 10–16.

Steiner, G. 1932. Die arktischen Mermithiden, Gordioiden und Nectonematoiden. *Fauna Arktika* 6: 161–174.

Sugiyama, K. 1958. Effects of the parasitism by a nematode on a grasshopper, *Oxya japonica*. I. Effects of the parasitism on wing length, pronotal length and genitals. *Zool. Mag.* 65: 382–385.

Vandel, A. 1927. Modification dèterminès par un Nèmatode du genre "Mermis" chez les ouvriès et les soldats de la fourmi *Pheidole pallidula*, Nyl. *Bull. Biol. France et Belg.* 61: 38–48.

Vandel, A. 1930. La production chez la fourmi *Pheidole pallidula*, sous l'action de parasites du genre *Mermis. C. R. Acad. Sci.* 190: 770–772.

Vandel, A. 1934. Le cycle èvolutif d'*Hexamermis* sp. pàrasite de la fourmi *Pheidole pallidula. Ann. Sci. Natur. Ser. Bot. Zool.* 47: 47–58.

Weaver, C. R., and King, O. R. 1954. Meadow Spittlebug. *Ohio Agr. Exp. Stat. Bull.* 741.

Webster, J. M. 1972. Nematodes and biological control. In *Economic Nematology*. Webster, J. M. ed. Academic Press, London, pp. 469–496.

Webster, J. M. 1980. Biocontrol: The potential of entomophilic nematodes in insect management. *J. Nematol.* 12: 270–278.

Webster, J. M., and Thong, C. H. S. 1984. Nematode parasites of Orthopterans. In *Plant and Insect Nematodes*. Nickle, W. R. ed. Marcel Dekker, New York, pp. 697–726.

Welch, H. E. 1963. *Amphimermis bogongae* sp. nov. and *Hexamermis cavicola* sp. nov. from the Australian bogong moth, *Agrotis infusa* (Boisd.), with a review of the genus, *Amphimermis* Kaburaki and Imamura, 1932. *Parasitology* 53: 55–62.

Wheeler, W. M. 1907. The polymorphism of ants, with an account of some singular abnormalities due to parasitism. *Bull. Am. Mus. Natur. Hist.* 23: 1–93.

Wheeler, W. M. 1928. *Mermis* parasitism and intercastes among ants. *J. Exp. Zool.* 50: 165–237.

Wouts, W. M. 1981. *Hexamermis truncata* (Rudolphi, 1809) new combination, the valid name for *Hexamermis albicans* (von Siebold, 1848). *Syst. Parasitol.* 3: 127–128.

Wouts, W. M. 1984. Nematode parasites of Lepidopterans. In *Plant and Insect Nematodes*. Nickle, W. R. ed. Marcel Dekker, New York, pp. 655–696.

22

Sphaerularioid Nematodes of Importance in Agriculture

MICHEL REMILLET* *University of Cairo, Cairo, Egypt*

CHRISTIAN LAUMOND *Institut National de la Recherche Agronomique, Antibes, France*

I. INTRODUCTION

Sphaerularioid nematode parasites in the body cavity of insects are represented by 208 species distributed in 29 genera and 6 families excluding phoretic nematodes. Six genera of Sphaerularioids are monospecific and 10 genera have 2–5 species.

Earlier contributors to the taxonomy and biology of the Sphaerularioids include Dufour (1837), Leuckart (1884), von Linstow (1890, 1893), zur Strassen (1892), Fuchs (1913, 1915, 1929, 1937, 1938), Cobb (1920, 1921), T. Goodey (1930, 1931), Bovien (1932, 1937, 1944), Filipjev (1934), Chitwood (1935), Chitwood and Chitwood (1937), and Christie (1938, 1941). From the comprehensive works published by Rühm (1954, 1956, 1960) and Wachek (1955) in Germany, studies on the taxonomy of Sphaerularioids developed in several countries, particularly the USA: Nickle (1963a, 1967a, 1984), Massey (1974), Poinar (1975, 1977, 1979); USSR: Slankis (1969a,b, 1972, 1974), Slobodyanyuk (1975a,b, 1976, 1984, 1986), Blinova and Gurando (1977), Blinova and Korenchenko (1986), Rubzov (1981, 1982); India: Reddy and Rao (1980, 1981, 1987); France: Remillet and Van Waerebeke (1975, 1976, 1978), Laumond and Beaucournu (1978), Laumond and Mauleon (1982), Laumond and Bonifassi (1991), Deunff and Launay (1984), Deunff et al. (1985), Launay et al. (1983), Launay and Deunff (1984, 1990).

Identification of the species is difficult because available keys are based on various combinations of morphological, biological, life history characters (Nickle, 1967a; Remillet, 1988). Placement of species at the genus level was greatly facilitated by the recent comprehensive generic diagnoses of Siddiqi (1986a). A most important advancement in the systematics and biology of Sphaerularioids has been the discovery of alternation of generations. This phenomenon, first described by Bovien (1937) with *Heterotylenchus aberrans* as an exception, is now known to occur in 11 genera. Alternation of generations, however, is not restricted to the insect nematodes. Siddiqi (1986a) presents evidence that this phenome-

*_Present affiliation_: Institut Français de Recherche Scientifique pour le Développement en Coopération, Paris, France

non also occurs in *Stictylus* Thorne, 1941, and *Sphaerulariopsis* Wachek, 1955 (both genera synonimized with *Prothallonema* Christie, 1938), in *Iotonchium* Cobb, 1920, in *Fungiotonchium* Siddiqi, 1986a and in the Paurodontidae Thorne, 1941.

The complex host-parasite relationships of Sphaerularioids remain little known. Survival and perpetuation of the nematode is insured by the annual rate of parasitism (rarely exceeding 25%), the host's fecundity reduction (but complete sterility is rare), the dissemination of juvenile nematodes by living adult insects, the adaptation of the length of the free-living period of infective females, and the synchronization with the host larval development period. Free-living or plant parasitic generations allow the survival of the nematodes in the absence of hosts.

These highly specialized adaptations lead to a high degree of specificity between the nematode and insect species, as evidenced by the ratio of 908 species of Sphaerularioids parasite to 335 species in 132 genera of insects (53 families, 6 orders) and 6 genera of Acarina (Appendix 1, Appendix 2). Sometimes one species is specific to one host. This specificity and the complex balance maintained between hosts and parasites are limiting factors in the use of Sphaerularioids in biocontrol programs.

II. TAXONOMY

The adopted classification is as follows:

Superfamily: Sphaerularioidea Lubbock, 1861 (Poinar, 1975)
Family: Allantonematidae Pereira (1932) in Chitwood and Chitwood, 1937
      Parasitylenchidae Siddiqi, 1986
      Iotonchiidae Goodey, 1953
      Sphaerulariidae Lubbock, 1861
      Fergusobiidae Goodey, 1963
      Phaenopsitylenchidae Blinova and Korenchenko, 1986

All the insect parasitic Tylenchida are placed in the superfamily Sphaerularioidea.

The family Allantonematidae, with a single heterosexual generation cycle, represents the basic type of Sphaerularioidea.

Parasitylenchidae are quite similar to Allantonematidae in their morphology. They are characterized by the presence of a secondary heterosexual or parthenogenetic generation.

Iotonchiidae also have a primary heterosexual generation alternating with a secondary parthenogenetic generation, but they have a characteristic morphology (L-shaped spicules) which may indicate a different phylogenetic origin (Siddiqi, 1986a).

Sphaerulariidae have hypertrophied or everted uterus and a single heterosexual generation cycle similar to that of Allantonematidae. The merger of the Sphaerulariidae (previously of Aphelenchid status) with the Allantonematids (Nickle, 1967a) was discussed by Poinar (1975) because the position of the substylet dorsal gland opening was unclear. According to Maggenti (1981) and Siddiqi (1986a), this family is considered as belonging to the Sphaerularioidea.

Members of the Fergusobiidae and Phaenopsitylenchidae present both parasitic and free-living generations. They were previously placed in the family Neotylenchidae (Poinar, 1975, 1977; Maggenti, 1981; Siddiqi, 1986a) because the free-living generations resemble *Deladenus* in many respects. Meanwhile, the parasitic heterosexual generation is morphologically perfectly similar to that of the other members of Sphaerularioidea. On the other hand, these two families have their free-living generations that must obligatorily alternate with the parasitic generation to complete the life cycle. It is therefore proposed here to place the Fergusobiidae and Phaenopsitylenchidae in the Sphaerularioidea with the other true insect parasitic nematodes.

Superfamily: Sphaerularioidea Lubbock, 1861 (Poinar, 1975)

Diagnosis (modified after Siddiqi, 1986a): Tylenchida. Obligate insect or mite parasitism (except for *Beddingia* parasitizing siricids, which can establish successive generations of fungal feeding populations). Female of several genera di–, tri–, tetramorphic according to according to feeding habits. Both entomoparasitic and free-living generations found in several genera. In entomoparasitic generation, only the adult female is parasitic in insect or mite hemocele. Several types of life cycles possible: one or two parasitic heterosexual generations, alternation of parasitic heterosexual and parasitic parthenogenetic generations, alternation of parasitic heterosexual and free-living heterosexual or parthenogenetic generations. Hemocele-inhabiting female usually has microvilli on body surface and canallike formation in body wall indicating absorption of food from general body surface. Cuticle smooth or finely annulated. Phasmids absent. Stylet generally under 20 µm long (hypertrophied in infective female and gravid parasitic female). Esophagus usually may not be divisible into corpus, isthmus, and basal region. No muscular median bulb (*Fergusobia* contains a pseudovalve in the metacorporal region). Three esophageal glands, but only two reported for *Sphaerularia* and *Tripius*, contained in a basal bulb or extending over intestine. Orifice of dorsal esophageal gland close to or at some distance behind stylet (in Sphaerulariidae, the position of the orifice is unclear).

Females: monodelphic, prodelphic. Preadult entomoparasitic female (infective female) with hypertrophied stylet (= pseudostylet) and esophagus, small vulva, elongated uterus serving as storage for sperm often numbering several hundred, usually still within the larval cuticle. In gravid parasitic female, hypertrophied stylet present but difficult to observe, sperm stored in membranous pouches localized in anterior region of elongated uterus. Uterus often with several eggs and/or juveniles, sometimes everted and hypertrophied to lead an independent life (= uterium of *Sphaerularia*). Ovary single, outstretched, reflexed at tip or coiled; a rachis may be present. Vagina may be tuboid and strongly muscular. Vulva a large transverse slit, oval, or small porelike, located posteriorly usually at over 85% of body length.

Males: with or without a stylet. Esophagus as in female, or rarely degenerated. Testis single, outstretched or with tip reflexed. Several thousand round or ameboid sperm produced by forms of entomoparasitic generation, much less by plant parasitic or mycetophagous forms. Spicules small (usually under 30 µm long); paired, arcuate, cephalated or in Iotonchiidae large, robust, and angular. Gubernaculum simple, fixed, may be lacking. Bursa simple, may be absent.

Family: Allantonematidae Pereira (1932)
in Chitwood and Chitwood, 1937

Diagnosis Sphaerularioidea. Only one heterosexual generation known. The male, never or rarely parasitic, has a very short life in the environment and dies soon after mating. The partially free-living fertilized female also has a short free life before penetrating a new host. Entomoparasitic female develops into obese, sausage-shaped, elongate saclike or oval, mature parasitic female with little mobility; most of the body cavity is occupied by reproductive organs. Partially free-living female thin, very active; stylet strong; three esophageal glands, elongated, well developed; ovary reduced to few cells, uterus elongated, packed with minute sperm. Male similar to partially free-living female but with shorter esophageal glands; testis outstretched; spicule arcuate; gubernaculum rarely absent; bursa present or absent.

Subfamilies

Parasitic female round, oval, or elongate saclike, never dorsally curved; vulva a transversal slit or indistinct: Allantonematinae Pereira (1932) (emended in Chitwood, 1935) in Chitwood and Chitwood, 1937.

Parasitic female small to large, dorsally curved, vulva prominent, deeply cleft: Contortylenchinae Rühm, 1956.

Allantonematinae Pereira (1932) (emended in Chitwood, 1935) in Chitwood and Chitwood, 1937

Type genus: *Allantonema* Leuckart, 1884

Key to genera [modified after Poinar (1977) and Siddiqi (1986a)]

1. Partially free-living females apparently lacking stylet . . . .
. . . . *Bradynema* zur Strassen, 1892
   Partially free-living females possessing stylet 2
2. Partially free-living males apparently lacking stylet; esophagus completely degenerate 3
   Partially free-living males possessing stylet; esophagus normal or partially degenerate 4
3. Partially free-living forms generally with clavate tails; parasitic female small, round to oval, can accommodate one or two eggs; parasites of thrips . . . .
. . . . *Thripinema* Siddiqi, 1986
   Partially free-living forms without clavate tails; parasitic female large, generally tuboid, can accommodate a number of eggs or juveniles; not parasites of thrips. .
. . . . *Howardula* Cobb, 1921
4. Parasitic female round, oval; or bean-shaped; lip region of partially free-living forms offset: . . . . *Allantonema* Leuckart, 1884
   Parasitic female elongate, sausage-shaped; lip region of partially free-living forms not offset 5
5. Excretory pore at anterior margin of nerve ring or more anterior . . . .
. . . . *Metaparasitylenchus* Wachek, 1955 (Nickle, 1967)
   Excretory pore posterior to nerve ring 6
6. Bursa absent . . . . *Protylenchus* Wachek, 1955
   Bursa present 7
7. Body surface of parasitic female wavy, with constrictions and swellings; stylet sunken into body . . . . *Sulphuretylenchus* Rühm, 1956 (Nickle, 1967)
   Body surface of parasitic female neither wavy nor with constrictions and swellings; stylet in normal position (except occasionally in *Parasitylenchoides*) 8
8. Excretory pore in partially free-living forms at 105–125 µm from anterior end; stylet of partially free-living female distinctly knobbed . . . .
. . . . *Parasitylenchoides* Wachek, 1955
   Excretory pore in partially free-living forms at less than 100 µm from anterior end
. . . . 9
9. Partially free-living forms with stylet distinctly knobbed; parasites of staphilinid beetles . . . . *Proparasitylenchus* Wachek, 1955 (Nickle, 1967)
   Partially free-living forms with stylet with basal thickenings, without well-developed knobs . . . . *Neoparasitylenchus* Nickle, 1967

Contortylenchinae Rühm, 1956

Type genus: *Contortylenchus* Rühm, 1956

Key to genera (modified after Siddiqi, 1986a)

1. Stylet with basal knobs or thickenings; parasites of Coleoptera 2
   Stylet without basal knobs or thickenings; parasites of Siphonaptera ....
   .... *Spilotylenchus* Launay, Deunff, and Bain, 1983
2. Parasitic female under 0.8 mm long; *a* = under 8; parasites of Scolytidae ....
   .... *Bovienema* Nickle, 1963
   Parasitic female over 0.8 mm long; *a* = over 10 3
3. Parasitic female strongly dorsally arcuate; parasites of Scolytidae ....
   .... *Contortylenchus* Rühm, 1956
   Parasitic female slightly dorsally arcuate or straight; parasites of Cerambycidae ..
   .... *Aphelenchulus* Cobb, 1920

Remarks on *Prothallonema* Christie, 1938: *Prothallonema* may be a valid genus with the only new combination *Prothallonema intermedium* proposed by Siddiqi (1986a). This genus is supposed to present an alternation of an insect parasitic heterosexual generation and a free-living generation (Khan, 1957a, b). More studies are needed because males, entomoparasitic forms, and host remain unknown. The female of the free-living generation was described as *Stictylus intermedius* by Geraert et al. (1984). The infective female of the partially free-living generation was described as *Prothallonema dubium* by Christie (1938) and as *Howardula dubia* by Nickle (1965). Heterosexual generation is not known. The synonymization of *Stictylus* and *Sphaerulariopsis* with *Prothallonema* as proposed by Siddiqi (1986a) seems to be hypothetical. Insect host unknown.

Parasitylenchidae Siddiqi, 1986

Diagnosis (modified after Siddiqi, 1986a): Sphaerularioidea. Two or three types of adult present in host's body cavity. Primary heterosexual generation alternating with a secondary heterosexual or parthenogenetic generation. Rarely, supposed nonfeeding parthenogenetic female of the secondary generation occurs in the environment (*Heteromorphotylenchus*). Primary heterosexual female not curving spirally when relaxed. Female stylet generally under 18 μm long, distinctly knobbed. Orifice of dorsal gland close to or farther behind stylet base. Excretory pore generally anterior to nerve ring. Vulva less than two body widths from anus, lips not modified; vagina poorly muscular. Postvulval uterine sac absent. Female tail short, subcylindroid or conoid. Male may occur in host's body cavity (*Parasitylenchus*). Spicules slender, ventrally arcuate, about 20 μm or less long (but 27–30 μm long in *Parasitylenchus macrobursatus*). Bursa enveloping tail, may be absent (*Heterotylenchus*). Gubernaculum present or absent.

Key to subfamilies of Parasitylenchidae (after Siddiqi, 1986a)

1. Copulation takes place in host's body cavity ....
   .... Parasitylenchinae Siddiqi, 1986
   Copulation does not take place in host's body cavity 2
2. Parthenogenetic female and its eggs found in host's body cavity ....
   .... Heterotylenchinae Siddiqi, 1986
   Parthenogenetic female and its eggs found in environment ....
   .... Heteromorphotylenchinae Siddiqi, 1986

Parasitylenchinae Siddiqi, 1986

Diagnosis (after Siddiqi, 1986a): Parasitylenchidae. Five distinct adult forms, with two types of alternating heterosexual generations. Three types of adult in host's body cavity—primary heterosexual female, secondary heterosexual female, and male. Copulation takes place in host's body cavity. Primary heterosexual female sausage– or spindle-shaped, produces heterosexual forms which multiply in the host's body cavity. The latter produce male and female juveniles that leave host for further development and copulation in the environment.

Type genus: *Parasitylenchus* Micoletzky, 1922
No other genus.

Heterotylenchinae Siddiqi, 1986

Diagnosis (after Siddiqi, 1986a): Parasitylenchidae. Four distinct adult forms, with alternation of heterosexual and parthenogenetic generations. Two types of adults in host's body cavity, a primary gamogenetic female and a secondary parthenogenetic female. Adult male not found in host's body cavity. Primary heterosexual female produces only female juveniles that develop to parthenogenetic females, which in turn produce male and female juveniles that quit the host for further development and copulation.

Type genus: *Heterotylenchus* Bovien, 1937

Key to genera of Heterotylenchinae

1. Parasitic females dorsally curved 2
   Parasitic females not dorsally curved 3
2. Both parasitic heterosexual and parthenogenetic females dorsally curved, with ventral side turned outward; parasites of Siphonaptera ....
   .... *Psyllotylenchus* Poinar and Nelson, 1973
   Only parasitic females of parthenogenetic generation with ventral side turned outward. Parasites of Siphonaptera ....
   .... *Incurvinema* Deunff, Launay and Beaucournu, 1985
3. Cephalic lip areas and papillae in parasitic females prominent; male with six paired caudal "papillae" ....
   .... *Paregletylenchus* Slobodyanyuk, 1984
   Cephalic lip areas and papillae in parasitic females not prominent; male without caudal "papillae" 4
4. Parasitic heterosexual female with elongate-shaped body; male without gubernaculum; parasites of Diptera ....
   .... *Heterotylenchus* Bovien, 1937
   Parasitic heterosexual female with sausage-shaped body; male with gubernaculum. Parasites of Coleoptera ....
   .... *Wachekitylenchus* Slobodyanyuk, 1986

Heteromorphotylenchinae Siddiqi, 1986

Diagnosis (modified after Siddiqi, 1986a): Parasitylenchidae. Four distinct adult forms with alternation of heterosexual and parthenogenetic generations; parthenogenetic generation mostly in free life. Parasitic female in insect hemocele produces juveniles, only of female sex, which have full development in insect hemocele and leave the host via the intestine just before the last molt. They molt and develop into adult parthenogenetic females which live on their food reserve, do not feed, and produce only a small number of eggs. The eggs produced by parthenogenetic females hatch and the juveniles develop into male and preadult female which mate. The mode of feeding is unknown. Preadult fertilized female invades host's larva or nymph.

Type genus: *Heteromorphotylenchus* Remillet and Van Waerebeke, 1978
No other genus.

Iotonchiidae Goodey, 1953 (Skarbilovich, 1959)

Diagnosis (modified after Siddiqi, 1986a): Sphaerularioidea. Adult parasitic females known only for *Paraiotonchium*, in which two types of female occur in host's body cavity—a primary heterosexual female curving ventrally and spirally and having vulva at more than two body widths from anus, and a secondary parthenogenetic female. Alternation of heterosexual and parthenogenetic generations present. Partially free-living forms usually with marked sexual dimorphism in anterior region in which male esophagus is degenerate and stylet degenerate or absent. Male cephalic region tri– or tetralobed, usually asymmetrical. Female stylet generally over 18 μm long, usually without basal knobs; indistinct basal knobs or thickenings may be present. Excretory pore opposite or behind nerve ring. Vulva more than two body widths in front of anus, with or without anterior lip flap. Vagina strongly muscular. No postvulval uterine sac. Tail generally elongate-conoid, or filiform. Spicules robust, angular, L-shaped or of an aberrant form, in two parts, proximal part broad and cephalated, distal part slender, with rounded or spined tip. Large postanal genital papillae may be present. Bursa small, adanal, or large, completely enveloping tail.

Genus: *Paraiotonchium* Slobodyanyuk, 1975
Other genera:
Iotonchium Cobb, 1920 (type genus)
Fungiotonchium Siddiqi, 1986
Remarks: entomoparasitic forms known only in *Paraiotonchium*.

Sphaerulariidae Lubbock, 1861 (Skarbilovich, 1947)

Diagnosis: Sphaerularioidea. Only one heterosexual generation known. Male rarely found in host. Parasitic female in host's hemocele, with hypertrophied uterus, generally more or less everted and which may lead independent life. Partially free-living female with stylet well developed, base split into three flanges or knobbed; two or three esophageal glands, orifice of dorsal gland at the base of stylet, not always distinct; uterus very long in fertilized female. Male similar to female; bursa present or absent.

Type genus: *Sphaerularia* Dufour, 1837

Key to genera

1. Parasitic female with uterus hypertrophied but not everted. Juvenile nematodes mature to adults and may reproduce within the body cavity of the female ......
   .... *Scatonema* Bovien, 1932
   Parasitic female with uterus more or less everted 2
2. Parasitic female with partially everted uterus not larger than body; male without bursa .... *Tripius* Chitwood, 1935
   Parasitic female with completely everted uterus much larger than body; male with bursa 3

3. Stylet knobbed; everted uterus surface smooth; bursa distinct, completely
    enveloping tail; parasites of bark beetles and hymenopterous parasitoids . . . .
    . . . . *Sphaerulariopsis* Wachek, 1955
    Stylet not knobbed; everted uterus surface with numerous large rounded elevations;
    bursa indistinct; parasites of bumblebee queens and their hymenopterous parasitoids
    . . . . *Sphaerularia* Dufour, 1837

Fergusobiidae Goodey, 1963 (Siddiqi and Goodey, 1964)

Diagnosis: Sphaerularioidea. Alternation of generations. One parasitic heterosexual generation in insect hemocele. Several plant parasitic parthenogenetic generations in galls. Cephalic framework hexaradiate. Stylet knobbed. Anterior portion of esophagus swollen, nonmuscular, with valvelike structure situated anteriorly. Esophagointestinal junction anterior
to nerve ring. Esophageal glands bulboid. Gubernaculum absent. Bursa well developed.

Type genus: *Fergusobia* Currie, 1937 (Christie, 1941)
No other genus.

Phaenopsitylenchidae Blinova and Korenchenko, 1986

Diagnosis: Sphaerularioidea. Alternation of generations. One parasitic heterosexual generation. One or several free-living heterosexual or parthenogenetic generations. Parasitic
generation: partially free-living female with a strong stylet, three esophageal glands well
developed, ovary reduced, excretory pore in front of hemizonid; male with bursa and minute sperm. Free-living generation: female with fine stylet, three esophageal glands but two
are reduced, ovary normally developed; male similar to that of parasitic generation but with
large sperm.

Type genus: *Beddingia* Blinova and Korenchenko, 1986

Key to genera

Parasitic forms with only one generation in insect host, with heterosexual female and juveniles; free-living forms with several successive generations possible; two types of males,
one heterosexual free-living female fungus-feeding, one partially free-living female: *Beddingia* Blinova and Korenchenko, 1986. Parasitic forms with only one generation in insect
host, with heterosexual female and juveniles; free-living forms with only one generation of
parthenogenetic female; male fungus-feeding and partially free-living female: *Phaenopsitylenchus* Blinova and Korenchenko, 1986.

III. LIFE CYCLES

Allantonematidae (Fig. 1A)

The fertilized slender infective female enters the body cavity of the host's larva by penetrating through the cuticle. The female then develops into a swollen mature parasitic female
that deposits eggs into the adult host hemocele. The juveniles develop to the third or fourth
stage before escape through the insect's intestine or reproductive system. Once in the host's
environment, the nematodes mature, mate, and the fertilized infective female enters a new
host. This cycle has only one heterosexual generation. Both parasitic forms and free-living
forms are found. After Siddiqi (1986a), the free-living forms are only partially free-living,
as they do not feed and multiply in the host environment.

Parasitylenchidae (Fig. 1B, C, D)

The entering female develops into normal heterosexual female that produces eggs which develop:

Into adults of a second heterosexual generation, which mate and oviposit while still within the host's body cavity; the eggs develop into juveniles which eventually leave the host (Parasitylenchinae) (Fig. 1B)

Into parthenogenetic females that produce eggs inside the host's body cavity; eggs develop into males and females in the environment (Heterotylenchinae) (Fig. 1C).

Into parthenogenetic females that leave the host and deposit eggs in the environment; the juveniles develop as males and females (Heteromorphotylenchinae) (Fig. 1D).

In these three cases, the remainder of the cycle is similar to the basic type Allantonematidae.

Iotonchiidae (Fig. 1E)

The life cycle is similar to that of Heterotylenchinae. Sometimes the juveniles produced by the parthenogenetic females develop as males and females before leaving the host.

Sphaerulariidae (Fig. 1A)

The life cycle is very similar to that of Allantonematidae. In *Scatonema*, meanwhile, juveniles may develop into adults before leaving the host.

Fergusobiidae (Fig. 1F)

The entering female develops into heterosexual female that produces eggs into the host body cavity. The juveniles that leave the host develop into parthenogenetic females that probably feed on plant tissues. These females may give rise to successive free-living generations of parthenogenetic females in the plant and eventually produce males and infective females, the latter of which entering a new host.

Phaenopsitylenchidae (Fig. 1G)

In *Phaenopsitylenchus*, the life cycle is similar to that of *Heteromorphotylenchus*, but with a parthenogenetic female of Neotylenchid type.

In *Beddingia* parasite of Curculionids (Fig. 1H), the juveniles that leave the host develop into males and females of Neotylenchid type that probably feed on plant tissues. This free-living generation may give rise to successive free-living generations. Juveniles of each free-living generation give in part males and infective females, the latter of which entering a new host.

In *Beddingia* parasite of Siricids (Fig. 1I), the juveniles that leave the host can establish successive generations of fungal feeding populations of Neotylenchid type. They can also have a classical insect parasitic cycle. Each of these life cycles can continue indefinitely without the intervention of the other. These nematodes could be considered as facultative insect parasites (Poinar, 1975).

IV. HOSTS AND BIOLOGY

Allantonema Leuckart, 1884
syn. *Tylenchomorphus* Fuchs, 1915

Diagnosis: Allantonematinae
Hosts: Coleoptera

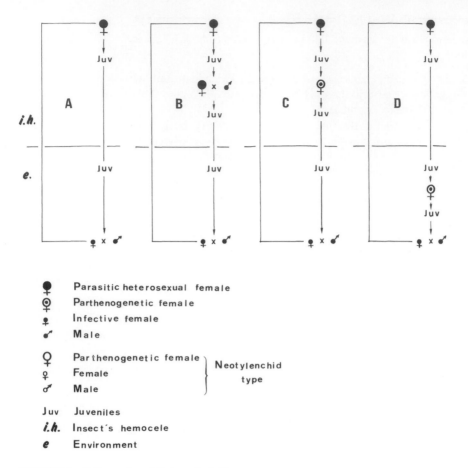

FIGURE 1   Life cycles of Sphaerularioidea.

Type species: *Allantonema mirabile* Leuckart, 1884 in *Geotrupes* (Germany), *Hylobius*
        (Germany, Denmark, USSR)
Other species: *A. bathycapsulatum* Slankis, 1974 in *Crypturgus* (USSR)
A. matthesi Wachek, 1955 in *Ochthebius* (Germany)
A. morosum (Fuchs, 1929) Filipjev, 1934 in *Hylastes* (Germany, USSR), *Tomicus* (Canada)
A. orthotomici Massey, 1974 in *Orthotomicus* (USA)
A. paramorosum Massey, 1974 in *Hylastes* (USA)
A. philonthi Wachek, 1955 in *Philonthus* (Germany)
A. sylvaticum V. Linstow, 1893 in *Geotrupes* (Germany)
Species *inquirendae*: *A. muscae* Roy and Mukherjee, 1937 in *Musca* (India) *A. stricklandi*
        Roy and Mukherjee, 1937 in *Musca* (India)

Biology

One heterosexual generation (Fig. 1A). In *A. philonthi* (Wachek, 1955), the juveniles, after
the second molt, leave the host through the intestine and anus. One molt takes place in the
environment. In laboratory conditions, the development into young adults is 10–13 days for
the male, 15–18 days for the female, but probably faster according with the observations of

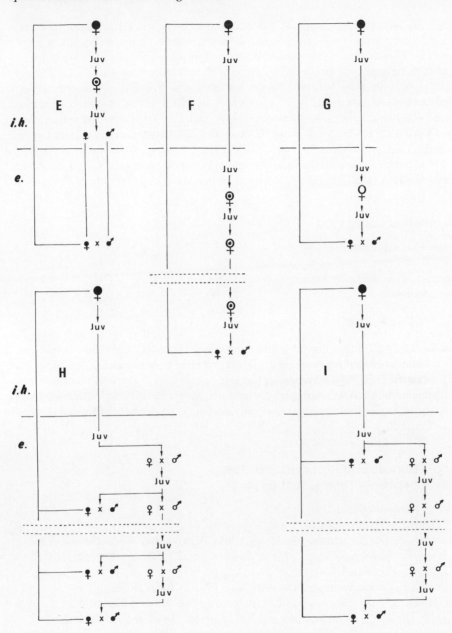

Bovien (1937) who observed adults in laboratory after 5–6 days. After mating, the male dies. The fertilized fourth stage infective female enters a pupa of the new host (*A. mirabile* parasitize *Hylobius* at the larval stage) and molts for last time. The female grows considerably, becomes bean-shaped and is incorporated with numerous host's tracheae. The percentage of parasitism is low (4% on 239 insects examined). Two or three parasitic females are found in one host, maximum 12. In *A. matthesi*, up to 150 juveniles with one parasitic female and 400 juveniles with four females were found. The larval development has been described by Wülker (1923) for *A. mirabile*. Juveniles of this species escape from the host like juveniles of *A. philonthi*.

The effect of *A. mirabile* on the host seems to be negligible contrary to the parasitism of *Ochthebius* by *A. matthesi* which sometimes brings about degeneration of the gonads.

Aphelenchulus Cobb, 1920

Diagnosis: Contortylenchinae
Hosts: Coleoptera
Type species: *Aphelenchulus mollis* Cobb, 1920 in *Cyllene* (USA)
No other species.

Biology

One heterosexual generation (Fig. 1A). Biology not well known (Cobb, 1920). One to eight parasitic females are found per host. Up to 16,000 juveniles were counted in one beetle, but it is few compared to the 700,000 juveniles found in one another Cerambycid beetle parasitized by an undetermined Allantonematidae in Madagascar (Remillet and Van Waerebeke, 1975). According to the biology of the host, the free-living stages of *A. mollis* should have a relatively long life.

Beddingia Blinova and Korenchenko, 1986
syn. *Deladenus* Thorne, 1941 (in part)

Diagnosis: Phaenopsitylenchidae
Hosts: Coleoptera, Hymenoptera
Type species: *Beddingia wilsoni* (Bedding, 1968) Blinova and Korenchenko, 1986 in *Megarhyssa, Pseudorhyssa, Rhyssa, Sirex* (Australia, Canada, Europe, India, New Zealand, Pakistan, USA)
Other species: *B. barisii* Laumond and Bonifassi, 1991 in *Baris* (France)
B. canii (Bedding, 1974) Blinova and Korenchenko, 1986 in *Sirex* (Canada)
B. imperialis (Bedding, 1974) Blinova and Korenchenko, 1986 in *Sirex* (Pakistan)
B. nevexii (Bedding, 1974) Blinova and Korenchenko, 1986 in *Sirex, Urocerus, Xeris* (USA)
B. proximus (Bedding, 1974) Blinova and Korenchenko, 1986 in *Sirex* (USA)
B. rudyi (Bedding, 1974) Blinova and Korenchenko, 1986 in *Sirex* (Turkey), *Urocerus, Xeris* (Europe, Japan)
B. siricidicola (Bedding, 1968) Blinova and Korenchenko, 1986 in *Ibalia, Sirex, Xeris* (Belgium, Hungary, North Africa, Japan, Australia, New Zealand), *Serropalpus* (Australia)
Remarks: *Deladenus durus* (Cobb, 1922) Thorne, 1941 should have entomoparasitic form (Baujard, 1979).

Biology

Beddingia in Siricids

Alternation (not obligatory) of one parasitic generation and several heterosexual free-living mycetophagous generations (Fig. 1I). Five to 20 parasitic females (maximum 100) were counted in the body cavity of infected insects. Ovoviviparous females release juveniles into the host hemocele. Up to 51,510 juveniles with nine parasitic females were observed; generally 750 to 2,000 juveniles are present with less than 10 parasitic females. Juveniles invade ovaries of mature siricid females and enter into the eggs. All eggs usually contain juvenile nematodes (50–200). Juveniles exit the host during the oviposition which in fact may be a "nemaposition." The adult stage is reached after 5 days. The adults are of *Deladenus* (Neotylenchoid) type or of Sphaerularioid type depending on environmental conditions. The two types have completely different life cycles: each cycle can continue indefinitely without the intervention of the other. The *Deladenus* type feeds on suitable fungus *Amylostereum areolatum* or *A. chailletii*; after copulation, the females lay 300–800 eggs in 2–3 weeks in laboratory cultures; hatching occurs after 3–4 days (22°C) and juveniles can develop in *Deladenus* or Sphaerularioid generation. The Sphaerularioid type has a classical cycle. The infective females are at fourth stage. The penetration takes 4–60 min and occurs more often ventrally, at the posterior end of the host larva. The last molt (abnormal molt) occurs within the host (Bedding, 1972) and microvilli develop on the outside of the nematode (Riding, 1970). The development of the gonad of the parasitic female begins only at the commencement of host pupation, which may occur up to 2 years after nematode entry. Females deposit juveniles into the host hemocele before the adult insect emerges for the wood. Generally host larvae are parasitized without damage. It appears that parasitism in males is a "dead end" for nematodes (Bedding, 1972). Parasitized host females are more often sterile with *B. canii, B. nevexii, B. proximus, B. rudyi* (Bedding, 1974). In other species, nematodes are deposited with the eggs. *Beddingia siricidicola* is used in the biological control of *Sirex noctilio* and has been introduced successfully into the northern Tasmania and Victoria regions in Australia (Bedding, 1984).

Beddingia in Curculionids

Alternation (obligatory) of one parasitic heterosexual generation and several heterosexual free-living generations (possibly phytoparasitic) (Figs. 1H, 2). Five to 90 parasitic heterosexual females lay eggs in the hemocele of the *Baris*, at the end of summer (Fig. 3). The first molt takes place in the egg and the second molt in the host's hemocele. In spring, juveniles leave host at the third stage through the genital or intestinal tract. They probably feed on tissues of colza and molt twice. Males (with large spermatozoa) and females are of *Deladenus* type. This free-living generation produces larvae which give in part a new free-living generation, in part males (with minute spermatozoa) and fourth-stage infective females of Sphaerularioid type. After mating, the infective females enter a new host's larva and molt a short time after. Mature parasitic females produce juveniles which develop into third stage. This stage stays during the winter within the adult host diapausing in the soil. The free-living females lay only 10–15 eggs. The number of free-living generations may be in relation with the temperature. It seems that two or three free-living generations succeed one another, so one or two entomoparasitic generations may occur yearly (Laumond, 1970; Laumond and Bonifassi, 1991).

This parasitism reduces significantly the fecundity and the longevity of the *Baris* females.

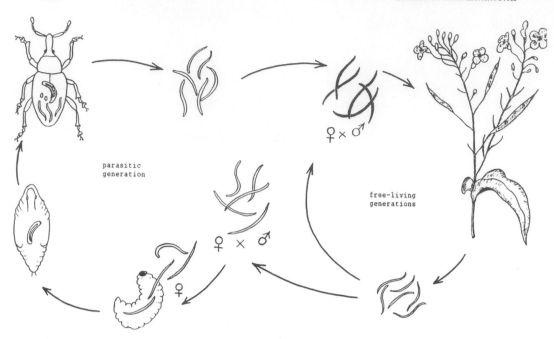

FIGURE 2   Life cycle of *Beddingia barisii*. (Modified after Slobodyanyuk, 1984.)

FIGURE 3   Parasitism of *Baris caerulescens* by *Beddingia barisii*: parasitic females extracted from the body cavity of the host.

Bovienema Nickle, 1963

Diagnosis: Contortylenchinae
Hosts: Coleoptera
Type species: *Bovienema tomici* (Bovien, 1937) Nickle, 1963 in *Pityogenes* (Denmark,
　　Germany, USA, USSR)
Other species: *B. cylindricum* (Slankis, 1967) Siddiqi, 1986 in *Ips* (USSR)
B. gifuchsi (Fuchs, 1938) Siddiqi, 1986 in *Cryphalus* (Germany), *Ips* (Czechoslovakia),
　　Pityogenes (Germany, USSR), *Scolytus, Xyleborus* (France)
B. pityophthori (Massey, 1974) Siddiqi, 1986 in *Pityophthorus* (USA)

Biology

One heterosexual generation (Fig. 1A). The parasitic females lay eggs in the body cavity of
the host beetle. One to three females per host are present in *Pityogenes fossifrons* (Leconte)
parasitized by *B. tomici*, two to five in *P. bidentatus* (Herbst) parasitized by *B. tomici*, two to
11 in *Ips duplicatus* parasitized by *B. cylindricum*. The eggs hatch and the juveniles grow in
size until the fourth stage. Then they leave the host by way of the intestinal tract and anus.
Sometimes the last molt takes place in the rectum (Bovien, 1937), but normally outside the
host. After the molt, young adults mate, the males die, and the inseminated infective females
enter new host larvae (Nickle, 1963a,b).

　　The rate of parasitism is very variable, the maximum is given by Slankis (1967) in
Ips parasitized by *B. cylindricum* (38.5%). Bovien (1937) observed 30–60% of parasitism in
P. bidentatus parasitized by *B. tomici* in beetles obtained from logs kept in laboratory.
Rühm (1956) mentioned 25–30% of parasitism by *B. tomici* in *P. bidentatus* and *P. quad-
ridens* (Hartig) in Germany.

Bradynema zur Strassen, 1892

Diagnosis: Allantonematinae
Hosts: Coleoptera, Diptera, Hemiptera, Siphonaptera
Type species: *Bradynema rigidum* (von Siebold, 1836) zur Strassen, 1892 in *Aphodius*
　　(Czechoslovakia, Denmark, Germany, UK, USA)
Other species: *B. bibionis* Wachek, 1955 in *Bibio* (Germany), *Philia* (Germany)
B. gerridis Poisson, 1933 in *Gerris* (France)
B. kurochkini Rubzov and Tshumakova, in Rubzov, 1981 in *Ctenophthalmus* (USSR)
B. nepae Poisson, 1933 in *Nepa* (France)
B. strasseni Wülker, 1923 in *Rhagium* (Germany), *Spondylis* (USSR)
B. trixagi Wachek, 1955 in *Throscus* (Germany)
B. veliae Poisson, 1933 in *Velia* (France)

Biology

One heterosexual generation (Figs. 1A, 4). Up to 15 parasitic females of *B. trixagi* were
found in *Throscus dermestoides* L. (Wachek, 1955) and 50 *B. veliae* females in *Velia cur-
rens* Fabr. (Poisson, 1933). These ovoviviparous females produce numerous juveniles; up
to 3000 were found in a male of *T. dermestoides*. The larval development of *Bradynema
rigidum* was described by Wülker (1923). Juveniles leave the host via the rectum or are
deposited with the insect eggs. In this latter case, ovaries are more or less destroyed. Larvae
of *B. trixagi* leave the host by genital tract and develop into adults after 3–4 days (males) and
4–7 days (females). Larvae of *B. rigidum* develop into adults after 8 days (males) and 7–10
days (females). The free-living adults mate after 8–14 days and survive 12–18 days (males)
to 40 days (females). Then infective females enter new host larvae.

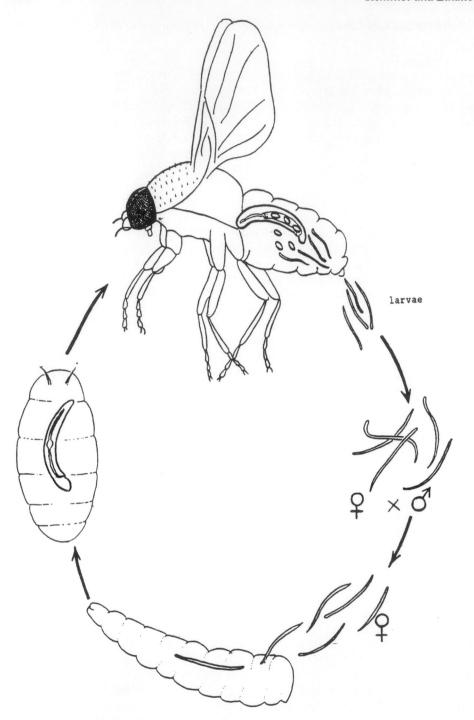

larvae

♀ × ♂

♀

FIGURE 4   Life cycle of *Bradynema*. (After Slobodyanyuk, 1984; courtesy of Slobodyanyuk.)

The rate of parasitism was 42% in *Bibio* (*n* = 65) parasitized by *B. bibionis* (Wachek, 1955).

Contortylenchus Rühm, 1956

Diagnosis: Contortylenchinae
Hosts: Coleoptera
Type species: *Contortylenchus diplogaster* (V. Linstow, 1890) Rühm, 1956 in *Ips* (Austria, Czechoslovakia, France, Germany, Switzerland, USSR)
Other species: *C. acuminati* Rühm, 1956 in *Ips* (France, Germany, USSR)
C. amitini Rühm, 1956 in *Ips* (Germany)
C. brevicomi(Massey, 1957) Rühm, 1960 in *Dendroctonus* (Canada, USA), *Ips* (USA)
C. bullus Massey, 1974 in *Hylurgops* (USA)
C. cribicolli Massey, 1974 in *Ips* (USA)
C. cryphali Rühm, 1956 in *Cryphalus* (Germany)
C. cunicularii (Fuchs, 1929) Rühm, 1956 in *Hylastes* (Germany, USSR)
C. elongatus (Massey, 1960) Nickle, 1963 in *Ips* (USA)
C. grandicolli (Massey, 1957) Rühm, 1960 in *Ips* (Australia, USA)
C. laricis (Fuchs, 1929) Rühm, 1956 in *Ips* (Germany, USSR)
C. orthotomici Massey, 1974 in *Orthotomicus* (USA)
C. proximus Kakuliya, 1967 in *Ips* (USSR)
C. pseudodiplogaster Slankis, 1969 in *Ips* (USSR)
C. rarus Slankis, 1967 in *Hylurgops* (USSR)
C. reversus (Thorne, 1935) Rühm, 1956 in *Dendroctonus, Ips* (Canada, USA)
C. spirus (Massey, 1957) Rühm, 1960 in *Ips* (USA)
C. terebranus Massey, 1974 in *Dendroctonus* (USA)

Biology

One heterosexual generation (Fig. 1A). Generally one to three parasitic females are present in the body cavity of parasitized Scolytidae. Meanwhile, 11 females of *C. acuminati* were observed in *Ips* (Rühm, 1956) and 25–29 females of *C. reversus* in *Dendroctonus engelmanni* Hopk. (Massey, 1956). The females lay eggs in host's hemocele. From 2000 juveniles of *C. reversus* (Furniss, 1967) to 7500 juveniles of *C. elongatus* (Nickle, 1963c) were released respectively in one parasitized host. *Contortylenchus brevicomi* females do not commence oviposition until the beetle host reaches maturity (MacGuidwin et al., 1980). According to Gibb et al. (1986), functioning of the reproductive system of *C. grandicolli* (in terms of egg deposition) is inhibited in larval and pupal hosts. The first molt occurs in the egg, the two further in the body cavity (Thong and Webster, 1972). The same larval development occurs in *C. elongatus* (Nickle, 1963b, c) but in *C. reversus* the three molts occur in the host hemocele (Thong and Webster, 1973). Juveniles at fourth stage leave the host via the digestive or reproductive tract. They develop into adults after about a week and mate. The infective females may survive about 1 week. (*C. reversus*) before penetrating a new host larva or pupa.

The rate of parasitism is often low, from 3% (*C. acuminati*) to 5% (*C. diplogaster*), or very high, from 41.3% (*C. elongatus*) to 73% (*C. reversus*). Lieutier (1981) noted that the rate of parasitism is generally the same for females and males. Massey (1960) studied the variations of parasitism with years. The adult scolytids infected by *Contortylenchus* present a lateness in emergence, about 3–4 days for the two yearly generations of *Ips sexdentatus* (Boern.) in France (Lieutier, 1981). Some parasitized species of scolytids are partially sterilized (Massey, 1957) such as *Ips sexdentatus* (50% of sterilization) according to Lieutier

(1981). Massey (1960) observed 52% reduction in the number of eggs laid by females of *Ips confusus* (Lec.) infected by *C. elongatus*. This author also observed considerable difference in egg galleries length constructed by infected individuals as compared with the uninfected.

Fergusobia Currie, 1937 (Christie, 1941)
syn.   *Anguillulina (Fergusobia*, Currie, 1937)
Dorsalla Jairajpuri, 1966 (Siddiqi, 1986)

Diagnosis: Fergusobiidae

Hosts: Diptera

Type species: *Fergusobia tumifaciens* (Currie, 1937) Wachek, 1955 in *Fergusonina* (Australia)

Other species: *F. indica* (Jairajpuri, 1962) Siddiqi, 1986, host unknown (India)

F. jambophila Siddiqi, 1986 in *Fergusonina* (India)

F. magna Siddiqi, 1986, host unknown (Australia)

Biology

Alternation of one parasitic heterosexual generation and several free-living parthenogenetic generations (Fig. 1F). The biology of *Fergusobia* is very particular but remains insufficiently known (Jairajpuri, 1966; Siddiqi, 1986b). Two to seven parasitic females are found per host but never in adult male flies. Eggs are laid in the body cavity of the fly. The larval development needs more study. The juveniles migrate into oviduct of the fly which deposits both insect eggs and nematodes into *Eucalyptus* tissues where nematodes develop into parthenogenetic females (Fisher and Nickle, 1968). According to Currie (1937), 1–50 juveniles are deposited with each insect egg. The larval development of the free-living generation is also not well known. It is assumed that the juveniles feed on tissues of the gall and may cause proliferation of plant tissues (in this case, parthenogenetic generations should be considered as plant parasitic). The parthenogenetic females of *F. tumifaciens* are oviparous. That of *F. jambophila* are ovoviviparous and one well-developed juvenile (118 µm long) was observed inside uterus (Siddiqi, 1986b). Several generations of parthenogenetic females may feed in galls and multiply before hatching of fly eggs; which takes 6 weeks. The resulting juveniles develop first into males and later the females appear. After mating, the inseminated infective females enter host larvae; thereby up to 100% of larvae are parasitized. During host pupation, the females grow rapidly. They lay eggs shortly after emergence of the fly from the gall. The parasitized female fly is partially sterile, which secures the durability of nematodes.

Heteromorphotylenchus Remillet and Van Waerebeke, 1978

Diagnosis: Heteromorphotylenchinae

Hosts: Coleoptera

Type species: *Heteromorphotylenchus stelidotae* Remillet and Van Waerebeke, 1978 in *Stelidota* (Madagascar)

Other species: *H. carpophili* Remillet and Van Waerebeke, 1978 in *Carpophilus* (Madagascar)

Biology

Alternation of one parasitic heterosexual generation and one free-living parthenogenetic generation (Fig. 1D). Inseminated infective females enter host larvae or pupae. The number of parasitic females per host is one or two (maximum 24 in *H. carpophili*). Mature parasitic

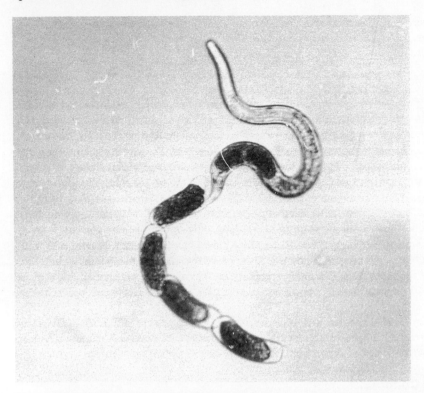

FIGURE 5   Parthenogenetic female of *Heteromorphotylenchus* laying eggs (eggs, remarkably large, remain in the last molt).

females lay eggs. After hatching, larvae develop into fourth larval stage parthenogenetic females, in about 1 month. Several thousand larvae (*H. carpophili*) can be found in the host's body cavity. In very old infestations (3 or 4 months), encapsulation reactions are observed. One month after the imaginal molt of the host (*Stelidota*), the fourth larval stage parthenogenetic females begin to enter through the intestine and escape via the anus. The last molt occurs in the host's body cavity or out of the insect. Six to 14 eggs are laid by each parthenogenetic female, and can be retained in the last molt cast (Fig. 5). In the case of *H. stelidotae*, development of the embryo and the three first larval stages takes place in the egg in less than 1 week. After hatching, it results directly in one adult male or a fourth larval stage female. After the last molt and mating, the infective female enters a new host. In the case of *H. carpophili*, only one molt is seen in the egg laid by the parthenogenetic female. After hatching, the second stage develops into adults. The way of feeding is unknown. Parthenogenetic females are able to survive during one month after laying eggs. Young heterosexual adults have a short free-living life. The male dies after mating and the female soon enters a new host (Remillet and Van Waerebeke, 1978).

In *Stelidota* and *Carpophilus*, ovaries of the host females are reduced and the sterility is total.

Heterotylenchus Bovien, 1937

Diagnosis: Heterotylenchinae
Hosts: Diptera

Type species: *Heterotylenchus aberrans* Bovien, 1937 in *Delia* (Denmark)
Other species: *H. hyderabadensis* Reddy and Rao, 1980 in *Morellia* (India)
H. simplex Slobodyanyuk, 1975 in *Morellia* (USSR)

Biology

Alternation of one heterosexual generation and one parthenogenetic generation, both in the insect (Fig. 1C). In the body cavity of the infested host, two types of females, larvae and a great number of eggs, are found. Heterosexual females, one to four per host (*H. aberrans*), grow and become mature during the imaginal molt of the host. The fertility is low. Only one egg is present in the uterus. The juveniles develop until fourth stage within the egg (this kind of evolution is also mentioned concerning the development of the juveniles into fourth stage within the eggs of parthenogenetic females of *Heteromorphotylenchus stelidota* Remillet and Van Waerebeke, 1978). After hatching, one molt occurs (*H. aberrans*), giving adult parthenogenetic females. In uterus of these females, only one egg is also present. Eggs of heterosexual females are bigger than those of parthenogenetic females. Juveniles born of parthenogenetic females are "nemaposited" by the host female. The infestation of male flies is a dead end for nematodes. One molt occurs in the environment and mating takes place within 24 hr. The male soon dies but the infective female may live for a week before infecting a new host larva.

   Bovien (1937) observed 10% of parasitism in 30 maggots, 9–25% in adult males and females of *Delia (Hylemyia) antiqua* (Meig.) in Denmark. *H. aberrans* has not effect on male but ovaries are atrophied on female. The sterility is partial or total.

Howardula Cobb, 1921
syn. *Acarinocola* Warren, 1941
Tylenchinema Goodey, 1930

Diagnosis: Allantonematinae
Hosts: Coleoptera, Diptera, Siphonaptera, Acarina
Type species: *Howardula benigna* Cobb, 1921 in *Acalymma, Diabrotica, Epitrix* (USA, USSR)
Other species: *H. acarinora* Wachek, 1955 in *Parasitus, Poecilochirus* (Germany)
H. acris Remillet and Van Waerebeke, 1976 in *Dactylosternum* (?) (Madagascar)
H. albopunctata Yatham and Rao, 1980 in *Sepsis* (India)
H. aoronymphia Welch, 1959 in *Drosophila* (UK)
H. apioni Poinar, Laumond, and Bonifassi, 1980 in *Apion* (France)
H. colaspidis Elsey, 1979 in *Colaspis* (USA)
H. dominicki Elsey, 1977 in *Epitrix* (USA)
H. husseyi Richardson, Hesling, and Riding, 1977 in *Megaselia* (UK)
H. madecassa Remillet and Van Waerebeke, 1975 in *Carpophilus* (Madagascar)
H. marginatus Reddy and Rao, 1981 in *Copromyza* (India)
H. oscinellae (Goodey, 1930) Wachek, 1955 in *Oscinella* (Denmark, Germany, UK, USSR)
H. phyllotretae Oldham, 1933 in *Longitarsus* (USSR), *Phyllotreta* (France, Germany, UK, USA)
H. prima Rubzov and Tshumakova in Rubzov, 1981 in *Ctenophthalmus* (USSR)
H. stenoloba Rubzov and Tshumakova in Rubzov, 1981 in *Ctenophthalmus* (USSR)
H. truncati Remillet and Van Waerebeke, 1975 in *Carpophilus* (India, Farquhar and Providence, British Indian Ocean Territory)
Species *inquirendae*: *H. claviger* (Warren, 1941) Wachek, 1955 in *Cosmolaelaps, Hypoaspis* (UK)

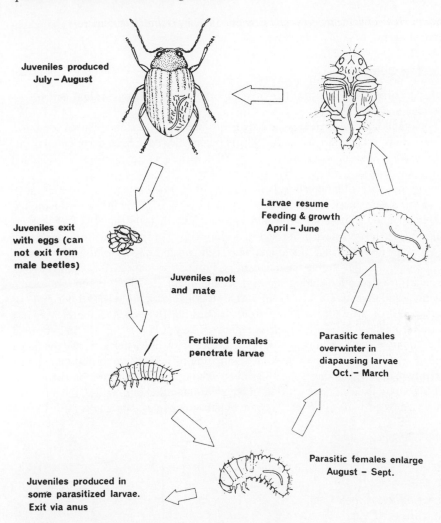

FIGURE 6   Life cycle of *Howardula colaspidis* and *Colaspis brunnea* in northeastern North Carolina. (After Esley, 1979.)

H. cuneifer (Warren, 1941) Wachek, 1955 in *Cosmolaelaps* (UK)
H. hirsuta (Warren, 1941) Wachek, 1955 in *Haemogamasus* (UK)
H. terribilis (Warren, 1941) Wachek, 1955 in *Euryparasitus* (UK)

Biology

One heterosexual generation (Figs. 1A, 6). The genus *Howardula* is represented by 20 species parasites of Insecta and Acarina. Parasitized insects belong to orders Coleoptera, Diptera, and Siphonaptera. Such a host range is uncommon in Sphaerularioids. This genus probably requires a revision, and recently Sidiqqi (1986a) transferred *H. aptini* parasite of *Thrips* to a new genus *Thripinema*. The nematodes described by Warren (1941) in Acarina are probably not part of the genus *Howardula* (Yatham and Rao, 1980; Siddiqi, 1986a).

Howardula parasitizes some insect pests of crops and thus presents great interest as a potential biocontrol agent.

Howardula in Coleoptera

Howardula parasitizes four families. The most important parasitism cases are found among Chrysomelidae and Apionidae.

Chrysomelidae: *Howardula* is the only genus in the Allantonematidae to parasitize natural populations of the family (Poinar, 1988). Four species have been described parasitizing six genera and 22 species of Chrysomelids. All these insects are injurious to crops such as corn, rice, soybean, sugar beet, tobacco, turnip, etc.

One to three parasitic females (maximum 28) of *H. phyllotretae* are found per host. Juveniles of this species may be encountered (up to 13,000), not only in the abdomen, but also in the thorax and the head of *Phyllotreta undulata* Kutsch. (Oldham, 1933). The parasitic females release juveniles when the beetles become active in spring; 80% of juveniles (*n* = 1000) are female juveniles in *H. dominicki*. In *H. benigna* and *H. colaspidis*, the juveniles are deposited with the eggs during the oviposition. From 50 juveniles (*H. benigna*) to 812 (*H. colaspidis*) are deposited during the egg laying. Juveniles of *H. dominicki* escape through genital and digestive tracts. From 300 to 400 juveniles are released per male or female beetle; 59% of those survived for 5–6 days, none after 13 days; 90% are released in a 10-day period from day 4 to 14 after adult emergence (Elsey, 1977a) (Fig. 7). An important point with *H. dominicki* is its ability to produce and release juveniles in host larva; the range of release is 3–339 before the larva dies. Dissemination of juveniles by foredoomed host larvae is rarely mentioned among Sphaerularioids, although probably common (Elsey and Pitts, 1976). Two molts take place in 48 hr in the environment. After mating, the infective female enters a new host larva or pupa. The rate of parasitism by *H. benigna* is 20%. Over a

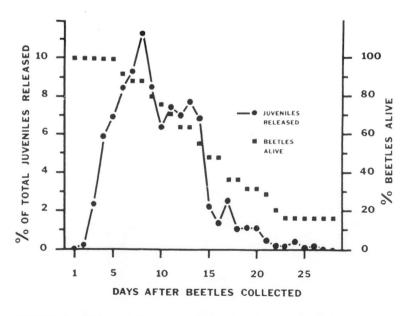

FIGURE 7   Daily variation in *Howardula* juveniles released from a cohort of 25 parasitized tobacco flea beetles held at 27°C and daily reduction in survival of the cohort. (After Esley, 1977b.)

2 year period, parasitism by *H. colaspidis* declined from 23% to 3%. *Howardula dominicki* may parasitize up to 56% of *Epitrix hirtipennis* (Melsh.) adults and 60% of the larvae (Elsey, 1977b). Besides the larval mortality, *Howardula* may castrate the host female by suppressing ovarian development.

Apionidae: *Howardula apioni* is parasitic in *Apion carduorum* Kirby, insect pest of young artichoke plants in the Mediterranean area (Poinar et al., 1980). The three larval stages as well the pupal stage of the insect are infested. The rate of parasitism fluctuates from 2 to 13% over an 8-year period, depending on the time and place of sampling. Third-stage juveniles leave the adult beetle via the digestive tract. In the tissue of the stems of artichoke, they mature in 1–2 weeks (in laboratory conditions at 20–22°C), molt twice to reach the adult stage, and mate. The inseminated infective females enter the larval stages of *Apion*. When the *Apion* adults leave the artichoke for their diapause, infective females begin their development into parasitic females. In this case of parasitism, it seems that the third and fourth stages feed plant tissue to achieve their development. Further studies on the behavior and impact of the parasitism on *Apion* populations would be of great interest.

Howardula in Diptera

Five families of Diptera are parasitized by four different species of *Howardula*. The parasitic females are found in the body cavity of the host. Up to 26 females of *H. husseyi* were reported in *Megaselia halterata* Wood (Richardson et al., 1977). In this species, 2200 eggs and larvae were produced by nine gravid females. Most of the larvae are liberated in the first 4 days after adult's emergence (Richardson and Chanter, 1979). Juveniles may leave the host through the intestinal tract as in *H. aoronymphia* and *H. oscinellae* (Goodey, 1930), or through the genital tract like in *H. husseyi*. In the case of *H. husseyi*, very few nematodes are liberated by female flies. The larvae released from host are the second-stage larvae which penetrate the ovaries, enter the oviduct, and are liberated during the oviposition of the host female. Outside three molts are visible at the same time for the male, and two for the female of *H. husseyi*. Thus the fourth-stage female is fertilized and enters a new host. After mating, males of *H. oscinellae* remain alive about 2 weeks and females a maximum of 4 weeks (Goodey, 1931). After penetration, the infective female of *H. husseyi* molts and the cuticle is replaced by microvilli (Riding, 1970). The parasitism of male phorids by *H. husseyi* is a dead end for the nematodes. Riding and Hague (1974) reported a rate of parasitism by *H.husseyi* reaching 60–75%. It is important to note that dead nematodes were found in parasitized flies without evidence of encapsulation or melanization. The nematodes reduce significantly the fecundity and the longevity in phorids (Richardson and Chanter, 1979) and delay the emergence (Richardson and Hesling, 1977) (Fig. 8).

Incurvinema Deunff, Launay, and Beaucournu, 1985

Diagnosis: Heterotylenchinae
Hosts: Siphonaptera
Type species: *Incurvinema helicoides* Deunff, Launay, and Beaucournu, 1985 in
      Rhadinopsylla (France)
No other species.

Biology

Alternation of one heterosexual generation and one parthenogenetic generation, both parasitic in the insect (Fig. 1C). The biology of *Incurvinema* is not well known. In laboratory experiments, the juveniles which leave the host give males after 1 week.

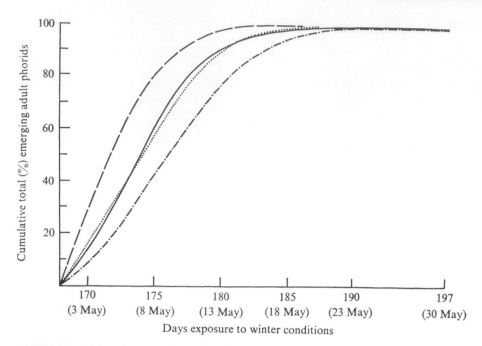

FIGURE 8   Effect of parasitism by *Howardula husseyi* on the emergence of overwintered male and female *Megaselia halterata.* _ _ _ _ _ , unparasitized females; . . . . . , parasitized females; ____, unparasitized males; _ . _ . _ ., parasitized males. (After Richardson and Hesling, 1977; courtesy of Richardson.)

Incurvinema helicoides parasitizes fleas of mole nests in France. This species is known only in *Rhadinopsylla pentacantha* (Rothschild). The intensity of the infestation is very important; the abdominal volume of the parasitized host increases. The males are castrated and the ovaries are reduced in the females (Deunff et al., 1985).

Metaparasitylenchus Wachek, 1955 (Nickle, 1967)
syn. *Parasitylenchus (Metaparasitylenchus* Wachek, 1955)

Diagnosis: Allantonematinae
Hosts: Coleoptera
Type species: *Metaparasitylenchus telmatophili* (Wachek, 1955) Nickle, 1967 in
      Telmatophilus (Germany)
Other species: *M. boopini* (Wachek, 1955) Siddiqi, 1986 in *Carpelimus* (Germany)
M. cossoni (Wülker, 1929) Nickle, 1967 in *Cossonus* (Germany)
M. cryptophagi (Wachek, 1955) Nickle, 1967 in *Cryptophagus* (Germany)
M. guadeloupensis Laumond and Mauleon, 1982 in *Hexacolus* (Guadeloupe)
M. helmidis (Wachek, 1955) Nickle, 1967 in *Helmis, Limnius, Riolus* (Germany)
M. leperisini (Massey, 1974) Siddiqi, 1986 in *Leperisinus* (USA)
M. mycetophagi (Wachek, 1955) Nickle, 1967 in *Mycetophagus* (Germany)
M. oschei (Rühm, 1956) Nickle, 1967 in *Rhizophagus* (Germany)
M. rhizophagi (Wachek, 1955) Nickle, 1967 in *Rhizophagus* (Germany)
M. strangaliae (Wachek, 1955) Nickle, 1967 in *Strangalia* (Germany)
M. tetropii (Wachek, 1955) Nickle, 1967 in *Tetropium* (Germany)

Biology

One heterosexual generation (Fig. 1A). The biology of all these species is noticeably the same. Generally one to three females parasitize the body cavity of the host, 11–16 maximum, but 123 females of *Metaparasitylenchus mycetophagi* have been counted in *Mycetophagus piceus* F. (Wachek, 1955). More than 2000 juveniles were found in the host (*Cryptophagus umbratus* Er.). Juveniles at third stage leave the host by genital or digestive tract and undergo a third molt in free life, after 6–20 days. The fertilized females could survive 1 or 2 months before infesting new hosts, such a relatively long period of the free life being uncommon among Allantonematidae. After copulation, the males die. The infective females penetrate young larvae or pupae of Coleoptera and molt for the last time.

This genus includes 12 species, parasitizing eight families of Coleoptera. The rate of parasitism is very variable, from 1.4–4% (*M. helmidis*) to 50% in populations of *Hexacolus guyanensis* Scheld. parasitized by *M. guadeloupensis* (Laumond and Mauleon, 1982). The effect of the nematodes is not well known. They may produce lesions in digestive tract and reduce the fecundity of hosts.

Neoparasitylenchus Nickle, 1967
syn. subgen. *Neoparasitylenchus* Nickle, 1963

Diagnosis: Allantonematinae
Hosts: Coleoptera
Type species: *Neoparasitylenchus cryphali* (Fuchs, 1914) Nickle, 1967 in *Cryphalus, Ips* (Germany)
Other species: *N. amvlocercus* Poinar and Caylor, 1974 in *Conophthorus* (USA)
N. avulsi (Massey, 1958) Nickle, 1967 in *Dendroctonus, Ips* (USA)
N. betulae (Rühm, 1956) Nickle, 1967 in *Scolytus* (Germany, USSR)
N. brachydorus Slankis, 1974 in *Crypturgus* (USSR)
N. caveocaudatus Slankis, 1974 in *Dryocoetes* (USSR)
N. chalcographi (Fuchs, 1938) Nickle, 1967 in *Pityogenes* (Germany, USSR)
N. cinerei (Fuchs, 1929) Nickle, 1967 in *Blastophagus* (France), *Crypturgus* (France, Germany)
N. coronatus (Massey, 1974) Siddiqi, 1986 in *Hylurgops* (USA)
N. hylastis (Wülker, 1923) Nickle, 1967 in *Hylastes* (Germany, New Zealand, USSR)
N. ipinius (Massey, 1974) Siddiqi, 1986 in *Ips* (USA)
N. ligniperdae (Fuchs, 1929) Nickle, 1967 in *Hylurgus* (Germany)
N. notati (Fuchs, 1929) Siddiqi, 1986 in *Pissodes* (Germany)
N. oriundus (Massey, 1974) Siddiqi, 1986 in *Ips* (USA)
N. orthotomici (Rühm, 1960) Nickle, 1967 in *Ips* (Germany)
N. ovarius (Massey, 1958) Nickle, 1967 in *Ips* (USA)
N. parasitus (Massey, 1974) Siddiqi, 1986 in *Polygraphus* (USA)
N. pessoni (Rühm and Chararas, 1957) Nickle, 1967 in *Dryocoetes* (France)
N. pityophthori (Rühm, 1956) Nickle, 1967 in *Pityophthorus* (Germany)
N. poligraphi (Fuchs, 1938) Nickle, 1967 in *Polygraphus* (Germany)
N. raphidophorus Slankis, 1974 in *Crypturgus* (USSR)
N. rugulosi (Schvester, 1957) Nickle, 1967 in *Scolytus* (France, USA)
N. scolyti (Oldham, 1930) Nickle, 1967 in *Hylesinus, Leperisinus, Phloeotribus* (France), *Scolytus* (UK)
N. scrutillus (Massey, 1964) Siddiqi, 1986 in *Scolytus* (USA)
N. senicus (Massey, 1974) Siddiqi, 1986 in *Pityophthorus* (USA)

N. wuelkeri (Rühm, 1956) Nickle, 1967 in *Dendroctonus* (Germany)
N. xylebori (Schvester, 1950) Nickle, 1967 in *Xyleborus* (France)

Biology

One heterosexual generation (Fig. 1A). All species of *Neoparasitylenchus* (except *N. notati*) are parasites of Scolytidae. After reaching the third stage, the juveniles produced by the parasitic females leave the host and molt twice in the environment. Adults mate and the infective females enter a new host. One to five gravid females are found per host; the maximum indicated by Schvester (1950) was 27 females of *N. xylebori* in *Xyleborus dispar* F. Observations using the technique of hanging drop (0.75% Ringer's solution) showed that the third-stage juveniles of *N. amvlocercus* reached the adult stage after molting twice (Poinar and Caylor, 1974). In case of *N. rugulosi*, the nematodes exit from the head end of the dead beetle as fourth-stage juveniles (Nickle, 1971). The specificity of this parasitism is very high with 27 species of *Neoparasitylenchus* parasitizing 19 genera of Scolytidae.

The rate of parasitism varies according to species, but generally reaches high percentages: average of 48% in *N. rugulosi* but maximum possible of 75% (Schvester, 1957), 50% in *N. betulae* (Rühm, 1956), about 60% in *N. amvlocercus* (Poinar and Caylor, 1974) and *N. scolyti* (Oldham, 1930). This parasitism brings about atrophy of the reproductive organs of hosts parasitized by *N. amvlocercus, N. hylastis* (Wülker, 1923), *N. scolyti*, and *N. xylebori*. Nickle (1971) reported that females of *Scolytus rugulosus* infected with *N. rugulosi* were sterilized and formed aberrant breeding galleries in the trees.

Paraiotonchium Slobodyanyuk, 1975
syn. *Heterotylenchus* Bovien, 1937 (in part)

Diagnosis: Iotonchiidae
Hosts: Diptera
Type species: *Paraiotonchium autumnale* (Nickle, 1967) Slobodyanyuk, 1975 in *Hydrotaea, Morellia* (Czechoslovakia), *Musca* (Canada, Czechoslovakia, USA, USSR), *Orthellia* (USA)
Other species: *P. crassirostris* (Yatham and Rao, 1981) Siddiqi, 1986 in *Musca, Stomoxys* (India)
P. nicholasi Slobodyanyuk, 1975 in *Musca* (Australia)
P. osiris Slobodyanyuk, 1976 in *Musca* (USSR)
P. xanthomelas (Reddy and Rao, 1987) n. comb. in *Musca* (India)

Remarks: Laumond and Lyon (1975) discovered an undetermined Iotonchiid parasitic in *Helophilus* (Diptera, Syrphidae), with a life cycle very similar to that of *Paraiotonchium* (Fig. 9). Nicholas and Hugues (1970) mentioned the record of *P. autumnale* (?) in *Scopeuma stercorarium* L. (Diptera, Scatophagidae) in Great Britain.

Biology

Alternation of one heterosexual generation and one parthenogenetic generation, both parasitic in insect (Figs. 1E, 10). This life cycle is very similar to that of *Heterotylenchus* of Diptera. Among the insects parasitized by this genus, *Musca autumnalis* De Geer, the face fly of cattle and horses, is a dipterous pest of economic importance. Many detailed studies have been conducted on this subject since Stoffolano and Nickle's first report (1966) and the study of the synchronization between *M. autumnalis* and *P. autumnale* life cycles (Stoffolano, 1967). Krafsur et al. (1983) counted 1–50 heterosexual females in one host; Nickle (1967b) reported up to 24 parthenogenetic females per host; Slobodyanyuk (1975b) re-

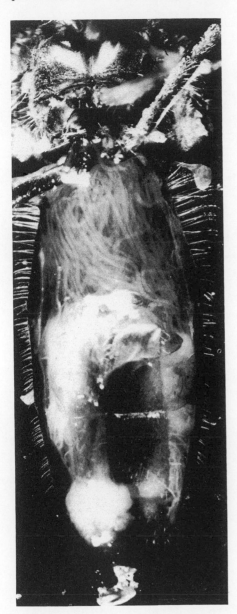

FIGURE 9 Males and females of the heterosexual generation of a Iotonchiid in a dissected *Helophilus* sp.

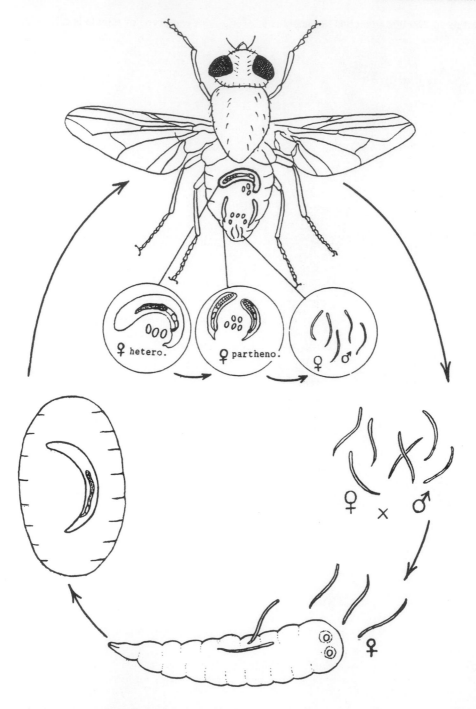

FIGURE 10   Life cycle of *Paraiotonchium*. (After Slobodyanyuk, 1984; courtesy of Slobodyanyuk.)

ported up to 120,000 juveniles per host. The fecundity of each type of female is different: the heterosexual female lays only 15– 24 brownish eggs whereas the parthenogenetic female may produce 250–300 white eggs. The larval development was described by Slobodyanyuk (1980). All molts occur in the host with the possibility that the last one occurs outside in *P. autumnale* and in *P. xanthomelas* (Reddy and Rao, 1987). Three molts were observed in the adult host parasitized by *P. crassirostris* (Yatham and Rao, 1981). Males and unmated females of *P. autumnale* were observed in ovaries at sites normally occupied by eggs (Nickle, 1967). The parasitized female fly releases juveniles during the oviposition which is often a "nemaposition." The free life of released juveniles is short, 4–6 days for males and 10 days for females (*P. crassirostris*).

Paraiotonchium seems to parasitize only the order Diptera. The incidence of parasitism by *P. autumnale* fluctuates with many factors such as sampling method, season, geographic localization etc. (Kaya and Moon, 1978). The rate of parasitism may reach 84% if flies are captured on the surface of manure pats (Thomas and Puttler, 1970; Thomas et al., 1972) (Fig. 11), but Nickle (1967) gave an average rate of 23%. It is as yet unclear what effects *P. autumnale* has on the host other than castration of the females (Geden and Stoffolano, 1984). The behavior of *Musca autumnalis* is changed. Kaya et al. (1979) reported that after the nematodes had invaded the ovaries, female flies frequented not the cattle but fresh cow pats to ovoposit eggs and nematodes, becoming "terminal dung feeders." Sterility of face flies is reported by Nappi (1973). Partial or total sterility of Australian bush flies results from *P. nicholasi* infestation. Complete sterility was also observed in *Musca* and *Stomoxys* parasitized by *P. crassirostris* in India (Yatham and Rao, 1981). Infested males are capable of mating and inseminating females but it is probably a dead end for the nematode. Attempts to introduce *P. autumnale* in natural populations of face flies in USA were not really successful (Nickle, 1978) and more studies on host-parasite relationships are needed (Thomas et al., 1972; Geden and Stoffolano, 1984).

Parasitylenchoides Wachek, 1955

Diagnosis: Allantonematinae
Hosts: Coleoptera
Type species: *Parasitylenchoides steni* Wachek, 1955 in *Stenus* (Germany)
Other species: *P. ditomae* Wachek, 1955 in *Ditoma* (Germany)
P. koerneri Wachek, 1955 in *Anotylus* (Germany)
P. paederi Wachek, 1955 in *Paederus* (Germany)
P. paromali Wachek, 1955 in *Paromalus* (Germany)
P. rheocharae Wachek, 1955 in *Aleochara* (Germany)
P. sciodrepae Wachek, 1955 in *Sciodrepa* (Germany)
P. wichmanni Wachek, 1955 in *Paromalus, Plegaderus* (Germany)

Biology

One heterosexual generation (Fig. 1A). One to four infective females, when fertilized, enter into pupa of a new host, through intersegmental membranes. They molt once in some species and develop into parasitic females in host's body cavity. These females lay eggs or juveniles in the hemocele. Up to 5500 larvae were recorded in *P. sciodrepae* infestation. After two molts, the juveniles pass through the rectum and leave the host via the anus. If larvae are reared in hanging drop, the free-living period is variable. Young adults were obtained after one molt in 14–20 days for *P. ditomae* and after two molts in 4–5 days for *P. paederi*. Infective females of *P. ditomae* stay alive 54 days before penetrating a new host.

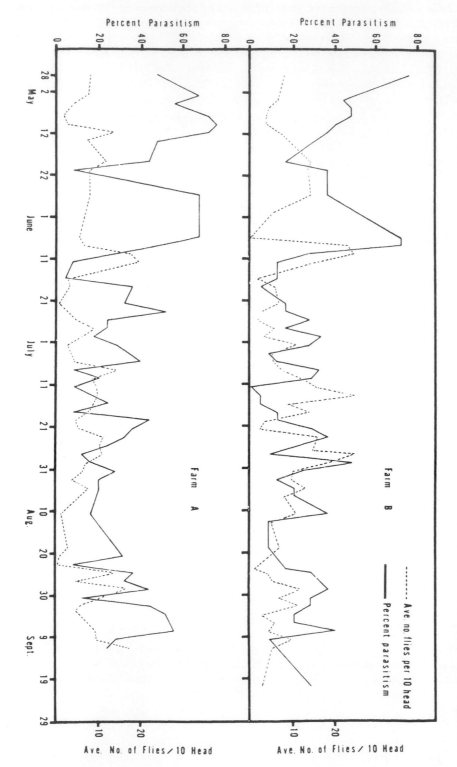

FIGURE 11 Seasonal parasitism of the face fly by the nematode *Paraiotonchium autumnale* and average face fly populations at two farms in central Missouri, 1970. (After Thomas et al., 1972.)

Parasitylenchoides parasitizes two families of predaceous beetles found in Germany (Wachek, 1955). The incidence of parasitism in beetles varies from 2% (530 *Paederus* examined) with *P. paederi* up to 54% (99 *Ditoma* examined) with *P. ditomae*. Sterility is sometimes observed in host females but never in males.

Parasitylenchus Micoletzky, 1922
syn. *Parasitylenchus* (*Parasitylenchus* Wachek, 1955)
Polymorphotylenchus (*Thylakolenchus* Rühm, 1956)
Vastotylenchus Slankis, 1967

Diagnosis: Parasitylenchinae
Hosts: Coleoptera, Diptera
Type species: *Parasitylenchus dispar* (Fuchs, 1915) Micoletzky, 1922 in *Cryphalus, Ips, Scolytus* (Europe, USSR)
Other species: *P. aculeatus* Slankis, 1972 in *Ips* (USSR)
P. curvidentis (Fuchs, 1914) Micoletzky, 1922 in *Ips, Pityokteines* (Germany, USSR)
P. diplogenus Welch, 1959 in *Drosophila* (UK)
P. klimenkorum Korenchenko, 1987 in *Orthotomicus* (USSR)
P. macrobursatus Blinova and Gurando, 1977 in *Blastophagus* (USSR)

Remarks: *Parasitylenchus coccinellinae* Iperti and Van Waerebeke, 1968, in fact, does not present alternation of morphologically different generations and would be placed near *Parasitylenchoides*: in *Adalia, Adonia, Harmonia, Semiadalia* (France).

Biology

Two heterosexual generations, both parasitic in insect (Figs. 1B, 12). Only the biology of *P. diplogenus* is well known. Inseminated infective females enter host larvae through the cuticle (Korenchenko, 1987). The number of primary heterosexual females is one to three (up to eight in laboratory infections). These females are oviparous or ovoviviparous and produced juveniles which give 10–20 females of the secondary generation after two molts, and males after one molt. If the number of gravid females increases, their length decreases. Secondary heterosexual females are oviparous and produce 100–200 eggs; up to 4000 juveniles can be counted in the host. The third molt occurs in host's gonads, the fourth out of the insect. After mating, the infective females enter a new host. The larval development was described by Welch (1959). The incidence of the parasitism (2–7%) is dependent on host species, geographic origin, and season. The parasitism increases in autumn and decreases in spring. *Parasitylenchus diplogenus* damages the genital and intestinal tract. The female gonads are almost completely destroyed by the emergence of the nematode larvae through the genital system.

The biology of the other species is not well known. Concerning the effect of the parasitism on hosts, Fuchs (1915) observed a reduction in the fat body and gonads of beetles infected by *P. dispar*; infested female bark beetles deposited less than half the number of eggs of healthy beetles. Rühm (1956) noted that 14% of the population of *Pityokteines curvidens* infected by *P. curvidentis* was sterile or had very reduced gonads, with a rate of parasitism between 15 and 25%.

Paregletylenchus Slobodyanyuk, 1984

Diagnosis: Heterotylenchinae
Hosts: Diptera

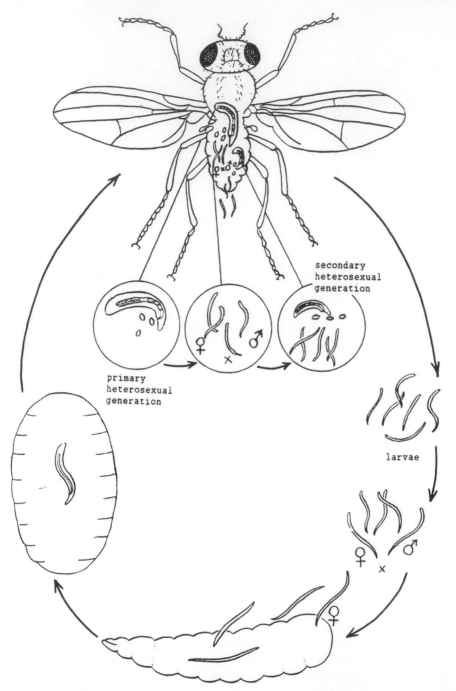

FIGURE 12 Life cycle of *Parasitylenchus*. (After Slobodyanyuk, 1984; courtesy of Slobodyanyuk.)

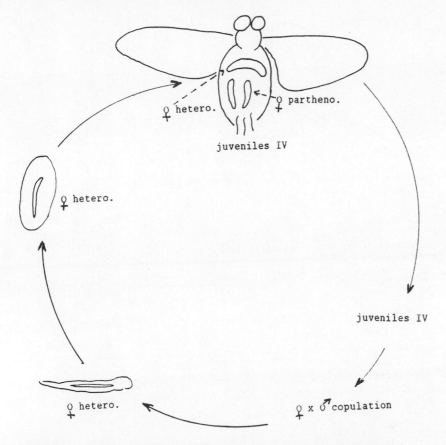

FIGURE 13 Life cycle of *Paregletylenchus*. (Slobodyanyuk, pers. commun.)

Type species: *Paregletylenchus papillatus* Slobodyanyuk, 1984 in *Paregle* (USSR)
No other species.

Biology

Alternation of one heterosexual generation and one parthenogenetic generation, both parasitic in insect (Figs. 1C, 13). One to three heterosexual females are found per host. Infective fertilized females penetrate new hosts through the cuticle (penetration not observed). When mature, the females produce about 50–60 eggs, laid in the host hemocele. Juveniles develop into parthenogenetic females. These females produce and lay about 250–300 eggs in the host hemocele; up to 20,000 juveniles were counted in host's body cavity. At the fourth stage, juvenile females leave the host through the genital tract of host female or digestive tract of host male. The fourth-stage male begins molting during the migration but the last molt of the juvenile female takes place only in the host's environment. Under laboratory conditions the last molt occurs within 24 hr. then mating takes place. Free-living infective females survive 4 days and males 3 days.

The incidence of nematode infection by *P. papillatus* is low: 1.9% in 474 insects dissected.

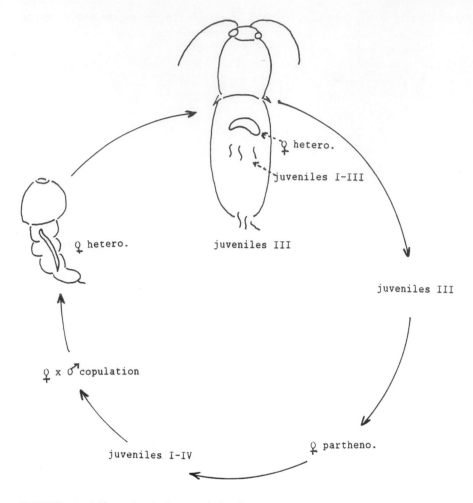

FIGURE 14   Life cycle of *Phaenopsitylenchus*. (Slobodyanyuk, pers. commun.)

Phaenopsitylenchus Blinova and Korenchenko, 1986

Diagnosis: Phaenopsitylenchidae
Hosts: Coleoptera
Type species: *Phaenopsitylenchus laricis* Blinova and Korenchenko, 1986 in *Phaenops*
        (USSR)
No other species

Biology

Alternation of one parasitic heterosexual generation and one free-living parthenogenetic
generation (Figs. 1G, 14). The heterosexual generation—parasitic female, male, and infective female—is typical of the superfamily Sphaerularioidea; the parthenogenetic female is
rather typical of the genus *Deladenus* (Neotylenchidae). Few data on the biology of *P.
laricis* were published by Blinova and Korenchenko (1986). In the host's body cavity, para-

sitic females are more or less encapsuled in the fat body. In the capsule, the eggs deposited by the females, then the first and second larval stages, are accumulate. Second stages escape in the host's hemocele and molt. The third stages develop during the flight of adult *Phaenops guttulata*. They enter into the oviducts of the females and are deposited under the bark of the tree (*Larix dahurica*) attacked by the insect. They stay and develop in parthenogenetic females only when a specific microflora is present in the host's galleries. The parthenogenetic females lay eggs in galleries. Juveniles develop into males and fourth-stage females of the heterosexual generation. After mating, young infective females enter larvae of *P. guttulata*.

Proparasitylenchus Wachek, 1955 (Nickle, 1967)
syn. *Parasitylenchus (Proparasitylenchus* Wachek, 1955)

Diagnosis: Allantonematinae
Hosts: Coleoptera
Type species: *Proparasitylenchus platystethi* (Wachek, 1955) Nickle, 1967 in *Platystethus* (Germany)
Other species: *P. athetae* (Wachek, 1955) Nickle, 1967 in *Atheta* (*Nehemitropia*) (Germany)
P. medonis (Wachek, 1955) Nickle, 1967 in *Medon* (Germany)
P. myrmedoniae (Wachek, 1955) Nickle, 1967 in *Zyras* (Germany)
P. oxyteli (Wachek, 1955) Nickle, 1967 in *Anotylus* (Germany)
P. trogophloei (Wachek, 1955) Nickle, 1967 in *Carpelimus* (Germany)

Biology

One heterosexual generation (Fig. 1A). Fertilized young females enter host's larvae or pupae. Generally one or two parasitic females are present per host, maximum 11 (*P. oxyteli*). In case of *P. medonis*, 2400 larvae were counted in host's body cavity. Two molts occur in the environment. Juveniles molt to adults after 4–6 days (male) or 5–8 days (female), but after 11 days with *P. athetae*. After molting, the longevity of adults may be 18 days (*P. myrmedoniae*).

The incidence of parasitism is low (0.5–8%). The sterility of the host was observed in *Anotylus* infested by *P. oxyteli* (Wachek, 1955).

Protylenchus Wachek, 1955

Diagnosis: Allantonematinae
Hosts: Coleoptera
Type species: *Protylenchus heteroceri* Wachek, 1955 in *Heterocerus* (Germany)
Other species: *Protylenchus anobii* (Wachek, 1955) Siddiqi, 1986 in *Anobium* (Germany)

Biology

One heterosexual generation (Fig. 1A). One to eight oviparous or ovoviviparous parasitic females (maximum 18) were found in host's body cavity, with up to 4500 juveniles (*P. heteroceri*). After two molts juveniles pass through the rectum and leave the host via the anus. They undergo one molt after 5 days for the male and 7 days for the female, then mate. The free-living infective females may survive for 30–40 days (*P. heteroceri*) before penetrating host's larvae or pupae. After penetration, the female molts for the last time.

The rate of parasitism was 4.5% in *Anobium pertinax* (L.) and *A. striatum* Oliv. and up to 23% in *Heterocerus marginatus* (Fabr.) and *H. fenestratus* (Thunb.)

Psyllotylenchus Poinar and Nelson, 1973
syn. *Aphanitylenchus* Rubzov, 1981

Diagnosis: Heterotylenchinae

Hosts: Siphonaptera

Type species: *Psyllotylenchus viviparus* Poinar and Nelson, 1973 in *Catallagia, Diamanus, Monopsyllus, Opisodasys* (USA)

Other species: *P. acuticapitus* Rubzov and Tshumakova in Rubzov, 1981 in *Neopsylla* (USSR)

P. acuticaudus Rubzov and Tshumakova in Rubzov, 1981 in *Ceratophyllus* (USSR)

P. acuticephalus Rubzov and Tshumakova in Rubzov, 1981 in *Ceratophyllus* (USSR)

P. caspius Rubzov and Samurov in Rubzov, 1981 in *Coptopsylla, Nosopsyllus* (USSR)

P. ceratophyllus Rubzov and Samurov in Rubzov, 1981 in *Ceratophyllus* (USSR)

P. chabaudi Deunff and Launay, 1984 in *Nosopsyllus* (France)

P. crassus (Rubzov and Tshumakova in Rubzov, 1981) Siddiqui, 1986 in *Ceratophyllus* (USSR)

P. curvans Rubzov and Kotty, 1981 in Rubzov, 1981 in *Megabothris* (USSR)

P. cuspidatus (Rubzov and Tshumakova in Rubzov, 1981) Siddiqui, 1986 in *Ceratophyllus* (USSR)

P. ioffi Rubzov and Morozov in Rubzov, 1981 in *Coptopsylla* (USSR)

P. kozlovi Rubzov and Tshumakova in Rubzov, 1981 in *Neopsylla, Nosopsyllus* (USSR)

P. larviparus Rubzov and Tshumakova in Rubzov, 1981 in *Neopsylla* (USSR)

P. latifrons Rubzov and Tshumakova in Rubzov, 1981 in *Neopsylla* (USSR)

P. longicauda (Rubzov and Tshumakova in Rubzov, 1981) Siddiqui, 1986 in *Neopsylla* (USSR)

P. macrocephalus Rubzov and Morozov in Rubzov, 1981 in *Coptopsylla* (USSR)

P. morozovi Rubzov, 1981 in *Ceratophyllus* (USSR)

P. neopsyllus Rubzov and Tshumakova in Rubzov, 1981 in *Neopsylla* (USSR)

P. pawlowskyi (Kurochkin, 1960) Poinar and Nelson, 1973 in *Ceratophyllus, Coptopsylla, Xenopsylla* (USSR)

P. rectangulatus Rubzov, 1982 in *Ceratophyllus* (USSR)

P. samurovi Rubzov, 1981 in *Neopsylla* (USSR)

P. tenuis Rubzov and Tshumakova in Rubzov, 1981 in *Neopsylla* (USSR)

P. tesquorae Rubzov and Nikulshin in Rubzov, 1981 in *Citellophilus* (*Ceratophyllus*) (USSR)

P. tiflovi Rubzov and Tshumakova in Rubzov, 1981 in *Neopsylla* (USSR)

P. zassuchini Rubzov and Morozov in Rubzov, 1981 in *Coptopsylla* (USSR)

Remarks: The species described by Rubzov need more studies. New genera and new species are to be published (Deunff, pers. commun., 1989; Slobodyanyuk, pers. commun., 1989).

Biology

Alternation of one heterosexual generation and one parthenogenetic generation, both parasitic in insect (Fig. 1C). Only the biology of *P. viviparus* is well known. In host's hemocele, the ovoviviparous heterosexual female of *P. chabaudi* produces juveniles of large size but in small number. These juveniles develop into parthenogenetic females also ovoviviparous, but producing numerous juveniles (Deunff and Launay, 1984). Juveniles of *P. viviparus* mature to the third or fourth stage before leaving the host via the intestine or the reproductive system (Poinar and Nelson, 1973). After one or two molts and mating, the males die and the fertilized infective females enter a new host.

Psyllotylenchus viviparus may partly castrate males but *P. pawlowskyi* always castrates male fleas (Kurochkin, 1960). Deunff and Launay (1984) emphasized the fact that the nature of soil may have an important effect on free-living stages of *Psyllotylenchus*: the fleas captured in sandy areas are more frequently parasitized than the fleas coming from regions with clayey soils.

Scatonema Bovien, 1932

Diagnosis: Sphaerulariidae
Hosts: Diptera
Type species: *Scatonema wuelkeri* Bovien, 1932 in *Scatopse* (Denmark, Germany)
No other species

Biology

One heterosexual generation (Fig. 1A). This species possesses the unique character in Sphaerularioids of having the juveniles develop to the adult stage and then mate within the uterus of the mature parasitic female, in the host's hemocele (Poinar, 1975). One to four (sometimes up to 20) parasitic females are found in the host's body cavity. Mature females have their uterus growing considerably but without eversion. About 15 days after penetration, the uterus is filled with all larval stages and sometimes adults. The number of molts is unknown; only one molt is reported by Bovien (1932). When males occur in the host, they mate but do not leave the body cavity. Juveniles or females escape during the oviposition of the host female after passing through the genital tract. Juveniles in host male should escape after its death. The free-living longevity is short, about 5 days in eggs clusters of *Scatopse* on manure. Infective females penetrate a new host larva, pupa, or imago through the body wall.

Up to 25% of *Scatopse* larvae were infested in laboratory experiments, but in the case of early parasitized larvae, they die and young infective females escape. In nature, the incidence of parasitism was 5–10%.

Sphaerularia Dufour, 1837

Diagnosis: Sphaerulariidae
Hosts: Hymenoptera
Type species: *Sphaerularia bombi* Dufour, 1837 in *Bombus* (Canada, France, Germany, Netherlands, New Zealand, Norway, Sweden, USA), *Psithyrus* (France, Germany, UK), *Vespula* (UK)
No other species

Biology

One heterosexual generation (Figs. 1A, 15). The biology of this species is not yet well known. Up to 42 gravid females were found in the host (Pouvreau, 1966) and 72 in *Bombus terrestris* L. by Poinar and van der Laan (1972) observed two molts in the egg. The nematodes that reach the third-stage juveniles in the host and are deposited via the anus. They enter the soil and become adult in approximately 2 months, which is very uncommon among Sphaerulariids. The way of infestation is still hypothetical. Infective females penetrate the queens during hibernation. Males and workers, which do not hibernate, are never parasitized. Infested queens seem to be unable to build nests (Pouvreau, 1966) and have a disoriented behavior favorable to the dispersal and survival of the parasite (Lundberg and Svensson, 1975).

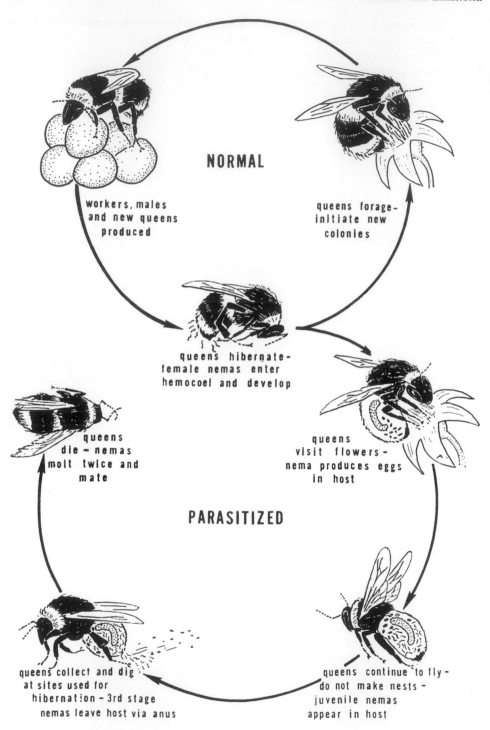

FIGURE 15   Normal and *Sphaerularia*-parasitized cycles of *Bombus* spp. in the Netherlands. (After Poinar and Van der Laan, 1972; courtesy of G. O. Poinar.)

The incidence of parasitism is more important when captures are made on hibernation areas than on flowers. The parasitism also varies with climatic conditions. According to Pouvreau (1962), *S. bombi* reduces corpora allata of the host, which lead to degeneration of nutritive cells and oocytes.

Sphaerulariopsis Wachek, 1955

Diagnosis: Sphaerulariidae
Hosts: Coleoptera, Hymenoptera
Type species: *Sphaerulariopsis stammeri* Wachek, 1955 in *Ernobius* (Germany)
Other species: *S. dendroctoni* (Massey, 1956) Nickle, 1963 in *Dendroctonus* (Canada, USA)
S. hastatus (Khan, 1957) Nickle, 1963 in *Dendroctonus* (Canada), *Coeloides* (Canada)
S. piceae (Fuchs, 1929) Nickle, 1967 in *Pissodes* (Germany)
S. pini (Fuchs, 1929) Nickle, 1963 in *Pissodes* (Germany)
S. piniphili (Fuchs, 1929) Nickle, 1963 in *Pissodes* (Germany)
S. ungulacaudus (Khan, 1957) Nickle, 1963 in *Dendroctonus* (Canada)

Biology

One heterosexual generation (Fig. 1A). Meanwhile, Rühm (1956) and Siddiqi (1986a) claimed that alternation of generations should exist. The adult females, which evert their uterus, and juveniles are free inside the body cavity of the host. Up to 50 females of *S. dendroctoni* were recorded from *Dendroctonus engelmanni* Hopk, and hundreds of juveniles and eggs were present in host's body cavity (Massey, 1956). Juveniles of *S. stammeri* leave host through the rectum, via anus, and molt twice in the environment. After mating, males die and the young inseminated infective females enter a new host at larval stage (Wachek, 1955). Massey (1956) observed males in host's body cavity of *S. dendroctoni*.
The annual incidence of parasitism by *S. dendroctoni* was 35.4% on *Dendroctonus* (Massey, 1956) but the level of parasitism could reach 76% in beetles on an individual tree. The egg production of *Dendroctonus engelmanni* was reduced to 63% in infested insects.

Spilotylenchus Launay, Deunff, and Bain, 1983

Diagnosis: Contortylenchinae
Hosts: Siphonaptera
Type species: *Spilotylenchus arthuri* Launay, Deunff, and Bain, 1983 in *Spilopsyllus* (France)
Other species: *S. beaucournui* Launay and Deunff, 1984 in *Spilopsyllus* (France)
S. laplandicus (Rubzov and Darskaja in Rubzov, 1981) Launay, Deunff, and Bain, 1983 in *Ceratophyllus (Megabothris)* (USSR)
S. megabothridis (Laumond and Beaucournu, 1978) Launay, Deunff, and Bain, 1983 in *Megabothris* (France)

Remarks: Launay and Deunff (1990) described *S. maisonabei*, parasitic in *Spilopsyllus cuniculi* (Dale) in France (in press); two other *Spilopsyllus* were found in *Ctenophthalmus* from Zaïre and *Coptopsylla* from Tunisia.

Biology

One heterosexual generation (Fig. 1A). Some data were reported from *S. megabothridis* (Laumond and Beaucournu, 1978) parasitizing the flea *Megabothris turbidus* (Rothschild) in France and from *S. beaucournui* (Launay and Deunff, 1984). One to four parasitic females are found in host's body cavity. They are ovoviviparous and release juveniles in host's hemocele. The four larval stages seem to be present in hemocele. Juveniles leave host via intestinal or genital tract. The last molt occurs in the environment in the case of females but was not observed in the case of males. The way of penetration in a new host is unknown. With *S. megabothridis* and *S. arthuri* (Launay et al., 1983) the parasitism has a blocking effect on the oogenesis in female fleas but no damage was noted in parasitized male fleas.

Sulphuretylenchus Rühm, 1956 (Nickle, 1967)
syn. *Parasitylenchus* (*Sulphuretylenchus* Rühm, 1956)

Diagnosis: Allantonematinae
Hosts: Coleoptera
Type species: *Sulphuretylenchus sulphureus* (Fuchs, 1938) Nickle, 1967 in *Pityogenes* (Germany, USSR)
Other species: *S. elongatus* (Massey, 1958) Nickle, 1967 in *Ips, Scolytus* (USA)
S. escherichi (Rühm, 1956) Nickle, 1967 in *Dryocoetes* (Germany)
S. fuchsi (Rühm, 1956) Nickle, 1967 in *Polygraphus* (Germany, USSR)
S. grosmannae (Rühm, 1954) Nickle, 1967 in *Pityogenes* (Germany, USSR)
S. kleinei (Rühm, 1956) Nickle, 1967 in *Hylastes* (Germany, USSR)
S. nopimingi (Tomalak, Welch, and Galloway, 1989) in *Pityokteines* (Canada)
S. pilifronus (Massey, 1958) Nickle, 1967 in *Ips* (USA)
S. posteruteri (Tomalak, Welch, and Galloway, 1989) in *Ips* (Canada)
S. pseudoundulatus (Tomalak, Welch, and Galloway, 1989) in *Polygraphus* (Canada)
S. pugionifer Slankis, 1969 in *Hylastes* (USSR)
S. stipatus (Massey, 1966) Siddiqi, 1986 in *Dendroctonus* (USA)
S. undulatus (Massey, 1974) Siddiqi, 1986 in *Pseudohylensinus* (USA)

Biology

One heterosexual generation (Figs. 1A, 16). The biology of *S. elongatus* (Massey, 1958) Nickle, 1967, is relatively well known. One to eight parasitic females with up to 2500 juveniles were counted in one individual (Massey, 1964). One mature parasitic female of *S. elongatus* may lay up to 642 eggs (Ashraf and Berryman, 1970a). The juveniles penetrate the gut of the host and are deposited via anus in the beetle larval galleries. Some juveniles may be released from dead larvae or dead adults. All stages of *Scolytus ventralis* Leconte are parasitized. The number of molts was not mentioned but Rühm (1954) reported that the infective female of *S. grosmannae* (Rühm, 1954) was the fourth stage. Copulation takes place in the larval galleries where males and young females of *S. stipatus* occur (Massey, 1966). The infective fertilized females enter a new host larva through the cuticle, in about 2 hr.

This genus, with 10 species described, seems to be parasitic only in Scolytidae. Female bark beetles are sterilized by heavy infection. Massive cellular necrosis in the internal organs are reported by Ashraf and Berryman (1970b). The emergence of adults is delayed by the parasitism and the attack behavior is aberrant. The incidence of parasitism reached 70% in individual tree bark beetle population (Massey, 1964).

Thripinema Siddiqi, 1986
syn. *Howardula* Cobb, 1921 (in part)

Diagnosis: Allantonematinae
Hosts: Thysanoptera

Type species: *Thripinema reniraoi* (Reddy, Nickle, and Rao, 1982) Siddiqi, 1986 in
     Megaluriothrips (India)
Other species: *T. aptini* (Sharga, 1932) Siddiqi, 1986 in *Aptinothrips* (UK)
T. nicklewoodi (Nickle and Wood, 1964) Siddiqi, 1986 in *Frankliniella* (Canada, USA),
     Taeniothrips (Canada)

Biology

One heterosexual generation (Fig. 1A). In *T. aptini* parasite of *Aptinothrips rufus* Gmelin in
UK, a maximum of 215 eggs, 255 juveniles, and 8 females were counted per host, the maximum of eggs and juveniles occurring between April and July. The parasitism in thrip males
is unknown. Juveniles in the host's hemocele are mainly females. Young male and female
juveniles at fourth stage enter the intestine and leave the host via anus (Sharga, 1932) but,
according to Reddy et al. (1982), juveniles of *T. reniraoi* emerge through the ovipositor of
the thrips. The number of molts is not exactly known. Lysaght (1936) mentioned one molt
after the penetration of the female into the host's body cavity but, according to Nickle and
Wood (1964), the nematodes leave the host as fully developed, partially free-living forms.
Males and females are found in the gall produced by the thrips on blueberry leaves. The
fertilized infective females enter host's nymphs.

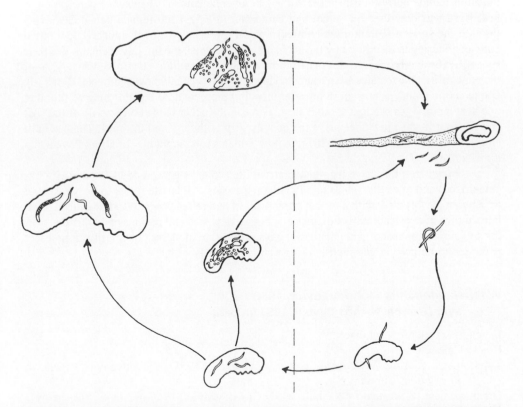

FIGURE 16   Life cycle of *Sulphuretylenchus elongatus*. (Left) parasitic. (Right) free-living. (After
Ashraf and Berryman, 1970a.)

Heavily infected host females are sterile. The rate of parasitism fluctuates considerably throughout the localities and seasons; the lowest parasitism is in spring and the highest in summer (Lysaght, 1937). *T. nicklewoodi* parasitized up to 71% of one sample of thrips from New Brunswick, Canada (Nickle and Wood, 1964) and *T. reniraoi* occurred in 63% of *Megaluriothrips* in India (Reddy et al., 1982).

Tripius Chitwood, 1935
syn. *Asconema* Leuckart, 1886
Atractonema Leuckart, 1886 (Leuckart, 1887)
Proatractonema Bovien, 1944

Diagnosis: Sphaerulariidae
Hosts: Diptera
Type species: *Tripius gibbosus* (Leuckart, 1886) Chitwood, 1935 in *Cecidomyia* (Germany)
Other species: *T. sciarae* (Bovien, 1944) Wachek, 1955 in *Bradysia* (UK), *Sciara* (Denmark)

Biology

One heterosexual generation (Fig. 1A). The biology of *T. sciarae* (Bovien, 1944) has been particularly well studied by Bovien (1944) in *Sciara* larvae, by Poinar (1965) and Poinar and Doncaster (1965) in *Bradysia paupera* Tuom. The infective stage of this nematode is the adult female but still ensheathed in the last juvenile cuticle. The way of penetration in host larva was described by Poinar and Doncaster (1965): adult female nematodes expel through the stylet a secretion which breaks down the front part of the unshed, last larval cuticle; adhering to their hosts by the adhesive mass formed, the infective females use their spears and probably enzymes for penetration which occurs within 10 min. Seven days after the penetration, the females are mature and begin to lay eggs into host hemocele. Generally one parasitic female is present in the adult host but more than 20 females may be found in one host larva. Eggs hatch in about 3 days. Three molts occur in the body cavity of the host and the fourth-stage juveniles leave the host by way of the gut and the anus. The last molt and then mating occur within 48 hr. Infective females remain viable for about 2 weeks in moist soil.

Parasitized flies have lengthened larval life (45 days instead of 17–18 days for unparasitized) and often die, releasing juvenile nematodes. All larvae found to be parasitized by Bovien (1944) died, with a rate of parasitism in nature reaching 30%. Parasitized adult female flies ovoposit normally but, instead of eggs, deposit small packets of nematodes over the soil surface or near organic matter. *Tripius sciarae* should have potentiality as an insect pathogen to be used in glasshouse against common gnats.

Wachekitylenchus Slobodyanyuk, 1986
syn. *Heterotylenchus* Bovien, 1937 (in part)

Diagnosis: Heterotylenchinae
Hosts: Coleoptera
Type species: *Wachekitylenchus bovieni* (Wachek, 1955) Slobodyanyuk, 1986 in *Bembidion* (Germany)
Other species: *W. stammeri* (Wachek, 1955) Slobodyanyuk, 1986 in *Clivina* (Germany)
W. wuelkeri (Wachek, 1955) Slobodyanyuk, 1986 in *Bembidion* (Germany)

Remarks: The parthenogenetic female of *W. wuelkeri* is strongly different from that of *W. bovieni* and *W. stammeri* but very similar to the parthenogenetic female of *Heterotylenchus simplex*, so *W. wuelkeri* seems to be a link between the two genera.

Biology

Alternation of one heterosexual generation and one parthenogenetic generation, both parasitic in the insect (Fig. 1C). The fertilized infective female of the heterosexual generation enters a new host pupa through intersegmental membranes. When mature, the parasitic female lays 40–60 eggs (*W. bovieni*) or 300–400 eggs (*W. wuelkeri*) in the host's hemocele. All juveniles produced develop into parthenogenetic females which lay 150–200 juveniles (*W. bovieni*) or 20–25 eggs (*W. wuelkeri*) in the hemocele. Curiously, the respective fecundity of the two types of females is reversed comparing the two species, but the same final level of 6000–10,000 juveniles is obtained in the two cases. The juveniles molt twice and escape from the host through the intestinal or genital tract. In the environment, they undergo two molts within 1 or 2 weeks. After the copulation, the males die but the fertilized females may live for 20–30 days before infecting a new host. The three species of this genus are parasitic in ground beetles (Carabidae) (Wachek, 1955). Rates of parasitism are maximum in August: *W. bovieni*, 22.5%, *W. stammeri*, 6%, *W. wuelkeri*, 11%. The parasitism causes reduction of size in ovaries.

V. CONCLUSION

It is apparent from this review that Sphaerularioids more frequently parasitize forest insects, about 53% of the species among 13 genera, than crop insects, about 12% among 8 genera. Of the forest insects, the members of the Scolytidae are the most highly developed parasites (Kaya, 1984). The nematode parasites of crop insects, belonging mainly to *Heterotylenchus, Howardula, Paraiotonchium*, and *Thripinema*, parasitize Diptera, Coleoptera, and Thysanoptera. The exact role of Sphaerularioids in the natural regulation of insect populations remains largely unknown. In fact, our knowledge of this group of entomophilic nematodes is still superficial if compared to that of Steinernematidae, Heterorhabditidae, and Mermithidae. For example, Geden and Stoffolano (1984) showed for Diptera how our knowledge of parasitism by nematodes is out of proportion to the number of insect species. Some important discoveries are recent, such as life cycles with alternation of different generations and new host records. The first mention of parasitism of Siphonaptera by Sphaerularioids is from Poinar and Nelson (1973); at the present time, three genera with 31 species are described. Sphaerularioids are unknown in Orthoptera and Lepidoptera.

Biological control of insects using Sphaerularioids presently is a neglected field of research, except for *Paraiotonchium, Howardula*, and *Beddingia. Paraiotonchium autumnale* was experimented against the face fly *Musca autumnalis* in the USA and *H. husseyi* against the mushroom phorid *Megaselia halterata* in Great Britain. *Beddingia wilsoni* is used at the present time in Australia to control the wood wasp *Sirex noctilio*. We hope that this review will encourage more scientists to work in this important and challenging field of research.

APPENDIX 1 Nematodes and Their Insect Hosts

Nematoda: genus	Order	Family	Genus
Allantonema	Coleoptera	Curculionidae	*Hylobius*
		Hydrophilidae	*Ochthebius*
		Scarabaeidae	*Geotrupes*
		Scolytidae	*Crypturgus, Hylastes, Orthotomicus, Tomicus (Myelophilus)*
		Staphylinidae	*Philonthus*
Aphelenchulus	Coleoptera	Cerambycidae	*Cyllene*
Beddingia	Coleoptera	Curculionidae	*Baris*
		Melandryidae	*Serropalpus*
	Hymenoptera	Cynipidae	*Ibalia*
		Ichneumonidae	*Megarhyssa, Pseudorhyssa, Rhyssa*
		Siricidae	*Sirex, Urocerus, Xeris*
Bovienema	Coleoptera	Scolytidae	*Cryphalus, Ips, Pityogenes, Pityophthorus, Scolytus, Xyleborus (Anisandrus)*
Bradynema	Coleoptera	Cerambycidae	*Rhagium, Spondylis*
		Scarabaeidae	*Aphodius*
		Trixagidae	*Throscus*
	Diptera	Bibionidae	*Bibio, Philia*
	Hemiptera	Gerridae	*Gerris*
		Nepidae	*Nepa*
		Veliidae	*Velia*
	Siphonaptera	Ctenophthalmidae	*Ctenophthalmus*

Contortylenchus	Coleoptera	Scolytidae	*Cryphalus, Dendroctonus, Hylastes, Hylurgops, Ips, Orthotomicus*
Fergusobia	Diptera	Agromyzidae	*Fergusonina*
Heteromorphotylenchus	Coleoptera	Nitidulidae	*Carpophilus, Stelidota*
Heterotylenchus	Diptera	Anthomyiidae	*Delia (Hylemyia)*
		Muscidae	*Morellia*
Howardula	Coleoptera	Apionidae	*Apion*
		Chrysomelidae	*Acalymma, Colaspis, Diabrotica, Epitrix, Longitarsus, Phyllotreta*
		Nitidulidae	*Carpophilus*
		Hydrophilidae	*Dactylosternum*
	Diptera	Chloropidae	*Oscinella*
		Drosophilidae	*Drosophila*
		Phoridae	*Megaselia*
		Sepsidae	*Sepsis*
		Sphaeroceridae	*Copromyza*
	Siphonaptera	Ctenophthalmidae	*Ctenophthalmus*
	Acarina	Gamasida	*Cosmolaelaps, Euryparasitus, Haemogamasus, Hypoaspis, Parasitus, Poecilochirus*
Incurvinema	Siphonaptera	Hystrichopsyllidae	*Rhadinopsylla*
Metaparasitylenchus	Coleoptera	Cerambycidae	*Strangalia, Tetropium*
		Cryptophagidae	*Cryptophagus, Telmatophilus*
		Curculionidae	*Cossonus*
		Dryopidae	*Helmis, Limnius, Riolus*
		Mycetophagidae	*Mycetophagus*
		Rhizophagidae	*Rhyzophagus*

APPENDIX 1   (Continued)

Nematoda: genus	Order	Family	Genus
(*Metaparasitylenchus*)	(Coleoptera)	Scolytidae	*Hexacolus (Scolytodes), Leperisinus*
		Staphylinidae	*Carpelimus*
Neoparasitylenchus	Coleoptera	Curculionidae	*Pissodes*
		Scolytidae	*Anisandrus (Xyleborus), Blastophagus, Conophthorus, Cryphalus, Crypturgus, Dendroctonus, Dryocoetes, Hylastes, Hylensinus, Hylurgops, Hylurgus, Ips, Leperesinus, Phloeoiribus, Pityogenes, Pityophthorus, Polygraphus, Scolytus, Xyleborus*
Paraiotonchium	Diptera	Muscidae	*Hydrotaea, Morellia, Musca, Orthellia, Stomoxys*
		Scatophagidae	*Scopeuma*
		Syrphidae	*Helophilus*
Parasitylenchoides	Coleoptera	Colydiidae	*Bitoma (Ditoma)*
		Histeridae	*Paromalus, Plegaderus*
		Silphidae	*Sciodrepa*
		Staphylinidae	*Aleochara, Anotylus (Oxytelus), Paederus, Stenus*
Parasitylenchus	Coleoptera	Coccinellidae	*Adalia, Adonia, Harmonia, Semiadalia*
		Scolytidae	*Blastophagus, Conophthorus, Cryphalus, Ips, Orthotomicus, Pityok-teines, Scolytus*
	Diptera	Drosophilidae	*Drosophila*

Nematode	Order	Family	Insect genera
Paregletylenchus	Diptera	Anthomyiidae	*Paregle*
Phaenopsitylenchus	Coleoptera	Buprestidae	*Phaenops*
Proparasitylenchus	Coleoptera	Staphylinidae	*Anotylus (Oxytelus), Carpelimus, Medon, Nehemitropia, (Atheta) Platystethus, Zyras*
Protylenchus	Coleoptera	Anobiidae	*Anobium*
		Heteroceridae	*Heterocerus*
Psyllotylenchus	Siphonaptera	Ceratophyllidae	*Ceratophyllus, (Callopsylla, Citellophilus), Diamanus, Megabothris, Monopsyllus, Nosopsyllus, Opisodasys*
		Coptopsyllidae	*Coptopsylla*
		Hystrichopsyllidae	*Catallagia, Neopsylla, Rhadinopsylla*
		Pulicidae	*Xenopsylla*
Scatonema	Diptera	Scatopsidae	*Scatopse*
Sphaerularia	Hymenoptera	Apidae	*Bombus, Psithyrus*
		Vespidae	*Vespa (Vespula)*
Sphaerulariopsis	Coleoptera	Anobiidae	*Ernobius*
		Curculionidae	*Pissodes*
		Scolytidae	*Dendroctonus*
	Hymenoptera	Braconidae	*Coeloides*
Spilotylenchus	Siphonaptera	Ceratophyllidae	*Ceratophyllus (Megabothris),*
		Coptopsyllidae	*Coptopsylla*
		Ctenophthalmidae	*Ctenophthalmus*
		Pulicidae	*Spilopsyllus*

APPENDIX 1   (Continued)

Nematoda: genus	Order	Family	Genus
Sulphuretylenchus	Coleoptera	Scolytidae (Ipidae)	*Dendroctonus, Dryocoetes, Hylastes, Ips, Pityogenes, Pityokteines, Polygraphus, Pseudohylesinus, Scolytus*
Thripinema	Thysanoptera	Thripidae	*Aptinothrips, Frankliniella, Megaluriothrips, Taeniothrips*
Tripius	Diptera	Cecidomyiidae	*Cecidomyia (Dasyneura)*
		Mycetophilidae	*Bradysia*
		Sciaridae	*Sciara*
Wachekitylenchus	Coleoptera	Carabidae	*Bembidion, Clivina*

APPENDIX 2 Genera List of Insects Parasitized by Identified Sphaerularioids

Order	Family	Genus
Coleoptera	Anobiidae	*Anobium, Ernobius*
	Apionidae	*Apion*
	Buprestidae	*Phaenops*
	Carabidae	*Bembidion, Clivina*
	Cerambycidae	*Cyllene (Megacyllene), Rhagium, Spondylis, Strangalia, Tetropium*
	Chrysomelidae	*Acalymma, Colaspis, Diabrotica, Epitrix, Longitarsus, Phyllotreta*
	Coccinellidae	*Adalia, Adonia, Harmonia, Semiadalia*
	Colydiidae	*Bitoma (Ditoma)*
	Cryptophagidae	*Cryptophagus, Telmatophilus*
	Curculionidae	*Baris, Cossonus, Hylobius, Pissodes*
	Dryopidae	*Helmis, Limnius (Latelmis), Riolus*
	Heteroceridae	*Heterocerus*
	Histeridae	*Paromalus, Plegaderus*
	Hydrophilidae	*Dactylosternum, Ochthebius*
	Melandryidae	*Serropalpus*
	Mycetophagidae	*Mycetophagus*
	Nitidulidae	*Carpophilus, Stelidota*
	Rhizophagidae	*Rhizophagus*
	Scarabaeidae	*Aphodius, Geotrupes*
	Scolytidae	*Anisandrus (Xyleborus), Blastophagus, Conophthorus, Cryphalus, Crypturgus, Dendroctonus, Dryocoetes, Hexacolus, Hylastes, Hylesinus, Hylurgops, Hylurgus, Ips, Leperisinus, Orthotomicus, Phloeotribus, Pityogenes, Pityokteines, Pityophthorus, Polygraphus, Pseudohylesinus, Scolytus, Tomicus (Myelophilus), Xyleborus*

APPENDIX 2 (Continued)

Order	Family	Genus
	Silphidae	*Sciodrepa*
	Staphylinidae	*Aleochara, Anotylus, Carpelimus (Trogophloeus), Medon, Nehemitropia (Atheta), Oxytelus, Paederus, Philonthus, Platystethus, Stenus, Zyras (Myrmedonia)*
	Trixagidae	*Throscus (Trixagus)*
Diptera	Agromyzidae	*Fergusonina*
	Anthomyiidae	*Delia (Hylemyia), Paregle*
	Bibionidae	*Bibio, Philia*
	Cecidomyiidae	*Cecidomyia*
	Chloropidae	*Oscinella*
	Drosophilidae	*Drosophila*
	Muscidae	*Hydrotaea, Morellia, Musca, Orthellia, Stomoxys*
	Mycetophilidae	*Bradysia*
	Phoridae	*Megaselia*
	Scatophagidae	*Scopeuma*
	Scatopsidae	*Scatopse*
	Sciaridae	*Sciara*
	Sepsidae	*Sepsis*
	Sphaeroceridae	*Copromyza*
	Syrphidae	*Helophilus*

Order	Family	Genera
Hemiptera	Gerridae	*Gerris (Aquarius)*
	Nepidae	*Nepa*
	Veliidae	*Velia*
Hymenoptera	Apidae	*Bombus, Psithyrus*
	Braconidae	*Coeloides*
	Cynipidae	*Ibalia*
	Ichneumonidae	*Megarhyssa, Pseudorhyssa, Rhyssa*
(Hymenoptera)	Siricidae	*Sirex, Urocerus, Xeris*
	Vespidae	*Vespa (Vespula)*
Siphonaptera	Ceratophyllidae	*Ceratophyllus (Callopsylla, Citellophilus), Diamanus, Megabothris, Monopsyllus, Nosopsyllus, Opisodasys*
	Coptopsyllidae	*Coptopsylla*
	Ctenophthalmidae	*Ctenophthalmus*
	Hystrichopsyllidae	*Catallagia, Neopsylla, Rhadinopsylla,*
	Pulicidae	*Spilopsyllus, Xenopsylla*
Thysanoptera	Thripidae	*Aptinothrips, Frankliniella, Megaluriothrips, Taeniothrips*
Acarina	Gamasida	*Cosmolaelaps, Euryparasitus, Haemogamasus, Hypoaspis, Parasitus, Poecilochirus*

REFERENCES

Ashraf, M., and Berryman, A. A. 1970a. Biology of *Sulphuretylenchus elongatus* (Nematoda: Sphaerulariidae), and its effect on its host, *Scolytus ventralis* (Coleoptera: Scolytidae). *Can. Entomol.* 102: 197–213.

Ashraf, M., and Berryman, A. A. 1970b. Histopathology of *Scolytus ventralis* (Coleoptera: Scolytidae) infected by *Sulphuretylenchus elongatus* (Nematoda: Sphaerulariidae). *Ann. Entomol. Soc. Am.* 63: 924–930.

Baujard, P. 1979. *Etudes nématofaunistiques en forêt landaise.* Thèse Sciences, Bordeaux.

Bedding, R. A. 1968. *Deladenus wilsoni* n. sp. and *D. siricidicola* n. sp. (Neotylenchidae) entomophagous-mycetophagous nematodes parasitic in siricid woodwasps. *Nematologica* 14: 515–525.

Bedding, R. A. 1972. Biology of *Deladenus siricidicola* (Neotylenchidae) an entomophagous-mycetophagous nematode parasitic in siricid woodwasps. *Nematologica* 18: 482–493.

Bedding, R. A. 1974. Five new species of *Deladenus* (Neotylenchidae), entomophagous-mycetophagous nematodes parasitic in siricid woodwasps. *Nematologica* 20: 204–225.

Bedding, R. A. 1984. Nematode parasites of Hymenoptera. In *Plant and Insect Nematodes*. Nickle, W. R. ed. Marcel Dekker, New York, pp. 755–795.

Blinova, S. L., and Gurando, E. V. 1977. *Parasitylenchus macrobursatus* sp. n. (Nematoda: Sphaerulariidae) and taxonomy of the genus *Parasitylenchus* Micoletzky 1922. *Zashchita Lesa. Vypusk. L.* 2: 24–28 (in Russian).

Blinova, S. L., and Korenchenko, E. A. 1986. *Phaenopsitylenchus laricis* gen. and sp. n. (Nematoda: Phaenopsitylenchidae fam. n.) parasite of *Phaenops guttulata* and remarks on taxonomy of nematodes of the superfamily Sphaerularioidea. *Acad. Nauk. SSSR Trudy Gel'mint. Lab.* 34: 14–23 (in Russian).

Bovien, P. 1932. On a new nematode, *Scatonema wülkeri* gen. and spec. n. parasitic in the body-cavity of *Scatopse fuscipes* Meig. (Diptera: Nematocera). *Vid. Meddel. Dansk Naturh. Forening, Kobenhavn* 94: 13–32.

Bovien, P. 1937. Some types of association between nematodes and insects. *Vid. Meddel. Dansk Naturh. Forening Kobenhavn* 101.

Bovien, P. 1944. *Proatractonema sciarae* n. g., n. sp., a parasitic nematode from the body cavity of a dipterous larva. *Vid. Meddel. Dansk Naturh. Forening Kobenhavn* 108: 1–14.

Chitwood, B. G. 1935. Nomenclatorial notes, I. *Proc. Helm. Soc. Wash.* 2: 51–54.

Chitwood, B. G., and Chitwood, M. B. 1937. *An Introduction to Nematology*. University Park Press, Baltimore.

Christie, J. R. 1938. Two nematodes associated with decaying citrus fruit. *Proc. Helm. Soc. Wash.* 5: 29–33.

Christie, J. R. 1941. Life history (zooparasitica). Parasites of invertebrates. In *An Introduction to Nematology*. Christie, J.R. ed. Babylon, New York, pp. 246–266.

Cobb, N. A. 1920. One hundred new nemas (Type species of 100 new genera). *Contrib. Sci. Nematol.* 9: 217–343.

Cobb, N. A. 1921. *Howardula benigna*, a nemic parasite of the cucumber-beetle (*Diabrotica*). *Science* 54: 667–670.

Currie, G. A. 1937. Galls on eucalyptus trees. A new type of association between flies and nematodes. *Proc. Linn. Soc. N.S.W.* 62: 147–174.

Deunff, J. 1984. *Les parasites de Siphonaptères. Etude de la Systématique, de la Biologie et du Pouvoir Pathogène des Tylenchides (Nematodea) dans une perspective de lutte biologique.* Thèse Pharmacie, Rennes.

Deunff, J., and Launay, H. 1984. *Psyllotylenchus chabaudi*, n. sp. (Nematodea, Tylenchida: Allantonematidae) parasite de *Nosopsyllus fasciatus* (Bosc.) (Siphonaptera: Ceratophyllidae). *Ann. Parasitol. Hum. Comp.* 59: 263–270.

Deunff, J., Launay, H., and Beaucournu, J. C. 1985. *Incurvinema helicoides* n. gen., n. sp. Nematodea, Tylenchida: Allantonematidae parasite de *Rhadinopsylla pentacantha* (Rothschild, 1897) (Siphonaptera: Hystrichopsyllidae). *Ann. Parasitol. Hum. Comp.* 60: 739–746.

Dufour, L. 1837. Recherches sur quelques entozoaires et larves parasites des insectes Orthoptères et Hyménoptères. *Ann. Sci. Natur. Zool.* 7: 5–20.

Elsey, K. D. 1977a. Parasitism of some economically important species of Chrysomelidae by nematodes of the genus *Howardula*. *J. Invert. Pathol.* 29: 384–385.

Elsey, K. D. 1977b. Dissemination of *Howardula* sp. nematodes by adult tobacco flea beetles (Coleoptera: Chrysomelidae). *Can. Entomol.* 109: 1283–1285.

Elscy, K. D. 1977c. *Howardula dominicki* n. sp. infesting the tobacco flea beetle in North Carolina. *J. Nematol.* 9: 338–342.

Elsey, K. D. 1979. *Howardula colaspidis* (Allantonematidae) n. sp. a new parasite of the grape *Colaspis* (Coleoptera: Chrysomelidae). *Nematologica* 25: 54–61.

Elsey, K. D., and Pitts, J. M. 1976. Parasitism of the tobacco flea beetle by a Sphaerulariid nematode *Howardula* sp. *Env. Entomol.* 5: 707–711.

Filipjev, I. N. 1934. The classification of the free-living nematodes and their relation to the parasitic nematodes. *Smithsonian Misc. Coll.* 89: 1–63.

Fisher, J. M., and Nickle, W. R. 1968. On the classification and life history of *Fergusobia curriei* (Sphaerulariidae: Nematoda). *Proc. Helm. Soc. Wash.* 35: 40–46.

Fuchs, G. 1913. Über Parasiten und andere biologisch an die Borkenkäfer gebundene Nematoden. *Verhandl. Ges. Deutsch. Naturf. u. érzte (85 Vers., Wien) Teil* 2: 689–692.

Fuchs, G. 1915. Die Naturgeschichte der Nematoden und einiger anderer Parasiten, 1. des *Ips typographus* L. 2. des *Hylobius abietis* L. *Zool. Jahrb.* 38: 109–222.

Fuchs, G. 1929. Die Parasiten einiger Rüssel-und Borkenkäfer. *Z. Parasitenkd* 2: 248–285.

Fuchs, G. 1937. Neue parasitische und halbparasitische Nematoden bei Borkenkäfern und einige andere Nematoden. I. *Zool. Jahrb.* 70: 291–380.

Fuchs, G. 1938. Neue parasiten und halbparasiten bei Borkenkäfern und einige andere Nematoden. II, III, u. IV. *Zool. Jahrb.* 71: 123–190.

Furniss, M. M. 1967. Nematode parasites of the Douglas-fir beetle in Idaho and Utah. *J. Econ. Entomol.* 60: 1323–1326.

Geden, C. J., and Stoffolano, J. G. 1984. Nematode parasites of other dipterans. In *Plant and Insect Nematodes*. Nickle, W. R., ed. Marcel Dekker, New York, pp. 849–898.

Geraert, E., Raski, D. J., and Choi, Y. E. 1984. A study of *Stictylus intermedius* n. comb. with a review of the genus (Nematoda: Tylenchida). *Nematologica* 30: 161–171.

Gibb, K. S., and Fisher, J. M. 1986. Co-ordination of the life-cycle of *Contortylenchus grandicolli* (Nematoda: Allantonematidae) with that of its host *Ips grandicollis* (Scolytidae). *Nematologica* 32: 222–233.

Goodey, T. 1930. One remarkable new nematode, *Tylenchinema oscinellae* gen. et sp. n., parasitic in the frit-fly, *Oscinella frit* L., attacking oats. *Phil. Trans. Roy. Soc. London* 218: 315–343.

Goodey, T. 1931. Further observations on *Tylenchinema oscinellae* Goodey, 1930, a nematode parasite of the frit-fly. *J. Helminthol.* 9: 157–174.

Iperti, G., and Van Waerebeke, D. 1968. Description, biologie et importance d'une nouvelle espèce d'Allantonematidae (Nématode) parasite des coccinelles aphidiphages: *Parasitylenchus coccinellinae* n. sp., *Entomophaga* 13: 107–119.

Jairajpuri, M. S. 1966. *Dorsalla indica* (Jairajpuri, 1962) n. comb. (Nematoda: Neotylenchidae). *Labdev. J. Sci. Tech. India* 4: 142–143.

Kaya, H. K. 1984. Nematode parasites of Bark Beetles. In *Plant and Insect Nematodes*. Nickle, W. R., ed. Marcel Dekker, New York, pp. 727–754.

Kaya, H. K., and Moon, R. D. 1978. The nematode *Heterotylenchus autumnalis* and face fly *Musca autumnalis*: A field study in northern California. *J. Nematol* 10: 333–341.

Kaya, H. K., Moon, R. D., and Witt P. L. 1979. Influence of the nematode, *Heterotylenchus autumnalis* on the behavior of face fly, *Musca autumnalis*. *Env. Entomol.* 8: 537–540.

Khan, M. A. 1957a. *Sphaerularia bombi* Duf. (Nematoda: Allantonematidae) infesting bumble bees and *Sphaerularia hastata* sp. nov. infesting bark beetles in Canada. *Can. J. Zool.* 35: 519–523.

Khan, M. A. 1957b. *Sphaerularia ungulacauda* sp. nov. (Nematoda: Allantonematidae) from the Douglas fir beetle, *Dendroctonus pseudotsugae* Hopk., with key to *Sphaerularia* species (emended). *Can. J. Zool.* 35: 635–639.

Korenchenko, E. A. 1987. *Parasitylenchus klimenkorum* sp. n. (Nematoda: Allantonematidae) a parasite of *Orthotomicus laricis* (Coleoptera, Ipidae). *Parazitologiya* 21: 567–576 (in Russian).

Krafsur, E. S., Church, C. J., Elvin, M. K., and Ernst, C. M. 1983. Epizootiology of *Heterotylenchus autumnalis* (Nematoda) among face flies (Diptera: Muscidae) in central Iowa, USA. *J. Med. Entomol.* 20: 318–324.

Kurochkin, Y. V. 1960. The nematode *Heterotylenchus pawlowskyi* sp. n., castrating flea-vectors of plague. *Dokl. Akad. Nauk SSSR* 135: 1281–1284 (in Russian)

Laumond, C. 1970. Hétérogonie et adaptations morphologiques chez un Sphaerulariidae (Nematoda) parasite de *Baris caerulescens*. *C. R. Acad. Sci. Paris* 271: 1575–1577.

Laumond, C., and Lyon J. P. 1975. Un nématode nouveau (Allantonematidae) parasite du genre *Helophilus* (Diptera: Syrphidae). *Acta Tropica* 32: 334–339.

Laumond, C., and Beaucournu, J. C. 1978. *Neoparasitylenchus megabothridis* n. sp. (Tylenchida: Allantonematidae) parasite de *Megabothris turbidus* (Siphonaptera: Ceratophyllidae); observations sur les Tylenchides de puces dans le Sud-Ouest de l'Europe. *Ann. Parasitol.* (Paris) 53: 291–302.

Laumond, C., and Mauleon, H. 1982. *Metaparasitylenchus guadeloupensis* n. sp. (Tylenchida, Allantonematidae) parasite d'*Hexacolus guyanensis* (Coleoptera, Scolytidae) en Guadeloupe. *Rev. Nematol.* 5: 65–69.

Laumond, C., and Bonifassi, E. 1991. *Beddingia barisii* n. sp. (Sphaerularioidea: Phaenopsitylenchidae), a nematode parasite of *Baris caerulescens* (Coleoptera: Curculionidae). *Nematologica* (in press).

Launay, H., Deunff, J., and Bain, O. 1983. *Spilotylenchus arthuri*, gen. n., sp. n. (Nematodea, Tylenchida: Allantonematidae), parasite of *Spilopsyllus cuniculi* (Dale, 1878) (Siphonaptera: Pulicidae). *Ann. Parasitol. Hum. Comp.*, 58: 141–150.

Launay, H., and Deunff, J. 1984. Un *Spilotylenchus* nouveau (Nematodea, Tylenchida: Allantonematidae) parasite de *Spilopsyllus cuniculi* (Dale) (Siphonaptera: Pulicidae), Siphonaptère inféodé au lapin de garenne. *Ann. Parasitol. Hum. Comp.* 59: 415–420.

Launay, H., and Deunff, J. 1990. *Spilotylenchus maisonabei* n. sp. (Nematoda: Allantonematidae) parasite de *Spilopsyllus cuniculi* (Dale, 1878) (Siphonaptera: Pulicidae), puce oïoxéne du lapin de garenne. *Rev. Nématol.* (in press).

Leuckart, R. 1884. Über einen neuen heterogenen Nematoden. *Tagebl. 57, Verhandl. Ges. Deutsch. Naturf. u. Ärzte,* 320.

Lieutier, F. 1981. Influence des nématodes parasites sur l'essaimage du scolyte *Ips sexdentatus* (Boern.). Action régulatrice du froid. *Acta Oecologica/Oecol. Appl.* 2: 357–368.

Linstow von, O. F. B. 1890. Über *Allantonema* und *Diplogaster*. *Zentralbl. Bakteriol.* 8: 489–493.

Linstow von, O. F. B. 1893. Über *Allantonema sylvaticum*. *Zentralbl. Bakteriol.* 14: 169–173.

Lundberg, H., and Svensson, B. G. 1975. Studies on the behavior of *Bombus* Latr. species (Hym., Apidae) parasitized by *Sphaerularia bombi* Dufour (Nematoda) in an alpine area. *Norw. J. Entomol.* 22: 129–134.

Lysaght, A. 1936. A note on the adult female of *Anguillulina aptini* (Sharga), a nematode parasitizing *Aptinothrips rufus* Gmelin. *Parasitology* 28: 290–292.

Lysaght, A. 1937. An ecological study of a Thrips (*Aptinothrips rufus*) and its nematode parasite (*Anguillulina aptini*). *J. Anim. Ecol.* 6: 169–192.

MacGuidwin, A. E., Smart, G. C., and Allen, G. E. 1980. Redescription and life history of *Contortylenchus brevicomi*, a parasite of the southern pine beetle *Dendroctonus frontalis*. *J. Nematol.* 12: 207–212.

MacGuidwin, A. E., Smart, G. C., Wilkinson, R. C., and Allen, G. E. 1980. Effect of the nematode *Contortylenchus brevicomi* on gallery construction and fecundity of the southern pine beetle. *J. Nematol.* 12: 278–282.

Maggenti, A. 1981. *General Nematology*. Springer-Verlag, New York.

Massey, C. L. 1956. Nematode parasites and associates of the Engelmann spruce beetle (*Dendroctonus engelmanni* Hopk.). *Proc. Helm. Soc. Wash.* 23: 14–24.

Massey, C. L. 1957. Four new species of *Aphelenchulus* (Nematoda) parasitic in bark beetles in the United States. *Proc. Helm. Soc. Wash.* 24: 29–34.

Massey, C. L. 1958. Four new species of *Parasitylenchus* (Nematoda) from scolytid beetles. *Proc. Helm. Soc. Wash.* 25: 26–30.

Massey, C. L. 1960. Nematode parasites and associates of the California five-spined engraver, *Ips confusus* (Lec.). *Proc. Helm. Soc. Wash.* 27: 14–22.

Massey, C. L. 1964. The nematode parasites and associates of the fir engraver beetle, *Scolytus ventralis* LeConte, in New Mexico. *J. Insect Pathol.* 6: 133–155.

Massey, C. L. 1966. The nematode parasites and associates of *Dendroctonus adjunctus* (Coleoptera: Scolytidae) in New Mexico. *Ann. Entomol. Soc. Am.* 59: 424–440.

Massey, C. L. 1974. *Biology and Taxonomy of Nematode Parasites and Associates of Bark Beetles in the United States*. Agr. Handbook 446, Forest Service, USDA, Washington, D.C.

Nappi, A. J. 1973. Effects of parasitization by the nematode *Heterotylenchus autumnalis* on mating and oviposition in the host, *Musca autumnalis*. *J. Parasitol.* 59: 963–969.

Nicholas, W. L., and Hugues, R. D. 1970. *Heterotylenchus* sp. (Nematoda: Sphaerulariidae), a parasite of the Australian bush fly, *Musca vetustissima*. *J. Parasitol.* 56: 116–122.

Nickle, W. R. 1963a. *Bovienema* (Nematoda: Allantonematidae), a new genus parasitizing bark beetles of the genus *Pityogenes* Bedel, with notes on other endoparasitic nematodes of scolytids. *Proc. Helm. Soc. Wash.* 30: 256–262.

Nickle, W. R. 1963b. The endoparasitic nematodes of California bark beetles with descriptions of *Bovienema* n. g. and *Neoparasitylenchus* n. subg. and with the presentation of new information on the life history of *Contortylenchus elongatus* n. comb. Ph.D. thesis, Univ. California.

Nickle, W. R. 1963c. Notes on the genus *Contortylenchus* Rühm, 1956, with observations on the biology and life history of *C. elongatus* (Massey, 1960) n. comb., a parasite of a bark beetle. *Proc. Helm. Soc. Wash.* 30: 218–223.

Nickle, W. R. 1965. On the status of *Prothallonema dubium* Christie, 1938 (Nematoda: Allantonematidae). *Nematologica* 11: 44.

Nickle, W. R. 1967a. On the classification of the insect parasitic nematodes of the Sphaerulariidae Lubbock, 1861 (Tylenchoidea: Nematoda). *Proc. Helm. Soc. Wash.* 34: 72–94.

Nickle, W. R. 1967b. *Heterotylenchus autumnalis* sp. n. (Nematoda: Sphaerulariidae), a parasite of the face fly, *Musca autumnalis* De Geer. *J. Parasit.* 53: 398–401.

Nickle, W. R. 1971. Behavior of the shothole borer, *Scolytus rugulosus*, altered by the nematode parasite *Neoparasitylenchus rugulosi*. *Ann. Entomol. Soc. Am.* 64: 751.

Nickle, W. R. 1978. Taxonomy of nematodes that parasitize insects, and their use as biological control agents. In *Biosystematics in Agriculture*. Romberger, J. A., ed. Osmum & Co., Allenheld, pp. 37–51.

Nickle, W. R., and Welch, H. E. 1984. History, development, and importance of insect nematology. In *Plant and Insect Nematodes*. Nickle, W. R. ed. Marcel Dekker, New York, pp. 627–653.

Nickle, W. R., and Wood, G. W. 1964. *Howardula aptini* (Sharga, 1932) parasitic in blueberry thrips in New Brunswick. *Can. J. Zool.* 42: 843–846.

Oldham, J. N. 1930. On the infestation of elm bark beetles (Scolytidae) by a nematode, *Parasitylenchus scolyti* n. sp. *J. Helminthol.* 8: 239–248.

Oldham, J. N. 1933. On *Howardula phyllotretae* n. sp., a nematode parasite of flea beetles (Chrysomelidae: Coleoptera), with some observations on its incidence. *J. Helminthol.* 11: 119–136.

Poinar, G. O., Jr. 1965. The bionomics and parasitic development of *Tripius sciarae* (Bovien) (Sphaerulariidae: Aphelenchoidea), a nematode parasite of sciarid flies (Sciaridae: Diptera). *Parasitolgy* 55: 559–569.

Poinar, G. O., Jr. 1975. *Entomogenous nematodes: A Manual and Host List of Insect-Nematode Associations.* E. J. Brill, Leiden, The Netherlands.

Poinar, G. O., Jr. 1977. *CIH Keys to the Groups and Genera of Nematode Parasites of Invertebrates.* Commonwealth Agricultural Bureaux, Farnham Royal, UK.

Poinar, G. O., Jr. 1979. *Nematodes for Biological Control of Insects.* CRC Press, Boca Raton, Florida.

Poinar, G. O., Jr. 1988. Nematode parasites of Chrysomelidae. In *Biology of Chrysomelidae.* Jolivet, P., Petitpierre, E., and Hsiao, T. H., eds. Kluwer, Dordrecht, pp. 433–448.

Poinar, G. O., Jr., and Caylor, J. N. 1974. *Parasitylenchus amvlocercus* sp. n. (Tylenchida: Nematodea) from *Conophthorus monophyllae* (Scolytidae: Coleoptera) in California with a synopsis of the nematode genera found in bark beetles. *J. Invert. Pathol.* 24: 112–119.

Poinar, G. O., Jr., and Doncaster, C. C. 1965. The penetration of *Tripius sciarae* (Sphaerulariidae: Aphelenchoidea) into its insect host, *Bradysia paupera* Tuom. (Mycetophilidae: Diptera). *Nematologica* 11: 73–78.

Poinar, G. O., Jr., and Van der Laan, P. A. 1972. Morphology and life history of *Sphaerularia bombi. Nematologica* 18: 239–252.

Poinar, G. O., Jr., and Nelson, B. C. 1973. *Psyllotylenchus viviparus* n. gen., n. sp. (Nematodea: Tylenchida: Allantonematidae) parasitizing fleas (Siphonaptera) in California. *J. Med. Entomol.* 10: 349–354.

Poinar, G. O., Jr., Laumond, C., and Bonifassi, E. 1980. *Howardula apioni* sp. n. (Allantonematidae: Nematoda), a parasite of *Apion carduorum* Kirby (Curculionidae: Coleoptera) in southern France. *Proc. Helm. Soc. Wash.* 47: 218–223.

Poisson, R. 1933. Trois nouvelles espèces de nématodes de la cavité générale d'hémiptères aquatiques. *Ann. Parasitol. Hum. Comp.* 11: 463–466.

Pouvreau, A. 1962. Contribution á l'étude de *Sphaerularia bombi* (Nematoda, Tylenchida), parasite des reines de bourdons. *Ann. Abeille.* 5: 181–193.

Pouvreau, A. 1966. Sur quelques ennemis des bourdons. Proc. 2nd Int. Symp. Pollinisation, London 1964. *Bee World* 47: 173–177.

Reddy, Y. N., and Rao, P. N. 1980. Studies on *Heterotylenchus hyderabadensis* sp. n. parasitic in the fly *Morellia hortensis* (Wiedemann). *Proc. Ind. Acad. Parasitol.* 1: 51–55.

Reddy, Y. N., and Rao, P. N. 1981. Description of morphology, biology, life history of a new species *Howardula marginatis* sp. n. from flies. *Riv. Parassit.* 42: 127–135.

Reddy, Y. N., and Rao, P. N. 1987. Studies on *Heterotylenchus xanthomelas* sp. n. parasitic in *Musca xanthomelas* Wiedemann (Muscidae: Diptera). *Ind. J. Nematol.* 17: 180–183.

Reddy, Y. N., Nickle, W. R., and Rao, P. N. 1982. Studies on *Howardula aptini* (Nematoda: Sphaerulariidae) parasitic in *Megaluriothrips* sp. in India. *Ind. J. Nematol.* 12: 1–5.

Remillet, M. 1988. Insect-parasitic nematodes resembling plant-parasitic forms. In *Nematode Identification and Expert System Technology.* Fortuner, R. ed. Plenum Press, New York, pp. 283–289.

Remillet, M., and Van Waerebeke, D. 1975. Description et cycle biologique de *Howardula madecassa* n. sp. et *Howardula truncati* n. sp. (Nematoda: Sphaerulariidae) parasites de *Carpophilus* (Coleoptera: Nitidulidae). *Nematologica* 21: 192–206.

Remillet, M., and Van Waerebeke, D. 1976. Description et cycle biologique de *Howardula acris* n. sp. (Nematoda: Sphaerulariidae) parasite d'Hydrophilidae (Coleoptera). *Cah. ORSTOM, Sér. Biol.* 11: 219–224.

Remillet, M., and Van Waerebeke, D. 1978. Description et cycle biologique de *Heteromorphotylenchus stelidotae* n. g., n. sp. et de *Heteromorphotylenchus carpophili* n. sp. (Nematoda: Allantonematidae). *Nematologica* 24: 222–238.

Richardson, P. N., and Chanter, D. O. 1979. Phorid fly (Phoridae: *Megaselia halterata*) longevity and the dissemination of nematodes (Allantonematidae: *Howardula husseyi*) by parasitized females. *Ann. Appl. Biol.* 93: 1–11.

Richardson, P. N., and Hesling, J. J. 1977. Studies on an overwintering population of the mushroom phorid *Megaselia halterata* (Diptera: Phoridae) parasitized by *Howardula husseyi* (Nematoda: Allantonematidae). *Ann. Appl. Biol.* 86: 321–327.

Richardson, P. N., Hesling, J. J., and Riding, I. L. 1977. Life cycle and description of *Howardula husseyi* n. sp. (Tylenchida: Allantonematidae), a nematode parasite of the mushroom phorid *Megaselia halterata* (Diptera: Phoridae). *Nematologica* 23: 217–231.

Riding, I. L. 1970. Microvilli on the outside of a nematode. *Nature* 226: 179–180.

Riding, I. L., and Hague, N. G. M. 1974. Some observations on a Tylenchid nematode *Howardula* sp. parasitizing the mushroom phorid *Megaselia halterata* (Phoridae, Diptera). *Ann. Appl. Biol.* 78: 205– 211.

Rubzov, I. A. 1981. Parasites and enemies of fleas. "Nauka," Leningrad.

Rubzov, I. A. 1982. A new species of nematodes (Tylenchida: Allantonematidae) parasitic in fleas. *Parazitologiya* 16: 237–238.

Rühm, W. 1954. Einige neue, ipidenspezifische Nematodenarten. *Zool. Anz.* 153: 221–242.

Rühm, W. 1956. Die Nematoden der Ipiden. *Parasitol. Schriftenr.*

Rühm, W. 1960. Ein Beitrag zur Nomenklatur und Systematik einiger mit Scolytiden vergesellschafteter Nematodenarten. *Zool. Anz.* 164: 201–213.

Schvester, D. 1950. Sur un nématode du groupe des *Parasitylenchus dispar* Fuchs, parasite nouveau du Xylebore disparate (*Xyleborus dispar* F.). *Ann. Epiphyt.* 1: 47–53.

Schvester, D. 1957. Contribution à l'étude des coléoptères scolytides. *Ann. Epiphyt.* 8: 1–162.

Sharga, U. S. 1932. A new nematode, *Tylenchus aptini* n. sp., parasite of Thysanoptera (Insecta: *Aptinothrips rufus* Gmelin). *Parasitology* 24: 268–279.

Siddiqi, M. R. 1986a. *Tylenchida: Parasites of Plants and Insects*. Commonw. Inst. Parasitol., Slough, UK.

Siddiqi, M. R. 1986b. A review of the Nematode genus *Fergusobia* Currie (Hexatylina) with descriptions of *F. jambophila* n. sp. and *F. magna* n. sp. In *Plant Parasitic Nematodes of India*. Swarup, G., and Dasgupta, D. R. eds. pp. 264–278.

Slankis, A. 1967. *Contortylenchus cylindricus* sp. n. and *Contortylenchus rarus* sp. n. (Tylenchida: Contortylenchidae) parasites of bark beetles and taxonomic notes on the genus *Contortylenchus* Rühm, 1956. *Trudy Gel'mint. Lab.* 18: 111–118 (in Russian).

Slankis, A. 1969a. *Contortylenchus pseudodiplogaster* n. sp. from the bark beetle *Ips sexdentatus*. *Mater. Nauch. Konf. Voes. Obshch. Gel'mint.* 2: 302–305 (in Russian).

Slankis, A. 1969b. *Sulphuretylenchus pugionifer* n. sp. (Nematoda: Allantonematidae) from the bark beetle *Hylastes cunicularius* Er. *Trudy Gel'mint. Lab.* 20: 156–157 (in Russian).

Slankis, A. 1972. *Parasitylenchus dispar* and *Parasitylenchus aculeatus* sp. n. *Zool. Zh.* 51: 1731–1737 (in Russian).

Slankis, A. 1974. Four new species of nematodes (Nematoda: Sphaerulariidae) from bark beetles (Coleoptera: Ipidae). *Parazitologiya* 8: 57–62 (in Russian).

Slobodyanyuk, O. V. 1975a. *Heterotylenchus simplex* n. sp. (Nematoda: Sphaerulariidae) parasitic in *Morellia simplex*. *Parazitologiya* 9: 127–134 (in Russian).

Slobodyanyuk, O. V. 1975b. Erection of *Paraiotonchium* n. g. (Nematoda: Sphaerulariidae) and description of the type-species, *P. autumnalis* (Nickle, 1967) n. comb. *Trudy Gel'mint. Lab.* 25: 156–168 (in Russian).

Slobodyanyuk, O. V. 1976. *Paraiotonchium osiris* (Iotonchiinae: Tylenchida) new nematode species in *Musca osiris* Wd. *Parazitologiya* 10: 30–39 (in Russian).

Slobodyanyuk, O. V. 1980. Post-embryonic development of the nematode *Paraiotonchium autumnalis* (Tylenchida: Sphaerulariidae). Translated from Russian. Sonin, M. D. ed. 1987 Amerind Pub. Co. Ltd., New Delhi, pp. 156–179.

Slobodyanyuk, O. V. 1984. *Entomopathogenic Nematodes of Diptera. Order Tylenchida*. "Nauka," Moscow.

Slobodyanyuk, O. V. 1986. Revision of the genus *Heterotylenchus* Bovien, 1937 and erection of the genus *Wachekitylenchus* gen. n. (Allantonematidae: Tylenchida). *Parazitologiya* 20: 466–475.

Stoffolano, J. G., Jr., 1967. The synchronization of the life cycle of diapausing face flies, *Musca autumnalis*, and of the nematode, *Heterotylenchus autumnalis*. *J. Invert. Pathol.* 9:395–397.

Stoffolano, J. G., Jr., and Nickle, W. R. 1966. Nematode parasite (*Heterotylenchus* sp.) of face fly in New York State. *J. Econ. Entomol.* 59: 221–222.

Strassen, O. K. L. 1892. *Bradynema rigidum* v. Sieb. *Z. Wiss. Zool.* 54: 655–747.

Thomas, G. D., and Puttler, B. 1970. Seasonal parasitism of the face fly by the nematode *Heterotylenchus autumnalis* in Central Missouri, 1968. *J. Econ. Entomol.* 63: 1922–1923.

Thomas, G. D., Puttler, B., and Morgan, C. E. 1972. Further studies of field parasitism of the face fly by the nematode *Heterotylenchus autumnalis* in central Missouri, with notes on the gonadotrophic cycles of the face fly. *Env. Entomol.* 1: 759–763.

Thong, C. H. S., and Webster, J. M. 1972. A redescription of the bark beetle nematode *Contortylenchus brevicomi*: synonym *Contortylenchus barberus* (Nematoda: Sphaerulariidae). *J. Nematol.* 4: 213–216.

Thong, C. H. S., and Webster, J. M. 1973. Morphology and the post-embryonic development of the bark beetle nematode *Contortylenchus reversus* (Sphaerulariidae). *Nematologica* 19: 159–168.

Thorne, G. 1935. Nemic parasites and associates of the mountain pine beetle (*Dendroctonus monticolae*) in Utah. *J. Agr. Res.* 51: 131–144.

Thorne, G. 1941. Some nematodes of the family Tylenchidae which do not possess a valvular median esophageal bulb. *Great Basin Naturalist* 2: 37–85.

Wachek, F. 1955. Die entoparasitischen Tylenchiden. *Parasitol. Schriftenr.* 3: 1–119.

Warren, E. 1941. On the occurence of nematodes in the haemocoel of certain gamasid mites. *Ann. Natal. Mus.* 10: 79–94.

Welch, H. E. 1959. Taxonomy, life cycle, development, and habits of two new species of Allantonematidae (Nematoda) parasitic in drosophilid flies. *Parasitology* 49: 83–103.

Wülker, G. 1923. Über Fortpflanzung und Entwicklung von *Allantonema* und verwandten Nematoden. *Ergebn. Fortschr. Zool.* 5: 389–507.

Yatham, N. R., and Rao, P. N. 1980. Description and biology of *Howardula albopunctata* sp. n. from the two species of *Sepsis* (Sepsidae: Diptera). *Proc. Ind. Acad. Parasitol.* 1: 42–46.

Yatham, N. R., and Rao, P. N. 1981. Studies on *Heterotylenchus crassirostris* sp. n. parasitic in *Musca crassirostris* Stein and *Stomoxys calcitrans* L. *Ind. J. Nematol.* 11: 19–24.

Index

Printed and bound by CPI Group (UK) Ltd, Croydon, CR0 4YY

30/10/2024

01781230-0001